Catalogue of Meteorites

Catalogue of Meteorites

With special reference to those represented
in the collection of the
British Museum (Natural History)

by
A. L. Graham, A. W. R. Bevan and R. Hutchison

Fourth Edition (Revised and Enlarged)

The University of Arizona Press
Tucson, Arizona

ISBNs

British Museum (Natural History) 0 565 00 941 9
University of Arizona Press 0.8165 0912 3

Fourth edition 1985

British Museum (Natural History), Cromwell Road,
London SW7 5BD

The *Catalogue of Meteorites* has been published simultaneously in the United Kingdom by the British Museum (Natural History). It is available exclusively in North America from The University of Arizona Press, 1615 East Speedway Boulevard, Tucson, Arizona, 85719, U.S.A.

Library of Congress Catalog Card Number: 84–48681

ISBN 0–8165–0912–3

British Library Cataloguing in Publication Data

British Museum (Natural History)
Catalogue of meteorites with special reference to those represented in the collection of the British Museum (Natural History)
4th ed. (rev. and enl.)
1. Meteorites——Catalogs and collections
I. Title II. Graham, A. L. III. Bevan, A. W. R.
IV. Hutchison, R.
523.5'1'0216 QB755

ISBN 0–565–00941–9

Printed and bound in Great Britain by
The Garden City Press Limited
Letchworth, Hertfordshire SG6 1JS

Contents

Preface

In 1881 Lazarus Fletcher, recently appointed as Keeper of Minerals, published a Guide to the Catalogue of Meteorites — a slim, 40 page volume, priced at 2d — which described the nature of meteorites and listed the 361 specimens then in the Museum's collections. This little book ran to 11 editions over the years up to the First World War. Dr. G. T. Prior, Fletcher's successor as Keeper, continued and greatly advanced the study of meteorites. He compiled the first Catalogue of Meteorites, published in 1923, which, besides listing those meteorites in the Museum's collection, attempted to include the names of all meteorites known at that time with the location of the main mass of each. Fletcher and Prior thus laid the foundation of what has become an international standard work of reference.

Prior retired in 1927, and the main thrust of meteorite research in the Museum passed first to Dr. L. J. Spencer and then to Dr. Max H. Hey, who maintained Prior's practice of using meticulous chemical analyses in seeking deeper insights into the nature of meteorites. Hey was responsible for producing the second appendix to Prior's Catalogue in 1940, and the second and third editions of the Catalogue published in 1953 and 1966, respectively. Hey improved the Catalogue by including additional information in the form of classification and topographical lists, the geographical co-ordinates of falls, summaries of certain chemical determinations, and a catalogue of tektites. Although Hey retired in 1969, he continued to visit the Museum regularly and his long experience and encyclopaedic knowledge were available to his successors in the production of the appendix to the third edition (1977). This fourth edition of the Catalogue is dedicated to Max Hey who died on 24 January 1984 in his eightieth year.

The format of the previous editions has been retained. The authors have incorporated more chemical data in entries for individual meteorites but the summary of the minor and trace element contents of meteorites has been omitted. This edition contains nearly one third more information than the third edition and its appendix.

We are indebted to many friends and colleagues for information and corrections, and this help is gratefully acknowledged.

Department of Mineralogy
May 1984

A. C. BISHOP
Keeper

Introduction

This edition incorporates and expands the information published in the third edition (1966) and its appendix (1977). It includes the names of all well-authenticated meteorites known up to January 1984, even if no material has been preserved. Since the publication of the third edition there have been significant changes in the science of meteoritics. Several excellent texts on the subject are now available and any attempt to summarize their more general aspects would be out of place in a catalogue. This introduction will therefore be confined to a short account of the nomenclature and classification of meteorites and the system of entries adopted.

Meteorites are named from their place of find or fall, which is usually the nearest inhabited place to the actual site. Meteorite names are labels, of no intrinsic significance. To avoid confusion it is desirable that the name, once given, should remain unchanged though the name of the place of fall may be altered or even though the name was originally given in error. Studies of geographical distribution are best served by the latitude and longitude, and each entry includes these wherever possible. The rules for naming newly recovered meteorites have been standardized by the Nomenclature Committee of the Meteoritical Society and are published in the Society's journal (Meteoritics, 1980, *15*, p. 102). Existing names have not been changed to agree with these rules. Where two or more distinct meteorites have been recovered from one locality and there are no further suitable place-names available, lower case letters, in brackets, follow the locality name to distinguish the meteorites, for example Tulia (a) and Tulia (b).

The recording of synonyms has been given special attention; some names may cover two or three distinct falls, and the accepted name for one meteorite may be a synonym of two or more others (for example, Saratov, Washington County, Salem). In the past distinct falls with the same accepted name have been distinguished, where possible, as "iron," "stone," or by amplifying the name of the place of fall (for example, Colby (Kansas) and Colby (Wisconsin); or where neither of these methods was applicable, by adding the date of fall or find to the name (for example Duel Hill (1854) and Duel Hill (1873)). Simple numbering (Barratta no. 1, no. 2, etc.) is reserved wherever possible to distinguish specimens of the same fall. Technical advances have helped to eliminate doubt in identifying individuals as "paired", that is those that possibly belong to the same fall. Some individuals or fragments have been shown conclusively to be parts of the same fall (i.e. synonymous) but others to belong to distinct falls. For meteorites from countries using the Cyrillic alphabet, the Latin spelling most commonly used in the literature is retained for the older falls, while names of more modern falls, for which no accepted Latin form can be said to exist, are transliterated according to the British Standard system. Older Chinese names have retained their Wade-Giles Latinization while the more recently reported falls have been named using the Pin-yin scheme. None of the falls recorded in the classical Chinese literature has been included. A recent compilation of these has been published by R. Xuan and X. Xia (Geochimica, 1982, p. 422). These records were previously collected by De Guignes (*Voyages à Peking, etc.*, Paris, 1808), by Abel-Remusat (*Mélanges asiatiques*, Paris, 1825), and by E. Biot (Catalogues des étoiles filantes et autre meteores observés en Chine. Mémoires de l'Academie des Sciences de l'Institute de France, 1848); see also P. A. Kesselmeyer (*Über den Ursprung der Meteorsteine*, Frankfurt-a.-M., 1860). A few falls mentioned in the ancient Greek or Roman literature have also been omitted because in no case is the evidence of meteoritic nature adequate.

The recent discovery of large numbers of meteoritic specimens in Antarctica has posed problems both in the traditional methods of nomenclature and also in the preparation of this catalogue. The names given to these specimens consist of a locality and number, the first two digits of which refer to the December year of the field season of recovery. It has proved difficult to assess the number of individual falls which are represented by the recovered material and consequently not all the individually numbered specimens have been included. The Catalogue contains individual entries for each locality and for each expedition season in Antarctica. Further, all the achondrites, enstatite and carbonaceous chondrites, stony-irons and irons recovered have been included. For some expeditions and localities it has proved necessary to restrict the number of ordinary chondrite entries, for example the 1977–78 season in the Allan Hills area during which 300 specimens were recovered and the 1979–80 season in the Yamato area in which over 3300 specimens were found. The ordinary chondrites recovered by these two expeditions are included here only if the individual mass found is greater than 500g. This weight was chosen in these cases to give the widest coverage of material that is likely to be available for research in reasonable amounts. In other areas with smaller collections the weight limit is much lower; for example for the 1979–80 expedition in the Reckling Peak locality, which recovered 15 specimens, all masses over 70g are included.

In all cases, the latitude and longitude are appended: where possible these are the co-ordinates of the actual site of fall or find, but more commonly they are those of the name-place. In many instances the name-place is a small village not to be found on the maps available, and then the co-ordinates of the nearest place shown in the Times Atlas are cited, or approximate values are given.

The date and time of fall are given for observed falls; for the most part these times are probably local times though this is not often explicitly stated in the literature. Where the time of fall is reported in Universal Time (Greenwich Mean Time) this is indicated (U.T.).

No attempt has been made towards a complete bibliography; the list of repositories aims at recording the whereabouts of specimens exceeding 5 per cent. of the total known mass, but this has not been consistently adhered to, nor is there any reason to believe that the records are complete. The literature cited is intended to cover details of fall or find, good descriptions and chemical analyses.

The chondritic meteorites have been classified on a chemical and textural basis by W. R. Van Schmus and J. A. Wood (Geochimica et Cosmochimica Acta, 1967, *31*, p. 747) and a chemical and structural classification of the irons is given by J. T. Wasson (*Meteorites*, Springer-Verlag, Berlin, 1974) and by V. F. Buchwald (*Handbook of Iron Meteorites*, Universities of California and Arizona State, Berkley, 1975). The carbonaceous chondrites are classified according to the scheme of W. R. Van Schmus and J. M. Hayes (Geochimica et Cosmochimica Acta, 1974, *38*, p. 47) and J. T. Wasson (*loc. cit.*). Where chemical or mineralogical data relevant to the classification are available they are included in the entry for that meteorite, for example Ni, Ga, Ge, Ir contents of irons and total Fe and/or olivine composition for stones.

The system of classification adopted here includes that of G. T. Prior (Mineralogical Magazine, 1920, *19*, p. 51) as modified by B. Mason (*Meteorites*, New York: Wiley, 1962), B. Mason and H. B. Wilk (Geochimica et Cosmochimica Acta, 1964, *28*, p. 533), with that of the carbonaceous chondrites and irons as modified by J. T. Wasson (*loc. cit.*). The three main divisions of irons, stony-irons and stones are subdivided as follows:

IRONS are classified according to chemistry. Currently thirteen chemical groups of irons are recognized on their bulk contents of the elements Ni, Ga, Ge and Ir (*IAB, IC, IIAB, IIC, IID, IIE, IIF, IIIAB, IIICD, IIIE, IIIF, IVA, IVB*). In addition, structural types are recognized which generally, but not exclusively, parallel bulk Ni content:

ataxites (D), have Ni contents normally greater than 16 weight % and often sub-microscopic structures;

octahedrites (O), have generally 7–12 weight % Ni; they are further subdivided according to mean true band-width of their kamacite lamellae into coarsest (Ogg) >3.3mm, coarse (Og) 1.3–3.3mm, medium (Om) 0.5–1.3mm, fine (Of) 0.2–0.5mm, finest (Off) <0.2mm continuous, and plessitic (Opl) <0.2mm spindles; and

hexahedrites (H), have Ni contents of 5–6 weight % and consist of crystals of kamacite.

Anomalous and ungrouped irons are those having structures or chemistries which do not conform to the above groups.

STONY-IRONS are usually divided into three classes:

pallasites, consisting of grains of olivine (often single crystals and up to 1cm across) embedded in a matrix of nickel–iron;

mesosiderites, a variable group consisting of approximately equal amounts of metal and silicates, often with much troilite; the metal does not form a continuous network as in the pallasites, and the silicates are mainly hypersthene and plagioclase; and

lodranites, consisting of approximately equal amounts of metal, olivine and pyroxene.

Some pallasites exhibit a considerable volume of silicate-free metal, and some irons carry considerable amounts of silicate inclusions.

STONY meteorites are divided into two major classes according to the presence or absence of the peculiar structures known as chondrules, and these classes are further sub-divided into groups according to the composition of the silicates.

The CHONDRITES, besides containing chondrules, all carry a variable amount of free metal, except for the carbonaceous chondrites, in some of which not only is metal absent, but the chondrules may be absent or ill-developed. Chondrites are divided into five groups:

enstatite chondrites, (E–group), with an enstatite close to pure $MgSiO_3$ as the dominant silicate, carry large amounts of metal (13–25 weight %) which is low in nickel and may contain appreciable silicon;

olivine-bronzite chondrites, (H–group), contain roughly equal amounts of an olivine with 15–20 mol. % Fe_2SiO_4 and bronzite, together with about 16–21 % free metal with a nickel content of 7–12%;

olivine-hypersthene chondrites, (L–group), contain an olivine with 21–26 mol. % Fe_2SiO_4 and hypersthene, together with about 7–12 % free metal;

olivine-hypersthene chondrites, amphoterites, (LL–group), contain an olivine with 27–32 mol. % Fe_2SiO_4, and less than about 7% free metal which is Ni- and Co-rich; and

carbonaceous chondrites, (C–group), contain little or no free metal and form a more heterogeneous group than those above. They are sub-divided into types CI (I for Ivuna), CM (M for Mighei), CV (V for Vigarano) and CO (O for Ornans). CI is equivalent to the type I of H. B. Wiik (Geochimica et Cosmochimica Acta, 1956, *9*, p. 279), CM is equivalent to his type II and CV and CO are together equivalent to his type III. The differences in chemistry between these are sufficiently large to warrant their consideration as groups in their own right. The 'super-group' of carbonaceous chondrites is retained here since significant aspects of the chemistry of these meteorites require that they are closely related (for example, oxygen isotope systematics).

In the older Tschermak-Brezina classification, considerable importance was attached to the colour and physical structure of meteorites; these characteristics appear generally to be of less significance than the chemical composition. Some of the appropriate adjectives (black, veined, brecciated) have been retained because they give some indication of the degree of shock suffered by the meteorites, or the retention or otherwise of rare gases.

The ACHONDRITES are a heterogeneous class, lacking the chondrules which characterize the commoner class of stony meteorites. They are generally more coarsely crystallized than the chondrites, and nearer in chemical and mineralogical composition and in structure to terrestrial igneous rocks. The achondrites are usually divided into two sub-classes, one of which is calcium-poor with less than 3% CaO, the other being calcium-rich with 5% or more CaO.

The CALCIUM-POOR ACHONDRITES are further sub-divided according to their mineralogical composition into:

enstatite achondrites (aubrites, AUB), consisting of an enstatite very near pure $MgSiO_3$, together with small amounts of accessory minerals;

hypersthene achondrites (diogenites, ADIO), consisting of a fairly iron-rich orthopyroxene with small amounts of accessory minerals; and

olivine-pigeonite achondrites (ureilites, AURE), a group containing minor carbon as finely divided graphite and as diamond.

The CALCIUM-RICH ACHONDRITES are, in the main, basaltic achondrites (*eucrites*, AEUC and *howardites*, AHOW). They consist of plagioclase (An_{80-97}) and pyroxene, dominantly pigeonite in the eucrites and hypersthene in the howardites. In a minor group, the shergottites, plagioclase is more sodic (about An_{50}) and present as a glassy pseudomorph, maskelynite.

Some stones defy classification into the above classes and have, in the past, been assigned to classes containing one to three individuals. Here the designation anomalous (ANOM) is used for such stones, ACANOM for ungrouped achondrites, for example Nakhla, and CHANOM for ungrouped chondrites, for example Kakangari.

In this edition the tabulation of the minor and trace elements in meteorites has been omitted. These data are now available in other publications, for example; B. Mason (ed.), *Handbook of Elemental Abundances in Meteorites*, Gordon and Breach, 1971; B. Mason, Cosmochemistry, Part 1, Meteorites. In *Data of Geochemistry* (ed. M. Fleischer), 6th edn., U.S. Geological Survey Professional Paper 440–B–1, 1979.

The Catalogue now contains entries for 2611 reasonably authenticated meteorites of which 1435 are represented in the collection of the British Museum (Natural History). A further 173 doubtful falls are also listed.

Appended to the catalogue of meteorites is a list of known or suggested meteorite craters. The list of prepared sections of meteorites in this Museum and in other institutions has also been included, as has the list of tektites in the British Museum (Natural History) Collection.

The Catalogue is now maintained as a computer file and Dr D. B. Williams is thanked for the help he has given in its production. In spite of careful checking there will inevitably be mistakes and omissions; corrections and new information are welcomed for incorporation in any future edition.

Summary of Classified Meteorites

The following table is a summary of the numbers of meteorites currently classified and is derived from the main body of the catalogue. In using these data for statistical purposes it should be noted that there is only limited inclusion of the chondritic meteorites found in Antarctica since 1977.

Class	*Total*	*Falls*	*Finds*	*BM(NH)*
IRONS				
IAB	107	6	101	73
IC	11	0	11	10
IIAB	68	5	63	42
IIC	7	0	7	5
IID	15	3	12	13
IIE	14	1	13	8
IIF	5	1	4	1
IIIAB	197	8	189	140
IIICD	21	2	19	11
IIIE	13	0	13	9
IIIF	6	0	6	4
IVA	56	3	53	32
IVB	12	0	12	12
IRANOM	78	5	73	50
Unclassified	115	8	107	11
TOTAL IRONS	725	42	683	421
STONY-IRONS				
Mesosiderites	32	6	26	23
Pallasites	39	3	36	24
Lodranites	2	1	1	1
TOTAL STONY-IRONS	73	10	63	48
STONES				
Achondrites				
Anomalous	10	4	6	6
Aubrites	11	9	2	10
Diogenites	15	9	6	8
Eucrites	55	25	30	22
Howardites	24	18	6	17
Ureilites	17	4	13	10
Total	132	69	63	73
Chondrites				
Enstatite (E-group)	24	13	11	18
Anomalous	7	3	4	5
Olivine-bronzite (H-group)				
H (no type)	82	26	56	27
H3	25	8	17	12
H4	157	50	107	83
H5	263	117	146	169
H6	154	75	79	99
Total H-group	681	276	405	390
Olivine-hypersthene (L-group)				
L (no type)	66	24	42	19
L3	21	8	13	10
L4	53	21	32	29
L5	105	53	52	73
L6	424	213	211	260
Total L-group	669	319	350	391
Amphoterites (LL-group)				
LL (no type)	5	2	3	1
LL3	14	10	4	7
LL4	10	7	3	5
LL5	20	15	5	14
LL6	47	32	15	30
Total LL-group	96	66	30	57
Carbonaceous (C-group)				
C (no type)	2	0	2	0
CI	5	5	0	2
CM2	33	18	15	13
CO3	10	5	5	7
CV3	14	6	8	7
C4	1	0	1	1
C5	2	1	1	2
Total C-group	67	35	32	32
Stones (no class)	137	72	65	0
Total chondrites	1681	784	897	893
TOTAL STONES	1813	853	960	966
Total authenticated meteorites	2611	905	1706	1435
Doubtful				
Unclassified	98	–	–	0
Irons	22	13	9	0
Stones	51	40	11	0
Stony-irons	2	1	1	0
Total	173	54	21	0
GRAND TOTAL	2784	959	1727	1435

A Summary of the Geographical Listing

Country	*Total*	*Falls*	*Finds*	*Doubtful*	*BM(NH)*
Afghanistan	2	1	0	1	0
Algeria	18	6	12	0	9
Angola	3	2	1	0	2
Argentina	54	20	32	2	28
Australia	218	12	204	2	104
Austria	7	3	2	2	1
Bangladesh	8	8	0	0	8
Belgium	7	4	0	3	3
Bolivia	2	0	2	0	1
Brazil	48	21	22	5	20
Bulgaria	5	5	0	0	2
Burma	3	3	0	0	2
Cambodia	2	2	0	0	1
Cameroun	2	2	0	0	0
Canada	49	9	37	3	21
Central African Republic	1	1	0	0	0
Chad	2	1	1	0	1
Chile	36	0	35	1	27
China	68	35	32	1	2
Colombia	1	0	1	0	1
Costa Rica	1	1	0	0	1
Cuba	1	0	1	0	0
Czechoslovakia	24	16	8	0	18
Denmark	5	3	1	1	2
Egypt	9	2	5	2	4
England	21	10	0	11	8
Ethiopia	5	5	0	0	1
Finland	12	5	7	0	6
France	69	60	4	5	54
Germany (East and West)	61	28	13	20	28
Greece	6	1	0	5	1
Greenland	3	0	3	0	3
Guatemala	1	0	1	0	1
Honduras	2	0	1	1	1
Hungary	11	6	1	4	2
India	124	111	7	6	94
Indonesia	17	15	1	1	11
Iran	3	2	0	1	2
Iraq	4	2	1	1	3
Ireland (Republic and Northern)	7	7	0	0	6
Italy	47	27	3	17	19
Ivory Coast	1	0	0	1	0
Jamaica	1	0	1	0	1
Japan	49	31	10	8	15
Jordan	1	1	0	0	1
Kenya	3	2	0	1	2
Korea	4	3	1	0	0
Lebanon	1	0	0	1	0
Libya	2	0	2	0	1
Malawi	5	4	1	0	3
Mauritania	4	2	2	0	3
Mexico	71	14	55	2	49
Mongolia	10	4	4	2	0
Morocco	2	1	1	0	1
Namibia	13	2	11	0	3
Netherlands	5	3	0	2	3
New Caledonia	1	1	0	0	0
New Zealand	8	1	7	0	6
Niger	8	6	2	0	2
Nigeria	12	10	2	0	9
Norway	11	8	3	0	5
Oman	7	0	7	0	7
Pakistan	15	14	0	1	13
Papua New Guinea	2	2	0	0	0
Paraguay	2	1	0	1	0
Peru	1	0	1	0	1
Philippines	4	3	1	0	3
Poland	23	9	7	7	9
Portugal	8	5	2	1	2
Romania	8	7	1	0	6
Rwanda	1	1	0	0	0
Saudi Arabia	25	4	19	2	9
Scotland	6	3	0	3	3
Somalia	3	2	1	0	2
South Africa	46	21	23	2	32
South Yemen	1	1	0	0	0
Spain	32	22	3	7	14
Sri Lanka	1	1	0	0	0
Sudan	6	6	0	0	6
Swaziland	1	1	0	0	1
Sweden	15	9	4	2	6
Switzerland	7	3	1	3	3
Syria	3	2	0	1	1
Tanzania	8	7	1	0	3
Thailand	2	2	0	0	1
Tibet	1	1	0	0	0
Tunisia	2	2	0	0	1
Turkey	13	8	1	4	2
Uganda	3	3	0	0	2
Upper Volta	6	6	0	0	2
U.S.A.	920	116	795	9	583
U.S.S.R.	176	97	75	4	90
Venezuela	1	1	0	0	1
Vietnam	3	3	0	0	0
Wales	3	2	0	1	2
Yugoslavia	12	10	2	0	8
Zaire	6	5	1	0	2
Zambia	1	1	0	0	1
Zimbabwe	1	1	0	0	1

Classified List of Meteorites

In the following list the known meteorites have been divided into irons, stony-irons and stones and then further sub-divided by chemical group. Doubtful falls and pseudometeorites are listed separately. An asterisk (*) following the name indicates that the meteorite is represented in the BM(NH) collection. An obelisk (†) following the name indicates an observed fall.

Iron

Adzhi-Bogdo (iron)
Agua Blanca *
Alatage
al-Ghanim (iron)
Allan Hills A80104
Allan Hills A81013
Allan Hills A81014
Ameca-Ameca
Answer
Anyujskij
Arabella
ar-Rakhbah
Bagnone
Benares (b)
Bilibino
Blue Tier
Botetourt County
Campo de Pucara (iron)
Castray River
Cleburne
Clinton
Coldwater (iron) *
Colorado Springs
Czestochowa Rakow I
Czestochowa Rakow II
Dadin
Dellys
Dor el Gani
Duan
Dunganville *
Ellisras
El Simbolar *
El Timbu
Espiritu Santo
Fengzhen
Fillmore
Fort Stockton
Fujian
Fukue
Fuling
Gerzeh
Guadaloupe County
Guangyuan
Guin
Guizhou
Hatfield
Hejing
Holliday
Jalandhar †
King Solomon
Komagome †
Krzadka
Kumdah
Lafayette (iron)
Laguna Manantiales *
Lancaster County
Landor
Lasher Creek
Las Salinas
Lebedinnyi
Lefroy
Los Sauces
Lujan
Lusk *
Majorca †
Manlai
Mariaville †
Mar'inka
Masua *
Midland
Milnesand (iron)
Minnesota (iron)
Mrirt
Murchison Downs
Naifa *
Nashville (iron)
Nejed (no. 2)
New Mexico
Nochtuisk *
Norin-Shibir
Northampton
Nyaung † *
Opava
Palinshih †
Parma Canyon
Patos de Minas (octahedrite)
Patti †
Paulding County
Pei Xian
Red Willow
Saginaw
Sakauchi †
Salina
Saotome
Selcany
Shohaku
South African Railways
South Dahna
Southern Michigan
Suwa
Teocaltiche
Tianlin
Un-named
Ur *
Vicenice
Wei-hui-fu (a)
Wei-hui-fu (b)
Wietrzno-Bobrka
Wu-chu-mu-ch'in
Xiquipilco no. 2
Zapata County
Zhigansk
Zhongxiang

Anomalous / ungrouped iron

Alikatnima
Allan Hills A77255
Babb's Mill (Blake's Iron)
Babb's Mill (Troost's Iron) *
Bacubirito *
Barbacena *
Bellsbank *
Bocaiuva
Britstown *
Butler *
Cambria *
Chebankol
Cowra *
Cruz del Aire
Deep Springs *
Dehesa *
De Hoek
Denver City *
Dermbach
Dorrigo
El Qoseir
Elton
Emsland
Garden Head
Glenormiston *
Glen Rose (iron)
Grand Rapids *
Guffey
Gun Creek *
Hammond *
Illinois Gulch *
Kendall County *
Kingston *
Klondike (Gay Gulch)
Kofa *
La Caille *
La Primitiva *
Laurens County *
Lazarev
Leshan
Lime Creek *
Linville *
Livingston (Tennessee)
Mbosi *
Morradal *
Mount Magnet *
Mundrabilla *
Muzaffarpur † *
Nedagolla † *
New Baltimore *
N'Goureyma † *
Nordheim
Petropavlovsk *
Piedade do Bagre *
Piñon *
Prambanan *
Quesa †
Rafrüti *
Redfields *
Reed City *
Santa Catharina *
Santiago Papasquiero *
Shingle Springs *
Sombrerete
Soper *

Soroti † *
South Byron *
Tishomingo
Tombigbee River *
Tucson *
Twin City
Ventura
Victoria West *
Washington County
Waterville *
Yamato 75031
Ysleta
Zacatecas (1792) *

I
Alexander County
Cuba
Monturaqui
Niagara *
Sierra Blanca *

I?
Tabarz *

IA
Allan Hills A76002 *
Allan Hills A77283
Balfour Downs
Bischtübe *
Black Mountain *
Bogou †
Bohumilitz *
Bolivia *
Burgavli
Burkett *
California
Campo del Cielo *
Cañon Diablo *
Casey County *
Comanche (iron)
Cookeville *
Coolac *
Copiapo *
Cosby's Creek *
Cranbourne *
Deelfontein
Deport *
Duel Hill (1873) *
Dungannon *
Ellicott *
Gladstone (iron) *
Goose Lake *
Haniet-el-Beguel *
Harlowton
Hasparos
Hope
Idaho
Itapuranga
Jenkins
Jenny's Creek *
Kaalijarv *
Landes *
Leeds *
Lexington County *
Linwood *
Magura *
Mayerthorpe
Mazapil † *
Misteca *
Moctezuma *
Morasko *
Mount Ayliff *
Nagy-Vázsony *
Neptune Mountains
New Leipzig *
Nieder Finow
Odessa (iron) *
Ogallala *
Oscuro Mountains *
Osseo *
Ozren *
Pan de Azucar *
Paracutu
Pooposo *
Purgatory Peak A77006
Reckling Peak A80226
Rosario *
Sardis *
Sarepta *
Seeläsgen *
Seligman *
Seymour *
Shrewsbury *
Silver Crown *
Smithville *
Southern Arizona
Surprise Springs *
Tacoma
Thoreau
Toluca *
Udei Station † *
Vaalbult *
Waldron Ridge *
Wichita County *
Yardea
Yardymly †
Yenberrie *
Yongning
Youndegin *

IA-ANOM
Annaheim *
Bahjoi † *
Karee Kloof *
Mertzon
Morrill *
Pine River *
Zenda

IA?
Morden

IB
Bitburg *
Colfax *
Four Corners *
Mesa Verde Park
Oktibbeha County *
Persimmon Creek
Pitts †
San Cristobal *

IB-ANOM
Woodbine *

IC
Arispe *
Bendegó *
Chihuahua City *
Etosha
Mount Dooling *
Murnpeowie *
St. Francois County *
Santa Rosa *
Union County *

IC-ANOM
Nocoleche *
Winburg *

IIA
Allan Hills A78100
Avce †
Bennett County *
Bingera *
Boguslavka † *
Braunau † *
Bruno *
Calico Rock *
Carver
Cedartown
Chesterville *
Chico Mountains
Cincinnati *
Coahuila *
Cosmo Newberry
Edmonton (Canada) *
El Mirage
Forsyth County *
Gressk *
Hex River Mountains *
Holland's Store *
Indian Valley *
Keen Mountain
Kopjes Vlei *
Lick Creek *
Locust Grove *
Lombard *
Mayodan
Mejillones
Murphy *
Negrillos *
North Chile *
Okahandja
Okano † *
Patos de Minas (hexahedrite)
Pima County
Pirapora *
Richland
San Francisco del Mezquital *
Scottsville *
Sierra Gorda
Siratik *
Smithonia *
Uwet *
Walker County *
Wathena
Yamato 75105
Yarroweyah *

IIAB
Squaw Creek

IIB
Ainsworth *
Alkhamasin
Derrick Peak A78001
El Burro *
Iredell
Jerslev
Lake Murray
Mount Joy *
Navajo *
Nenntmannsdorf *
North Portugal
Old Woman
Sandia Mountains *
Santa Luzia *
São Julião de Moreira *
Sikhote-Alin † *
Silver Bell
Smithsonian Iron *

Summit *

IIC

Ballinoo *
Cratheús (1950)
Kumerina *
Perryville *
Salt River *
Unter-Mässing
Wiley *

IID

Alt Bela
Bridgewater *
Brownfield (iron) *
Carbo *
Elbogen † *
Hraschina †
Losttown *
Mount Ouray *
Needles *
N'Kandhla † *
Puquios *
Richa *
Rodeo *
Wallapai *

IID-ANOM

Arltunga *

IIE

Arlington *
Barranca Blanca *
Colomera
Egvekinot
Elga
Garhi Yasin † *
Lonaconing *
Seymchan
Techado
Tobychan
Verkhne Dnieprovsk *
Weekeroo Station *

IIE-ANOM

Kodaikanal *
Netschaëvo *

IIF

Corowa
Del Rio
Dorofeevka
Monahans *
Repeev Khutor †

IIIA

Aggie Creek
Aliskerovo
Angelica
Asarco Mexicana
Aswan
Augusta County *
Avoca (Western Australia) *
Bagdad
Bartlett
Bear Lodge *
Benedict *
Billings *
Bluewater
Boxhole *
Briggsdale *
Cabin Creek † *
Cacaria *
Canton *
Canyon City *
Caperr *
Cape York *
Carthage *
Casas Grandes *
Casimiro de Abreu
Chambord
Charcas *
Chilkoot *
Chulafinnee *
Costilla Peak *
Cowell *
Cumpas *
Dalton *
Davis Mountains *
Denton County *
Dexter *
Dimitrovgrad *
Drum Mountains
Duketon *
Durango *
El Sampal
Elyria *
Emmitsburg *
Fort Pierre *
Franceville *
Frankfort (iron) *
Glasgow *
Gnowangerup
Greenbrier County *
Guilford County *
Guixi
Gundaring *
Haig *
Harriman (Om)
Hayden Creek *
Henbury *
Ider *
Ilimaes (iron) *
Iron Creek *
Itutinga
Ivanpah *
Jackson County *
Jianshi †
Joel's Iron *
Juncal *
Juromenha †
Kalkaska
Kapunda
Kayakent † *
Kenton County *
Kifkakhsyagan
Kyancutta *
Lanton *
La Porte *
Lenarto *
Liangcheng
Livingston (Montana) *
Loreto
Lucky Hill *
Madoc *
Maldyak
Manitouwabing
Mapleton *
Marshall County *
Merceditas *
Milly Milly *
Moorumbunna *
Morito *
Nazareth (iron) *
Norfolk *
Norfork † *
Nova-Petropolis
Nuleri
Picacho *
Plymouth *
Point of Rocks (iron)
Providence
Puente del Zacate *
Quartz Mountain *
Quinn Canyon
Rancho de la Pila (1882) *
Rateldraai *
Red River *
Red Rock
Roebourne *
Rowton † *
Ruff's Mountain *
Russel Gulch *
Sacramento Mountains *
Samelia † *
San Angelo *
Sanclerlandia
Sandtown *
Santa Apolonia *
Savannah
Schwetz *
Seneca Falls *
Sierra Sandon *
Slaghek's Iron
Spearman *
Ssyromolotovo *
Susuman *
Tamarugal *
Tamentit *
Tendo
Thule *
Thunda *
Tonganoxie *
Toubil River *
Trenton *
Uegit
Uwharrie
Veliko-Nikolaevsky Priisk *
Verissimo
Verkhne Udinsk *
View Hill *
Wabar *
Waingaromia *
Welland *
Willamette *
Yarri
York (iron)
Youanmi *

IIIA-ANOM

Murfreesboro *
Zerhamra *

IIIA?

Lismore
Withrow
Wooster *

IIIAB

Karasburg

IIIB

Apoala *
Aprel'sky
Asheville *
Augustinovka *
Bald Eagle *
Baquedano *
Bear Creek *
Bella Roca *
Brainard

Buenaventura
Campbellsville *
Chupaderos *
Cleveland *
El Capitan *
Floydada
Grant *
Hopper
Ilinskaya Stanitza *
Joe Wright Mountain *
Kinsella
Knowles *
Kouga Mountains *
Kuga
Los Reyes *
Luis Lopez *
Mount Edith *
Narraburra *
Norristown
Orange River (iron) *
Oroville *
Owens Valley *
Rancho Gomelia
Roper River
Sam's Valley *
Sanderson *
Smith's Mountain *
Tambo Quemado *
Tepla *
Thurlow *
Tieraco Creek
Turtle River *
Wimberley
Wolf Creek *
Wonyulgunna *
Zacatecas (1969)

IIIB-ANOM
Delegate *
Treysa † *

IIIB?
Norquín
Pierceville (iron) *

IIICD
Algoma *
Anoka
Ballinger *
Carlton *
Corrizatillo
Dayton
Föllinge
Freda
Havana
Jaralito
Lamesa *
Mungindi *
Nantan
Pittsburg *
Tazewell *
Wedderburn

IIICD-ANOM
Edmonton (Kentucky)
Hassi-Jekna † *
Magnesia † *
Quarat al Hanish *
Zaffra *

IIIE
Armanty
Burlington *
Cachiyuyal *
Colonia Obrera
Coopertown *
Kokstad *
Paloduro
Paneth's Iron *
Rhine Villa *
Shangdu
Staunton *
Tanokami Mountain *
Willow Creek *

IIIF
Clark County *
Klamath Falls
Moonbi *
Nelson County *
Oakley (iron)
St. Genevieve County *

IVA
Allan Hills A78252
Altonah *
Bishop Canyon *
Bodaibo *
Boogaldi *
Bristol *
Bushman Land
Charlotte † *
Cratheús (1931) *
Duchesne *
Fuzzy Creek
Gibeon *
Guanghua
Harriman (Of)
Hill City
Huangling
Huizopa *
Iron River
Jamestown *
Jonesboro
La Grange *
Mantos Blancos *
Maria Elena (1935) *
Mart *
Millarville *
Mount Sir Charles
Muonionalusta *
New Westville
Ningbo †
Novorybinskoe
Obernkirchen *
Otchinjau *
Para de Minas
Puerta de Arauco
Putnam County *
Rembang † *
Rica Aventura
San Francisco Mountains *
Seneca Township *
Serrania de Varas *
Shirahagi *
Signal Mountain
Smithland *
Social Circle
Western Arkansas *
Wood's Mountain *
Yanhuitlan *
Yingde
Yudoma

IVA-ANOM
Chinautla *
Duel Hill (1854) *
Longchang
São João Nepomuceno
Steinbach *

IVA?
Alvord
Cranberry Plains

IVB
Cape of Good Hope *
Hoba *
Iquique *
Klondike (Skookum Gulch) *
Kokomo *
Santa Clara *
Tawallah Valley *
Ternera *
Tlacotepec *
Warburton Range *
Weaver Mountains *

IVB-ANOM
Chinga *

Iron?
Bulls Run †
El Chiflón

Not reported
Paso Rio Mayo

Stone
Adzhi-Bogdo (stone) †
Aiken
al-Jimshan
Andhara †
Arrabury
as-Sanam
Barcelona (stone) †
Barntrup †
Belville †
Bethel
Boolka
Bronco
Brownell
Cacak †
Caldwell
Calivo †
Campo Perrando
Castel Berardenga †
Castrovillari †
Caswell County †
Changxing
Chetrinahatti †
Clarendon
Colonia Suiza †
Crocker's Well
Dharwar †
Divnoe
Dora (stone)
Dowa †
Dunhua †
Elida
Ferguson †
Ferintosh
Fort Flatters †
Fünen †
Fuyang †
Gue†a †
Gyokukei †
Hachi-oji †
Hamilton (Texas)
Harrison Township
Hart Camp

Hatford †
Hechi
Hereford
Holetta †
Hoxie †
Ibrisim †
Idalia
Jiapigou
Jilong
Julesburg
Kamyshla
Khenteisky †
Kijima (1906) †
Konovo †
Krutikha †
Kushiike †
La Charca †
Lichtenberg †
Loot
Los Martinez †
Lost Draw
Lucerne Valley
Luray
Malaga †
Matad
Meerut †
Milnesand (stone)
Minamino †
Minnichhof †
Mount Vaisi †
Mtola †
Mühlau
Myersville
Myhee Caunta †
Natal
Needmore
New Moore
Nicorps †
Novy-Ergi †
Novy-Projekt †
Oasis State Park
Oliva-Gandia †
Olton
Ortenau †
Owrucz †
Palahatchie †
Pep
Pettiswood †
Phulmari
Picote †
Pine Bluffs
Plainview (c)
Pony Creek
Portugal †
Po-wang Chen †
Preobrazhenka
Ratyn †
Richland Springs †
Ridgecrest
Rivolta de Bassi †
Rockhampton †
Rodach †
Round Top
Sabetmahet †
Sagan †
St. Vrain
Sarratola
Selden
Severny Kolchim
Shiraiwa
Simbirsk (of P. Partsch)
Soheria †
Sopot †
Stolzenau †
Stratton
Stretchleigh †
Taicang
Thurman
Tounkin †
Tres Estacas
Truckton
Tulung Dzong †
Two Buttes (b)
Unkoku †
Valdinoce †
Villarrica †
Wauneta
Wellman (b)
Wolamo †
Wuzhi †
Yamato 6905
Yoshiki †
Zhuanghe †

Achondrite

Anomalous

Allan Hills A77005
Allan Hills A81005
Angra dos Reis (stone) † *
Brachina
Chassigny † *
Governador Valadares *
Lafayette (stone) *
Nakhla † *
Peckelsheim †
Tierra Blanca *

Aubrite

Allan Hills A78113
Aubres † *
Bishopville † *
Bustee † *
Cumberland Falls † *
Khor Temiki † *
Mayo Belwa † *
Norton County † *
Peña Blanca Spring † *
Pesyanoe † *
Shallowater *

Diogenite

Aioun el Atrouss † *
Allan Hills A77256
Elephant Moraine A79002
Ellemeet † *
Garland †
Ibbenbüren † *
Johnstown † *
Manegaon † *
Roda † *
Shalka † *
Tatahouine † *
Yamato 6902
Yamato 74013
Yamato 75032
Yamato 75299

Eucrite

Adalia
Allan Hills A76005 *
Allan Hills A78040
Allan Hills A78132
Allan Hills A79017
Allan Hills A80102
Allan Hills A81001
Allan Hills A81009
Allan Hills A81011
Béréba † *
Bouvante
Brient †
Cachari *
Chervony Kut †
Elephant Moraine A79001
Elephant Moraine A79004
Elephant Moraine A79005
Elephant Moraine A79006
Emmaville † *
Haraiya † *
Ibitira †
Jonzac † *
Juvinas † *
Kirbyville †
Lakangaon † *
Macibini †
Medanitos † *
Millbillillie † *
Moama *
Moore County † *
Nagaria † *
Nobleborough † *
Nuevo Laredo *
Padvarninkai † *
Palo Blanco Creek
Pasamonte † *
Pecora Escarpment 82501
Pecora Escarpment 82502
Peramiho †
Pomozdino
Reckling Peak A80204
Reckling Peak A80224
Serra de Magé † *
Shergotty † *
Sioux County † *
Stannern † *
Thiel Mountains 82403
Vetluga †
Yamato 74159
Yamato 74356
Yamato 74450
Yamato 790122
Yamato 790260
Yamato 790266
Zagami † *

Howardite

Allan Hills A77302
Allan Hills A78006
Bholghati † *
Białystok † *
Binda *
Bununu †
Chaves † *
Erevan †
Frankfort (stone) † *
Jodzie † *
Kapoeta † *
Le Teilleul † *
Luotolax † *
Malvern † *
Mässing † *
Melrose (b) *
Molteno † *
Muckera
Pavlovka † *

Petersburg † *
Washougal † *
Yamato 7308
Yurtuk † *
Zmenj †

Ureilite

Allan Hills A77257
Allan Hills A78019
Dingo Pup Donga *
Dyalpur † *
Goalpara *
Hajmah (a) *
Haverö † *
Kenna *
Lahrauli † *
Nilpena *
North Haig *
Novo-Urei † *
Reckling Peak A80239
Yamato 74123
Yamato 74130
Yamato 74659
Yamato 790981

Chondrite

Anomalous

Acapulco † *
Allan Hills A77081
Carlisle Lakes (a)
Kakangari † *
Mount Morris (Wisconsin) *
Pontlyfni † *
Winona *

C

Yamato 75003

C?

Bench Crater

CI

Alais † *
Ivuna †
Orgueil † *
Revelstoke †
Tonk †

CM2

Adelaide
Allan Hills A77306
Allan Hills 82100
Al Rais † *
Banten †
Belgica 7904
Bells †
Boriskino † *
Cochabamba
Cold Bokkeveld † *
Crescent †
Erakot †
Essebi † *
Haripura †
Kaidun †
Kivesvaara
Lookout Hill *
Mighei † *
Murchison † *
Murray † *
Nawapali † *
Niger (C2)
Nogoya † *
Pollen † *
Renazzo † *
Santa Cruz † *
Yamato 74641
Yamato 74642
Yamato 74662
Yamato 75260
Yamato 790003
Yamato 791824
Yamato 793321

CO3

Allan Hills A77003
Allan Hills A77307
Allan Hills 82101
Colony *
Felix † *
Isna *
Kainsaz † *
Lancé † *
Ornans † *
Warrenton † *

CV3

Allan Hills A80133
Allan Hills A81003
Allende † *
Arch (a)
Bali †
Efremovka *
Grosnaja † *
Kaba † *
Leoville *
Mokoia † *
Reckling Peak A80241
Tibooburra
Vigarano † *
Yamato 6903

C4

Coolidge *

C5

Karoonda † *
Mulga (west) *

E3

Qingzhen †
Yamato 6901

E4

Abee † *
Adhi Kot † *
Indarch † *
Kota-Kota *
Parsa †
Saint-Sauveur † *
South Oman *

E4-5

Bethune *

E5

Reckling Peak A80259
St. Mark's † *
Yilmia *

E6

Allan Hills A81021
Atlanta *
Blithfield *
Daniel's Kuil † *
Happy Canyon *
Hvittis † *
Jajh deh Kot Lalu † *
Khairpur † *
North West Forrest (E6)
Pillistfer † *
Ufana † *

H

Abo
Aguas Calientes
Akhricha
Akwanga †
Anton
Avilez †
Baandee
Bethlehem † *
Brunflo
Canyon
Chail † *
Comanche (stone)
Cranfills Gap
Crosbyton
De Kalb
Dispatch
Distrito Quebracho † *
Djermaia †
Duncanville
Ekh Khera † *
Eunice *
Felt *
Gail
Gao (Upper Volta) †
Gerona
Granada Creek
Gualeguaychú † *
Happy (a)
Hassayampa
Hinojo
Hugo (stone) *
Ishinga †
Kaffir (b) *
Kalvesta
Kangean †
Kanzaki
Karval *
Kasamatsu †
Lider (b)
Lunan †
Malampaka †
Maricopa
Massenya *
Mayday
Mayfield *
McLean *
Mjelleim †
Moshesh
Mount Morris (New York)
Nallah
Naruna
Nazareth (stone) *
Nio †
Okabe † *
Oshkosh
Otomi †
Ouallen *
Pavel †
Phuoc-Binh †
Piquetberg †
Plainview (1950)
Rolla (1941) *
Rosebud
St. Ann *
San Carlos *
Sasagase † *
Shafter Lake
Simmern † *
Sone †
Springer
Takenouchi † *
Tatum

Tokio (a) *
Tostado *
Tromøy †
Uvalde
Valdavur † *
Venus
Wellman (d)
Wray *
Yamato 6907
Yorktown (Texas) *

H3

Allan Hills A78084
Bremervörde † *
Brownfield (1937) *
Clovis (no.1)
Fleming *
Florence †
Frenchman Bay *
Grady (1937) *
Laundry East
Outpost Nunatak A80301
Prairie Dog Creek *
Roosevelt *
Sharps †
Snyder
Suwahib (Buwah) *
Tieschitz † *
Willaroy
Yamato 75028

H3-4

Dhajala † *
Gorlovka †
Happy Draw
Raguli
Tulia (a) *

H3-5

Luponnas † *

H3-6

Hainaut †

H4

Adrian *
Akbarpur † *
Alexandrovsky †
Allan Hills A77004
Allan Hills A77208
Allan Hills A77224
Allan Hills A77225
Allan Hills A77226
Allan Hills A77232
Allan Hills A77262
Allan Hills A81041
Ankober †
Arch (b) *
Aurora *
Avanhandava †
Bath † *
Beaver Creek † *
Bielokrynitschie † *
Birni N'konni †
Bledsoe *
Broken Bow *
Bushnell *
Cashion *
Cedar (Texas) *
Chernyi Bor †
Chico Hills *
Conquista †
Coonana *
Correo
Culbertson *
Cullison *
Cushing
Dimmitt *
Edjudina
Edmonson (b) *
Ekeby †
Elephant Moraine 82602
Elm Creek *
Enigma
Erofeevka *
Eustis
Farmville †
Fayetteville †
Feid Chair †
Forest Vale † *
Garnett *
Glasatovo † *
Gobabeb
Grady (c) *
Grüneberg † *
Gruver *
Guenie †
Gursum †
Gütersloh † *
Hale Center (no. 2) *
Hammond Downs *
Hat Creek *
Hobbs *
Holly *
Holyoke *
Huntsman
Imperial
Jiddat al Harasis *
Kabakly
Kabo † *
Kalaba † *
Kamiomi
Kargapole
Kesen † *
Kiffa † *
Kimbolton
Kittakittaooloo
Lakeview
La Villa *
Lixna † *
Lone Star *
Lone Tree *
Los Lunas
Marilia † *
Markovka
Menow † *
Meteorite Hills A78001
Metsäkylä
Millen
Miller (Kansas) *
Monroe † *
Mosquero
Motta di Conti † *
Mulga (south)
Namib Desert
Nassirah †
Ness County (1938) *
Nikolaevka †
North West Forrest (H)
Noventa Vicentina †
Numakai †
Ochansk † *
Oczeretna *
Odessa (stone) *
Orimattila
Petropavlovka
Phu Hong †
Phum Sambo † *
Polujamki
Portales (a) *
Portales (c) *
Quenggouk † *
Ranchapur † *
Ransom *
Reckling Peak A78002
Reckling Peak A78004
Reid
Rolla (1939) *
Romero *
Ste. Marguerite † *
St. Louis †
Salaices *
San Emigdio *
São Jose do Rio Preto †
Schaap-Kooi *
Seagraves
Selma *
Seminole *
Sena † *
Seneca *
Seres † *
Sete Lagoas †
Silverton (Texas)
Skiff
Slavetic † *
Sylacauga †
Tafoya (a)
Tobe
Tysnes Island † *
Udaipur †
Ulysses *
Ute Creek *
Villedieu
Weldona *
Wellman (c) *
Wellman (e) *
Weston † *
Willowdale *
Wingellina
Witchelina *
Woodward County *
Wynella *
Yamato 7301
Yamato 74155
Yamato 74364
Yamato 790269
Ybbsitz
Yonozu † *

H4-5

Dawn (b)
Little River (b) *
Yamato 74193

H4-6

Muleshoe *

H5

Abajo *
Achilles *
Acme *
Adams County *
ad-Dahbubah
Agen † *
Alamogordo *
Alessandria † *
Allan Hills A77182
Allan Hills A77294

Allan Hills A79025
Allan Hills A79026
Allan Hills A79029
Allan Hills A81015
Allan Hills A81019
Allan Hills A81020
Allan Hills A81042
Allegan † *
Ambapur Nagla † *
Anlung †
Anthony *
Arbol Solo †
Arroyo Aguiar † *
Ashmore *
Assisi † *
Barbotan † *
Barnaul †
Barwise *
Beardsley † *
Beddgelert † *
Bir Hadi
Boaz (stone)
Bogoslovka
Bolshaya Korta *
Bonita Springs *
Borodino † *
Bowden
Breitscheid † *
Brownfield (1964) *
Burdett *
Bur-Gheluai † *
Burnabbie *
Cañellas † *
Cangas de Onis † *
Carcote *
Carichic *
Caroline *
Casilda *
Castalia † *
Cee Vee *
Centerville †
Cereseto † *
Chamberlin *
Changde †
Channing *
Cocklebiddy *
Colby (Kansas) *
Coldwater (stone) *
Collescipoli † *
Cope *
Cosina † *
Cottonwood
Covert *
Cronstad † *
Cross Roads † *
Cuero *
Dalhart *
Darmstadt † *
Dimboola
Dokachi † *
Doyleville *
Dresden (Kansas) *
Dumas (a)
Dundrum † *
Ehole † *
Eichstädt † *
Elba *
Elephant Moraine A79007
Elkhart *
El Perdido
Elsinora *
Enshi †
Épinal † *
Eva *
Faith *
Farley *
Faucett *
Favars † *
Fenbark
Fenghsien-Ku †
Ferguson Switch *
Forest City † *
Forrest (a)
Franklin *
Garrison *
Geidam †
Gilgoin *
Gnadenfrei † *
Grayton
Gross-Divina † *
Grzempach †
Gumoschnik † *
Gunnadorah
Haviland (stone) *
Haviland (b) *
Hawk Springs *
Hedeskoga †
Heredia † *
Hessle † *
Hickiwan
Higashi-koen † *
Horace *
Howe *
Hugoton *
Idutywa † *
Isthilart † *
Itapicuru-Mirim † *
Jiange †
Jilin † *
Kaee † *
Kangra Valley † *
Kansas City (1903) *
Kearney *
Kennard
Kerilis † *
Kielpa *
Kilbourn † *
Kissij *
Krasnyi Klyuch †
Kutais †
Laborel † *
La Colina † *
La Escondida
Laundry Rockhole
Leighton † *
Leon
Lillaverke † *
Limerick † *
Logan *
Lost City † *
Macau † *
Malotas † *
Mardan † *
Marsland *
Meadow (a)
Meester-Cornelis † *
Menindee Lakes 002
Menindee Lakes 005
Merua † *
Miami *
Mianchi †
Miller (Arkansas) † *
Misshof † *
Mooresfort † *
Mornans † *
Morro do Rocio *
Morven *
Nadiabondi †
Nammianthal † *
Nardoo (no.1)
Nuevo Mercurio † *
Nullarbor
Nullagine
Oesede †
Ohaba † *
Oldfield River
Olmedilla de Alarcón † *
Orlovka *
Oro Grande
Palolo Valley †
Panhandle
Pantar † *
Patricia
Pavlodar (stone) †
Penokee
Plains *
Plainview (1917) *
Pleasanton *
Pokhra † *
Poonarunna
Pribram † *
Pulsora † *
Pułtusk † *
Raco † *
Rancho de la Presa † *
Rawlinna (stone)
Reckling Peak A79004
Reckling Peak A79014
Rhineland
Richardton † *
Riverton
Rolla (1936)
Rose City † *
Rush County *
Saline *
Salt Lake City *
San José
Schenectady †
Scott City *
Scurry *
Searsmont † *
Seibert *
Selakopi †
Seldebourak †
Seth Ward
Shields *
Shuangyang †
Sierra County
Sindhri † *
Sitathali † *
Ställdalen † *
Stonington *
Sungach †
Sutton *
Sweetwater *
Tabor † *
Tafoya (b)
Texline *
Thackaringa
Timochin † *

Tomhannock Creek *
Toulon
Travis County *
Two Buttes (a) *
Uberaba † *
Ucera † *
Udipi † *
Utzenstorf † *
Vengerovo † *
Verkhne Tschirskaia † *
Villa Coronado *
Wairarapa Valley
Wellington *
Wellman (a) *
Wessely † *
West Forrest
Whitman *
Wikieup
Wilbia *
Wildara *
Wiluna † *
Witklip Farm † *
Wynyard
Xingyang †
Xinyi
Yamato 6906
Yamato 6908
Yamato 74001
Yamato 74079
Yamato 74115
Yamato 74371
Yamato 74647
Yamato 791209
Yambo †
Yangchiang †
Yatoor † *
Zaisan †
Zebrak † *
Zhovtnevyi † *
Zsadany † *

H5-6

Glanggang † *
Novosibirsk
Smyer

H6

Aarhus † *
Abbott
Akron (1940) *
Allan Hills A76006 *
Allan Hills A76008 *
Allan Hills A77258
Allan Hills A77271
Allan Hills A77288
Allan Hills A78115
Allan Hills A79002
Allan Hills A79016
Andura † *
Archie † *
Belly River
Belmont
Benld †
Benoni † *
Bjelaja Zerkov † *
Blansko † *
Boerne *
Bond Springs
Buckleboo
Butsura † *
Cacilandia
Canon City † *
Canyonlands
Cape Girardeau † *
Capilla del Monte † *
Cavour *
Cedar (Kansas) *
Charsonville † *
Charwallas † *
Chiang Khan † *
Chitado †
Cobija *
Coomandook
Cortez *
Dawn (a)
De Cewsville †
Densmore (1950) *
Desuri † *
Djati-Pengilon † *
Donga Kohrod † *
Doroninsk † *
Dresden (Ontario) †
Dumas (b)
Dwaleni † *
Ellis County *
Erxleben † *
Esnandes † *
Estacado *
Etter *
Florey *
Fluvanna (b) *
Forsbach † *
Galapian †
Georgetown
Gladstone (stone) *
Gopalpur † *
Great Bear Lake
Great Bend
Guareña † *
Haven *
Hungen † *
Ichkala †
Ijopega †
Indio Rico *
Ingalls *
Ipiranga †
Jamkheir † *
Jeedamya
Judesegeri † *
Kadonah † *
Kaldoonera Hill
Kappakoola *
Kernouve † *
Khetri † *
Kikino † *
Killeter † *
Kimble County *
Kingai † *
Klein-Wenden † *
Krider *
Kulp †
Lancon † *
Little River (a) *
Littlerock
Lumpkin † *
Maridi † *
Menindee Lakes 006
Meteorite Hills A78006
Mills *
Modoc (1948) *
Morland *
Morton *
Moti-ka-nagla † *
Mount Baldr *
Mount Browne † *
Mulga (north)
Nanjemoy † *
Naoki † *
Naragh † *
Nulles † *
Oakley (stone) *
Ogi † *
Ovid *
Ozona *
Patora † *
Pickens County *
Pipe Creek *
Pirthalla † *
Plantersville †
Queen's Mercy † *
Rabbit Flat
Ras Tanura †
Reckling Peak A79003
Reckling Peak A80201
Reckling Peak A80231
Rochester † *
Rowena
St. Germain-du-Pinel † *
Salles † *
San Juan Capistrano †
Seguin
Seoni † *
Shupiyan † *
Sinnai †
Somesbar
Supuhee † *
Suwahib ('Ain Sala) *
Taiga (stone)
Tell *
Thal † *
Tirupati † *
Tjabe † *
Tomatlan † *
Torrington † *
Toulouse † *
Trenzano † *
Tugalin-Bulen †
Vago † *
Vernon County † *
Visuni † *
Vulcan *
Wayside *
West Reid
Wilmot *
Yamato 6904
Yamato 74014
Yamato 74094
Yamato 74418
Yamato 74459
Yamato 74640
Zvonkov †

LL

Holman Island
Meru † *
Yalgoo
Yamato 790964

LL?

Rampurhat †

LL3

Allan Hills A76004 *
Allan Hills A77304
Allan Hills A81030

Allan Hills A81031
Bishunpur † *
Chainpur † *
Krymka † *
Manych †
Ngawi † *
Parnallee † *
Piancaldoli †
St. Mary's County †
Semarkona †

LL3-6

Bhola † *

LL4

Bo Xian †
El Paso
Hamlet † *
Kelly *
Savtschenskoje † *
Soko-Banja † *
Witsand Farm † *
Yamato 74442

LL4-5

Salem †

LL4-6

Sevilla †

LL5

Aldsworth † *
Alta'ameem † *
Forrest Lakes
Guidder †
Hunter *
Innisfree †
Khanpur † *
Krähenberg † *
Lazbuddie *
North Reid
Nyirábrany †
Oberlin *
Olivenza † *
Paragould † *
Parambu †
Perth † *
Pevensey *
Siena † *
Tuxtuac † *
Umbala † *

LL6

Appley Bridge † *
Arcadia *
Athens †
Bandong † *
Bates Nunataks A78004
Beeler *
Benares (a) † *
Benton † *
Bison *
Bloomington †
Boelus
Borgo San Donino † *
Caratash † *
Cherokee Springs † *
Chicora †
Dahmani †
Dhurmsala † *
Dongtai †
Douar Mghila † *
Ensisheim † *
Galim †
Jelica † *
Karakol † *
Karatu † *
Lake Labyrinth *
Manbhoom † *
Mangwendi † *
Mellenbye *
Min-Fan-Zhun †
Näs
Niger (LL6) †
Oberon Bay
Okniny † *
Ottawa † *
Oubari *
Reckling Peak A80235
St. Lawrence *
St. Mesmin † *
Saint-Séverin † *
Tomakovka †
Uden † *
Vavilovka † *
Vishnupur † *
Yamato 74646
Yamato 75258
Yukan † *
Zerga

L

Aguila Blanca †
Alberta †
Albin (stone)
Allen
Asco † *
Aus
Benthullen
Bou Hadid
Bradford Woods †
Çanakkale † *
Coronel Arnold
Crab Hole *
Dale Dry Lake
Delhi †
Doolgunna
Ethiudna
Forestburg
Hardtner *
Hartley
Jerome (Idaho)
Juarez *
Junction
Kisvarsany †
Leander
Lider (a)
Linum † *
Lost Lake *
Madhipura †
Marlow
McAddo *
Mike †
Minnesota (stone)
Mission
Montferre † *
Moriarty
Mulletiwu †
Muroc
Nainital
Nan Yang Pao †
Newsom *
Ofehértó †
Oufrane
Pampa de Agua Blanca
Park
Phillips County (stone) † *
Ploschkovitz † *
Portis
Post
Puente-Ladron
Puerto Libertad
Rio Bunge
Rogers
Rosamond Dry Lake *
Sazovice †
Schellin † *
Sidney
Stratford †
Strathmore † *
Taonan †
Tomita † *
Trysil †
Twentynine Palms
Ulmiz †
West Point *
Wilburton *
York (stone) *

L3

Allan Hills A77011
Andreevka †
Bovedy † *
Carraweena *
Hallingeberg † *
Inman *
Ioka
Khohar † *
Mezö-Madaras † *
Monte Colina
Moorabie
Ragland
Rakity
Reckling Peak A79008
Yamato 74191
Yambo no. 2

L3-4

Mafra †

L3-6

Dubrovnik † *
Grady (1933) *
Hedjaz † *
Umbarger *

L4

Albareto † *
Allan Hills A77230
Allred
ash-Shalfah
Atarra † *
Awere † *
Bald Mountain † *
Barratta *
Beyrout †
Bjurböle † *
Botschetschki †
Clohars † *
Cynthiana † *
Dalgety Downs *
Daoura
Delphos (a)
Floyd *
Fremont Butte *
Goodland *
Grassland *
Harding County *
Hardwick *
Hökmark †
Ikhrarene

Jerome (Kansas) *
Kediri † *
Kendleton †
Kramer Creek *
Lanzenkirchen †
Laundry West
Maria Linden †
McKinney *
Mossgiel *
Naretha
Nikolskoe † *
Nora Creina
North East Reid
Okechobee
Rio Negro † *
Rock Creek
Rupota †
Santa Barbara † *
Saratov † *
Slobodka † *
Suwahib (Adraj) *
Tennasilm † *
Vera *
Vigo Park
Yamato 75097
Yamato 75108
Yamato 75110
Yamato 75271
Yocemento *

L5

Adelie Land *
Allan Hills A81017
Almelo Township *
Arapahoe *
Armel *
Arriba *
Assam *
as-Su'aydan
Ausson † *
Banswal † *
Beaver *
Beenham *
Beuste † *
Black Moshannan Park †
Blackwell †
Bluff *
Borkut † *
Briscoe *
Cardanumbi
Chajari †
Chandakapur † *
Chervettaz † *
Cilimus † *
Claytonville *
Crumlin † *
Domanitch †
Elenovka † *
Elephant Moraine A79009
Ellerslie
Ergheo † *
Farmington † *
Farnum *
Finney *
Fluvanna (a)
Friona
Fukutomi † *
Ghubara *
Grefsheim †
Gretna *
Grier (b)
Guibga † *
Hale Center (no. 1) *
Hendersonville *
Hildreth *
Homestead † *
Honolulu † *
Indianola *
Jhung † *
Kaufman
Khmelevka †
Kingfisher *
Knyahinya † *
Kulak † *
Kybunga
La Lande *
Lishui †
Little Piney † *
Loyola
Lua † *
Lubbock *
Mabwe-Khoywa †
Malakal † *
Marion (Kansas) *
Marmande † *
Melrose (a) *
Messina † *
Mirzapur † *
Molina † *
Monte das Fortes †
Monte Milone † *
Muddoor † *
Nagy-Borové † *
New Almelo *
Oak
Ohuma † *
Pampanga † *
Peck's Spring *
Pohlitz † *
Reliegos † *
Renca † *
Richmond † *
River
Roy (1933) *
Rushville
Ryechki †
St. Peter
Sevrukovo † *
Shelburne † *
South Plains
Summerfield *
Suwanee Spring
Tadjera † *
Taiban *
Tané † *
Tarbagatai
Timmersoi *
Tjerebon † *
Tsarev
Umm Ruaba † *
Vilna †
Vincent
Wittekrantz † *
Yorktown (New York) † *

L5-6

Genichesk
Hajmah (c) *

L6

Abernathy *
Achiras † *
Aguada † *
Air † *
Akaba † *
Akron (1961) *
Alamosa
Aleppo † *
Alfianello † *
al-Ghanim (stone)
Allan Hills A76001 *
Allan Hills A76003 *
Allan Hills A76007 *
Allan Hills A76009 *
Allan Hills A77001
Allan Hills A77050
Allan Hills A77105
Allan Hills A77231
Allan Hills A77251
Allan Hills A77269
Allan Hills A77270
Allan Hills A77272
Allan Hills A77280
Allan Hills A77281
Allan Hills A77282
Allan Hills A77296
Allan Hills A77297
Allan Hills A77305
Allan Hills A78043
Allan Hills A78050
Allan Hills A78103
Allan Hills A78105
Allan Hills A78112
Allan Hills A78126
Allan Hills A78130
Allan Hills A78251
Allan Hills A79033
Allan Hills A80101
Allan Hills A81016
Allan Hills A81026
Allan Hills A81027
Amber *
Amherst
Andover † *
Andryushki
Angers † *
Anson *
Apt † *
Artracoona *
Ashdon † *
Atemajac †
Atoka †
Atwood *
Aumale † *
Aumieres † *
Aztec † *
Bachmut † *
Bakhardok
Baldwyn †
Bansur †
Baroti † *
Barwell † *
Bates Nunataks A78001
Bates Nunataks A78002
Bath Furnace † *
Baxter † *
Bayard
Beaver-Harrison
Belle Plaine *
Berdyansk
Berlanguillas † *
Bhagur † *
Bherai † *

Billygoat Donga
Blaine Lake
Blanket † *
Bocas † *
Bori † *
Bowesmont *
Brady
Brewster *
Bruderheim † *
Burrika
Bursa † *
Buschhof † *
Cabezo de Mayo † *
Cadell
Calliham *
Cartoonkana
Castine † *
Catherwood
Chandpur † *
Chantonnay † *
Château Renard † *
Chico *
Chitenay †
Circle Back *
Clareton *
Clovis (no.2) *
Cockarrow Creek
Cockburn
Cocunda
Colby (Wisconsin) † *
Concho *
Coolamon
Coon Butte *
Coorara
Cotesfield *
Cranganore † *
Credo *
Dandapur † *
Danville † *
Davy
Deal † *
Demina † *
De Nova *
Densmore (1879) *
Denver †
Diep River † *
Dix *
Dolgovoli † *
Dosso †
Drake Creek † *
Durala † *
Duruma † *
Duwun †
Dwight *
Edmonson (a) *
Elephant Moraine A79003
Elephant Moraine A79010
Eli Elwah *
Ella Island *
Erie
Fisher † *
Forksville † *
Forrest (b) *
Forsyth † *
Frankel City *
Franklinville
Futtehpur † *
Galatia *
Gambat † *
Garraf *
Gifu † *
Girgenti † *
Git-Git † *
Gomez
Granes †
Grant County *
Grossliebenthal † *
Guangrao †
Gurram Konda † *
Hajmah (b) *
Hamilton (Queensland) *
Harleton † *
Harrison County † *
Harrisonville *
Haskell *
Hayes Center *
Hermitage Plains *
Hesston
High Possil † *
Hinojal
Holbrook † *
Homewood
Hotse †
Inner Monglia †
Isoulane-n-Amahar *
Jackalsfontein † *
Jartai †
Jemlapur † *
Johnson City *
Junan †
Kagarlyk †
Kakowa † *
Kalumbi † *
Kamalpur † *
Kamsagar † *
Kandahar (Afghanistan) † *
Kaptal-Aryk † *
Karagai
Karewar † *
Karkh † *
Karloowala † *
Kermichel *
Keyes *
Kharkov † *
Kheragur † *
Kiel † *
Kinley *
Koraleigh *
Krasnoi-Ugol † *
Kress *
Kukschin †
Kuleschovka † *
Kulnine *
Kunashak † *
Kusiali † *
Kuttippuram † *
Kuznetzovo † *
Kyle *
Kyushu † *
La Bécasse † *
Ladder Creek *
L'Aigle † *
Lake Bonney
Lake Brown *
Lake Grace
Laketon *
Lakewood *
Lalitpur † *
Långhalsen †
Launton † *
Lavrentievka †
Lawrence *
Leedey † *
Leeuwfontein † *
Leikanger
Leonovka †
Le Pressoir † *
Les Ormes † *
Lesves † *
Lincoln County *
Lissa † *
Lockney *
Long Island *
Loomis *
Loop *
Louisville †
Lucé † *
Lundsgård † *
Lutschaunig's Stone *
Machinga † *
Madiun †
Madrid † *
Mainz *
Makarewa *
Mamra Springs †
Marion (Iowa) † *
Maryville †
Mascombes † *
Mauerkirchen † *
Mauritius † *
Maziba †
Menindee Lakes 001
Menindee Lakes 003
Menindee Lakes 004
Mern † *
Meteorite Hills A78002
Meteorite Hills A78003
Meteorite Hills A78028
Meuselbach † *
Mezel † *
Mhow † *
Middlesbrough † *
Milena † *
Minas Gerais *
Mocs † *
Modoc (1905) † *
Montlivault † *
Monze † *
Moorleah †
Moradabad † *
Mosca
Motpena *
Muizenberg *
Muraid † *
Muroc Dry Lake
Nagai †
Nakhon Pathom †
Nardoo (no.2)
Narellan † *
Nashville (stone) *
Nazareth (c)
Neenach *
Nejo † *
Nerft † *
Ness County (1894) *
New Concord † *
Niger (L6) †
Nogata †
Norcateur *
Noyan-Bogdo †

Oesel † *
Ojuelos Altos †
Oldenburg (1930) †
Orvinio † *
Oterøy †
Otis *
Ovambo †
Oviedo †
Pacula † *
Pampa del Infierno *
Pannikin
Paranaiba †
Patrimonio † *
Pavlograd † *
Peace River † *
Peetz *
Perpeti † *
Pervomaisky † *
Pierceville (stone) *
Pinto Mountains *
Pirgunje † *
Plainview (d)
Pnompehn †
Point of Rocks (stone)
Portales (b) *
Potter *
Prambachkirchen †
Pricetown † *
Putinga † *
Quincay † *
Rakovka † *
Ramnagar † *
Ramsdorf † *
Rangala † *
Reager
Reckling Peak A78001
Reckling Peak A79001
Reckling Peak A79002
Red Deer Hill
Renqiu †
Rewari † *
Rich Mountain †
Roy (1934) *
Ruhobobo †
Rush Creek *
St. Caprais-de-Quinsac † *
St.-Chinian †
St. Christophe-la-Chartreuse †
St. Denis Westrem † *
St. Francis Bay
St. Michel † *
Salla
San Pedro Springs
Santa Isabel † *
Sauguis † *
Schönenberg † *
Sediköy † *
Segowlie † *
Seminole Draw (a) *
Shaw *
Shibayama
Shikarpur † *
Shytal † *
Silverton (New South Wales) *
Sinai † *
Ski † *
Sleeper Camp
Slobodka (of P. Partsch)
Smith Center *
Springfield *
Springlake
Stavropol † *
Sublette
Success †
Suchy Dul †
Sultanpur † *
Tambakwatu †
Tarfa *
Tathlith † *
Tauk † *
Tauti †
Temple *
Tenham † *
Thomson
Tiberrhamine *
Tilden † *
Tillaberi †
Torreon de Mata *
Tourinnes-la-Grosse † *
Trifir
Troup †
Tryon *
Tuan Tuc †
Tulia (b) *
Tuzla *
Ularring
Umm Tina *
Usti nad Orlici † *
Utrecht † *
Valdinizza † *
Valkeala
Valley Wells *
Varpaisjärvi *
Virba † *
Vouillé † *
Waconda *
Waldo *
Walltown *
Walters †
Wardswell Draw *
Webb
Wethersfield (1971) †
Wethersfield (1982) †
Wickenburg (stone) *
Willard *
Willowbar
Wold Cottage † *
Woolgorong † *
Xi Ujimgin †
Yamato 6909
Yamato 7304
Yamato 7305
Yamato 74077
Yamato 74080
Yamato 74118
Yamato 74190
Yamato 74354
Yamato 74362
Yamato 74445
Yamato 74454
Yamato 74605
Yamato 75102
Yandama *
Yayjinna
Zaborzika † *
Zabrodje † *
Zavetnoe †
Zavid † *
Zemaitkiemis † *
Zomba † *

Stone?

Tamir-Tsetserleg
Urasaki †

Stony-iron

Lodranite

Lodran † *
Yamato 791493

Mesosiderite

Allan Hills A77219
Allan Hills A81059
Barea †
Bencubbin *
Bondoc *
Budulan *
Chinguetti *
Clover Springs *
Crab Orchard *
Dalgaranga *
Dyarrl Island †
Emery *
Estherville † *
Hainholz *
Łowicz † *
Mincy *
Morristown *
Mount Padbury *
Patwar † *
Pinnaroo *
Simondium *
Um-Hadid
Vaca Muerta *
Veramin † *
Weatherford *
Weiyuan
Yamato 75274

Mesosiderite?

Donnybrook

Mesosiderite, anomalous

Enon *
Horse Creek *
Mount Egerton *
Reckling Peak A79015

Pallasite

Admire *
Ahumada *
Albin (pallasite) *
Argonia
Barcis
Bendock
Brahin *
Brenham *
Cold Bay *
Dora (pallasite) *
Eagle Station *
El Rancho Grande
Esquel *
Finmarken *
Giroux
Glorieta Mountain *
Huckitta *
Imilac *
Itzawisis
Krasnojarsk *
Lipovsky
Marburg
Marjalahti † *
Molong *
Mount Dyrring *
Mount Vernon † *

Newport *
Pavlodar (pallasite) *
Phillips County (pallasite) *
Port Orford
Rawlinna (pallasite)
Santa Rosalia
Somervell County
South Bend *
Springwater *
Sterling
Thiel Mountains *
Yamato 74044
Zaisho †

Pallasite?

Singhur

Doubtful

Aegospotami †
Aglar †
Ahmedabad †
Aidin †
Alaschan
Aphaca
Atlantic Ocean †
Augsburg (952) †
Barcelona (iron) †
Basingstoke †
Beeston †
Carpentras †
Cassandra
Cecil Plains
Civitavecchia †
Crevalcore †
Curvello
Daoukro
Delphi
Denmark †
Dordrecht (iron) †
Dordrecht (stone) †
Dorpat †
Eastman
East Norton †
Emesa
Ephesus
Friedland †
Gaj †
Gao (Mali) †
Gdynia †
Glarus †
Glastonbury †
Glen Osborne
Harrogate †
Ida †
Italy (956) †
Italy (963) †
Kerulensky
Kis-Gyor
Kloster Schefftlar †
Kuli-schu †
Kunersdorf †
Larissa †
Limousin †
Loch Tay †
Lowell †
Macatuba †
Magdeburg †
Malpas †
Marsala †
Massa-lubrense †
McAlester
Mikolawa †
Milan †
Minden †
Mixbury †
Mons †
Nagareyama
Nanseiki †
North Africa (481) †
North Africa (1020) †
Novellara †
Ofen †
Oldenburg (1368) †
Oldisleben †
Omoa
Oriang †
Ostrzeszów †
Piedmont †
Pölau †
Prince George †
Prince of Wales's Straits †
Provence †
Pulrose †
Pulter
Quedlinburg †
Rio Mocoreta †
Ruschany †
San Cerre de Mallorca †
Sano
São Sebastião da Boa Vista †
Skåne-Tranås †
Sotik †
Soweida †
Steiermark †
Terranova di Sibari †
Thebes
Thrace †
Torgau †
Tregnie †
Turin †
Velikoi-Ustyug †
Viterbo †
Warsaw I
Woodbridge †
Yorkshire †

Iron

Alwal †
Bechuanaland
Bezerros †
Caparrosa
Juzjanan †
La Rinconada †
Lausanne †
Lucania †
Meers
Meissen †
Monte Alto
Naunhof †
Oborniki
Ogasawara
Otasawian
Paint Creek †
Pfullingen †
Pomorze
Sakurayama †
Tarapaca
Tyumen †

Iron?

Birgi †

Stone

Arabia
Arundu †
Augsburg (1528) †
Avoca (Texas)
Bheenwal †
Brussels †
Calce †
Cardiff †
Charleroi †
Cholula
Copinsay †
Crema †
Ermendorf †
Fabriano †
Ferrette †
Glady
Hyderabad (1898) †
Iowa City †
Java †
Jubila del Agua †
Kaaba
Klagenfurt †
Kochi †
Masanderan †
Miskolcz †
Montrose
Mulino †
Nörten †
Orange River (stone) †
Ponta Grossa †
Roa †
Schleusingen †
Simonod †
Swindnica Gorna
Tiree
Trevlac
Walkringen †
Yakushima †

Eucrite

Constantinople †

H

San Luis
Ultuna

Stone?

Atsuma †
Hanau †
Hyderabad (1936) †
Ipacaray †
Kandahar (India) †
Kansas City (1876) †
Niederreissen †
Palenca de Baixo †
Pentolina †
Shiriya †

Stony-iron

Pallasite

Australia
Mineo †

Pseudometeorite

Aba
Allport
Alto Verde
Bamenda
Baydon
Bleckenstad
Carlstadt †

Ceylon
Chaco Santafecino
Colston Bassett
Cormeilles
Eaton †
El Marplatense
Estes Park
Fair Play
Grazac †
Groslée
Halstead
Hautes Alpes
Horsham
Keilce
Koso-sho
Kurumi
Langeac
Le Gould's Stone
Leroy
Lodi
Loerbeek
London
Macquarie River
Mar del Plata
Menabilly
Mikomotojima
Nauheim
Newtown
Ovifak
Scriba
South Dixon
Sumampa
Takysie Lake
Tandil
Trentino
Tule
Villanueva del Fresno
Yaddlethorpe
Yafee Mountains
Zenith

Iron
Agricultural College
Angara
Brianza †
Calumet County
Dowerin
Kirkland †
Nova Lima
South Strafford
Wolfsegg

Stone
Legnano †
Orebro †
Thatcher

List of Meteorites arranged Geographically

In the following list of meteorites arranged according to country of fall or find, names in italics indicate doubtful falls or pseudometeorites. An asterisk (*) following the name indicates that the meteorite is represented in the BM(NH) collection. An obelisk (†) following the name indicates an observed fall.

Afghanistan

Juzjanan †
Kandahar (Afghanistan) † *

Algeria

Akhricha
Bou Hadid
Daoura
Dellys
Fort Flatters †
Haniet-el-Beguel *
Isoulane-n-Amahar *
Ouallen *
Oufrane
Seldebourak †
Tadjera † *
Tamentit *
Tiberrhamine *
Zerhamra *

Alger

Aumale † *

Algerian Sahara

Ikhrarene

Constantine

Feid Chair †

El Golea

Hassi-Jekna † *

Angola

Cunene

Chitado †
Ehole † *
Otchinjau *

Antarctica

Adelie Land *
Allan Hills A77001
Allan Hills A78006
Allan Hills A79002
Allan Hills A80101
Allan Hills A81001
Allan Hills 82100
Bates Nunataks A78001
Belgica 7901
Belgica 7904
Derrick Peak A78001
Elephant Moraine A79001
Elephant Moraine 82602
Lazarev
Meteorite Hills A78001
Mount Baldr *
Neptune Mountains
Outpost Nunatak A80301
Pecora Escarpment 82501
Purgatory Peak A77006
Reckling Peak A78001
Reckling Peak A79001
Reckling Peak A80201
Thiel Mountains *
Thiel Mountains 82403
Yamato 6901
Yamato 6902
Yamato 6903
Yamato 6904
Yamato 6905
Yamato 6906
Yamato 6907
Yamato 6908
Yamato 6909
Yamato 7301
Yamato 7304
Yamato 7305
Yamato 7308
Yamato 74001
Yamato 75003
Yamato 790001
Yamato 8001
Yamato 81001

Argentina

Bahia Blanca

El Perdido

Buenos Aires province

Cachari *
El Marplatense
Hinojo
Indio Rico *
Juarez *
La Colina † *
Lujan
Mar del Plata
Palermo †
San Carlos *
Tandil

Catamarca

Aguas Calientes
Campo de Pucara (iron)
Medanitos † *

Chaco

Campo del Cielo *
Pampa del Infierno *
Tres Estacas

Chubut

Caperr *
El Sampal
Esquel *
Paso Rio Mayo

Cordoba

Achiras † *
Aguada † *
Aguila Blanca †
Belville †
Capilla del Monte † *
Casilda *
El Simbolar *
Mojigasta

Corrientes

Rio Mocoreta †

Entre Rios

Chajari †
Distrito Quebracho † *
Gualeguaychú † *
Hinojal
Isthilart † *
Nogoya † *

La Rioja

Agua Blanca *
El Chiflón
Los Sauces
Puerta de Arauco

Mendoza

Alto Verde

Neuquen

Dadin
Norquín

Rio Negro

Colonia Suiza †

San Luis

Arbol Solo †
Renca † *
San Luis

Santa Cruz

Laguna Manantiales *

Santa Fe

Arroyo Aguiar † *
Campo Perrando
Coronel Arnold
El Timbu
Santa Isabel † *
Tostado *
Vera *

Santiago del Estero

Malotas † *
Sumampa

Tucuman

Raco † *

Australia

New South Wales
Barratta *
Binda *
Bingera *
Boogaldi *
Boolka
Cartoonkana
Coolac *
Coolamon
Corowa
Cowra *
Delegate *
Dorrigo
Eli Elwah *
Elsinora *
Emmaville † *
Forest Vale † *
Gilgoin *
Hermitage Plains *
Koraleigh *
Macquarie River
Menindee Lakes 001
Menindee Lakes 002
Menindee Lakes 003
Menindee Lakes 004
Menindee Lakes 005
Menindee Lakes 006
Moama *
Molong *
Moonbi *
Moorabie
Morden
Mossgiel *
Mount Browne † *
Mount Dyrring *
Mungindi *
Nardoo (no.1)
Nardoo (no.2)
Narellan † *
Narraburra *
Nocoleche *
Pevensey *
Rowena
Silverton (New South Wales) *
Thackaringa
Tibooburra
Willaroy
Yandama *

Northern Territory
Alikatnima
Arltunga *
Bond Springs
Boxhole *
Henbury *
Huckitta *
Mount Sir Charles
Rabbit Flat
Roper River
Tawallah Valley *
Yenberrie *

Queensland
Answer
Arrabury
Cecil Plains
Ellerslie
Gladstone (iron) *
Glenormiston *
Hamilton (Queensland) *
Hammond Downs *
King Solomon
Le Gould's Stone
Rockhampton †
Tenham † *
Thunda *
Wynella *

South Australia
Adelaide
Artracoona *
Brachina
Buckleboo
Cadell
Caroline *
Carraweena *
Cockburn
Cocunda
Coomandook
Coonana *
Cowell *
Crocker's Well
Ethiudna
Glen Osborne
Kaldoonera Hill
Kappakoola *
Kapunda
Karoonda † *
Kielpa *
Kittakittaooloo
Kyancutta *
Kybunga
Lake Bonney
Lake Labyrinth *
Monte Colina
Moorumbunna *
Motpena *
Muckera
Murnpeowie *
Nilpena *
Nora Creina
Nullarbor
Pinnaroo *
Poonarunna
Rhine Villa *
Vincent
Weekeroo Station *
Wilbia *
Witchelina *
Yardea

Tasmania
Blue Tier
Castray River
Lefroy
Moorleah †

Victoria
Bendock
Cranbourne *
Dimboola
Horsham
Kulnine *
Lismore
Murchison † *
Oberon Bay
Wedderburn
Yarroweyah *

Western Australia
Avoca (Western Australia) *
Baandee
Balfour Downs
Ballinoo *
Bencubbin *
Billygoat Donga
Burnabbie *
Burrika
Cardanumbi
Carlisle Lakes (a)
Cockarrow Creek
Cocklebiddy *
Coorara
Cosmo Newberry
Crab Hole *
Credo *
Dalgaranga *
Dalgety Downs *
Dingo Pup Donga *
Donnybrook
Doolgunna
Dowerin
Duketon *
Edjudina
Fenbark
Forrest (a)
Forrest (b) *
Forrest Lakes
Frenchman Bay *
Gnowangerup
Gundaring *
Gunnadorah
Haig *
Jeedamya
Johnny's Donga
Korrelocking
Kumerina *
Lake Brown *
Lake Grace
Landor
Laundry East
Laundry Rockhole
Laundry West
Lookout Hill *
Mellenbye *
Millbillillie † *
Milly Milly *
Mount Dooling *
Mount Edith *
Mount Egerton *
Mount Magnet *
Mount Padbury *
Mulga (north)
Mulga (south)
Mulga (west) *
Mundrabilla *
Murchison Downs
Nallah
Naretha
North East Reid
North Haig *
North Reid
North West Forrest (E6)
North West Forrest (H)
Nuleri
Nullagine
Oak
Oldfield River
Pannikin

Rawlinna (pallasite)
Rawlinna (stone)
Redfields *
Reid
River
Roebourne *
Sleeper Camp
Tieraco Creek
Ularring
Un-named
Warburton Range *
Webb
West Forrest
West Reid
Wildara *
Wiluna † *
Wingellina
Wolf Creek *
Wonyulgunna *
Woolgorong † *
Yalgoo
Yarri
Yayjinna
Yilmia *
Youanmi *
Youndegin *

Austria

Carinthia
Klagenfurt †

Nieder Osterreich
Lanzenkirchen †
Ybbsitz

Ober-Österreich
Mauerkirchen † *
Prambachkirchen †
Wolfsegg

Steiermark
Steiermark †

Tyrol
Mühlau

Bangladesh

Chittagong
Patwar † *
Perpeti † *

Dacca
Dokachi † *
Muraid † *

East Bengal
Pirgunje † *

Khulna
Bhola † *

Rajshahi
Gopalpur † *

Tangail
Shytal † *

Belgium
Brussels †
Charleroi †
Hainaut †
Lesves † *
Mons †
St. Denis Westrem † *
Tourinnes-la-Grosse † *

Bolivia

Ouro
Pooposo *

Bolivia?
Cochabamba

Brazil
Angra dos Reis (stone) † *
Cratheús (1950)
São Sebastião da Boa Vista †

Bahia
Bendegó *
Monte Alto

Ceara
Cratheús (1931) *
Parambu †

Goias
Cacilandia
Itapuranga
Sanclerlandia
Santa Luzia *
Verissimo

Maranhão
Itapicuru-Mirim † *

Mato Grosso
Paranaiba †

Minas Gerais
Barbacena *
Bocaiuva
Conquista †
Curvello
Governador Valadares *
Ibitira †
Itutinga
Minas Gerais *
Nova Lima
Paracutu
Para de Minas
Patos de Minas (hexahedrite)
Patos de Minas (octahedrite)
Patrimonio † *
Piedade do Bagre *
Pirapora *
São João Nepomuceno
Sete Lagoas †
Uberaba † *

Parana
Ipiranga †
Ponta Grossa †
Rio Negro † *

Pernambuco
Bezerros †
Serra de Magé † *

Rio de Janeiro
Casimiro de Abreu

Rio Grande do Norte
Macau † *

Rio Grande do Sul
Nova-Petropolis
Putinga † *
Santa Barbara † *

Santa Catarina
Mafra †
Santa Catharina *

Santa Catharina
Morro do Rocio *

São Paulo
Avanhandava †
Marilia † *
São Jose do Rio Preto †

Bulgaria
Rasgrad †
Virba † *

Gabrovo
Gumoschnik † *

Sliven
Konovo †

Veliko Turnovo
Pavel †

Burma
Quenggouk † *

Kayah
Mabwe-Khoywa †

Upper Burma
Nyaung † *

Cambodia
Phum Sambo † *
Pnompehn †

Cameroun
Galim †

Nord province
Guidder †

Sud-ouest province
Bamenda

Canada

Alberta
Abee † *
Belly River
Bruderheim † *
Edmonton (Canada) *
Ferintosh

Innisfree †
Iron Creek *
Kinsella
Mayerthorpe
Millarville *
Otasawian
Peace River † *
Skiff
Vilna †
Vulcan *

British Columbia
Beaver Creek † *
Prince George †
Revelstoke †
Takysie Lake

Manitoba
Giroux
Homewood
Riverton

New Brunswick
Benton † *

North-west Territories
Great Bear Lake
Holman Island
Klondike (Gay Gulch)
Klondike (Skookum Gulch) *

Ontario
Blithfield *
De Cewsville †
Dresden (Ontario) †
Madoc *
Manitouwabing
Midland
Osseo *
Shelburne † *
Thurlow *
Welland *

Quebec
Chambord
Eastman
Leeds *

Saskatchewan
Annaheim *
Blaine Lake
Bruno *
Catherwood
Fillmore
Garden Head
Kinley *
Red Deer Hill
Springwater *
Wynyard

Central African Republic
Bali †

Chad
Djermaia †
Massenya *

Chile
Dehesa *

Antofagasta
Cobija *
Las Salinas
Maria Elena (1935) *
Monturaqui
North Chile *
Pampa de Agua Blanca
Rica Aventura
San Cristobal *
Sierra Gorda
Sierra Sandon *

Atacama
Baquedano *
Barranca Blanca *
Cachiyuyal *
Carcote *
Copiapo *
Corrizatillo
Ilimaes (iron) *
Imilac *
Joel's Iron *
Juncal *
Lutschaunig's Stone *
Mantos Blancos *
Mejillones
Merceditas *
Pan de Azucar *
Puquios *
Serrania de Varas *
Slaghek's Iron
Ternera *
Vaca Muerta *

Tarapaca
Iquique *
La Primitiva *
Negrillos *
Tamarugal *
Tarapaca

China
Koso-sho
Kuli-schu †

Anhui
Bo Xian †
Fuyang †
Po-wang Chen †

Fujian
Fujian

Gansu
Nan Yang Pao †
Weiyuan

Guangdong
Yangchiang †
Yingde

Guangxi
Duan
Hechi
Nantan
Tianlin
Yongning

Guizhou
Anlung †
Guizhou
Qingzhen †

Hebei
Renqiu †

Henan
Mianchi †
Wuzhi †
Xingyang †

Hubei
Enshi †
Guanghua
Huangling
Jianshi †
Zhongxiang

Hunan
Changde †
Wei-hui-fu (a)
Wei-hui-fu (b)

Jiangsu
Dongtai †
Fenghsien-Ku †
Lishui †
Min-Fan-Zhun †
Pei Xian
Taicang
Xinyi

Jiangxi
Guixi
Yukan † *

Jilin
Dunhua †
Jiapigou
Jilin † *
Shuangyang †
Taonan †

Liaoning
Palinshih †
Zhuanghe †

Nei Monggol
Fengzhen
Inner Monglia †
Liangcheng
Shangdu
Wu-chu-mu-ch'in
Xi Ujimgin †

Ningxia
Jartai †

Shandong
Guangrao †
Hotse †
Junan †
Nanseiki †

Shanghai
Changxing

Sichuan
Fuling
Guangyuan

Jiange †
Leshan
Longchang

Taiwan
Jilong

Xinjiang
Alatage
Armanty
Hejing

Yunan
Lunan †

Zhejiang
Ningbo †

Colombia

Boyaca
Santa Rosa *

Costa Rica
Heredia † *

Cuba
Cuba

Czechoslovakia
Selcany
Suchy Dul †
Usti nad Orlici † *

Bohemia
Lissa † *

Jihocesky
Bohumilitz *
Tabor † *

Jihomoravsky
Blansko † *
Stannern † *
Tieschitz † *
Vicenice
Wessely † *

Moravia
Alt Bela
Sazovice †

Severomoravsky
Opava

Sevrocesky
Ploschkovitz † *

Slovensko
Lenarto *
Magura *

Stredocesky
Pribram † *
Zebrak † *

Stredoslovensky
Gross-Divina † *
Nagy-Borové † *

Vychodocesky
Braunau † *

Zapadocesky
Elbogen † *

Zapadocesky Kraj
Tepla *

Denmark
Aarhus † *
Denmark †
Fünen †
Jerslev
Mern † *

Egypt
Aswan
El Qoseir
Gerzeh
Isna *
Nakhla † *
Quarat al Hanish *
Sinai † *
Soweida †
Thebes

England
Allport
London
Yorkshire †

Berkshire
Hatford †

Cheshire
Malpas †

Cornwall
Menabilly

Devonshire
Stretchleigh †
Tregnie †

Essex
Ashdon † *
Halstead

Gloucestershire
Aldsworth † *

Hampshire
Basingstoke †

Lancashire
Appley Bridge † *

Leicestershire
Barwell † *
East Norton †

Lincolnshire
Yaddlethorpe

Nottinghamshire
Beeston †
Colston Bassett

Oxfordshire
Launton † *
Mixbury †

Shropshire
Rowton † *

Somerset
Glastonbury †

Suffolk
Woodbridge †

Wiltshire
Baydon

Yorkshire
Harrogate †
Middlesbrough † *
Wold Cottage † *

Ethiopia
Ankober †
Gursum †
Holetta †
Nejo † *
Wolamo †

Finland
Abo
Bjurböle † *
Haverö † *
Hvittis † *
Kivesvaara
Luotolax † *
Metsäkylä
Orimattila
St. Michel † *
Salla
Valkeala
Varpaisjärvi *

France
Hautes Alpes

Ain
Groslée *
Luponnas † *
Simonod †

Alpes Maritimes
La Caille *
Mount Vaisi †

Alsace
Ensisheim † *
Ferrette †

Ardeche
Juvinas † *

Aube
St. Mesmin † *

Aude
Granes †
Montferre † *

Aveyron
Favars † *

Basses-Pyrénées
Beuste † *
Sauguis † *

Bouches du Rhone
Lancon † *

Charente
Saint-Séverin † *

Charente Maritime
Esnandes † *
Jonzac † *

Correze
Mascombes † *

Corsica
Asco † *

Cote-d'Or
Villedieu

Cotes-du-Nord
Kerilis † *

Doubs
Ornans † *

Drome
Aubres † *
Bouvante
Laborel † *
Mornans † *

Finistére
Clohars † *

Gard
Alais † *

Gers
Barbotan † *

Gironde
St. Caprais-de-Quinsac † *

Haute Garonne
Ausson † *
Saint-Sauveur † *
Toulouse † *

Haute Loire
Grazac †
Langeac

Haute Marne
Chassigny † *

Haute Vienne
Limousin †

Herault
St.-Chinian †

Ille-et-Vilaine
St. Germain-du-Pinel † *

Indre
La Bécasse † *

Indre-et-Loir
Le Pressoir † *

Loir-et-Cher
Chitenay †
Lancé † *
Montlivault † *

Loiret
Charsonville † *
Château Renard † *

Lot-et-Garonne
Agen † *
Galapian †
Marmande † *

Lozere
Aumieres † *

Maine et Loire
Angers † *

Manche
Le Teilleul † *

Morbihan
Kermichel *
Kernouve † *

Nord
Ste. Marguerite † *

Normandy
Nicorps †

Orne
L'Aigle † *

Pas de Calais
Aire-sur-la-Lys †

Puy de Dome
Mezel † *

Rhone
Salles † *

Sarthe
Lucé † *

Tarn-et-Garonne
Orgueil † *

Val d'Oise
Cormeilles

Var
Provence †

Vaucluse
Apt † *
Carpentras †

Vendée
Chantonnay † *
St. Christophe-la-Chartreuse †

Vienne
Quincay † *
Vouillé † *

Vosges
Épinal † *

Yonne
Les Ormes † *

Germany
Kunersdorf †

Baden-Württemberg
Ortenau †
Pfullingen †

Bayern
Augsburg (952) †
Augsburg (1528) †
Eichstädt † *
Kloster Schefftlar †
Mässing † *
Rodach †
Schönenberg † *
Unter-Mässing

Dresden
Ermendorf †
Meissen †
Nenntmannsdorf *

Erfurt
Klein-Wenden † *
Niederreissen †
Tabarz *

Frankfurt
Friedland †
Nieder Finow

Gera
Pohlitz † *

Halle
Oldisleben †
Quedlinburg †

Hessen
Breitscheid † *
Darmstadt † *
Hanau †
Hungen † *
Marburg
Nauheim
Treysa † *

Karl-Marx-Stadt
Pölau †
Steinbach *

Leipzig
Naunhof †
Torgau †

Magdeburg
Magdeburg †

Niedersachsen
Benthullen
Bremervörde † *
Emsland

Nörten †
Obernkirchen *
Oesede †
Oldenburg (1368) †
Oldenburg (1930) †
Stolzenau †

Nordrhein-Westfalen
Barntrup †
Forsbach † *
Gütersloh † *
Hainholz *
Ibbenbüren † *
Minden †
Pecklesheim †
Ramsdorf † *

Potsdam
Linum † *
Menow † *

Rheinland-Pfalz
Bitburg *
Krähenberg † *
Mainz *
Simmern † *

Sachsen-Anhalt
Erxleben † *

Schleswig-Holstein
Kiel † *

Suhl
Dermbach
Schleusingen †

Thuringen
Meuselbach † *

Greece
Aegospotami †
Cassandra
Delphi
Larissa †
Seres † *
Thrace †

Greenland
Cape York *
Ella Island *
Ovifak
Thule *

Guatemala
Chinautla *

Holland
Dordrecht (iron) †
Dordrecht (stone) †
Ellemeet † *
Loerbeek
Uden † *
Utrecht † *

Honduras
Omoa
Rosario *

Hungary
Kaba † *
Kis-Gyor
Kisvarsany †
Mike †
Mikolawa †
Minnichhof †
Miskolcz †
Nagy-Vázsony *
Nyirábrany †
Ofehértó †
Ofen †

India
Jemlapur † *
Kandahar (India) †

Andhra Pradesh
Alwal †
Gurram Konda † *
Hyderabad (1898) †
Hyderabad (1936) †
Nedagolla † *
Punganaru †
Tirupati † *
Yatoor † *

Assam
Assam *
Goalpara *

Bihar
Andhara †
Butsura † *
Madhipura †
Muzaffarpur † *
Parsa †
Ranchapur † *
Segowlie † *
Shergotty † *
Shikarpur † *
Soheria †

Delhi state
Delhi †

Gujarat
Bherai † *
Dhajala † *
Myhee Caunta †

Haryana
Charwallas † *
Pirthalla † *
Rewari † *

Himachal Pradesh
Baroti † *
Dhurmsala † *
Kangra Valley † *

Karnataka
Chetrinahatti †
Dharwar †
Judesegeri † *
Kamsagar † *
Muddoor † *
Udipi † *

Kashmir
Shupiyan † *

Kerala
Cranganore † *
Kuttippuram † *

Madhya Pradesh
Bori † *
Donga Kohrod † *
Erakot †
Lakangaon † *
Oriang †
Patora † *
Pulsora † *
Sarratola
Semarkona †
Seoni † *
Sitathali † *

Maharashtra
Andura † *
Bhagur † *
Chandakapur † *
Jamkheir † *
Kalumbi † *
Manegaon † *
Naoki † *
Phulmari
Singhur

Orissa
Bholghati † *
Nawapali † *

Punjab
Durala † *
Jalandhar †
Umbala † *

Rajasthan
Bansur †
Bheenwal †
Desuri † *
Haripura †
Khetri † *
Lua † *
Moti-ka-nagla † *
Rangala † *
Samelia † *
Tonk †
Udaipur †

Tamil Nadu
Kakangari † *
Kodaikanal *
Nammianthal † *
Parnallee † *
Valdavur † *

Uttar Pradesh
Akbarpur † *
Ambapur Nagla † *
Atarra † *
Bahjoi † *
Banswal † *
Benares (a) † *
Benares (b)
Bishunpur † *
Bustee † *

Chail † *
Chainpur † *
Chandpur † *
Dandapur † *
Dyalpur † *
Ekh Khera † *
Futtehpur † *
Haraiya † *
Kadonah † *
Kaee † *
Kamalpur † *
Khanpur † *
Kheragur † *
Khohar † *
Kusiali † *
Lahrauli † *
Lalitpur † *
Meerut †
Merua † *
Mhow † *
Mirzapur † *
Moradabad † *
Nagaria † *
Nainital
Pokhra † *
Ramnagar † *
Sabetmahet †
Sultanpur † *
Supuhee † *

West Bengal
Manbhoom † *
Rampurhat †
Shalka † *
Vishnupur † *

Indonesia

Java
Bandong † *
Banten †
Cilimus † *
Djati-Pengilon † *
Glanggang † *
Java †
Kediri † *
Madiun †
Meester-Cornelis † *
Ngawi † *
Prambanan *
Rembang † *
Selakopi †
Tambakwatu †
Tjabe † *
Tjerebon † *

Kangean Island
Kangean †

Iran
Masanderan †
Naragh † *
Veramin † *

Iraq
Ahmedabad †
Alta'ameem † *
Tauk † *
Ur *

Ireland
Dundrum † *
Pettiswood †

County Antrim
Crumlin † *

County Limerick
Limerick † *

County Tipperary
Mooresfort † *

County Tyrone
Killeter † *

Isle of Man
Pulrose †

Italy
Aglar †
Trentino

Ancona
Fabriano †

Basilicata
Lucania †

Calabria
Castrovillari †
Terranova di Sibari †

Campania
Massa-lubrense †

Emilia
Crevalcore †
Novellara †
Renazzo † *
Vigarano † *

Friuli
Barcis

Lazio
Civitavecchia †
Orvinio † *
Viterbo †

Lombardy
Alfianello † *
Crema †
Lodi
Milan †
Rivolta de Bassi †
Trenzano † *
Valdinizza † *

Marche
Monte Milone † *

Milano
Brianza †

Parma
Borgo San Donino † *

Piemonte
Alessandria † *
Cereseto † *
Motta di Conti † *
Piedmont †
Turin †

Romagna
Albareto † *
Valdinoce †

Sardinia
Masua *
Sinnai †

Sicily
Girgenti † *
Marsala †
Messina † *
Mineo †
Patti †

Tuscany
Bagnone
Castel Berardenga †
Pentolina †
Piancaldoli †
Siena † *

Umbria
Assisi † *
Collescipoli † *
Narni †

Veneto
Legnano †
Noventa Vicentina †
Vago † *

Vicenza
Calce †

Ivory Coast
Daoukro

Jamaica
Lucky Hill *

Japan

Hokkaido
Atsuma †
Numakai †

Honshu
Aba
Gifu † *
Hachi-oji †
Ida †
Kamiomi
Kasamatsu †
Kesen † *
Kijima (1906) †
Komagome †
Kuga
Kurumi
Kushiike †
Mikomotojima
Minamino †
Nagai †
Nagareyama
Nio †

Ogasawara
Okabe † *
Okano † *
Otomi †
Sakauchi †
Sakurayama †
Sano
Saotome
Sasagase † *
Shibayama
Shirahagi *
Shiraiwa
Shiriya †
Sone †
Suwa
Takenouchi † *
Tané † *
Tanokami Mountain *
Tendo
Tomita † *
Urasaki †
Yonozu † *
Yoshiki †

Kyushu
Fukue
Fukutomi † *
Higashi-koen † *
Kanzaki
Kyushu † *
Nogata †
Ogi † *
Yakushima †

Shikoku
Kochi †
Zaisho †

Jordan
Akaba † *

Kenya
Duruma † *
Meru † *
Sotik †

Korea
Gyokukei †
Shohaku

Choeure-namdo
Duwun †

Lebanon
Aphaca

Libya
Dor el Gani
Oubari *

Malawi
Kota-Kota *

Central province
Dowa †

Mzimba district
Mtola †

Southern province
Machinga † *
Zomba † *

Mali
Gao (Mali) †
N'Goureyma † *
Siratik *
Trifir

Mauritania
Aioun el Atrouss † *
Chinguetti *
Kiffa † *

Adrar
Zerga

Mexico
Chihuahua City *

Baja California
Loreto
Santa Rosalia
Signal Mountain

Chihuahua
Abajo *
Ahumada *
Allende † *
Buenaventura
Carichic *
Casas Grandes *
Chupaderos *
Huizopa *
Morito *
Salaices *
Sierra Blanca *
Torreon de Mata *
Tule
Villa Coronado *

Coahuila
Coahuila *
El Burro *
Puente del Zacate *

Durango
Avilez †
Bella Roca *
Cacaria *
Colonia Obrera
Durango *
Jaralito
Rancho de la Pila (1882) *
Rancho Gomelia
Rodeo *
San Francisco del Mezquital *
Santa Clara *
Santiago Papasquiero *

Guanajuato
Cosina † *
La Charca †

Guerrero
Acapulco † *
Caparrosa

Hidalgo
Pacula † *

Jalisco
Atemajac †
Teocaltiche
Tomatlan † *

Mexico State
Ameca-Ameca

Mexico state
Los Reyes *

Mexico State
Toluca *

Mexico state
Xiquipilco no. 2

Michoacan
Espiritu Santo
Rancho de la Presa † *

Nuevo Leon
Cruz del Aire

Oaxaca
Apoala *
Misteca *
Yanhuitlan *

Puebla
Cholula
Tlacotepec *

San Luis Potosi
Bocas † *
Charcas *

Sinaloa
Bacubirito *

Sonora
Arispe *
Carbo *
Cumpas *
Moctezuma *
Puerto Libertad

Tamaulipas
Nuevo Laredo *
San José
Santa Cruz † *

Tlaxcala
Santa Apolonia *

Zacatecas
La Escondida
Mazapil † *
Nuevo Mercurio † *
Sombrerete
Tuxtuac † *
Zacatecas (1792) *
Zacatecas (1969)

Mongolia
Adzhi-Bogdo (iron)
Adzhi-Bogdo (stone) †

Alaschan

Arhangay Aymag
Tamir-Tsetserleg

Choybalsan aymag
Kerulensky
Khenteisky †
Matad

Omnogov aymag
Manlai
Noyan-Bogdo †

Ovor-Hangay aymag
Tugalin-Bulen †

Moon

Oceanus Procellarum
Bench Crater

Morocco
Douar Mghila † *
Mrirt

Namibia
Aus
Bushman Land
Etosha
Gibeon *
Gobabeb
Hoba *
Itzawisis
Karasburg
Namib Desert
Okahandja
Ovambo †
St. Francis Bay
Witsand Farm † *

New Caledonia
Nassirah †

New Zealand
Morven *

Canterbury
View Hill *

County Southland
Makarewa *

East Coast province
Waingaromia *

North Island
Kimbolton
Mokoia † *

Wellington
Wairarapa Valley

Westland
Dunganville *

Niger
Air † *
Birni N'konni †
Dosso †
Niger (C2)
Niger (L6) †
Niger (LL6) †
Tillaberi †
Timmersoi *

Nigeria
Akwanga †
Bununu †
Geidam †
Git-Git † *
Kabo † *
Karewar † *
Mayo Belwa † *
Ohuma † *
Richa *
Udei Station † *
Uwet *
Zagami † *

Northern Ireland

County Londonderry
Bovedy † *

Norway
Finmarken *
Grefsheim †
Leikanger
Mjelleim †
Morradal *
Oterøy †
Pollen † *
Ski † *
Tromøy †
Trysil †
Tysnes Island † *

Oman
Ghubara *
Jiddat al Harasis *
South Oman *
Tarfa *

Jiddat al Harasis
Hajmah (a) *
Hajmah (b) *
Hajmah (c) *

Pakistan
Karloowala † *

Baluchistan
Karkh † *
Kulak † *

Kashmir
Arundu †

North-west Frontier
Mardan † *
Thal † *

Punjab
Adhi Kot † *
Jhung † *
Lodran † *

Sind
Gambat † *
Garhi Yasin † *
Jajh deh Kot Lalu † *
Khairpur † *
Sindhri † *
Visuni † *

Papua-New Guinea
Dyarrl Island †
Ijopega †

Paraguay
Villarrica †

Asuncion
Ipacaray †

Peru

Ayacucho
Tambo Quemado *

Philippines

Luzon
Bondoc *
Pampanga † *

Mindanao
Pantar † *

Panay
Calivo †

Poland
Białystok † *
Czestochowa Rakow I
Czestochowa Rakow II
Gaj †
Gdynia †
Gnadenfrei † *
Grüneberg † *
Grzempach †
Keilce
Krzadka
Łowicz † *
Morasko *
Oborniki
Ostrzeszów †
Pomorze
Pułtusk † *
Ratyn †
Sagan †
Schellin † *
Schwetz *
Seeläsgen *
Swindnica Gorna
Warsaw I
Wietrzno-Bobrka

Portugal
Chaves † *
Juromenha †
Monte das Fortes †
North Portugal
Palenca de Baixo †
Portugal †

Minho
São Julião de Moreira *

Traz-os-Montes
Picote †

Romania
Kakowa † *
Sopot †
Tuzla *
Zsadany † *

Harghita
Mezö-Madaras † *

Oradea
Tauti †

Transylvania
Mocs † *
Ohaba † *

Rwanda

Ruhengeri Prefecture
Ruhobobo †

Saudi Arabia
ad-Dahbubah
al-Ghanim (iron)
al-Ghanim (stone)
al-Jimshan
Alkhamasin
Al Rais † *
Arabia
ar-Rakhbah
ash-Shalfah
as-Sanam
as-Su'aydan
Bir Hadi
Hedjaz † *
Kaaba
Kumdah
Naifa *
Ras Tanura †
South Dahna
Suwahib (Adraj) *
Suwahib ('Ain Sala) *
Suwahib (Buwah) *
Tathlith † *
Um-Hadid
Umm Tina *
Wabar *
Yafee Mountains

Scotland
Copinsay †
Loch Tay †
Tiree

Perthshire
Perth † *
Strathmore † *

Strathclyde
High Possil † *

Somalia
Bur-Gheluai † *
Ergheo † *
Uegit

South Africa
Bechuanaland
Maria Linden †

Cape Province
Bellsbank *
Bowden
Britstown *
Cape of Good Hope *
Cold Bokkeveld † *
Daniel's Kuil † *
Deelfontein
De Hoek
Diep River † *
Hex River Mountains *
Idutywa † *
Jackalsfontein † *
Karee Kloof *
Kokstad *
Kopjes Vlei *
Kouga Mountains *
Molteno † *
Moshesh
Mount Ayliff *
Muizenberg *
Orange River (stone) †
Piquetberg †
Queen's Mercy † *
Rateldraai *
St. Mark's † *
Schaap-Kooi *
Simondium *
South African Railways
Vaalbult *
Victoria West *
Wittekrantz † *

Natal
Bulls Run †
Macibini †
Natal
N'Kandhla † *

Orange Free State
Cronstad † *
Malvern † *
Orange River (iron) *
Winburg *

Transvaal
Benoni † *
Ellisras
Leeuwfontein † *
Lichtenberg †
Witklip Farm † *

South Korea
Unkoku †

South Yemen
Kaidun †

Spain
Barcelona (iron) †
Barcelona (stone) †
Barea †
Berlanguillas † *
Cabezo de Mayo † *
Cañellas † *
Cangas de Onis † *
Colomera
Cuenta
Garraf *
Gerona
Guareña † *
Jubila del Agua †
La Rinconada †
Los Martinez †
Madrid † *
Majorca †
Molina † *
Nulles † *
Ojuelos Altos †
Oliva-Gandia †
Olivenza † *
Olmedilla de Alarcón † *
Oviedo †
Pulter
Quesa †
Reliegos † *
Roa †
Roda † *
San Cerre de Mallorca †
Sena † *
Sevilla †
Villanueva del Fresno

Sri Lanka
Mulletiwu †

Sudan
Kapoeta † *
Khor Temiki † *
Kingai † *
Malakal † *
Maridi † *
Umm Ruaba † *

Swaziland
Dwaleni † *

Sweden
Bleckenstad
Brunflo
Carlstadt †
Ekeby †
Föllinge
Hallingeberg † *
Hedeskoga †
Hessle † *
Hökmark †
Långhalsen †
Lillaverke † *
Lundsgård † *
Muonionalusta *
Näs
Orebro †
Skåne-Tranås †
Ställdalen † *
Ultuna

Switzerland
Chervettaz † *
Glarus †
Lausanne †
Rafrüti *
Ulmiz †
Utzenstorf † *
Walkringen †

Syria
Aleppo † *
Beyrout †
Emesa

Tanzania
Ishinga †
Ivuna †
Karatu † *
Malampaka †
Mbosi *
Peramiho †
Rupota †
Ufana † *

Thailand
Chiang Khan † *
Nakhon Pathom †

Tibet
Tulung Dzong †

Tunisia
Dahmani †
Tatahouine † *

Turkey
Adalia
Aidin †
Birgi †
Bursa † *
Çanakkale † *
Caratash † *
Constantinople †
Domanitch †
Ephesus
Ibrisim †
Kayakent † *
Magnesia † *
Sediköy † *

Uganda
Awere † *
Maziba †
Soroti † *

Upper Volta
Béréba † *
Bogou †
Gao (Upper Volta) †
Guenie †
Guibga † *
Nadiabondi †

U.S.A.

Alabama
Athens †
Carver
Chulafinnee *
Danville † *
Felix † *
Frankfort (stone) † *
Guin
Ider *
Leighton † *
Lime Creek *
Selma *
Summit *
Sylacauga †
Tombigbee River *
Walker County *

Alaska
Aggie Creek
Chilkoot *
Cold Bay *
Prince of Wales's Straits †

Arizona
Bagdad
Cañon Diablo *
Clover Springs *
Coon Butte *
Cottonwood
El Mirage
Gun Creek *
Hassayampa
Hickiwan
Holbrook † *
Kofa *
Maricopa
Navajo *
Pima County
San Francisco Mountains *
Seligman *
Silver Bell
Southern Arizona
Tucson *
Wallapai *
Weaver Mountains *
Wickenburg (stone) *
Wikieup
Winona *

Arkansas
Cabin Creek † *
Calico Rock *
Fayetteville †
Hatfield
Hope
Joe Wright Mountain *
Miller (Arkansas) † *
Newport *
Norfork † *
Paragould † *
Sandtown *
Success †
Western Arkansas *

California
Canyon City *
Dale Dry Lake
Goose Lake *
Imperial
Ivanpah *
Littlerock
Lucerne Valley
Muroc
Muroc Dry Lake
Needles *
Neenach *
Old Woman
Oroville *
Owens Valley *
Pinto Mountains *
Red Rock
Ridgecrest
Rosamond Dry Lake *
San Emigdio *
San Juan Capistrano †
Shingle Springs *
Somesbar
Surprise Springs *
Twentynine Palms
Valley Wells *
Ventura

Colorado
Adams County *
Akron (1940) *
Akron (1961) *
Alamosa
Arapahoe *
Armel *
Arriba *
Atwood *
Bear Creek *
Bethune *
Bishop Canyon *
Briggsdale *
Canon City † *
Colorado Springs
Cope *
Cortez *
De Nova *
Denver †
Doyleville *
Eaton †
Elba *
Ellicott *
Erie
Estes Park
Fleming *
Franceville *
Fremont Butte *
Georgetown
Granada Creek
Guffey
Holly *
Holyoke *
Horse Creek *
Hugo (stone) *
Idalia
Johnstown † *
Julesburg
Karval *
Kelly *
Kramer Creek *
Lafayette (iron)
Lincoln County *
Lost Lake *
Mesa Verde Park
Mosca
Mount Ouray *
Newsom *
Ovid *
Peetz *
Phillips County (pallasite) *
Rush Creek *
Russel Gulch *
Seibert *
Shaw *
Springfield *
Sterling
Stonington *
Stratton
Thatcher
Thurman
Tobe

Truckton
Two Buttes (a) *
Two Buttes (b)
Washington County
Wauneta
Weldona *
Wiley *
Wray *

Connecticut
Newtown
Stratford †
Weston † *
Wethersfield (1971) †
Wethersfield (1982) †

Florida
Bonita Springs *
Eustis
Grayton
Okechobee

Georgia
Canton *
Cedartown
Dalton *
Enigma
Forsyth † *
Holland's Store *
Locust Grove *
Losttown *
Lumpkin † *
Millen
Norristown
Paulding County
Pickens County *
Pitts †
Putnam County *
Sardis *
Smithonia *
Social Circle
Thomson
Twin City
Union County *

Hawaii
Honolulu † *
Palolo Valley †

Idaho
Hayden Creek *
Idaho
Jerome (Idaho)
Oakley (iron)
Parma Canyon

Illinois
Benld †
Bloomington †
Havana
South Dixon
Tilden † *
Toulon
Woodbine *

Indiana
Hamlet † *
Harrison County † *
Kokomo *
Lafayette (stone) *
La Porte *
Plymouth *
Rochester † *
Rush County *
Rushville
South Bend *
Trevlac

Iowa
Alvord
Estherville † *
Forest City † *
Homestead † *
Iowa City †
Lone Tree *
Mapleton *
Marion (Iowa) † *

Kansas
Achilles *
Admire *
Almelo Township *
Anson *
Anthony *
Argonia
Beardsley † *
Beeler *
Belle Plaine *
Bison *
Brenham *
Brewster *
Brownell
Burdett *
Caldwell
Cedar (Kansas) *
Colby (Kansas) *
Coldwater (iron) *
Coldwater (stone) *
Coolidge *
Covert *
Cullison *
Densmore (1879) *
Densmore (1950) *
Dispatch
Dresden (Kansas) *
Dwight *
Elkhart *
Ellis County *
Elm Creek *
Elyria *
Farmington † *
Franklinville
Galatia *
Garnett *
Goodland *
Grant County *
Great Bend
Gretna *
Hardtner *
Harrison Township
Haven *
Haviland (stone) *
Haviland (b) *
Hesston
Hill City
Horace *
Hoxie †
Hugoton *
Ingalls *
Inman *
Jerome (Kansas) *
Johnson City *
Kalvesta
Ladder Creek *
Lawrence *
Leon
Leoville *
Little River (a) *
Little River (b) *
Long Island *
Luray
Marion (Kansas) *
Mayday
Mayfield *
Miller (Kansas) *
Modoc (1905) † *
Modoc (1948) *
Morland *
Nashville (stone) *
Ness County (1894) *
Ness County (1938) *
New Almelo *
Norcateur *
Norton County † *
Oakley (stone) *
Oberlin *
Otis *
Ottawa † *
Park
Penokee
Phillips County (stone) † *
Pierceville (iron) *
Pierceville (stone) *
Pleasanton *
Portis
Prairie Dog Creek *
Ransom *
Reager
Rolla (1936)
Rolla (1939) *
Rolla (1941) *
St. Peter
Saline *
Scott City *
Seguin
Selden
Seneca *
Shields *
Smith Center *
Sublette
Tonganoxie *
Ulysses *
Waconda *
Waldo *
Wathena
Wilburton *
Willowdale *
Wilmot *
Yocemento *
Zenith

Kentucky
Bath Furnace † *
Campbellsville *
Casey County *
Clark County *
Cumberland Falls † *
Cynthiana † *
Eagle Station *
Edmonton (Kentucky)

Frankfort (iron) *
Franklin *
Glasgow *
Kenton County *
La Grange *
Louisville †
Marshall County *
Mount Vernon † *
Murray † *
Nelson County *
Providence
Salt River *
Scottsville *
Smithland *
Walltown *

Louisiana
Atlanta *

Maine
Andover † *
Castine † *
Nobleborough † *
Searsmont † *

Maryland
Emmitsburg *
Lonaconing *
Nanjemoy † *
St. Mary's County †

Massachusetts
Lowell †
Northampton

Michigan
Allegan † *
Grand Rapids *
Iron River
Kalkaska
Reed City *
Rose City † *
Seneca Township *
Southern Michigan

Minnesota
Anoka
Arlington *
Fisher † *
Hardwick *
Minnesota (iron)
Minnesota (stone)
Turtle River *

Mississippi
Baldwyn †
Oktibbeha County *
Palahatchie †

Missouri
Archie † *
Baxter † *
Billings *
Butler *
Cape Girardeau † *
De Kalb
Fair Play
Faucett *
Harrisonville *
Jenkins
Kansas City (1876) †
Kansas City (1903) *
Lanton *
Little Piney † *
Mincy *
Perryville *
St. Francois County *
St. Genevieve County *
St. Louis †
Seymour *
Warrenton † *

Montana
Harlowton
Illinois Gulch *
Livingston (Montana) *
Lombard *

Nebraska
Ainsworth *
Amherst
Arcadia *
Bayard
Benedict *
Boelus
Brainard
Broken Bow *
Bushnell *
Cotesfield *
Culbertson *
Dix *
Farnum *
Hayes Center *
Hildreth *
Huntsman
Indianola *
Kearney *
Kennard
Lancaster County
Linwood *
Loomis *
Mariaville †
Marsland *
Morrill *
Ogallala *
Potter *
Red Willow
St. Ann *
Sidney
Sioux County † *
Sutton *
Tryon *
Whitman *
York (iron)
York (stone) *

Nevada
Quartz Mountain *
Quinn Canyon

New Jersey
Deal † *

New Mexico
Abbott
Acme *
Alamogordo *
Arabella
Arch (a)
Arch (b) *
Aurora *
Aztec † *
Beenham *
Bethel
Bluewater
Boaz (stone)
Chico *
Chico Hills *
Clovis (no.1)
Clovis (no.2) *
Correo
Costilla Peak *
Delphos (a)
Dora (pallasite) *
Dora (stone)
El Capitan *
Elida
El Rancho Grande
Farley *
Floyd *
Four Corners *
Garrison *
Gladstone (stone) *
Glorieta Mountain *
Grady (1933) *
Grady (1937) *
Grady (c) *
Grant *
Grier (b)
Hasparos
Hobbs *
Kenna *
Kingston *
Krider *
Lakewood *
La Lande *
Los Lunas
Luis Lopez *
Malaga †
Melrose (a) *
Melrose (b) *
Mills *
Milnesand (iron)
Milnesand (stone)
Moriarty
Mosquero
New Mexico
Oasis State Park
Oro Grande
Oscuro Mountains *
Palo Blanco Creek
Pasamonte † *
Picacho *
Piñon *
Point of Rocks (iron)
Point of Rocks (stone)
Portales (a) *
Portales (b) *
Portales (c) *
Puente-Ladron
Ragland
Rogers
Roy (1933) *
Roy (1934) *
Sacramento Mountains *
St. Vrain
Sandia Mountains *
Sierra County
Suwanee Spring
Tafoya (a)

Tafoya (b)
Taiban *
Tatum
Techado
Thoreau
Ute Creek *
Willard *

New York
Bethlehem † *
Burlington *
Cambria *
Lasher Creek
Mount Morris (New York)
Schenectady †
Scriba
Seneca Falls *
South Byron *
Tomhannock Creek *
Yorktown (New York) † *

North Carolina
Alexander County
Asheville *
Bald Mountain † *
Black Mountain *
Bridgewater *
Castalia † *
Caswell County †
Colfax *
Cross Roads † *
Deep Springs *
Duel Hill (1854) *
Duel Hill (1873) *
Farmville †
Ferguson †
Guilford County *
Hendersonville *
Lick Creek *
Linville *
Mayodan
Monroe † *
Moore County † *
Murphy *
Nashville (iron)
Persimmon Creek
Rich Mountain †
Smith's Mountain *
Uwharrie
Wood's Mountain *

North Dakota
Bowesmont *
Freda
Jamestown *
New Leipzig *
Niagara *
Richardton † *

Ohio
Cincinnati *
Dayton
Enon *
New Concord † *
New Westville
Paint Creek †
Pricetown † *
Wooster *

Oklahoma
Amber *
Atoka †
Beaver *
Blackwell †
Cashion *
Colony *
Crescent †
Cushing
Eva *
Felt *
Hunter *
Keyes *
Kingfisher *
Knowles *
Lake Murray
Leedey † *
Logan *
Lost City † *
Marlow
McAlester
Meers
Roosevelt *
Soper *
Springer
Tishomingo
Walters †
Weatherford *
Willowbar
Woodward County *
Zaffra *

Oregon
Klamath Falls
Mulino †
Port Orford
Salem †
Sam's Valley *
Willamette *

Pennsylvania
Bald Eagle *
Black Moshannan Park †
Bradford Woods †
Chicora †
Mount Joy *
New Baltimore *
Pittsburg *
Shrewsbury *

South Carolina
Bishopville † *
Cherokee Springs † *
Chesterville *
Forsyth County *
Laurens County *
Lexington County *
Ruff's Mountain *

South Dakota
Bath † *
Bennett County *
Cavour *
Centerville †
Emery *
Faith *
Fort Pierre *
Harding County *
Mission

Tennessee
Babb's Mill (Blake's Iron)
Babb's Mill (Troost's Iron) *
Bristol *
Carthage *
Charlotte † *
Cleveland *
Clinton
Cookeville *
Coopertown *
Cosby's Creek *
Crab Orchard *
Drake Creek † *
Harriman (Of)
Harriman (Om)
Jackson County *
Jonesboro
Livingston (Tennessee)
Maryville †
Morristown *
Murfreesboro *
Petersburg † *
Savannah
Smithville *
Tazewell *
Waldron Ridge *

Texas
Abernathy *
Adrian *
Aiken
Allen
Allred
Anton
Ashmore *
Avoca (Texas)
Ballinger *
Bartlett
Barwise *
Bells †
Blanket † *
Bledsoe *
Bluff *
Boerne *
Brady
Briscoe *
Bronco
Brownfield (iron) *
Brownfield (1937) *
Brownfield (1964) *
Burkett *
Calliham *
Canyon
Carlton *
Cedar (Texas) *
Cee Vee *
Chamberlin *
Channing *
Chico Mountains
Circle Back *
Clarendon
Claytonville *
Cleburne
Comanche (iron)
Comanche (stone)
Concho *
Cranfills Gap
Crosbyton
Cuero *
Dalhart *

Davis Mountains *
Davy
Dawn (a)
Dawn (b)
Del Rio
Denton County *
Denver City *
Deport *
Dexter *
Dimmitt *
Dumas (a)
Dumas (b)
Duncanville
Edmonson (a) *
Edmonson (b) *
El Paso
Elton
Estacado *
Etter *
Eunice *
Ferguson Switch *
Finney *
Florence †
Florey *
Floydada
Fluvanna (a)
Fluvanna (b) *
Forestburg
Fort Stockton
Frankel City *
Friona
Fuzzy Creek
Gail
Glady
Glen Rose (iron)
Gomez
Grassland *
Gruver *
Guadaloupe County
Hale Center (no. 1) *
Hale Center (no. 2) *
Hamilton (Texas)
Happy (a)
Happy Canyon *
Happy Draw
Harleton † *
Hart Camp
Hartley
Haskell *
Hereford
Holliday
Howe *
Iredell
Junction
Kaffir (b) *
Kaufman
Kendall County *
Kendleton †
Kimble County *
Kirbyville †
Kress *
Kyle *
Laketon *
Lakeview
Lamesa *
La Villa *
Lazbuddie *
Leander
Lider (a)
Lider (b)
Lockney *
Lone Star *
Loop *
Lost Draw
Lubbock *
Mart *
McAddo *
McKinney *
McLean *
Meadow (a)
Mertzon
Miami *
Monahans *
Morton *
Muleshoe *
Myersville
Naruna
Nazareth (c)
Nazareth (iron) *
Nazareth (stone) *
Needmore
New Moore
Nordheim
Odessa (iron) *
Olton
Ozona *
Paloduro
Panhandle
Patricia
Peck's Spring *
Peña Blanca Spring † *
Pep
Pipe Creek *
Plains *
Plainview (1917) *
Plainview (1950)
Plainview (c)
Plainview (d)
Plantersville †
Pony Creek
Post
Red River *
Rhineland
Richland
Richland Springs †
Rock Creek
Romero *
Rosebud
Round Top
Saginaw
St. Lawrence *
San Angelo *
Sanderson *
San Pedro Springs
Scurry *
Seagraves
Seminole *
Seminole Draw (a) *
Seth Ward
Shafter Lake
Shallowater *
Silverton (Texas)
Smyer
Snyder
Somervell County
South Plains
Spearman *
Springlake
Squaw Creek
Summerfield *
Sweetwater *
Tell *
Temple *
Texline *
Tierra Blanca *
Tokio (a) *
Travis County *
Troup †
Tulia (a) *
Tulia (b) *
Umbarger *
Uvalde
Venus
Vigo Park
Wardswell Draw *
Wayside *
Wellington *
Wellman (a) *
Wellman (b)
Wellman (c) *
Wellman (d)
Wellman (e) *
West Point *
Wichita County *
Wimberley
Yorktown (Texas) *
Ysleta
Zapata County

Utah
Altonah *
Beaver-Harrison
Canyonlands
Drum Mountains
Duchesne *
Garland †
Ioka
Salina
Salt Lake City *

Vermont
South Strafford

Virginia
Augusta County *
Botetourt County
Cranberry Plains
Dungannon *
Forksville † *
Hopper
Indian Valley *
Keen Mountain
Norfolk *
Richmond † *
Sharps †
Staunton *

Washington
Kirkland †
Tacoma
Washougal † *
Waterville *
Withrow

West Virginia
Greenbrier County *
Jenny's Creek *
Landes *
Montrose

Wisconsin
Algoma *
Angelica
Belmont
Calumet County
Colby (Wisconsin) † *
Hammond *
Kilbourn † *
Mount Morris (Wisconsin) *
Oshkosh
Pine River *
Trenton *
Vernon County † *
Zenda

Wyoming
Albin (pallasite) *
Albin (stone)
Bear Lodge *
Clareton *
Hat Creek *
Hawk Springs *
Lusk *
Pine Bluffs
Silver Crown *
Torrington † *
Willow Creek *

U.S.A.?
California

USSR

Altay region
Markovka
Polujamki
Rakity

Armenian SSR
Erevan †

Astrakhan region
Repeev Khutor †

Azerbaydzhan SSR
Indarch † *
Kulp †
Yardymly †

Bashkirskaya ASSR
Krasnyi Klyuch †

Belorussiya SSR
Brahin *
Chernyi Bor †
Gressk *
Ruschany †
Zabrodje † *
Zmenj †

Buryat ASSR
Norin-Shibir
Tarbagatai

Checheno-Ingushskaya ASSR
Grosnaja † *

Chita
Budulan *

Estonian SSR
Dorpat †
Kaalijarv *
Oesel † *
Pillistfer † *
Tennasilm † *

Federated SSR
Aliskerovo
Angara
Anyujskij
Aprel'sky
Barnaul †
Bilibino
Bodaibo *
Boguslavka † *
Bolshaya Korta *
Boriskino † *
Borodino † *
Brient †
Burgavli
Chebankol
Demina † *
Dorofeevka
Doroninsk † *
Elga
Ichkala †
Ilinskaya Stanitza *
Karagai
Karakol † *
Kargapole
Khmelevka †
Kifkakhsyagan
Kikino † *
Krasnoi-Ugol † *
Krasnojarsk *
Krutikha †
Kunashak † *
Kuznetzovo † *
Lavrentievka †
Lebedinnyi
Lipovsky
Maldyak
Manych †
Netschaëvo *
Nochtuisk *
Novosibirsk
Novy-Ergi †
Ochansk † *
Orlovka *
Pavlovka † *
Pervomaisky † *
Pesyanoe † *
Petropavlovka
Petropavlovsk *
Preobrazhenka
Raguli
Rakovka † *
Saratov † *
Sarepta *
Seymchan
Sikhote-Alin † *
Simbirsk (of P. Partsch)
Slobodka † *
Slobodka (of P. Partsch)
Stavropol † *
Sungach †
Susuman *
Taiga (stone)
Timochin † *
Toubil River *
Tounkin †
Tsarev
Tunguska †
Tyumen †
Velikoi-Ustyug †
Veliko-Nikolaevsky Priisk *
Vengerovo † *
Verkhne Tschirskaia † *
Verkhne Udinsk *
Vetluga †
Yudoma
Zavetnoe †

Kalinin
Glasatovo † *

Karelia ASSR
Marjalahti † *

Kazakh SSR
Bogoslovka
Nikolaevka †

Kazakhstan SSR
Bischtübe *
Efremovka *
Erofeevka *
Mamra Springs †
Novorybinskoe
Zaisan †

Kazan SSR
Pavlodar (pallasite) *
Pavlodar (stone) †

Kirghizian Republic
Kaptal-Aryk † *

Komi ASSR
Pomozdino

Krasnodar district
Kutais †

Kursk
Sevrukovo † *

Kuybyshev region
Kamyshla

Latvian SSR
Buschhof † *
Lixna † *
Misshof † *
Nerft † *

Lithuanian SSR
Jodzie † *
Novy-Projekt †
Padvarninkai † *
Zemaitkiemis † *

Magadan
Egvekinot

Mordovsky ASSR
Novo-Urei † *

Moscow
Agricultural College
Nikolskoe † *

Perm region
Severny Kolchim

Russian SSR
Ssyromolotovo *

Tatar Republic
Kainsaz † *
Kissij *

Turkmen SSR
Kabakly

Turkmenistan
Bakhardok

Turvinskaya
Chinga *

Ukraine
Alexandrovsky †
Andreevka †
Andryushki
Augustinovka *
Bachmut † *
Berdyansk
Bielokrynitschie † *
Bjelaja Zerkov † *
Borkut † *
Botschetschki †
Chervony Kut †
Divnoe
Dolgovoli † *
Elenovka † *
Genichesk
Gorlovka †
Grossliebenthal † *
Kagarlyk †
Kharkov † *
Knyahinya † *
Krymka † *
Kukschin †
Kuleschovka † *
Leonovka †
Mar'inka
Mighei † *
Oczeretna *
Odessa (stone) *
Okniny † *
Owrucz †
Pavlograd † *
Ryechki †
Savtschenskoje † *
Tomakovka †
Vavilovka † *
Verkhne Dnieprovsk *
Yurtuk † *
Zaborzika † *
Zhovtnevyi † *
Zvonkov †

Yakutsk ASSR
Tobychan
Zhigansk

Venezuela
Ucera † *

Vietnam
Phu Hong †
Phuoc-Binh †
Tuan Tuc †

Wales
Cardiff †

Gwynedd
Beddgelert † *
Pontlyfni † *

Yugoslavia
Avce †
Cacak †
Dimitrovgrad *
Dubrovnik † *
Gueťa †
Hraschina †
Jelica † *
Milena † *
Ozren *
Slavetic † *
Soko-Banja † *
Zavid † *

Zaire
Alberta †
Essebi † *
Kalaba † *
Lusaka †
Yambo †
Yambo no. 2

Zambia

Southern Province
Monze † *

Zimbabwe

Mashonaland
Mangwendi † *

Catalogue of Meteorites

The names of meteorites are printed in roman type; those in italics are synonyms. M.A. refers to Mineralogical Abstracts. Specimens in the British Museum (Natural History) collection are indicated by their register numbers (in square brackets); their weights are given in grams, kilograms or tonnes (1000kg).

Aarhus 56°11′ N., 10°14′ E.
Jutland, Denmark
Fell 1951, October 2, 1813 hrs
Synonym(s): *Arhus*
Stone. Olivine-bronzite chondrite (H6).
Reported, Nord. Astron. Tidsskr., 1951, p.150, 1953, p.56, A. Garboe and K. Callisen, Medd. Dansk Geol. Foren., 1952, **12**, p.284 [M.A. 12-247]. Two stones, 300g and 420g, were recovered; the first stone broke into four pieces, V.F. Buchwald and S. Munck, Analecta geol., 1965, **1**, p.1 [M.A. 17-593]. Olivine Fa_{18}, B. Mason, Geochimica et Cosmochimica Acta, 1963, **27**, p.1011. Analysis, 25.75 % total iron, H.B. Wiik, Geochimica et Cosmochimica Acta, 1956, **9**, p.279.
520g Copenhagen, Univ. Geol. Mus.; 2.9g New York, Amer. Mus. Nat. Hist.; Thin section, Washington, U.S. Nat. Mus.;
Specimen(s): [1972,491], 44.3g. partly crusted end piece off the smaller stone

Aba 35°57′ N., 140°24′ E.
Inashiki, Ibaraki, Honshu, Japan
Pseudometeorite..
A small stone is said to have fallen, 28 April 1927, 0900hrs, and was described and figured, I. Yamamoto and S. Murayama, Pop. Astron., Northfield, Minnesota, 1951, **59**, p.431 [M.A. 11-529]. Not meteoritic, S. Murayama, Nat. Sci. Mus. Tokyo, 1953, **20**, p.152 [M.A. 13-80].

Abajo 26°48′ N., 105°25′ W. approx.
Chihuahua, Mexico
Found 1982
Stone. Olivine-bronzite chondrite (H5).
A crusted individual weighing 319g and a fragment of 12g were found within the strewnfield of the Allende fall, near Cienega de Ceniceros de Abajo, W. Zeitschel, letter of 8 January, 1983 in Min. Dept. BM(NH). Olivine $Fa_{18.9}$, A.L. Graham, priv. comm., 1983.
Specimen(s): [1983,M.2], 319g.

Abakan v Toubil River.

Abancay v St. Genevieve County.

Abbe error for Abee.

Abbott 36°18′ N., 104°17′ W.
Colfax County, New Mexico, U.S.A.
Found 1951, between 1951 and 1960
Stone. Olivine-bronzite chondrite (H6).
Listed, L. LaPaz, Cat. Coll. Inst. Meteor. Univ. New Mexico, 1965. Abbott is about 12 miles from Farley (*q.v.*), 20 miles from Tafoya and Chico (*q.q.v.*). Olivine Fa_{19}, B. Mason, Geochimica et Cosmochimica Acta, 1967, **31**, p.1100. Description, polymict H3 to H6, 25.35 % total iron, R.V. Fodor et al., New Mexico Geol. Soc. Spec. Publ. no. 6, 1976, p.206.
9.5kg Albuquerque, Univ. New Mexico;

Abdel Malek v Nakhla.

Abee 54°13′ N., 113°0′ W.
Alberta, Canada
Fell 1952, June 10, 1105 hrs
Stone. Enstatite chondrite (E4), black, brecciated.
Seen to fall near Abee, N. of Edmonton Alberta. A stone of 107kg was recovered from a hole 2ft. to 3ft. in diameter, 6ft. deep and inclined at 25 degrees to the vertical, P.M. Millman, J. Roy. Astron. Soc. Canada, 1953, **47**, p.32 [M.A. 12-358]. Description, analysis, 30.35 % total iron, K.R. Dawson et al., Geochimica et Cosmochimica Acta, 1960, **21**, p.127. Mineralogy, K. Keil, J. Geophys. Res., 1968, **73**, p.6945. Trace element abundances, J.C. Laul et al., Geochimica et Cosmochimica Acta, 1973, **37**, p.329, C.M. Binz et al., Geochimica et Cosmochimica Acta, 1974, **38**, p.1579. Analysis, 32.52 % total iron, H. von Michaelis et al., Earth planet. Sci. Lett., 1969, **5**, p.387. Consortium study, age etc, K. Marti, Earth planet. Sci. Lett., 1983, **62**, p.116 et seq.
100kg Ottawa, Geol. Surv. Canada; 2.95kg Washington, U.S. Nat. Mus.; 1.86kg Chicago, Field Mus. Nat. Hist.; 1.3kg Tempe, Arizona State Univ.; 1.5kg New York, Amer. Mus. Nat. Hist.;
Specimen(s): [1970,169], 257g.

Abel v Cranbourne.

Aberdeen v Bath.

Aberden v Bath.

Abernathy 33°51′ N., 101°48′ W.
Lubbock County, Texas, U.S.A.
Found 1941
Stone. Olivine-hypersthene chondrite (L6).
A mass of 2914g was found in a rock garden near Hale Center, Texas; it had been picked up 2 miles north of Abernathy, H.H. and A.D. Nininger, The Nininger Collection of Meteorites, Winslow, Arizona, 1950, p.26. Olivine Fa_{23}, B. Mason, Geochimica et Cosmochimica Acta, 1963, **27**, p.1011.
784g Tempe, Arizona State Univ.; 510g Fort Worth, Texas, Monnig Colln.; 485g Chicago, Field Mus. Nat. Hist.; 83g Washington, U.S. Nat. Mus.;
Specimen(s): [1959,842], 788g.

Abert Iron v Toluca.

Abo 60°26′ N., 22°18′ E.
Turku, Finland
Found 1840, known before this year
Stone. Olivine-bronzite chondrite (H).
Olivine Fa_{19}, B. Mason, Geochimica et Cosmochimica Acta, 1963, **27**, p.1011.
1g Paris, Mus. d'Hist. Nat.;

Acapulco 16°53′ N., 99°54′ W.
El Quemado Colony, Acapulco, Guerrero, Mexico
Fell 1976, August 11, 1100 hrs
Synonym(s): *El Quemado*
Stone. Chondrite, anomalous (CHANOM).
A single mass of 1914g was collected from a crater about 30 cm in diameter, Meteor. Bull., 1978 (55), Meteoritics, 1978, **13**, p.327. Full description; chondritic composition, achondritic texture; analysis, 27.5 % total iron, olivine $Fa_{11.9}$, H. Palme et al., Geochimica et Cosmochimica Acta, 1981, **45**, p.727. An unusual meteorite, similar to Allan Hills A77081. Oxygen isotopes, T.K. Mayeda and R.N. Clayton, Geochimica et Cosmochimica Acta, 1980 (Suppl. 14), p.1145.
36g Chicago, Field Mus. Nat. Hist.;
Specimen(s): [1978,M.26], 40.9g.

Accalana v Carraweena.

Achilles 39°46′36″ N., 100°48′48″ W.
Rawlins County, Kansas, U.S.A.
Found 1924, recognized 1950
Stone. Olivine-bronzite chondrite (H5), veined.
A mass the original weight of which is estimated at 16kg was found but was not recognised as a meteorite until 1950. Described and figured, H.O. Stockwell and R.A. Morley, Pop. Astron., Northfield, Minnesota, 1951, **59**, p.429, Contr. Meteoritical Soc., **5**, p.72 [M.A. 11-529]. Olivine Fa_{20}, B. Mason, Geochimica et Cosmochimica Acta, 1963, **27**, p.1011.
12.02kg Hutchinson, Kansas, H.O. Stockwell's Coll.; 546g Washington, U.S. Nat. Mus.; 181g Los Angeles, Univ. of California; 178g Tempe, Arizona State Univ.; 138g New York, Amer. Mus. Nat. Hist.;
Specimen(s): [1960,106], 272g. two fragments

Achiras 33°10′ S., 64°57′ W.
Dept. Rio Cuarto, Cordoba, Argentina
Fell 1902, at night
Stone. Olivine-hypersthene chondrite (L6).
A fragment of 780 grams is described, figured and analysed, 23.07 % total iron, J. Olsacher et al., Bol. Acad. Nac. Cienc. Cordoba, 1951, **39**, p.261 [M.A. 11-530]. Olivine Fa_{25}, B. Mason, Geochimica et Cosmochimica Acta, 1963, **27**, p.1011. The fall took place at night, letter of F.E. Wickman of 26 June, 1972 in Min. Dept. BM(NH)., L.M. Villar, Cienc. Investig., 1968, **24**, p.302 suggests that the time of fall is 1920 January 15 2000hrs. and the same as for Aguilla Blanca and Capilla del Monte. Capilla del Monte appears to be an H group chondrite, the others L group, and part of Villar's data appear to be wrong.
48g New York, Amer. Mus. Nat. Hist.; 6.8g Washington, U.S. Nat. Mus.;
Specimen(s): [1963,530], 31g.; [1962,178], 2.5g.

Acme 33°38′ N., 104°16′ W.
Chaves County, New Mexico, U.S.A.
Found 1947
Stone. Olivine-bronzite chondrite (H5).
A mass of 75kg plus fragments was found, H.H. and A.D. Nininger, The Nininger Collection of Meteorites, Winslow, Arizona, 1950, p.26., Ward's Nat. Sci. Estab. Cat. FM-2, 1949, p.22. Olivine Fa_{19}, B. Mason, Geochimica et Cosmochimica Acta, 1963, **27**, p.1011.
41kg Tempe, Arizona State Univ.; 60g Los Angeles, Univ. of California;
Specimen(s): [1950,141], 26g.; [1959,1009], 32432g.

Adalia 36°54′ N., 30°41′ E.
Konia, Turkey
Found 1883, known in this year
Synonym(s): *Konia*
Stone. Achondrite, Ca-rich. Eucrite (AEUC).
Mentioned, S. Meunier, Meteorites, 1884, p.295. Analysis, B. Mason et al., Smithson. Contrib. Earth Sci., 1979 (22), p.30.
0.95g Paris, Mus. d'Hist. Nat.;

Adams County v Mount Joy.

Adams County 39°58′ N., 103°46′ W.
Colorado, U.S.A.
Found 1928
Stone. Olivine-bronzite chondrite (H5), brecciated.
A badly oxidised stone of 5.7kg was found in section 17, township 1 south, range 57 W. in Adams County, H.H. Nininger, Am. J. Sci., 1931, **22**, p.414 [M.A. 5-12]. Olivine Fa_{19}, B. Mason, Geochimica et Cosmochimica Acta, 1963, **27**, p.1011. Contains unequilibrated lithic fragments, R.V. Fodor et al., Meteoritics, 1980, **15**, p.41.
Main mass, Denver, Colorado School of Mines; 193g Tempe, Arizona State Univ.; 569g Washington, U.S. Nat. Mus.; 250g Chicago, Field Mus. Nat. Hist.; 458g Harvard Univ.; 83g Denver, Mus. Nat. Hist.;
Specimen(s): [1931,274], 350g. and fragments, 29g; [1959, 843], 161.5g.

Adams Diggings v Bluewater.

Adana v Ibrisim.

Adare v Limerick.

Adargas v Chupaderos.

ad-Dahbubah 19°50′ N., 51°15′ E.
Rub'al Khali, Saudi Arabia
Found 1961, March
Synonym(s): *Dahbubah*
Stone. Olivine-bronzite chondrite (H5).
A weathered mass of about 200lb was found, and broke up on moving, the largest fragment being about 90lb, analysis, A.D. Holm, Am. J. Sci., 1962, **260**, p.303. Olivine Fa_{19}, J.T. Wasson, Meteorites, Springer-Verlag, 1974, p.268.
40kg Washington, U.S. Nat. Mus.;

Adelaide
South Australia, Australia
Found 1972, recognized in this year
Stone. Carbonaceous chondrite, type II (CM2).
A single mass of 2031g was in the possession of the South Australian Department of Mines for several years prior to being recognized as a meteorite. Description, with an analysis, R. Davy et al., Meteoritics, 1978, **13**, p.121. Discussion, relationship to Bench Crater (*q.v.*), M.J. Fitzgerald and J.B. Jones, Meteoritics, 1977, **12**, p.443.
1.93kg Adelaide, Geol. Surv. South Austr.; 38g Adelaide, Univ.;

Adelaide v Rhine Villa.

Adelie Land 67°11′ S., 142°23′ E.
Antarctica
Found 1912
Synonym(s): *Antarctic*
Stone. Olivine-hypersthene chondrite (L5).
A stone of 2.2lb was found on snow about 20 miles west of Cape Denison during the Australasian Antarctic Expedition of 1911-1914, D.Mawson, The Home of the Blizzard, 1915, **2**, p.11. Description, with an analysis, F.L. Stillwell, Australasian Antarctic Exp., 1911-1914 Sci. Rep. Ser. A, 1923, **4, Geology** (1), p.1-13. Olivine Fa_{24}, B. Mason, Geochimica et Cosmochimica Acta, 1963, **27**, p.1011. Some chondrules contain chromite, P. Ramdohr, Geochimica et Cosmochimica Acta, 1967, **31**, p.1961.
Main mass, Adelaide, South Australian Mus.; 69g Washington, U.S. Nat. Mus.; 2.4g New York, Amer. Mus. Nat. Hist.;
Specimen(s): [1930,441], 43g. plus 0.5g fragments

Adhi Kot 32°6′ N., 71°48′ E.
Nurpur, Shahpur district, Punjab, Pakistan
Fell 1919, May 1, 1200 hrs
Stone. Enstatite chondrite (E4).
A stone of 4239g fell after detonations from NW. and appearance of smoke; description, G.V. Hobson, Rec. Geol. Surv. India, 1927, **60**, p.128, B. Mason, Geochimica et Cosmochimica Acta, 1966, **30**, p.23. Detailed mineralogy, classification as an enstatite chondrite, type 1, K. Keil, J. Geophys. Res., 1968, **73**, p.6945. Trace element abundances, J.C. Laul et al., Geochimica et Cosmochimica Acta, 1973, **37**, p.329.
4kg Calcutta, Mus. Geol. Surv. India; 65g New York, Amer. Mus. Nat. Hist.; 22g Washington, U.S. Nat. Mus.;
Specimen(s): [1974,M.16], 28.63g. part crusted fragment

Admire 38°42′ N., 96°6′ W. approx.
Lyon County, Kansas, U.S.A.
Found 1881
Synonym(s): *Illinois*
Stony-iron. Pallasite (PAL).
A mass of 12 to 15lb was ploughed up in 1881, and other masses later, making a total weight of over 80kg; description, G.P. Merrill, Proc. U.S. Nat. Mus., 1902, **24**, p.907. Analysis of metal, 10.7 %Ni, 20.3 ppm.Ga, 39.2 ppm.Ge, 0.017 ppm.Ir, J.T. Wasson and S.P. Sedwick, Nature, 1969, **222**, p.22. Olivine $Fa_{12.0}$, P.R. Buseck and J.I. Goldstein, Bull. Geol. Soc. Amer., 1969, **80**, p.2141.
29kg Albuquerque, Univ. of New Mexico; 22kg Washington, U.S. Nat. Mus.; 14kg Chicago, Field Mus. Nat. Hist.; 5.9kg Tempe, Arizona State Univ.; 760g New York, Amer. Mus. Nat. Hist.; 476g Los Angeles, Univ. of California; 3.6kg Univ. of Michigan; 1.5kg Bloomfield Hills, Michigan, Cranbrook Inst. Sci.; 1kg Lansing, Michigan, Michigan State College; 1.8kg Glasgow, Hunterian Mus.; 477g Prague, Nat. Mus; 559g Yale Univ.;
Specimen(s): [1950,377], 3850g. slice, with 58g fragments; [85741], 1036g. slice and 299g fragments; [1959,1049], 3526g. and fragments totalling 623.5g

Adon v Bendigó.

Adraj v Suwahib (Adraj).

Adrar v Chinguetti.

Adrian 35°9′ N., 102°43′ W.
Deaf Smith County, Texas, U.S.A.
Found 1936
Stone. Olivine-bronzite chondrite (H4), veined.
A stone of 6.3kg was found in 1936, and a second of 16.3kg later, A.D. Nininger, Pop. Astron., Northfield, Minnesota, 1937, **45**, p.449 [M.A. 7-62]. Listed, H.H. and A.D. Nininger, The Nininger Collection of Meteorites, Winslow, Arizona, 1950, p.26. Mentioned, W. Wahl, letter of May 23, 1950 in Min. Dept. BM(NH). Olivine Fa_{19}, B. Mason, Geochimica et Cosmochimica Acta, 1963, **27**, p.1011.
7.8kg Tempe, Arizona State Univ.; 140g Chicago, Field Mus. Nat. Hist.; 129g New York, Amer. Mus. Nat. Hist.;
Specimen(s): [1938,39], 838g. slice; [1959,1008], 8391g.

Adzhi-Bogdo (iron) 44°52′ N., 95°25′ E.
Gobi Altay, Mongolia, [Аджи-Богдо II]
Found, about the middle of the 19th century
Synonym(s): *Adzhi-Bogdo II*
Iron.
A large mass about 70×50×45cm was found on the Adzhi-Bogdo Ridge. Several fragments were sent to Ulan-Bator and shown to be meteoritic, G.G. Vorobyev and O. Namnandorzh, Метеоритика, 1958 (16), p.134. The possible identity with Armanty, L.J. Spencer [M.A. 14-129] is very unlikely, as Armanty is much larger.
582kg Ulan-Bator, State Mus.;

Adzhi-Bogdo (stone) 44°50′ N., 95°10′ E.
Gobi Altay, Mongolia, [Аджи-Богдо I]
Fell 1949, October 30, end of the month
Synonym(s): *Adzhi-Bogdo I, Hobdo, Khobdo, Kobdo*
Stone. Chondrite, brecciated.
A mass of 910gm was found, and several more pieces later; the locality was first given as Khobdo, 48°0′N., 91°43′E., L.G. Kvasha, Метеоритика, 1954, **11**, p.81, 1956, **14**, p.14 [M.A. 13-52, -359] and later corrected, G.G. Vorobyev and O. Namnandorzh, Метеоритика, 1958 (16), p.134 [M.A. 14-129].
775g Ulan-Bator, State Mus.; 64g Moscow, Acad. Sci.;

Adzhi-Bogdo I v Adzhi-Bogdo (stone).

Adzhi-Bogdo !I v Agzhi-Bogdo (iron).

Aegospotami
Thrace, Greece
Fell 465, B.C., approx.
Doubtful..
A meteorite is believed to have fallen, E.F.F. Chladni, Die Feuer-Meteore, Wien, 1819, p.177, G.A. Wainwright, J. Egypt. Archaeol., 1930, **16**, p.35.

Aeriotopos v Bear Creek.

Agen 44°13′ N., 0°37′ E.
Lot-et-Garonne, France
Fell 1814, September 5, 1200 hrs
Synonym(s): *Galapian (in part), Monclar-d'Agenais*
Stone. Olivine-bronzite chondrite (H5), veined, xenolithic.
A shower of stones, of total weight of about 30kg the largest weighing about 9kg fell after the appearance of cloud and detonations, H.F. Boudon de Saint-Amans, Ann. Chim., 1814, **92**, p.25., L.W. Gilbert, Ann. Phys. (Gilbert), 1814, **48**, p.340, 395, 402. Olivine Fa_{20}, B. Mason, Geochimica et Cosmochimica Acta, 1963, **27**, p.1011. Xenolithic, R.A. Binns, Geochimica et Cosmochimica Acta, 1968, **32**, p.299.

9.1kg Paris, Mus. d'Hist. Nat.; 1.7kg Rome, Vatican Colln; 250g Chicago, Field Mus. Nat. Hist.; 200g Vienna, Naturhist. Mus.; 163g Washington, U.S. Nat. Mus.; 17.5g Budapest, Nat. Hist. Mus.; 79g Prague, Nat. Mus.; 18g Berlin, Humbolt Univ.; 17g Tempe, Arizona State Univ.;
Specimen(s): [19971], 31g.

Agen v Barbotan.

Agen v Galapian.

Aggie Creek 64°53′ N., 163°10′ W.
Seward peninsula, Alaska, U.S.A.
Found 1942
Synonym(s): *Fairbanks*
Iron. Octahedrite, medium (1.2mm) (IIIA).
A mass of about 43kg was raised by a gold dredge in Aggie Creek. Described, with analysis, 8.54 %Ni, E.P. Henderson, Am. Miner., 1949, **34**, p.229 [M.A. 10-521]. The 75g Fairbanks iron, found before 1947 is part of Aggie Creek according to B. Mason and E.P. Henderson. It has been suggested that this and six other siderites of almost identical composition may be fragments of one original cosmic mass, fallen at different times and places from the same recurrent meteorite swarm, C.P. Olivier, Am. Miner., 1949, **34**, p.232 [M.A. 10-521]. Analysis, 8.48 %Ni, 20.5 ppm.Ga, 39.9 ppm.Ge, 0.46 ppm.Ir, E.R.D. Scott et al., Geochimica et Cosmochimica Acta, 1973, **37**, p.1957.
1kg Washington, U.S. Nat. Mus.; 75g Chicago, Field Mus. Nat. Hist., Fairbanks mass;

Aglar
45°47′ N., 13°22′ E. co-ordinates those of Aquileia, near Trieste
East coast of Adriatic(=Aquileia), Italy
Fell 1112
Doubtful..
Several stones fell, black and hard as iron, E.F.F. Chladni, Die Feuer-Meteore, Wien, 1819, p.197.

Agpalilik v Cape York.

Agra v Kadonah.

Agra v Kheragur.

Agra v Nagaria.

Agram v Hraschina.

Agricultural Academy v Agricultural College.

Agricultural College
Petrovskoye Razumovskoye, Moscow, USSR
Synonym(s): *Agricultural Academy, Petrovskoie-Rasumovskoye, Rousoumousky*
Pseudometeorite. Iron.
A specimen of 180.1g thus labelled is in New York, (Amer. Mus. Nat. Hist.). Its history is unknown, C.A. Reeds, Bull. Am. Mus. Nat. Hist., 1937, **73** (6), p.531 [M.A. 7-61]. A fragment of 725 gm labelled Rousoumouski in the La Plata Mus. acquired in 1905 from Ward of Chicago, is believed to belong to the same mass, M.M. Radice, Notas Mus. La Plata, 1949, **14** (55), p.221 [M.A. 11-440]. There is no record of this mass in Russia, and its history and place of find are obscure, Meteor. Bull., 1959 (13). Considered to be wrought iron, V.F. Buchwald, Iron Meteorites, Univ. of California, 1975, p.248.

Agrigento v Girgenti.

Agua Blanca v Charcas.

Agua Blanca 28°55′ S., 66°57′ W.
Pinchas, dept. Castro Barras, La Rioja, Argentina
Found 1938, before this year
Iron. Octahedrite, medium.
A mass of 49kg was found. Described, with two analyses, 7.42 %Ni, 4.51 %Ni, E.H. Ducloux and R.G. Loyarte, Notas Mus. La Plata, 1939, **4** (8), p.339 [M.A. 7-544].
Main mass, Inca Hausi Mus., La Rioja;
Specimen(s): [1962,169], 2.5g.

Aguada 31°36′ S., 65°14′ W.
Pocho, Cordoba, Argentina
Fell 1930, September
Synonym(s): *Quebrada de la Aguada*
Stone. Olivine-hypersthene chondrite (L6).
A meteorite fell in the Quebrada de la Aguada. Two pieces totalling 1620g are described, with an analysis, J. Olsacher et al., Bol. Acad. Nac. Cienc. Cordoba, 1951, **39**, p.261 [M.A. 11-531]. The fall took place at night, F.E. Wickman, letter of 26 June 1972, in Min. Dept. BM(NH). Olivine Fa_{25}, B. Mason, Geochimica et Cosmochimica Acta, 1963, **27**, p.1011.
Main mass, Cordoba, Argentina, Min. Geol. Mus. of the Facult. Cienc. Exact., Fisic. y Natural.; 74g New York, Amer. Mus. Nat. Hist.; 8g Washington, U.S. Nat. Mus.;
Specimen(s): [1962,172], 3.5g.

Aguas Calientes 25°30′ S., 68°24′ W. approx.
Catamarca, Argentina
Found 1971
Stone. Olivine-bronzite chondrite (H).
Listed, Cat. Meteorites Max-Planck-Inst., Heidelberg, 1979. Classification, co-ordinates, T. Kirsten, priv. comm., 1982.
257g Heidelberg, Max-Planck-Inst.;

Aguila Blanca 30°52′ S., 64°33′ W.
Rio Dolores, Punilla dept., Cordoba, Argentina
Fell 1920, January 15, 2000 hrs
Stone. Olivine-hypersthene chondrite (L).
Two stones were collected, but only one, of 1440g, is preserved. Described, with analysis (the separation of the silicate and metal phases seems to have been incomplete), E.H. Ducloux, Notas Mus. La Plata, 1939, **4** (9), p.353 [M.A. 7-545], L.M. Villar, Cienc. Investig., 1968, **24**, p.302 gives time of fall, (1920, January 15, 2000), the same as for Achiras and Capilla del Monte. Capilla del Monte is probably an olivine-bronzite chondrite, whereas the other two are probably olivine-hypersthene chondrites; in part at least Villar's data are wrong.
Main mass, Buenos Aires, Mus. de Cienc. Nat.;

Ahmedabad 32°5′ N., 44°20′ E.
near Kufah, Iraq
Fell 892, A.D.
Doubtful..
A number of black stones are said to have fallen, and some were brought to Baghdad, E.F.F. Chladni, Die Feuer-Meteore, Wien, 1819, p.192. The evidence of meteoritic character is not conclusive.

Ahnighito v Cape York.

Ahumada 30°42′ N., 105°30′ W. approx.
Chihuahua, Mexico
Found 1909
Stony-iron. Pallasite (PAL).
A mass of 116lb. was found 60 miles east of Ahumada; described, O.C. Farrington, Field Mus. Nat. Hist. Geol. Ser., 1914, **5** (178), p.1. Olivine $Fa_{11.5}$, P.R. Buseck and J.I. Goldstein, Bull. Geol. Soc. Amer., 1969, **80**, p.2141. Analysis of metal, 8.0 %Ni, 21.4 ppm.Ga, 49.0 ppm.Ge, 0.057 ppm.Ir, J.T. Wasson and S.P. Sedwick, Nature, 1969, **222**, p.22.
45.9kg Chicago, Field Mus. Nat. Hist., main mass; 834g Washington, U.S. Nat. Mus.; 711g Tempe, Arizona State Univ.; 3.4kg New York, Amer. Mus. Nat. Hist.; 665g Budapest, Nat. Mus.; 340g Vienna, Naturhist. Mus.;
Specimen(s): [1913,177], 604g. slice; [1912,112], 47g.

Aidin 37°50′ N., 27°48′ E.
Turkey
Fell 1340
Doubtful..
One stone is said to have fallen, G. von Boguslawski, Ann. Phys. Chem. (Poggendorff), 1854, **4** (suppl.), p.10. The evidence is not conclusive.

Aigla v L'Aigle.

Aigle v L'Aigle.

Aiken 34°12′ N., 101°30′ W. approx.
Floyd County, Texas, U.S.A.
Found 1936
Stone. Chondrite?.
One stone of 956g was found, A.D. Nininger, Pop. Astron., Northfield, Minnesota, 1940, **48**, p.555, Contr. Soc. Res. Meteorites, 1940, **2**, p.227 [M.A. 8-54].

Aimi v Gifu.

Ain v Luponnas.

Ain v Simonod.

Ainsa Iron v Tucson.

Ain Sala v Suwahib ('Ain Sala).

Ainsworth 42°36′ N., 99°48′ W. approx.
Brown County, Nebraska, U.S.A.
Found 1907
Synonym(s): *Central Missouri, Dacotah, Dakota, Ponca Creek*
Iron. Octahedrite, coarsest (5mm) (IIB).
A mass of 23.5lb was found; described, with an analysis, 6.49 %Ni, E.E. Howell, Am. J. Sci., 1908, **25**, p.105. Analysis, 5.9 %Ni, 55.7 ppm.Ga, 144 ppm.Ge, 0.023 ppm.Ir, J.T. Wasson, Geochimica et Cosmochimica Acta, 1969, **33**, p.859. Heavily shocked and annealed, paired with Central Missouri and Ponca Creek, V.F. Buchwald, Iron Meteorites, Univ. of California, 1975, p.249. Central Missouri, a 55lb mass found in about 1855; description, H.L. Preston, Am. J. Sci., 1900, **9**, p.285.
6.7kg Washington, U.S. Nat. Mus., 1.7kg Ainsworth, 5kg Central Missouri; 4.5kg New York, Amer. Mus. Nat. Hist., 3.5kg Ainsworth, 1kg Central Missouri; 2.5kg Tempe, Arizona State Univ.; 450g Vienna, Naturhist. Mus.; 0.4kg Michigan Univ.; 347g Budapest, Nat. Mus.; 3208g Chicago, Field Mus. Nat. Hist., 2.5kg Central Missouri, 708g Ainsworth; 1075g Harvard Univ., 1kg Central Missouri, 75g Ainsworth; 558g Los Angeles, Univ. of California, Central Missouri;
Specimen(s): [1908,184], 536g. slice in three pieces, 308g, 219g, 9g.; [35963], 224g. of Ponca Creek; [84554], 988g. slice, of Central Missouri

Aioun el Atrouss 16°23′53″ N., 9°34′13″ W.
Gounquel, Mauritania
Fell 1974, April 17
Stone. Achondrite, Ca-poor. Diogenite (ADIO).
After a fireball and a sonic boom, meteoritic material was recovered by tribesmen from three separate sites in sandy desert terrain. The total mass recovered is not reported, I.S.M. Lomena et al., Meteoritics, 1976, **11**, p.51.
443g Washington, U.S. Nat. Mus.; 104g Chicago, Field Mus. Nat. Hist.; 111g Tempe, Arizona State Univ.; 95g New York, Amer. Mus. Nat. Hist.;
Specimen(s): [1977,M.9], 35.1g.

Air 19°5′ N., 8°23′ E.
Agadez, Niger
Fell 1925
Synonym(s): *Iferouane*
Stone. Olivine-hypersthene chondrite (L6).
A stone (20kg in fragments) was reported to have fallen in 1925, 15km SE of Zoarika and 70km WSW of Iferouane (coordinates as given above), another, (3 pieces, 4kg) was found at Kori Zilalet 50km SSE of Iferouane, 95km from the other; they are attributed to the same fall. Description, A. Lacroix, C. R. Acad. Sci. Paris, 1935, **200**, p.1641 [M.A. 6-103]. Olivine $Fa_{23.3}$, 21.78 % total iron, R.T. Dodd and E. Jarosewich, Meteoritics, 1980, **15**, p.69.
22.9kg Paris, Mus. d'Hist. Nat., main mass;
Specimen(s): [1972,234], 31.75g. an irregular fragment with a little crust

Aire-sur-la-Lys 50°40′ N., 2°20′ E. approx.
Pas de Calais, France
Fell 1769, towards the end of the year
No specimen now preserved, A. Lacroix, Bull. Mus. d'Hist. Nat. Paris, 1927, **33**, p.421. Listed, E.F.F. Chladni, Die Feuer-Meteore, Wien, 1819, p.252.

Akaba 29°31′ N., 35°3′ E.
Jordan
Fell 1949, September 21
Stone. Olivine-hypersthene chondrite (L6).
A bedouin travelling in Saudi Arabia and camping near the frontier of Jordan saw a flaming object fall towards him that appeared to break into three, but he was only able to find one piece. He broke off about one third; the other two thirds, a mass of 779g was purchased by Prof. F.A. Paneth (letter of F.A. Paneth of June 17 1950 in Min. Dept. Brit. Mus. Nat. Hist.), F.A. Paneth, Geochimica et Cosmochimica Acta, 1950, **1**, p.70. Olivine Fa_{24}, B. Mason, Geochimica et Cosmochimica Acta, 1963, **27**, p.1011.
650g Mainz, Max-Planck-Inst.; 8g Washington, U.S. Nat. Mus.; 6g Berlin, Humboldt Univ.; 4g Yale Univ.;
Specimen(s): [1950,339], 70g.

Akbarpoor v Akbarpur.

Akbarpur 29°43′ N., 77°57′ E.
Saharanpur district, Uttar Pradesh, India
Fell 1838, April 18, 0800 hrs
Synonym(s): *Saharanpur, Akbarpoor*
Stone. Olivine-bronzite chondrite (H4), brecciated.
A stone weighing about 4lb fell, after detonations, Peshkar of Munglour, copy of letter of April 28, 1838, in Min. Dept. BM(NH). Olivine Fa_{19}, B. Mason, Geochimica et Cosmochimica Acta, 1963, **27**, p.1011.
30g Vienna, Naturhist. Mus.; 7g Chicago, Field Mus. Nat. Hist.; 8g Berlin, Humboldt Univ.;
Specimen(s): [15646], 1040g. the nearly complete stone, and fragments, 3g

Akershuus v Ski.

Akhicha v Akhricha.

Akhricha 28°27′45″ N., 1°1′30″ E.
Algeria
Found 1968
Synonym(s): *Akhicha, Daiet el Akhricha*
Stone. Olivine-bronzite chondrite (H).
One almost complete stone, and a small fragment 0.5m from it, were found about 15km SSW. of the centre of Daiet el Akricha, 145km ENE. of the city of Adrar (Touat Valley), western Tademait Plateau, Algerian Sahara. The total weight of the meteorite was 1760g and it was found on Quaternary gravel overlying the Cretaceous plateau. The stone is partly oxidised and is evidently old, Meteor. Bull., 1971 (50), Meteoritics, 1971, **6**, p.117.
1817g Paris, Mus. d'Hist. Nat.;

Akpohon v Cape York.

Akron (1940) 40°9′ N., 103°10′ W.
Washington County, Colorado, U.S.A.
Found 1940
Stone. Olivine-bronzite chondrite (H6).
A mass of 4 to 7kg was found, but only 407g were preserved, A.D. Nininger, Pop. Astron., Northfield, Minnesota, 1940, **48**, p.555, Contr. Soc. Res. Meteorites, **2**, p.227 [M.A. 8-54]. A second stone of 643g found in 1954 is also an olivine-bronzite chondrite and is probably part of the same fall. Olivine Fa_{19}, B. Mason, Geochimica et Cosmochimica Acta, 1963, **27**, p.1011. H.H. Nininger, Meteoritics, 1970, **5**, p.215 suggests that the two stones may well be from different falls.
253g Tempe, Arizona State Univ., from the first stone.; Main mass, Tempe, Arizona State Univ., of the second (1954) stone.;
Specimen(s): [1959,844], 118.5g. from the first stone.

Akron (1961) 40°9′ N., 103°10′ W.
Washington County, Colorado, U.S.A.
Found 1961
Synonym(s): *Akron no. 3*
Stone. Olivine-hypersthene chondrite (L6).
A mass of 4kg was found, B. Mason, letter of March 4 1964 in Min. Dept. BM(NH). Reported, Meteor. Bull., 1964 (32). A distinct fall, olivine Fa_{25}, B. Mason, Geochimica et Cosmochimica Acta, 1967, **31**, p.1100.
1.3kg Mainz, Max-Planck-Inst.; 70g Washington, U.S. Nat. Mus.; 42g New York, Amer. Mus. Nat. Hist.;
Specimen(s): [1965,189], 21.7g.

Akron no. 3 v Akron (1961).

Akwanga 8°55′ N., 8°26′ E.
Plateau Province, Nigeria
Fell 1959, July 2
Stone. Olivine-bronzite chondrite (H).
The fall was heard but not seen. A mass of 3kg was found 2 miles from Gaji village in a hole 18 inches deep. Report, with an analysis, 26.13 % total iron, W.N. MacLeod, Rec. Geol. Surv. Nigeria for 1959, 1962, p.15 [M.A. 15-535]. Mentioned, R.R.E. Jacobson, letter of March 24 1962, in Min. Dept. BM(NH). Olivine Fa_{19}, B. Mason, Geochimica et Cosmochimica Acta, 1963, **27**, p.1011.
Main mass, Kaduna, Nigeria, Geol. Surv. Mus.; 0.9g New York, Amer. Mus. Nat. Hist.;

Alabama v Carver.

Alabama v Lime Creek.

Alabama v Walker County.

Alais 44°7′ N., 4°5′ E.
Gard, France
Fell 1806, March 15, 1700 hrs
Synonym(s): *Allais, Valence*
Stone. Carbonaceous chondrite, type I (CI).
Two stones, of about 4 and 2kg respectively, fell after detonations, one at Saint Etienne de Lolm and the other at Valence, M.M. Pages and L.A. d'Hombres-Firmas, J. Phys. Chim. Hist. Nat., 1806, **62**, p.440. Analysis, J.J. Berzelius, Ann. Phys. Chem. (Poggendorff), 1834, **33**, p.113. Analysis by H.B. Wiik, 17.76 % total iron, B. Mason, Space Sci. Rev., 1962, **1**, p.621 [M.A. 16-640]. Very little preserved. Type I classification, B. Mason, Meteoritics, 1971, **6**, p.59. Trace element data, G.W. Kallemeyn and J.T. Wasson, Geochimica et Cosmochimica Acta, 1981, **45**, p.1217.
45g Paris, Mus. d'Hist. Nat.; 14g Chicago, Field Mus. Nat. Hist.; 6.2g Rome, Vatican Colln; 4.5g Tempe, Arizona State Univ.; 3g Ottawa, Mus. Geol. Surv. Canada.; 1.9g Tübingen, Univ.;
Specimen(s): [61329], 10.5g.; [33964], 2.5g.

Alamogordo 32°54′ N., 105°56′ W.
Otero County, New Mexico, U.S.A.
Found 1938
Synonym(s): *Sacramento Peak*
Stone. Olivine-bronzite chondrite (H5).
Many fragments were found, totalling 13.6kg., at 32°26′N., 105°44′W., A.D. Nininger, Pop. Astron., Northfield, Minnesota, 1939, **47**, p.211. Corrected co-ordinates, H.H. and A.D. Nininger, The Nininger Collection of Meteorites, Winslow, Arizona, 1950, p.27. Olivine Fa_{19}, B. Mason, Geochimica et Cosmochimica Acta, 1963, **27**, p.1011.
2040g Tempe, Arizona State Univ.; 905g Washington, U.S. Nat. Mus.; 365g New York, Amer. Mus. Nat. Hist.; 84g Los Angeles, Univ. of California; 574g Chicago, Field Mus. Nat. Hist.; 642g Albuquerque, Univ. of New Mexico;
Specimen(s): [1949,73], 84.4g.; [1959,838], 1623g. four fragments, 740g, 554g, 256.5g, 73g.

Alamosa 37°28′ N., 105°52′ W.
Alamosa County, Colorado, U.S.A.
Found 1937
Stone. Olivine-hypersthene chondrite (L6).
One stone of 1.8kg was found, A.D. Nininger, Pop. Astron., Northfield, Minnesota, 1939, **47**, p.211, The Nininger Collection of Meteorites, Winslow, Arizona, 1950, p.27. Olivine Fa_{25}, B. Mason, Geochimica et Cosmochimica Acta, 1963, **27**, p.1011.
1784g Washington, U.S. Nat. Mus., main mass;

Alandroal v Juromenha.

Alaschan
Mongolia
Doubtful..
Mentioned, F.A. Berwerth, Fortschr. Min. Krist. Petr., 1912, **2**, p.230. Description, P.K. Kovlov, Notiz. Isv. Russ. Geogr. Soc. St. Petersburg (for 1907), 1908, **43**, p.215-217 as an extraordinary meteor, but no meteorite was found.

Alastoeva v Djati-Pengilon.

Alastoewa v Djati-Pengilon.

Alatage 42°20′ N., 93° E.
Xinjiang, China
Found 1959, April
Synonym(s): *Hami*
Iron.
A single mass of 37.5kg was found, no further references, D. Bian, Meteoritics, 1981, **16**, p.115.
37.5kg Changchun, School of Metallurgy and Geology Mus.;

Alatyr v Novo-Urei.

Albacher Mühle v Bitburg.

Alba Julia v Ohaba.

Albany County v Bethlehem.

Albarello v Albareto.

Albareto 44°39′ N., 11°1′ E.
Modena, Romagna, Italy
Fell 1766, July, 1700 hrs, in the middle of the month
Synonym(s): *Albarello, Alboreto, Alboretto, Modena*
Stone. Olivine-hypersthene chondrite (L4).
A stone of about 2kg fell, after detonations, D. Troili, Della caduta di un sasso dall'aria, Modena, 1766, E.F.F. Chladni, Die Feuer-Meteore, Wien, 1819, p.250. Analysis, P. Maissen, Gazzetta Chimica Italiana, 1880, **10**, p.20. New analysis, P. Gallitelli, Periodico Miner., 1939, **10**, p.345 [M.A. 7-540], B. Baldanza, Min. Mag., 1965, **35**, p.214. Co-ordinates from G.R. Levi-Donati (priv. comm.). Olivine Fa_{24}, B. Mason, Geochimica et Cosmochimica Acta, 1963, **27**, p.1011.
605g Modena Univ.; 145g Rome, Univ.; 80g Vienna, Naturhist. Mus.; 13g Tempe, Arizona State Univ.; 6g Chicago, Field Mus. Nat. Hist.; 3g Prague, Nat. Mus.; 3g Berlin, Humboldt Univ.; 1.7g Washington, U.S. Nat. Mus.;
Specimen(s): [55387], 50.5g.; [35728], 1g.

Alberta 2° N., 22°40′ E. approx.
Bumba, Zaire
Fell 1949, November 13, 1400 hrs
Stone. Olivine-hypersthene chondrite (L).
One stone of 625g fell at Alberta, 12 km from Bumba. Described, with analysis, G. Haine and G. Viseur, Bull. Serv. Geol. Congo Belge, 1954, **5**, p.29 [M.A. 12-611]. Reported, Meteor. Bull., 1958 (8).

Albert Iron v Toluca.

Albin (pallasite) 41°30′ N., 104°6′ W.
Laramie County, Wyoming, U.S.A.
Found 1915, recognized 1935
Stony-iron. Pallasite (PAL).
A mass of 83lb was found 5 miles north of Albin in 1915 and recognized as meteoritic in 1935. The olivine crystals are very clear, and up to 37mm across. Description, H.H. Nininger, The Mines Mag., Golden, Colorado, 1937, **27**, p.16 [M.A. 7-70]. Analysis of metal, 10.4 %Ni, 16.8 ppm.Ga, 29.4 ppm.Ge, 0.015 ppm.Ir, J.T. Wasson and S.P. Sedwick, Nature, 1969, **222**, p.22. Olivine $Fa_{12.5}$, P.R. Buseck and J.I. Goldstein, Bull. Geol. Soc. Amer., 1969, **80**, p.2141.
21kg Tempe, Arizona State Univ.; 3.2kg Washington, U.S. Nat. Mus.; 1.7kg New York, Amer. Mus. Nat. Hist.; 400g Vienna, Naturhist. Mus.; 346g Chicago, Field Mus. Nat. Hist.;
Specimen(s): [1959,1010], 5357g.

Albin (stone) 41°25′ N., 104°6′ W.
Laramie County, Wyoming, U.S.A.
Found 1949, recognized in this year
Stone. Olivine-hypersthene chondrite (L).
A mass of 15.4kg was found, H.H. and A.D. Nininger, The Nininger Collection of Meteorites, Winslow, Arizona, 1950, p.28. Olivine Fa_{22}, B. Mason, Geochimica et Cosmochimica Acta, 1963, **27**, p.1011.
8g Tempe, Arizona State Univ.;

Alboreto v Albareto.

Alboretto v Albareto.

Albuquerque v Cañon Diablo.

Albuquerque v Glorieta Mountain.

Aldsworth 51°47′ N., 1°47′ W.
Cirencester, Gloucestershire, England
Fell 1835, August 4, 1630 hrs
Synonym(s): *Cirencester*
Stone. Olivine-hypersthene chondrite, amphoterite (LL5), veined.
One stone of 1.5lb and a shower of smaller stones fell 0.5 miles from Aldsworth, after detonations and appearance of a fireball at Cirencester, T.C. Brown, Rep. Brit. Assn., 1857, **27**, p.140. Description, with analysis, 20.24 % total iron, olivine Fa_{28}, R. Hutchison, Acta Geophys. Polon., 1973, **21**, p.203.
15g Vienna, Naturhist. Mus.; 5.5g Harvard Univ.; 4g Chicago, Field Mus. Nat. Hist.; 1.4g New York, Amer. Mus. Nat. Hist.;
Specimen(s): [61308], 498g. main mass

Aleksandrovskii v Alexandrovsky.

Aleksandrovskii Khutor v Alexandrovsky.

Alekseevka v Bachmut.

Aleksinac v Soko-Banja.

Aleppo 36°14′ N., 37°8′ E.

Syria
Fell 1873, approx.
Synonym(s): *Haleb, Tirnova*
Stone. Olivine-hypersthene chondrite (L6), brecciated.

According to Dr. Halid Edhem Bey's labels sent by Dr. L. Eger with the BM(NH). specimen, the meteorite is called Haleb (not Tirnova, as given earlier by mistake), and Haleb is identical with Aleppo. A specimen had been offered to the Paris Museum by Dr. Halid Edhem Bey, by mistake as from Tirnova, Roumelia, S. Meunier, C. R. Acad. Sci. Paris, 1893, **117**, p.257. The stone weighed about 7lb. Olivine Fa_{24}, B. Mason, Geochimica et Cosmochimica Acta, 1963, **27**, p.1011.

581g Vienna, Naturhist. Mus.; 477g Budapest, Nat. Mus.; 298g Prague, Bohemian Mus.; 162g Washington, U.S. Nat. Mus.; 131g New York, Amer. Mus. Nat. Hist.; 19.5g Chicago, Field Mus. Nat. Hist.; 9g Berlin, Humboldt Univ.;

Specimen(s): [70349], 66g. and fragments, 7g

Alessandria 44°53′ N., 8°45′ E.

Piemonte, Italy
Fell 1860, February 2, 1145 hrs
Synonym(s): *Alexandria, Allessandria, Piedmont, San Giuliano, San Giuliano di Alessandria, San Giuliano Vecchio, Santa Giulietta, Thal von San Giuliano Vecchio*
Stone. Olivine-bronzite chondrite (H5), veined.

Several stones are said to have fallen, about seven in number, weighing from 300g to 1kg each, at San Giuliano Vecchio, G. Missaghi, Il Nuovo Cimento, Pisa, 1861, **13**, p.272. The co-ordinates are of San Giuliano Vecchio, G.R. Levi-Donati (priv. comm.). Mentioned, B. Baldanza, Min. Mag., 1965, **35**, p.214. Brief description, with an analysis, olivine $Fa_{18.3}$, 28.32 % total iron, G.R. Levi-Donati and G.P. Sighinolfi, Meteoritics, 1977, **12**, p.291, abs.

256g Turin, Univ.; 78g Vienna, Naturhist. Mus.; 37g Washington, U.S. Nat. Mus.; 70g Chicago, Field Mus. Nat. Hist.; 21g Prague, Nat. Mus.; 8.7g Tempe, Arizona State Univ.; 12g Berlin, Humboldt Mus.; 8g New York, Amer. Mus. Nat. Hist.;

Specimen(s): [1920,280], 137g.; [35332], 9g.; [33296], thin section

Alexander v Gibeon.

Alexander County 35°45′ N., 81°15′ W. approx.

North Carolina, U.S.A.
Found 1875, before this year
Synonym(s): *Cedar Creek*
Iron. Octahedrite, coarse (2.0mm) (I).

Description, with an analysis, 5.86 %Ni, S.C.H. Bailey, J. Elisha Mitchell Sci. Soc., 1891, **8**, p.17. Intensely shocked, V.F. Buchwald, Iron Meteorites, Univ. of California, 1975, p.255.

193g, were in Bailey's collection in 1891; 60g New York, Amer. Mus. Nat. Hist.; 12g Washington, U.S. Nat. Mus.; 1g Tempe, Arizona State Univ.;

Alexandria v Alessandria.

Alexandrovsky 50°57′ N., 31°49′ E.

Nyezhin district, Ukraine, USSR, [Александровский Хутор]
Fell 1900, July 8, 1600 hrs
Synonym(s): *Aleksandrovskii, Aleksandrovskii Khutor, Alexandrovskii Khutor, Alexandrowskij Chutor, Alexandrovsky Khutor, Njeschin, Nyezhin*
Stone. Olivine-bronzite chondrite (H4).

A stone of 9251g fell, L.A. Kulik, Метеоритика, 1941, **1**, p.73 [M.A. 9-294], E.L. Krinov, Астрон. Журнал, 1945, **22**, p.303 [M.A. 9-297], E.L. Krinov, Каталог Метеоритов Акад. Наук СССР, Москва, 1947 [M.A. 10-511]. Analysis, 27.37 % total iron, M.I. D'yakonova and V.Ya. Kharitonova, Метеоритика, 1961, **21**, p.52. Olivine Fa_{18}, B. Mason, Geochimica et Cosmochimica Acta, 1963, **27**, p.1011.

9172g Moscow, Acad. Sci., main mass;

Alexandrovskii Khutor v Alexandrovsky.

Alexandrovsky Khutor v Alexandrovsky.

Alexandrowskij Chutor v Alexandrovsky.

Alexejevka v Bachmut.

Alexinatz v Soko-Banja.

Alfenas v Patrimonio.

Alfianello 45°16′ N., 10°9′ E.

Brescia, Lombardy, Italy
Fell 1883, February 16, 1500 hrs
Synonym(s): *Brescia, Cremona, Verolanuova*
Stone. Olivine-hypersthene chondrite (L6).

A stone of about 228kg fell, after detonations, L. Bombicci, Atti R. Accad. Lincei Roma, Cl. Sci. Fis. Mat. Nat., 1882-3, **14**, p.675. Description and analysis, H. von Foullon, Sitzungsber. Akad. Wiss. Wien, Math.-naturwiss. Kl., 1883, **88**, p.433. Mentioned, B. Baldanza, Min. Mag., 1965, **35**, p.214, G.R. Levi-Donati, Atti Mem. Acad. Sci. Lett. Arti Modena, 1955, **13**, p.160 [M.A. 14-50]. Olivine Fa_{24}, B. Mason, Geochimica et Cosmochimica Acta, 1963, **27**, p.1011. Lightly shocked, G.R. Levi-Donati, Meteoritics, 1971, **6**, p.225. Analysis, 21.80 % total iron, V.Ya. Kharitonova, Метеоритика, 1968, **28**, p.138. 21.25 % total iron, H. von Michaelis et al., Earth planet. Sci. Lett., 1969, **5**, p.387.

12.7kg Berlin, Humboldt Univ.; 7.5kg Rome, Univ.; 5kg Budapest, Nat. Mus.; 6.5kg Chicago, Field Mus. Nat. Hist.; 2kg Bologna,; 5kg Brescia,; 1.5kg Copenhagen, Univ. Geol. Mus.; 3kg Milan,; 1.5kg New York, Amer. Mus. Nat. Hist.; 1kg Paris, Mus. d'Hist. Nat.; 7kg Tartu,; 3.5kg Rome, the Vatican; 1.5kg Vienna, Naturhist. Mus.; 756g Harvard Univ.; 680g Washington, U.S. Nat. Mus.; 169g Dublin, Nat. Mus.;

Specimen(s): [55240], 2457g. and 194g and fragments 47.5g

al-Ghanim (iron) 19°50′ N., 54°5′ E.

Rub'al Khali, Saudi Arabia
Found 1960
Synonym(s): *Ghanim*
Iron. Octahedrite?.

About 500g of fragments, mostly badly oxidised, were recovered. Described, with an analysis, 8.2 %Ni, D.A. Holm, Am. J. Sci., 1962, **260**, p.303.

500g Washington, U.S. Nat. Mus., fragments;

al-Ghanim (stone) 19°42′ N., 53°58′ E.
Rub'al Khali, Saudi Arabia
Found 1960, May
Synonym(s): *Ghanim*
Stone. Olivine-hypersthene chondrite (L6).
An oxidised mass of 780g was found. Later two more fragments were recovered, one of 1kg and one of 1975g in the same area. Analyses of all three specimens suggest that they are parts of the same fall, D.A. Holm, Am. J. Sci., 1962, **260**, p.303. Classification, olivine Fa_{25}, B. Mason, priv. comm., 1980.
3.7kg Washington, U.S. Nat. Mus., main mass;

Algoma 44°39′ N., 87°28′ W.
Kewaunee County, Wisconsin, U.S.A.
Found 1887
Iron. Octahedrite, medium (0.60mm) (IIICD).
A mass of 9lb., discoid in shape, was ploughed up. Described, with an analysis, 10.62 %Ni, W.H. Hobbs, Bull. Geol. Soc. Amer., 1903, **14**, p.97. Chemically anomalous, 10.75 %Ni, 17.9 ppm.Ga, 38.3 ppm.Ge, 0.35 ppm.Ir, E.R.D. Scott et al., Geochimica et Cosmochimica Acta, 1973, **37**, p.1957. Description, V.F. Buchwald, Iron Meteorites, Univ. of California, 1975, p.255.
Main mass, Wisconsin Univ.; 22g Washington, U.S. Nat. Mus.; 10g Chicago, Field Mus. Nat. Hist.; 4g Berlin, Humboldt Univ.;
Specimen(s): [86139], 18g. four fragments

al-Hadida v Wabar.

Alice Springs v Huckitta.

Alikatnima 23°20′ S., 134°7′ E.
Northern Territory, Australia
Found 1931
Iron. Ataxite, Ni-rich (IRANOM).
Two pieces of 20 and 15lb in the South Austr. Mus. Adelaide a third in Central Australia, A.R. Alderman, Rec. S. Austr. Mus., 1936, **5**, p.537 [M.A. 7-71]. Listed, D.W.P. Corbett, Rec. S. Austr. Mus., 1968, **15**, p.767. Chemically anomalous, 13.8 %Ni, 0.28 ppm.Ga, 0.14 ppm.Ge, 4.2 ppm.Ir, D.J. Malvin et al., priv. comm., 1983.
14.9kg Adelaide, S. Austr. Mus., includes the main mass of 9.07kg; 32g Washington, U.S. Nat. Mus.;

Aliskerovo 67°53′ N., 167°30′ E.
Magadan region, Federated SSR, USSR
Found 1977, July 10
Iron. Octahedrite, medium (1.2mm) (IIIA).
A single mass of 58.4kg was found at a depth of 7-8 metres, Meteor. Bull., 1978 (55), Meteoritics, 1978, **13**, p.328. Description and analysis, 9.25 %Ni, S.G. Zhelnin et al., Метеоритика, 1980, **39**, p.54. Further analysis, 8.8 % Ni, 19 ppm.Ga, 45 ppm.Ge, 0.34 ppm.Ir, G.M. Kolesov et al., Метеоритика, 1982, **40**, p.45.

al-Jimshan 20°42′ N., 52°50′ E.
Rub'al Khali, Saudi Arabia
Found 1955
Synonym(s): *Jimshan*
Stone.
No details are available for this probably stony meteorite, D.A. Holm, Am. J. Sci., 1962, **260**, p.303. See ar-Rakhbah.

Alkhamasin 20°36′ N., 44°53′ E.
Wadi al Dawasir, Saudi Arabia
Found 1973
Iron. Octahedrite, coarsest (IIB).
A mass of 1200kg which had been known for some time by natives was recognized as meteoritic in 1973. It was found about 25 km NE. of Al Khamasin and removed to King Saud University, Riyadh. Reported, with an analysis, Meteor. Bull., 1983 (61), Meteoritics, 1983, **18**, p.77. Brief report, analysis, 6.15 %Ni, 56 ppm.Ga, 160 ppm.Ge, 0.07 ppm.Ir, A.A. Almohandis and R.S. Clarke, Jr., Meteoritics, 1983, **18**, p.abs.
1200kg Riyadh, King Saud Univ.;

Allahabad v Chail.

Allahabad v Futtehpur.

Allais v Alais.

Allan Hills A76001 76°45′ S., 159°20′ E. approx.
Found 1977, January
Synonym(s): *Allan Hills No. 1, Allan Nunatak No. 1*
Stone. Olivine-hypersthene chondrite (L6).
A single mass of 20.151kg was found, olivine $Fa_{24.5}$. This is one of nine meteorites, see following entries, found in the Allan Hills area during the 1976-1977 field season in Antarctica. The specimens were numbered in the sequence of their recovery and are so named, E. Olsen et al., Meteoritics, 1978, **13**, p.209, W.A. Cassidy et al., Science, 1977, **198**, p.727. Briefly reported, Meteor. Bull., 1979 (56), Meteoritics, 1979, **14**, p.161.
Specimen(s): [1978,M.1], 598.1g.

Allan Hills A76002 76°45′ S., 159°22′34″ E.
Synonym(s): *Allan Hills No. 2, Allan Nunatak No. 2*
Iron. Octahedrite, coarsest (IA).
A single mass of 1510g was found. Analysis and classification, 7.0 %Ni, 92.4 ppm.Ga, 423 ppm.Ge, 2.3 ppm.Ir, A. Kracher et al., Geochimica et Cosmochimica Acta, 1980, **44**, p.773. Chemically identical to Allan Hills A77250, 10.5kg, Allan Hills A77263, 1.66kg, Allan Hills A77289, 2186g and Allan Hills A77290, 3784g, R.S. Clarke, Jr. et al., Meteoritics, 1980, **15**, p.273, abs.
Specimen(s): [1978,M.2], 118.1g.

Allan Hills A76003 76°44′11″ S., 159°20′47″ E.
Synonym(s): *Allan Hills No. 3, Allan Nunatak No. 3*
Stone. Olivine-hypersthene chondrite (L6).
Three masses, totalling 10.495kg, were found. Olivine $Fa_{24.6}$. Contains cristobalite-pyroxene inclusion, E.J. Olsen et al., Earth planet. Sci. Lett., 1981, **56**, p.82.
Specimen(s): [1978,M.3], 312.4g.

Allan Hills A76004 76°46′30″ S., 159°21′5″ E.
Synonym(s): *Allan Hills No. 4, Allan Nunatak No. 4*
Stone. Olivine-hypersthene chondrite, amphoterite (LL3).
A single mass of 305g was found olivine $Fa_{0\text{-}34}$.
Specimen(s): [1978,M.4], 17.5g.

Allan Hills A76005 76°37′46″ S., 159°15′42″ E.
Synonym(s): *Allan Hills A81006, Allan Hills A81007, Allan Hills A81008, Allan Hills A81010*
Stone. Achondrite, Ca-rich. Eucrite (AEUC).
A single mass of 1425g was found. Description, polymict; analysis, L. Grossman et al., Geochimica et Cosmochimica Acta, 1981, **45**, p.1267.
Specimen(s): [1978,M.5], 73.8g.

Allan Hills A76006 76°38′59″ S., 159°14′42″ E.

Synonym(s): *Allan Hills No. 6, Allan Nunatak No.6*
Stone. Olivine-bronzite chondrite (H6).
A single mass of 1137g was found, olivine $Fa_{18.3}$.
Specimen(s): [1978,M.6], 70g.

Allan Hills A76007 76°40′36″ S., 159°13′14″ E.

Synonym(s): *Allan Hills No. 7, Allan Nunatak No. 7*
Stone. Olivine-hypersthene chondrite (L6).
A single mass of 410g was found, olivine $Fa_{24.4}$.
Specimen(s): [1978,M.7], 13.9g.

Allan Hills A76008 76°40′44″ S., 159°10′24″ E.

Synonym(s): *Allan Hills No. 8, Allan Nunatak No. 8*
Stone. Olivine-bronzite chondrite (H6).
A single mass of 1150g was found, olivine $Fa_{19.2}$.
Specimen(s): [1978,M.8], 62.1g.

Allan Hills A76009 76°42′26″ S., 159°7′43″ E.

Synonym(s): *Allan Hills No. 9, Allan Nunatak No. 9*
Stone. Olivine-hypersthene chondrite (L6).
Thirty-three masses were found, totalling 407kg, olivine $Fa_{24.1}$.
Specimen(s): [1978,M.9], 200.6g.

Allan Hills A77001 76°45′ S., 159°20′ E. approx.

Victoria Land, Antarctica
Found 1977, December 1977 and January 1978
Stone. Olivine-hypersthene chondrite (L6).
A mass of 252g was found, olivine Fa_{25}, Antarctic Meteorite Newsletter, 1981, **4**, p.9. Over 300 masses, mainly chondritic and representing a large number of individual falls, were found in the Allan Hills area during the 1977-1978 field season in Antarctica. The specimens range in weight up to 10.5kg and were numbered in the sequence of recovery and are so named. Listed here are those ordinary chondrite specimens of mass greater than 500g, achondrites and irons. For a fuller listing of the recovered material and references see, Meteor. Bull., 1980 (57), Meteoritics, 1980, **15**, p.95, U.B. Marvin and B. Mason, Smithson. Contrib. Earth Sci., 1980 (23).

Allan Hills A77003

Stone. Carbonaceous chondrite, type III (CO3).
A mass of 780g was found, olivine $Fa_{4\text{-}48}$, 24.51 % total iron. Classification, E.R.D. Scott et al., Meteoritics, 1981, **16**, p.385, abs.

Allan Hills A77004

Synonym(s): *Allan Hills A77191, Allan Hills A77192, Allan Hills A77233, Allan Hills A81044, Allan Hills A81048*
Stone. Olivine-bronzite chondrite (H4).
A mass of 2.23kg was found (Allan Hills A77004). Other masses thought to be synonymous are Allan Hills A77191, 642g; Allan Hills A77192, 845g; Allan Hills A77233, 4087g.

Allan Hills A77005

Stone. Achondrite, anomalous (ACANOM).
A single mass of 482g was found. Anomalous achondrite, olivine Fa_{28}. Description, H.Y. McSween, Jr., et al., Earth planet. Sci. Lett., 1979, **45**, p.275. Analysis, RE data, M.-S. Ma et al., Geochimica et Cosmochimica Acta, 1981 (suppl. 16), p.1349. Petrogenesis, chronology, C.-Y. Shih et al., Geochimica et Cosmochimica Acta, 1982, **46**, p.2323.

Allan Hills A77009 v Allan Hills A78084.

Allan Hills A77011

Synonym(s): *Allan Hills A77015, Allan Hills A77052, Allan Hills A77115, Allan Hills A77166, Allan Hills A77167, Allan Hills A77214, Allan Hills A77241, Allan Hills A77249, Allan Hills A77260, Allan Hills A78038, Allan Hills A79045, Allan Hills A81024*
Stone. Olivine-hypersthene chondrite (L3).
An individual of 291.5g was found. Classification, pairing with 33 other specimens [only those of over 100g included here], contains graphite/magnetite aggregates, olivine $Fa_{4\text{-}36}$, S.G. McKinley et al., Geochimica et Cosmochimica Acta, 1981 (suppl. 16), p.1039. Analysis, of Allan Hills A77214 specimen, 20.77 % total iron, E. Jarosewich, Smithson. Contrib. Earth Sci., 1980 (23), p.48.

Allan Hills A77015 v Allan Hills A77011.

Allan Hills A77050

Stone. Olivine-hypersthene chondrite (L6).
A fragment weighing 1.04kg was found.

Allan Hills A77052 v Allan Hills A77011.

Allan Hills A77081

Stone. Chondrite, anomalous (CHANOM).
A mass of 8.59g was recovered. Equilibrated; achondritic structure; mineralogically chondritic, olivine Fa_{11}, Antarctic Meteorite Newsletter, 1981, **4**, p.20. Very similar to Acapulco (*q.v.*). Brief petrographic description, H. Takeda and K. Yanai, Meteoritics, 1980, **15**, p.373, abs. Discussion, trace element analysis, gas retention ages, L. Schultz et al., Earth planet. Sci. Lett., 1982, **61**, p.23.

Allan Hills A77105

Stone. Olivine-hypersthene chondrite (L6).
A mass of 9.41kg was found.

Allan Hills A77115 v Allan Hills A77011.

Allan Hills A77166 v Allan Hills A77011.

Allan Hills A77167 v Allan Hills A77011.

Allan Hills A77182

Stone. Olivine-bronzite chondrite (H5).
A mass of 1.12kg was found.

Allan Hills A77191 v Allan Hills A77004.

Allan Hills A77192 v Allan Hills A77004.

Allan Hills A77208

Stone. Olivine-bronzite chondrite (H4).
A mass of 1.73kg was found.

Allan Hills A77214 v Allan Hills A77011.

Allan Hills A77215 v Allan Hills A77011.

Allan Hills A77219

Stony-iron. Mesosiderite (MES).
A mass of 637g was found. It contains rare olivine, 36.64 % total iron. Description, mineralogy, W.N. Agosto et al., Geochimica et Cosmochimica Acta, 1980 (Suppl. 14), p.1027.

Allan Hills A77224
Stone. Olivine-bronzite chondrite (H4).
A mass of 787g was found.

Allan Hills A77225
Stone. Olivine-bronzite chondrite (H4).
A mass of 5.87kg was found.

Allan Hills A77226
Stone. Olivine-bronzite chondrite (H4).
A mass of 15.23kg was found.

Allan Hills A77230
Stone. Olivine-hypersthene chondrite (L4).
A mass of 2.47kg was found.

Allan Hills A77231
Stone. Olivine-hypersthene chondrite (L6).
A mass of 9.27kg was found.

Allan Hills A77232
Stone. Olivine-bronzite chondrite (H4).
A mass of 6.49kg was found.

Allan Hills A77233 v Allan Hills A77004.

Allan Hills A77241 v Allan Hills A77011.

Allan Hills A77249 v Allan Hills A77011.

Allan Hills A77250 v Allan Hills A76002.

Allan Hills A77251
Stone. Olivine-hypersthene chondrite (L6).
A mass of 1.31kg was found.

Allan Hills A77255
Iron. Ataxite, Ni-rich (IRANOM).
A mass of 765g was found. Analysis, 12.23 %Ni, 0.6 ppm.Ga, 12.0 ppm.Ir, R.S. Clarke, Jr. et al., Meteoritics, 1980, **15**, p.273, abs.

Allan Hills A77256
Stone. Achondrite, Ca-poor. Diogenite (ADIO).
A mass of 676g was found. Analysis, 12.49 % total iron.

Allan Hills A77257
Stone. Achondrite, Ca-poor. Ureilite (AURE).
A mass of 1,99kg was found, olivine $Fa_{9\text{-}23}$, 10.55 % total iron.

Allan Hills A77258
Stone. Olivine-bronzite chondrite (H6).
A mass of 597g was found.

Allan Hills A77260 v Allan Hills A77011.

Allan Hills A77262
Stone. Olivine-bronzite chondrite (H4).
A mass of 862g was found.

Allan Hills A77263 v Allan Hills A76002.

Allan Hills A77269
Stone. Olivine-hypersthene chondrite (L6).
A mass of 1.04kg was found.

Allan Hills A77270
Stone. Olivine-hypersthene chondrite (L6).
A mass of 589g was found.

Allan Hills A77271
Stone. Olivine-bronzite chondrite (H6).
A mass of 610g was found.

Allan Hills A77272
Stone. Olivine-hypersthene chondrite (L6).
A mass of 674g was found.

Allan Hills A77280
Stone. Olivine-hypersthene chondrite (L6).
A mass of 3226g was recovered. Very similar to Allan Hills A77273 (492g) and Allan Hills A77277 (142.7g).

Allan Hills A77281
Stone. Olivine-hypersthene chondrite (L6).
A mass of 1231g was recovered. Very similar to Allan Hills A77280.

Allan Hills A77282
Stone. Olivine-hypersthene chondrite (L6).
A mass of 4127g was recovered. Very similar to Allan Hills A77280 and Allan Hills A77281.

Allan Hills A77283
Iron. Octahedrite, coarse (1.8mm) (IA).
A mass of 10.51kg was found. It is carbon-rich, containing diamonds and graphite; retains an undistorted heat-affected zone; not a crater-forming meteorite. Diamonds therefore pre-terrestrial, R.S. Clarke, Jr. et al., Nature, 1981, **291**, p.396. Analysis, 7.33 %Ni, 69.0 ppm.Ga, 230 ppm.Ge, 2.24 ppm.Ir, R.S. Clarke, Jr. et al., Meteoritics, 1980, **15**, p.273, abs.

Allan Hills A77288
Stone. Olivine-bronzite chondrite (H6).
A mass of 1880g was found.

Allan Hills A77289 v Allan Hills A76002.

Allan Hills A77290 v Allan Hills A76002.

Allan Hills A77294
Stone. Olivine-bronzite chondrite (H5).
A mass of 1351g was found.

Allan Hills A77296
Stone. Olivine-hypersthene chondrite (L6).
A mass of 963g was found.

Allan Hills A77297
Stone. Olivine-hypersthene chondrite (L6).
A mass of 952g was found.

Allan Hills A77302
Stone. Achondrite, Ca-rich. Howardite (AHOW).
A mass of 235.5g was found. Petrology, classification, T.C. Labotka and J.J. Papike, Geochimica et Cosmochimica Acta, 1980 (Suppl. 14), p.1103.

Allan Hills A77304
Stone. Olivine-hypersthene chondrite, amphoterite (LL3).
A mass of 650g was found.

Allan Hills A77305
Stone. Olivine-hypersthene chondrite (L6).
A mass of 6444g was found.

Allan Hills A77306
Synonym(s): *Allan Hills A78261, Allan Hills A81002, Allan Hills A81004*
Stone. Carbonaceous chondrite, type II (CM2).
A mass of 19.9g was found. Trace element data, G.W. Kallemeyn and J.T. Wasson, Geochimica et Cosmochimica Acta, 1981, **45**, p.1217.

Allan Hills A77307
Stone. Carbonaceous chondrite, type III (CO3).
A mass of 181.3g was found. Classification, E.R.D. Scott et al., Meteoritics, 1981, **16**, p.385, abs.

Allan Hills A78006 76°45′ S., 159°30′ E. approx.
Victoria Land, Antarctica
Found 1978, between December 1978 and January 1979
Stone. Achondrite, Ca-rich. Howardite (AHOW).
A mass of 8.0g was found, Antarctic Meteorite Newsletter, 1981, **4**, p.75. Approximately 260 masses, mainly chondritic and representing a large number of individual falls, were found during the 1978-1979 field season in Antarctica. The masses have been sequentially numbered and are so named, listing, U.B. Marvin and B. Mason, Smithson. Contrib. Earth Sci., 1982 (24). Listed here are those ordinary chondrites of mass greater than 500g, achondrites and the iron.

Allan Hills A78019
Synonym(s): *Allan Hills A78262*
Stone. Achondrite, Ca-poor. Ureilite (AURE).
A mass of 30.3 g was found, similar in all respects to Allan Hills A78262, 26.1g, Antarctic Meteorite Newsletter, 1981, **4**, p.100. Petrography, olivine $Fa_{23.3}$, J.L. Berkley and J.H. Jones, J. Geophys. Res., 1982, **87** (suppl.), p.A353.

Allan Hills A78038 v Allan Hills A77011.

Allan Hills A78040
Stone. Achondrite, Ca-rich. Eucrite (AEUC).
A mass of 211g was found.

Allan Hills A78043
Stone. Olivine-hypersthene chondrite (L6).
A mass of 680g was found, similar to Allan Hills A78045, 396g, olivine Fa_{25}, Antarctic Meteorite Newsletter, 1981, **4**, p.78.

Allan Hills A78050
Stone. Olivine-hypersthene chondrite (L6).
A mass of 1045g was found, olivine Fa_{23}, Antarctic Meteorite Newsletter, 1981, **8**, p.80.

Allan Hills A78084
Synonym(s): *Allan Hills A77009, Allan Hills A81022*
Stone. Olivine-bronzite chondrite (H3).
A mass of 14.28kg was found, olivine Fa_{18}, pyroxene Fs8-24, Antarctic Meteorite Newsletter, 1981, **4**, p.84.

Allan Hills A78100
Iron. Hexahedrite (IIA).
A mass of 85g was found. Analysis, 5.5 %Ni, 61 ppm.Ga, 27 ppm.Ir, W. Daode et al., Lunar Sci. Conf., 1982, p.139, abs.

Allan Hills A78103
Synonym(s): *Allan Hills A78104*
Stone. Olivine-hypersthene chondrite (L6).
A mass of 589g was found, similar to Allan Hills A78104, 672g, in all respects, olivine Fa_{24}, Antarctic Meteorite Newsletter, 1981, **4**, p.85.

Allan Hills A78104 v Allan Hills A78103.

Allan Hills A78105
Stone. Olivine-hypersthene chondrite (L6).
A mass of 941g was found, olivine Fa_{23}.

Allan Hills A78112
Synonym(s): *Allan Hills A78114*
Stone. Olivine-hypersthene chondrite (L6).
A mass of 2485g was found, similar in all respects to Allan Hills A78114, 808g, olivine Fa_{25}, Antarctic Meteorite Newsletter, 1981, **4**, p.90.

Allan Hills A78113
Stone. Achondrite, Ca-poor. Aubrite (AUB).
A fragment of 298.6g was found. Mineralogy, T.R. Watters et al., Meteoritics, 1980, **15**, p.386, abs.

Allan Hills A78114 v Allan Hills A78112.

Allan Hills A78115
Stone. Olivine-bronzite chondrite (H6).
A mass of 847g was found, olivine Fa_{18}.

Allan Hills A78126
Synonym(s): *Allan Hills A78130, Allan Hills A78131*
Stone. Olivine-hypersthene chondrite (L6).
A mass of 606g was found, similar in all respects to Allan Hills A78130, 2733g and Allan Hills A78131, 268g, olivine Fa_{25}, Antarctic Meteorite Newsletter, 1981, **4**, p.92.

Allan Hills A78130
Stone. Olivine-hypersthene chondrite (L6).
A mass of 2733g was found.

Allan Hills A78132
Synonym(s): *Allan Hills A78158, Allan Hills A78165*
Stone. Achondrite, Ca-rich. Eucrite (AEUC).
A mass of 656g was found.

Allan Hills A78158 v Allan Hills A78132.

Allan Hills A78165 v Allan Hills A78132.

Allan Hills A78251
Stone. Olivine-hypersthene chondrite (L6).
A mass of 1312g was found, olivine Fa_{23}.

Allan Hills A78252
Iron. Octahedrite, medium (0.5mm) (IVA).
A single mass of 2789g was found. Analysis, 9.33 %Ni, 2.5 ppm.Ga, 0.45 ppm.Ir, R.S. Clarke, Jr. et al., Meteoritics, 1980, **15**, p.273, abs.

Allan Hills A78261 v Allan Hills A77306.

Allan Hills A78262 v Allan Hills A78019.

Allan Hills A79002 76°45′ S., 159°30′ E. approx.
Victoria Land, Antarctica
Found 1979, between December 1979 and January 1980
Stone. Olivine-bronzite chondrite (H6).
A mass of 222g was found, olivine Fa_{18}, U.B. Marvin and B. Mason, Smithson. Contrib. Earth Sci., 1982 (24). This is one of about 55 masses which were found in the Allan Hills region during the 1979-1980 field season in Antarctica. Most are chondritic and weigh less than 100g, listed here are those ordinary chondrite masses of over 200g and the sole achondrite. For a fuller listing, Antarctic Meteorite Newsletter, 1981, **4**, p.115, U.B. Marvin and B. Mason, Smithson. Contrib. Earth Sci., 1982 (24).

Allan Hills A79003 v Allan Hills A81031.

Allan Hills A79016
Stone. Olivine-bronzite chondrite (H6).
A mass of 1146g was found, olivine Fa_{17}.

Allan Hills A79017
Stone. Achondrite, Ca-rich. Eucrite (AEUC).
A mass of 310g was found.

Allan Hills A79025
Stone. Olivine-bronzite chondrite (H5).
A mass of 1208g was found, olivine Fa_{17}.

Allan Hills A79026
Stone. Olivine-bronzite chondrite (H5).
A mass of 572g was found, olivine Fa_{18}.

Allan Hills A79029
Stone. Olivine-bronzite chondrite (H5).
A mass of 505g was found, olivine Fa_{18}.

Allan Hills A79033
Stone. Olivine-hypersthene chondrite (L6).
A mass of 208g was found, olivine Fa_{24}.

Allan Hills A79045 v Allan Hills A77011.

Allan Hills A80101 76°45′ S., 159°30′ E. approx.
Victoria Land, Antarctica
Found 1980, December 1980-January 1981
Synonym(s): *Allan Hills A80103, Allan Hills A80105, Allan Hills A80107, Allan Hills A80108, Allan Hills A80110, Allan Hills A80112, Allan Hills A80113, Allan Hills A80114, Allan Hills A80115, Allan Hills A80116, Allan Hills A80117, Allan Hills A80119, Allan Hills A80120, Allan Hills A80125*
Stone. Olivine-hypersthene chondrite (L6).
A mass of 8725g was found, olivine Fa_{24}. Probably synonymous masses of over 400g are Allan Hills A80103, 535g; Allan Hills A80105, 445g. These are part of the material found by the 1980-1981 American Expedition to Antarctica, that from the Allan Hills region is named Allan Hills A80101 to Allan Hills A80133, Antarctic Meteorite Newsletter, 1982, **5** (no. 1). Included here are all ordinary chondrite masses of over 400g and all other types found.

Allan Hills A80102
Stone. Achondrite, Ca-rich. Eucrite (AEUC).
A stone of 471.2g was found, plagioclase An_{87}.

Allan Hills A80103 v Allan Hills A80101.

Allan Hills A80104
Iron. Ataxite.
A single mass of 882g was found.

Allan Hills A80105 v Allan Hills A80101.

Allan Hills A80107 v Allan Hills A80101.

Allan Hills A80108 v Allan Hills A80101.

Allan Hills A80110 v Allan Hills A80101.

Allan Hills A80112 v Allan Hills A80101.

Allan Hills A80113 v Allan Hills A80101.

Allan Hills A80114 v Allan Hills A80101.

Allan Hills A80115 v Allan Hills A80101.

Allan Hills A80116 v Allan Hills A80101.

Allan Hills A80117 v Allan Hills A80101.

Allan Hills A80119 v Allan Hills A80101.

Allan Hills A80120 v Allan Hills A80101.

Allan Hills A80125 v Allan Hills A80101.

Allan Hills A80133
Stone. Carbonaceous chondrite, type III (CV3).
A mass of 3.6g was found, olivine $Fa_{0.5-35}$.

Allan Hills A81001
Allan Hills region, Antarctica
Found 1981, between December 1981 and January 1982
Stone. Achondrite, Ca-rich. Eucrite (AEUC).
A single mass of 52.9g was found in the Allan Hills region during the 1981-1982 field season in Antarctica. During this season a total of 373 specimens were found; 6 achondrites, 4 carbonaceous chondrites, two irons and 361 ordinary chondrites. Listed here are those ordinary chondrites of mass over 500g and all other types found. Listed, with some descriptions, Antarctic Meteorite Newsletter, 1982, **6** (1).

Allan Hills A81002 v Allan Hills A77306.

Allan Hills A81003
Found 1981
Stone. Carbonaceous chondrite, type III (CV3).
A single mass of 10.1g was found, olivine Fa_{0-60}, Antarctic Meteorite Newsltter, 1982, **6** (1).

Allan Hills A81004 v Allan Hills A77306.

Allan Hills A81005
Stone. Achondrite, anomalous (ACANOM).
A mass of 31.4g was found by the American Antarctic Expedition during the 1981-82 field season in the Allan Hills region. Anorthositic breccia, olivine Fa_{11-40}, plagioclase An_{97}, Antarctic Meteorite Newsletter, 1982, **5** (4). Lunar origin, D. Bogard, ed., Geophys. Res. Lett., 1983, **10**, p.773 et seq.

Allan Hills A81006 Allan Hills A76005

Allan Hills A81007 v Allan Hills A76005.

Allan Hills A81008 v Allan Hills A76005.

Allan Hills A81009
Synonym(s): *Allan Hills A81012*
Stone. Achondrite, Ca-rich. Eucrite (AEUC).
A single mass of 229g was found; paired with Allan Hills A81012, 36.6g.

Allan Hills A81010 v Allan Hills A76005.

Allan Hills A81011
Stone. Achondrite, Ca-rich. Eucrite (AEUC).
A single mass of 405g was found.

Allan Hills A81012 v Allan Hills A81009.

Allan Hills A81013
Iron. Hexahedrite.
A mass of 17.72kg was found.

Allan Hills A81014
Iron. Octahedrite, fine.
A mass of 188g was found.

Allan Hills A81015
Stone. Olivine-bronzite chondrite (H5).
A mass of 5489g was found, olivine Fa_{19}.

Allan Hills A81016
Stone. Olivine-hypersthene chondrite (L6).
A mass of 3850g was found, olivine Fa_{25}.

Allan Hills A81017
Synonym(s): *Allan Hills A81018, Allan Hills A81023*
Stone. Olivine-hypersthene chondrite (L5).
A mass of 1434g was found, olivine Fa_{25}. Paired with Allan Hills A81018, 2236g; Allan Hills A81023, 413g.

Allan Hills A81018 v Allan Hills A81017.

Allan Hills A81019
Stone. Olivine-bronzite chondrite (H5).
A mass of 1051g was found, olivine Fa_{19}.

Allan Hills A81020
Stone. Olivine-bronzite chondrite (H5).
A mass of 1352g was found olivine Fa_{19}.

Allan Hills A81021
Stone. Enstatite chondrite (E6).
A mass of 695g was found.

Allan Hills A81022 v Allan Hills A78084.

Allan Hills A81023 v Allan Hills A81017.

Allan Hills A81024 v Allan Hills A77011.

Allan Hills A81025 v Allan Hills A79003.

Allan Hills A81026
Stone. Olivine-hypersthene chondrite (L6).
A mass of 515g was found, olivine Fa_{25}.

Allan Hills A81027
Synonym(s): *Allan Hills A81028, Allan Hills A81029*
Stone. Olivine-hypersthene chondrite (L6).
A mass of 3835g was found, olivine Fa_{25}. Paired with Allan Hills A81028, 80g and Allan Hills A81029, 153g.

Allan Hills A81028 v Allan Hills A81027.

Allan Hills A81029 v Allan Hills A81027.

Allan Hills A81030
Stone. Olivine-hypersthene chondrite, amphoterite (LL3).
A mass of 1851g was found, olivine Fa_{1-49}.

Allan Hills A81031
Synonym(s): *Allan Hills A79003, Allan Hills A81032*
Stone. Olivine-hypersthene chondrite, amphoterite (LL3).
A mass of 1594g was found. Paired with Allan Hills A79003, 3g and Allan Hills A81032, 726g.

Allan Hills A81032 v Allan Hills A81031.

Allan Hills A81041
Stone. Olivine-bronzite chondrite (H4).
A mass of 728g was found, olivine Fa_{18}.

Allan Hills A81042
Stone. Olivine-bronzite chondrite (H5).
A mass of 534g was found, olivine Fa_{19}.

Allan Hills A81044 v Allan Hills A77004.

Allan hills A81048 v Allan Hills A77004.

Allan Hills A81059
Synonym(s): *Allan Hills A81098*
Stony-iron. Mesosiderite (MES).
A very weathered mass of 539g was found, olivine Fa_{28}.

Allan Hills A81098 v Allan Hills A81059.

Allan Hills 82100
Antarctica
Found 1982, between December 1982 and January 1983
Stone. Carbonaceous chondrite, type II (CM2).
A single mass of 24.3g was found, the first of a total of 113 meteorite specimens (106 chondritic and 7 achondritic) found during the 1982-1983 field season in the Allan Hills, Elephant Nunatak, Pecora Escarpment and Thiel Mountains regions of Antarctica. Listed, Antarctic Meteorite Newsletter, 1983, **6** (3).

Allan Hills 82101
Stone. Carbonaceous chondrite, type III (CO3).
A mass of 29.1g was found.

Allegan — 42°32′ N., 85°53′ W.
Allegan County, Michigan, U.S.A.
Fell 1899, July 10, 0800 hrs
Stone. Olivine-bronzite chondrite (H5).
A stone of about 70lb fell, after detonations, on Thomas Hill on the Saugatuck Road, H.A. Ward, Am. J. Sci., 1899, **8**, p.412. Description, analysis, G.P. Merrill, Proc. Washington Acad. Sci., 1900, **2**, p.41. Nature of the metal, H.C. Urey and T. Mayeda, Geochimica et Cosmochimica Acta, 1959, **17**, p.113. Mineralogy, analysis, 28.54 % total iron, E.

Jarosewich and B. Mason, Geochimica et Cosmochimica Acta, 1969, **33**, p.411, B. Mason and A.D. Maynes, Proc. U.S. Nat. Mus., 1967, **124** (3624). Charged particle track study, P. Pellas et al., Meteoritics, 1973, **8**, p.418. Olivine Fa_{18}, B. Mason, Geochimica et Cosmochimica Acta, 1963, **27**, p.1011. Pyroxene structure, J.R. Ashworth, Earth planet. Sci. Lett., 1980, **46**, p.167.

18kg Washington, U.S. Nat. Mus.; 1.1kg Fort Worth, Texas, Monnig Colln.; 861g Chicago, Field Mus. Nat. Hist.; 460g Rome, Vatican Colln; 369g Vienna, Naturhist. Mus.; 162g Tempe, Arizona State Univ.; 112g Berlin, Humboldt Univ.; 95g New York, Amer. Mus. Nat. Hist.; 89g Yale Univ.; 60g Prague, Nat. Mus.; 500g Michigan Univ.; 500g Paris, Mus. d'Hist. Nat.; 295g Harvard Univ.;

Specimen(s): [84553], 759g. and fragments, 4g; [1920,281], 153g. four pieces, and fragments, 42g

Alleghany Mountains v Greenbrier County.

Allegheny County v Pittsburgh.

Allen 33°6′ N., 96°42′ W. approx.

Collin County, Texas, U.S.A.
Found 1923, approx., recognized 1938
Stone. Olivine-hypersthene chondrite (L), veined.

One stone of 1.4kg was found, A.D. Nininger, Pop. Astron., Northfield, Minnesota, 1939, **47**, p.211.

1.4kg Fort Worth, Texas, Monnig Colln.;

Allen County v Scottsville.

Allende 26°58′ N., 105°19′ W.

Chihuahua, Mexico
Fell 1969, February 8, 0105 hrs
Synonym(s): *Pueblito de Allende, Qutrixpileo*
Stone. Carbonaceous chondrite, type III (CV3).

A shower of stones fell after a bolide was seen. More than two tons were probably collected. One individual weighed 100-110kg but had fragmented on impact. The strewnfield is estimated to be more than 150 square km, E.A. King et al., Science, 1969, **163**, p.928. Comprehensive description with analysis, 23.85 % total iron, R.S. Clarke, Jr. et al., Smithson. Contrib. Earth Sci., 1970 (5)., R.S. Clarke, Jr., Meteoritics, 1970, **5**, p.189, D.P. Elston, Meteoritics, 1970, **5**, p.195, R.M. Housley and M. Blander, Meteoritics, 1970, **5**, p.203. X-ray fluorescence analysis, T.S. McCarthy and L.H. Ahrens, Earth planet. Sci. Lett., 1972, **14**, p.97. Rb-Sr measurments., C.M. Gray et al., Icarus, 1973, **20**, p.213. Grossular in inclusions, L.H. Fuchs, Meteoritics, 1974, **9**, p.11. Ti isotopic anomalies, S. Niemeyer and G.W. Lugmair, Earth planet. Sci. Lett., 1981, **53**, p.211 [M.A. 81-4290]. Trace element data, G.W. Kallemeyn and J.T. Wasson, Geochimica et Cosmochimica Acta, 1981, **45**, p.1217.

380kg Washington, U.S. Nat. Mus.; 55kg Chicago, Field Mus. Nat. Hist.; 34kg Los Angeles, Univ. of California; 10kg Copenhagen, Univ. Geol. Mus.; 6.7kg Fort Worth, Texas, Monnig Colln.; 3kg Harvard Univ.;

Specimen(s): [1969,147], 1232g. complete stone; [1969,148], 327g. fragments; [1981,M.5], 67.8g. individual; [1982,M.10], 7.5g. individual

Allessandria v Alessandria.

Allport

Derbyshire, England
Pseudometeorite..

The material supposed to have fallen after a detonating fireball had passed in August or September, 1827 or 1828, B. Powell, Rep. Brit. Assn., 1850, p.89, was clearly not meteoritic.

Allred 33°6′ N., 102°57′ W.

Yoakum County, Texas, U.S.A.
Found
Stone. Olivine-hypersthene chondrite (L4).

A single mass of 6.57kg was found 16 km SW. of Plains, Yoakum County, Meteor. Bull., 1979 (56), Meteoritics, 1979, **14**, p.165. Olivine $Fa_{25.5}$, J.T. Wasson, priv. comm.

Main mass, Los Angeles, Univ. of California;

Almelo Township 39°36′ N., 100°7′ W.

New Almelo, Norton County, Kansas, U.S.A.
Found 1949
Stone. Olivine-hypersthene chondrite (L5).

One stone and two smaller pieces, total weight 3.18kg, were found in a church yard, and presented to Fort Hays, Kansas State College Mus. The material is probably the same as New Almelo (examination by R. Hutchison), E.A. King, letter of 19 Nov. 1971, in Min. Dept. BM(NH). Not an olivine-pigeonite chondrite as reported, Meteor. Bull., 1969 (47), Meteoritics, 1970, **5**, p.104.

Specimen(s): [1970,4], Thin section

Alpine v Chico Mountains.

Al Rais 24°25′ N., 39°31′ E.

Medina, Saudi Arabia
Fell 1957, December 10
Synonym(s): *Rais*
Stone. Carbonaceous chondrite, type II (CM2).

A total weight of 160g fell near the city of Medina, Saudi Arabia, B. Mason, Meteorites, Wiley, 1962, p.96. Analysis, 23.78 % total iron, B. Mason, Space Sci. Rev., 1963, **1**, p.621 [M.A. 16-640].

132g Washington, U.S. Nat. Mus., main mass;

Specimen(s): [1971,289], 14g. a sawn, partly crusted fragment and 0.25g of fragments

Alta v Finmarken.

Alta'ameem 35°16′24″ N., 44°12′56″ E. approx.

Humira, Iraq
Fell 1977, August 20, 1030-1100 hrs
Stone. Olivine-hypersthene chondrite, amphoterite (LL5).

After detonations, a mass of about 6kg was recovered. It had broken into fragments on impact. The place of fall is about 100 metres north of the village of Humira, Meteor. Bull., 1978 (55), Meteoritics, 1978, **13**, p.329. Mineralogy and chemistry, olivine Fa_{27}, 20.3 % total iron, K.S. Al-Bassan, Meteoritics, 1978, **13**, p.257.

Main mass, Baghdad, Ministry of Education, Dept. of Sci.;

Specimen(s): [1978,M.22], 50g.

Altai v Barnaul.

Altai v Demina.

Alt Bela 49°46′ N., 18°15′ E.

Ostrava, Moravia, Czechoslovakia
Found 1898
Synonym(s): *Alt Biela, Stara Bela*
Iron. Octahedrite, medium (0.7mm) (IID).

A mass of about 4kg was said to have fallen at the beginning of the nineteenth century; described, with an analysis, 12.89 %Ni, F. Smycka, Jahresber. Real-gymm. Mährisch-Ostrau,

1899., Zeits. Kryst. Min., 1901, **34**, p.707., Verhand. Naturf. Ver. Brünn, 1900, **38**, p.29. Classification and analysis, 10.04 %Ni, 75.0 ppm.Ga, 84 ppm.Ge, 16 ppm.Ir, A. Kracher et al., Geochimica et Cosmochimica Acta, 1980, **44**, p.773. Description, V.F. Buchwald, Iron Meteorites, Univ. of California, 1975, p.256.

2.7kg Prague, Bohemian Mus.; 428g Vienna, Naturhist. Mus.; 410g Vienna, Naturhist. Mus.; 19g Chicago, Field Mus. Nat. Hist.; 10g Washington, U.S. Nat. Mus.;

Alt Biela v Alt Bela.

Alte v Finmarken.

Alten v Finmarken.

Alton v Covert.

Altona v Altonah.

Altonah 40°34′ N., 110°29′ W.
Duchesne County, Utah, U.S.A.
Found 1932, about
Synonym(s): *Altona*
Iron. Octahedrite, fine (0.3mm) (IVA).

Known weight 21.5kg;, H.H. Nininger, Our Stone Pelted Planet, 1933, The Mines Mag., Golden, Colorado, 1933, **23** (8), p.6, A.D. Nininger, Pop. Astron., Northfield, Minnesota, 1939, **47**, p.211. Partial analysis and determinations of Au, Pd and Ga, 8.55 %Ni, 0.44 %Co, E. Goldberg et al., Geochimica et Cosmochimica Acta, 1951, **2**, p.1. Mentioned, H.H. and A.D. Nininger, The Nininger Collection of Meteorites, Winslow, Arizona, 1950, p.28. Almost certainly not identical to Duchesne (or Mount Tabby), V.F. Buchwald, Iron Meteorites, Univ. of California, 1975, p.258. Analysis, distinct from Duchesne (*q.v.*), 8.17 %Ni, 2.33 ppm.Ga, 0.123 ppm.Ge, 1.5 ppm.Ir, R. Schaudy et al., Icarus, 1972, **17**, p.174.

10.5kg Washington, U.S. Nat. Mus.; 4.2kg Tempe, Arizona State Univ.; 1.2kg Harvard Univ.;

Specimen(s): [1959,970], 3940g. and 74g sawings

Alto Verde 33°7′ S., 68°20′ W.
Mendoza, Argentina
Pseudometeorite..

The fall of a large meteor on 14 January 1898 was observed, A. Michaut, Misc. Anal. Soc. Cient. Argentina, 1898, **45**, p.363 and a specimen was recovered, but this has been identified by B. Mason as earthy hematite, L.O. Giacomelli, letter of 20 January 1964 in Min. Dept. BM(NH).

Alva v Woodward County.

Alvord 43°19′20″ N., 96°17′20″ W.
Lyon County, Iowa, U.S.A.
Found 1976, June 15, approx.
Iron. Octahedrite (IVA?).

A single mass of 17.5kg was found on a farm 1.3 miles SE. of Alvord, Meteor. Bull., 1978 (55), Meteoritics, 1978, **13**, p.329.

Alwal 17°22′ N., 78°28′ E.
Hyderabad, Andhra Pradesh, India
Fell 1901, approx.
Synonym(s): *Boenpalli, Hyderabad*
Doubtful. Iron.

A mass of about the size of a tennis ball is said to have fallen, but has been lost and the description is inadequate, M.A.R. Khan, Pop. Astron., Northfield, Minnesota, 1936, **44**, p.565. Listed by Coulson and by Leonard.

Alzhi-Bogdo error for Adzhi-Bogdo.

Amakaken v Caperr.

Amalia Farm v Gibeon.

Amana v Homestead.

Amates v Toluca.

Ambala v Umbala.

Amba Madapur v Ambapur Nagla.

Ambapur Nagla 27°40′ N., 78°15′ E.
Aligarh district, Uttar Pradesh, India
Fell 1895, May 27, 0100 hrs
Synonym(s): *Amba Madapur, Nagla, Wagla*
Stone. Olivine-bronzite chondrite (H5).

After the appearance of a luminous meteor moving from east to west, a stone of about 14lb., broken into two pieces, was found, T.H. Holland, letter of 15 April, 1896, with abstract of report of the fall by C.E. Crawford, in Min. Dept. BM (NH). There is no village of this name in tahsil Sikandra Rao; possibly the place of fall was Amba Madapur., C.A. Silberrad, Min. Mag., 1932, **23**, p.298. Analysis, 27.63 % total iron, olivine $Fa_{18.7}$, R. Hutchison et al., Proc. Roy. Soc. London, 1981, **A374**, p.159.

3kg Calcutta, Mus. Geol. Surv. India; 511g Vienna, Naturhist. Mus.; 86g New York, Amer. Mus. Nat. Hist.; 45g Budapest, Nat. Hist. Mus.; 45g Washington, U.S. Nat. Mus.; 29.6g Chicago, Field Mus. Nat. Hist.; 14g Prague, Nat. Mus.;

Specimen(s): [81117], 917g. and fragments, 6g; [81118], 143g.

Amber 37°10′ N., 97°53′ W.
Grady County, Oklahoma, U.S.A.
Found 1934, possible year, recognized 1955
Synonym(s): *Chickasha*
Stone. Olivine-hypersthene chondrite (L6), black.

A stone of 4532g was found, possibly in 1934, but was not recognized as meteoritic until 1955. Probably a very old fall, and may belong to the same fall as Cashion or Kingfisher, F.C. Leonard, Classif. Cat. Meteor., 1956, p.79, Publ. Astron. Soc. Pacific, 1956, **68** (405), p.547, Meteoritics, 1956, **1**, p.490 [M.A. 14-130]. The stone was sectioned at the Univ. of Oklahoma, but the finder was in possession of all the slices, E.P. Henderson, letter of 30 November 1956 in Min. Dept. BM(NH). Distinct from Cashion (H4) and from Kingfisher (L5). Olivine Fa_{24}, B. Mason, Geochimica et Cosmochimica Acta, 1963, **27**, p.1011.

2139g Los Angeles, Univ. of California, main mass; 1318g Fort Worth, Texas, Monnig Colln.;

Specimen(s): [1968,185], 48g. a crusted piece

Ameca-Ameca 20°35′ N., 104°4′ W.
Mexico State, Mexico
Found 1889, known before this year
Iron.
A small nodule in the Mexican Nat. Mus., Mexico City; described, A. Castillo, Cat. Meteorites Mexique, Paris, 1889, p.3. Perhaps identical with Toluca, L. Fletcher, Min. Mag., 1890, **9**, p.168. This mass may be lost, V.F. Buchwald, Iron Meteorites, Univ. of California, 1975, p.260.
1.2g Chicago, Field Mus. Nat. Hist.;

Amherst 40°48′ N., 99°12′ W. approx.
Buffalo County, Nebraska, U.S.A.
Found 1947, known in this year
Stone. Olivine-hypersthene chondrite (L6).
Two masses, approximately 8kg and 500g were in S.H. Perry's collection, Meteorite Coll. of S.H. Perry, Adrian, Michigan, 1947, p.14, and are now in Washington, U.S. Nat. Mus. Olivine Fa_{25}, B. Mason, Geochimica et Cosmochimica Acta, 1963, **27**, p.1011.
8.47kg Washington, U.S. Nat. Mus., main masses of both stones.;

Amun v Thebes.

Anderson v Brenham.

Andhara 26°35′ N., 85°34′ E.
Muzaffarpur district, Bihar, India
Fell 1880, December 2, 1600 hrs
Stone.
A stone weighing about 6lb was seen to fall and was made an object of worship in a temple built over the place of the fall, L.L. Fermor, Rec. Geol. Surv. India, 1907, **35** (2), p.92.

Andover 44°37′ N., 70°45′ W.
Oxford County, Maine, U.S.A.
Fell 1898, August 5, 0730 hrs
Stone. Olivine-hypersthene chondrite (L6).
A stone of about 7lb fell, after detonations, H.A. Ward, Proc. Rochester Acad. Sci., 1902, **4**, p.79. Olivine Fa_{25}, B. Mason, Geochimica et Cosmochimica Acta, 1963, **27**, p.1011.
2823g Washington, U.S. Nat. Mus., main mass; 90g Chicago, Field Mus. Nat. Hist.; 15g Vienna, Naturhist. Mus.;
Specimen(s): [86762], 19.7g.

Andreevka 48°42′ N., 37°30′ E.
Slaviansky district, Donetsk region, Ukraine, USSR, [Андреевка]
Fell 1969, August 7, 1900 hrs
Stone. Olivine-hypersthene chondrite (L3).
A single mass of about 600g was recovered after it had fallen through the slate roof of a house, Meteor. Bull., 1976 (54), Meteoritics, 1976, **11**, p.69. Description, analysis, 23.6 % total iron, E.G. Osadchy et al., Метеоритика, 1982, **40**, p.25.
150g Donetz, Mus.; 43g Moscow, Acad. Sci.;

Andrioniskis v Padvarninkai.

Andriushki v Andryushki.

Andrjuschki v Andryushki.

Andronishkis v Padvarninkai.

Andronishkyai v Padvarninkai.

Andryushki 49°39′ N., 29°38′ E.
Near Berdichev, Ukraine, USSR, [Андрюшки]
Found 1898
Synonym(s): *Andriushki, Andrjuschki*
Stone. Olivine-hypersthene chondrite (L6).
Two small pieces of a chondritic stone (11.1g and 15.2g) were preserved in the Acad. Sci. USSR, Moscow, under the name Andryushki; little is known of their history and they may be part of a fall already described under some other name, E.L. Krinov, Астрон. Журнал, 1945, **22**, p.303 [M.A. 9-297], Каталог Метеоритов Акад. Наук СССР, Москва, 1947 [M.A. 10-511]. Listed, I.S. Astapowitsch, Trans. Roy. Astron. Soc. Canada, 1938, **32**, p.195 [M.A. 7-172]. Olivine Fa_{23}, B. Mason, Geochimica et Cosmochimica Acta, 1967, **31**, p.1100.
14.9g Moscow, Acad. Sci.;

Andura 20°53′ N., 76°52′ E.
Akola district, Berar, Maharashtra, India
Fell 1939, August 9, evening
Stone. Olivine-bronzite chondrite (H6).
A stone of 17.9kg fell, M.A.R. Khan, Hyderabad Acad. Studies, 1950, **12** [M.A. 11-441]. Mentioned, P. Chatterjee, letter of 25 March 1952 in Min. Dept. BM(NH). Olivine Fa_{20}, B. Mason, Geochimica et Cosmochimica Acta, 1967, **31**, p.1100.
17.7kg Calcutta, Mus. Geol. Surv. India; 9g Washington, U.S. Nat. Mus.; 1g New York, Amer. Mus. Nat. Hist.;
Specimen(s): [1974,M.19], 54.65g. a sawn, crusted fragment.

Angara v Ssyromolotovo.

Angara 58°41′ N., 94°14′ E.
Yeniseisk, Federated SSR, USSR, [Ангара]
Found 1885
Synonym(s): *Borovaya, Borowaja, Muroshna, Murozhnaya, Muroznaja, Uderei, Vorova, Worowo*
Pseudometeorite. Iron.
Three small pieces were found; one of 65.5g from the Borovaya river, 58°41'N., 94°14'E.; one of 103.8g from the Murozhna river 58°41'N., 94°14'E., near its junction with the Borovaya, and one of 15.2g from the Uderei mine, 58°43'N., 94°26'E. All three places are quite close together in the Angara valley, but the specimens are listed separately, E.L. Krinov, Каталог Метеоритов Акад. Наук СССР, Москва, 1947 [M.A. 10-511]. The whole of the material is in the Acad. Sci. Moscow; it contains no nickel and is not meteoritic, A.N. Zavaritsky and L.G. Kvasha, Метеориты СССР, Москва, 1952, p.238. Further references, M.H. Hey, Catalogue of Meteorites, Brit. Mus. Nat. Hist., 1966, p.21.

Angela v La Primitiva.

Angelica 44°15′ N., 88°15′ W.
Shawano County, Wisconsin, U.S.A.
Found 1916
Iron. Octahedrite, medium (1.2mm) (IIIA).
A mass of 14.8kg was ploughed up. Described, with a poor analysis, R.N. Buckstaff, Trans. Wisconsin Acad. Sci. Arts Letters, 1943, **35**, p.99 [M.A. 9-302]. Classification and analysis, 7.42 %Ni, 18.3 ppm.Ga, 34.9 ppm.Ge, 9.3 ppm.Ir, E.R.D. Scott et al., Geochimica et Cosmochimica Acta, 1973, **37**, p.1957. Additional data, V.F. Buchwald, Iron Meteorites, Univ. of California, 1975, p.260.
Main mass, Oshkosh, Wisconsin, Public Mus.; 331g

Washington, U.S. Nat. Mus.; 110g New York, Amer. Mus. Nat. Hist.; 84g Chicago, Field Mus. Nat. Hist.;

Angers 47°28′ N., 0°33′ W.
Maine et Loire, France
Fell 1822, June 3, 2015 hrs
Stone. Olivine-hypersthene chondrite (L6), veined.
After appearance of luminous meteor and detonations, several stones fell, the largest about 900g, L.W. Gilbert, Ann. Phys. (Gilbert), 1822, **71**, p.345, A. Lacroix, Bull. Mus. d'Hist. Nat. Paris, 1927, **33**, p.433. Olivine Fa_{25}, B. Mason, Geochimica et Cosmochimica Acta, 1963, **27**, p.1011.
Main mass, Angers, Mus.; 136g Paris, Mus. d'Hist. Nat.; 28g Chicago, Field Mus. Nat. Hist.; 19g New York, Amer. Mus. Nat. Hist.; 6g Berlin, Humboldt Univ.;
Specimen(s): [63924], 14g.; [48761], 8g.

Angra dos Reis (iron) v Pirapora.

Angra dos Reis (stone) 22°58′ S., 44°19′ W.
Rio de Janeiro, Brazil
Fell 1869, January, 0500 hrs, latter half of month
Stone. Augite achondrite (ACANOM).
A stone of about 1.5kg fell; described, with an analysis, G. Tschermak, Tschermaks Min. Petr. Mitt., 1887, **8**, p.341, 1888, **9**, p.423. Mineral compositions, olivine Fa_{46}, R. Hutchison, Nature, 1972, **240** (Phys. Sci.), p.58. Neutron activation analysis, A.A. Smales et al., Geochimica et Cosmochimica Acta, 1970 (Suppl. 1), p.1575. Very low Sr87/86 ratio, H.G. Sanz et al., Geochimica et Cosmochimica Acta, 1970, **34**, p.1227. Consortium study, K. Keil, Earth planet. Sci. Lett., 1977, **35**, p.271.
101g Rio de Janeiro, Mus. Nac.; 8.5g Washington, U.S. Nat. Mus.; 2g Berlin, Humboldt Univ.; 15g Paris, Mus. d'Hist. Nat.; 6.4g Chicago, Nat. Hist. Mus.; 7g Vienna, Naturhist. Mus.; 6g Budapest, Nat. Mus.;
Specimen(s): [63233], 5g.

Anighito v Cape York.

Anjela v La Primitiva.

Ankober 9°32′ N., 39°43′ E.
Bolede, Ethiopia
Fell 1942, July 7, 1100 hrs
Stone. Olivine-bronzite chondrite (H4).
One stone, said to weigh about 6500g fell at Basso, near Ankober, Meteor. Bull., 1957 (5)., B. Mason, letters of 28 April 1964 and 11 February 1965 in Min. Dept. BM(NH). Olivine Fa_{18}, B. Mason, Geochimica et Cosmochimica Acta, 1967, **31**, p.1100. Analysis, 27.47 % total iron, R.S. Clarke, Jr. et al., Smithson. Contrib. Earth Sci., 1975 (14), p.63.
6.9kg Washington, U.S. Nat. Mus., entire mass;

Anlong v Anlung.

Anlung 25°9′ N., 105°11′ E.
Guizhou, China
Fell 1971, May 2
Synonym(s): *Anlong*
Stone. Olivine-bronzite chondrite (H5).
Full description and analysis, 26.98 % total iron, Meteorite Lab. etc., Geochim., 1974, p.105 [M.A. 75-1269]. K-Ar age 3280 m.y., exposure age 3.0 my., Geochim., 1973, p.272 [M.A. 74-2348]. A mass of 2.5kg fell, co-ordinates, D. Bian, Meteoritics, 1981, **16**, p.115.
2.5kg Guiyang, Acad. Sin. Inst. Geochem.;

Annaheim 52°20′ N., 104°52′ W.
Saskatchewan, Canada
Found 1916
Iron. Octahedrite, coarse (1.4mm) (IA-ANOM).
A crescent-shaped mass of 11.84kg(?) was found in 1916 six miles north of Annaheim; it may possibly have fallen in 1914 on January 21 at 1430 hrs., as a fireball was seen and detonations were heard on that date. Described, with an analysis, 7.84 %Ni, R.A.A. Johnston and H.V. Ellsworth, Trans. Roy. Soc. Canada, 1921, **15** (4), p.69 [M.A. 1-406]. Classification and analysis, 7.74 %Ni, 79.8 ppm.Ga, 302 ppm.Ge, 3.5 ppm.Ir, J.T. Wasson, Icarus, 1970, **12**, p.407. Description, the mass weighed over 13.5kg, V.F. Buchwald, Iron Meteorites, Univ. of California, 1975, p.262.
12.4kg Ottawa, Mus. Geol. Surv. Canada;
Specimen(s): [1924,1094], 359g.

Annapolis v Nanjemoy.

Anoka 45°12′ N., 93°26′ W.
Anoka County, Minnesota, U.S.A.
Found 1961, approx.
Iron. Octahedrite, fine (0.34mm) (IIICD).
A mass of 1.108kg was found, Meteor. Bull., 1964 (32). Description, analysis and co-ordinates, G.I Huss et al., Meteoritics, 1966, **3**, p.58. Classification and new analysis, 11.95 %Ni, 17.2 ppm.Ga, 15.7 ppm.Ge, 0.16 ppm.Ir, J.T. Wasson and R. Schaudy, Icarus, 1971, **14**, p.59. Structural classification, V.F. Buchwald, Iron Meteorites, Univ. of California, 1975, p.263.
168g Mainz, Max-Planck Inst., main mass; 39g New York, Amer. Mus. Nat. Hist.; 132g Washington, U.S. Nat. Mus.; 181g Tempe, Arizona State Univ.;

Anson 39°20′ N., 97°33′42″ W.
Sumner County, Kansas, U.S.A.
Found 1972, recognized in this year
Stone. Olivine-hypersthene chondrite (L6).
A total weight of 3.9kg was found by a farmer, olivine Fa_{23}, Meteor. Bull., 1979 (56), Meteoritics, 1979, **14**, p.165.
3.8kg Mainz, Max-Planck-Inst.;
Specimen(s): [1976,M.14], 116g.

Answer 21°39′35″ S., 140°54′25″ E.
Selwyn, Queensland, Australia
Found 1970, June 16
Iron. Ataxite.
A single mass of 11.09kg was found near the Answer mine south of Selwyn, Meteor. Bull., 1976 (54), Meteoritics, 1976, **11**, p.71., B.R. Houston, Queensland Govt. Min. J., 1971, **72**, p.482.

Antarctic v Adelie Land.

Anthony 37°5′ N., 98°3′ W.
Harper County, Kansas, U.S.A.
Found 1919
Stone. Olivine-bronzite chondrite (H5).
A much oxidised stone of about 20kg was ploughed up 6.5 miles east and 1 mile south of Anthony; described, with an analysis, G.P. Merrill, Proc. Nat. Acad. Sci., 1924, **10**, p.306. Olivine Fa_{19}, B. Mason, Geochimica et Cosmochimica Acta, 1963, **27**, p.1011.
9kg Washington, U.S. Nat. Mus.; 51g Tempe, Arizona State Univ.; 21g Chicago, Field Mus. Nat. Hist.;
Specimen(s): [1924,325], 221g. slice; [1959,811], 61.5g. slice

Anticoli Corradi v Orvinio.

Antifona v Collescipoli.

Antofagasta v Mantos Blancos.

Antofagasta v San Cristobal.

Antofagasta v Imilac.

Anton 33°46′57″ N., 102°10′52″ W.
Hockley County, Texas, U.S.A.
Found 1965
Stone. Olivine-bronzite chondrite (H).
A single stone weighing 41.8kg was ploughed up, G.I Huss, letter of 26 October, 1983 in Min. Dept. BM(NH).
41.5kg Denver, Amer. Meteorite Colln.; 287g Fort Worth, Texas, Monnig Colln.;

Anyujskij 66°54′ N., 164°12′ E.
Near Anyujskij, Magadan region, Federated SSR, USSR
Found 1981, July 19
Iron.
A single mass of about 100kg was found at a depth of 4.5 m., Meteor. Bull., 1982 (60), Meteoritics, 1982, **17**, p.93.

Aphaca 34°4′ N., 35°52′ E.
Byblos, Lebanon
Doubtful..
A cult object worshipped in classical times at Aphaca (=Afqua) was perhaps a meteorite, G.A. Wainwright, Zeits. Ägypt. Sprache, 1935, **71**, p.41.

Apoala 17°42′ N., 97°0′ W. approx.
Oaxaca, Mexico
Found 1889
Iron. Octahedrite, medium (0.65mm) (IIIB).
A mass of about 85kg was found; described, E. Cohen, Meteoritenkunde, 1905, **3**, p.384. Some specimens, including B.M.86068, have been heated by man to 900C. The main mass in Mexico City (Mus. Inst. Geol.) appears to be undamaged, V.F. Buchwald, Iron Meteorites, Univ. of California, 1975, p.266. Classification and analysis, 9.39 % Ni, 18.4 ppm.Ga, 35.7 ppm.Ge, 0.016 ppm.Ir, E.R.D. Scott et al., Geochimica et Cosmochimica Acta, 1973, **37**, p.1957.
Main mass, Mexico, Mus. Inst. Geol.; 189g Tempe, Arizona State Univ., part of Durango, V.F. Buchwald; 456g Berlin, Humboldt Univ.; 324g Washington, U.S. Nat. Mus.; 2kg Chicago, Field Mus. Nat. Hist.;
Specimen(s): [86068], 281g. has been heated, V.F. Buchwald.; [1959,969], 208g. and 13.5g of sawings, almost certainly NOT Apoala; it is probably part of the 164kg Durango mass, V.F. Buchwald, letter of 14 August 1972 in Min. Dept. Brit. Mus. (Nat. Hist.)

Apollonia v Santa Apollonia.

Appley Bridge 53°35′ N., 2°43′ W.
Wigan, Lancashire, England
Fell 1914, October 13, 2045 hrs
Synonym(s): *Lancashire, Wigan*
Stone. Olivine-hypersthene chondrite, amphoterite (LL6), veined.
After appearance of a luminous fireball and detonations, a stone of about 33lb. was found next day at Halliwell Farm, W.F. Denning, Nature, 1914, **94**, p.258, W.C. Jenkins, Nature, 1914, **94**, p.505. Described and analysed, 20.08 % total iron, B. Mason and H.B. Wiik, Geochimica et Cosmochimica Acta, 1964, **28**, p.533. Olivine Fa_{31}, B. Mason, Geochimica et Cosmochimica Acta, 1963, **27**, p.1011. U-Pb measurements., N.H. Gale et al., Naturwiss., 1979, **66**, p.419. Ni and Co contents of metal, D.W. Sears and H.J. Axon, Meteoritics, 1976, **11**, p.97.
850g Edinburgh, Royal Scottish Mus.; 545g Washington, U.S. Nat. Mus.; 214g New York, Amer. Mus. Nat. Hist.;
Specimen(s): [1920,40], 8.69kg. main mass with 1.71kg end piece and fragments 23g; [1920,41], 4g. powder

Aprelskij v Aprel'sky.

Aprel'sky 53°18′ N., 126°7′ E.
Amur region, Federated SSR, USSR, [Апрельский]
Found 1969
Synonym(s): *Aprelskij*
Iron. Octahedrite, medium (0.7mm) (IIIB).
A mass of 54.6kg was recovered from gold-bearing deposits, Meteor. Bull., 1969 (48), Meteoritics, 1970, **5**, p.108. Classification and analysis, 10.07 %Ni, 19.4 ppm.Ga, 37 ppm.Ge, 0.054 ppm.Ir, E.R.D. Scott and J.T. Wasson, Geochimica et Cosmochimica Acta, 1976, **40**, p.103.
54.6kg Moscow, Acad. Sci.;

Apt 43°52′ N., 5°23′ E.
Vaucluse, France
Fell 1803, October 8, 1030 hrs
Synonym(s): *Saurette, Vaucluse*
Stone. Olivine-hypersthene chondrite (L6), veined.
A stone of about 3.2kg fell, after detonations, A. Laugier, Ann. Phys. (Gilbert), 1804, **16**, p.72. Analysis, olivine $Fa_{24.2}$, 21.58 % total iron, R.T. Dodd and E. Jarosewich, Meteoritics, 1980, **15**, p.69.
2kg Paris, Mus. d'Hist. Nat.; 297g Vienna, Naturhist. Mus.; 51g Rome, Vatican Colln; 34g Chicago, Field Mus. Nat. Hist.; 15g Berlin, Humboldt Univ.; 10g Budapest, Nat. Hist. Mus.;
Specimen(s): [35167], 36.7g.; [1960,329], 36g.

Arabella
Lincoln County, New Mexico, U.S.A., Co-ordinates not reported
Found 1955, before this year
Iron. Octahedrite, medium.
Total known weight 1kg; listed, Cat. Huss Coll. Meteor., 1976, p.3.
36.5g Mainz, Max-Planck-Inst.;

Arabia
Saudi Arabia
Doubtful. Stone.
A stone in the Mus. Nac. Rio de Janeiro is labelled Arabia, E. de Oliveira, Anais Acad. Brasil. Cienc., 1931, **3**, p.52 [M.A. 5-14]. It is probably part of the Hedjaz stone, W. Campbell Smith, Min. Mag., 1932, **23**, p.47.

Aragon v Cedartown.

Arapahoe 38°48′ N., 102°12′ W. approx.
Cheyenne County, Colorado, U.S.A.
Found 1940
Stone. Olivine-hypersthene chondrite (L5), black.
A mass of 19.083kg was found, H.H. and A.D. Nininger, The Nininger Collection of Meteorites, Winslow, Arizona, 1950, p.29, Ward's Nat. Sci. Estab., Meteorite Price List, 1950. Olivine Fa_{23}, B. Mason, Geochimica et Cosmochimica Acta, 1963, **27**, p.1011. Old I-Xe age, R.J. Drozd and F.

Podosek, Earth planet. Sci. Lett., 1976, **31**, p.15.
5.4kg Tempe, Arizona State Univ.; 422g Washington, U.S. Nat. Mus.; 343g Denver, Mus. Nat. Hist.; 53g Los Angeles, Univ. of California.; 1.5kg Chicago, Field Mus. Nat. Hist.; 52g New York, Amer. Mus. Nat. Hist.;
Specimen(s): [1950,222], 15.4g. two pieces; [1959,827], 6670g. in two pieces, a slice of 1055g and a piece of 5615g.

Arbol Solo 33° S., 66° W. approx.

Socoscora district, San Luis, Argentina
Fell 1954, September 11, 2100 hrs
Stone. Olivine-bronzite chondrite (H5).

A shower of stones fell, but not much has been preserved, Meteor. Bull., 1964 (32). Possibly the olivine-bronzite chondrite examined by B. Mason under the name San Luis (*q.v.*) may be a specimen of Arbol Solo.
600g Mendoza, Cuyo Univ.; 165g L.O. Giacomelli's collection.;

Arcadia 41°25′ N., 99°6′ W.

Valley County, Nebraska, U.S.A.
Found 1937
Stone. Olivine-hypersthene chondrite, amphoterite (LL6).

A mass of 19.4kg was found in NW. 1/4, sect. 11, township 18, range 16, Valley County, A.D. Nininger, Pop. Astron., Northfield, Minnesota, 1939, **47**, p.211. Mentioned, E.P. Henderson, letter of 3 June 1939 in Min. Dept. BM(NH). Listed, H.H. and A.D. Nininger, The Nininger Collection of Meteorites, Winslow, Arizona, 1950, p.29. Olivine Fa_{29}, B. Mason, Geochimica et Cosmochimica Acta, 1963, **27**, p.1011.
6kg Tempe, Arizona State Univ.; 1.7kg Washington, U.S. Nat. Mus.; 388g Chicago, Field Mus. Nat. Hist.; 125g Los Angeles, Univ. of California; 63g New York, Amer. Mus. Nat. Hist.;
Specimen(s): [1953,6], 5.4g.; [1959,995], 5004g. and fragments, 8.5g

Arch v Arch (a).

Arch (a) 34°9′ N., 103°13′12″ W.

Roosevelt County, New Mexico, U.S.A.
Found 1972, spring
Stone. Carbonaceous chondrite, type III (CV3).

A single mass of 985.8g was found 8km NW. of Arch, T.V.V. King et al., Meteoritics, 1977, **12**, p.276, abs., Meteor. Bull., 1978 (55), Meteoritics, 1978, **13**, p.330.
49g Mainz, Max-Planck-Inst.;

Arch (b) 34°12′21″ N., 103°13′20″ W.

Roosevelt County, New Mexico, U.S.A.
Found 1979, approx.
Stone. Olivine-bronzite chondrite (H4).

A mass of 1.5kg was ploughed up 9.6 km ENE. of Portales, G.I Huss, priv. comm., 1982. Olivine Fa_{20}, A.L. Graham, priv. comm., 1983.
Specimen(s): [1983,M.19], 42g.

Archie 38°30′ N., 94°18′ W. approx.

Cass County, Missouri, U.S.A.
Fell 1932, August 10, 1630 hrs
Stone. Olivine-bronzite chondrite (H6).

Seven stones were found, totalling 11lb. 2.5oz., the largest being 8lb. Described, E.S. Haynes, Pop. Astron., Northfield, Minnesota, 1935, **43**, p.181 [M.A. 6-104]. Brief report, H.H. Nininger, Pop. Astron., Northfield, Minnesota, 1936, **44**, p.93 [M.A. 6-399]. Mentioned, H.H. Nininger, The Mines Mag., Golden, Colorado, 1933, **23** (8), p.6, C.C. Wylie, Pop. Astron., Northfield, Minnesota, 1933, **41**, p.55 [M.A. 5-404], C.C. Wylie, Pop. Astron., Northfield, Minnesota, 1935, **43**, p.637. Olivine Fa_{20}, B. Mason, Geochimica et Cosmochimica Acta, 1963, **27**, p.1011. One of the seven stones narrowly missed a man's head. A shower of minute fragments over an area six miles long was reported but none of this material was preserved, H.H. and A.D. Nininger, The Nininger Collection of Meteorites, Winslow, Arizona, 1950, p.29. Calculation of the orbit, C.C. Wylie, Pop. Astron., Northfield, Minnesota, 1948, **56**, p.273 [M.A. 10-401].
3.76kg Washington, U.S. Nat. Mus.; 139g Tempe, Arizona State Univ.; 57g Los Angeles, Univ. California; 46g New York, Amer. Mus. Nat. Hist.; 22oz. Michigan Univ.; 97g Chicago, Field Mus. Nat. Hist.;
Specimen(s): [1959,813], 199g.

Arenazzo v Renazzo.

Argonia 37°16′ N., 97°46′ W.

Sumner County, Kansas, U.S.A.
Found 1940, before this year
Stony-iron. Pallasite (PAL).

A mass estimated at 34kg was found, but only 84g have been preserved, A.D. Nininger, Pop. Astron., Northfield, Minnesota, 1940, **48**, p.555, Contr. Soc. Res. Meteorites, **2**, p.227 [M.A. 8-54], H.H. and A.D. Nininger, The Nininger Collection of Meteorites, Winslow, Arizona, 1950, p.29. Olivine $Fa_{13.5}$, P.R. Buseck and J.I. Goldstein, Bull. Geol. Soc. Amer., 1969, **80**, p.2141.
25g Tempe, Arizona State Univ.; Main mass, Friends Univ.;

Arhus v Aarhus.

Aricarie v Washington County.

Arickaree v Washington County.

Arickarie v Washington County.

Arispe 30°20′ N., 109°59′ W.

Sonora, Mexico
Found 1896
Synonym(s): *Arizpe, Moctezuma (of F. Bewerth), Noon*
Iron. Octahedrite, coarse (2.9mm) (IC).

A mass of about 272lb was found in 1898 about 15 miles NW. of Arispe and two other masses of 116lb and 20lb had been found in 1896 about 25 miles NW. of Arispe; the largest mass was described, H.A. Ward, Proc. Rochester Acad. Sci., 1902, **4**, p.79, A.F. Wunsche, Proc. Colorado Sci. Soc., 1903, **7**, p.67. Description of the smaller masses, O.C. Farrington, Field Mus. Nat. Hist. Geol. Ser., 1914, **5** (178), p.2. Analysis, 6.77 %Ni, E. Goldberg et al., Geochimica et Cosmochimica Acta, 1951, **2**, p.1. Classification and analysis, 6.54 %Ni, 50.3 ppm.Ga, 243 ppm.Ge, 9.7 ppm.Ir, E.R.D. Scott and J.T. Wasson, Geochimica et Cosmochimica Acta, 1976, **40**, p.103. Contains 117g per tonne Pt metals, V.M. Goldschmidt, Proc. Roy. Inst. Gt. Britain, 1929, **26**, p.73, Die Naturwiss., 1930, **18**, p.999 [M.A. 5-7]. Another mass of 122kg had been in use as an anvil, H.H. and A.D. Nininger, The Nininger Collection of Meteorites, Winslow, Arizona, 1950, p.29. Description, V.F. Buchwald, Iron Meteorites, Univ. of California, 1975, p.269.
178kg Washington, U.S. Nat. Mus.; 137.2kg Tempe, Arizona State Univ.; 13.8kg Vienna, Naturhist. Mus.; 74.6kg Fort Worth, Texas, Monnig Colln.; 42kg Chicago, Field Mus. Nat. Hist.; 6kg New York, Amer. Mus. Nat. Hist.; 3.1kg Mainz, Max-Planck-Inst.; 1.8kg Michigan

State Univ.; 1kg Mexico City, Inst. Geol.; 2.8kg Albuquerque, Univ. of New Mexico.; 610g Yale Univ.; 587g Tübingen, Univ.;
Specimen(s): [1959,1051], 39.908kg.; [86425], 1846g. a slice

Arizona v Cañon Diablo.

Arizona v Tucson.

Arizpe v Arispe.

Arlington 44°36′ N., 94°6′ W. approx.
Sibley County, Minnesota, U.S.A.
Found 1894
Synonym(s): *Sibley County*
Iron. Octahedrite, medium (0.8mm) (IIE).
A mass of about 19.7lb was found 2.5 miles NE. of Arlington; described, with an analysis, 8.60 %Ni, N.H. Winchell, Amer. Geologist, 1896, **18**, p.267. Classification and analysis, 8.42 %Ni, 21.8 ppm.Ga, 64.9 ppm.Ge, 5.8 ppm.Ir, E.R.D. Scott and J.T. Wasson, Geochimica et Cosmochimica Acta, 1976, **40**, p.103.
6.1kg Washington, U.S. Nat. Mus., main mass; 164g Chicago, Field Mus. Nat. Hist.; 56g Tempe, Arizona State Univ.; 55g Berlin, Humboldt Univ.; 1kg New York, Amer. Mus. Nat. Hist.;
Specimen(s): [1921,438], 101g. end slice; [83393], 56g.

Arltunga also see Cranbourne.

Arltunga 23°20′ S., 134°40′ E.
Northern Territory, Australia
Found 1908, September
Iron. Ataxite (IID-ANOM).
A mass of 40 lb was found 2 miles south of the Government Cyanide Works, C. Anderson, Rec. Austr. Mus., 1913, **10**, p.54, L.L. Smith, Am. J. Sci., 1910, **30**, p.264.. Apparently a recent fall, probably within a few months of the find. The appearence of an etched surface suggests an ataxite, but under high magnification a micro-octahedral structure is seen. Described, with an analysis, 10.22 %Ni, D. Mawson, Trans. Roy. Soc. South Austr., 1934, **58**, p.1 [M.A. 6-15]. Classification and new analysis, 9.64 %Ni, 76.8 ppm.Ga, 81.5 ppm.Ge, 17 ppm.Ir, E.R.D. Scott and J.T. Wasson, Geochimica et Cosmochimica Acta, 1976, **40**, p.103. Mentioned, J.F. Lovering and L.G. Parry, Geochimica et Cosmochimica Acta, 1962, **26**, p.361. Structural description, V.F. Buchwald, Iron Meteorites, Univ. of California, 1975, p.273., H.J. Axon, Min. Mag., 1968, **36**, p.1139.
Main mass, Adelaide, South Austr. Mus.; 157g Sydney, Australian Mus.; 979g Washington, U.S. Nat. Mus.;
Specimen(s): [1937,245], 320g. and a fragment, 48g.

Armanty 47° N., 88° E. approx.
Xinjiang, China
Found 1898, before this year
Synonym(s): *Kumisch Choi Cha, Kumys-Tyuya, Mungen Dusch, Wushike, Xinjiang*
Iron. Octahedrite, medium (IIIE).
The main mass (20 tons) of this little known meteorite remains at the place of find; five pieces weighing 74.92g were acquired by the Acad. Sci. USSR. in 1939, L.A. Kulik, Метеоритика, 1941, **2**, p.123 [M.A. 10-174], E.L. Krinov, Каталог Метеоритов Акад. Наук СССР, Москва, 1947 [M.A. 10-511]. Description, analysis, 9.92 %Ni, O.S. Vyalov, Метеоритика, 1949, **5**, p.23 [M.A. 12-106]. Analysis, 9.09 %Ni, M.I. D'yakonova, Метеоритика, 1958, **16**, p.180, Метеоритика, 1959, **17**, p.96 [M.A. 15-534]. Listed, with references, D. Bian, Meteoritics, 1981, **16**, p.115. Classification and analysis, 9.1 %Ni, 16.2 ppm.Ga, 31.5 ppm.Ge, 0.23 ppm.Ir, A. Kracher et al., Geochimica et Cosmochimica Acta, 1980, **44**, p.773.
Main mass, Wulumuqi, Exhibition Center; 42g Moscow, Acad. Sci.;

Armel 39°46′ N., 102°8′ W.
Yuma County, Colorado, U.S.A.
Found 1967
Stone. Olivine-hypersthene chondrite (L5).
A stone of 9.2kg was ploughed up, Meteor. Bull., 1970 (49), Meteoritics, 1970, **5**, p.174. Olivine Fa_{23} X-Ray diffraction method, M.J. Frost, Min. Dept. BM(NH).
3.2kg Mainz, Max-Planck-Inst.; 489g Denver, Mus. Nat. Hist.; 301g Washington, U.S. Nat. Mus.; 168g Tempe, Arizona State Univ.; 63g New York, Amer. Mus. Nat. Hist.;
Specimen(s): [1969,2], 51g. a slice

Aroos v Yardymly.

Arrabury 26°30′ S., 141°5′ E.
Queensland, Australia
Found 1960, approx.
Stone.
A mass of 5.3kg, found on Arrabury station close to the border with South Australia, was purchased by the South Australian Mus., Meteor. Bull., 1983 (61), Meteoritics, 1983, **18**, p.77.
55g Washington, U.S. Nat. Mus.; Specimen, Adelaide, South Austr. Mus.;

ar-Rakhbah 20°36′ N., 52°30′ E.
Rub'al Khali, Saudi Arabia
Found 1955
Synonym(s): *Rakhbah*
Iron.
A completely weathered iron meteorite, B. Mason, letter of 23 December 1968 in Min. Dept. BM(NH). Not a stone as suggested, D.A. Holm, Am. J. Sci., 1962, **260**, p.303.
395g Washington, U.S. Nat. Mus.;

Arriba 39°18′ N., 103°15′ W.
Lincoln County, Colorado, U.S.A.
Found 1936
Stone. Olivine-hypersthene chondrite (L5), brecciated.
One stone of 15kg in three pieces was found, and later a further four stones, 9.9kg, 2.9kg, 2.8kg, and 0.5kg, A.D. Nininger, Pop. Astron., Northfield, Minnesota, 1937, **45**, p.449 [M.A. 7-62]. Olivine Fa_{25}, B. Mason, Geochimica et Cosmochimica Acta, 1963, **27**, p.1011. The 0.5kg mass is regarded as a distinct fall (Arriba no. 2), A.D. Nininger, Pop. Astron., Northfield, Minnesota, 1937, **45**, p.449 [M.A. 7-62], F.C. Leonard, Pop. Astron., Northfield, Minnesota, 1947, **55**, p.381. Not listed separately, H.H. and A.D. Nininger, The Nininger Collection of Meteorites, Winslow, Arizona, 1950, p.29.
13.3kg Tempe, Arizona State Univ.; 5kg Denver, Mus. Nat. Hist.; 1.1kg Washington, U.S. Nat. Mus.; 247g Los Angeles, Univ. of California; 169g New York, Amer. Mus. Nat. Hist.; 2kg Chicago, Field Mus. Nat. Hist.;
Specimen(s): [1937,1653], 587g. end slice.; [1959,781], 8720g. an almost complete stone.

Arroyo Aguiar 31°25′ S., 60°40′ W.
La Capital department, Santa Fe, Argentina
Fell 1950, Summer
Stone. Olivine-bronzite chondrite (H5).
One stone of 7.45kg was seen to fall, and was found soon after at a depth of 50 cm, J.L. Benet, El Meteorito de Arroyo Aguiar, Publ. Univ. Nac. del Litoral, Cuidad de Santa Fe, Argentina, 1961, L.O. Giacomelli, letter of 25 June 1962 in Min. Dept. BM(NH). Reported, Meteor. Bull., 1962 (25). Olivine Fa_{20}, B. Mason, Geochimica et Cosmochimica Acta, 1967, **31**, p.1100.
Main mass, Santa Fe, Museo Facultad Ingenieria Quimica.; 2.9g Washington, U.S. Nat. Mus.;
Specimen(s): [1964,62], 8.5g.

Artracoona 29°4′ S., 139°55′ E.
South Australia, Australia
Found 1914
Stone. Olivine-hypersthene chondrite (L6).
A complete weathered stone, 20.81kg, was found 8 miles north of the old Carraweena Head Station and 6 miles west of Artracoona Hill. It is distinct from Carraweena and Accalana, D. Heymann, Geochimica et Cosmochimica Acta, 1965, **29**, p.1203. Description, analysis, A.W. Kleeman, Trans. Roy. Soc. South Austr., 1936, **60**, p.73 [M.A. 7-71]. Olivine $Fa_{25.6}$, B. Mason, Rec. Austr. Mus., 1974, **29**, p.169. Analysis, 21.19 % total iron, M.J. Fitzgerald, Ph.D. Thesis, Univ. of Adelaide, 1979, p.23.
20kg Adelaide, Univ., main mass; 500g Adelaide, South Austr. Mus.; 375g Washington, U.S. Nat. Mus.; 5.7g New York, Amer. Mus. Nat. Hist.;
Specimen(s): [1966,492], 85g. a fragment.

Arundu 35°52′ N., 75°20′ E.
Baltistan, Kashmir, Pakistan
Fell 1936, September
Doubtful. Stone.
A stone was given to Mr. J.B. Auden by the Tahsildar of Skardu, but was lost and the description is inadequate, A.M. Heron, Rec. Geol. Surv. India, 1938, **73**, p.28.

Arva v Magura.

Asarco Mexicana
Not known,
Found
Iron. Octahedrite, medium (1.1mm) (IIIA).
Classification and analysis, 8.32 %Ni, 21.2 ppm.Ga, 44.4 ppm.Ge, 0.23 ppm.Ir, E.R.D. Scott et al., Geochimica et Cosmochimica Acta, 1973, **37**, p.1957.
Specimen, Washington, U.S. Nat. Mus.;

Asco 42°27′ N., 9°2′ E.
Corsica, France
Fell 1805, November
Stone. Olivine-hypersthene chondrite (L).
Original weight not known, P. Partsch, Die Meteoriten, Wien, 1843, p.64. Olivine Fa_{26}, B. Mason, Geochimica et Cosmochimica Acta, 1963, **27**, p.1011.
18g Vienna, Naturhist. Mus.; 6g Berlin, Humboldt Univ.; 2g New York, Amer. Mus. Nat. Hist.; 0.9g Tübingen, Univ.; 5g Budapest, Nat. Mus.; 9g Chicago, Field Mus. Nat. Hist.;
Specimen(s): [35169], less than 1g.

Ashburton Downs v Dalgety Downs.

Ashdon 52°3′ N., 0°18′ E.
Saffron Walden, Essex, England
Fell 1923, March 9, 1300 hrs
Stone. Olivine-hypersthene chondrite (L6).
A stone of about 1300g was seen to fall; described and partially analysed, G.T. Prior, Min. Mag., 1923, **20**, p.131. Olivine Fa_{25}, B. Mason, Geochimica et Cosmochimica Acta, 1963, **27**, p.1011.
Specimen(s): [1923,484], 1241g. the nearly complete stone, and fragments 11.5g.

Asheville v Black Mountain.

Asheville v Duel Hill (1854).

Asheville 35°35′ N., 82°32′ W.
Buncombe County, North Carolina, U.S.A.
Found 1839
Synonym(s): *Baird's Farm, Baird's Plantation, Buncombe County*
Iron. Octahedrite, medium (0.6mm) (IIIB).
A mass about the size of a man's head was found 6 miles north of Baird's Farm, C.U. Shepard, Am. J. Sci., 1839, **36**, p.81, 1847, **4**, p.79. Analysis, 9.07 %Ni, M.I. D'yakonova, Метеоритика, 1958, **16**, p.130. Description, V.F. Buchwald, Iron Meteorites, Univ. of California, 1975, p.274.
276g Tempe, Arizona State Univ.; 271g Vienna, Naturhist. Mus.; 36g Yale Univ.; 13g Berlin, Humboldt Univ.; 6g Washington, U.S. Nat. Mus.; 5g Los Angeles, Univ. of California;
Specimen(s): [33749], 70g. fragments; [34610], 28g.; [24003], 13g. fragments; [34375], 4.5g.

Ashfork v Cañon Diablo.

Ashmore 32°54′ N., 102°17′ W.
Gaines County, Texas, U.S.A.
Found 1969
Synonym(s): *Loop*
Stone. Olivine-bronzite chondrite (H5).
A single individual of 55.4kg was discovered during deep ploughing approximately 2.5 miles north and 1.5 miles west of Ashmore, at a depth of 18 inches; partial analysis, 27.5 % total iron, olivine $Fa_{18.8}$, J.R. Craig et al., Meteoritics, 1971, **6**, p.33. At one time thought to be part of the Loop meteorite. Described, with analysis, 27.24 % total iron, W.B. Bryan and G. Kullerud, Meteoritics, 1975, **10**, p.41.
17.7kg Mainz, Max-Planck-Inst.; 328g Fort Worth, Texas, Monnig Colln.; 187g Copenhagen, Univ. Geol. Mus.; 130g Vienna, Naturhist. Mus.; 100g New York, Amer. Mus. Nat. Hist.; 16g Washington, U.S. Nat. Mus., fragments; Fragments, West Texas Mus., Texas Tech. Univ.;
Specimen(s): [1971,287], 72.7g. a crusted part slice.

ash-Shalfah 21°52′30″ N., 49°43′10″ E.
Rub'al Khali, Saudi Arabia
Found 1961, Summer
Synonym(s): *ash-Shalfar, Shalfa*
Stone. Olivine-hypersthene chondrite (L4).
A much weathered mass of 935g was recovered, D.A. Holm, Am. J. Sci., 1962, **260**, p.303. Classification, B. Mason, priv. comm., 1980.
935g Washington, U.S. Nat. Mus., the entire mass;

ash-Shalfar v ash-Shalfah.

Assam 26° N., 92° E. approx.
Assam, India
Found 1846, known in this year
Stone. Olivine-hypersthene chondrite (L5), brecciated.
Three pieces, together weighing about 6 lb, were found in Calcutta in the "Coal and Iron Committee's" collection, and were probably obtained from Assam, H. Piddington, J. Asiatic Soc. Bengal, 1846, **15**, p.Proc. XLVI, LXXVI. Description, W. von Haidinger, Sitzungsber. Akad. Wiss. Wien, Math.-naturwiss. Kl., 1860, **41**, p.752. Olivine Fa_{24}, B. Mason, Geochimica et Cosmochimica Acta, 1963, **27**, p.1011. Gas-rich, L. Schultz et al., Earth planet. Sci. Lett., 1971, **12**, p.119.
301g Calcutta, Mus. Geol. Surv. India; 188g Vienna, Naturhist. Mus.; 41g Paris, Mus. d'Hist. Nat.; 21g Tempe, Arizona State Univ.;
Specimen(s): [33760], 357g.; [34798], 180g.; [96256], Three thin sections

as-Sanam 22°45′ N., 51°10′ E.
Rub'al Khali, Saudi Arabia
Found 1954
Stone.
"Several large chunks of rock" were found, one of which 20×20×15cm on analysis compares closely to the average of many stony meteorites", D.A. Holm, Am. J. Sci., 1962, **260**, p.303. The analysis suggests a very severely oxidised stone.

Assisi 43°2′ N., 12°33′ E.
Perugia, Umbria, Italy
Fell 1886, May 24, 0700 hrs
Synonym(s): *Bettona, Perugia, Torre, Torre Assisi, Torre presso Assisi*
Stone. Olivine-bronzite chondrite (H5).
A stone of about 2kg was seen to fall at Tordandrea, G.R. Levi-Donati, Min. Mag., 1967, **36**, p.595, F. Burragato, Periodico Miner., 1967, **36**, p.463. Initial report, G. Bellucci, Il meteorito di Assisi, Perugia, Tipographia di Vincenzo Santucci, Perugia, 1887. Olivine Fa_{18}, B. Mason, Geochimica et Cosmochimica Acta, 1963, **27**, p.1011.
227g Turin, Univ.; 200g Vienna, Naturhist. Mus.; 114.3g Chicago, Field Mus. Nat. Hist.; 114g Rome, Univ.; 29g Washington, U.S. Nat. Mus.; 23g Berlin, Humboldt Univ.; 18g Prague, Nat.Mus.; 5g Tempe, Arizona State Univ.; 4g New York, Amer. Mus. Nat. Hist.; 107g Paris, Mus. d'Hist. Nat.;
Specimen(s): [63621], 146.5g.; [92565], 3g.

as-Su'aydan 21°15′15″ N., 55°18′18″ E.
Rub'al Khali, Saudi Arabia
Found 1960, December 14
Stone. Olivine-hypersthene chondrite (L5).
An oxidised mass of 5855g was found. Reported with an analysis, D.A. Holm, Am. J. Sci., 1962, **260**, p.303. Olivine Fa_{26}, B. Mason, priv. comm., 1981.
5.53kg Washington, U.S. Nat. Mus., main mass;

Aswam error for Aswan.

Aswan 23°59′10″ N., 32°37′25″ E.
Egypt
Found 1955
Iron. Octahedrite, medium (1.2mm) (IIIA).
One mass of 12kg was found on a road. Described, E.M. El Shazly, Egypt J. Geol., 1958, **2**, p.71 [M.A. 14-409]. Classification and analysis, 8.21 %Ni, 20 ppm.Ga, 41.8 ppm.Ge, 0.22 ppm.Ir, E.R.D. Scott et al., Geochimica et Cosmochimica Acta, 1973, **37**, p.1957.
11g Washington, U.S. Nat. Mus.;

Atacama v Cachiyuyal.

Atacama v Dehesa.

Atacama v Imilac.

Atacama Desert v Copiapo.

Atacama Desert v Ilimaes (iron).

Atacama Desert v Joel's Iron.

Atacama Desert v Lutschaunig's Stone.

Atacama Desert v Slaghek's Iron.

Atarra 25°15′15″ N., 80°37′30″ E.
Manikpur, Banda district, Uttar Pradesh, India
Fell 1920, December 23, 1735 hrs
Synonym(s): *Turra*
Stone. Olivine-hypersthene chondrite (L4), black.
After detonations and appearance of trail of smoke from north to south three stones of total weight 1280g fell at Turra village, 4 miles from Atarra railway station; description, G.V. Hobson, Rec. Geol. Surv. India, 1927, **60**, p.131. Mineralogy, analysis, 21.44 % total iron, olivine $Fa_{24.9}$, A. Dube et al., Smithson. Contrib. Earth Sci., 1977 (19), p.71.
617g Calcutta, Mus. Geol. Surv. India; 40g Tempe, Arizona State Univ.; 37g Washington, U.S. Nat. Mus.;
Specimen(s): [1925,443], 74g.

Atemajac 20°4′ N., 103°40′ W.
Sierra de Topalpo, Jalisco, Mexico
Fell 1896, February 26
Stone. Olivine-hypersthene chondrite (L6).
Description, analysis, J.C. Haro, Bol. Inst. Geol. Mexico, 1931 (50), p.44. Olivine Fa_{23}, B. Mason, Geochimica et Cosmochimica Acta, 1963, **27**, p.1011.
8g Washington, U.S. Nat. Mus.; 5g Tempe, Arizona State Univ.; 44g Mexico City, Inst. Geol.; 32g Chicago, Field Mus. Nat. Hist.;

Athens 34°45′ N., 87°0′ W.
Limestone County, Alabama, U.S.A.
Fell 1933, July 11, 0930 hrs
Stone. Olivine-hypersthene chondrite, amphoterite (LL6), brecciated.
One stone of 265g fell; description, C.C. Wylie and S.H. Perry, Pop. Astron., Northfield, Minnesota, 1933, **41**, p.468 [M.A. 5-404]. Also with an unsatisfactory analysis, S.H. Perry and C.C. Wylie, Pop. Astron., Northfield, Minnesota, 1935, **43**, p.331 [M.A. 6-104]. Olivine Fa_{31}, B. Mason and H.B. Wiik, Geochimica et Cosmochimica Acta, 1964, **28**, p.533.
206g Washington, U.S. Nat. Mus., main mass;

Atlanta 31°48′ N., 92°45′ W.
Winn County, Louisiana, U.S.A.
Found 1938
Stone. Enstatite chondrite (E6).
A fragment of 5.5kg was found in SE. 1/4, section 14, township 9 N., range 4W., A.D. Nininger, Pop. Astron., Northfield, Minnesota, 1939, **47**, p.211. Analysis by H.B. Wiik, 28.95 % total iron, B. Mason, Geochimica et Cosmochimica Acta, 1966, **30**, p.23. XRF analysis, 24.81 % total iron, H. von Michaelis et al., Earth planet. Sci. Lett.,

1969, **5**, p.387. Mineralogy, K. Keil, J. Geophys. Res., 1968, **73**, p.6945.
1.2kg Tempe, Arizona State Univ.; 141g New York, Amer. Mus. Nat. Hist.; 267g Chicago, Field Mus. Nat. Hist.;
Specimen(s): [1959,1001], 1568g. and fragments, 32.5g.

Atlantic Ocean 30°50′ N., 70°25′ W.
Fell 1809, June 19, 2300 hrs
Doubtful..
Many stones fell into the sea, and one of 170g on the deck of a ship, south of Rhode Island, U.S.A., but this was lost and the description is inadequate, E.F.F. Chladni, Die Feuer-Meteore, Wien, 1819, p.290, F. Berwerth, Fortschr. Min. Krist. Petr., 1912, **2**, p.230.

Atobe v Gifu.

Atoka 34°19′ N., 96°9′ W.
Atoka County, Oklahoma, U.S.A.
Fell 1945, September 17
Stone. Olivine-hypersthene chondrite (L6).
Fragments totalling 475g (the largest 330g) were in Washington, U.S. Nat. Mus., Rep. U.S. Nat. Mus., 1947, p.43. Olivine Fa_{24}, B. Mason, Geochimica et Cosmochimica Acta, 1963, **27**, p.1011.
952g Fort Worth, Texas, Monnig Colln.; 432g Washington, U.S. Nat. Mus.;

Atsuma 43° N., 141° E.
Yufutsu, Hokkaido, Japan
Fell 1935, September 3
Doubtful. Stone?.
A meteorite was seen to fall in the Atsumamura river, and a stone of 10.69kg was recovered, described and analysed, S. Nakao, J. Geogr., Tokyo Geogr. Soc., 1936, **48**, p.485 [M.A. 6-395]. Report describes a block of gabbro; the meteorite, if any, was never found, S. Murayama, letter of 6 April 1962 in Min. Dept. BM(NH).

Atwood 40°31′ N., 103°16′ W.
Logan County, Colorado, U.S.A.
Found 1963, recognized in this year
Stone. Olivine-hypersthene chondrite (L6).
Three fragments of total weight 2.1kg of one stone were ploughed up, G.I. Huss, letter of 20 September 1970, in Min. Dept. BM(NH). Olivine Fa_{25}, B. Mason, Geochimica et Cosmochimica Acta, 1967, **31**, p.1100.
560g Mainz, Max-Planck-Inst.; 202g Denver, Mus. Nat. Hist.; 161g Washington, U.S. Nat. Mus.; 49g New York, Amer. Mus. Nat. Hist.; 37g Tempe, Arizona State Univ.;
Specimen(s): [1965,409], 34.5g. crusted part slice

Aubres 44°23′ N., 5°10′ E.
Nyons, Drome, France
Fell 1836, September 14, 1500 hrs
Synonym(s): *Nyons*
Stone. Achondrite, Ca-poor. Aubrite (AUB).
A stone of about 800g was seen to fall, J.R. Gregory, Geol. Mag., 1887, **4** (Decade III), p.552. Analysis of enstatite, A.M. Reid and A.J. Cohen, Geochimica et Cosmochimica Acta, 1967, **31**, p.661.
24g Budapest, Nat. Hist. Mus.; 9g Berlin, Humboldt Univ.; 2g Chicago, Field Mus. Nat. Hist.; 11g Paris, Mus. d'Hist. Nat.;
Specimen(s): [63552], 440g. the greater part of the stone.

Auburn v Tombigbee River.

Aucamville v Toulouse.

Augsburg (952) 48°22′ N., 10°53′ E.
Bayern, Germany
Fell 952, A.D.
Doubtful..
A large stone fell, E.F.F. Chladni, Die Feuer-Meteore, Wien, 1819, p.193, Ann. Phys. (Gilbert), 1814, **47**, p.105, but the evidence is not conclusive.

Augsburg (1528) 48°22′ N., 10°53′ E.
Bayern, Germany
Fell 1528, June 29
Doubtful. Stone.
One stone fell, E.F.F. Chladni, Die Feuer-Meteore, Wien, 1819, p.212, but the evidence is not conclusive.

Augusta v Castine.

Augusta County also see Staunton.

Augusta County 38°10′ N., 79°5′ W. approx.
Virginia, U.S.A.
Found 1858, or 1859, recognized 1877
Synonym(s): *Staunton No. 4 and No. 6*
Iron. Octahedrite, medium (1.2mm) (IIIA).
The 69kg and 7kg masses previously assigned to the Staunton find differ significantly from the other Staunton masses, which are coarse octahedrites of group IIIE. It was therefore proposed that the two masses be renamed Augusta County, V.F. Buchwald, Iron Meteorites, Univ. of California, 1975, p.277. Analysis, 8.19 %Ni, 18.7 ppm.Ga, 35.3 ppm.Ge, 9.2 ppm.Ir, E.R.D. Scott et al., Geochimica et Cosmochimica Acta, 1973, **37**, p.1957.
4.84kg Vienna, Naturhist. Mus., according to V.F. Buchwald (loc. cit.); 4.83kg Washington, U.S. Nat. Mus., V.F. Buchwald (loc. cit.); 2.84kg Chicago, Field Mus. Nat. Hist., two masses, 1.58kg and 1.26 kg V.F. Buchwald (loc. cit.); 6.57kg Budapest, Nat. Hist. Mus., V.F. Buchwald (loc. cit.); 2.74kg Harvard Univ., V.F. Buchwald (loc. cit.);
Specimen(s): [54820], 1293g. and fragments, 12g.

Augustinovka 48°4′ N., 35°5′ E.
Ekaterinoslav, Ukraine, USSR, [Августиновка]
Found 1890
Synonym(s): *Augustinowka, Avgustinovka, Ekaterinoslav, Jekaterinoslav, Verkhne Dnieprovsk (in part)*
Iron. Octahedrite, medium (0.8mm) (IIIB).
A mass of about 400kg was found in loess, V.F. Alexejev, Zap. Imp. Miner. Obshch., 1893, **30**, p.475, S. Meunier, C. R. Acad. Sci. Paris, 1893, **116**, p.1151, A. von Kupffer, Ann. Naturhist. Hofmus. Wien, 1911, **25**, p.436. Classification and analysis, 9.56 %Ni, 17.8 ppm.Ga, 37.6 ppm.Ge, 0.03 ppm.Ir, E.R.D. Scott et al., Geochimica et Cosmochimica Acta, 1973, **37**, p.1957. Identical with most [but not all] of Verkhne Dnieprovsk specimens, V.F. Buchwald, Iron Meteorites, Univ. of California, 1975, p.279.
Main mass, Leningrad,; 2.3kg Vienna, Naturhist. Mus.; 1.9kg Rome, Vatican Colln; 1.2kg Chicago, Field Mus. Nat. Hist.; 500g Moscow, Acad. Sci.; 300g New York, Amer. Mus. Nat. Hist.; 129g Berlin, Humboldt Univ.;
Specimen(s): [83956], 869g. a slice and fragments 56.5g.

Augustinowka v Augustinovka.

Aukoma v Pillistfer.

Aumale 36°10′ N., 3°40′ E.
Sour el Ghozlane, Alger, Algeria
Fell 1865, August 25, 1100-1200 hrs
Synonym(s): *Senhadja*
Stone. Olivine-hypersthene chondrite (L6), veined.
Two stones, each of about 25kg fell about 3 miles apart, one in " tribe" of Senhadja, the other in "tribe" of Ouled Sidi Salem, G.A. Daubrée, C. R. Acad. Sci. Paris, 1866, **62**, p.72. Olivine Fa_{25}, B. Mason, Geochimica et Cosmochimica Acta, 1963, **27**, p.1011.
8.3kg Paris, Mus. d'Hist. Nat.; 336g Chicago, Field Mus. Nat. Hist.; 81g Washington, U.S. Nat. Mus.; 60g New York, Amer. Mus. Nat. Hist.; 49g Prague, Nat. Mus.; 42g Berlin, Humboldt Univ.;
Specimen(s): [1920,282], 58g.; [84190], 13g. and fragments, 1g.; [41108], 9g.

Aumieres 44°20′ N., 3°14′ E.
Lozere, France
Fell 1842, June 3, 2100 hrs
Synonym(s): *Berrias, Lozere, St. Georges-de-Lévéjac*
Stone. Olivine-hypersthene chondrite (L6), veined.
A single stone of about 2kg fell after appearance of a luminous meteor, J. de Malbos, C. R. Acad. Sci. Paris, 1842, **14**, p.917. Olivine Fa_{24}, B. Mason, Geochimica et Cosmochimica Acta, 1963, **27**, p.1011.
1.4kg Paris, Mus. d'Hist. Nat.; 182g Rome, Vatican Colln; 45g Chicago, Field Mus. Nat. Hist.; 45g New York, Amer. Mus. Nat. Hist.; 34g Berlin, Humboldt Univ.; 21g Tempe, Arizona State Univ.; 18g Prague, Nat. Mus.; 16g Washington, U.S. Nat. Mus.;
Specimen(s): [71575], 40.7g.; [34596], 0.5g.

Aurora 36°20′ N., 105°3′ W.
Colfax County, New Mexico, U.S.A.
Found 1938
Stone. Olivine-bronzite chondrite (H4).
One stone of 1kg was found, A.D. Nininger, Pop. Astron., Northfield, Minnesota, 1939, **47**, p.211. Olivine $Fa_{18.3}$, D.E. Lange and K. Keil, Meteoritics, 1976, **11**, p.315, abs.
247g Tempe, Arizona State Univ.; 105g Chicago, Field Mus. Nat. Hist.;
Specimen(s): [1959,845], 228g.

Aus 26°40′ S., 16°15′ E.
Namibia
Found, year not reported
Stone. Olivine-hypersthene chondrite (L).
Listed, Cat. Meteorites Max-Planck-Inst., Heidelberg, 1979.
30.2g Heidelberg, Max-Planck-Inst.;

Aussen error for Ausson.

Ausson 43°5′ N., 0°35′ E.
Haute Garonne, France
Fell 1858, December 9, 0730 hrs
Synonym(s): *Aussen, Aussun, Clarac, Montrejeau*
Stone. Olivine-hypersthene chondrite (L5).
Two stones, weighing about 9kg and 41kg respectively, fell, the first near Ausson and the other near Clarac, about 3 miles distant, F. Petit, C. R. Acad. Sci. Paris, 1858, **47**, p.1053. Analysis (doubtful, no CaO reported), G. Chancel and A. Moitessier, C. R. Acad. Sci. Paris, 1859, **48**, p.267 and 479. Olivine Fa_{24}, B. Mason, Geochimica et Cosmochimica Acta, 1963, **27**, p.1011.
2.5kg Paris, Mus. d'Hist. Nat.; 539g Berlin, Humboldt Univ.; 453g Tübingen Univ.; 437g New York, Amer. Mus. Nat. Hist.; 349g Tempe, Arizona State Univ.; 223g Washington, U.S. Nat. Mus.; 122g Prague, Nat. Mus.; 176g Budapest, Nat. Hist. Mus.; 1kg Edinburgh, Royal Scot. Mus.; 1kg Vienna, Naturhist. Mus.; 593g Moscow, Acad. Sci.; 1.1kg Toulouse, Mus. d'Hist. Nat.; 860g Toulouse, Fac. Sci. Univ.; 374g Chicago, Field Mus. Nat. Hist.;
Specimen(s): [31987], 334g. Ausson stone; [90273], 110g. from Clarac stone; [1920,283], 110g.; [55531], 32g.; [1950, 391], 97g. in four fragments.

Aussun error for Ausson.

Austin v Denton County.

Austin v Wichita County.

Australia
Found 1880
Doubtful. Stony-iron. Pallasite (PAL).
116g. in Harvard Univ. labelled as above by J.L. Smith, O.W. Huntington, Proc. Amer. Acad. Arts and Sci., 1888, **23**, p.99. Discredited, possibly a fragment of Mount Dyrring, T. Hodge-Smith, Mem. Austr. Mus., 1939 (7), p.34.
116g Harvard Univ.; 21g New York, Amer. Mus. Nat. Hist.;

Authon v Lancé.

Avanhandava 21°27′37″ S., 49°57′3″ W.
São Paulo, Brazil
Fell 1952
Stone. Olivine-bronzite chondrite (H4).
One mass of some 30 cm diameter fell, but was broken up, and only 9.33kg are known to have been preserved, W.S. Curvello, letter of February 1960 in Min. Dept. BM(NH). Full description, with an analysis, 27.15 % total iron, olivine $Fa_{17.3}$, W. Paar et al., Rev. Brasil. Geocienc., 1976, **6**, p.201.
1.59kg Avanhandava, Municipal Prefecture; 7.74kg Lins, Colegio Estadual and Escuola Normal, Lins is close to Avanhandava; 162g Washington, U.S. Nat. Mus.;

Avce 46° N., 13°30′ E. approx.
Isonzo Valley, Gorizia, Yugoslavia
Fell 1908, March 31, 0845 hrs
Synonym(s): *Avec, Avse, Gorizia, Isonzo, Isonzothal*
Iron. Hexahedrite (IIA).
A mass of 1230g fell, after detonations, F.A. Berwerth, Anz. Akad. Wiss. Wien, Math-naturwiss. Kl., 1908, **45**, p.298, B. Baldanza, Min. Mag., 1965, **35**, p.214. In 1908 Avce was in Austrian territory but it is now in Yugoslav territory, G.R. Levi-Donati, letter of 30 April 1969 in Min. Dept. BM(NH). Analysis, 5.49 %Ni, 58.1 ppm.Ga, 182 ppm.Ge, 57 ppm.Ir, J.T. Wasson, Meteorites, Springer-Verlag, 1974, p.299. Description, V.F. Buchwald, Iron Meteorites, Univ. of California, 1975, p.281.
1.1kg Vienna, Naturhist. Mus., main mass;

Avec error for Avce.

Avgustinovka v Augustinovka.

Avilez 25° N., 103°30′ W. approx.
Cuencame, Durango, Mexico
Fell 1855, June
Stone. Olivine-bronzite chondrite (H).
Several stones are said to have fallen, but only fragments, including a piece of 146g, were preserved, F. Wöhler, Nachr.

Gessell. Wiss. Göttingen, 1867, p.57, L. Häpke, Abhand. Naturwiss. Ver. Bremen, 1884, **8**, p.515. Olivine Fa_{19}, B. Mason, Geochimica et Cosmochimica Acta, 1963, **27**, p.1011.

142g Göttingen, Univ.; 6g Chicago, Field Mus. Nat. Hist.; 5g Prague, Nat. Mus.;

Avoca v Tulia.

Avoca (Texas) 32°54′ N., 99°48′ W. approx.

Jones County, Texas, U.S.A.
Found 1924
Doubtful. Stone.

One stone of 7.1kg was found, H.H. Nininger, Our Stone Pelted Planet, 1933. Not listed, B. Mason, Meteorites, Wiley, 1962, p.242.

Avoca (Western Australia) 30°51′ S., 122°19′ E.

Western Australia, Australia
Found 1966, January, or February
Iron. Octahedrite, medium (0.9mm) (IIIA).

A mass of 37.85kg was found 6 miles north of Avoca Downs Station Homestead, co-ordinates, G.J.H. McCall, First Suppl. Cat. West. Austr. Met. Coll., 1968. See, with an analysis, 8.65 %Ni, 0.52 %Co, G.J.H. McCall, Min. Mag., 1968, **36**, p.859. New analysis, 9.21 %Ni, 20.6 ppm.Ga, 44.5 ppm.Ge, 0.30 ppm.Ir, J.T. Wasson, Meteorites, Springer-Verlag, 1974, p.303. Description, V.F. Buchwald, Iron Meteorites, Univ. of California, 1975, p.282.

Main mass, Perth, West. Austr. Mus.; 1.1kg Kalgoolie, West. Austr. School of Mines.; 360g Mrs.J. Warren (private coll.);

Specimen(s): [1967,254], 225g. polished slice; [1968,277], 754g. part slice; [1968,278], 344g. part slice.

Avse v Avce.

Awere 2°43′ N., 32°50′ E. approx.

Omoro County, Uganda
Fell 1968, July 12, 0300 hrs, approx.
Stone. Olivine-hypersthene chondrite (L4).

A stone of 134g fell at Mukungu Parwec, Awere. Another 'larger' stone was reputedly seen to fall in grass but has not been found, olivine Fa_{25}, R. Hutchison, Meteoritics, 1971, **6**, p.53. The co-ordinates given are those of Awere. Reported, Meteor. Bull., 1972 (51), Meteoritics, 1972, **7**, p.215.

Main mass, Entebbe, Uganda, Geol. Surv. and Mines Dept.;

Specimen(s): [1970,28], 5.25g.

Aztec v Holbrook.

Aztec 36°48′ N., 108°0′ W.

San Juan County, New Mexico, U.S.A.
Fell 1938, February 1, 1700 hrs
Stone. Olivine-hypersthene chondrite (L6).

One stone of 2.83kg fell, A.D. Nininger, Pop. Astron., Northfield, Minnesota, 1940, **48**, p.555., Contr. Soc. Res. Meteorites, 1942, **2**, p.227 [M.A. 8-54]. Olivine Fa_{23}, B. Mason, Geochimica et Cosmochimica Acta, 1963, **27**, p.1011.

2.75kg Chicago, Field Mus. Nat. Hist.; 35.9g Tempe, Arizona State Univ.;

Specimen(s): [1959,767], 41.5g. fragment.

Aztek error for Aztec.

Baandee 31°37′ S., 118°2′ E.

Western Australia, Australia
Found 1967, possibly earlier, possibly fell 1961 or 1962
Stone. Olivine-bronzite chondrite (H).

A single completely crusted stone of 256.3g was found by a farmer while ploughing adjacent to Hunters Dam on land unit 13929, 4.25 miles ESE. of Baandee Railway Station. s.g. 3.33, G.J.H. McCall, 2nd. Suppl. to West. Austr. Mus. Spec. Publ. no. 3, 1972., G.J.H. McCall, letter of 17 February 1969 in Min. Dept. BM(NH). Corrected co-ordinates, Meteor. Bull., 1975 (53), Meteoritics, 1975, **10**, p.139.

188.5g Perth, West. Austr. Mus., main mass and also 42.3g;

Babb's Mill (Blake's Iron) 36°18′ N., 82°53′ W.

Greene County, Tennessee, U.S.A.
Found 1876
Synonym(s): *Babb's Mill, Blake's Iron, Greene County*
Iron. Ataxite (IRANOM).

A mass of about 300lb. was ploughed up 10 miles north of Greenville in 1876, W.P. Blake, Am. J. Sci., 1886, **31**, p.41. Full description, V.F. Buchwald, Iron Meteorites, Univ. of California, 1975, p.284. Analysis, 11.8 %Ni, 0.203 ppm.Ga, 0.029 ppm.Ge, 1.7 ppm.Ir, R. Schaudy et al., Icarus, 1972, **17**, p.174. This mass is distinct from Babb's Mill (Troost's Iron) (*q.v.*).

129kg Vienna, Naturhist. Mus., main mass;

Babb's Mill (Troost's Iron) 36°18′ N., 82°53′ W.

Greene County, Tennessee, U.S.A.
Found 1842
Synonym(s): *Babb's Mill, Troost's Iron*
Iron. Ataxite, Ni-rich (IRANOM).

A mass of 14lb. was ploughed up 10 miles north of Greenville in 1842, G. Troost, Am. J. Sci., 1845, **49**, p.342. Full description, V.F. Buchwald, Iron Meteorites, Univ. of California, 1975, p.285. Analysis, 17.7 %Ni, 18.6 ppm.Ga, 41 ppm.Ge, 35 ppm.Ir, E.R.D. Scott et al., Geochimica et Cosmochimica Acta, 1973, **37**, p.1957. The 6.25kg and 2.75kg masses are distinct from Babb's Mill (Blake's Iron) (*q.v.*). The larger mass has been heated and forged.

1kg Harvard Univ., heated.; 63g Copenhagen, Min. Mus., heated; 52g New York, Amer. Mus. Nat. Hist., undamaged material.; 1.85kg Tempe, Arizona State Univ., undamaged material; 33g Tübingen Univ., undamaged material.;

Specimen(s): [18490], 1989g. a slice and 69.5g, also a polished mount. This is forged material.

Babb's Mill v Babb's Mill (Blake's Iron).

Babb's Mill v Babb's Mill (Troost's Iron).

Bachmut 48°36′ N., 38°0′ E.

Ekaterinoslav, Ukraine, USSR, [Бахмут]
Fell 1814, February 15, 1200 hrs
Synonym(s): *Alekseevka, Alexejevka, Bakhmut, Ekaterinoslav, Scholakov*
Stone. Olivine-hypersthene chondrite (L6).

A stone of 18kg fell, after detonations, F. von Giese, Ann. Phys. (Gilbert), 1815, **50**, p.117. Analysis, A. Kuhlberg, Arch. Naturk. Liv.-Ehst.-u. Kurlands, Ser. 1, Min. Wiss., Dorpat, 1867, **4**, p.18. The place of fall is incorrectly mapped, P.M. Millman, Trans. Roy. Astron. Soc. Canada, 1938, **32**, p.199. There is some confusion between the Bachmut and the Pavlograd falls. Three falls seem to be represented, R. Ganapathy and E. Anders, Очерки Современной Геохимии и Аналитической

Химии Наука, Москва, 1972, p.72. Olivine Fa_{24}, B. Mason, Geochimica et Cosmochimica Acta, 1963, **27**, p.1011. Analysis, 22.37 % total iron, O.A. Kirova et al., Метеоритика, 1978, **37**, p.87.
8kg Kharkov, Univ., main mass; 44g Moscow, Acad. Sci.; 1.6kg Vienna, Naturhist. Mus.;
Specimen(s): [46012], 35g. was obtained from J.R. Gregory, labelled Ekaterinoslav and so may be Pavlograd, since Ganapathy and Anders (loc. cit) established that a similarly labelled specimen from the same source has a radiation age identical to that of the Moscow specimen of Pavlograd; [35153], 2g.; [35162], 2g.; [34614], 1.5g.

Bacubirito 26°12′ N., 107°50′ W.
Sinaloa, Mexico
Found 1863
Synonym(s): *El Ranchito, Ranchito, Sinaloa*
Iron. Octahedrite, finest (0.08mm) (IRANOM).
A huge mass, 12 ft long and estimated to weigh 22 tons, was found on the farm El Ranchito, 7 miles south of Bacubirito, V.F. Buchwald, Iron Meteorites, Univ. of California, 1975, p.289. Details of recovery, H.A. Ward, Proc. Rochester Acad. Sci., 1902, **4**, p.67, L. Fletcher, Min. Mag., 1890, **9**, p.151. Analysis, 9.4 %Ni, E. Cohen, Meteoritenkunde, 1905, **3**, p.281. Chemically anomalous, new analysis and classification, 9.62 %Ni, 17.7 ppm.Ga, 31.9 ppm.Ge, 4.9 ppm.Ir, E.R.D. Scott et al., Geochimica et Cosmochimica Acta, 1973, **37**, p.1957.
Main mass, Centro Civico Constitution, Culiacan, Sinalao, removed to here in 1959 from original find site.; 997g Washington, U.S. Nat. Mus.; 303g Tübingen, Univ.; 164g Tempe, Arizona State Univ.; 98g Yale Univ.; 1.75kg Chicago, Field Mus. Nat. Hist.; 493g Harvard Univ.;
Specimen(s): [84235], 525.5g. and polished fragment, 51.5g.; [86070], 470g.; [84236], 61.5g.

Badger v Sacramento Mountains.

Baffin's Bay v Cape York.

Bagdad 34°32′ N., 113°25′ W.
Mohave County, Arizona, U.S.A.
Found 1959, spring
Iron. Octahedrite, medium (1.1mm) (IIIA).
One iron of 2.2kg was found on the surface of the desert, 7.8 %Ni, C.B. Moore and S.L. Tackett, J. Arizona Acad. Sci., 1963, **2**, p.191, Meteor. Bull., 1962 (25). Description, V.F. Buchwald, Iron Meteorites, Univ. of California, 1975, p.292. Analysis, 8.01 %Ni, 19.8 ppm.Ga, 39.7 ppm.Ge, 6.8 ppm.Ir, E.R.D. Scott et al., Geochimica et Cosmochimica Acta, 1973, **37**, p.1957.
Main mass, Tempe, Arizona State Univ.; 38g Copenhagen, Univ. Geol. Mus.;

Bagnone 44°19′ N., 9°59′ E.
Massa Carrara, Tuscany, Italy
Found 1904, or 1905, recognized 1967
Iron. Octahedrite, medium.
One mass of about 48kg was found on a hill near Bagnone, at the beginning of the 20th century. Described, with analysis, 8.46 %Ni, S. Bonatti et al., Atti Soc. Tosc. Sci. Nat., Mem. Ser. A, 1970, **77**, p.123.
Main mass, Pisa, Univ. Inst. Min. Pet.;

Bahia v Bendegó.

Bahjoi 28°29′ N., 78°30′ E.
Moradabad district, Uttar Pradesh, India
Fell 1934, July 23, 2130 hrs
Synonym(s): *Hapur, Moradabad District*
Iron. Octahedrite, coarse (1.5mm) (IA-ANOM).
A mass of 10.322kg fell at Chandankati Muazam, 1.5 miles north of Bahjoi (co-ordinates above). Description, M.S. Krishnan, Rec. Geol. Surv. India, 1936, **71**, p.144 [M.A. 6-393]. The locality was at first incorrectly given as Hapur, Meerut district, M.A.R. Khan, Meteors and meteoric iron in India, Secunderabad, 1934 [M.A. 6-102]. Description, V.F. Buchwald, Iron Meteorites, Univ. of California, 1975, p.294. Analysis, 7.95 %Ni, 62.7 ppm.Ga, 272 ppm.Ge, 1.9 ppm.Ir, J.T. Wasson, Icarus, 1970, **12**, p.407.
9.5kg Calcutta, Geol. Surv. India.; 497g Washington, U.S. Nat. Mus.; 10g Tempe, Arizona State Univ.;
Specimen(s): [1951,328], 80.5g.

Baird's Farm v Asheville.

Baird's Plantation v Asheville.

Bakhardok 38°36′ N., 58°0′ E.
Karakum desert, Turkmenistan, USSR
Found 1978, September
Stone. Olivine-hypersthene chondrite (L6).
A single mass of 4.12kg was found half-buried in sand in the Karakum desert, about 48km SW. of Bakhardok, Meteor. Bull., 1981 (59), Meteoritics, 1981, **16**, p.193. Brief description, R.L. Khotinok, Метеоритика, 1982, **40**, p.6. Analysis, 19.65 % total iron, olivine $Fa_{25.4}$, M.A. Nazarov et al., Метеоритика, 1982, **41**, p.57.

Bakhmut v Bachmut.

Bald Eagle 41°17′ N., 77°3′ W. approx.
Williamsport, Lycoming County, Pennsylvania, U.S.A.
Found 1891
Synonym(s): *Park Hotel, Susquehanna, Williamsport*
Iron. Octahedrite, medium (0.8mm) (IIIB).
A mass of about 7 lb was found on the east side of Bald Eagle Mountain, H.A. Ward, Proc. Rochester Acad. Sci., 1902, **4**, p.79, 86. Described and figured, R.W. Stone and E.M. Starr, Bull. Pa. Geol. Surv., 1967 (Rep. G2, 1932, rev.), p.13, 33. Analysis, 9.25 %Ni, 18.1 ppm.Ga, 37.1 ppm.Ge, 0.018 ppm.Ir, E.R.D. Scott et al., Geochimica et Cosmochimica Acta, 1973, **37**, p.1957. Location of site, description, V.F. Buchwald, Iron Meteorites, Univ. of California, 1975, p.295.
Main mass, Lewisburg, Pa., Bucknell Univ.; 300g Chicago, Field Mus. Nat. Hist.; 51g Berlin, Humboldt Univ.; 6g Tempe, Arizona State Univ.;
Specimen(s): [1927,5], 15g. filings

Bald Mountain 35°58′ N., 82°29′ W.
Yancy County, North Carolina, U.S.A.
Fell 1929, July 9, afternoon
Stone. Olivine-hypersthene chondrite (L4), veined.
Two stones fell, 3.7kg together, H.H. Nininger, Our Stone Pelted Planet, 1933. Olivine Fa_{22}, B. Mason, Geochimica et Cosmochimica Acta, 1963, **27**, p.1011.
306g Chicago, Field Mus. Nat. Hist.; 222g Washington, U.S. Nat. Mus.; 93g Tempe, Arizona State Univ.;
Specimen(s): [1959,814], 122g.

Baldohn v Misshof.

Baldwyn 34°30′ N., 88°40′ W.

Lee County, Mississippi, U.S.A.
Fell 1922, February 2, in the daytime
Stone. Olivine-hypersthene chondrite (L6), veined.
After humming noise to the NW., a stone of 345g fell about 10 ft. from a negro on a farm about 1.5 miles NW. of Baldwyn; described, G.P. Merrill, Proc. U.S. Nat. Mus., 1925, **67** (6), p.1. Reported, L.C. Glenn, Am. J. Sci., 1925, **9**, p.488. Olivine Fa_{25}, B. Mason, Geochimica et Cosmochimica Acta, 1963, **27**, p.1011.
Main mass, Nashville, Vanderbilt Univ.; 0.5g Washington, U.S. Nat. Mus.;

Balfour Downs 22°45′ S., 120°50′ E.

Western Australia, Australia
Found 1962
Iron. Octahedrite, medium (1.3mm) (IA).
A mass of 2.4kg was found on Balfour Downs Station, B. Mason, letters of 4 March and 28 April 1964 in Min. Dept. BM(NH). Analysis, 8.39 %Ni, H.B. Wiik, Geochimica et Cosmochimica Acta, 1965, **29**, p.1003, Spec. Publ. West. Austr. Mus., 1965 (3), p.20. Classification and analysis, 8.39 %Ni, 56.4 ppm.Ga, 194 ppm.Ge, 2 ppm.Ir, J.T. Wasson, Icarus, 1970, **12**, p.407. Description, V.F. Buchwald, Iron Meteorites, Univ. of California, 1975, p.297.
900g Washington, U.S. Nat. Mus.; 754g New York, Amer. Mus. Nat. Hist.; 140g Perth, West. Austr. Mus.; 140g Sydney, Austr. Mus.;

Bali 5°23′ N., 16°23′ E.

Central African Republic
Fell 1907, November 22 or 23, 1030 hrs
Stone. Carbonaceous chondrite, type III (CV3).
Obtained from the Bali mission station on the Lobaye [Bali] river. Listed, F. Berwerth, Fortschr. Min. Krist. Petr., 1912, **2**, p.230. Partial analysis, 23.34 % total iron, 15.8 %Si, 14.8 %Mg, B. Mason, Meteoritics, 1971, **6**, p.59. Mineralogy, G. Hoinkes and G. Kurat, Meteoritics, 1975, **10**, p.416, abs. Trace element data, G.W. Kallemeyn and J.T. Wasson, Geochimica et Cosmochimica Acta, 1981, **45**, p.1217.
815g Vienna, Naturhist. Mus., includes main mass, 569g.; 154g Washington, U.S. Nat. Mus.; 5g Budapest, Nat. Mus.; 5g Paris, Mus. d'Hist. Nat.;

Balin v Palinshih.

Ballinee v Ballinoo.

Ballinger 31°46′ N., 99°59′ W.

Runnels County, Texas, U.S.A.
Found 1927
Iron. Octahedrite, coarse (2.6mm) (IIICD).
One mass of 1250g in the Nininger Collection in 1933. Described, with analysis, 6.54 %Ni, H.H. Nininger, J. Geol., 1929, **37**, p.88 [M.A. 4-119]. Further analysis, 6.19 %Ni, 84.5 ppm.Ga, 326 ppm.Ge, 2.1 ppm.Ir, J.T. Wasson, Icarus, 1970, **12**, p.407. Description and classification, V.F. Buchwald, Iron Meteorites, Univ. of California, 1975, p.298.
459g Washington, U.S. Nat. Mus.; 89g Harvard Univ.; 73g New York, Amer. Mus. Nat. Hist.; 123g Chicago, Field Mus. Nat. Hist.; 58g Tempe, Arizona State Univ.; 16g Los Angeles, Univ. of California;
Specimen(s): [1959,915], 56.5g. thin slice and 2g sawings.

Ballinoo 27°42′ S., 115°46′ E.

Murchison River, Western Australia, Australia
Found 1892
Synonym(s): *Ballinee, Ballinos, Mount Erin*
Iron. Octahedrite, plessitic (IIC).
A mass of 93 lb was found 10 miles south of Ballinoo, T. Cooksey, Rec. Austr. Mus., 1897, **3**, p.55, H.A. Ward, Am. J. Sci., 1898, **5**, p.136. Described, with an analysis, 9.87 % Ni, E. Cohen, Sitzungsber. Akad. Wiss. Berlin, 1898, p.19, Meteoritenkunde, 1905, **3**, p.284. Classification and new analysis, 9.72 %Ni, 39.0 ppm.Ga, 94.4 ppm.Ge, 9.0 ppm.Ir, J.T. Wasson, Geochimica et Cosmochimica Acta, 1969, **33**, p.859. Shock and subsequent re-heating history, V.F. Buchwald, Iron Meteorites, Univ. of California, 1975, p.300. Listed, Spec. Publ. West. Austr. Mus., 1965 (3), p.20.
11kg Chicago, Field Mus. Nat. Hist.; 2.25kg Harvard Univ.; 1.5kg Washington, U.S. Nat. Mus.; 3.5kg New York, Amer. Mus. Nat. Hist.; 1.9kg Vienna, Naturhist. Mus.; 599g Paris, Mus. d'Hist. Nat.; 357g Sydney, Australian Mus.; 327g Budapest, Nat. Mus.; 2.5kg Perth, Western Austr. Mus.;
Specimen(s): [81732], 3160g.; [1920,285], 365g. an etched slice, and a polished mount.

Ballinos v Ballinoo.

Ballua v Butsura.

Bambuk v Siratik.

Bamenda 5°55′ N., 10°15′ E.

Sud-ouest province, Cameroun
Pseudometeorite..
A mass described, M.D.W. Jeffreys, Man, London, 1955, **55**, p.167 [M.A. 13-81] as meteoritic is a boulder of maghemite, M.H. Hey, Min. Mag., 1961, **32**, p.910.

Bancoorah v Shalka.

Bandera County v Pipe Creek.

Bandhya v Duketon.

Bandong 6°55′ S., 107°36′ E.

Java, Indonesia
Fell 1871, December 10, 1330 hrs
Stone. Olivine-hypersthene chondrite, amphoterite (LL6).
Six stones, of total weight about 11.5kg, fell, after detonations; described, with an analysis, G.A. Daubrée, C. R. Acad. Sci. Paris, 1872, **75**, p.1676. Contains recrystallised xenoliths, olivine $Fa_{29.4}$, R.A. Binns, Geochimica et Cosmochimica Acta, 1968, **32**, p.299.
8kg Bandung, Geol. Mus.; 2kg Paris, Mus. d'Hist. Nat.; 250g Prague, Nat. Mus.; 113g Vienna, Naturhist. Mus.; 56g Washington, U.S. Nat. Mus.; 27g New York, Amer. Mus. Nat. Hist.; 26g Chicago, Field Mus. Nat. Hist.;
Specimen(s): [1920,286], 44g.; [48760], 11.5g.; [1980,M.15], 30.5g.

Bandya v Duketon.

Banja v Soko-Banja.

Banjaca v Jelica.

Bankura v Shalka.

Bansur 27°42′ N., 76°20′ E. approx.
Rajasthan, India
Fell 1892
Stone. Olivine-hypersthene chondrite (L6).
About 15kg fell, K. Gopalan and M.N. Rao, Meteoritics, 1976, **11**, p.131.
0.5g Washington, U.S. Nat. Mus.; Specimen, Jaipur, Mus.;

Banswal 30°24′ N., 78°12′ E.
Dehra Dun district, Uttar Pradesh, India
Fell 1913, January 12, 1800 hrs
Stone. Olivine-hypersthene chondrite (L5).
After a brilliant luminous meteor (moving NW. to SE.), and detonations, a single stone fell, but only about 14g of fragments were recovered, J.C. Brown, Rec. Geol. Surv. India, 1913, **43** (3), p.237. Olivine Fa_{24}, B. Mason, Geochimica et Cosmochimica Acta, 1963, **27**, p.1011.
7.6g Calcutta, Mus. Geol. Surv. India; 5.7g Washington, U.S. Nat. Mus.;
Specimen(s): [1951,331], 2.2g.

Banten 6°20′ S., 106°0′ E. approx.
Java, Indonesia
Fell 1933, May 24, 0400-0500 hrs
Stone. Carbonaceous chondrite, type II (CM2).
Four individuals, totalling 629g, were recovered, Meteor. Bull., 1980 (57), Meteoritics, 1980, **15**, p.96. Description and mineral analyses, olivine Fa_{0-70}, K. Fredriksson et al., Meteoritics, 1979, **14**, p.400. Correct date of fall 24 May, 1933, S. Darsoprajitno, priv. comm., 1983.
560g Bandung, Geol. Mus.; 60g Washington, U.S. Nat. Mus.;

Baquedano 23°18′ S., 69°53′ W.
Atacama, Chile
Found 1932, before this year
Iron. Octahedrite, medium (1.2mm) (IIIB).
A mutilated mass of 22kg was found near the railway junction at Baquedano. Described, with an analysis, 8.82 % Ni, C. Palache and F.A. Gonyer, Am. Miner., 1932, **17**, p.357 [M.A. 5-158]. May be a transported mass belonging to one of the known falls in this region, L.J. Spencer, Min. Abs. [M.A. 5-158]. Classification and new analysis, 8.76 % Ni, 20.4 ppm.Ga, 43.1 ppm.Ge, 0.10 ppm.Ir, E.R.D. Scott et al., Geochimica et Cosmochimica Acta, 1973, **37**, p.1957. Structural study, distinct fall, V.F. Buchwald, Iron Meteorites, Univ. of California, 1975, p.302.
20.2kg Harvard Univ.; 880g Washington, U.S. Nat. Mus.;
Specimen(s): [1967,380], 99g. an etched slice

Baratta v Barratta.

Barbacena 21°13′ S., 43°56′ W.
Minas Gerais, Brazil
Found 1918, March 20
Iron. Octahedrite, fine (0.12mm) (IRANOM).
Two oxidised masses, 6140g and 2886g were found, E. de Oliveira, Anais Acad. Brasil. Cienc., 1931, **3**, p.33 [M.A. 5-14]. Described, with an analysis, 10.5 %Ni, W.S. Curvello, Bol. Mus. Nac. Rio de Janeiro, 1951 (geol. no. 14) [M.A. 11-530]. Classification and analysis, 10.9 %Ni, 12.8 ppm.Ga, 1.16 ppm.Ge, 2.9 ppm.Ir, A. Kracher et al., Geochimica et Cosmochimica Acta, 1980, **44**, p.773.
Main masses, Ouro Preto, Escuola de Minas; 57g Rio de Janeiro, Mus. Nac.;
Specimen(s): [1938,467], 33g. two fragments

Barber County v Nashville (stone).

Barbotan 43°57′ N., 0°3′ W.
Gers, France
Fell 1790, July 24, 2100 hrs
Synonym(s): *Agen, Bordeaux, Landes, Roquefort*
Stone. Olivine-bronzite chondrite (H5), veined.
A shower of stones, the largest of 9kg, fell after detonations and appearance of a fireball travelling from south to north, -. Baudin, Ann. Phys. (Gilbert), 1803, **13**, p.346. Description, H. Pfahler, Tschermaks Min. Petr. Mitt., 1893, **13**, p.353. Olivine Fa_{19}, B. Mason, Geochimica et Cosmochimica Acta, 1967, **31**, p.1100.
774g Tübingen Univ.; 354g Washington, U.S. Nat. Mus.; 88g New York, Amer. Mus. Nat. Hist.; 121g Tempe, Arizona State Univ.; 618g Vienna, Naturhist. Mus.; 500g Paris, Mus. d'Hist. Nat.; 449g Chicago, Field Mus. Nat. Hist.;
Specimen(s): [56548], 515g. a complete stone; [18582], 144.7g.; [90244], 122g.; [1920,287], 5g. and fragments, 1g.; [1935,53], 0.5g.

Barcelona v Cañellas.

Barcelona v Nulles.

Barcelona (iron) 41°22′ N., 2°10′ E.
Barcelona, Spain
Fell 1850, September
Doubtful..
A mass of iron, sp. gr. 8.12, is said to have fallen, R.P. Greg, Rep. Brit. Assn., 1860, p.90 but the fall is not mentioned, M. Faura y Sans, Meteoritos caidos en la Peninsula Iberica, Tortosa, 1922 and must be regarded as very doubtful.

Barcelona (stone) 41°22′ N., 2°10′ E.
Barcelona, Spain
Fell 1704, December 25, 1700 hrs
Stone. Chondrite?.
A black stone, grey internally, fell, E.F.F. Chladni, Ann. Phys. Chem. (Poggendorff), 1826, **8**, p.46., M. Faura y Sans, Meteoritos caidos en la Peninsula Iberica, Tortosa, 1922, p.6 reports this meteorite as falling into the sea; while one or more masses may have fallen in the sea, the evidence collected by Chladni shows that one or more stones were preserved, for a time at least.

Barcis 46°6′ N., 12°21′ E.
Pordenone, Friuli, Italy
Found 1950
Stony-iron. Pallasite (PAL).
Recognized among moraine deposits of a glacial valley near Barcis, a village in the Friuli region. The find was sawn into two parts, one part of 1650g was deposited in the Museum of the Scuola Mineraria of Agordo, and later lent to the Italian Centre for Meteorite Studies. The total mass is not known, Suppl. to Cat. Italian Centre for Meteorite Studies, Univ. Perugia, 1971.

Bardsley v Beardsley.

Bare v Mocs.

Barea 42°23′ N., 2°30′ W.
Logrono, Spain
Fell 1842, July 4
Synonym(s): *Logrono, Varea*
Stony-iron. Mesosiderite (MES).
A stone of about 7 lb fell, R.P. Greg, Phil. Mag., 1854, **8**, p.460. Description, S. Meunier, C. R. Acad. Sci. Paris, 1872, **75**, p.1547. Analysis and mineralogy, B. Mason and E. Jarosewich, Min. Mag., 1973, **39**, p.204.
1.7kg Madrid,; 132g Washington, U.S. Nat. Mus.; 11.5g Chicago, Field Mus. Nat. Hist.; 85g Tempe, Arizona State Univ.;

Barnaul 52°44′ N., 84°5′ E.
Tomsk, Federated SSR, USSR, [Барнаул]
Fell 1904, May 22, 2330 hrs
Synonym(s): *Altai, Teleutskoe Osero, Teleutskoe Ozero, Teleutskoje Osero*
Stone. Olivine-bronzite chondrite (H5).
After the appearance of a brilliant meteor and detonations many small stones fell near Barnaul, but only six were recovered, V.N. Mamontov, Trudy geol. miner. Muz., 1909, **3**, p.107. The total recovered weight is 23.2g, and the heaviest fragment 9.3g; all the material is in the Acad. Sci. Moscow, E.L. Krinov, Астрон. Журнал, 1945, **22**, p.303 [M.A. 9-297], Каталог Метеоритов АИад. Наук СССР, Москва, 1947 [M.A. 10-511]. Classification, olivine Fa_{18}, A.Ya. Skripnik and L.G. Kvasha, Метеоритика, 1980, **39**, p.38.
21.2g Moscow, Acad. Sci.;

Barntrup 52°0′ N., 9°6′ E.
Lippe, Nordrhein-Westfalen, Germany
Fell 1886, May 28, 1430 hrs
Synonym(s): *Krähenholz, Lippe*
Stone. Chondrite, veined.
A small stone of 17g fell, after detonations, L. Häpke, Abhand. Naturwiss. Ver. Bremen, 1889, **11**, p.323.
9.5g Detmold,; 6g Vienna, Naturhist. Mus.;

Baroti 31°37′ N., 76°48′ E.
Bilaspur, Simla Hill States, Himachal Pradesh, India
Fell 1910, September 15, 1000 hrs
Stone. Olivine-hypersthene chondrite (L6).
One stone of about 10 lb fell, G. de P. Cotter, Rec. Geol. Surv. India, 1912, **42**, p.273. Description, analysis, 23.18 % total iron, G.T. Prior, Min. Mag., 1913-14, **17**, p.22, 132. Olivine Fa_{25}, B. Mason, Geochimica et Cosmochimica Acta, 1963, **27**, p.1011.
42g Chicago, Field Mus. Nat. Hist.; 6.6g Washington, U.S. Nat. Mus.;
Specimen(s): [1912,555], 858g.; [1912,556], 6.5g. two fragments

Barraba v Bingera.

Barranca Bjanca v Barranca Blanca.

Barranca Blanca 28°5′ S., 69°20′ W.
Come Caballo Pass, Atacama, Chile
Found 1855
Synonym(s): *Barranca Bjanca, San Francisco Pass*
Iron. Anomalous (IIE).
A mass of about 12kg was found between Copiapo and Catamarca; described, with an analysis, 8.01 %Ni, L. Fletcher, Min. Mag., 1889, **8**, p.262. This iron is sulphide-rich. Unusual structure and cooling history, H.J. Axon and D. Faulkner, Min. Mag., 1970, **37**, p.898. Trace element analysis, A.A. Smales et al., Geochimica et Cosmochimica Acta, 1967, **31**, p.673. New classification and analysis, 8.07 %Ni, 22.1 ppm.Ga, 63.9 ppm.Ge, 4.9 ppm.Ir, E.R.D. Scott and J.T. Wasson, Geochimica et Cosmochimica Acta, 1976, **40**, p.103. Description, V.F. Buchwald, Iron Meteorites, Univ. of California, 1975, p.304.
161g Tempe, Arizona State Univ.; 84g Washington, U.S. Nat. Mus.; 73g New York, Amer. Mus. Nat. Hist.; 67g Vienna, Naturhist. Mus.; 27g Harvard Univ.; 16g Chicago, Field Mus. Nat. Hist.;
Specimen(s): [41187], 11610g. the main mass and filings, 112.5g.

Barratta 35°18′ S., 144°34′ E.
Deniliquin, County Townsend, New South Wales, Australia
Found 1845
Synonym(s): *Baratta, Deniliquin*
Stone. Olivine-hypersthene chondrite (L4), black.
Five stones, weighing respectively about 145lb, 31lb, 48lb, 48lb, and 175lb. were found at different times. The first stone, described, A. Liversidge, Trans. Roy. Soc. New South Wales, 1872, **6**, p.97 was found in 1845; the second and third stones in 1889, and these were also described, A. Liversidge, J. and Proc. Roy. Soc. New South Wales, 1902, **36**, p.350. The fourth stone was acquired by H.A. Ward in 1902, Cat. Ward-Coonley Coll. Meteorites, Chicago, 1904, p.35. Listed, T. Hodge-Smith, Mem. Austr. Mus., 1939 (7), p.12, 30 [M.A. 7-380]. Olivine Fa_{22-32}, R.T. Dodd and W.R. Van Schmus, J. Geophys. Res., 1965, **70**, p.3801. Described, with an analysis, 21.07 % total iron, B. Mason and H.B. Wiik, Am. Mus. Novit., 1966 (2273). Listed, B. Mason, Rec. Austr. Mus., 1974, **29**, p.169.
Main masses, Sydney, Australian Mus., main masses of the first, second and third stones.; 88.7kg Chicago, Field Mus. Nat. Hist., main masses of the fourth and fifth stones.; 4.2kg Washington, U.S. Nat. Mus.; 2kg New York, Amer. Mus. Nat. Hist.; 1.3kg Tempe, Arizona State Univ.; 1.1kg Vienna, Naturhist. Mus.; 549g Budapest, Nat. Mus.; 529g Ottawa, Geol. Surv. Canada; 1.8kg Harvard Univ.;
Specimen(s): [86428], 2678g. of the fifth stone.; [1959,770], 1110g.; [1927,1284], 349g.; [1927,1282], 194g. four pieces; [1927,1286], 66.5g. two pieces.; [83339], 46.5g. of the first stone.; [1927,1287], 19g. fragments.; [56493], Five thin sections

Barringer v Cañon Diablo.

Bartlet v Bartlett.

Bartlett Meteorite v Tucson.

Bartlett 30°50′ N., 97°30′ W.
Bell County, Texas, U.S.A.
Found 1938, recognized in this year
Synonym(s): *Bartlet*
Iron. Octahedrite, medium (1.1mm) (IIIA).
One mass of 8.59kg was recognized, A.D. Nininger, Pop. Astron., Northfield, Minnesota, 1939, **47**, p.211, E.P. Henderson, letter of 3 June 1939, in Min. Dept. BM(NH). The mass, which was ploughed up 5 miles W. of Bartlett is described, with an analysis, 8.88 %Ni, F.M. Bullard, Am. Miner., 1940, **25**, p.497 [M.A. 8-60]. New analysis and classification, 8.68 %Ni, 20.6 ppm.Ga, 46.0 ppm.Ge, 0.64 ppm.Ir, E.R.D. Scott et al., Geochimica et Cosmochimica Acta, 1973, **37**, p.1957. Description, V.F. Buchwald, Iron Meteorites, Univ. of California, 1975, p.306.
6.8kg Austin, Univ. of Texas.; 670g Washington, U.S. Nat. Mus.;

Barwell 52°34′ N., 1°20′ W.

Leicestershire, England
Fell 1965, December 24, 1620 hrs
Stone. Olivine-hypersthene chondrite (L6).

A shower of stones fell after a fireball was seen, with sonic effects; one fragment was seen to fall and shatter and more material was recovered over the following few weeks; the total mass recovered is at least 44kg. Described with analysis, 21.45 % total iron, E.A. Jobbins et al., Min. Mag., 1966, **35**, p.881. Fall described, H.G. Miles and A.J. Meadows, Nature, 1966, **210**, p.983. U-Pb data, D.M. Unruh et al., Geochimica et Cosmochimica Acta, 1979 (Suppl. 11), p.1011. Plateau Ar-Ar age, G. Turner et al., Geochimica et Cosmochimica Acta, 1978 (Suppl. 10), p.989. Analysis, 21.71 % total iron, A.A. Moss et al., Min. Mag., 1967, **36**, p.101.

4.1kg London, Geol. Mus.; 2.2kg Leicester, Mus.; 1.5kg Washington, U.S. Nat. Mus.; 1.5kg Los Angeles, Univ. of California; 560g Armagh Planetarium; 412g Coventry, Mus.;

Specimen(s): [1966,56], 4702g. two masses 3015g and 1687g with fragments, 144g; [1966,57], 2396g.; [1966,58], 427.5g. four fragments; [1966,59], 3343g. fragments; [1966,60], 2845g. crusted piece; [1966,62], 548g. two crusted fragments; [1966,63], 7739g. the largest fragment found.; [1966,64], 3438g. crusted piece; [1966,65], 2205g. crusted piece; [1966, 66], 36g. and fragments, 9g.; [1966,68], 48.9g. two crusted fragments; [1966,67], 11.6g. fragments; [1966,286], 382g. crusted fragment; [1971,21], 2422g. coke with meteoritic fragments; [1972,5], 12g. fragments and dust.

Barwise 34°0′ N., 101°30′ W.

Floyd County, Texas, U.S.A.
Found 1950, recognized 1965
Stone. Olivine-bronzite chondrite (H5).

A stone of 10.4kg was ploughed up, Meteor. Bull., 1967 (40), Meteoritics, 1970, **5**, p.93. Partial analysis, REE abundances, 25.1-33.9 % total iron, N. Nakamura and A. Masuda, Meteoritics, 1973, **8**, p.149. Olivine Fa_{18}, M.J. Frost, priv. comm.

5kg Mainz, Max-Planck-Inst., main mass; 290g Washington, U.S. Nat. Mus.; 112g Harvard Univ.; 86g Chicago, Field Mus. Nat. Hist.; 82g Tempe, Arizona State Univ.; 78g New York, Amer. Mus. Nat. Hist.;

Specimen(s): [1967,386], 88g. a slice

Basedow Range v Henbury.

Basingstoke 51°16′ N., 1°5′ W.

Hampshire, England
Fell 1806, May 17
Doubtful..

One stone of 2.5 lb is said to have fallen on the highway but nothing was preserved and the description is inadequate, E.F.F. Chladni, Die Feuer-Meteore, Wien, 1819, p.280, R.P. Greg, Rep. Brit. Assn., 1860, p.63, T.M. Hall, Min. Mag., 1879, **3**, p.8.

Bassein v Quenggouk.

Basti v Bustee.

Basti v Lahrauli.

Bates County v Butler.

Bates Nunataks A78001 80°15′ S., 153 30° E.

100km SW. of Darwin Glacier, Antarctica
Found 1978, between December 1978 and January 1979
Stone. Olivine-hypersthene chondrite (L6).

A mass of 160.7g was found, olivine Fa_{24}. One of about six masses found during the 1978-1979 field season in Antarctica near the Bates Nunataks. They were numbered in the sequence of recovery and are so named Bates Nunataks A78001 to Bates Nunataks A78006. Description of some of this material, Antarctic Meteorite Newsletter, 1981, **4**, p.100.

Bates Nunataks A78002

Stone. Olivine-hypersthene chondrite (L6).

Two pieces which fitted together and totalling 4301g in weight, were found, olivine Fa_{24}.

Bates Nunataks A78004

Stone. Olivine-hypersthene chondrite, amphoterite (LL6).

A mass of 1079g was recovered, olivine Fa_{30}.

Batesville v Joe Wright Mountain.

Bath 45°25′ N., 98°19′ W.

Brown County, South Dakota, U.S.A.
Fell 1892, August 29, 1600 hrs
Synonym(s): *Aberdeen, Aberden*
Stone. Olivine-bronzite chondrite (H4), brecciated.

One stone of about 46.7lb fell after detonations, A.E. Foote, Am. J. Sci., 1893, **45**, p.64. Microscopic description, G.P. Merrill, Mem. Nat. Acad. Sci. Washington, 1919, **14** (4), p.1. New description and analysis, 25.41 % total iron, B. Mason and H.B. Wiik, Am. Mus. Novit., 1966 (2272). Olivine Fa_{19}, B. Mason, Geochimica et Cosmochimica Acta, 1963, **27**, p.1011.

3kg Chicago, Field Mus. Nat. Hist., approx. weight; 3kg Vienna, Naturhist. Mus.; 2.8kg New York, Amer. Mus. Nat. Hist.; 2kg Harvard Univ.; 1kg Washington, U.S. Nat. Mus.; 675g Rome, Vatican Colln; 462g Tübingen Univ.; 325g Prague, Nat. Mus.; 560g Paris, Ecole des Mines; 247g Budapest, Nat. Hist. Mus.;

Specimen(s): [71526], 2112g. and fragments, 2g.

Bath Furnace 38°15′ N., 83°45′ W. approx.

Bath County, Kentucky, U.S.A.
Fell 1902, November 15, 1845 hrs
Stone. Olivine-hypersthene chondrite (L6).

A stone of about 13lb was seen to fall, after detonations and appearance of a luminous meteor, H.A. Ward, Am. J. Sci., 1903, **15**, p.316. Two other stones, of about 0.5lb and 177lb respectively, were found later, A.H. Miller, Science, 1903, **18**, p.243. The smallest stone was described, O.C. Farrington, Field Mus. Nat. Hist. Geol. Ser., 1907, **3**, p.111 and the largest, H.A. Ward, Proc. Rochester Acad. Sci., 1905, **4**, p.193. Olivine Fa_{24}, B. Mason, Geochimica et Cosmochimica Acta, 1963, **27**, p.1011.

83kg Chicago, Field Mus. Nat. Hist., includes the largest stone; 250g Hanau, Zeitschel Colln.; 388g Washington, U.S. Nat. Mus.; 75g Vienna, Naturhist. Mus.; 60g Tempe, Arizona State Univ.; 49g New York, Amer. Mus. Nat. Hist.;

Specimen(s): [86427], 1007g.

Bathurst v Cowra.

Batsura v Butsura.

Battle River v Iron Creek.

Baxter 36°45′ N., 93°30′ W. approx.
Stone County, Missouri, U.S.A.
Fell 1916, January 18, 0900 hrs
Stone. Olivine-hypersthene chondrite (L6).
One stone, of 611g, fell through the roof of a house, H.H. Nininger, Science, 1938, **87**, p.234, Pop. Astron., Northfield, Minnesota, 1938, **46**, p.407 [M.A. 7-177]. Olivine Fa_{24}, B. Mason, Geochimica et Cosmochimica Acta, 1963, **27**, p.1011.
244g Tempe, Arizona State Univ.;
Specimen(s): [1959,846], 273.5g.

Bayard 41°49′ N., 103°22′ W.
Scotts Bluff County, Nebraska, U.S.A.
Found 1982
Stone. Olivine-hypersthene chondrite (L6).
A single mass of 75kg was found, A.W. Struempler, letter of 20 October, 1983 in Min. Dept. BM(NH). Olivine $Fa_{24.5}$, A.L. Graham, priv. comm., 1983.

Baydon 51°31′ N., 1°36′ W.
Wiltshire, England
Pseudometeorite..
The supposed iron meteorite, said to have fallen at Baydon on May 12, 1825, E.F.F. Chladni, Ann. Phys. Chem. (Poggendorff), 1826, **8**, p.49 is from the description, clearly a pyrite nodule with a crystalline surface, largely oxidised.

Bazar v Butsura.

Beaconsfield v Cranbourne.

Bear Creek 39°36′ N., 105°18′ W.
Jefferson County, Colorado, U.S.A.
Found 1866
Synonym(s): *Aeriotopos, Bear River, Colorado, Denver, Denver County, Jefferson, Jefferson County*
Iron. Octahedrite, medium (0.60mm) (IIIB).
A mass of about 500lb was found about 25 or 30 miles from Denver; description, C.U. Shepard and J. Henry, Am. J. Sci., 1866, **42**, p.250, 286, E. Cohen, Meteoritenkunde, 1905, **3**, p.299. Analysis, 10.14 %Ni, E. Goldberg et al., Geochimica et Cosmochimica Acta, 1951, **2**, p.1. BM.1959,973 has been intensely shocked, H.J. Axon, Prog. Materials Sci., 1968, **13**, p.221. Further analysis and classification, 9.8 %Ni, 18.4 ppm.Ga, 32.8 ppm.Ge, 0.019 ppm.Ir, E.R.D. Scott et al., Geochimica et Cosmochimica Acta, 1973, **37**, p.1957. Description, V.F. Buchwald, Iron Meteorites, Univ. of California, 1975, p.307.
160kg Tempe, Arizona State Univ.; 32.2kg Denver, Mus. Nat. Hist.; 326g Fort Worth, Texas, Monnig Colln.; 275g Washington, U.S. Nat. Mus.; 156g Yale Univ.; 148g Harvard Univ.; 146.7g Chicago, Field Mus. Nat. Hist.; 116g New York, Amer. Mus. Nat. Hist.; 107g Los Angeles, Univ. of California;
Specimen(s): [41030], 35g. and a fragment, 4g; [40878], 7g.; [1959,973], 4270g. a slice, and fragments, 59g., a polished mount, and filings, 25g.

Beardsley 39°48′ N., 101°12′ W.
Rawlins County, Kansas, U.S.A.
Fell 1929, October 15, 2330 hrs
Synonym(s): *Bardsley*
Stone. Olivine-bronzite chondrite (H5).
60 stones were recovered, from 9285g to 70g in weight, totalling 16kg. Described, H.H. Nininger, Am. Miner., 1932, **17**, p.563., W.A. Waldschmidt, Am. Miner., 1932, **17**, p.565 [M.A. 5-299]. Nature of the metal, H.C. Urey and T. Mayeda, Geochimica et Cosmochimica Acta, 1959, **17**, p.113. Olivine Fa_{20}, B. Mason, Geochimica et Cosmochimica Acta, 1963, **27**, p.1011. Analysis, 26.64 % total iron, H. von Michaelis et al., Earth planet. Sci. Lett., 1969, **5**, p.387.
4.2kg Tempe, Arizona State Univ.; 945g Washington, U.S. Nat. Mus.; 860g Michigan Univ.; 443g Chicago, Field Mus. Nat. Hist.; 262g Mainz, Max-Planck-Inst.; 193g Los Angeles, Univ. of California; 158g New York, Amer. Mus. Nat. Hist.; 71g Yale Univ.;
Specimen(s): [1932,7], 294g.; [1959,996], 4827g.

Bear Lodge 44°30′ N., 104°12′ W.
Crook County, Wyoming, U.S.A.
Found 1931
Iron. Octahedrite, medium (1.2mm) (IIIA).
One mass of 48.5kg was found during road repairs on North Redwater Creek, Bear Lodge Mts., 12 miles NE. of Sundance. Described, with analysis, 8.12 %Ni, C.C. O'Harra, Science, 1932, **76**, p.34 [M.A. 5-157]. Listed, S.H. Perry, Meteorite Coll. of S.H. Perry, Adrian, Michigan, 1947, p.2. Classification and analysis, 7.67 %Ni, 19.3 ppm.Ga, 38.7 ppm.Ge, 4.5 ppm.Ir, A. Kracher et al., Geochimica et Cosmochimica Acta, 1980, **44**, p.773. Description, V.F. Buchwald, Iron Meteorites, Univ. of California, 1975, p.310.
Main mass, South Dakota School of Mines; 3.1kg Washington, U.S. Nat. Mus.; 313g Tempe, Arizona State Univ.;
Specimen(s): [1959,943], 537.5g. a slice, and sawings, 45.5g.

Bear River v Bear Creek.

Beaufort v Orange River (stone).

Beaugency v Charsonville.

Beaver 36°48′ N., 100°32′ W.
Beaver County, Oklahoma, U.S.A.
Found 1940, approx., recognized 1981
Stone. Olivine-hypersthene chondrite (L5).
A single mass of 25.628kg was found in use as a door-stop in the county jail in Beaver. It had been there for about 40 years, Meteor. Bull., 1983 (61), Meteoritics, 1983, **18**, p.77.
1066g Sedona, Arizona, Westcott Colln.; 829g Fort Worth, Texas, Monnig Colln.; 705g Vienna, Naturhist. Mus.; 532g Geneva, Nat. Hist. Mus.; 400g Toronto, Roy. Ontario Mus.;
Specimen(s): [1983,M.46], 104g.

Beaver Creek 51°10′ N., 117°20′ W.
West Kootenay district, British Columbia, Canada
Fell 1893, May 26, 1530 hrs
Stone. Olivine-bronzite chondrite (H4).
One stone fell weighing about 31lb., E.E. Howell, Science, 1893, p.41. Described, with an analysis, E.E. Howell and G.P. Merrill, Am. J. Sci., 1894, **47**, p.430. Analysis, olivine $Fa_{18.8}$, 28.31 % total iron, R. Hutchison et al., Proc. Roy. Soc. London, 1981, **A374**, p.159. I-Xe age, J. Jordon et al., Meteoritics, 1978, **13**, p.506.
Main mass, was in Howell's collection in 1897.; 2.75kg New York, Amer. Mus. Nat. Hist.; 2kg Chicago, Field Mus. Nat. Hist.; 700g Washington, U.S. Nat. Mus.; 448g Budapest, Nat. Mus.; 405g Harvard Univ.; 400g Vienna, Naturhist. Mus.; 212g Rome, Vatican Colln; 116g Ottawa, Geol. Surv. Canada; 80g Yale Univ.; 104g Tempe, Arizona State Univ.; 26g Los Angeles, Univ. of California;
Specimen(s): [73646], 661g.; [1920,289], 36g.

Beaver-Harrison 38°29'3" N., 113°8'11" W.
Beaver County, Utah, U.S.A.
Found 1979, July 24
Stone. Olivine-hypersthene chondrite (L6).
A single mass of 925g was found on the surface at the abandoned Beaver-Harrison mine, Beaver Lake Mountains. Description, olivine $Fa_{25.0}$, K. Keil et al., Meteoritics, 1981, **16**, p.13.

Bécasse v La Bécasse.

Bechuanaland 25° S., 24° E.
South Africa
Found 1888, known in this year?
Doubtful. Iron.
A meteorite is mentioned, with the above particulars, F. Berwerth, Ann. Naturhist. Hofmus. Wien, 1903, **18**, p.51. Listed, E.A. Wülfing, Die Meteoriten in Samml., Tübingen, 1897, p.23 's Anhang' but there is no further entry in the appendix. It is not mentioned in later catalogues, nor is the weight of the mass stated or it's whereabouts; it can only be regarded as doubtful.

Beddgelert 53°1' N., 4°6' W.
Gwynedd, Wales
Fell 1949, September 21, 0147 hrs
Stone. Olivine-bronzite chondrite (H5), black.
One mass of 794g fell through the roof of an hotel; described, with age determination and partial analysis, K.F. Chackett et al., Geochimica et Cosmochimica Acta, 1950, **1**, p.3. Reported, F.A. Paneth, Nature, 1949, **164**, p.990. Olivine Fa_{19}, B. Mason, Geochimica et Cosmochimica Acta, 1963, **27**, p.1011.
374g Mainz, Max-Planck-Inst., ex Durham Univ.; 22g New York, Amer. Mus. Nat. Hist.; 17g Washington, U.S. Nat. Mus.; 15g Nat. Mus. Wales; 13g Yale Univ.;
Specimen(s): [1949,259], 377.5g.; [1971,103], 0.6g. small, crusted fragments

Beeler 38°32' N., 100°13' W.
Ness County, Kansas, U.S.A.
Found 1924, and probably in 1894
Synonym(s): *Caster Farm, Castor Brothers' Farm, Castor Farm, Ness County (in part)*
Stone. Olivine-hypersthene chondrite, amphoterite (LL6).
A stone of 8.64kg was ploughed up, Meteor. Bull., 1969 (47), Meteoritics, 1970, **5**, p.104. The Kansada stone previously assigned to the Ness County find of 1894 is weathered brecciated and probably identical to Beeler. Description and location, W.F. Read, Meteoritics, 1972, **7**, p.417. Olivine Fa_{30}, J.T. Wasson, Meteorites, Springer-Verlag, 1974, p.283.
Main mass, Los Angeles, Univ. of California;
Specimen(s): [83489], 2008g. about a quarter of the Kansada stone.; [1970,3], 4g. of Beeler, a polished mount.

Beenham 36°13' N., 103°39' W.
Union County, New Mexico, U.S.A.
Found 1937
Stone. Olivine-hypersthene chondrite (L5), veined.
Many masses and fragments were found, totalling 44.4kg, in township 25 range 30E., Union County, E.P. Henderson, letter of 12 May and 3 June, 1939 in Min. Dept. BM(NH). Described, R.M. Leard, Pop. Astron., Northfield, Minnesota, 1939, **47**, p.385 [M.A. 7-376]. Listed, H.H. and A.D. Nininger, The Nininger Collection of Meteorites, Winslow, Arizona, 1950, p.32. Olivine $Fa_{23.6}$, D.E. Lange and K. Keil, Meteoritics, 1976, **11**, p.315, abs.. Post formational history, P.R. Buseck, Geochimica et Cosmochimica Acta, 1967, **31**, p.1583.
4kg Washington, U.S. Nat. Mus., complete stone; 1.9kg Tempe, Arizona State Univ.; 750g Chicago, Nat. Hist. Mus.; 864g Alberquerque, Univ.of New Mexico; 73g New York, Amer. Mus. Nat. Hist.;
Specimen(s): [1949,74], 37.7g.; [1959,1007], 2993g.

Beeston 52°56' N., 1°13' W.
Nottinghamshire, England
Fell 1780, April 11, 2100 hrs
Doubtful..
Several stones are said to have fallen after a prominent meteor was seen, but nothing has been preserved and the description is inadequate, E.F.F. Chladni, Die Feuer-Meteore, Wien, 1819, p.256. Listed, R.P. Greg, Rep. Brit. Assn., 1860, p.60, T.M. Hall, Min. Mag., 1879, **3**, p.7.

Behar v Shergotty.

Beirut v Beyrout.

Belaia Tserkov v Bjelaja-Zerkov.

Belaja-Zerkov v Bjelaja-Zerkov.

Belaja Zerkow v Bjelaja-Zerkov.

Belaya Gora v Saratov.

Belfast v Bovedy.

Belgica 7901
Belgica Mountains, Antarctica
Found 1979
Five specimens, all chondritic, one of them carbonaceous, were found by the Japanese Antarctic Expedition during the 1979-1980 field season.

Belgica 7904
Belgica Mountains, Antarctica
Found 1979
Stone. Carbonaceous chondrite, type II (CM2).
A single, nearly complete stone of 1234g was recovered in the Belgica Mountains region during the 1979-1980 Japanese Antarctic Expedition, Meteorites News, Tokyo, 1982, **1**, p.26.

Belgorod v Sevrukovo.

Belgradjek v Virba.

Bella Roca 24°54' N., 105°24' W.
Sierra de San Francisco, Santiago Papasquiaro, Durango, Mexico
Found 1888, known in this year
Synonym(s): *La Bella Roca, La Bella Rocka, Papasquiaro*
Iron. Octahedrite, medium (0.7mm) (IIIB).
A mass of about 73lb was found, L. Fletcher, Min. Mag., 1890, **9**, p.155. Described and analysed, 9.78 %Ni, J.E. Whitfield, Am. J. Sci., 1889, **37**, p.439, E. Cohen, Meteoritenkunde, 1905, **3**, p.374. Classification and new analysis, 10.06 %Ni, 16.7 ppm.Ga, 31.1 ppm.Ge, 0.014 ppm.Ir, E.R.D. Scott et al., Geochimica et Cosmochimica Acta, 1973, **37**, p.1957. Description, V.F. Buchwald, Iron Meteorites, Univ. of California, 1975, p.311.
12.5kg Vienna, Naturhist. Mus.; 19.7kg Tübingen Univ.;

2kg New York, Amer. Mus. Nat. Hist.; 1.9kg Harvard Univ.; 666g Washington, U.S. Nat. Mus.; 665g Tempe, Arizona State Univ.; 680g Chicago, Field Mus. Nat. Hist.; 431g Ottawa, Mus. Geol. Surv. Canada; 405g Yale Univ.; 274g Prague, Nat. Mus.; 207g Budapest, Nat. Hist. Mus.;
Specimen(s): [64206], 3505g. a slice, and troilite, 20g

Belle Plaine 37°19′ N., 97°15′ W.

Sumner County, Kansas, U.S.A.
Found 1950
Stone. Olivine-hypersthene chondrite (L6).

Three stones were found within a two mile strip SE. of Belle Plaine; the first was found in 1950, the second a few years later and the third in 1963, the weights are 26.9kg, 28kg, and 23.9kg respectively, E. Cillerman and W.E. Hill, Meteoritics, 1967, **3**, p.147. Mineralogy, olivine Fa_{25}, J. Huth and F. Begemann, Meteoritics, 1983, **18**, p.abs. A fourth mass, 17.6kg was found in 1981, W. Zeitschel, letter of 27 January, 1982 in Min. Dept. BM(NH).

54.3kg Fort Worth, Texas, Monnig Colln., the first and second stones; 22kg Los Angeles, Univ. of California, of the third stone; 23g New York, Amer. Mus. Nat. Hist., of the second stone; 90g Washington, U.S. Nat. Mus.;
Specimen(s): [1974,M.4], 1582g. of the third stone, a slice.

Bells 33°36′ N., 96°28′ W.

Grayson County, Texas, U.S.A.
Fell 1961, September 9
Stone. Carbonaceous chondrite, type II (CM2).

A detonating fireball was heard on September 9 over NE. Texas; one fragment was picked up next day, another six weathered fragments later. Total weight about 10oz., O.E. Monnig, Meteoritics, 1963, **2**, p.67. Reported, Meteor. Bull., 1962 (25), Meteor. Bull., 1963 (28).

283g Fort Worth, Texas, Monnig Colln.; 4.4g New York, Amer. Mus. Nat. Hist.; 0.5g Washington, U.S. Nat. Mus.;

Bellsbank 28°5′ S., 24°5′ E.

Barkly West district, Cape Province, South Africa
Found 1955
Iron. Hexahedrite (IRANOM).

One mass of 38kg was found 2ft below the surface in dolomitic strata 27 miles NNW. of Barkly West, D. Groenewald, Trans. Geol. Soc. S. Africa, 1959, **62**, p.75 [M.A. 15-38]. Analysis, 4.5 %Ni, A.A. Moss, priv. comm., 1964. Classification and new analysis confirming low Ni content, 4.13 %Ni, 39.2 ppm.Ga, 54.6 ppm.Ge, 0.15 ppm.Ir, E.R.D. Scott et al., Geochimica et Cosmochimica Acta, 1973, **37**, p.1957. Description, V.F. Buchwald, Iron Meteorites, Univ. of California, 1975, p.313.

343g Washington, U.S. Nat. Mus.; Main mass, Pretoria, Geol. Surv. Mus.;
Specimen(s): [1961,403], 524g. a slice

Belley v Groslée.

Belly River 49°30′ N., 113° W. approx.

Alberta, Canada
Found 1943, winter of 1943-44
Stone. Olivine-bronzite chondrite (H6), veined.

One complete stone was found, of 7.9kg, on the eastern bank of the Belly River, L. LaPaz, Meteoritics, 1953, **1**, p.106. Reported, P.M. Millman, J. Roy. Astron. Soc. Canada, 1953, **47**, p.162 [M.A. 12-358]. Description, analysis, 25.89 % total iron, B. Mason and H.B. Wiik, Am. Mus. Novit., 1967 (2280). Olivine Fa_{20}, B. Mason, Geochimica et Cosmochimica Acta, 1963, **27**, p.1011.

1518g Ottawa, Geol. Surv. Canada; 31g New York, Amer. Mus. Nat. Hist.; 33.6g Alberquerque, Univ. of New Mexico;

Belmont v Simonod.

Belmont 42°44′ N., 90°21′ W.

Lafayette County, Wisconsin, U.S.A.
Found 1958, in the spring
Stone. Olivine-bronzite chondrite (H6), veined.

One mass of 25.3kg was found; sp.gr. 3.56, W.A. Broughton and L. LaPaz, Wisconsin Acad. Review, 1962, **9**, p.155. Mentioned, W.A. Broughton, letters of 27 March 1963, Meteor. Bull., 1963 (26). Olivine Fa_{18}, B. Mason, Geochimica et Cosmochimica Acta, 1967, **31**, p.1100.

1965g Albuquerque, Univ. of New Mexico;

Belokrinice v Bielokrynitschie.

Belokrinich'e v Bielokrynitschie.

Belokrinitschje v Bielokrynitschie.

Belostok v Białystok.

Belville 32°20′ S., 64°52′ W.

Unión, Cordoba, Argentina
Fell 1937, December
Stone. Chondrite.

Report with figures, L.M. Villar, Cienc. Investig., 1968, **24**, p.302.

Bemdego v Bendegó.

Benares (a) 25°22′ N., 82°55′ E.

Varanasi, Uttar Pradesh, India
Fell 1798, December 19, 2000 hrs
Synonym(s): *Krakhut*
Stone. Olivine-hypersthene chondrite, amphoterite (LL6).

Many stones fell at Krakhut, about 14 miles from Benares, one of about 2lb through a roof, after the appearance of a luminous fireball and detonations, E. Howard, Phil. Trans. Roy. Soc. London, 1802, p.168, 175, Ann. Phys. (Gilbert), 1812, **41**, p.453. Olivine Fa_{28}, B. Mason, Geochimica et Cosmochimica Acta, 1963, **27**, p.1011.

1274g Edinburgh, Roy. Scottish Mus.; 662g Vienna, Naturhist. Mus.; 286.8g Tübingen Univ.; 180g Copenhagen, Univ. Geol. Mus.; 88g Washington, U.S. Nat. Mus.; 66g Calcutta, Indian Mus.; 110g Budapest, Nat. Mus.; 77g Tempe, Arizona State Univ.; 59g Chicago, Field Mus. Nat. Hist.; 40g Prague, Nat. Mus.;
Specimen(s): [61310], 190g. a complete stone; [90247], 106g.; [61311], 136g.; [1935,52], 0.4g.

Benares (b) 25°20′ N., 83° E. approx.

Varanasi, Uttar Pradesh, India
Iron.

A specimen of iron labelled Benares is in Lucknow Mus. A small fragment sent to Min. Dept. BM(NH) to confirm its meteoritic nature. It has been artificially heated, A.W.R. Bevan, priv. comm., 1982.

Bench Crater

Apollo 12 landing site, Oceanus Procellarum, Moon
Stone. Carbonaceous chondrite, anomalous (C?).

A fragment found in soil sample 12037 brought to the earth from the moon by the Apollo 12 mission has a matrix similar to that of CI chondrites, H.Y. McSween, Jr., Earth planet. Sci. Lett., 1976, **31**, p.193 [M.A. 77-2021].

Bencubbin 30°45′ S., 117°47′ E.
Western Australia, Australia
Found 1930, July 30
Synonym(s): *Mandiga, North Mandiga*
Stony-iron. Mesosiderite (MES).
One mass of 54kg was found 12 miles NW. of Bencubbin, described, with analysis, E.S. Simpson and D.G. Murray, Min. Mag., 1932, **23**, p.33. Analysis, A.J. Easton and J.F. Lovering, Geochimica et Cosmochimica Acta, 1963, **27**, p.753. Mentioned, A.D. Nininger, Pop. Astron., Northfield, Minnesota, 1940, **48**, p.558, Contr. Soc. Res. Meteorites, **2**, p.227 [M.A. 8-54]. Another mass of 142lb was found in 1959 or 1960 at Mandinga and is considered to be part of the Bencubbin fall, G.J.H. McCall, letters of 23 July and 9 August 1963 in Min. Dept. BM(NH)., Spec. Publ. West. Austr. Mus., 1965 (3), p.21. A similarity with the Patwar meteorite has been suggested; a new class of enstatite-olivine stony-irons is suggested to cover these two anomalous meteorites, G.J.H. McCall, Min. Mag., 1965, **35**, p.476, G.J.H. McCall, Min. Mag., 1966, **36**, p.726. Posslbly includes Korrelocking, G.J.H. McCall, letter of 14 June 1972 in Min. Dept. BM(NH). Shocked and heated, A.V. Jain and M.E. Lipschutz, Nature, 1973, **242** (Phys. Sci.), p.26. The silicate has variable composition, G.J.H. McCall, 2nd. Suppl. to West. Austr. Mus. Spec. Publ. no. 3, 1972, p.5. Formation history, G.W. Kallemeyn et al., Geochimica et Cosmochimica Acta, 1978, **42**, p.507.
24kg Perth, West. Austr. Mus., 12kg of each mass; 1062g Kalgoorlie, West. Austr. School of Mines, of the second mass; 7kg Washington, U.S. Nat. Mus.; 435g New York, Amer. Mus. Nat. Hist.; 190g Vienna, Naturhist. Mus.; 143g Sydney, Australian Mus.;
Specimen(s): [1931,536], 4258g. and fragments, 20g

Bendegó 10°7′ S., 39°12′ W.
Monte Santo, Bahia, Brazil
Found 1784
Synonym(s): *Adon, Bahia, Bemdego, Bendengó, Bendigo, Benegó, Sergipe, Wollaston's Iron*
Iron. Octahedrite, coarse (1.8mm) (IC).
A large mass of about 5 tons was found near the rivulet called the Bendegó, A.F. Mornay and W.H. Wollaston, Phil. Trans. Roy. Soc. London, 1816, **106**, p.270, 281. It was removed to Rio de Janeiro in 1888, and was described and analysed, 6.8 %Ni, O.A. Derby, Rev. Mus. Nac. Rio de Janeiro, 1896, **9**, p.89. The mass was heated in a large fire. The original weight is given by O.A. Derby (loc.cit.) as 5360kg on the railroad scales, O.A. Derby, Am. J. Sci., 1888, **36**, p.158. Mentioned, H.E. de Araujo, Rev. Soc. Brazil. Chim., 1931, **2**, p.365 [M.A. 5-153], A. Betim Paes Leme, Anais Acad. Brasil. Cienc., 1935, **7**, p.177 [M.A. 6-205], E. de Oliveira, Ann. Acad. Brasil. Sci., 1931, **3**, p.33 [M.A. 5-14]. Analysis, 6.39 %Ni, 54 ppm.Ga, 234 ppm.Ge, 0.20 ppm.Ir, J.T. Wasson, Icarus, 1970, **12**, p.407. Bandwidth, description, V.F. Buchwald, Iron Meteorites, Univ. of California, 1975, p.315.
Main mass, Rio de Janeiro,; 2.75kg Chicago, Field Mus. Nat. Hist.; 1.37kg La Plata Mus.; 1.5kg Paris, Mus. d'Hist. Nat.; 2648g Tübingen Univ.; 2.4kg Vienna, Naturhist. Mus.; 1980g Washington, U.S. Nat. Mus.; 901g Copenhagen, Univ. Geol. Mus.; 750g New York, Amer. Mus. Nat. Hist.; 901g Tempe, Arizona State Univ.; 573g Budapest, Nat. Mus.;
Specimen(s): [19962], 2220g.; [66585], 863g. and a fragment, 2.5g; [1911,718], 516g.; [1964,711], 6.5g.

Bendengó v Bendegó.

Bendigo v Bendegó.

Bendoc v Bendock.

Bendock 37°9′ S., 148°55′ E.
County Croajingolong, Victoria, Australia
Found 1898
Synonym(s): *Bendoc*
Stony-iron. Pallasite (PAL).
A mass of about 60lb was found; described and analysed, 7.81 %Ni, J.C.H. Mingaye, Ann. Rep. Dept. Mines New South Wales, 1899 (for 1898), p.21, Rep. Austr. Assn. Adv. Sci., 1903 (for 1902), p.162.

Benedict 41° N., 97°32′ W.
York County, Nebraska, U.S.A.
Found 1970
Iron. Octahedrite, medium (1.1mm) (IIIA).
A 16.38kg oriented individual was found on land which had been levelled for irrigation, about 18 miles from the find site of the York iron (*q.v.*), G.I. Huss, letter of 11 July, 1975. Although superficially similar to York, structural study by V.F. Buchwald indicates that Benedict is richer in Ni and P and so is distinct from York, V.F. Buchwald, letter of 13 June 1975 in Min. Dept. BM(NH). Classification and analysis, 8.62 %Ni, 21.4 ppm.Ga, 45.4 ppm.Ge, 0.18 ppm.Ir, A. Kracher et al., Geochimica et Cosmochimica Acta, 1980, **44**, p.773.
9.8kg Mainz, Max-Planck-Inst.; 836g Copenhagen, Univ. Geol. Mus.; 602g Fort Worth, Texas, Monnig Colln.;
Specimen(s): [1975,M.2], 115.7g. a crusted, etched part slice.

Benegó error for Bendegó.

Benld 39°5′ N., 89°9′ W.
Macoupin County, Illinois, U.S.A.
Fell 1938, September 29, 0900 hrs
Stone. Olivine-bronzite chondrite (H6), veined.
One mass of 1770.5g fell through a garage roof. Description, B.H. Wilson, Pop. Astron., Northfield, Minnesota, 1938, **46**, p.458, Science, 1939, **89**, p.34 [M.A. 7-273], H.W. Nichols, Sci. Monthly, New York, 1939, **49**, p.235 [M.A. 7-379], S.K. Roy and R.K. Wyant, Field Mus. Nat. Hist. Geol. Ser., 1951, **7** (11), p.145 [M.A. 11-530]. Analysed by R.K. Wyant (op. cit.). Olivine Fa_{20}, B. Mason, Geochimica et Cosmochimica Acta, 1963, **27**, p.1011.
Main mass, Chicago, Field Mus. Nat. Hist., the nearly complete stone; 6.3g Washington, U.S. Nat. Mus.;

Bennett County 43°30′ N., 101°15′ W.
South Dakota, U.S.A.
Found 1934
Synonym(s): *Norris*
Iron. Hexahedrite (IIA).
One mass of 89kg was found at the head of Black Pipe Creek, 35 miles NE. of Martin, Bennett County. Described, with an analysis, 5.25 %Ni, C.C. O'Harra, Science, 1935, **81**, p.72 [M.A. 6-105]. Reported, Meteor. Bull., 1961 (22). New analysis, 5.28 %Ni, 59.1 ppm.Ga, 179 ppm.Ge, 41 ppm.Ir, J.T. Wasson, Geochimica et Cosmochimica Acta, 1969, **33**, p.859. Description, V.F. Buchwald, Iron Meteorites, Univ. of California, 1975, p.318.
Main mass, South Dakota School of Mines; 8.1kg Washington, U.S. Nat. Mus.; 1183g Tempe, Arizona State Univ.;
Specimen(s): [1959,932], 922g. a slice, and a fragment, 15g and sawings, 13g.

Benoni 26°10′ S., 28°25′ E.
Transvaal, South Africa
Fell 1943, July 25, 1735 hrs
Synonym(s): *Groenewald, Groenewald-Benoni*
Stone. Olivine-bronzite chondrite (H6).
A stone of 3880g fell 55 paces from Mr.J.W. Groenewald at Benoni, central Transvaal. Description, analysis, T.W. Gevers et al., Trans. Geol. Soc. S. Africa, 1945, **48**, p.83 [M.A. 10-175]. Gas-rich, J. Zähringer, Geochimica et Cosmochimica Acta, 1968, **32**, p.209. Olivine Fa_{18}, R. Hutchison, priv. comm. Also described, C. Frick and E.C.I. Hammerbeck, Bull. Geol. Surv. S. Africa, 1973 (57), p.11.
14.4g Tempe, Arizona State Univ.; Main mass, Univ. of Witwatersrand, Geol. Mus.;
Specimen(s): [1970,336], 32.3g. a sawn, part crusted fragment

Benthullen 53°3′ N., 8°6′ E.
Near Oldenburg, Niedersachsen, Germany
Found 1951
Stone. Olivine-hypersthene chondrite (L), brecciated.
A stone of about 17kg was found in a peat bog; it may be from the Oldenburg fall of 1368 (*q.v.*). The stone is partly crusted and in a good state of preservation. The find site is about six km N. of Bissel and Beverbruch, where the stones of the Oldenburg (1930) fall landed (*q.v.*) and so Benthullen is nearer the town of Oldenburg than the fall of that name, olivine Fa_{25}, P. Ramdohr and A. El Goresy, Meteoritics, 1974, **9**, p.397, abs.

Benton 45°57′ N., 67°33′ W.
Porten Settlement, New Brunswick, Canada
Fell 1949, January 16, 2000 hrs
Stone. Olivine-hypersthene chondrite, amphoterite (LL6).
Two stones, 1500g and 1340g, fell, H.H. and A.D. Nininger, The Nininger Collection of Meteorites, Winslow, Arizona, 1950, p.33, P.M. Millman, J. Roy. Astron. Soc. Canada, 1953, **47**, p.29, 92, 165 [M.A. 12-358]. Olivine $Fa_{30.5}$, K. Fredriksson et al., Origin and Distribution of the Elements, ed. L.H. Ahrens, Pergamon, 1968, p.457.
1200g Ottawa, Geol. Surv. Canada, the second stone; 124g New York, Amer. Mus. Nat. Hist.; 225g Washington, U.S. Nat. Mus.; 15.3g Tempe, Arizona State Univ.;
Specimen(s): [1963,532], 49.5g. three fragments, 31g, 12.5g, and 6g.

Berar v Chandakapur.

Beraun v Zebrak.

Berdiansk v Berdyansk.

Berdjansk v Berdyansk.

Berdjansk v Pavlograd.

Berdyansk 46°45′ N., 36°49′ E.
Taurida, Ukraine, USSR, [Бердянск]
Found 1843, known in this year
Synonym(s): *Berdiansk, Berdjansk, Kavkaz, Kawkas*
Stone. Olivine-hypersthene chondrite (L6), veined.
A mass of 2.25kg was described, M. Hiriakov and A.A. Inostrantzev, Geol. För. Förh. Stockholm, 1878, **4**, p.72, I.S. Astapowitsch, Trans. Roy. Astron. Soc. Canada, 1938, **32**, p.195 [M.A. 7-172]. The fall is probably a prehistoric one, as the mass was found in a tumulus. Shortly described and figured, E.L. Krinov, Астрон. Журнал, 1945, **22**, p.303 [M.A. 9-297], Каталог Метеоритов Акад. Наук СССР, Москва, 1947 [M.A. 10-511]. Olivine Fa_{25}, B. Mason, Geochimica et Cosmochimica Acta, 1963, **27**, p.1011. Analysis, 21.82 % total iron, V.Ya. Kharitonova, Метеоритика, 1965, **26**, p.146.
2kg Moscow, Acad. Sci.;

Béréba 11°39′ N., 3°39′ W.
Upper Volta
Fell 1924, June 27, 1530 hrs
Synonym(s): *Haute Volta*
Stone. Achondrite, Ca-rich. Eucrite (AEUC).
A stone of about 18kg fell after detonations; description, A. Lacroix, C. R. Acad. Sci. Paris, 1925, **181**, p.745, Nouv. Arch. Mus. d'Hist. Nat. Paris, 1926, **1**, p.15-58. Rare gas content, J. Zähringer, Geochimica et Cosmochimica Acta, 1968, **32**, p.1421. Rb-Sr study, J.L. Birck and C.J. Allegre, Earth planet. Sci. Lett., 1978, **39**, p.37 [M.A. 78-4757].
303g Paris, Mus. d'Hist. Nat.; 123g Chicago, Field Mus. Nat. Hist.; 4g Tempe, Arizona State Univ.; 15g Washington, U.S. Nat. Mus.;
Specimen(s): [1970,337], 10.3g.; [1972,235], 45g. two crusted fragments.

Berg Emir v Krasnojarsk.

Berlanguillas 41°41′ N., 3°48′ W.
Burgos, Spain
Fell 1811, July 8, 2000 hrs
Synonym(s): *Burgos*
Stone. Olivine-hypersthene chondrite (L6), veined.
Three stones, one of 2.75kg, fell, L.W. Gilbert, Ann. Phys. (Gilbert), 1812, **40**, p.116, **41**, p.452. Olivine Fa_{25}, B. Mason, Geochimica et Cosmochimica Acta, 1963, **27**, p.1011.
1kg Paris, Mus. d'Hist. Nat.; 198g Vienna, Naturhist. Mus.; 25g Chicago, Field Mus. Nat. Hist.; 16g Tübingen, Univ.; 14g Yale Univ.; 7.5g New York, Amer. Mus. Nat. Hist.; 6g Modena, Univ.;
Specimen(s): [90259], 16.5g.; [44133], 8.3g.

Berrias v Aumieres.

Beshkalia v Samelia.

Beshki v Samelia.

Besouros v Bezerros.

Bethany v Gibeon.

Bethel 34°14′ N., 103°23′ W.
Roosevelt County, New Mexico, U.S.A.
Found 1968
Stone. Chondrite.
Two masses are listed, Bethel (a) 18.3g, and Bethel (b) 56.8g, Cat. Huss Coll. Meteorites, 1976, p.5.
Both masses, Mainz, Max-Planck-Inst.;

Bethlehem 42°32′ N., 73°50′ W.
Albany County, New York, U.S.A.
Fell 1859, August 11, 0730 hrs
Synonym(s): *Albany County, Troy*
Stone. Olivine-bronzite chondrite (H).
After the appearance of a luminous meteor, and detonations, a small stone about the size of a pigeon's egg was seen to fall, D.A. Wells, Proc. Boston Soc. Nat. Hist., 1859, **7**, p.176. Description, C.U. Shepard, Am. J. Sci., 1860, **30**, p.206. Olivine Fa_{19}, B. Mason, Geochimica et Cosmochimica

Acta, 1963, **27**, p.1011.
8g Albany (New York State), Museum; 1g Ottawa, Mus. Geol. Surv. Canada; 2.5g New York, Amer. Mus. Nat. Hist.; 1g Chicago, Field Mus. Nat. Hist.;
Specimen(s): [34593], less than 1g, all fragments.

Bethune 39°18′ N., 102°25′ W.
Kit Carson County, Colorado, U.S.A.
Found 1941
Stone. Enstatite chondrite (E4-5).
39.5g were in H.H. Nininger's collection, H.H. and A.D. Nininger, The Nininger Collection of Meteorites, Winslow, Arizona, 1950, p.34. Mineralogy, 67g recovered, B. Mason, Geochimica et Cosmochimica Acta, 1966, **30**, p.23.
14.6g New York, Amer. Mus. Nat. Hist.; 14.3g Tempe, Arizona State Univ.;
Specimen(s): [1959,847], 25g. a weathered fragment.

Bétréchies v Hainaut.

Bettona v Assisi.

Bettrechies v Hainaut.

Beuste 43°13′ N., 0°14′ W.
Pau, Basses-Pyrénées, France
Fell 1859, May, afternoon
Synonym(s): *Bueste, Pau*
Stone. Olivine-hypersthene chondrite (L5), brecciated.
Two fragments of about 1.5kg and 0.5kg were found, about 700m apart, G.A. Daubrée, C. R. Acad. Sci. Paris, 1873, **76**, p.315. Olivine Fa_{25}, B. Mason, Geochimica et Cosmochimica Acta, 1963, **27**, p.1011.
420g Pau, Mus.; 400g Paris, Ecole des Mines; 212g Paris, Mus. d'Hist. Nat.; 87g Budapest, Nat. Mus.; 77g Vienna, Naturhist. Mus.; 37g Chicago, Field Mus. Nat. Hist.; 3g Washington, U.S. Nat. Mus.;
Specimen(s): [64340], 33.7g.

Beverbruch v Oldenburg (1930).

Beverbrück v Oldenburg (1930).

Bewitched Burgrave v Elbogen.

Beyrout 33°53′ N., 35°30′ E.
Syria
Fell 1921, December 31, 1545 hrs
Synonym(s): *Beirut*
Stone. Olivine-hypersthene chondrite (L4).
Fell through the roof of a hut near the Univ. of St. Joseph, Beyrout. Original weight 1100g, only two small fragments preserved, A. Lacroix, C. R. Acad. Sci. Paris, 1929, **188**, p.949 [M.A. 4-117.]. Olivine Fa_{26}, B. Mason, Geochimica et Cosmochimica Acta, 1963, **27**, p.1011.
51g Paris, Mus. d'Hist. Nat.;

Bezerros 8°18′ S., 36°6′ W.
Pernambuco, Brazil
Fell 1915, May 9
Synonym(s): *Besouros*
Doubtful. Iron.
Original weight estimated at 20 tons, H. Michel, Fortschr. Min. Krist. Petr., 1922, **7**, p.258. Another account gives the supposed weight as about 1 ton, E. de Oliveira, Anais Acad. Brasil. Cienc., 1931, **3**, p.33,[M.A. 5-15]. Fall almost certainly mythical, C.C. Wylie, Pop. Astron., Northfield, Minnesota, 1930, **38**, p.446.

Bhagur 20°53′ N., 74°50′ E.
West Khandesh district, Maharashtra, India
Fell 1877, November 27, 1800 hrs
Synonym(s): *Dhulia, Khandesh district*
Stone. Olivine-hypersthene chondrite (L6), veined.
After appearance of a luminous meteor over Dhule and other places in the Khandesh district, a stone, of unrecorded weight, fell near the village of Bhagur, E. Cordeaux, J. Bombay Branch Roy. Asiatic Soc., 1878, **14** (36, Abstr. of Soc. Proc., pp. iii-vi). Olivine Fa_{25}, B. Mason, Geochimica et Cosmochimica Acta, 1963, **27**, p.1011.
7g Vienna, Naturhist. Mus.; 2.5g Calcutta, Mus. Geol. Surv. India; 2.5g Chicago, Field Mus. Nat. Hist.;
Specimen(s): [64511], 6g.

Bhawalpur v Khairpur.

Bheenwal 25°9′ N., 73°4′ E.
Erinpura, Sirohi, Rajasthan, India
Fell 1865, August 20, 1330 hrs
Synonym(s): *Erinpoorah*
Doubtful. Stone.
A stone of 3.25lb was seen to fall, but has apparently been lost, and the description is inadequate, A.L. Coulson, Mem. Geol. Surv. India, 1940, **75**, p.1 [M.A. 8-54], Rep. Brit. Assn., 1866, **36**, p.132.

Bherai 20°50′ N., 71°28′ E.
Junagarh, Kathiawar, Gujarat, India
Fell 1893, April 28, 0800 hrs
Synonym(s): *Jafferabad, Kathiawar*
Stone. Olivine-hypersthene chondrite (L6), veined.
After appearance of a luminous meteor, and detonations, a stone of less than 0.25lb fell near the village of Kovaya, between Bherai and Jafferabad, J.W. Judd, Nature, 1893, **49**, p.32. Olivine Fa_{24}, B. Mason, Geochimica et Cosmochimica Acta, 1963, **27**, p.1011. Detailed description and micrometric analysis, P.K. Ghosh, Rec. Geol. Surv. India, 1941, **75** (Proof paper no. 14) [M.A. 9-297].
8.5g Calcutta, Mus. Geol. Surv. India;
Specimen(s): [76802], 17.5g.

Bhola 22°41′ N., 90°39′ E.
Bakarganj district, Khulna, Bangladesh
Fell 1940, March 27
Stone. Olivine-hypersthene chondrite, amphoterite (LL3-6), brecciated.
Three pieces, of 841g, 184g, and 22g, were recovered, M.A.R. Khan, Hyderabad Acad. Studies, 1950, **12** [M.A. 11-441]. Mentioned, P. Chatterjee, letter of 25 March 1952 in Min. Dept. BM(NH). Brecciated, pyroxene Fs4-25, olivine $Fa_{27.8}$, F. Wlotzka et al., Geochimica et Cosmochimica Acta, 1983, **47**, p.743.
841g Calcutta, Mus. Geol. Surv. India.; 116g Washington, U.S. Nat. Mus.;
Specimen(s): [1983,M.8], 16g.

Bholgati v Bholghati.

Bholghati 22°5′ N., 86°54′ E.
Deoli pargana, Maurbhanj, Orissa, India
Fell 1905, October 29, 0830 hrs
Synonym(s): *Bholgati*
Stone. Achondrite, Ca-rich. Howardite (AHOW).
Two stones of about 2lb and 3.5lb fell, after detonations, L.L. Fermor, Rec. Geol. Surv. India, 1907, **35**, p.83. Figured, M.B. Duke and L.T. Silver, Geochimica et Cosmochimica Acta, 1967, **31**, p.1637. Radiation and gas

retention ages, R. Ganapathy and E. Anders, Geochimica et Cosmochimica Acta, 1969, **33**, p.775. Analysis, B. Mason et al., Smithson. Contrib. Earth Sci., 1979 (22), p.30.
740g Calcutta, Mus. Geol. Surv. India, nearly complete stone; 44.6g New York, Amer. Mus. Nat. Hist.; 1.6g Washington, U.S. Nat. Mus.;
Specimen(s): [1915,140], 27g. from the 2lb stone.

Bhur-Gheluai v Bur-Gheluai.

Bhurtpur v Kheragur.

Bhurtpur v Moti-ka-nagla.

Bialacerkiew v Bjelaja Zerkov.

Białystok 53°6′ N., 23°12′ E.
Grodno, Poland
Fell 1827, October 5, 0930 hrs
Synonym(s): *Belostok, Bielostok, Fasti, Jaski, Jasly, Jaśli, Knaasta, Knasta, Knyszyn, Kuasti-Knasti, Kwasli, Phasti*
Stone. Achondrite, Ca-rich. Howardite (AHOW).
A shower of stones fell, of which four, of total weight 4kg were found, the largest weighing about 2kg, Ann. Chim. Phys., 1828, **39**, p.421, A. Göbel, Bull. Acad. Sci. St.-Petersbourg, 1867, **11**, p.270. Figured, M.B. Duke and L.T. Silver, Geochimica et Cosmochimica Acta, 1967, **31**, p.1637. Listed as a eucrite, B. Mason, Geochimica et Cosmochimica Acta, 1967, **31**, p.107 but as a howardite, B. Mason, Smithson. Contrib. Earth Sci., 1975 (14), p.71. Analysis, B. Mason et al., Smithson. Contrib. Earth Sci., 1979 (22), p.30.
120g Kiev Univ.; 91g Moscow, Acad. Sci.; 101g Budapest, Nat. Mus.; 73g Berlin, Humboldt Univ.; 59g Vienna, Naturhist. Mus.; 16g Washington, U.S. Nat. Mus.; 4.5g Chicago, Field Mus. Nat. Hist.;
Specimen(s): [54640], 3.5g.; [35154], less than a gram; [96257], thin section

Bibandi v Dokachi.

Bielokrynitschie 50°8′ N., 27°10′ E.
Zaslavl, Volhynia, Ukraine, USSR, [Белокриничье]
Fell 1887, January 1, 1800 hrs
Synonym(s): *Belokrinice, Belokrinich'e, Belokrinitschje, Bjelokrynitschie*
Stone. Olivine-bronzite chondrite (H4), veined.
After the appearance of a fireball moving SW. to NE., and detonations, several stones were seen to fall of which eight were found, the largest weighing about 2kg, Y.I. Simashko, Cat. Meteorites, 1891, p.51. Description, G.P. Merrill, Mem. Nat. Acad. Sci. Washington, 1919, **14** (4), p.1. Olivine Fa_{20}, B. Mason, Geochimica et Cosmochimica Acta, 1963, **27**, p.1011.
400g Vienna, Naturhist. Mus., main mass; 57g Prague, Nat. Mus.; 56g Washington, U.S. Nat. Mus.; 49g Budapest, Nat. Mus.; 29g New York, Amer. Mus. Nat. Hist.; 28g Moscow, Acad. Sci.; 257g Chicago, Field Mus. Nat. Hist.; 99g Paris, Mus. d'Hist. Nat.; 44g Stockholm,; 31g Dorpat,; 35g Tempe, Arizona State Univ.;
Specimen(s): [66213], 47g.; [1920,291], 5.3g.

Bielostok v Białystok.

Bierbele v Bjurböle.

Big Skookum v Klondike (Skookum Gulch).

Bilibino 67°18′ N., 160°48′ E.
Bilibino district, Magadan region, Federated SSR, USSR
Found 1981, June 20
Iron. Octahedrite, medium.
A single mass of about 1000kg was found, Meteor. Bull., 1982 (60), Meteoritics, 1982, **17**, p.93. Description, analysis, 5.8 %Ni, G.F. Pavlov et al., Метеоритика, 1983, **42**, p.49.

Billings 37°4′ N., 93°33′ W.
Christian County, Missouri, U.S.A.
Found 1903
Synonym(s): *Christian County*
Iron. Octahedrite, medium (1.2mm) (IIIA).
A mass of about 54lb resembling an axe was found about 4 miles east of Billings; described, with an analysis, 7.38 %Ni, H.A. Ward and O.C. Farrington, Am. J. Sci., 1905, **19**, p.240. Classification and new analysis, 7.77 %Ni, 19.5 ppm.Ga, 37.4 ppm.Ge, 3.7 ppm.Ir, E.R.D. Scott et al., Geochimica et Cosmochimica Acta, 1973, **37**, p.1957. Description, V.F. Buchwald, Iron Meteorites, Univ. of California, 1975, p.319.
12.75kg Chicago, Field Mus. Nat. Hist.; 2.2kg New York, Amer. Mus. Nat. Hist.; 742g Vienna, Naturhist. Mus.; 531g Harvard Univ.; 510g Tempe, Arizona State Univ.; 434g Washington, U.S. Nat. Mus.; 305g Budapest, Nat.Hist. Mus.; 100g Berlin, Humboldt Univ.;
Specimen(s): [1950,52], 590g. slice and fragments, 22g.

Billygoat Donga 30°23′ S., 126°20′ E. approx.
Western Australia, Australia
Found 1962
Stone. Olivine-hypersthene chondrite (L6).
Three masses were found 5 miles N. of Billygoat Donga and about 21 miles NE. of Sleeper Camp; total weight unknown; two masses lost. Remaining stone very weathered. Another stone of 142g was found in 1963 two miles S. of Billygoat Donga but has been lost. Description, G.J.H. McCall and W.H. Cleverly, Min. Mag., 1968, **36**, p.691., G.J.H. McCall and W.H. Cleverly, J. Roy. Soc. West. Austr., 1970, **53**, p.69. Additional finds in 1970 and 1971, within the Mulga (north) strewnfield (*q.v.*), 392.4g and 358.2g, G.J.H. McCall, 2nd. Suppl. to West. Austr. Mus. Spec. Publ. no. 3, 1972, p.5. The three weathered specimens of total weight 129g found in December 1963 are assigned to Mulga (south) (*q.q.v.*). Olivine $Fa_{25.1}$, B. Mason, Rec. Austr. Mus., 1974, **29**, p.169. Preserved material comprises three masses.
0.6g Washington, U.S. Nat. Mus.; Two masses, Perth, West. Austr. Mus., found in 1970 and 1971; 142g Kalgoorlie, West. Austr. School of Mines, mass found in 1962;

Billygoat Donga III v Mulga (south).

Binda 34°19′ S., 149°23′ E.
Crookwell, County King, New South Wales, Australia
Found 1912
Stone. Achondrite, Ca-rich. Howardite (AHOW).
A mass of 12lb was found; it possibly fell on the night of May 25, 1912, when a luminous meteor was seen and detonations heard; description, with an analysis, C. Anderson, Rec. Austr. Mus., 1913, **10**, p.49. Analysis of metal, 2.0 to 3.8 %Ni, 0.3 to 0.9 %Co, J.F. Lovering, Nature, 1964, **203**, p.70. May have genetic significance in relation to eucrites and diogenites; analysis, T.S. McCarthy et al., Earth planet. Sci. Lett., 1973, **18**, p.433. Radiation and

gas retention ages, R. Ganapathy and E. Anders, Geochimica et Cosmochimica Acta, 1969, **33**, p.775. Rb/Sr age, 3450 m.y., J.L. Birck and C.J. Allegre, Nature, 1979, **282**, p.288 [M.A. 81-1749].
2.6kg Sydney, Austr. Mus., Main mass; 228g Washington, U.S. Nat. Mus.; 30g New York, Amer. Mus. Nat. Hist.; 143g Paris, Mus. d'Hist. Nat.;
Specimen(s): [1980,M.24], 6.46g.

Bingara v Bingera.

Bingera 29°53′ S., 150°34′ E.
County Murchison, New South Wales, Australia
Found 1880
Synonym(s): *Barraba, Bingara, Warialda*
Iron. Hexahedrite (IIA).
A pear-shaped mass of about 0.5lb was found; described and analysed, 4.39 %Ni, A. Liversidge, J. and Proc. Roy. Soc. New South Wales, 1882 (1883), **16**, p.35. On etching shows a granular structure, E. Cohen, Meteoritenkunde, 1905, **3**, p.233. In 1924 another mass of 6.4kg was found about 9 miles N. of Bingera, G.W. Card, letters of 21 October 1924 and 25 January 1926 in Min. Dept. BM(NH). Another analysis, 4.73 %Ni, H.P. White, Ann. Rep. Dep. Mines, New South Wales (for 1924), 1925, p.104. Further analysis, 5.58 %Ni, 59.7 ppm.Ga, 185 ppm.Ge, 3.2 ppm.Ir, J.T. Wasson, Geochimica et Cosmochimica Acta, 1969, **33**, p.859. Identical with Barraba and Warialda, V.F. Buchwald, Iron Meteorites, Univ. of California, 1975, p.320.
220g Sydney, Mining and Geol. Mus., main mass of the 0.5lb stone, approx. weight.; 85g Vienna, Naturhist. Mus., from the 0.5lb mass.; Main masses, Sydney, Austr. Mus., of the 6.4kg mass, and of the Warialda masses; 268g New York, Amer. Mus. Nat. Hist., from the 6.4kg mass.; 197g Washington, U.S. Nat. Mus.;
Specimen(s): [1925,62], 212g. end piece from the 6.4kg mass; [1916,3], 15g. a slice from the 0.5lb mass; [1916,2], 107g. slice, of Barraba; [1921,675], 36g. slice, of Warialda

Birgi 38°14′ N., 28°6′ E.
Smyrna, Turkey
Fell 1332, approx.
Doubtful. Iron?
A mass of about 1cwt., perhaps an iron, is said to have fallen, M.A.R. Khan, Nature, 1944, **154**, p.465.

Bir Hadi 19°25′ N., 51°2′ E.
Rub'al Khali, Saudi Arabia
Found 1958
Stone. Olivine-bronzite chondrite (H5).
A fragment of 525g was found, described, (incorrectly as a stony-iron), D.A. Holm, Am. J. Sci., 1962, **260**, p.303. Olivine Fa_{19}, B. Mason, Geochimica et Cosmochimica Acta, 1967, **31**, p.1100.
525g Washington, U.S. Nat. Mus., main mass;

Birni N'konni 13°46′ N., 5°18′ E.
Niger
Fell 1923, April, or May
Stone. Olivine-bronzite chondrite (H4).
A piece of 560g was found in 1932 in possession of the natives; two other pieces had been lost. Description, A. Lacroix, C. R. Acad. Sci. Paris, 1932, **194**, p.1533 [M.A. 5-299]. Listed, A.D. Nininger, Pop. Astron., Northfield, Minnesota, 1940, **48**, p.557, A.D. Nininger, Contr. Soc. Res. Meteorites, **2**, p.227 [M.A. 8-54]. Olivine Fa_{18}, B. Mason, Geochimica et Cosmochimica Acta, 1963, **27**, p.1011.
535g Paris, Mus. d'Hist. Nat.;

Birüssa v Veliko-Nikolaevsky Priisk.

Biryusa v Veliko-Nilkolaevsky Priisk.

Bischopville v Bishopville.

Bischtübe 51°57′ N., 62°12′ E.
Nikolaev, Turgai, Kazakhstan SSR, USSR, [Бисштюбе]
Found 1888
Synonym(s): *Bischtuebe, Bishtiube, Bistjube, Nikolaev, Nikolajev, Turgai, Turgaj*
Iron. Octahedrite, coarse (1.8mm) (IA).
Three masses, of about 32kg, 16kg, 0.25kg respectively, were ploughed up, E.D. Kislakovsky, Bull. Soc. Nat. Moscou, 1890, **4** (2), p.187. Description, A.N. Zavaritsky, Метеоритика, 1954 (11), p.64 [M.A. 13-48]. New analysis, 7.88 %Ni, 68.4 ppm.Ga, 238 ppm.Ge, 1.9 ppm.Ir, J.T. Wasson, Icarus, 1970, **12**, p.407. Description, V.F. Buchwald, Iron Meteorites, Univ. of California, 1975, p.322.
24kg Leningrad, Mus. Mining Inst.; 5.7kg Bonn, Univ. Mus.; 3.3kg Washington, U.S. Nat. Mus.; 2.8kg Vienna, Naturhist. Mus.; 266g Tübingen Univ.; 361g Yale Univ.; 174g Berlin, Humboldt Univ.; 13g Moscow, Acad. Sci.; 2.5kg Chicago, Field Mus. Nat. Hist.; 270g Prague, Nat. Mus.; 402g New York, Amer. Mus. Nat. Hist.;
Specimen(s): [84372], 1595g. a slice and fragments, 45g; [1920,290], 263g. a slice; [67039], 36g. a slice.

Bischtuebe v Bischtübe.

Bishop Canyon 38° N., 108°30′ W. approx.
San Miguel County, Colorado, U.S.A.
Found 1912
Iron. Octahedrite, fine (0.30mm) (IVA).
A mass of 19lb was found 4 miles west of Bishop Canyon; description, O.C. Farrington, Field Mus. Nat. Hist. Geol. Ser., 1914, **5** (Publ. 178), p.3. Analysis, 7.85 %Ni, O.C. Farrington, letter of 15 November 1926 in Min. Dept. BM (NH). New analysis, 7.5 %Ni, 2.20 ppm.Ga, 0.110 ppm.Ge, 2.6 ppm.Ir, R. Schaudy et al., Icarus, 1972, **17**, p.174. Description and bandwidth, V.F. Buchwald, Iron Meteorites, Univ. of California, 1975, p.324.
6kg Chicago, Field Mus. Nat. Hist.; 588g Hanau, Zeitschel Colln.; 426g Denver, Mus. Nat. Hist.; 226g Washington, U.S. Nat. Mus.; 131g Tempe, Arizona State Univ.;
Specimen(s): [1926,472], 529g. a slice

Bishopville 34°10′ N., 80°17′ W.
Lee County, South Carolina, U.S.A.
Fell 1843, March 25
Synonym(s): *Bischopville, Sumter County*
Stone. Achondrite, Ca-poor. Aubrite (AUB).
A stone of about 13lb fell after detonations, C.U. Shepard, Am. J. Sci., 1848, **6**, p.411. Analysis, G.P. Merrill, Mem. Nat. Acad. Sci. Washington, 1916, **14** (1), p.12. Description, G. Tschermak, Sitzungsber. Akad. Wiss. Wien, Math.-naturwiss. Kl., 1883, **88**, p.363, 367. Analysis of enstatite, A.M. Reid and A.J. Cohen, Geochimica et Cosmochimica Acta, 1967, **31**, p.661.
1.5kg Tempe, Arizona State Univ.; 569g Tübingen, Univ.; 630g Washington, U.S. Nat. Mus.; 217g Berlin, Humboldt Univ.; 200g Yale Univ.; 77g Vienna, Naturhist. Mus.; 33g Tempe, Arizona State Univ.; 50g Budapest, Nat. Mus.; 40g Paris, Mus. d'Hist. Nat.; 72g Chicago, Field Mus. Nat. Hist.; 56g Harvard Univ.;
Specimen(s): [20795], 460g. and fragments, 17g.

Bishtiube v Bischtübe.

Bishunpur 25°23′ N., 82°36′ E.
Mirzapur district, Uttar Pradesh, India
Fell 1895, April 26, 1500 hrs
Synonym(s): *Parjabatpur*
Stone. Olivine-hypersthene chondrite, amphoterite (LL3), black.
Four stones fell, after detonations, but only two were recovered, one of 942g at Bishunpur, and the other of 97g at Parjabatpur a mile distant, E.A. Wülfing, Die Meteoriten in Samml., Tübingen, 1897, p.31. See also a translation of police report dated 7 May 1895 in Min. Dept. BM(NH). The place of fall is now in Benares State, C.A. Silberrad, Min. Mag., 1932, **23**, p.301. Unequilibrated, R.T. Dodd et al., Geochimica et Cosmochimica Acta, 1967, **31**, p.921. Analysis, 19.97 % total iron, E. Jarosewich, Geochimica et Cosmochimica Acta, 1966, **30**, p.1261. LL-group classification, R.T. Dodd and E. Jarosewich, Meteoritics, 1979, **14**, p.380, abs.
376g Calcutta, Mus. Geol. Surv. India; 82g Vienna, Naturhist. Mus.; 42g Washington, U.S. Nat. Mus.; 14g Tempe, Arizona State Univ.; 4.5g New York, Amer. Mus. Nat. Hist.; 1g Chicago, Field Mus. Nat. Hist.;
Specimen(s): [80339], 359g. of Bishunpur, and fragments, 8g.; [80340], 31g. of Parjabatpur

Bison 38°18′24″ N., 99°42′36″ W.
Rush County, Kansas, U.S.A.
Found 1958
Stone. Olivine-hypersthene chondrite, amphoterite (LL6).
A mass of about 3kg was found about 6.8km NE. of Bison, G.I Huss, priv. comm., 1982. Heavily shocked breccia, olivine $Fa_{31.9}$, A.L. Graham, priv. comm., 1983.
Specimen(s): [1983,M.20], 148g.

Bissempore v Shalka.

Bistjube v Bischtübe.

Bitburg 49°58′ N., 6°32′ E.
Trier, Rheinland-Pfalz, Germany
Found 1805, known before this year
Synonym(s): *Albacher Mühle, Bitsburg, Eifel, Trier*
Iron. Iron with silicate inclusions (IB).
A mass of about 1.5 tons, most of which had been smelted in a furnace, was seen in 1805, Col. Gibbs, Am. Mineral. J., 1814, **1**, p.219, E.F.F. Chladni, Ann. Phys. (Gilbert), 1819, **60**, p.242. Only a little of the unaltered material has been preserved. Classification, figure and analysis, 12.4 %Ni, 34.8 ppm.Ga, 140 ppm.Ge, 0.46 ppm.Ir, E. Rambaldi et al., Min. Mag., 1974, **39**, p.595. Metallography, V.F. Buchwald, Iron Meteorites, Univ. of California, 1975, p.326.
3.5kg Bonn, Univ. Mus.; 2kg Tübingen, Univ.; 1.0kg Paris, Mus. d'Hist. Nat.; 956g Chicago, Field Mus. Nat. Hist.; 749g Budapest, Nat. Hist. Mus.; 378g Tempe, Arizona State Univ.; 168g Washington, U.S. Nat. Mus.; 85g Vienna, Naturhist. Mus.; 79g New York, Amer. Mus. Nat. Hist.; 69g Tempe, Arizona State Univ.; 64g Yale Univ.;
Specimen(s): [13406], 1318g. and fragments, 7g; [90218], 171g. oxidised; [33933], 51.5g.; [33198], 22.5g.; [33924a], 15g.; [33920], 12g.; [25463], 2.5g.

Bithur v Futtehpur.

Bitsburg v Bitburg.

Bjelaja Zerkov 49°47′ N., 30°10′ E.
Kiev, Ukraine, USSR, [Белая Церков]
Fell 1796, January 15
Synonym(s): *Belaia Tserkov, Belaja Zerkov, Belaja Zerkow, Bialacerkiew, Kiev*
Stone. Olivine-bronzite chondrite (H6).
A fairly big stone fell with the usual phenomena, A. Stoikovitz, Ann. Phys. (Gilbert), 1809, **31**, p.307. Olivine Fa_{20}, B. Mason, Geochimica et Cosmochimica Acta, 1963, **27**, p.1011.
1350g Kiev, Ukraine Acad. Sci., main mass; 118g Vienna, Naturhist. Mus.; 90g Budapest, Nat. Mus.; 28g Moscow, Acad. Sci.; 17g Berlin, Humboldt Univ.; 10g Washington, U.S. Nat. Mus.; 7g Chicago, Field Mus. Nat. Hist.;
Specimen(s): [54638], 7g. two pieces; [43195], 1g.

Bjelokrynitschie v Bielokrynitschie.

Bjorböle error for Bjurböle.

Bjurboele v Bjurböle.

Bjurböle 60°24′ N., 25°48′ E.
Borga, Nyland, Finland
Fell 1899, March 12, 2230 hrs
Synonym(s): *Bierbele, Bjorböle, Bjurboele*
Stone. Olivine-hypersthene chondrite (L4), friable.
One stone fell through the sea-ice and broke into fragments, the largest of which weighed 80kg, the total weight being about 330kg; described and analysed, W. Ramsay and L.H. Borgström, Bull. Comm. Geol. Finlande, 1902 (12), p.1. Olivine Fa_{26}, B. Mason, Geochimica et Cosmochimica Acta, 1963, **27**, p.1011. Further analysis, 20.71 % total iron, H. von Michaelis et al., Earth planet. Sci. Lett., 1969, **5**, p.387. Chondrule composition, L.S. Walter, Meteorite Research, ed. P.M. Millman, D. Reidel, Dordrecht-Holland, 1969, p.191. Further analysis, 18.95 % total iron, A.J. Easton and C.J. Elliott, Meteoritics, 1977, **12**, p.409.
215kg Helsinki, Univ.; Large piece, Stockholm, Riksmus.; 5.7kg Chicago, Field Mus. Nat. Hist.; 4kg Washington, U.S. Nat. Mus.; 2.1kg Paris, Mus. d'Hist. Nat.; 2kg Vienna, Naturhist. Mus.; 1939g Moscow, Acad. Sci.; 448g Tempe, Arizona State Univ.; 1.1kg Oxford, Univ. Mus.; 370g Tübingen, Univ.; 3.9kg Oslo, Min.-Geol. Mus.; 649g Berlin, Humboldt Univ.; 175g Yale Univ.; 613g Prague, Nat. Mus.; 500g Copenhagen, Univ.; 540g Harvard Univ.; 794g New York, Amer. Mus. Nat. Hist.;
Specimen(s): [1926,494], 89g. two fragments; [1927,11], 588g. and fragments 6g; [86138], 150g.; [1906,30], thin section

Blaauw-Kapel v Utrecht.

Black Moshannan Park 40°55′ N., 78°5′ W.
Center County, Pennsylvania, U.S.A.
Fell 1941, July 10, 0600-0630 hrs
Synonym(s): *Black Moshannon Park*
Stone. Olivine-hypersthene chondrite (L5).
A number of stones fell, one of 523.86g was recovered at the time of the fall, F.J. Keeley, Notulae Naturae, Acad. Nat. Sci. Philadelphia, 1942 (99) [M.A. 8-374]. Another ten specimens together weighing 181.5g have since been recovered, the original specimen now weighs 458.3g, R.W. Stone and E.M. Starr, Bull. Pa. Geol. Surv., 1967 (Rep. G2, 1932, rev.), p.30, 34. Olivine Fa_{24}, B. Mason, Geochimica et Cosmochimica Acta, 1963, **27**, p.1011.
610g Philadelphia, Acad. Nat. Sci.; 29g Washington, U.S. Nat. Mus.;

Black Moshannon Park v Black Moshannan Park.

Black Mountain 35°35′ N., 82°21′ W.
15 miles east of Asheville, Buncombe County, North Carolina, U.S.A.
Found 1839, about
Synonym(s): *Asheville, Buncombe County*
Iron. Octahedrite, coarse (2.6mm) (IA).
A piece of about 21oz was described, C.U. Shepard, Am. J. Sci., 1847, **4**, p.82. Analysis, 6.73 %Ni, 98 ppm.Ga, 460 ppm.Ge, 2.2 ppm.Ir, J.T. Wasson, Meteorites, Springer-Verlag, 1974, p.297. Description, V.F. Buchwald, Iron Meteorites, Univ. of California, 1975, p.327.
358g Tempe, Arizona State Univ.; 54g New York, Amer. Mus. Nat. Hist.; 18g Washington, U.S. Nat. Mus.; 76g London, Geol. Mus.; 44g Budapest, Nat. Mus.; 84g Vienna, Naturhist. Mus.; 33g Berlin, Humboldt Univ.; 22g Calcutta, Mus. Geol. Surv. India;
Specimen(s): [34578], 53g.

Blackwell 36°50′ N., 97°20′ W.
Kay County, Oklahoma, U.S.A.
Fell 1906, May, 2100 hrs
Stone. Olivine-hypersthene chondrite (L5).
A mass of 2381g fell, and was used as a doorstop for 28 years, E.P. Henderson, letter of 3 June 1939 in Min. Dept. BM(NH). Olivine Fa_{26}, B. Mason, Geochimica et Cosmochimica Acta, 1967, **31**, p.1100.
2362g Washington, U.S. Nat. Mus.;

Blaine Lake 52°46′12″ N., 106°53′48″ W.
Blaine Lake, Saskatchewan, Canada
Found 1974, summer
Stone. Olivine-hypersthene chondrite (L6).
A single mass of 1896.4g was found 7 km SSW. of Blaine Lake. Similar to Red Deer Hill (*q.v.*) which was found 80 km to the ENE. and may be paired with Blaine Lake, Meteor. Bull., 1978 (55), Meteoritics, 1978, **13**, p.330.
Main mass, Ottawa, Nat. Meteorite Collection;

Blake's Iron v Babb's Mill (Blake's Iron).

Blanket 31°50′ N., 98°50′ W.
Brown County, Texas, U.S.A.
Fell 1909, May 30, 2230 hrs
Stone. Olivine-hypersthene chondrite (L6).
Several stones fell, the total weight is not known, H.H. and A.D. Nininger, The Nininger Collection of Meteorites, Winslow, Arizona, 1950, p.34. Olivine Fa_{24}, B. Mason, Geochimica et Cosmochimica Acta, 1963, **27**, p.1011.
2.9kg Chicago, Field Mus. Nat. Hist., main masses of three of the stones; 1815g Washington, U.S. Nat. Mus.; 84g Tempe, Arizona State Univ.;
Specimen(s): [1959,848], 79g.

Blansko 49°22′ N., 16°38′ E.
Brno, Jihomoravsky, Czechoslovakia
Fell 1833, November 25, 1830 hrs
Stone. Olivine-bronzite chondrite (H6), veined.
A shower of stones fell, after appearance of fireball and detonations and eight, weighing altogether 350g, were found some days later, F. von Reichenbach, Neues Jahrb. Min., 1834, p.125, Ann. Phys. Chem. (Poggendorff), 1865, **124**, p.213. Analysis, J.J. Berzelius, Ann. Phys. Chem. (Poggendorff), 1834, **33**, p.8. Olivine Fa_{19}, B. Mason, Geochimica et Cosmochimica Acta, 1963, **27**, p.1011. Another stone of about 120g was found in 1866, Ann. Phys. Chem. (Poggendorff), **136**, p.446. Microprobe analysis of Ni-Cu alloy, E. Olsen, Meteoritics, 1973, **8**, p.259.
88g Tübingen, Univ.; 69g Vienna, Naturhist. Mus.; 26g Berlin, Humboldt Univ.; 19g Prague, Nat. Mus.; 9g Chicago, Field Mus. Nat. Hist.;
Specimen(s): [35170], 0.75g.

Blaauw-Kapel v Utrecht.

Bleckenstad 58° N., 16° E.
Ostergötland, Sweden
Found 1925, April 11
Pseudometeorite..
A meteor was observed, leaving a trail of smoke. Stones are said to have fallen, and fragments of a white, porous limestone were picked up, differing from the local rocks. The possibly meteoritic nature of this material has been the subject of considerable discussion, A. Hadding, Geol. För. Förh. Stockholm, 1943, **65**, p.15, 316, N. Zenzen, Geol. För. Förh. Stockholm, 1942, **64**, p.61, N. Zenzen, Geol. För. Förh. Stockholm, 1943, **65**, p.313 [M.A. 9-292], F.C. Cross, Pop. Astron., Northfield, Minnesota, 1947, **55**, p.96 [M.A. 10-171]. Pseudometeorite, F.E. Wickman and A. Uddenberg-Anderson, Geol. För. Förh. Stockholm, 1982, **104**, p.57.

Bledsoe 33°35′30″ N., 103°1′36″ W.
Cochran County, Texas, U.S.A.
Found 1970
Stone. Olivine-bronzite chondrite (H4).
One weathered specimen of 30.5kg was found at a depth of 18 to 20 inches while deep ploughing, G.I Huss, letter of 16 March 1972 in Min. Dept. BM(NH). Reported, Meteor. Bull., 1974 (52), Meteoritics, 1974, **9**, p.108.
20.5kg Mainz, Max-Planck-Inst.; 445g Copenhagen, Univ. Geol. Mus.; 99g New York, Amer. Mus. Nat. Hist.; 97g Vienna, Naturhist. Mus.;
Specimen(s): [1972,497], 108.9g. a crusted slice

Blendija v Soko-Banja.

Blithfield 45°30′ N., 77° W. approx.
Renfrew County, Ontario, Canada
Found 1910
Stone. Enstatite chondrite (E6).
A stone of 1830g was found, R.A.A. Johnston, letter of 8 July 1920 in Min. Dept. BM(NH). Described, with an analysis, R.A.A. Johnston and M.F. Connor, Trans. Roy. Soc. Canada, 1922, **16** (4), p.187 [M.A. 2-259]. Mineralogy, K. Keil, J. Geophys. Res., 1968, **73**, p.6945. Analysis, 19.66 % total iron, H. von Michaelis et al., Earth planet. Sci. Lett., 1969, **5**, p.387.
722g Ottawa, Mus. Geol. Surv. Canada; 466g Chicago, Field Mus. Nat. Hist.; 146g New York, Amer. Mus. Nat. Hist.; 129g Washington, U.S. Nat. Mus.;
Specimen(s): [1924,185], 1g. fragment, 102g missing.; [1978, M.10], 12g.

Bloody Basin v Cañon Diablo.

Bloomington 40°28′48″ N., 89°0′15″ W.
McClean County, Illinois, U.S.A.
Fell 1938, summer, between 2100 and 2200 hrs
Stone. Olivine-hypersthene chondrite, amphoterite (LL6).
Two interlocking fragments of total weight 67.8g were found the morning after they were thought to have fallen on the back porch of 301 Howard Street, Bloomington, Meteor. Bull., 1975 (53), Meteoritics, 1975, **10**, p.133. Description, E. Olsen and H. Nelson, Trans. Illinois Acad. Sci., 1975, **68**, p.403.
Main mass, in the possession of the finder, Rev.H. Cox; 27g Chicago, Field Mus. Nat. Hist.;

Blount County v Summit.

Blue Tier 41°11′ S., 148°2′ E.
County Dorset, Tasmania, Australia
Found 1890
Synonym(s): *Tasmania*
Iron. Octahedrite, medium.
A mass of about 3lb was found, C. Anderson, Rec. Austr. Mus., 1913, **10**, p.56. The Chicago specimen (29g) is probably mislabelled Black Mountain (*q.v.*), V.F. Buchwald, Iron Meteorites, Univ. of California, 1975, p.328.
Main mass, Launceston, Queen Victoria Mus.;

Bluewater 35°16′ N., 107°58′ W.
Cibola County, New Mexico, U.S.A.
Found 1946, recognized in this year
Synonym(s): *Adams Diggings*
Iron. Octahedrite, medium (1.0mm) (IIIA).
Listed, L. LaPaz, Cat. Coll. Inst. Meteoritics, Univ. of New Mexico, 1965. Classification and analysis, 8.12 %Ni, 19.8 ppm.Ga, 40.6 ppm.Ge, 2.6 ppm.Ir, A. Kracher et al., Geochimica et Cosmochimica Acta, 1980, **44**, p.773. Cibola Co. was created from part of Valencia Co. in 1981.
538g Albuquerque, Univ. of New Mexico;

Bluff 29°54′ N., 96°48′ W. approx.
Fayette County, Texas, U.S.A.
Found 1878, approx.
Synonym(s): *Fayette County, La Grange*
Stone. Olivine-hypersthene chondrite (L5), brecciated.
A stone of about 320lb was found 3 miles SW. of La Grange; described, J.E. Whitfield and G.P. Merrill, Am. J. Sci., 1888, **36**, p.113 with an analysis (doubtful since no alkalies were reported). A second stone of 17lb 1oz was found in 1896 and a third stone of 30lb was found before or during 1917. Analysis, 22.11 % total iron, B. Mason and H.B. Wiik, Am. Mus. Novit., 1967 (2280). Olivine Fa_{25}, B. Mason, Geochimica et Cosmochimica Acta, 1963, **27**, p.1011.
26.9kg Chicago, Field Mus. Nat. Hist.; 8.5kg Washington, U.S. Nat. Mus.; 17kg Vienna, Naturhist. Mus.; 8kg Harvard Univ.; 15kg New York, Amer. Mus. Nat. Hist.; 5.5kg Michigan Univ.; 683.5g Tübingen, Univ.; 417g Oxford, Univ. Mus.; 368g Paris, Mus. d'Hist. Nat.; 338g Copenhagen, Univ. Geol. Mus.; 2.8g Budapest, Nat. Hist. Mus.; 17lb Austin, Texas, Bureau of Economic Geology, the second stone; 30lb Fort Worth, Texas Observers Collection, the third stone;
Specimen(s): [64204], 12401g. and fragments, 6g; [1964,573], thin section

Boaz (iron) v Hope.

Boaz (stone) 33°39′6″ N., 103°42′30″ W.
Chaves County, New Mexico, U.S.A.
Found 1968
Stone. Olivine-bronzite chondrite (H5).
Listed, a mass of 1.8kg, Cat. Huss Coll. Meteorites, 1976, p.8.
1.7kg Mainz, Max-Planck-Inst.;

Bobrik v Kharkov.

Bocaiuva 17°10′ S., 43°50′ W. approx.
Minas Gerais, Brazil
Found
Iron. Ataxite (IRANOM).
An iron of this name is in Belo Horizonte, C.B. Gomes, letter of 15 October, 1981. Brief description, silicate mineralogy, analysis, 8.49 %Ni, 19.5 ppm.Ga, 178 ppm.Ge, 2.9 ppm.Ir, W.S. Curvello et al., Meteoritics, 1983, **18**, p.abs.
Main mass, Belo Horizonte, Univ. Minas Gerais;
Specimen, Rio de Janeiro, Mus. Nac.;

Bocas 23° N., 102° W.
San Luis Potosi, Mexico
Fell 1804, November 24
Synonym(s): *Hacienda de Bocas, Ramos, San Luis Potosi*
Stone. Olivine-hypersthene chondrite (L6).
Original weight and details of fall unknown; small fragments of the stone were said to be preserved in the School of Engineers in Mexico, A. Castillo, Cat. Meteorites Mexique, Paris, 1889, p.13, H.J. Burkart, Verh. Naturh. Ver. Preuss. Rheinl., 1865, **22**, p.71. Olivine Fa_{26}, B. Mason, Geochimica et Cosmochimica Acta, 1963, **27**, p.1011.
9g Paris, Mus. d'Hist. Nat.; 10g Mexico City, Inst. Geol.; 2g Washington, U.S. Nat. Mus.; 1g Chicago, Field Mus. Nat. Hist.; 1g Berlin, Humboldt Univ.;
Specimen(s): [92564], 33g. labelled Ramos (Mexico); [40768], less than a gram

Bochechki v Botschetschki.

Bodaibo 57°51′ N., 114°12′ E.
Irkutsk, Federated SSR, USSR, [Бодайбо]
Found 1907
Iron. Octahedrite, fine (0.30mm) (IVA).
A specimen in the Geol. Mus. Acad. Sci., Leningrad of the iron found at Bodaibo on the Vitim River was described and analysed, N.T. Belaiew, Trudy geol. miner. Muz., 1914, **8**, p.129. Brief description, A.N. Zavaritsky, Метеоритика, 1954 (11), p.64 [M.A. 13-48]. Classification and new analysis, 7.91 %Ni, 1.96 ppm.Ga, 0.111 ppm.Ge, 1.7 ppm.Ir, R. Schaudy et al., Icarus, 1972, **17**, p.174. Description, V.F. Buchwald, Iron Meteorites, Univ. of California, 1975, p.328.
11.6kg Moscow, Acad. Sci.; 261g Washington, U.S. Nat. Mus.; 152g Tempe, Arizona State Univ.;
Specimen(s): [1970,27], 18.2g. a fragment

Boelus 41°6′ N., 98°36′ W.
Howard County, Nebraska, U.S.A.
Found 1941
Stone. Olivine-hypersthene chondrite, amphoterite (LL6).
A stone weighing 730g was found, Rep. U.S. Nat. Mus., 1942, p.56. Olivine Fa_{30}, B. Mason and H.B. Wiik, Geochimica et Cosmochimica Acta, 1964, **28**, p.533.
675g Washington, U.S. Nat. Mus.;

Boenpalli v Alwal.

Boerne 29°48′ N., 98°48′ W. approx.
Kendall County, Texas, U.S.A.
Found 1932
Stone. Olivine-bronzite chondrite (H6).
One mass of 1.1kg was found, A.D. Nininger, Pop. Astron., Northfield, Minnesota, 1937, **45**, p.449 [M.A. 7-62]. Mentioned, F.C. Leonard, Univ. New Mexico Publ., Albuquerque, 1946 (meteoritics ser. no. 1), p.45. Listed, H.H. and A.D. Nininger, The Nininger Collection of Meteorites, Winslow, Arizona, 1950, p.34. Olivine Fa_{20}, B.

Mason, Geochimica et Cosmochimica Acta, 1963, **27**, p.1011.
537g Washington, U.S. Nat. Mus., main mass; 274g Fort Worth, Texas, Monnig Colln.; 45g Tempe, Arizona State Univ.;
Specimen(s): [1959,849], 37.5g. thin slice

Bogota v Santa Rosa.

Bogoslovka 52°30′ N., 68°48′ E. approx.
Molotov district, Akmolinsk region, Kazakh SSR, USSR, [Богословка]
Found 1948, August 5, probably fell in 1942
Stone. Olivine-bronzite chondrite (H5).
One stone of 2.18kg was found, E.L. Krinov, Метеоритика, 1949 (6), p.106. Analysis, 26.97 % total iron, M.I. D'yakonova and V.Ya. Kharitonova, Метеоритика, 1961, **21**, p.52. Olivine Fa_{19}, B. Mason, Geochimica et Cosmochimica Acta, 1963, **27**, p.1011.
2.1kg Moscow, Acad. Sci.;

Bogou 12°30′ N., 0°42′ E. approx.
Upper Volta
Fell 1962, August 14, 1000 hrs
Iron. Octahedrite, coarse (1.9mm) (IA).
A mass of 8.8kg was recovered from a hole 20 to 30cm in radius and 50cm deep, Meteor. Bull., 1962 (25), New Scientist, 1962 (December 27), p.713. The place of fall is reported to be "near the village of Bogou, approximately 130km NE. of Feda N'Gourma, Upper Volta, 12°52'N., 0°48'E". The cited latitude and longitude are about 60km to the NNE., and near the village of Bartibogou; 130km NE. of Feda N'Gourma would be well inside the state of Niger and on the Niger river, approximately 110km ENE. of Feda N'Gourma is Botou, 12°42'N., 1°59'E. There is evidently some confusion, and the exact place of fall remains in doubt. According to one eye-witness report in Orloff the time of fall was "noon", F.E. Wickman, letter of 14 May 1973 in Min. Dept. BM(NH). Analysis, 7.15 %Ni, 77.4 ppm.Ga, 301 ppm.Ge, 1.4 ppm.Ir, J.T. Wasson, Icarus, 1970, **12**, p.407. Co-ordinates and description, V.F. Buchwald, Iron Meteorites, Univ. of California, 1975, p.330.
3.1kg Washington, U.S. Nat. Mus.;

Boguslavka 44°33′ N., 131°38′ E.
220km N. of Vladivostok, Federated SSR, USSR, [Богуславка]
Fell 1916, October 18, 1147 hrs
Synonym(s): *Boguslawka*
Iron. Hexahedrite (IIA).
Two masses of 199kg and 57kg fell, H. Backlund, Bull. Acad. Sci. Petrograd, 1916, **10**, p.1817, Geol. För. Förh. Stockholm, 1917, **39**, p.105. Analysis, 5.48 %Ni, E.P. Henderson, Am. Miner., 1941, **26**, p.546 [M.A. 8-195]. From the shape of the two known masses, it is concluded that a third mass remains to be found, E.L. Krinov, Метеоритика, 1946, **3**, p.59 [M.A. 10-174]. Further analysis, 5.45 %Ni, 60.5 ppm.Ga, 180 ppm.Ge, 24 ppm.Ir, J.T. Wasson, Geochimica et Cosmochimica Acta, 1969, **33**, p.859. Description, V.F. Buchwald, Iron Meteorites, Univ. of California, 1975, p.333.
201kg Moscow, Acad. Sci.; 185g Washington, U.S. Nat. Mus.; 244g Copenhagen, Univ. Geol. Mus.;
Specimen(s): [1956,318], 418g.

Boguslawka v Boguslavka.

Bohumilice v Bohumilitz.

Bohumilitz 49°3′ N., 13°46′ E.
Vimperk (=Winterberg), Jihocesky, Czechoslovakia
Found 1829
Synonym(s): *Bohumilice, Prachin, Vyskovice*
Iron. Octahedrite, coarse (1.9mm) (IA).
A mass of about 52kg was found near Bohumilitz Castle, Verh. Ges. vaterl. Mus. Prag, 1830, **8**, p.15, 26, Edinburgh J. Sci., 1830, **3** (6, new ser.), p.310. A second mass of 962g was found near Bohumilitz in 1899, and a third mass of 5850g was found in 1925 at Vyskovice. References, V.F. Buchwald, Iron Meteorites, Univ. of California, 1975, p.334. This meteorite may have fallen on 1 January, 1770, J.V. Zelikzo, Prehled mineralu jiznich Czech, Vodnany, Bohemia, 1936 [M.A. 6-392] but is too heavily weathered for this to be the case, V.F. Buchwald (op. cit.). Classification and analysis, 7.37 %Ni, 75.3 ppm.Ga, 264 ppm.Ge, 1.8 ppm.Ir, J.T. Wasson, Icarus, 1970, **12**, p.407.
44kg Prague, Nat. Mus.; 4.7kg Vienna, Naturhist. Mus.; 1.5kg Paris, Mus. d'Hist. Nat.; 1.5kg Chicago, Field Mus. Nat. Hist.; 1.3kg Berlin, Humboldt Univ.; 634g Budapest, Nat. Hist. Mus.; 325g Washington, U.S. Nat. Mus.; 466g Tempe, Arizona State Univ.; 248g New York, Amer. Mus. Nat. Hist.;
Specimen(s): [1927,155], 244g. a slice; [35726], 97g.; [1920, 292], 44g. etched slice; [34613], 21g.

Bois de Fontaine v Charsonville.

Boise City v Keyes.

Boisfontaine v Charsonville.

Bokkeveld v Cold Bokkeveld.

Bolivia
Locality not known
Found, date unknown
Iron. Octahedrite, coarse (2.7mm) (IA).
A mass of 21.25kg in the Cranfield Collection of minerals was described, with an analysis, 5.63 %Ni, G.P. Merrill, Proc. U.S. Nat. Mus., 1927, **72** (4), p.1. Further analysis, 6.6 %Ni, 85.4 ppm.Ga, 393 ppm.Ge, 1.5 ppm.Ir, J.T. Wasson, Icarus, 1970, **12**, p.407. Distinct from Pooposa (*q.v.*), A. Kracher et al., Geochimica et Cosmochimica Acta, 1980, **44**, p.773.
20kg Washington, U.S. Nat. Mus., main mass; 106g Tempe, Arizona State Univ.; 44g Los Angeles, Univ. of California;
Specimen(s): [1959,926], 105g. a slice and 5g sawings.

Bolshaja Korta v Bolshaya Korta.

Bolshaya Korta 57°38′ N., 83°22′ E.
Narym, Novosibirsk, Federated SSR, USSR, [Большая Корта]
Found 1939, May 10
Synonym(s): *Bolshaja Korta*
Stone. Olivine-bronzite chondrite (H5).
The mass is estimated to have weighed 2 to 2.5kg but only 1593g were recovered, B.S. Mitropolsky, C. R. (Doklady) Acad. Sci. URSS, 1940, **28**, p.120 [M.A. 8-373]. Partial description, A.A. Onosovskaya, C. R. (Doklady) Acad. Sci. URSS, 1940, **28**, p.122 [M.A. 8-192], L.A. Kulik, Метеоритика, 1941, **2**, p.123 [M.A. 10-174], E.L. Krinov, Каталог Метеоритов АИад. Наук СССР, Москва, 1947 [M.A. 10-511]. Olivine Fa_{19}, B. Mason, Geochimica et Cosmochimica Acta, 1963, **27**, p.1011. Perhaps part of Ichkala, V.F. Buchwald, letter in

Min. Dept. BM(NH). Analysis, 27.32 % total iron, V.Ya. Kharitonova, Метеоритика, 1965, **26**, p.146.
1592g Moscow, Acad. Sci.;
Specimen(s): [1956,324], 24.5g.

Bolson de Mapimi v Coahuila.

Bomba v Vaca Muerta.

Bonanza Iron v Coahuila.

Bondoc 12°20′ N., 122°52′ E. approx.
Bondoc Peninsula, Luzon, Philippines
Found 1956
Synonym(s): *Bondoc Peninsula*
Stony-iron. Mesosiderite (MES).
A single mass of 888.6kg and a few smaller masses were found, G.I Huss, letter of 16 March 1972 in Min. Dept. BM (NH). A specimen was received by H.H. Nininger in 1962, Meteor. Bull., 1962 (25), H.H. Nininger, Science, 1963, **139**, p.345. Metallography, B.N. Powell, Geochimica et Cosmochimica Acta, 1969, **33**, p.789. Silicate mineralogy, B.N. Powell, Geochimica et Cosmochimica Acta, 1971, **35**, p.5. Unshocked, A.V. Jain and M.E. Lipschutz, Nature, 1973, **242** (Phys. Sci.), p.26. Analysis of metal, 7.3 %Ni, 15.6 ppm.Ga, 48 ppm.Ge, 3.8 ppm.Ir, J.T. Wasson et al., Geochimica et Cosmochimica Acta, 1974, **38**, p.135. Fission track age, E.A. Carver and E. Anders, Geochimica et Cosmochimica Acta, 1976, **40**, p.467.
558.8kg Tempe, Arizona State Univ.; 28kg Washington, U.S. Nat. Mus.;
Specimen(s): [1964,650], 32g.; [1966,523], 34.1g. weathered fragment.

Bondoc Peninsula v Bondoc.

Bond Springs see also Mount Sir Charles.

Bond Springs 23°30′ S., 133°50′ E.
Northern Territory, Australia
Found 1898, before this year
Stone. Olivine-bronzite chondrite (H6).
A small, complete stone weighing 6.18g was found near Bond Springs. Description, G. Baker and A.B. Edwards, Mem. Nat. Mus. Melbourne, 1941 (12), p.49 [M.A. 8-197].

Bonita Springs 26°16′ N., 81°45′ W.
Lee County, Florida, U.S.A.
Found 1938, in the summer, recognized in 1956
Stone. Olivine-bronzite chondrite (H5).
One stone of 41.8kg was found, Meteor. Bull., 1957 (5). Analysis, 27.06 % total iron, olivine Fa_{18}, E. Jarosewich, Geochimica et Cosmochimica Acta, 1966, **30**, p.1261.
34kg Washington, U.S. Nat. Mus.; 663g Adelaide, S. Austr. Mus.; 546g Chicago, Field Mus. Nat. Hist.; 373g New York, Amer. Mus. Nat. Hist.; 303g Tempe, Arizona State Univ.; 601g Harvard Univ.;
Specimen(s): [1966,45], 355g. slice, and fragments, 5g

Boogaldi 31°9′ S., 149°7′ E.
Coonabarabran, County Baradine, New South Wales, Australia
Found 1900
Synonym(s): *Bugaldi*
Iron. Octahedrite, fine (0.42mm) (IVA).
A pear-shaped mass of about 4.5lb was found about 2 miles from Boogaldi, R.T. Baker, J. and Proc. Roy. Soc. New South Wales, 1900, **34**, p.81. Described and analysed, 8.05 %Ni, A. Liversidge, J. and Proc. Roy. Soc. New South Wales, 1902, **36**, p.341. Further analysis, 8.79 %Ni, 2.26 ppm.Ga, 0.132 ppm.Ge, 0.56 ppm.Ir, R. Schaudy et al., Icarus, 1972, **17**, p.174. Description, V.F. Buchwald, Iron Meteorites, Univ. of California, 1975, p.337.
Main mass, Sydney, Australian Mus.; 79g Washington, U.S. Nat. Mus.; 28g Vienna, Naturhist. Mus.;
Specimen(s): [86924], 68g. a slice and fragments, 7g.; [1927, 1270], 137g.; [1927,1271], 16g. and filings, 17g.

Boolka 30°4′ S., 141°4′ E.
New South Wales, Australia
Found 1968, July
Stone. Chondrite.
One very weathered and fissured mass of about 6-8 lbs. was found approximately 100 miles N. of Broken Hill and within a mile of Moorabbie (*q.v.*), D.H. McColl, letters of 12 June 1969, 25 January 1972 and 6 June 1972 in Min. Dept. BM (NH).
Main mass, with finder, Mr.L. Russell.;

Borcut v Borkut.

Bordeaux v Barbotan.

Borgo San Donnino v Borgo San Donino.

Borgo San Donino 44°52′ N., 10°3′ E.
Parma, Italy
Fell 1808, April 19, 1200 hrs
Synonym(s): *Borgo San Donnino, Casignano, Cella di Costamezzana, Cusignano, Gabiano, Parma, Piacenza, Pieve-di-Casignano, Varano, Varano de' Marchesi, Vignabona, Villa di Cella*
Stone. Olivine-hypersthene chondrite, amphoterite (LL6), brecciated.
Several stones fell, after detonations, the largest weighing about 1kg, L.W. Gilbert, Ann. Phys. (Gilbert), 1808, **29**, p.209. Listed, B. Baldanza, Min. Mag., 1965, **35**, p.214. Olivine Fa_{29}, B. Mason, Geochimica et Cosmochimica Acta, 1963, **27**, p.1011. The correct spelling of the town is Borgo San Donnino; the town is now called Fidenza, G.R. Levi-Donati, priv.comm., 1969 in Min. Dept. BM(NH). Further references, F. Burragato, Periodico Miner., 1967, **36**, p.463, B. Baldanza et al., Meteoritics, 1970, **5**, p.137.
477g Parma, Univ.; 264g Vienna, Naturhist. Mus.; 405g Paris, Mus. d'Hist. Nat.; 36g Tempe, Arizona State Univ.; 14g Berlin, Humboldt Univ.; 10g Chicago, Field Mus. Nat. Hist.;
Specimen(s): [19975], 7.5g. two pieces.

Bori 21°57′ N., 78°2′ E.
Betul district, Madhya Pradesh, India
Fell 1894, May 9, 1600 hrs
Stone. Olivine-hypersthene chondrite (L6), veined.
A stone of about 19lb fell, J. Coggin Brown, Mem. Geol. Surv. India, 1916, **43** (2), p.173. Olivine Fa_{25}, B. Mason, Geochimica et Cosmochimica Acta, 1963, **27**, p.1011.
5kg Calcutta, Mus. Geol. Surv. India; 482g Chicago, Field Mus. Nat. Hist.; 695g Paris, Mus. d'Hist. Nat.; 400g Vienna, Naturhist. Mus.; 339g New York, Amer. Mus. Nat. Hist.; 136g Washington, U.S. Nat. Mus.; 55g Prague, Nat. Mus.;
Specimen(s): [77431], 1262.5g.

Boriskino 54°14′ N., 52°29′ E.
Orenburg, Federated SSR, USSR, [Старое Борискино]
Fell 1930, April 20, 1330 hrs
Synonym(s): *Staroe Boriskino, Staroye Boriskino*
Stone. Carbonaceous chondrite, type II (CM2).
Two stones were recovered, 1165.6g together, three others were destroyed, E.L. Krinov, C. R. (Doklady) Acad. Sci. URSS, 1931, p.262, Trudy miner. Inst., 1933 (2), p.69 [M.A. 6-393]. The total known mass is 1342g. Described, with an analysis by O.A. Alekseeva, L.G. Kvasha, Метеоритика, 1948, **4**, p.83 [M.A. 10-516]. Listed, E.L. Krinov, Каталог Метеоритов Акад. Наук СССР, Москва, 1947 [M.A. 10-511], A.D. Nininger, Pop. Astron., Northfield, Minnesota, 1940, **48**, p.559, A.D. Nininger, Contr. Soc. Res. Meteorites, **2**, p.227 [M.A. 8-54]. Further analysis, 22.15 % total iron, H.B. Wiik, Geochimica et Cosmochimica Acta, 1956, **9**, p.279. Mentioned, B. Mason, Space Sci. Rev., 1963, **1**, p.621 [M.A. 16-640].
961g Moscow, Acad. Sci., three pieces of one stone and one of the other.; 2g New York, Amer. Mus. Nat. Hist.;
Specimen(s): [1965,394], 0.6g. fragments

Borkut 48°9′ N., 24°17′ E.
Marmaros, Ukraine, USSR
Fell 1852, October 13, 1500 hrs
Synonym(s): *Borcut, Marmaros, Marmoros*
Stone. Olivine-hypersthene chondrite (L5).
A stone of about 7kg fell, after detonations; described, with an analysis, F. Leydolt, Sitzungsber. Akad. Wiss. Wien, Math.-naturwiss. Kl., 1856, **20**, p.398. Olivine Fa_{26}, B. Mason, Geochimica et Cosmochimica Acta, 1963, **27**, p.1011. Elliptical chondrules are orientated, W. Keidel, Beitr. Miner. Petrogr., 1965, **11**, p.487 [M.A. 17-686].
3kg Tübingen, Univ., minimum weight; 202g Budapest, Nat. Hist. Mus.; 190g Vienna, Naturhist. Mus.; 115g Harvard Univ.; 87g Chicago, Field Mus. Nat. Hist.; 40g Berlin, Humboldt Univ.;
Specimen(s): [35168], 40g.

Borodino 55°28′ N., 35°52′ E.
Moscow, Federated SSR, USSR, [Бородино]
Fell 1812, September 5, 0100 hrs
Synonym(s): *Borordino, Kolocha, Kolotscha, Stonitsa, Stonitza*
Stone. Olivine-bronzite chondrite (H5), brecciated.
The fall of a stone of about 500g is said to have been observed by a soldier on guard before the battle of Borodino, E.A. Wülfing, Die Meteoriten in Samml., Tübingen, 1897, p.40., Y.I. Simashko, letter of 20 June 1892 in Min. Dept. BM(NH). The time of fall, O.C. Farrington, Cat. Meteor. Field Mus. Nat. Hist. Chicago, 1916 is not substantiated from other well-known sources, A. Yavnel, letters of late 1971-1972 in Min. Dept. BM(NH). Olivine Fa_{20}, B. Mason, Geochimica et Cosmochimica Acta, 1963, **27**, p.1011.
Main mass, Leningrad, Mus. Mining Inst.; 124g Moscow, Acad. Sci.; 2g Vienna, Naturhist. Mus.; 1g Chicago, Field Mus. Nat. Hist.; 1g Berlin, Humboldt Univ.;
Specimen(s): [1911,140], 1.5g.

Borordino v Borodino.

Borovaya v Angara.

Borowaja v Angara.

Bosna v Ozren.

Bosnia v Ozren.

Botetourt County 37°30′ N., 79°45′ W. approx.
Virginia, U.S.A.
Found 1850
Iron. Ataxite, Ni-rich.
A large mass, 'not easily transported on horseback', was found, C.U. Shepard, Am. J. Sci., 1866, **42**, p.250. Fragments referred to Botetourt County have been described, with an analysis, 17 %Ni, E. Cohen, Meteoritenkunde, 1905, **3**, p.114. The large mass appears to have been lost and may not have been meteoritic, E.A. Wülfing, Die Meteoriten in Samml., Tübingen, 1897, p.297. Reheated, has a structure similar to Babb's Mill (Troost's Iron), V.F. Buchwald, Iron Meteorites, Univ. of California, 1975, p.338.
2.1g Tempe, Arizona State Univ.; Specimen, Washington, U.S. Nat. Mus.;

Botschetschki 51°20′ N., 33°53′ E.
Putivl, Ukraine, USSR, [Бочечки]
Fell 1823, at end of year
Synonym(s): *Bochechki, Koursk, Kursk, Pulivl, Putiwl*
Stone. Olivine-hypersthene chondrite (L4).
A stone of 614g was acquired by the Mus. of Acad. Sci. of Petrograd in 1824, G. von Blöde, Bull. Acad. Sci. St.-Petersbourg, 1848, **6**, p.5, A. Göbel, Bull. Acad. Sci. St.-Petersbourg, 1876, **11**, p.256. Listed, E.L. Krinov, Каталог Метеоритов АИад. Наук СССР, Москва, 1947 [M.A. 10-511], Астрон. Журнал, 1945, **22**, p.303 [M.A. 9-297]. Olivine Fa_{26}, B. Mason, Geochimica et Cosmochimica Acta, 1963, **27**, p.1011.
519g Moscow, Lomonossov Inst., Acad. Sci.; 3g Vienna, Naturhist. Mus.;

Bottcher Island v Mauritius.

Bou Hadid 28°18′45″ N., 0°14′ E.
Algeria
Found 1969
Synonym(s): *Moungar Bou Hadid*
Stone. Olivine-hypersthene chondrite (L).
One fragment of 374g, apparently of a larger mass, was found 8km NW. of Moungar Bou Hadid, 62km NE. of the city of Adrar, Touat Valley, Western Tademait Plateau. It was found on Quaternary gravel overlying the Cretaceous plateau and from its state of oxidation, appears to be part of a very old fall, Meteor. Bull., 1971 (50), Meteoritics, 1971, **6**, p.118.
373g Paris, Mus. d'Hist. Nat.;

Boulia v Glenormiston.

Bourbon-Vendée v Chantonnay.

Bourbon-Vendée v St. Christophe la Chartreuse.

Bourdeaux v Mornans.

Bouvante 44°55′20″ N., 5°16′4″ E.
Bouvante-le-Haut, Drome, France
Found 1978, July 30
Synonym(s): *Bouvante-le-Haut*
Stone. Achondrite, Ca-rich. Eucrite (AEUC).
A single crusted stone weighing 8.3kg was found 100m. SE. of Bouvante Lake, Meteor. Bull., 1980 (57), Meteoritics, 1980, **15**, p.96. Description, petrology, 15.1 % total iron, M. Christophe Michel-Levy et al., Meteoritics, 1980, **15**, p.272, abs.
8.3kg Paris, Mus. d'Hist. Nat.;

Bouvante-le-Haut v Bouvante.

Bovedy 54°34′ N., 6°20′ W.
County Londonderry, Northern Ireland
Fell 1969, April 25, 2122 hrs
Synonym(s): *Belfast, Kilrea, Sprucefield*
Stone. Olivine-hypersthene chondrite (L3).
A fireball was observed over the British Isles moving from SE. to NW. Three days later a specimen, in two pieces totalling 513g, was found at Sprucefield, County Antrim, after it had fallen through an asbestos roof. A second stone fell on a farm at Bovedy, County Londonderry, about 60km NW. of Sprucefield, and weighed 4.95kg, I.G. Meighan and P.S. Doughty, Nature, 1969, **223**, p.24. Reported, Meteor. Bull., 1969 (46), Meteoritics, 1970, **5**, p.103. The stone is unequilibrated, it contains glassy chondrules and anorthositic glass, olivine Fa_{24}, R. Hutchison and A.L. Graham, Nature, 1975, **255**, p.471. Description, analysis, 22.48 % total iron, A.L. Graham et al., Geochimica et Cosmochimica Acta, 1976, **40**, p.529.
Main mass, Armagh Observatory; Specimens, Belfast, Queen's Univ. and Ulster Mus. (jointly);
Specimen(s): [1971,1], 41.2g. a slice and fragments, 3.3g of Bovedy; [1971,102], thin section of the Sprucefield stone; [1972,233], 14.15g. from the Sprucefield stone; [1975,M.12], 63.47g. two fragments

Bowden 33°0′ S., 26°24′ E. approx.
Cape Province, South Africa
Found 1907
Synonym(s): *Waterfall*
Stone. Olivine-bronzite chondrite (H5).
A mass, of estimated original weight 603g, was found on Waterfall farm, Bowden, 25 miles NNW. of Grahamstown, and was probably a recent fall, E.D. Mountain, S. African J. Sci., 1934, **31**, p.225 [M.A. 6-103]. Description, analysis, E.D. Mountain, Rec. Albany Mus., Grahamstown, 1935, **4**, p.248 [M.A. 6-207]. Olivine Fa_{18}, B. Mason, Geochimica et Cosmochimica Acta, 1963, **27**, p.1011. The mass mentioned, F. Berwerth, Fortschr. Min. Krist. Petr., 1912, **2**, p.240 under the name of Waterfall is clearly identical with that later described under the name Bowden.
Main mass, Grahamstown, Albany Mus.; 12.4g Los Angeles, Univ. of California; 6.5g Washington, U.S. Nat. Mus.;

Bowesmont 48°41′ N., 97°10′ W.
Pembina County, North Dakota, U.S.A.
Found 1962
Stone. Olivine-hypersthene chondrite (L6).
A mass of 5lb was found, Cat. Meteor. Arizona State Univ., 1964, p.353, Meteor. Bull., 1964 (32). Olivine Fa_{25}, B. Mason, Geochimica et Cosmochimica Acta, 1967, **31**, p.1100. A mass of 1.3kg, found in 1972, may belong to this fall, G.I Huss, Cat. Huss Coll. Meteorites, 1976, p.6.
344g Mainz, Max-Planck-Inst.; 116g Washington, U.S. Nat. Mus.; 96g New York, Amer. Mus. Nat. Hist.; 121g Tempe, Arizona State Univ.;
Specimen(s): [1965,190], 115g. slice, and a fragment, 13.5g

Boxhole 22°37′ S., 135°12′ E.
Plenty River, Northern Territory, Australia
Found 1937, June
Synonym(s): *Hart Range*
Iron. Octahedrite, medium (1.0mm) (IIIA).
A meteorite crater was recognised in June 1937 on Boxhole station and a number of iron fragments and shale balls were found, C.T. Madigan, Trans. Roy. Soc. South Austr., 1937, **61**, p.187 [M.A. 7-72]. Later several larger masses were found, including one of 82kg. The material is very similar to the Henbury irons. This find includes Hart Range, J.R. de Laeter, J. Roy. Soc. West. Austr., 1973, **56**, p.123. Analysis, 7.64 %Ni, 18.1 ppm.Ga, 37.2 ppm.Ge, 8.2 ppm.Ir, E.R.D. Scott et al., Geochimica et Cosmochimica Acta, 1973, **37**, p.1957. Bandwidth, description, V.F. Buchwald, Iron Meteorites, Univ. of California, 1975, p.338.
178kg Adelaide, South Austr. Mus.; 9kg New York, Amer. Mus. Nat. Hist.; 7.3kg Washington, U.S. Nat. Mus.; 3.3kg Sydney, Austr. Mus.; 3.1kg Tempe, Arizona State Univ.; 1kg Tübingen, Univ.;
Specimen(s): [1937,1652], 82kg.; [1938,405], 1329g. a slice and fragments, 7.5g and filings, 44g.; [1938,403], 97.5g.; [1938,406], 79.5g.; [1938,407], 365g. shale ball; [1938,408], 817g. shale ball.

Bo Xian 33°50′ N., 115°50′ E.
Anhui, China
Fell 1977, October 20, 1430 hrs
Stone. Olivine-hypersthene chondrite, amphoterite (LL4).
Two masses, 7.5kg and 5.5kg, were recovered, D. Bian, Meteoritics, 1981, **16**, p.118. Partial analysis, 19.35 % total iron, W. Yi and D. Wang, Meteoritics, 1981, **16**, p.abs. Classification, olivine $Fa_{28.4}$, J.T. Wasson, letter of 8 November, 1982 in Min. Dept. BM(NH).
5.5kg Guiyang, Inst. Geochemistry;

Boyett v Cross Roads.

Brachina 31°18′ S., 138°23′ E.
near Brachina, South Australia, Australia
Found 1974, May 26
Stone. Achondrite, Ca-poor (ACANOM).
Two stones totalling 202.85g were found. Description and analysis, olivine Fa_{33}, 20.70 % total iron, J.E. Johnson et al., Rec. S. Austr. Mus., 1977, **17**, p.309. Oxygen isotopic data, not a chassignite, R. Clayton and T. Mayeda, Earth planet. Sci. Lett., 1983, **62**, p.1.
202g Adelaide, South Austr. Mus.;

Bradford Woods 40°30′ N., 80°5′ W.
Alleghany County, Pennsylvania, U.S.A.
Fell 1886, September, 1200 hrs, approx. time, recognized 1946
Stone. Olivine-hypersthene chondrite (L).
A mass of 762g fell on the Deutihl property, 2.5 miles SW. of Bradford Woods. Described and figured, R.W. Stone and E.M. Starr, Bull. Pa. Geol. Surv., 1967 (Rep. G2, 1932, rev.), p.8, 33.
Entire mass, in the possession of R. Hillman, Baden, Penn.;

Bradley County v Cleveland.

Brady 31°3′ N., 99°28′ W.
McCulloch County, Texas, U.S.A.
Found 1937
Stone. Olivine-hypersthene chondrite (L6).
One stone of 927g was found 9 miles SW. of Brady, A.D. Nininger, Pop. Astron., Northfield, Minnesota, 1939, **47**, p.211. Mentioned, E.P. Henderson, letter of 3 June, 1939 in Min. Dept. BM(NH). Listed, H.H. and A.D. Nininger, The Nininger Collection of Meteorites, Winslow, Arizona, 1950, p.35. Olivine Fa_{25}, B. Mason, Geochimica et Cosmochimica Acta, 1963, **27**, p.1011.
573g Tempe, Arizona State Univ., main mass; 67g Chicago, Field Mus. Nat. Hist.; 60g Washington, U.S. Nat. Mus.;

Bragin v Brahin.

Brahin 52°30′ N., 30°20′ E.

Minsk, Belorussiya SSR, USSR, [Брагин]
Found 1810
Synonym(s): *Bragin, Komarinsky, Kruki, Krukov, Minsk, Rocicky, Rokicky, Rokitskii*
Stony-iron. Pallasite (PAL).

Two masses of about 80kg and 20kg respectively were found in 1810, Ann. Phys. (Gilbert), 1819, **63**, p.32 and a third of 183kg later, P.I. Grishchinsky, Ann. Geol. Min. Russ., 1911, **13**, p.72. Analysis of metal, 8.38 %Ni, L.L. Ivanov, Ann. Geol. Min. Russ., 1911, **13**, p.111. Olivine $Fa_{11.5}$, P.R. Buseck and J.I. Goldstein, Bull. Geol. Soc. Amer., 1969, **80**, p.2141. Areal analysis gave olivine 37.18% by weight, P.N. Chirvinsky, Izvest. Don Polytech. Inst. Novocherkassk, 1929, **11**, p.194 [M.A. 4-260]. There has been considerable confusion over the number and weights of the known masses of this fall. Eight masses reported, totalling 633kg, E.L. Krinov, Каталог Метеоритов Акад. Наук СССР, Москва, 1947 [M.A. 10-511]. A number of masses are reported, since discredited by Krinov (loc. cit.), P.I. Grishchinsky, Mem. Soc. Nat. Kiev, 1928, **27** (3), p.8 [M.A. 4-420]. Eleven masses listed, total weight 823kg, including masses of 73kg found in 1952, 12.7kg found in 1968 and 21.8kg found in 1979, G.I. Emel'yanov, Метеоритика, 1982, **41**, p.117.

100kg Moscow, Acad. Sci., the two masses found in 1810; 66.38kg Moscow, Acad. Sci., from Kutzovka; 80kg Kiev, Univ., two masses found near Kolyban; 183kg Kiev, Univ., found at Kruki in 1890-92; 16kg Minsk, Acad. Sci., found in the Komarinsk district before 1937; 12.39kg Minsk, Inst. of Geol., found in 1968; 3.25kg Vienna, Naturhist. Mus.; 308g Berlin, Humboldt Univ.; 225g Prague, Nat. Mus.; 169g Tempe, Arizona State Univ.; 76g Chicago, Field Mus. Nat. Hist.; 75g New York, Amer. Mus. Nat. Hist.;

Specimen(s): [33965], 22g.

Brainard 41°9′12″ N., 96°57′18″ W.

Butler County, Nebraska, U.S.A.
Found 1978, recognized in this year
Iron. Octahedrite, medium (1.0mm) (IIIB).

A single mass of 138.3kg was found 5.6 km SE. of Brainard. It had been ploughed up in a field some years before 1978, Meteor. Bull., 1982 (60), Meteoritics, 1982, **17**, p.93. Analysis, 9.1 %Ni, 22.5 ppm.Ga, 45.2 ppm.Ge, 0.058 ppm.Ir, D.J. Malvin et al., priv. comm., 1983.

Main mass, Denver, Amer. Meteorite Lab.;

Brambanan v Prambanan.

Bramudor v Tomatlan.

Brandenburg v Linum.

Brandenburg v Seelasgen.

Branibor v Seelasgen.

Brasil v Minas Gerais.

Braunau 50°36′ N., 16°18′ E. approx.

Trutnov, Vychodocesky, Czechoslovakia
Fell 1847, July 14, 0345 hrs
Synonym(s): *Broumov, Hauptmannsdorf*
Iron. Hexahedrite (IIA).

Two masses, of about 22kg and 17kg fell at Hauptmannsdorf after detonations, and appearance of a luminous cloud, C.C. Beinert, Ann. Phys. Chem. (Poggendorff), 1847, **72**, p.70. Structural description, N.A. Neumann, Haidinger's Naturwiss. Abhand. Wien, 1850, **3** (2), p.45. Analysis, 5.21 %Ni, E. Cohen, Meteoritenkunde, 1905, **3**, p.207. Historical note, K. Tucek, Casopis Narod. Mus. Praha, 1947, **116**, p.1 [M.A. 10-398]. The smaller mass was intact in 1897 and in the abbey of Braunau, the larger has been distributed. Description, shock re-heating, V.F. Buchwald, Iron Meteorites, Univ. of California, 1975, p.340, H.J. Axon, Min. Mag., 1968, **36**, p.1139. Further analysis, 5.49 %Ni, 61.5 ppm.Ga, 183 ppm.Ge, 12 ppm.Ir, J.T. Wasson, Geochimica et Cosmochimica Acta, 1969, **33**, p.859.

18kg Prague, Nat. Mus., includes entire smaller mass; 2.5kg Vienna, Naturhist. Mus.; 1.5kg Berlin, Humboldt Univ.; 916g Tübingen, Univ.; 569g Harvard Univ.; 453g Chicago, Field Mus. Nat. Hist.; 241g Washington, U.S. Nat. Mus.; 222g New York, Amer. Mus. Nat. Hist.;

Specimen(s): [33954], 519g.; [90217], 32g.

Brazos v Red River.

Brazos v Wichita County.

Brazos River v Wichita County.

Brazzaville v Massenya.

Breece v Grant.

Breitenbach v Steinbach.

Breitscheid 50°40′1″ N., 8°11′1″ E.

Dillkreis, Hessen, Germany
Fell 1956, August 11, 1530 hrs
Stone. Olivine-bronzite chondrite (H5), veined, xenolithic.

At least 14 fragments were found 30 min. after the fall, and the original weight is estimated at 1kg. Description, F. Paneth, Angew. Chem., 1957, **69**, p.714. Analysis, F. Paneth et al., Geochimica et Cosmochimica Acta, 1959, **17**, p.315. Olivine Fa_{19}, B. Mason, Geochimica et Cosmochimica Acta, 1963, **27**, p.1011. Xenolithic, R.A. Binns and F. Wlotzka, Meteoritics, 1977, **12**, p.177, abs.

Main mass, Mainz, Max-Planck-Inst.; 24g New York, Amer. Mus. Nat. Hist.; 19g Tübingen, Univ.; Thin section, Washington, U.S. Nat. Mus.;

Specimen(s): [1965,29], 4.6g.

Bremervörde 53°24′ N., 9°6′ E. approx.

Niedersachsen, Germany
Fell 1855, May 13, 1700 hrs
Synonym(s): *Gnarrenburg, Stade*
Stone. Olivine-bronzite chondrite (H3), brecciated.

Several stones, five at least, fell after detonations, near the village of Gnarrenburg; the total weight was about 7.25kg and the largest stone weighed about 2.75kg, Ann. Phys. Chem. (Poggendorff), 1856, **98**, p.609. Description, with analysis, 26.52 % total iron, B. Mason and H.B. Wiik, Am. Mus. Novit., 1967 (2280). Unequilibrated, R.T. Dodd et al., Geochimica et Cosmochimica Acta, 1967, **31**, p.921. Contains equilibrated fragments, analysis, 24.76 % total iron,

olivine $Fa_{17.4}$, R. Hutchison et al., Proc. Roy. Soc. London, 1981, **A374**, p.159.
2.8kg Göttingen, Univ.; 1kg Clausthal, Bergakad.; 521g Calcutta, Mus. Geol. Surv. India; 339g Vienna, Naturhist. Mus.; 254g New York, Amer. Mus. Nat. Hist.; 283g Berlin, Humboldt Univ.; 86g Chicago, Field Mus. Nat. Hist.; 91g Tempe, Arizona State Univ.;
Specimen(s): [33910], 711g. and 46.5g, also fragments, 6g.; [33739], 14g.; [1920,305], 2g.

Brenham 37°36′ N., 99°12′ W. approx.
Kiowa County, Kansas, U.S.A.
Found 1882
Synonym(s): *Anderson, Hamilton County, Haviland, Haviland Township, Hopewell Mounds, Kiowa, Kiowa County, Little Miami Valley, Turner Mound*
Stony-iron. Pallasite (PAL).
About twenty masses were found, weighing together about 2000lb individual masses weighing from 1oz to 466lb. Described, with an analysis of metal, 10.35 %Ni, G.F. Kunz, Am. J. Sci., 1890, **40**, p.312. Many other small masses, weighing mostly less than 450g were found in 1892, R. Hay, Am. J. Sci., 1892, **43**, p.80. Partially to completely oxidised fragments and limonitic masses ("meteorodes") were found in a swampy depression which is probably a meteorite crater, Haviland Crater, associated with this fall, H.H. Nininger, Trans. Kansas Acad. Sci., 1929, **32**, p.63, H.H. Nininger, Am. Miner., 1938, **23**, p.536 [M.A. 7-177]. A mass of 740lb was found in 1947, 0.75 miles WNW. of the crater, also others of 474lb., 381lb., 357lb., 220lb., and 110lb at depths of between one and two feet, and more recently, at a depth of 5ft 3in, a mass of 1000lb. This has been described, incorrectly, as the world's largest pallasite, (compare, for example Huckitta, 2 tons). Metallography, H.J. Axon and E.D. Yardley, Min. Mag., 1969, **37**, p.275. Prehistoric transport; analysis of metal, 11.1 %Ni, 26.2 ppm.Ga, 70.8 ppm.Ge, 0.041 ppm.Ir, J.T. Wasson and S.P. Sedwick, Nature, 1969, **222**, p.22. Olivine $Fa_{12.5}$, P.R. Buseck and J.I. Goldstein, Bull. Geol. Soc. Amer., 1969, **80**, p.2141.
281kg Fort Worth, Texas, Monnig Colln.; 200kg New York, Amer. Mus. Nat. Hist., approximate mass; 170kg Michigan Univ.; 302kg Chicago, Field Mus. Nat. Hist.; 109.8kg Tempe, Arizona State Univ.; 89kg Washington, U.S. Nat. Mus.; 61kg Harvard Univ.; 55kg Yale Univ.; 27.5kg Los Angeles, Univ. of California; 7kg Vienna, Naturhist. Mus.; 3kg Budapest, Nat. Mus.;
Specimen(s): [66202], 1502g.; [68725], 506g.; [1930,437], seven limonitic nodules (" meteorodes"); [1959,1054], 12270g. slice; [1960,492], 15.5g.; [1959,171], 6.5g.; [1959, 1055], 100kg. meteorodes and meteorode fragments; [1962, 141], nickeliferous nodules from soil near crater.; [1969,172], 26g. polished slice.

Brescia v Alfianello.

Brescia v Trenzano.

Brewster 39°15′ N., 101°20′ W.
Thomas County, Kansas, U.S.A.
Found 1940, recognized in this year
Stone. Olivine-hypersthene chondrite (L6).
One mass of 17kg was found, W. Wahl, letter of 23 May 1950 in Min. Dept. BM(NH)., H.H. and A.D. Nininger, The Nininger Collection of Meteorites, Winslow, Arizona, 1950, p.36. Olivine Fa_{24}, B. Mason, Geochimica et Cosmochimica Acta, 1963, **27**, p.1011.
6.8kg Tempe, Arizona State Univ.; 986g Washington, U.S. Nat. Mus.;
Specimen(s): [1959,778], 5871g. and slices of 456g, 429g, 404g, and 388g, and fragments, 17g.

Brianza 45°45′ N., 9°15′ E. approx.
Milano, Italy
Fell 1760
Pseudometeorite. Iron.
A small fragment of 14g in the Univ. of Parma is labelled as belonging to this fall, B. Baldanza, Min. Mag., 1965, **35**, p.214, L.W. Gilbert, Ann. Phys. (Gilbert), 1822, **72**, p.329. Manufactured iron, G.R. Levi-Donati, priv. comm., 1982.

Bridgewater 35°43′ N., 81°52′ W.
Burke County, North Carolina, U.S.A.
Found 1890
Synonym(s): *Bridgewater Station, Burke County, Fairweather*
Iron. Octahedrite, medium (0.65mm) (IID).
A mass of 30lb was found by a ploughman 2 miles from Bridgewater Station; description, with an analysis, 9.94 %Ni, G.F. Kunz, Am. J. Sci., 1890, **40**, p.320. Classification and analysis, 9.8 %Ni, 81.0 ppm.Ga, 82.0 ppm.Ge, 10 ppm.Ir, J.T. Wasson, Meteorites, Springer-Verlag, 1974, p.301. Description, V.F. Buchwald, Iron Meteorites, Univ. of California, 1975, p.342.
8.25kg Vienna, Naturhist. Mus.; 0.5kg Budapest, Nat. Mus.; 172g Prague, Nat. Mus.; 115g New York, Amer. Mus. Nat. Hist.; 102g Chicago, Field Mus. Nat. Hist.;
Specimen(s): [77094], 51g. slice

Bridgewater Station v Bridgewater.

Brient 52°8′ N., 59°19′ E.
Chkalov (=Orenburg), Federated SSR, USSR, [Бриент]
Fell 1933, April 19, 2030 hrs
Stone. Achondrite, Ca-rich. Eucrite (AEUC).
A mass of 5 or 6kg fell at Brient, in the valley of the Suvunduk, a tributary of the Ural river, but only 219g were preserved, E.L. Krinov, Астрон. Журнал, 1945, **22**, p.303 [M.A. 9-297], Каталог Метеоритов Акад. Наук СССР, Москва, 1947 [M.A. 10-511], L.A. Kulik, Метеоритика, 1941, **1**, p.73 [M.A. 9-294]. Analysis, M.I. D'yakonova and V.Ya. Kharitonova, Метеоритика, 1961, **21**, p.52. Listed, I.S. Astapowitsch, Trans. Roy. Astron. Soc. Canada, 1938, **32**, p.195 [M.A. 7-172]. Incorrect co-ordinates, P.M. Millman, Trans. Roy. Astron. Soc. Canada, 1938, **32**, p.198.
194g Moscow, Acad. Sci.;

Briggsdale 40°40′ N., 104°19′ W.
Weld County, Colorado, U.S.A.
Found 1949, recognized in this year
Iron. Octahedrite, medium (1.3mm) (IIIA).
A mass of 2230g was found, H.H. Nininger, letter of 19 October 1949 in Min. Dept. BM(NH)., H.H. and A.D. Nininger, The Nininger Collection of Meteorites, Winslow, Arizona, 1950, p.36. Analysis, 8.17 %Ni, 20.1 ppm.Ga, 40.7 ppm.Ge, 0.72 ppm.Ir, E.R.D. Scott et al., Geochimica et Cosmochimica Acta, 1973, **37**, p.1957. Shocked and annealed, V.F. Buchwald, Iron Meteorites, Univ. of California, 1975, p.343.
1109g Tempe, Arizona State Univ.; 153g Denver, Mus. Nat. Hist.; 60g Washington, U.S. Nat. Mus.;
Specimen(s): [1959,945], 902g. and sawings, 34g.

Briscoe 34°21′ N., 101°24′ W.
Briscoe County, Texas, U.S.A.
Found 1940
Stone. Olivine-hypersthene chondrite (L5).
A stone of 1773g was found 16 miles E. of Kress, Briscoe Co., H.H. and A.D. Nininger, The Nininger Collection of Meteorites, Winslow, Arizona, 1950, p.36. Olivine Fa_{26}, B. Mason, Geochimica et Cosmochimica Acta, 1963, **27**, p.1011.
579g Tempe, Arizona State Univ.; 260g Washington, U.S. Nat. Mus.; 54g New York, Amer. Mus. Nat. Hist.;
Specimen(s): [1959,812], 758g.

Briscoe County v Briscoe.

Bristol 36°34′ N., 82°11′ W.
Sullivan County, Tennessee, U.S.A.
Found 1937
Iron. Octahedrite, fine (0.3mm) (IVA).
One mass of 20kg was found 12 miles SE. of Bristol, A.D. Nininger, Pop. Astron., Northfield, Minnesota, 1939, **47**, p.211., E.P. Henderson, letter of 3 June 1939 in Min. Dept. BM(NH). Analysis, and determination of Au, Ga and Pd, 8.15 %Ni, E. Goldberg et al., Geochimica et Cosmochimica Acta, 1951, **2**, p.1. Further analysis, 7.9 %Ni, 2.13 ppm.Ga, 0.124 ppm.Ge, 1.6 ppm.Ir, R. Schaudy et al., Icarus, 1972, **17**, p.174. Description, " possible connection with Harriman (Of)", V.F. Buchwald, Iron Meteorites, Univ. of California, 1975, p.344.
2.1kg Tempe, Arizona State Univ.; 1.1kg Harvard Univ.; 443g Chicago, Field Mus. Nat. Hist.; 193g New York, Amer. Mus. Nat. Hist.; 139g Washington, U.S. Nat. Mus.; 217g Michigan Univ.; 69g Los Angeles, Univ. of California;
Specimen(s): [1959,977], 3973g. a corner piece, and also a slice, 465g and fragments, 5g; [1955,226], 13.2g. fragments.

Britstown 30°35′ S., 23°33′ E.
Cape Province, South Africa
Found 1910, before this year
Iron. Octahedrite, plessitic (IRANOM).
A mass of about 544g was brought from South Africa to Germany by an engineer before 1910; description, C. Palache, Am. J. Sci., 1926, **12**, p.139. Description, structural classification, V.F. Buchwald, Iron Meteorites, Univ. of California, 1975, p.346. Classification and analysis, 19.5 % Ni, 39.6 ppm.Ga, 183 ppm.Ge, 2.0 ppm.Ir, A. Kracher et al., Geochimica et Cosmochimica Acta, 1980, **44**, p.773.
19g Washington, U.S. Nat. Mus.;
Specimen(s): [1927,87], 17.5g. slice

Broken Bow 41°26′ N., 99°42′ W.
Custer County, Nebraska, U.S.A.
Found 1937
Stone. Olivine-bronzite chondrite (H4).
One stone of 6.8kg was found 12 miles S. and 3 miles W. of Broken Bow, E.P. Henderson, letter of 12 May 1939 in Min. Dept. BM(NH)., A.D. Nininger, Pop. Astron., Northfield, Minnesota, 1939, **47**, p.211. Olivine Fa_{20}, B. Mason, Geochimica et Cosmochimica Acta, 1963, **27**, p.1011.
2.7kg Tempe, Arizona State Univ.; 283g Washington, U.S. Nat. Mus.; 287g Chicago, Field Mus. Nat. Hist.;
Specimen(s): [1959,993], 2530g.

Bronco 33°17′ N., 102°58′ W.
Yoakum County, Texas, U.S.A.
Found 1971
Stone. Chondrite.
Five small stones were found, totalling 972g, the largest 590g, Meteor. Bull., 1980 (58), Meteoritics, 1980, **15**, p.235.
590g Los Angeles, Univ. of California; 344g Mainz, Max-Planck-Inst.;

Brookville v Rushville.

Broumov v Braunau.

Brownell 38°38′ N., 99°45′ W. approx.
Ness County, Kansas, U.S.A.
Found 1972
Stone. Chondrite.
Briefly mentioned as found near Brownell, co-ordinates as above. Differs from Ness County, Franklinville and Wellmanville specimens, W.F. Read, Meteoritics, 1972, **7**, p.417.

Brownfield (b) v Brownfield (1964).

Brownfield (iron) 33°13′ N., 102°11′ W.
Terry County, Texas, U.S.A.
Found 1960, approx., recognized 1966
Synonym(s): *Brownfield (1966)*
Iron. Octahedrite, medium (0.8mm) (IID).
A mass of 1626g was ploughed up, Meteor. Bull., 1967 (40), Meteoritics, 1970, **5**, p.94., G.I. Huss, letter of 18 March 1970 in Min. Dept. BM(NH). Analysis, 10.32 %Ni, 77.8 ppm.Ga, 85.2 ppm.Ge, 10 ppm.Ir, J.T. Wasson, Meteorites, Springer-Verlag, 1974, p.301.
424g Mainz, Max-Planck-Inst., main mass; 106g Los Angeles, Univ. of California; 63g Tempe, Arizona State Univ.;
Specimen(s): [1967,385], 28.5g. polished and etched slice.

Brownfield no. 2 v Brownfield (1964).

Brownfield (1937) 33°13′ N., 102°11′ W.
Terry County, Texas, U.S.A.
Found 1937
Stone. Olivine-bronzite chondrite (H3).
Two stones, of 1oz and 15oz, were found in 1937, A.D. Nininger, Pop. Astron., Northfield, Minnesota, 1937, **45**, p.449 [M.A. 7-62]. Five additional stones have been found, of 31lb 8oz, 17lb, 22lb 6oz, 12lb 5oz, and 6lb. The total weight now recovered is 90.1lb (40.96kg.), G.I. Huss, letters of 9 April and 5 May 1971 in Min. Dept. BM(NH). Unequilibrated, mean composition, olivine $Fa_{18.5}$, R. Hutchison, pers. comm. 26.03 % total iron, A.J. Easton, unpublished analysis.
25kg Mainz, Max-Planck-Inst.; 1.1kg Washington, U.S. Nat. Mus.; 277g Tübingen, Univ.; 223g Harvard Univ.; 190g Los Angeles, Univ. of California; 172g Tempe, Arizona State Univ.;
Specimen(s): [1959,850], 200g.; [1971,116], 21g. fragments; [1971,121], 493g. part slice

Brownfield (1964) 33°13′ N., 102°11′ W.

Terry County, Texas, U.S.A.
Found 1964
Synonym(s): *Brownfield no. 2, Brownfield (b)*
Stone. Olivine-bronzite chondrite (H5).

An incomplete stone of 4.1kg was found, G.I. Huss, letter of 18 March 1970 in Min. Dept. BM(NH). The stone is an equilibrated chondrite, olivine Fa_{17}, R. Hutchison, pers. comm. A Brownfield (c) stone of 1550g is also listed, Cat. Huss Coll. Meteorites, 1976, p.7.

2.4kg Mainz, Max-Planck-Inst.; 226g Copenhagen, Univ. Geol. Mus.; 168g Washington, U.S. Nat. Mus.; 101g Tempe, Arizona State Univ.;

Specimen(s): [1965,410], 72g.

Brownfield (1966) v Brownfield (iron).

Bruce v Cranbourne.

Bruderheim 53°54′ N., 112°53′ W.

Alberta, Canada
Fell 1960, March 4, 0806 hrs, U.T.
Stone. Olivine-hypersthene chondrite (L6).

A bolide was observed by many witnesses and about 303kg of fragments of the shower were recovered over an area 3km across, Meteor. Bull., 1960 (18). Description, analysis, H. Baadsgaard et al., J. Geophys. Res., 1961, **66**, p.3574. Analysis quoted, H. König, Geochimica et Cosmochimica Acta, 1964, **28**, p.1697. Olivine Fa_{24}, B. Mason, Geochimica et Cosmochimica Acta, 1963, **27**, p.1011. Trace element distribution, R.O. Allen, Jr. and B. Mason, Geochimica et Cosmochimica Acta, 1973, **37**, p.1435. Al26 content, P.J. Cressy, Jr., Meteoritics, 1970, **5**, p.190, abs.. U-Pb measurements, N.H. Gale et al., Earth planet. Sci. Lett., 1980, **48**, p.311 [M.A. 81-0692].

129kg Alberta Univ., main mass; 31kg Ottawa, Mus. Geol. Surv. Canada, largest individual mass; 18.8kg Mainz, Max-Planck-Inst.; 12.5kg Toronto, Roy. Ontario Mus.; 4kg Tempe, Arizona State Univ., approx.; 1199g Chicago, Field Mus. Nat. Hist., complete individual; 5.5kg Washington, U.S. Nat. Mus.; 1.08kg Los Angeles, Univ. of California; 838g Alburqueque, Univ. of New Mexico; 391g New York, Amer. Mus. Nat. Hist.; 305g Copenhagen, Univ. Geol. Mus.; 272g Yale Univ.;

Specimen(s): [1967,256], 466g. an individual and fragments, 30g; [1968,276], 66g. an individual

Brule v Ogallala.

Brunflo 63°7′ N., 14°17′ E.

Rödbrottet quarry, near Brunflo, Sweden
Found 1980, recognized in this year
Stone. Olivine-bronzite chondrite (H).

Chondrule structures were recognized in a block of Ordovician limestone (approx. age 450 m.y.). The silicate minerals have been replaced by calcite, barite and a Cr-V-'phengite'. Only original mineral preserved is chromite, the composition of which suggests an H-group classification, P. Thorslund and F.E. Wickman, Nature, 1981, **289**, p.285.

Bruno 52°16′ N., 105°21′ W.

Saskatchewan, Canada
Found 1931
Iron. Hexahedrite (IIA).

A mass of 13kg., only slightly weathered, was found at Bruno. Described with analysis, 5.79 %Ni, H.H. Nininger, Am. J. Sci., 1936, **31**, p.209 [M.A. 6-398]. Analysis, 5.41 % Ni, 61.3 ppm.Ga, 185 ppm.Ge, 37 ppm.Ir, J.T. Wasson, Geochimica et Cosmochimica Acta, 1969, **33**, p.859. Description of structure, surface features, V.F. Buchwald, Iron Meteorites, Univ. of California, 1975, p.347.

12.4kg Tempe, Arizona State Univ.; 17g Washington, U.S. Nat. Mus.;

Specimen(s): [1959,916], 58g. and sawings, 10.5g

Brununu v Bununu.

Brusa v Domanitch.

Brussels 50°52′ N., 4°22′ E.

Belgium
Fell 1520, before this year
Doubtful. Stone.

One stone is said to have fallen, E.F.F. Chladni, Die Feuer-Meteore, Wien, 1819, p.208 but the evidence is not conclusive.

Bubuowly v Supuhee.

Bückeberg v Obernkirchen.

Buckleboo 32°57′20″ S., 136°8′15″ E.

South Australia, Australia
Found 1963, approx.
Stone. Olivine-bronzite chondrite (H6).

A single mass of 992g was ploughed up 12 km SW. of Buckleboo railway siding, Meteor. Bull., 1983 (61), Meteoritics, 1983, **18**, p.78.

Specimen, Adelaide, South Austr. Mus.;

Budetin v Gross-Divina.

Budulan 50°34′ N., 114°54′ E. approx.

Buryat National District, Chita, USSR, [Будулан]
Found 1962, summer
Stony-iron. Mesosiderite (MES).

One mass of 100kg was found, Meteor. Bull., 1963 (26). Analysis of metal, 7.48 %Ni, 15.0 ppm.Ga, 58 ppm.Ge, 4.4 ppm.Ir, J.T. Wasson et al., Geochimica et Cosmochimica Acta, 1974, **38**, p.135. No evidence of severe shock, A.V. Jain and M.E. Lipschutz, Nature, 1973, **242** (Phys. Sci.), p.26.

91kg Moscow, Acad. Sci.; 520g Perth, West. Austr. Mus.; 215g Tempe, Arizona State Univ.; 97g Washington, U.S. Nat. Mus.; 63g Vienna, Naturhist. Mus.;

Specimen(s): [1972,490], 36.64g. an irregular fragment.

Buenaventura 29°48′ N., 107°33′ W.

Chihuahua, Mexico
Found 1969, approx.
Iron. Octahedrite, medium (0.8mm) (IIIB).

A single mass of 113.6kg was purchased by the Univ. of California. It was found 5 miles SW. of Buenaventura, Meteor. Bull., 1978 (55), Meteoritics, 1978, **13**, p.332. Classification and analysis, 9.92 %Ni, 17.4 ppm.Ga, 34.5 ppm.Ge, 0.011 ppm.Ir, A. Kracher et al., Geochimica et Cosmochimica Acta, 1980, **44**, p.773.

113kg Los Angeles, Univ. of California;

Buen Huerto v North Chile.

Buenos Aires v El Perdido.

Bueste v Beuste.

Buey Muerto v North Chile.

Bugaldi v Boogaldi.

Bulloah v Butsura.

Bulls Run
Natal, South Africa, Locality not known
Fell
Iron?.
It was reported that a 2.25kg meteorite which fell on the Bulls Run Farm, was donated to the Geology Department, Univ. of Pretoria. The mass has not been located, C. Frick and E.C.I. Hammerbeck, Bull. Geol. Surv. S. Africa, 1973 (57), p.12.

Buncombe County v Ashville.

Buncombe County v Black Mountain.

Bununu 10°1′ N., 9°35′ E.
Bauchi, Nigeria
Fell 1942, April, 1000 hrs
Synonym(s): *Brununu*
Stone. Achondrite, Ca-rich. Howardite (AHOW).
One stone of 357g fell 20 miles S of Bununu and 50 miles S. of Bauchi, north Central Nigeria, F.C. Leonard, Classif. Cat. Meteor., 1956, p.8, 15, 63, Meteor. Bull., 1957 (5). Description and analysis, B. Mason, Geochimica et Cosmochimica Acta, 1967, **31**, p.107., E. Jarosewich, Geochimica et Cosmochimica Acta, 1967, **31**, p.1103. Analyses of glass spheres, C. Desnoyers and D.Y. Jerome, C. R. Acad. Sci. Paris, 1974, **278** (D), p.3275., A.F. Noonan, Meteoritics, 1974, **9**, p.233.
331g Washington, U.S. Nat. Mus., main mass;

Bunzlau v Lissa.

Bunzlau v Ploschkovitz.

Bur v Bur-Gheluai.

Burdett 38°14′ N., 99°32′ W.
Pawnee County, Kansas, U.S.A.
Found 1940, approx., recognized 1966
Stone. Olivine-bronzite chondrite (H5).
One stone of 8.6kg was recognised as meteoritic, having lain in a farmyard near Burdett for many years, Meteor. Bull., 1967 (40), Meteoritics, 1970, **5**, p.93. Full description and analysis, olivine Fa_{18}, 26.42 % total iron, R.V. Fodor et al., Chem. Erde, 1971, **30**, p.103.
5kg Mainz, Max-Planck-Inst.; 203g Washington, U.S. Nat. Mus.; 127g Chicago, Field Mus. Nat. Hist.; 106g Harvard Univ.; 74g Tempe, Arizona State Univ.;
Specimen(s): [1966,520], 61g. a slice and fragments, 2g.

Burgavii error for Burgavli.

Burgavli 66°24′ N., 137°28′ E.
Yakutsk, Federated SSR, USSR, [Бургавли]
Found 1941
Iron. Octahedrite, coarse (2.6mm) (IA).
One mass of 24.9kg was found and was in the Kolymsk geological museum, E.L. Krinov, Астрон. Журнал, 1945, **22**, p.303 [M.A. 9-297], Каталог Метеоритов Акад. Наук СССР, Москва, 1947 [M.A. 10-511]. Description, G.P. Vdovykin, Метеоритика, 1965, **25**, p.134. Analysis, 6.71 %Ni, 95.8 ppm.Ga, 519 ppm.Ge, 1.1 ppm.Ir, J.T. Wasson, Icarus, 1970, **12**, p.407. Description, V.F. Buchwald, Iron Meteorites, Univ. of California, 1975, p.349.
20.8kg Moscow, Acad. Sci.; 130g Washington, U.S. Nat. Mus.;

Burggraf v Elbogen.

Bur-Gheluai 5° N., 48° E. approx.
Bur-Hagaba district, Somalia
Fell 1919, October 16, 0800 hrs
Synonym(s): *Bhur-Gheluai, Bur, Bur-Hagaba*
Stone. Olivine-bronzite chondrite (H5), xenolithic.
A shower of over 100 stones fell, of total weight about 120kg, the largest of 15.4kg, A. Neviani, letter of 20 October 1921 in Min. Dept. BM(NH)., A. Neviani, Boll. Soc. Geol. Ital., 1921, **40**, p.209. Description, the stones fell after the appearance of a fireball travelling from north to south and after detonations, A. Neviani, Boll. Soc. Geol. Ital., 1922, **41**, p.1. Further description and analysis, 28.45 % total iron, B. Mason and A.D. Maynes, Proc. U.S. Nat. Mus., 1967, **124** (3624). Illustrated description., G.R. Levi-Donati, Meteoritics, 1968, **4**, p.23. Xenolithic, R.A. Binns, Geochimica et Cosmochimica Acta, 1968, **32**, p.299. Olivine Fa_{19}, B. Mason, Geochimica et Cosmochimica Acta, 1963, **27**, p.1011. A mass of 2.5kg found in the Bur district in 1939, described, C. Lippi-Boncambi, Boll. Soc. Geol. Ital., 1949, **67**, p.218 [M.A. 11-269] is a chondrite and evidently to be referred to this fall.
78kg Rome, Univ., includes a 15.4kg individual; 17.7kg Bologna, Min. Mus. Univ.; 3.5kg Washington, U.S. Nat. Mus.; 1.5kg Chicago, Field Mus. Nat. Hist.; 1.5kg Chicago, Field Mus. Nat. Hist.; 1.1kg New York, Amer. Mus. Nat. Hist.; 649g Paris, Mus. d'Hist. Nat.; 843g Harvard Univ.;
Specimen(s): [1922,10], 1907g.; [1922,11], 26g.

Burgos v Berlanguillas.

Burguon v Chandakapur.

Bur-Hagaba v Bur-Gheluai.

Burke County v Bridgewater.

Burke County v Linville.

Burkett 32°2′ N., 99°15′ W.
Coleman County, Texas, U.S.A.
Found 1913
Iron. Octahedrite, coarse (2.0mm) (IA).
A mass of about 8.4kg was found about 3.5 miles S. of Burkett, and was later cut into six pieces, L.W. MacNaughton, Am. Mus. Novit., 1926 (207), p.2. Analysis, 6.87 %Ni, 87.2 ppm.Ga, 368 ppm.Ge, 2.1 ppm.Ir, J.T. Wasson, Icarus, 1970, **12**, p.407. Description, V.F. Buchwald, Iron Meteorites, Univ. of California, 1975, p.350.
6.58kg New York, Amer. Mus. Nat. Hist., main masses; 1173g Washington, U.S. Nat. Mus.; 100g Tempe, Arizona State Univ.;
Specimen(s): [1959,933], 75.5g. and sawings, 6g.

Burlington 42°45′ N., 75°11′ W.

Otsego County, New York, U.S.A.
Found 1819, before this year
Synonym(s): *Cooperstown, Otsego County*
Iron. Octahedrite, medium (1.3mm) (IIIE).

A mass of about 150lb was ploughed up, but only about 12lb have been preserved; described, with an analysis, B. Silliman, Am. J. Sci., 1844, **46**, p.401. Further analysis, 8.15 %Ni, 16.9 ppm.Ga, 34.9 ppm.Ge, 0.45 ppm.Ir, E.R.D. Scott et al., Geochimica et Cosmochimica Acta, 1973, **37**, p.1957. Most of the find was forged into agricultural implements, and the remainder has been heated, V.F. Buchwald, Iron Meteorites, Univ. of California, 1975, p.351.

1.5kg Washington, U.S. Nat. Mus.; 130g Vienna, Naturhist. Mus.; 114g Chicago, Field Mus. Nat. Hist.; 723g Yale Univ.; 114g Berlin, Humboldt Univ.; 759g Tübingen, Univ.; 68g New York, Amer. Mus. Nat. Hist.; 57g Tempe, Arizona State Univ.;

Specimen(s): [14621], 175g.; [90223], 115g.

Burnabbie 32°3′ S., 126°10′ E.

Western Australia, Australia
Found 1965
Stone. Olivine-bronzite chondrite (H5).

One weathered, incompletely crusted mass of 2.3kg was found 4 miles E. of Cocklebiddy Tank, Eyre Highway. Three chips of 187.3g, 21.5g, and 5.6g were recovered nearby, their distribution suggesting approach on an eastwards path, G.J.H. McCall, First Suppl. to West. Austr. Mus. Spec. Publ. no. 3, 1968, p.8. Co-ordinates, G.J.H. McCall and W.H. Cleverly, J. Roy. Soc. West. Austr., 1970, **53**, p.69. Olivine $Fa_{18.5}$, B. Mason, Rec. Austr. Mus., 1974, **29**, p.169.

2.3kg Kalgoorlie, West. Austr. School of Mines; 176g Perth, West. Austr. Mus.; 65g Washington, U.S. Nat. Mus.;

Specimen(s): [1968,285], 37g. crusted slice

Burracoppin v Lake Brown.

Burrika 31°58′ S., 125°50′ E.

Western Australia, Australia
Found 1966
Stone. Olivine-hypersthene chondrite (L6).

A weathered fragment of 20.4g was found, G.J.H. McCall, First Suppl. to West. Austr. Mus. Spec. Publ. no. 3, 1968, p.9. Co-ordinates, G.J.H. McCall and W.H. Cleverly, J. Roy. Soc. West. Austr., 1970, **53**, p.69. Olivine Fa_{25}, B. Mason, Rec. Austr. Mus., 1974, **29**, p.169.

Main mass, Kalgoorlie, West. Austr. School of Mines; 0.6g Washington, U.S. Nat. Mus.; Thin section, Perth, West. Austr. Mus.;

Bursa 40°12′ N., 29°14′ E.

Turkey
Fell 1946
Stone. Olivine-hypersthene chondrite (L6).

One mass of 25kg was recovered, D. Heymann, Sci. Rep. Faculty Sci. Ege Univ., 1966 (28) data on He, Ne, and Ar are also given. Reported, Meteor. Bull., 1967 (39), Meteoritics, 1970, **5**, p.90. Olivine Fa_{24}, B. Mason, Geochimica et Cosmochimica Acta, 1967, **31**, p.1100.

Main mass, Izmir, Ege Univ.; 207g Tempe, Arizona State Univ.; 68g Chicago, Field Mus. Nat. Hist.;

Specimen(s): [1968,8], 93g. crusted fragment.

Buschhof 46°27′ N., 25°47′ E.

Zemgale, Latvian SSR, USSR
Fell 1863, June 2, 0730 hrs
Synonym(s): *Bushhof, Scheikahr Stattan, Scheikaher-Statten*
Stone. Olivine-hypersthene chondrite (L6), veined.

A stone of about 5kg fell, after detonations, G. Rose, Ann. Phys. Chem. (Poggendorff), 1863, **120**, p.619. Described, with an analysis, C. Grewingk and C. Schmidt, Arch. Naturk. Liv.-Ehst.-u. Kurlands, Ser. 1, Min. Wiss., Dorpat, 1864, **3**, p.452, 473. Olivine Fa_{24}, B. Mason, Geochimica et Cosmochimica Acta, 1963, **27**, p.1011.

1.3kg Tartu, Univ.; 828g Vienna, Naturhist. Mus.; 449g Sarajevo, Nat. Mus.; 288g Moscow, Acad. Sci.; 156g New York, Amer. Mus. Nat. Hist.; 99g Prague, Nat. Mus.; 74g Berlin, Humboldt Univ.; 73g Chicago, Field Mus. Nat. Hist.; 55g Tempe, Arizona State Univ.;

Specimen(s): [36269], 98g.; [1920,293], 63g.

Bushhof v Buschhof.

Bushman Land 30° S., 20° E. approx.

Namibia
Found 1933, approx.
Iron. Octahedrite, fine (0.33mm) (IVA).

Description, V.F. Buchwald, Iron Meteorites, Univ. of California, 1975, p.353. The co-ordinates are very approximate and are in fact outside Namibia. Reported, Meteor. Bull., 1975 (53), Meteoritics, 1975, **10**, p.141. Distinct from Gibeon (*q.v.*), analysis, 8.83 %Ni, 2.08 ppm.Ga, 0.134 ppm.Ge, 0.98 ppm.Ir, R. Schaudy et al., Icarus, 1972, **17**, p.174.

2.8kg Washington, U.S. Nat. Mus.;

Bushnell 41°14′ N., 103°54′ W.

Kimball County, Nebraska, U.S.A.
Found 1939
Stone. Olivine-bronzite chondrite (H4).

One stone of 1240g was found in Section 25, township 15, range 59, A.D. Nininger, Pop. Astron., Northfield, Minnesota, 1940, **48**, p.555., A.D. Nininger, Contr. Soc. Res. Meteorites, **2**, p.227 [M.A. 9-54], H.H. and A.D. Nininger, The Nininger Collection of Meteorites, Winslow, Arizona, 1950, p.37. Olivine Fa_{19}, B. Mason, Geochimica et Cosmochimica Acta, 1963, **27**, p.1011.

636g Tempe, Arizona State Univ.; 56g New York, Amer. Mus. Nat. Hist.; 50g Washington, U.S. Nat. Mus.;

Specimen(s): [1959,851], 212g. and fragments, 1g

Businski v Milena.

Bustee 26°47′ N., 82°50′ E.

between Gorakhpur and Fyzabad, Basti district, Uttar Pradesh, India
Fell 1852, December 2, 1000 hrs
Synonym(s): *Basti, Goruckpur*
Stone. Achondrite, Ca-poor. Aubrite (AUB).

A stone of about 3-4lb fell, after detonations; described, with an analysis, N.S. Maskelyne, Phil. Trans. Roy. Soc. London, 1870, **160**, p.193. The stone is brecciated and contains osbornite, oldhamite and diopside. Analysis of enstatite, A.M. Reid and A.J. Cohen, Geochimica et Cosmochimica Acta, 1967, **31**, p.661.

78g Washington, U.S. Nat. Mus.; 13g Vienna, Naturhist. Mus.; 11g Yale Univ.; 7.7g Tempe, Arizona State Univ.;

Specimen(s): [32100], 1267g. the greater part of the stone

Butcher Iron v Coahuila.

Buther Iron v Coahuila.

Butler 38°18′ N., 94°22′ W.

Bates County, Missouri, U.S.A.
Found 1874, before this year
Synonym(s): *Bates County*
Iron. Octahedrite, plessitic (0.15mm) (IRANOM).

A mass of about 90lb was ploughed up 8 miles SW. of Butler, G.C. Broadhead, Am. J. Sci., 1875, **10**, p.401. Description, A. Brezina, Sitzungsber. Akad. Wiss. Wien, Math.-naturwiss. Kl., 1881, **82**, p.348. Analysis, 15.31 %Ni, M.I. D'yakonova, Метеоритика, 1958, **16**, p.180 [M.A. 14-49]. Classification and new analysis, 15.2 %Ni, 87.1 ppm.Ga, 1970 ppm.Ge, 1.2 ppm. Ir, J.T. Wasson, Icarus, 1970, **12**, p.407. Description, some material damaged during cutting, V.F. Buchwald, Iron Meteorites, Univ. of California, 1975, p.356.

12.7kg Harvard Univ.; 3.6kg Paris, Mus. d'Hist. Nat.; 3kg Vienna, Naturhist. Mus.; 500g Washington, U.S. Nat. Mus.; 970g Yale Univ.; 495g Budapest, Nat. Mus.; 504g Chicago, Field Mus. Nat. Hist.; 539g Tempe, Arizona State Univ.; 336g New York, Amer. Mus. Nat. Hist.;

Specimen(s): [53292], 315g.; [67222], 25g.; [1959,978], 263g. and filings, 15g.

Butsura 27°5′ N., 84°5′ E.

Champaran district, Bihar, India
Fell 1861, May 12, 1200 hrs
Synonym(s): *Ballua, Batsura, Bazar, Bulloah, Chireya, Chiriya, Gurkpur, Gorukhur, Gorukhpur, Gorukpur, Gurukpur, Jataha Bazar, Piprassi, Quatahar, Qutahar Bazar*
Stone. Olivine-bronzite chondrite (H6).

Five stones fell, after detonations, two of 5 and 7oz, respectively at Bulloah, one of 11lb at Piprassi, a much larger mass at Qutahar Bazar, and one of 8.75lb at Chireya; all could be fitted together showing that they formed part of one mass, N.S. Maskelyne and V. von Lang, Phil. Mag., 1863, **25**, p.50. Maskelyne gives the weight of the Qutahar Bazar mass as 13lb (6kg); this is probably an error for 43lb (19.5kg), since close on 18kg of this mass can be accounted for in collections. The total known mass of the fall is accordingly about 29kg. Mineral and bulk analysis, olivine Fa_{19}, 28.68 % total iron, R. Hutchison et al., Proc. Roy. Soc. London, 1981, **A374**, p.159. The places of fall are Piprasi, Ballua, Chiriya, and Jataha Bazar, near Bagaha, 27°5'N., 84°5'E., all are now in Gorakhpur district, Uttar Pradesh, except Bagaha and Piprasi, C.A. Silberrad, Min. Mag., 1932, **23**, p.294.

7.4kg Calcutta, Mus. Geol. Surv. India, 4.9kg of Qutahar Bazar 2.1kg of Chireya, and smaller pieces.; 500g Vienna, Naturhist. Mus.; 147g Tempe, Arizona State Univ.; 92g Yale Univ.; 88g Berlin, Humboldt Univ.; 53g New York, Amer. Mus. Nat. Hist.; 39g Chicago, Field Mus. Nat. Hist.;

Specimen(s): [34794], 12980g. main mass of Qutahar Bazar, and fragments, 2.5g.; [34796], 5080g. main mass of Piprassi; [34795], 706g. of Chireya, and fragment, 29g.; [34797], 158g. of Bulloah.

Buwah v Suwahib (Buwah).

Buwaj Suwaihib v Suwahib (Buwah).

Cabarras County v Monroe.

Cabarrus County v Monroe.

Cabaya v Gibeon.

Cabell error for Cadell.

Cabeza de Mayo v Cabezo de Mayo.

Cabezo de Mayo 37°59′ N., 1°10′ W.

Murcia, Spain
Fell 1870, August 18, 0615 hrs
Synonym(s): *Cabeza de Mayo, Cabezzo de Mayo, Murcia*
Stone. Olivine-hypersthene chondrite (L6).

A stone of 25kg fell, after detonations; report and analysis (doubtful, no CaO reported), J.M. Solano y Eulate, Anal. Soc. Españ. Hist. Nat. Madrid, 1872, **1**, p.77. Partial analysis, 22.01 % total iron, W. Wahl and H.B. Wiik, Geochimica et Cosmochimica Acta, 1951, **1**, p.123. Olivine Fa_{24}, B. Mason, Geochimica et Cosmochimica Acta, 1963, **27**, p.1011.

184g Madrid, Mus. Cienc. Nat.; 210g Washington, U.S. Nat. Mus.; 162g Chicago, Field Mus. Nat. Hist.; 47.9g Budapest, Nat. Hist. Mus.; 36g Tempe, Arizona State Univ.; 27g Paris, Mus. d'Hist. Nat.;

Specimen(s): [1920,354], 61g.; [54636], 3.5g.

Cabezzo de Mayo v Cabezo de Mayo.

Cabin Creek 35°30′ N., 93°30′ W. approx.

Johnson County, Arkansas, U.S.A.
Fell 1886, March 27, 1500 hrs
Synonym(s): *Johnson County*
Iron. Octahedrite, medium (1.1mm) (IIIA).

A mass of about 107lb fell, after detonations, about 6 miles E. of Cabin Creek; described, with an analysis, 6.6 %Ni, G.F. Kunz, Am. J. Sci., 1887, **33**, p.494. Description, V.F. Buchwald, Iron Meteorites, Univ. of California, 1975, p.359. Classification and analysis 8.2 %Ni, 21.1 ppm.Ga, 39.6 ppm.Ge, 0.7 ppm.Ir, A. Kracher et al., Geochimica et Cosmochimica Acta, 1980, **44**, p.773.

47.4kg Vienna, Naturhist. Mus., main mass; 34g Washington, U.S. Nat. Mus.;

Specimen(s): [67453], 5g.

Cacak v Guèa.

Cacak v Jelica.

Cacak 43°50′20″ N., 20°20′ E.

Sjbija, Yugoslavia
Fell 1919, June 6, 2030 hrs
Stone. Chondrite.

One stone of 212g was found after a notable bolide, Zap. Srpsk. Geol., 1920-1922, Meteor. Bull., 1963 (27), M. Ramovic, Cat. Meteor. Coll. Yugoslavia, 1965, p.63.

Whole mass, Belgrade, Nat. Hist. Mus., mislaid;

Cacaria 24°30′ N., 104°48′ W. approx.

Durango, Mexico
Found 1867
Iron. Octahedrite, medium (1.2mm) (IIIA).

A mass of 41.4kg was used as an anvil in Durango and was later removed to the Mexican National Museum, A. Castillo, Cat. Meteorites Mexique, Paris, 1889, p.5, L. Fletcher, Min. Mag., 1890, **9**, p.154. Description, analysis, 7.7 %Ni, E. Cohen, Meteoritenkunde, 1905, **3**, p.400. H.A. Ward analysed a specimen from the main mass in Mexico City (Nat. Mus.), and obtained 12.06%Ni. Further description, H.H. Nininger, Anal. Inst. Biologia Mexico, 1931, **2**, p.181 [M.A. 5-159], J.C. Haro, Bol. Inst. Geol. Mexico, 1931 (50), p.35. The mass has been heated and forged, V.F. Buchwald,

Iron Meteorites, Univ. of California, 1975, p.362. Distinct from Rancho de la Pila. New analysis, 7.56 %Ni, 19.1 ppm.Ga, 35.6 ppm.Ge, 8.7 ppm.Ir, E.R.D. Scott et al., Geochimica et Cosmochimica Acta, 1973, **37**, p.1957.
38kg Mexico City, Mus. de Chopo, main mass, approx. weight; 2kg Chicago, Field Mus. Nat. Hist.; 165g Washington, U.S. Nat. Mus.; 71g Tempe, Arizona State Univ.;
Specimen(s): [84883], 296g. a slice from the mass in the Mexican Nat. Mus., and polished piece, 8g

Cachari 36°24′ S., 59°30′ W.
Azul, Buenos Aires province, Argentina
Found 1916, May
Stone. Achondrite, Ca-rich. Eucrite (AEUC).
A stone of 23.5kg was found at a depth of 1.5m near Cachari railway station, M. Kantor, Rev. Mus. La Plata, 1920, **25**, p.118. Description, analysis, E.H. Ducloux, Rev. fac. cienc. quim. Univ. nac. La Plata, 1928, **5**, p.13 [M.A. 4-120]. Description, A. Lacroix, Nouv. Arch. Mus. d'Hist. Nat. Paris, 1926, **1**, p.56. Shock melting age 3.0 Ga, D.D. Bogard et al., Meteoritics, 1981, **16**, p.296.
16.8kg La Plata Mus., main mass; 625g Paris, Mus. d'Hist. Nat.; 1.65kg Washington, U.S. Nat. Mus.;
Specimen(s): [1962,167], 3g.; [1970,338], 6.7g. a sawn fragment and fragments, 1.25g.

Cachinal v Vaca Muerta.

Cachiyuyal 25° S., 69°30′ W. approx.
Atacama, Chile
Found 1874
Synonym(s): *Atacama*
Iron. Octahedrite, medium (1.3mm) (IIIE).
A mass of about 2.5kg was found about 20 leagues from the coast, I. Domeyko, C. R. Acad. Sci. Paris, 1875, **81**, p.597. Analysis, 7.88 %Ni, 16.9 ppm.Ga, 30.3 ppm.Ge, 3.1 ppm.Ir, E.R.D. Scott et al., Geochimica et Cosmochimica Acta, 1973, **37**, p.1957. Discussion of co-ordinates, description, V.F. Buchwald, Iron Meteorites, Univ. of California, 1975, p.365.
1kg Santiago, Mus., main mass; 723g Chicago, Field Mus. Nat. Hist.; 460g Vienna, Naturhist. Mus.; 350g Paris, Mus. d'Hist. Nat.; 11g New York, Amer. Mus. Nat. Hist.;
Specimen(s): [71570], 23g. slice.

Cacilandia
Goias, Brazil, Co-ordinates not reported
Found, date not known
Stone. Olivine-bronzite chondrite (H6).
Mentioned, without full details, H. Hintenberger et al., Z. Natur., 1965, **20A** (8), p.983. Olivine Fa_{20}, J.T. Wasson, Meteorites, Springer-Verlag, 1974, p.269.
Specimen, Rio de Janeiro, Mus. Nacion.;

Cacova v Kakowa.

Cadell 34°4′ S., 139°45′ E.
E. side of Murray River, South Australia, Australia
Found 1910, April
Stone. Olivine-hypersthene chondrite (L6).
A stone of 7.25lb was found 3 miles from Morgan, C. Anderson, Rec. Austr. Mus., 1913, **10**, p.57. Olivine $Fa_{25.5}$, B. Mason, Rec. Austr. Mus., 1974, **29**, p.169. XRF analysis, 21.56 % total iron, M.J. Fitzgerald, Ph.D. Thesis, Univ. of Adelaide, 1979, p.23.
2.81kg Adelaide, South Austr. Mus., includes 2.27kg main mass; 71g Washington, U.S. Nat. Mus.;

Caille v La Caille.

Calce 45°30′ N., 11°30′ E. approx.
Vicenza, Italy
Fell 1635, July 7
Doubtful. Stone.
A stone of 11oz is said to have fallen, E.F.F. Chladni, Die Feuer-Meteore, Wien, 1819, p.224 but the evidence is not conclusive.

Calchaqui v Vera.

Caldiero v Vago.

Calderilla v Imilac.

Caldwell 37°2′ N., 97°37′30″ W.
Sumner County, Kansas, U.S.A.
Found 1961, recognized in this year
Stone. Chondrite.
A mass of 12.9kg was found, Cat. Huss Coll. Meteorites, 1976, p.8.
53g Mainz, Max-Planck-Inst.;

Calico Rock 36°5′ N., 92°9′ W.
Izard County, Arkansas, U.S.A.
Found 1938, recognized 1964
Iron. Hexahedrite (IIA).
A mass of 7.275kg was found, Meteor. Bull., 1965 (33). Analysis, 5.45 %Ni, 57.3 ppm.Ga, 185 ppm.Ge, 8.6 ppm.Ir, J.T. Wasson, Geochimica et Cosmochimica Acta, 1969, **33**, p.859. Description, some plastic deformation, V.F. Buchwald, Iron Meteorites, Univ. of California, 1975, p.368.
643g Tempe, Arizona State Univ.; 539g Washington, U.S. Nat. Mus.; 73g Los Angeles, Univ. of California; 94g New York, Amer. Mus. Nat. Hist.; Main mass, Los Angeles, Griffith Observatory;
Specimen(s): [1966,44], 110g.

California
U.S.A.?
Iron. Octahedrite? (IA).
Listed, without details of location, but with an analysis, 7.7 %Ni, 66.7 ppm.Ga, 253 ppm.Ge, 1.9 ppm.Ir, J.T. Wasson, Meteorites, Springer-Verlag, 1974, p.298.

Calivo 11°45′ N., 122°20′ E.
Panay, Philippines
Fell 1916, May 26, afternoon hrs
Stone. Achondrite?, brecciated.
A mass of 2400g fell in Tinigao, Calivo. Described, M. Selga, Publ. Manila Observ., 1930, **1** (9), p.3 [M.A. 4-421] with unsatisfactory analyses by G.O. Opiana and L. Ocampo.

Callac v Kerilis.

Calliham 28°25′ N., 98°15′ W.
McMullen County, Texas, U.S.A.
Found 1958
Synonym(s): *Callihan*
Stone. Olivine-hypersthene chondrite (L6).
A mass of 40kg was found, B. Mason, Meteorites, Wiley, 1962, p.242. Olivine Fa_{23}, B. Mason, Geochimica et Cosmochimica Acta, 1963, **27**, p.1011. Co-ordinates reported are of a site in Live Oak County, about 5 miles SW. of Three Rivers.
11.6kg Mainz, Max-Planck-Inst.; 3.9kg Tempe, Arizona

State Univ.; 255g Harvard Univ.; 230g Washington, U.S. Nat. Mus.; 162g Tübingen, Univ.; 876g Copenhagen, Univ.; 145g Ottawa, Geol. Surv. Canada; 125g Perth, West. Austr. Mus.; 110g Chicago, Field Mus. Nat. Hist.; 82g New York, Amer. Mus. Nat. Hist.;
Specimen(s): [1959,1002], 2.04kg.

Callihan error for Calliham.

Calumet County 44° N., 88°15′ W. approx.
Wisconsin, U.S.A.
Found 1934, before this year
Pseudometeorite. Iron.
A mass about the size of a baseball was found but is of manufactured iron, B. Mason, letter to W.F. Read, 16 April 1968, copy in Min. Dept. BM(NH). Reported, Meteor. Bull., 1969 (47), Meteoritics, 1970, **5**, p.106.

Camaro v Messina.

Camaro Superiore v Messina.

Cambria 43°12′ N., 78°48′ W. approx.
Niagara County, New York, U.S.A.
Found 1818
Synonym(s): *Lockport*
Iron. Octahedrite, fine (0.48mm) (IRANOM).
A mass of 36lb was turned up by the plough, B. Silliman, Am. J. Sci., 1845, **48**, p.388. Analysis, 10.65 %Ni, C. Rammelsberg, Monatsber. Akad. Wiss. Berlin, 1870, p.444. Meteorite chemically anomalous, 10.17 %Ni, 11.1 ppm.Ga, 1.52 ppm.Ge, 0.88 ppm.Ir, J.T. Wasson and R. Schaudy, Icarus, 1971, **14**, p.59. Description, oriented troilite, plastic deformation, V.F. Buchwald, Iron Meteorites, Univ. of California, 1975, p.369.
3kg Yale Univ.; 801g Tübingen, Univ.; 340g Tempe, Arizona State Univ.; 294g Vienna, Naturhist. Mus.; 204g Berlin, Humboldt Univ.; 351g Washington, U.S. Nat. Mus.; 209g New York, Amer. Mus. Nat. Hist.; 175g Chicago, Field Mus. Nat. Hist.;
Specimen(s): [19005], 4951g. the largest mass preserved, and two polished mounts.

Campamento Dadin v Dadin.

Campbellsville 37°22′ N., 85°22′ W.
Taylor County, Kentucky, U.S.A.
Found 1929, June
Iron. Octahedrite, medium (1.3mm) (IIIB).
One mass of 34lb was found, D.M. Young, letters of 6 March and 4 April, 1935 in Min. Dept. BM(NH). Described, with an analysis, D.M. Young, Pop. Astron., Northfield, Minnesota, 1939, **47**, p.382 [M.A. 7-376]. Analysis, 8.65 % Ni, 20.4 ppm.Ga, 43.8 ppm. Ge., 0.09 ppm.Ir, E.R.D. Scott et al., Geochimica et Cosmochimica Acta, 1973, **37**, p.1957. Description, epsilon structure, V.F. Buchwald, Iron Meteorites, Univ. of California, 1975, p.372.
9.5kg Washington, U.S. Nat. Mus., main mass; 768g Fort Worth, Texas, Monnig Colln.; 460g Harvard Univ.; 265g Tempe, Arizona State Univ.; 250g Miami, Univ.;
Specimen(s): [1935,891], 416g. and a fragment, 7.7g.

Campo del Cielo 27°28′ S., 60°35′ W.
Gran Chaco Gualamba, Chaco, Argentina
Found 1576
Synonym(s): *Chaco Gualamba, Charata, El Abipon, El Hacha, El Mataco, El Mocoví, El Patio, El Perdido, El Rosario, El Taco, El Toba, El Tonocoté, Gancedo, Gran Chaco Gualamba, Gran Chaco (iron), Gran Chaco I, Gran Chaco II, La Perdida, Los Guanacos, Mesón de Fierro, Otumpa, Nihuá, Pinalta, Pozo del Cielo, Runa Pocito, San Jago del Estero, Santiago del Estero, Silva, Tucuman, Wöhler's Iron*
Iron. Octahedrite, coarse (3.0mm), with silicate inclusions (IA).
A mass estimated at about 15 tons was found, Don Rubin de Celis, Phil. Trans. Roy. Soc. London, 1788, **78**, p.37, 183, L. Fletcher, Min. Mag., 1889, **8**, p.229. In 1813 a mass of about 1400lb was brought to Buenos Aires and was later given to Sir Woodbine Parish who presented it to the British Museum, Phil. Trans. Roy. Soc. London, 1834, **125**, p.53. Full description, E. Cohen, Meteoritenkunde, 1905, **3**, p.35. Wöhler's Iron, a 114g mass from an unknown locality in Wöhler's collection, was referred to Otumpa by A. Brezina; described and analysed, 7.38 %Ni, F. Wöhler, Ann. Chem. Pharm. Leipzig, 1852, **81**, p.252. Locality description, A. Alvarez, El Meteorito del Chaco, Buenos Aires, 1926, **149**, p.1055. The large mass found by Don Rubin de Celis in 1783 appears to have been recorded by Hernan Mexia de Miraval as early as 1576. The mass of about 1000kg, of which the 1400lb mass in the British Museum (Nat. Hist.) is the greater part, was found in 1803 at Runa Pocito, Campo del Cielo, Santiago del Estero. Some pieces of this mass have been heated and forged, V.F. Buchwald, Iron Meteorites, Univ. of California, 1975, p.373. Other masses have been found in the Campo del Cielo, near the boundary between Santiago del Estero and Chaco Nacional; viz., in 1913, pieces estimated to weigh about 1500kg at Pozo del Cielo (Hoyo Rubin de Celis), Chaco Nacional; in 1923, an immense mass of 4210kg, known as "El Toba", at "El Rosario", 21km SE. of Gancedo, Santiago del Estero; in 1924, an implement of 2500g at Gancedo, Chaco Nacional; and in 1925 a large mass of 732kg, known as "El Mocovi", at "Los Guanacos", Chaco Nacional. These masses have a similar Ni content as the Runa Pocito mass. Other masses are El Abipon, 1936, 460kg; El Mataco, 1937, 1000kg; El Tonocoté, 1931, 850kg; Nihuá, 1948, 15kg; and Pinaltá, 1937, 9kg, L.O. Giacomelli, letters of 1960 to 1962 in Min. Dept. BM(NH). A further mass of 900kg was found in about 1937, E. Zamboni, letter of 11 January 1938 in Min. Dept. BM(NH)., may refer to the El Mataco mass. The El Mocoví mass is described and analysed, 5.57 %Ni, E.H. Ducloux, Rev. fac. cienc. quim. Univ. nac. La Plata, 1928, **5** (2), p.8 [M.A. 4-120]. Classification and new analysis of metal, 6.62 %Ni, 90.0 ppm.Ga, 392 ppm.Ge, 3.2 ppm.Ir, J.T. Wasson, Icarus, 1970, **12**, p.407. Description and analysis of the Odessa-type silicates, T.E. Bunch et al., Contr. Miner. Petrol., 1970, **25**, p.297, F. Wlotzka and E. Jarosewich, Smithson. Contrib. Earth Sci., 1977 (19), p.104. Discussion, B. Mason, Min. Mag., 1967, **36**, p.120. A mass of 18 tons was found at a depth of 5 meters in crater 10, W.A. Cassidy, Meteoritics, 1970, **5**, p.187. Comprehensive, fully illustrated description, V.F. Buchwald, Iron Meteorites, Univ. of California, 1975, p.373.
1530g La Plata Mus., of "Pozo del Cielo"; "El Rosario", Buenos Aires, Mus. Argentino Cient. Nat.; "Gancedo", Buenos Aires, Mus. Argentino Cient. Nat.; Named masses, Buenos Aires, Mus. Argentino Cient. Nat., the masses known as "El Abipon", "El Mocoví", "Pinaltá", "Los Guanacos"; 350g Buenos Aires, Mus. Argentino Cient. Nat., of Pozo del Cielo; Named masses, Buenos Aires,

Direcc. Geol. Min., the masses known as "El Tonocoté", "Nihuá"; 2.25kg Paris, Mus. d'Hist. Nat.; 1kg Copenhagen, Univ.; 500g Budapest, Nat. Mus.; 8.7kg Albuquerque, Univ. of New Mexico; 530g Chicago, Field Mus. Nat. Hist.; 196kg Washington, U.S. Nat. Mus., two slices from a mass of 1993.8kg found in 1962 at El Taco; 4.7kg Tempe, Arizona State Univ.;
Specimen(s): [61313], 634kg. main mass of Otumpa; [90232], 302g.; [1964,66], 78g. and fragments,42.5g.; [1911,717], 48g. fragments; [1964,49], 152g. of El Toba; [1962,131], 52g. fragments, of El Toba; [1964,60], 16g. fragments, of El Toba; [1962,164], 5.5g. of El Toba; [1964,61], 4g. of El Toba; [1927,980], 80g. a completely weathered fragment of Pozo del Cielo; [35164], 26g. of Wöhler's Iron; [63877], 4.5g. of Wöhler's Iron; [1962,171], 14.5g. of El Mocovi; [1962,175], 9.5g. of Gran Chaco

Campo del Puchara v Imilac.

Campo del Puchara v Imilac.

Campo de Pucara (iron) 27°40′ S., 67°7′ W. approx.
Andalgala district, Catamarca, Argentina
Found 1879
Iron. Hexahedrite.
In addition to the small pallasite specimen thought to be Imilac (*q.v.*) a hexahedrite mass of 41g is listed from the same locality, L.M. Villar, Cienc. Investig., 1968, **24**, p.302.

Campo Perrando
Santa Fe, Argentina, Co-ordinates not reported
Stone. Achondrite.
Named, with no information, L.M. Villar, Cienc. Investig., 1968, **24**, p.302.

Camp Verde v Cañon Diablo.

Cañada de Hierro v Tucson.

Çanakkale 39°48′ N., 26°36′ E.
Bayramiç, Turkey
Fell 1964, July, end of the month
Synonym(s): *Cannakale*
Stone. Olivine-hypersthene chondrite (L).
Several pieces, the largest weighing 4kg, were collected, Meteor. Bull., 1965 (33). Olivine Fa_{25}, B. Mason, Geochimica et Cosmochimica Acta, 1967, **31**, p.1100.
400g Izmir, Univ.; 35g Tempe, Arizona State Univ.; 10g Oslo, Min.-Geol. Mus.;
Specimen(s): [1966,2], 4.5g.

Canara v Udipi.

Cancã v Paranaiba.

Can-Can v Paranaiba.

Cañellas 41°15′ N., 1°40′ E.
Barcelona, Spain
Fell 1861, May 14, 1330 hrs
Synonym(s): *Barcelona, Canyelles, Vilanova de Sitjes, Villanova, Villa Nueva*
Stone. Olivine-bronzite chondrite (H5), brecciated.
Several stones fell, after detonations, near Villa Nueva, but most were lost, or broke into fragments the largest of which weighed only 18oz, R.P. Greg, Phil. Mag., 1861, **22**, p.107, M. Faura y Sans, Butll. Centr. Excurs. Catalunya, 1921, **31** (322), p.277. Olivine Fa_{17}, B. Mason, Geochimica et Cosmochimica Acta, 1963, **27**, p.1011. Figured, with an analysis, M. Faura y Sans, Meteoritos caidos en la Peninsula Iberica, Tortosa, 1922, p.19.
500g Madrid, Mus. Cienc. Nat.; 237g Paris, Mus. d'Hist. Nat.; 121g Barcelona,; 38g Tempe, Arizona State Univ.; 9g Chicago, Field Mus. Nat. Hist.; 6.5g Washington, U.S. Nat. Mus.; 7g Berlin, Humboldt Univ.;
Specimen(s): [35781], 1.5g.

Canemorto v Orvinio.

Caney Fork v Carthage.

Caney Fork v Smithville.

Cangas de Onis 43°23′ N., 5°9′ W.
Asturias, Spain
Fell 1866, December 6, 1100 hrs
Synonym(s): *Elgueras, Oviedo*
Stone. Olivine-bronzite chondrite (H5).
A shower of stones fell, after detonations, the largest stone weighing about 11kg; described and analysed, J.R. de Luanco, Anal. Soc. Españ. Hist. Nat. Madrid, 1874, **3**, p.69. Olivine Fa_{18}, B. Mason, Geochimica et Cosmochimica Acta, 1963, **27**, p.1011.
3kg Seville, Univ.; 1.47kg Paris, Mus. d'Hist. Nat.; 14.8kg Madrid, Mus. Cienc. Nat.; 1103g Washington, U.S. Nat. Mus.; 15kg Orviedo Univ.; 731g Harvard Univ.; 132g Budapest, Nat. Hist. Mus.;
Specimen(s): [54813], 89g.; [1925,900], 204g.

Cannakale v Çanakkale.

Canoncito v Glorieta Mountain.

Canon City 38°28′13″ N., 105°14′29″ W.
Fremont County, Colorado, U.S.A.
Fell 1973, October 27, approx. 1800 hrs
Stone. Olivine-bronzite chondrite (H6).
One stone of 1.4kg fell through the roof of a garage, 3.18km N. of Canon City Post Office; it broke into four major fragments of 559g, 531g, 74g, and 53g. A meteor was observed between 1800 and 1815 hours that evening, Meteor. Bull., 1974 (52), Meteoritics, 1974, **9**, p.101. Ar isotopes, E.L. Firman and R.W. Stoenner, Meteoritics, 1979, **14**, p.1.
295g Mainz, Max-Planck-Inst.; 55g Washington, U.S. Nat. Mus.;
Specimen(s): [1977,M.6], 11.1g.

Cañon Diablo 35°3′ N., 111°2′ W.
Coconino County, Arizona, U.S.A.
Found 1891
Synonym(s): *Albuquerque, Arizona, Ashfork, Barringer, Bloody Basin, Camp Verde, Canyon Diablo, Canyon Diablo no. 2, Canyon Diablo no. 3, Canyon Diablo (1936), Canyon Diablo (1949), Colorado River, Cut Off, Diablo Canyon, Ehrenberg, Elden, Fair Oaks, Fossil Springs, Ganado, Helt Township, Houck, La Paz, Las Vegas, Moab, Monument Rock, Mount Elden, Oildale, Palisades Park, Panamint Range, Pulaski County, Rifle, Roswell, Schertz, Wickenberg (iron), Winsloe, Winslow*
Iron. Octahedrite, coarse (2.0mm) (IA).
Numerous masses, of total weight over 30 tons and ranging from minute fragments to pieces of over 1000lb, have been found in the vicinity of a crater like elevation known as Coon Butte, "Crater Mountain" or " Meteor Crater" 10

miles SE. of Cañon Diablo, H.H. Nininger, Pop. Astron., Northfield, Minnesota, 1949, **57**, p.17 [M.A. 10-522]. Metallography, J.O. Lord, Pop. Astron., Northfield, Minnesota, 1941, **49**, p.492 [M.A. 8-375]. Contains diamond, H.H. Nininger, Pop. Astron., Northfield, Minnesota, 1939, **47**, p.504, C.J. Ksanda and E.P. Henderson, Am. Miner., 1939, **24**, p.677. The presence of diamond in BM 83025 has been confirmed by X-ray examination, the same mass contains numerous lamellae of cohenite. Two estimates of the mass of the Cañon Diablo fall are 25000 tons, and 5000 to 15000 tons, C.C. Wylie, Pop. Astron., Northfield, Minnesota, 1943, **51**, p.97, 220 [M.A. 8-378], L. LaPaz, Pop. Astron., Northfield, Minnesota, 1943, **51**, p.339 [M.A. 9-287]. For further details of the very large crater, Meteor Crater, see crater listing. The composition of many of the synonomous irons has been determined, 6.98 %Ni, 81.8 ppm.Ga, 324 ppm. Ge., 1.9 ppm.Ir, J.T. Wasson, Meteorites, Springer-Verlag, 1974, p.297. Further references, D.M. Barringer, Proc. Acad. Nat. Sci. Philadelphia, 1905, **57**, p.861, B.C. Tilghman, Proc. Acad. Nat. Sci. Philadelphia, 1905, **57**, p.887, D. Heymann et al., J. Geophys. Res., 1966, **71**, p.619 [M.A. 17-685], H.J. Axon, Prog. Materials Sci., 1968, **13**, p.185, R.R. Jaeger and M.E. Lipschutz, Geochimica et Cosmochimica Acta, 1968, **32**, p.773. Contains krinovite, E. Olsen and L. Fuchs, Science, 1968, **161**, p.786 [M.A. 20-324]. Comprehensive description, transported masses, history, V.F. Buchwald, Iron Meteorites, Univ. of California, 1975, p.381.

639.1kg Barringer Meteorite Company, the largest mass; 485kg Christchurch, New Zealand, Canterbury Mus., single mass; 1748kg Chicago, Field Mus. Nat. Hist.; 600kg New York, Amer. Mus. Nat. Hist.; 301kg Budapest, Nat. Hist. Mus.; 213kg Philadelphia, Acad. Nat. Sci.; 1675kg Washington, U.S. Nat. Mus.; 380kg Paris, Mus. d'Hist. Nat.; 690kg Tempe, Arizona State Univ.; 200kg Vienna, Naturhist. Mus.; 14kg Harvard Univ.; 70kg Prague, Bohemian Mus.; 193kg Stockholm, Riksmus., single mass; 123kg Collection of Mr. Stevan Celebonovic, includes a 113kg mass.; 24.7kg Bonn, Univ. Mus.; 2.5kg Madrid, Mus. Cienc. Nat.; 25.8kg Schönenwerd, Bally Mus.; 67.5kg Los Angeles, Univ. of California; 2.5kg Ottawa, Mus. Geol. Surv. Canada;

Specimen(s): [1959,1052], 98.66kg.; [68578], 65.17kg.; [1959, 1016], 21.77kg. of Rifle; [77188], 13.63kg.; [67592], 3008g.; [83025], 1524g.; [1949,282], 1756g. a complete individual; [1949,283], 1595g. and fragments, 10g.; [1949,284], 1284g.; [1949,285], 1700g. fragments of iron shale; [1950,388], 25g.; [1959,918], 18g.; [1959,936], 25g. of Monument Rock; [68248], 128g. oxidised fragments; [69138], 29g.; [1934,53], 436.5g.; [1953,9], 59.5g.; [1933,321], 532g. iron shale; [1929, 1499], 73g. graphitic fragments of Mount Elden; [1926,491], 188g. seven shale "balls"; [1960,489], 67g.; [1962,137], 250g. slice; [1959,940], 50.5g. of Wickenburg (iron), and sawings, 2.5g; [1959,934], 671g. of Camp Verde; [1950,251], 34g.; [1959,938], 27.5g. and sawings, 2.5g, of Palisades Park; [1959,982], 20.5g. and sawings, 2.5g, of Roswell; [1982,M.1], 81g.

Canton 34°12′ N., 84°29′ W.

Cherokee County, Georgia, U.S.A.
Found 1894
Synonym(s): *Cherokee County, Cherokee Mills*
Iron. Octahedrite, medium (1.1mm) (IIIA).

A mass of 15.5lb was ploughed up about 5 miles SW. of Canton; described, with an analysis, 6.7 %Ni, E.E. Howell, Am. J. Sci., 1895, **50**, p.252. Another analysis, 7.69 %Ni, H.B. Wiik and B. Mason, Geochimica et Cosmochimica Acta, 1965, **29**, p.1003. Further analysis, 7.58 %Ni, 18.6 ppm.Ga, 35.9 ppm.Ge, 9.3 ppm.Ir, E.R.D. Scott et al., Geochimica et Cosmochimica Acta, 1973, **37**, p.1957. Description, history, distinct from Losttown (*q.v.*), V.F. Buchwald, Iron Meteorites, Univ. of California, 1975, p.379.

262g Chicago, Field Mus. Nat. Hist.; 877g New York, Amer. Mus. Nat. Hist.; 409g Washington, U.S. Nat. Mus.; 150g Berlin, Humboldt Mus.; 82g Yale Univ.;

Specimen(s): [80684], 310.5g. slice, and fragments, 5.5g.

Canyelles v Cañellas.

Cany Fork v Smithville.

Canyon 34°58′ N., 101°56′ W.

Randall County, Texas, U.S.A.
Found 1957, recognized in this year
Stone. Olivine-bronzite chondrite (H).

Listed, L. LaPaz, Cat. Coll. Inst. Meteor. Univ. New Mexico, 1965. Olivine Fa_{18}, B. Mason, Geochimica et Cosmochimica Acta, 1967, **31**, p.1100.

120.9g Albuquerque, Univ. of New Mexico;

Canyon City 40°54′ N., 123°6′ W. approx.

Trinity County, California, U.S.A.
Found 1875
Synonym(s): *Trinity County*
Iron. Octahedrite, medium (1.0mm) (IIIA).

A mass of about 19lb was found 3 miles NE. of Canyon City; described with an analysis, 7.85 %Ni, H.A. Ward, Am. J. Sci., 1904, **17**, p.383. Further analysis, 7.58 %Ni, 19.8 ppm.Ga, 36.8 ppm.Ge, 11 ppm.Ir, E.R.D. Scott et al., Geochimica et Cosmochimica Acta, 1973, **37**, p.1957. Description, epsilon structure, V.F. Buchwald, Iron Meteorites, Univ. of California, 1975, p.380.

4.5kg Chicago, Field Mus. Nat. Hist.; 668g Tempe, Arizona State Univ.; 284g Washington, U.S. Nat. Mus.; 163g Los Angeles, Univ. of California; 801g New York, Amer. Mus. Nat. Hist.; 355g Berlin, Humboldt Univ.; 320g Harvard Univ.;

Specimen(s): [86944], 193g.; [1959,946], 387g. and sawings, 9g.

Canyon Diablo v Cañon Diablo.

Canyon Diablo (1936) v Cañon Diablo.

Canyon Diablo (1949) v Cañon Diablo.

Canyon Diablo no. 2 v Cañon Diablo.

Canyon Diablo no. 3 v Cañon Diablo.

Canyonlands 38°11′ N., 109°53′ W.

San Juan County, Utah, U.S.A.
Found 1961
Stone. Olivine-bronzite chondrite (H6), brecciated.

A partly crusted stone of 1.52kg was found near the confluence of the Green and the Colorado Rivers in Canyonlands National Park, Meteor. Bull., 1975 (53), Meteoritics, 1975, **10**, p.141. Description, contains maskelynite, olivine $Fa_{19.6}$, D.E. Lange et al., Meteoritics, 1974, **9**, p.271.

186g Tempe, Arizona State Univ.;

Caparrosa 17°30′ N., 99°30′ W.
SW. of Chilpanzingo, Guerrero, Mexico
Found 1858, before this year
Synonym(s): *Chilpanzingo, Rincon de Caparrosa*
Doubtful. Iron.
A nodule of 341g is said to have been found in a piece of copper pyrites, A. Castillo, Cat. Meteorites Mexique, Paris, 1889, p.1. Referred to Toluca by A. Brezina.
5g Mexico City, Inst. Geol.;

Cape v Cape of Good Hope.

Cape Girardeau 37°16′ N., 89°35′ W.
Cape Girardeau County, Missouri, U.S.A.
Fell 1846, August 14, 1500 hrs
Stone. Olivine-bronzite chondrite (H6).
A stone of about 5lb fell with a loud report 7.5 miles south of Cape Girardeau; described, with an analysis, E.S. Dana and S.L. Penfield, Am. J. Sci., 1886, **32**, p.229. Olivine Fa_{19}, B. Mason, Geochimica et Cosmochimica Acta, 1963, **27**, p.1011.
1478g Yale Univ.; 98g Rome, Vatican Colln; 93g Vienna, Naturhist. Mus.; 57g Chicago, Field Mus. Nat. Hist.; 35g New York, Amer. Mus. Nat. Hist.; 7.6g Tempe, Arizona State Univ.; 4.9g Budapest, Nat. Mus.;
Specimen(s): [64342], 76.2g. and fragments, 1.2g.

Cape Iron v Cape of Good Hope.

Capeisen v Cape of Good Hope.

Cape of Good Hope 33°30′ S., 26° E. Approx.
Cape Province, South Africa
Found 1793
Synonym(s): *Cape, Capeisen, Cape Iron, Capland, Great Fish River, Joremenyseg-fok, Kapeisen*
Iron. Ataxite, Ni-rich (IVB).
A mass of about 300lb was found between Sunday River and Bushman River (W. of Great Fish River), J. Barrow, Account of Travels into the interior of Southern Africa, London, 1801, **1**, p.226, A. von Dankelmann, J.H. Voight's Mag. Naturkunde, 1805, **10**, p.3. Described, with an analysis, 15.67 %Ni, E. Cohen, Meteoritenkunde, 1905, **3**, p.138. Partial analysis, and determination of Ga, Au and Pd, 16.48 %Ni, E. Goldberg et al., Geochimica et Cosmochimica Acta, 1951, **2**, p.1. The sword made of the Cape Iron and presented to the Emperor of Russia, J. Sowerby, Exotic Mineralogy, London, 1811-1820, **2**, p.138, could not be traced in 1937, M.A.R. Khan, Nature, 1944, **154**, p.465. Structural description, H.J. Axon and P.L. Smith, Min. Mag., 1972, **38**, p.736. Further analysis, 16.92 %Ni, 0.198 ppm.Ga, 0.059 ppm.Ge, 36 ppm.Ir, R. Schaudy et al., Icarus, 1972, **17**, p.174. Description, parts of the mass were removed and forged, V.F. Buchwald, Iron Meteorites, Univ. of California, 1975, p.407.
67kg Budapest, Nat. Mus.; 947g Vienna, Naturhist. Mus.; 701g Berlin, Humboldt Univ.; 306g Paris, Mus. d'Hist. Nat.; 209g Washington, U.S. Nat. Mus.; 208g Chicago, Field Mus. Nat. Hist.; 127g Los Angeles, Univ. of California; 119g New York, Amer. Mus. Nat. Hist.; 157g Tempe, Arizona State Univ.; 101g Harvard Univ.; 55g Cape Town, South African Mus.;
Specimen(s): [54646], 273g. slice; [46975], 18g.; [90221], 16.5g.; [15143], 4.5g.; [33746], 1g.; [1935,47], 1.44g. two hammered flakes, 1g and 0.44g.; [1949,68], less than 1g., hammered flake.

Cape Province v Deelfontein.

Caperr 45°17′ S., 70°29′ W.
Rio Senguerr, Chubut, Argentina
Found 1869, known in this year
Synonym(s): *Amakaken, Caperr Aiken*
Iron. Octahedrite, medium (1.0mm) (IIIA).
A large mass weighing 251lb was seen by Dr. F.P. Moreno in 1896 who secured it for the La Plata Museum; described and analysed, 9.33 %Ni, L. Fletcher, Min. Mag., 1899, **12**, p.167, L.M. Villar, Cienc. Investig., 1968, **24**, p.302 gives the date of discovery as 1896, April 4, slightly different co-ordinates for the find site. It is clear that the existence of this mass was known for at least 25 years before 1896, M.M. Radice, Cat. Meteor. La Plata Mus., 1959, **5**, p.29. Further analysis, 8.58 %Ni, 21 ppm.Ga, 45.3 ppm.Ge, 0.24 ppm.Ir, E.R.D. Scott et al., Geochimica et Cosmochimica Acta, 1973, **37**, p.1957. Description, epsilon structure, V.F. Buchwald, Iron Meteorites, Univ. of California, 1975, p.409.
113kg La Plata, Mus., main mass; 383g Chicago, Field Mus. Nat. Hist.; 174g Vienna, Naturhist. Mus.; 36g Washington, U.S. Nat. Mus.;
Specimen(s): [1906,53], 219g. slice and fragments, 25g.; [84134], 20g. and filings, 7.5g.

Caperr Aiken v Caperr.

Cape York 76°8′ N., 64°56′ W.
West Greenland, Greenland
Found 1818, known in this year
Synonym(s): *Agpalilik, Ahnighito, Akpohon, Anighito, Baffin's Bay, Davis' Strait, Melville, Melville Bay, North Star Bay, Northumberland Island, Ross's Iron, Savik, Saviksue, Sowallick Mountains, Swallik*
Iron. Octahedrite, medium (1.2mm) (IIIA).
Knives of iron with bone handles were given to Capt. John Ross in 1818 by the Esquimos of Prince Regent's Bay, John Ross, Voyage of Discovery in Baffin's Bay, London, 1819, p.102-118. Three large masses, purported to weigh respectively about 59 tons, 3 tons, and 896lb and named "The Tent" or Ahnighito, "The Woman," and "The Dog", were shown to Lieut. R.E. Peary, by whom they were later transported to New York, R.E. Peary, Northward over the Great Ice, London, 1898, **2**, p.145, 553, 600. The weight of Ahnighito was long in doubt but is now known to be 31 metric tons, B. Mason, Meteorites, Wiley, 1962, p.27. The nickel content varies in different parts of the meteorite, 7.34 to 7.86 %Ni, V.M. Goldschmidt, Proc. Roy. Inst. Gt. Britain, 1929, **26**, p.73, V.M. Goldschmidt, Die Naturwiss., 1930, **18**, p.999 [M.A. 5-7]. Another mass of 3.5 tons was found by Esquimos on a small peninsula called Savik in the Prince Regent Bay; it was removed to Copenhagen, Univ. Min. Geol. Mus., in 1925; description, O.B. Bøggild, Meddel. Grønland, 1927, **74**, p.11 [M.A. 3-535]. The mass of 48kg found in 1955 and reported, F.C. Leonard, Meteoritics, 1965, **1**, p.305 [M.A. 13-82] as a new mass of Cape York is a distinct fall, Thule (*q.v.*), V.F. Buchwald, Geochimica et Cosmochimica Acta, 1961, **25**, p.95. Three further masses have been reported, one of 7.8kg found at Saveqarfik in 1961 (Savik II) and one of about 15 tons found at Agpalilik, V.F. Buchwald, Geochimica et Cosmochimica Acta, 1964, **28**, p.125. The third was of 250g, was found on Northumberland Island, 77°30′N., 71°W., but is believed to be a transported fragment from the Melville Bay area, V.F. Buchwald, Naturens Verden, 1964, p.33. The Agpalilik and the 7.8kg Savik masses are in Copenhagen, V.F. Buchwald and S. Munck, Analecta geol., 1965, **1**, p.1 [M.A. 17-593]. Analysis, 7.58 %Ni, 19.2 ppm.Ga, 36.0 ppm.Ge, 5.0 ppm.Ir, E.R.D. Scott et al., Geochimica et Cosmochimica Acta, 1973, **37**, p.1957. Comprehensive history, description, oriented troilite, artificial cold working, V.F. Buchwald, Iron Meteorites,

Univ. of California, 1975, p.410.
34tons New York, Amer. Mus. Nat. Hist., the three masses collected by Peary, approx weight.; 23.5tons Copenhagen, Univ. Geol. Mus., most of Agpalilik and Savik I masses; 8.1kg Washington, U.S. Nat. Mus.; 1kg Prague, Nat. Mus.; 1.25kg Paris, Mus. d'Hist. Nat.; 225g Tempe, Arizona State Univ.;
Specimen(s): [87561], Esquimo knife; [87562] Esquimo knife, similar to 87561, one figured in Ross's Voyage of Discovery.; [1926,515], 1637g. slice of the North Star Bay iron; [1972,4], 880g. part-slice of Agpalilik.

Capilla del Monte 30°53′ S., 64°33′ W.
Cordoba, Argentina
Fell 1934
Synonym(s): *Corboda, Cordoba*
Stone. Olivine-bronzite chondrite (H6).
One mass of 750g, L.O. Giacomelli, letters of 30 March 1960, 30 October 1962 in Min. Dept. BM(NH). Olivine Fa_{19}, B. Mason, Geochimica et Cosmochimica Acta, 1963, **27**, p.1011. The classification is given, probably erroneously, as olivine-hypersthene chondrite, and the date of fall the same as that of Achiras, L.M. Villar, Cienc. Investig., 1968, **24**, p.302.
Main mass, Mus. Nat. Inst. Geol., Argentina; 8g New York, Amer. Mus. Nat. Hist.;
Specimen(s): [1962,132], 8g.; [1964,68], 22.8g.

Capitan Range v El Capitan Range.

Capland v Cape of Good Hope.

Capland v Hex River Mountains.

Caralue v Cowell.

Caracoles v Imilac.

Caranzalillo v Corrizatillo.

Caranzatillo v Corrizatillo.

Caratash 38°30′ N., 27° E. approx.
Izmir, Turkey
Fell 1902, August 22, 2000 hrs
Synonym(s): *Karatash*
Stone. Olivine-hypersthene chondrite, amphoterite (LL6).
After appearance of a fireball and detonations, a stone "about as big as a melon" fell, A.S. Anastassiadhis, Bull. Soc. Astron. France, 1902, **16**, p.486, letter of 22 January 1903 in Min. Dept. BM(NH). The main mass was sent to Constantinople. In thin section this stone presents the same characters as Manbhoom, G.T. Prior, priv. comm. Olivine Fa_{29}, B. Mason, Geochimica et Cosmochimica Acta, 1963, **27**, p.1011.
Specimen(s): [86521], 8g.

Caraweena v Carraweena.

Carbo 29°40′ N., 111°30′ W.
Sonora, Mexico
Found 1923
Iron. Octahedrite, medium (0.9mm) (IID).
A mass of 1000lb was found on the Alamo ranch, 40 miles W. of Carbo. Described and analysed, 8.68 %Ni, C. Palache and F.A. Gonyer, Am. Miner., 1930, **15**, p.388 [M.A. 4-423]. The mass was either split in half late in its lifetime or eroded away on one side since its fall, J.H. Hoffman and N.O. Nier, Geochimica et Cosmochimica Acta, 1959, **17**, p.32-36. Further analysis, 10.02 %Ni, 70.0 ppm.Ga, 87.2 ppm.Ge, 13 ppm.Ir, J.T. Wasson, Geochimica et Cosmochimica Acta, 1969, **33**, p.859. Description, references to noble gas and isotopic analyses, V.F. Buchwald, Iron Meteorites, Univ. of California, 1975, p.426.
465kg Harvard Univ., main mass; 24kg Washington, U.S. Nat. Mus.; 3.79kg Tempe, Arizona State Univ.; 641g New York, Amer. Mus. Nat. Hist.;
Specimen(s): [1959,947], 4057g. and sawings, 26.5g

Carcoar v Cowra.

Carcote 24° S., 69° W. approx.
western Cordilleras, Atacama, Chile
Found 1888, known in this year
Stone. Olivine-bronzite chondrite (H5).
Original weight unknown; a piece of 79g was sent to Antofagasta as a "silver ore"; description, F. von Sandberger, Neues Jahrb. Min., 1889, **2**, p.173. Analysis, W. Will and J. Pinnow, Ber. Deutsch. Chem. Gesell. Berlin, 1890, **23**, p.345. Nature of the metal, H.C. Urey and T. Mayeda, Geochimica et Cosmochimica Acta, 1959, **17**, p.113. Olivine Fa_{20}, B. Mason, Geochimica et Cosmochimica Acta, 1963, **27**, p.1011.
208g Würzburg Univ.; 80g Vienna, Naturhist. Mus.; 1g Chicago, Field Mus. Nat. Hist.;
Specimen(s): [67454], 2.5g.

Cardanumbi 32°10′30″ S., 125°38′ E.
Western Australia, Australia
Found 1966
Stone. Olivine-hypersthene chondrite (L5).
A single, completely crusted stone of 6.4g was found, First Suppl. to West. Austr. Mus. Spec. Publ. no. 3, 1968, p.9. Olivine $Fa_{25.7}$, B. Mason, Rec. Austr. Mus., 1974, **29**, p.169.
Main mass, Kalgoorlie, West. Austr. School of Mines; 0.6g Washington, U.S. Nat. Mus.; Thin section, Perth, West. Austr. Mus.;

Cardiff 51°28′ N., 3°10′ W.
Glamorgan, Wales
Fell 1842, before this year
Doubtful. Stone.
A stone is said to have fallen, G. von Boguslawski, Ann. Phys. Chem. (Poggendorff), 1854, **4 (suppl.)**, p.366 but the evidence is very unsatisfactory.

Carevar v Cowra.

Cariani v Vigarano.

Caribrod v Dimitrovgrad.

Carichic 27°56′ N., 107°3′ W.
Near Casas Grandes, Chihuahua, Mexico
Found 1983
Stone. Olivine-bronzite chondrite (H5).
Three very weathered fragments totalling 17kg were found, W. Zeitschel, letter of 12 September, 1983. Olivine Fa_{19}, A.L. Graham, priv. comm., 1983.
Specimen(s): [1983,M.7], 570g.

Carleton Iron v Tucson.

Carlisle Lakes (a) 29°10′ S., 127°05′ E. approx.
Carlisle Lakes, Western Australia, Australia
Found 1977, December 27
Stone. Chondrite, anomalous (CHANOM).
A stone of 49.5g was found 6km S.of Carlisle Lakes on the northern fringe of the Nullarbor Plain. It is one of five meteorites found in the area, the others are normal H-group chondrites. The stone does not resemble Mulga (west) (*q.v.*), olivine Fa_{37}, Meteor. Bull., 1980 (57), Meteoritics, 1980, **15**, p.97. Unequilibrated, R.A. Binns and G.D. Pooley, Meteoritics, 1979, **14**, p.349, abs.

Carlsburg v Ohaba.

Carlstadt 59°22′ N., 13°31′ E.
Sweden
Fell 1822, September 10
Pseudometeorite.
Several stones are said to have fallen, E.F.F. Chladni, Ann. Phys. (Gilbert), 1823, **75**, p.230, 244, but the evidence suggests that the occurrence was an earth tremor, P. Geijer, Arkiv Min. Geol., 1964, **3**, p.465.

Carlton 31°55′ N., 98°2′ W.
Hamilton County, Texas, U.S.A.
Found 1887
Synonym(s): *Carlton-County, Carlton-Hamilton, Corlton-Hamilton, Eroth County (false), False Eroth County, Hamilton County*
Iron. Octahedrite, fine (0.2mm) (IIICD).
A mass of 179lb was ploughed up 5 miles S. of Carlton; described with an analysis, 12.77 %Ni, E.E. Howell, Am. J. Sci., 1890, **40**, p.223. Description and partial analysis for Ga, Au and Pd, E. Goldberg et al., Geochimica et Cosmochimica Acta, 1951, **2**, p.1. Classification and new analysis, 13.0 % Ni, 11.4 ppm.Ga, 8.59 ppm.Ge, 0.076 ppm.Ir, J.T. Wasson and R. Schaudy, Icarus, 1971, **14**, p.59. Description, polycrystalline, V.F. Buchwald, Iron Meteorites, Univ. of California, 1975, p.428. Analysis of included silicates, A. Kracher and G. Kurat, Meteoritics, 1977, **12**, p.282, abs.
9.25kg Chicago, Field Mus. Nat. Hist.; 7.5kg Vienna, Naturhist. Mus.; 5kg New York, Amer. Mus. Nat. Hist.; 3.2kg Harvard Univ.; 5.5kg Michigan Univ.; 3.5kg Paris, Mus. d'Hist. Nat.; 800g Rome, Vatican Colln; 1.4kg Washington, U.S. Nat. Mus.; 1.2kg Tempe, Arizona State Univ.; 335g Los Angeles, Univ. of California; 177g Perth, West. Austr. Mus.; 371g Tübingen, Univ.;
Specimen(s): [65970], 6021g. two slices, and fragments, 20g.; [1959,163], 11.5g.

Carlton-County v Carlton.

Carlton-Hamilton v Carlton.

Carnarvon v Kopjes Vlei.

Carnavelpattu v Mulletiwu.

Caroline 37°59′ S., 140°59′ E.
South Australia, Australia
Found 1941, before this year
Stone. Olivine-bronzite chondrite (H5), veined.
A stone of approximately 800g was found on an aboriginal camp site. Description, F.L. Stillwell, Mem. Nat. Mus. Melbourne, 1941 (12), p.41 [M.A. 8-197]. Co-ordinates, olivine $Fa_{20.9}$, B. Mason, Rec. Austr. Mus., 1974, **29**, p.169.
750g Melbourne, Nat. Mus. Victoria, main mass; 33g Adelaide, South Austr. Mus.; 13g Washington, U.S. Nat. Mus.;
Specimen(s): [1983,M.33], 2g.

Carpentras 44°3′ N., 5°3′ E.
Vaucluse, France
Fell 1738, October 18, 1630 hrs
Doubtful..
Many stones are said to have fallen near Carpentras and Champfort, but nothing has been preserved and the description is inadequate, E.F.F. Chladni, Die Feuer-Meteore, Wien, 1819, p.241.

Carraweena 29°14′ S., 139°56′ E.
South Australia, Australia
Found 1914
Synonym(s): *Accalana, Caraweena*
Stone. Olivine-hypersthene chondrite (L3).
A stone of 28.8kg was found about 6 miles SW. of Carraweena. Analysed and described, A.R. Alderman, Rec. S. Austr. Mus., 1936, **5**, p.537 [M.A. 7-70]. Rare gas study shows that Accalana and Carraweena are identical, D. Heymann, Geochimica et Cosmochimica Acta, 1965, **29**, p.1203. The Accalana specimen was known before 1917 the original weight being 6.25lb, A.R. Alderman, Rec. S. Austr. Mus., 1936, **5**, p.537 [M.A. 7-71]. Co-ordinates, B. Mason, Rec. Austr. Mus., 1974, **29**, p.169. Described, with analysis, 19.83 % total iron, B. Mason and H.B. Wiik, Am. Mus. Novit., 1966 (2273). Unequilibrated, R.T. Dodd et al., Geochimica et Cosmochimica Acta, 1967, **31**, p.921. Monte Coline (*q.v.*) may be part of this fall.
26kg Adelaide, South Austr. Mus., includes main mass, 21.77kg, and main mass of Accalana, 2.5kg; 386g New York, Amer. Mus. Nat. Hist.; 44g Washington, U.S. Nat. Mus.; 20g Tempe, Arizona State Univ.;
Specimen(s): [1937,246], 521g.

Carrisalillo v Vaca Muerta.

Carrizalillo v Corrizatillo.

Carrizalillo v Vaca Muerta.

Carroll County v Eagle Station.

Carsie v Strathmore.

Carsons Well v Needles.

Carthage 36°16′ N., 85°59′ W.
Smith County, Tennessee, U.S.A.
Found 1840
Synonym(s): *Caney Fork, Carthago, Caryfort, Coney Fork, Karthago, Smith County (Tennessee)*
Iron. Octahedrite, medium (1.3mm) (IIIA).
A mass of about 280lb was found; description, G. Troost, Am. J. Sci., 1846, **2**, p.356. Analysis, 8.24 %Ni, 21.5 ppm. Ga., 43.7 ppm.Ge, 0.57 ppm.Ir, E.R.D. Scott et al., Geochimica et Cosmochimica Acta, 1973, **37**, p.1957. Description, some specimens, and possibly the whole mass, were artificially heated, V.F. Buchwald, Iron Meteorites, Univ. of California, 1975, p.431.
64kg Tübingen, Univ.; 12.3kg Harvard Univ.; 6kg Budapest, Nat. Mus.; 2kg Prague, Nat. Mus.; 1.75kg Paris, Mus. d'Hist. Nat.; 1.4kg Washington, U.S. Nat. Mus.; 500g Chicago, Field Mus. Nat. Hist.;
Specimen(s): [20793], 24.61kg. single piece

Carthago v Carthage.

Cartoonkana 29°45′ S., 141°2′ E.
Big Plain, Lake Stewart station, New South Wales, Australia
Found 1914, before this year
Stone. Olivine-hypersthene chondrite (L6).
A stone of 290g was obtained by the South Australian Mus. in 1914 from Yandama station. It is petrologically similar to Yandama (*q.v.*) and the two could belong to the same fall, A.R. Alderman, Rec. S. Austr. Mus., 1936, **5**, p.537 [M.A. 7-70]. Co-ordinates olivine $Fa_{25.7}$, B. Mason, Rec. Austr. Mus., 1974, **29**, p.169. Analysis, 20.04 % total iron, M.J. Fitzgerald, Ph.D. Thesis, Univ. of Adelaide, 1979, p.23.
Main mass, Adelaide, South Austr. Mus.; 1.2g New York, Amer. Mus. Nat. Hist.; thin section, Washington, U.S. Nat. Mus.;

Carver 32°0′ N., 86°0′ W. approx.
Alabama, U.S.A.
Found
Synonym(s): *Alabama*
Iron. Ataxite (IIA).
Exact locality unknown, L. LaPaz, Cat. Coll. Inst. Meteor. Univ. New Mexico, 1965. Alabama is a synonym of both Lime Creek, a nickel-rich ataxite and Walker County, a hexahedrite. Named Carver and described, L. LaPaz, Univ. New Mexico Publ. in Meteoritics no. 6, 1969, p.155. Analysis 5.5 %Ni, 59 ppm.Ga, 184 ppm.Ge, 12 ppm.Ir, A. Kracher et al., Geochimica et Cosmochimica Acta, 1980, **44**, p.773.
9.5kg Albuquerque, Univ. of New Mexico, approximate weight;

Caryfort v Carthage.

Caryfort v Smithville.

Casale v Cereseto.

Casale v Motta di Conti.

Casale Monferrato v Cereseto.

Casale Piemonte v Motta di Conti.

Casas Grandes 30°24′ N., 107°48′ W. approx.
Chihuahua, Mexico
Found 1867, recognized in this year
Synonym(s): *Chihuahua*
Iron. Octahedrite, medium (1.2mm) (IIIA).
A mass of 3407lb was found in an ancient tomb, E.G. Tarayre, Arch. Comm. Sci. Mexique, Paris, 1867, **3**, p.348. As implied, L. Fletcher, Min. Mag., 1890, **9**, p.119 this mass was presented to the Smithsonian Institution in 1876. Description, W. Tassin, Proc. U.S. Nat. Mus., 1902, **25**, p.69. Analysis, 7.74 %Ni, G.P. Merrill, Am. J. Sci., 1913, **35**, p.514. Further analysis, 7.77 %Ni, 19.9 ppm.Ga, 37.4 ppm.Ge, 5.1 ppm.Ir, E.R.D. Scott et al., Geochimica et Cosmochimica Acta, 1973, **37**, p.1957. Historical note, O.E. Monnig, Pop. Astron., Northfield, Minnesota, 1939, **47**, p.152 [M.A. 7-380]. Description, historical details, V.F. Buchwald, Iron Meteorites, Univ. of California, 1975, p.433.
1317kg Washington, U.S. Nat. Mus.; 10kg Chicago, Field Mus. Nat. Hist., approx. weight; 5kg Michigan Univ.; 1.7kg Harvard Univ.; 820g New York, Amer. Mus. Nat. Hist.; 762g Tempe, Arizona State Univ.; 618g Vienna, Naturhist. Mus.; 217g Berlin, Humboldt Univ.;
Specimen(s): [85481], 989g. slice

Caserio Ucera v Ucera.

Casey County 37°15′ N., 85°0′ W. approx.
Casey County, Kentucky, U.S.A.
Found 1877, known in this year
Iron. Octahedrite, coarse (2.2mm) (IA).
Original weight unknown. First mentioned, J.L. Smith, Am. J. Sci., 1877, **14**, p.246. Description, A. Brezina, Sitzungsber. Akad. Wiss. Wien, Math.-naturwiss. Kl., 1880, **82** (1), p.351. Analysis, 6.96 %Ni, 81.7 ppm.Ga, 317 ppm.Ge, 1.1 ppm.Ir, J.T. Wasson, Icarus, 1970, **12**, p.407. Part of the mass has been heated and forged, V.F. Buchwald, Iron Meteorites, Univ. of California, 1975, p.435.
167g Harvard Univ.; 92g Washington, U.S. Nat. Mus.; 71g Chicago, Field Mus. Nat. Hist.; 65g Vienna, Naturhist. Mus.; 21g New York, Amer. Mus. Nat. Hist.;
Specimen(s): [53294], 45.5g.

Cashion 35°50′52″ N., 97°42′12″ W.
Kingfisher County, Oklahoma, U.S.A.
Found 1936
Synonym(s): *Cushion*
Stone. Olivine-bronzite chondrite (H4), black.
A partly crusted stone of 5896.8g was ploughed up on a farm 2.25 miles NW. of Cashion, H.O. Stockwell and R.A. Morley, Earth Sci. Digest, 1953, **7**, p.20 [M.A. 12-360], F.C. Leonard, Geochimica et Cosmochimica Acta, 1963, **27**, p.261. Olivine Fa_{18}, B. Mason, Geochimica et Cosmochimica Acta, 1963, **27**, p.1011.
497g Copenhagen, Univ. Geol. Mus.; 85g Los Angeles, Univ. of California; 72g New York, Amer. Mus. Nat. Hist.; 71g Washington, U.S. Nat. Mus.; 52g Yale Univ.;
Specimen(s): [1960,103], 139.5g.; [1960,104], 152g. three fragments

Casignano v Borgo San Donino.

Casilda 33°6′ S., 61°8′ W.
Cordoba, Argentina
Found 1937
Stone. Olivine-bronzite chondrite (H5).
The co-ordinates are for Casilda, near which one 5.25kg stone was found, Meteor. Bull., 1974 (52), Meteoritics, 1974, **9**, p.108. Description, with an analysis, A.J. Toselli and M. de Brootkorb, Acta Geol. Lilloana, 1972, **12**, p.133.
Main mass, Inst. Geol. Univ. Nacional de Tucuman; 203g Washington, U.S. Nat. Mus.;
Specimen(s): [1983,M.43], 12g.

Casimiro de Abreu 22°28′ S., 42°13′ W.
Rio de Janeiro, Brazil
Found 1947
Iron. Octahedrite, medium (1.3mm) (IIIA).
A mass of 24kg was found on the Andorinhas estate. Described, with an analysis, 8.27 %Ni, W.S. Curvello, Bol. Mus. Nac. Rio de Janeiro, 1950 (geol. no. 11) [M.A. 11-446]. Further analysis, 8.4 %Ni, 20.9 ppm.Ga, 41.0 ppm.Ge, 0.25 ppm.Ir, E.R.D. Scott et al., Geochimica et Cosmochimica Acta, 1973, **37**, p.1957. Description, cosmic heating and recrystallisation, V.F. Buchwald, Iron Meteorites, Univ. of California, 1975, p.436.
212g Washington, U.S. Nat. Mus.; 180g Tempe, Arizona State Univ.; Main mass, Rio de Janeiro, Mus. Nac.;

Cassandra 39°59′ N., 23°24′ E.
Macedonia, Greece
Doubtful..
A cult object worshipped at Cassandra may have been a meteorite, G.A. Wainwright, J. Egypt. Archaeol., 1930, **16**, p.35.

Castalia 36°5′ N., 78°4′ W.
Nash County, North Carolina, U.S.A.
Fell 1874, May 14, 1430 hrs
Synonym(s): *Nash County*
Stone. Olivine-bronzite chondrite (H5), brecciated, xenolithic.
After detonations, "a dozen or more" stones fell over an area of 10 by 3 miles but only three, of 5.5kg, 1kg, and 800g, were found. Described, analysis, J.L. Smith, Am. J. Sci., 1875, **10**, p.147. The analysis shows no CaO and is doubtful, W. Wahl, Geochimica et Cosmochimica Acta, 1950, **1**, p.28. Further analysis, 25.83 % total iron, H.B. Wiik, Comm. Phys.-Math. Soc. Sci. Fenn., 1969, **34**, p.135. Xenolithic, olivine Fa_{19}, R.A. Binns, Geochimica et Cosmochimica Acta, 1968, **32**, p.299.
5.5kg Vienna, Naturhist. Mus., the 5.5kg stone.; 12g Harvard Univ.; 189g Chicago, Field Mus. Nat. Hist.; 176g Yale Univ.; 89g New York, Amer. Mus. Nat. Hist.; 26g Tempe, Arizona State Univ.;
Specimen(s): [50804], 27g.

Castel Beradengo error for Castel Berardenga.

Castel Berardenga 43°21′ N., 11°30′ E.
Tuscany, Italy
Fell 1791, May 17, 0500 hrs
Stone.
Several stones fell, but nothing has been preserved; however it is clear from the description that they were true meteorites - they had a black crust and contained metallic iron, D.A. Soldani, Atti Accad. Sci. Siena, 1808, **9**, p.1, E.F.F. Chladni, Die Feuer-Meteore, Wien, 1819, p.260, B. Baldanza, Min. Mag., 1965, **35**, p.214. The co-ordinates given are those of Castelnuovo Berardenga, the modern name of Castel Berardenga, G.R. Levi-Donati, priv. comm. 30 April 1969 in Min. Dept. BM(NH).

Caster Farm v Beeler.

Castine 44°23′ N., 68°45′ W.
Hancock County, Maine, U.S.A.
Fell 1848, May 20, 0400 hrs
Synonym(s): *Augusta*
Stone. Olivine-hypersthene chondrite (L6), veined.
A stone of about 3oz was seen to fall after the appearance of a fireball and detonations; described, with an analysis of the metal, C.U. Shepard, Am. J. Sci., 1848, **6**, p.251. Olivine Fa_{25}, B. Mason, Geochimica et Cosmochimica Acta, 1963, **27**, p.1011.
42g Chicago, Field Mus. Nat. Hist.; 25g Tempe, Arizona State Univ.; 15g Yale Univ.; 2.5g Budapest, Nat. Mus.;
Specimen(s): [34592], 2.5g.

Castor Brothers' Farm v Beeler.

Castor Farm v Beeler.

Castray River 41°30′ S., 145°25′ E.
Tributary of the Heazlewood River, Russell County, Tasmania, Australia
Found 1899
Iron.
Three pieces, each of about 3.3g were found, W.F. Petterd, Proc. Roy. Soc. Tasmania, 1900-1(1902), p.48. Meteoritic nature unconfirmed, V.F. Buchwald, Iron Meteorites, Univ. of California, 1975, p.438.
3.3g Launceston, Tasmania, Queen Victoria Mus., main mass;

Castrovillari 39°48′ N., 16°12′ E.
Cosenza, Calabria, Italy
Fell 1583, January 9
Stone.
A stone, 33lb, fell but nothing has been preserved, D.A. Soldani, Atti Accad. Sci. Siena, 1808, **9**, p.1, E.F.F. Chladni, Die Feuer-Meteore, Wien, 1819, p.219, E.F.F. Chladni, Phil. Mag., 1826, **67**, p.3, 179, B. Baldanza, Min. Mag., 1965, **35**, p.214.

Caswell County 36°30′ N., 79°15′ W. approx.
North Carolina, U.S.A.
Fell 1810, January 30, 1400 hrs
Stone. Chondrite?
A stone of 1360g fell, and was described as resembling Weston; it is not known whether any material is preserved. Described, Bishop Madison, Med. Philos. Register, 1814, **1 (2nd. edition)**, p.118. Listed, F.P. Venable, List of Meteorites of N. Carolina, J. Elisha Mitchell Sci. Soc., 1890, **7**, p.33. References listed, E.P. Henderson, letter of 3 June 1939 in Min. Dept. BM(NH).

Catamarca v Imilac.

Catamarka v Imilac.

Catherwood 51°57′54″ N., 107°26′18″ W.
Saskatchewan, Canada
Found 1965, possibly 1964
Stone. Olivine-hypersthene chondrite (L6).
A 3.92kg stone was found, Meteor. Bull., 1974 (52), Meteoritics, 1974, **9**, p.109. Contains ringwoodite and majorite, olivine $Fa_{24.7}$, L.C. Coleman, Canadian Min., 1977, **15**, p.97 [M.A. 79-2718].
3.2kg Ottawa, Geol. Surv. Canada;

Catorce v Charcas.

Catozze v Charcas.

Cavour 44°20′ N., 98°2′ W.
Beadle County, South Dakota, U.S.A.
Found 1943, before this year
Stone. Olivine-bronzite chondrite (H6), brecciated.
An 8lb mass, at first thought to be an iron, was found on a farm 11 miles SW. of Cavour. In 1941 and 1942 two other masses, 8lb 6oz and 9lb, were found nearby, and later four more masses, three of which weigh 12.5lb, 8lb, and 1lb. A 5lb mass examined by J.D. Buddhue proved to be a dark, much oxidised and very friable stone, W.J. Lindsey, The Mineralogist, Portland, Oregon, 1943, **11**, p.316., W.J. Lindsey, Pop. Astron., Northfield, Minnesota, 1945, **53**, p.40, Contr. Soc. Res. Meteorites, 1945, **3**, p.119, 178 [M.A. 9-301]. Analysis, J.D. Buddhue, Pop. Astron., Northfield, Minnesota, 1948, **56**, p.385 [M.A. 10-403]. Olivine Fa_{19}, B.

Mason, Geochimica et Cosmochimica Acta, 1963, **27**, p.1011.
10kg Washington, U.S. Nat. Mus.; 3.9kg Albuquerque, Univ. of New Mexico; 729g Fort Worth, Texas, Monnig Colln.; 301g Chicago, Field Mus. Nat. Hist.;
Specimen(s): [1953,7], 42.5g.

Cecil Plains 27°31′ S., 151°11′ E.
Queensland, Australia
Doubtful..
A supposed meteorite has been listed under this name, G.W. Card, Ms. list of Australian Meteorites in Min. Dept. BM (NH)., but no further information has been received and the report is discredited, T. Hodge-Smith, Mem. Austr. Mus., 1939 (7), p.34.

Cedar (Kansas) 39°42′ N., 98°59′ W.
Smith County, Kansas, U.S.A.
Found 1937
Synonym(s): *Smith County (Kansas)*
Stone. Olivine-bronzite chondrite (H6), black.
A stone of 4.9kg was found, A.D. Nininger, Pop. Astron., Northfield, Minnesota, 1939, **47**, p.214, Ann. Rep. U.S. Nat. Mus., 1938, p.108, E.P. Henderson, letter of 3 June 1939 in Min. Dept. BM(NH)., H.H. and A.D. Nininger, The Nininger Collection of Meteorites, Winslow, Arizona, 1950, p.42. Olivine Fa_{18}, B. Mason, Geochimica et Cosmochimica Acta, 1963, **27**, p.1011. Co-ordinates, B. Mason, Meteorites, Wiley, 1962, p.232.
1.46kg Tempe, Arizona State Univ.; 191g Washington, U.S. Nat. Mus.; 77g Chicago, Field Mus. Nat. Hist.;
Specimen(s): [1938,304], 128.8g.; [1959,1005], 1545g.

Cedar (Texas) 29°54′ N., 96°48′ W. approx.
Fayette County, Texas, U.S.A.
Found 1890
Stone. Olivine-bronzite chondrite (H4), veined.
Three stones of 16.5lb, 12lb, and 2.5lb were found, and are said to differ from Bluff (*q.v.*) in microscopic characters, G.P. Merrill, Proc. U.S. Nat. Mus., 1918, **54**, p.557. A fourth stone, of 25.5lb was found in about 1906. Analysis, 22.11 % total iron, B. Mason and H.B. Wiik, Am. Mus. Novit., 1967 (2280). Olivine Fa_{18}, B. Mason, Geochimica et Cosmochimica Acta, 1963, **27**, p.1011. The main mass of the 16.5lb stone is in Chicago, Field Mus. Nat. Hist. labelled 'Bluff' in their 1965 catalogue.
500g Tempe, Arizona State Univ.; 2.18kg Washington, U.S. Nat. Mus., main mass of the 12lb stone; 2.5lb Waco, Texas, Baylor Univ., probable location of main mass of 2.5lb stone; 25.5lb. Austin, Texas, Bureau of Econ. Geol.;
Specimen(s): [1924,441], 215.7g. slice

Cedar Creek v Alexander County.

Cedartown 34°1′ N., 85°16′ W.
Polk County, Georgia, U.S.A.
Found 1898, before this year
Synonym(s): *Aragon, Cedertown*
Iron. Hexahedrite (IIA).
A somewhat rusted mass of 11.3kg was ploughed up before 1898 about 62 miles WNW. of Atlanta. It was recovered from the ashes of a house that had been burnt down, and though exposed to a temperature estimated at 560C. showed no alteration of its microstructure. Described, with an analysis, 5.48 %Ni, S.H. Perry, Smithsonian Misc. Coll., 1946, **104** (23) [M.A. 10-177]. Analysis, 5.36 %Ni, 63.2 ppm.Ga, 181 ppm.Ge, 8.2 ppm.Ir, J.T. Wasson, Meteorites, Springer-Verlag, 1974, p.299. Includes Aragon, effects of cosmic heating, V.F. Buchwald, Iron Meteorites, Univ. of California, 1975, p.438.
9940g Washington, U.S. Nat. Mus., main mass; 318g Chicago, Field Mus. Nat. Hist., includes Aragon; 271g New York, Amer. Mus. Nat. Hist.; 245g Harvard Univ.; 102g Michigan Univ.;

Cedertown v Cedartown.

Cee Vee 34°12′ N., 100°29′ W.
Cottle County, Texas, U.S.A.
Found 1959, or 1960, recognized 1965
Stone. Olivine-bronzite chondrite (H5).
One mass of 3.6kg was ploughed up in a field, Meteor. Bull., 1967 (40), Meteoritics, 1970, **5**, p.94.
1422g Mainz, Max-Planck-Inst., main mass; 169.8g Arcadia, California, Colln. of H.J. Meyers; 130g Washington, U.S. Nat. Mus.; 67g Los Angeles, Univ. of California; 13g Chicago, Field Mus. Nat. Hist.;
Specimen(s): [1966,521], 80g. slice

Cella di Costamezzana v Borgo San Donino.

Centerville 43°12′ N., 96°55′ W.
Turner County, South Dakota, U.S.A.
Fell 1956, February 29, at night
Synonym(s): *McMurchie*
Stone. Olivine-bronzite chondrite (H5).
A mass of 45.6g fell through the roof of a shed, H. Martin, Proc. South Dakota Acad. Sci., 1958, **37**, p.135, Meteor. Bull., 1963 (26). Olivine Fa_{19}, B. Mason, Geochimica et Cosmochimica Acta, 1967, **31**, p.1100.
Main mass, Rapid City, South Dakota, South Dakota School of Mines and Technology; 0.7g Tempe, Arizona State Univ.;

Cento v Renazzo.

Central Missouri v Ainsworth.

Central Wyoming v Willow Creek.

Cereseto 45°5′ N., 8°18′ E.
Alessandria, Piemonte, Italy
Fell 1840, July 17, 0730 hrs
Synonym(s): *Casale, Casale Monferrato, Mailand, Milan, Offiglia, Ottiglio, Pastrona, Piedmont*
Stone. Olivine-bronzite chondrite (H5), brecciated.
A stone of about 5kg fell, after detonations, and appearance of a fireball, Ann. Phys. Chem. (Poggendorff), 1840, **50**, p.668. Listed, B. Baldanza, Min. Mag., 1965, **35**, p.214. Olivine Fa_{19}, B. Mason, Geochimica et Cosmochimica Acta, 1963, **27**, p.1011. The co-ordinates are of Cereseto, G.R. Levi-Donati, priv. comm., 29 April 1969 in Min. Dept. BM (NH). Mentioned, F. Burragato, Periodico Miner., 1967, **36**, p.463.
3.4kg Turin Univ.; 2.02kg Perugia Univ. (Italian Center for Meteorite Studies); 258g Vienna, Naturhist. Mus.; 182g Budapest, Nat.Hist. Mus.; 65g Washington, U.S. Nat. Mus.; 27g Prague, Nat. Mus.; 16g Berlin, Humboldt Univ.;
Specimen(s): [33960], 102.5g.; [33297], 21.7g.

Cerralvo v Coahuila.

Cerro Cosina v Cosina.

Cerro la Bomba v Vaca Muerta.

Cerros del Buei Muerto v North Chile.

Cerros del Buen Muerto v North Chile.

Cerros del Buey Muerto v North Chile.

Cesena v Valdinoce.

Ceylon v Mulletiwu.

Ceylon

Pseudometeorite..

Two pieces of iron of 23g and 21g in the Museum of Practical Geology, London, labelled "Ceylon, found 1869 in the hands of the natives, working into charms as 'Lightning Iron'," are referred to, E.A. Wülfing, Die Meteoriten in Samml., Tübingen, 1897, p.389. Contains no nickel (G.T. Prior).

Chachak v Jelica.

Chaco Gualambo v Campo del Cielo.

Chaco Santafecino

Pseudometeorite..

A mass of limonite in the Nat. Mus., Buenos Aires, is described and shown to be of terrestrial origin, E.H. Ducloux, Rev. fac. cienc. quim. Univ. nac. La Plata, 1928, **5** (1), p.77 [M.A. 4-121].

Chaharwala v Charwallas.

Chail 25°22′ N., 81°40′ E.

Allahabad, Uttar Pradesh, India

Fell 1814, November 5, 1630 hrs

Synonym(s): *Allahabad*

Stone. Olivine-bronzite chondrite (H), brecciated.

Nineteen stones, some of them up to 30lb in weight, were said to have fallen in the Doab, after detonations, Phil. Mag., 1815, **46**, p.155. Olivine Fa_{18}, B. Mason, Geochimica et Cosmochimica Acta, 1963, **27**, p.1011.

Specimen(s): [63874], 0.5g. the only known specimen

Chainpur 25°51′ N., 83°29′ E.

Azamgarh district, Uttar Pradesh, India

Fell 1907, May 9, 1330 hrs

Stone. Olivine-hypersthene chondrite, amphoterite (LL3).

Several stones appear to have fallen after detonations near villages on the borders of Ghazipur and Azamgarh districts, and about 18lb of fragments were recovered, including one from Chainpur weighing about 12lb, G. de P. Cotter, Rec. Geol. Surv. India, 1912, **42**, p.268. Description and analysis, 19.78 % total iron, K. Keil et al., Am. Mus. Novit., 1963 (2173) [M.A. 17-56]. Classification and variation in olivine composition, R.T. Dodd et al., Geochimica et Cosmochimica Acta, 1967, **31**, p.921, R.T. Dodd, Geochimica et Cosmochimica Acta, 1969, **33**, p.161. Chemical variation between chondrules, L.S. Walter and R.T. Dodd, Meteoritics, 1972, **7**, p.341. Xe retention age, F.A. Podosek, Geochimica et Cosmochimica Acta, 1970, **34**, p.341. H/D ratios, F. Robert et al., Nature, 1979, **282**, p.785. Preferred orientation of chondrules, P.M. Martin and A.A. Mills, Earth planet. Sci. Lett., 1980, **51**, p.18.

8kg Calcutta, Mus. Geol. Surv. India; 160g New York, Amer. Mus. Nat. Hist.; 114g Washington, U.S. Nat. Mus.; 36g Tempe, Arizona State Univ.;

Specimen(s): [1915,86], 347.5g. and fragments, 3g

Chajari 30°47′ S., 58°3′ W.

Federacion department, Entre Rios, Argentina

Fell 1933, November 29, 1300 hrs

Synonym(s): *Entre Rios I*

Stone. Olivine-hypersthene chondrite (L5).

One mass of 18.3kg was recovered, L.O. Giacomelli, letters of 30 March, 19 Sept. 1960 and 25 June, 30 Oct. 1962 in Min. Dept. BM(NH). Reported, Meteor. Bull., 1961 (21). Classification and analysis, M.E. Teruggi and L.O. Giacomelli, Rev. Mus. La Plata, 1968, **6** (sec. geol.), p.189.

11kg L.O. Giacomelli's collection, main mass; 100g Buenos Aires, Mus. Dir. Geol. Min.;

Chamberlin 36°12′ N., 102°27′ W.

Dallam County, Texas, U.S.A.

Found 1941

Stone. Olivine-bronzite chondrite (H5).

One mass of 2.4kg was found, W. Wahl, letter of 23 May 1950 in Min. Dept. BM(NH)., H.H. and A.D. Nininger, The Nininger Collection of Meteorites, Winslow, Arizona, 1950, p.42. Olivine Fa_{18}, B. Mason, Geochimica et Cosmochimica Acta, 1963, **27**, p.1011.

792g Tempe, Arizona State Univ.; 362g Fort Worth, Texas, Monnig Colln.; 271g Chicago, Field Mus. Nat. Hist.; 153g Los Angeles, Univ. of California; 98g Washington, U.S. Nat. Mus.;

Specimen(s): [1959,835], 765g.

Chambord 48°27′ N., 72°4′ W.

Lake St. John County, Quebec, Canada

Found 1904

Iron. Octahedrite, medium (0.9mm) (IIIA).

A mass of 6.6kg was found, R.A.A. Johnston, The Ottawa Naturalist, 1906, **20**, p.51. Analysis, 7.53 %Ni, 18.4 ppm.Ga, 35.0 ppm.Ge, 10 ppm.Ir, E.R.D. Scott et al., Geochimica et Cosmochimica Acta, 1973, **37**, p.1957. Shock hardened, epsilon structure, V.F. Buchwald, Iron Meteorites, Univ. of California, 1975, p.441.

6.47kg Ottawa, Geol. Surv. Canada; 18g Chicago, Field Mus. Nat. Hist.;

Chañaral v Ilimaes (iron).

Chañaral v Merceditas.

Chañaral v Vaca Muerta.

Chañarlino v Merceditas.

Chandakapur 20°16′ N., 76°1′ E.

Buldhana, Maharashtra, India

Fell 1838, June 6, 1200 hrs

Synonym(s): *Berar, Burguon, Denulguon*

Stone. Olivine-hypersthene chondrite (L5), brecciated.

Three stones, weighing respectively about 11lb, 7.5lb and 1lb, fell, after detonations, at the villages Chandakapur, Denulgaon, and Burguon, (Burgaon-Dewalgaon), C.A. Silberrad, Min. Mag., 1932, **23**, p.295. Description, analysis, H.L. Bowman, Min. Mag., 1910, **15**, p.350. Olivine Fa_{24}, B. Mason, Geochimica et Cosmochimica Acta, 1963, **27**, p.1011.

3.5kg Edinburgh, Roy. Scot. Mus.; 1839g Oxford, Univ.; 1.46kg Vienna, Naturhist. Mus.; 197g Tübingen, Univ.; 83g Chicago, Field Mus. Nat. Hist.; 126g Tempe, Arizona State Univ.; 41g Calcutta, Mus. Geol. Surv. India;

Specimen(s): [16354], 486g. and fragments, 15g; [16868], 225g. and powder, 10g.

Chandpur 27°17′ N., 79°3′ E.

Mainpuri district, Uttar Pradesh, India
Fell 1885, April 6, evening hrs
Synonym(s): *Mainpuri*
Stone. Olivine-hypersthene chondrite (L6), veined.

After detonations, a stone of about 2.5lb was heard to fall and was found next day 5 miles NW. of Mainpuri, H.B. Medlicott, Rec. Geol. Surv. India, 1885, **18**, p.148. Olivine Fa_{24}, B. Mason, Geochimica et Cosmochimica Acta, 1963, **27**, p.1011.

384g Calcutta, Mus. Geol. Surv. India; 88g Vienna, Naturhist. Mus.; 59g Paris, Mus. d'Hist. Nat.;
Specimen(s): [56444], 477g. about half the stone

Changanorein v Cranganore.

Changde 29°05′ N., 111°45′ E.

Changde, Hunan, China
Fell 1977, March 11, 1154 hrs
Stone. Olivine-bronzite chondrite (H5).

After a fireball and detonations, a shower of stones, totalling 1.81kg, fell within an elliptical area 6km by 4km. The largest individual weighed 900g, Meteor. Bull., 1978 (55), Meteoritics, 1978, **13**, p.332. Listed, D. Wang et al., Geochimica, 1977, p.288, D. Bian, Meteoritics, 1981, **16**, p.115. Analysis, 28.21 % total iron, W. Daode et al., Geochemistry, 1982, **1**, p.186.

900g Guiyang, Acad. Sin. Inst. Geochem.;

Changxing 31°20′ N., 121°40′ E.

Shanghai, China
Found 1964
Synonym(s): *Changxing Dao*
Stone.

Total mass recovered was 26.9kg, D. Bian, Meteoritics, 1981, **16**, p.115.

Changxing Dao v Changxing.

Channing 35°43′ N., 102°17′ W.

Hartley County, Texas, U.S.A.
Found 1936
Stone. Olivine-bronzite chondrite (H5), black.

One stone was found in fragments, total 15.3kg, A.D. Nininger, Pop. Astron., Northfield, Minnesota, 1937, **45**, p.449 [M.A. 7-62]. Olivine Fa_{19}, B. Mason, Geochimica et Cosmochimica Acta, 1963, **27**, p.1011.

2.5kg Tempe, Arizona State Univ.; 639g Chicago, Field Mus. Nat. Hist.; 1530g Washington, U.S. Nat. Mus.; 63g New York, Amer. Mus. Nat. Hist.;
Specimen(s): [1937,388], 45g.; [1959,826], 2434g.

Chantonnay 46°41′ N., 1°3′ E.

Vendée, France
Fell 1812, August 5, 0200 hrs
Synonym(s): *Bourbon-Vendée, La Rochelle*
Stone. Olivine-hypersthene chondrite (L6), brecciated.

A stone of 31.5kg fell, after appearance of a fireball and detonations, -.Cavoleau, Ann. Phys. (Gilbert), 1819, **63**, p.228. Description, G. Tschermak, Sitzungsber. Akad. Wiss. Wien, Math.-naturwiss. Kl., 1874, **70**, p.465. Analysis, C. Rammelsberg, Z. Deutsch. Geol. Ges., 1870, **22**, p.889. Further bulk and mineral analysis, olivine $Fa_{23.9}$, 21.13 % total iron, R.T. Dodd and E. Jarosewich, Meteoritics, 1981, **16**, p.93.

2.75kg Vienna, Naturhist. Mus.; 2.1kg Paris, Mus. d'Hist. Nat.; 465g Paris, Ecole des Mines; 371g Washington, U.S. Nat. Mus.; 244g Berlin, Humboldt Univ.; 185g Moscow, Acad. Sci.; 359g Tempe, Arizona State Univ.; 181g Tübingen, Univ.; 114g Copenhagen, Min. Mus. Univ.; 160g New York, Amer. Mus. Nat. Hist.; 99g Chicago, Field Mus. Nat. Hist.; 96g Oxford, Univ.;
Specimen(s): [90261], 582g.; [33905], 531g.; [33906], 239g.; [33190], 48.5g.

Charata v Campo del Cielo.

Charca v La Charca.

Charcas 23°5′ N., 101°1′ W.

San Luis Potosi, Mexico
Found 1804, known in this year
Synonym(s): *Agua Blanca, Catorce, Catozze, Descubridora, Poblazon, San Luis Potosi, Venagas*
Iron. Octahedrite, medium (1.1mm) (IIIA).

A mass of over 780kg was mentioned, F.T. Sonnenschmid, Tablas Mineralogicas, Mexico, 1804, p.288. It was standing at the corner of the churchyard at Charcas, and was said to have been brought from San José del Sitio, 12 leagues (50km) distant; in 1886 it was removed to Paris, L. Fletcher, Min. Mag., 1890, **9**, p.157, 160. Description, G.A. Daubrée, C. R. Acad. Sci. Paris, 1867, **64**, p.633, 636. Descubridora, 576kg, was found between 1780 and 1783 at about 23°40′N., 100°51′W., H.J. Burkart, Neues Jahrb. Min., 1874, p.22. A further 41.5kg was found in 1855 at Catorce, G.F. Kunz, Am. J. Sci., 1887, **33**, p.233. Some fragments, including specimen no. 119 in Prague (Nat. Mus.), nos. 1355 and 146 in Washington, U.S. Nat. Mus., no. A770 in Vienna, Naturhist. Mus. have been hammered and heated to 900C. The main mass in Paris (700kg) and specimen no. 356.1x in Tempe, Arizona State Univ. are probably undamaged, V.F. Buchwald, Iron Meteorites, Univ. of California, 1975, p.442. Synonymy with Descubridora, analysis, 7.86 %Ni, 19.4 ppm.Ga, 41.4 ppm.Ge, 1.9 ppm.Ir, E.R.D. Scott et al., Geochimica et Cosmochimica Acta, 1973, **37**, p.1957.

700kg Paris, Mus. d'Hist. Nat.; 39kg Chicago, Field Mus. Nat. Hist., includes 33kg of Descubridora; 509g Los Angeles, Univ. of California, of Descubridora; 366g Budapest, Nat. Hist. Mus.; large mass, Mexico City, Nat. Mus., main mass of Descubridora; 41kg Vienna, Naturhist. Mus., of Descubridora (Catorce); 3kg Washington, U.S. Nat. Mus., of Descubridora; 3kg New York, Amer. Mus. Nat. Hist., of Descubridora; 1.5kg Harvard Univ., of Descubridora; 1.5kg Mexico City, Inst. Geol., of Descubridora; 668g Prague, Nat. Mus., of Descubridora; 673g Tempe, Arizona State Univ.;
Specimen(s): [68957], 295g.; [41109], 38.5g.; [1959,948], 721g. slice, and sawings, 12.5g; [85075], 4324g. slice, of Descubridora, and fragments, 83g; [64203], 29g. of Descubridora

Charkov v Kharkov.

Charleroi 50°24′ N., 4°28′ E.

Belgium
Fell 1634, October 27, 0800 hrs
Doubtful. Stone.

Several stones, weighing up to 8lb, fell; nothing is preserved, E.F.F. Chladni, Die Feuer-Meteore, Wien, 1819, p.223 and the evidence is inadequate.

Charles County v Nanjemoy.

Charleston v Jenny's Creek.

Charlotte v Monroe.

Charlotte 36°10′ N., 87°20′ W.
Dickson County, Tennessee, U.S.A.
Fell 1835, July 31, 1400-1500 hrs, or August 1
Synonym(s): *Dickson County*
Iron. Octahedrite, fine (0.3mm) (IVA).
A drop-shaped mass of about 9.5lb fell, after detonation and a vivid light, G. Troost, Am. J. Sci., 1845, **49**, p.337. Analysis, 8.01 %Ni, J.L. Smith, Am. J. Sci., 1875, **10**, p.349. Description, E. Cohen, Meteoritenkunde, 1905, **3**, p.320. Occluded gases analysed, A.W. Wright, Am. J. Sci., 1876, **11**, p.257. Analysis, 8.04 %Ni, 2.24 ppm.Ga, 0.118 ppm.Ge, 1.5 ppm.Ir, R. Schaudy et al., Icarus, 1972, **17**, p.174. Partly reheated and rapidly cooled, possibly during atmospheric flight, V.F. Buchwald, Iron Meteorites, Univ. of California, 1975, p.447.
2kg Harvard Univ., main mass, anterior portion; 182g Washington, U.S. Nat. Mus.; 177g Tempe, Arizona State Univ.; 168g Vienna, Naturhist. Mus.; 58g New York, Amer. Mus. Nat. Hist.;
Specimen(s): [50808], 62g.; [20199], 15.5g.

Charsonville 47°56′ N., 1°34′ E.
Meung, Loiret, France
Fell 1810, November 23, 1330 hrs
Synonym(s): *Beaugency, Bois de Fontaine, Boisfontaine, Chartres, La Touanne, Meung sur Loire, Orléans, Touanne*
Stone. Olivine-bronzite chondrite (H6), veined.
Three stones fell, after detonations, but only two were found, one of 18kg and the other of 9kg, P.M.S. Bigot de Morogues, Ann. Phys. (Gilbert), 1811, **37**, p.349, L.N. Vauquelin, Ann. Phys. (Gilbert), 1812, **40**, p.83. Bois de Fontaine and Chartres, J.R. Gregory, Geol. Mag., 1886, **3**, p.357, L. Fletcher, Min. Mag., 1889, **8**, p.146 are identical with Charsonville. Olivine Fa_{18}, B. Mason, Geochimica et Cosmochimica Acta, 1963, **27**, p.1011.
4.5kg Paris, Mus. d'Hist. Nat.; 640g Vienna, Naturhist. Mus.; 309g New York, Amer. Mus. Nat. Hist.; 93g Tempe, Arizona State Univ.; 83g Washington, U.S. Nat. Mus.; 78g Chicago, Field Mus. Nat. Hist.; 53g Los Angeles, Univ. of California; 49g Berlin, Humboldt Univ.;
Specimen(s): [1920,294], 153g. of Charsonville; [42516], 36.5g. of Charsonville; [90257], 3g. of Charsonville; [56542], 1163g. of Bois de Fontaine; [53598], 85.5g. of Bois de Fontaine; [52151], 1.5g. of Bois de Fontaine; [63875], 20g. of Chartres; [96258], thin section

Chartres v Charsonville.

Charwallas 29°29′ N., 75°30′ E.
Hisar district, Haryana, India
Fell 1834, June 12, 0800 hrs
Synonym(s): *Chaharwala*
Stone. Olivine-bronzite chondrite (H6).
A stone of about 26.5lb fell, after detonations, -. Parsons, J. Asiatic Soc. Bengal, 1834, **3**, p.413 where the date of fall is given doubtfully as June 8, whereas in the Cat. of the Indian Mus. Calcutta, it is given as June 12. The place of fall is Chaharwala, co-ordinates as above, C.A. Silberrad, Min. Mag., 1932, **23**, p.295. Only a few grams known in collections. Olivine Fa_{19}, B. Mason, Geochimica et Cosmochimica Acta, 1963, **27**, p.1011.
60g Tempe, Arizona State Univ.; 19g Vienna, Naturhist. Mus.; 9g Budapest, Nat. Mus.;
Specimen(s): [25461], 30.5g.; [34602], 7g.; [96259], Thin section

Chassigny 47°43′ N., 5°22′ E.
Haute Marne, France
Fell 1815, October 3, 0800 hrs
Synonym(s): *Langres, Shassini*
Stone. Achondrite, Ca-poor. Chassignite (ACANOM).
A stone (or perhaps several) fell after detonations, the total weight being about 4kg, -. Pistollet, Ann. Phys. (Gilbert), 1818, **58**, p.171. It contains chondrules. Chemical analysis and mineralogical description, E. Jérémine et al., Bull. Soc. franc. Min. Crist., 1962, **85**, p.262 [M.A. 16-171], Метеоритика, 1960, **18**, p.66. XRF analysis, 21.34 % total iron, T.S. McCarthy et al., Meteoritics, 1974, **9**, p.215. Analyses of minerals and trace element analysis, olivine Fa_{32}, B. Mason et al., Meteoritics, 1976, **11**, p.21. A cumulate dunite, with kaersutite bearing melt inclusions, R.J. Floran et al., Geochimica et Cosmochimica Acta, 1978, **42**, p.1213.
344g Paris, Mus. d'Hist. Nat.; 23g Paris, Ecole des Mines; 100g Vienna, Naturhist. Mus.; 27g Rome, Vatican Colln; 23g Washington, U.S. Nat. Mus.; 18g Oslo, Min.-Geol. Mus.; 13g Berlin, Humboldt Univ.; 7g Tempe, Arizona State Univ.;
Specimen(s): [19972], 32g.; [33991], 6g.

Château Renard 47°56′ N., 2°55′ E.
Montargis, Loiret, France
Fell 1841, June 12, 1330 hrs
Synonym(s): *Trigueres, Triguerre*
Stone. Olivine-hypersthene chondrite (L6), veined.
A stone of about 30kg fell, after detonations and appearance of a fireball, -. Delavaux, C. R. Acad. Sci. Paris, 1841, **12**, p.1190. Metal composition, G.T. Prior, Min. Mag., 1919, **18**, p.354. Analysis, 22.20 % total iron, B. Mason and H.B. Wiik, Am. Mus. Novit., 1961 (2069). Olivine Fa_{25}, B. Mason, Geochimica et Cosmochimica Acta, 1963, **27**, p.1011. Contains maskelynite, R.A. Binns, Nature, 1967, **213**, p.1111.
1.5kg Paris, Mus. d'Hist. Nat.; 2.4kg Paris, Ecole des Mines; 1kg Tübingen, Univ.; 837g Vienna, Naturhist. Mus.; 727g Washington, U.S. Nat. Mus.; 614g Budapest, Nat. Mus.; 438g Berlin, Humboldt Univ.; 341g Copenhagen, Univ.; 258g Chicago, Field Mus. Nat. Hist.;
Specimen(s): [16355], 3199g.; [1920,295], 60g.; [46971], 66g.

Chatillens v Chervettaz.

Chattooga County v Holland's Store.

Chautonnay error for Chantonnay.

Chaves 41°56′ N., 7°28′ W.
Traz-os-Montes, Portugal
Fell 1925, May 3
Stone. Achondrite, Ca-rich. Howardite (AHOW).
A stone of 2670g fell, R. de Serpa Pinto, Rev. sismol. geofis. Coimbra, 1932, **3**, p.45 [M.A. 5-297], E. Jérémine, Bolm. Soc. geol. Port., 1953-1954, **11**, p.127 includes an analysis. Mentioned, E. Jérémine, Comm. Serv. Geol. Portugal, 1954, **35** [M.A. 12-609].
38g Harvard Univ.; 0.5g New York, Amer. Mus. Nat. Hist.; Thin section, Washington, U.S. Nat. Mus.;
Specimen(s): [1933,89], 3.49g. and fragments, 0.6g.; [1954, 104], 11.5g. two fragments.

Chebankol 53°40′ N., 88°0′ E.
Novosibirsk, Federated SSR, USSR, [Чебанкол]
Found 1938
Synonym(s): *Tchebankol*
Iron. Octahedrite, coarse (2.5mm) (IRANOM).
An oxidised mass of 126.6kg and fragments totalling 1111g were found 130cm down in alluvium near the Chebankol stream, a tributary of the Maly Kondon, L.A. Kulik, Метеоритика, 1941, **2**, p.123 [M.A. 10-174], E.L. Krinov, Метеоритика, 1948, **4**, p.97 [M.A. 10-517], E.L. Krinov, Астрон. Журнал, 1945, **22** (5), p.303 [M.A. 9-297], E.L. Krinov, Каталог Метеоритов Акад. Наук СССР, Москва, 1947 [M.A. 10-511]. Chemically anomalous, analysis, 8.80 %Ni, 21.8 ppm.Ga, 52.5 ppm.Ge, 0.11 ppm.Ir, E.R.D. Scott et al., Geochimica et Cosmochimica Acta, 1973, **37**, p.1973. Description, metal rich in dissolved carbon, related to group IIIE, V.F. Buchwald, Iron Meteorites, Univ. of California, 1975, p.449.
122kg Moscow, Acad. Sci.; 418g Perth, West. Austr. Mus.; 57g Washington, U.S. Nat. Mus.;

Chernyi Bor 53°42′ N., 30°6′ E. approx.
Mogilёv district, Belorussiya SSR, USSR, [Черный Бор]
Fell 1964, summer, 0530 hrs, U.T., approx.
Stone. Olivine-bronzite chondrite (H4).
After detonations, two fragments, together weighing about 6kg were seen to fall into trees and about 50m apart. One specimen was not recovered and the other was broken into small fragments, Meteor. Bull., 1972 (51), Meteoritics, 1972, **7**, p.216. Analysis, 27.89 % total iron, O.A. Kirova and M.I. D'yakonova, Метеоритика, 1976, **35**, p.43.
282g Moscow, Acad. Sci., and 5.5g in two fragments.; 20g Mogilёv, Nat. Hist. Mus.;

Cherokee County v Canton.

Cherokee County v Losttown.

Cherokee Mills v Canton.

Cherokee Springs 35°2′ N., 81°53′ W.
Spartenburg County, South Carolina, U.S.A.
Fell 1933, July 1, 0942 hrs
Stone. Olivine-hypersthene chondrite, amphoterite (LL6).
Two stones fell, 5.7kg and 2.7kg, 6.5 miles apart. Description, S. H. Perry, Science, 1933, **78**, p.312 [M.A. 5-405]. Description and analysis, olivine Fa_{28}, 20.38 % total iron, E. Jarosewich and B. Mason, Geochimica et Cosmochimica Acta, 1969, **33**, p.411.
7885g Washington, U.S. Nat. Mus., main masses of both stones; 50g Tempe, Arizona State Univ.;
Specimen(s): [1959,775], 52g. and fragments, 1g

Cherson v Grossliebenthal.

Cherson v Savtschenskoje.

Cherson v Vavilovka.

Chervettaz 46°33′ N., 6°49′ E.
Palézieux, Vaud, Switzerland
Fell 1901, November 30, 1400 hrs
Synonym(s): *Chatillens, Palézieux*
Stone. Olivine-hypersthene chondrite (L5).
After appearance of luminous meteor and detonations, a stone of about 0.75kg fell and was found in the forest of Chervettaz; description, M. Lugeon, Bull. Soc. Vaud. Sci. Nat., 1904, **40**, p.1. Olivine Fa_{26}, B. Mason, Geochimica et Cosmochimica Acta, 1963, **27**, p.1011.
Main mass, Lausanne, Mus. Géol.; 26g Chicago, Field Mus. Nat. Hist.;
Specimen(s): [86761], 27g.

Chervonnyi Kut v Chervony Kut.

Chervonyi Kut v Chervony Kut.

Chervony Kut 50°50′ N., 34°0′ E.
Sumy region, Ukraine, USSR, [Червоный Кут]
Fell 1939, June 23, 1300 hrs
Synonym(s): *Chervonnyi Kut, Chervonyi Kut, Tchervony Koot, Tschervony Kut*
Stone. Achondrite, Ca-rich. Eucrite (AEUC).
One mass fell, the weight variously given as 1800g (Kulik), 1734g (Chirvinsky and Sokolova), and 1692g (Krinov). The time of fall is given as 0400 hrs and the place is stated to be in Spivakovsk district, L.A. Kulik, Метеоритика, 1941, **2**, p.123 [M.A. 10-174]. Time given as 1300 hrs., the place as Talalaevskoi district; description, with an analysis, P.N. Chirvinsky and A.I. Sokolova, Метеоритика, 1946, **3**, p.37 [M.A. 10-173]. Listed, E.L. Krinov, Метеоритика, 1945, **22**, p.303 [M.A. 9-297], E.L. Krinov, Каталог Метеоритов Акад. Наук СССР, Москва, 1947 [M.A. 10-511].
1036g Moscow, Acad. Sci.;

Chester County v Chesterville.

Chesterville 34°42′ N., 81°12′ W. approx.
Chester County, South Carolina, U.S.A.
Found 1849, before this year
Synonym(s): *Chester County*
Iron. Hexahedrite (IIA).
A mass of 16.5kg was ploughed up a few years before 1849, C.U. Shepard, Am. J. Sci., 1849, **7**, p.449. About half the mass was forged into horseshoes, etc and the whole was probably heated. Description, artificial heat alteration, V.F. Buchwald, Iron Meteorites, Univ. of California, 1975, p.450. Described, with an analysis, 5.50 %Ni, E. Cohen, Meteoritenkunde, 1905, **3**, p.62. Further analysis, 5.61 %Ni, 58.9 ppm.Ga, 178 ppm.Ge, 1.8 ppm.Ir, J.T. Wasson, Meteorites, Springer-Verlag, 1974, p.300.
2.55kg Tempe, Arizona State Univ.; 738g Vienna, Naturhist. Mus.; 270g Tübingen, Univ.; 144g Chicago, Field Mus. Nat. Hist.; 751g Yale Univ.; 336g Berlin, Humboldt Univ.; 139g Tempe, Arizona State Univ.;
Specimen(s): [24001], 2058g. largest piece preserved; [90224], 86g.

Chetrinahatti 14°30′ N., 76°30′ E. approx.
Challakere, Chitradurga, Karnataka, India
Fell 1880, September 6, 2210 hrs
Stone.
A specimen weighing 72g is in the Govt. Mus., Bangalore; Challakere is 14°19'N., 76°40'E.

Chiang Khan 17°54′ N., 101°38′ E.
Loei, Thailand
Fell 1981, November 17, 0530 hrs
Stone. Olivine-bronzite chondrite (H6).
After a bright fireball and detonations, thirty-one pieces, totalling 367g, the largest 51.3g, were recovered from the town of Chiang Khan on the Thailand-Laos border, Meteor. Bull., 1982 (60), Meteoritics, 1982, **17**, p.94. Classification,

olivine Fa_{18}, R.S. Clarke, Jr., letter of 18 December, 1981 in Min. Dept. BM(NH).
12g Washington, U.S. Nat. Mus.;
Specimen(s): [1983,M.38], 22g.

Chiapikou v Jiapigou.

Chiara v Trenzano.

Chichimeguilas v Mazapil.

Chichimequillas v Mazapil.

Chickasha v Amber.

Chico 36°30′ N., 104°12′ W.
Colfax County, New Mexico, U.S.A.
Found 1954, January
Stone. Olivine-hypersthene chondrite (L6).
A weathered stone of 231lb was found; the chondrules are up to 2cm in diameter, L. LaPaz, Meteoritics, 1954, **1**, p.182 [M.A. 13-53]. Silicate composition suggests an amphoterite, olivine Fa_{27}, B. Mason and H.B. Wiik, Geochimica et Cosmochimica Acta, 1964, **28**, p.533. L6 stone, olivine Fa_{24}, D.E. Lange and K. Keil, Meteoritics, 1976, **11**, p.315, abs. Analysis, 19.06 % total iron, V.Ya. Kharitonova, Метеоритика, 1969, **29**, p.91.
100kg Albuquerque, Univ. of New Mexico, main mass; 44g Los Angeles, Univ. of California;
Specimen(s): [1973,M.2], 59.5g. partly crusted piece; [1973, M.3], 49.6g. partly crusted piece

Chico Hills 36°0′ N., 104°30′ W. approx.
Colfax County, New Mexico, U.S.A.
Found 1951, between 1951 and 1960
Stone. Olivine-bronzite chondrite (H4).
Listed, L. LaPaz, Cat. Coll. Inst. Meteor. Univ. New Mexico., 1965. Classification, olivine $Fa_{19.5}$, D.E. Lange and K. Keil, Meteoritics, 1976, **11**, p.315, abs.
5.75kg Albuquerque, Univ. of New Mexico;
Specimen(s): [1973,M.1], 270.5g. part-crusted fragment

Chico Mountains 29°0′ N., 103°15′ W. approx.
Brewster County, Texas, U.S.A.
Found 1915, known in this year
Synonym(s): *Alpine*
Iron. Hexahedrite (IIA).
Original weight said to have been about 2 tons; described, with an analysis, 5.62 %Ni, G.P. Merrill, Proc. U.S. Nat. Mus., 1922, **61** (4), p.1. Further analysis, 5.50 %Ni, 59.3 ppm.Ga, 176 ppm.Ge, 6.2 ppm.Ir, J.T. Wasson, Meteorites, Springer-Verlag, 1974, p.299. Description, shock-recrystallised, V.F. Buchwald, Iron Meteorites, Univ. of California, 1975, p.453.
176g Washington, U.S. Nat. Mus.;

Chicora 40°56′ N., 79°44′ W.
Butler County, Pennsylvania, U.S.A.
Fell 1938, June 24, 1800 hrs
Stone. Olivine-hypersthene chondrite, amphoterite (LL6).
Two stones, of 242g and 61g, fell, A.D. Nininger, Pop. Astron., Northfield, Minnesota, 1939, **47**, p.214, E.P. Henderson, letter of 3 June 1939 in Min. Dept. BM(NH). The mass found fell some miles short of the calculated point of impact and the main mass may yet be found. From the smoke trail and energy considerations the total mass is calculated at 519 tons. In 1940 two more fragments were found 400ft from the original find site, and are now in Washington, U.S. Nat. Mus. Described, with an analysis, F.W. Preston et al., Proc. U.S. Nat. Mus., 1941, **90**, p.387 [M.A. 8-193]. Olivine Fa_{29}, B. Mason and H.B. Wiik, Geochimica et Cosmochimica Acta, 1964, **28**, p.533. Description, R.W. Stone and E.M. Starr, Bull. Pa. Geol. Surv., 1967 (Rep. G2, 1932, rev.), p.28, 34.
275g Washington, U.S. Nat. Mus.; 45g in possession of Adam Garing of Chicora; 8g Pittsburg, Mellon Inst.;

Chihuahua v Casas Grandes.

Chihuahua (of H.H. Nininger, 1931) v Chihuahua City.

Chihuahua City 28°40′ N., 106°7′ W.
Mexico
Found 1929, recognized in this year
Synonym(s): *Chihuahua (of H.H. Nininger, 1931)*
Iron. Chemically and structurally anomalous (IC).
A mass of 11kg is described, with an analysis, 6.96 %Ni and a mass of 43kg is mentioned, H.H. Nininger, Am. J. Sci., 1931, **22**, p.69 [M.A. 5-14]. Described with determination of Ga, Au, Pd and partial analysis, 6.91 %Ni, E. Goldberg et al., Geochimica et Cosmochimica Acta, 1951, **2**, p.1. Classification and new analysis, 6.68 %Ni, 52.7 ppm.Ga, 212 ppm.Ge, 0.11 ppm.Ir, J.T. Wasson, Icarus, 1970, **12**, p.407. Description, recrystallised, composed of polycrystalline Fe/Ni with cohenite, V.F. Buchwald, Iron Meteorites, Univ. of California, 1975, p.454.
15.8kg Tempe, Arizona State Univ.; 19kg Chihuahua City, Chamber of Mines; 2.6kg Fort Worth, Texas, Monnig Colln.; 2kg Chicago, Field Mus. Nat. Hist.; 2kg Harvard Univ.; 363g Washington, U.S. Nat. Mus.;
Specimen(s): [1959,1011], 18086g.

Chilcat v Chilkoot.

Chilcoot v Chilkoot.

Chilcoot Inlet v Chilkoot.

Chile v Dehesa.

Chile v Maria Elena (1935).

Chile v North Chile.

Chile v Tamarugal.

Chile v Vaca Muerta.

Chilkat v Chilkoot.

Chilkoot 59°20′ N., 136°0′ W.
Portage Bay, Alaska, U.S.A.
Found 1881, known in this year
Synonym(s): *Chilcat, Chilcoot, Chilcoot Inlet, Chilkat*
Iron. Octahedrite, medium (1.0mm) (IIIA).
A mass of 43kg was obtained from Indians, who said it had been seen to fall about 100 years ago, Ann. Rep. California State Mining Bureau, Sacramento, 1884, p.262, O.C. Farrington, Mem. Nat. Acad. Sci. Washington, 1915, **13**, p.122. Determination of Ga, Au and Pd, with an analysis, 7.89 %Ni, E. Goldberg et al., Geochimica et Cosmochimica Acta, 1951, **2**, p.1. Further analysis, 7.77 %Ni, 20.0 ppm.Ga, 39.3 ppm.Ge, 1.8 ppm.Ir, E.R.D. Scott et al.,

Geochimica et Cosmochimica Acta, 1973, **37**, p.1957. Description, pre-terrestrial plastic deformation, V.F. Buchwald, Iron Meteorites, Univ. of California, 1975, p.457.
Main mass, San Francisco, California State Mining Bureau; 5.5kg Washington, U.S. Nat. Mus.; 234g Mainz, Max-Planck-Inst.; 155g Harvard Univ.; 62g Chicago, Field Mus. Nat. Hist.;
Specimen(s): [1924,444], 39g. and filings, 5g

Chilpanzingo v Caparrosa.

Chinautla 14°30′ N., 90°30′ W. approx.
Guatemala
Found 1902
Synonym(s): *Guatemala*
Iron. Octahedrite, fine (0.4mm) (IVA-ANOM).
A mass of 5.72kg was found; described and analysed, 9.05 % Ni, S. Meunier, C. R. Acad. Sci. Paris, 1902, **134**, p.755. Further analysis, 9.54 %Ni, 2.08 ppm.Ga, 0.112 ppm.Ge, 0.12 ppm.Ir, R. Schaudy et al., Icarus, 1972, **17**, p.174. Description, some pre-terrestrial plastic deformation, V.F. Buchwald, Iron Meteorites, Univ. of California, 1975, p.459.
206g Vienna, Naturhist. Mus.; 204g Budapest, Nat. Hist. Mus.; 162g Washington, U.S. Nat. Mus.; 133g Paris, Mus. d'Hist. Nat.; 121g Chicago, Field Mus. Nat. Hist.; 95g Harvard Univ.; 94g New York, Amer. Mus. Nat. Hist.;
Specimen(s): [1924,16], 972.5g. corner piece, and a slice, 198g.; [1913,434], 143g. a slice, and fragments, 3g.

Chinga 51°3′30″ N., 94°24′ E.
Tanna Tuva, Turvinskaya, USSR, [Чинге]
Found 1913
Synonym(s): *Chinge, Tannuola, Tannu-Ola, Tchinge, Tschinga, Tschinge, Tuva, Urgailyk-Chinge*
Iron. Ataxite, Ni-rich (IVB-ANOM).
Many masses, individuals up to 20kg, of total weight about 80kg, were found in the Chinga stream, a tributary of the Upper Yenisei from the Tanno-ola Mts. Recognised as meteoritic, G. Pehrmann, Acta Acad. Aboensis Math. et Phys., 1923, **3** (1) [M.A. 6-103]. The place of fall is incorrectly mapped, P.M. Millman, Trans. Roy. Astron. Soc. Canada, 1938, **32**, p.199. Analysis, 16.56 %Ni, M.I. D'yakonova, Метеоритика, 1958 (16), p.180. Further analysis, 16.38 %Ni, 0.181 ppm.Ga, 0.082 ppm.Ge, 3.6 ppm.Ir, R. Schaudy et al., Icarus, 1972, **17**, p.174. Structural description, H.J. Axon and P.L. Smith, Min. Mag., 1972, **38**, p.736. Many of the specimens are strongly oxidised, and some have the torn appearance characteristic of fragments from meteoritic craters such as Cañon Diablo and Henbury. It is thought probable that this fall was associated with a crater that has disappeared through erosion, or perhaps has simply passed unnoticed, but fieldwork in 1963 failed to find any crater, B.I. Vronskii and I.T. Zotkin, Метеоритика, 1968, **28**, p.125. Fragments of metal with 16.37%Ni, found in the alluvium of the upper Argolik river, Tuva district are now attributed to the Chinga fall, A.L. Dodin, Зап. сесиюз. Мин. ОбчИ., 1952, **81** (2), p.64 [M.A. 12-106]. Description, co-ordinates, V.F. Buchwald, Iron Meteorites, Univ. of California, 1975, p.461.
68kg Moscow, Acad. Sci., includes the largest mass; 1.6kg Washington, U.S. Nat. Mus.; 316g New York, Amer. Mus. Nat. Hist.; 315g Tempe, Arizona State Univ.;
Specimen(s): [1956,317], 154g.

Chinge v Chinga.

Chinguetti 20°15′ N., 12°41′ W.
Dhar Adrar, Mauritania
Found 1920
Synonym(s): *Adrar*
Stony-iron. Mesosiderite (MES).
An enormous mass, said to be 100m long and 45m high, with other smaller masses, was found in the desert of Adrar, about 45km to the SW. of Chinguetti. Description, A. Lacroix, C. R. Acad. Sci. Paris, 1924, **179**, p.309. There are doubts about the existence of the huge mass, T. Monod, Méharées, Paris, 1937, p.145, 160, 188., L. LaPaz and J. LaPaz, Meteoritics, 1954, **1**, p.187 [M.A. 13-54]. Mesosiderite, E. Jarosewich and B. Mason, Geochimica et Cosmochimica Acta, 1969, **33**, p.411. Analysis of metal, 9.6 %Ni, 15.2 ppm.Ga, 55.2 ppm.Ge, 6.2 ppm.Ir, J.T. Wasson et al., Geochimica et Cosmochimica Acta, 1974, **38**, p.135. See also Aouelloul in the list of meteorite craters. Corona structures, references, C.E. Nehru et al., Geochimica et Cosmochimica Acta, 1980, **44**, p.1103.
3.9kg Paris, Mus. d'Hist. Nat.; 409g Washington, U.S. Nat. Mus.;
Specimen(s): [1980,M.25], 36.1g.

Chireya v Butsura.

Chiriya v Butsura.

Chitado 17°21′ S., 13°58′ E.
Cunene, Angola
Fell 1966, October 20
Stone. Olivine-bronzite chondrite (H6).
A shower of meteorites, of unknown total weight, fell near Chitado, within an area of about 6 square km. Several stones, weighing several hundred grams each, and a few of some kilograms, were collected, Meteor. Bull., 1975 (53), Meteoritics, 1975, **10**, p.134. Description, with an analysis, M. Portugal Ferreira et al., Estudos Notas Trab. Serv. Fom. Min., 1971, **20**, p.209 [M.A. 73-2791].
Specimen, Luanda, Serv. Geol. Minas.;

Chitenay 47°28′15″ N., 0°58′36″ W.
Chitenay, Loir-et-Cher, France
Fell 1978, February 21, 1956 hrs
Stone. Olivine-hypersthene chondrite (L6).
After a notable bolide a stone of about 4kg fell. It was recovered next morning from a hole in a lawn, Meteor. Bull., 1979 (56), Meteoritics, 1979, **14**, p.166. Brief description, olivine $Fa_{23.4}$, M. Christophe Michel-Levy, Bull. Min., 1978, **101**, p.580 [M.A. 79-2726].

Chmelewka v Khmelevka.

Cholula v Yanhuitlan.

Cholula 15°5′ N., 98°20′ W.
Puebla, Mexico
Found
Doubtful. Stone.
An aerolite, shaped like a frog, is said to have been housed in the temple on top of the pyramid of Cholula, Z. Nuttall, Archaeol. Ethnol. Papers, Peabody Mus., Harvard, 1901, **2**, p.270.

Christian County v Billings.

Chukri v Merua.

Chulafinnee 33°30′ N., 85°40′ W.
Cleburne County, Alabama, U.S.A.
Found 1873
Synonym(s): *Cleburne County*
Iron. Octahedrite, medium (1.1mm) (IIIA).
A mass of about 35.75lb was ploughed up, W.E. Hidden, Am. J. Sci., 1880, **19**, p.370. Analysis, 7.37 %Ni, J.B. Mackintosh, Am. J. Sci., 1880, **20**, p.74. Description, A. Brezina, Sitzungsber. Akad. Wiss. Wien, Math.-naturwiss. Kl., 1882, **84**, p.281. Further analysis, 7.47 %Ni, 17.8 ppm.Ga, 33.7 ppm.Ge, 5.5 ppm.Ir, J.T. Wasson, Meteorites, Springer-Verlag, 1974, p.301. Description, the whole mass has been artifically heated to about 800C, V.F. Buchwald, Iron Meteorites, Univ. of California, 1975, p.464.
11.75kg Vienna, Naturhist. Mus., main mass; 1kg Budapest, Nat. Mus.; 119g Washington, U.S. Nat. Mus.; 116g Chicago, Field Mus. Nat. Hist.; 103g Harvard Univ.; 101g Berlin, Humboldt Univ.; 83g New York, Amer. Mus. Nat. Hist.;
Specimen(s): [54276], 60g.

Chupaderos 27°0′ N., 105°6′ W. approx.
Jimenez, Chihuahua, Mexico
Found 1852, known for centuries, first mentioned in this year
Synonym(s): *Adargas, Concepcion, Cuernavaca, Huejuquilla, Jimenez, Las Adargas, Morelos, Rio Florido, Sierra de las Adargas, Valle de Allende, Valle de San Bartolome*
Iron. Octahedrite, medium (0.65mm) (IIIB).
Two large masses of 14114kg and 6767kg respectively were found about 16 miles from Jimenez (formerly Huejuquilla), A. Castillo, Cat. Meteorites Mexique, Paris, 1889, p.8, L. Fletcher, Min. Mag., 1890, **9**, p.148. The two masses were removed in 1891 to the School of Mines, Mexico City. Adargas is another mass of Chupaderos, and Cuernavaca a fragment of Chupaderos. Details of Adargas, L. Fletcher, Min. Mag., 1890, **9**, p.140 and of Cuernavaca, H.A. Ward, Proc. Rochester Acad. Sci., 1902, **4**, p.81. Older analyses of these masses, for example of Chupaderos, 8.76 %Ni, E. Cohen and E. Weinschenk, Ann. Naturhist. Hofmus. Wien, 1891, **6**, p.147. Further analysis, 9.7 % Ni, 17.2 ppm.Ga, 29.6 ppm.Ge, 0.020 ppm.Ir, E.R.D. Scott et al., Geochimica et Cosmochimica Acta, 1973, **37**, p.1957. Full description, localities, history, V.F. Buchwald, Iron Meteorites, Univ. of California, 1975, p.465.
14114kg Mexico City, near the Palace of Bellas Artes in Tacuba no. 5; 6767kg Mexico City, near the Palace of Bellas Artes in Tacuba no. 5; 3325kg Mexico City, near the Palace of Bellas Artes in Tacuba no. 5, of Adargas; Main mass, Mexico City, Nat. Mus., of Cuernavaca; 28kg Chicago, Field Mus. Nat. Hist., of Chupaderos, 10kg; of Cuernavaca, 17.5kg; of Adargas, 340g; 1.2kg Tempe, Arizona State Univ.; 2.3kg Yale Univ.; 2.5kg Washington, U.S. Nat. Mus.; 1.3kg Harvard Univ., of Cuernavaca; 1.23kg Vienna, Naturhist. Mus., 660g of Chupaderos, 570g of Cuernavaca; 817g Tübingen, Univ.; 786g Paris, Mus. d'Hist. Nat.;
Specimen(s): [85074], 1087g. slice; [80111], 47.3g. of Adargas; [1959,941], 192.5g. and sawings, 8g. of Adargas; [86069], 1024g. of Cuernavaca.

Chuvashskie Kissy v Kissij.

Chuvashskii Kissy v Kissij.

Cilimus 6°57′ S., 108°6′ E.
Sampora, Java, Indonesia
Fell 1979, May 7, 0930 hrs
Stone. Olivine-hypersthene chondrite (L5).
After a fireball and detonations, a stone of 1.6kg fell in a garden in the village of Sampora. Cilimus is the nearest post office, Meteor. Bull., 1980 (57), Meteoritics, 1980, **15**, p.97. Description, shocked, olivine $Fa_{24.0}$, J. Miller et al., Meteoritics, 1981, **16**, p.69.
Main mass, Bandong, Geol. Mus.; 260g Washington, U.S. Nat. Mus.;
Specimen(s): [1980,M.16], 45.6g.

Cilli v Stannern.

Cincinnati 39°7′ N., 84°30′ W.
Hamilton County, Ohio, U.S.A.
Found 1870
Iron. Hexahedrite (IIA).
A piece of 28g in Munich, (Bavarian State Mus.) was described, with an analysis, 5.43 %Ni, E. Cohen, Meteoritenkunde, 1905 (3), p.51. The iron was found in 1870 (not 1898 as at first stated) and was acquired by Mr. Hoseus after whose death two small slices came into the possession of Dr. Engelmann and later the Basel Mus., C. Wendler, letter of 9 June 1927 in Min. Dept. BM(NH). Photomicrographs show a peculiar recrystallisation of the kamacite along Neumann lines, M. von Schwarz and H. Bauer, Zentr. Miner., 1936 (Abt. A), p.207 [M.A. 6-390]. Described, with an analysis, 5.33 %Ni, E.P. Henderson and S.H. Perry, Proc. U.S. Nat. Mus., 1958, **107**, p.339. Reheated, probably artificially, V.F. Buchwald, Iron Meteorites, Univ. of California, 1975, p.476. Classification and analysis, 5.44 %Ni, 56 ppm.Ga, 178 ppm.Ge, 20 ppm.Ir, A. Kracher et al., Geochimica et Cosmochimica Acta, 1980, **44**, p.773.
165g Vienna, Naturhist. Mus.; 106g New York, Amer. Mus. Nat. Hist.; 63g Chicago, Field Mus. Nat. Hist.; 38g Harvard Univ.; 32g Washington, U.S. Nat. Mus.;
Specimen(s): [1927,1012], 51.4g.

Circle Back 34°1′41″ N., 102°40′48″ W.
Bailey County, Texas, U.S.A.
Found 1977
Stone. Olivine-hypersthene chondrite (L6).
A single mass of 2268g was found, G.I Huss, priv. comm., 1982. Olivine $Fa_{26.2}$, A.L. Graham, priv. comm., 1983.
Specimen(s): [1983,M.21], 44g.

Cirencester v Aldsworth.

Civitavecchia 42°5′ N., 11°47′ E.
Lazio, Italy
Fell 1855, October 17
Doubtful..
After the usual phenomena, a mass is said to have been seen to fall within a few feet of a ship in the harbour, but as nothing was recovered the evidence is not fully conclusive, R.P. Secchi, Cosmos, 1856, **9** (1), p.421, Giachetti, Nuovo Cimento, 1856, **4**, p.312.

Claiborne v Lime Creek.

Claiborne v Walker County.

Claiborne County v Tazewell.

Clairborne v Walker County.

Clarac v Ausson.

Clarendon v Paloduro.

Clarendon 34°54'18" N., 100°54'48" W.
Donley County, Texas, U.S.A.
Found 1979
Stone. Chondrite.
A single mass was ploughed up but broke into pieces; less than half, weighing 1964g, was recovered, G.I Huss, letter of 26 October, 1983, 1981 in Min. Dept. BM(NH).
Main mass, Denver, Amer. Meteorite Lab.;

Clareton 43°42' N., 104°42' W.
Weston County, Wyoming, U.S.A.
Found 1931, known between 1931 and 1937
Stone. Olivine-hypersthene chondrite (L6).
A stone of 1050g was described, H.H. Nininger, The Mines Mag., Golden, Colorado, 1937, **27**, p.16 [M.A. 7-69]. Olivine Fa_{25}, B. Mason, Geochimica et Cosmochimica Acta, 1963, **27**, p.1011.
500g Tempe, Arizona State Univ.; 208g Washington, U.S. Nat. Mus.;
Specimen(s): [1959,819], 50.5g. slice

Clark County 38°0' N., 84°10' W.
Kentucky, U.S.A.
Found 1937, before this year
Iron. Octahedrite, medium (1.0mm) (IIIF).
One mass of 11.3kg was found, A.D. Nininger, Pop. Astron., Northfield, Minnesota, 1939, **47**, p.211., D.M. Young, Pop. Astron., Northfield, Minnesota, 1939, **47**, p.382 [M.A. 7-376], E.P. Henderson, letter of 3 June 1939 in Min. Dept. BM(NH). Listed, H.H. and A.D. Nininger, The Nininger Collection of Meteorites, Winslow, Arizona, 1950, p.44. Determination of Ga, Au, Pd and also a partial analysis, 7.02 %Ni, E. Goldberg et al., Geochimica et Cosmochimica Acta, 1951, **2**, p.1. Classification and analysis, 6.79 %Ni, 6.92 ppm.Ga, 0.993 ppm.Ge, 6.2 ppm.Ir, E.R.D. Scott and J.T. Wasson, Geochimica et Cosmochimica Acta, 1976, **40**, p.103. Description, references to trace element analyses, V.F. Buchwald, Iron Meteorites, Univ. of California, 1975, p.476.
2.28kg Lexington, Univ. of Kentucky; 2kg Washington, U.S. Nat. Mus.; 1.8kg Tempe, Arizona State Univ.;
Specimen(s): [1959,949], 2711g. and fragments, 99g

Clarke County v Lime Creek.

Claytonville 34°20'48" N., 101°29'24" W.
Swisher County, Texas, U.S.A.
Found 1964, recognized 1966
Stone. Olivine-hypersthene chondrite (L5).
One stone of 10.5kg was ploughed up in a field, Meteor. Bull., 1970 (49), Meteoritics, 1970, **5**, p.174. Olivine Fa_{24}, R. Hutchison, priv. comm.
6.0kg Mainz, Max-Planck-Inst., main mass; 198g Washington, U.S. Nat. Mus.; 69g Copenhagen, Univ. Geol. Mus.; 50g New York, Amer. Mus. Nat. Hist.;
Specimen(s): [1970,33], 172g. crusted part-slice and chip, 0.7g

Claywater v Vernon County.

Claywater Stone v Vernon County.

Cleburne 32°19' N., 97°25' W.
Johnson County, Texas, U.S.A.
Found 1907
Iron. Octahedrite, fine.
A mass of 6.8kg, found in 1907, was recognized as meteoritic in 1934, A.D. Nininger, Pop. Astron., Northfield, Minnesota, 1937, **45**, p.449 [M.A. 7-62]. Mentioned, E.P. Henderson, letter of 12 May 1939 in Min. Dept. BM(NH).
6.5kg Fort Worth, Texas, Monnig Colln.;

Cleburne County v Chulafinnee.

Cléguérec v Kernouve.

Clegueric v Kernouve.

Cleveland 34°53' N., 84°47' W.
Bradley County, Tennessee, U.S.A.
Found 1860
Synonym(s): *Bradley County, East Tennessee, Lea Iron, Philadelphia Iron, Whitfield, Whitfield County*
Iron. Octahedrite, medium (1.0mm) (IIIB).
A mass of 254lb was found; described and analysed, 8.06 % Ni, F.A. Genth, Proc. Acad. Nat. Sci. Philadelphia, 1886, p.366. Further analysis, 8.85 %Ni, 21.0 ppm.Ga, 41.9 ppm.Ge, 0.094 ppm.Ir, E.R.D. Scott et al., Geochimica et Cosmochimica Acta, 1973, **37**, p.1957. Full description, the 13lb (Whitfield County) mass, formerly attributed to Dalton (*q.v.*), is part of Cleveland, V.F. Buchwald, Iron Meteorites, Univ. of California, 1975, p.478.
930g Philadelphia, Mus. Acad. Sci., slices; 3.9kg Vienna, Naturhist. Mus., includes 2.9kg, main mass, of "Whitfield County"; 620g Chicago, Field Mus. Nat. Hist.; 500g Tempe, Arizona State Univ.; 819g Washington, U.S. Nat. Mus.; 478g New York, Amer. Mus. Nat. Hist.; 429g Harvard Univ., of "Whitfield County";
Specimen(s): [64205], 209g. slice.; [53291], 133g. of "Whitfield County"

Clinton 36°5' N., 84°12' W.
Anderson County, Tennessee, U.S.A.
Found 1950, known in this year
Iron. Octahedrite?.
Total known weight 7.7kg, H.H. and A.D. Nininger, The Nininger Collection of Meteorites, Winslow, Arizona, 1950, p.45.
10g Tempe, Arizona State Univ., oxidized fragments;

Clohars
Fouesnant, Quimper, Finistére, France
Fell 1822, June 21
Stone. Olivine-hypersthene chondrite (L4).
A fragment of 6g, with label of locality and date of fall, was acquired by the Mus. d'Hist. Nat., Paris in 1897, S. Meunier, C. R. Acad. Sci. Paris, 1897, **124**, p.1543. Olivine Fa_{25}, B. Mason, Geochimica et Cosmochimica Acta, 1963, **27**, p.1011. Doubtful provenance, A. Lacroix, Bull. Mus. d'Hist. Nat. Paris, 1927, **33**, p.445.
Specimen(s): [1960,330], 43.5g. almost complete stone

Clois error for Clovis.

Clondike v Klondike (Gay Gulch).

Clondike v Klondike (Skookum Gulch).

Clover Springs 34°27′ N., 111°22′ W.

Gila County, Arizona, U.S.A.
Found 1954, recognized in this year
Stony-iron. Mesosiderite (MES).

A mass of 7700g was found about 13 miles SW. of Clover Springs, Meteor. Bull., 1957 (5), Cat. Meteor. Arizona State Univ., 1964, p.701. Analysis of metal, 6.10 %Ni, 11.0 ppm.Ga, 42 ppm.Ge, 2.1 ppm.Ir, J.T. Wasson, Geochimica et Cosmochimica Acta, 1974, **38**, p.135. Corona structure, mineral analyses, C.E. Nehru et al., Geochimica et Cosmochimica Acta, 1980, **44**, p.1103.

1kg Washington, U.S. Nat. Mus.; 921g Mainz, Max-Planck-Inst.; 450g Tempe, Arizona State Univ.;

Specimen(s): [1959,994], 2419g.

Clovis (b) v Clovis (no. 2).

Clovis (no.1) 34°18′ N., 103°8′5″ W.

Curry County, New Mexico, U.S.A.
Found 1961
Stone. Olivine-bronzite chondrite (H3).

A mass of 283kg was found 6 miles SE. of Clovis. Distinct from Grady (1933) and from Melrose (a), which are L-group chondrites; analysis, 23.33 % total iron, E. Jarosewich, Geochimica et Cosmochimica Acta, 1966, **30**, p.1261. It is more oxidised and has less total iron than Grady (1933) (*q.v.*). Unequilibrated, R.T. Dodd et al., Geochimica et Cosmochimica Acta, 1967, **31**, p.921. Contains Ca-Al-Ti-rich inclusions, A.F. Noonan, Meteoritics, 1975, **10**, p.51.

283kg Washington, U.S. Nat. Mus., main mass; 500g Tempe, Arizona State Univ.;

Clovis (no.2) 34°18′ N., 103°8′ W.

Curry County, New Mexico, U.S.A.
Found 1963, recognized in this year
Synonym(s): *Clois no.2, Clovis (b)*
Stone. Olivine-hypersthene chondrite (L6).

A second mass of 13.3kg is distinct from Clovis (no.1), Meteor. Bull., 1964 (32). Total known weight 41.3kg, listed (as a bronzite chondrite in error), Cat. Huss Coll. Meteorites, 1976, p.11. Olivine Fa_{24}, B. Mason, Geochimica et Cosmochimica Acta, 1967, **31**, p.1100.

32kg Mainz, Max-Planck-Inst.; 642g Tempe, Arizona State Univ.; 180g Vienna, Naturhist. Mus.; 171g New York, Amer. Mus. Nat. Hist.;

Specimen(s): [1964,651], 102g.

Cmien v Zmenj.

Coahuila 28°42′ N., 102°44′ W.

Coahuila, Mexico
Found 1837, known in this year
Synonym(s): *Bolson de Mapimi, Bonanza Iron, Butcher Iron, Buther Iron, Cerralvo, Couch Iron, Fort Duncan, Fort Dunkan, Hacienda de Potosi, Lupton's Iron, Maverick County, Nuevo Leon, Potosi, Saltillo, Sancha (Sanchez) Estate, Santa Rosa, Santillo*
Iron. Hexahedrite (IIA).

Fourteen masses (Bonanza Iron), some said to be of 2000lb to 3000lb were seen by E.M. Hamilton in 1866, C.U. Shepard, Am. J. Sci., 1866, **42**, p.347, C.U. Shepard, Am. J. Sci., 1867, **43**, p.384. Eight pieces (Butcher Iron) from 290lb to 654lb in weight, totalling 4000lb were removed to the U.S.A. by H.B. Butcher in 1868, J.L. Smith, Am. J. Sci., 1869, **47**, p.383. Another mass of Butcher Iron of 192lb was seen by N.T. Lupton in 1879 at Santa Rosa, Am. J. Sci., 1885, **29**, p.232. The Sancha (Sanchez) Estate mass of 252lb was found in use as an anvil at Saltillo by D.N. Couch in 1853, J.L. Smith, Am. J. Sci., 1855, **19**, p.160. A mass of 97lb was found at Fort Duncan, Maverick County Texas in 1882 by C.C. Cusick, W.E. Hidden, Am. J. Sci., 1886, **32**, p.304. The probable identity of all these masses was suggested, A. Brezina, Ann. Naturhist. Hofmus. Wien, 1886, **1**, p.26., L. Fletcher, Min. Mag., 1890, **9**, p.107. The Smithsonian Iron (*q.v.*) of about 6lb has been referred to Coahuila, E. Cohen, Meteoritenkunde, 1905, **3**, p.190 but is distinct as also is Richland (*q.v.*). Analysis, 5.59 %Ni, E.P. Henderson, Am. J. Sci., 1941, **239**, p.407 [M.A. 8-195]. Determination of Co, Au, Ga, Pd and analysis, 5.63 %Ni, E. Goldberg et al., Geochimica et Cosmochimica Acta, 1951, **2**, p.1. Further analysis, 5.49 %Ni, 57.6 ppm.Ga, 178 ppm.Ge, 16 ppm.Ir,, J.T. Wasson, Geochimica et Cosmochimica Acta, 1969, **33**, p.859. Full description, shower part of a single kamacite crystal, V.F. Buchwald, Iron Meteorites, Univ. of California, 1975, p.480.

819kg Harvard Univ., of Butcher Iron; and 865g of Sanchez Estate; 220kg Vienna, Naturhist. Mus., from various masses; 114kg Washington, U.S. Nat. Mus., main mass of Sanchez Estate; 259kg Paris, Mus. d'Hist. Nat.; 15kg Budapest, Nat. Mus.; 9kg Chicago, Field Mus. Nat. Hist.; 7kg Stockholm, Riksmus; 8kg New York, Amer. Mus. Nat. Hist.; 3.2kg Yale Univ.; 2.2kg Berlin, Humboldt Univ.; 1kg Michigan Univ.; 580g Oxford, Univ.; 746g Tempe, Arizona State Univ.; 652g Prague, Nat. Mus.; 1kg Copenhagen, Univ.;

Specimen(s): [49517], 243.5kg. of Butcher Iron, one of the original masses; [54242], 1780g. of Butcher Iron; [53295], 525.5g. and a polished slice, 26.5g, and sawings, 12g, all of Butcher Iron; [43052], 9g. of Butcher Iron; [62849], 4500g. of Fort Duncan; [61921], 572g. of Sancha Estate; [35276], 9g. of Santa Rosa; [34587], 7g. of Santa Rosa; [90230], Microprobe mount, P.313 of Santa Rosa; [41031], 5g. of Bonanza Iron; [1959,161], 41g.; [1959,162], 8g.

Cobaya v Gibeon.

Cobija 22°34′ S., 70°15′ W.

Pampa of Santa Barbara, Antofagasta, Chile
Found 1892
Synonym(s): *Lampa, Santiago de Chile*
Stone. Olivine-bronzite chondrite (H6).

An ellipsoidal mass of 3690g was found, H.A. Ward, Proc. Rochester Acad. Sci., 1906, **4**, p.229. A second mass of 6.25lb obtained from H.A. Ward was described, O.C. Farrington, Field Mus. Nat. Hist. Geol. Ser., 1922, **3**, p.115 under the name of Lampa. The Cobija mass was analysed, E.D. Mountain, Min. Mag., 1926, **21**, p.87. Lampa was also partially analysed and its suspected identity with Cobija proved. Olivine Fa_{19}, B. Mason, Geochimica et Cosmochimica Acta, 1963, **27**, p.1011.

1.25kg Chicago, Field Mus. Nat. Hist., of Cobija; 1.4kg Chicago, Field Mus. Nat. Hist., of Lampa; 1kg Santaigo de Chile, School of Mines, approx. weight, of Cobija; 252g New York, Amer. Mus. Nat. Hist.; 605g Vienna, Naturhist. Mus.; 202g Washington, U.S. Nat. Mus.; 136g Berlin, Humboldt Univ.;

Specimen(s): [1905,441], 269g. end piece, and fragments, 12g, of Cobija; [1907,133], 252g. slice, of Cobija; [1910,120], 181g. and fragments, 16.5g, of Lampa.

Cobija v Joel's Iron.

Cochabamba
Bolivia?
Found
Stone. Carbonaceous chondrite, type II (CM2).
A partly crusted fragment weighing 85g and labelled 'Cochabamba, Chile', is in the Naturhist. Mus., Vienna. The provenance of the mass is unknown. It has a carbonaceous matrix, Ca-Al-rich inclusions, and olivine of variable composition, G. Kurat and A. Kracher, Meteoritics, 1975, **10**, p.432, abs. Matrix contains cronstedtite, W.F. Muller et al., Tschermaks Min. Petr. Mitt., 1979, **26**, p.293. Trace element data, G.W. Kallemeyn and J.T. Wasson, Geochimica et Cosmochimica Acta, 1981, **45**, p.1217.
85g Vienna, Naturhist. Mus., only mass known;

Cochin China v Tuan Tuc.

Cocina v Cosina.

Cockarrow Creek 26°40′ S., 120°10′ E.
Western Australia, Australia
Found 1970
Stone. Olivine-hypersthene chondrite (L6).
Two interlocking pieces, of 221g and 208g, were found about 8 miles SW. of Wiluna, G.J.H. McCall, 2nd. Suppl. to West. Austr. Mus. Spec. Publ. no. 3, 1972, p.9. Olivine $Fa_{25.4}$, B. Mason, Rec. Austr. Mus., 1974, **29**, p.169.
420g Perth, West. Austr. Mus.; 3.2g Washington, U.S. Nat. Mus.;

Cockburn 32°8′ S., 141°2′ E.
South Australia, Australia
Found 1946
Stone. Olivine-hypersthene chondrite (L6).
An incomplete stone of 10.13g was found near the Cockburn-Egebek road where it crosses Ophara Creek, about 6 miles SE. of Cockburn. Further searches resulted in additional finds, including an individual of 2.207kg, the total weight recovered being 2.46kg, Meteor. Bull., 1975 (53), Meteoritics, 1975, **10**, p.142. Description, J.E. Johnson and D.H. McColl, Trans. Roy. Soc. South Austr., 1967, **91**, p.37 [M.A. 20-47]. Co-ordinates, classification. olivine $Fa_{25.1}$, B. Mason, Rec. Austr. Mus., 1974, **29**, p.169. Discussion and analysis, (no. 3 stone) 20.61 % total iron, M.J. Fitzgerald, Ph.D. Thesis, Univ. of Adelaide, 1979, p.23.
1.8kg Washington, U.S. Nat. Mus., main mass;

Cocke County v Cosby's Creek.

Cocklebiddy 31°55′ S., 126°15′ E.
Western Australia, Australia
Found 1949
Stone. Olivine-bronzite chondrite (H5).
A mass of 19.5kg was found 2.6 miles WNW. of Nallah Nallah Rockhole, G.J.H. McCall and W.H. Cleverly, Min. Mag., 1968, **36**, p.691, G.J.H. McCall, First Suppl. to West. Austr. Mus. Spec. Publ. no. 3, 1968, p.3. Olivine $Fa_{17.2}$, B. Mason, Rec. Austr. Mus., 1974, **29**, p.169.
16.8kg Perth, West. Austr. Mus.; 138g New York, Amer. Mus. Nat. Hist.; 119g Washington, U.S. Nat. Mus.;
Specimen(s): [1968,284], 255g. crusted end piece.

Cocunda 32°49′ S., 134°48′ E.
Eyre Peninsula, South Australia, Australia
Found 1945
Stone. Olivine-hypersthene chondrite (L6).
A mass of 500g was found, 0.75 mile W. of Cocunda rocks, B. Mason, Geochimica et Cosmochimica Acta, 1963, **27**, p.1022, D.W.P. Corbett, Rec. S. Austr. Mus., 1968, **15**, p.767., L. Greenland and J.F. Lovering, Geochimica et Cosmochimica Acta, 1965, **29**, p.843. Olivine $Fa_{25.8}$, B. Mason, Rec. Austr. Mus., 1974, **29**, p.169. Analysis, 20.04 % total iron, M.J. Fitzgerald, Ph.D. Thesis, Univ. of Adelaide, 1979, p.23.
Main mass, Adelaide, South Austr. Mus.; 0.8g New York, Amer. Mus. Nat. Hist.; Thin section, Washington, U.S. Nat. Mus.;

Colby (Kansas) 39°25′ N., 101°3′ W.
Thomas County, Kansas, U.S.A.
Found 1940
Stone. Olivine-bronzite chondrite (H5).
One mass of 2.4kg was found, W. Wahl, letter of 23 May 1950 in Min. Dept. BM(NH). Listed, H.H. and A.D. Nininger, The Nininger Collection of Meteorites, Winslow, Arizona, 1950, p.45. Olivine Fa_{18}, B. Mason, Geochimica et Cosmochimica Acta, 1963, **27**, p.1011.
1.05kg Tempe, Arizona State Univ.; 108g Chicago, Field Mus. Nat. Hist.;
Specimen(s): [1959,832], 1000g. and fragments, 3g

Colby (Wisconsin) 44°54′ N., 90°17′ W.
Clark County, Wisconsin, U.S.A.
Fell 1917, July 4, 1820 hrs
Stone. Olivine-hypersthene chondrite (L6), veined.
After the appearance of a luminous meteor with a trail of black smoke, moving from NW. to SE., and detonations, two stones, of about 150lb and 80lb respectively, fell about 0.5 mile apart and were broken into pieces, R. N. Buckstaff, letter of 22 December 1921 in Min. Dept. BM(NH)., H.L. Ward, Science, 1917, **46**, p.262. Description, with an analysis (no CaO reported), G.P. Merrill, Proc. U.S. Nat. Mus., 1925, **67** (2), p.1. Further analysis, 24.38 % total iron, H. von Michaelis et al., Earth planet. Sci. Lett., 1969, **5**, p.387. Olivine Fa_{25}, B. Mason, Geochimica et Cosmochimica Acta, 1963, **27**, p.1011.
Main masses, Milwaukee Mus.; 2kg Washington, U.S. Nat. Mus.; 10kg Chicago, Field Mus. Nat. Hist.; 822g New York, Amer. Mus. Nat. Hist.; 500g Ottawa, Mus. Geol. Surv. Canada; 225g Tempe, Arizona State Univ.;
Specimen(s): [1922,792], 1620g.; [1922,9], 28.5g. from the larger stone

Cold Bay 55°11′ N., 162°33′ W.
Alaska, U.S.A.
Found 1921, June
Stony-iron. Pallasite (PAL).
Fragments totalling 320g were found, G.P. Merrill, Proc. U.S. Nat. Mus., 1922, **61** (4). Olivine Fa_{20}, P.R. Buseck and J.I. Goldstein, Bull. Geol. Soc. Amer., 1969, **80**, p.2141. Belongs to the Eagle Station sub-group, analysis of metal, 13.6 %Ni, 6.2 ppm.Ga, 113 ppm.Ge, 5.8 ppm.Ir, E.R.D. Scott, Geochimica et Cosmochimica Acta, 1977, **41**, p.349.
287g Washington, U.S. Nat. Mus.;
Specimen(s): [1924,443], 33.5g.

Cold Bokeveld v Cold Bokkeveld.

Cold Bokevelt v Cold Bokkeveld.

Cold Bokkaveld v Cold Bokkeveld.

Cold Bokkeveld 33°8′ S., 19°23′ E.

Cape Province, South Africa
Fell 1838, October 13, 0900 hrs
Synonym(s): *Bokkeveld, Cold Bokkaveld, Cold Bokeveld, Cold Bokevelt, Koul Bokkeveld*
Stone. Carbonaceous chondrite, type II (CM2).

Many stones, the largest of about 4.5lb, fell, after appearance of a fireball and detonations, T. Maclear and -. Watermeyer, Phil. Trans. Roy. Soc. London, 1839, **129**, p.83, T. Maclear, Phil. Trans. Roy. Soc. London, 1840, **130**, p.177. Analysis, 20.85 % total iron, H.B. Wiik, Geochimica et Cosmochimica Acta, 1956, **9**, p.279. C and H isotopes, G. Boato, Geochimica et Cosmochimica Acta, 1954, **6**, p.209. Carbon compounds present, G. Mueller, Geochimica et Cosmochimica Acta, 1952, **4**, p.1. Distribution of carbon in silicates, J. Orcel and B. Alpern, C. R. Acad. Sci. Paris, 1967, **265, ser. D**, p.897 [M.A. 20-308]. Partial analysis, 19.49 % total iron, H. von Michaelis et al., Earth planet. Sci. Lett., 1969, **5**, p.387. Trace element data, G.W. Kallemeyn and J.T. Wasson, Geochimica et Cosmochimica Acta, 1981, **45**, p.1217.

1.8kg Paris, Mus. d'Hist. Nat.; 680g Vienna, Naturhist. Mus.; 800g Edinburgh, Roy. Scottish Mus.; 245g Cape Town, South African Mus.; 130g Moscow, Acad. Sci.; 86g Budapest, Nat. Hist. Mus.; 75g Yale Univ.; 59g Chicago, Field Mus. Nat. Hist.; 50g Tübingen, Univ.; 41g Tempe, Arizona State Univ.; 37g Washington, U.S. Nat. Mus.;

Specimen(s): [13989], 610g. and fragments, 48g; [1727], 245.5g. nearly complete stone; [19002], 60g. fourteen individuals, and fragments, 4.5g; [1907,176], 10g. fragments; [1944,72], 4.5g. fragments; [82818], 4.8g.; [1935,46], 2.24g.; [33907], 2g.; [1964,579], 16.5g. of treated powder; [1964,705], 79g. and fragments, 58g, residues from extraction of hydrocarbons; [96260], Thin section

Coldwater (iron) 37°16′ N., 99°20′ W.

Comanche County, Kansas, U.S.A.
Found 1923
Iron. Octahedrite.

An almost completely oxidized, limonitic mass of 18.4kg was found in 1923. Described and analysed, 4.32 %Ni, E.V. Shannon, Proc. U.S. Nat. Mus., 1927, **72** (21) [M.A. 3-534]. Some slices still show traces of the Widmanstätten structure, H.H. and A.D. Nininger, The Nininger Collection of Meteorites, Winslow, Arizona, 1950, p.139. Reported, H.H. Nininger, Trans. Kansas Acad. Sci., 1928, **31**, p.87 [M.A. 5-155], H.H. Nininger and G.A. Muilenburg, J. Geol., 1931, **39**, p.592 [M.A. 5-13], H.H. Nininger, The Mines Mag., Golden, Colorado, 1933, **23**, p.6 [M.A. 6-405] these accounts differ somewhat.

4.7kg Washington, U.S. Nat. Mus.; 2.3kg New York, Amer. Mus. Nat. Hist.; 1.9kg Tempe, Arizona State Univ.; 3.75kg Chicago, Field Mus. Nat. Hist.;

Specimen(s): [1959,834], 1550g. oxidised

Coldwater (stone) 37°16′ N., 99°20′ W.

Comanche County, Kansas, U.S.A.
Found 1924
Stone. Olivine-bronzite chondrite (H5).

Two weathered stones were brought to G.P. Merrill by H.H. Nininger, G.P. Merrill, letter of 18 October 1926 in Min. Dept. BM(NH). Possibly referable to the meteor of 9 November 1923, H.H. Nininger, Trans. Kansas Acad. Sci., 1928, **31**, p.91, 94 [M.A. 5-155, -156], H.H. Nininger and G.A. Muilenburg, J. Geol., 1931, **39**, p.592 [M.A. 5-13]. More material has since been found and the total known weight is now 11kg. Listed, H.H. and A.D. Nininger, The Nininger Collection of Meteorites, Winslow, Arizona, 1950, p.45. Olivine Fa_{17}, B. Mason, Geochimica et Cosmochimica Acta, 1963, **27**, p.1011.

5.4kg Washington, U.S. Nat. Mus.; 3.5kg Chicago, Field Mus. Nat. Hist.; 2.7kg Tempe, Arizona State Univ.; 770g Mainz, Max-Planck-Inst.;

Specimen(s): [1928,473], 86g. and fragments, 4g; [1959,1006], 2170g.

Colfax 35°18′ N., 81°44′ W.

Rutherford County, North Carolina, U.S.A.
Found 1880
Synonym(s): *Ellenboro, Rutherford County*
Iron. Octahedrite, medium (0.6mm) (IB).

A mass of about 5lb was ploughed up near Ellenboro; described and analysed, 10.37 %Ni, L.G. Eakins, Am. J. Sci., 1890, **39**, p.395. Further analysis, 10.84 %Ni, 52.8 ppm.Ga, 153 ppm.Ge, 1.5 ppm.Ir, J.T. Wasson, Icarus, 1970, **12**, p.407. The mass has been artificially reheated and hammered, V.F. Buchwald, Iron Meteorites, Univ. of California, 1975, p.486.

900g Chicago, Field Mus. Nat. Hist.; 743g Harvard Univ.; 312g Washington, U.S. Nat. Mus.; 186g New York, Amer. Mus. Nat. Hist.;

Specimen(s): [84193], 44g. slice, and fragment, 4g.

Collantifona v Collescipoli.

Colle Antifona v Collescipoli.

Collescipoli 42°32′ N., 12°37′ E.

Terni, Umbria, Italy
Fell 1890, February 3, 1330 hrs
Synonym(s): *Antifona, Collantifona, Colle Antifona, Spoleto, Terni*
Stone. Olivine-bronzite chondrite (H5).

After appearance of fireball followed by detonations, a stone of about 5kg fell, G. Terrenzi, Riv. Ital. Sci. Nat. Siena, 1890, **10** (3), p.25. Analysis, G.P. Merrill, Mem. Nat. Acad. Sci. Washington, 1916, **14**, p.13. Further analysis, 26.88 % total iron, H.B. Wiik, Geochimica et Cosmochimica Acta, 1956, **9**, p.279. Mentioned, B. Baldanza, Min. Mag., 1965, **35**, p.214. Olivine Fa_{19}, B. Mason, Geochimica et Cosmochimica Acta, 1963, **27**, p.1011. Co-ordinates, G.R. Levi-Donati, priv. comm., 1969.

1.2kg Budapest, Nat. Mus.; 380g Vienna, Naturhist. Mus.; 303g Milan, Mus.; 171g New York, Amer. Mus. Nat. Hist.; 127g Washington, U.S. Nat. Mus.; 88g Chicago, Field Mus. Nat. Hist.;

Specimen(s): [1920,299], 151.5g.; [85834], 229g.

Collin County v McKinney.

Colombia v Santa Rosa.

Colomera 37°26′ N., 3°39′ W.

Granada, Spain
Found 1912
Synonym(s): *Granada*
Iron. Octahedrite, medium, with silicate inclusions (IIE).

Weight 134kg, described and analysed, 7.16 %Ni, J. Dorronsoro and F. Moreno Martin, Anal. Soc. Españ. Fis. Quim., 1934, **32**, p.1111 [M.A. 6-392]. The 'Granada' fall listed separately by Coulson and by Leonard is merely a synonym of Colomera. Chemically and structurally anomalous; analysis of metal, 7.86 %Ni, 28.4 ppm.Ga, 74.6 ppm.Ge, 7.9 ppm.Ir, E.R.D. Scott et al., Geochimica et Cosmochimica Acta, 1973, **37**, p.1957. Description of silicate inclusions, T.E. Bunch et al., Contr. Mineral. Petrol., 1970, **25**, p.297. Rb/Sr age, 4610 my., H.G. Sanz et al.,

Geochimica et Cosmochimica Acta, 1970, **34**, p.1227. Description, kamacite recrystallised, V.F. Buchwald, Iron Meteorites, Univ. of California, 1975, p.488.
Main mass, Madrid, Nat. Mus.; 133g Washington, U.S. Nat. Mus.;

Colonia Obrera 24°1′ N., 104°40′ W.
Durango, Mexico
Found 1973
Synonym(s): *Colonia Obrerra*
Iron. Octahedrite, coarse (1.4mm) (IIIE).
A single mass of 12.2kg was found, C.F. Lewis and C.B. Moore, Cat. Meteor. Arizona State Univ., 1976, p.189. Classification and analysis, 8.62 %Ni, 17.4 ppm.Ga, 37 ppm.Ge, 0.055 ppm.Ir, A. Kracher et al., Geochimica et Cosmochimica Acta, 1980, **44**, p.773.
12kg Tempe, Arizona State Univ.; 62g Chicago, Field Mus. Nat. Hist.;

Colonia Obrerra v Colonia Obrera.

Colonia Suiza 41°11′ S., 71°23′ W. approx.
San Carlos de Bariloche, Rio Negro, Argentina
Fell 1936, October 9, before this date
Stone.
A 250g, partly crusted stone was found in the collection of the Argentine Mus. Nat. Sci., Buenos Aires. It was reportedly part of a fall in Lago Moreno, and was retrieved from the water, S. Rivas and T.E. Rivas, Mus. Argentino Cienc. Nat., Rev. Minera, 1973, **30**, p.44.
Main mass, Buenos Aires, Mus. Argentino Cienc. Nat.;

Colony 35°21′ N., 98°41′ W.
Washita County, Oklahoma, U.S.A.
Found 1975, approx.
Stone. Carbonaceous chondrite, type III (CO3).
A single mass of 3912g was caught in the tines of a cotton cultivator, J. Westcott, letter of 20 February, 1983 in Min. Dept. BM(NH).
Specimen(s): [1983,M.2], 1154g.; [1983,M.45], 10.5g.

Colorado v Bear Creek.

Colorado v Russel Gulch.

Colorado (of A. Brezina) v Trenton.

Colorado River v Cañon Diablo.

Colorado Springs 38°50′ N., 104°50′ W. approx.
El Paso County, Colorado, U.S.A.
Found 1950, approx.
Iron.
A small iron was found by a Mr. J.H. Alexander and identified by R.M. Pearl. The specimen was later mislaid, R.M. Pearl, letter of 11 September 1973 in Min. Dept. BM (NH).

Colston Bassett
Nottinghamshire, England
Pseudometeorite..
A boulder lying in the churchyard has been identified as cinnabar, probably from Peru. Its meteoritic origin appears to be 'purely a fabrication of local legend', W.A.S. Sarjeant, The Mercian Geologist, 1971, **4**, p.41.

Comanche (iron) 32°1′ N., 98°42′ W.
Comanche County, Texas, U.S.A.
Found 1940
Iron. Octahedrite, coarse (1.5mm) (IA).
A mass of 19.7kg was found, F.C. Leonard, Classif. Cat. Meteor., 1956, p.8, 46, 64. Classification and analysis, 8.1 % Ni, 73.9 ppm.Ga, 269 ppm.Ge, 2.2 ppm.Ir, J.T. Wasson, Icarus, 1970, **12**, p.407. Description, co-ordinates, V.F. Buchwald, Iron Meteorites, Univ. of California, 1975, p.493.
550g Washington, U.S. Nat. Mus.;

Comanche (stone) 31°54′ N., 98°30′ W.
Comanche County, Texas, U.S.A.
Found 1956, or 1957
Stone. Olivine-bronzite chondrite (H).
A mass of about 2kg was found, B. Mason, letters of 4 March and 28 April 1964 in Min. Dept. BM(NH). Olivine Fa_{19}, B. Mason, Geochimica et Cosmochimica Acta, 1967, **31**, p.1100.
2.35kg Fort Worth, Texas, Monnig Colln.;

Commune des Ormes v Les Ormes.

Conceicão Aparecida v Patrimonio.

Concepcion v Chupaderos.

Concepcion v Nogoya.

Concho 32°0′ N., 101°30′ W. approx.
Glasscock County, Texas, U.S.A.
Found 1939, recognized in this year
Stone. Olivine-hypersthene chondrite (L6).
One much weathered stone of 93.5kg, A.D. Nininger, Pop. Astron., Northfield, Minnesota, 1940, **48**, p.555, A.D. Nininger, Contr. Soc. Res. Meteorites, 1940, **2**, p.227 [M.A. 8-54]. Olivine Fa_{23}, B. Mason, Geochimica et Cosmochimica Acta, 1963, **27**, p.1011.
758g New York, Amer. Mus. Nat. Hist.; 138g Washington, U.S. Nat. Mus.; 129g Tempe, Arizona State Univ.;
Specimen(s): [1963,31], 126g.

Coney Fork v Carthage.

Conquista 19°51′ S., 47°33′ W. approx.
Corrego do Lageado, Conquista, Minas Gerais, Brazil
Fell 1965, December, 0600 hrs, early in the month
Stone. Olivine-bronzite chondrite (H4).
The fall was accompanied by a loud noise; many fragments were said to have been distributed around the impact site but only a single mass of 20.35kg appears to have been preserved. Description, with an analysis, olivine $Fa_{17.2}$, 25.83 % total iron, K. Keil et al., Meteoritics, 1978, **13**, p.177.
20.3kg Belo Horizonte, Mus. Hist. Nat.; 87g Sao Paulo, Inst. Geocienc.; 49g Albuquerque, Univ. of New Mexico; 5.2g Washington, U.S. Nat. Mus.;

Constantia v Diep River.

Constantinopel v Constantinople.

Constantinople 41°3′ N., 28°55′ E.
Istanbul, Turkey
Fell 1805, June
Synonym(s): *Constantinopel*
Doubtful. Stone. Achondrite, Ca-rich. Eucrite (AEUC).
Several stones are said to have fallen on the shambles in Constantinople, E.F.F. Chladni, Die Feuer-Meteore, Wien, 1819, p.278. The fragment of 6g acquired by the Vienna Mus. in 1832 probably belongs to Stannern, G. Tschermak, Tschermaks Min. Petr. Mitt., 1872, p.85, P. Partsch, Die Meteoriten, Wien, 1843, p.26.
0.3g Tübingen, Univ., perhaps Stannern;

Cook
The determination of a number of trace elements in two unidentified meteorites (Cook and Cook 6089) from the Cook collection is reported, L. Greenland and J.F. Lovering, Geochimica et Cosmochimica Acta, 1965, **29**, p.843. A magnetic study refers to Cook 6064, F.D. Stacy et al., J. Geophys. Res., 1961, **66**, p.1523.

Cookeville 36°10′ N., 85°31′ W.
Putman County, Tennessee, U.S.A.
Found 1913, about
Iron. Octahedrite, coarse (IA).
A mass of about 5lb was found; described, with an analysis, 6.38 %Ni, G.P. Merrill, Proc. U.S. Nat. Mus., 1916, **51**, p.325. Further analysis, 6.4 %Ni, 91.4 ppm.Ga, 395 ppm.Ge, 2.1 ppm.Ir, J.T. Wasson, Icarus, 1970, **12**, p.407. Paired with Smithville (*q.v.*), V.F. Buchwald, Iron Meteorites, Univ. of California, 1975, p.494.
450g Washington, U.S. Nat. Mus.; 252g New York, Amer. Mus. Nat. Hist.; 216g Chicago, Field Mus. Nat. Hist.;
Specimen(s): [1920,116], 135.6g. slice

Coolac 34°58′ S., 148°7′30″ E.
County Harden, New South Wales, Australia
Found 1874
Iron. Octahedrite, coarse (2.1mm) (IA).
A mass of 19.28kg was found 3 miles W. of Coolac, and was used for many years as a fire stop. Described, with an erroneous analysis, 4.72 %Ni, T. Hodge-Smith, Rec. Austr. Mus., 1937, **20**, p.130 [M.A. 7-71]. Analysis, 6.78 %Ni, E.P. Henderson, Pop. Astron., Northfield, Minnesota, 1951, **59**, p.205 [M.A. 11-446]. Further analysis, 7.4 %Ni, 91.5 ppm.Ga, 423 ppm.Ge, 2.4 ppm.Ir, J.T. Wasson, Meteorites, Springer-Verlag, 1974, p.298. Description, properties unaffected by heating, except hardness of kamacite, V.F. Buchwald, Iron Meteorites, Univ. of California, 1975, p.494.
510g Sydney, Austr. Mus., main mass missing?; 2.75kg Washington, U.S. Nat. Mus.; 453g Tübingen, Univ.; 225g New York, Amer. Mus. Nat. Hist.; 221g Chicago, Field Mus. Nat. Hist.;
Specimen(s): [1965,212], 150.5g. and fragments, 1g; [1966, 52], 2939g. and 87g and pieces, 219g.

Coolamon 34°49′ S., 147°8′ E.
New South Wales, Australia
Found 1921, between 1920 and 1922
Stone. Olivine-hypersthene chondrite (L6).
A complete stone of 393g was found by G. Eisenhauer while ploughing in his property "Bonnie Doon", B. Mason, letters of 4 March, 28 April 1964 in Min. Dept. BM(NH). Reported, Meteor. Bull., 1965 (33), L. Greenland and J.F. Lovering, Geochimica et Cosmochimica Acta, 1965, **29**, p.843. Contains ringwoodite, olivine $Fa_{25.9}$, B. Mason, Rec. Austr. Mus., 1974, **29**, p.169.
Main mass, Sydney, Austr. Mus.; Thin section, Washington, U.S. Nat. Mus.;

Coolidge 38°2′ N., 101°59′ W.
Hamilton County, Kansas, U.S.A.
Found 1937
Stone. Carbonaceous chondrite, type IV (C4).
One mass of 4.5kg was found "at 36°47′N., 102°W.", A.D. Nininger, Pop. Astron., Northfield, Minnesota, 1939, **47**, p.211 the latitude here given is incorrect. Mentioned, F.C. Leonard, Univ. New Mexico Publ., Albuquerque, 1946 (meteoritics ser. no. 1), p.37, H.H. and A.D. Nininger, The Nininger Collection of Meteorites, Winslow, Arizona, 1950, p.45. Olivine Fa_{14}, B. Mason, Geochimica et Cosmochimica Acta, 1963, **27**, p.1011. Analysis, 23.68 % total iron, E. Jarosewich, Geochimica et Cosmochimica Acta, 1966, **30**, p.1261. XRF analysis, 24.00 % total iron, T.S. McCarthy and L.H. Ahrens, Earth planet. Sci. Lett., 1972, **14**, p.97. Classification, W.R. Van Schmus, Meteorite Research, ed. P.M. Millman, D. Reidel, Dordrecht-Holland, 1969, p.480, B. Mason, Meteoritics, 1971, **6**, p.59.
1.3kg Tempe, Arizona State Univ.; 228g New York, Amer. Mus. Nat. Hist.; 185g Chicago, Field Mus. Nat. Hist.; 203g Washington, U.S. Nat. Mus.; Specimen, Perth, West. Austr. Mus.;
Specimen(s): [1938,305], 107.6g. and fragments, 20.2g; [1959, 854], 1179g. slice, and fragments, 10.5g

Coomandook 35°25′ S., 139°45′ E.
South Australia, Australia
Found 1939
Synonym(s): *Coomandook I, Coomandook II, Ki-Ki*
Stone. Olivine-bronzite chondrite (H6).
One mass (Coomandook I) of 1.1kg (B. Mason) or 0.91kg, D.H. McColl, letter of 30 March 1972 in Min. Dept. BM (NH)., was found in 1939, 8 miles due E. of Coomandook. A second mass (Coomandook II, or Ki-Ki) of 4.18kg was found in January 1967 4.75 miles NE. of Ki-Ki, and less than 10 miles from the original discovery, D.W.P. Corbett, Rec. S. Austr. Mus., 1968, **15**, p.767. Olivine $Fa_{18.8}$, B. Mason, Rec. Austr. Mus., 1974, **29**, p.169. XRF analysis, 24.82 % total iron, M.J. Fitzgerald, Ph.D. Thesis, Univ. of Adelaide, 1979, p.23.
5.28kg Adelaide, South Austr. Mus., main masses of both stones; 11g Washington, U.S. Nat. Mus.;

Coomandook I v Coomandook.

Coomandook II v Coomandook.

Coonana 29°51′ S., 140°42′ E.
South Australia, Australia
Found 1962
Stone. Olivine-bronzite chondrite (H4).
A single mass of 6.8kg was found 5.5km west of Coonana Bore. XRF analysis, 24.87 % total iron, M.J. Fitzgerald, Ph.D. Thesis, Univ. of Adelaide, 1979, p.23, 201. Mentioned, D.H. McColl, letter of 30 March 1972 in Min. Dept. BM (NH). Olivine $Fa_{19.2}$, B. Mason, Rec. Austr. Mus., 1974, **29**, p.169.
5kg Adelaide, South Austr. Mus., includes main mass, 4.36kg; 17g New York, Amer. Mus. Nat. Hist.;
Specimen(s): [1968,4], 20.5g. fragments

Coon Butte 35°0′ N., 111°0′ W. approx.
Coconino County, Arizona, U.S.A.
Found 1905
Stone. Olivine-hypersthene chondrite (L6), brecciated.
A stone of 2.75kg was found about a mile W. of Coon Butte (=Meteor Crater); described and partially analysed, J.W. Mallet, Am. J. Sci., 1906, **21**, p.347. Olivine Fa_{24}, B. Mason,

Geochimica et Cosmochimica Acta, 1963, **27**, p.1011.
627g Tempe, Arizona State Univ.; 308g Chicago, Field Mus. Nat. Hist.; 200g Washington, U.S. Nat. Mus.;
Specimen(s): [1925,15], 172.8g. slice

Cooperstown v Burlington.

Coopertown 36°26′ N., 87°0′ W.
Robertson County, Tennessee, U.S.A.
Found 1860, known in this year
Synonym(s): *Robertson County*
Iron. Octahedrite, coarse (1.5mm) (IIIE).
A mass of 37lb was sent to J.L. Smith in 1860 and analysed by him, 9.12 %Ni, J.L. Smith, Am. J. Sci., 1861, **31**, p.266. Relative orientation of the kamacite and rhabdite, O.B. Bøggild, Meddel. Grønland, 1927, **74**, p.28. Classification and new analysis, 8.47 %Ni, 17.0 ppm.Ga, 34.9 ppm.Ge, 0.51 ppm.Ir, E.R.D. Scott et al., Geochimica et Cosmochimica Acta, 1973, **37**, p.1957. Description, swollen kamacite lamellae, epsilon structure, V.F. Buchwald, Iron Meteorites, Univ. of California, 1975, p.496.
2.1kg Tempe, Arizona State Univ.; 960g Washington, U.S. Nat. Mus.; 860g Vienna, Naturhist. Mus.; 815g Yale Univ.; 677g New York, Amer. Mus. Nat. Hist.; 358g Tempe, Arizona State Univ.; 213g Chicago, Field Mus. Nat. Hist.; 172g Berlin, Humboldt Univ.; 54g Harvard Univ.;
Specimen(s): [40879], 92g.; [35407], 77g. and fragments, 4g.

Coorara 30°27′ S., 126°6′ E.
Nullarbor Plain, Western Australia, Australia
Found 1966
Stone. Olivine-hypersthene chondrite (L6).
A broken, weathered mass of 92.4g was found about 0.4 miles S. of the Dingo Pup Donga recovery site (*q.v.*). Eight small fragments, together weighing 24.3g., were found about 0.2 miles NW. of the Dingo Pup Donga site, G.J.H. McCall and W.H. Cleverly, J. Roy. Soc. West. Austr., 1970, **53**, p.69, Meteor. Bull., 1976 (54), Meteoritics, 1976, **11**, p.72. Coorara contains veinlets with ringwoodite and majorite, olivine $Fa_{25.9}$, B. Mason, Rec. Austr. Mus., 1974, **29**, p.169.
80g Perth, West. Austr. Mus., main mass, approx. weight; 11.5g Kalgoorlie, West. Austr. School of Mines; 7.6g Washington, U.S. Nat. Mus.;

Cope 39°40′ N., 102°50′ W.
Washington County, Colorado, U.S.A.
Found 1934, recognized 1937
Stone. Olivine-bronzite chondrite (H5).
Eleven (?eight) stones, totalling 12kg were found in sect.26, township 5, range 49, in Washington County, and sect.35, township 5 range 49, in Kit Carson County, A.D. Nininger, Pop. Astron., Northfield, Minnesota, 1939, **47**, p.211. Mentioned, E.P. Henderson, letter of 3 June 1939 in Min. Dept. BM(NH). Olivine Fa_{18}, B. Mason, Geochimica et Cosmochimica Acta, 1963, **27**, p.1011.
5.1kg Washington, U.S. Nat. Mus., four stones; 4.9kg Denver, Mus. Nat. Hist.; 1.38kg Tempe, Arizona State Univ.; 451g Chicago, Field Mus. Nat. Hist.;
Specimen(s): [1952,140], 30g.; [1959,853], 1067.5g.

Copiapo 27°18′ S., 70°24′ W. approx.
Atacama, Chile
Found 1863
Synonym(s): *Atacama Desert, Deesa (of G.A. Daubrée), Dehese (of G.A. Daubrée), Desert of Atacama, Sierra de Deesa*
Iron. Octahedrite, coarse (1.5mm), with silicate inclusions (IA).
Numerous masses were brought to Copiapo since 1863; some, owing to interchange of labels by G. Daubrée, have been supposed to come from the Sierra de Deesa, L. Fletcher, Min. Mag., 1889, **8**, p.255. Described, with an analysis, 8.7 %Ni, G.A. Daubrée, C. R. Acad. Sci. Paris, 1868, **66**, p.571. Analysis of metal, 7.01 %Ni, 69.8 ppm.Ga, 252 ppm.Ge, 2.5 ppm.Ir, J.T. Wasson, Icarus, 1970, **12**, p.407. Mentioned, B. Mason, Min. Mag., 1967, **36**, p.120. Silicate mineralogy, T.E. Bunch et al., Contr. Mineral. Petrol., 1970, **25**, p.297. About 23kg are known in collections and G. Daubrée had a mass of 7kg. Description, V.F. Buchwald, Iron Meteorites, Univ. of California, 1975, p.497.
8.9kg Paris, Mus. d'Hist. Nat.; 3.4kg Berlin, Humboldt Univ.; 1kg Vienna, Naturhist. Mus.; 295g Harvard Univ.; 195g Chicago, Field Mus. Nat. Hist.;
Specimen(s): [45951], 450g. four pieces, and filings, 1.3g; [41110], 12.5g.; [40225], 2.5g.

Copinsay 58°54′ N., 2°41′ W.
Orkney, Scotland
Fell 1676
Doubtful. Stone.
One stone fell into a boat, E.F.F. Chladni, Die Feuer-Meteore, Wien, 1819, p.237 but the evidence is not conclusive.

Corboda v Capilla del Monte.

Cordoba v Capilla del Monte.

Corizatillo error for Corrizatillo.

Corlton-Hamilton v Carlton.

Cormeilles 48°59′ N., 2°13′ E.
Val d'Oise, France
Pseudometeorite..
Fragments of limonite-coated metal, found in 1944 in the Lambert quarry on the surface of limestone and immediately below a bed of Oligocene marl, are described as possibly meteoritic, J. Tricart and A. Cailleux, C. R. Acad. Sci. Paris, 1947, **225**, p.131 but chemical analysis shows only 0.01% Ni and some 3.5% C, and their meteoritic origin seems very improbable.

Coro v Ucera.

Coronel Arnold 33°4′ S., 61°0′ W.
San Lorenzo, Santa Fe, Argentina
Found 1962, known in this year
Stone. Olivine-hypersthene chondrite (L).
A 450g mass is listed, L.M. Villar, Cienc. Investig., 1968, **24**, p.302. May have been confused with the El Timbu iron (*q.v.*).

Corowa 36°0′ S., 146°22′ E.
New South Wales, Australia
Found 1964, before this year
Iron. Octahedrite, plessitic, chemically anomalous (IIF).
A mass of 25lb was found, described, with an analysis, 13.4 %Ni, G. Baker et al., Geochimica et Cosmochimica Acta, 1964, **28**, p.1377. Classification and new analysis, 13.13 % Ni, 10.1 ppm.Ga, 159 ppm.Ge, 0.77 ppm.Ir, J.T. Wasson, Geochimica et Cosmochimica Acta, 1969, **33**, p.859. Description; a unique iron, V.F. Buchwald, Iron Meteorites, Univ. of California, 1975, p.499.
Specimen, Melbourne, Nat. Mus. Victoria;

Corrego Areado v Patos de Minas (hexahedrite).

Correo 34°57′ N., 107°10′ W.
Valencia County, New Mexico, U.S.A.
Found 1979, July, or August
Stone. Olivine-bronzite chondrite (H4).
About 35 fragments, totalling about 700g were found, olivine $Fa_{18.6}$, Meteor. Bull., 1980 (58), Meteoritics, 1980, **15**, p.235. Description, A.E. Rubin et al., Meteoritics, 1981, **16**, p.9.
400g Albuquerque, Univ. of New Mexico; 89g Tempe, Arizona State Univ.;

Corrizalillo v Corrizatillo.

Corrizatillo 26°2′ S., 70°20′ W. see text
Copiapo, Atacama, Chile
Found 1884, known in this year
Synonym(s): *Caranzalillo, Caranzatillo, Carrizalillo, Corrizalillo*
Iron. Octahedrite, coarse (1.9mm) (IIICD).
A mass of 1328g was sent to Oslo Univ. from Chile by L. Sundt. It was provisionally referred to Copiapo, E.A. Wülfing, Die Meteoriten in Samml., Tübingen, 1897, p.87. The reputed place of find is possibly Carrizalillo 28°10′S., 69°46′W., but almost certainly Carrizalillo, 26°2′S., 70°20′W. The former is nearer to Copiapo. Since "Carrizalillo" is a discredited synonym of Vaca Muerta, the name "Corrizatillo" is retained. This is the meteorite in which Pt metals were sought, V.M. Goldschmidt, Proc. Roy. Inst. Gt. Britain, 1929, **26**, p.73. Analysis, 6.4 %Ni, 81.2 ppm.Ga, 328 ppm.Ge, 1.1 ppm.Ir, R. Schaudy and J.T. Wasson, Chem. Erde, 1970, **30**, p.287. Provenance, paired with Pan de Azucar (*q.v.*), V.F. Buchwald, Iron Meteorites, Univ. of California, 1975, p.961.
1.27kg Oslo, Min.-Geol. Mus.;

Corston v Strathmore.

Cortez 37°21′ N., 108°41′ W.
Montezuma County, Colorado, U.S.A.
Found 1940
Stone. Olivine-bronzite chondrite (H6), veined.
A mass of 715.6g was found, H.H. and A.D. Nininger, The Nininger Collection of Meteorites, Winslow, Arizona, 1950, p.46. Olivine Fa_{19}, B. Mason, Geochimica et Cosmochimica Acta, 1963, **27**, p.1011.
155g Tempe, Arizona State Univ.; 116g Harvard Univ.; 54g Washington, U.S. Nat. Mus.; 29g Denver, Mus. Nat. Hist.;
Specimen(s): [1959,852], 250g.

Cosby's Creek 35°47′ N., 83°15′ W.
Cocke County, Tennessee, U.S.A.
Found 1837, known before this year
Synonym(s): *Cocke County, East Tennessee, Sevier County, Wilson County*
Iron. Octahedrite, coarse (2.5mm) (IA).
Two masses, one said to have weighed 2000lb and the other 112lb, were known before 1837, G. Troost, Am. J. Sci., 1840, **38**, p.250, C.U. Shepard, Am. J. Sci., 1842, **43**, p.354, C.U. Shepard, Am. J. Sci., 1847, **4**, p.74. The larger mass was forged into various articles, V.F. Buchwald, Iron Meteorites, Univ. of California, 1975, p.500. Distinct from Waldron Ridge (*q.v.*) and Greenbrier County (*q.v.*). Analysis, 6.57 %Ni, 91.5 ppm. Ga., 431 ppm.Ge, 2.9 ppm.Ir, J.T. Wasson, Icarus, 1970, **12**, p.407.
17.2kg Harvard Univ.; 12.5kg Tübingen, Univ.; 3kg Chicago, Field Mus. Nat. Hist.; 2.9kg Oxford, Univ. Mus.; 1.68kg Paris, Mus. d'Hist. Nat.; 1.1kg Tempe, Arizona State Univ.; 1.2kg Yale Univ.; 1kg Copenhagen, Univ.; 1.3kg Washington, U.S. Nat. Mus.; 644g New York, Amer. Mus. Nat. Hist.;
Specimen(s): [16865], 25.85kg. one of the largest masses; [16866], 24.61kg. one of the largest masses preserved; [46976], 30g.; [1920,298], 21.5g.; [33204], 18.5g.; [33924], 35g.; [1964,710], 29g. fragments; [19938], 52g. graphite fragments

Cosina 21°10′ N., 100°52′ W.
Dolores Hidalgo, Guanajuato, Mexico
Fell 1844, January, 1100 hrs
Synonym(s): *Cerro Cosina, Cocina, Dolores Hidalgo, Loma de la Cosina*
Stone. Olivine-bronzite chondrite (H5).
A stone of about 1.2kg was seen to fall, after detonations and appearance of a fireball, O. Büchner, Ann. Phys. Chem. (Poggendorff), 1866, **129**, p.351, A. Castillo, Cat. Météorites Mexique, Paris, 1889, p.12. Olivine Fa_{19}, B. Mason, Geochimica et Cosmochimica Acta, 1963, **27**, p.1011.
121g Paris, Mus. d'Hist. Nat.; 57g Vienna, Naturhist. Mus.; 23g Berlin, Humboldt Univ.; 5g Chicago, Field Mus. Nat. Hist.;
Specimen(s): [40767], 38.2g.

Cosmo Newberry 27°57′ S., 122°53′ E.
Western Australia, Australia
Found 1980
Iron. Hexahedrite (IIA).
A single mass of 2156g was found 4 km north of Cosmo Newberry homestead, 5.64 %Ni, 62 ppm.Ga, 190 ppm.Ge, 19 ppm.Ir, J.R. de Laeter, Meteoritics, 1982, **17**, p.135.
2.15kg Perth, West. Austr. Mus.;

Cosona v Siena.

Cossipore v Manbhoom.

Costal de Garraf v Garraf.

Costa Rica v Heredia.

Costilla v Costilla Peak.

Costilla County v Costilla Peak.

Costilla Peak 36°50′ N., 105°14′ W.
Taos County, New Mexico, U.S.A.
Found 1881
Synonym(s): *Costilla, Costilla County*
Iron. Octahedrite, medium (1.0mm) (IIIA).
A mass of about 35.5kg was found; described, with an analysis, 7.71 %Ni, R.C. Hills, Proc. Colorado Sci. Soc., 1895, **5**, p.121. Determination of Ga, Au, Pd and partial analysis, 7.56 %Ni, E. Goldberg et al., Geochimica et Cosmochimica Acta, 1951, **2**, p.1. Further analysis, 7.49 % Ni, 18.7 ppm.Ga, 33.6 ppm.Ge, 14 ppm.Ir, E.R.D. Scott et al., Geochimica et Cosmochimica Acta, 1973, **37**, p.1957.
9.5kg Chicago, Field Mus. Nat. Hist.; 3.5kg New York, Amer. Mus. Nat. Hist.; 2.1kg Tempe, Arizona State Univ.; 1.5kg Vienna, Naturhist. Mus.; 1.8kg Washington, U.S. Nat. Mus.; 1.5kg Harvard Univ.; 500g Paris, Mus. d'Hist. Nat., approx. weight; 469g Tübingen, Univ.; 250g Prague, Nat. Mus.;
Specimen(s): [77097], 1595g.

Coterfield error for Cotesfield.

Cotesfield 41°22′ N., 98°38′ W.
Howard County, Nebraska, U.S.A.
Found 1928
Stone. Olivine-hypersthene chondrite (L6).
An oxidised stone of 1160g was found at Cotesfield. Description, H.H. Nininger, Am. Miner., 1933, **18**, p.56 [M.A. 5-300]. Olivine Fa_{23}, B. Mason, Geochimica et Cosmochimica Acta, 1963, **27**, p.1011.
Main mass, Denver, Colorado Mus. Nat. Hist.; 77g Chicago, Field Mus. Nat. Hist.; 33g Tempe, Arizona State Univ.; 17g Washington, U.S. Nat. Mus.; 17g Harvard Univ.;
Specimen(s): [1959,855], 27g. thin slice

Cottonwood 34°50′ N., 112°1′ W.
Yavapai County, Arizona, U.S.A.
Found 1955
Stone. Olivine-bronzite chondrite (H5).
A mass of about 800g was found, B. Mason, Meteorites, Wiley, 1962, p.228. Olivine Fa_{18}, B. Mason, Geochimica et Cosmochimica Acta, 1963, **27**, p.1011.
754g Tempe, Arizona State Univ., main mass; 40g New York, Amer. Mus. Nat. Hist.;

Couch Iron v Coahuila.

Covert 39°12′ N., 98°47′ W.
Osborne County, Kansas, U.S.A.
Found 1896, before this year
Synonym(s): *Alton*
Stone. Olivine-bronzite chondrite (H5), veined.
One stone, known before 1896, was recognised as a meteorite in 1929, and search between 1929 and 1931 yielded nine more stones, total weight 37kg, the largest being of 13kg, all weathered. Description, H.H. Nininger and G.A. Muilenburg, J. Geol., 1931, **39**, p.592 [M.A. 5-13]. Total weight now known is 61kg, H.H. and A.D. Nininger, The Nininger Collection of Meteorites, Winslow, Arizona, 1950, p.46. Olivine Fa_{18}, B. Mason, Geochimica et Cosmochimica Acta, 1963, **27**, p.1011.
19.2kg Tempe, Arizona State Univ.; 1.5kg Washington, U.S. Nat. Mus.; 1.36kg Yale Univ.; 798g Chicago, Field Mus. Nat. Hist.;
Specimen(s): [1931,223], 1618.5g.; [1959,1042], 3700g. half a stone.

Cowell 33°18′ S., 136°1′ E.
South Australia, Australia
Found 1932
Synonym(s): *Caralue*
Iron. Octahedrite, medium (1.2mm) (IIIA).
A mass of 5.72kg was found about 90 miles NW. of the town of Cowell and close to the location of the Kyancutta mass, M.J. Fitzgerald, letter of 11 December 1972 in Min. Dept. BM(NH). Analysis, 8.2 %Ni, 21 ppm.Ga, 38 ppm.Ge, S.J.B. Reed, Meteoritics, 1972, **7**, p.257. This mass almost certainly should be assigned to the Kyancutta fall (*q.v.*).
Specimen(s): [1966,491], 96.5g. slice

Cowley County v Wilmot.

Cowra 33°51′ S., 148°41′ E.
County Bathurst, New South Wales, Australia
Found 1888
Synonym(s): *Bathurst, Carcoar, Carevar*
Iron. Octahedrite, plessitic (IRANOM).
A mass of 12.25lb was found ot the top of Battery Mountain, G.W. Card, Rec. Geol. Surv. New South Wales, 1897, **5**, p.51. Analysis, chemically anomalous, 12.94 %Ni, 73.5 ppm.Ga, 12.3 ppm.Ge, 14 ppm.Ir, J.T. Wasson and R. Schaudy, Icarus, 1971, **14**, p.59. XRF analysis, S.J.B. Reed, Meteoritics, 1972, **7**, p.257. Structure, H.J. Axon and P.L. Smith, Min. Mag., 1972, **38**, p.736. Description, large skeletal schreibersite crystals, V.F. Buchwald, Iron Meteorites, Univ. of California, 1975, p.505.
Main mass, Sydney, Australian Mus.; 678g New York, Amer. Mus. Nat. Hist.; 210g Vienna, Naturhist. Mus.; 99g Budapest, Nat. Hist. Mus.; 87g Chicago, Field Mus. Nat. Hist.; 80g Prague, Nat. Mus.; 61g Washington, U.S. Nat. Mus.;
Specimen(s): [68206], 102g.; [68205], 78g. and a polished mount; [68211], 56.5g. filings

Coya Norte v North Chile.

Crab Hole 30°24′ S., 127°26′ E.
Donga, Western Australia, Australia
Found 1980, in the spring
Stone. Olivine-hypersthene chondrite (L).
Somewhat weathered material. Total weight recovered not reported. Mentioned, D. Heinlein, letter of 6 December, 1980 in Brit. Mus. (Nat. Hist.).
Specimen, Washington, U.S. Nat. Mus.; 25g Chicago, Field Mus. Nat. Hist.;
Specimen(s): [1981,M.8], 34.7g.

Crab Orchard 35°50′ N., 84°55′ W.
Rockwood, Cumberland County, Tennessee, U.S.A.
Found 1887
Synonym(s): *Cumberland County, Powder Mill Creek, Rockwood*
Stony-iron. Mesosiderite (MES).
About five masses, of total weight about 107lb, the largest weighing 85lb, were found 8.5 miles W. of Rockwood furnace, E.E. Howell, Science, 1887, **10**, p.107. Description, G.F. Kunz, Am. J. Sci., 1887, **34**, p.476. Analysis of metal, 7.02 %Ni, 13.3 ppm.Ga, 49.5 ppm.Ge, 2.9 ppm.Ir, J.T. Wasson et al., Geochimica et Cosmochimica Acta, 1974, **38**, p.135. Silicate mineralogy, olivine Fa_{28-51}, B.N. Powell, Geochimica et Cosmochimica Acta, 1971, **35**, p.5. Corona structure, mineralogy, C.E. Nehru et al., Geochimica et Cosmochimica Acta, 1980, **44**, p.1103.
8.4kg Chicago, Field Mus. Nat. Hist.; 7kg Vienna, Naturhist. Mus.; 3kg New York, Amer. Mus. Nat. Hist.;

2kg Washington, U.S. Nat. Mus.; 1kg Glasgow, Hunterian Mus.; 982g Rome, Vatican Colln; 800g Michigan Univ.; 714g Mainz, Max-Planck-Inst.; 670g Harvard Univ.; 618g Tempe, Arizona State Univ.; 500g Budapest, Nat. Mus.; 328g Yale Univ.; 244g Tübingen, Univ.;
Specimen(s): [63547], 1122.5g. slice, and pieces, 29g; [1959, 174], 10g. fragments; [1964,575], Thin section

Cranberry Plains 37°14′ N., 80°44′ W.

Poplar Hill, Giles County, Virginia, U.S.A.
Found 1852
Synonym(s): *Poplar Hill, Poplar Camp*
Iron. Octahedrite, fine (0.4mm) (IVA?).
Little or nothing is known of its history, E. Cohen, Meteoritenkunde, 1905, **3**, p.346. Description, specimens heated artificially to about 1000C, V.F. Buchwald, Iron Meteorites, Univ. of California, 1975, p.507.
29g Harvard Univ.; 22g Yale Univ.; 16g Paris, Mus. d'Hist. Nat.; 7.9g Washington, U.S. Nat. Mus.; 7g New York, Amer. Mus. Nat. Hist.; 11.7g Tempe, Arizona State Univ.;

Cranbourne 38°6′ S., 145°18′ E.

Melbourne, Victoria, Australia
Found 1854
Synonym(s): *Abel, Beaconsfield, Bruce, Dandenong, Langwarrin, Melbourne, Pakenham, Victoria, Western Point district, Western Port district, Yarra Yarra River*
Iron. Octahedrite, coarse (2.2mm) (IA).
Two large masses, the Bruce mass of 3.5 tons (no.1) and the Abel mass of about 1.5 tons (no.2), were found nearly four miles apart, W. Haidinger, Sitzungsber. Akad. Wiss. Wien, Math.-naturwiss. Kl., 1861, **44** (2), p.31. The mass of 1.5 tons said to have been found at Dandenong appears to be identical with the Abel mass. A mass of 165lb was found in 1876 at Beaconsfield (no.9), six miles from Cranbourne. A mass of 15lb (no.3) was found between 1854 and 1869 but is now lost. Another mass, of 18 cwt. (no.10) was found in 1886 about 5 miles SE. of Langwarrin railway station. In 1923, four more masses, of 1.25 tons, (no.4), 7 cwt (no.5), 3 cwt (no.7), and 52lb (no.8), were found and in 1928 a mass of 89lb was found 3 miles W. of Pakenham (no.6). The above numbers are assigned and earlier, in part inaccurate, data are corrected, A.B. Edwards and G. Baker, Mem. Nat. Mus. Melbourne, 1944 (14), p.23 [M.A. 9-299], T. Hodge-Smith, Mem. Austr. Mus., 1939 (7), p.15, 29, F.C. Leonard, Pop. Astron., Northfield, Minnesota, 1947, **55**, p.497. An eleventh mass, of 750kg, was found at Pearsedale and is now in Washington, U.S. Nat. Mus., Ann. Rep. U.S. Nat. Mus., 1939, p.56. Nothing is known of its history. The large mass of 3.5 tons was described and analysed, 7.74 %Ni, W. Flight, Phil. Trans. Roy. Soc. London, 1882, **173**, p.885. The Beaconsfield mass was described, with an analysis, 7.34 % Ni, E. Cohen, Sitzungsber. Akad. Wiss. Berlin, 1897, **46**, p.1035. Full description of Cranbourne, R.H. Walcott, Mem. Nat. Mus. Melbourne, 1915 (6). Fragments found in Abel's Collection of Minerals with the label "Yarra Yarra River - Date 1858", probably came from one of the Cranbourne masses. Langwarrin was described and analysed, 6.24 %Ni, R.H. Walcott, Mem. Nat. Mus. Melbourne, 1915 (6), p.36. The four masses found in 1923 were described, with analyses of nos. 4, 5, and 7 and a full description of four distinct varieties of schreibersite found in these masses, A.B. Edwards and G. Baker, Mem. Nat. Mus. Melbourne, 1944 (14), p.23 [M.A. 9-299]. The rust contains 0.1% Ge, and also Cu, Zn, Ga, Pb, As, Sn, Mo, Ag, Au and Pt metals, V.M. Goldschmidt, Z. Physikal. Chem., 1930, **146** (Abt. A), p.404 [M.A. 4-427], V.M. Goldschmidt, Proc. Roy. Inst. Gt. Britain, 1929, **26**, p.73. More recent analysis, 6.80 %Ni, 85.4 ppm.Ga, 358 ppm.Ge, 1.8 ppm.Ir, J.T. Wasson, Icarus, 1970, **12**, p.407. Description, listing of masses and find sites, V.F. Buchwald, Iron Meteorites, Univ. of California, 1975, p.508.
1.5tons Melbourne, Nat. Mus., Abel mass; 18cwt Melbourne, Nat. Mus., Langwarrin mass; 89lb Melbourne, Nat. Mus., Pakenham mass; 1.4tons Melbourne, Nat.Mus., No. 4 mass; 7cwt Melbourne, Victoria Geol. Surv. Mus., No. 5 mass; 52lb Melbourne, Victoria Geol. Surv. Mus., No. 8 mass; 3cwt Melbourne, Geol. Dept. Univ., No. 7 mass; 750kg Washington, U.S. Nat. Mus., Pearsedale mass; 3kg Budapest, Nat. Mus.; 2.6kg Chicago, Field Mus. Nat. Hist.; 4kg Sydney, Australian Mus.; 2.5kg Vienna, Naturhist. Mus.; 1kg Harvard Univ.; 500g Prague, Nat. Mus., of Beaconsfield; 662g Paris, Mus.d'Hist.Mat.;
Specimen(s): [55532], 3.5tons. of Cranbourne, the largest mass, and a piece, 3190g; [61307], 6.5g. of Cranbourne; [92566], 21g. of Cranbourne; [1959,168], 13g.; [1964,712], 10.5g. of Cranbourne; [92569], 122g. of "Yarra Yarra River"; [63884], 98g. of " Yarra Yarra River"; [63880], 62.5g. of "Yarra Yarra River"; [92568], 44g. of "Yarra Yarra River"; [92567], 35g. of "Yarra Yarra River"; [63882], 12.5g. of "Yarra Yarra River"; [63881], 12.5g. of "Yarra Yarra River"; [1922,161], 71g. of Beaconsfield, a slice; [1970, 308(a)], 1490g. crusted fragment with two polished faces, one with large troilite nodules; [1970,308(b)], 1379g. crusted fragment with two polished faces; [1970,308(c)], 105g. crusted slice; [1970,308(d)], 11g. sawings

Cranfills Gap 31°45′ N., 97°45′ W.

Bosque County, Texas, U.S.A.
Found 1940, about
Stone. Olivine-bronzite chondrite (H).
One stone of about 6kg was found, Meteor. Bull., 1967 (39), Meteoritics, 1970, **5**, p.90. Olivine Fa_{18}, B. Mason, Geochimica et Cosmochimica Acta, 1967, **31**, p.1100.
24g Fort Worth, Texas, Monnig Colln.;

Cranganore 10°12′ N., 76°16′ E.

Trichur, Kerala, India
Fell 1917, July 3, 1245 hrs
Synonym(s): *Changanorein, Kranganur*
Stone. Olivine-hypersthene chondrite (L6).
After detonations, some six stones fell within a radius of 3 miles at Cranganur, at the mouth of the Palliport river, 14 miles N. of Cochin. The total weight of pieces obtained was 1460g, the largest weighing 713g, described, H. Walker, Rec. Geol. Surv. India, 1924, **55**, p.139. Listed, C.A. Silberrad, Min. Mag., 1932, **23**, p.301, H.H. Hayden, Rec. Geol. Surv. India, 1918, **49** (1), p.8. Mineralogy, analysis, 21.49 % total iron, olivine $Fa_{24.7}$, A. Dube et al., Smithson. Contrib. Earth Sci., 1977 (19), p.71.
1.1kg Calcutta, Mus. Geol. Surv. India; 27g Tempe, Arizona State Univ.; 10g Washington, U.S. Nat. Mus.;
Specimen(s): [1925,444], 248.9g.

Cratheús (1931) 5°15′ S., 40°30′ W. approx.

Ceara, Brazil
Found 1914, known in this year
Iron. Octahedrite, fine (0.30mm) (IVA).
A mass of 27.5kg was purchased by the Brazilian Geol. Surv. in 1914. Description, with an analysis, 7.41 %Ni, E. de Oliveira, Anais Acad. Brasil. Cienc., 1931, **3**, p.33 [M.A. 5-14]. Classification and analysis, 7.72 %Ni, 2.19 ppm.Ga, 0.108 ppm.Ge, 2.3 ppm.Ir, R. Schaudy et al., Icarus, 1972, **17**, p.174. Cosmically reheated, V.F. Buchwald, Iron Meteorites, Univ. of California, 1975, p.511.
27.5kg Rio de Janeiro, Ser. Geol.; 74g Canberra, Austr. Nat. Univ.; 42g Washington, U.S. Nat. Mus.;
Specimen(s): [1931,254], 47g.

Cratheús (1950)
Brazil, Co-ordinates not known
Found 1909?
Iron. Octahedrite, plessitic (0.06mm) (IIC).
No information available, distinct from Cratheús (1931), V.F. Buchwald, Iron Meteorites, Univ. of California, 1975, p.512. Classification and analysis, 8.97 %Ni, 36.3 ppm.Ga, 91.4 ppm.Ge, 9.5 ppm.Ir, J.T. Wasson, Meteorites, Springer-Verlag, 1974, p.300. Mentioned, W.S. Curvello, Bol. Mus. Nac. Rio de Janeiro, 1950 (geol. no. 10) [M.A. 11-445].
347g Rio de Janeiro, Mus. Nat. Hist.; 20g Colln. of W.S. Curvello;

Crawford County v Mincy.

Credo 30°22′ S., 120°44′ E.
Western Australia, Australia
Found 1967
Stone. Olivine-hypersthene chondrite (L6).
One crusted stone of 10.82kg was found near Credo station, 50 miles NW. of Kalgoorlie, G.J.H. McCall and W.H. Cleverly, Min. Mag., 1969, **37**, p.281. Olivine $Fa_{24.9}$, B. Mason, Rec. Austr. Mus., 1974, **29**, p.169.
10.3kg Kalgoorlie, Western Austr. School of Mines; 260g Perth, Western Austr. Mus.; 212g with the finder, P.J. Howell;
Specimen(s): [1968,286], 50g. crusted part-slice

Crema 45°21′ N., 9°42′ E.
Cremona, Lombardy, Italy
Fell 1511, September 4
Doubtful. Stone.
Many stones, some of at least 100lb fell near the river Adda, E.F.F. Chladni, Die Feuer-Meteore, Wien, 1819, p.209, E.F.F. Chladni, Phil. Mag., 1826, **67**, p.3, 179 but nothing has been preserved. The stones are said to have killed birds, sheep, and a man. Listed, B. Baldanza, Min. Mag., 1965, **35**, p.214. It has been suggested that the Crema fireball is depicted in Raphael's painting the " Madonna di Foligno", in the Vatican Art Gallery, C.M. Bottley, letters of 3 and 7 August 1968 in Min. Dept. BM(NH).

Cremona v Alfianello.

Crescent 35°57′ N., 97°35′ W.
Logan County, Oklahoma, U.S.A.
Fell 1936, August 17, 1907 hrs
Stone. Carbonaceous chondrite, type II (CM2).
Two stones were found, 72.7g and 5.7g. Short description, O.E. Monnig and R. Brown, Pop. Astron., Northfield, Minnesota, 1936, **44**, p.568 [M.A. 6-397].
7g Fort Worth, Texas, Monnig Colln.; Thin section, Washington, U.S. Nat. Mus.;

Crevalcore 44°43′ N., 11°10′ E.
Bologna, Emilia, Italy
Fell 1596, March 1, 1700-1800 hrs
Doubtful..
A shower of stones is said to have fallen, E.F.F. Chladni, Die Feuer-Meteore, Wien, 1819, p.220 but the evidence does not appear conclusive.

Crocker's Well 32°1′ S., 139°47′ E.
South Australia, Australia
Found 1956
Synonym(s): *Crocker Well*
Stone. Chondrite.
One fragment of 3.8g was found, 0.25 miles N. of East Crocker uranium prospect, D.W.P. Corbett, Rec. S. Austr. Mus., 1968, **15**, p.767. Possibly related to Ethiudna (*q.v.*) which was found 6.5km away, M.J. Fitzgerald, Ph.D. Thesis, Univ. of Adelaide, 1979, p.203.
3.8g Adelaide, South Austr. Mus., main mass;

Crocker Well v Crocker's Well.

Crockett County v Ozona.

Cronstad 27°42′ S., 27°18′ E. approx.
Orange Free State, South Africa
Fell 1877, November 19, 1600 hrs
Synonym(s): *Geluksfontein, Kroonstad*
Stone. Olivine-bronzite chondrite (H5), veined.
After detonations, a shower of stones fell, the largest of about 6lb, but few were found; described and analysed, G.T. Prior, Min. Mag., 1916, **18**, p.10. Olivine Fa_{18}, B. Mason, Geochimica et Cosmochimica Acta, 1963, **27**, p.1011. Another analysis, 26.65 % total iron, H. von Michaelis et al., Earth planet. Sci. Lett., 1969, **5**, p.387.
Two stones, Bloemfontain, Mus.; 122g Berlin, Humboldt Univ.; 58g Cape Town, South African Mus.; 46g Chicago, Field Mus. Nat. Hist.; 12g Washington, U.S. Nat. Mus.;
Specimen(s): [55428], 787g. and fragments, 17g; [52149], 287.5g.

Crosbyton 33°40′ N., 101°16′ W.
Crosby County, Texas, U.S.A.
Found 1963, known before this year
Stone. Olivine-bronzite chondrite (H).
Olivine Fa_{17}, B. Mason, Geochimica et Cosmochimica Acta, 1963, **27**, p.1011.
1.3g New York, Amer. Mus. Nat. Hist.;

Cross Roads 35°38′ N., 78°8′ W.
Boyett, Wilson County, North Carolina, U.S.A.
Fell 1892, May 24, 0500 hrs
Synonym(s): *Boyett, Wilson County*
Stone. Olivine-bronzite chondrite (H5).
A stone of about 167g was seen to fall, after detonations, E.E. Howell, Am. J. Sci., 1893, **46**, p.67. Olivine Fa_{18}, B. Mason, Geochimica et Cosmochimica Acta, 1963, **27**, p.1011.
49g New York, Amer. Mus. Nat. Hist.; 26g Vienna, Naturhist. Mus.; 18g Chicago, Field Mus. Nat. Hist.; 9g Harvard Univ.;
Specimen(s): [73647], 9.8g.

Cross Timbers v Red River.

Crow Creek v Silver Crown.

Crumlin 54°37′ N., 6°13′ W.
County Antrim, Ireland
Fell 1902, September 13, 1030 hrs
Stone. Olivine-hypersthene chondrite (L5).
A stone of 9lb 5.5oz was seen to fall, after detonations, L. Fletcher, Nature, 1902, **66**, p.577. Description, analysis, L. Fletcher, Min. Mag., 1921, **19**, p.149. Further analysis, 21.56 % total iron, E. Jarosewich, priv. comm., 1976. Olivine Fa_{24},

B. Mason, Geochimica et Cosmochimica Acta, 1963, **27**, p.1011.
140g Armagh Observatory; 99g Edinburgh, Roy. Scottish Mus.; 38g Calcutta, Mus. Geol. Surv. India; 35g Washington, U.S. Nat. Mus.; 31g Vienna, Naturhist. Mus.; 13g Chicago, Field Mus. Nat. Hist.;
Specimen(s): [86115], 3572g. and pieces, 4g

Cruz del Aire 26°30′ N., 100°0′ W. approx.
Sabina Hidalgo, Nuevo Leon, Mexico
Found 1911
Iron. Octahedrite, fine (0.5mm) (IRANOM).
A mass of 15.2kg was found, possibly identical with Coahuila, L.W. MacNaughton, Am. Mus. Novit., 1926 (207), p.2. Now known to be distinct from Coahuila. A second mass of 7815g was found in 1930 in Cerro Chico de Santa Clara, Sabina Hidalgo. Described, with an analysis, 7.62 %Ni, R.E.S. Heineman, Am. J. Sci., 1932, **24**, p.465 [M.A. 5-300]. Analysis, chemically anomalous, 9.00 %Ni, 38.2 ppm.Ga, 186 ppm.Ge, 5.9 ppm.Ir, J.T. Wasson, Meteorites, Springer-Verlag, 1974, p.307. Description, unusual microstructures caused by shock followed by reheating, V.F. Buchwald, Iron Meteorites, Univ. of California, 1975, p.513.
Main mass, New York, Amer. Mus. Nat. Hist., found 1911; 5.8kg Tucson, Arizona College of Mines, the second mass; 655g Tempe, Arizona State Univ.; 351g Washington, U.S. Nat. Mus.; 217g Chicago, Field Mus. Nat. Hist.;

Csillagfalva v Knyahinya.

Cuanta v Cuenta.

Cuba 22° N., 80° W. approx.
Cuba
Found 1871, described in this year
Iron. Octahedrite, medium (1.3mm) (I).
A mass of about 1.5kg in the Madrid Mus. was described and analysed, J.M. Solano y Eulate, Anal. Soc. Españ. Hist. Nat. Madrid, 1872, **1**, p.183. Description, transported Toluca?, V.F. Buchwald, Iron Meteorites, Univ. of California, 1975, p.514.
23g Washington, U.S. Nat. Mus.; 2.6g Chicago, Field Mus. Nat. Hist.;

Cubertson error for Culbertson.

Cuenta
Spain
Synonym(s): *Cuanta*
Determinations of trace elements were made on a meteorite under the name "Cuenta, Spain", G. von Hevesy and K. Würstlin, Z. anorg. Chem., 1934, **216**, p.305,312 [M.A. 6-86]., W.H. Pinson et al., Geochimica et Cosmochimica Acta, 1953, **4**, p.251. This was probably a specimen of Olmedilla de Alarcón, Cuenca, Spain.

Cuerco error for Cuero.

Cuernavaca v Chupaderos.

Cuero 29°1′ N., 97°17′ W.
De Witt County, Texas, U.S.A.
Found 1936, May 16
Stone. Olivine-bronzite chondrite (H5), veined.
One mass of 46.5kg was found 4-5ft down on Chisholm Creek, 3 miles S. of Cuero, A.D. Nininger, Pop. Astron., Northfield, Minnesota, 1937, **45**, p.449 [M.A. 7-62]. The mass is dark green, and contains much water and ferric iron, suggesting the presence of serpentinous weathering products. Described, with an analysis, V.E. Barnes, Univ. Texas Publ., 1940 (3945), p.613 [M.A. 8-59]. Olivine Fa_{19}, B. Mason, Geochimica et Cosmochimica Acta, 1963, **27**, p.1011.
11g Mainz, Max-Planck-Inst.;
Specimen(s): [1961,392], 675g.

Culbertson 40°14′ N., 100°50′ W.
Hitchcock County, Nebraska, U.S.A.
Found 1913
Synonym(s): *Cubertson*
Stone. Olivine-bronzite chondrite (H4).
A stone of 13lb was turned up while planting corn, L.W. MacNaughton, Am. Mus. Novit., 1926 (207), p.1., F.C. Leonard, Univ. New Mexico Publ., Albuquerque, 1946 (meteoritics ser. 1), p.40. Olivine Fa_{18}, B. Mason, Geochimica et Cosmochimica Acta, 1963, **27**, p.1011.
Main mass, New York, Amer. Mus. Nat. Hist.; 84g Tempe, Arizona State Univ.;
Specimen(s): [1959,818], 66g. and fragments, 2g

Cullison 37°37′ N., 98°55′ W.
Pratt County, Kansas, U.S.A.
Found 1911
Stone. Olivine-bronzite chondrite (H4), brecciated.
A stone of about 10kg was found in 1911, but is said to have fallen 22 December 1902; described, with an analysis, G.P. Merrill, Proc. U.S. Nat. Mus., 1913, **44**, p.325. Olivine Fa_{18}, B. Mason, Geochimica et Cosmochimica Acta, 1963, **27**, p.1011.
2.5kg Washington, U.S. Nat. Mus.; 442g New York, Amer. Mus. Nat. Hist.; 300g Vienna, Naturhist. Mus.; 267g Budapest, Nat. Hist. Mus.; 250g Tempe, Arizona State Univ.; 151g Chicago, Field Mus. Nat. Hist.; 202g Prague, Nat. Mus.;
Specimen(s): [1912,315], 2540g. and fragments, 10g.

Cumberland County v Crab Orchard.

Cumberland Falls 36°50′ N., 84°21′ W.
Whitley County, Kentucky, U.S.A.
Fell 1919, April 9, 1200 hrs
Stone. Achondrite, Ca-poor. Aubrite (AUB).
Several stones fell, the largest, which broke into fragments, estimated to weigh 31lb, after the appearance of a fireball and detonations, A.M. Miller, Science, 1919, **49**, p.541. Description, analysis, G.P. Merrill, Proc. U.S. Nat. Mus., 1920, **57**, p.97. Contains a unique type of chondritic inclusion. Discussion, with an analysis of an inclusion, 19.5 % total iron, R.A. Binns, Meteorite Research, ed. P.M. Millman, D. Reidel, Dordrecht-Holland, 1969, p.696. These inclusions of forsterite chondrite material may be related to Kakangari (*q.v.*), A.L. Graham et al., Min. Mag., 1977, **41**, p.201. Further inclusion analysis, 19.0 % total iron, E. Jarosewich, Geochimica et Cosmochimica Acta, 1967, **31**, p.1103. Analysis of enstatite, A.M. Reid and A.J. Cohen, Geochimica et Cosmochimica Acta, 1967, **31**, p.661. Chondritic inclusion petrology, C.W. Neal and M.E. Lipschutz, Geochimica et Cosmochimica Acta, 1981, **45**, p.2091.
8.64kg Washington, U.S. Nat. Mus., main masses; 3kg New York, Amer. Mus. Nat. Hist.; 372g Tempe, Arizona State Univ.; 3.7kg Chicago, Field Mus. Nat. Hist.; 324g Copenhagen, Univ. Geol. Mus.; 164g Paris, Mus. d'Hist. Nat.; 163g Berlin, Humboldt Univ.;
Specimen(s): [1920,106], 358g.; [1920,107], 256.5g.

Cumpas 30°0′ N., 109°40′ W.
Sonora, Mexico
Found 1903
Iron. Octahedrite, medium (1.2mm) (IIIA).
A mass of 63lb was found 25 miles E. of Cumpas; described, with an analysis, 6.00 %Ni, C. Palache, Am. J. Sci., 1926, **12**, p.141. Not identical with Moctezuma (*q.v.*). New analysis, 7.94 %Ni, 21.0 ppm.Ga, 42.8 ppm.Ge, 2.7 ppm.Ir, E.R.D. Scott et al., Geochimica et Cosmochimica Acta, 1973, **37**, p.1957. Description, epsilon structure, V.F. Buchwald, Iron Meteorites, Univ. of California, 1975, p.514.
23kg Harvard Univ.; 2212g Washington, U.S. Nat. Mus.;
Specimen(s): [1927,84], 407g. slice, and fragments 59g.

Currant Creek v Guffey.

Curvello 18°48′ S., 44°37′ W.
Minas Gerais, Brazil
Doubtful..
A meteorite was seen to fall at 1845hrs, April 11, 1833, near Curvello, P. Claussen, Bull. Acad. Roy. Belg., 1841, **8** (1), p.341 and a piece of 85g(?) was recovered and placed in the Nat. Mus. Rio de Janeiro. A piece of 218g was found in the museum without labels, and believed to be Curvello; this proved to be artificial iron and the fall was discredited, A. Brezina, Jahrb. Geol. Reichsanst. Wien, 1885, **35**, p.221, O.A. Derby, Revista do Observatorio, Rio de Janeiro, 1888, **3**, p.35. It is not possible that Piedade do Bagre (*q.v.*) really represents this fall, V.F. Buchwald, Iron Meteorites, Univ. of California, 1975, p.973 as has been suggested, L.J. Spencer, Min. Mag., 1930, **22**, p.271.

Cushing 35°58′ N., 96°46′ W.
Payne County, Oklahoma, U.S.A.
Found 1932, in the spring
Stone. Olivine-bronzite chondrite (H4).
One stone of 567g was found in SE. 1/4, section 14, township 18, range 5 east, E.P. Henderson, letter of 3 June 1939 in Min. Dept. BM(NH). Olivine Fa_{19}, B. Mason, Geochimica et Cosmochimica Acta, 1963, **27**, p.1011.
513g Washington, U.S. Nat. Mus., main mass; 2.9g New York, Amer. Mus. Nat. Hist.;

Cushion v Cashion.

Cusignano v Borgo San Donino.

Cut Off v Cañon Diablo.

Cuvasskije Kissy v Kissij.

Cynthiana 38°24′ N., 84°15′ W.
Harrison County, Kentucky, U.S.A.
Fell 1877, January 23, 1600 hrs
Synonym(s): *Harrison County, Robinson Station*
Stone. Olivine-hypersthene chondrite (L4).
After the appearance of a fireball and detonations, a stone of about 6kg fell nine miles from Cynthiana; described and analysed, J.L. Smith, Am. J. Sci., 1877, **14**, p.224. Composition of metal, G.T. Prior, Min. Mag., 1919, **18**, p.353. Description, analysis, 19.2 % total iron, B. Mason and A.D. Maynes, Proc. U.S. Nat. Mus., 1967, **124** (3624). Equilibrated, olivine $Fa_{25.6}$, R.T. Dodd et al., Geochimica et Cosmochimica Acta, 1967, **31**, p.921.
2.8kg Harvard Univ.; 686g Paris, Mus. d'Hist. Nat.; 643g Washington, U.S. Nat. Mus.; 404g New York, Amer. Mus. Nat. Hist.;
Specimen(s): [53288], 136g. and powder, 9g.

Czartorya v Zaborzika.

Czestochowa Rakow I 50°48′ N., 19°7′ E.
Katowice (Stalinogrod), Poland
Found
Iron. Octahedrite, finest to nickel-rich ataxite.
A bracelet, bracelet no. 3, found in a grave dated at 700-500 B.C., is composed of meteoritic iron. It is oval, about 73×61mm, and 6.3mm thick. Analysis, 18.25 %Ni, 0.05 % C, 0.05 %Cu, 0.052 %P, J. Pokrzywnicki, Acta Geophys. Polon., 1971, **19**, p.235.
Bracelet, Czestochowa Mus., Cat. no. 9:60;

Czestochowa Rakow II 50°48′ N., 19°7′ E.
Katowice (Stalinogrod), Poland
Found
Iron. Octahedrite, finest to nickel-rich ataxite.
Another bracelet (bracelet no. 4) found in a grave dated at 700-500 B.C. is composed of meteoritic iron, it is oval, 70×50mm and 4.5mm thick. Analysis, 12.47 %Ni, 0.05 % C, 0.05 %Cu, 0.052 %P, J. Pokrzywnicki, Acta Geophys. Polon., 1971, **19**, p.235.
Bracelet, Czestochowa Mus., Cat. no. 294:61.;

Dabra v Lua.

Dabrowa Łuzycka v Sagan.

Dacca v Shytal.

Dacotah v Ainsworth.

Dadin 38°55′ S., 69°12′ W.
Neuquen, Argentina
Found 1949, known before this year
Iron. Octahedrite, coarse to medium.
A mass of 37.3kg was found near Campamento Dadín, 12km from the railway station Plaza Huincul. Described with an analysis, 6.7 %Ni, E.H. Ducloux, Notas Mus. La Plata, 1949, **14** (geol. no. 54), p.177 [M.A. 11-445]. Original weight probably 27.3kg, V.F. Buchwald, Iron Meteorites, Univ. of California, 1975, p.516.
27.2kg La Plata, Mus., main mass; 22g Harvard Univ.;

Dahbubah v ad-Dahbubah.

Dahmani 35°37′ N., 8°50′ E. approx.
Kasserine, Tunisia
Fell 1981, May
Stone. Olivine-hypersthene chondrite, amphoterite (LL6).
A single stone of 18kg fell near a village not far from Dahmani and was recovered by soldiers, olivine Fa_{31}, Meteor. Bull., 1982 (60), Meteoritics, 1982, **17**, p.94.
Main mass, Tunis, Geol. Surv.;

Daiet el Akhricha v Akhricha.

Dakhin Paiksha v Dokachi.

Dakota v Ainsworth.

Dale Dry Lake 34°2′ N., 115°54′ W.

San Bernardino County, California, U.S.A.
Found 1957
Stone. Olivine-hypersthene chondrite (L).

A mass of 300g was found, B. Mason, Meteorites, Wiley, 1962, p.229. The stone was found about 2 miles N. of the Virginia Dale Mine, Bull. Calif. Div. Mines Geol., 1966, **189**, p.232. Olivine Fa_{24}, B. Mason, Geochimica et Cosmochimica Acta, 1963, **27**, p.1011.

34g Tempe, Arizona State Univ.;

Dalgaranga 27°43′ S., 117°15′ E.

Western Australia, Australia
Found 1923
Stony-iron. Mesosiderite (MES).

A number of small fragments of iron were found in and around a crater 70 feet across and 11 feet deep, but only one of 40g was preserved in E.S. Simpson's collection. Described, with a partial analysis, 8.63 %Ni, E.S. Simpson, Min. Mag., 1938, **25**, p.157. In 1960 search around the crater yielded 207 fragments with a total weight of 1.1kg, the largest weighing 57g, of which about half were oxidised irons and half mesosiderite. Within the crater 280 fragments, totalling about 20lb were collected, all of which proved to be highly oxidised mesosiderites, H.H. Nininger and G.I. Huss, Min. Mag., 1960, **32**, p.619, G.J.H. McCall, Min. Mag., 1965, **35**, p.476, G.J.H. McCall, Spec. Publ. West. Austr. Mus., 1965 (3), p.28. Analysis of metal, 8.8 %Ni, 15.5 ppm.Ga, 56 ppm.Ge, 4.2 ppm.Ir, J.T. Wasson et al., Geochimica et Cosmochimica Acta, 1974, **38**, p.135. Olivine corona structure, C.E. Nehru et al., Geochimica et Cosmochimica Acta, 1980, **44**, p.1103.

4.3kg Mainz, Max-Planck-Inst.; 641g Tempe, Arizona State Univ.; 159g Washington, U.S. Nat. Mus.; 86g Perth, West. Austr. Mus.; 41g New York, Amer. Mus. Nat. Hist.; 26g Vienna, Naturhist. Mus.;

Specimen(s): [1961,394-401], 55.6g. eight fragments

Dalgety Doun error for Dalgety Downs.

Dalgety Downs 25°20′ S., 116°11′ E.

Gascoyne District, Western Australia, Australia
Found 1941, known in this year
Synonym(s): *Ashburton Downs, Dalgetty Downs, Dalgety Doun*
Stone. Olivine-hypersthene chondrite (L4).

A mass, broken into fragments totalling 217.7kg., was found 6 miles S. of Dalgety Downs, H. Bowley, Ann. Rep. Dep. Mines West. Austr., for 1942, 1944, p.76. In 1963 further masses to a total of 214kg were found by B. Mason and E.P. Henderson, and shown to be chondritic; a mass of 1.8kg said to have come from Ashburton Downs is part of the same fall, G.J.H. McCall, letter of 9 August 1963, in Min. Dept. BM(NH). Mentioned, G.J.H. McCall, J. Roy. Soc. West. Austr., 1966, **49**, p.52 [M.A. 18-272]. Another recovery of numerous fragments of about 40.9kg was made in 1964 by W.H. Cleverly, these are now in Kalgoorlie, (Western Austr. School of Mines), G.J.H. McCall, First Suppl. to West. Austr. Mus. Spec. Publ. no. 3, 1968, p.4. Olivine $Fa_{25.2}$, B. Mason, Rec. Austr. Mus., 1974, **29**, p.169. Analysis, 22.61 % total iron, A.J. Easton and C.J. Elliott, Meteoritics, 1977, **12**, p.409.

125kg Washington, U.S. Nat. Mus.; 18.9kg New York, Amer. Mus. Nat. Hist., includes a single mass of 11kg; 439g Tempe, Arizona State Univ.; 42kg Kalgoorlie, West. Austr. School of Mines; 5kg Perth, West. Austr. Mus.; 917g Chicago, Field Mus. Nat. Hist.; 724g Harvard Univ.; 401g Los Angeles, Univ. of California;

Specimen(s): [1964,747], 1440g. two masses of Ashburton Downs; [1964,581], 69g. of Dalgety Downs; [1964,582], 19.5g. fragments, of Dalgety Downs

Dalhart 36°2′36″ N., 102°24′30″ W.

Dallam County, Texas, U.S.A.
Found 1968
Stone. Olivine-bronzite chondrite (H5).

One stone of 4.4kg was found on grassland, G.I. Huss, letter of 23 April 1970, in Min. Dept. BM(NH). Olivine Fa_{18}, R. Hutchison, priv. comm.

2.3kg Mainz, Max-Planck-Inst.; 95g Vienna, Naturhist. Mus.; 69g New York, Amer. Mus. Nat. Hist.; 65g Washington, U.S. Nat. Mus.; 50g Los Angeles, Univ. of California;

Specimen(s): [1970,31], 26.5g. a part-slice

Dalton 34°48′ N., 84°59′ W.

Whitfield County, Georgia, U.S.A.
Found 1879
Iron. Octahedrite, medium (1.1mm) (IIIA).

In 1879 a rounded mass of 117lb was ploughed up 14 miles NE. of Dalton, C.U. Shepard, Am. J. Sci., 1883, **26**, p.336. Analysis, 7.57 %Ni, G.P. Merrill, Proc. U.S. Nat. Mus., 1916, **51**, p.447. Further analysis, 7.35 %Ni, 18.4 ppm.Ga, 33.1 ppm.Ge, 9.6 ppm.Ir, E.R.D. Scott et al., Geochimica et Cosmochimica Acta, 1973, **37**, p.1957. A 13lb mass "Whitfield County", found in 1877, was originally assigned to Dalton. Merrill (op. cit.) doubted this relationship. This mass is now assigned to Cleveland (*q.v.*); description, V.F. Buchwald, Iron Meteorites, Univ. of California, 1975, p.516.

49kg Washington, U.S. Nat. Mus., main mass; 500g Michigan Univ.; 429g Harvard Univ.; 91g Chicago, Field Mus. Nat. Hist.; 76g New York, Amer. Mus. Nat. Hist.;

Specimen(s): [61995], 131.5g.

Damaraland v Gibeon.

Dandapur 26°55′ N., 83°58′ E.

Gorakhpur district, Uttar Pradesh, India
Fell 1878, September 5, 1700 hrs
Synonym(s): *Goruckpur*
Stone. Olivine-hypersthene chondrite (L6), veined.

Two stones, of about 6.5lb and 5lb 14oz respectively, fell 300 paces apart, after the appearance of a moving cloud and detonations, H. Fraser, Proc. Asiatic Soc. Bengal, 1878, p.175. Olivine Fa_{25}, B. Mason, Geochimica et Cosmochimica Acta, 1963, **27**, p.1011.

2kg Calcutta, Mus. Geol. Surv. India, of the larger stone; 254g Paris, Mus. d'Hist. Nat.; 229g New York, Amer. Mus. Nat. Hist.; 212g Harvard Univ.; 183g Vienna, Naturhist. Mus.; 65g Chicago, Field Mus. Nat. Hist.; 57g Washington, U.S. Nat. Mus.;

Specimen(s): [53321], 2377g. main mass of the smaller of the two stones, and a fragment, 4g.

Dandenong v Cranbourne.

Dangtai v Dongtai.

Daniel's Kuil 28°12′ S., 24°34′ E.

Griqualand West, Cape Province, South Africa
Fell 1868, March 20
Synonym(s): *Griqualand*
Stone. Enstatite chondrite (E6).

A stone of 2lb 5oz was seen to fall by a native, J.R. Gregory, Geol. Mag., 1868, **5**, p.531. Description, analysis, G.T. Prior, Min. Mag., 1916, **18**, p.13. The stone contains oldhamite and daubréelite. References, B. Mason, Geochimica et Cosmochimica Acta, 1966, **30**, p.23, K. Keil, J. Geophys. Res., 1968, **73**, p.6945. Trace element abundances, C.M. Binz et al., Geochimica et Cosmochimica Acta, 1974, **38**, p.1579. Further analysis, 37.19 % total iron, A.J. Easton and C.J. Elliott, Meteoritics, 1977, **12**, p.409.

52g Calcutta, Mus. Geol. Surv. India; 33g New York, Amer. Mus. Nat. Hist.; 24g Cape Town, S. African Mus.; 24g Forth Worth, Texas, Monnig Colln.; 18g Vienna, Naturhist. Mus.; 13g Chicago, Field Mus. Nat. Hist.; 11g Harvard Univ.; 6.1g Budapest, Nat. Hist. Mus.;

Specimen(s): [42388], 222g.; [42502], 105g.; [42507], 18.5g. fragments; [42503], thin section; [96261], thin section

Dangtu v Po-wang Chen.

Dantu v Po-wang Chen.

Danville 34°24′ N., 87°4′ W.

Morgan County, Alabama, U.S.A.
Fell 1868, November 27, 1700 hrs
Stone. Olivine-hypersthene chondrite (L6), veined, brecciated.

Several stones appear to have fallen after detonations, but only one, of about 4.5lb, was recovered; described and analysed, J.L. Smith, Am. J. Sci., 1870, **49**, p.90. Olivine Fa_{23}, B. Mason, Geochimica et Cosmochimica Acta, 1963, **27**, p.1011.

84g Harvard Univ.; 41g Vienna, Naturhist. Mus.; 11g Yale Univ.; 5g Chicago, Field Mus. Nat. Hist.; 5g New York, Amer. Mus. Nat. Hist.; 3.1g Budapest, Nat. Hist. Mus.;

Specimen(s): [47237], 22g.; [43056], 4g.

Daohe v Nan Yang Pao.

Daoukro 7°5′ N., 3°58′ W.

Ivory Coast
Doubtful..

A copper meteorite is supposed to have fallen north of Daoukro, J.M. Saul, letter of 6 June 1969, in Min. Dept. BM(NH).

Daoura 29°35′ N., 3°45′ W.

Hamada de la Daoura, Algeria
Found 1952, February
Stone. Olivine-hypersthene chondrite (L4).

Two fragments of 20g and 15g were found on Pliocene sediments halfway between Taberbala and Hassi Bou Lahdam, in the Hamada de la Daoura. Description, E. Jérémine, C. R. Soc. Géol. France, 1953, p.47 [M.A. 12-610]. Olivine Fa_{23}, B. Mason, Geochimica et Cosmochimica Acta, 1963, **27**, p.1011.

Both fragments, Paris, Mus. d'Hist. Nat.;

Darmstadt 49°52′ N., 8°39′ E.

Hessen, Germany
Fell 1804, before this year
Stone. Olivine-bronzite chondrite (H5), veined.

A stone of about 100g fell, after detonations, G.A. Suckow, Mineralogie, Leipzig, 1804, **2**, p.649. It contains intergrowths of troilite and Fe-Ni metal, R. Vogel, Chem. Erde, 1966, **25**, p.142 [M.A. 20-308]. Olivine Fa_{19}, B. Mason, Geochimica et Cosmochimica Acta, 1963, **27**, p.1011.

50g Heidelberg Univ.; 24g Prague, Nat. Mus.; 16g Göttingen,; 8.8g Budapest, Nat. Mus.; 4.2g Mainz, Max-Planck-Inst.;

Specimen(s): [35277], 1.5g.

Daugarvpils v Lixna.

Davidson County v Drake Creek.

Davis Mountains 30°45′ N., 104°15′ W. approx.

Jeff Davis County, Texas, U.S.A.
Found 1903
Synonym(s): *Toyah*
Iron. Octahedrite, medium (0.9mm) (IIIA).

A mass of 1520lb was found; described, with an analysis by H.W. Nichols, 7.40 %Ni, O.C. Farrington, Field Mus. Nat. Hist. Geol. Ser., 1914, **5** (178), p.4. New analysis, 7.29 %Ni, 17.4 ppm.Ga, 33.7 ppm.Ge, 14 ppm.Ir, E.R.D. Scott et al., Geochimica et Cosmochimica Acta, 1973, **37**, p.1957. Description, shocked, partly annealed, V.F. Buchwald, Iron Meteorites, Univ. of California, 1975, p.519.

693.5kg Chicago, Field Mus. Nat. Hist., main mass; 117g Washington, U.S. Nat. Mus.;

Specimen(s): [1914,67], 41g.

Davis Strait v Cape York.

Davy 29°6′ N., 97°36′ W. approx.

De Witt County, Texas, U.S.A.
Found 1940, recognized in this year
Stone. Olivine-hypersthene chondrite (L6).

One mass and fragments, totalling about 45kg were found, A.D. Nininger, Pop. Astron., Northfield, Minnesota, 1940, **48**, p.555, A.D. Nininger, Contr. Soc. Res. Meteorites, 1942, **2**, p.227 [M.A. 8-54]. Olivine Fa_{23}, B. Mason, Geochimica et Cosmochimica Acta, 1963, **27**, p.1011.

51kg Fort Worth, Texas, Monnig Colln.; 0.6g New York, Amer. Mus. Nat. Hist.;

Dawn (a) 34°51′36″ N., 102°7′12″ W.

Randall County, Texas, U.S.A.
Found 1981, recognized in this year
Stone. Olivine-bronzite chondrite (H6).

A single mass of 7682g was found, G.I Huss, priv. comm., 1983. Olivine $Fa_{18.8}$, A.L. Graham, priv. comm., 1983.

7.36kg Fort Worth, Texas, Monnig Colln.;

Dawn (b) 34°51′36″ N., 102°7′12″ W.

Randall County, Texas, U.S.A.
Found 1981, recognized in this year
Stone. Olivine-bronzite chondrite (H4-5).

A mass of 257g was found, olivine $Fa_{18.7}$, A.L. Graham, priv. comm., 1983.

238g Fort Worth, Texas, Monnig Colln.;

Dawson v Klondike (Skookum Gulch).

Dayton 39°45′ N., 84°10′ W.

Montgomery County, Ohio, U.S.A.
Found 1892, or 1893, reported 1951
Iron. Octahedrite, finest (0.04mm) (IIICD).

A mass of 26.3kg was noted, with an analysis, 18.10 %Ni, E.P. Henderson and S.H. Perry, Geochimica et Cosmochimica Acta, 1954, **6**, p.223. Listed, F.C. Leonard, Classif. Cat. Meteor., 1956, Cat. Meteor. Arizona State Univ., 1964, p.1101. Classification and new analysis, 17.02 %Ni, 5.16 ppm.Ga, 3.52 ppm.Ge, 0.028 ppm.Ir, J.T. Wasson and R. Schaudy, Icarus, 1971, **14**, p.59. Although reported in Dayton, the place of find is unknown, description, V.F. Buchwald, Iron Meteorites, Univ. of California, 1975, p.520.

24kg Washington, U.S. Nat. Mus.; 486g Chicago, Field Mus. Nat. Hist.; 324g Tempe, Arizona State Univ.; 237g Sydney, Austr. Mus.; 200g Mainz, Max-Planck-Inst.;

Deal 40°15′ N., 74°0′ W.
Long Branch, Monmouth County, New Jersey, U.S.A.
Fell 1829, August 15, 0030 hrs
Synonym(s): *Monmouth County, New Jersey*
Stone. Olivine-hypersthene chondrite (L6).
Several stones appear to have fallen after the appearance of a fireball and detonations, but only one, about 3 inches long, was found on a farm 5 miles SW. of Long Branch, R. Vaux and T. McEuen, J. Acad. Nat. Sci., Philadelphia, 1829, **6**, p.181. Description, F.J. Keeley, Proc. Acad. Nat. Sci. Philadelphia, 1920, **72**, p.358 [M.A. 1-405]. Olivine Fa_{25}, B. Mason, Geochimica et Cosmochimica Acta, 1963, **27**, p.1011.
20g Philadelphia, Acad. Nat. Sci.; 5g Paris, Mus. d'Hist. Nat.; 3g Washington, U.S. Nat. Mus.; 1g Tempe, Arizona State Univ.;
Specimen(s): [34597], less than a gram, fragments

Death Valley v Plainview (1917).

Debnevo v Gumoschnik.

Debreczen v Kaba.

De Calb County v Smithville.

Decatur County v Prairie Dog Creek.

De Cewsville 43°0′ N., 80°0′ W. approx.
Haldimand County, Ontario, Canada
Fell 1887, January 21, 1400 hrs
Synonym(s): *Talbot Road*
Stone. Olivine-bronzite chondrite (H6).
A stone of 340g was seen to fall in Talbot Road in the village of De Cewsville, E.E. Howell, Proc. Rochester Acad. Sci., 1890, **1**, p.92. Olivine Fa_{18}, B. Mason, Geochimica et Cosmochimica Acta, 1963, **27**, p.1011.
337g Vienna, Naturhist. Mus., main mass, and 1.7g fragments; 0.9g New York, Amer. Mus. Nat. Hist.;

Deelfontein 30°11′ S., 23°16′ E.
Cape Province, South Africa
Found 1932, approx., recognized 1966
Synonym(s): *Cape Province*
Iron. Octahedrite, coarse (1.8mm) (IA).
Date of find and co-ordinates, C. Frick and E.C.I. Hammerbeck, Bull. Geol. Surv. S. Africa, 1973 (57), p.15. A mass of 28kg was found, but thought to be part of a larger mass, Meteor. Bull., 1967 (41), Meteoritics, 1970, **5**, p.95. Described, with an analysis, M.L. Comerford et al., Meteoritics, 1968, **4**, p.7. Classification and another analysis, 7.11 %Ni, 83.1 ppm.Ga, 306 ppm.Ge, 1.4 ppm.Ir, J.T. Wasson, Icarus, 1970, **12**, p.407. Description, different co-ordinates and place of find, V.F. Buchwald, Iron Meteorites, Univ. of California, 1975, p.522.
Main mass, Pretoria, Geol. Surv. Mus.; Specimen, Univ. of Missouri Mus.; 105g Washington, U.S. Nat. Mus.;

Deep River v Diep River.

Deep Springs 36°30′ N., 79°45′ W. approx.
Rockingham County, North Carolina, U.S.A.
Found 1846, recognized 1889
Synonym(s): *Rockingham County*
Iron. Ataxite, Ni-rich (IRANOM).
A mass of 11.5kg, found at Deep Springs farm, was said to have been seen to fall, F.P. Venable, Am. J. Sci., 1890, **40**, p.161. Described, with an analysis by J. Fahrenhorst, 13.44 %Ni, E. Cohen, Meteoritenkunde, 1905, **3**, p.112. Description and structure, H.J. Axon and P.L. Smith, Min. Mag., 1972, **38**, p.736. New analysis, 13.2 %Ni, 0.405 ppm.Ga, 0.108 ppm.Ge, 9.4 ppm.Ir, R. Schaudy et al., Icarus, 1972, **17**, p.174. Description, terrestrially old, V.F. Buchwald, Iron Meteorites, Univ. of California, 1975, p.525.
Main mass, Raleigh, North Carolina, State Mus.; 700g New York, Amer. Mus. Nat. Hist.; 430g Chicago, Field Mus. Nat. Hist.; 456g Tempe, Arizona State Univ.; 318g Washington, U.S. Nat. Mus.; 306g Berlin, Humboldt Univ.;
Specimen(s): [84437], 162g. slice, and fragments, 12g

Deesa (of G.A. Daubrée) v Copiapo.

Dehesa (of G.A. Daubrée) v Copiapo.

Dehesa 33°30′ S., 70°30′ W. approx.
Santiago, Chile
Found 1866
Synonym(s): *Atacama, Chile, Sierra de Deesa*
Iron. Ataxite, Ni-rich (IRANOM).
Original weight unknown, for the 7kg mass mentioned by I. Domeyko as sent by Signor Ludere appears to have been of Copiapo. A specimen of about 280g, obtained from Domeyko, was wrongly described by Daubrée, owing to an interchange of labels, as having been found in an unspecified locality of Chile, G.A. Daubrée, C. R. Acad. Sci. Paris, 1868, **66**, p.572. True locality, I. Domeyko, Mineralojia, 3rd. ed., 1879, p.134., L. Fletcher, Min. Mag., 1889, **8**, p.256. Description, E. Cohen, Meteoritenkunde, 1905, **3**, p.116. Analysis, 11.97 %Ni, F. Berwerth, Tschermaks Min. Petr. Mitt., 1917, **34**, p.272. Classification, Paris specimen no. 378 is the only undoubtedly authentic material, V.F. Buchwald, Iron Meteorites, Univ. of California, 1975, p.527. Classification and analysis, 11.8 %Ni, 3.0 ppm.Ga, 174 ppm.Ge, 32 ppm.Ir, A. Kracher et al., Geochimica et Cosmochimica Acta, 1980, **44**, p.773.
221g Paris, Mus. d'Hist. Nat.; 235g Tempe, Arizona State Univ., not Dehesa, V.F. Buchwald.; 60g Bordeaux,;
Specimen(s): [56473], 2g.

De Hoek 29°23′ S., 23°6′ E.
Hay District, Cape Province, South Africa
Found 1960, possibly fell 1959, January 2
Iron. Anomalous (IRANOM).
A 3.83kg mass was found on the De Hoek farm, a portion of Lanyon Vale 376, 29km ESE. of Niekerkshoop and almost 48km NE. of Prieska. In 1967 a second mass, of 17.1kg, was found approximately 800m SE. of the first find, C. Frick and E.C.I. Hammerbeck, Cat. S. African and SW. African Meteorites, Bull. Geol. Surv. S. Africa, 1973 (57), p.15. Chemically and structurally anomalous, analysis, 9.95 %Ni, 0.236 ppm.Ga, 0.079 ppm.Ge, 0.27 ppm.Ir, E.R.D. Scott and J.T. Wasson, Geochimica et Cosmochimica Acta, 1976, **40**, p.103. Mentioned, F.W. Schumann, Ann. Geol. Surv. S. Africa, 1967, **6**, p.99. Description, C. Frick and E.A. Viljoen, S. African J. Sci., 1973, **69**, p.310.
17.2kg Pretoria, Geol. Surv. Mus., main mass;

De Kalb 30°35′30″ N., 94°55′19″ W.
Buchanan County, Missouri, U.S.A.
Found 1969
Stone. Olivine-bronzite chondrite (H).
A stone of about 575g was found after ploughing. Although the discovery site is only 3 miles west of the known area of the Faucett meteorites (*q.v.*), it is thought to represent a distinct fall, W.F. Read, Meteoritics, 1971, **6**, p.105.
Main mass, Los Angeles, Univ. of California;

DeKalb County v Smithville.

Delegate 37°0′ S., 149°2′ E.
County Wellesley, New South Wales, Australia
Found 1904, about
Iron. Octahedrite, medium (0.9mm) (IIIB-ANOM).
A mass of 61lb of boomerang shape was found on the SE. side of Sawpit Creek; described and analysed, 9.25 %Ni, J.C.H. Mingaye, Rec. Geol. Surv. New South Wales, 1916, **9**, p.159. New analysis, 9.5 %Ni, 20.3 ppm.Ga, 41.7 ppm.Ge, 1.6 ppm.Ir, E.R.D. Scott et al., Geochimica et Cosmochimica Acta, 1973, **37**, p.1957. Description, weathered, shock-hardened, V.F. Buchwald, Iron Meteorites, Univ. of California, 1975, p.528.
Main mass, Sydney, Austr. Mus.; 359g Chicago, Field Mus. Nat. Hist.; 202g Washington, U.S. Nat. Mus.; 130g New York, Amer. Mus. Nat. Hist.;
Specimen(s): [1915,145], 181g. slice

Delhi 28°34′ N., 77°15′ E.
Delhi state, India
Fell 1897, October 18, 1930 hrs
Stone. Olivine-hypersthene chondrite (L).
After the appearance of a brilliant meteor, and detonations, two stones each of about 1lb fell 5 miles SSW. of Delhi, but only a fragment of 0.8g was secured, L.L. Fermor, Rec. Geol. Surv. India, 1907, **35** (2), p.90. Olivine Fa_{24}, B. Mason, Geochimica et Cosmochimica Acta, 1967, **31**, p.1100.
0.8g Calcutta, Mus. Geol. Surv. India, only mass;

Dellys 36°55′ N., 3°55′ E.
Algeria
Found 1865, known before this year
Iron. Octahedrite, medium (1.0mm).
A piece of 76g was acquired by the Mus. d'Hist. Nat., Paris, and was described, G.A. Daubrée, C. R. Acad. Sci. Paris, 1866, **62**, p.78. Brief account, possibly part of Tamentit (*q.v.*) but no analysis is available, V.F. Buchwald, Iron Meteorites, Univ. of California, 1975, p.529.
76g Paris, Mus. d'Hist. Nat.; 9g Vienna, Naturhist. Mus.; 3g Chicago, Field Mus. Nat. Hist.;

Del Parque v Imilac.

Delphi 38°29′ N., 22°30′ E.
Greece
Doubtful..
The sacred "stone of Cronos" at Delphi was probably a meteorite, G.A. Wainwright, Ann. Serv. Antiq. Egypte, 1929, **28**, p.175, Palestine Exploration Fund, 1934 (January), p.32, Zeits. Ägypt. Sprache, 1935, **71**, p.41. There is not really sufficient evidence for this.

Delphos (a) 34°5′ N., 103°31′30″ W.
Roosevelt County, New Mexico, U.S.A.
Found 1968
Stone. Olivine-hypersthene chondrite (L4).
A total of 697g was found; a Delphos (b) stone, 547.7g is also listed, Cat. Huss Coll Meteorites, 1976, p.15.
679g Mainz, Max-Planck-Inst., of Delphos (a); 547g Mainz, Max-Planck-Inst., of Delphos (b); 3.26g Albuquerque, Univ. of New Mexico;

Del Rio 29°22′ N., 100°58′ W.
Val Verde County, Texas, U.S.A.
Found 1965
Iron. Ataxite, Ni-rich (IIF).
A mass of 3.346kg and five fragments together weighing 250g were found 4.5 miles W. of Del Rio, E.P. Henderson and E.A. King, Meteoritics, 1977, **12**, p.1. Analysis, 11.34 % Ni, 9.19 ppm.Ga, 98.6 ppm.Ge, 19 ppm.Ir, J.T. Wasson, Meteorites, Springer-Verlag, 1974, p.307. Description, V.F. Buchwald, Iron Meteorites, Univ. of California, 1975, p.529.
Main mass, privately held;

Demina 51°28′ N., 84°46′ E.
Biysk district, Altai govt., Federated SSR, USSR, [Демина]
Fell 1911, September 6, 1530 hrs
Synonym(s): *Altai, Tomsk, Tomskii*
Stone. Olivine-hypersthene chondrite (L6).
After detonations and the appearance of a fireball moving from SW. to NE., a 16.4kg stone, and perhaps others, was seen to fall; described, with an analysis by A.P. Kalishev, M.A. Ussov, Petrographische Untersuchung, Tomsk, 1916, P.N. Chirvinsky, Centralblatt Min., 1923, p.553, abs. Analysis, 22.86 % total iron, M.I. D'yakonova and V.Ya. Kharitonova, Метеоритика, 1961, **21**, p.52. One mass of 1189g fell, E.L. Krinov, Астрон. Журнал, 1945, **22**, p.303 [M.A. 9-297], Каталог Метеоритов Акад. Наук СССР, Москва, 1947 [M.A. 10-511]. Olivine Fa_{23}, B. Mason, Geochimica et Cosmochimica Acta, 1963, **27**, p.1011.
1126g Moscow, Acad. Sci., in six pieces, the largest weighing 794g.;
Specimen(s): [1956,320], 42g.

Deniliquin v Barratta.

Denmark
Denmark
Fell 1076
Doubtful..
A stone or stones is said to have fallen, G. von Boguslawski, Ann. Phys. Chem. (Poggendorff), 1854, **4** (suppl.), p.9, but the evidence is not conclusive.

De Nova 39°51′ N., 102°57′ W.
Washington County, Colorado, U.S.A.
Found 1940
Stone. Olivine-hypersthene chondrite (L6), veined.
A mass of 12.7kg was found, W. Wahl, letter of 23 May 1950, in Min. Dept. BM(NH)., H.H. and A.D. Nininger, The Nininger Collection of Meteorites, Winslow, Arizona, 1950, p.48. Olivine Fa_{23}, B. Mason, Geochimica et Cosmochimica Acta, 1963, **27**, p.1011.
5.8kg Tempe, Arizona State Univ.; 632g Chicago, Field Mus. Nat. Hist.; 460g Denver, Mus. Nat. Hist.;
Specimen(s): [1959,990], 4784g.

Densmore (1879) 39°39′ N., 99°41′ W.
Norton County, Kansas, U.S.A.
Found 1879, recognized 1939
Stone. Olivine-hypersthene chondrite (L6).
Three fragments totalling 37.2kg were found by F.M. Kendrick, 3 miles E. and 1.5 miles S. of Densmore, A.D. Nininger, Pop. Astron., Northfield, Minnesota, 1940, **48**, p.555, Contr. Soc. Res. Meteorites, **2**, p.227 [M.A. 8-54]. An additional specimen of about 34lb was found in about 1950, the finder sold it to the American Meteor. Lab., Denver in 1965, W.F. Read, Trans. Kansas Acad. Sci., 1970, **73**, p.292. Nature of the metal, H.C. Urey and T. Mayeda, Geochimica et Cosmochimica Acta, 1959, **17**, p.113. Olivine Fa_{23}, B. Mason, Geochimica et Cosmochimica Acta, 1963, **27**, p.1011.
42kg Washington, U.S. Nat. Mus., includes main mass; 5.6kg Mainz, Max-Planck-Inst.; 119g Harvard Univ.; 102g New York, Amer. Mus. Nat. Hist.;
Specimen(s): [1967,387], 51g. part-slice, and fragments, 2g.

Densmore (1950) 39°33′57″ N., 99°38′50″ W.
Graham County, Kansas, U.S.A.
Found 1950
Stone. Olivine-bronzite chondrite (H6).
One mass of 11.3kg was found about 7 miles SE. of Densmore, Meteor. Bull., 1969 (47), Meteoritics, 1970, **5**, p.106, W.F. Read, Trans. Kansas Acad. Sci., 1970, **73**, p.292.
Main mass, Los Angeles, Univ. of California; Specimen, Lawrence, Kansas, Kansas Geol. Surv.; 92g Tempe, Arizona State Univ.; 13.9g Moscow, Acad. Sci.;
Specimen(s): [1971,99], 1547g. sawn, crusted fragment, and fragments, 1.4g.

Denton County 33° N., 97° W.
Denton County, Texas, U.S.A.
Found 1856, known since this year
Synonym(s): *Austin*
Iron. Octahedrite, medium (1.1mm) (IIIA).
The original mass was said to have been about 40lb, but only about 12lb was found by G.G. Shumard in the possession of a blacksmith at McKinney, Collin County in 1860, B.F. Shumard, Trans. St. Louis Acad. Sci., 1860, **1**, p.623. This mass has been heated to about 1000C and many specimens have been hammered and forged, V.F. Buchwald, Iron Meteorites, Univ. of California, 1975, p.530. Analysis, 8.21 %Ni, 19.7 ppm.Ga, 42.7 ppm. Ge., 0.28 ppm.Ir, E.R.D. Scott et al., Geochimica et Cosmochimica Acta, 1973, **37**, p.1957.
5.6kg Austin, Texas, State Mus., reputed place of repose, lost according to Smithsonian Institution; 200g Vienna, Naturhist. Mus.; 161g Harvard Univ.; 66g Yale Univ.; 44g Chicago, Field Mus. Nat. Hist.; 26g Göttingen, Univ.;
Specimen(s): [35412], 122g.

Denulguon v Chandakapur.

Denver 39°46′57″ N., 104°55′50″ W.
Denver County, Colorado, U.S.A.
Fell 1967, July 11-15
Stone. Olivine-hypersthene chondrite (L6).
One stone of 230g was found on a warehouse roof on 17 July 1967; it had fallen during the preceding week, Meteor. Bull., 1968 (43), Meteoritics, 1970, **5**, p.98. Described, with an analysis, 21.1 % total iron, B. Mason and E. Jarosewich, Science, 1968, **160**, p.878.
208g Washington, U.S. Nat. Mus., main mass;

Denver v Bear Creek.

Denver City 33°4′17″ N., 102°48′1″ W.
Yoakum County, Texas, U.S.A.
Found 1975
Iron. Octahedrite, fine (0.2mm) (IRANOM).
A single mass of 26.1kg was found during ploughing about 12 km NNE. of Denver City, Texas. Description, classification and analysis, 8.40 %Ni, 1.03 ppm.Ga, 0.5 ppm.Ge, 5.2 ppm.Ir, E.R.D. Scott et al., Meteoritics, 1977, **12**, p.425.
26kg Los Angeles, Univ. of California;
Specimen(s): [1978,M.13], 419.8g.

Denver County v Bear Creek.

Deoli v Ranchapur.

Deport 33°31′ N., 95°18′ W.
Red River County, Texas, U.S.A.
Found 1926
Synonym(s): *Red River County*
Iron. Octahedrite, coarse (1.3mm) (IA).
Masses of 2.7kg, 1.3kg, and 0.9kg were found in 1926 one mile E. of Deport; at least 23 more masses totalling a further 10kg were found later, E.P. Henderson, letter of 12 May 1939, in Min. Dept. BM(NH). Described and analysed, 7.9 %Ni, C. Palache and F.A. Gonyer, Am. Miner., 1932, **17**, p.357 [M.A. 5-158]. Further analysis, 8.11 %Ni, 69.9 ppm.Ga, 255 ppm.Ge, 2.2 ppm.Ir, J.T. Wasson, Icarus, 1970, **12**, p.407. Description, V.F. Buchwald, Iron Meteorites, Univ. of California, 1975, p.531.
9.6kg Fort Worth, Texas, Monnig Colln.; 4.45kg Washington, U.S. Nat. Mus.; 1.1kg Austin, Texas, Univ. of Texas; 1kg Michigan Univ.; 3.9kg Harvard Univ.; 482g Mainz, Max-Planck-Inst.; 396g New York, Amer. Mus. Nat. Hist.; 342g Los Angeles, Univ. of California; 204g Tempe, Arizona State Univ.; 314g Chicago, Field Mus. Nat. Hist.;
Specimen(s): [1959,935], 1048g. a complete mass

Deretschin v Ruschany.

Dermbach 50°43′ N., 10°7′ E.
near Dermbach, Suhl, Germany
Found 1924
Iron. Ataxite, Ni-rich (IRANOM).
A 1.5kg mass was found, consisting of Ni/Fe (69%), troilite (25%), and schreibersite (6%) by volume, G. Hoppe, Chem. Erde, 1976, **35**, p.305 [M.A. 77-3273]. Analysis, 42.1 %Ni, 4.66 ppm.Ga, 0.144 ppm.Ge, 0.029 ppm.Ir, E.R.D. Scott and J.T. Wasson, Geochimica et Cosmochimica Acta, 1976, **40**, p.103. Reported, Meteor. Bull., 1978 (55), Meteoritics, 1978, **13**, p.333.
1.5kg Dermbach, Mus.;

Derrick Peak A78001 80°5′ S., 156°25′ E.
Victoria Land, Antarctica
Found 1978, between December 1978 and January 1979
Iron. Octahedrite, coarsest (5-8mm) (IIB).
16 masses of iron, totalling 320kg, were found during the 1978-1979 field season in Antarctica and are named Derrick Peak A78001 to Derrick Peak A78016. The major masses are Derrick Peak A78009, 138kg; Derrick Peak A78008, 59.4kg; Derrick Peak A78005, 18.4kg; Derrick Peak A78001, 15.1kg; Derrick Peak A78002, 7.2kg. The masses appear to be similar, all have large protruding schreibersites, 6.64 % Ni, R.S. Clarke, Jr., Meteoritics, 1982, **17**, p.129. Find-site description, P.J.J. Kamp and D.J. Lowe, Meteoritics, 1982, **17**, p.119.

122kg Washington, U.S. Nat. Mus.; 120kg Tokyo, Nat. Inst. Polar Res.; 72kg Hamilton, N.Z., Waikato Univ.;

Descubridora v Charcas.

Desert of Atacama v Copiapo.

De Sotoville v Tombigbee River.

Desuri 25°44′ N., 73°37′ E.

Pali district, Rajasthan, India
Fell 1962, July 18
Stone. Olivine-bronzite chondrite (H6).
A stone of about 25.4kg was found in a pit nearly 61cm deep. Branches of some shrubs in the vicinity were destroyed. Description, P.K. Venkataraman, Indian Minerals, 1962, **16** (3), p.304, Meteor. Bull., 1964 (29). Olivine Fa_{20}, B. Mason, Geochimica et Cosmochimica Acta, 1967, **31**, p.1100.
21.9kg Calcutta, Geol. Surv. India;
Specimen(s): [1974,M.20], 49.6g. 3 fragments

Devica v Soko Banja.

Device v Soko Banja.

Dexter 33°49′ N., 97°0′ W.

Cooke County, Texas, U.S.A.
Found 1889
Iron. Octahedrite, medium (1.1mm) (IIIA).
A mass of 1724g was found; it may have fallen about a year before, C.A. Reeds, Bull. Am. Mus. Nat. Hist., 1937, **73** (6), p.517 [M.A. 7-61]. Description, H.H. Uhlig, Geochimica et Cosmochimica Acta, 1954, **6**, p.282, Geochimica et Cosmochimica Acta, 1955, **7**, p.34. Classification, analysis, 7.67 %Ni, 20.5 ppm.Ga, 40.9 ppm.Ge, 1.2 ppm.Ir, E.R.D. Scott et al., Geochimica et Cosmochimica Acta, 1973, **37**, p.1957. An old fall, shock hardened, epsilon structure, V.F. Buchwald, Iron Meteorites, Univ. of California, 1975, p.532.
1276g New York, Amer. Mus. Nat. Hist., main mass; 212g Washington, U.S. Nat. Mus.;
Specimen(s): [1963,533], 79g.

Dhajala 22°22′40″ N., 71°25′38″ E.

Surendranager district, Gujarat, India
Fell 1976, January 28, 2100 hrs, U.T.
Stone. Olivine-bronzite chondrite (H3-4).
Following a fireball accompanied by a bright flash and two loud detonations a shower of meteorites fell; about 45kg of fragments have been recovered, Meteor. Bull., 1978 (55), Meteoritics, 1978, **13**, p.334. Pb isotopic data, J.H. Chen and G.R. Tilton, Meteoritics, 1977, **12**, p.193, abs. Description and analysis, 27.10 % total iron, S.P. Das Gupta et al., Min. Mag., 1978, **42**, p.493. Ar isotopes, E.L. Firman and R.W. Stoenner, Meteoritics, 1979, **14**, p.1.
33.3kg Calcutta, Mus. Geol. Surv. India; Specimen, Ahmadabad, Physical Research Labs.;
Specimen(s): [1976,M.12], 73g. and fragments, 2g

Dharmsala v Dhurmsala.

Dharwar 14°53′ N., 75°36′ E.

Dharwar district, Karnataka, India
Fell 1848, February 15, 1300 hrs
Stone. Chondrite?.
A stone of 4lb fell near the village of Negloor, and was sent to Bombay Geogr. Soc., H. Giraud, Edinburgh New Phil. J., 1849, **47**, p.54. The place of fall is Naglur, district Dharwar, C.A. Silberrad, Min. Mag., 1932, **23**, p.295.

Dhenagur v Kheragur.

Dhulia v Bhagur.

Dhurmsala 32°14′ N., 76°28′ E.

Kangra district, Himachal Pradesh, India
Fell 1860, July 14, 1415 hrs
Synonym(s): *Dharmsala*
Stone. Olivine-hypersthene chondrite, amphoterite (LL6).
Several stones, the largest estimated at about 329lb, fell, after detonations and the appearance of a fireball, J. Asiatic Soc. Bengal, 1860 (1861), **29**, p.410. Analysis (doubtful, no CaO reported), S. Haughton, Phil. Mag., 1866, **32**, p.266. Partial analysis, 18.33 % total iron, W. Wahl and H.B. Wiik, Geochimica et Cosmochimica Acta, 1951, **1**, p.123. Olivine $Fa_{27.2}$, K. Fredriksson et al., Origin and Distribution of the Elements, ed. L.H. Ahrens, Pergamon, 1968, p.457.
3.8kg Turin Univ.; 3kg Chicago, Field Mus. Nat. Hist.; 1.5kg Vienna, Naturhist. Mus.; 1.3kg Calcutta, Mus. Geol. Surv. India; 1.4kg Budapest, Nat. Mus.; 700g Edinburgh, Roy. Scot. Mus.; 681g Washington, U.S. Nat. Mus.; 530g Harvard Univ.; 520g New York, Amer. Mus. Nat. Hist.; 241g Copenhagen, Univ.; 235g Prague, Nat.Mus.; 218g Paris, Mus. d'Hist. Nat.; 216g Dublin, Trinity College; 201g Berlin, Humboldt Univ.; 185g Tempe, Arizona State Univ.; 115g Oxford, Univ. Mus;
Specimen(s): [33762], 8559g. and fragments, 5.5g; [33763], 3784g. and fragments, 13g; [1964,563], 228g.; [96262], Six thin sections

Diablo Canyon v Cañon Diablo.

Diamantina v Thunda.

Dickson County v Charlotte.

Diep River 33°45′ S., 18°34′ E. approx.

Cape Province, South Africa
Fell 1906, November 4, 1630 hrs, possible time and date
Synonym(s): *Constantia, Deep River, Hermitage*
Stone. Olivine-hypersthene chondrite (L6).
Fell on farm Hermitage, F. Berwerth, Fortschr. Min. Krist. Petr., 1916, **5**, p.271, E.L. Gill, letter of 6 March 1926, in Min. Dept. BM(NH). The fall formerly listed as Constantia, a stone of about 2lb which fell through the iron roof of a house, J.P. Maclear, letter of 28 November 1906, Miss M. Wilman, letter of 14 November 1907, in Min. Dept. BM (NH)., appears to be only an early reference to the fall that subsequently became known as Diep River; the date and general description agree and the Diep River runs through the vale of Constantia. Partial analysis, 21.59 % total iron, H. von Michaelis et al., Earth planet. Sci. Lett., 1969, **5**, p.387. Olivine Fa_{25}, B. Mason, Geochimica et Cosmochimica Acta, 1963, **27**, p.1011.
20.25oz Cape Town, South African Mus., main mass; 57g Vienna, Naturhist. Mus.;
Specimen(s): [1950,253], 44g.

Dimboola 36°30′ S., 142°2′ E.

Victoria, Australia
Found 1944
Stone. Olivine-bronzite chondrite (H5).
About 16kg were found, B. Mason, Geochimica et Cosmochimica Acta, 1963, **27**, p.1022, L. Greenland and J.F. Lovering, Geochimica et Cosmochimica Acta, 1965, **29**,

p.843. Olivine $Fa_{19.4}$, B. Mason, Rec. Austr. Mus., 1974, **29**, p.169.
16kg Adelaide, S. Austr. Mus., main mass; 484g New York, Amer. Mus. Nat. Hist.; 46g Washington, U.S. Nat. Mus.;

Dimitrovgrad 43°2′47″ N., 22°51′50″ E.
Yugoslavia
Found 1949, recognized 1956
Synonym(s): *Caribrod, Dimitrovgradsko gvozde, Pirot*
Iron. Octahedrite, medium (1.1mm) (IIIA).
A mass of 100kg was found, M. Ramovic, Geol. Glasnik, 1958, **4**, p.273, Meteor. Bull., 1960 (18). Listed, M. Ramovic, Cat. Meteor. Coll. Yugoslavia, 1965, p.51. Analysis, 7.64 % Ni, 20.3 ppm.Ga, 40.2 ppm.Ge, 3 ppm.Ir, E.R.D. Scott et al., Geochimica et Cosmochimica Acta, 1973, **37**, p.1957. Description, references, shock hardened matrix, V.F. Buchwald, Iron Meteorites, Univ. of California, 1975, p.533.
99.7kg Belgrade, Serbian Nat. Mus., main mass; 219g Washington, U.S. Nat. Mus.;
Specimen(s): [1959,221], 12g.

Dimitrovgradsko gvozde v Dimitrovgrad.

Dimmitt 34°35′ N., 102°10′ W.
Castro County, Texas, U.S.A.
Found 1942, about
Stone. Olivine-bronzite chondrite (H4), brecciated.
At least 21 stones, totalling 13.5kg, were found; the fall may perhaps be identical with Tulia (*q.v.*), H.H. and A.D. Nininger, The Nininger Collection of Meteorites, Winslow, Arizona, 1950, p.130. Analysis, 24.8 % total iron, V.Ya. Kharitonova, Метеоритика, 1969, **29**, p.91. Olivine Fa_{20}, B. Mason, Geochimica et Cosmochimica Acta, 1963, **27**, p.1011. Breccia, contains H5 and LL-group clasts, A.E. Rubin et al., Meteoritics, 1981, **16**, p.382, abs. 364 specimens are included under the Dimmitt name in the Monnig collection and totalling 177kg. These may be from more than a single fall, Cat. Monnig Colln. Meteorites, 1983, p.7.
177kg Fort Worth, Texas, Monnig Colln.; 11.7kg Mainz, Max-Planck-Inst.; 6.28kg Washington, U.S. Nat. Mus.; 1kg Michigan Univ.; 2.8kg Tempe, Arizona State Univ.; 1.1kg Chicago, Field Mus. Nat. Hist.; 709g Harvard Univ.; 563g New York, Amer. Mus. Nat. Hist.; 290g Bloomingfield, Michigan, Cranbrook Inst. Sci.; 93g Los Angeles, Univ. of California;
Specimen(s): [1959,856], 96.5g.

Dinagepur v Pirgunje.

Dingo Pup Donga 30°26′ S., 126°6′ E.
Western Australia, Australia
Found 1965
Stone. Achondrite, Ca-poor. Ureilite (AURE).
A single stone of 122.7g was found, G.J.H. McCall, First Suppl. to West. Austr. Mus. Spec. Publ. no. 3, 1968, p.10. Analysis, 15.63 % total iron, G.J.H. McCall and W.H. Cleverly, Min. Mag., 1968, **36**, p.691. Contains diamond, G.P. Vdovykin, Space Sci. Rev., 1970, **10**, p.483. Mineralogy, petrology, J.L. Berkley et al., Geochimica et Cosmochimica Acta, 1980, **44**, p.1579.
Main mass, Kalgoorlie, West. Austr. School of Mines; 19g Perth, West. Austr. Mus.; 4g Washington, U.S. Nat. Mus.;
Specimen(s): [1968,339], 3.38g. part-slice

Dirranbandi v Wynella.

Dispatch 39°30′ N., 98°32′ W.
Smith County, Kansas, U.S.A.
Found 1956
Stone. Olivine-bronzite chondrite (H).
220g were found, B. Mason, Meteorites, Wiley, 1962, p.233. Olivine Fa_{18}, B. Mason, Geochimica et Cosmochimica Acta, 1963, **27**, p.1011.
215g Tempe, Arizona State Univ., main mass; 38g Chicago, Field Mus. Nat. Hist.; 5.4g New York, Amer. Mus. Nat. Hist.;

Distrito Quebracho 31°53′ S., 60°28′ W.
Parana dept., Entre Rios, Argentina
Fell 1957, March 13, 2000 hrs, approx. time
Synonym(s): *Quebracho*
Stone. Olivine-bronzite chondrite (H).
A partially crusted 400g stone was found next morning in a kitchen garden; described, with an analysis, C.A. Gordillo, Mus. Entre Rios, Direcc. Prensa, 1959 (Publ. no. 1), Meteor. Bull., 1962 (24). Olivine Fa_{18}, B. Mason, Geochimica et Cosmochimica Acta, 1967, **31**, p.1100.
Main mass, Parana City, Entre Rios Mus.; 1.8g New York, Amer. Mus. Nat. Hist.;
Specimen(s): [1964,69], 5g.

Divnoe 45°42′ N., 43°42′ E.
Near Divnoe, Stavropol region, Ukraine, USSR
Found 1981
Stone.
A single mass of 12.7kg was found in a field after grass-cutting, Meteor. Bull., 1983 (61), Meteoritics, 1983, **18**, p.78.

Dix 41°14′ N., 103°29′ W.
Kimball County, Nebraska, U.S.A.
Found 1927, recognized 1938
Stone. Olivine-hypersthene chondrite (L6).
Three fragments were found, totalling 44kg, in NW. 1/4, sect. 36, township 13, range 54, Kimball County, A.D. Nininger, Pop. Astron., Northfield, Minnesota, 1939, **47**, p.212, E.P. Henderson, letter of 3 June 1939, in Min. Dept. BM(NH).. Listed, H.H. and A.D. Nininger, The Nininger Collection of Meteorites, Winslow, Arizona, 1950, p.48. Olivine Fa_{23}, B. Mason, Geochimica et Cosmochimica Acta, 1963, **27**, p.1011.
16.5kg Tempe, Arizona State Univ.; 8.9kg Washington, U.S. Nat. Mus.; 830g Chicago, Field Mus. Nat. Hist.;
Specimen(s): [1959,1012], 17.771kg.

Djati-Pengilon 7°30′ S., 111°30′ E. approx.
Ngawi district, Java, Indonesia
Fell 1884, March 19, 1630 hrs
Synonym(s): *Alastoeva, Alastoewa*
Stone. Olivine-bronzite chondrite (H6).
A stone of about 166kg fell, after detonation and appearance of a fireball, R.D.M. Verbeek, Jaarb. Mijnwezen Nederland. Oost-Indie, 1886, **15**, p.145, includes an analysis by J.W. Retgers. The mass fell into the river Alastoeva, near Djati-Pengilon, K. Tucek, Cat. Meteor. Coll. Nat. Mus. Prague, 1968. Phosphate minerals, W.R. Van Schmus and P.H. Ribbe, Geochimica et Cosmochimica Acta, 1969, **33**, p.637. Olivine Fa_{20}, B. Mason, Geochimica et Cosmochimica Acta, 1963, **27**, p.1011.
156kg Bandung, Geol. Mus.; 727g Paris, Mus. d'Hist. Nat.; 480g Berlin, Humboldt Univ.; 464g Washington, U.S. Nat. Mus.; 377g Vienna, Naturhist. Mus.; 256g Harvard Univ.; 240g Yale Univ.;
Specimen(s): [63052], 465g.; [1964,564], 77.5g.; [1980,M.17], 16.72g. fragments

Djaul Island v Dyarrl Island.

Djermaia 12°44′ N., 15°3′ E.
Chad
Fell 1961, February 25
Stone. Olivine-bronzite chondrite (H), xenolithic, gas rich.
About 1kg was recovered. Noble gas data, L. Schultz and P. Signer, Meteoritics, 1977, **12**, p.359, abs. Track densities, J.C. Lorin and P. Pellas, Meteoritics, 1977, **12**, p.299, abs. Date of fall, olivine Fa_{19}, B. Mason, Geochimica et Cosmochimica Acta, 1963, **27**, p.1011.
83g Paris, Mus. d'Hist. Nat.;

Dluga Wola v Dolgovoli.

Doab v Futtehpur.

Doab v Kadonah.

Docco v Dosso.

Dog Creek v Prairie Dog Creek.

Dokachi 23°30′ N., 90°20′ E.
Munshiganj, Dacca, Bangladesh
Fell 1903, October 22, 1900 hrs
Synonym(s): *Bibandi, Dakhin Paiksha, Hariya, Kolapara, Paiksha, Rana*
Stone. Olivine-bronzite chondrite (H5), veined.
After appearance of fireball and detonations, a shower of over a hundred stones fell in several villages; twenty-four stones were recorded, the largest weighing 1571g and the total weight being about 3838g, L.L. Fermor, Rec. Geol. Surv. India, 1907, **35**, p.68. Described, with an analysis, H.L. Bowman, Min. Mag., 1911, **16**, p.35. Olivine Fa_{19}, B. Mason, Geochimica et Cosmochimica Acta, 1963, **27**, p.1011.
1.84kg Calcutta, Mus. Geol. Surv. India; 570g Oxford, Univ. Mus., includes five complete stones; 218g Chicago, Field Mus. Nat. Hist.; 164g Tempe, Arizona State Univ.; 78g Paris, Mus. d'Hist. Nat.; 16g Washington, U.S. Nat. Mus.;
Specimen(s): [1906,134], 593.5g. complete stone, Dokachi, (240 A.2 of Fermor's report); [1906,135], 27g. Dakhin Paiksha, (240 C.4 of Fermor's report).

Dolgaia v Dolgovoli.

Dolgaia Volia v Dolgovoli.

Dolgaja v Dolgovoli.

Dolgaja Volja v Dolgovoli.

Dolgaja Wolja v Dolgovoli.

Dolgovoli 50°45′ N., 25°18′ E.
Luck, Ukraine, USSR, [Долгая воля]
Fell 1864, June 26, 0700 hrs
Synonym(s): *Dluga Wola, Dolgaia, Dolgaia Volia, Dolgaja, Dolgaja Volja, Dolgaja Wolja*
Stone. Olivine-hypersthene chondrite (L6).
A stone of about 1.6kg fell after detonations, O. Büchner, Ann. Phys. Chem. (Poggendorff), 1865, **124**, p.591. Listed, E.L. Krinov, Астрон. Журнал, 1945, **22**, p.303 [M.A. 9-297], E.L. Krinov, Каталог Метеоритов Акад. Наук СССР, Москва, 1947 [M.A. 10-511]. Olivine Fa_{25}, B. Mason, Geochimica et Cosmochimica Acta, 1963, **27**, p.1011.
773g Kiev, Geol. Mus. Acad. Sci. Ukraine; 115g Moscow, Acad. Sci.; 102g Vienna, Naturhist. Mus.; 59g Paris, Mus. d'Hist. Nat.; 6g Berlin, Humboldt Univ.; 6g Chicago, Field Mus. Nat. Hist.;
Specimen(s): [43197], 3g.

Dolores Hidalgo v Cosina.

Domanitch 40° N., 29° E. approx.
Carakewy, Brusa, Turkey
Fell 1907, February 1
Synonym(s): *Brusa*
Stone. Olivine-hypersthene chondrite (L5).
Two large and two small stones were, in 1914, in the Museum at Brusa, A. Russell, letter of 8 November 1923, in Min. Dept. BM(NH). Olivine Fa_{24}, B. Mason, Geochimica et Cosmochimica Acta, 1963, **27**, p.1011.
442g Paris, Mus. d'Hist. Nat.; 5.7g Tempe, Arizona State Univ.;

Doña Inez v Vaca Muerta.

Donas v Gibeon.

Donga Karaod v Donga Kohrod.

Donga Kohrod 21°52′ N., 82°27′ E.
Bilaspur district, Madhya Pradesh, India
Fell 1899, September 23, 1500 hrs
Synonym(s): *Donga Karaod*
Stone. Olivine-bronzite chondrite (H6).
A small stone of about 0.5lb fell, Director, Geol. Surv. India, letter of 2 October 1901, in Min. Dept. BM(NH)., Cat. Meteorites, Calcutta Mus., 1901, p.4. The place of fall is Donga Karaod, C.A. Silberrad, Min. Mag., 1932, **23**, p.295. Olivine Fa_{19}, B. Mason, Geochimica et Cosmochimica Acta, 1963, **27**, p.1011.
90g Calcutta, Mus. Geol. Surv. India;
Specimen(s): [85671], 39.75g.

Dongning v Nantan.

Dongtai 32°55′ N., 120°47′ E.
Dongtai County, Jiangsu, China
Fell 1970, January 20, 2000 hrs
Synonym(s): *Dangtai*
Stone. Olivine-hypersthene chondrite, amphoterite (LL6).
A single mass of 5.5kg was recovered; references, D. Bian, Meteoritics, 1981, **16**, p.117. Analysis, 19.98 % total iron, D. Wang and Z. Ouyang, Geochimica, 1979, p.120, in Chinese [M.A. 80-2085].
5.5kg Nanjing, Purple Mountain Observatory;

Donguz v Saratov.

Donnybrook 33°37′ S., 115°55′ E. approx.
Thompson Brook, near Donnybrook, Western Australia, Australia
Found 1918, recognized 1971
Stony-iron. Mesosiderite? (MES?).
Two fragments totalling 414g were found in the mineral collection of the Univ. of Western Australia, catalogued as magnetite; they had been found in stream gravels, Meteor. Bull., 1974 (52), Meteoritics, 1974, **9**, p.110. The meteorite appears to be of unusual composition, R.A. Binns, letters of

19 April, 31 May, 9 June 1972, in Min. Dept. BM(NH).
414g Perth, Univ. West. Austr.;

Doolgunna 25°56′ S., 119°18′ E.
Meekatharra district, Western Australia, Australia
Found 1967
Synonym(s): *Meekatharra*
Stone. Olivine-hypersthene chondrite (L).
A fragment of 20g was found in a claypan near the southern boundary of Doolgunna station; described, W.N. MacLeod, Ann. Rep. Geol. Surv. West. Austr. for 1967, 1968, p.68 [M.A. 20-46]. Listed, G.J.H. McCall, 2nd. Suppl. to West. Austr. Mus. Spec. Publ. no. 3, 1972, p.11.
14g Perth, Geol. Surv. West. Austr., main mass;

Dooralla v Durala.

Doornport v Winburg.

Dora (pallasite) 33°55′36″ N., 103°57′18″ W.
Roosevelt County, New Mexico, U.S.A.
Found 1955, approx. year, recognized 1967
Stony-iron. Pallasite (PAL).
A 7.6kg mass was ploughed up 10 or 12 years prior to recognition and left by a fence in a farmyard, G.I. Huss, letter of 16 March 1973, in Min. Dept. BM(NH). Analysis of metal, 11.7 %Ni, 16.7 ppm.Ga, 33.4 ppm.Ge, 0.093 ppm.Ir, E.R.D. Scott, Geochimica et Cosmochimica Acta, 1977, **41**, p.349. Olivine $Fa_{12.9}$, E.R.D. Scott, Geochimica et Cosmochimica Acta, 1977, **41**, p.693.
3.5kg Mainz, Max-Planck-Inst., main mass; 1.25kg Tempe, Arizona State Univ.; 37g New York, Amer. Mus. Nat. Hist.;
Specimen(s): [1973,M.4], 32.8g. part-slice

Dora (stone) 33°55′24″ N., 103°21′ W.
Roosevelt County, New Mexico, U.S.A.
Found 1970
Stone. Chondrite.
Listed, total known weight 1.57kg, Cat. Huss Coll. Meteorites, 1976, p.16.
1480g Mainz, Max-Planck-Inst.;

Dordrecht (iron) 51°48′ N., 4°39′ E.
Holland
Fell 1650, August 6
Doubtful..
One mass is said to have fallen, E.F.F. Chladni, Die Feuer-Meteore, Wien, 1819, p.228, but the evidence is not conclusive.

Dordrecht (stone) 51°48′ N., 4°39′ E.
Holland
Fell 1808, a few years before this date
Doubtful..
One stone is said to have fallen, E.F.F. Chladni, Die Feuer-Meteore, Wien, 1819, p.275, but the evidence is not adequate.

Dor el Gani 26°57′ N., 16°2′ E.
Dor el Gani, Libya
Found 1972
Iron. Octahedrite.
A single mass of 2575g was found in the southern part of Dor el Gani, analysis, 7.05 %Ni, A. Mücke and E. Klitsch, Chem. Erde, 1976, **35**, p.169 [M.A. 77-2044]. Reported, Meteor. Bull., 1978 (55), Meteoritics, 1978, **13**, p.334.

Dores do Campo v Uberaba.

Dores dos Campos Formosos v Uberaba.

Dorofeevka 53°20′ N., 70°4′ E.
Kokchetav district, Omsk, Federated SSR, USSR, [Дорофеевка]
Found 1910
Iron. Octahedrite, plessitic (0.09mm) (IIF).
A mass of about 12.6kg was found by the expedition under L.A. Kulik that was sent to Siberia in 1921 by the Russian Academy of Sciences, P.N. Chirvinsky, Centralblatt Min., 1922, p.587, L.A. Kulik, Bull. Acad. Sci. Russ., 1922, **16**, p.391. Listed, E.L. Krinov, Астрон. Журнал, 1945, **22**, p.303 [M.A. 9-297], Каталог Метеоритов Акад. Наук СССР, Москва, 1947 [M.A. 10-511], A.N. Zavaritsky, Метеоритика, 1954 (11), p.64 [M.A. 13-48]. Analysis, 11.57 %Ni, M.I. D'yakonova, Метеоритика, 1958, **16**, p.180 [M.A. 14-49]. Classification, new analysis, 11.26 %Ni, 9.10 ppm.Ga, 124 ppm.Ge, 23 ppm.Ir, J.T. Wasson, Meteorites, Springer-Verlag, 1974, p.307. Description, unshocked, V.F. Buchwald, Iron Meteorites, Univ. of California, 1975, p.534.
12.39kg Moscow, Acad. Sci.;

Doroninsk 51°12′ N., 112°18′ E.
Irkutsk, Federated SSR, USSR, [Доронинск]
Fell 1805, April 6, 1700 hrs
Synonym(s): *Irkutsk*
Stone. Olivine-bronzite chondrite (H6), brecciated.
Two stones of about 3.25kg and 1kg fell, after appearance of cloud and detonations, A. Stoikovitz, Ann. Phys. (Gilbert), 1809, **31**, p.308. Two stones fell, weighing 3891g together, E.L. Krinov, Астрон. Журнал, 1945, **22**, p.303 [M.A. 9-297], Каталог Метеоритов Акад. Наук СССР, Москва, 1947 [M.A. 10-511]. Olivine Fa_{19}, B. Mason, Geochimica et Cosmochimica Acta, 1963, **27**, p.1011. Analysis, 28.07 % total iron, M.I. D'yakonova, Метеоритика, 1972, **31**, p.119.
1563g Moscow, Acad. Sci.; 215g Tübingen, Univ.; 96g Tempe, Arizona State Univ.; 74g Berlin, Humboldt Univ.; 61g Vienna, Naturhist. Mus.; 49g Chicago, Field Mus. Nat. Hist.;
Specimen(s): [33971], 5.5g.; [56470], 1.43g.

Dorpat 58°24′ N., 26°36′ E. approx.
Estonian SSR, USSR
Fell 1704, July 20, evening hrs
Doubtful..
Part of a letter of Field-marshal B.P. Sheremetev to F.A. Golovin has been interpreted by Yu. Simashko as describing an east-west travelling fireball, Падение Авухъ Метеоритовъ ъ Историческия Зпохи 1704 г близъ Дерпга и 1812 г.И. Бородино, St. Petersburg, 1892.

Dorrigo 30°17′ S., 152°40′ E.
New South Wales, Australia
Found 1948, before this year
Iron. Octahedrite, plessitic (0.15mm) (IRANOM).
A much weathered mass was found among weathered basalt boulders whilst breaking them for road-making. A 17lb mass and fragments were recovered, but the mass has since exuded much lawrencite, resulting in further decomposition, R.O. Chalmers, Austr. Mus. Mag., 1948, **9**, p.263 [M.A. 10-521]. Classification, description, V.F. Buchwald, Iron Meteorites, Univ. of California, 1975, p.536. Analysis, 16.3 %Ni, 17 ppm.Ga, 1.9 ppm.Ge, 11 ppm.Ir, D.J. Malvin et al., priv.

comm., 1983.
8.5kg Sydney, Austr. Mus.; 39g Washington, U.S. Nat. Mus., fragments;

Dosso 13°3′ N., 3°10′ E.
Niger
Fell 1962, February 19, 1230 hrs, approx. time
Synonym(s): *Docco*
Stone. Olivine-hypersthene chondrite (L6).
Two pieces were picked up by the villagers of Dosso, 30km from the village of Hanam Tombo. One fragment of 1025g was taken to the Mus. d'Hist. Nat. Paris, the second piece was retained by the local authorities, Le Niger newspaper, 1962, March 5, E. Jérémine, letter of 18 September 1962, in Min. Dept. BM(NH)., Meteor. Bull., 1962 (25). Olivine Fa_{25}, B. Mason, Geochimica et Cosmochimica Acta, 1963, **27**, p.1011.
872g Paris, Mus. d'Hist. Nat.;

Douar Maghila v Douar Mghila.

Douar Mghila 32°20′ N., 6°18′ W.
Morocco
Fell 1932, August 20, or 22, 1500 hrs
Synonym(s): *Douar Maghila, Douar M'Guila*
Stone. Olivine-hypersthene chondrite, amphoterite (LL6).
About 40 stones fell near Douar Mghila; only two were recovered one of 846g and one of 315g. Description, A. Lacroix, C. R. Acad. Sci. Paris, 1933, **197**, p.368 [M.A. 5-403], E. Jérémine, Notes Mém. Serv. Geol. Maroc, 1951, **5**, p.201 [M.A. 12-610], includes an analysis. Olivine $Fa_{29.8}$, K. Fredriksson et al., Origin and Distribution of the Elements, ed. L.H. Ahrens, Pergamon, 1968, p.457.
643g Paris, Mus. d'Hist. Nat., the larger stone; Specimen, Rabat, Mus. de l'Inst.;
Specimen(s): [1966,281], 103.3g. and fragments, 3g

Douar M'Guila v Douar Mghila.

Dowa 13°40′ S., 33°55′ E.
Central province, Malawi
Fell 1976, March 25, 1300 hrs
Stone.
A stone of 1lb 6oz fell with a thunderous noise at Chinguno village, about 3 km NE. of Dowa. A second, smaller stone, fell at Moya village, 3 km north of Dowa. The present whereabouts of these stones is not known, M.J. Crow, Geol. Surv. Malawi, 1982 (report MJC/32).

Dowerin 31°12′ S., 117°4′ E.
Western Australia, Australia
Found 1932, before this date
Pseudometeorite. Iron.
Many small fragments were found near Dowerin, one fragment, (0.35g) in E.S. Simpson's collection, E.S. Simpson, Min. Mag., 1938, **25**, p.158. Discredited, J.R. de Laeter, J. Roy. Soc. West. Austr., 1973, **56**, p.65.

Doyleville 38°25′ N., 106°35′ W.
Gunnison County, Colorado, U.S.A.
Found 1887
Stone. Olivine-bronzite chondrite (H5).
One stone, 112g, in H.H. Nininger's collection in 1933, H.H. Nininger, The Mines Mag., Golden, Colorado, 1933, **23** (8), p.6 [M.A. 5-405]. Listed, H.H. and A.D. Nininger, The Nininger Collection of Meteorites, Winslow, Arizona, 1950, p.48. Olivine Fa_{20}, B. Mason, Geochimica et Cosmochimica Acta, 1963, **27**, p.1011.
34g Tempe, Arizona State Univ.; 17g Washington, U.S. Nat. Mus.;
Specimen(s): [1959,857], 46g.

Drake Creek 36°24′ N., 86°30′ W. approx.
Nashville, Sumner County, Tennessee, U.S.A.
Fell 1827, May 9, 1600 hrs
Synonym(s): *Davidson County, Nashville, Sumner County*
Stone. Olivine-hypersthene chondrite (L6), veined, brecciated.
Five stones, the largest of 11.5lb, were seen to fall, after the appearance of a cloud and detonations, B. Silliman, Am. J. Sci., 1830, **18**, p.378. Analysis, E.H. von Baumhauer, Ann. Phys. Chem. (Poggendorff), 1845, **66**, p.498. Olivine Fa_{25}, B. Mason, Geochimica et Cosmochimica Acta, 1963, **27**, p.1011.
2.2kg Leiden, Rijksmus. Geol. Min.; 1.1kg Harvard Univ.; 717g Tübingen, Univ.; 430g Yale Univ.; 152g Washington, U.S. Nat. Mus.; 119g Chicago, Field Mus. Nat. Hist.;
Specimen(s): [20200], 9.5g. two fragments; [14697], 8g.; [90265], 0.5g.; [96263], Thin section

Drazkov v Pribram.

Dresden (Kansas) 38°48′ N., 100°25′ W.
Decatur County, Kansas, U.S.A.
Found 1953, recognized 1965
Stone. Olivine-bronzite chondrite (H5).
One stone of 6.76kg was turned up by a bulldozer during building of a country road, Meteor. Bull., 1967 (40), Meteoritics, 1970, **5**, p.94. Olivine Fa_{18}, M.J. Frost, priv. comm.
3.7kg Mainz, Max-Planck-Inst.; 224g Copenhagen, Univ. Geol. Mus.; 212g Washington, U.S. Nat. Mus.; 156g Harvard Univ.; 94g Tempe, Arizona State Univ.;
Specimen(s): [1967,388], 96g. part-slice, and fragments, 2g.

Dresden (Ontario) 42°31′12″ N., 82°15′36″ W.
Ontario, Canada
Fell 1939, June 11, 2056 hrs
Stone. Olivine-bronzite chondrite (H6).
Three stones, totalling 47.7kg were recovered. The phenomena of fall are described, P.M. Millman, Publ. Am. Astron. Soc., 1940, **10**, p.23 [M.A. 8-192]. Further description, W.G. Colgrove, J. Roy. Astron. Soc. Canada, 1939, **33**, p.301, E.G. Pleva and W.G. Colgrove, J. Roy. Astron. Soc. Canada, 1939, **33**, p.303 [M.A. 7-543]. Date and time of fall given as 2049hrs, July 12, A.D. Nininger, Pop. Astron., Northfield, Minnesota, 1940, **48**, p.557, Contr. Soc. Res. Meteorites, **2**, p.227 [M.A. 8-54]. Mentioned, P.M. Millman, The Sky, New York, 1939, **3** (12), p.27. Olivine Fa_{20}, B. Mason, Geochimica et Cosmochimica Acta, 1963, **27**, p.1011.
1758g Ottawa, Mus. Geol. Surv. Canada; 27g Washington, U.S. Nat. Mus.;

Drevdalen v Trysil.

Drum Mountains 39°30′ N., 112°54′ W.
Millard County, Utah, U.S.A.
Found 1944, September 24
Iron. Octahedrite, medium (1.2mm) (IIIA).
A weathered mass of 529kg was found on the surface of basalt. Described, with an analysis, 8.59 %Ni, E.P. Henderson and S.H. Perry, Smithsonian Misc. Coll., 1948, **110** (12) [M.A. 10-520], Contr. Soc. Res. Meteorites, 1945, **3**, p.185. Analysis, 8.23 %Ni, 20.4 ppm.Ga, 41.8 ppm.Ge, 0.64 ppm.Ir, E.R.D. Scott et al., Geochimica et Cosmochimica

Acta, 1973, **37**, p.1957. Description, shock deformation, V.F. Buchwald, Iron Meteorites, Univ. of California, 1975, p.538.
522kg Washington, U.S. Nat. Mus., main mass; 1kg Chicago, Field Mus. Nat. Hist.; 500g Michigan Univ.; 652g Tempe, Arizona State Univ.; 550g Calcutta, Mus. Geol. Surv. India;

Duan 23°54′ N., 108°6′ E.

Guangxi, China
Found
Iron.
A single mass, weight not reported, was found, D. Bian, Meteoritics, 1981, **16**, p.122.
Main mass, Guiyang, Acad. Sin. Inst. Geochem.;

Dubrovnik 42°27′30″ N., 18°26′30″ E.

Molunat, Yugoslavia
Fell 1951, February 20, 1400 hrs
Synonym(s): *Herceg, Hercegnovi, Mocila, Molunat*
Stone. Olivine-hypersthene chondrite (L3-6), brecciated.
One mass of 1.9kg was found, M. Ramovic, Geol. Glasnik, 1958, **4**, p.273, Meteor. Bull., 1960 (18). The stone is a polymict breccia with some unequilibrated fragments, olivine Fa_{21-25}, G. Kurat et al., Fortschr. Min., 1974, **52**, p.44.
1.79kg Dubrovnik Mus., main mass; 44g Vienna, Naturhist. Mus.; 5g Sarajevo, Nat. Mus.;
Specimen(s): [1959,222], 7.1g.

Dubrow v Sagan.

Duchesen v Duchesne.

Duchesne 40°23′ N., 110°52′ W.

Duchesne County, Utah, U.S.A.
Found 1906
Synonym(s): *Duchesen, Duchesne County, Mount Tabby*
Iron. Octahedrite, fine (0.3mm.) (IVA).
A 50lb mass. Described, with an analysis, 9.20 %Ni, H.H. Nininger, J. Geol., 1929, **37**, p.83 [M.A. 4-118], The Mines Mag., Golden, Colorado, 1933, **23** (8), p.6 [M.A. 5-405]. Mount Tabby material is from Duchesne (Ni=9.37%, P=0.18%), but Altonah (*q.v.*) is almost certainly different, V.F. Buchwald, Iron Meteorites, Univ. of California, 1975, p.540. Further analysis, 9.32 %Ni, 2.19 ppm.Ga, 0.125 ppm.Ge, 0.42 ppm.Ir, R. Schaudy et al., Icarus, 1972, **17**, p.174. The Ir content is significantly lower than that of Altonah.
8.6kg Univ. of Utah; 2.78kg Washington, U.S. Nat. Mus.; 2.5kg Tempe, Arizona State Univ.; 1kg Chicago, Field Mus. Nat. Hist.; 1kg Harvard Univ.; 354g Paris, Mus. d'Hist. Nat.; 245g New York, Amer. Mus. Nat. Hist.;
Specimen(s): [1928,472], 366g.

Duchesne County v Duchesne.

Duck Creek v Mount Edith.

Duel Hill (1854) 35°51′ N., 82°42′ W.

Walnut Mts., Madison County, North Carolina, U.S.A.
Found 1854
Synonym(s): *Asheville, Jewel Hill, Jewell Hill, Madison County*
Iron. Octahedrite, fine (0.3mm) (IVA-ANOM).
Two masses of about 40lb and 8lb were found in 1856 and 1854 respectively, F.P. Venable, List of Meteorites of North Carolina, J. Elisha Mitchell Sci. Soc., 1890, **7**, p.44, J.L. Smith, Am. J. Sci., 1860, **30**, p.240. Analysis, 9.8 %Ni, E. Cohen, Meteoritenkunde, 1905, **3**, p.354. New analysis, 10.39 %Ni, 1.93 ppm.Ga, 0.111 ppm.Ge, 0.64 ppm.Ir, R. Schaudy et al., Icarus, 1972, **17**, p.174. Description, shocked, epsilon structure, V.F. Buchwald, Iron Meteorites, Univ. of California, 1975, p.543.
1.2kg Tempe, Arizona State Univ.; 138g Washington, U.S. Nat. Mus.; 139g Harvard Univ.; 104g Paris, Mus. d'Hist. Nat.; 99g Berlin, Humboldt Univ.; 75g New York, Amer. Mus. Nat. Hist.;
Specimen(s): [40877], 87g.; [32049], 43g.

Duel Hill (1873) 35°51′ N., 82°42′ W.

Walnut Mts., Madison County, North Carolina, U.S.A.
Found 1873
Synonym(s): *Jewel Hill, Madison County*
Iron. Octahedrite, coarse (2.4mm) (IA).
A mass of about 25lb was found on a hillside, F.P. Venable, List of Meteorites of North Carolina, J. Elisha Mitchell Sci. Soc., 1890, **7**, p.45. See also, with an analysis, 5.17 %Ni, B.S. Burton, Am. J. Sci., 1876, **12**, p.439. New analysis, 6.51 %Ni, B. Mason and H.B. Wiik, Geochimica et Cosmochimica Acta, 1965, **29**, p.1003. Description, V.F. Buchwald, Iron Meteorites, Univ. of California, 1975, p.544. Classification and analysis, 6.6 %Ni, 84 ppm.Ga, 426 ppm.Ge, 4.3 ppm.Ir, A. Kracher et al., Geochimica et Cosmochimica Acta, 1980, **44**, p.773.
4.2kg Yale Univ., main mass; 1.5kg New York, Amer. Mus. Nat. Hist.; 1.2kg Vienna, Naturhist. Mus.; 219g Harvard Univ.; 190g Chicago, Field Mus. Nat. Hist.;
Specimen(s): [55123], 12g.; [1974,M.21], 50g. polished part-slice in a resin mount, approx. weight.

Dugo Polje v Soko Banja.

Duketon 27°30′ S., 122°22′ E. approx.

Nuleri Land district, Western Australia, Australia
Found 1948, October
Synonym(s): *Bandhya, Bandya, Duketown*
Iron. Octahedrite, medium (1.0mm) (IIIA).
A mass of 260.7lb was found 10 miles N. of Duketon, which is 70 miles N. of Laverton, J.M. Lindsay, letter of 12 September, 1957. in Min. Dept. BM(NH). Mentioned, M.J. Frost, J. Roy. Soc. West. Austr., 1958, **41**, p.55, K. Keil, Fortschr. Min., 1960, **38**, p.252. Listed, Spec. Publ. West. Austr. Mus., 1965 (3), p.32. Analysis, 7.25 %Ni, M.J. Frost, J. Roy. Soc. West. Austr., 1965, **48**, p.128. Further analysis, 7.52 %Ni, 19.8 ppm.Ga, 38.1 ppm.Ge, 4 ppm.Ir, E.R.D. Scott et al., Geochimica et Cosmochimica Acta, 1973, **37**, p.1957.
Main mass, Perth, Univ. Western Australia; 505g Los Angeles, Univ. of California;
Specimen(s): [1982,M.4], 186g.

Duketown v Duketon.

Dumas (a) 35°54′ N., 101°54′ W. approx

Moore County, Texas, U.S.A.
Found 1956, February
Stone. Olivine-bronzite chondrite (H5), veined.
A weathered mass of 46.05kg was found at the side of a road 14 miles SW. of Dumas. Described and analysed, L. LaPaz, Meteoritics, 1956, **1**, p.470 [M.A. 13-362]. Olivine Fa_{16}, B. Mason, Geochimica et Cosmochimica Acta, 1967, **31**, p.1100.
120g Albuquerque, Univ. of New Mexico;

Dumas (b) 35°55′30″ N., 101°53′42″ W.
Moore County, Texas, U.S.A.
Found 1980
Stone. Olivine-bronzite chondrite (H6).
A single fragment of 2284g was found after being struck by a mowing machine. Petrographically distinct from Dumas (a), contains fewer chondrules, smaller metal grains, G.I Huss, letter of 7 December, 1982 in Min. Dept. BM(NH).
61g Chicago, Field Mus. Nat. Hist.;

Dünaburg v Lixna.

Duncanville 32°38′ N., 96°52′ W.
Dallas County, Texas, U.S.A.
Found 1961
Stone. Olivine-bronzite chondrite (H).
A mass of 17.8kg was found, B. Mason, Meteorites, Wiley, 1962, p.242. Olivine Fa_{19}, B. Mason, Geochimica et Cosmochimica Acta, 1963, **27**, p.1011.
9g New York, Amer. Mus. Nat. Hist.;

Dundrum 52°33′ N., 8°2′ W.
County Tipperary, Ireland
Fell 1865, August 12, 1900 hrs
Synonym(s): *Tipperary*
Stone. Olivine-bronzite chondrite (H5).
A stone of 4lb 14.5oz was seen to fall, after detonations; described and analysed, S. Haughton, Proc. Roy. Irish Acad., 1867, **9**, p.336. Olivine Fa_{19}, B. Mason, Geochimica et Cosmochimica Acta, 1963, **27**, p.1011.
Main mass, Dublin, Trinity College; 69g Yale Univ.; 18g Vienna, Naturhist. Mus.;
Specimen(s): [67745], 241g.

Dungannon 36°51′ N., 82°27′ W.
Scott County, Virginia, U.S.A.
Found 1922, or 1923
Synonym(s): *Nickelsville*
Iron. Octahedrite, coarse (2.0mm) (IA).
A mass of about 13kg, broken by a plough into two pieces of 5lb and 23lb respectively, was found on Copper Ridge, 3 miles SE. of Dungannon. Described, with an analysis by J.E. Whitfield, 7.44 %Ni, G.P. Merrill, Proc. U.S. Nat. Mus., 1923, **62** (18). Description, recrystallised, V.F. Buchwald, Iron Meteorites, Univ. of California, 1975, p.547. Analysis, 6.9 %Ni, 78.5 ppm.Ga, 330 ppm.Ge, 2.1 ppm.Ir, J.T. Wasson, Icarus, 1970, **12**, p.407.
7.2kg Chicago, Field Mus. Nat. Hist.; 1kg Washington, U.S. Nat. Mus.; 554g New York, Amer. Mus. Nat. Hist.; 144g Tempe, Arizona State Univ.;
Specimen(s): [1924,440], 75g. etched slice

Dunganville 42°33′ S., 171°21′ E.
Greymouth, Westland, New Zealand
Found 1976, January
Iron. Octahedrite, coarse.
A single mass and fragments of weathering crust, together totalling 54kg in weight, were found in a creek bed, Meteor. Bull., 1978 (55), Meteoritics, 1978, **13**, p.335.
Specimen(s): [1978,M.26], 182g.

Dungo Polje v Soko-Banja.

Dunhua 43°20′ N., 128°15′ E.
Dunhua County, Jilin, China
Fell 1976, July 9
Stone.
Listed, D. Bian, Meteoritics, 1981, **16**, p.118.
Specimen, Jilin, Bureau of Sci. and Tech.;

Dun-le-Poëlier v La Bécasse.

Durala 30°18′ N., 76°38′ E.
Patiala district, Punjab, India
Fell 1815, February 18, 1200 hrs
Synonym(s): *Dooralla, Surala*
Stone. Olivine-hypersthene chondrite (L6), veined.
A stone of about 29lb fell, after detonations, C. Bird, Phil. Mag., 1820, **56**, p.146. The stone is said to have fallen at Durala, 30° 21'N. 76°41'E., 17 miles from Ambala, Patiala State; there is no such village, and the place of fall is probably Surala kalan or Surala khurd, 30°18'N., 76°38'E., 13 miles SE. of Rajpura, Patiala State, C.A. Silberrad, Min. Mag., 1932, **23**, p.301. Olivine Fa_{25}, B. Mason, Geochimica et Cosmochimica Acta, 1963, **27**, p.1011.
82g Fort Worth, Texas, Monnig Colln.; 78g Chicago, Field Mus. Nat. Hist.; 59g Washington, U.S. Nat. Mus.; 42g Vienna, Naturhist. Mus.; 36g Harvard Univ.;
Specimen(s): [32097], 12.0kg. main mass, and pieces, 46.1g.

Durango
Durango, Mexico, Not known
Found 1804, probable date
Synonym(s): *Pseudo-Apoala*
Iron. Octahedrite, medium (1.2mm) (IIIA).
A mass of 164kg is listed, J.C. Haro, Bol. Inst. Geol. Mexico, 1931 (50), p.80. Weight given as 167kg and date of find as 1904, H.H. Nininger, Our Stone Pelted Planet, 1933. Distinct from Morito (*q.v.*). There is confusion over material labelled Durango from differing sources, co-ordinates of original location not known. Shocked and recrystallised, V.F. Buchwald, Iron Meteorites, Univ. of California, 1975, p.550. Some specimens labelled 'Apoala' are identical to Durango. Analysis, 8.00 %Ni, 20.4 ppm.Ga, 40.2 ppm.Ge, 1.0 ppm.Ir, E.R.D. Scott et al., Geochimica et Cosmochimica Acta, 1973, **37**, p.1957.
164kg Mexico City, Inst. Geol., main mass; 283g Chicago, Field Mus. Nat. Hist.; 145g Tempe, Arizona State Univ., labelled Apoala;
Specimen(s): [1959,950], 91.5g. and sawings, 1.5g; [1959,969], 208g. and sawings, 13.5g. Previously labelled Apoala

Durango (of P. Partsch) v Rancho de la Pila.

Duruma 4° S., 39°30′ E. approx.
Mombasa, Kenya
Fell 1853, March 6
Synonym(s): *Turuma*
Stone. Olivine-hypersthene chondrite (L6), veined.
A stone of 577g was seen to fall, after detonations, R.P. Greg, Phil. Mag., 1862, **24**, p.538, O. Büchner, Die Meteoriten in Samml., Leipzig, 1863, p.85. Olivine Fa_{25}, B. Mason, Geochimica et Cosmochimica Acta, 1963, **27**, p.1011. The main mass, 507g, formerly in the Munich Mus. was lost in a bombing raid in 1944.
60g Budapest, Nat. Mus.; 22g Vienna, Naturhist. Mus.; 4g Chicago, Field Mus. Nat. Hist.;
Specimen(s): [35730], less than 0.5g; [83672], thin section

Duwun 33°26′ N., 127°16′ E.
Kohung-gun, Choeure-namdo, Korea
Fell 1943, November 23, 1547 hrs
Stone. Olivine-hypersthene chondrite (L6).
A single stone of 2117g is reported to have fallen. Description, analysis, 22.8 % total iron, olivine $Fa_{23.4}$, H. Yabuki and M. Shima, Bull. Nat. Sci. Mus. Tokyo, Ser. E, 1980, **3**, p.1.

Dwalene v Dwaleni.

Dwaleni 27°12′ S., 31°19′ E.
near Nhlangano, Swaziland
Fell 1970, October 12, 1030 hrs
Synonym(s): *Dwalene*
Stone. Olivine-bronzite chondrite (H6), veined, brecciated.
Three fragments of 2.37kg, 510g and 350g were collected from an area of about 12 sq.km, 6km SE. of the village of Dwaleni, after explosions had been heard. The specimens were found buried to a depth of about 15 cm in moist soil, Meteor. Bull., 1971 (50), Meteoritics, 1971, **6**, p.113. Gas-rich, M.A. Reynolds et al., Trans. Amer. Geophys. Union, EOS, 1971, **52**, p.269, abs. Olivine $Fa_{19.5}$, R. Hutchison, priv. comm.
115g Washington, U.S. Nat. Mus.;
Specimen(s): [1971,96], 15.6g. crusted fragments

Dwight 38°51′ N., 96°35′ W.
Geary County, Kansas, U.S.A.
Found 1940
Stone. Olivine-hypersthene chondrite (L6).
A mass of 4.1kg was found, H.H. and A.D. Nininger, The Nininger Collection of Meteorites, Winslow, Arizona, 1950, p.49. Olivine Fa_{25}, B. Mason, Geochimica et Cosmochimica Acta, 1963, **27**, p.1011.
1.6kg Tempe, Arizona State Univ.; 435g Chicago, Field Mus. Nat. Hist.; 174g Washington, U.S. Nat. Mus.; 58g New York, Amer. Mus. Nat. Hist.;
Specimen(s): [1959,859], 1374g.

Dyalpur 26°15′ N., 82° E. approx.
Sultanpur district, Uttar Pradesh, India
Fell 1872, May 8
Synonym(s): *Oude, Pura Dayal, Purwa Dayalpur, Sultanpur*
Stone. Achondrite, Ca-poor. Ureilite (AURE).
A stone of about 10oz fell, F. Fedden, Cat. Meteor., Indian Mus., Calcutta, 1880, p.171. This stone is said to have fallen at Dyalpur, 26° 14'N., 82°12'E., approx., but there is no place of this name in the district; the place of fall is probably one of the hamlets Pura Dayal, Pura Dayal Singh, or Purwa Dayalpur, C.A. Silberrad, Min. Mag., 1932, **23**, p.209. Noble gas content and gas retention age, about 1600 m.y., E. Mazor et al., Geochimica et Cosmochimica Acta, 1970, **34**, p.781. Analysis, 11.31 % total iron, H.B. Wiik, Meteoritics, 1972, **7**, p.553. Discussion, G.P. Vdovykin, Space Sci. Rev., 1970, **10**, p.483. Texture, mineralogy, J.L. Berkley et al., Geochimica et Cosmochimica Acta, 1980, **44**, p.1579.
19.5g Moscow, Acad. Sci.; 14g Calcutta, Mus. Geol. Surv. India; 14g Vienna, Naturhist. Mus.; 4.5g Chicago, Field Mus. Nat. Hist.;
Specimen(s): [51185], 209g. and fragments, 1.5g

Dyarrl Island 3° S., 151° E.
New Ireland, Papua-New Guinea
Fell 1933, January 31, 1630 hrs
Synonym(s): *Djaul Island, New Ireland*
Stony-iron. Mesosiderite (MES).
A fragment weighing 6oz was recovered; description, T. Hodge-Smith, Austr. Mus. Mag., 1933, **5**, p.56 [M.A. 6-105]. Analysis and mineralogy, B. Mason and E. Jarosewich, Min. Mag., 1973, **39**, p.204. Belongs to sub-group 1, R.J. Floran et al., Geochimica et Cosmochimica Acta, 1978 (Suppl. 10), p.1083.
8g Washington, U.S. Nat. Mus.; Main mass, Sydney, Austr. Mus.;

Dzemajtkemis v Zemaitkiemis.

Eagle v Eagle Station.

Eagle Station 38°37′ N., 84°58′ W.
Carroll County, Kentucky, U.S.A.
Found 1880
Synonym(s): *Eagle, Carroll County*
Stony-iron. Pallasite (PAL).
A mass of about 80lb was found about 0.75 miles from Eagle Station, G.F. Kunz, Am. J. Sci., 1887, **33**, p.228, contains an analysis. Also analysed, G.T. Prior, Min. Mag., 1918, **18**, p.178. Analysis of metal, 15.37 %Ni, 4.54 ppm.Ga, 75.3 ppm.Ge, 10 ppm. Ir., J.T. Wasson, Meteorites, Springer-Verlag, 1974, p.307. Olivine Fa_{20}, P.R. Buseck and J.I. Goldstein, Bull. Geol. Soc. Amer., 1969, **80**, p.2141. References, belongs to the Eagle Station sub-group of pallasites, E.R.D. Scott, Geochimica et Cosmochimica Acta, 1977, **41**, p.693.
18kg Vienna, Naturhist. Mus., minimum weight; 2.2kg Paris, Mus. d'Hist. Nat.; 521g Budapest, Nat. Hist. Mus.; 345g Washington, U.S. Nat. Mus.; 290g Tempe, Arizona State Univ.; 245g Chicago, Field Mus. Nat. Hist.;
Specimen(s): [64407], 701.5g. slice

Easter v Summerfield.

Easter Essendy v Strathmore.

Eastman
Quebec, Canada
Doubtful..
Mentioned, without details, J.A. Douglas, Paper Geol. Surv. Canada, 1968 (1A), p.109.

East Mount Magnet v Mount Magnet.

East Norton 52°33′ N., 0°50′ W.
Leicestershire, England
Fell 1803, July 4
Doubtful..
A large stone is said to have fallen and done much damage, Phil. Mag., 1803, **16**, p.191, E.F.F. Chladni, Die Feuer-Meteore, Wien, 1819, p.272, but the evidence is not conclusive, T.M. Hall, Min. Mag., 1879, **3**, p.8.

East Tennessee v Cleveland.

East Tennessee v Cosby's Creek.

East Tennessee v Jonesboro.

East Tennessee v Morristown.

East Tennessee v Tazewell.

Eaton 40°31′ N., 104°41′ W.
Weld County, Colorado, U.S.A.
Fell 1931, May 10, 1715 hrs
Pseudometeorite..
A copper nugget weighing 29.457g is reported to have been observed to fall, H.H. Nininger, Pop. Astron., Northfield, Minnesota, 1943, **51**, p.273, Contr. Soc. Res. Meteorites, 1943, **3**, p.85 [M.A. 9-300], The Nininger Collection of Meteorites, Winslow, Arizona, 1950, p.49. This copper 'meteorite' is discredited, P.R. Buseck et al., Geochimica et Cosmochimica Acta, 1973, **37**, p.1249. It contains 33% Zn, small Pb inclusions and less than 0.1% Ni. It is similar to yellow brass, and distinct from native terrestrial copper and meteoritic copper, the latter having 0.4% to 2.4% Ni.

Eau Claire v Hammond.

Echigo v Yonozu.

Echo v Salt Lake City.

Ector County v Odessa (iron).

Ector County No. 3 v Odessa (iron).

Eddy County v Sacramento Mountains.

Edjudina 29°35′11″ S., 122°10′54″ E.

Edjudina Station, Leonora District, Western Australia, Australia
Found 1969
Synonym(s): *Ejudina*
Stone. Olivine-bronzite chondrite (H4).

One crusted mass of 4.48kg was found by the Geological Survey of Western Australia at Edjudina station, about 80 miles N. of Kalgoorlie, Meteor. Bull., 1975 (53), Meteoritics, 1975, **10**, p.144, R.A. Binns, letter of 19 April 1972, in Min. Dept. BM(NH).. Listed, G.J.H. McCall, 2nd. Suppl. to West. Austr. Mus. Spec. Publ. no. 3, 1972, p.12. Description, J.D. Lewis, Ann. Rep. Geol. Surv. West. Austr., 1971, p.62. Olivine Fa_{19}, B. Mason, Rec. Austr. Mus., 1974, **29**, p.169.

Main mass, Perth, Geol. Surv. West. Austr.; 10g Perth, West. Austr. Mus.; 0.8g Washington, U.S. Nat. Mus.;

Edmonson (a) 34°17′ N., 101°50′ W.

Hale County, Texas, U.S.A.
Found 1955, recognized 1965
Stone. Olivine-hypersthene chondrite (L6).

One fractured mass of 12kg was ploughed up in a field, Meteor. Bull., 1967 (40), Meteoritics, 1970, **5**, p.94.

5.8kg Mainz, Max-Planck-Inst.; 281g Washington, U.S. Nat. Mus.; 186g Chicago, Field Mus. Nat. Hist.; 141g Harvard Univ.; 110g Vienna, Naturhist. Mus.; 104g Tempe, Arizona State Univ.; 89g New York, Amer. Mus. Nat. Hist.;

Specimen(s): [1966,522], 102g. slice, and fragments, 4.9g

Edmonson (b) 34°17′ N., 101°50′ W. approx.

Hale County, Texas, U.S.A.
Found 1981, February
Stone. Olivine-bronzite chondrite (H4).

A mass of 14.402kg was brought to the surface by ploughing, Meteor. Bull., 1983 (61), Meteoritics, 1983, **18**, p.78.

914g Sedona, Arizona, Westcott Colln.; 204g Vienna, Naturhist. Mus.;

Specimen(s): [1982,M.16], 38g.

Edmonton (Canada) 53°35′ N., 113°30′ W.

Alberta, Canada
Found 1939, recognized in this year
Iron. Hexahedrite (IIA).

One mass of 7.34kg was found, A.D. Nininger, Pop. Astron., Northfield, Minnesota, 1940, **48**, p.557, Contr. Soc. Res. Meteorites, 1941, **2**, p.227 [M.A. 8-54]. Analysis, 5.37 %Ni, 60.4 ppm.Ga, 172 ppm.Ge, 33 ppm.Ir, J.T. Wasson, Meteorites, Springer-Verlag, 1974, p.299. Description, V.F. Buchwald, Iron Meteorites, Univ. of California, 1975, p.554.

Main mass, Edmonton, Univ. of Alberta; 584g Washington, U.S. Nat. Mus.;

Specimen(s): [1967,255], 65g. polished fragment

Edmonton (Kentucky) 37°2′ N., 85°38′ W.

Metcalfe County, Kentucky, U.S.A.
Found 1942
Iron. Octahedrite, fine (0.32mm) (IIICD-ANOM).

A rusted mass of 10.2kg was ploughed up. Described, with an analysis, 12.57 %Ni, E.P. Henderson and S.H. Perry, Smithsonian Misc. Coll., 1947, **107** (13) [M.A. 10-402]. Further analysis and determination of Ga, Pd and Au, 12.66 %Ni, E. Goldberg et al., Geochimica et Cosmochimica Acta, 1951, **2**, p.1. Classification and new analysis, 12.7 %Ni, 25.4 ppm.Ga, 34.6 ppm.Ge, 0.55 ppm.Ir, E.R.D. Scott et al., Geochimica et Cosmochimica Acta, 1973, **37**, p.1957. Description, V.F. Buchwald, Iron Meteorites, Univ. of California, 1975, p.555.

8.28kg Washington, U.S. Nat. Mus.; 163g Harvard Univ.; 151g Chicago, Field Mus. Nat. Hist.;

Edo v Hachi-oji.

Efremovka 52°30′ N., 77° E. approx.

Pavlodar district, Kazakhstan SSR, USSR, [Ефремовка]
Found 1962, July
Stone. Carbonaceous chondrite, type III (CV3).

A mass of 21kg was found, Meteor. Bull., 1962 (25). Analysis, 25.98 % total iron, L.G. Kvasha and V.Ya. Kharitonova, Метеоритика, 1966, **27**, p.153. Another analysis, 22.49 % total iron, E. Jarosewich and B. Mason, Geochimica et Cosmochimica Acta, 1969, **33**, p.411. Further analysis, 21.05 % total iron, L.H. Ahrens et al., Meteoritics, 1973, **8**, p.133. Classification, W.R. Van Schmus, Meteorite Research, ed. P.M. Millman, D. Reidel, Dordrecht-Holland, 1969, p.480.

19.6kg Moscow, Acad. Sci.; 269g Washington, U.S. Nat. Mus.; 185g New York, Amer. Mus. Nat. Hist.; 172g Chicago, Field Mus. Nat. Hist.;

Specimen(s): [1971,196], 54g.

Eggenfelden v Mässing.

Egvekinot 66°48′ N., 178°12′ E. approx.

Egvekinot, Chukotka, Magadan, USSR, [Эгвекинот]
Found 1970, July
Iron. Octahedrite, fine (0.2mm) (IIE).

Fragments, totalling 8-10kg, were found in alluvium in a stream valley. Two fragments, of 2275g and 1260g, were received by the Committee on Meteorites of the Academy of Sciences, Moscow in 1975. Description and analysis, 13.5 % Ni, N.I. Zaslavskaya et al., Метеоритика, 1977, **36**, p.53. Classification, further analysis, 13.3 %Ni, 19 ppm.Ga, 70 ppm.Ge, 0.05 ppm.Ir, A.A. Yavnel and G.M. Kolesov, Метеоритика, 1977, **36**, p.162.

7.0kg Moscow, Acad. Sci.;

Egyptian Meteorite v Sinai.

Ehole 17°18′ S., 15°50′ E. approx.

Cunene, Angola
Fell 1961, August 31, 1645 hrs, approx. time
Stone. Olivine-bronzite chondrite (H5).

Two masses fell at Ehole, near the village of Namacunda. One mass of 2.4kg was sent to tne Smithsonian Astrophysical Observatory, and 1.53kg from this mass was transferred to the U.S. National Museum, Meteor. Bull., 1962 (23). Analysis, 27.35 % total iron, E. Jarosewich, Geochimica et Cosmochimica Acta, 1966, **30**, p.1261. Olivine Fa_{19}, B. Mason, Geochimica et Cosmochimica Acta, 1963,

27, p.1011.
1.5kg Washington, U.S. Nat. Mus.; 75g Harvard Univ.;
Specimen(s): [1966,46], 30.5g.

Ehrenberg v Cañon Diablo.

Eibenstock v Steinbach.

Eichstädt 48°54′ N., 11°13′ E.

Middle Franconia, Bayern, Germany
Fell 1785, February 19, 1215 hrs
Synonym(s): *Eichstätt, False Wittens, Wittens, Wittmess*
Stone. Olivine-bronzite chondrite (H5).

A stone of about 3kg was seen to fall, after detonations, in the district of Wittmess, 5 miles from Eichstädt, J. Pickel, von Moll's Ann. Berg- u. Hüttenk, Salzburg, 1805, **3**, p.251. Described, with an analysis, C.W. Gümbel, Sitzungsber. Akad. Wiss. München, Math.-phys. Kl., 1878, **8**, p.25. Olivine Fa_{20}, B. Mason, Geochimica et Cosmochimica Acta, 1963, **27**, p.1011. The specimens previously in Münich and Budapest have been destroyed.

293g Zürich, Univ.; 130g Vienna, Naturhist. Mus.; 15g Berlin, Humboldt Univ.; 14g Chicago, Field Mus. Nat. Hist.; 5g New York, Amer. Mus. Nat. Hist.;
Specimen(s): [84188], 33g.; [54639], 14g.; [83671], thin section

Eichstätt v Eichstädt.

Eifel v Bitburg.

Ejudina v Edjudina.

Ekaterinoslav v Augustinovka.

Ekaterinoslav v Bachmut.

Ekaterinoslav v Pavlograd.

Ekaterinoslav v Verkhne Dnieprovsk.

Ekeby 56°2′ N., 13°0′ E.

Skåne, Sweden
Fell 1939, April 5, 0612 hrs, U.T.
Synonym(s): *Lund*
Stone. Olivine-bronzite chondrite (H4).

After three detonations, a stone of 3310g fell; a fragment of 26g fell eight metres away. Described, with an analysis, A. Hadding, Geol. För. Förh. Stockholm, 1940, **62**, p.148 [M.A. 8-57]. Mentioned, A. Corlin, Pop. Astron., Northfield, Minnesota, 1939, **47**, p.436. Olivine Fa_{19}, B. Mason, Geochimica et Cosmochimica Acta, 1963, **27**, p.1011.

2.5kg Lund, Univ.; 18g Washington, U.S. Nat. Mus.;

Ekh Khera 28°16′ N., 78°47′ E.

Bisauli tahsil, Budaun district, Uttar Pradesh, India
Fell 1916, April 5, 0230 hrs
Stone. Olivine-bronzite chondrite (H), veined.

After detonations and the appearance of a fireball on April 5, a stone of about 840g was found on April 21; described, H. Walker, Rec. Geol. Surv. India, 1916, **47**, p.276. Olivine Fa_{18}, B. Mason, Geochimica et Cosmochimica Acta, 1967, **31**, p.1100.

803g Calcutta, Mus. Geol. Surv. India; 1g New York, Amer. Mus. Nat. Hist.;
Specimen(s): [1974,M.17], 20.17g. part-crusted fragment

El Abipon v Campo del Cielo.

Elandsburg v Gibeon.

Elba 39°50′ N., 103°13′ W.

Washington County, Colorado, U.S.A.
Found 1966, recognized 1967
Stone. Olivine-bronzite chondrite (H5).

One half of an oriented individual was ploughed up. It weighed 4.174kg, Meteor. Bull., 1972 (51), Meteoritics, 1972, **7**, p.221.

2kg Mainz, Max-Planck-Inst.; 171g Denver, Mus. Nat. Hist.; 139g Chicago, Field Mus. Nat. Hist.; 105g Copenhagen, Univ. Geol. Mus.; 94g Washington, U.S. Nat. Mus.; 61g Los Angeles, Univ. of California;
Specimen(s): [1971,117], 51.2g. and a fragment, 0.4g

Elberton v Smithonia.

Elbogen 50°11′ N., 12°44′ E.

Zapadocesky, Czechoslovakia
Fell 1400, possible date, recognized 1811 or 1812
Synonym(s): *Bewitched Burgrave, Burggraf, Ellbogen, Loket*
Iron. Octahedrite, medium (0.75mm) (IID).

A mass of about 107kg was preserved for centuries at the Rathhaus of Elbogen and was known as 'the bewitched burgrave'; mentioned in 1785, J. Schaller, Topographie Böhmen, 1785, **2**, p.6. Recognized as meteoritic, K.A. Neumann, Ann. Phys. (Gilbert), 1812, **42**, p.197. Analysis, 8.52 %Ni, J.J. Berzelius, Ann. Phys. Chem. (Poggendorff), 1834, **33**, p.135. Part, if not all, of the mass has been artificially heated and some forged; description, history, V.F. Buchwald, Iron Meteorites, Univ. of California, 1975, p.557. For notes on the history of the mass, K. Tucek, Casopis Národního Musea, Praha, 1947, **116**, p.1 [M.A. 10-398]. Analysis, 10.2 %Ni, 74.5 ppm.Ga, 87 ppm.Ge, 14 ppm.Ir, J.T. Wasson, Meteorites, Springer-Verlag, 1974, p.301.

79kg Vienna, Naturhist. Mus.; 14kg Elbogen, Rathaus; 6kg Prague, Univ.; 6.7kg Prague, Nat. Mus.; 323g Budapest, Nat. Hist. Mus.; 225g Berlin, Humboldt Univ.; 169g Harvard Univ.; 146g Tübingen, Univ.; 94g Washington, U.S. Nat. Mus.; 60g Chicago, Field Mus. Nat. Hist.; 46g New York, Amer. Mus. Nat. Hist.;
Specimen(s): [90219], 55g.; [33201], 37g.; [1922,162], 16.5g.

El Burro 29°20′ N., 101°50′ W.

Coahuila, Mexico
Found 1939
Iron. Octahedrite, coarsest (10mm) (IIB).

A 79lb mass was found near El Burro hills. Described, with an analysis, 6.02 %Ni, E.P. Henderson, Am. Miner., 1941, **26**, p.655 [M.A. 8-376]. A mass that was in H.H. Nininger's collection said to come from 'El Burro, Val Verde County, Texas', H.H. and A.D. Nininger, The Nininger Collection of Meteorites, Winslow, Arizona, 1950, p.49, is clearly part of this fall. The El Burro range is just across the Rio Grande from Val Verde County. Classification and new analysis, 5.95 %Ni, 58.1 ppm.Ga, 167 ppm.Ge, 0.059 ppm.Ir, J.T. Wasson, Meteorites, Springer-Verlag, 1974, p.300. Description, V.F. Buchwald, Iron Meteorites, Univ. of California, 1975, p.560.

34kg Washington, U.S. Nat. Mus., main mass; 1kg Tempe, Arizona State Univ.; 315g New York, Amer. Mus. Nat. Hist.;
Specimen(s): [1959,927], 676.5g. and sawings, 24.5g.

El Capitan 33°30′ N., 105°30′ W. approx.
Lincoln County, New Mexico, U.S.A.
Found 1893
Synonym(s): *Capitan Range*
Iron. Octahedrite, medium (1.1mm) (IIIB).
A mass of about 27.5kg was found in 1893, and may have fallen in July 1882, as a fireball was seen to fall behind the El Capitan Range; described, with an analysis by H.N. Stokes, 8.40 %Ni, E.E. Howell, Am. J. Sci., 1895, **50**, p.253. New analysis, 8.56 %Ni, 21.5 ppm.Ga, 45.1 ppm.Ge, 0.11 ppm.Ir, E.R.D. Scott et al., Geochimica et Cosmochimica Acta, 1973, **37**, p.1957. The main mass was in the Howell collection in 1895. Heavily corroded, not a recent fall; description, V.F. Buchwald, Iron Meteorites, Univ. of California, 1975, p.562.
2.1kg Chicago, Field Mus. Nat. Hist.; 6.3kg New York, Amer. Mus. Nat. Hist.; 4.8kg Washington, U.S. Nat. Mus.; 540g Harvard Univ.; 495g Tübingen, Univ.; 306g Berlin, Humboldt Univ.; 252g Yale Univ.; 100g Tempe, Arizona State Univ.;
Specimen(s): [80683], 956g. slice

El Chañaralino v Merceditas.

El Chiflón 29°26′ S., 66°50′ W. approx.
La Rioja, Argentina
Found
Iron?. Hexahedrite?.
Listed, without details, L.M. Villar, Cienc. Investig., 1968, **24**, p.302.

Elden v Cañon Diablo.

Eldorado County v Shingle Springs.

El Dorado County v Shingle Springs.

Eldorado Creek v Klondike (Skookum Gulch).

Elenovka 47°50′ N., 37°40′ E.
Donetsk region, Ukraine, USSR, [Еленовка]
Fell 1951, October 17, 1630 hrs
Synonym(s): *Jelenowka, Yelenovka*
Stone. Olivine-hypersthene chondrite (L5).
Eight stones and fragments fell near Elenovka railway station, and 54.64kg were collected, S.S. Fonton, Метеоритика, 1954 (11), p.169, L.G. Kvasha, Метеоритика, 1954, **11**, p.76, E.L. Krinov, Метеоритика, 1955, **12**, p.29 [M.A. 13-51]. Analysis, 21.01 % total iron, G.V. Baryshnikova and A.K. Lavrukhina, Метеоритика, 1979, **38**, p.37. Olivine Fa_{25}, B. Mason, Geochimica et Cosmochimica Acta, 1963, **27**, p.1011. K-Ar age, 4310-4670 m.y., P. Bochsler et al., Meteorite Research, ed. P.M. Millman, D. Reidel, Dordrecht-Holland, 1969, p.857.
43kg Moscow, Acad. Sci.; 379g Tempe, Arizona State Univ.; 219g Washington, U.S. Nat. Mus.;
Specimen(s): [1956,166], 263g.

Elephant Moraine A79001 76°15′ S., 156°30′ E.
Near Reckling Peak, Antarctica
Found 1979, between Dec. 1979 and Jan. 1980
Stone. Achondrite, Ca-rich. Eucrite (shergottite) (AEUC).
A single mass of 794.2g was recovered. An unbrecciated shergottite consisting of clinopyroxene, Fs25-38, maskelynite (An60-55), and rare olivines (Fo77-73), Antarctic Meteorite Newsletter, 1981, **4**, p.133. Petrography, analysis, H.Y. McSween and E. Jarosewich, Geochimica et Cosmochimica Acta, 1983, **47**, p.1501. This is one of about twelve specimens recovered from area named Elephant Moraine during the 1979-1980 field season in Antarctica. See following entries.

Elephant Moraine A79002
Stone. Achondrite, Ca-poor. Diogenite (ADIO).
A single mass of 2843g was found, o-pyroxene Fs22, olivine $Fa_{24.5}$, Antarctic Meteorite Newsletter, 1981, **4**, p.134.

Elephant Moraine A79003
Stone. Olivine-hypersthene chondrite (L6).
A mass of 435.6g was found, olivine Fa_{24}.

Elephant Moraine A79004
Stone. Achondrite, Ca-rich. Eucrite (AEUC).
A single mass of 390g was recovered.

Elephant Moraine A79005
Synonym(s): *Elephant Moraine A79011*
Stone. Achondrite, Ca-rich. Eucrite (AEUC), polymict.
A mass of 450g was found. Includes Elephant Moraine A79011, 86.4g, U.B. Marvin and B. Mason, Smithson. Contrib. Earth Sci., 1982 (24), p.42.

Elephant Moraine A79006
Synonym(s): *Elephant Moraine 82600*
Stone. Achondrite, Ca-rich. Eucrite (AEUC), becciated.
A single mass of 716g was recovered. Classification, petrology, S.B. Simon et al., Meteoritics, 1982, **17**, p.149.

Elephant Moraine A79007
Stone. Olivine-bronzite chondrite (H5).
A mass of 199.9g was found, olivine Fa_{18}.

Elephant Moraine A79009
Stone. Olivine-hypersthene chondrite (L5).
A stone of 140g was found, olivine Fa_{24}.

Elephant Moraine A79010
Stone. Olivine-hypersthene chondrite (L6).
A stone of 287.3g was found, olivine Fa_{24}.

Elephant Moraine A79011 v Elephant Moraine A79005.

Elephant Moraine 82600 v Elephant Moraine A79006.

Elephant Moraine 82602
Antarctica
1982, between December 1982 and January 1983
Stone. Olivine-bronzite chondrite (H4).
A single mass of 1824g was found by an American party during the 1982-1983 field season in Antarctica, Antarctic Meteorite Newsletter, 1984, **7** (no.1), p.17.

Elga 64°42′ N., 141°12′ E. approx.
Yakutia, Federated SSR, USSR, [Эльга]
Found 1959, August 28
Iron. Octahedrite with silicate inclusions (IIE).
One mass of 28.8kg was found, Meteor. Bull., 1960 (16). Coordinates, E.L. Krinov, Метеоритика, 1962, **22**, p.125. Classification, analysis, 7.98 %Ni, 24.1 ppm.Ga, 72.4 ppm.Ge, 4 ppm.Ir, E.R.D. Scott et al., Geochimica et Cosmochimica Acta, 1973, **37**, p.1957. Inclusion mineralogy, Eu.G. Osadchii et al., Geochimica et Cosmochimica Acta, 1981 (suppl. 16), p.1049.

10.5kg Novosibirsk, Siberian Branch, Acad. Sci.; 790g Moscow, Acad. Sci.;

El Garganitello v Tomatlan.

Elgueras v Cangas de Onis.

El Hacha v Campo del Cielo.

Elida 33°47′ N., 103°34′36″ W.
Roosevelt County, New Mexico, U.S.A.
Found 1968
Stone. Chondrite.
Three Elida meteorites are listed; Elida (a) of 936.8g, Elida (b) of 537.8g (found at 33°55′N., 103°31′6″W.) and Elida (c) of 385.5g (found at 33°47′24″N., 103°33′48″W.), Cat. Huss Coll. Meteorites, 1976, p.16. Elida (c) is an H5 chondrite, D.E. Lange et al., Cat. Meteor. Univ. New Mexico, 1980, p.18.
Main masses of all three stones, Mainz, Max-Planck-Inst.; 2.8g Albuquerque, Univ. of New Mexico;

Eli Elwah 34°30′ S., 144°43′ E.
Hay, New South Wales, Australia
Found 1888
Synonym(s): *Hay*
Stone. Olivine-hypersthene chondrite (L6).
A mass of about 33.5lb was found 15 miles W. of Hay, H.C. Russell, J. and Proc. Roy. Soc. New South Wales, 1888 (1889), **22**, p.341. Analysis, A. Liversidge, J. and Proc. Roy. Soc. New South Wales, 1903, **36**, p.356. Amount and composition of Ni-Fe determined, G.T. Prior, Min. Mag., 1919, **18**. Olivine Fa_{23}, B. Mason, Geochimica et Cosmochimica Acta, 1963, **27**, p.1011.
170g Harvard Univ.; 84g Chicago, Field Mus. Nat. Hist.; 55g Washington, U.S. Nat. Mus.; 29g Tempe, Arizona State Univ.;
Specimen(s): [1908,284], 13.29kg. and two pieces, 101g; [1927,1273], 41g.; [1927,1275], 46g.; [1927,1274], 5g.

El Inca v Tamarugal.

Elisabethpol v Indarch.

Elisabethpol v Mighei.

Elizavetpol v Mighei.

Elkhart 37°1′ N., 101°53′ W.
Morton County, Kansas, U.S.A.
Found 1936
Stone. Olivine-bronzite chondrite (H5).
A mass of 573g was found, H.H. and A.D. Nininger, The Nininger Collection of Meteorites, Winslow, Arizona, 1950, p.49. Olivine Fa_{18}, B. Mason, Geochimica et Cosmochimica Acta, 1963, **27**, p.1011.
422g Tempe, Arizona State Univ., main mass;
Specimen(s): [1959,858], 110g. and fragments, 3.5g

Ella Island 72°53′ N., 25°7′ W.
Solitairebught, Greenland
Found 1971
Stone. Olivine-hypersthene chondrite (L6).
A stone of about 3kg was found on a glacial surface 600m NW. of Ella Island station, north Greenland. A 715g fragment from the mass was donated to the Department of Geology, Univ. of Iowa. Description, microprobe analyses of minerals, olivine Fa_{23}, J.H. Carman and G.R. McCormick, Meteoritics, 1975, **10**, p.1. A second stone, 4.5kg, was found in 1975, V.F. Buchwald and P. Graff-Petersen, Analecta geol., 1976 (4), p.33.
5.7kg Copenhagen, Univ. Geol. Mus.; 157g Chicago, Field Mus. Nat. Hist.;
Specimen(s): [1976,M.10], 132g. from the second stone

Ellbogen v Elbogen.

Ellemeet 51°45′ N., 4° E. approx.
Isle of Schouwen, Zeeland, Holland
Fell 1925, August 28, 1130 hrs
Stone. Achondrite, Ca-poor. Diogenite (ADIO).
After the appearance of a fireball and detonation, at least two stones fell, one of 970g and the other of 500g. Described, with an analysis, W. Nieuwenkamp, Versl. Afdeel. Natuurk. Konink. Akad. Wetensch. Amsterdam, 1927, **36**, p.625 [M.A. 3-393]. The 500g stone was left where it fell and in 1927 had decomposed to a powdery mass.
918g Utrecht, Univ. Min. Geol. Inst.; Thin section, Washington, U.S. Nat. Mus.;
Specimen(s): [1927,984], 15g.

Ellenboro v Colfax.

Ellerslie 28°54′ S., 146°46′ E.
Tego, Maranoa, Queensland, Australia
Found 1905, known in this year
Stone. Olivine-hypersthene chondrite (L5).
One stone of 22.5lb was found about 80 miles N. of Bourke, New South Wales, just across the Queensland border, T. Hodge-Smith, Mem. Austr. Mus., 1939 (7), p.17 [M.A. 7-380]. Olivine Fa_{25}, B. Mason, Geochimica et Cosmochimica Acta, 1963, **27**, p.1011.
Main mass, Melbourne, Nat. Mus.; 28g Washington, U.S. Nat. Mus.;

Ellicott 38°48′28″ N., 104°34′10″ W.
El Paso County, Colorado, U.S.A.
Found 1960, between 1960 and 1964, recognized 1973
Iron. Octahedrite, medium (0.78mm) (IA).
Two masses of about 19lb and 15.5lb, (totalling 15.7kg), were found in a dry creek bed, R.M. Pearl, letter of 11 September 1973, in Min. Dept. BM(NH)., Meteor. Bull., 1975 (53), Meteoritics, 1975, **10**, p.145. Distinct from Franceville (*q.v.*), although found only 1.2km away; description and analysis, 8.0 %Ni, 57 ppm.Ga, 252 ppm.Ge, 3.4 ppm.Ir, E.J. Olsen et al., Meteoritics, 1974, **9**, p.263.
15kg In the possession of Prof. R.M. Pearl Colorado Springs, both masses; 534g Chicago, Field Mus. Nat. Hist.; 447g Mainz, Max-Planck-Inst.;
Specimen(s): [1976,M.1], 61.8g. part-slice

Ellis County 38°47′ N., 99°20′ W. approx.
Kansas, U.S.A.
Found 1948, or 1949
Stone. Olivine-bronzite chondrite (H6).
A single mass of 4.692kg was ploughed up about 6.5 miles south of Hays, Ellis County, E.J. Olsen and G.I Huss, Earth Sci., 1976 (September-October). Reported, Meteor. Bull., 1979 (56), Meteoritics, 1979, **14**, p.166.
348g Hanau, Zeitschel Colln.; 116g Chicago, Field Mus. Nat. Hist.;
Specimen(s): [1981,M.6], 267g. slice

Ellisras 23°50′ S., 27°55′ E.
Near Ellisras village, Transvaal, South Africa
Found 1970, approx., recognized 1981
Iron. Octahedrite, medium.
A single mass of 2.066kg was found on Ventershoek 579 LQ farm, about 27 km SE. of Ellisras village. A similar but larger specimen was reported to have been found in the same area two years earlier but its present whereabouts are not known, Meteor. Bull., 1983 (61), Meteoritics, 1983, **18**, p.79.
2kg Pretoria, Univ. Geol. Dept.;

El Marplatense 37°59′ S., 57°38′ W.
Buenos Aires province, Argentina
Pseudometeorite..
A mass reported to have fallen on a garage roof on November 16, 1947 is a sedimentary rock, M.E. Teruggi, Rev. Mus. Munic. Cienc. Nat. Mar del Plata, 1957, **1** (3), L.O. Giacomelli, letters of 30 March, 25 June, 30 October 1962, in Min. Dept. BM(NH).

El Mataco v Campo del Cielo.

Elm Creek 38°30′ N., 96°12′ W. approx.
Admire, Lyon County, Kansas, U.S.A.
Found 1906
Stone. Olivine-bronzite chondrite (H4).
A mass of 7kg was ploughed up 3 miles NNE. of Admire, K.S. Howard, Am. J. Sci., 1907, **23**, p.379. Described, with an analysis, G.P. Merrill, Mem. Nat. Acad. Sci. Washington, 1916, **14** (Mem. 1), p.15. The analysis shows no alkalies and is doubtful, W. Wahl, Geochimica et Cosmochimica Acta, 1950, **1**, p.28. Olivine Fa_{18}, B. Mason, Geochimica et Cosmochimica Acta, 1963, **27**, p.1011. Gas-rich, H. Hintenberger et al., Z. Natur., 1965, **20A**, p.983.
1kg Washington, U.S. Nat. Mus.; 764g Chicago, Field Mus. Nat. Hist.; 511g Vienna, Naturhist. Mus.; 485g Harvard Univ.; 324g Berlin, Humboldt Univ.; 259g New York, Amer. Mus. Nat. Hist.; 65g Yale Univ.; 53g Tempe, Arizona State Univ.;
Specimen(s): [1907,444], 845g. and fragments, 25g.

El Mirage 33°41′ N., 112°16′ W.
Maricopa County, Arizona, U.S.A.
Found 1972
Iron. Hexahedrite (IIA).
A mass of iron, 598g, was found in the desert at approximately Sect. 16, township 4N, range 1E., Meteor. Bull., 1975 (53), Meteoritics, 1975, **10**, p.145. Classification and analysis, 5.63 %Ni, 57.7 ppm.Ga, 185 ppm.Ge, 6.1 ppm.Ir, A. Kracher et al., Geochimica et Cosmochimica Acta, 1980, **44**, p.773.
548g Tempe, Arizona State Univ., main mass; 23g Chicago, Field Mus. Nat. Hist.;

Elmo v Joe Wright Mountain.

El Mocovi v Campo del Cielo.

El Morito v Morito.

El Nakhla el Baharia v Nakhla.

El Paso 31°47′ N., 106°14′ W. approx.
El Paso County, Texas, U.S.A.
Found 1950
Stone. Olivine-hypersthene chondrite, amphoterite (LL4).
A single stone of 276g was found on the surface among sand dunes about 40km E. of El Paso. The composition of the olivine is consistent with classification as an amphoterite; description and analysis, olivine Fa_{28}, W.S. Strain and J.M. Hoffer, Meteoritics, 1971, **6**, p.15.
Main mass, El Paso, Univ. of Texas;

El Patio v Campo del Cielo.

El Perdido 38°41′ S., 61°6′ W.
between Irene and Dorrego, Bahia Blanca, Argentina
Found 1905
Synonym(s): *Buenos Aires*
Stone. Olivine-bronzite chondrite (H5).
A stone of about 30.25kg was ploughed up, M. Kantor, Rev. Mus. La Plata, 1920, **25**, p.97, Centralblatt Min., 1906, p.716. Has been considered identical with Indio Rico,(*q.v.*), but is distinct, E. Fossa-Mancini, Notas Mus. La Plata, 1947, **12** (geol. no. 45), p.109 [M.A. 10-403]. Olivine Fa_{19}, B. Mason, Geochimica et Cosmochimica Acta, 1963, **27**, p.1011.
25kg La Plata, Mus., main mass; 700g New York, Amer. Mus. Nat. Hist.; 221g Vienna, Naturhist. Mus.; 77g Harvard Univ.; 39g Washington, U.S. Nat. Mus.;

El Perdido v Campo del Cielo.

El Qoseir 26°17′ N., 34°15′ E.
Egypt
Found 1921
Synonym(s): *Kosseir*
Iron. Ataxite, Ni-rich (IRANOM).
A mass of 2405g was picked up 15km S. of Qoseir (=Kosseir), on the Red Sea coast. Described, with an analysis, 15.17 %Ni, E.L. Krinov, C. R. (Doklady) Acad. Sci. URSS, 1945, **47**, p.497 [M.A. 9-298], Каталог Метеоритов Акад. Наук СССР, Москва, 1947 [M.A. 10-511]. Chemically anomalous, analysis, 13.19 %Ni, 6.15 ppm.Ga, 11.7 ppm.Ge, 5.5 ppm.Ir, J.T. Wasson and R. Schaudy, Icarus, 1971, **14**, p.59.
2290g Moscow, Acad. Sci., main mass;

El Quemado v Acapulco.

El Ranchito v Bacubirito.

El Rancho Grande 37° N., 105° W. approx.
Colfax County, New Mexico, U.S.A.
Found 1954
Stony-iron. Pallasite (PAL).
Listed, L. LaPaz, Cat. Coll. Inst. Meteor. Univ. New Mexico, 1965. Co-ordinates from unpublished list of J.T. Wasson, December 1971.
7.9g Albuquerque, Univ. of New Mexico;

El Rodeo v Rodeo.

El Rosario v Campo del Cielo.

El Sampal 44°32′ S., 70°22′ W.
Nueva Lubuka, Chubut, Argentina
Found 1973
Iron. Octahedrite, medium (0.95mm) (IIIA).
An individual of 142kg was found, partly buried in soil, about 20km E. of the town of Nueva Lubuka, Meteor. Bull., 1975 (53), Meteoritics, 1975, **10**, p.146. Brief description and analysis, 8.9 %Ni, 0.48 %Co, 0.28 %P, C.F. Lewis et al., Meteoritics, 1974, **9**, p.365, abs. Classification and analysis, 8.8 %Ni, 20.0 ppm.Ga, 39.6 ppm.Ge, 0.58 ppm.Ir, A. Kracher et al., Geochimica et Cosmochimica Acta, 1980, **44**, p.773.
142kg Tempe, Arizona State Univ., main mass; 2.25kg Chicago, Field Mus. Nat. Hist.;

Elsass v Ensisheim.

El Simbolar 30°38′ S., 64°53′ W.
Cruz del Eje, Cordoba, Argentina
Found 1938
Iron. Octahedrite, coarse.
A mass of 40kg was found; described, with an analysis, 6.29 %Ni, J. Olsacher, Bol. Fac. Cienc. Univ. Nac. Cordoba, 1939, **2**, p.79 [M.A. 12-255].
Specimen(s): [1962,177], 0.5g.

Elsinora 29°27′ S., 143°36′ E.
Delalah County, New South Wales, Australia
Found 1922
Stone. Olivine-bronzite chondrite (H5), veined.
A stone of about 12 × 6 inches was found on Elsinora station; two fragments, totalling 1797g have been preserved. Described and analysed, T. Hodge-Smith, Rec. Austr. Mus., 1929, **17**, p.50 [M.A. 4-262]. Olivine Fa_{17}, B. Mason, Geochimica et Cosmochimica Acta, 1963, **27**, p.1011. One of the two Nardoo stones (*q.v.*) is possibly part of the Elsinora fall.
1028kg Sydney, Austr. Mus., main mass; 87g Washington, U.S. Nat. Mus.;
Specimen(s): [1930,440], 290g.

El Taco v Campo del Cielo.

El Timbu 33°7′ S., 60°58′ W.
San Lorenzo department, Santa Fe, Argentina
Found 1942
Synonym(s): *Santa Fe I*
Iron.
A mass of approximately 500kg was found, L.O. Giacomelli, letters of 30 March, 1960 25 June and 30 October 1962, in Min. Dept. BM(NH)., Meteor. Bull., 1962 (24).
Main mass, Buenos Aires, Dir. Geol. Min.;

El Toba v Campo del Cielo.

Elton 33°43′ N., 100°50′ W.
Dickens County, Texas, U.S.A.
Found 1936, approx., recognized 1938
Iron. Anomalous (IRANOM).
A mass of 1.9kg was found, A.D. Nininger, Pop. Astron., Northfield, Minnesota, 1939, **47**, p.212, E.P. Henderson, letter of 3 June 1939, in Min. Dept. BM(NH). Classification and analysis, 6.97 %Ni, 46.3 ppm.Ga, 165 ppm.Ge, 0.053 ppm.Ir, J.T. Wasson, Icarus, 1970, **12**, p.407.
Main mass, Austin, Univ. of Texas; Specimen, Houston, Johnson Space Center;

El Tonocoté v Campo del Cielo.

El Tlahi v Hedjaz.

El Tule v Tule.

Elwell v Johnstown.

Elyria 38°16′48″ N., 97°21′54″ W.
McPherson County, Kansas, U.S.A.
Found 1971, recognized in this year
Iron. Octahedrite, medium (IIIA).
One mass of 10.9kg, encased in a 4-5mm layer of oxide, was ploughed up, Meteor. Bull., 1974 (52), Meteoritics, 1974, **9**, p.110. Mentioned, G.I Huss, letter of 16 March, 1972, in Min. Dept. BM(NH).
5.4kg Mainz, Max-Planck-Inst.; 299g Fort Worth, Texas, Monnig Colln.; 294g Copenhagen, Univ. Geol. Mus.; 234g Washington, U.S. Nat. Mus.; 146g Chicago, Field Mus. Nat. Hist.; 66g New York, Amer. Mus. Nat. Hist.;
Specimen(s): [1972,134], 145.6g. crusted part-slice.

Emery 43°33′48″ N., 97°35′0″ W.
Hanson County, South Dakota, U.S.A.
Found 1962, approx., recognized 1968
Stony-iron. Mesosiderite (MES).
One mass of 16.7kg was found in a pile of stones cleared from a field. It lay in a farmyard from about 1962 until 1968, G.I. Huss, letter of 7 January, 1971, in Min. Dept. BM(NH)., Meteor. Bull., 1972 (51), Meteoritics, 1972, **7**, p.222. Analysis and mineralogy, B. Mason and E. Jarosewich, Min. Mag., 1973, **39**, p.204. Analysis of metal, 7.0 %Ni, 11.5 ppm.Ga, 47.1 ppm.Ge, 2.4 ppm.Ir, J.T. Wasson et al., Geochimica et Cosmochimica Acta, 1974, **38**, p.135. Olivine coronas, composition, references, C.E. Nehru et al., Geochimica et Cosmochimica Acta, 1980, **44**, p.1103.
7.6kg Mainz, Max-Planck-Inst.; 1kg Washington, U.S. Nat. Mus.; 430g Chicago, Field Mus. Nat. Hist.; 80g Los Angeles, Univ. of California;
Specimen(s): [1970,310], 139.3g. slice

Emesa 34°44′ N., 36°42′ E.
Homs, Syria
Doubtful..
The sacred object at Emesa (the modern Homs) was probably meteoritic, G.A. Wainwright, Ann. Serv. Antiq. Egypte, 1929, **28**, p.175, Palestine Explor. Fund, quarterly statement, 1934 (January), p.32, but the evidence is not conclusive.

Emir v Krasnojarsk.

Emmaville 29°28′ S., 151°37′ E.
County Gough, New South Wales, Australia
Fell 1900
Stone. Achondrite, Ca-rich. Eucrite (AEUC).
A stone of 127g was found, C. Anderson, Rec. Austr. Mus., 1913, **10**, p.58. Only 99g preserved, T. Hodge-Smith, Mem. Austr. Mus., 1939 (7), p.17. Analysis, B. Mason et al., Smithson. Contrib. Earth Sci., 1979 (22), p.27.
99g Sydney, Austr. Mus.; Thin-section, Washington, U.S. Nat. Mus.;
Specimen(s): [1968,188], two polished thin-sections, P293 and P294

Emmet County v Estherville.

Emmetsburg v Emmitsburg.

Emmitsburg 39°43′ N., 77°18′ W.
Frederick County, Maryland, U.S.A.
Found 1854
Synonym(s): *Emmetsburg, Frederick County, Maryland*
Iron. Octahedrite, medium (1.0mm) (IIIA).
A mass of about 1lb was found, and passed into the possession of Dr. J.R. Chilton of New York, from whom S.C.H. Bailey obtained specimens, S.C.H. Bailey, letter of 7 January, 1885, in Min. Dept. BM(NH). Description, some specimens attributed to this fall are mis-labelled, V.F. Buchwald, Iron Meteorites, Univ. of California, 1975, p.564.
31g New York, Amer. Mus. Nat. Hist.; 41g Chicago, Field Mus. Nat. Hist.; 10g Harvard Univ.; 9g Vienna, Naturhist. Mus.; 7g Washington, U.S. Nat. Mus.;
Specimen(s): [56158], 6.5g. obtained from S.C.H. Bailey

Emsland 53°6′ N., 7°12′ E.
Niedersachsen, Germany
Found 1940
Iron. Octahedrite, medium (0.90mm) (IRANOM).
A mass of 19kg was found near the village of Brahe on the Ems. One side shows shows natural Widmanstetter figures, evidently due to etching by soil acids. Described, with an analysis, 8.71 %Ni, R. Vogel, Chem. Erde, 1943, **15**, p.52 [M.A. 9-292]. Further analysis, 9.40 %Ni, 2.90 ppm.Ga, 35.0 ppm.Ge, 2.9 ppm.Ir, E.R.D. Scott et al., Geochimica et Cosmochimica Acta, 1973, **37**, p.1957. Description, V.F. Buchwald, Iron Meteorites, Univ. of California, 1975, p.566.
18.2kg Göttingen, Univ.; 253g Los Angeles, Univ. of California; 109g Fort Worth, Texas, Monnig Colln.; 82g Copenhagen, Univ. Geol. Mus.;

Encantada v Putinga.

Enigma 31°20′ N., 82°19′ W.
Berrien County, Georgia, U.S.A.
Found 1967, first reported in this year
Stone. Olivine-bronzite chondrite (H4).
Reported, Meteor. Bull., 1967 (39), Meteoritics, 1970, **5**, p.91. Olivine Fa_{17}, B. Mason, Geochimica et Cosmochimica Acta, 1967, **31**, p.1100.
94g Washington, U.S. Nat. Mus.;

Enon 39°52′ N., 83°57′ W.
Clarke County, Ohio, U.S.A.
Found 1883, approx., recognized 1938
Stony-iron. Mesosiderite, anomalous (MESANOM).
A mass of 763g, A.D. Nininger, Pop. Astron., Northfield, Minnesota, 1939, **47**, p.212. Partial description, H.H. Nininger, Pop. Astron., Northfield, Minnesota, 1942, **50**, p.563, Contr. Soc. Res. Meteorites, 1942, **3**, p.61 [M.A. 8-374]. Although it is texturally similar to the mesosiderites, the plagioclase is sodic oligoclase and not calcic bytownite, as in most mesosiderites, T.E. Bunch et al., Contr. Mineral. Petrol., 1970, **25**, p.297. Ar-Ar age of silicates, 4.59 b.y., S. Niemeyer, Geochimica et Cosmochimica Acta, 1983, **47**, p.1007.
366g Tempe, Arizona State Univ., main mass; 44g Washington, U.S. Nat. Mus.;
Specimen(s): [1959,986], 40g. slice

Enshi 30°18′ N., 109°30′ E.
Bajiao, Hubei, China
Fell 1974, December 26, 1530 hrs
Stone. Olivine-bronzite chondrite (H5).
Several stones fell, totalling over 8kg, the largest piece recovered weighing 4.5kg, D. Bian, Meteoritics, 1981, **16**, p.118. Analysis, 29.84 % total iron, D. Wang and Z. Ouyang, Geochimica, 1979, p.120, in Chinese [M.A. 80-2085].
Main mass, Beijing, Acad. Sin. Inst. Geol.;

Ensigahara v Kyushu.

Ensisheim 47°52′ N., 7°21′ E.
Alsace, France
Fell 1492, November 16, 1130 hrs
Synonym(s): *Elsass*
Stone. Olivine-hypersthene chondrite, amphoterite (LL6), brecciated.
A stone of about 127kg fell, after detonations, and was preserved for a long time in the parish church of Ensisheim, Ann. Phys. (Gilbert), 1804, **18**, p.279, Ann. Phys. Chem. (Poggendorff), 1864, **121**, p.333, Ann. Phys. Chem. (Poggendorff), 1865, **122**, p.182. Analysis, G.T. Prior, Min. Mag., 1921, **19**, p.169. Olivine Fa_{28}, B. Mason, Geochimica et Cosmochimica Acta, 1963, **27**, p.1011. Xenolithic, R.A. Binns, Geochimica et Cosmochimica Acta, 1968, **32**, p.299. A chromite-plagioclase chondrule is figured, P. Ramdohr, Geochimica et Cosmochimica Acta, 1967, **31**, p.1961.
55.75kg Ensisheim, Rathaus; 9.79kg Paris, Mus. d'Hist. Nat.; 905g Berlin, Humboldt Univ.; 458g Washington, U.S. Nat. Mus.; 660g Vienna, Naturhist. Mus.; 316g Tübingen, Univ.; 209g Tempe, Arizona State Univ.; 17.5g Budapest, Nat. Hist. Mus.;
Specimen(s): [90241], 441g.; [1920,300], 10g.; [1938,409], 238g.

Entre Rios I v Chajari.

Entre Rios II v Hinojal.

Ephesus 37°57′ N., 27°23′ E.
Turkey
Doubtful..
The famous sacred object, 'Diana of the Ephesians,' may have been a meteorite, G.A. Wainwright, Palestine Explor. Fund, quarterly statement, 1934 (January), p.32, but the evidence is not conclusive. The object was not a meteorite, C.C. Wylie and J.R. Naidu, Pop. Astron., Northfield, Minnesota, 1936, **44**, p.514 [M.A. 6-388].

Épinal 48°11′ N., 6°28′ E.
Vosges, France
Fell 1822, September 13, 0700 hrs
Synonym(s): *La Baffe*
Stone. Olivine-bronzite chondrite (H5).
A stone of "about the size of a 6lb cannon ball" was seen to fall, after detonations, during a " thunder storm", in the parish of La Baffe, 6 miles from Épinal, -. Parisot, Ann. Phys. (Gilbert), 1822, **72**, p.323. Olivine Fa_{19}, B. Mason, Geochimica et Cosmochimica Acta, 1963, **27**, p.1011.
210g Paris, Mus. d'Hist. Nat.; 19g Chicago, Field Mus. Nat. Hist.; 16g Vienna, Naturhist. Mus.; 9g Berlin, Humboldt Univ.;
Specimen(s): [35780], 1.5g.

Erakot 19°2′ N., 81°53′30″ E.
Jagdalpur tahsil, Bastar district, Madhya Pradesh, India
Fell 1940, June 22, 1700 hrs
Synonym(s): *Erokote*
Stone. Carbonaceous chondrite, type II (CM2).
One stone of 113g was recovered, P. Chatterjee, letter of 25 March 1952, in Min. Dept. BM(NH). Reported, M.A.R. Khan, Hyderabad Acad. Studies, 1950, **12** [M.A. 11-441].

Description and analysis, 22.42 % total iron, B. Mason and H.B. Wiik, Am. Mus. Novit., 1962 (2115) [M.A. 16-639].
91g Calcutta, Mus. Geol. Surv. India; 2.3g New York, Amer. Mus. Nat. Hist.;

Erevan 40°18′ N., 44°30′ E.

Erevan, Armenian SSR, USSR, [Ереван]
Fell 1911, or 1912
Stone. Achondrite, Ca-rich. Howardite (AHOW).
A single mass weighing 107.2g was recovered soon after its fall near the town of Erevan, R.L. Khotinok, Метеоритика, 1978, **37**, p.74. Reported, Meteor. Bull., 1978 (55), Meteoritics, 1978, **13**, p.335.
55g Moscow, Acad. Sci.;

Ergheo 1°10′ N., 44°10′ E.

Brava, Somalia
Fell 1889, July, before the full moon
Stone. Olivine-hypersthene chondrite (L5).
A stone of about 20kg was seen to fall, but was left for five years lying in the ground; described, with an analysis, E. Artini and G. Melzi, Esplorazione Commerciale, December, 1898, Neues Jahrb. Min., 1900, **1**, p.357. Shock-metamorphism, B. Baldanza and G.R. Levi-Donati, Min. Mag., 1971, **38**, p.197. Olivine Fa_{25}, B. Mason, Geochimica et Cosmochimica Acta, 1963, **27**, p.1011.
1.08kg Washington, U.S. Nat. Mus.; 637g Vienna, Naturhist. Mus.; 529g Chicago, Field Mus. Nat. Hist.; 559g Yale Univ.; 384g Prague, Nat. Mus.; 303g Tempe, Arizona State Univ.; 210g Rome, Univ. Min. Mus.; 178g New York, Amer. Mus. Nat. Hist.; 148g Dublin, Nat. Mus.; 155g Harvard Univ.; 105g Berlin, Humboldt Univ.;
Specimen(s): [85838], 915g. and a fragment, 3.7g; [1920,301], 115g.

Erie 40°1′54″ N., 105°3′24″ W.

Weld County, Colorado, U.S.A.
Found 1965, July 15, approx.
Stone. Olivine-hypersthene chondrite (L6).
A mass of 3.3kg was found at a site with the above co-ordinates, just over the border into Boulder County, G. Goles, letter of October, 1965, in Min. Dept. BM(NH)., Meteor. Bull., 1965 (34). Olivine Fa_{24}, B. Mason, Geochimica et Cosmochimica Acta, 1967, **31**, p.1100.
Main mass, San Diego, Univ. of California; 1.2kg Washington, U.S. Nat. Mus.; 1kg Tempe, Arizona State Univ.; 209g Denver, Mus. Nat. Hist.;

Erinpoorah v Bheenwall.

Ermendorf 51°18′ N., 13°33′ E.

Grossenhain, Dresden, Germany
Fell 1677, May 28, afternoon
Doubtful. Stone.
Many small stones are said to have fallen, blue and green, soluble in acid, and containing much copper, E.F.F. Chladni, Die Feuer-Meteore, Wien, 1819, p.237 evidence is not conclusive. Mentioned, F.A. Paneth, Geochimica et Cosmochimica Acta, 1951, **1**, p.117. The latitude and longitude cited are those of Grossenhain; Ermendorf was not located.

Erofeevka 51°52′ N., 70°21′ E.

North Kazahk, Kazakhstan SSR, USSR, [Ерофеевка]
Found 1937, may have fallen in 1925, February 8 or 9.
Synonym(s): *Erofeewka*
Stone. Olivine-bronzite chondrite (H4), black.
A stone of 1771.9g was found, P.L. Dravert, J. Roy. Astron. Soc. Canada, 1937, **31**, p.364 [M.A. 7-67], P.L. Dravert, Метеоритика, 1941, **1**, p.43 [M.A. 9-293], E.L. Krinov, Каталог Метеоритов АИад. Наук СССР, Москва, 1947 [M.A. 10-511]. Analysis, 27.15 % total iron, V.Ya. Kharitonova, Метеоритика, 1965, **26**, p.146. Olivine Fa_{17}, B. Mason, Geochimica et Cosmochimica Acta, 1963, **27**, p.1011.
1637g Moscow, Acad. Sci.;
Specimen(s): [1971,197], 13.28g.

Erofeewka v Erofeevka.

Erokote v Erakot.

Eroth County (false) v Carlton.

Erxleben 52°13′ N., 11°15′ E.

Magdeburg, Sachsen-Anhalt, Germany
Fell 1812, April 15, 1600 hrs
Synonym(s): *Erx Uben, Magdeburg*
Stone. Olivine-bronzite chondrite (H6).
A stone of about 2.25kg fell, after detonations, between Magdeburg and Helmstedt, J.F.L. Hausmann and G.U.A. Vieth, Ann. Phys. (Gilbert), 1812, **40**, p.450. Analysis, F. Stromeyer, Ann. Phys. (Gilbert), 1812, **42**, p.105. Chondrule, with chromite-rich rim is figured, P. Ramdohr, Geochimica et Cosmochimica Acta, 1967, **31**, p.1961. Olivine Fa_{19}, B. Mason, Geochimica et Cosmochimica Acta, 1963, **27**, p.1011.
297g Göttingen, Univ.; 194g Tempe, Arizona State Univ.; 103g Berlin, Humboldt Univ.; 87g Vienna, Naturhist. Mus.; 49g Chicago, Field Mus. Nat. Hist.; 44g Tübingen, Univ.; 31g Washington, U.S. Nat. Mus.;
Specimen(s): [33733], 31.5g.

Erx Uben v Erxleben.

Eschigo v Yonozu.

Esna v Isna.

Esnandes 46°15′ N., 1°6′ W.

Charente Maritime, France
Fell 1837, August
Synonym(s): *La Rochelle*
Stone. Olivine-bronzite chondrite (H6).
A stone of about 1.5kg fell, L'Institut, Paris, 1837, **5**, p.334. Very little preserved in collections. Olivine Fa_{18}, B. Mason, Geochimica et Cosmochimica Acta, 1963, **27**, p.1011.
42g Vienna, Naturhist. Mus.; 23g Chicago, Field Mus. Nat. Hist.; 3.2g Tempe, Arizona State Univ.;
Specimen(s): [63983], 2.75g.

Espiritu Santo 20°0′ N., 102°11′ W.

Michoacan, Mexico
Found
Iron. Octahedrite, fine (0.65mm).
The specimen in Chicago was chiselled from some larger mass, almost certainly Chupaderos, V.F. Buchwald, Iron Meteorites, Univ. of California, 1975, p.475.

54g Chicago, Field Mus. Nat. Hist.; 35g New York, Amer. Mus. Nat. Hist.;

Esquel 42°54′ S., 71°20′ W.

Chubut, Argentina
Found 1951, before this year
Stony-iron. Pallasite (PAL).

A mass of 1500kg was found embedded in the soil, L.O. Giacomelli, Meteoritos hallados en la Patagonia, Argentina Austral, 1962, **34**, p.14, letter of 30 September 1963, in Min. Dept. BM(NH). Olivine Fa_{12}, P.R. Buseck and J.I. Goldstein, Bull. Geol. Soc. Amer., 1969, **80**, p.2141. Analysis of metal, 8.50 %Ni, 21.5 ppm.Ga, 55.5 ppm.Ge, 0.023 ppm.Ir, E.R.D. Scott, Geochimica et Cosmochimica Acta, 1977, **41**, p.349.

Main mass, Buenos Aires, in the possession of the finder; 9.8g New York, Amer. Mus. Nat. Hist.;

Specimen(s): [1964,65], 39g.; [1964,63], 1.5g. olivine fragments.

Essendy v Strathmore.

Essebi 2°53′ N., 30°50′ E.

Aru, Haut Zaire, Zaire
Fell 1957, July 28, 0720 hrs
Stone. Carbonaceous chondrite, type II (CM2).

A stone fell in the region of Essebi, about 25km W. of Aru. Described, with an analysis, F. Corin, Bull. Acad. Roy. Sci. Outre-mer, Brussels, 1960, **6**, p.954 [M.A. 16-53]. Analysis, 20.24 % total iron, H.B. Wiik, Comm. Phys.-Math. Soc. Sci. Fenn., 1969, **34**, p.135.

Several fragments, Brussels, Roy. Acad. Sci. Outre-mer; 61g Washington, U.S. Nat. Mus.;

Specimen(s): [1967,417], 0.2g. powder; [1973,M.37], 0.1g.

Estacade v Estacado.

Estacado 33°54′ N., 101°45′ W.

Hale County, Texas, U.S.A.
Found 1883
Synonym(s): *Estacade, Llano Estacado*
Stone. Olivine-bronzite chondrite (H6).

A stone of 290kg was found in 1883, twelve miles S. of Hale Center; described, with an analysis (with improbably high alkalies), K.S. Howard and J.M. Davidson, Am. J. Sci., 1906, **22**, p.55. Further analysis, 27.88 % total iron, B. Mason and H.B. Wiik, Am. Mus. Novit., 1963 (2154) [M.A. 16-639]. Olivine Fa_{19}, B. Mason, Geochimica et Cosmochimica Acta, 1963, **27**, p.1011. Thermoluminescence and pre-atmospheric shape, D.W. Sears, Earth planet. Sci. Lett., 1975, **26**, p.97.

123kg New York, Amer. Mus. Nat. Hist.; 118.5kg Chicago, Field Mus. Nat. Hist.; 18kg Harvard Univ.; 12.8kg Tempe, Arizona State Univ.; 10kg Washington, U.S. Nat. Mus.; 8.7kg Rome, Vatican Colln; 5.58kg Paris, Mus. d'Hist. Nat.; 4kg Sydney, Austr. Mus.; 2.25kg Berlin, Humboldt Univ.; 376g Dublin, Nat. Mus.;

Specimen(s): [1906,259], 17103g. large slice; [1950,390], 235g. three slices, 149g, 81.5g, 4.5g.

Estes Park 40°23′ N., 105°31′ W.

Larimer County, Colorado, U.S.A.
Pseudometeorite..

Titaniferous magnetite sand, H.B. Van Valkenburg and J.E. Sellers, Univ. Colorado Studies, Boulder, Colorado, 1929, **17**, p.17 [M.A. 5-161].

Estherville 43°25′ N., 94°50′ W.

Emmet County, Iowa, U.S.A.
Fell 1879, May 10, 1700 hrs
Synonym(s): *Emmet County, Iowa, Jowa, Peery Meteor, Perry Meteor*
Stony-iron. Mesosiderite (MES).

Several large masses, of total weight of over 700lb, the two largest weighing about 437lb and 151lb respectively, and hundreds of small fragments of nickel-iron, fell, after detonations and the appearance of a brilliant fireball, S.F. Peckham, Am. J. Sci., 1879, **18**, p.77, C.U. Shepard, Am. J. Sci., 1879, **18**, p.186. Description, analysis, J.L. Smith, Am. J. Sci., 1880, **19**, p.459. Mentioned, G.P. Merrill, Proc. U.S. Nat. Mus., 1920, **58**, p.363. Popular account of the fall, B.H. Wilson, The Palimpsest, Iowa, 1928, **9**, p.317 [M.A. 4-117]., G.S. Dillé, Proc. Iowa Acad. Sci., 1928, **35**, p.225 [M.A. 5-157]. Texture, metallography and silicate mineralogy, B.N. Powell, Geochimica et Cosmochimica Acta, 1969, **33**, p.789, B.N. Powell, Geochimica et Cosmochimica Acta, 1971, **35**, p.5. Analysis of metal, 9.0 %Ni, 9.0 ppm.Ga, 52.3 ppm.Ge, 2.7 ppm.Ir, J.T. Wasson et al., Geochimica et Cosmochimica Acta, 1974, **38**, p.135. The location of the 150lb mass is unknown. A 125lb mass, labelled 'Estherville', and formerly in the Univ. of Minnesota, is now in Washington (U.S. Nat. Mus.); cutting showed it to be "clearly a Cañon Diablo type meteorite", R.S. Clarke, Jr., letter of 14 December 1971, in Min. Dept. BM(NH). The 437lb mass was divided between the London, Paris and Vienna museums. Olivine coronas, references, C.E. Nehru et al., Geochimica et Cosmochimica Acta, 1980, **44**, p.1103.

55kg Paris, Mus. d'Hist. Nat.; 40kg Yale Univ.; 16.8kg Harvard Univ.; 23kg Vienna, Naturhist. Mus., approx. weight, includes a 21kg mass; 9.5g Chicago, Field Mus. Nat. Hist.; 6.5kg Budapest, Nat. Mus.; 4.5kg Berlin, Humboldt Univ.; 4kg Washington, U.S. Nat. Mus.; 3kg New York, Amer. Mus. Nat. Hist.;

Specimen(s): [53764], 104kg. part of the largest mass, also a slice, 10896g, and pieces, 515g.; [65575], 409g. 42 small fragments; [1918,3-19], 187g. 17 small fragments; [53786], 21.5g.; [53787], 10g. two fragments, 7g and 3g.; [1927,1293], 5g.; [1959,172], 8g.

Estland v Oesel.

Esu v Maridi.

Ethiudna 31°59′ S., 139°46′ E.

County Lytton, South Australia, Australia
Found 1977, February 20
Stone. Olivine-hypersthene chondrite (L).

A single mass of 74.318kg was found 6.8km NNW. of Ethiudna Hill and NW. of Plumbago station. Possibly related to Crocker's Well (*q.v.*), Meteor. Bull., 1980 (57), Meteoritics, 1980, **15**, p.97.

74.3kg, held by the finder;

Etosha 18°30′ S., 16° E. approx.

Ovamboland, Namibia
Found
Iron. Octahedrite, medium (IC).

A mass of 110.7kg was found near the Etosha Game Reserve, NW. of Tsumeb, and presented to the Department of Geology, Univ. of Potchefstroom, C. Frick and E.C.I. Hammerbeck, Bull. Geol. Surv. S. Africa, 1973 (57), p.16, Meteor. Bull., 1975 (53), Meteoritics, 1975, **10**, p.146. Has cohenite inclusions with unusual morphologies and pencil-shaped troilite nodules; classification and analysis, 6.85 %Ni, 48.9 ppm.Ga, 217 ppm.Ge, 0.10 ppm.Ir, E.R.D. Scott and

J.T. Wasson, Geochimica et Cosmochimica Acta, 1976, **40**, p.103.
Main mass, Potchefstroom Univ.; 203g Pretoria, Geol. Surv. Mus.;

Etter 35°59′ N., 101°54′ W.
Moore County, Texas, U.S.A.
Found 1965, recognized 1966
Stone. Olivine-bronzite chondrite (H6).
Four individuals and one fragment, of total weight 153.3kg, were ploughed up, Meteor. Bull., 1975 (53), Meteoritics, 1975, **10**, p.147, G.I. Huss, letter of 16 March 1973, in Min. Dept. BM(NH).
130kg Mainz, Max-Planck-Inst.; 873g Chicago, Field Mus. Nat. Hist.; 464g Copenhagen, Univ. Geol. Mus.; 188g New York, Amer. Mus. Nat. Hist.; 185g Vienna, Naturhist. Mus.;
Specimen(s): [1972,498], 209g. crusted, part-slice.

Et-Tlahi v Hedjaz.

Eunice 34°28′12″ N., 101°45′30″ W.
Swisher County, Texas, U.S.A.
Found 1961
Stone. Olivine-bronzite chondrite (H).
A stone of 2.3kg was ploughed up in a wheat field, G.I. Huss, letter of 8 February 1974, in Min. Dept. BM(NH).
1.6kg Mainz, Max-Planck-Inst.;
Specimen(s): [1974,M.2], 27.6g. crusted, part-slice

Eustis 28°50′ N., 81°41′ W.
Lake County, Florida, U.S.A.
Found 1918
Stone. Olivine-bronzite chondrite (H4).
A stone of 502g was ploughed up, G.P. Merrill, Am. J. Sci., 1918, **44**, p.64. Olivine Fa_{18}, B. Mason, Geochimica et Cosmochimica Acta, 1963, **27**, p.1011.
480g Washington, U.S. Nat. Mus., main mass;

Eva 36°12′ N., 101°54′30″ W.
Texas County, Oklahoma, U.S.A.
Found 1965, or 1966, recognized 1970
Stone. Olivine-bronzite chondrite (H5), polymict breccia.
One fragment of 6.7kg, probably representing less than half the parent mass, was found, G.I. Huss, letters of 7 January and 9 April 1971, in Min. Dept. BM(NH)., Meteor. Bull., 1972 (51), Meteoritics, 1972, **7**, p.223. Polymict, olivine Fa_{19}, R.V. Fodor and K. Keil, Earth planet. Sci. Lett., 1976, **29**, p.1.
2.7kg Mainz, Max-Planck-Inst.; 300g Washington, U.S. Nat. Mus.; 248g New York, Amer. Mus. Nat. Hist.;
Specimen(s): [1970,309], 73.8g. slice

Ezel v Oesel.

Fabriano 43°19′ N., 12°56′ E.
S. Anatolio, Ancona, Italy
Fell 1776, January, in the afternoon hrs, or 1777, or February
Doubtful. Stone.
Several stones similar to those of the Siena shower are said to have fallen, D.A. Soldani, Atti Accad. Sci. Siena, 1808, **9**, p.1, E.F.F. Chladni, Die Feuer-Meteore, Wien, 1819, p.255, but the evidence is not conclusive. Listed, B. Baldanza, Min. Mag., 1965, **35**, p.214.

Faha v Limerick.

Fairbanks v Aggie Creek.

Fairfield County v Weston.

Fair Oaks v Cañon Diablo.

Fair Play 37°40′ N., 93°33′ W.
Polk County, Missouri, U.S.A.
Found 1928
Synonym(s): *Fairplay*
Pseudometeorite..
91kg known. This supposed mesosiderite has been shown to be white cast iron, S.H. Perry, Pop. Astron., Northfield, Minnesota, 1946, **54**, p.49 [M.A. 9-300].

Fairplay v Fair Play.

Fairweather v Bridgewater.

Faith 45°20′ N., 102°5′ W.
Perkins County, South Dakota, U.S.A.
Found 1952, recognized 1967
Stone. Olivine-bronzite chondrite (H5).
A large single stone weighing 105kg was found 23 miles NNE. of Faith. Olivine Fa_{18}, Cat. Meteor. Arizona State Univ., 1970, p.37, C.B. Moore, letter of 10 July 1970, in Min. Dept. BM(NH).
83kg Tempe, Arizona State Univ.; 723g Chicago, Field Mus. Nat. Hist.; 467g New York, Amer. Mus. Nat. Hist.; 434g Washington, U.S. Nat. Mus.;
Specimen(s): [1968,180], 76g. slice

False Eroth County v Carlton.

False Kalamaroo v Grand Rapids.

False Wittens v Eichstädt.

False Yanhuitlan v Misteca.

Farley 36°20′ N., 104°3′ W.
Colfax County, New Mexico, U.S.A.
Found 1936
Stone. Olivine-bronzite chondrite (H5), veined.
Two stones 19.4kg together, A.D. Nininger, Pop. Astron., Northfield, Minnesota, 1937, **45**, p.449 [M.A. 7-62]. Listed, H.H. and A.D. Nininger, The Nininger Collection of Meteorites, Winslow, Arizona, 1950, p.51. Olivine $Fa_{18.7}$, D.E. Lange and K. Keil, Meteoritics, 1976, **11**, p.315, abs.
2.2kg Tempe, Arizona State Univ.; 930g Washington, U.S. Nat. Mus.; 700g Albuquerque, Univ. of New Mexico; 620g Chicago, Field Mus. Nat. Hist.; 132g New York, Amer. Mus. Nat. Hist.;
Specimen(s): [1937,382], 457g.; [1937,393], 0.5g.; [1959,810], 1930g. thick slice

Farmington 39°45′ N., 97°2′ W.
Washington County, Kansas, U.S.A.
Fell 1890, June 25, 1300 hrs
Synonym(s): *Washington, Washington County*
Stone. Olivine-hypersthene chondrite (L5), black, brecciated.
After the appearance of a fireball, and detonations, a stone of 188lb was seen to fall, and another of 9lb was found, F.H. Snow, Science, 1890, **16**, p.38. See also, with an analysis, G.F. Kunz and E. Weinschenk, Tschermaks Min. Petr. Mitt., 1891, **12**, p.177, Am. J. Sci., 1892, **43**, p.65. More

recent description and analysis, 21.22 % total iron, P.R. Buseck et al., Geochimica et Cosmochimica Acta, 1966, **30**, p.1. Magnetic properties, M.W. Rowe, Meteoritics, 1975, **10**, p.23. A fragment of an Apollo asteroid?, B.J. Levin et al., Icarus, 1976, **28**, p.307. U-Pb isotope study, M.C.B. Abranches et al., Earth planet. Sci. Lett., 1980, **46**, p.311 [M.A. 80-4728].

26.2kg Chicago, Field Mus. Nat. Hist.; 11.7kg Budapest, Nat. Mus.; 6kg New York, Amer. Mus. Nat. Hist.; 2.6kg Washington, U.S. Nat. Mus.; 2.6kg Paris, École des Mines; 3kg Harvard Univ.; 1.68kg Tempe, Arizona State Univ.; 1.3kg Vienna, Naturhist. Mus.; 424g Ottawa, Mus. Geol. Surv. Canada; 225g Prague, Nat. Mus.;

Specimen(s): [66200], 737g.; [67016], 60.5g. slice.; [1964,571], thin section

Farmville — 35°33′ N., 77°32′ W.

Pitt County, North Carolina, U.S.A.
Fell 1934, December 4
Stone. Olivine-bronzite chondrite (H4).

Two stones fell, 56kg together, A.D. Nininger, Pop. Astron., Northfield, Minnesota, 1937, **45**, p.449 [M.A. 7-62]. Olivine Fa_{18}, B. Mason, Geochimica et Cosmochimica Acta, 1963, **27**, p.1011.

5.9kg Washington, U.S. Nat. Mus., complete stone; Main mass, Raleigh, N. Carolina, North Carolina State Mus.;

Farnum — 40°15′ N., 100°14′ W.

Lincoln County, Nebraska, U.S.A.
Found 1937
Stone. Olivine-hypersthene chondrite (L5).

One stone of 4.2kg was found midway between Farnum and Ingham in township 9, range 26 W., Lincoln County, A.D. Nininger, Pop. Astron., Northfield, Minnesota, 1939, **47**, p.212, E.P. Henderson, letter of 3 June 1939, in Min. Dept. BM(NH). Olivine Fa_{25}, B. Mason, Geochimica et Cosmochimica Acta, 1963, **27**, p.1011.

1.5kg Tempe, Arizona State Univ.; 150g Washington, U.S. Nat. Mus.;

Specimen(s): [1959,861], 1480g. and fragments, 1.4g

Fasti v Białystok.

Fatehpur v Futtehpur.

Faucett — 39°37′ N., 94°52′ W.

Buchanan County, Missouri, U.S.A.
Found 1966
Synonym(s): *Fawcett*
Stone. Olivine-bronzite chondrite (H5).

Following the discovery of the first 12lb stone in 1966, Meteor. Bull., 1966 (37), at least seven others have been found, totalling over 100kg in weight. They are internally fresh and may have fallen in the summer of 1907, W.F. Read, Meteoritics, 1969, **4**, p.244. Olivine Fa_{18}, B. Mason, Geochimica et Cosmochimica Acta, 1967, **31**, p.1100. Description, analysis, 26.59 % total iron, E.A. King et al., Meteoritics, 1977, **12**, p.13.

88kg Los Angeles, Univ. of California, most of the masses; 6.2kg Washington, U.S. Nat. Mus.; 1.1kg Moscow, Acad. Sci.; 414g Tempe, Arizona State Univ.; 130g Vienna, Naturhist. Mus.;

Specimen(s): [1969,145], 76g. a sawn piece.

Favars — 44°23′ N., 2°49′ E.

Laissac, Aveyron, France
Fell 1844, October 21, 0645 hrs
Synonym(s): *Gaillac, Laissac*
Stone. Olivine-bronzite chondrite (H5).

After detonations, a stone of 1.5kg was found, A. Boisse, L'Institut, Paris, 1844, **12**, p.399. Olivine Fa_{18}, B. Mason, Geochimica et Cosmochimica Acta, 1963, **27**, p.1011.

335g Paris, Mus. d'Hist. Nat.; 29g Chicago, Field Mus. Nat. Hist.;

Specimen(s): [56466], 5g.; [34598], 0.2g.

Fawcett v Faucett.

Fayette County v Bluff.

Fayetteville v Petersburg.

Fayetteville — 36°3′ N., 94°10′ W.

Washington County, Arkansas, U.S.A.
Fell 1934, December 26, 1158 hrs
Stone. Olivine-bronzite chondrite (H4).

A stone of 2.25kg was dug out immediately after its fall, another of 108g was found a few days later in a new hole, 3.5 miles away, D.P. Richardson, Pop. Astron., Northfield, Minnesota, 1935, **43**, p.384 [M.A. 6-104]. Olivine Fa_{19}, B. Mason, Geochimica et Cosmochimica Acta, 1963, **27**, p.1011. Gas-rich; analysis of light fraction, 29.5 % total iron, and of dark, 25.5 % total iron, rare gas contents, O. Müller and J. Zähringer, Earth planet. Sci. Lett., 1966, **1**, p.25.

Specimens of both stones, Fayetteville, Univ. of Arkansas; 161g Washington, U.S. Nat. Mus.; 26g Tempe, Arizona State Univ.;

Fehrbellin v Linum.

Feid Chair — 36°53′ N., 8°27′ E.

La Calle, Constantine, Algeria
Fell 1875, August 16, midday hrs
Synonym(s): *La Calle*
Stone. Olivine-bronzite chondrite (H4), brecciated.

A stone of 380g fell, G.A. Daubrée, C. R. Acad. Sci. Paris, 1877, **84**, p.70. Olivine Fa_{17}, B. Mason, Geochimica et Cosmochimica Acta, 1963, **27**, p.1011.

25g Paris, Mus. d'Hist. Nat.;

Fekete v Mezö-Madaras.

Felix — 32°32′ N., 87°10′ W.

Perry County, Alabama, U.S.A.
Fell 1900, May 15, 1130 hrs
Stone. Carbonaceous chondrite, type III (CO3).

After the appearance of a fireball and detonations, a stone of about 7lb was found; described, with an analysis, G.P. Merrill, Proc. U.S. Nat. Mus., 1902, **24**, p.193. Amount and composition of nickel-iron determined, G.T. Prior, Min. Mag., 1919, **18**, p.353. New analysis, 25.94 % total iron, H.B. Wiik, Geochimica et Cosmochimica Acta, 1956, **9**, p.279. H and C isotopes, G. Boato, Geochimica et Cosmochimica Acta, 1954, **6**, p.209. Further analysis, 24.81 % total iron, T.S. McCarthy and L.H. Ahrens, Earth planet. Sci. Lett., 1972, **14**, p.97. Trace element data, G.W. Kallemeyn and J.T. Wasson, Geochimica et Cosmochimica Acta, 1981, **45**, p.1217.

1286g Washington, U.S. Nat. Mus., main mass; 150g Paris, Mus. d'Hist. Nat.; 72g Chicago, Field Mus. Nat. Hist.; 16g Vienna, Naturhist. Mus.;

Specimen(s): [1919,89], 29g. in two pieces, 20g and 9g.; [1971,290], 92.3g. crusted slice.

Felt 36°33′ N., 102°47′ W.

Cimarron County, Oklahoma, U.S.A.
Found 1970
Stone. Olivine-bronzite chondrite (H).

Two stones of 1.8kg and 3.6kg were ploughed up in one field, G.I. Huss, letter of 16 March 1973, in Min. Dept. BM (NH). Olivine Fa_{19}, A.L. Graham, priv. comm., 1978.

3.0kg Mainz, Max-Planck-Inst.; 140g Vienna, Naturhist. Mus.; 53g New York, Amer. Mus. Nat. Hist.;

Specimen(s): [1973,M.17], 78g. a crusted slice.

Fenbark 30°26′25″ S., 121°15′25″ E.

Western Australia, Australia
Found 1968
Stone. Olivine-bronzite chondrite (H5), brecciated.

A brecciated stone of 1.861kg was found by nickel prospectors, 25 miles NNW. of Kalgoorlie and 30 miles from Credo. It is badly weathered, cracked and decomposed, G.J.H. McCall and W.H. Cleverly, Min. Mag., 1969, **37**, p.281. Olivine $Fa_{19.6}$, B. Mason, Rec. Austr. Mus., 1974, **29**, p.169.

1.368kg Kalgoorlie, West. Austr. School of Mines, Main mass, and a slice, 14.5g.; Three specimens, one to each of the finders; 38g Perth, West. Austr. Mus.; 4g Washington, U.S. Nat. Mus.;

Feng v Fenghsien-Ku.

Fenghsien-Ku 34°36′ N., 116°45′ E.

Tang-shan prefecture, Jiangsu, China
Fell 1924, October 5, 1822 hrs
Synonym(s): *Feng, Fengxian, Kiangsu*
Stone. Olivine-bronzite chondrite (H5).

After the appearance of a luminous meteor, passing from SW. to NE., and detonations, stones fell near the town of Fe†ng-Hsien; a conical stone weighing 82g is described, H.T. Lee, Bull. Geol. Soc. China, 1925, **4**, p.273. Two stones figured, and the time given as 1722 hrs, H.T. Chang, Mem. Geol. Surv. China, 1927 (ser. B, no. 2). Olivine Fa_{18}, B. Mason, Geochimica et Cosmochimica Acta, 1963, **27**, p.1011.

Fengxian v Fenghsien-Ku.

Fengzhen 40°30′ N., 113° E. approx.

Fengzhen County, Nei Monggol, China
Found
Iron. Octahedrite, medium.

A 458g mass was found. Listed, references, D. Bian, Meteoritics, 1981, **16**, p.122.

458g Beijing, Acad. Sin. Inst. Geol.;

Ferguson 36°6′ N., 81°25′ W.

Wilkes County, North Carolina, U.S.A.
Fell 1889, July 18
Synonym(s): *Haywood County*
Stone. Chondrite.

After detonations, a stone of about 0.5lb fell, it was sent to G.F. Kunz and was lost in New York City, G.F. Kunz, Am. J. Sci., 1890, **40**, p.320.

Ferguson Switch 34° N., 101°30′ W. approx.

Hale County, Texas, U.S.A.
Found 1937
Stone. Olivine-bronzite chondrite (H5).

A mass of 1.3kg was found, H.H. and A.D. Nininger, The Nininger Collection of Meteorites, Winslow, Arizona, 1950, p.51. Olivine Fa_{17}, B. Mason, Geochimica et Cosmochimica Acta, 1963, **27**, p.1011.

737g Tempe, Arizona State Univ.; 536g Mainz, Max-Planck-Inst.; 50g New York, Amer. Mus. Nat. Hist.;

Specimen(s): [1959,860], 290g. and fragments, 14.5g

Ferintosh 52°48′ N., 112°59′ W.

Alberta, Canada
Found 1965
Stone. Chondrite.

One stone of 2.201kg was found, Meteor. Bull., 1966 (35).

Main mass, Edmonton, Univ. of Alberta;

Ferrara v Renazzo.

Ferrara v Trenzano.

Ferrara v Vigarano.

Ferrette 47°30′ N., 7°19′ E.

Haut-Rhin, Alsace, France
Fell 1947, November 2, 0200 hrs, approx. time
Doubtful. Stone.

A mass of about 7kg fell a few yards from Ferrette railway station, and was found at daybreak. It was broken up and only about 4.3kg were preserved, L'Alsace (newspaper), 1947 (November 12), Bull. Soc. Astron. France, 1948, **62**, p.27. The main mass, once in the Ferrette Mus., is apparently no longer there. It was borrowed and then lost some years ago. The specimen is said to be of sandstone and not meteoritic, M. Christophe-Michel-Levy, priv. comm.

Fezzan v Oubari.

Fillmore 52°2′48″ N., 107°14′ W.

Saskatchewan, Canada
Found 1916, approx., recognized 1968
Iron. Octahedrite.

A specimen of 113g, possibly found near Fillmore, was acquired by a teacher. In 1968 it was recognized as meteoritic by the Geol. Surv. Canada, Meteor. Bull., 1971 (50), Meteoritics, 1971, **6**, p.118.

12.8g Ottawa, Geol. Surv. Canada;

Filomena v North Chile.

Finmark v Finmarken.

Finmarken 70° N., 24° E.

Finmark, Norway
Found 1902
Synonym(s): *Alta, Alte, Alten, Finmark, Finnmarken*
Stony-iron. Pallasite (PAL).

A mass of about 77.5kg was found, E. Cohen, Mitt. Naturwiss. Ver. Neu-Vorp. u. Rügen, Greifswald, 1903, **35**, p.1. Olivine Fa_{13}, P.R. Buseck and J.I. Goldstein, Bull. Geol. Soc. Amer., 1969, **80**, p.2141. Analysis of metal, 10.7 %Ni, 18.7 ppm.Ga, 43.7 ppm.Ge, 1.8 ppm.Ir, J.T. Wasson and S.P. Sedwick, Nature, 1969, **222**, p.22. Texture, references, E.R.D. Scott, Geochimica et Cosmochimica Acta, 1977, **41**, p.693.

5.9kg Vienna, Naturhist. Mus.; 3.65kg Chicago, Field Mus. Nat. Hist.; 2.2kg Schönenwerd, Bally Mus.; 1.5kg Budapest, Nat. Mus.; 977g Washington, U.S. Nat. Mus.; 874g Berlin, Humboldt Univ.; 607g Copenhagen, Univ. Geol. Mus.; 390g Dublin, Nat. Mus.; 341g Harvard Univ.; 200g Moscow, Acad. Sci.;
Specimen(s): [86755], 1289g. a slice, and fragments, 15g; [1964,565], 74.5g. slice.

Finney 34°16′ N., 101°34′ W.
Hale County, Texas, U.S.A.
Found 1962
Stone. Olivine-hypersthene chondrite (L5).
A mass of 10.7kg was found, Cat. Meteor. Arizona State Univ., 1976, p.109. Olivine Fa_{24}, B. Mason, Geochimica et Cosmochimica Acta, 1967, **31**, p.1100. Reported, Meteor. Bull., 1964 (32).
2.3kg Mainz, Max-Planck-Inst.; 277g Washington, U.S. Nat. Mus.; 113g New York, Amer. Mus. Nat. Hist.; 71g Tempe, Arizona State Univ.;
Specimen(s): [1964,652], 176g.

Finnmarken v Finmarken.

Finsterhölzelries v Teplá.

Fischer v Fisher.

Fisher 47°49′ N., 96°51′ W.
Polk County, Minnesota, U.S.A.
Fell 1894, April 9, 1600 hrs
Synonym(s): *Fischer, Polk County*
Stone. Olivine-hypersthene chondrite (L6), veined.
In April detonations were heard and in June two stones were found, one of 9.25lb and a larger one, which was broken and mostly lost; later, in 1895, two other small pieces were found, and in 1898 another stone of about 3lb. Described, with an analysis, G.P. Merrill, Proc. U.S. Nat. Mus., 1915, **48**, p.503. This analysis shows no alkalies and is doubtful, W. Wahl, Geochimica et Cosmochimica Acta, 1950, **1**, p.28. Olivine Fa_{23}, B. Mason, Geochimica et Cosmochimica Acta, 1963, **27**, p.1011.
6.5kg Washington, U.S. Nat. Mus.; 1.67kg New York, Amer. Mus. Nat. Hist.; 9.25lb Minnesota Univ.; 736g Yale Univ.; 371g Harvard Univ.; 461g Chicago, Field Mus. Nat. Hist.; 347g Prague, Nat. Mus.; 188g Paris, Mus. d'Hist. Nat.; 83g Vienna, Naturhist. Mus.;
Specimen(s): [81069], 372.5g.; [81070], 226g.

Fish River v Gibeon.

Fisterhölzelries v Teplá.

Fiume Potenza v Monte Milone.

Fleming 40°41′ N., 102°49′ W.
Logan County, Colorado, U.S.A.
Found 1940
Stone. Olivine-bronzite chondrite (H3), black, brecciated.
One mass of 1.75kg was found, A.D. Nininger, Pop. Astron., Northfield, Minnesota, 1940, **48**, p.555, Contr. Soc. Res. Meteorites, 1942, **2**, p.227 [M.A. 8-54]. Listed, H.H. and A.D. Nininger, The Nininger Collection of Meteorites, Winslow, Arizona, 1950, p.51, W. Wahl, Geochimica et Cosmochimica Acta, 1952, **2**, p.91. Olivine Fa_{18}, B. Mason, Geochimica et Cosmochimica Acta, 1963, **27**, p.1011.
860g Denver, Mus. Nat. Hist.; 472g Tempe, Arizona State Univ.; 68g Washington, U.S. Nat. Mus.;
Specimen(s): [1959,823], 91g. thin slice

Florence 30°50′ N., 97°46′ W.
Williamson County, Texas, U.S.A.
Fell 1922, January 21, 2000 hrs
Stone. Olivine-bronzite chondrite (H3), brecciated.
One stone of 3640g fell. Described, with an analysis, J.T. Lonsdale, Am. Miner., 1927, **12**, p.398 [M.A. 3-537]. The analysis shows more K than Na and is doubtful, W. Wahl, Geochimica et Cosmochimica Acta, 1950, **1**, p.28.
42g Tempe, Arizona State Univ.; Part of main mass, Univ. of Texas;

Florey 32°31′21″ N., 102°42′18″ W.
Andrews County, Texas, U.S.A.
Found 1978
Stone. Olivine-bronzite chondrite (H6).
A single stone of 110.8kg was found 12-15 miles SSW. of Seminole, G.I Huss, priv. comm., 1983. Olivine $Fa_{19.4}$, A.L. Graham, priv. comm., 1983.
Specimen(s): [1983,M.23], 838g.

Flows v Monroe.

Floyd 34°11′36″ N., 103°34′30″ W.
Roosevelt County, New Mexico, U.S.A.
Found 1966, recognized 1967
Stone. Olivine-hypersthene chondrite (L4).
An individual of 13kg was found under a fence by an arrowhead hunter, G.I Huss, letter of 7 January 1971, in Min. Dept. BM(NH). Reported, Meteor. Bull., 1972 (51), Meteoritics, 1972, **7**, p.223. Olivine $Fa_{23.2}$, A.L. Graham, priv. comm., 1981.
8.0kg Mainz, Max-Planck-Inst.; 453g Copenhagen, Univ. Geol. Mus.; 305g New York, Amer. Mus. Nat. Hist.; 256g Washington, U.S. Nat. Mus.; 256g Chicago, Field Mus. Nat. Hist.; 50g Los Angeles, Univ. of California;
Specimen(s): [1971,101], 66g. polished slice, and fragments, 2.3g.

Floydada 33°59′ N., 101°17′ W.
Floyd County, Texas, U.S.A.
Found 1912, known before this year
Iron. Octahedrite, medium (0.9mm) (IIIB).
One mass of 12.5kg was recognized in 1938, from SW. 1/4, sect. 8, block D5, Floyd County, A.D. Nininger, Pop. Astron., Northfield, Minnesota, 1939, **47**, p.212. Listed, H.H. and A.D. Nininger, The Nininger Collection of Meteorites, Winslow, Arizona, 1950, p.52. Analysis, 9.1 %Ni, 21.1 ppm.Ga, 41.4 ppm.Ge, 0.013 ppm.Ir, D.J. Malvin et al., priv. comm., 1983.
12.5kg Fort Worth, Texas, Texas Observers Collection; 20.7g Tempe, Arizona State Univ.;

Floyd County v Indian Valley.

Floyd Mountain v Indian Valley.

Fluvanna (a) 32°48′ N., 101°7′ W.
Scurry County, Texas, U.S.A.
Found 1967, approx.
Stone. Olivine-hypersthene chondrite (L5).
More than seven pieces totalling over 9.5kg were found, the largest 6.3kg. Some have since been lost, Meteor. Bull., 1983 (61), Meteoritics, 1983, **18**, p.79.
147g Chicago, Field Mus. Nat. Hist.; 82g Watchung, Dupont Colln.;

Fluvanna (b) 32°54′12″ N., 101°9′46″ W.
Scurry County, Texas, U.S.A.
Found 1976
Stone. Olivine-bronzite chondrite (H6).
Many fragments totalling 4.11kg were found, G.I Huss, priv. comm., 1982. Olivine $Fa_{20.5}$, A.L. Graham, priv. comm., 1983.
Specimen(s): [1983,M.22], 57g.

Foellinge v Föllinge.

Föllinge 63°44′ N., 14°51′ E.
Ottsjön, Jämtland, Sweden
Found 1932, September
Synonym(s): *Foellinge, Ottsjö, Ottsjön*
Iron. Octahedrite, finest (0.040mm) (IIICD).
One piece of 400g was ploughed up. Described, with a preliminary analysis, F.E. Wickman, Pop. Astron. Tidskr., 1951, p.63 [M.A. 12-356]. Mentioned, S. Hjelmqvist, letter of March 1958, in Min. Dept. BM(NH). Reported, Meteor. Bull., 1961 (20), V.F. Buchwald, Analecta geol., 1967 (2), V.F. Buchwald, Geochimica et Cosmochimica Acta, 1967, **31**, p.1559. Analysis, 18.13 %Ni, 4.02 ppm.Ga, 3.15 ppm.Ge, 0.072 ppm.Ir, J.T. Wasson and R. Schaudy, Icarus, 1971, **14**, p.59. One of the smallest individual irons found; a well-preserved drop-like mass with crust, V.F. Buchwald, Iron Meteorites, Univ. of California, 1975, p.567.
Main mass, Stockholm, Riksmuseet; 17.9g Chicago, Field Mus. Nat. Hist.; 14g Washington, U.S. Nat. Mus.;

Folersville v Staunton.

Fomatlan v Tomatlan.

Forest v Forest City.

Forestburg 33°30′ N., 97°39′ W.
Montague County, Texas, U.S.A.
Found 1957
Stone. Olivine-hypersthene chondrite (L).
26.6kg were found, B. Mason, Meteorites, Wiley, 1962, p.243. Olivine Fa_{24}, B. Mason, Geochimica et Cosmochimica Acta, 1963, **27**, p.1011.
Main mass, Fort Worth, Texas, Childrens Mus.;

Forest City 43°15′ N., 93°40′ W.
Winnebago County, Iowa, U.S.A.
Fell 1890, May 2, 1715 hrs
Synonym(s): *Forest, Forestcity, Iowa, Jowa, Kossuth County, Leland, Winnebago County*
Stone. Olivine-bronzite chondrite (H5), brecciated.
After appearance of a brilliant fireball moving from W. to E. and detonations, a shower of stones fell over an area of 2 miles by 1 mile; the shower consisted of five large stones of 80lb, 66lb, 10lb, 5lb, 4lb and over 500 small ones. The total weight preserved is about 122kg, J. Torrey and E.H. Barbour, Am. J. Sci., 1890, **39**, p.521, Amer. Geologist, 1891, **8**, p.67, G.F. Kunz, Am. J. Sci., 1890, **40**, p.318, H.A. Newton, Am. J. Sci., 1890, **39**, p.522. Olivine Fa_{19}, B. Mason, Geochimica et Cosmochimica Acta, 1963, **27**, p.1011. H and C isotopes, G. Boato, Geochimica et Cosmochimica Acta, 1954, **6**, p.209. Nature of the metal, H.C. Urey and T. Mayeda, Geochimica et Cosmochimica Acta, 1959, **17**, p.113. Bi, Tl and Hg analysis, W.D. Ehmann and J.R. Huizenga, Geochimica et Cosmochimica Acta, 1959, **17**, p.125. Recent analysis, 27.2 % total iron, B. Mason and H.B. Wiik, Am. Mus. Novit., 1965 (2220).
43.36kg New York, Amer. Mus. Nat. Hist., includes the largest stone recovered, now 33.97kg; 36kg Washington, U.S. Nat. Mus.; 28kg Yale Univ.; 17.8kg Chicago, Field Mus. Nat. Hist.; 2.5kg Lansing, Michigan, Michigan State College; 1.2kg Budapest, Nat. Mus.; 1.4kg Tempe, Arizona State Univ.; 1.1kg Vienna, Naturhist. Mus.; 940g Harvard Univ.;
Specimen(s): [66201], 2265g. complete stone; [65972], 265g. five complete stones and three fragments; [1960,495], 28.5g.

Forest Vale 33°21′ S., 146°51′30″ E.
County Dowling, New South Wales, Australia
Fell 1942, August 7, 1500 hrs
Stone. Olivine-bronzite chondrite (H4).
Five fragments were found fitting together to form an almost complete stone of about 26kg. The largest fragment weighed 10.49kg and another stone of 1.9kg was found about 12 chains distant, and a third is said to have been broken up. Description, T. Hodge-Smith, Austr. J. Sci., 1943, **5**, p.154 [M.A. 9-298]. Olivine Fa_{17}, B. Mason, Geochimica et Cosmochimica Acta, 1963, **27**, p.1011. Further description and analysis, 27.45 % total iron, A.F. Noonan et al., Smithson. Contrib. Earth Sci., 1972 (9), p.57.
147g Washington, U.S. Nat. Mus.; 91g New York, Amer. Mus. Nat. Hist.;
Specimen(s): [1970,30], 7.2g.

Forksville 36°47′ N., 78°5′ W.
Mecklenburg County, Virginia, U.S.A.
Fell 1924, July 16, 1745 hrs
Stone. Olivine-hypersthene chondrite (L6).
After detonations four stones, weighing 2250g, 1850g, 1114g and 853g, were seen to fall. Described, with an analysis, G.P. Merrill, Proc. U.S. Nat. Mus., 1927, **70** (21). Microprobe analyses of whitlockite and chlor-apatite, W.R. Van Schmus and P.H. Ribbe, Geochimica et Cosmochimica Acta, 1969, **33**, p.637. Olivine Fa_{26}, B. Mason, Geochimica et Cosmochimica Acta, 1963, **27**, p.1011.
1995g Washington, U.S. Nat. Mus.; 3.36kg Univ. of Virginia; 85g Tempe, Arizona State Univ.;
Specimen(s): [1926,490], 110g. from the 853g stone.

Forli v Valdinoce.

Formosas v Uberaba.

Forrestcity v Forest City.

Forrest (a) 30°49′ S., 128°13′ E.
Nullarbor Plain, Western Australia, Australia
Found 1967
Stone. Olivine-bronzite chondrite (H5).
A highly weathered mass of 97.7g with a remnant of fusion crust was found 6-7 miles ENE. of Forrest Station on the Trans-Australian railway, G.J.H. McCall and W.H. Cleverly, J. Roy. Soc. West. Austr., 1970, **53**, p.69. Classification, olivine $Fa_{19.6}$, B. Mason, Rec. Austr. Mus., 1974, **29**, p.169.
Main mass, Kalgoorlie, West. Austr. School of Mines; 2.5g Perth, West. Austr. Mus., and thin section; 1.7g Washington, U.S. Nat. Mus.;

Forrest (b) 30°59′ S., 127°53′ E.
Western Australia, Australia
Found 1980, October
Synonym(s): *South Forrest*
Stone. Olivine-hypersthene chondrite (L6).
A number of fragments, totalling about 26kg, were found about 20 miles SW. of Forrest, olivine $Fa_{26.0}$, Meteor. Bull., 1983 (61), Meteoritics, 1983, **18**, p.79.

410g Chicago, Field Mus. Nat. Hist.;
Specimen(s): [1981,M.1], 518g.

Forrest Lakes 29°25′ S., 129°30′ E. approx.

Western Australia, Australia
Found 1948, before this year
Synonym(s): *Loongana (stone)*
Stone. Olivine-hypersthene chondrite, amphoterite (LL5).
A small weathered stone was found 120 miles NE. of Reid Station (on the Trans-Australian railway) near Forrest Lakes, Rep. Director Govt. Chem. Lab. West. Austr., 1948, p.20, G.J.H. McCall, Min. Mag., 1965, **35**, p.241. A mass of 500g was found, D.H. McColl, letter of 30 March 1972, in Min. Dept. BM(NH). Classification, olivine $Fa_{27.5}$, B. Mason, Rec. Austr. Mus., 1974, **29**, p.169.
208g Perth, Govt. Chem. Lab.; 38g Perth, West. Austr. Mus.; 33g New York, Amer. Mus. Nat. Hist.; Thin section, Washington, U.S. Nat. Mus.;

Forsbach 50°57′ N., 7°19′ E.

Hoffnungstal, Cologne, Nordrhein-Westfalen, Germany
Fell 1900, June 12, 1400 hrs
Stone. Olivine-bronzite chondrite (H6).
A stone of over 240g was seen to fall, R. Brauns, Verh. Naturh. Ver. Preuss. Rheinl., 1918, **75**, p.129. Olivine Fa_{19}, B. Mason, Geochimica et Cosmochimica Acta, 1963, **27**, p.1011.
Nearly complete stone, Bonn, Univ. Mus.; 8.5g Vienna, Naturhist. Mus.;
Specimen(s): [1924,326], 7.5g.

Forsyth 33°1′ N., 83°58′ W.

Monroe County, Georgia, U.S.A.
Fell 1829, May 8, 1530 hrs
Synonym(s): *Monroe County*
Stone. Olivine-hypersthene chondrite (L6), veined.
A stone of 36lb fell, after detonations, B. Silliman, Am. J. Sci., 1830, **18**, p.388. Olivine Fa_{23}, B. Mason, Geochimica et Cosmochimica Acta, 1963, **27**, p.1011. Repository of the main mass not known.
150g Tempe, Arizona State Univ.; 132g Yale Univ.; 88g Vienna, Naturhist. Mus.; 89g New York, Amer. Mus. Nat. Hist.; 84g Chicago, Field Mus. Nat. Hist.; 59g Tübingen, Univ.; 57g Harvard Univ.; 19g Berlin, Humboldt Univ.;
Specimen(s): [90266], 50.5g.; [33947], 21.5g.

Forsyth v Mincy.

Forsyth County 36°6′ N., 80°12′ W.

South Carolina, U.S.A.
Found 1891, approx. year
Iron. Hexahedrite (IIA).
A mass of about 50lb was ploughed up "about three years before 1895", E.A. de Schweinitz, Am. J. Sci., 1896, **1**, p.208. Description; may be from Forsyth County, Georgia; a single hexahedrite crystal cosmically shocked and reheated, V.F. Buchwald, Iron Meteorites, Univ. of California, 1975, p.569. Described, with an analysis, 5.55 %Ni, E. Cohen, Sitzungsber. Akad. Wiss. Berlin, 1897, p.386. Newer analysis, 5.50 %Ni, 60.8 ppm.Ga, 176 ppm.Ge, 31 ppm.Ir, J.T. Wasson, Meteorites, Springer-Verlag, 1974, p.299. Repository of the main mass unknown.
0.9kg New York, Amer. Mus. Nat. Hist.; 1121g Washington, U.S. Nat. Mus.; 638g Yale Univ.; 536g Chicago, Field Mus. Nat. Hist.; 514g Vienna, Naturhist. Mus.; 504g Berlin, Humboldt Univ.; 480g Raleigh, N. Carolina State Mus.; 358g Michigan Univ.; 339g Tübingen, Univ.;
Specimen(s): [82774], 308g.; [1920,302], 218g.

Fort Duncan v Coahuila.

Fort Dunkan v Coahulia.

Forte Pierre v Fort Pierre.

Fort Flatters 28°15′ N., 7° E. approx.

Eastern Erg, Algeria
Fell 1944, June 23
Stone.
A stone was seen to fall about 35km NE. of Fort Flatters, 28°4′N., 6°38′E., and is preserved in the Musée Saharienne, Ouargla, Algeria, L.F. Brady, letter of 26 January, 1953, in Min. Dept. BM(NH).

Fort Pierre 44°21′ N., 100°23′ W.

Stanley County, South Dakota, U.S.A.
Found 1856
Synonym(s): *Forte Pierre, Nebraska*
Iron. Octahedrite, medium (1.1mm) (IIIA).
A mass of 35lb was found about 20 miles NW. of Fort Pierre, between Council Bluff and Fort Union; analysis, 7.18 %Ni, N. Holmes, Trans. St. Louis Acad. Sci., 1860, **1**, p.711, C.U. Shepard, Am. J. Sci., 1860, **30**, p.204. The mass was artifically heated to about 800C. by the finders, V.F. Buchwald, Iron Meteorites, Univ. of California, 1975, p.572. Classification and analysis, 7.46 %Ni, 18.3 ppm.Ga, 35.9 ppm.Ge, 7.4 ppm.Ir, A. Kracher et al., Geochimica et Cosmochimica Acta, 1980, **44**, p.773.
4.5kg St. Louis, Mus. Acad. Sci.; 1829g New York, Amer. Mus. Nat. Hist.; 1.7kg Vienna, Naturhist. Mus.; 356g Yale Univ.; 337g Tempe, Arizona State Univ.; 184g Harvard Univ.; 99g Washington, U.S. Nat. Mus.; 64g Chicago, Field Mus. Nat. Hist.; 12g Berlin, Humboldt Univ.;
Specimen(s): [77187], 1882g.; [35415], 68g.

Fort Stockton 30°55′ N., 103°4′ W.

Pecos County, Texas, U.S.A.
Found 1952
Iron. Octahedrite, medium.
A single mass of 4.7kg was found, Cat. Monnig Colln., 1983.
4.7kg Fort Worth, Texas, Monnig Colln.;

Fossil Springs v Cañon Diablo.

Foum Tatahouine v Tatahouine.

Four Corners 37°0′ N., 109°3′ W.

San Juan County, New Mexico, U.S.A.
Found 1924, approx.
Iron. Octahedrite, medium (0.8mm), with silicate inclusions (IB).
A mass of about 25kg was found 15 miles SE. from the common corner of Colorado, New Mexico, Arizona and Utah; described, with an analysis of iron and insoluble silicate, 9.57 %Ni, G.P. Merrill, Proc. Nat. Acad. Sci., 1924, **10**, p.312. Mentioned, B. Mason, Min. Mag., 1967, **36**, p.120. Analysis of metal, 8.90 %Ni, 48.7 ppm.Ga, 179 ppm.Ge, 2.0 ppm.Ir, J.T. Wasson, Icarus, 1970, **12**, p.407. Silicate inclusions of the Copiapo type, T.E. Bunch et al., Contr. Mineral. Petrol., 1970, **25**, p.297. K-Ar age of inclusions, about 4500 m.y., D. Bogard et al., Earth planet. Sci. Lett., 1968, **3**, p.275. Description, polycrystalline parent taenite has decomposed to kamacite plus taenite, V.F. Buchwald, Iron

Meteorites, Univ. of California, 1975, p.573.
7.25kg Washington, U.S. Nat. Mus., main mass; 2.5kg Canberra, Austr. Nat. Univ.; 1.5kg Harvard Univ.; 1917g New York, Amer. Mus. Nat. Hist.; 1850g Chicago, Field Mus. Nat. Hist.; 1226g Harvard Univ.; 509g Tempe, Arizona State Univ.; 414g Copenhagen, Univ. Geol. Mus.; 98g Berlin, Humboldt Univ.;
Specimen(s): [1924,331], 1132g. slice, and fragments, 20g

Fouta Senegalais v Siratik.

Framington v Farmington.

Franceville 38°49′ N., 104°37′ W.

El Paso County, Colorado, U.S.A.
Found 1890
Iron. Octahedrite, medium (1.0mm) (IIIA).
A mass of 41lb 6.5oz was found; described, with an analysis, 8.05 %Ni, H.L. Preston, Proc. Rochester Acad. Sci., 1902, **4**, p.75. More recent analysis, 8.2 %Ni, 20.4 ppm.Ga, 42.4 ppm.Ge, 0.38 ppm.Ir, E.R.D. Scott et al., Geochimica et Cosmochimica Acta, 1973, **37**, p.1957. Has been shocked at 400-750kb and is distinct from Ellicott (*q.v.*), E.J. Olsen et al., Meteoritics, 1974, **9**, p.263. Description, V.F. Buchwald, Iron Meteorites, Univ. of California, 1975, p.575.
4.6kg Vienna, Naturhist. Mus.; 2.25kg Harvard Univ.; 863g Chicago, Field Mus. Nat. Hist.; 656g New York, Amer. Mus. Nat. Hist.; 397g Tempe, Arizona State Univ.; 355g Budapest, Nat. Hist. Mus.; 319g Prague, Nat. Mus.; 297g Washington, U.S. Nat. Mus.;
Specimen(s): [86424], 772g. slice

Francfort v Frankfort (iron).

Franckfort v Frankfort (iron).

Frankel City 32°20′ N., 102°44′42″ W.

Andrews County, Texas, U.S.A.
Found 1977
Stone. Olivine-hypersthene chondrite (L6).
A single mass of 4.7kg was ploughed up, G.I Huss, letter of 7 December, 1982 in Min. Dept. BM(NH).
Specimen(s): [1983,M.44], 3g.

Frankfort (iron) 38°12′ N., 84°50′ W.

Franklin County, Kentucky, U.S.A.
Found 1866
Synonym(s): *Francfort, Franckfort, Franklin County*
Iron. Octahedrite, medium (1.2mm) (IIIA).
A mass of 24lb was found on a hill 8 miles SW. of Frankfort. Described, with an analysis, 8.53 %Ni, J.L. Smith, Am. J. Sci., 1870, **49**, p.331. New analysis, 7.85 %Ni, 20.2 ppm.Ga, 40.4 ppm.Ge, 1.8 ppm.Ir, E.R.D. Scott et al., Geochimica et Cosmochimica Acta, 1973, **37**, p.1957. Mass heavily corroded, probably found in Anderson County; shocked but not annealed, V.F. Buchwald, Iron Meteorites, Univ. of California, 1975, p.576.
5.5kg Harvard Univ.; 1.6kg Washington, U.S. Nat. Mus.; 37g Yale Univ.; 33g Tempe, Arizona State Univ.;
Specimen(s): [40888], 149.5g.; [40889], 66.5g.

Frankfort (stone) 34°29′ N., 87°50′ W.

Franklin County, Alabama, U.S.A.
Fell 1868, December 5, 1500 hrs
Synonym(s): *Franklin County*
Stone. Achondrite, Ca-rich. Howardite (AHOW).
A stone of about 650g was seen to fall, after detonations, 4 miles S. of Frankfort; described, with an analysis, by, G.J. Brush, Am. J. Sci., 1869, **48**, p.240. New description, and analysis, 13.96 % total iron, B. Mason and H.B. Wiik, Am. Mus. Novit., 1966 (2272). Petrology, M.B. Duke and L.T. Silver, Geochimica et Cosmochimica Acta, 1967, **31**, p.1637.
22g Harvard Univ.; 90g Yale Univ.; 64g Vienna, Naturhist. Mus.; 21g Chicago, Field Mus. Nat. Hist.; 19g New York, Amer. Mus. Nat. Hist.;
Specimen(s): [47238], 24.5g.; [43055], 5.75g.; [43685], 0.5g.

Franklin 36°43′ N., 86°34′ W.

Simpson County, Kentucky, U.S.A.
Found 1921
Stone. Olivine-bronzite chondrite (H5), black.
A mass of 9062g was found by farmers ploughing a field, but was not recognised as meteoritic until 1956, F.C. Leonard, Publ. Astron. Soc. Pacific, 1957, **69**, p.459 [M.A. 14-130]. Olivine Fa_{19}, B. Mason, Geochimica et Cosmochimica Acta, 1963, **27**, p.1011.
7.959kg Los Angeles, Univ. of California; 207g Washington, U.S. Nat. Mus.; 95g Tempe, Arizona State Univ.;
Specimen(s): [1966,488], 66.5g. a sawn piece, and fragment, 1.2g.

Franklin County v Frankfort (iron).

Franklin County v Frankfort (stone).

Franklinville 38°30′ N., 100° W. approx.

Ness County, Kansas, U.S.A.
Found 1888, approx.
Stone. Olivine-hypersthene chondrite (L6).
A number of stones were found in the Franklinville area, SW. of Ness City, O.C. Farrington, Field Mus. Nat. Hist. Geol. Ser., 1902 (64), p.300. There is some confusion about the number of distinct falls in this area, W.F. Read, Meteoritics, 1972, 7, p.417. Possibly part of Ness County (*q.v.*).

Fransfontein v Gibeon.

Fraserburg v Jackalsfontein.

Freda 46°23′ N., 101°14′ W.

Grant County, North Dakota, U.S.A.
Found 1919
Iron. Ataxite, Ni-rich (IIICD).
An oriented mass of 268g, showing marked flowlines, was ploughed up. Described, with an analysis, 23.49 %Ni, E.P. Henderson and S.H. Perry, Proc. U.S. Nat. Mus., 1942, **92** (3134), p.21 [M.A. 8-375]. Mentioned, W.D. Brentnall and H.J. Axon, J. Iron Steel Inst., 1962, **200**, p.947. Analysis, 22.57 %Ni, 2.09 ppm.Ga, 2.24 ppm.Ge, 0.021 ppm.Ir, J.T. Wasson and R. Schaudy, Icarus, 1971, **14**, p.59. Description; interior not affected by heat during atmospheric flight (vide Brentnall and Axon), V.F. Buchwald, Iron Meteorites, Univ. of California, 1975, p.577.
255g Washington, U.S. Nat. Mus., main mass;

Frederick County v Emmitsburg.

Fremont Butte 38°30′ N., 105°30′ W. approx.

Colorado, U.S.A.
Found 1963, recognized in this year
Stone. Olivine-hypersthene chondrite (L4).
One weathered mass of 6.647kg was recovered from a farm near Fremont Butte in 1963. Description, G.I. Huss et al.,

Meteoritics, 1966, **3**, p.73. Chromite in chondrules, P. Ramdohr, Geochimica et Cosmochimica Acta, 1967, **31**, p.1961. Olivine Fa_{24}, B. Mason, Geochimica et Cosmochimica Acta, 1967, **31**, p.1100.

2.6kg Mainz, Max-Planck-Inst.; 370g Denver, Mus. Nat. Hist.; 162g Washington, U.S. Nat. Mus.; 137g Harvard Univ.; 131g Los Angeles, Univ. of California; 93g Chicago, Field Mus. Nat. Hist.;

Specimen(s): [1965,411], 120g.

Frenchman Bay 30°36′30″ S., 115°10′ E.

Western Australia, Australia
Found 1964
Stone. Olivine-bronzite chondrite (H3).

One weathered mass of 8.8kg was found among limestone outcrops and sand dunes SSE. of Frenchman Bay. The stone is roughly pear-shaped with maximum dimensions 10 × 7.75 × 4.25 inches and a coating of iron shale about 0.75 inch thick and composed mainly of limonite, G.J.H. McCall, J. Roy. Soc. West. Austr., 1966, **49**, p.45. Olivine $Fa_{18.5}$, B. Mason, Rec. Austr. Mus., 1974, **29**, p.169.

Almost complete stone, Perth, West. Austr. Mus.; 82g Washington, U.S. Nat. Mus.; 69g New York, Amer. Mus. Nat. Hist.;

Specimen(s): [1968,282], 28g. a part-slice; [1968,283], 3.4g.

Friedeburg v Friedland.

Friedland

Frankfurt, Germany
Fell 1304, October 1
Synonym(s): *Friedeburg*
Doubtful..

Several stones, black and hard as iron, are said to have fallen at a place given by some authors as Friedland (Brandenburg, 52°7′N., 14°15′E.) and by some others as Friedeburg (on the Saale, 51°37′N., 11°45′E.), E.F.F. Chladni, Die Feuer-Meteore, Wien, 1819, p.200. The evidence is not conclusive.

Friedrichsau v Grzempach.

Friona 34°32′24″ N., 102°34′6″ W.

Parmer County, Texas, U.S.A.
Found 1981, recognized in this year
Stone. Olivine-hypersthene chondrite (L5).

Two stones, totalling 19.6kg, were picked up in a ploughed field, Meteor. Bull., 1983 (61), Meteoritics, 1983, **18**, p.79.

144g Chicago, Field Mus. Nat. Hist.;

Fujian 26° N., 119° E. approx.

Fujian, China
Found
Iron.

A single mass of unreported weight, is listed without further references, D. Bian, Meteoritics, 1981, **16**, p.115. Co-ordinates from map figure, D. Wang and Z. Ouyang, Geochimica, 1979, p.120 [M.A. 80-2085].

Specimen, Nanjing, Geol. Mus.;

Fukue 32°42′ N., 128°51′ E.

Goto Islands, Nagasaki prefecture, Kyushu, Japan
Found 1849, January, or February
Synonym(s): *Hukue, Hukuye*
Iron. Octahedrite, medium.

Said to have fallen in January or February 1849, at Fukue village, Fukue Island, but there is no credible information on the matter, K. Jimbo, Beitr. Miner. Japan, 1906 (2), p.33. According to I. Yamamoto, this iron was found in February 1842, I. Yamamoto, Kwasan Observ. Bull., 1935, p.306, P.M. Millman, Trans. Roy. Astron. Soc. Canada, 1938, **32**, p.197 [M.A. 7-173]. Date of fall, and classification, S. Murayama, letter of 6 April 1962, in Min. Dept. BM(NH).

8g Tokyo, Imp. Mus.;

Fukutomi 33°11′ N., 130°12′ E.

Kinoshima, Saga, Kyushu, Japan
Fell 1882, March 19, 1300 hrs
Synonym(s): *Hiokomo, Hukutomi, Kijima, Kinejima*
Stone. Olivine-hypersthene chondrite (L5), veined.

Two stones fell, after detonations, weighing 7.20kg and 4.42kg, K. Jimbo, Beitr. Miner. Japan, 1906 (2), p.41. Mentioned, S. Murayama, letter of 6 April 1962, in Min. Dept. BM(NH). Olivine Fa_{24}, B. Mason, Geochimica et Cosmochimica Acta, 1963, **27**, p.1011. Description, analysis, 22.78 % total iron, M. Shima et al., Bull. Nat. Sci. Mus. Tokyo, Ser. E, 1979, **2**, p.17.

Main masses of the two stones, Tokyo, Imp. Mus.; 175g Chicago, Field Mus. Nat. Hist.; 28g Tempe, Arizona State Univ.; 27g Budapest, Nat. Mus.; 20g Vienna, Naturhist. Mus.;

Specimen(s): [1905,71], 207g. and fragments, 3.2g; [68212], 3.61g.

Fuling 29°42′ N., 107°24′ E.

Sichuan, China
Found 1960, before this year
Iron.

A single mass of unreported weight is in Beijing (Peking), D. Bian, Meteoritics, 1981, **16**, p.121.

Specimen, Beijing, Acad. Sin. Inst. Geol.;

Fulton County v Rochester.

Fünen 55°20′ N., 10°20′ E.

Denmark
Fell 1654, March 30, 0800 hrs
Synonym(s): *Fyen, Fyn, Fynen, Ørsted*
Stone.

Many stones fell, light coloured with a thick black crust, and with small metallic patches inside, E.F.F. Chladni, Die Feuer-Meteore, Wien, 1819, p.228. This was clearly a genuine meteorite, but nothing is now preserved. Account of the fall, A. Garboe, Medd. Dansk Geol. Foren., 1952, **12**, p.296, K. Callinsen, Medd. Dansk Geol. Foren., 1952, **12**, p.297 [M.A. 12-246].

Furnas County v Norton County.

Fürstenburg v Menow.

Futtehpur 25°57′ N., 80°49′ E.

Allahabad district, Uttar Pradesh, India
Fell 1822, November 30, 1800 hrs
Synonym(s): *Allahabad, Bithur, Doab, Fatehpur, Shahpur*
Stone. Olivine-hypersthene chondrite (L6), veined.

After appearance of a fireball and detonations, several stones, weighing from 1lb to 4lb each, fell at Futtehpur and Bithur, about 70 miles NW. of Allahabad, J. Tytler, Edinburgh J. Sci., 1828, **8**, p.171, W. von Haidinger, J. Asiatic Soc. Bengal, 1859, **28**, p.259. Olivine Fa_{24}, B. Mason, Geochimica et Cosmochimica Acta, 1963, **27**, p.1011.

778g Calcutta, Mus. Geol. Surv. India; 492g Vienna, Naturhist. Mus., two stones; 97g Paris, Mus. d'Hist. Nat.; 450g Edinburgh, Roy. Scot. Mus.; 93g Chicago, Field Mus. Nat. Hist.; 53g Harvard Univ.; 36g Yale Univ.; 50g Tempe, Arizona State Univ.;

Specimen(s): [34800], 798g.; [33757], 444g. and fragments, 4g; [34805], 136g. Bithur

Fuyang 32°54′ N., 115°48′ E.
Fuyang County, Anhui, China
Fell 1977, June
Stone.
Three stones were recovered, totalling 2-3kg, D. Bian, Meteoritics, 1981, **16**, p.115.
Main mass, Hefei, Univ. of Sci. and Tech.;

Fuzzy Creek 31°36′40″ N., 99°54′15″ W.
Runnels County, Texas, U.S.A.
Found, year not reported
Iron. Anomalous (IVA).
Listed, without details, analysis, 11.8 %Ni, 2.28 ppm.Ga, 0.138 ppm.Ge, 0.18 ppm.Ir, D.J. Malvin et al., priv. comm., 1983.
2.6kg Fort Worth, Texas, Monnig Colln.;

Fyen v Fünen.

Fyn v Fünen.

Fynen v Fünen.

Gabiano v Borgo San Donino.

Gail 32°42′ N., 101°36′ W.
Borden County, Texas, U.S.A.
Found 1948
Stone. Olivine-bronzite chondrite (H).
An almost complete, weathered stone of 4.65kg was found in three pieces in a farmyard in the NW. corner of Borden County, close to the former settlement of Tredway, O.E. Monnig, letter of 28 November, 1948 to E.P. Henderson, copy in Min. Dept. BM(NH). Olivine Fa_{18}, B. Mason, Geochimica et Cosmochimica Acta, 1963, **27**, p.1011.
4.67kg Fort Worth, Texas, Monnig Colln.;

Gaillac v Favars.

Gaj 52°31′35″ N., 21°25′45″ E. approx.
Warsaw, Poland
Fell 1927, June 10-15, 1030 hrs, approx. time
Doubtful..
A meteorite was seen to fall into a marsh near Wolomin, NE. of Warsaw, J. Pokrzywnicki, Studia Geol. Polon., 1964, **15**, p.122.

Galapian 44°18′ N., 0°24′ E.
Agen, Lot-et-Garonne, France
Fell 1826, August, or May
Synonym(s): *Agen*
Stone. Olivine-bronzite chondrite (H6), veined.
A large stone is said to have fallen in the month of August, 1826, B. de Ferussac, Bull. Sci. Natur., Paris, 1827, **11**, p.420. A piece of 39g said to have fallen on May 19, 1826, (the same date as Pavlograd), is in Paris (Mus. d'Hist. Nat.), S. Meunier, Cat. Coll. Meteorites Mus. d'Hist. Nat., Paris, 1909, p.30. Olivine Fa_{20}, B. Mason, Geochimica et Cosmochimica Acta, 1963, **27**, p.1011.
5g Chicago, Field Mus. Nat. Hist.; 2g Vienna, Naturhist. Mus.;

Galapian v Agen.

Galatia 38°38′30″ N., 98°53′ W.
Barton County, Kansas, U.S.A.
Found 1971, August 8
Stone. Olivine-hypersthene chondrite (L6).
A single mass of 23.9kg was found in 1971 and identified as meteoritic in 1976. Description, classification and mineral data, olivine $Fa_{24.9}$, W.R. Van Schmus et al., Meteoritics, 1978, **13**, p.267. Reported, Meteor. Bull., 1979 (56), Meteoritics, 1979, **14**, p.161.
16.2kg Los Angeles, Univ. of California; 3.8kg Mrs. C. DeWald; 1kg Barton County Community College; 400g Univ. of Kansas; 88g Chicago, Field Mus. Nat. Hist.;
Specimen(s): [1978,M.14], 2.49kg.

Galim 7°3′ N., 12°26′ E.
Adamoua, Cameroun
Fell 1952, November 13, 1700 hrs
Stone. Olivine-hypersthene chondrite, amphoterite (LL6), veined, brecciated.
Two fragments, 19g and 9g, were recovered from a fall observed in the region of Galim, 120km W. of Ngaoundere, Adamoua region, and were sent to Paris, E. Jérémine, C. R. Acad. Sci. Paris, 1953, **237**, p.1740 [M.A. 12-610]. The fireball was seen over Adamawa province, Nigeria, R. Walls, letter of 24 August, 1953 in Min. Dept. BM(NH). Reported, Meteor. Bull., 1958 (8). Three small pieces were found at Tignere, W.N. MacLeod and R. Walls, Rec. Geol. Surv. Nigeria, 1958, p.25., J. Orcel and E. Jérémine, 83 Congr. Soc. Sav., 1958, p.211 [M.A. 15-39]. Discussion, olivine Fa_{29}, B. Mason and H.B. Wiik, Geochimica et Cosmochimica Acta, 1964, **28**, p.533.
28g Paris, Mus. d'Hist. Nat., the two pieces found at Galim.;

Galleguillos v Ternera.

Galleguitos v Ternera.

Gallipoli v Henbury.

Gambat 27°21′ N., 68°32′ E.
Khairpur district, Sind, Pakistan
Fell 1897, September 15
Stone. Olivine-hypersthene chondrite (L6), veined.
A stone of about 14lb fell, H.H. Hayden, letter of 24 March, 1898 in Min. Dept. BM(NH). Olivine Fa_{23}, B. Mason, Geochimica et Cosmochimica Acta, 1963, **27**, p.1011.
3.75kg Calcutta, Mus. Geol. Surv. India; 438g Vienna, Naturhist. Mus.; 400g Paris, Mus. d'Hist. Nat.; 90g Budapest, Nat. Mus.; 77g Oxford, Univ. Mus.; 72g Berlin, Humboldt Univ.; 72g Chicago, Field Mus. Nat. Hist.; 43g New York, Amer. Mus. Nat. Hist.; 30g Tempe, Arizona State Univ.;
Specimen(s): [83864], 1752g. and 84.6g

Ganado v Cañon Diablo.

Gancedo v Campo del Cielo.

Gao (Mali) 16°18′ N., 0° W. approx.
Soudan, Mali
Fell 1932, January 19, 2100 hrs
Doubtful..
A meteorite is said to have fallen, but nothing appears to have been recovered and the fall must be regarded as doubtful, F.C. Leonard, Contr. Soc. Res. Meteorites, 1945, **3**, p.224 [M.A. 9-289].

Gao (Upper Volta) 11°39′ N., 2°11′ W.
Upper Volta
Fell 1960, March 5, 1700 hrs, approx. time
Stone. Olivine-bronzite chondrite (H).
At least 16 stones fell 60km north of Leo, near the Upper Volta-Ghana border; the largest stone weighed 2.5kg olivine Fa_{18}, B. Mason, Geochimica et Cosmochimica Acta, 1967, **31**, p.1100.

Gara Ikhrarene v Ikharene.

Garden Head 49°49′ N., 108°27′36″ W.
Saskatchewan, Canada
Found 1944, early autumn
Iron. Octahedrite, plessitic (0.1mm) (IRANOM).
A mass of 1.296kg was found and is in the Canadian National Meteorite Collection, Meteor. Bull., 1964 (32). Classification and analysis, 16.96 %Ni, 10.7 ppm.Ga, 16.6 ppm.Ge, 0.12 ppm.Ir, J.T. Wasson and R. Schaudy, Icarus, 1971, **14**, p.59. Description, spindle width, V.F. Buchwald, Iron Meteorites, Univ. of California, 1975, p.580.
1256g Ottawa, Geol. Surv. Canada;

Gargantillo v Tomatlan.

Garhi Yasin 27°53′ N., 68°32′ E.
Shikarpur taluk, Sukkur district, Sind, Pakistan
Fell 1917, January, during darkness hrs
Iron. Octahedrite, medium (1.1mm) with silicate inclusions (IIE).
After the appearance of a 'ball of fire', and detonations, a mass of 380g was seen by a Sikh to fall and was given to E.L. Moysey next day, E.L. Moysey, letter of 2 October, 1924 in Min. Dept. BM(NH). Classification and analysis, 8.3 %Ni, 20.9 ppm.Ga, 65 ppm.Ge, 4.6 ppm.Ir, E.R.D. Scott and J.T. Wasson, Geochimica et Cosmochimica Acta, 1976, **40**, p.103. Description, thoroughly annealed, V.F. Buchwald, Iron Meteorites, Univ. of California, 1975, p.580.
Specimen(s): [1924,832], 337g. and two pieces, 14.5g, and filings, 10g

Garland 41°41′ N., 112°8′ W.
Box Elder County, Utah, U.S.A.
Fell 1950, in the summer
Stone. Achondrite, Ca-poor. Diogenite (ADIO).
One stone of 102g was recovered, B. Mason, Meteorites, Wiley, 1962, p.244, B. Mason, Am. Mus. Novit., 1963 (2155). The time of fall was about 1100hrs., F.E. Wickman, letter of 14 May, 1973 in Min. Dept. BM(NH). Unequilibrated, breccia, microprobe analyses, orthopyroxene Fs20-42, J.V. Heyse, Meteoritics, 1975, **10**, p.413, abs.
102g Washington, U.S. Nat. Mus., main mass.;

Garnet v Garnett.

Garnett 38°16′ N., 95°15′ W.
Anderson County, Kansas, U.S.A.
Found 1938, recognized in this year
Synonym(s): *Garnet*
Stone. Olivine-bronzite chondrite (H4).
One stone of 4788g was found in sect. 22, township 21, range 20, Anderson Co., A.D. Nininger, Pop. Astron., Northfield, Minnesota, 1939, **47**, p.212, E.P. Henderson, letter of 3 June, 1939 in Min. Dept. BM(NH). This weathered stone was ploughed up in 1937. Described, with an analysis, H.H. Nininger, Pop. Astron., Northfield, Minnesota, 1941, **49**, p.326 [M.A. 8-193]. Olivine Fa_{20}, B. Mason, Geochimica et Cosmochimica Acta, 1963, **27**, p.1011. Further bulk analysis, 24.40 % total iron, A.J. Easton and C.J. Elliott, Meteoritics, 1977, **12**, p.409.
1.4kg Tempe, Arizona State Univ.; 331g Harvard Univ.; 286g Washington, U.S. Nat. Mus.; 250g Chicago, Field Mus. Nat. Hist.;
Specimen(s): [1959,862], 742g. and fragments, 12g

Garraf 41°16′ N., 1°55′ E.
Barcelona, Spain
Found 1905
Synonym(s): *Costal de Garraf*
Stone. Olivine-hypersthene chondrite (L6).
A stone of about 8.8kg was found, M. Faura y Sans, Butll. Centr. Excurs. Catalunya, 1921, **31** (322), p.270. Description, with an analysis, M. Faura y Sans, Meteoritos caidos en la Peninsula Iberica, Tortosa, 1922, p.58. Olivine Fa_{24}, B. Mason, Geochimica et Cosmochimica Acta, 1963, **27**, p.1011. The Costal de Garraf is a stretch of coast from Sitjes, 41°15'N., 1°50'E., to Castell de Fels, 41°17'N., 2°0'E.
4.9kg Barcelona, Collegi Comercial de la Bonanova and Mus. Cienc. Nat.; 365g Paris, Mus. d'Hist. Nat.; 323g Tempe, Arizona State Univ.; 185g Harvard Univ.; 70g New York, Amer. Mus. Nat. Hist.; 68g Chicago, Field Mus. Nat. Hist.;
Specimen(s): [1927,1050], 41.6g.

Garrett County v Lonaconing.

Garrison 33°51′48″ N., 103°20′48″ W.
Roosevelt County, New Mexico, U.S.A.
Found 1969
Stone. Olivine-bronzite chondrite (H5).
One stone of 792.8g was found during ploughing in 1969. A second stone weighing 4323g was subsequently found 3 miles from the find site of the first stone, olivine Fa_{18}, Meteor. Bull., 1979 (56), Meteoritics, 1979, **14**, p.166.
5kg Mainz, Max-Planck-Inst.;
Specimen(s): [1977,M.2], 40.2g.

Garz v Schellin.

Gawler Range v Yardea.

Gay Gulch v Klondike (Gay Gulch).

Gdynia 54°35′ N., 18°40′ E. approx.
Gdansk, Poland
Fell 1959, January 21, 0500 hrs, approx. time
Doubtful..
Following detonations, many people witnessed a meteorite falling into the sea near Gdynia. Soon after the incident a diver examined the sea bed and in 1960 a dredge was used, but no meteorite was found, J. Pokrzywnicki, Studia Geol. Polon., 1964, **15**, p.124.

Geidam 12°55′ N., 11°55′ E.
Bornu province, Nigeria
Fell 1950, July 6
Synonym(s): *Seidam*
Stone. Olivine-bronzite chondrite (H5).
A stone of 725g fell at Garau in the Zajibiri village unit of Geidam district, W.N. MacLeod and R. Walls, Rec. Geol. Surv. Nigeria, 1958, p.24, 1959, p.18. It closely resembles Akwanga (*q.v.*). Description and analysis, 27.2 % total iron, B. Mason and H.B. Wiik, Am. Mus. Novit., 1965 (2220).
Main mass, Geol. Surv. Nigeria; 41g New York, Amer. Mus. Nat. Hist.; 5.5g Washington, U.S. Nat. Mus.;

Geluksfontein v Cronstad.

Genichesk 46°42′ N., 32°36′ E.

Kherson district, Ukraine, USSR, [Геническ]
Found 1927, August 10-15
Stone. Olivine-hypersthene chondrite (L5-6).

A single mass weighing 531g was found on the surface of a cultivated field, 3-4km west of the town of Genichesk. One or two weeks previously a fireball as bright as the full moon was observed. It fragmented into sparks but no sound was heard, Meteor. Bull., 1978 (55), Meteoritics, 1978, **13**, p.336. Description, R.L. Khotinok, Метеоритика, 1978, **37**, p.74. Description and analysis, 22.80 % total iron, O.A. Kirova et al., Метеоритика, 1980, **39**, p.49.

449g Moscow, Acad. Sci., includes main mass, 240g;

Gent v St. Denis Westrem.

Georgetown 39°42′ N., 105°42′ W.

Clear Creek County, Colorado, U.S.A.
Found
Stone. Olivine-bronzite chondrite (H6).

One stone of 687g is in Washington, U.S. Nat. Mus. olivine Fa_{19}, B. Mason, Geochimica et Cosmochimica Acta, 1967, **31**, p.1100.

687g Washington, U.S. Nat. Mus.;

Gera v Pohlitz.

Gerano v Orvinio.

Gerona 41°58′ N., 2°49′ E.

Catalonia, Spain
Found 1899
Synonym(s): *Girona*
Stone. Olivine-bronzite chondrite (H), brecciated.

A stone of about 148g was found in the environs of Gerona, M. Faura y Sans, Butll. Centr. Excurs. Catalunya, 1921, **31** (322), p.270. Described, with an analysis, S. Calderon and J.R. Mourelo, Acta R. Soc. Esp. Hist. Nat., 1900, **29**, p.70. Figured, M. Faura y Sans, Meteoritos caidos en la Peninsula Iberica, Tortosa, 1922, p.55. Olivine Fa_{19}, B. Mason, Geochimica et Cosmochimica Acta, 1963, **27**, p.1011.

85g Madrid, Mus. Cienc. Nat.; 55g Barcelona, Dr. M. Cazurro's collection; 10g Tempe, Arizona State Univ.; 1g Chicago, Field Mus. Nat. Hist.;

Gerzeh 29°19′ N., 31°17′ E.

Egypt
Found
Iron.

Iron beads found in 1911 in predynastic graves at Gerzch, 50 miles south of Cairo, G.A. Wainwright, Revue archeolgie, 1912, **1**, p.255 were found, C.H. Desch, Rep. Brit. Assn., 1928, **98**, p.440, to contain 7.5%Ni, and are certainly of meteoritic origin, G.A. Wainwright, J. Egypt. Archaeol., 1932, **18**, p.3 [M.A. 5-151]. Mentioned, W.M. Flinders Petrie et al., The Labyrinth Gerzeh and Mazguneh, London, 1912. The three iron beads in the University College, London, Collection are oxidised. Description, V.F. Buchwald, Iron Meteorites, Univ. of California, 1975, p.583.

Gettysburg v Mount Joy.

Ghanim v al-Ghanim (iron).

Ghanim v al-Ghanin (stone).

Ghazeepore v Mhow.

Ghent v St. Denis Westrem.

Ghoordha v Moti-ka-nagla.

Ghubara 19°13′40″ N., 56°8′34″ E.

Oman
Found 1954
Synonym(s): *Ghubara (1954), Gnubara*
Stone. Olivine-hypersthene chondrite (L5), black, xenolithic.

Found on the surface of the desert, the stones are fresh internally, and the crust only slightly weathered. See letters from the Iraq Petroleum Co. of March, 1957 and September, 1958, in Min. Dept. BM(NH). Reported, Meteor. Bull., 1959 (13). Mentioned, G.P. Vdovykin, Метеоритика, 1964, **25**, p.134. Xenolithic, analyses of xenolith and host; the host is unequilibrated, with olivine Fa21.6 to Fa26.5, but olivine of xenoliths is Fa24. Host analysis, 21.33 % total iron, R.A. Binns, Geochimica et Cosmochimica Acta, 1968, **32**, p.299. See also under Jiddat al Harasis (*q.v.*).

1.7kg Chicago, Field Mus. Nat. Hist.; 1.5kg Los Angeles, Univ. of California; 303g New York, Amer. Mus. Nat. Hist.; 216g Mainz, Max-Planck-Inst.;

Specimen(s): [1954,207], 16kg. approx.; [1956,242], 23kg. approx.; [1958,805], 101.6kg. complete stone.; [1967,416], thin section

Ghubara (1954) v Ghubara.

Ghubara (1962) v Jiddat al Harasis.

Gibb's Meteorite v Red River.

Gibeon 25°30′ S., 18° E. approx.

Great Nama Land, Namibia
Found 1836, known before this year
Synonym(s): *Alexander, Amalia Farm, Bethany, Cabaya, Coamus Farm, Cobaya, Damaraland, Donas, Fish River, Fransfontein, Gibon, Goamus Farm, Great Fish River, Great Namaqualand, Gröndorn, Grossnamaqualand, Haruchas, Kameelhaar, Kamkas, Keetmanshoop, Kinas Putts, Lichtenfels, Lion River, Löwenfluss, Mukerop, Namaqualand, Nico, Springbok River, Tsess, Wild*
Iron. Octahedrite, fine (0.30mm) (IVA).

Large masses were reported near the E. bank of the Great Fish River and three days journey NE. from Bethany, J.E. Alexander, J. Roy. Geogr. Soc. London, 1838, **8**, p.24. Other large masses have since been found; description, V.F. Buchwald, Iron Meteorites, Univ. of California, 1975, p.584. Structure, H.J. Axon and P.L. Smith, Min. Mag., 1970, **37**, p.888. Trace element analysis, A.A. Smales et al., Geochimica et Cosmochimica Acta, 1967, **31**, p.673. Analysis, 7.68 %Ni, 1.97 ppm.Ga, 0.111 ppm.Ge, 2.4 ppm.Ir, R. Schaudy et al., Icarus, 1972, **17**, p.174.

880kg Cape Town, South African Mus., includes main mass, 650kg.; 12613kg Windhoek, Mus.; 730kg Pretoria, Geol. Mus.; 540kg Fort Worth, Texas, Monnig Colln.; 253kg Bonn, Univ. Mus.; 500kg Yale Univ.; 305kg Berlin, Humboldt Univ.; 424kg Hamburg, Mus.; 556kg Frankfurt am Main, Senk. Naturfors. Ges. Mus.; 263kg Harvard Univ.; 123kg Copenhagen, Univ. Geol. Mus.; 29.2kg Washington, U.S. Nat. Mus.; 19kg Prague, Nat. Mus.; 236.5kg Harvard Univ.; 173kg Tempe, Arizona State Univ.; 138kg Stuttgart, Nat.-Kab.; 500kg Vienna, Naturhist. Mus.; 53kg Chicago, Field Mus. Nat. Hist.; 21.6kg New York, Amer. Mus. Nat. Hist.; 11kg Rio de

Janiero, Serv. Geol.; 15.8kg Budapest, Nat. Mus.; 6kg Dublin, Nat. Mus.; 68kg Ottawa, Mus. Geol. Surv. Canada;
Specimen(s): [1930,422], 136kg. of Gibeon; [1929,1563], 195g. complete individual of Gibeon; [1909,4], 5130g. of Gibeon; [1910,753], 6610g. slice, of Goamus; [1921,386], 1208g. an end slice, of Goamus; [85891], 4320g. slice, of Mukerop; [1921,190], 123.5g. of Mukerop; [1959,158], an engraved paper knife, of Mukerop; [54729], 1436g. of Great Namaqualand; [1941,1], 1422g. of Kameelhaar, and filings, 11g; [1941,2], 755g. and fragments, 18g, of Kamkas; [32048], 350g. of Lion River; [63885], 9g. of Springbok River; [1911, 716], 2g. of Great Fish River, presented by J.E. Alexander to Geol. Soc., London, in 1838; [1959,159], 92g.

Gibon v Gibeon.

Gifu 35°32′ N., 136°53′ E.
Yamagata, Gifu prefecture, Honshu, Japan
Fell 1909, July 24, 0544 hrs
Synonym(s): *Aimi, Atobe, Hachiman, Hiromi, Itumi, Iwa, Izumi, Kamiariti, Kitano, Mino, Ooyada, Oyada, Takano, Taromaru, Umehara, Yawata*
Stone. Olivine-hypersthene chondrite (L6).
After detonations over 100 stones fell at Aimi and other villages near Gifu, and 27 were recovered, the largest weighing about 4kg, T. Wakimizu, Beitr. Miner. Japan, 1912 (4), p.145, S. Murayama, letter of 6 April, 1962, in Min. Dept. BM(NH). See report, which includes an analysis, A. Miyashiro, Jap. J. Geol. Geogr., 1962, **33**, p.125 [M.A. 17-56]. Olivine Fa_{25}, B. Mason, Geochimica et Cosmochimica Acta, 1963, **27**, p.1011.
7.4kg Tokyo, Nat. Sci. Mus.; 18g Washington, U.S. Nat. Mus.; 11g Tempe, Arizona State Univ.; 7g Vienna, Naturhist. Mus.; 1.6g New York, Amer. Mus. Nat. Hist.;
Specimen(s): [1973,M.27], 1.1g. fragments; [1978,M.16], 16.9g.

Gilgoin 30°23′ S., 147°12′ E.
Brewarrina, County Clyde, New South Wales, Australia
Found 1889
Synonym(s): *Gilgoin Station*
Stone. Olivine-bronzite chondrite (H5).
Seven stones, (nos. 1-7), weighing respectively, about 67.5lb, 74.2lb, 55.2lb, 37lb, 26.5lb, 16lb, 21.7lb, were found scattered over an area of about 4 square miles about 40 miles SE. of Brewarrina, H.C. Russell, J. and Proc. Roy. Soc. New South Wales, 1893, **27**, p.361, C. Anderson, Rec. Austr. Mus., 1913, **10** (5), p.58. Analyses, A. Liversidge, J. and Proc. Roy. Soc. New South Wales, 1902, **36**, p.354, J.C.H. Mingaye, Rec. Geol. Surv. New South Wales, 1916, **9**, p.166. Nature of the metal, H.C. Urey and T. Mayeda, Geochimica et Cosmochimica Acta, 1959, **17**, p.113. Olivine Fa_{17}, B. Mason, Geochimica et Cosmochimica Acta, 1963, **27**, p.1011. A further mass of 11.9kg was found in 1920 and is in the Austr. Mus.
75kg Sydney, Austr. Mus.; 13.3kg Tempe, Arizona State Univ.; 3kg New York, Amer. Mus. Nat. Hist.; 4.28kg Washington, U.S. Nat. Mus.; 2kg Yale Univ.; 800g Budapest, Nat. Mus.; 459g Prague, Nat. Mus.; 800g Ottawa, Mus. Geol. Surv. Canada;
Specimen(s): [86131], 1951g.; [1915,151], 261.5g. of the no. 7 stone; [1927,1276], 150g.; [1927,1277], 147g.; [1927,1278], 5.5g.; [1927,1279], 14.5g. fragments and powder

Gilgoin Station v Gilgoin.

Gilpin County v Russell Gulch.

Gindorcha v Indarch.

Girgenti 37°19′ N., 13°34′ E.
Sicily, Italy
Fell 1853, February 10, 1830 hrs, U.T.
Synonym(s): *Agrigento, Girghenti*
Stone. Olivine-hypersthene chondrite (L6), veined.
Three stones, one of about 7lb, fell, R.P. Greg, Phil. Mag., 1862, **24**, p.538. See also, with an analysis, G. vom Rath, Ann. Phys. Chem. (Poggendorff), 1869, **138**, p.541. Account of the fall, M.H. Hey, Am. J. Sci., 1951, **249**, p.249 [M.A. 11-440]. For a description and new analysis, 23.51 % total iron, G.R. Levi-Donati and E. Jarosewich, Meteoritics, 1972, **7**, p.109.
8.9kg Washington, U.S. Nat. Mus.; 2.5kg Milan, Mus.; 2.2kg Chicago, Field Mus. Nat. Hist.; 1kg Rome, Univ.; 411g Paris, Mus. d'Hist. Nat.; 474g Berlin, Humboldt Univ.; 423g Budapest, Nat. Mus.;
Specimen(s): [1948,255], 467.5g.; [1920,303], 121g.; [55241], 118g.; [46387], 55.5g.; [54762], 41.5g.; [90271], 1.5g.

Girger v Khor Temiki.

Girghenti v Girgenti.

Girona v Gerona.

Giroux 49°37′ N., 96°33′ W.
Manitoba, Canada
Found 1954
Stony-iron. Pallasite (PAL).
One mass of 4.275kg was found 3 miles NE. of Giroux, F.C. Leonard, Classif. Cat. Meteor., 1956, p.8, 43, 66, Meteor. Bull., 1957 (5), 1958 (8). Olivine Fa_{11}, P.R. Buseck and J.I. Goldstein, Bull. Geol. Soc. Amer., 1969, **80**, p.2141. Analysis of metal, 9 %Ni, 22.6 ppm.Ga, 50.8 ppm.Ge, 0.03 ppm.Ir, J.T. Wasson, Meteorites, Springer-Verlag, 1974, p.306. References, metal composition, 10.3 %Ni, E.R.D. Scott, Geochimica et Cosmochimica Acta, 1977, **41**, p.693.
2.3kg Washington, U.S. Nat. Mus.; 2kg Toronto, Roy. Ontario Mus.;

Git-Git 9°36′ N., 9°55′ E. approx.
Baram, Bauchi district, Nigeria
Fell 1947, January 9, 0700 hrs, approx. time.
Stone. Olivine-hypersthene chondrite (L6).
Two stones, of 280g and about 200g, were seen to fall, R. Walls, letters of 10 January, 1950, 15 February, 1951, 14 March, 20 April, 1951 in Min. Dept. BM(NH)., Rec. Geol. Surv. Nigeria, 1958, p.23. Olivine Fa_{24}, B. Mason, Geochimica et Cosmochimica Acta, 1963, **27**, p.1011.
105g Kaduna, Geol. Surv. Nigeria;
Specimen(s): [1951,263], 265g. and fragments, 2g.; [1951, 115], Thin section

Gjilatelke v Mocs.

Gladstone (iron) 23°54′ S., 151°18′ E.
Portcurtis, Queensland, Australia
Found 1915
Synonym(s): *Queensland, South Queensland*
Iron. Octahedrite, coarse (2.8mm) (IA).
A mass of 736.6kg was found 4 miles S. of Gladstone. Described, with an analysis, 6.4 %Ni, H.C. Richards, Mem. Queensland Mus., 1930, **10**, p.65 [M.A. 4-426]. A second mass of 24.1kg was found in 1940, N.A.H. Simmonds, Geol. Surv. Queensland, publ. no 320, 1964, Meteor. Bull., 1964

(32). Another analysis, 6.53 %Ni, 93.7 ppm.Ga, 418 ppm.Ge, 3.0 ppm.Ir, J.T. Wasson, Icarus, 1970, **12**, p.407. Description, a shower-producing coarse octahedrite, includes Queensland, V.F. Buchwald, Iron Meteorites, Univ. of California, 1975, p.593.

660kg Chicago, Field Mus. Nat. Hist., main mass; 21kg Washington, U.S. Nat. Mus.; 16kg New York, Amer. Mus. Nat. Hist.; 24kg Brisbane, Geol. Surv. Queensland; 13kg Sydney, Austr. Mus.; 6kg Harvard Univ.; 646g Tempe, Arizona State Univ.;

Specimen(s): [1930,101], 7.29kg. and fragments, 78.5g

Gladstone (stone) 36°18′ N., 104°0′ W.

Union County, New Mexico, U.S.A.
Found 1936
Stone. Olivine-bronzite chondrite (H6), black, veined.

Three stones were found just over the border in Colfax Co., weight 57.3kg, A.D. Nininger, Pop. Astron., Northfield, Minnesota, 1937, **45**, p.449 [M.A. 7-62]. Classification as H4, olivine $Fa_{18.7}$, D.E. Lange and K. Keil, Meteoritics, 1976, **11**, p.315.

45.5kg Tempe, Arizona State Univ.; 3.25kg Washington, U.S. Nat. Mus.; 2kg Albuquerque, Univ. of New Mexico; 1.26kg Harvard Univ.; 892g Los Angeles, Univ. of California; 800g Chicago, Field Mus. Nat. Hist.; 330g New York, Amer. Mus. Nat. Hist.;

Specimen(s): [1937,381], 451g.; [1959,824], 3872g. in two pieces, a slice of 1631g and an end piece of 2241g.

Glady 32°42′ N., 100°48′ W. approx.

Curry County, Texas, U.S.A.
Found 1946, known in this year
Doubtful. Stone.

Reported, F.C. Leonard, Univ. New Mexico Publ., Albuquerque, 1946 (meteoritics ser. no. 1), p.46. A place of this name has not been located and the lat. and long. quoted are those of Scurry Co., Texas; there is no Curry Co. in Texas. Probably a mistake for Grady, (*q.v.*), Curry County, New Mexico.

Glanggang 7°15′ S., 107°42′ E.

Near Bandung, Java, Indonesia
Fell 1939, September 26
Stone. Olivine-bronzite chondrite (H5-6).

A single stone of 1303g is said to have fallen on the above date, which is also the reported date of fall of Selakopi (*q.v.*). Reported, olivine $Fa_{18.4}$, Meteor. Bull., 1982 (60), Meteoritics, 1982, **17**, p.94. Description, K. Fredriksson and G.S. Peretsman, Meteoritics, 1982, **17**, p.77.

Specimen(s): [1982,M.21], 1.4g.

Glarus 47°3′ N., 9°4′ E.

Switzerland
Fell 1674, October 6
Doubtful..

Two large stones are said to have fallen, E.F.F. Chladni, Die Feuer-Meteore, Wien, 1819, p.237, but the evidence is not conclusive.

Glasatovo 57°21′ N., 37°37′ E.

Kashin, Tver govt., Kalinin, USSR, [Глазатово], [Кашин]
Fell 1918, February 27, 1245 hrs
Synonym(s): *Kaschin, Kashin*
Stone. Olivine-bronzite chondrite (H4), veined.

A stone of over 150kg fell, after detonations, 2 versts from Kashin, L.A. Kulik, Bull. Acad. Sci. Russ., 6, 1918, **12**, p.1089, P.N. Chirvinsky, Centralblatt Min., 1928, p.327 [M.A. 4-119]. Analysis, 27.40 % total iron, M.I. D'yakonova and V.Ya. Kharitonova, Метеоритика, 1961, **21**, p.52. Olivine Fa_{18}, B. Mason, Geochimica et Cosmochimica Acta, 1963, **27**, p.1011.

121.056kg Moscow, Acad. Sci., includes main mass, 121.05kg.;

Specimen(s): [1956,323], 110g. and fragments, 1.6g

Glasgow 37°1′ N., 85°55′ W.

Barren County, Kentucky, U.S.A.
Found 1922
Iron. Octahedrite, medium (1.1mm) (IIIA).

Two masses were found, of about 25lb and 20lb; described, with an analysis, G.P. Merrill, Am. J. Sci., 1922, **4**, p.329, 1923, **5**, p.63. Classification and new analysis, 7.6 %Ni, 20.6 ppm.Ga, 38.8 ppm.Ge, 5.0 ppm.Ir, E.R.D. Scott et al., Geochimica et Cosmochimica Acta, 1973, **37**, p.1957. Description, shock-hardened, V.F. Buchwald, Iron Meteorites, Univ. of California, 1975, p.595.

4047g Washington, U.S. Nat. Mus.; 3.5kg New York, Amer. Mus. Nat. Hist.; 265g Harvard Univ.; 138g Chicago, Field Mus. Nat. Hist.;

Specimen(s): [1935,892], 206g. and fragments, 32g.

Glasgow v High Possil.

Glasotovo error for Glasatovo.

Glastonbury 51°8′ N., 2°43′ W.

Somerset, England
Fell 1806, May 17
Doubtful..

A stone of 2.5lb is said to have fallen, E.J. Harris, Diss., Göttingen, 1859, p.67, quoting Monthly Mag. for April 1811. Compare, R.P. Greg, Phil. Mag., 1854, **8**, p.458. There is also a hazy separate report of a fall in July or August, 1816, Phil. Mag., 1816, **48**, p.235.

Glen Helen v Henbury.

Glenormiston 22°54′ S., 138°43′ E.

Boulia, Queensland, Australia
Found 1925
Synonym(s): *Boulia*
Iron. Structurally and chemically anomalous (IRANOM).

A mass of 90lb was found at Glenormiston. Described, with an analysis, 8.71 %Ni, H.C. Richards, Mem. Queensland Mus., 1930, **10**, p.65 [M.A. 4-426]. It is described as composed of an irregular aggregate of kamacite grains averaging 6mm across, with some interstitial taenite and plessite, showing well-marked Neumann lines but no Widmanstetter figures. Classification and new analysis, 7.45 %Ni, 16.9 ppm. Ga., 76.8 ppm.Ge, 2.7 ppm.Ir, E.R.D. Scott et al., Geochimica et Cosmochimica Acta, 1973, **37**, p.1957.

Main mass, Brisbane, Queensland Mus.;

Specimen(s): [1983,M.34], 6.9g.

Glen Osborne

South Australia, Australia
Doubtful..

Undescribed; was in possession of Sir D. Mawson. Discredited, A.R. Alderman, Rec. S. Austr. Mus., 1936, **5**, p.537 [M.A. 7-70].

Glen Rose (iron) 32°15′ N., 97°43′ W.
Somervell County, Texas, U.S.A.
Found 1934, approx. year
Iron. Octahedrite, fine (IRANOM).
A mass of about 24.5lb was recognized as a meteorite in 1936, A.D. Nininger, Pop. Astron., Northfield, Minnesota, 1937, **45**, p.449 [M.A. 7-62]. The locality is near Cleburne (*q.v.*), V.F. Buchwald, Iron Meteorites, Univ. of California, 1975, p.597. Classification, analysis, 9.4 %Ni, 0.74 ppm.Ga, 0.305 ppm.Ge, 3.9 ppm.Ir, D.J. Malvin et al., priv. comm., 1983.
9.9kg Fort Worth, Texas, Monnig Colln.;

Glen Rose (stone) v Rosebud.

Glindorcha v Indarch.

Glorieta v Glorieta Mountain.

Glorieta Mountain 35°36′ N., 105°48′ W.
Canoncito, Santa Fe County, New Mexico, U.S.A.
Found 1884
Synonym(s): *Albuquerque, Canoncito, Glorieta, Pojoaque, Rio Arriba, Santa Fe, Santa Fe County, Trinity County*
Stony-iron. Pallasite (PAL).
Several masses, of total weight about 320lb, the largest weighing 148.5lb were found, G.F. Kunz, Am. J. Sci., 1885, **30**, p.235, 1886, **32**, p.311. Another mass of 2.5kg was found in Rio Arriba Co. New Mexico and described, R.C. Hills, Proc. Colorado Sci. Soc., 1914, **11**, p.1-4. Description, review of older analyses, V.F. Buchwald, Iron Meteorites, Univ. of California, 1975, p.597. The Santa Fe iron is identical to Glorieta Mountain. Recent analysis. 12.04 %Ni, 13.2 ppm.Ga, 10.7 ppm.Ge, 0.014 ppm.Ir, J.T. Wasson, Meteorites, Springer-Verlag, 1974, p.307. Olivine Fa_{13}, P.R. Buseck and J.I. Goldstein, Bull. Geol. Soc. Amer., 1969, **80**, p.2141. Anomalous main-group pallasite, E.R.D. Scott, Geochimica et Cosmochimica Acta, 1977, **41**, p.693.
60.7kg Vienna, Naturhist. Mus.; 30kg New York, Amer. Mus. Nat. Hist.; 11kg Washington, U.S. Nat. Mus.; 10.5kg Paris, Mus. d'Hist. Nat.; 9.75kg Berlin, Humboldt Univ.; 7.8kg Albuquerque, Univ. of New Mexico; 4.5kg Chicago, Field Mus. Nat. Hist.; 2.6kg Yale Univ.; 2kg Dublin, Nat. Mus.; 1.25kg Harvard Univ.; 691g Tempe, Arizona State Univ.; 400g Ottawa, Mus. Geol. Surv. Canada; 327g Los Angeles, Univ. of California; 250g Prague, Nat. Mus.;
Specimen(s): [56541], 1472g.; [1906,149], 977g. a mass of 944g and a slice, 33g.; [1953,2], 105g.; [62350], 61.5g.; [56569], 56.5g.; [1959,975], 17.5g. and sawings, 0.5g, of Santa Fe

Glorietta Mountain error for Glorieta Mountain.

Gnadenfrei 50°40′ N., 16°46′ E.
between Reichenbach (=Dzierzoniów) and Frankenstein, Poland
Fell 1879, May 17, 1600 hrs
Synonym(s): *Pilawa Gorna, Schobergrund*
Stone. Olivine-bronzite chondrite (H5), brecciated.
Two stones of about 1kg and 0.75kg fell, after detonations. Described, with an analysis, J.G. Galle and A. von Lasaulx, Monatsber. Akad. Wiss. Berlin, 1879, p.750. Amount and composition of Ni/Fe determined, G.T. Prior, Min. Mag., 1919, **18**, p.353. Olivine Fa_{18}, B. Mason, Geochimica et Cosmochimica Acta, 1963, **27**, p.1011.
750g Breslau Univ.; 87g Vienna, Naturhist. Mus.; 66g New York, Amer. Mus. Nat. Hist.; 24g Chicago, Field Mus. Nat. Hist.; 13g Berlin, Humboldt Univ.;
Specimen(s): [55409], 40g.

Gnarrenburg v Bremervörde.

Gnowangerup 34°0′ S., 118°6′ E.
Western Australia, Australia
Found 1976
Iron. Octahedrite, medium (1.0mm) (IIIA).
A single mass of 33.6kg was found during road building 9 km SE. of Gnowangerup. Description, analysis, 8.59 %Ni, 20 ppm.Ga, 44 ppm.Ge, 1 ppm.Ir, J.R. de Laeter, Meteoritics, 1982, **17**, p.135.
33.6kg Perth, West. Austr. Mus.;

Gnubara error for Ghubara.

Goalpara 26°10′ N., 90°36′ E.
Assam, India
Found 1868
Stone. Achondrite, Ca-poor. Ureilite (AURE).
A stone of about 6lb was found amongst other specimens sent to the Calcutta Mus. by the Rajah of Goalpara, W. von Haidinger, Sitzungsber. Akad. Wiss. Wien, Math.-naturwiss. Kl., 1869, **59** (2), p.665. Analysis, 16.4 % total iron, H.B. Wiik, Meteoritics, 1972, **7**, p.553. XRF analysis, 16.94 % total iron, T.S. McCarthy et al., Meteoritics, 1974, **9**, p.215. Olivine composition variable, R. Hutchison and R.F. Symes, Meteoritics, 1972, **7**, p.23. Rare gases, E. Mazor et al., Geochimica et Cosmochimica Acta, 1970, **34**, p.781. Carbon polymorphs, G.P. Vdovykin, Space Sci. Rev., 1970, **10**, p.483. Mineralogy and petrology, J.L. Berkley et al., Geochimica et Cosmochimica Acta, 1980, **44**, p.1579.
434g Calcutta, Mus. Geol. Surv. India; 163g Washington, U.S. Nat. Mus.; 153g Vienna, Naturhist. Mus.; 99g Harvard Univ.;
Specimen(s): [51187], 830g. and fragments, 34.5g; [43058], 51g.

Goamus Farm v Gibeon.

Gobabeb 23°33′ S., 15°2′ E.
Namibia
Found 1969
Stone. Olivine-bronzite chondrite (H4).
A mass of 23kg, plus a number of smaller fragments, in total about 27kg, were found on the flanks of a large sand dune about 8 miles SSE. of Gobabeb, Meteor. Bull., 1975 (53), Meteoritics, 1975, **10**, p.148. Description and analyses of minerals, contains equilibrated silicates and unequilibrated spinels, olivine Fa_{18}, R.F. Fudali and A.F. Noonan, Meteoritics, 1975, **10**, p.31.
25g Washington, U.S. Nat. Mus.;

Goldbach's Iron v Yanhuitlan.

Gomez 33°10′53″ N., 102°24′5″ W.
Terry County, Texas, U.S.A.
Found 1974, before this year
Stone. Olivine-hypersthene chondrite (L6).
A single mass of 47kg was found, olivine Fa_{26}, Meteor. Bull., 1980 (58), Meteoritics, 1980, **15**, p.236. Description, P.F. Sipiera et al., Meteoritics, 1980, **15**, p.201.

Goodland 39°21′ N., 101°40′ W.
Sherman County, Kansas, U.S.A.
Found 1923
Stone. Olivine-hypersthene chondrite (L4).
Two fragments fitting together, weighing 3628g, were recognized as meteoritic in 1938, from SE. 1/4, sect. 1,

township 10, range 41, Sherman County, A.D. Nininger, Pop. Astron., Northfield, Minnesota, 1939, **47**, p.212., E.P. Henderson, letter of 3 June, 1939 in Min. Dept. BM(NH). Analysis, 21.49 % total iron, E. Jarosewich, Geochimica et Cosmochimica Acta, 1967, **31**, p.1103. Slightly unequilibrated, olivine Fa_{25}, R.T. Dodd et al., Geochimica et Cosmochimica Acta, 1967, **31**, p.921. Contains calcic plagioclase, An85, R. Hutchison and A.L. Graham, Nature, 1975, **255**, p.471.

1.4kg Tempe, Arizona State Univ.; 174g Washington, U.S. Nat. Mus.; 142g Chicago, Field Mus. Nat. Hist.; 77g Los Angeles, Univ. of California; 60g New York, Amer. Mus. Nat. Hist.;

Specimen(s): [1959,863], 800g.

Goose Lake 41°58′48″ N., 120°32′30″ W.

Modoc County, California, U.S.A.
Found 1938, October 13
Iron. Octahedrite, medium (1.2mm) (IA).

A mass of 1169.5kg was found on lava beds near Goose Lake, E.P. Henderson, letter of 3 June, 1939 in Min. Dept. BM(NH). Mentioned, F.C. Leonard, Pop. Astron., Northfield, Minnesota, 1940, **48**, p.432 [M.A. 8-61], E.G. Linsley, Calif. J. Mines Geol., 1939, **35**, p.308 [M.A. 7-543]. Description, E.P. Henderson and S.H. Perry, Proc. U.S. Nat. Mus., 1958, **107**, p.339 [M.A. 14-130]. It has been suggested that Goose Lake may be a 'throw-out' of the Cañon Diablo impact, C.S. Cook and C.P. Butler, Nature, 1965, **206**, p.704. Recent analysis, 8.00 %Ni, 67.2 ppm.Ga, 305 ppm.Ge, 2.3 ppm.Ir, J.T. Wasson, Icarus, 1970, **12**, p.407. The Ga content is just significantly lower than the range for Cañon Diablo (*q.v*). Structure, H.J. Axon and R. Rieche, Nature, 1967, **215**, p.379. Full description, V.F. Buchwald, Iron Meteorites, Univ. of California, 1975, p.601.

1165kg Washington, U.S. Nat. Mus.; 1kg Ann Arbor, Univ. of Michigan; 543g Calcutta, Mus. Geol. Surv. India; 517g Chicago, Field Mus. Nat. Hist.; 325g Moscow, Acad. Sci.;

Specimen(s): [1959,951], 72g. and sawings, 2g.

Gopalpur 24°14′ N., 89°3′ E.

Pabna district, Rajshahi, Bangladesh
Fell 1865, May 23, 1800 hrs
Stone. Olivine-bronzite chondrite (H6).

A stone of about 3.5lb was seen to fall, Proc. Asiatic Soc. Bengal, 1865, p.94. Described, with an analysis, G. Tschermak, Tschermaks Min. Petr. Mitt., 1872, p.95. Olivine Fa_{20}, B. Mason, Geochimica et Cosmochimica Acta, 1963, **27**, p.1011.

1217g Calcutta, Mus. Geol. Surv. India; 159 g Vienna, Naturhist. Mus.;

Specimen(s): [41019], 134g. and fragments, 0.4g

Gorakhpur v Pokhra.

Gorizia v Avce.

Gorlovka 48°17′ N., 38°5′ E. approx.

Donetsk region, Ukraine, USSR, [Горловка]
Fell 1974, July 17, 1730 hrs, U.T.
Stone. Olivine-bronzite chondrite (H3-4).

Fragments totalling 3618g were recovered in the town of Gorlovka, Meteor. Bull., 1976 (54), Meteoritics, 1976, **11**, p.70. Description, with an analysis, 27.63 % total iron, O.A. Kirova et al., Метеоритика, 1977, **36**, p.46.

3.4kg Moscow, Acad. Sci.;

Goruckpur v Bustee.

Goruckpur v Butsura.

Goruckpur v Dandapur.

Goruckpur v Supuhee.

Gorukhpur v Butsura.

Gorukhur v Butsura.

Gosnells v Mount Dooling.

Gostkowo v Pułtusk.

Gotha v Tabarz.

Governador Valadares 18°51′ S., 41°57′ W.

Minas Gerais, Brazil
Found 1958
Stone. Achondrite, Ca-rich. Nakhlite (ACANOM).

A well preserved individual of 158g was found, but little is known of its history. Description, analysis, olivine Fa_{67}, F. Burragato et al., Meteoritics, 1975, **10**, p.374, abs. Petrology, distinct from Nakhla and Lafayette, J.L. Berkley et al., Geochimica et Cosmochimica Acta, 1980 (Suppl. 14), p.1089.

Main mass, Rome, Univ. Min. Mus.;

Specimen(s): [1975,M.16], 6.4g. a sawn, partly crusted fragment.

Grady (no. 1) v Grady (1933).

Grady (no. 2) v Grady (1937).

Grady (no. 3) v Grady (c).

Grady (1933) 34°48′ N., 103°19′ W.

Curry County, New Mexico, U.S.A.
Found 1933
Synonym(s): *Grady (no. 1)*
Stone. Olivine-hypersthene chondrite (L3-6).

One stone of 4.23kg was found in 1933, another of 9.3kg in 1937, A.D. Nininger, Pop. Astron., Northfield, Minnesota, 1937, **45**, p.449 [M.A. 7-62]. The two stones, at first attributed to one fall, are definitely distinct, H.H. and A.D. Nininger, The Nininger Collection of Meteorites, Winslow, Arizona, 1950, p.54. Olivine Fa_{25}, B. Mason, Geochimica et Cosmochimica Acta, 1963, **27**, p.1011.

1.7kg Washington, U.S. Nat. Mus.; 101g Albuquerque, Univ. of New Mexico; 22g Tempe, Arizona State Univ.;

Specimen(s): [1959,837], 20g.

Grady (1937) 34°48′ N., 103°19′ W.

Curry County, New Mexico, U.S.A.
Found 1937
Synonym(s): *Grady (no. 2)*
Stone. Olivine-bronzite chondrite (H3), veined.

This 9.3kg stone is quite distinct from the 4.23kg Grady (1933) stone. Analysis, 26.23 % total iron, E. Jarosewich, Geochimica et Cosmochimica Acta, 1967, **31**, p.1103. Olivine Fa_{15}, B. Mason, Geochimica et Cosmochimica Acta, 1963, **27**, p.1011. The total iron is higher than in Clovis (no. 1) (*q.v.*). Chondrule with chromite-rich rim figured, P. Ramdohr, Geochimica et Cosmochimica Acta, 1967, **31**, p.1961.

2.2kg Tempe, Arizona State Univ.; 1610g Albuquerque,

Univ. of New Mexico; 772g Washington, U.S. Nat. Mus.; 520g Chicago, Field Mus. Nat. Hist.; 449g Harvard Univ.; 330g New York, Amer. Mus. Nat. Hist.;
Specimen(s): [1959,1004], 2682g.

Grady (c) 34°48′ N., 103°19′ W.

Curry County, New Mexico, U.S.A.
Found 1970, recognized in this year
Synonym(s): *Grady (no. 3)*
Stone. Olivine-bronzite chondrite (H4).

Total known weight 5.7kg, G.I Huss, letter of 16 March, 1973 in Min. Dept. BM(NH). The stone is apparently different from Grady (1933) and from Grady (1937), D.E. Lange et al., Cat. Meteor. Univ. New Mexico, 1980, p.26.
3.55kg Mainz, Max-Planck-Inst.; 100g Vienna, Naturhist. Mus.; 60g Albuquerque, Univ. of New Mexico;
Specimen(s): [1972,499], 52.6g. a slice, broken and crusted.

Graham County v Morland.

Granada v Colomera.

Granada Creek 37°58′ N., 102°20′ W.

Prowers County, Colorado, U.S.A.
Found 1975, in the spring
Stone. Olivine-bronzite chondrite (H).

A single mass of 479g was found in a field, Meteor. Bull., 1978 (55), Meteoritics, 1978, **13**, p.336.

Gran Chaco (iron) v Campo del Cielo.

Gran Chaco (pallasite) v Imilac.

Gran Chaco I v Campo del Cielo.

Gran Chaco II v Campo del Cielo.

Gran Chaco Gualamba v Campo del Cielo.

Grand Rapids 42°58′ N., 85°46′ W.

Kent County, Michigan, U.S.A.
Found 1883
Synonym(s): *False Kalamazoo, Kalamazoo, Walker Township*
Iron. Octahedrite, medium (0.55mm) (IRANOM).

A mass of 114lb was found, I.R. Eastman, Am. J. Sci., 1884, **28**, p.299. Chemically anomalous, 9.26 %Ni, 17.9 ppm.Ga, 13.7 ppm.Ge, 14 ppm.Ir, J.T. Wasson and R. Schaudy, Icarus, 1971, **14**, p.59. Description, V.F. Buchwald, Iron Meteorites, Univ. of California, 1975, p.606.
11.2kg Chicago, Field Mus. Nat. Hist.; 8.37kg New York, Amer. Mus. Nat. Hist.; 3526g Washington, U.S. Nat. Mus.; 1.8kg Vienna, Naturhist. Mus.; 1.5kg Tempe, Arizona State Univ.; 1008g Harvard Univ.; 697g Yale Univ.;
Specimen(s): [68724], 997.5g. includes a slice of 973g; [1920, 304], 333g. slice; [61923], 112g.

Granes 42°54′ N., 2°15′ E.

Quillion, Aude, France
Fell 1964, November 13, 1700 hrs
Stone. Olivine-hypersthene chondrite (L6).

A mass of 9kg fell; brief description, olivine Fa_{24}, M. Christophe Michel-Levy, Bull. Min., 1978, **101**, p.580 [M.A. 79-2726].
Main mass, Montpelier Univ.; 900g Paris, Mus. d'Hist. Nat.;

Granite Peak v Wiluna.

Grant 35°10′ N., 107°53′ W.

Cibola County, New Mexico, U.S.A.
Found 1929, about
Synonym(s): *Breece, San Rafael*
Iron. Octahedrite, medium (0.80mm) (IIIB).

A conical mass of 480kg was found in the Zuni Mts., 45 miles S. of Grants. Described, with an analysis, 8.58 %Ni, E.P. Henderson, Pop. Astron., Northfield, Minnesota, 1934, **42**, p.511 [M.A. 6-14]. Analysis, 9.35 %Ni, E.P. Henderson, Am. J. Sci., 1941, **239**, p.407 [M.A. 8-195]. More recent analysis, Further analysis, 9.24 %Ni, 19.8 ppm.Ga, 37.0 ppm.Ge, 0.040 ppm.Ir, E.R.D. Scott et al., Geochimica et Cosmochimica Acta, 1973, **37**, p.1957. Synonymous with Breece and San Raphael, V.F. Buchwald, Iron Meteorites, Univ. of California, 1975, p.607. Cibola Co. was created from part of Valencia Co. in 1981.
481kg Washington, U.S. Nat. Mus.; 43kg Chicago, Field Mus. Nat. Hist., includes 'Breece';
Specimen(s): [1959,1047], 32.5g.; [1959,944], 344g. and sawings, 6g, of Breece

Grant County 37°28′ N., 101°26′ W.

Kansas, U.S.A.
Found 1936
Stone. Olivine-hypersthene chondrite (L6).

Several fragments, totalling 2.3kg, were found, E.P. Henderson, letters of 12 May and 3 June 1939 in Min. Dept. BM(NH). Mentioned, A.D. Nininger, Pop. Astron., Northfield, Minnesota, 1939, **47**, p.212. Olivine Fa_{26}, B. Mason, Geochimica et Cosmochimica Acta, 1963, **27**, p.1011.
696g Washington, U.S. Nat. Mus.; 469g Tempe, Arizona State Univ.; 147g New York, Amer. Mus. Nat. Hist.;
Specimen(s): [1959,815], 455g. and fragments, 1.5g

Grants error for Grant.

Grasse v La Caille.

Grassland 33°7′ N., 101°35′ W.

Lynn County, Texas, U.S.A.
Found 1964
Stone. Olivine-hypersthene chondrite (L4).

One badly weathered stone of 4.4kg was found lying on the ground after the first rain following the terracing of a field, G.I Huss, letter of 20 September, 1970 in Min. Dept. BM (NH). Olivine Fa_{23}, B. Mason, Geochimica et Cosmochimica Acta, 1967, **31**, p.1100.
2kg Mainz, Max-Planck-Inst.; 114g Washington, U.S. Nat. Mus.; 83g New York, Amer. Mus. Nat. Hist.; 80g Tempe, Arizona State Univ.;
Specimen(s): [1966,272], 108g. a slice, and fragments, 1g.

Graves v Moore County.

Grayton 30°18′42″ N., 86°10′ W.

Grayton Beach, Walton County, Florida, U.S.A.
Found 1983
Stone. Olivine-bronzite chondrite (H5).

A single mass of 10.9kg was found near Grayton Beach by H. Povenmire, G.I Huss, letter of 26 October, 1983 in Min. Dept. BM(NH). Olivine $Fa_{18.6}$, A.L. Graham, priv. comm., 1983.

Grazac 45°11′ N., 4°17′ E.
Haute Loire, France
Fell 1885, August 10, 0400 hrs
Synonym(s): *Montpelagry, Tarn*
Pseudometeorite..
After detonations, about 20 stones are said to have fallen between the villages of Grazac and Montpelagry, the largest weighing about 600g, A. Caraven-Cachin, C. R. Acad. Sci. Paris, 1887, **104**, p.1813. Not meteoritic, A. Lacroix, Bull. Mus. d'Hist. Nat. Paris, 1927, **33**, p.446 [M.A. 3-534].

Great Bear Lake 66° N., 120° W.
North-west Territories, Canada
Found 1936, June
Stone. Olivine-bronzite chondrite (H6).
One stone was found by an Indian on the ice of Great Bear Lake; the original weight is not known, and the main mass has been lost, E.P. Henderson, letters of 12 May and 3 June, 1939 in Min. Dept. BM(NH). Olivine Fa_{19}, B. Mason, Geochimica et Cosmochimica Acta, 1963, **27**, p.1011.
40g Washington, U.S. Nat. Mus.;

Great Bend 38°23′42″ N., 98°54′57″ W.
Barton County, Kansas, U.S.A.
Found 1983
Stone. Olivine-bronzite chondrite (H6).
A single mass of 28.77kg was found, G.I Huss, letter of 26 October, 1983 in Min. Dept. BM(NH). Olivine $Fa_{19.4}$, A.L. Graham, priv. comm., 1984.
28.77kg Denver, Amer. Meteorite Lab.;

Great Fish River v Cape of Good Hope.

Great Fish River v Gibeon.

Great Namaqualand v Gibeon.

Greeley County v Ladder Creek.

Greenbrier County 37°50′ N., 80°19′ W.
West Virginia, U.S.A.
Found 1880
Synonym(s): *Alleghany Mountains, White Sulphur Springs*
Iron. Octahedrite, medium (1.0mm) (IIIA).
A mass of 11lb was found near the top of Alleghany Mountains, 3 miles N. of White Sulphur Springs; described and analysed, 7.11 %Ni, L. Fletcher, Min. Mag., 1887, **7**, p.183. Chemically distinct from Crosby's Creek (*q.v.*) and Waldron Ridge (*q.q.v.*), J.T. Wasson, Icarus, 1970, **12**, p.407. Classification and analysis, 7.38 %Ni, 18.1 ppm.Ga, 33.3 ppm.Ge, 10 ppm.Ir, E.R.D. Scott et al., Geochimica et Cosmochimica Acta, 1973, **37**, p.1957. The mass has been heated and forged, V.F. Buchwald, Iron Meteorites, Univ. of California, 1975, p.616.
473g Washington, U.S. Nat. Mus.; 64g Harvard Univ.;
Specimen(s): [55239], 1780g. and fragments, 30g.

Greene County v Babb's Mill (Blake's Iron).

Grefsheim 60°40′ N., 11°0′ E.
Lake Mjosa, Norway
Fell 1976, January 25-31
Stone. Olivine-hypersthene chondrite (L5).
Fragments totalling 45.5g which could be fitted together into a complete stone, with fusion crust, were found on the ice covering Lake Mjosa, 1.6km from the shore at Grefsheim, olivine $Fa_{23.7}$, Meteor. Bull., 1979 (56), Meteoritics, 1979, **14**, p.167. Description, W.L. Griffin and K.A. Jorgensen, Meteoritics, 1979, **14**, p.117 [M.A. 81-1743].
45.5g Oslo, Min.-Geol. Mus.;

Grenade v Toulouse.

Gressk 53°14′ N., 27°20′ E.
Minsk, Belorussiya SSR, USSR, [Гресск]
Found 1955, July
Synonym(s): *Hressk*
Iron. Hexahedrite (IIA).
One mass of 303kg was found, V.G. Fesenkov, Метеоритика, 1958, **16**, p.5 [M.A. 14-45]. Analysis, 5.8 %Ni, M.I. D'yakanova, Метеоритика, 1958, **16**, p.179 [M.A. 14-49]. Another analysis, 5.61 %Ni, 62.1 ppm.Ga, 177 ppm.Ge, 7.7 ppm.Ir, J.T. Wasson, Geochimica et Cosmochimica Acta, 1969, **33**, p.859. Uranium content, L.I. Genaeva et al., Метеоритика, 1972, **31**, p.137. Shock melted troilite, mildly reheated, V.F. Buchwald, Iron Meteorites, Univ. of California, 1975, p.617.
Main mass, Minsk, Inst. Geol. Sci.; 1.9kg Washington, U.S. Nat. Mus.; 1.3kg Tempe, Arizona State Univ.; 1.17kg Moscow, Acad. Sci.; 1kg Yale Univ.; 984g Calcutta, Mus. Geol. Surv. India; 892g Albuquerque, Univ. of New Mexico; 785g Berlin, Humboldt Univ.; 590g Prague, Nat. Mus.; 538g Chicago, Field Mus. Nat. Hist.; 500g Perth, West. Austr. Mus.;
Specimen(s): [1972,319], 74.6g. internal slice

Gretna 39°56′ N., 99°13′ W.
Phillips County, Kansas, U.S.A.
Found 1912
Stone. Olivine-hypersthene chondrite (L5).
A stone, broken into three pieces, was found 12 miles N. of Gretna. Two of the pieces weigh 36kg and 22.7kg and fit together, indicating the loss of a third fragment, probably about 23kg. Described and figured, H.H. Nininger, Trans. Kansas Acad. Sci., 1936, **39**, p.172. Mentioned, A.D. Nininger, Pop. Astron., Northfield, Minnesota, 1937, **45**, p.449 [M.A. 7-62]. Olivine Fa_{25}, B. Mason, Geochimica et Cosmochimica Acta, 1963, **27**, p.1011. May be synonymous with Phillips County (stone) (*q.v.*).
32kg Washington, U.S. Nat. Mus., main mass of the 36kg piece; Large mass, Hays, Kansas, Fort Hays State College, main mass of the 22.7kg piece; 3kg Michigan Univ.;
Specimen(s): [1959,864], 19.8g. and fragments, 1g

Grier (b) 34°16′ N., 103°23′ W. approx.
Curry County, New Mexico, U.S.A.
Found 1969
Stone. Olivine-hypersthene chondrite (L5).
A single mass of 929g was found, Meteor. Bull., 1980 (58), Meteoritics, 1980, **15**, p.236. Description, classification, olivine $Fa_{25.5}$, K. Fredriksson et al., Meteoritics, 1981, **16**, p.129. A distinct Grier stone is also mentioned, G.I Huss, Cat. Huss Coll. Meteorites, 1976, p.20.
145g Washington, U.S. Nat. Mus.;

Grimma v Naunhof.

Grimma v Steinbach.

Griqualand v Daniel's Kuil.

Griqualand v Orange River (iron).

Groenewald v Benoni.

Groenewald-Benoni v Benoni.

Gröndorn v Gibeon.

Grootfontein v Hoba.

Groslée 45°42′ N., 5°37′ E.
Belley, Ain, France
Found 1827
Synonym(s): *Belley*
Pseudometeorite..
2g in Paris, Mus. d'Hist. Nat., labelled doubtful, is artificial iron, A. Lacroix, Bull. Mus. d'Hist. Nat. Paris, 1927, **33**, p.445. The BM(NH) specimens are not meteoritic.
Specimen(s): [34164], 50g. two oxidized fragments.

Grosnaia v Grosnaja.

Grosnaja 43°40′ N., 45°23′ E.
Mekensk, Terek, Checheno-Ingushskaya ASSR, USSR, [Гросная]
Fell 1861, June 28, 1900 hrs
Synonym(s): *Grosnaia, Groznaia, Groznaja, Groznaya, Mikenskoi, Mikenskoj, Mikentskaia, Mikentskaia, Mikentskaja, Terek*
Stone. Carbonaceous chondrite, type III (CV3).
A shower of stones fell, after detonations, but only one of about 3.5kg was recovered, as the rest fell into the river Terek. Described, with an analysis, G. Tschermak, Tschermaks Min. Petr. Mitt., 1878, **1**, p.153. Another analysis, 24.8 % total iron, B. Mason, Space Sci. Rev., 1963, **1**, p.621 [M.A. 16-640]. One stone of 3251g fell, E.L. Krinov, Астрон. Журнал, 1945, **22**, p.303 [M.A. 9-297]. Classification, W.R. Van Schmus, Meteorite Research, ed. P.M. Millman, D. Reidel, Dordrecht-Holland, 1969, p.480.
1840g Moscow, Acad. Sci., main mass; 534g Vienna, Naturhist. Mus.; 47g Chicago, Field Mus. Nat. Hist.; 42g Berlin, Humboldt Univ.; 25g New York, Amer. Mus. Nat. Hist.;
Specimen(s): [63624], 146g. and fragments, 3.5g; [35217], 6.2g.

Gross-Divina 49°16′ N., 18°43′ E.
Sillein, Trencsén, Stredoslovensky, Czechoslovakia
Fell 1837, July 24, 1130 hrs
Synonym(s): *Budetin, Nagydévény, Nagy-Divina, Nagy-Diwina, Velka-Divina*
Stone. Olivine-bronzite chondrite (H5).
A stone of about 10.5kg fell, D. Zipser, Neues Jahrb. Min., 1840, p.89, G. von Boguslawski, Ann. Phys. Chem. (Poggendorff), 1854, **4** (suppl.), p.356. Olivine Fa_{19}, B. Mason, Geochimica et Cosmochimica Acta, 1963, **27**, p.1011.
10kg Budapest, Nat. Mus.; 215g Paris, Mus. d'Hist. Nat.; 66g Vienna, Naturhist. Mus.;
Specimen(s): [56465], 3g.; [35171], minute fragment.

Grossliebenthal 46°21′ N., 30°35′ E.
Odessa, Kherson, Ukraine, USSR, [Гросслибенталь]
Fell 1881, November 19, 0630 hrs
Synonym(s): *Cherson, Odesa, Odessa*
Stone. Olivine-hypersthene chondrite (L6), veined.
After the appearance of a fireball, a stone of about 8kg was found, another stone fell at 'Sitschawska' and was broken up and lost, G.A. Daubrée, C. R. Acad. Sci. Paris, 1884, **98**, p.323. Analysis, P.G. Melikov and C. Schwalbe, Ber. Deutsch. Chem. Gesell. Berlin, 1893, **26**, p.234. Further analysis, 22.51 % total iron, M.I. D'yakonova, Метеоритика, 1972, **31**, p.119. Mentioned, L.A. Kulik, Метеоритика, 1941, **1**, p.73 [M.A. 9-294], E.L. Krinov, Астрон. Журнал, 1945, **22**, p.303 [M.A. 9-297]. Olivine Fa_{25}, B. Mason, Geochimica et Cosmochimica Acta, 1963, **27**, p.1011.
6.42kg Moscow, Acad. Sci.; 174g Vienna, Naturhist. Mus.; 80g Prague, Nat. Mus.; 75g Paris, Mus. d'Hist. Nat.; 54g New York, Amer. Mus. Nat. Hist.; 49g Washington, U.S. Nat. Mus.; 46g Berlin, Humboldt Univ.;
Specimen(s): [1920,306], 79g.; [63928], 52g.; [54637], 8.5g.

Grossnamaqualand v Gibeon.

Groznaia v Grosnaja.

Groznaja v Grosnaja.

Groznaya v Grosnaja.

Gruever v Gruver.

Grünberg v Grüneberg.

Grüneberg 51°56′ N., 15°30′ E.
Silesia, Poland
Fell 1841, March 22, 1530 hrs
Synonym(s): *Grünberg, Heinrichau, Schloin, Seifersdorf, Seifersholz, Wilkanowko, Zielena Gora, Zielona Gora*
Stone. Olivine-bronzite chondrite (H4), veined.
A stone of about 1kg fell near Heinrichau, after detonations, Ann. Phys. Chem. (Poggendorff), 1841, **52**, p.495, J. Pokrzywnicki, Urania, 1955, **26**, p.166. Olivine Fa_{17}, B. Mason, Geochimica et Cosmochimica Acta, 1963, **27**, p.1011.
750g Berlin, Humboldt Univ.; 102g Chicago, Field Mus. Nat. Hist.; 108g Tübingen, Univ.; 63g Wroclaw Univ.; 16g Vienna, Naturhist. Mus.; 16g Washington, U.S. Nat. Mus.;
Specimen(s): [35179], 21g.; [90267], 9.5g.

Gruver 36°20′ N., 101°24′ W.
Hansford County, Texas, U.S.A.
Found 1934
Synonym(s): *Gruever*
Stone. Olivine-bronzite chondrite (H4).
Masses totalling 11.1kg were found, A.D. Nininger, Pop. Astron., Northfield, Minnesota, 1937, **45**, p.449 [M.A. 7-62]. Another mass was recovered in 1966. Olivine Fa_{19}, B. Mason, Geochimica et Cosmochimica Acta, 1963, **27**, p.1011. A main mass is in the Gillespie collection.
3.5kg Fort Worth, Texas, Monnig Colln.; 1.6kg Tempe, Arizona State Univ.; 1.7kg Mainz, Max-Planck-Inst., the mass found in 1966; 403g Calcutta, Mus. Geol. Surv. India; 1.25kg Washington, U.S. Nat. Mus.; 783g Chicago, Field Mus. Nat. Hist.; 258g Los Angeles, Univ. of California; 256g Copenhagen, Univ. Geol. Mus.;
Specimen(s): [1937,389], 5.5g.; [1959,865], 1248g.

Grzempach 52°52′ N., 16°38′ E.
Poznan, Poland
Fell 1910, September 3, 1500 hrs
Synonym(s): *Friedrichsau, Grzempy*
Stone. Olivine-bronzite chondrite (H5).
A mass of 690g fell at Grzempach (=Grzempy, =Friedrichsau), near Czarnków, F. Slavik and L.J. Spencer, Min. Mag., 1928, **21**, p.478, A. Laszkiewicz, Mineralogia,

Warszawa, 1936, p.195 [M.A. 6-385], J. Pokrzywnicki, Acta Geophys. Polon., 1956, **4**, p.21 [M.A. 13-79]. Mineralogy, olivine Fa_{20}, M. Christophe Michel-Levy, Bull. Acad. Sci. Pol., 1976, **24** (ser. sci. terre), p.63.
640g Warsaw, Polish Acad. Sci., main mass;

Grzempy v Grzempach.

Guadaloupe County 29°30′ N., 98° W. approx.
Texas, U.S.A.
Found 1950, known before this year
Iron. Octahedrite, medium (1.2mm).
A small specimen of 20.5g, formerly in the Nininger collection, is probably a piece of the Cañon Diablo fall, H.H. and A.D. Nininger, The Nininger Collection of Meteorites, Winslow, Arizona, 1950, p.55. Not Cañon Diablo, possibly part of Sanderson, V.F. Buchwald, Iron Meteorites, Univ. of California, 1975, p.1393.
20.5g Tempe, Arizona State Univ.;

Guadalupe v Rancho de la Pila (1882).

Gualalupe (of Karavinsky) v Rancho de la Pila (1882).

Gualeguaychú 33° S., 58°37′ W.
Entre Rios, Argentina
Fell 1932, October
Stone. Olivine-bronzite chondrite (H).
A mass of 22kg was found on the ranch La Constancia, dept. Gualeguaychú. Description, with an analysis, E.H. Ducloux, Anal. Mus. Argentino Cienc. Nat., 1942, **40**, p.123 [M.A. 10-179]. Co-ordinates and date of fall, L.M. Villar, Cienc. Investig., 1968, **24**, p.302. Olivine Fa_{19}, B. Mason, Geochimica et Cosmochimica Acta, 1963, **27**, p.1011.
Specimen(s): [1962,166], 15g.

Guanghua 32°24′ N., 110°42′ E.
Wudang Mountains, Hubei, China
Found 1932, approx.
Synonym(s): *Guanghua Wudan Shan*
Iron. Octahedrite, fine (0.3mm) (IVA).
A mass weighing over 190kg was found, D. Bian, Meteoritics, 1981, **16**, p.121. Classification, analysis, 7.7 % Ni, 1.71 ppm.Ga, 0.091 ppm.Ge, 2.8 ppm.Ir, D.J. Malvin et al., priv. comm., 1983.
Specimen, Wuhan, Hubei Bureau of Geology;

Guanghua Wudan Shan v Guanghua.

Guangrao 37°6′ N., 118°24′ E.
Guangrao County, Shandong, China
Fell 1980, June 21, 1615 hrs
Stone. Olivine-hypersthene chondrite (L6).
A single stone weighing 1.9kg fell. Brief description, analysis, 22.38 % total iron, olivine $Fa_{24.3}$, H. Wei and W. Daode, Geochimica, 1982, p.321.
1.8kg Guiyang, Acad. Sin. Inst. Geochem.;

Guangyuan 32°24′ N., 105°54′ E.
Sichuan, China
Found 1965
Iron.
A single mass was found, no further details are reported, D. Bian, Meteoritics, 1981, **16**, p.122.

Guareña 38°44′ N., 6°1′ E.
Badajos, Spain
Fell 1892, July 20, 1030 hrs
Stone. Olivine-bronzite chondrite (H6).
Two stones, one of about 32kg and the other 7kg, fell, after detonations about 5 miles apart. Described, S. Calderón and F. Quiroga, Anal. Soc. Españ. Hist. Nat. Madrid, 1893, **22**, p.127, E. Cohen, Ann. Naturhist. Hofmus. Wien, 1896, **11**, p.36. Analysis, 27.74 % total iron, olivine Fa_{19}, E. Jarosewich and B. Mason, Geochimica et Cosmochimica Acta, 1969, **33**, p.411. Rb-Sr age, 4,560 m.y., G.J. Wasserburg et al., Earth planet. Sci. Lett., 1969, **7**, p.33. I-Xe correlation, J. Jordon et al., Meteoritics, 1979, **14**, p.434.
29.4kg Madrid, Mus. Cienc. Nat., main mass; 7kg Badajos, La Comision de Monumentos, approx. weight, the smaller stone.; 416g Washington, U.S. Nat. Mus.; 180g Paris, Mus. d'Hist. Nat.; 54g Tempe, Arizona State Univ.;
Specimen(s): [85426], 56.6g.

Guatemala v Chinautla.

Guc v Guêa.

Guêa 43°46′ N., 20°14′ E.
Serbia, Yugoslavia
Fell 1891, September 28, 1700 hrs
Synonym(s): *Cacak, Guc, Gutsch*
Stone.
A piece of 1915g was preserved at Belgrade, E.A. Wülfing, Die Meteoriten in Samml., Tübingen, 1897, p.135, in 1897, but was "taken away by the enemy during the war", P.S. Pavlović, letter of 12 April 1924, in Min. Dept. BM(NH).

Guenie 11°45′ N., 2°9′ W.
Upper Volta
Fell 1960, April
Stone. Olivine-bronzite chondrite (H4).
A shower of stones fell though the number and total weight are not recorded, olivine Fa_{19}, Meteor. Bull., 1980 (57), Meteoritics, 1980, **15**, p.167. Brief report, M. Bourot et al., Meteoritics, 1975, **10**, p.368, abs. Petrography, M. Christophe Michel-Levy, Earth planet. Sci. Lett., 1981, **54**, p.67. Co-ordinates, fall site 60 km from Leo and 140 km from Ouagadougou, M. Christophe Michel-Levy, priv. comm., 1982.
3.6kg Paris, Mus. d'Hist. Nat.;

Guernsey County v New Concord.

Guetersloh v Gütersloh.

Guffey 38°46′ N., 105°31′ W.
Park County, Colorado, U.S.A.
Found 1907
Synonym(s): *Currant Creek, Park Creek*
Iron. Ataxite (IRANOM).
A pear-shaped mass of 682lb was found on Currant Creek in Freemont County, 22 miles SW. of Cripple Creek; described, as Currant Creek, with an analysis, 10 %Ni, W.P. Headden, Proc. Colorado Sci. Soc., 1908, **9**, p.79. Also described, as Guffey, with an analysis, 10.5 %Ni, E.O. Hovey, Am. Mus. J., 1909, **9**, p.237. Classification and analysis, 9.94 %Ni, 0.146 ppm.Ga, 0.082 ppm.Ge, 5.0 ppm.Ir, R. Schaudy et al., Icarus, 1972, **17**, p.174. Description of ataxitic structure, V.F. Buchwald, Iron Meteorites, Univ. of California, 1975, p.618.
309kg New York, Amer. Mus. Nat. Hist., main mass, minimum weight; 8.2kg Washington, U.S. Nat. Mus.;

Guibga 13°30′ N., 0°41′ W.
near Pissila, Upper Volta
Fell 1972, February 26
Synonym(s): *Upper Volta*
Stone. Olivine-hypersthene chondrite (L5).
A stone of approximate dimensions 10 × 5 × 5cm fell at the village of Guibga, about 44km NE. of Pissila, Upper Volta. The total weight is not known. A 288g fragment was sent to the U.S. Geol. Surv. by M. Ph. Ouedraogo in November, 1972. This specimen was transferred to the U.S. Nat. Mus. for study, Meteor. Bull., 1974 (52), Meteoritics, 1974, **9**, p.102.
250g Washington, U.S. Nat. Mus.;
Specimen(s): [1974,M.5], 19.4g. a sawn, crusted fragment.

Guidder 9°55′ N., 13°59′ E.
Nord province, Cameroun
Fell 1949, January 7, 1845 hrs
Stone. Olivine-hypersthene chondrite, amphoterite (LL5).
A shower of stones fell, 6 to 8km SE. of Guidder, of which at least two, of 476g and 492g, were recovered. Described, with an analysis, E. Jérémine and A. Sandréa, Geochimica et Cosmochimica Acta, 1953, **4**, p.83. Analysis, 18.98 % total iron, B. Mason and H.B. Wiik, Geochimica et Cosmochimica Acta, 1964, **28**, p.535. Olivine $Fa_{27.8}$, K. Fredriksson et al., Origin and Distribution of the Elements, ed. L.H. Ahrens, Pergamon, 1968, p.457.
743g Paris, Mus. d'Hist. Nat.;

Guilford County 35°34′ N., 79°50′ W. approx.
North Carolina, U.S.A.
Found 1822, known before this year
Synonym(s): *Randolf County*
Iron. Octahedrite, medium (1.1mm) (IIIA).
Two pieces of iron, one of about 2lb, from Randolph County, and the other of 7oz from Guilford County (10 to 15 miles distant), were found in a collection of North Carolina minerals and rocks formed by Prof. Olmsted. The smaller piece is said to have come from a mass weighing 28lb some of which had been worked by a blacksmith into horse-shoe nails, C.U. Shepard, Am. J. Sci., 1841, **40**, p.369. Less than 100g known in collections. Perhaps a fragment of Uwharrie (*q.v.*), V.F. Buchwald, Iron Meteorites, Univ. of California, 1975, p.619.
21g Tempe, Arizona State Univ.; 20g Yale Univ.; 15g Chicago, Field Mus. Nat. Hist.; 8g Vienna, Naturhist. Mus.;
Specimen(s): [24004], 15g. of Guilford County; [90231a], 2g. of Randolph County; [15542], 1.5g. of Randolph County

Guin 33°58′ N., 87°55′ W.
Marion County, Alabama, U.S.A.
Found 1969
Iron. Octahedrite, coarse.
A single mass of 34.5kg was found during hay mowing, J. Westcott, letter of 20 February, 1983 in Min. Dept. BM (NH).

Guixi 28°17′ N., 117°11′ E.
Jiangxi, China
Found
Iron. Octahedrite, medium (1.2mm) (IIIA).
A single mass of 220kg is in Beijing Planetarium. Date of find not reported, D. Bian, Meteoritics, 1981, **16**, p.122. Listed, co-ordinates, A.S. Walker, Meteoritics, 1980, **15**, p.253. Classification, analysis, 8.1 %Ni, 20 ppm.Ga, 39.7 ppm.Ge, 2.4 ppm.Ir, D.J. Malvin et al., priv. comm., 1983.
220kg Beijing, Planetarium;

Guizhou 25°24′ N., 107°30′ E.
Guizhou, China
Iron.
A single mass of unreported weight is listed, D. Bian, Meteoritics, 1981, **16**, p.122.
Specimen, Guiyang, Acad. Sin. Inst. Geochem.;

Gumashnik v Gumoschnik.

Gumoschnik 42°54′ N., 24°42′ E. approx.
Trojan, Gabrovo, Bulgaria
Fell 1904, April 28, 1820 hrs
Synonym(s): *Debnevo, Gumashnik, Humoshnik, Trojan, Troyan*
Stone. Olivine-bronzite chondrite (H5).
After detonations, five or six stones were found, the largest weighing 3.8kg and the total 5.7kg. Described, with an analysis, G. Bontschew, Neues Jahrb. Min., 1912, **2**, p.354, abs.. Further analysis, 25.52 % total iron, D.I. Dimov, Annu. Univ. Sofia, Fac. Geol. Geogr., 1974, **66** (Livre 1, geol.), p.209. Olivine Fa_{19}, B. Mason, Geochimica et Cosmochimica Acta, 1963, **27**, p.1011.
253g Sophia Univ.; 69g Moscow, Acad. Sci.; 59g Vienna, Naturhist. Mus.;
Specimen(s): [1924,407], 5.75g.

Gumuruh v Bandong.

Gun Creek 34°0′ N., 111°0′ W. approx.
Gila County, Arizona, U.S.A.
Found 1909
Iron. Octahedrite, medium (0.8mm) (IRANOM).
A mass of about 50lb was found about 70 miles NE. of Globe. Description, with an analysis, 6.5 %Ni, C. Palache, Am. J. Sci., 1926, **12**, p.146. Further analysis, 8.38 %Ni, 22.4 ppm.Ga, 69.7 ppm.Ge, 0.052 ppm.Ir, E.R.D. Scott et al., Geochimica et Cosmochimica Acta, 1973, **37**, p.1957. Description, high phosphide content, V.F. Buchwald, Iron Meteorites, Univ. of California, 1975, p.621.
2.9kg Harvard Univ.; 1204g Washington, U.S. Nat. Mus.; 42g Tempe, Arizona State Univ.;
Specimen(s): [1927,85], 200g. slice, and fragments, 10g; [1927,85a], 84.5g. slice

Gundaring 33°18′ S., 117°40′ E.
Western Australia, Australia
Found 1937, May
Iron. Octahedrite, coarse (1.4mm) (IIIA).
One mass of 248lb, found in 1937, was thought to have fallen in April, 1930. Description, with an analysis, 8.18 % Ni, E.S. Simpson, Min. Mag., 1938, **25**, p.158. Listed, Spec. Publ. West. Austr. Mus., 1965 (3), p.34. Classification and new analysis, 8.27 %Ni, 20.0 ppm.Ga, 43.9 ppm.Ge, 0.31 ppm.Ir, E.R.D. Scott et al., Geochimica et Cosmochimica Acta, 1973, **37**, p.1957. Description, corrosion indicates it fell long before 1930, V.F. Buchwald, Iron Meteorites, Univ. of California, 1975, p.622.
Main mass, Perth, West. Austr. Mus.; 237g Washington, U.S. Nat. Mus.;
Specimen(s): [1954,105], 168g.

Gunnadorah 31°0′ S., 125°56′ E.
Nullarbor Plain, Western Australia, Australia
Found 1968
Stone. Olivine-bronzite chondrite (H5).
One weathered fragment, 19.7g, was found 168m on a bearing 08° from the 837 mile post on the Trans-Australian Railway, between Rawlinna and Haig stations and close to

the boundary of the Gunnadorah pastoral station, G.J.H. McCall and W.H. Cleverly, J. Roy. Soc. West. Austr., 1970, **53**, p.69. Classification, olivine Fa_{19}, B. Mason, Rec. Austr. Mus., 1974, **29**, p.169.
Main mass, Kalgoorlie, West. Austr. School of Mines; Thin section, and chip, Perth, West. Austr. Mus.; 1.2g Washington, U.S. Nat. Mus.;

Gúrdha v Moti-ka-nagla.

Gurkpur v Butsura.

Gurram Konda 13°47′ N., 78°36′ E.
Chittoor district, Andhra Pradesh, India
Fell 1814
Synonym(s): *Kadapa*
Stone. Olivine-hypersthene chondrite (L6).
Specimen 51188 in the British Museum, Natural History, collection is the only known fragment. It was one (numbered H. 62) of a collection of minerals sent to the East India Company's Museum in 1818 by Dr. B. Heyne, who had obtained it from Mr. Skinner at Chittoor, who "had received it from the Punganur Rajah residing near Gurram Konda, where it fell in the year 1814" (list of minerals sent by Dr. Heyne, in Min. Dept. BM(NH).). Possibly a fragment of the lost Punganaru fall (*q.v.*). Olivine Fa_{24}, B. Mason, Geochimica et Cosmochimica Acta, 1963, **27**, p.1011.
Specimen(s): [51188], 8g.

Gursum 9°22′ N., 42°25′ E.
Harer region, Ethiopia
Fell 1981, February 10, 1330 hrs
Stone. Olivine-bronzite chondrite (H4).
After a fireball and a sonic boom, three fragments totalling 34.65kg were recovered about 5 km west of the village of Gursum, Meteor. Bull., 1983 (61), Meteoritics, 1983, **18**, p.80. Brief report, G.P. Sighinolfi et al., Meteoritics, 1983, **18**, p.abs.

Gurukpur v Butsura.

Gütersloh 51°55′ N., 8°23′ E.
Nordrhein-Westfalen, Germany
Fell 1851, April 17, 2000 hrs
Synonym(s): *Guetersloh*
Stone. Olivine-bronzite chondrite (H4), brecciated.
After appearance of a fireball followed by detonations, next day a stone of about 1kg, and a year later a smaller stone almost decomposed, were found, Ann. Phys. Chem. (Poggendorff), 1852, **87**, p.500. Olivine Fa_{20}, B. Mason, Geochimica et Cosmochimica Acta, 1963, **27**, p.1011.
750g Berlin, Humboldt Univ., min. weight; 87g Vienna, Naturhist. Mus.; 40g Tübingen, Univ.; 22g Washington, U.S. Nat. Mus.;
Specimen(s): [33738], 92g. and fragments, 5g.

Gutsch v Gűea.

Gwarzo v Kabo.

Gyokukei 35° N., 127°30′ E. approx.
Keisyo-hokudo, Korea
Fell 1930, March 17
Synonym(s): *Zindoo*
Stone. Chondrite.
A mass of 1.32kg fell, A.D. Nininger, Pop. Astron., Northfield, Minnesota, 1940, **48**, p.558, Contr. Soc. Res. Meteorites, 1942, **2**, p.227 [M.A. 8-54], Meteor. Bull., 1959 (12).

Gyulateke v Mocs.

Gyulatelke v Mocs.

Hachiman v Gifu.

Hachi-oji 35°39′ N., 139°20′ E.
Tokyo Fu, Honshu, Japan
Fell 1817, December 29
Synonym(s): *Edo, Hati-ozi*
Stone. Chondrite.
References, J. Geol. Soc. Tokyo, 1895, **2**, p.246, K. Jimbo, Beitr. Miner. Japan, 1906 (2), p.52, I. Yamamoto, Kwasan Observ. Bull., 1935, p.306 [M.A. 7-173]., P.M. Millman, Trans. Roy. Astron. Soc. Canada, 1938, **32**, p.197 [M.A. 7-173], S. Murayama, Nat. Sci. Mus. Tokyo, 1953, **20**, p.141 [M.A. 13-80].
Less than 0.2g, Tokyo, Nat. Sci. Mus.;

Hacienda Concepcion v Chupaderos.

Hacienda de Bocas v Bocas.

Hacienda de Potosi v Coahuila.

Hacienda di Mani v Toluca.

Haig 31°23′ S., 125°38′ E.
Western Australia, Australia
Found 1951
Synonym(s): *Rawlinna (iron)*
Iron. Octahedrite, medium (0.9mm) (IIIA).
A mass of 480kg was in the Western Australian Museum in 1954, L. Glauert, Nature, 1954, **174**, p.65. Description, and analysis, 7.4 %Ni, Spec. Publ. West. Austr. Mus., 1965 (3), p.35. A mass of 22.8kg and other material together totalling 23.3kg, all found in 1965, interlock with the main mass, W.H. Cleverly, J. R. Soc. West. Austr., 1968, **51**, p.76. Classification and analysis, 7.24 %Ni, 18.8 ppm.Ga, 33.2 ppm.Ge, 10 ppm.Ir, E.R.D. Scott et al., Geochimica et Cosmochimica Acta, 1973, **37**, p.1957. Description, highly shocked, V.F. Buchwald, Iron Meteorites, Univ. of California, 1975, p.624.
480kg Perth, West. Austr. Mus., main mass; 22.88kg Kalgoorlie, West. Austr. School of Mines; 450g Washington, U.S. Nat. Mus.;
Specimen(s): [1968,280], 100g.

Hainant error for Hainaut.

Hainaut 50°19′ N., 3°44′ E.
Bettrechies, Belgium
Fell 1934, November 26, 2000-2100 hrs
Synonym(s): *Bétrèchies, Bettrechies, Lille*
Stone. Olivine-bronzite chondrite (H3-6), brecciated.
Fell in the commune of Bettrechies, dept. Nord, 60m east of the Belgian frontier; probable weight 15-20kg, but only 9kg were recovered. Description, and analysis, M. Lecompte, Mém. Mus. R. Hist. Nat. Belgique, 1935 (66) [M.A. 6-101]. Further analysis, 27.02 % total iron, H.B. Wiik, Geochimica et Cosmochimica Acta, 1956, **9**, p.279. Olivine Fa_{19}, B. Mason, Geochimica et Cosmochimica Acta, 1963, **27**, p.1011.
8425g Lille, Mus. Gosselet; 712g Paris, Mus. d'Hist. Nat.; 922g Brussels,;

Hainholz 52°17′ N., 8°55′ E.

Minden, Nordrhein-Westfalen, Germany
Found 1856
Synonym(s): *Paderborn*
Stony-iron. Mesosiderite (MES).

A mass of 16.5kg was found, F. Wöhler, Ann. Phys. Chem. (Poggendorff), 1857, **100**, p.342. Analysis, G.T. Prior, Min. Mag., 1918, **18**, p.158. Analysis of metal, 8.12 %Ni, 15.1 ppm.Ga, 59 ppm.Ge, 3.7 ppm.Ir, J.T. Wasson et al., Geochimica et Cosmochimica Acta, 1974, **38**, p.135. Silicate mineralogy, B.N. Powell, Geochimica et Cosmochimica Acta, 1971, **35**, p.5. Corona structure, classification, references, C.E. Nehru et al., Geochimica et Cosmochimica Acta, 1980, **44**, p.1103.

6.5kg Tübingen, Univ.; 2.25kg Chicago, Field Mus. Nat. Hist.; 930g Vienna, Naturhist. Mus.; 458g Budapest, Nat. Mus.; 382g Washington, U.S. Nat. Mus.; 380g Amherst College; 355g Tempe, Arizona State Univ.; 294g Berlin, Humboldt Univ.; 291g New York, Amer. Mus. Nat. Hist.; 104g Moscow, Acad. Sci.; 90g Prague, Nat. Mus.;

Specimen(s): [33911], 298g.; [90233], 153g.

Hajmah (a) 19°55′ N., 56°15′ E.

near Hajmah, Jiddat al Harasis, Oman
Found 1958, before this year
Stone. Achondrite, Ca-poor. Ureilite (AURE).

A single stone of 596g was found during oil prospecting in the Hajmah area before 1958. The co-ordinates given are those of Haima oil bore, now called Hajmah, R. Hutchison, Meteoritics, 1977, **12**, p.263, abs. Reported, Meteor. Bull., 1978 (55), Meteoritics, 1978, **13**, p.337.

5g Chicago, Field Mus. Nat. Hist.;

Specimen(s): [1980,M.19], 596g.

Hajmah (b) 19°55′ N., 56°15′ E. approx.

near Hajmah, Jiddat al Harasis, Oman
Found 1958, before this year
Stone. Olivine-hypersthene chondrite (L6).

A single mass of 890.6g was found during oil prospecting in the Hajmah area. Texturally distinct from Hajmah (c) and compositionally distinct from Hajmah (a), olivine $Fa_{25.5}$, Meteor. Bull., 1980 (57), Meteoritics, 1980, **15**, p.98.

Specimen(s): [1980,M.20], 890g.

Hajmah (c) 19°55′ N., 56°15′ E. approx.

near Hajmah, Jiddat al Harasis, Oman
Found 1958, before this year
Stone. Olivine-hypersthene chondrite (L5-6).

Two masses, 1065g and 67g, were found during oil prospecting in the Hajmah area. They are texturally very similar and believed to be part of the same fall, but distinguished from the Hajmah (b) stone of very similar composition found in the same area, olivine $Fa_{25.6}$, Meteor. Bull., 1980 (57), Meteoritics, 1980, **15**, p.98.

Specimen(s): [1980,M.21], 1065g.; [1980,M.22], 67g.

Hakata v Higashi-koen.

Haleb v Aleppo.

Hale Center (no. 1) 34°3′ N., 101°45′ W.

Hale County, Texas, U.S.A.
Found 1936
Stone. Olivine-hypersthene chondrite (L5).

Two stones, of 695g and 610g were found, and were at first supposed to belong to the same fall, A.D. Nininger, Pop. Astron., Northfield, Minnesota, 1937, **45**, p.449 [M.A. 7-62]. Later they were recognized as distinct falls; the 610g stone is reckoned as no. 1, H.H. and A.D. Nininger, The Nininger Collection of Meteorites, Winslow, Arizona, 1950, p.55. Olivine Fa_{25}, B. Mason, Geochimica et Cosmochimica Acta, 1963, **27**, p.1011.

188g Washington, U.S. Nat. Mus.; 26g Tempe, Arizona State Univ.;

Specimen(s): [1959,866], 26g. slice

Hale Center (no. 2) 34°3′ N., 101°45′ W.

Hale County, Texas, U.S.A.
Found 1936
Stone. Olivine-bronzite chondrite (H4).

Two stones were found, of 695g and 610g, and were thought at first to belong to the same fall, A.D. Nininger, Pop. Astron., Northfield, Minnesota, 1937, **45**, p.449 [M.A. 7-62]. Later they were recognized as distinct falls; the 695g stone is reckoned as no. 2, H.H. and A.D. Nininger, The Nininger Collection of Meteorites, Winslow, Arizona, 1950, p.56. Olivine Fa_{17}, B. Mason, Geochimica et Cosmochimica Acta, 1963, **27**, p.1011. The main mass was divided between Wichita Univ. and the Bingham Collection.

324g Washington, U.S. Nat. Mus.; 24g Tempe, Arizona State Univ.;

Specimen(s): [1959,867], 22g. slice

Hale County v Plainview (1917).

Hallingeberg 57°49′ N., 16°14′ E.

Västervik, Kalmar, Sweden
Fell 1944, February 1, 1400 hrs
Stone. Olivine-hypersthene chondrite (L3).

One stone fell at Hallingeberg, 25km NNW. of Västervik. Fragments totalling 1456g (the largest 685g) were collected, A. Hadding, Geol. För. Förh. Stockholm, 1944, **61**, p.314 [M.A. 9-291]. Analysis, 21.97 % total iron, E. Jarosewich, Geochimica et Cosmochimica Acta, 1966, **30**, p.1261. Unequilibrated, olivine Fa_{0-30}, R.T. Dodd et al., Geochimica et Cosmochimica Acta, 1967, **31**, p.921. Petrology of chondrules, R.T. Dodd, Contr. Mineral. Petrol., 1974, **47**, p.97.

1165g Lund Univ., includes main mass, 685g; 144g Washington, U.S. Nat. Mus.; 37g New York, Amer. Mus. Nat. Hist.;

Specimen(s): [1974,M.24], 3.15g. fragments

Halstead

Essex, England
Pseudometeorite..

A stone is said to have fallen on March 12, 1731, R.P. Greg, Rep. Brit. Assn., 1860, p.56, T.M. Hall, Min. Mag., 1879, **3**, p.7, but the original account, A. Vievar, Phil. Trans., 1739, **41**, p.288, does not tally with the fall of a meteorite, but rather suggests a small whirlwind. A similar phenomenon was also observed three days later at Springfield, also in Essex, S. Shepheard, Phil. Trans., 1739, **41**, p.289.

Hamblen County v Morristown.

Hamersley v Roebourne.

Hamersley Range v Roebourne.

Hami v Alatage.

Hamilton v Hamilton (Queensland).

Hamilton (Queensland) 28°29′ S., 148°15′ E.
near Dirranbandi, Queensland, Australia
Found 1966
Synonym(s): *Hamilton*
Stone. Olivine-hypersthene chondrite (L6), veined.
One 68kg stone was found almost buried in mud in the flood plain of the Balonne River, Meteor. Bull., 1968 (42), Meteoritics, 1970, **5**, p.96. Classification, olivine $Fa_{25.3}$, B. Mason, Rec. Austr. Mus., 1974, **29**, p.169.
Main mass, Armidale, Univ. of New England; 275g Tempe, Arizona State Univ.; 201g Washington, U.S. Nat. Mus.;
Specimen(s): [1968,187], 516g. thick slice

Hamilton (Texas) 31°35′36″ N., 98°15′ W.
Hamilton County, Texas, U.S.A.
Found 1965, recognized in this year
Stone. Chondrite.
A mass of about 2.7kg was found, Cat. Huss Coll. Meteorites, 1976, p.21.
52.2g Mainz, Max-Planck-Inst.;

Hamilton County v Brenham.

Hamilton County v Carlton.

Hamilton Hotel v Henbury.

Hamilton Station v Hamilton (Queensland).

Hamlet 41°23′ N., 86°36′ W.
Starke County, Indiana, U.S.A.
Fell 1959, October 13, 0905 hrs
Stone. Olivine-hypersthene chondrite, amphoterite (LL4).
One stone of 2.045kg struck a house and was found in a street 30 minutes after its fall, Meteor. Bull., 1960 (17). A second stone of 1.66kg was found in 1963. Analysis, 20.09 % total iron, H.B. Wiik, Comm. Phys.-Math. Soc. Sci. Fenn., 1969, **34**, p.135. Olivine Fa_{27}, B. Mason, Geochimica et Cosmochimica Acta, 1963, **27**, p.1011.
910g Chicago, Field Mus. Nat. Hist.; 1.5kg Washington, U.S. Nat. Mus.;
Specimen(s): [1971,291], 82.5g. crusted, part-slice

Hammersley v Roebourne.

Hammersley Range v Roebourne.

Hammond 44°55′ N., 92°26′ W.
St. Croix County, Wisconsin, U.S.A.
Found 1884
Synonym(s): *Eau Claire, St. Croix County, St. Croix River, Wisconsin*
Iron. Octahedrite, medium (0.60mm) (IRANOM).
A mass of over 53lb was ploughed up, D. Fisher, Am. J. Sci., 1887, **34**, p.381. Described, with an analysis, 7.34 %Ni, E. Cohen, Meteoritenkunde, 1905, **3**, p.406. New analysis, chemically anomalous, 8.07 %Ni, 26.2 ppm.Ga, 58.4 ppm.Ge, 0.098 ppm.Ir, E.R.D. Scott et al., Geochimica et Cosmochimica Acta, 1973, **37**, p.1957. Part of this fall has been heated and forged, V.F. Buchwald, Iron Meteorites, Univ. of California, 1975, p.625. Mechanical and thermal metamorphism pre-dates atmospheric entry, H.J. Axon, Prog. Materials Sci., 1968, **13**, p.183.
23.1kg Yale Univ., includes main mass, 22.7kg; 842g New York, Amer. Mus. Nat. Hist.; 535g Vienna, Naturhist. Mus.; 378g Tempe, Arizona State Univ.; 293g Washington, U.S. Nat. Mus.; 240g Paris, Mus. d'Hist. Nat.; 132g Chicago, Field Mus. Nat. Hist.;
Specimen(s): [67447], 62g.

Hammond Downs 25°28′ S., 142°48′ E.
South Gregory, Queensland, Australia
Found 1950, recognized as distinct from Tenham 1972
Stone. Olivine-bronzite chondrite (H4).
A stone of 27kg was found by Mr. H.M. Hammond about 7 miles north of Ingella homestead, north of the Tenham strewnfield. The stone was sent with some Tenham stones to the Smithsonian Institution where it was recognized as a new find. There it was sawn and distributed. Analysis, 23.8 % total iron, olivine $Fa_{18.8}$, B. Mason, Meteoritics, 1973, **8**, p.1. Reported, Meteor. Bull., 1974 (52), Meteoritics, 1974, **9**, p.111.
14.455kg Washington, U.S. Nat. Mus.; 6.673kg New York, Amer. Mus. Nat. Hist.; 5.581kg Sydney, Austr. Mus.;
Specimen(s): [1983,M.31], 14.5g.

Hanakai v Kyushu.

Hanakita v Kyushu.

Hanau 50°6′ N., 8°56′ E.
Hessen, Germany
Fell 1877, August 21?
Doubtful. Stone?.
A hot stone the size of a pea was picked up, weight 0.37g. No trace in local collections, record doubtful. A contemporary newspaper account is reprinted, Ber. Senckenb. Naturforsch. Gesell. (Natur und Museum), Frankfurt-a-M., 1928, **58**, p.333 [M.A. 4-418].

Haniet-el-Beguel 32°29′ N., 4°24′ E.
Oued Mzab, Algeria
Found 1888
Synonym(s): *Ouaregla, Teniet-el-Beguel*
Iron. Octahedrite, coarse (1.6mm) (IA).
A mass of about 2kg was found 5 metres deep in digging a well; description, G.A. Daubrée, C. R. Acad. Sci. Paris, 1889, **108**, p.930. Classification and analysis, 8.3 %Ni, 67.4 ppm.Ga, 271 ppm.Ge, 2.4 ppm.Ir, E.R.D. Scott and J.T. Wasson, Geochimica et Cosmochimica Acta, 1976, **40**, p.103. Description, V.F. Buchwald, Iron Meteorites, Univ. of California, 1975, p.628.
1935g Paris, Mus. d'Hist. Nat., main mass; 11g Chicago, Field Mus. Nat. Hist.;
Specimen(s): [1980,M.26], 14.2g.

Happy (a) 35°40′48″ N., 101°59′36″ W.
Swisher County, Texas, U.S.A.
Found 1971
Stone. Olivine-bronzite chondrite (H).
A mass of 2.86kg was found. Two other unclassified chondrites, Happy (b) 2504g, and Happy (c) 884g, are also listed, Cat. Huss Coll. Meteorites, 1976, p.21.
Main masses of the three stones, Mainz, Max-Planck-Inst.;

Happy Canyon 34°48′6″ N., 101°34′ W.
Armstrong County, Texas, U.S.A.
Found 1971
Stone. Enstatite chondrite (E6).
A single mass of about 16.3kg was ploughed up 2 km NW. of Wayside, Texas. It was cut and a section removed; the whereabouts of this section is not certain, Meteor. Bull., 1976 (54), Meteoritics, 1976, **11**, p.75. Description and analysis, 20.99 % total iron, E. Olsen et al., Meteoritics,

1977, **12**, p.109.
9.3kg Mainz, Max-Planck-Inst.; 73g Fort Worth, Texas, Monnig Colln.; 61g Chicago, Field Mus. Nat. Hist.;
Specimen(s): [1976,M.15], 76.7g.

Happy Draw
Randall County, Texas, U.S.A.
Found 1981
Stone. Olivine-bronzite chondrite (H3-4).
Two stones are listed, one of 224g, the other of 120g, Cat. Monnig Colln. Meteorites, 1983. Both are very oxidised, olivine $Fa_{7.1\text{-}27.4}$, A.L. Graham, priv. comm., 1983.
303g Fort Worth, Texas, Monnig Colln., main masses of both stones;

Hapur v Bahjoi.

Haraiya 26°48′ N., 82°32′ E.
Basti district, Uttar Pradesh, India
Fell 1878, August or September, afternoon
Stone. Achondrite, Ca-rich. Eucrite (AEUC).
After detonations a stone of about 1kg was seen to fall, L.L. Fermor, Rec. Geol. Surv. India, 1907, **35** (2), p.90. Trace element contents in plagioclase and pyroxene, R.O. Allen, Jr. and B. Mason, Geochimica et Cosmochimica Acta, 1973, **37**, p.1435. Analysis, B. Mason et al., Smithson. Contrib. Earth Sci., 1979 (22), p.30.
562g Calcutta, Mus. Geol. Surv. India; 66g Tempe, Arizona State Univ.; 56g New York, Amer. Mus. Nat. Hist.; 2.2g Washington, U.S. Nat. Mus.;
Specimen(s): [1983,M.14], 9.1g.

Harding County 45°30′ N., 103°30′ W. approx.
South Dakota, U.S.A.
Found 1941
Stone. Olivine-hypersthene chondrite (L4).
A mass of 3075g was found, H.H. and A.D. Nininger, The Nininger Collection of Meteorites, Winslow, Arizona, 1950, p.56. Olivine Fa_{26}, B. Mason, Geochimica et Cosmochimica Acta, 1963, **27**, p.1011.
34g Tempe, Arizona State Univ.;
Specimen(s): [1959,868], 38.5g. slice

Hardtner 37°4′ N., 98°39′42″ W.
Barber County, Kansas, U.S.A.
Found 1972, recognized in this year
Stone. Olivine-hypersthene chondrite (L).
A 13kg individual was ploughed up in a field which had been terraced, Meteor. Bull., 1975 (53), Meteoritics, 1975, **10**, p.149.
10.2kg Mainz, Max-Planck-Inst.; 225g Copenhagen, Univ. Geol. Mus.;
Specimen(s): [1974,M.23], 123.5g.

Hardwick 43°48′ N., 96°10′ W.
Battle Plain Township, Rock County, Minnesota, U.S.A.
Found 1937
Stone. Olivine-hypersthene chondrite (L4).
One mass of 7.8kg was found, A.D. Nininger, Pop. Astron., Northfield, Minnesota, 1939, **47**, p.212. Olivine Fa_{24}, B. Mason, Geochimica et Cosmochimica Acta, 1963, **27**, p.1011.
2.8kg Tempe, Arizona State Univ.; 1.09kg Washington, U.S. Nat. Mus.; 279g Los Angeles, Univ. of California; 280g Chicago, Field Mus. Nat. Hist.;
Specimen(s): [1959,1035], 2912.5g. about half of the mass

Haripura 28°23′ N., 75°47′ E.
Jhunjhun district, Rajasthan, India
Fell 1921, January 17, 2100 hrs
Stone. Carbonaceous chondrite, type II (CM2).
After detonations and appearance of light, a stone fell and broke into two pieces: description, G.V. Hobson, Rec. Geol. Surv. India, 1927, **60**, p.136. According to another account, a shower of stones fell, one of which weighed 315g, C.A. Silberrad, Min. Mag., 1932, **23**, p.301. Analysis, 21.23 % total iron, H.B. Wiik, Geochimica et Cosmochimica Acta, 1956, **9**, p.279.
281g Calcutta, Mus. Geol. Surv. India; Specimen, Jaipur Mus.;

Hariya v Dokachi.

Harleton 32°40.5′ N., 94°30.7′ W.
Harrison County, Texas, U.S.A.
Fell 1961, May 30, 2230 hrs
Stone. Olivine-hypersthene chondrite (L6).
One stone of 8.36kg was recovered a few minutes after its fall. It penetrated the sandy soil to a depth of 75cm, Meteor. Bull., 1961 (22). Description, analysis, 22.6 % total iron, olivine $Fa_{24.6}$, R.S. Clarke, Jr. et al., Smithson. Contrib. Earth Sci., 1977 (19), p.61.
4400g Washington, U.S. Nat. Mus.; 508g Chicago, Field Mus. Nat. Hist.;
Specimen(s): [1966,47], 38g.

Harliton error for Harleton.

Harlowton 46°26′ N., 109°50′ W.
Wheatland County, Montana, U.S.A.
Found 1975
Iron. Octahedrite, medium (1.0mm) (IA).
A single mass of about 4kg was found, C.F. Lewis and C.B. Moore, Cat. Meteor., Arizona State Univ., 1976, p.176. Classification and analysis, 8.82 %Ni, 61.5 ppm.Ga, 222 ppm.Ge, 2.5 ppm.Ir, A. Kracher et al., Geochimica et Cosmochimica Acta, 1980, **44**, p.773.
2.89kg Tempe, Arizona State Univ.; 247g Chicago, Field Mus. Nat. Hist.;

Harriman (Of) 35°57′ N., 84°34′ W.
Roane County, Tennessee, U.S.A.
Found 1947, before this year
Iron. Octahedrite, fine (0.30mm) (IVA).
A 67lb mass was found; description, V.F. Buchwald, Iron Meteorites, Univ. of California, 1975, p.630. Analysis, 7.96 %Ni, 2.21 ppm.Ga, 0.130 ppm.Ge, 2.3 ppm.Ir, R. Schaudy et al., Icarus, 1972, **17**, p.174.
28kg Fort Worth, Texas Observers Colln.; Polished sections, Washington, U.S. Nat. Mus.;

Harriman (Om) 35°57′ N., 84°34′ W.
Roane County, Tennessee, U.S.A.
Found 1938, before this year
Iron. Octahedrite, medium (0.95mm) (IIIA).
A mass of about 13kg, Rep. U.S. Nat. Mus., 1947, p.43, Meteorite Coll. of S.H. Perry, Adrian, Michigan, 1947, p.6. Analysis, 7.41 %Ni, 19.2 ppm.Ga, 36.8 ppm.Ge, 10 ppm.Ir, E.R.D. Scott et al., Geochimica et Cosmochimica Acta, 1973, **37**, p.1957. Description, shocked with epsilon structure, V.F. Buchwald, Iron Meteorites, Univ. of California, 1975, p.631.
12.3kg Washington, U.S. Nat. Mus., main mass; 121g New York, Amer. Mus. Nat. Hist.;

Harrison County 38°15′ N., 86°10′ W. approx.
Indiana, U.S.A.
Fell 1859, March 28, 1600 hrs
Stone. Olivine-hypersthene chondrite (L6), brecciated.
A shower of stones fell, after detonations, within an area of four square miles, but only four stones, of total weight about 1.5lb, were found: described and analysed, J.L. Smith, Am. J. Sci., 1859, **28**, p.409. Olivine Fa_{26}, B. Mason, Geochimica et Cosmochimica Acta, 1963, **27**, p.1011.
73g Harvard Univ.; 320g Tempe, Arizona State Univ.; 19g Washington, U.S. Nat. Mus.; 17g Yale Univ.; 11g Copenhagen, Univ. Geol. Mus.; 9g Vienna, Naturhist. Mus.;
Specimen(s): [35402], 30g.; [34612a], 4g.

Harrison County v Cynthiana.

Harrison Township 38°20′ N., 101°42′ W.
Greenley County, Kansas, U.S.A.
Found 1945
Stone. Chondrite.
A single mass of 3263g was found, Cat. Monnig Colln. Meteorites.
3260g Fort Worth, Texas, Monnig Colln.;

Harrisonville 38°39′ N., 94°20′ W.
Cass County, Missouri, U.S.A.
Found 1933, April 9
Stone. Olivine-hypersthene chondrite (L6), veined.
14 stones, totalling 12.9kg, were found, H.H. Nininger, The Mines Mag., Golden, Colorado, 1933, **23**, p.6 [M.A. 5-405], A.D. Nininger, Pop. Astron., Northfield, Minnesota, 1937, **45**, p.449 [M.A. 7-62]. Listed, F.C. Leonard, Univ. New Mexico Publ., Albuquerque, 1946 (meteoritic ser. no. 1), p.40, W. Wahl, letter of 23 May, 1950, in Min. Dept. BM (NH)., H.H. and A.D. Nininger, The Nininger Collection of Meteorites, Winslow, Arizona, 1950, p.56. Olivine Fa_{24}, B. Mason, Geochimica et Cosmochimica Acta, 1963, **27**, p.1011.
15.4kg Tempe, Arizona State Univ.; 10kg Washington, U.S. Nat. Mus.; 7kg Los Angeles, Univ. of California; 2.5kg Michigan Univ.; 1.8kg Yale Univ.; 0.5kg Bloomfield Hills, Michigan, Cranbrook Inst. Sci.; 1.5kg Chicago, Field Mus. Nat. Hist.;
Specimen(s): [1937,1655], 133g.; [1959,997], 5949g. a complete stone; [1959,998], 2420g. a complete stone

Harrogate 53°59′ N., 1°33′ W.
Yorkshire, England
Fell 1842, August 5, 1700 hrs
Doubtful..
A stone is said to have fallen, but was regarded as very doubtful, R.P. Greg, Rep. Brit. Assn., 1869, p.80.

Hart Camp 33°59′54″ N., 102°10′36″ W.
Lamb County, Texas, U.S.A.
Found 1970
Stone. Chondrite.
Listed, total known weight 1.2kg, Cat. Huss Coll. Meteorites, 1976, p.22.
1147g Mainz, Max-Planck-Inst.;

Hartford v Marion (Iowa).

Hartley 35°56′30″ N., 102°9′36″ W.
Hartley County, Texas, U.S.A.
Found 1967
Stone. Olivine-hypersthene chondrite (L).
Listed, total known weight 2.1kg, Cat. Huss Coll. Meteorites, 1976, p.22.
1853g Mainz, Max-Planck-Inst.;

Hart Range v Boxhole.

Haruchas v Gibeon.

Harvard University v Vaca Muerta.

Haskell 33°9′ N., 99°45′ W.
Haskell County, Texas, U.S.A.
Found 1909
Stone. Olivine-hypersthene chondrite (L6).
A mass of 36kg was found, H.H. and A.D. Nininger, The Nininger Collection of Meteorites, Winslow, Arizona, 1950, p.56. Olivine Fa_{23}, B. Mason, Geochimica et Cosmochimica Acta, 1963, **27**, p.1011.
15kg Tempe, Arizona State Univ.; 994g Fort Worth, Texas, Monnig Colln.; 820g Chicago, Field Mus. Nat. Hist.;
Specimen(s): [1959,1013], 18824g. half of the mass

Hasparos 34° N., 105°30′ approx.
Lincoln County, New Mexico, U.S.A.
Found 1935
Iron. Octahedrite, coarse (2.4mm) (IA).
A single mass of 12.88kg was found on the northern slope of the Capitan Mountains and became available for research in 1982, Meteor. Bull., 1983 (61), Meteoritics, 1983, **18**, p.80. Classification, analysis, 6.4 %Ni, 94.7 ppm.Ga, 486 ppm.Ge, 5.6 ppm.Ir, D.J. Malvin et al., priv. comm, 1983.
Specimen, Los Angeles, Univ. of California;

Hassayampa 33°45′ N., 112°40′ W. approx.
Maricopa County, Arizona, U.S.A.
Found 1963, before this year
Stone. Olivine-bronzite chondrite (H).
Main mass in O.E. Monnig's collection. Olivine Fa_{18}, B. Mason, Geochimica et Cosmochimica Acta, 1963, **27**, p.1011.
16kg Fort Worth, Texas, Monnig Colln.;

Hassi-Iekna v Hassi-Jekna.

Hassi-Jekna 28°57′ N., 0°49′ E.
Oued Mequiden, El Golea, Algeria
Fell 1890, some years before
Synonym(s): *Hassi-Iekna*
Iron. Octahedrite, fine (0.47mm) (IIICD-ANOM).
A pear-shaped mass of 1.25kg fell, after detonations: described and analysed, S. Meunier, C. R. Acad. Sci. Paris, 1892, **115**, p.531. Further analysis, 10.50 %Ni, 27.4 ppm.Ga, 69.6 ppm.Ge, 0.18 ppm.Ir, E.R.D. Scott et al., Geochimica et Cosmochimica Acta, 1973, **37**, p.1957. Description, carbon-rich, haxonite well developed, V.F. Buchwald, Iron Meteorites, Univ. of California, 1975, p.632.
1250g Paris, Mus. d'Hist. Nat., main mass;
Specimen(s): [1980,M.27], 11.5g.

Hässle v Hessle.

Hastings County v Madoc.

Hataya v Shiraiwa.

Hat Creek 42°55′ N., 104°25′ W.
Niabrara County, Wyoming, U.S.A.
Found 1939
Stone. Olivine-bronzite chondrite (H4), veined.
One mass of 8.9kg was found, A.D. Nininger, Pop. Astron., Northfield, Minnesota, 1940, **48**, p.556, A.D. Nininger, Contr. Soc. Res. Meteorites, **2**, p.227, W. Wahl, letter of 23 May, 1950 in Min. Dept. BM(NH)., H.H. and A.D. Nininger, The Nininger Collection of Meteorites, Winslow, Arizona, 1950, p.56. Olivine Fa_{18}, B. Mason, Geochimica et Cosmochimica Acta, 1963, **27**, p.1011.
191g Tempe, Arizona State Univ.;
Specimen(s): [1959,817], 346g. and fragments, 2g

Hatfield 34°29′ N., 94°27′ W.
Polk County, Arkansas, U.S.A.
Found 1941
Iron. Octahedrite, medium (1.15mm).
Total weight not known, H.H. and A.D. Nininger, The Nininger Collection of Meteorites, Winslow, Arizona, 1950, p.57. Brief description, V.F. Buchwald, Iron Meteorites, Univ. of California, 1975, p.634.
19.7g Tempe, Arizona State Univ.;

Hatford 51°39′ N., 1°31′ W.
Berkshire, England
Fell 1628, April 9, 1700 hrs
Stone.
At least three stones fell, at Hatford (24lb), Challow (26lb), and Balking (14lb). Although nothing has been preserved, it is quite clear from the description that this was a true meteorite, Gentleman's Mag., 1796, **66**, p.1007, T.W. Webb, Nature, 1870, **2**, p.212, T.M. Hall, Min. Mag., 1879, **3**, p.4.

Hati-ozi v Hachi-oji.

Hautes Alpes
France
Pseudometeorite..
Mentioned, pseudometeorite, C. Cailliatte, J. Observ., 1954, **37**, p.11.

Hauptmannsdorf v Braunau.

Haute Volta v Béréba.

Havana 40°20′ N., 90°3′ W.
Mason County, Illinois, U.S.A.
Found, prehistoric
Iron. Octahedrite, fine (0.35mm) (IIICD).
Two beads of meteoritic iron were found in an Indian mound, Rep. U.S. Nat. Mus., 1950, p.56, Meteorite Coll. of S.H. Perry, Adrian, Michigan, 1949, p.16. It is not related to Brenham or the Hopewellian pallasites, J.T. Wasson, letter of 1 August, 1972, in Min. Dept. BM(NH). Classification and analysis, 11.37 %Ni, 20.5 ppm.Ga, 21.6 ppm.Ge, 0.3 ppm.Ir, J.T. Wasson and R. Schaudy, Icarus, 1971, **14**, p.59. Description, bandwidth, artificially cold-worked and annealed, V.F. Buchwald, Iron Meteorites, Univ. of California, 1975, p.635.
12g Washington, U.S. Nat. Mus.;

Haven 37°58′ N., 97°45′ W.
Reno County, Kansas, U.S.A.
Found 1950
Stone. Olivine-bronzite chondrite (H6), veined.
A mass of 2948g was found, R.A. Morley, Earth Sci. Digest, 1952, **6**, p.33 [M.A. 12-254], E.P. Henderson, Am. Miner., 1955, **40**, p.937, Meteor. Bull., 1958 (8). Total known mass 6.1kg, Cat. Monnig Colln., 1983, p.10. Olivine Fa_{20}, B. Mason, Geochimica et Cosmochimica Acta, 1963, **27**, p.1011.
3.1kg Fort Worth, Texas, Monnig Colln.; 78g New York, Amer. Mus. Nat. Hist.; 23g Tempe, Arizona State Univ.; 21g Washington, U.S. Nat. Mus.;
Specimen(s): [1971,295], 24.7g. crusted slice

Haverö 60°14′44″ N., 22°3′43″ E.
Finland
Fell 1971, August 2, 1545 hrs
Stone. Achondrite, Ca-poor. Ureilite (AURE).
A stone of 1544g fell into an outbuilding of a farm, making holes at three different levels, indicating that its path was from WSW. and 8 degrees from the vertical, K.J. Neuvonen et al., Meteoritics, 1972, **7**, p.515. Various papers on this meteorite, including an analysis by H.B. Wiik, 14.9 % total iron, Meteoritics, 1972, **7**, p.515. Primordial gases, H.W. Weber et al., Earth planet. Sci. Lett., 1976, **29**, p.81. Mineralogy and petrology, J.L. Berkley et al., Geochimica et Cosmochimica Acta, 1980, **44**, p.1579.
Main mass, Turku, Univ.;
Specimen(s): [1975,M.8], polished thin section

Haviland v Brenham.

Haviland (stone) 37°37′ N., 99°6′ W.
Kiowa County, Kansas, U.S.A.
Found 1937
Stone. Olivine-bronzite chondrite (H5).
One stone of 1035g was found in NE. 1/4, sect. 27, township 28, range 17W., Kiowa Co., A.D. Nininger, Pop. Astron., Northfield, Minnesota, 1939, **47**, p.212, E.P. Henderson, letter of 3 June, 1939, in Min. Dept. BM(NH). Olivine Fa_{19}, B. Mason, Geochimica et Cosmochimica Acta, 1963, **27**, p.1011.
422g Tempe, Arizona State Univ.; 84g Washington, U.S. Nat. Mus.;
Specimen(s): [1959,869], 503g. and fragments, 1.5g

Haviland (b) 37°35′48″ N., 99°7′30″ W.
Kiowa County, Kansas, U.S.A.
Found 1976, recognized in this year
Stone. Olivine-bronzite chondrite (H5).
A single stone of 2092g was ploughed up about 2.8 km SW. of Haviland, G.I Huss, priv. comm., 1982 in Min. Dept. BM (NH). Probably distinct from Haviland (stone), olivine $Fa_{18.5}$, A.L. Graham, priv. comm. 1982 in Min. Dept. BM (NH). Differs from Haviland (stone), rare gas contents distinct, L. Schultz, letter of 11 February, 1983 in Min. Dept. BM(NH).
Specimen(s): [1982,M.23], 29g.

Haviland Township v Brenham.

Hawaii v Honolulu.

Hawk Springs 41°47′ N., 104°17′ W.
Goshen County, Wyoming, U.S.A.
Found 1935, December 23
Stone. Olivine-bronzite chondrite (H5), black.
Three fragments, fitting together and of total weight 367g, were found. Described, H.H. Nininger, The Mines Mag., Golden, Colorado, 1937, **27**, p.16 [M.A. 7-69]. Olivine Fa_{19}, B. Mason, Geochimica et Cosmochimica Acta, 1963, **27**, p.1011.
91g Tempe, Arizona State Univ.; 88g Washington, U.S. Nat. Mus.; 54g Chicago, Field Mus. Nat. Hist.;
Specimen(s): [1959,829], 110g. and fragments, 1g

Hay v Eli Elwah.

Hay v Pevensey.

Hayami v Tané.

Hayden Creek 45° N., 114° W. approx.
Lemhi County, Idaho, U.S.A.
Found 1895, before this year
Iron. Octahedrite, medium (1.0mm) (IIIA).
A mass of 9.5oz, (270g), was found by a gold prospector at the bottom of a twelve-foot shaft, W.E. Hidden, Am. J. Sci., 1900, **9**, p.367. Description, shock-hardened; possibly paired with Livingston (Montana), V.F. Buchwald, Iron Meteorites, Univ. of California, 1975, p.637. Analysis, 7.58 %Ni, 18.4 ppm.Ga, 36.6 ppm.Ge, 8.1 ppm.Ir, E.R.D. Scott and J.T. Wasson, Geochimica et Cosmochimica Acta, 1976, **40**, p.103.
68g Chicago, Field Mus. Nat. Hist.; 54g Vienna, Naturhist. Mus.; 14g New York, Amer. Mus. Nat. Hist.;
Specimen(s): [84438], 77g.

Hayes Center 40°31′ N., 101°2′ W.
Hayes County, Nebraska, U.S.A.
Found 1941
Stone. Olivine-hypersthene chondrite (L6), black.
A mass of 4.5kg was found in sect. 36, township 8N, range 32W, H.H. and A.D. Nininger, The Nininger Collection of Meteorites, Winslow, Arizona, 1950, p.57. Olivine Fa_{23}, B. Mason, Geochimica et Cosmochimica Acta, 1963, **27**, p.1011.
1.4kg Tempe, Arizona State Univ.; 102g Perth, West. Austr. Mus.; 442g Harvard Univ.; 82g New York, Amer. Mus. Nat. Hist.;
Specimen(s): [1959,1024], 1002g. thick slice

Haywood County v Ferguson.

Hebrides v Tiree.

Hechi 24°42′ N., 108°0′ E.
Hechi County, Guangxi, China
Found 1956
Stone.
Listed, without further details, D. Bian, Meteoritics, 1981, **16**, p.115.

Hedeskoga 55°28′ N., 13°47′ E.
Ystad, Sweden
Fell 1922, April 20, 1945 hrs
Stone. Olivine-bronzite chondrite (H5), veined.
After the appearance of a fireball, and detonations, A stone of 3.5kg fell 3km NNW. of Ystad. Described, with an analysis, A. Hadding, Geol. För. Förh. Stockholm, 1924, **46**, p.383, A. Corlin, Pop. Astron., Northfield, Minnesota, 1939, **46**, p.383. Olivine Fa_{19}, B. Mason, Geochimica et Cosmochimica Acta, 1963, **27**, p.1011.
2.5kg Lund Univ.; 40g Chicago, Field Mus. Nat. Hist.; 26g Washington, U.S. Nat. Mus.;

Hedjaz 27° N., 36° E. ?
Et-Tlahi, Saudi Arabia
Fell 1910, in the spring, at night hrs
Synonym(s): *El Tlahi, Et-Tlahi, Hejaz*
Stone. Olivine-hypersthene chondrite (L3-6), brecciated.
After detonations, four stones fell at Et-Tlahi, the largest weighing 4kg. Described, with an analysis, J. Couyat, C. R. Acad. Sci. Paris, 1912, **155**, p.916. A misprint in Couyat's paper reverses the percentage of troilite and metal. The analysis shows more K than Na and is doubtful, W. Wahl, Geochimica et Cosmochimica Acta, 1950, **1**, p.28. Couryat gives the place of fall as Et-Tlahi, near Deba in Madian; Deba is probably Dheba, 27°20'N., 35°40'E., in Midian. A polymict breccia of L3-L6 fragments, F. Kraut and K. Fredriksson, Meteoritics, 1971, **6**, p.284, abs. Olivine Fa_{24}, B. Mason, Geochimica et Cosmochimica Acta, 1963, **27**, p.1011. Analysis, 22.09 % total iron, B. Mason, priv. comm., 1981.
5.8kg Paris, Mus. d'Hist. Nat.; 133g Calcutta, Mus. Geol. Surv. India; 107g Washington, U.S. Nat. Mus.; 70g Copenhagen, Univ. Geol. Mus.;
Specimen(s): [1925,13], 80g. eight pieces

Heinrichau v Grüneberg.

Hejaz v Hedjaz.

Hejing 42°24′ N., 86°18′ E.
Xinjiang, China
Found 1965
Iron.
Listed, without details of mass recovered, D. Bian, Meteoritics, 1981, **16**, p.122.

Helt Township v Cañon Diablo.

Hembury v Henbury.

Henbury 24°34′ S., 133°10′ E.
Northern Territory, Australia
Found 1931
Synonym(s): *Basedow Range, Gallipoli, Glen Helen, Hamilton Hotel, Hembury, Nutwood Downs*
Iron. Octahedrite, medium (0.9mm) (IIIA).
Large numbers of fragments are found outside, and a few inside, the craters; about 1500lb had been collected by June 1933. Described, A.R. Alderman, Min. Mag., 1932, **23**, p.19. Further description and analysis, 7.54 %Ni, A.R. Alderman, Rec. S. Austr. Mus., 1932, **4**, p.561. See also, includes analyses of the glass and the local sandstone, L.J. Spencer, Min. Mag., 1933, **23**, p.387. Analysis, synonymy of Basedow Range and Nutwood Downs, 7.47 %Ni, 17.7 ppm.Ga, 33.7 ppm.Ge, 13 ppm.Ir, E.R.D. Scott et al., Geochimica et Cosmochimica Acta, 1973, **37**, p.1957. Description, deformation of metal, Gallipoli and Nutwood Downs transported Henbury, V.F. Buchwald, Iron Meteorites, Univ. of California, 1975, p.638, 1385, 1408.
133kg Adelaide, S. Austr. Mus.; 227kg Washington, U.S. Nat. Mus.; 174kg Tempe, Arizona State Univ.; 56kg Sydney, Austr. Mus.; 36kg New York, Amer. Mus. Nat. Hist.; 20kg Chicago, Field Mus. Nat. Hist.; 9.5kg Berlin, Humboldt Univ.; 9.4kg Mainz, Max-Planck-Inst.; 7.3kg Budapest, Nat. Mus.; 5.3kg Paris, Mus. d'Hist. Nat.; 2.2kg

Los Angeles, Univ. of California; 5kg Perth, Govt. Chem. Lab.; 2.1kg Harvard Univ.; 1.3kg Ottawa, Mus. Geol. Surv. Canada; 1kg Yale Univ.;

Specimen(s): [1932,1359], 123.8kg. pieces of 68kg, 50.8kg, and a slice, 5kg; [1932,1360], 54.4kg.; [1932,1361], 10.75kg.; [1932,1362], 2kg.; [1933,5], 77.5kg.; [1934,133], 21kg.; [1934, 134], 11.4kg.; [1934,135], 11kg.; [1932,1423], 12.1kg.; [1932, 1424], 15kg.; [1932,1425], 6.9kg.; [1931,488], 782g.; [1932, 98], 272.5g.; [1932,1363-1422], 3928g. 60 pieces; [1932,1426-1442]; [1932,1444]; [1932,1446-1477]; [1932,1479-1530], 29.8kg. 103 pieces; [1934,136-517]; [1934,519-539]; [1934, 541-546]; [1934,548-562]; [1934,564-606], 31.2kg. 467 pieces; [1934,658], 63.5g.; [1950,389], 262g.; [1932,1531-1535]; [1932,1549-1560]; [1932,1563] shale balls and shale; [1932, 1536-1548] silica glass; [1933,7-15] silica glass; [1934,659-661] silica glass; [1932,1561-1562] country rock; [1933,16-24] country rock

Hendersonville 35°19′ N., 81°28′ W.

Henderson County, North Carolina, U.S.A.
Found 1901
Stone. Olivine-hypersthene chondrite (L5).

A mass of about 13lb was found about 3 miles NW. of Hendersonville, and it possibly fell in 1876 when a meteor passed over the town, L.C. Glenn, Am. J. Sci., 1904, **17**, p.215. Described, with an analysis, G.P. Merrill, Proc. U.S. Nat. Mus., 1907, **32**, p.79. Amount and composition of metal, G.T. Prior, Min. Mag., 1919, **18**, p.353. Olivine Fa_{24}, B. Mason, Geochimica et Cosmochimica Acta, 1963, **27**, p.1011.

2695g Washington, U.S. Nat. Mus., main mass; 253g Tempe, Arizona State Univ.; 178g New York, Amer. Mus. Nat. Hist.;

Specimen(s): [1919,87], 123g. slice

Henry County v Hopper.

Henry County v Locust Grove.

Hentey v Khenteisky.

Herceg v Dubrovnik.

Hercegnovi v Dubrovnik.

Heredia 10°0′ N., 84°6′ W.

San José, Costa Rica
Fell 1857, April 1, at night hrs
Synonym(s): *Costa Rica, San José*
Stone. Olivine-bronzite chondrite (H5), brecciated.

After the appearance of a fireball and detonations, many stones were found, the largest weighing about 1kg. Described, with an analysis, I. Domeyko, Anal. Univ. Chile, Santiago, 1859, **16**, p.235. Listed, O. Büchner, Die Meteoriten in Samml., Leipzig, 1863, p.93. Olivine Fa_{18}, B. Mason, Geochimica et Cosmochimica Acta, 1963, **27**, p.1011.

422g Göttingen, Univ.; 70g Tübingen, Univ.; 24g Vienna, Naturhist. Mus.; 9.5g Chicago, Field Mus. Nat. Hist.; 7.5g New York, Amer. Mus. Nat. Hist.;

Specimen(s): [56981], 44g.; [34671], 6g.; [35152], 0.5g.

Hereford 33°47′42″ N., 102°14′30″ W.

Deaf Smith County, Texas, U.S.A.
Found 1970, recognized in this year
Stone. Chondrite.

Listed, total known weight 2276g, Cat. Huss Coll. Meteorites, 1976, p.22.

1116g Mainz, Max-Planck-Inst.;

Herlen v Kerulensky.

Hermadale v Hermitage Plains.

Hermidale v Hermitage Plains.

Hermitage v Diep River.

Hermitage Plains 31°44′ S., 146°24′ E.

County Canbelego, New South Wales, Australia
Found 1909
Synonym(s): *Hermadale, Hermidale*
Stone. Olivine-hypersthene chondrite (L6).

A mass of about 70lb was found 20 miles SE. of Canbelego, C.A. Anderson, Rec. Austr. Mus., 1913, **10** (5), p.60. Analysis, H.P. White, Rec. Geol. Surv. New South Wales, 1920, **9**, p.108. Olivine Fa_{24}, B. Mason, Geochimica et Cosmochimica Acta, 1963, **27**, p.1011.

Main mass, Sydney, Austr. Mus.; 749g Vienna, Naturhist. Mus.; 520g New York, Amer. Mus. Nat. Hist.; 500g Rome, Vatican Colln; 314g Harvard Univ.; 992g Ottawa, Mus. Geol. Surv. Canada; 216g Chicago, Field Mus. Nat. Hist.; 130g Washington, U.S. Nat. Mus.;

Specimen(s): [1912,86], 653g. and fragments, 7.5g.

Hessle 59°51′ N., 17°40′ E.

Uppsala, Sweden
Fell 1869, January 1, 1230 hrs
Synonym(s): *Hässle, Stockholm*
Stone. Olivine-bronzite chondrite (H5).

After detonations, a shower of stones, weighing from about 1.8kg to a few grams each, fell over an area of about 3 × 9 miles; some fell upon ice a few inches thick without breaking it, and powdery, carbonaceous matter was found in association with the stones, A.E. Nordenskiöld, Svenska Vetensk.-Akad. Handl. Stockholm, 1870, **8** (9), Ann. Phys. Chem. (Poggendorff), 1870, **141**, p.205. Analysis, G. Lindström, Övers. Vetensk.-Akad. Förh. Stockholm, 1869 (8), p.715. The analysis shows no K and is doubtful, W. Wahl, Geochimica et Cosmochimica Acta, 1950, **1**, p.28. The fall occurred 20km from Uppsala, F.E. Wickman, letter of 3 December, 1971 in Min. Dept. BM(NH). Olivine Fa_{19}, B. Mason, Geochimica et Cosmochimica Acta, 1963, **27**, p.1011.

7kg Stockholm, Riksmus.; 5kg Uppsala, Univ.; 1.8kg Washington, U.S. Nat. Mus.; 730g Tempe, Arizona State Univ.; 640g Paris, Mus. d'Hist. Nat.; 476g Harvard Univ.; 422g Chicago, Field Mus. Nat. Hist.; 122g New York, Amer. Mus. Nat. Hist.;

Specimen(s): [42467], 365g.; [42468], 331g. half a stone; [42470], 144.5g. complete stone; [42469], 23g. and fragments, 2.5g.; [1927,1288], 37.5g.; [42471], 16g. three stones; [42473], 1g. small stone; [42472], 1g. three stones.

Hesston 38°7′ N., 97°26′ W.

Harvey County, Kansas, U.S.A.
Found 1951, July
Stone. Olivine-hypersthene chondrite (L6).

A mass of 12.9kg was found, F.C. Leonard, Classif. Cat. Meteor., 1956, p.8, 20, 67, Meteor. Bull., 1958 (8). The main mass was in the collection of the late H.O. Stockwell. Olivine Fa_{24}, B. Mason, Geochimica et Cosmochimica Acta, 1963, **27**, p.1011.

98g Los Angeles, Univ. of California; 42g New York, Amer. Mus. Nat. Hist.;

Hex River Mountains 33°19′ S., 19°37′ E.
Cape Province, South Africa
Found 1882
Synonym(s): *Capland, Kapland*
Iron. Hexahedrite (IIA).
A mass of about 60kg was found, A. Brezina, Verh. Geol. Reichsanst. Wien, 1887, p.289. Analysis, 5.68 %Ni, E. Cohen, Meteoritenkunde, 1905, **3**, p.222. New analysis, 5.59 %Ni, 60.7 ppm.Ga, 181 ppm.Ge, 4.4 ppm.Ir, J.T. Wasson, Geochimica et Cosmochimica Acta, 1969, **33**, p.859. Structure, H.J. Axon and C.V. Waine, Min. Mag., 1972, **38**, p.725. Description, V.F. Buchwald, Iron Meteorites, Univ. of California, 1975, p.644.
37.7kg Vienna, Naturhist. Mus.; 2.5kg New York, Amer. Mus. Nat. Hist.; 2.5kg Budapest, Nat. Mus.; 582g Washington, U.S. Nat. Mus.; 683g Chicago, Field Mus. Nat. Hist.; 294g Cape Town, South African Mus.;
Specimen(s): [1921,19], 285.5g. slice; [77098], 245g. slice

Heze v Hotse.

Hickiwan 32°21′30″ N., 112°24′32″ W.
Pima County, Arizona, U.S.A.
Found 1974, March 23
Stone. Olivine-bronzite chondrite (H5).
A single individual weighing 1.928kg was found 7 km east of Hickiwan. Description, olivine $Fa_{19.4}$, D.E. Lange et al., Meteoritics, 1977, **12**, p.286, abs. Reported, Meteor. Bull., 1978 (55), Meteoritics, 1978, **13**, p.337.
0.93g Albuquerque, Univ. of New Mexico, and three thin sections;

Hidalgo v Pacula.

Higashi-koen 33°36′ N., 136°26′ E.
Fukuoka, Kyushu, Japan
Fell 1897, August 11
Synonym(s): *Hakata*
Stone. Olivine-bronzite chondrite (H5).
A stone of about 750g fell, K. Jimbo, Beitr. Miner. Japan, 1906 (2), p.50. Olivine Fa_{19}, B. Mason, Geochimica et Cosmochimica Acta, 1963, **27**, p.1011.
28g Vienna, Naturhist. Mus.; 15g Budapest, Nat. Mus.;
Specimen(s): [84043], 29.5g.

Highland County v Pricetown.

High Possil 55°54′ N., 4°14′ W.
Glasgow, Strathclyde, Scotland
Fell 1804, April 5, morning
Synonym(s): *Glasgow*
Stone. Olivine-hypersthene chondrite (L6).
After detonations, a stone of about 10lb was seen to fall, and broke into two pieces, one about 5cm long, the other measuring 15×10×10cm. Most of the stone was subsequently lost, A. Tilloch, Phil. Mag., 1804, **18**, p.371. Olivine Fa_{25}, B. Mason, Geochimica et Cosmochimica Acta, 1963, **27**, p.1011.
170g Glasgow, Hunterian Mus.; 26g Calcutta, Mus. Geol. Surv. India; 15g Vienna, Naturhist. Mus.; 3.5g Chicago, Field Mus. Nat. Hist.; 3g Tempe, Arizona State Univ.; 2.3g Budapest, Nat. Mus.;
Specimen(s): [19970], 91.5g.

Hildreth 40°20′ N., 99°2′ W.
Franklin County, Nebraska, U.S.A.
Found 1894
Stone. Olivine-hypersthene chondrite (L5).
One mass of 3.06kg was found and recognized as meteoritic in 1937, E.P. Henderson, letters of 12 May and 3 June, 1939 in Min. Dept. BM(NH)., A.D. Nininger, Pop. Astron., Northfield, Minnesota, 1939, **47**, p.212. Olivine Fa_{25}, B. Mason, Geochimica et Cosmochimica Acta, 1963, **27**, p.1011.
1kg Tempe, Arizona State Univ.; 172g Washington, U.S. Nat. Mus.;
Specimen(s): [1959,870], 1425g.

Hill City 39°22′ N., 99°51′ W.
Graham County, Kansas, U.S.A.
Found 1947, known about this year
Iron. Octahedrite, fine (0.38mm) (IVA).
A mass of 11.7kg was found, S.H. Perry, Meteorite Coll. of S.H. Perry, Adrian, Michigan, 1947, p.6. Analysis, with determinations of Au, Ga and Pd, 9.23 %Ni, E. Goldberg et al., Geochimica et Cosmochimica Acta, 1951, **2**, p.1. Classification and new analysis, 9.09 %Ni, 2.29 ppm.Ga, 0.144 ppm.Ge, 0.88 ppm.Ir, R. Schaudy et al., Icarus, 1972, **17**, p.174. Description, shock annealed, V.F. Buchwald, Iron Meteorites, Univ. of California, 1975, p.646.
10kg Washington, U.S. Nat. Mus.; 161g Chicago, Field Mus. Nat. Hist.; 115g New York, Amer. Mus. Nat. Hist.;

Hill's Stone v Travis County.

Hinojal 32°22′ S., 60°9′ W.
Partido de Victoria, Entre Rios, Argentina
Found 1927, between 1927 and 1934
Synonym(s): *Entre Rios II, Victoria*
Stone. Olivine-hypersthene chondrite (L6).
A stone estimated at 50kg was ploughed up and broken into fragments, most of which appear to have been lost, L.O. Giacomelli, letters of 30 March 1960, 25 June and 30 October 1962, and 30 September 1963, in Min. Dept. BM (NH). Reported, Meteor. Bull., 1964 (29). Olivine Fa_{26}, B. Mason, Geochimica et Cosmochimica Acta, 1967, **31**, p.1100.
150g Buenos Aires, Dir. Geol. Min.;

Hinojo 36°52′ S., 60°10′ W.
Partido de Olavarria, Buenos Aires province, Argentina
Found 1928, before this date
Stone. Olivine-bronzite chondrite (H).
A weathered stone of 1155g was found near Hinojo railway station. Described, with an analysis, E.H. Ducloux, Rev. fac. cienc. quim. Univ. nac. La Plata, 1928, **5** (2), p.1 [M.A. 4-119], Anal. Soc. Cient. Argentina, 1931, **112**, p.247 [M.A. 5-159]. Olivine Fa_{20}, B. Mason, Geochimica et Cosmochimica Acta, 1963, **27**, p.1011.
1.1kg La Plata Mus., main mass; 42g Paris, Mus. d'Hist. Nat.; 12g Harvard Univ.;

Hiokomo v Fukutomi.

Hiquipilco v Toluca.

Hiromi v Gifu.

Hishikari v Kyushu.

Hishugari v Kyushu.

Hisikari v Kyushu.

Hislugari v Kyushu.

Hizen v Ogi.

Hmeljevka v Khemlevka.

Hoba 19°35′ S., 17°55′ E.
Grootfontein, Namibia
Found 1920
Synonym(s): *Grootfontein, Hobart West, Hoba Wes, Hoba West*
Iron. Ataxite, Ni-rich (IVB).
A mass measuring over 9×9×3.2 feet, of very ductile iron was found on a farm 12 miles west of Grootfontein. Weight estimated from measurements at 60 metric tons, analysis, 16.24 %Ni, L.J. Spencer, Min. Mag., 1932, **23**, p.1. Structure, H.J. Axon and P.L. Smith, Min. Mag., 1972, **38**, p.736. Further analysis, 16.56 %Ni, 0.192 ppm.Ga, 0.049 ppm.Ge, 27 ppm.Ir, R. Schaudy et al., Icarus, 1972, **17**, p.174. The main mass remains at the place of fall. Description, V.F. Buchwald, Iron Meteorites, Univ. of California, 1975, p.647.
4203g Washington, U.S. Nat. Mus., includes 1.8kg iron shale; 4.2kg Philadelphia, Acad. Nat. Sci.; 2kg Chicago, Field Mus. Nat. Hist.; 1kg Paris, Mus. d'Hist. Nat.; 685g Harvard Univ.;
Specimen(s): [1930,976], 2251g. and pieces, 31.5g, and filings, 116g.; [1929,1963] iron shale; [1931,144] iron shale; [1971,22] iron shale; [1971,299] iron shale

Hobart West v Hoba.

Hoba Wes v Hoba.

Hoba West v Hoba.

Hobbs 32°44′ N., 103°6′ W.
Lea County, New Mexico, U.S.A.
Found 1933, recognized in this year
Stone. Olivine-bronzite chondrite (H4).
Known weight 2.3kg, H.H. Nininger, The Mines Mag., Golden, Colorado, 1933, **23** (8), p.6 [M.A. 5-405], H.H. and A.D. Nininger, The Nininger Collection of Meteorites, Winslow, Arizona, 1950, p.60. Olivine Fa_{19}, B. Mason, Geochimica et Cosmochimica Acta, 1963, **27**, p.1011. A Hobbs (b) stone, total known weight 1.6kg, is listed, Cat. Huss Coll. Meteorites, 1976, p.22. Hobbs (b) classified as H5, D.E. Lange et al., Cat. Meteor. Coll. Univ. New Mexico, 1980, p.22.
224g Washington, U.S. Nat. Mus.; 145g Chicago, Field Mus. Nat. Hist.; 65g Tempe, Arizona State Univ.;
Specimen(s): [1959,871], 99g. slice, and fragments, 3g

Hobdo v Adzhi-Bogdo (stone).

Hochscheid v Simmern.

Hoekmark v Hökmark.

Hofmeyer v Karee Kloof.

Hojsin v Pribram.

Hökmark 64°26′ N., 21°12′ E.
Västerbotten, Sweden
Fell 1954, June 9, 2031 hrs, U.T.
Synonym(s): *Hoekmark*
Stone. Olivine-hypersthene chondrite (L4).
After detonations, two small pieces, 108.8g and 196.7g, were seen to fall. The smaller one was picked up immediately and the bigger one was found next day, S. Hjelmqvist, letter of March 1958, in Min. Dept. BM(NH). Described and figured, D. Malmqvist, Bull. Geol. Inst. Univ. Uppsala, 1961, **40**, p.118 [M.A. 16-170]. Analysis, 21.00 % total iron, H.B. Wiik, Comm. Phys.-Math. Soc. Sci. Fenn., 1969, **34**, p.135. Olivine Fa_{24}, B. Mason, Geochimica et Cosmochimica Acta, 1963, **27**, p.1011.
Both stones, Stockholm, Riksmus.; 12g Washington, U.S. Nat. Mus.; 9g Chicago, Field Mus. Nat. Hist.;

Holata v Holetta.

Holbrook 34°54′ N., 110°11′ W.
Navajo County, Arizona, U.S.A.
Fell 1912, July 19, 1915 hrs
Synonym(s): *Aztec*
Stone. Olivine-hypersthene chondrite (L6).
After the appearance of a smoky trail in the sky, and detonations, a shower of stones fell, estimated to number 14000, of total weight about 481lb, with individuals weighing from 14.5lb to a few grains. Description, W.M. Foote, Am. J. Sci., 1912, **34**, p.437, G.P. Merrill, Smithson. Misc. Coll., 1912, **60** (2149). Analysis, 21.56 % total iron, B. Mason and H.B. Wiik, Geochimica et Cosmochimica Acta, 1961, **21**, p.276. Olivine Fa_{25}, B. Mason, Geochimica et Cosmochimica Acta, 1963, **27**, p.1011. XRF analysis, 19.8 % total iron, H. von Michaelis et al., Earth planet. Sci. Lett., 1969, **5**, p.387. Nature of the metal, H.C. Urey and T. Mayeda, Geochimica et Cosmochimica Acta, 1959, **17**, p.113.
25.7kg New York, Amer. Mus. Nat. Hist., 2050 stones; 15.5kg Ottawa, Mus. Geol. Surv. Canada, 1883 stones; 13.7kg Fort Worth, Texas, Monnig Colln.; 4kg Philadelphia, Acad. Nat. Sci.; 10.8kg Tempe, Arizona State Univ.; 10.7kg Chicago, Field Mus. Nat. Hist., 196 stones; 8.45kg Washington, U.S. Nat. Mus.; 3.7kg Harvard Univ.; 4.8kg Vienna, Naturhist. Mus.; 2.25kg Bonn, Univ. Mus.; 2.2kg Budapest, Nat. Mus.; 400kg Stockholm, Riksmus; 1kg Los Angeles, Univ. of California; 400g Edinburgh, Roy. Scot. Mus.; 1240g Dublin, Nat. Mus. (640g); Univ. Coll. (600g); 276g Prague, Nat. Mus.;
Specimen(s): [1912,653-660]; [1912,663-665]; [1912,668-678], 7590g. 98 mostly complete stones, largest mass 3120.5g, smallest 0.329g; [1962,138], 2g.

Holetta 9°4′ N., 38°25′ E.
Addis-Ababa, Ethiopia
Fell 1923, April 14, 1125 hrs
Synonym(s): *Holata, Walata*
Stone.
A stone of 1415g fell after detonations, H.M. Baghdassarian, Bull. Soc. Astron. France, 1924, **38**, p.81.

Holland's Store 34°22′ N., 85°26′ W.
Chattooga County, Georgia, U.S.A.
Found 1887
Synonym(s): *Chattooga County*
Iron. Hexahedrite (IIA).
A mass of 27lb was found and was afterwards broken up. Described, with an analysis, 4.97 %Ni, G.F. Kunz, Am. J. Sci., 1887, **34**, p.471. Another analysis, 5.35 %Ni, E. Cohen, Meteoritenkunde, 1905, **3**, p.239. More recent analysis, 5.35

%Ni, 60.9 ppm.Ga, 184 ppm.Ge, 20 ppm.Ir, J.T. Wasson, Meteorites, Springer-Verlag, 1974, p.299. Description, shocked and recrystallised, V.F. Buchwald, Iron Meteorites, Univ. of California, 1975, p.651.
3kg Vienna, Naturhist. Mus.; 247g Chicago, Field Mus. Nat. Hist.; 212g New York, Amer. Mus. Nat. Hist.; 92g Yale Univ.; 54g Washington, U.S. Nat. Mus.; 51g Berlin, Humboldt Univ.;
Specimen(s): [67445], 190g.

Holliday 33°45′ N., 98°50′ W.

Archer County, Texas, U.S.A.
Found 1950
Iron. Octahedrite.
A fragment of 10g was found, B. Mason, Meteorites, Wiley, 1962, p.243.
12g Fort Worth, Texas, Monnig Colln.;

Holly 38°4′ N., 102°6′ W.

Prowers County, Colorado, U.S.A.
Found 1937
Stone. Olivine-bronzite chondrite (H4).
Two stones were found, 299g together, E.P. Henderson, letters of 12 May and 3 June 1939, in Min. Dept. BM(NH)., A.D. Nininger, Pop. Astron., Northfield, Minnesota, 1939, **47**, p.212, H.H. and A.D. Nininger, The Nininger Collection of Meteorites, Winslow, Arizona, 1950, p.61. Olivine Fa_{18}, B. Mason, Geochimica et Cosmochimica Acta, 1963, **27**, p.1011.
88g Tempe, Arizona State Univ.; 37g Washington, U.S. Nat. Mus.; 30g Denver, Mus. Nat. Hist.; 19g New York, Amer. Mus. Nat. Hist.;
Specimen(s): [1959,872], 43.5g.

Holman Island 70°44′ N., 117°45′ W.

Victoria Island, North-west Territories, Canada
Found 1951, March
Stone. Olivine-hypersthene chondrite, amphoterite (LL).
A stone of 552g was found on sea ice in March, P.M. Millman, J. Roy. Astron. Soc. Canada, 1953, **47**, p.32, 92, 162 [M.A. 12-358], K.R. Dawson, Cat. Canad. Nat. Meteor. Colln., Geol. Surv. Canada, 1963 (63-37), p.31. Olivine Fa_{29}, B. Mason and H.B. Wiik, Geochimica et Cosmochimica Acta, 1964, **28**, p.533. Analyses of light and dark fractions, O. Müller and J. Zähringer, Earth planet. Sci. Lett., 1966, **1**, p.25.
174g Ottawa, Mus. Geol. Surv. Canada;

Holyoke 40°34′ N., 102°18′ W.

Phillips County, Colorado, U.S.A.
Found 1933, about
Stone. Olivine-bronzite chondrite (H4).
Recognized 1935. One stone of 5.5kg was found, A.D. Nininger, Pop. Astron., Northfield, Minnesota, 1937, **45**, p.449 [M.A. 7-62], H.H. and A.D. Nininger, The Nininger Collection of Meteorites, Winslow, Arizona, 1950, p.61. Olivine Fa_{19}, B. Mason, Geochimica et Cosmochimica Acta, 1963, **27**, p.1011. Contains inclusions of ?carbonaceous chondrite, P. Ramdohr, The opaque minerals in stony meteorites, Elsevier, New York, 1973, p.224. A 5.7kg mass is in Denver, possibly representing a second stone.
6kg Denver, Mus. Nat. Hist.; 1824g Tempe, Arizona State Univ.; 344g Chicago, Field Mus. Nat. Hist.; 260g New York, Amer. Mus. Nat. Hist.; 222g Washington, U.S. Nat. Mus.; 195g Michigan Univ.;
Specimen(s): [1959,873], 1765g. and fragments, 3.5g.

Homestead 41°48′ N., 91°52′ W.

Iowa County, Iowa, U.S.A.
Fell 1875, February 12, 2215 hrs
Synonym(s): *Amana, Iowa County, Jowa, Marengo, Sherlock, West Liberty*
Stone. Olivine-hypersthene chondrite (L5), brecciated.
After the appearance of a brilliant fireball, moving from south to north, and detonations, about 100 stones, weighing together about 500lb, the largest about 74lb, were found scattered over an area of about 18 square miles from Amana to Boltonville, N.R. Leonard, Am. J. Sci., 1875, **10**, p.357, G. Hinrichs, Popular Sci. Monthly, New York, 1875, **7**, p.588. Description, C.W. Gümbel, Sitzungsber. Akad. Wiss. München, Math.-phys. Kl., 1875, **5**, p.313. Analysis, G.T. Prior, Min. Mag., 1918, **18**, p.173. Olivine Fa_{24}, B. Mason, Geochimica et Cosmochimica Acta, 1963, **27**, p.1011.
35kg Yale Univ.; 19.1kg Fort Worth, Texas, Monnig Colln.; 17.5kg Chicago, Field Mus. Nat. Hist.; 9.1kg Harvard Univ.; 10.8kg New York, Amer. Mus. Nat. Hist.; 7.49kg Paris, Mus. d'Hist. Nat.; 6.5kg Vienna, Naturhist. Mus.; 3.3kg Tempe, Arizona State Univ.; 5.6kg Washington, U.S. Nat. Mus.; 3kg Copenhagen, Univ. Geol. Mus.; 2.36kg Berlin, Humboldt Univ.; 2kg Stockholm, Riksmus.; 3.5kg Moscow, Acad. Sci.; 3.8kg Budapest, Nat. Mus.;
Specimen(s): [48474], 3775g. obtained from G. Hinrichs; [50803], 127g. and fragments, 6g.; [1962,157], 60g. two fragments; [1971,300], 1.5g.

Homewood 49°30′30″ N., 97°49′ W.

Homewood, Manitoba, Canada
Found 1970
Stone. Olivine-hypersthene chondrite (L6).
A single stone of 325g was found 4 km east of Homewood, Manitoba. Description and analysis, olivine $Fa_{25.4}$, 21.60 % total iron, W.K. Mysyk et al., Meteoritics, 1979, **14**, p.207.
Specimen, Winnipeg, Mineral Mus. Univ. of Manitoba;

Honduras v Rosario.

Honolulu 21°18′ N., 157°52′ W.

Oahu, Hawaii, U.S.A.
Fell 1825, September 27, 1030 hrs
Synonym(s): *Hawaii, Sandwich Islands*
Stone. Olivine-hypersthene chondrite (L5), veined.
After detonations, several stones fell, two of which weighed about 1.5kg each, E. Hofmann, Karsten's Arch. Min. Berlin, 1829, **1**, p.311. Analysis, A. Kuhlberg, Arch. Naturk. Liv.-Ehst.-u. Kurlands, Ser. 1, Min. Wiss., Dorpat, 1867, **4**, p.14. Olivine Fa_{24}, B. Mason, Geochimica et Cosmochimica Acta, 1963, **27**, p.1011. Further analysis, 22.46 % total iron, M.I. D'yakonova, Метеоритика, 1968, **28**, p.131.
577g Yale Univ.; 549g Tartu, Univ.; 438g Moscow, Acad. Sci.; 261g Vienna, Naturhist. Mus.; 103g Rome, Vatican Colln; 63g Berlin, Humboldt Univ.; 44g Tempe, Arizona State Univ.; 35g Budapest, Nat. Mus.;
Specimen(s): [36609(7)], 42.5g.; [25460], 24g.; [34603], 14g.; [1920,307], 2.9g. powder

Hope 33°41′ N., 93°36′ W. approx.

Hempstead County, Arkansas, U.S.A.
Found 1955, some years before this
Synonym(s): *Boaz (iron)*
Iron. Octahedrite, coarse (2.1mm) (IA).
This is the 6.8kg mass listed as coming from Boaz, Alabama, F.C. Leonard, Classif. Cat. Meteor., 1956, p.8, 44, 63, B. Mason, Meteorites, Wiley, 1962, p.228. The main mass is in the collection of O.E. Monnig, Fort Worth, Texas, who has

informed the U.S. Nat. Mus. that it was allegedly found near Hope, Arkansas, but only became known when possessed by a teacher in Boaz, R.S. Clarke, Jr., letter of 10 March 1976, in Min. Dept. BM(NH). Classification and analysis, 6.77 % Ni, 85.7 ppm.Ga, 398 ppm.Ge, 0.59 ppm.Ir, J.T. Wasson, Icarus, 1970, **12**, p.407. Description, V.F. Buchwald, Iron Meteorites, Univ. of California, 1975, p.654.
4.3kg Fort Worth, Texas, Monnig Colln., main mass; 885g Washington, U.S. Nat. Mus.;

Hopewell Mounds v Brenham.

Hopper 36°33′ N., 79°47′ W.
Henry County, Virginia, U.S.A.
Found 1889
Synonym(s): *Henry County*
Iron. Octahedrite, medium (0.70mm) (IIIB).
A mass of about 4lb was found; described and analysed, 7.7 %Ni, F.P. Venable, Am. J. Sci., 1890, **40**, p.162. Co-ordinates, distinct from Smith's Mountain, V.F. Buchwald, Iron Meteorites, Univ. of California, 1975, p.660. Location of main mass not known.
47g Chicago, Field Mus. Nat. Hist.; 25g Washington, U.S. Nat. Mus.; 8g New York, Amer. Mus. Nat. Hist.;

Horace 38°21′ N., 101°47′ W.
Greeley County, Kansas, U.S.A.
Found 1940
Synonym(s): *Horace (no. 2) (9.2kg)*
Stone. Olivine-bronzite chondrite (H5).
Two stones were found and at first believed to be of the same fall, A.D. Nininger, Pop. Astron., Northfield, Minnesota, 1940, **48**, p.556. The larger stone, 9.2kg, is the bronzite chondrite and distinct from the L group stone previously known as Horace no. 1 (2.9kg). This and the Tribune stones are now assigned to Ladder Creek (*q.v*). Incorrectly reported, B. Mason, Meteoritics, 1969, **4**, p.240. The co-ordinates are of locality 3 in figure 1 of Mason. A further mass of 10.1kg is listed as a bronzite chondrite, Cat. Huss Coll. Meteorites, 1976, p.23.
10.1kg Mainz, Max-Planck-Inst.; 2.786kg Tempe, Arizona State Univ.; 1.1kg Chicago, Field Mus. Nat. Hist.; 464g Washington, U.S. Nat. Mus.;
Specimen(s): [1959,1036], 2584g. and a fragment, 61g, formerly listed as Horace no. 2.

Horace (no. 1) (2.9kg) v Ladder Creek.

Horace (no. 2) (9.2kg) v Horace.

Horace (no. 2) (in part) v Ladder Creek.

Horlick Mountains v Thiel Mountains.

Horowitz v Zebrak.

Horse Creek 37°35′ N., 102°46′ W.
Baca County, Colorado, U.S.A.
Found 1937
Stony-iron. Mesosiderite, anomalous (MESANOM).
A piece of 570g was found, A.D. Nininger, Pop. Astron., Northfield, Minnesota, 1937, **45**, p.449 [M.A. 7-62] Analysis, determination of Au, Ga and Pd, 5.87 %Ni, E. Goldberg et al., Geochimica et Cosmochimica Acta, 1951, **2**, p.1. The metal is structurally anomalous, analysis of metal, 5.75 %Ni, 47.5 ppm.Ga, 110 ppm.Ge, 2.5 ppm.Ir, J.T. Wasson, Meteorites, Springer-Verlag, 1974, p.307. Description of metal, perryite precipitated along 111 planes from a single kamacite crystal; metal has 2.5% Si; references, V.F. Buchwald, Iron Meteorites, Univ. of California, 1975, p.661.
140g Washington, U.S. Nat. Mus.; 47g Chicago, Field Mus. Nat. Hist.; 39g Tempe, Arizona State Univ.; 38g Harvard Univ.;
Specimen(s): [1959,919], 50g. two fragments, and sawings, 10g.

Horsham
Victoria, Australia
Pseudometeorite..
Mentioned, G. Baker, Meteoritics, 1953, **1**, p.92, Mem. Nat. Mus. Victoria, 1957, p.72 [M.A. 14-131].

Hotse 35°40′ N., 115°30′ E.
Heze, Shandong, China
Fell 1956, June 26
Synonym(s): *Heze, Hotsu, Shantung Hotse*
Stone. Olivine-hypersthene chondrite (L6).
Described and figured, with an analysis, 22.66 % total iron, C. Tzewen, Acta Geol. Sinica, 1966, **46**, p.64, Zentr. Miner., 1967 (Tiel 1), p.152. Three fragments were found, the largest weighing 180g, D. Bian, Meteoritics, 1981, **16**, p.117.

Hotsu v Hotse.

Houck v Cañon Diablo.

Howard County v Kokomo.

Howe 33°30′ N., 96°36′ W.
Grayson County, Texas, U.S.A.
Found 1938, recognized in this year
Stone. Olivine-bronzite chondrite (H5).
A single mass of 8.63kg was found, A.D. Nininger, Pop. Astron., Northfield, Minnesota, 1940, **48**, p.556, Contr. Soc. Res. Meteorites, 1940, **2**, p.227 [M.A. 8-54]. Olivine Fa_{19}, B. Mason, Geochimica et Cosmochimica Acta, 1963, **27**, p.1011.
2.7kg Tempe, Arizona State Univ.; 1.9kg Fort Worth, Texas, Monnig Colln.; 402g Chicago, Field Mus. Nat. Hist.;
Specimen(s): [1959,874], 2630g. a slab, and fragments, 9g.

Hoxie 39°21′ N., 100°27′ W. approx.
Sheridan County, Kansas, U.S.A.
Fell 1963, reported in this year
Stone. Chondrite.
A stone of 266.1g was seen to fall on a farm SW. of the town of Hoxie, W.E. Hill, Trans. Kansas Acad. Sci., 1963, **66**, p.270, Meteor. Bull., 1964 (29).

Hozowitz v Zebrak.

Hraschina 46°6′ N., 16°20′ E.
Zagreb, Croatia, Yugoslavia
Fell 1751, May 26, 1800 hrs
Synonym(s): *Agram, Hrascina, Hrasina, Zagrab, Zagreb, Zagrebacko zeljezo*
Iron. Octahedrite, medium (0.70mm) (IID).
After the appearance of a fireball, which divided into two parts with detonations, two masses of about 40kg and 9kg respectively were found to have fallen, W. von Haidinger, Sitzungsber. Akad. Wiss. Wien, Math.-naturwiss. Kl., 1859, **35**, p.361. Analysis, 10.6 %Ni, 74.5 ppm.Ga, 89.4 ppm.Ge, 13 ppm.Ir, J.T. Wasson, Meteorites, Springer-Verlag, 1974,

p.301. The smaller mass appears to have been lost. The specimen B.M.19963 is not authentic Hraschina. Description, cosmically deformed and annealed, V.F. Buchwald, Iron Meteorites, Univ. of California, 1975, p.664.

39kg Vienna, Naturhist. Mus., main mass of the larger iron.; 80g New York, Amer. Mus. Nat. Hist.;

Hrascina v Hraschina.

Hrasina v Hraschina.

Hressk v Gressk.

Hualapai v Wallapai.

Huanilla v Vaca Muerta.

Huangling 29°30′ N., 110° E. approx.

Hubei, China
Found
Iron. Octahedrite (IVA).

A single mass is in Beijing, Acad. Sci. Inst. Geochem., D. Bian, Meteoritics, 1981, **16**, p.122. Co-ordinates from map figure, D. Wang and Z. Ouyang, Geochimica, 1979, p.120 [M.A. 80-2085]. Classification, analysis, 7.8 %Ni, 1.8 ppm.Ga, 0.18 ppm.Ge, 2.8 ppm.Ir, D.J. Malvin et al., priv. comm., 1983.

Huckitta 22°22′ S., 135°46′ E.

Northern Territory, Australia
Found 1924
Synonym(s): *Alice Springs, Huckitte*
Stony-iron. Pallasite (PAL).

A mass of 1084g was found by H. Basedow in 1924 on the Burt Plains, Alice Springs, 23°33'S., 133°52'E. It was described, with an analysis, L.J. Spencer, Min. Mag., 1932, **23**, p.38. The troilite mentioned has since been found to be largely schreibersite. In July, 1937, the main mass of 1411.5kg of which Alice Springs is evidently a transported fragment, was found at Huckitta, 22°22'S., 135°46'E., surrounded by over 900kg of iron shale, C.T. Madigan, Trans. Roy. Soc. South Austr., 1937, **61**, p.187 [M.A. 7-73], C.T. Madigan, Min. Mag., 1939, **25**, p.353. Olivine $Fa_{12.5}$, P.R. Buseck and J.I. Goldstein, Bull. Geol. Soc. Amer., 1969, **80**, p.2141. Analysis of metal, 7.79 %Ni, 26.0 ppm.Ga, 65 ppm.Ge, 0.94 ppm.Ir, E.R.D. Scott, Geochimica et Cosmochimica Acta, 1977, **41**, p.349.

Main mass, Adelaide, S. Austr. Mus.; 4.3kg Fort Worth, Texas, Monnig Colln.; 2.2kg Tempe, Arizona State Univ.; 733g Mainz, Max-Planck-Inst.; 403g Washington, U.S. Nat. Mus.; 352g Chicago, Field Mus. Nat. Hist.;

Specimen(s): [1932,9], 1004g. the whole mass of Alice Springs, and fragments, 17.5g; [1940,1], 4.3g. two fragments

Huckitte v Huckitta.

Huejuquilla v Chupaderos.

Huejuquilla v Morito.

Huejuquilla v Sierra Blanca.

Huejuquilla v Tule.

Huesca v Roda.

Hugo (iron) v Soper.

Hugo (stone) 39°8′ N., 103°29′ W.

Lincoln County, Colorado, U.S.A.
Found 1936
Stone. Olivine-bronzite chondrite (H).

One stone of 80g was found, A.D. Nininger, Pop. Astron., Northfield, Minnesota, 1937, **45**, p.449 [M.A. 7-62], H.H. and A.D. Nininger, The Nininger Collection of Meteorites, Winslow, Arizona, 1950, p.62. Olivine Fa_{17}, B. Mason, Geochimica et Cosmochimica Acta, 1963, **27**, p.1011.

30g Tempe, Arizona State Univ.; 12g Los Angeles, Univ. of California; 7.2g Denver, Mus. Nat. Hist.;

Specimen(s): [1959,875], 31g.

Hugoton 37°12′ N., 101°21′ W.

Stevens County, Kansas, U.S.A.
Found 1927, recognized 1936
Stone. Olivine-bronzite chondrite (H5), black, brecciated.

A mass of 7.3kg, evidently only part of a larger mass, was ploughed up in 1927. Later the main mass of 325kg and fragments totalling 6.8kg were found. Described, with an analysis, H.H. Nininger, Trans. Kansas Acad. Sci., 1936, **39**, p.175., H.H. Nininger, J. Geol., 1936, **44** (5), p.66 [M.A. 6-390]. Another stone of 5.5kg was found later, H.H. Nininger, Pop. Astron., Northfield, Minnesota, 1937, **45**, p.449 [M.A. 7-62]. Olivine $Fa_{18.3}$, A.L. Graham, priv. comm., 1982.

314.8kg Tempe, Arizona State Univ., main mass; 2.5kg Washington, U.S. Nat. Mus.; 1.5kg Michigan Univ.; 1.25kg Chicago, Field Mus. Nat. Hist.; 500g Los Angeles, Univ. of California;

Specimen(s): [1937,1654], 66g.; [1959,1041], 1195.5g.; [1983, M.48], 86g.

Huittinen v Hvittis.

Huizopa 28°54′ N., 108°34′ W.

Temosachic, Guerrero district, Chihuahua, Mexico
Found 1907
Synonym(s): *Temosachic*
Iron. Octahedrite, fine (0.28mm) (IVA).

A mass of 108.5kg and four smaller masses of from 5 to 10kg were found in a ruin near Huizopa, 60 miles west of Temosachic, and were brought to Chihuahua for sale as silver ore, R. Tower, letters of 1908 in Min. Dept. BM(NH). quoting G. Griggs, Director State Mining Exhibition, Chihuahua, in 1908. The mass of 108.5kg was seen about 1932 in the State Mines Exhibit, Chihuahua City. Described, with an analysis, 7.90 %Ni, H.H. Nininger, The Mines Mag., Golden, Colorado, 1932, **22**, p.11 [M.A. 5-405]. Determination of Ga, Pd, Au and Ni, 7.81 %Ni, E. Goldberg et al., Geochimica et Cosmochimica Acta, 1951, **2**, p.1. New analysis, 7.48 %Ni, 2.22 ppm.Ga, 0.118 ppm.Ge, 2.2 ppm.Ir, R. Schaudy et al., Icarus, 1972, **17**, p.174. Description, severely deformed, some shock melting and recrystallisation, V.F. Buchwald, Iron Meteorites, Univ. of California, 1975, p.668.

65kg Chihuahua City, main mass, approx. weight; 3.5kg New York, Amer. Mus. Nat. Hist.; 6.3kg Mexico City, Inst. Geol.; 2.79kg Washington, U.S. Nat. Mus.; 2kg Michigan Univ.; 1.7kg Tempe, Arizona State Univ.; 1kg Los Angeles, Univ. of California; 250g Chicago, Field Mus. Nat. Hist.;

Specimen(s): [1908,239], 1422g. three pieces, and filings, 48g; [1959,1046], 1206.5g. slice, and sawings, 10g.

Hukue v Fukue.

Hukutomi v Fukutomi.

Hukuye v Fukue.

Humboldt Iron v Morito.

Humoshnik v Gumoschnik.

Hungen 50°18′ N., 8°55′ E.
Hessen, Germany
Fell 1877, May 17, 0700 hrs
Stone. Olivine-bronzite chondrite (H6), veined.
After detonations, a stone of 86g was seen to fall, and another of 26g was found later in a wood, 3 miles from Hungen, O. Buchner, Tschermaks Min. Petr. Mitt., 1877, p.313. Olivine Fa_{19}, B. Mason, Geochimica et Cosmochimica Acta, 1963, **27**, p.1011.
56g Giessen Univ.; 26g Vienna, Naturhist. Mus.; 21g Washington, U.S. Nat. Mus.; 11.7g Budapest, Nat. Mus.; 8g Chicago, Field Mus. Nat. Hist.;
Specimen(s): [54632], 5.5g.

Hunsrück v Simmern.

Hunter 36°33′36″ N., 97°40′ W.
Garfield County, Oklahoma, U.S.A.
Found 1962, approx., recognized 1971
Stone. Olivine-hypersthene chondrite, amphoterite (LL5).
A single mass of 74.6kg was unearthed by a bull-dozer during the terracing of a field. A second, smaller piece, was said to have been found, Meteor. Bull., 1979 (56), Meteoritics, 1979, **14**, p.167. Olivine $Fa_{26.8}$, o-pyroxene Fs22.7, A.L. Graham, priv. comm., 1981.
74kg Mainz, Max-Planck-Inst.; 309g Fort Worth, Texas, Monnig Colln.;
Specimen(s): [1976,M.16], 131g.

Huntsman 41°11′ N., 103° W.
Cheyenne County, Nebraska, U.S.A.
Found 1910
Stone. Olivine-bronzite chondrite (H4).
A single stone of 14.5kg was found, olivine Fa_{17}, Meteor. Bull., 1981 (59), Meteoritics, 1981, **16**, p.195.
1.34kg Tempe, Arizona State Univ.;

Hvittis 61°11′ N., 22°41′ E.
Abo, Finland
Fell 1901, October 21, 1200 hrs
Synonym(s): *Huittinen*
Stone. Enstatite chondrite (E6).
Fell after detonations; three days later a stone of 14kg was found. Described, with an analysis, L.H. Borgström, Bull. Comm. Geol. Finlande, 1903 (14). Classical analysis, 22.63 % total iron, H.B. Wiik, Comm. Phys.-Math. Soc. Sci. Fenn., 1969, **34**, p.135. XRF analysis, 23.49 % total iron, H. von Michaelis et al., Earth planet. Sci. Lett., 1969, **5**, p.387. Abundances of 13 trace elements, C.M. Binz et al., Geochimica et Cosmochimica Acta, 1974, **38**, p.1579.
Main mass, Helsinki Univ.; 656g Chicago, Field Mus. Nat. Hist.; 190g Washington, U.S. Nat. Mus.; 166g Berlin, Humboldt Univ.; 176g New York, Amer. Mus. Nat. Hist.; 125g Vienna, Naturhist. Mus.; 76g Yale Univ.;
Specimen(s): [86754], 125g.; [1906,29], thin section

Hyderabad v Alwal.

Hyderabad (1898) 17°22′ N., 78°28′ E.
Andhra Pradesh, India
Fell 1898, in the summer
Doubtful. Stone.
Nothing is preserved and the evidence is not conclusive, M.A.R. Khan, Contr. Soc. Res. Meteorites, 1936, **1**, p.57, A.L. Coulson, Mem. Geol. Surv. India, 1940, **75**, p.1 [M.A. 8-54], F.C. Leonard, Univ. New Mexico Publ., Albuquerque, 1946 (meteoritics ser. no. 1), p.6.

Hyderabad (1936) 17°22′ N., 78°28′ E.
Andhra Pradesh, India
Fell 1936, October 13
Doubtful. Stone?.
After a notable meteor, a shower of fine 'sand' was observed beating on tree leaves. It is believed that this was meteoritic but nothing was collected, M.A.R. Khan, Nature, 1944, **155**, p.53, M.A.R. Khan, Pop. Astron., Northfield, Minnesota, 1945, **53**, p.39.

Hyen v Mjelleim.

Iamysheva v Pavlodar (pallasite).

Iardymlinskii v Yardymly.

Ibbenbühren v Ibbenbüren.

Ibbenbüren 52°17′ N., 7°42′ E.
Nordrhein-Westfalen, Germany
Fell 1870, June 17, 1400 hrs
Synonym(s): *Ibbenbühren*
Stone. Achondrite, Ca-poor. Diogenite (ADIO).
Fell, after detonation and appearance of light. A stone of about 2kg was found two days later; described and analysed, G. vom Rath, Ann. Phys. Chem. (Poggendorff), 1872, **146**, p.463. Thermal history, H. Mori and H. Takeda, Earth planet. Sci. Lett., 1981, **53**, p.266.
1.88kg Berlin, Humboldt Univ., main mass; 18g Vienna, Naturhist. Mus.; 9g Washington, U.S. Nat. Mus.; 3g Chicago, Field Mus. Nat. Hist.;
Specimen(s): [46270], 1.8g.

Ibitira 20° S., 45° W. approx.
Martinho Campos, Minas Gerais, Brazil
Fell 1957, June 30, 1715 hrs
Synonym(s): *Martinho Campos, Monjolo Farm*
Stone. Achondrite, Ca-rich. Eucrite (AEUC), vesicular, unbrecciated.
One stone of 2.5kg was found in the village of Ibitira, near Martinho Campos, 19°22'S., 45°14'W. Described, with a preliminary analysis, V. Menezes, Sky and Telescope, 1957, **17**, p.10, Meteor. Bull., 1957 (6). Vesicular and unbrecciated, L.L. Wilkening and E. Anders, Geochimica et Cosmochimica Acta, 1975, **39**, p.1205. Rb-Sr age, 4520 m.y., J.L. Birck and C.J. Allegre, Earth planet. Sci. Lett., 1978, **39**, p.37 [M.A. 78-4757]. Mineralogy, I.M. Steele and J.V. Smith, Earth planet. Sci. Lett., 1976, **33**, p.67 [M.A. 77-3268].
2.5kg Belo Horizonte, Centro Estudos Astron. M.G.; 30g Sao Paulo, Inst. Geocienc.; 17.9g Albuquerque, Univ. of New Mexico; 1.6g Washington, U.S. Nat. Mus.;

Ibrisim 38° N., 35° E.
Seyham, Turkey
Fell 1949, late summer
Synonym(s): *Saricam, Adana*
Stone. Chondrite.
A stone fell at the village of Ibrisim (=Saricam), 65km north of the town of Adana, 36°55'N., 35°17'E. Described, with an unsatisfactory analysis, O. Bayramgil, Bull. Geol. Soc. Turkey, 1952, **3**, p.21 [M.A. 11-529].

Ichkala 58°12′ N., 82°56′ E.
Tomskaya, Federated SSR, USSR, [Ичкала]
Fell 1936, May 29, 1934 hrs
Synonym(s): *Itchkala, Itschkala*
Stone. Olivine-bronzite chondrite (H6).
Mentioned, I.S. Astapowitsch, Trans. Roy. Astron. Soc. Canada, 1938, **32**, p.195 [M.A. 7-172]. A stone fell at 58°12'N., 82°56'E., V.I. Vernadsky, letter of 5 August, 1939 in Min. Dept. BM(NH)., in the valley of the Ob. The time of fall is given as 1134hrs, L.A. Kulik, Метеоритика, 1941, **1**, p.73 [M.A. 9-294]. Listed, E.L. Krinov, Астрон. Журнал, 1945, **22**, p.303 [M.A. 9-297], Каталог Метеоритов Акад. Наук СССР, Москва, 1947 [M.A. 10-511]. Olivine Fa_{18}, B. Mason, Geochimica et Cosmochimica Acta, 1963, **27**, p.1011.
3972g Moscow, Acad. Sci., main mass.;

Ida 33°47′ N., 136°2′ E.
Mie, Honshu, Japan
Fell 1929, October 10
Doubtful..
Doubtful, I. Yamamoto, Kwasan Observ. Bull., 1935, p.306 [M.A. 7-173]. Listed, P.M. Millman, Trans. Roy. Astron. Soc. Canada, 1938, **32**, p.197 [M.A. 7-173].

Idaho
Idaho, U.S.A., Not located
Iron. Octahedrite (IA).
Exact locality not known. Chemically very similar to Cañon Diablo. Classification and analysis, 7.09 %Ni, 81 ppm.Ga, 321 ppm.Ge, 2.1 ppm.Ir, E.R.D. Scott and J.T. Wasson, Geochimica et Cosmochimica Acta, 1976, **40**, p.103.
Specimen, Washington, U.S. Nat. Mus.;

Idalia 39°41′42″ N., 102°17′42″ W.
Yuma County, Colorado, U.S.A.
Found 1968, recognized in this year
Stone. Chondrite.
Listed, total known weight 7.4kg, Cat. Huss Coll. Meteorites, 1976, p.23.
1677g Mainz, Max-Planck-Inst.;

Ider 34°41′ N., 85°39′ W.
De Kalb County, Alabama, U.S.A.
Found 1957
Iron. Octahedrite, medium (1.2mm) (IIIA).
A mass of 140kg was found, B. Mason, Meteorites, Wiley, 1962, p.228. Classification and analysis, 8.3 %Ni, 20.1 ppm.Ga, 40 ppm.Ge, 2.8 ppm.Ir, E.R.D. Scott et al., Geochimica et Cosmochimica Acta, 1973, **37**, p.1957. Heavily weathered, shock hardened, V.F. Buchwald, Iron Meteorites, Univ. of California, 1975, p.671.
90kg Washington, U.S. Nat. Mus.; 188g Perth, West. Austr. Mus.;
Specimen(s): [1966,49], 193g. slice

Idutywa 32°6′ S., 28°20′ E.
Transkei, Cape Province, South Africa
Fell 1956, February 1, 1815 hrs
Stone. Olivine-bronzite chondrite (H5).
A shower of stones fell, but only two, of 3009g and 448g respectively, were recovered. The larger one made a hole in the ground, the smaller did not. Described, with an analysis, E.D. Mountain, S. African J. Sci., 1956, **53**, p.73 [M.A. 13-360], letter of 9 February, 1960 in Min. Dept. BM(NH). Analysis, 26.2 % total iron, H. von Michaelis et al., Earth planet. Sci. Lett., 1969, **5**, p.387.
Main mass, Grahamstown, possession of Prof. E.D. Mountain; 448g East London Mus.; 50g Washington, U.S. Nat. Mus.;
Specimen(s): [1962,134], 54g.

Iferouane v Aïr.

Igast See separate entry after tektites

Iglau v Stannern.

Iharaota v Lalitpur.

Ihung v Jhung.

Ijopega 6°2′ S., 145°22′ E.
Asaro valley, Goroka, Papua-New Guinea
Fell 1975, March 3, 2145 hrs, U.T.
Stone. Olivine-bronzite chondrite (H6), brecciated.
A single stone of 7.33kg spiralled down emitting a smoke trail and accompanied by sonic phenomena. It made a hole in soft ground 25m from an eyewitness. The location is 8km NW. of Goroka in the eastern highlands of Papua-New Guinea. It is a brecciated stone, olivine $Fa_{19.9}$, 27.56 % total iron, A.L. Jaques et al., Meteoritics, 1975, **10**, p.289.
Half the stone, Ijopega, with the finders; Half the stone, Canberra, Austr. Nat. Univ.;

Ikharene v Ikhrarene.

Ikhrarene 28°36′15″ N., 1°2′30″ E.
Western Tademait Plateau, Algerian Sahara, Algeria
Found 1969
Synonym(s): *Gara Ikhrarene, Ikharene*
Stone. Olivine-hypersthene chondrite (L4).
One complete stone of 5465g was found half buried in a Quaternary sandy gravel overlying the Cretaceous plateau. The stone is oxidised and is part of a very old fall, Meteor. Bull., 1971 (50), Meteoritics, 1971, **6**, p.119.
5393g Paris, Mus. d'Hist. Nat.;

Ilimaë v Ilimaes (iron).

Ilimaes (iron) 26° S., 70° W. approx.
Taltal, Atacama, Chile
Found 1870, known since this year
Synonym(s): *Atacama Desert, Chañaral, Ilimaë*
Iron. Octahedrite, medium (1.1mm) (IIIA).
A mass of about 51kg was found at 'Ilimaë' (probably a misprint for Ilimaes), L. Fletcher, Min. Mag., 1889, **8**, p.260. Described, with an analysis, 7.82 %Ni, G. Tschermak, Denkschr. Akad. Wiss. Wien, Math.-naturwiss. Kl., 1872, **31**, p.187. New analysis, 8.1 %Ni, 21.1 ppm.Ga, 43.5 ppm.Ge, 0.17 ppm.Ir, J.T. Wasson, Meteorites, Springer-Verlag, 1974, p.302. Includes Chañaral, a 1.2kg mass found in 1884 which has the same structure and chemical

composition as Ilimaes (iron), V.F. Buchwald, Iron Meteorites, Univ. of California, 1975, p.675.
51kg Vienna, Naturhist. Mus., includes main mass 50.75kg; 1.2kg Santiago, School of Mines, main mass of Chañaral; 8.4g Chicago, Field Mus. Nat. Hist., of Chañaral;
Specimen(s): [67450], 39.5g.

Ilimaes (pallasite) v Imilac.

Il'inskaia Stanitsa v Ilinskaya Stanitza.

Ilinskaya Stanitza 51°14′ N., 57°23′ E.
Orenburg, Federated SSR, USSR, [Илъинская Станица]
Found 1915, before this year
Synonym(s): *Il'inskaia Stanitsa*
Iron. Octahedrite, medium (0.7mm) (IIIB).
One mass of 5.621kg was found, E.L. Krinov, Астрон. Журнал, 1945, **22**, p.303 [M.A. 9-297], Каталог Метеоритов Акад. Наук СССР, Москва, 1947 [M.A. 10-511]. Analysis, 9.40 %Ni, M.I. D'yakonova, Метеоритика, 1958, **16**, p.180. New analysis, 9.28 % Ni, 19.5 ppm.Ga, 39.2 ppm.Ge, 0.29 ppm.Ir, E.R.D. Scott et al., Geochimica et Cosmochimica Acta, 1973, **37**, p.1957. Description, contains Brezina lamellae and has been shocked, V.F. Buchwald, Iron Meteorites, Univ. of California, 1975, p.677.
4538g Moscow, Acad. Sci.; 184g Washington, U.S. Nat. Mus.;
Specimen(s): [1956,319], 37g.

Illinois v Admire.

Illinois Gulch 46°41′ N., 112°33′ W.
Powell County, Montana, U.S.A.
Found 1899
Synonym(s): *Ophir*
Iron. Ataxite, Ni-rich (IRANOM).
A mass of about 2.5kg was found about 4 feet below the surface, H.L. Preston, Am. J. Sci., 1900, **9**, p.201. Described, with an analysis, 12.67 %Ni, E. Cohen, Meteoritenkunde, 1905, **3**, p.83. Chemically anomalous, analysis, 11.68 %Ni, 2.80 ppm.Ga, 2.76 ppm.Ge, 5.3 ppm.Ir, J.T. Wasson and R. Schaudy, Icarus, 1971, **14**, p.59. Locality, co-ordinates; shocked, reheated and rapidly cooled, contains martensite, V.F. Buchwald, Iron Meteorites, Univ. of California, 1975, p.678.
603g Chicago, Field Mus. Nat. Hist.; 268g New York, Amer. Mus. Nat. Hist.; 88g Rome, Vatican Colln; 69g Berlin, Humboldt Univ.;
Specimen(s): [84785], 570g. and a fragment, 28g.

Imilac 24°12.2′ S., 68°48.4′ W.
Atacama Desert, Atacama, Chile
Found 1822, known in this year
Synonym(s): *Antofagasta, Atacama, Calderilla, Campo del Pucara, Campo del Puchara, Caracoles, Catamarca, Catamarka, Del Parque, Gran Chaco (pallasite), Ilimaes (pallasite), La Encantada, La Rioja, Ollague, Peine, Potosi, Salta, San Pedro, San Pedro de Atacama, Toconao*
Stony-iron. Pallasite (PAL).
Numerous masses weighing together several hundredweight, with individuals up to 450lb, were found in a valley to the SW. of Imilac, T. Allan, Trans. Roy. Soc. Edinburgh, 1831, **11**, p.223, L. Fletcher, Min. Mag., 1889, **8**, p.243. The specimen found in 1879 at Campo del Puchara, Argentina, was probably carried from Imilac, E. Cohen, Neues Jahrb. Min., 1887, **2**, p.45. Co-ordinates, transported masses, strewnfield, craters, metal deformation, V.F. Buchwald, Iron Meteorites, Univ. of California, 1975, p.1393. Olivine $Fa_{12.5}$, P.R. Buseck and J.I. Goldstein, Bull. Geol. Soc. Amer., 1969, **80**, p.2141. Analysis of metal, 9.9 %Ni, 21.1 ppm.Ga, 46.0 ppm.Ge, 0.071 ppm.Ir, J.T. Wasson and S.P. Sedwick, Nature, 1969, **222**, p.22. Texture, includes Antofagasta, Ollague, Salta, references, E.R.D. Scott, Geochimica et Cosmochimica Acta, 1977, **41**, p.693.
70kg Copiapo, Lyceo, main mass of Ilimaes (pallasite); 18.6kg Yale Univ.; 11.2kg Chicago, Field Mus. Nat. Hist., includes 10.5kg Ilimaes (pallasite); 10.5kg Oxford, Univ.; 19kg Washington, U.S. Nat. Mus.; 5.2kg Vienna, Naturhist. Mus., includes 1.46kg Ilimaes (pallasite); 5kg New York, Amer. Mus. Nat. Hist.; 5.1kg Paris, Mus. d'Hist. Nat.; 4kg Budapest, Nat. Mus.; 3.8kg Harvard Univ., of Ollague; 3.5kg Tempe, Arizona State Univ.; 3.8kg Berlin, Humboldt Univ.; 1.8kg Copenhagen, Univ. Geol. Mus.; 1.8kg Edinburgh, Roy. Scottish Mus.; 1kg Paris, Ecole des Mines; 550g Moscow, Acad. Sci.; 250g Prague, Nat. Mus.;
Specimen(s): [53322], 198.1kg. presented by G. Hicks in 1879; [90239], 9265g. presented by Sir Woodbine Parish; [90240], 2107g.; [27283], 1941g. presented by W. Bollaert; [1927,88], 240g. slice, and fragments, 3g, of Ollague; [1906, 52], 266.5g. of Ilimaes (pallasite); [40534], 235g.; [80669], 201.5g.; [33923], 169g.; [1911,720], 129.5g.; [33939], 102.5g.; [35789], 37.5g.; [27216a], 5.5g.; [27217], 1g.; [1959,173], 7g.

Imperial 32°52′ N., 115°35′ W.
Imperial County, California, U.S.A.
Found 1908, known in this year
Stone. Olivine-bronzite chondrite (H4).
Original mass unknown, F.C. Leonard, Pop. Astron., Northfield, Minnesota, 1947, **55**, p.381, Contr. Meteoritical Soc., 1942, **4**, p.58. Olivine Fa_{18}, B. Mason, Geochimica et Cosmochimica Acta, 1963, **27**, p.1011.
4g Washington, U.S. Nat. Mus.;

Inca v Tamarugal.

Inca v Vaca Muerta.

Indarch 39°45′ N., 46°40′ E.
Shusha, Elisavetpol, Azerbaydzhan SSR, USSR, [Индарх]
Fell 1891, April 7, 2010 hrs
Synonym(s): *Elisabethpol, Gindorcha, Glindorcha, Indarh, Indarkh, Schuscha, Suscha*
Stone. Enstatite chondrite (E4).
After detonations, and appearance of flame, a stone of about 27kg fell, and was found next morning, Y.I. Simashko, Cat. Meteorites, St.-Petersbourg, 1891, p.55, S. Meunier, C. R. Acad. Sci. Paris, 1897, **125**, p.894. Description, analysis, G.P. Merrill, Proc. U.S. Nat. Mus., 1915, **49**, p.109. Another analysis, 33.15 % total iron, H.B. Wiik, Geochimica et Cosmochimica Acta, 1954, **6**, p.209. Further analysis, 30.41 % total iron, H. von Michaelis et al., Earth planet. Sci. Lett., 1969, **5**, p.387. Trace element analysis, C.M. Binz et al., Geochimica et Cosmochimica Acta, 1974, **38**, p.1579.
12.4kg Chicago, Field Mus. Nat. Hist., main mass; 2840g Washington, U.S. Nat. Mus.; 288g Budapest, Nat. Mus.; 854g Harvard Univ.; 345g Vienna, Naturhist. Mus.; 100g Paris, Mus. d'Hist. Nat.; 524g Ottawa, Mus. Geol. Surv. Canada; 1.24kg Moscow, Acad. Sci.;
Specimen(s): [86948], 306g. and fragments, 7g; [1921,23], 87.5g.; [70350], 0.5g.; [68726], 8.5g.; [1980,M.2], 443g.

Indarh v Indarch.

Indarkh v Indarch.

Independence v Kenton County.

Independence County v Joe Wright Mountain.

Indianola 40°14′ N., 100°25′ W.
Red Willow County, Nebraska, U.S.A.
Found 1939
Stone. Olivine-hypersthene chondrite (L5), xenolithic.
One stone of 4kg was found in section 2, township 2, A.D. Nininger, Pop. Astron., Northfield, Minnesota, 1940, **48**, p.556, Contr. Soc. Res. Meteorites, 1942, **2**, p.227 [M.A. 8-54], The Nininger Collection of Meteorites, Winslow, Arizona, 1950, p.63. Olivine Fa_{24}, B. Mason, Geochimica et Cosmochimica Acta, 1963, **27**, p.1011. Xenolithic, R.A. Binns, Geochimica et Cosmochimica Acta, 1968, **32**, p.299.
1.74kg Tempe, Arizona State Univ.; 225g Washington, U.S. Nat. Mus.; 200g Chicago, Field Mus. Nat. Hist.;
Specimen(s): [1959,876], 1400g.

Indian Valley 36°56′ N., 80°30′ W.
Floyd County, Virginia, U.S.A.
Found 1887
Synonym(s): *Floyd County, Floyd Mountain, Radford Furnace*
Iron. Hexahedrite (IIA).
A mass of 31lb was ploughed up near the base of the south side of Floyd Mountain. Described, with an analysis, 5.56 % Ni, G.F. Kunz and E. Weinschenk, Am. J. Sci., 1892, **43**, p.424. Another analysis and determination of Ga, Au and Pd, 5.64 %Ni, E. Goldberg et al., Geochimica et Cosmochimica Acta, 1951, **2**, p.1. Shocked, partially recrystallised, possibly paired with Mayodan, V.F. Buchwald, Iron Meteorites, Univ. of California, 1975, p.679. Analysis, 5.48 %Ni, 61.4 ppm.Ga, 174 ppm.Ge, 11 ppm.Ir, J.T. Wasson, Geochimica et Cosmochimica Acta, 1969, **33**, p.859.
8kg Chicago, Field Mus. Nat. Hist., approx. weight; 700g Tempe, Arizona State Univ.; 578g Rome, Vatican Colln; 489g Washington, U.S. Nat. Mus.; 455g Berlin, Humboldt Univ.; 303g Vienna, Naturhist. Mus.; 179g Budapest, Nat. Mus.;
Specimen(s): [84192], 81.5g. slice; [1959,920], 628g. slice, and sawings, 20.5g.

Indio Rico 38°20′ S., 60°53′ W.
Buenos Aires province, Argentina
Found 1887
Stone. Olivine-bronzite chondrite (H6).
A stone of 15kg was found. Described, with an analysis, J.J.J. Kyle, Anal. Soc. Cient. Argentina, 1887, **24**, p.128. Is distinct from El Perdido. For a recalculation of the analysis and a review of the literature, E. Fossa-Mancini, Notas Mus. La Plata, 1947, **12** (geol. 46), p.143 [M.A. 10-403]. Olivine Fa_{18}, B. Mason, Geochimica et Cosmochimica Acta, 1963, **27**, p.1011.
10.6kg La Plata Mus.; 219g Vienna, Naturhist. Mus.; 214g Harvard Univ.; 57.7g Budapest, Nat. Mus.; 26g Berlin, Humboldt Univ.; 23g Prague, Nat. Mus.; 11g Chicago, Field Mus. Nat. Hist.;
Specimen(s): [85673], 1.5g.

Ingalls 37°50′ N., 100°27′ W.
Gray County, Kansas, U.S.A.
Found 1937
Stone. Olivine-bronzite chondrite (H6).
One fragment of 226g was found in SW. 1/4, sect. 22, township 26, range 29, Gray Co., A.D. Nininger, Pop. Astron., Northfield, Minnesota, 1939, **47**, p.212, The Nininger Collection of Meteorites, Winslow, Arizona, 1950, p.64. Olivine Fa_{20}, B. Mason, Geochimica et Cosmochimica Acta, 1963, **27**, p.1011.
120g Tempe, Arizona State Univ.;
Specimen(s): [1959,877], 113.5g.

Ingast error for Igast.

Inman 38°15′ N., 97°40′ W.
McPherson County, Kansas, U.S.A.
Found 1966
Stone. Olivine-hypersthene chondrite (L3).
One mass of 7.25kg was found during ploughing, Meteor. Bull., 1967 (41), Meteoritics, 1970, **5**, p.95. Description, olivine unequilibrated, 19.45 % total iron, K. Keil et al., Meteoritics, 1978, **13**, p.11.
Main mass, Kansas State Univ.; 14g Washington, U.S. Nat. Mus.;
Specimen(s): [1982,M.11], 152g.

Inner Monglia 41° N., 112° E. approx.
Nei Monggol, China
Fell 1963, probably
Synonym(s): *Nei Monggol*
Stone. Olivine-hypersthene chondrite (L6).
Described with an analysis, C. Tzewen, Acta Geol. Sinica, 1966, **46**, p.64, Zentr. Miner., 1967, p.152. Co-ordinates from map, analysis, 22.24 % total iron, D. Wang and Z. Ouyang, Geochimica, 1979, p.120, in Chinese [M.A. 80-2085]. Listed, 3kg fell in 1962, references, D. Bian, Meteoritics, 1981, **16**, p.115.
3kg Beijing, Acad. Sin. Inst. Geol.;

Innisfree 53°24′54″ N., 111°20′15″ W.
Alberta, Canada
Fell 1977, February 5
Stone. Olivine-hypersthene chondrite, amphoterite (LL5).
Six specimens and fragments altogether totalling 3.79kg were found at the site predicted from the study of fireball photographs taken on the night of February 5, Meteor. Bull., 1978 (55), Meteoritics, 1978, **13**, p.338. Description of discovery and orbit, I. Halliday et al., J. Roy. Astron. Soc. Canada, 1978, **72**, p.15. Cosmogenic radionuclides, trace element contents, L.A. Rancitelli and J.C. Laul, Meteoritics, 1977, **12**, p.346, abs. Fall dynamics, 4.58kg now known, I. Halliday et al., Meteoritics, 1981, **16**, p.153. Mineralogy, olivine $Fa_{27.1}$, D.G.W. Smith, Canadian Min., 1980, **18**, p.433.
4.03kg Ottawa, Geol. Surv. Canada;

Invercargill v Makariwa.

Ioka 40°15′ N., 110°5′ W.
Duchesne County, Utah, U.S.A.
Found 1931, recognized 1956
Stone. Olivine-hypersthene chondrite (L3).
A weathered stone of 31.5kg was ploughed up. Analysis, 20.77 % total iron, E. Jarosewich, Geochimica et Cosmochimica Acta, 1966, **30**, p.1261. Olivine Fa_{23}, B. Mason, Geochimica et Cosmochimica Acta, 1963, **27**, p.1011.
8.9kg Washington, U.S. Nat. Mus.; 2.1g New York, Amer. Mus. Nat. Hist.;

Ioluca v Toluca.

Iowa v Estherville.

Iowa v Forest City.

Iowa v Marion (Iowa).

Iowa City 41°39′ N., 91°31′ W.
Iowa, U.S.A.
Fell 1861, November 15, 2230 hrs
Doubtful. Stone.
Following the appearance of a fireball, and detonations, seven stones are reported to have fallen, J. Glaisher et al., Rep. Brit. Assn., 1877, **47**, p.98.

Iowa County v Homestead.

Ipacaray 25°18′ S., 57°16′ W.
Ipacaray Lake, Asuncion, Paraguay
Fell 1877, before this year
Doubtful. Stone?.
After detonations, a nearly spherical stone, two yards in diameter, is said to have fallen and penetrated the soil to a depth of two yards, Anal. Soc. Cient. Argentina, 1877, **3**, p.336.

Ipiranga 25°30′ S., 54°30′ W.
Lageado Ipiranga, Foz do Iguacu, Parana, Brazil
Fell 1972, December 27, 1030 hrs
Synonym(s): *Ipiringa, Lageado Ipiranga, Lajeado Ipiranga*
Stone. Olivine-bronzite chondrite (H6).
An individual of 2.65kg, and a number of smaller stones, totalling about 30 specimens, were collected after a fireball was seen travelling in a westerly direction over SW. Parana state. Detonations lasted several seconds and the length of the strewnfield was at least 40km, Meteor. Bull., 1974 (52), Meteoritics, 1974, **9**, p.102. Description, with an analysis, 26.27 % total iron, C.B. Gomes et al., Chem. Erde, 1978, **37**, p.265 [M.A. 79-1569].
3.2kg Rome, Mus. Min.; 2.6kg Curitiba, Observ. Astron. Parana; 234g Sao Paulo, Inst. Geocienc.; 19g Albuquerque, Univ. of New Mexico;

Ipiringa v Ipiranga.

Iquique 20°11′ S., 69°44′ W.
Pampa del Tamarugal, Tarapaca, Chile
Found 1871
Iron. Ataxite, Ni-rich (IVB).
A mass of 12.5kg was found embedded in nitrate, 30 miles east of Iquique, G. Rose, Am. J. Sci., 1874, **8**, p.398. Analysis, 15.99 %Ni, E.P. Henderson, Am. J. Sci., 1941, **239**, p.407 [M.A. 8-195]. Co-ordinates, L.M. Villar, Cienc. Investig., 1968, **24**, p.302. More recent analysis, 16.03 %Ni, 0.170 ppm.Ga, 0.051 ppm.Ge, 28 ppm.Ir, R. Schaudy et al., Icarus, 1972, **17**, p.174. Description, some artificial heating, V.F. Buchwald, Iron Meteorites, Univ. of California, 1975, p.682.
9.4kg Berlin, Humboldt Univ., main mass; 168g Washington, U.S. Nat. Mus.; 140g Moscow, Acad. Sci.; 87g New York, Amer. Mus. Nat. Hist.; 33g Vienna, Naturhist. Mus.; 16g Chicago, Field Mus. Nat. Hist.;
Specimen(s): [1972,320], 24.2g.

Irapuato v La Charca.

Iredell 31°58′ N., 97°52′ W.
Bosque County, Texas, U.S.A.
Found 1898
Iron. Octahedrite, coarsest (10mm) (IIB).
A mass of about 1.5kg was found on a sheep ranch, 5 or 6 miles SW. of Iredell, but only about 500g have been preserved, W.M. Foote, Am. J. Sci., 1899, **8**, p.415. The mass was broken or heated and forged; may be related to El Burro (*q.v.*), V.F. Buchwald, Iron Meteorites, Univ. of California, 1975, p.685. Classification and analysis, 6.0 %Ni, 58.1 ppm.Ga, 163 ppm.Ge, 0.07 ppm.Ir, J.T. Wasson, Meteorites, Springer-Verlag, 1974, p.300.
98g Washington, U.S. Nat. Mus.; 179g New York, Amer. Mus. Nat. Hist.; 92g Rome, Min. Mus.; 11g Chicago, Field Mus. Nat. Hist.;

Irkutsk v Doroninsk.

Irkutsk v Tounkin.

Iron Creek 53° N., 112° W.
Battle River, Alberta, Canada
Found 1869, known before this year
Synonym(s): *Battle River, Saskatchewan, Victoria*
Iron. Octahedrite, medium (1.1mm) (IIIA).
A mass of 386lb on a hill near Iron Creek had long been known to the Indians before its removal to Victoria College in about 1869, A.P. Coleman, Trans. Roy. Soc. Canada, 1886, **4** (3), p.97. Description, O.C. Farrington, Field Mus. Nat. Hist. Geol. Ser., 1907, **3** (6), p.113. Analysis, 7.72 %Ni, 20.2 ppm.Ga, 39.6 ppm.Ge, 3.3 ppm.Ir, E.R.D. Scott et al., Geochimica et Cosmochimica Acta, 1973, **37**, p.1957. Description, shocked and partly annealed, V.F. Buchwald, Iron Meteorites, Univ. of California, 1975, p.686.
Main mass, Toronto, Victoria College; 250g Chicago, Field Mus. Nat. Hist.; 122g Washington, U.S. Nat. Mus.; 117g Vienna, Naturhist. Mus.; 88g New York, Amer. Mus. Nat. Hist.;
Specimen(s): [63548], 79.5g.

Ironhannock Creek v Tomhannock Creek.

Iron River 46°5′ N., 88°34′ W.
Iron County, Michigan, U.S.A.
Found 1889, recognized 1965
Iron. Octahedrite, fine (0.3mm) (IVA).
A mass of 1420g was found, V.D. Chamberlain, letter of 16 September, 1965 in Min. Dept. BM(NH). Description and co-ordinates, analysis, 8.0 %Ni, V.D. Chamberlain, Meteoritics, 1971, **6**, p.161. Another analysis, 7.87 %Ni, 2.12 ppm.Ga, 0.118 ppm.Ge, 2.1 ppm.Ir, R. Schaudy et al., Icarus, 1972, **17**, p.174. Description, V.F. Buchwald, Iron Meteorites, Univ. of California, 1975, p.687.
Main mass, East Lansing, Michigan State Univ.; 44g Tempe, Arizona State Univ.; Specimen, Mainz, Max-Planck Inst.;

Irwin v Tucson.

Irwin-Ainsa Iron v Tucson.

Isaacs River v Le Gould's Stone.

Ishinga 8°56′ S., 33°48′ E.

Mbeya district, Tanzania
Fell 1954, October 8, 1500 hrs
Stone. Olivine-bronzite chondrite (H).

The stone, of about 1.3kg, buried itself to a depth of about 30 cm, J.R. Harpum, Rec. Geol. Surv. Tanganyika for 1957, 1965, **11**, p.59, Meteor. Bull., 1958 (10). Olivine Fa_{18}, B. Mason, Geochimica et Cosmochimica Acta, 1963, **27**, p.1011.

Main mass, Nzovwe, near Mbeya, St. Joseph's Middle School; 127g Dodoma, Geol. Surv. Tanzania; 13g New York, Amer. Mus. Nat. Hist.;

Isle de France v Mauritius.

Isna 24°50′ N., 31°40′ E.

Luxor, Egypt
Found 1970
Synonym(s): *Esna*
Stone. Carbonaceous chondrite, type III (CO3).

A 23kg individual was found about 100km SW. of Isna, on the Nile River near Luxor, Meteor. Bull., 1975 (53), Meteoritics, 1975, **10**, p.149. Description and analysis, 24.83 % total iron, R.L. Methot et al., Meteoritics, 1975, **10**, p.121.

Main mass, Cairo, Geol. Mus.; 107g Tempe, Arizona State Univ.; 31g Chicago, Field Mus. Nat. Hist.;
Specimen(s): [1975,M.13], 6g. part-slice; [1983,M.28], 7.6g.

Isonzo v Avce.

Isonzothal v Avce.

Isoulane-n-Amahar 27°7′45″ N., 8°40′2″ E.

Algeria
Found 1945, May 19
Stone. Olivine-hypersthene chondrite (L6), veined.

A somewhat weathered mass of 72kg was found at the co-ordinates given, NNE. of Fort Polinac. Described, with an analysis, E. Jérémine et al., C. R. Acad. Sci. Paris, 1956, **242**, p.2369 [M.A. 13-361]. Olivine Fa_{25}, B. Mason, Geochimica et Cosmochimica Acta, 1963, **27**, p.1011.

72kg Paris, Mus. d'Hist. Nat.;
Specimen(s): [1972,236], 124.25g.

Isthilart 31°11′ S., 57°57′ W.

Departmento Federacion, Entre Rios, Argentina
Fell 1928, November 12, 0730 hrs
Stone. Olivine-bronzite chondrite (H5).

Fell near the railway station at Isthilart. Two pieces, 3050g total weight. Described, with an analysis, E.H. Ducloux and F. Pastore, Rev. fac. cienc. quim. Univ. nac. La Plata, 1929, **6** (2), p.13 [M.A. 4-425]. Olivine Fa_{18}, B. Mason, Geochimica et Cosmochimica Acta, 1963, **27**, p.1011.
Specimen(s): [1962,176], 2g.

Italy (956)

Fell 956, A.D.
Doubtful..

A stone is said to have fallen, G. von Boguslawski, Ann. Phys. Chem. (Poggendorff), 1854, **4** (suppl.), p.8, but the evidence is not conclusive.

Italy (963)

Fell 963, A.D.
Doubtful..

A stone or stones reputedly fell, G. von Boguslawski, Ann. Phys. Chem. (Poggendorff), 1854, **4** (suppl.), p.8, but the evidence is not conclusive.

Itapicuru-Mirim 3°24′ S., 44°20′ W.

Maranhão, Brazil
Fell 1879, March
Stone. Olivine-bronzite chondrite (H5).

A stone of 2kg fell, O.A. Derby, Anais Acad. Brasil. Cienc., 1888, **49**, p.407. Mineralogy and analysis, olivine $Fa_{18.8}$, 29.58 % total iron, C.B. Gomes et al., Meteoritics, 1977, **12**, p.241, abs., C.B. Gomes et al., Anais Acad. Brasil. Cienc., 1977, **49**, p.407.

1.3kg Rio de Janeiro, Mus. Nac.; 220g Chicago, Field Mus. Nat. Hist.; 110g Sao Paulo, Inst. Geocienc.; 9.7g Washington, U.S. Nat. Mus.; 7g Vienna, Naturhist. Mus.;
Specimen(s): [63234], 6.25g.

Itapuranga 15°35′ S., 50°9′ W.

Curral de Pedro, Goias, Brazil
Found
Iron. Octahedrite, coarse (1.5mm) (IA).

A single mass of 628kg was found, date not reported, Meteor. Bull., 1980 (58), Meteoritics, 1980, **15**, p.237. Classification and analysis, 6.7 %Ni, 97 ppm.Ga, 478 ppm.Ge, 2.8 ppm.Ir, A. Kracher et al., Geochimica et Cosmochimica Acta, 1980, **44**, p.773. Brief description, D.P. Svisero et al., Bol. Inst. Geocienc. Univ. São Paulo, 1980, **11**, p.21.

Main mass, São Paulo, Mus. Inst. Geocienc.;

Itchkala v Ichkala.

Itschkala v Ichkala.

Itumi v Gifu.

Itutinga 21°20′ S., 44°40′ W.

Minas Gerais, Brazil
Found 1960, known before this year
Iron. Octahedrite, medium (1.0mm) (IIIA).

Brief description, V.F. Buchwald, Iron Meteorites, Univ. of California, 1975, p.688. Classification and analysis, 7.2 %Ni, 18.6 ppm.Ga, 36.0 ppm.Ge, 13 ppm.Ir, A. Kracher et al., Geochimica et Cosmochimica Acta, 1980, **44**, p.773.

Main mass, Ouro Preto, School of Mines; Specimens, Rio de Janeiro, Mus. Nat. Hist.; 24g Albuquerque, Univ. of New Mexico;

Itzawisis 26°16′ S., 18°11′ E.

Keetmanshoop, Namibia
Found 1946
Stony-iron. Pallasite (PAL).

A very fresh mass of about 350g was found 4 miles S. of Itzawisis railway siding and about 20 miles N. of Keetmanshoop. Described, with an analysis, H.J. Nel, Geol. Surv. Union South Africa, 1949 (Memoir 43), p.9 [M.A. 11-136]. Olivine Fa_{20}, P.R. Buseck and J.I. Goldstein, Bull. Geol. Soc. Amer., 1969, **80**, p.2141. Analysis of metal, 14.7 %Ni, 5.73 ppm.Ga, 85.9 ppm.Ge, 15 ppm.Ir, J.T. Wasson, Meteorites, Springer-Verlag, 1974, p.307. Eagle station group, references, E.R.D. Scott, Geochimica et Cosmochimica Acta, 1977, **41**, p.693.

69g Pretoria, Geol. Surv. Mus.;

Iudoma v Yudoma.

Iuknov v Timochin.

Iurtuk v Yurtuk.

Iurtyk v Yurtuk.

Ivanova v Pervomaisky.

Ivanpah 35°20′ N., 115°19′ W.
San Bernardino County, California, U.S.A.
Found 1880
Synonym(s): *San Bernardino County*
Iron. Octahedrite, medium (1.0mm) (IIIA).
A mass of about 128lb was found, C.U. Shepard, Am. J. Sci., 1880, **19**, p.381. Analysis, 7.34 %Ni, E. Cohen, Ann. Naturhist. Hofmus. Wien, 1892, **7**, p.149. More recent analysis, 7.51 %Ni, 21.1 ppm.Ga, 38.1 ppm.Ge, 3.8 ppm.Ir, E.R.D. Scott et al., Geochimica et Cosmochimica Acta, 1973, **37**, p.1957. Brief description, V.F. Buchwald, Iron Meteorites, Univ. of California, 1975, p.688.
Main mass, San Francisco, State Mining Bureau Mus.; 508g New York, Amer. Mus. Nat. Hist.; 3.1kg Washington, U.S. Nat. Mus.; 359g Harvard Univ.; 220g Chicago, Field Mus. Nat. Hist.; 139g Vienna, Naturhist. Mus.;
Specimen(s): [56161], 19g.; [64144*], 14g.

Ivuna 8°25′ S., 32°26′ E.
Tanzania
Fell 1938, December 16, 1730 hrs
Stone. Carbonaceous chondrite, type I (CI).
Two or three stones fell at Ivuna, near the W. shore of Rukwa, one of 704.5g was recovered, Ann. Rep. Geol. Div. Tanganyika Terr., 1940, p.22 [M.A. 8-58], F. Oates, Tanganyika Notes and Records, 1941, **12**, p.28 [M.A. 8-373]. Mentioned, J.R. Harpum, Rec. Geol. Surv. Tanganyika for 1961, 1965, **11**, p.54. Analysis, 19.01 % total iron, H.B. Wiik, Geochimica et Cosmochimica Acta, 1956, **9**, p.279.
Main mass, Dodoma, Tanzania Geol. Surv.; 122g Washington, U.S. Nat. Mus.; 66g New York, Amer. Mus. Nat. Hist.;

Iwa v Gifu.

Iwate v Kesen.

Iwati v Takenouchi.

Ixtlahuaca v Toluca.

Izumi v Gifu.

Jacala v Pacula.

Jackalsfontein 32°30′ S., 21°54′ E.
Beaufort West, Cape Province, South Africa
Fell 1903, April 22, 1130 hrs
Synonym(s): *Fraserburg*
Stone. Olivine-hypersthene chondrite (L6).
Two stones, one very large, fell after detonations, to the NW. of Uitkijk on the boundary of the farms Tamboersfontein and Jackalsfontein, L. Peringuey, letter of 12 August, 1919 in Min. Dept. BM(NH). Analysis, 21.57 % total iron, H. von Michaelis et al., Earth planet. Sci. Lett., 1969, **5**, p.387. Olivine Fa_{24}, B. Mason, Geochimica et Cosmochimica Acta, 1963, **27**, p.1011.
106lb Cape Town, South African Mus., main mass, and fragments, 1.5lb; 160g Vienna, Naturhist. Mus.; 30g Washington, U.S. Nat. Mus.; Specimen, Witwatersrand Univ.;
Specimen(s): [1908,431], 56g. four fragments of the smaller stone

Jackson County 36°25′ N., 85°30′ W. approx.
Tennessee, U.S.A.
Found 1846, known in this year
Iron. Octahedrite, medium (1.20mm) (IIIA).
A piece of about 1lb from a large mass, since lost, was described, G. Troost, Am. J. Sci., 1846, **2**, p.357 The mass was probably heated or forged; may be part of Carthage (*q.v.*), V.F. Buchwald, Iron Meteorites, Univ. of California, 1975, p.689. Chemically very similar to Carthage, 9.2 %Ni, 21.0 ppm.Ga, 46.7 ppm.Ge, 0.66 ppm.Ir, A. Kracher et al., Geochimica et Cosmochimica Acta, 1980, **44**, p.773.
116g Chicago, Field Mus. Nat. Hist.; 46g Washington, U.S. Nat. Mus.; 44g New York, Amer. Mus. Nat. Hist.;
Specimen(s): [33957], 91g.

Jafferabad v Bherai.

Jajh deh Kot Lalu 26°45′ N., 68°25′ E.
Faizganj taluk, Khairpur district, Sind, Pakistan
Fell 1926, May 2, 1700-1800 hrs
Synonym(s): *Jaj-Khot-Laulu*
Stone. Enstatite chondrite (E6), veined.
After detonations a stone fell and two pieces, of 753g and 220g, respectively, were recovered: described, G.V. Hobson, Rec. Geol. Surv. India, 1927, **60**, p.150. Mineral chemistry, K. Keil, J. Geophys. Res., 1968, **73**, p.6945. Analysis, 22.17 % total iron, B. Mason, Geochimica et Cosmochimica Acta, 1966, **30**, p.23.
538g Calcutta, Mus. Geol. Surv. India.; Smaller piece, Karachi Mus.; 23g New York, Amer. Mus. Nat. Hist.; 18g Washington, U.S. Nat. Mus.; 16g Tempe, Arizona State Univ.;
Specimen(s): [1928,479], 32g. and fragment, 1g

Jaj-Khot-Laulu v Jajh deh Kot Lalu.

Jalandhar 31° N., 75° E.
Punjab, India
Fell 1621, April 10
Iron., or stony-iron.
A mass of about 1967 grams fell in the reign of King Jahangir, and was forged into sword blades, H. Blockmann, Proc. Asiatic Soc. Bengal, 1869 (6), p.167, M.A.R. Khan, Meteors and meteoric iron in India, Secunderabad, 1934 [M.A. 6-102], W. Campbell Smith, Nature, 1935, **135**, p.39. Mentioned, C. Greville, Phil. Trans., 1803, p.200, Phil. Mag., 1803, **16**, p.294. The place of fall is Jalandhar (=Jullundur) pargana.

Jalisco v Tomatlan.

Jamaica v Lucky Hill.

Jamestown 46°37′ N., 98°30′ W.
Stutsman County, North Dakota, U.S.A.
Found 1885
Synonym(s): *Stutsman County*
Iron. Octahedrite, fine (0.26mm) (IVA).
A mass of about 4kg was found 15 to 20 miles SE. of Jamestown. Described, with analysis, 9.75 %Ni, O.W. Huntington, Proc. Amer. Acad. Arts and Sci., 1891, **25**, p.229. Analysis, 7.45 %Ni, 1.80 ppm.Ga, 0.093 ppm.Ge, 3.5 ppm.Ir, R. Schaudy et al., Icarus, 1972, **17**, p.174. Description of deformation, H.J. Axon and A.W.R. Bevan, Nature, 1976, **263**, p.302. Description, V.F. Buchwald, Iron Meteorites, Univ. of California, 1975, p.690.
581g Chicago, Field Mus. Nat. Hist.; 329g Washington, U.S. Nat. Mus.; 128g Berlin, Humboldt Univ.; 97g Vienna, Naturhist. Mus.; 91g New York, Amer. Mus. Nat. Hist.; 86g Harvard Univ.;
Specimen(s): [67215], 1311g. the largest piece preserved, and a fragment, 252g

Jamkheir 18°45′ N., 75°20′ E.
Ahmadnagar district, Maharashtra, India
Fell 1866, October 5, 1200 hrs
Stone. Olivine-bronzite chondrite (H6), brecciated.
Two stones of "Size of a wood apple" (about 4 inches in diameter) are said to have fallen after detonations, but only small fragments were recovered, W. D'Oyly, copy of letter of 14 February, 1867 to Government Secretary, Bombay, in Min. Dept. BM(NH). Olivine Fa_{19}, B. Mason, Geochimica et Cosmochimica Acta, 1963, **27**, p.1011.
2g Calcutta, Mus. Geol. Surv. India; 1g Chicago, Field Mus. Nat. Hist.;
Specimen(s): [40602], 15.5g. the largest piece preserved

Jamyscheva v Pavlodar (pallasite).

Jamyseva v Pavlodar (pallasite).

Janacera Pass v Vaca Muerta.

Japan v Ogi.

Jaralito 26°15′57″ N., 103°53′6″ W.
Jaralito, Durango, Mexico
Found 1977, September
Iron. Octahedrite, coarse (1.9mm) (IIICD).
A single mass of 11.138kg was found in a field, Meteor. Bull., 1980 (57), Meteoritics, 1980, **15**, p.98. Classification and analysis, 6.52 %Ni, 91.6 ppm.Ga, 376 ppm.Ge, 1.5 ppm.Ir, A. Kracher et al., Geochimica et Cosmochimica Acta, 1980, **44**, p.773.
Specimen, Los Angeles, Univ. of California;

Jardymlinsky v Yardymly.

Jarquera v Vaca Muerta.

Jarso v Nejo.

Jartai 39°42′ N., 105°48′ E.
Jilantai People's Commune, Ningxia, China
Fell 1979, March 15
Synonym(s): *Jilantai*
Stone. Olivine-hypersthene chondrite (L6).
A mass of 20.5kg is listed, references, D. Bian, Meteoritics, 1981, **16**, p.115. Analysis, 20.93 % total iron, Y. Yang and X. Yang, Geochimica, 1981, p.411.

Jaska v Slavetic.

Jaski v Białystok.

Jasłi v Białystok.

Jasly v Białystok.

Jataha Bazar v Butsura.

Java 7°30′ S., 110° E. approx.
Java, Indonesia
Fell 1421
Doubtful. Stone.
According to local records, a large stone fell, E.F.F. Chladni, Die Feuer-Meteore, Wien, 1819, p.202.
3g New York, Amer. Mus. Nat. Hist.;

Jeedamya 29°35′ S., 121°10′ E.
Menzies district, Western Australia, Australia
Found 1971
Stone. Olivine-bronzite chondrite (H6).
A single, oriented stone of 914g was found in a shallow depression (1cm) in soil, Meteor. Bull., 1972 (51), Meteoritics, 1972, **7**, p.224. Olivine Fa_{19}, G.J.H. McCall, 2nd. Suppl. to West. Austr. Mus. Spec. Publ. no. 3, 1972, p.15.
Entire mass, Perth, West. Austr. Mus.;

Jefferson v Bear Creek.

Jefferson City v Little Piney.

Jefferson County v Bear Creek.

Jekaterinoslav v Augustinovka.

Jekaterinoslav v Pavlograd.

Jelenowka v Elenovka.

Jelica 43°50′ N., 20°26′30″ E.
Serbia, Yugoslavia
Fell 1889, December 1, 1430 hrs
Synonym(s): *Banjaca, Cacak, Chachak, Jeliza, Jezevica, Piljusa*
Stone. Olivine-hypersthene chondrite, amphoterite (LL6), brecciated.
After detonations and appearance of light, a shower of stones fell over an area of 5 × 3 miles: the stones, of which 26 or more were found, varied in weight from 8.5kg to 70g and had a total weight of about 34kg, E. Doll, Jahr. Geol. Reichsanst. Wien, 1890, p.70, G. von Niessl, Verh. Naturf. Ver. Brünn, 1890, **29**, p.166. Analysis, S.M. Losanitsch, Ber. Deutsch. Chem. Gesell. Berlin, 1892, **25**, p.876. Olivine $Fa_{32.3}$, K. Fredriksson et al., Origin and Distribution of the Elements, ed. L.H. Ahrens, Pergamon, 1968, p.457. Analysis of lithic fragments, R.V. Fodor and K. Keil, Meteoritics, 1975, **10**, p.325. Ni and Co contents of metal, D. Sears and H.J. Axon, Meteoritics, 1976, **11**, p.97.
7.3kg Belgrade, Mus.; 5.4kg Budapest, Nat. Hist. Mus; 1.05kg Vienna, Naturhist. Mus.; 515g Paris, Mus. d'Hist. Nat.; 753g New York, Amer. Mus. Nat. Hist.; 200g Moscow, Acad. Sci.; 307g Chicago, Field Mus. Nat. Hist.; 231g Washington, U.S. Nat. Mus.; 136g Prague, Nat. Mus.; 114g Berlin, Humboldt Univ.;

Specimen(s): [65486], 1511g. a nearly complete stone; [65605], 207g. and fragments, 59g; [1920,308], 3g.

Jeliza v Jelica.

Jemlapur
India
Fell 1901, February
Stone. Olivine-hypersthene chondrite (L6).
A stone of about 1lb, labelled "Jemlapur, Muddera Tk (fell Feb./01)," was found in a curio dealer's shop in Beckenham, Kent, in 1913, but neither the locality nor the fall of a meteorite on that date can be traced, G.T. Prior, Nature, 1916, **97**, p.241. Olivine Fa_{25}, B. Mason, Geochimica et Cosmochimica Acta, 1963, **27**, p.1011.
Specimen(s): [1918,364], 464g. the nearly complete stone, and fragments, 2 grams.

Jenkins 36°49′28″ N., 93°45′40″ W.
Barry County, Missouri, U.S.A.
Found 1946, about, recognized 1965
Iron. Octahedrite, coarse (2.3mm) (IA).
One mass of 55.4kg was found 8km northwest of Jenkins, Meteor. Bull., 1966 (38), Meteoritics, 1970, **5**, p.89. Description, W.F. Read, Meteoritics, 1967, **3**, p.141. Analysis, 6.85 %Ni, 86.2 ppm.Ga, 353 ppm.Ge, 1.8 ppm.Ir, J.T. Wasson, Icarus, 1970, **12**, p.407. Description, inclusion-rich, possibly paired with Seymour (*q.v.*), V.F. Buchwald, Iron Meteorites, Univ. of California, 1975, p.691.
55kg Washington, U.S. Nat. Mus., main mass;

Jennies Creek v Jenny's Creek.

Jenny's Creek 37°54′ N., 82°23′ W.
Wayne County, West Virginia, U.S.A.
Found 1883
Synonym(s): *Charleston, Jennies Creek, Kanawha County, Kanwahoe County, Old Fork, Wayne County*
Iron. Octahedrite, coarse (2.2mm) (IA).
Three masses, of about 23lb, 2.5lb and 1lb respectively, were found in 1883-85, but only about 2lb have been preserved, G.F. Kunz, Proc. Amer. Assoc., 1885, **34**, p.246. Analysis, 7.0 %Ni, 84.6 ppm.Ga, 320 ppm.Ge, 2.3 ppm.Ir, J.T. Wasson, Icarus, 1970, **12**, p.407. Description, co-ordinates, V.F. Buchwald, Iron Meteorites, Univ. of California, 1975, p.693.
587g Vienna, Naturhist. Mus.; 228g New York, Amer. Mus. Nat. Hist.; 131g Chicago, Field Mus. Nat. Hist.; 67g Budapest, Nat. Mus.; 18g Tempe, Arizona State Univ.; 14g Washington, U.S. Nat. Mus.; 32g Harvard Univ.;
Specimen(s): [56918], 78g.; [1959,164], 0.9g.

Jerome (Idaho) 42°38′ N., 114°50′ W.
Jerome County, Idaho, U.S.A.
Found 1954
Stone. Olivine-hypersthene chondrite (L).
A mass of 6.8kg was found, B. Mason, Meteorites, Wiley, 1962, p.231. Olivine Fa_{24}, B. Mason, Geochimica et Cosmochimica Acta, 1963, **27**, p.1011.
189g Tempe, Arizona State Univ.; 6g New York, Amer. Mus. Nat. Hist.;

Jerome (Kansas) 38°46′ N., 100°44′ W.
Gove County, Kansas, U.S.A.
Found 1894
Stone. Olivine-hypersthene chondrite (L4).
A much oxidized mass of 62lb, and several smaller pieces weighing together 3.25lb, were found on the Smoky Hill River, 15 miles E. of Jerome. Description and analysis, H.S. Washington, Am. J. Sci., 1898, **5**, p.447. Olivine Fa_{19}, B. Mason, Geochimica et Cosmochimica Acta, 1963, **27**, p.1011.
30kg Yale Univ.; 166g Washington, U.S. Nat. Mus.; 64g Chicago, Field Mus. Nat. Hist.; 55g Paris, Mus. d'Hist. Nat.;
Specimen(s): [1923,1006], 6.5g.; [1973,M.22], 56.5g. crusted fragment

Jerslev 55°36′ N., 11°13′ E.
Sjaelland, Denmark
Found 1976, September
Iron. Octahedrite, coarsest (10mm) (IIB).
A single mass of 40kg (after cleaning) was found in soil by a farmer, Meteor. Bull., 1978 (55), Meteoritics, 1978, **13**, p.339. Analysis (classification as IIIB in error), 5.66 %Ni, 58.0 ppm.Ga, 168 ppm.Ge, 0.19 ppm.Ir, A. Kracher et al., Geochimica et Cosmochimica Acta, 1980, **44**, p.773. Classification, V.F. Buchwald, priv. comm., 1980.
40kg Copenhagen, Univ. Geol. Mus.;

Jewel Hill v Duel Hill (1854).

Jewel Hill v Duel Hill (1873).

Jewell Hill v Duel Hill (1854).

Jezevica v Jelica.

Jhang v Jhung.

Jharaota v Lalitpur.

Jhung 31°18′ N., 72°23′ E.
Lyallpur district, Punjab, Pakistan
Fell 1873, June, 1500 hrs
Synonym(s): *Ihung, Jhang, Jung Kot Divan*
Stone. Olivine-hypersthene chondrite (L5).
After detonations, four stones, of 6lb, 4lb, 2lb and 1lb, fell in villages, in a line from south to north, in the district of Jhang, A. Brandreth, letter of 9 December, 1875 accompanying a copy of a report of G. Lewis to the Royal Society, in Min. Dept. BM(NH). Olivine Fa_{25}, B. Mason, Geochimica et Cosmochimica Acta, 1963, **27**, p.1011. Lightly shocked, R.T. Dodd and E. Jarosewich, Earth planet. Sci. Lett., 1979, **44**, p.335.
760g Calcutta, Mus. Geol. Surv. India; 92g Paris, Mus. d'Hist. Nat.; 44g Harvard Univ.; 24g New York, Amer. Mus. Nat. Hist.; 24g Oslo, Min.-Geol. Mus.; 23g Chicago, Field Mus. Nat. Hist.; 22g Washington, U.S. Nat. Mus.;
Specimen(s): [51190], 1770g. a nearly complete stone, and pieces, 33.5g.

Jiange 31°55′ N., 104°55′ E.
Minyang, Sichuan, China
Fell 1964, October 9, 1900? hrs
Stone. Olivine-bronzite chondrite (H5).
Listed, references, D. Bian, Meteoritics, 1981, **16**, p.115. Analysis, 29.49 % total iron, D. Wang and Z. Ouyang, Geochimica, 1979, p.120, in Chinese [M.A. 80-2085].
200g Beijing, Acad. Sin. Inst. Geol., approx. weight;

Jianshi 30°48′30″ N., 109°30′ E.
Hubei, China
Fell 1890, approx.
Iron. Octahedrite, medium (1.0mm) (IIIA).
A single mass weighing over 600kg remains at the place of fall, D. Bian, Meteoritics, 1981, **16**, p.115. Classification, analysis, 8.6 %Ni, 21.3 ppm.Ga, 44.5 ppm.Ge, 0.4 ppm.Ir, D.J. Malvin et al., priv. comm., 1983.
Specimen, Beijing, Acad. Sinica Inst. Geol.;

Jiapigou 42°50′ N., 127°30′ E.
Jilin, China
Found 1880, before this year
Synonym(s): *Chiapikou*
Stone.
Several 10's of kilograms were recovered, D. Bian, Meteoritics, 1981, **16**, p.115.

Jiddat al Harasis 19°15′ N., 56°4′ E.
Oman
Found 1957, or 1958
Synonym(s): *Ghubara (1962)*
Stone. Olivine-bronzite chondrite (H4).
A weathered mass of 1270g and 89g of fragments, previously assumed to belong to the Ghubara find, were shown to be distinct. Further fragments found in 1968 at 19°30′N., 55°54′E., belong to the same fall. Olivine Fa_{17}, M.J. Frost, priv. comm.
Specimen(s): [1962,133], 1270g. and fragments, 74g.; [1968, 272], 171.5g. two stones

Jigalovka v Kharkov.

Jigalowka v Kharkov.

Jilin 44°0′ N., 126°30′ E.
Jilin, China
Fell 1976, March 8, 1500 hrs
Synonym(s): *Kirin*
Stone. Olivine-bronzite chondrite (H5).
After a fireball and several explosions, a shower of stones fell, totalling about 4 tonnes and with the largest individual weighing 1770kg, analysis, 28.6 % total iron, Joint Investigating Group, Sci. Sinica, 1977, **20**, p.502 [M.A. 78-1794], Joint Investigating Group, Geochimica, 1976, p.157 [M.A. 77-2037]. Description, olivine Fa_{18}, Meteorite Research Group, Acta Geol. Sin., 1976, p.176 [M.A. 77-3262].
Specimen(s): [1981,M.9], 54.2g.; [1981,M.10], 8.3g.

Jilong
Taiwan, China, Co-ordinates not reported
Found year not reported
Stone.
Listed, without details, W. Daode et al., Geochemistry, 1982, **1**, p.186.

Jimenez v Chupaderos.

Jimenez v Sierra Blanca.

Jimshan v al-Jimshan.

Jiquipilco v Toluca.

Jodzie 55°42′ N., 24°24′ E.
Panevezys, Kovno, Lithuanian SSR, USSR, [Йоджяй]
Fell 1877, June 17, 0430 hrs
Synonym(s): *Yodze*
Stone. Achondrite, Ca-rich. Howardite (AHOW).
A stone of unknown weight fell near the village of Jodzie and was lost except for a few fragments, Y.I. Simashko, letter of 16 November, 1891 in Min. Dept. BM(NH)., and his label with specimen (68214). Description, A. Brezina, Verh. Ges. Deut. Naturf. Arzte, 1893, **65**, p.159. Gas-rich, E. Mazor and E. Anders, Geochimica et Cosmochimica Acta, 1967, **31**, p.1441. Figured, M.B. Duke and L.T. Silver, Geochimica et Cosmochimica Acta, 1967, **31**, p.1637. Analysis, B. Mason et al., Smithson. Contrib. Earth Sci., 1979 (22), p.30.
21g Chicago, Field Mus. Nat. Hist., principal known mass; 1g Paris, Mus. d'Hist. Nat.; 1g Vienna, Naturhist. Mus.; 1g Berlin, Humboldt Univ.; 4.6g Moscow, Acad. Sci.;
Specimen(s): [68214], 1.5g.

Joel's Iron 24° S., 69° W. approx.
Atacama, Chile
Found 1858
Synonym(s): *Atacama Desert, Cobija*
Iron. Octahedrite, medium (1.1mm) (IIIA).
A mass of 1300g, found in 1858 in an unspecified part of the desert, was presented to the British Museum in 1863 by Mr. Lewis Joel, the British Vice-Consul at Cobija. Description and analysis, 8.8 %Ni, L. Fletcher, Min. Mag., 1889, **8**, p.263. Newer analysis, 8.4 %Ni, 22.6 ppm.Ga, 43.6 ppm.Ge, 0.26 ppm.Ir, E.R.D. Scott et al., Geochimica et Cosmochimica Acta, 1973, **37**, p.1957. Description, recrystallised and annealed, V.F. Buchwald, Iron Meteorites, Univ. of California, 1975, p.694.
43.5g Chicago, Field Mus. Nat. Hist.; 13g Harvard Univ.;
Specimen(s): [35782], 1199g. includes main mass 1144g, and filings, 35g.

Joe Wright Mountain 35°46′ N., 91°30′ W.
Independence County, Arkansas, U.S.A.
Found 1884
Synonym(s): *Batesville, Elmo, Independence County*
Iron. Octahedrite, medium (0.9mm) (IIIB).
A mass of 94lb was found 7 miles E. of Batesville, W.E. Hidden, Am. J. Sci., 1886, **31**, p.461. Analysis, 9.1 %Ni, 20.1 ppm.Ga, 35.5 ppm.Ge, 0.015 ppm.Ir, E.R.D. Scott et al., Geochimica et Cosmochimica Acta, 1973, **37**, p.1957. Description, V.F. Buchwald, Iron Meteorites, Univ. of California, 1975, p.697. See also Sandtown (*q.v.*).
33kg Vienna, Naturhist. Mus., main mass; 540g Washington, U.S. Nat. Mus.; 1.46kg Budapest, Nat. Mus.; 265g Chicago, Field Mus. Nat. Hist.;
Specimen(s): [56919], 372g. slice

Joe Wright Mountain II v Sandtown.

Johanngeorgenstadt v Steinbach.

Johnny's Donga 30°20′ S., 126°22′ E. approx.
Nullarbor Plain, Western Australia, Australia
Found 1965, before this year
A specimen from Johnny's Donga is reported as "lost", West. Austr. Mus. Spec. Publ. no. 3, 1965, p.25.

Johnson v Johnson City.

Johnson City 37°33′ N., 101°41′ W.
Stanton County, Kansas, U.S.A.
Found 1937
Synonym(s): *Johnson*
Stone. Olivine-hypersthene chondrite (L6).
Two stones were found, 10.4kg together, E.P. Henderson, letters of 12 May and 3 June, 1939 in Min. Dept. BM(NH). Listed, Ann. Rep. U.S. Nat. Mus., 1938, p.50, A.D. Nininger, Pop. Astron., Northfield, Minnesota, 1939, **47**, p.212, H.H. and A.D. Nininger, The Nininger Collection of Meteorites, Winslow, Arizona, 1950, p.64. Olivine Fa_{25}, B. Mason, Geochimica et Cosmochimica Acta, 1963, **27**, p.1011.
3.9kg Tempe, Arizona State Univ.; 2kg Washington, U.S. Nat. Mus.; 485g New York, Amer. Mus. Nat. Hist.;
Specimen(s): [1959,1037], 4039g. half of one of the masses

Johnson County v Cabin Creek.

Johnstown 40°21′ N., 104°54′ W.
Weld County, Colorado, U.S.A.
Fell 1924, July 6, 1620 hrs
Synonym(s): *Elwell, Jonstown, Weld County*
Stone. Achondrite, Ca-poor. Diogenite (ADIO).
After four explosions, twenty-seven stones fell near Johnstown. The total weight recovered was about 40.3kg, and the largest stone weighed about 23.5 kg. Description, E.O. Hovey et al., Am. Mus. Novit., 1925 (203). Analysis, B. Mason and E. Jarosewich, Meteoritics, 1971, **6**, p.241. Petrology, mineral chemistry, R.J. Floran et al., Geochimica et Cosmochimica Acta, 1981, **45**, p.2385. Thermal history, H. Mori and H. Takeda, Earth planet. Sci. Lett., 1981, **53**, p.266. Rb-Sr study, J.L. Birck and C.J. Allegre, Earth planet. Sci. Lett., 1981, **55**, p.116.
22kg New York, Amer. Mus. Nat. Hist., largest stone; 5.2kg Denver, Mus. Nat. Hist.; 1.6kg Chicago, Field Mus. Nat. Hist.; 1.5kg Tempe, Arizona State Univ.; 752g Harvard Univ.;
Specimen(s): [1926,492], 204g. nearly complete stone; [1926, 493], 150g. nearly complete stone; [1959,828], 996g. and fragments, 68g.

Jonesboro 36°18′ N., 82°28′ W.
Washington County, Tennessee, U.S.A.
Found 1891
Synonym(s): *East Tennessee*
Iron. Octahedrite, fine (IVA).
No known history; a piece of 30g that was in Ward's Natural Science Establishment in 1892, H.A. Ward, Cat. Meteorites, Rochester, New York, 1892, p.15 was sold to the Naturhist. Mus., Vienna. Description, E. Cohen, Meteoritenkunde, 1905, **3**, p.388. Possibly a weathered fragment of Duel Hill (1854), V.F. Buchwald, Iron Meteorites, Univ. of California, 1975, p.698.
28g Vienna, Naturhist. Mus., main mass;

Jonstown v Johnstown.

Jonzac 45°26′ N., 0°27′ W.
Charente Maritime, France
Fell 1819, June 13, 0545 hrs
Synonym(s): *Saintonge*
Stone. Achondrite, Ca-rich. Eucrite (AEUC).
After the appearance of a luminous meteor and detonations, a shower of stones fell, the two largest of which weighed 3kg and 2kg respectively, F. de Bellevue, J. Phys. Chim. Hist. Nat., 1821, **92**, p.136. Description, analysis, H. Michel, Tschermaks Min. Petr. Mitt., 1912, **31**, p.577 589, A. Lacroix, Arch. Mus. Hist. Nat. Paris, 1926, **1** (ser. 6), p.15 [M.A. 3-392]. Further analysis, B. Mason et al., Smithson. Contrib. Earth Sci., 1979 (22), p.30.
1097g Vienna, Naturhist. Mus.; 250g Paris, Mus. d'Hist. Nat.; 39g Rome, Vatican Colln; 14g Prague, Nat. Mus.; 7.5g Chicago, Field Mus. Nat. Hist.;
Specimen(s): [19974], 9g.; [1935,48], minute fragment.

Joremenyseg-fok v Cape of Good Hope.

Joundegin v Youndegin.

Joutnevy v Zhovtnevyi.

Jowa v Estherville.

Jowa v Forest City.

Jowa v Homestead.

Juarez 37°33′ S., 60°9′ W.
Buenos Aires province, Argentina
Found 1938, before this year
Stone. Olivine-hypersthene chondrite (L).
Two stones, together weighing 6.1kg, were found, Meteor. Bull., 1962 (24). Olivine Fa_{24}, B. Mason, Geochimica et Cosmochimica Acta, 1963, **27**, p.1011.
Main masses of both specimens, Buenos Aires, Mus. Nat. Hist.;
Specimen(s): [1962,163], 4.5g.

Jubila del Agua
Spain, Not located
Fell 1908, December
Doubtful. Stone.
Five stones are said to have fallen, F. Berwerth, Fortschr. Min. Krist. Petr., 1912, **2**, p.234. Not mentioned, M. Faura y Sans, Meteoritos caidos en la Peninsula Iberica, Tortosa, 1922.

Juchnow v Timochin.

Judesegeri 12°51′ N., 76°48′ E.
Tumkur district, Karnataka, India
Fell 1876, February 16, evening
Synonym(s): *Judesegiri, Judesgherry, Mysore*
Stone. Olivine-bronzite chondrite (H6).
After the appearance of a luminous meteor and detonations, a stone fell into the bed of the tank of Judesegeri village. It was broken up and only fragments weighing about 1.5lb were preserved, H.B. Medlicott, Proc. Asiatic Soc. Bengal, 1876, p.221. Olivine Fa_{18}, B. Mason, Geochimica et Cosmochimica Acta, 1963, **27**, p.1011.
300g Calcutta, Mus. Geol. Surv. India; 16g Vienna, Naturhist. Mus.; 16g Washington, U.S. Nat. Mus.;
Specimen(s): [51367], 114g.

Judesegiri v Judesegeri.

Judesgherry v Judesegeri.

Judoma v Yudoma.

Jukao v Min-Fan-Zhun.

Julesburg 39°58′30″ N., 102°16′ W.
Sedgewick County, Colorado, U.S.A.
Found 1983
Stone. Chondrite.
A single mass of 56.6kg was found in a landfill at Julesburg, G.I Huss, letters of 26 October, 14 December 1983 in Min. Dept. BM(NH).

Junan 35°12′ N., 118°48′ E.
Shandong, China
Fell 1976, May 15, 1100 hrs
Stone. Olivine-hypersthene chondrite (L6).
Listed, with references, 950g was recovered, D. Bian, Meteoritics, 1981, **16**, p.115.
950g Beijing, Acad. Sin. Inst. Geochem.;

Juncal 26° S., 69°15′ W.
Atacama, Chile
Found 1866
Iron. Octahedrite, medium (1.1mm) (IIIA).
A mass of 104kg was found between the Rio Juncal and the Salinas de Pedernal, analysis, 7.0 %Ni, G.A. Daubrée, C. R. Acad. Sci. Paris, 1868, **66**, p.568, L. Fletcher, Min. Mag., 1889, **8**, p.261. Further analysis, 8.05 %Ni, 20.5 ppm.Ga, 41.2 ppm.Ge, 1.8 ppm.Ir, E.R.D. Scott et al., Geochimica et Cosmochimica Acta, 1973, **37**, p.1957. Description, shock-hardened, V.F. Buchwald, Iron Meteorites, Univ. of California, 1975, p.698.
Main mass, Paris, Mus. d'Hist. Nat.; 871g Vienna, Naturhist. Mus.; 200g Budapest, Nat. Mus.; 140g New York, Amer. Mus. Nat. Hist.; 110g Chicago, Field Mus. Nat. Hist.;
Specimen(s): [68580], 70g.; [43202], 2.5g.

Junction 30°30′ N., 99°50′ W.
Kimble County, Texas, U.S.A.
Found 1932
Stone. Olivine-hypersthene chondrite (L).
A fragment of 241g, recognized as meteoritic in 1938, A.D. Nininger, Pop. Astron., Northfield, Minnesota, 1939, **47**, p.212. Olivine Fa_{24}, B. Mason, Geochimica et Cosmochimica Acta, 1963, **27**, p.1011.
40g Fort Worth, Texas, Monnig Colln.;

Jung Kot Divan v Jhung.

Juromenha 38°44′25″ N., 7°16′12″ W.
Alentejo, Portugal
Fell 1968, November 14, 1755 hrs, U.T.
Synonym(s): *Alandroal*
Iron. Structurally anomalous (IIIA).
A mass of 25.25kg was seen to fall, making a crater 60 cm deep at Herdade de Tenazes, near the village of Alandroal. The mass was said to have been incandescent when discovered and still warm when recovered next morning, C. Teixeira, Bolm. Soc. geol. Port., 1968, **16**, p.267. Analysis, 8.81 %Ni, 21.2 ppm.Ga, 40.3 ppm.Ge, 0.24 ppm.Ir, E.R.D. Scott et al., Geochimica et Cosmochimica Acta, 1973, **37**, p.1957. Ataxitic structure formed by cosmic heating to 1400C. followed by quenching, V.F. Buchwald, Iron Meteorites, Univ. of California, 1975, p.700.
Main mass, Lisbon, Centre of Geol. Studies, Faculty of Science; 16g Washington, U.S. Nat. Mus.;

Jurtuk v Yurtuk.

Juvinas 44°43′ N., 4°18′ E.
Libonnes, Entraigues, Ardeche, France
Fell 1821, June 15, 1500 hrs
Synonym(s): *Libonnez*
Stone. Achondrite, Ca-rich. Eucrite (AEUC).
After the appearance of a fireball and detonations, a stone of over 91kg fell near the village of Libonnes, L.W. Gilbert, Ann. Phys. (Gilbert), 1821, **69**, p.407, L.W. Gilbert, Ann. Chim. Phys., 1821, **17**, p.434. Analysis, G.P. Merrill, Mem. Nat. Acad. Sci. Washington, 1916, **14** (1), p.19. Mineralogy, analysis, M.B. Duke and L.T. Silver, Geochimica et Cosmochimica Acta, 1967, **31**, p.1637. U-Th-Pb systematics, M. Tatsumoto and D.M. Unruh, Meteoritics, 1975, **10**, p.500, abs., G. Manhes et al., Meteoritics, 1981, **16**, p.353, abs. Rb-Sr study, J.C. Birck et al., Earth planet. Sci. Lett., 1978, **39**, p.37 [M.A. 78-4757].
42kg Paris, Mus. d'Hist. Nat.; 2kg Tübingen, Univ., approx. weight; 985g Berlin, Humboldt Univ.; 680g Vienna, Naturhist. Mus.; 333g Tempe, Arizona State Univ.; 321g New York, Amer. Mus. Nat. Hist.; 317g Chicago, Field Mus. Nat. Hist.;
Specimen(s): [90263], 316g.; [458], 396.5g.; [1920,309], 5g. fragments; [90262], thin section

Juzjanan 34°22′ N., 62°8′ E. Co-ordinates of Herat
Herat, Afghanistan
Fell 1009, about this year
Doubtful. Iron., or stony-iron.
A large mass fell, S. de Sacy, Ann. Phys. (Gilbert), 1815, **50**, p.293, E.F.F. Chladni, Die Feuer-Meteore, Wien, 1819, p.194, E.J. Holmyard and D.C. Mandeville, Avicennae de congelatione et conglutatione lapidum, Paris, 1927 [M.A. 3-466]. The place of fall is also spelt Dschuzzan or Djouzdjan, and is in Khorasan.

Kaaba 21°25′ N., 39°54′ E.
Mecca, Saudi Arabia
Found 1772, mentioned in this year
Synonym(s): *Mecca*
Doubtful. Stone.
A stone built in the sanctuary of the Kaaba in Mecca is said to be possibly meteoritic, P. Partsch, Denkschr. Akad. Wiss. Wien, Math.-naturwiss. Kl., 1857, **13**, p.1. Full bibliography, E.A. Wülfing, Die Meteoriten in Samml., Tübingen, 1897, p.400. Mentioned, M.A.R. Khan, Pop. Astron., Northfield, Minnesota, 1938, **46**, p.403 [M.A. 7-176]. Reports of diffusion banding and other physical attributes suggest that it is an agate; may therefore be a pseudometeorite, R.S. Dietz and J. McHone, Meteoritics, 1974, **9**, p.173.

Kaali v Kaalijarv.

Kaalijarv 58°24′ N., 22°40′ E.
Saaremaa (=Oesel), Estonian SSR, USSR
Found 1937, July
Synonym(s): *Kaali, Sall*
Iron. Octahedrite, coarse (2.0mm) (IA).
Thirty rusted fragments, from 0.1g to 24g, total weight 100g, were found in the Kaalijarv craters after a long search, I.A. Reinwald, Natur u. Volk, Frankfurt-a-M., 1938, **68**, p.16 [M.A. 7-73]. Desribed, with an analysis, 8.32 %Ni, 6.45 % Ni, L.J. Spencer, Min. Mag., 1937, **25**, p.75. Description, Z.A. Yudin, Метеоритика, 1968, **28**, p.44. Description, impact deformation, V.F. Buchwald, Iron Meteorites, Univ. of California, 1975, p.704.
100g Tartu, Geol. Inst.; 9g New York, Amer. Mus. Nat. Hist.; 4.3g Washington, U.S. Nat. Mus.; 2.2g Tempe,

Arizona State Univ.; 88g Moscow, Acad. Sci.;
Specimen(s): [1938,134], 7.7g.; [1938,135], 4.9g.; [1938,136], 2.0g.

Kaande v Oesel.

Kaba 47°21′ N., 21°18′ E.
Debreczen, Hungary
Fell 1857, April 15, 2200 hrs
Synonym(s): *Debreczen*
Stone. Carbonaceous chondrite, type III (CV3).
After appearance of a fireball and detonations, a stone of about 3kg was found, J. von Török, Ann. Phys. Chem. (Poggendorff), 1858, **105**, p.329. Analysis, F. Wöhler, Sitzungsber. Akad. Wiss. Wien, Math.-naturwiss. Kl., 1858, **33**, p.205. The analysis shows an improbably high percentage of alumina, W. Wahl, Geochimica et Cosmochimica Acta, 1950, **1**, p.28. Detailed history, A. Hoffer, Debreczeni Szemle, 1928 [M.A. 4-117], K.I. Sztrokay, Neues Jahrb. Min. Abh., 1960, **94**, p.1284 [M.A. 15-286], K.I. Sztrokay et al., Acta Geol. Polon., 1961, **7**, p.57. Classification, W.R. Van Schmus and J.M. Hayes, Geochimica et Cosmochimica Acta, 1974, **38**, p.47.
2655g Debreczen, Reform College, main mass; 32g Vienna, Naturhist. Mus.; 8g Tübingen, Univ.;
Specimen(s): [35794], 98g.; [33969a], 4.5g.

Kabakly 39°46′ N., 62°31′ E.
Dejnan district, Turkmen SSR, USSR, [Кабаклы]
Found 1965
Stone. Olivine-bronzite chondrite (H4).
One stone of 71.6g was found at the bottom of a hole, 10-15 cm deep, in desert WNW. of the village of Kabakly, Meteor. Bull., 1969 (45), Meteoritics, 1970, **5**, p.102. Co-ordinates are of Kabakly.
58.6g Moscow, Acad. Sci.;

Kabala v Kalaba.

Kabo 11°51′ N., 8°13′ E. approx.
Kano State, Nigeria
Fell 1971, April 25, 1630 hrs
Synonym(s): *Gwarzo*
Stone. Olivine-bronzite chondrite (H4), xenolithic.
At least 4 stones, totalling about 13.4kg, were seen to fall at the end of a west-east trajectory, and near the village of Kabo. The largest stone weighed about 6kg but broke into two pieces on impact. Description, analysis, olivine $Fa_{18.9}$, 29.4 % total iron, R. Hutchison et al., Min. Mag., 1973, **39**, p.340.
Main mass, Kano, Military Governor's Office; Specimens, Kaduna, Geol. Surv. Nigeria, from half of the largest stone; 80g Washington, U.S. Nat. Mus.; Specimen, Univ. of Ibadan;
Specimen(s): [1972,131], 209.5g. three fragments, and smaller fragments, 6.4g.

Kadapa v Gurram Konda.

Kadonah 27°5′ N., 78°20′ E.
Agra district, Uttar Pradesh, India
Fell 1822, August 7, night hrs
Synonym(s): *Agra, Doab, Karondha*
Stone. Olivine-bronzite chondrite (H6), veined.
A large stone (weight not recorded) fell, after detonations, near the village of Kadonah and was found next morning, Rep. Brit. Assn., 1851, **20**, p.120. There is no village Kadonah in the Agra district; the probable place of fall is Karondha, tahsil Fatehabad, district Agra, C.A. Silberrad, Min. Mag., 1932, **23**, p.299. Olivine Fa_{17}, B. Mason, Geochimica et Cosmochimica Acta, 1963, **27**, p.1011.
4g Calcutta, Mus. Geol. Surv. India; 19g Chicago, Field Mus. Nat. Hist.;
Specimen(s): [35852], 39g.

Kaee 27°15′ N., 79°58′ E.
Sandi pargana, Hardoi district, Uttar Pradesh, India
Fell 1838, January 29
Synonym(s): *Kaikhai, Oude*
Stone. Olivine-bronzite chondrite (H5).
After detonations, a stone of about 0.5lb fell in the village of Kaee, N.S. Maskelyne, Phil. Mag., 1864, **28**, p.149. Olivine Fa_{19}, B. Mason, Geochimica et Cosmochimica Acta, 1963, **27**, p.1011. The place of fall is probably Kaikhai, Sandi pargana, C.A. Silberrad, Min. Mag., 1932, **23**, p.299.
4g Vienna, Naturhist. Mus.;
Specimen(s): [35605], 204g. main mass

Kaffir (b) 34°40′18″ N., 101°48′54″ W.
Swisher County, Texas, U.S.A.
Found 1966, recognized in this year
Stone. Olivine-bronzite chondrite (H).
A total weight of 3.5kg was found, Meteor. Bull., 1979 (56), Meteoritics, 1979, **14**, p.167. Olivine Fa_{18}, A.L. Graham, priv. comm., 1978. Another stone, found in 1965, of 1874.9g is listed as Kaffir (a), G.I Huss, Cat. Huss Coll. Meteorites, 1976, p.24.
1726g Mainz, Max-Planck Inst, of Kaffir (a); 2108g Mainz, Max-Planck Inst, of Kaffir (b);
Specimen(s): [1975,M.3], 66.8g.

Kagarlyk 49°52′ N., 30°50′ E.
Kiev district, Ukraine, USSR, [Кагарлык]
Fell 1908, June 30, 0700 hrs
Stone. Olivine-hypersthene chondrite (L6).
After the appearance of a fireball, followed by detonations, a stone of 1.9kg was found 60km from Kiev, L.A. Kulik, Bull. Soc. Russ. des Amis de l'Etude de l'Univers, Leningrad, 1926, **15**, p.173. Listed, P.N. Chirvinsky, Centralblatt Min., 1923, p.551, E.L. Krinov, Каталог Метеоритов Акад. Наук СССР, Москва, 1947 [M.A. 10-511]. Olivine Fa_{23}, B. Mason, Geochimica et Cosmochimica Acta, 1963, **27**, p.1011.
1886g Moscow, Acad. Sci., main mass;

Kahangarai v Kakangari.

Kahrapar v Khanpur.

Kaidun 15°0′ N., 48°18′ E.
Near Kaidun, [Khuraybah], South Yemen
Fell 1980, December 3, 0445 hrs, U.T.
Synonym(s): *Kaydun*
Stone. Carbonaceous chondrite, type II (CM2).
After a fireball had been seen travelling NW. to SE., a single mass of 841.6g was recovered from a small impact pit, Meteor. Bull., 1982 (60), Meteoritics, 1982, **17**, p.95. The fall date (Jan. 7, 1981) is incorrectly reported, SEAN Bull., 1981, **6** (4). Xe isotopes, fall date December 30, Yu.A. Shukolyukov et al., Геохимиа, 1983, p.54. CI classification, G.W. Kallemeyn and J.F. Kerridge, Meteoritics, 1983, **18**, p.abs.
Specimen, Moscow, Acad. Sci.;

Kaikhai v Kaee.

Kaïnsas v Kainsaz.

Kainsaz 55°26′ N., 53°15′ E.
Muslyumov, Tatar Republic, USSR, [Каинсаз]
Fell 1937, September 13, 1415 hrs
Synonym(s): *Kaïnsas*
Stone. Carbonaceous chondrite, type III (CO3).
15 pieces were seen to fall near Kainsaz, the largest 102.5kg, and totalling over 200kg. Briefly described, L.S. Selivanov, Nature, 1938, **142**, p.623, C. R. (Doklady) Acad. Sci. URSS, 1938, **20**, p.263 [M.A. 7-270]. The date of fall is given as September 14, I.S. Astapowitsch, Trans. Roy. Astron. Soc. Canada, 1938, **32**, p.195 [M.A. 7-172]. The place of fall incorrectly mapped, P.M. Millman, Trans. Roy. Astron. Soc. Canada, 1938, **32**, p.199. Analysis, M.I. D'yakanova, Метеоритика, 1964, **25**, p.129. Classification, W.R. Van Schmus and J.M. Hayes, Geochimica et Cosmochimica Acta, 1974, **38**, p.47. Another analysis, 25.56 % total iron, L.H. Ahrens et al., Meteoritics, 1973, **8**, p.133. Trace element data, G.W. Kallemeyn and J.T. Wasson, Geochimica et Cosmochimica Acta, 1981, **45**, p.1217.
101.9kg Moscow, Acad. Sci., main mass; 79kg Kazan Univ.; 126g Chicago, Field Mus. Nat. Hist.; 26g Washington, U.S. Nat. Mus.; 97g Perth, West. Austr. Mus.;
Specimen(s): [1972,232], 34.1g.; [1965,395], 0.9g.

Kakangari 12°23′ N., 78°31′ E.
Tiruppattur taluq, Dharmapuri district, Tamil Nadu, India
Fell 1890, June 4, 0800 hrs
Synonym(s): *Kahangarai, Kangankarai, Salem*
Stone. Chondrite, anomalous (CHANOM).
After detonations, two stones were seen to fall, one was broken up, the other weighed about 0.75lb, Nature, 1892, **45**, p.20. The place of fall is Kangankarai, 12°23'N., 78°31'E., C.A. Silberrad, Min. Mag., 1932, **23**, p.296. An unusual chondrite, rich in troilite and with numerous armoured chondrules. Chemical analysis, 22.47 % total iron, B. Mason and H.B. Wiik, Am. Mus. Novit., 1966 (2272). Chemically distinct, further analysis, 22.79 % total iron, olivine $Fa_{3-9.5}$, A.L. Graham et al., Min. Mag., 1977, **41**, p.201. Trace element data, A.M. Davis et al., Nature, 1977, **265**, p.230. Oxygen isotopic ratios, R.N. Clayton et al., Earth planet. Sci. Lett., 1976, **30**, p.10. Noble gas abundances, B. Srinivasan and E. Anders, Meteoritics, 1977, **12**, p.417.
163g Calcutta, Mus. Geol. Surv. India; 7g Chicago, Field Mus. Nat. Hist.; 7g New York, Amer. Mus. Nat. Hist.;
Specimen(s): [69062], 100g.

Kakova v Kakowa.

Kakowa 45°8′ N., 21°40′ E.
Oravita, Romania
Fell 1858, May 19, 0800 hrs
Synonym(s): *Cacova, Kakova*
Stone. Olivine-hypersthene chondrite (L6), veined.
After detonations, a stone of 577g was seen to fall, W. von Haidinger, Sitzungsber. Akad. Wiss. Wien, Math.-naturwiss. Kl., 1859, **34**, p.11. Amount and composition of Ni-Fe, G.T. Prior, Min. Mag., 1919, **18**, p.353. A chromite-feldspar intergrowth is figured, P. Ramdohr, Geochimica et Cosmochimica Acta, 1967, **31**, p.1961. Olivine Fa_{23}, B. Mason, Geochimica et Cosmochimica Acta, 1963, **27**, p.1011.
327g Vienna, Naturhist. Mus.; 9g Berlin, Humboldt Univ.; 4.9g Tempe, Arizona State Univ.;
Specimen(s): [35166], 150g.

Kakrapar v Khanpur.

Kalaba 6°50′ S., 29°30′ E. approx.
Baudouinville (Moba), Katanga, Zaire
Fell 1951, October 31, 1730 hrs
Synonym(s): *Kabala, Kapofi, Kasongo*
Stone. Olivine-bronzite chondrite (H4).
Two pieces were found, one of which was broken up; the other weighed 950g. A third piece is said to have fallen about 15km from Kalaba. Described, with an analysis, Beugnies and Rorive, Ann. Serv. Mines Katanga for 1952-53, 1954, **17**, p.79, Bull. Serv. Geol. Congo Belge, 1954 (5), p.50 [M.A. 12-611].
Main mass, Moba, Mus. Grand Séminaire; 11g Washington, U.S. Nat. Mus.;
Specimen(s): [1960,55], thin section

Kalak v Kulak.

Kalamazoo v Grand Rapids.

Kalambha v Kalumbi.

Kalambi v Kalumbi.

Kaldoonera Hill 32°37′ S., 134°51′ E.
South Australia, Australia
Found 1956, before this year
Stone. Olivine-bronzite chondrite (H6).
Two masses of total weight 7.49kg were found, D.W.P. Corbett, Rec. S. Austr. Mus., 1968, **15**, p.767. Co-ordinates of find site, classification, olivine $Fa_{18.7}$, B. Mason, Rec. Austr. Mus., 1974, **29**, p.169. Over 12kg now recovered, analysis, 24.78 % total iron, M.J. Fitzgerald, Ph.D. Thesis, Univ. of Adelaide, 1979, p.23.
Main mass, Adelaide, South Austr. Mus.; 478g New York, Amer. Mus. Nat. Hist.; 38g Mainz, Max-Planck-Inst.; 1.4g Washington, U.S. Nat. Mus.;

Kalkaska 44°38′49″ N., 85°8′12″ W.
Kalkaska County, Michigan, U.S.A.
Found 1947, or 1948, in the summer
Iron. Octahedrite, medium (1.0mm) (IIIA).
A mass of 20.72lb was found, V.D. Chamberlain, Meteoritics, 1965, **2**, p.361, Meteorites of Michigan, Michigan Geol. Surv. Bull., 1968 (5), p.28. Analysis, 7.39 % Ni, 18.1 ppm.Ga, 33.5 ppm.Ge, 11 ppm.Ir, E.R.D. Scott et al., Geochimica et Cosmochimica Acta, 1973, **37**, p.1957. Description; shock-hardened, V.F. Buchwald, Iron Meteorites, Univ. of California, 1975, p.707.
Main mass, East Lansing, Michigan State Univ.; 759g Washington, U.S. Nat. Mus.; 483g Tempe, Arizona State Univ.;

Kalumbi 17°50′ N., 73°59′ E.
Satara district, Maharashtra, India
Fell 1879, November 4
Synonym(s): *Kalambha, Kalambi*
Stone. Olivine-hypersthene chondrite (L6), veined.
A stone of about 10lb fell at Kalambi, J. Bombay Branch Roy. Asiatic Soc., 1880, **14**, p.lv. The place of fall is probably Kalambha, 17°50'N., 73°59'E., C.A. Silberrad, Min. Mag., 1932, **23**, p.299. Olivine Fa_{24}, B. Mason, Geochimica et Cosmochimica Acta, 1963, **27**, p.1011.
Main mass, Bombay, Mus. Roy. Asiatic Soc., probable location; 380g Paris, Mus. d'Hist. Nat.; 149g Vienna, Naturhist. Mus.;
Specimen(s): [64509], 28g.; [64510], 1g.

Kalvesta 38°5′ N., 100°15′ W.

Finney County, Kansas, U.S.A.
Found 1968, recognized in this year
Stone. Olivine-bronzite chondrite (H).

One stone of about 22lb was removed from a concrete slab porch in which it had been embedded, Meteor. Bull., 1969 (47), Meteoritics, 1970, **5**, p.105.

127g Fort Hays, Kansas State College;

Kamalpur 26°2′ N., 81°28′ E.

Rae Bareli district, Uttar Pradesh, India
Fell 1942, August 18
Stone. Olivine-hypersthene chondrite (L6).

One mass of 2770g fell, P. Chatterjee, letter of 25 March, 1952 in Min. Dept. BM(NH). Listed, M.A.R. Khan, Hyderabad Acad. Studies, 1950, **12** [M.A. 19-441]. Olivine Fa_{25}, B. Mason, Geochimica et Cosmochimica Acta, 1967, **31**, p.1100.

2.7kg Calcutta, Mus. Geol. Surv. India; 0.9g New York, Amer. Mus. Nat. Hist.;

Specimen(s): [1983,M.9], 30g.

Kameelhaar v Gibeon.

Kamiariti v Gifu.

Kamiomi 36°2.5′ N., 139°57.4′ E.

Sashima-gun, Ibaraki, Honshu, Japan
1913, March, 1500 hrs, late March or early April, between 1913 and 1916
Stone. Olivine-bronzite chondrite (H4).

After detonations a single stone of 448g was seen to fall into a rice field in the village of Kamiomi about 40km NE. of Tokyo. Kamiomi has recently been incorporated into the city of Iwai, Meteor. Bull., 1975 (53), Meteoritics, 1975, **10**, p.134. Analysis, olivine Fa_{19}, 27.33 % total iron, A. Okada et al., Meteoritics, 1979, **14**, p.177.

Main mass, in the possession of the finder, Mr. Y. Shimamura; Specimen, Tokyo, Nat. Sci. Mus.;

Kamkas v Gibeon.

Kamsagar 14°11′ N., 75°48′ E.

Shimoga district, Karnataka, India
Fell 1902, November 12, 1300 hrs
Stone. Olivine-hypersthene chondrite (L6).

A stone of 1293g fell, J.C. Brown, Rec. Geol. Surv. India, 1915, **45**, p.223. Olivine Fa_{24}, B. Mason, Geochimica et Cosmochimica Acta, 1963, **27**, p.1011.

1kg Calcutta, Mus. Geol. Surv. India; 9.6g Washington, U.S. Nat. Mus.;

Specimen(s): [1983,M.10], 19.6g.

Kamyk nad Vltavou v Pribram.

Kamyshla 54°0′ N., 52°12′ E.

Kuybyshev region, USSR
Found 1981, June
Stone. Chondrite.

A single mass of 1540g was found partly buried in the soil of an unploughed field, Meteor. Bull., 1983 (61), Meteoritics, 1983, **18**, p.80.

Kanawha County v Jenny's Creek.

Kandahar (Afghanistan) 31°36′ N., 65°47′ E.

Afghanistan
Fell 1959, November
Stone. Olivine-hypersthene chondrite (L6).

No details are available, Meteor. Bull., 1960 (19). Olivine Fa_{24}, B. Mason, Geochimica et Cosmochimica Acta, 1963, **27**, p.1011.

123g Washington, U.S. Nat. Mus.;

Specimen(s): [1963,942], 12.5g.

Kandahar (India)

India, Not located
Fell 1833, November, end of the month
Doubtful. Stone?.

A shower of stones are said to have fallen at an unidentified village of the name of Kandahar (not Kandahar, Afghanstan). The account is imperfect, and there seems to be some possibility that the fall may be identical with Kadonah, although the dates do not tally. Listed, A.L. Coulson, Mem. Geol. Surv. India, 1940, **75**, p.116 [M.A. 8-54], Malte Brun, Nouv. Ann. Voyages, geogr. et hist., 3, **2**, p.415, J.C. Brown, Mem. Geol. Surv. India, 1916, **43**, p.213.

Kangankarai v Kakangari.

Kangean 7° S., 115°30′ E. approx.

Kangean Island, Indonesia
Fell 1908, September 27, 1100 hrs, approx. time
Stone. Olivine-bronzite chondrite (H).

A stone fell into a water-covered rice field in the area of Kalisangka, near the town of Se Djabang, on Kangean Island. The fall was accompanied by sound effects. The stone was broken into pieces by local people; five pieces of total weight 1.63kg are preserved. Illustrated description, olivine Fa_{19}, E. Niggli, Geol. en Mijnbouwn, 1966, **45**, p.1. Reported, Meteor. Bull., 1976 (54), Meteoritics, 1976, **11**, p.70.

Main mass, Delft, Mus. Technical Inst.;

Kangra Valley 32°5′ N., 76°18′ E.

Himachal Pradesh, India
Fell 1897, about
Stone. Olivine-bronzite chondrite (H5).

A stone of about 400g, labelled "Seen to fall in the Kangra valley, north Punjab" was sent by Col. St. John Grant to W.N. Hartley, on July 21, 1897, W.N. Hartley, J. Chem. Soc. London, Trans., 1906, **89** (2), p.1566. Olivine Fa_{19}, B. Mason, Geochimica et Cosmochimica Acta, 1963, **27**, p.1011.

Specimen(s): [1907,22], 322g. the nearly complete stone

Kansada v Ness County (1894).

Kansas v Tonganoxie.

Kansas City (1876) 39°6′ N., 94°38′ W.

Missouri, U.S.A.
Fell 1876, June 25, 0900-1000 hrs
Doubtful. Stone?.

A small fragment, 1.75 inches in diameter by 0.33 inches thick, is said to have fallen on and partially pierced a tin roof; crusted on one side, looked like troilite on the other, J.D. Parker, Am. J. Sci., 1876, **12**, p.316. This fall is not mentioned by Wülfing (1897) nor in Farrington's Catalogue of North American Meteorites (1909), but neither confirmation nor disproof of its meteoritical nature has been traced, K. Servos, letter of 28 January, 1957 in Min. Dept. BM(NH).

Kansas City (1903) 39°6′ N., 94°38′ W.
Missouri, U.S.A.
Found 1903
Stone. Olivine-bronzite chondrite (H5).
A much oxidized stone of 36kg was found in Kansas City, some 6 feet below the surface, G.P. Merrill, Proc. U.S. Nat. Mus., 1919, **55**, p.95. Olivine Fa_{19}, B. Mason, Geochimica et Cosmochimica Acta, 1963, **27**, p.1011.
Main mass, Kansas City, Dyar Mus.; 283g Washington, U.S. Nat. Mus.; 21g New York, Amer. Mus. Nat. Hist.;
Specimen(s): [1949,182], 18g.

Kansu v Nan Yang Pao.

Kantarah v Sinai.

Kanwahoe County v Jenny's Creek.

Kanzaki 33°18′ N., 130°22′ E.
Saga, Kyushu, Japan
Found
Stone. Olivine-bronzite chondrite (H).
A small stone of 123.75g, of unknown history, was in the Museum of the Geol. Surv. Japan but is now lost. It resembled Kyushu and is possibly identical, K. Jimbo, Beitr. Miner. Japan, 1906 (2), p.32, 51, S. Murayama, letter of 6 April, 1962 in Min. Dept. BM(NH)., who considers it to be distinct, S. Murayama, Nat. Sci. Mus. Tokyo, 1953, **20**, p.129 [M.A. 13-80] where the weight is given as 483g.

Kapeisen v Cape of Good Hope.

Kapland v Hex River Mountains.

Kapoeta 4°42′ N., 33°38′ E.
Equatoria, Sudan
Fell 1942, April 22, 1900 hrs
Stone. Achondrite, Ca-rich. Howardite (AHOW).
A stone of 11.355kg fell on the Kapoeta-Nathalani road, G. Andrew, letters of 5 October, 1942 and 30 December, 1945 in Min. Dept. BM(NH)., K. Fredriksson and K. Keil, Geochimica et Cosmochimica Acta, 1963, **27**, p.717. Described, with an analysis, 13.96 % total iron, B. Mason and H.B. Wiik, Am. Mus. Novit., 1967 (2273). Contains crystals evenly irradiated by solar flare particles, P. Pellas et al., Nature, 1969, **223**, p.272. Contains carbonaceous chondrite fragments, L. Wilkening, Geochimica et Cosmochimica Acta, 1973, **37**, p.1985. Distribution of He, Ne, and Ar, G.H. Megrue, Meteorite Research, ed. P.M. Millman, D. Reidel, Dordrecht-Holland, 1969, p.922. Clast petrography, references, extended Ar-Ar ages of clasts, R.S. Rajan et al., Geochimica et Cosmochimica Acta, 1979, **43**, p.957.
Main mass, Khartoum, Sudan Geol. Surv.; 51g New York, Amer. Mus. Nat. Hist.; 49g Chicago, Field Mus. Nat. Hist.; 0.9g Washington, U.S. Nat. Mus.;
Specimen(s): [1946,140], 781g. and fragments, 7g; [1946,141], 80.3g. and fragments, 5.1g

Kapofi v Kalaba.

Kappakoola 33°15′ S., 135°32′ E.
Le Hunte County, South Australia, Australia
Found 1929, September
Stone. Olivine-bronzite chondrite (H6), brecciated.
A mass of 392.5g was found on sect. 11, hundred of Kappakoola, Eyre Peninsula. Description, L.J. Spencer, Min. Mag., 1936, **24**, p.357. Co-ordinates, M.J. Fitzgerald, Ph.D. Thesis, Univ. of Adelaide, 1979, p.214. The main mass was in Adelaide Univ., South Australia, but may be lost. Olivine $Fa_{19.6}$, B. Mason, Rec. Austr. Mus., 1974, **29**, p.169.
9g Adelaide, South Austr. Mus.; 0.7g Washington, U.S. Nat. Mus.;
Specimen(s): [1936,1027], 45g.

Kaptal-Aryk 42°27′ N., 73°22′ E.
Kalinin, Kirghizian Republic, USSR, [Каптал-Арык]
Fell 1937, May 12, 2045 hrs
Stone. Olivine-hypersthene chondrite (L6), veined.
A mass of about 3.5kg was seen to fall; short description, L.P. Maliuga, Nature, 1938, **142**, p.623, C. R. (Doklady) Acad. Sci. URSS, 1938, **20**, p.265 [M.A. 7-270]. Corrected longitude and classification, E.L. Krinov, Астрон. Журнал, 1945, **22**, p.303 [M.A. 9-297], Каталог Метеоритов Акад. Наук СССР, Москва, 1947 [M.A. 10-511]. Incorrect weight reported, A.D. Nininger, Pop. Astron., Northfield, Minnesota, 1940, **48**, p.559, Contr. Soc. Res. Meteorites, 1942, **2**, p.227 [M.A. 8-54]. Analysis, 20.47 % total iron, M.I. D'yakonova and V.Ya. Kharitonova, Метеоритика, 1961, **21**, p.52. Olivine Fa_{23}, B. Mason, Geochimica et Cosmochimica Acta, 1963, **27**, p.1011.
2854g Moscow, Acad. Sci., main mass and fragments, 63g;
Specimen(s): [1971,198], 11.6g.

Kapunda 34°19′ S., 138°56′ E. approx.
South Australia, Australia
Found 1965, approx., recognized 1972
Iron. Octahedrite, medium (0.9mm) (IIIA).
A mass of 542g was found by two girls and originally thought to be slag. After its recognition as a meteorite, the finders had forgotten the exact location of the find site. Classification, analysis, 8.05 %Ni, M.J. Fitzgerald, Ph.D. Thesis, Univ. of Adelaide, 1979, p.112, 215.
518g Adelaide, Univ.;

Karagai 51°7′ N., 57°55′ E.
Orenburg, Federated SSR, USSR, [Карагай]
Found 1900, between 1900 and 1937
Synonym(s): *Karagaj*
Stone. Olivine-hypersthene chondrite (L6).
One stone of 115g fell, L.A. Kulik, Метеоритика, 1941, **1**, p.73 [M.A. 9-294], E.L. Krinov, Астрон. Журнал, 1945, **22**, p.303 [M.A. 9-297], Каталог Метеоритов Акад. Наук СССР, Москва, 1947 [M.A. 10-511]., V.I. Vernadsky, letter of 5 August, 1939 in Min. Dept. BM(NH)., P.M. Millman, Trans. Roy. Astron. Soc. Canada, 1938, **32**, p.198, I.S. Astapowitsch, Trans. Roy. Astron. Soc. Canada, 1938, **32**, p.195 [M.A. 7-172]. Olivine Fa_{23}, B. Mason, Geochimica et Cosmochimica Acta, 1963, **27**, p.1011.
111g Moscow, Acad. Sci.;

Karagaj v Karagai.

Karakol 47°13′ N., 81°1′ E.
Ayagus, Semipalatinsk, Federated SSR, USSR, [Каракал]
Fell 1840, May 9, 1200 hrs
Synonym(s): *Kirghiz Steppes, Kirgisen Steppe, Kirgizskaia Step*
Stone. Olivine-hypersthene chondrite, amphoterite (LL6).
A stone of about 3kg fell, after detonations, by the river Karakol, E. von Eichwald, Arch. Wiss. Kunde Russl., 1847,

5, p.180, A. Göbel, Bull. Acad. Sci. St.-Petersbourg, 1867, **11**, p.264. It is reported that two stones fell, one of which was broken up by the natives, Y.I. Simashko, Cat. Meteorites, St.-Petersbourg, 1891, p.34. Olivine Fa_{28}, B. Mason, Geochimica et Cosmochimica Acta, 1963, **27**, p.1011.

2734g Moscow, Acad. Sci.; 30g Chicago, Field Mus. Nat. Hist.;

Specimen(s): [84189], 22g.; [63926], 2g.

Karand v Veramin.

Karang Modjo v Ngawi.

Karasburg 27°40′ S., 18°58′ E.

Warmbad district, Namibia
Found 1964, before this year
Iron. Octahedrite, medium (1.2mm) (IIIAB).

A small mass was found on Duurdrift Nord farm, in the Warmbad district, 45km NNE. of Karasburg. It had been reheated by the finder to 800-900C. Analysis, 8.68 %Ni, V.F. Buchwald, Iron Meteorites, Univ. of California, 1975, p.709.

11.5kg Johannesburg, Public Library, main mass; 4.5g Washington, U.S. Nat. Mus.;

Karatash v Caratash.

Karatu 3°30′ S., 35°35′ E.

Arusha, Tanzania
Fell 1963, September 11
Stone. Olivine-hypersthene chondrite, amphoterite (LL6).

A stone of 2.22kg fell at Tlae Daat, co-ordinates as above, between Bassoduwish and Karatu, R.W. Bartholomew, letter of 26 March, 1964 in Min. Dept. BM(NH). Analysis, 18.56 % total iron, E. Jarosewich, Geochimica et Cosmochimica Acta, 1966, **30**, p.1261. Olivine Fa_{30}, B. Mason, Geochimica et Cosmochimica Acta, 1967, **31**, p.1100.

Main mass, Dodoma, Geol. Surv. Tanzania; 57g Washington, U.S. Nat. Mus.;

Specimen(s): [1964,577], 153g.; [1964,578], thin sections (two)

Karavinsky Iron v Rancho de la Pila (1882).

Karee Kloof 31°36′ S., 25°48′ E.

Hofmeyer, Cape Province, South Africa
Found 1914, about
Synonym(s): *Hofmeyer*
Iron. Octahedrite, coarse (1.6mm) (IA-ANOM).

A mass of about 203lb was found, Director's Report for 1914, Port Elizabeth Museum, 1914, p.4. Described and analysed, 8.27 %Ni, G.T. Prior, Min. Mag., 1923, **20**, p.134. New analysis, 8.1 %Ni, 79.5 ppm.Ga, 355 ppm.Ge, 1.5 ppm.Ir, J.T. Wasson, Icarus, 1970, **12**, p.407. Anomalous bandwidth; troilite-graphite and cohenite minerals not seen, V.F. Buchwald, Iron Meteorites, Univ. of California, 1975, p.710.

92kg Port Elizabeth Mus., main mass; 192g Vienna, Naturhist. Mus.; 27g Washington, U.S. Nat. Mus.;

Specimen(s): [1922,350], 405g.; [1920,431], 25g. filings; [1922, 351], 14g.

Karewar 12°54′ N., 7°9′ E.

Ruma, Katsina province, Nigeria
Fell 1949, September 19, 1900 hrs, approx. time
Stone. Olivine-hypersthene chondrite (L6).

One stone of about 180g was seen to fall, and penetrated the ground to about 3 feet, R. Walls, letters of 10 January, 1950, and 20 April, 1951 in Min. Dept. BM(NH)., R. Walls, Rec. Geol. Surv. Nigeria, 1958, p.23. Olivine Fa_{24}, B. Mason, Geochimica et Cosmochimica Acta, 1963, **27**, p.1011.

Specimen(s): [1951,16], 140g.

Kargapole 55°53′ N., 64°18′ E.

Kurgan region, Federated SSR, USSR, [Каргаполье]
Found 1961, July
Stone. Olivine-bronzite chondrite (H4).

One mass of 21.8kg was found; description and analysis, 24.22 % total iron, L.N. Ovchinnikov and I.A. Yudin, Метеоритика, 1966, **27**, p.76. New analysis, 25.21 % total iron, V.Ya. Kharitonova, Метеоритика, 1969, **29**, p.91.

Main mass, Sverdlovsk, Urals Geol. Mus.; 10.38kg Moscow, Acad. Sci.;

Karkh 27°48′ N., 67°10′ E.

Jhalawan, Baluchistan, Pakistan
Fell 1905, April 27, 1300 hrs
Stone. Olivine-hypersthene chondrite (L6).

After the appearance of a meteor followed by detonations, two stones at least fell, one below the Sumbaji Hills and another in the Michara Hills; six pieces were recovered, of total weight about 22kg. The largest, from the Sumbaji Hills, weighed 14.5kg, L.L. Fermor, Rec. Geol. Surv. India, 1907, **35**, p.85. Mineralogy, analysis, 22.39 % total iron, olivine $Fa_{23.3}$, A. Dube et al., Smithson. Contrib. Earth Sci., 1977 (19), p.71.

21kg Calcutta, Mus. Geol. Surv. India; 121g Paris, Mus. d'Hist. Nat.; 49g Washington, U.S. Nat. Mus.; 30g Chicago, Field Mus. Nat. Hist.;

Specimen(s): [1915,85], 226.5g. 18 fragments

Karloowala 31°35′ N., 71°36′ E.

Mianwali district, Pakistan
Fell 1955, July 21, 1200 hrs
Synonym(s): *Karluwala*
Stone. Olivine-hypersthene chondrite (L6).

A loud noise was heard, and a cloud of dust seen at the spot where a mass of 6.5lb was later dug up, letters of Geol. Surv. Pakistan of 28 July, 29 September, 1956 in Min. Dept. BM(NH). Olivine Fa_{23}, B. Mason, Geochimica et Cosmochimica Acta, 1963, **27**, p.1011.

Main mass, Quetta, Geol. Surv. Pakistan;

Specimen(s): [1956,241], 70g.

Karlsburg v Ohaba.

Karluwala v Karloowala.

Karondha v Kadonah.

Karoona v Karoonda.

Karoonda 35°5′ S., 139°55′ E.
Buccleuch County, South Australia, Australia
Fell 1930, November 25, 2253 hrs
Synonym(s): *Karoona*
Stone. Carbonaceous chondrite (C5).
Numerous fragments, the largest 7lb, total 92lb, were found about two weeks after the fall, K. Grant and G.F. Dodwell, Nature, 1931, **127**, p.402 [M.A. 5-15]. Description, with doubtful analysis, D. Mawson, Trans. Roy. Soc. South Austr., 1934, **58**, p.1 [M.A. 6-15]. New analysis, and description, 25.55 % total iron, B. Mason and H.B. Wiik, Am. Mus. Novit., 1962 (2115) [M.A. 16-639]. Co-ordinates, M.J. Fitzgerald, Ph.D. Thesis, Univ. of Adelaide, 1979, p.216. Silicates homogeneous, olivine $Fa_{33.4}$, W.R. Van Schmus, Meteorite Research, ed. P.M. Millman, D. Reidel, Dordrecht-Holland, 1969, p.480. Classification, W.R. Van Schmus and J.M. Hayes, Geochimica et Cosmochimica Acta, 1974, **38**, p.47. XRF analysis, 23.86 % total iron, M.J. Fitzgerald, Meteoritics, 1979, **14**, p.109. Noble gases in minerals, B. Srinivasan et al., J. Geophys. Res., 1977, **82**, p.762.
6.4kg Adelaide, South Austr. Mus., includes main mass, 5.5kg; 1.3kg New York, Amer. Mus. Nat. Hist.; 151g Sydney, Austr. Mus.; 127g Mainz, Max-Planck-Inst.; 123g Washington, U.S. Nat. Mus.; 41g Tempe, Arizona State Univ.; 28g Chicago, Field Mus. Nat. Hist.;
Specimen(s): [1931,311], 106g. and fragments, 23g; [1931, 489], 125.5g. four pieces; [1973,M.25], 44g.

Karthago v Carthage.

Karval 38°43′ N., 103°31′ W.
Lincoln County, Colorado, U.S.A.
Found 1936
Stone. Olivine-bronzite chondrite (H).
One mass of 1104g was found, A.D. Nininger, Pop. Astron., Northfield, Minnesota, 1937, **45**, p.449 [M.A. 7-62], H.H. Nininger, Amer. Antiquity, 1938, **4**, p.39 [M.A. 7-272], H.H. and A.D. Nininger, The Nininger Collection of Meteorites, Winslow, Arizona, 1950, p.64. Olivine Fa_{19}, B. Mason, Geochimica et Cosmochimica Acta, 1963, **27**, p.1011.
418g Tempe, Arizona State Univ.; 194g Chicago, Field Mus. Nat. Hist.; 19.7g Denver, Mus. Nat. Hist.;
Specimen(s): [1959,878], 378.5g.

Kasamatsu 35°22′ N., 136°46′ E.
Hashima, Gifu, Honshu, Japan
Fell 1938, March 31
Synonym(s): *Kasamatu*
Stone. Olivine-bronzite chondrite (H).
One stone of 710g fell through the roof of a house, A.D. Nininger, Pop. Astron., Northfield, Minnesota, 1940, **48**, p.558, Contr. Soc. Res. Meteorites, **2**, p.227 [M.A. 8-54], S. Murayama, Nat. Sci. Mus. Tokyo, 1953, **20**, p.129 [M.A. 13-80], Meteoritics, 1955, **1**, p.300. Description, analysis, 29.13 % total iron, A. Miyashiro et al., Jap. J. Geol. Geogr., 1963, **34**, p.191. Olivine Fa_{18}, B. Mason, Geochimica et Cosmochimica Acta, 1967, **31**, p.1100.
Main mass, Hashima, in the house where it fell; Specimen, Tokyo, Nat. Sci. Mus.;

Kasamatu v Kasamatsu.

Kaschin v Glasatovo.

Kashin v Glasatovo.

Kasiali v Kusiali.

Kasongo v Kalaba.

Kathiawar v Bherai.

Kaufman 32°35′ N., 96°25′ W.
Kaufman County, Texas, U.S.A.
Found 1893
Stone. Olivine-hypersthene chondrite (L5), black.
One mass of 23kg was found 4 or 5 miles W. of Kaufman, Meteor. Bull., 1957 (5). Olivine Fa_{23}, B. Mason, Geochimica et Cosmochimica Acta, 1963, **27**, p.1011.
19.9kg Fort Worth, Texas, Monnig Colln.; 168g Washington, U.S. Nat. Mus.;

Kavkaz v Berdyansk.

Kawagaon v Naoki.

Kawkas v Berdyansk.

Kayakent 39°15′48″ N., 31°46′48″ E.
Turkey
Fell 1961, April
Iron. Octahedrite, medium (1.2mm) (IIIA).
A mass of 85kg was brought to Ege University in 1967, Meteor. Bull., 1967 (40), Meteoritics, 1970, **5**, p.92. One evening in April, 1961, villagers heard "a burst in the air." The following August the mass was found in a hole 30 cm deep, and kept in the village until 1967. It is well preserved. Analysis, 8.09 %Ni, A. Kizilirmak et al., Sci. Rep. Faculty Sci. Ege Univ. no. 68, 1969 (Astronomy No. 7). Another analysis, 8.32 %Ni, 19.9 ppm.Ga, 44.0 ppm.Ge, 1.1 ppm.Ir, E.R.D. Scott et al., Geochimica et Cosmochimica Acta, 1973, **37**, p.1957. Description; shock-hardened; may have fallen before 1961, V.F. Buchwald, Iron Meteorites, Univ. of California, 1975, p.711.
Main mass, Izmir, Ege Univ.; 150g Tempe, Arizona State Univ.; 135g Collection of K. Hindley; 112g Copenhagen, Univ. Geol. Mus.;
Specimen(s): [1969,9], 100g. and fragment, 4g.

Kaydun v Kaidun.

Kearney 40°41′ N., 99°4′ W.
Kearney County, Nebraska, U.S.A.
Found 1934, recognized in this year
Stone. Olivine-bronzite chondrite (H5).
Probably fell in June, 1932. One stone of 1101g, A.D. Nininger, Pop. Astron., Northfield, Minnesota, 1937, **45**, p.449 [M.A. 7-62], H.H. and A.D. Nininger, The Nininger Collection of Meteorites, Winslow, Arizona, 1950, p.65, 112. Two masses, 8.4kg and 0.5kg, since found, Rep. U.S. Nat. Mus., 1950, p.55. Olivine Fa_{20}, B. Mason, Geochimica et Cosmochimica Acta, 1963, **27**, p.1011.
1008g Washington, U.S. Nat. Mus., main mass; 37g Tempe, Arizona State Univ.;
Specimen(s): [1959,879], 40g. slice

Kediri 7°45′ S., 112°1′ E.
Java, Indonesia
Fell 1940, about
Stone. Olivine-hypersthene chondrite (L4).
About 1940 a shower of stones travelling north to south was seen to fall on the Soemboer Wadoek rubber plantation, 10km south of the town of Kediri. 70 fragments of unknown total weight were recovered. The largest stone, 3.304kg, was taken to Amsterdam where it was cut and distributed. The

co-ordinates given are those of Kediri, Meteor. Bull., 1975 (53), Meteoritics, 1975, **10**, p.135., Th. van Dijk, letter of 21 March, 1974 in Min. Dept. BM(NH).
2887g Tempe, Arizona State Univ., main mass; 175g Teyler Mus., Holland; 69g Chicago, Field Mus. Nat. Hist.;
Specimen(s): [1974,M.14], 65g. crusted part-slice

Keen Mountain 37°13′ N., 82°0′ W.
Buchanan County, Virginia, U.S.A.
Found 1950
Iron. Hexahedrite (IIA).
A mass of 14.75lb was found 30 ft from the crest of the S. face of Keen Mountain. Described, with an analysis, 5.65 % Ni, E.P. Henderson and S.H. Perry, Proc. U.S. Nat. Mus., 1958, **107**, p.393 [M.A. 14-130]. 3He and 4He content, J.H. Hoffmann and A.O. Nier, J. Geophys. Res., 1960, **65**, p.1063 [M.A. 15-536]. Description; shock-hardened and plastically deformed, V.F. Buchwald, Iron Meteorites, Univ. of California, 1975, p.715. More recent analysis, 5.38 %Ni, 62.0 ppm.Ga, 183 ppm.Ge, 12 ppm.Ir, J.T. Wasson, Meteorites, Springer-Verlag, 1974, p.299.
5.6kg Washington, U.S. Nat. Mus., main mass; 208g Tempe, Arizona State Univ.;

Keetmanshoop v Gibeon.

Keilce
Poland
Pseudometeorite..
A fragment of 14.55g of supposed meteorite from Keilce, in the Museum of Luov, Ukrainian SSR., is a limonite concretion, J. Pokrzywnicki, Acta Geophys. Polon., 1971, **19**, p.235, A.A. Yasinskaya, Метеоритика, 1968, **28**, p.171.

Keithick v Strathmore.

Kelly 40°28′ N., 103°2′ W.
Logan County, Colorado, U.S.A.
Found 1937
Stone. Olivine-hypersthene chondrite, amphoterite (LL4), brecciated.
One mass was found weighing 44.3kg, A.D. Nininger, Pop. Astron., Northfield, Minnesota, 1937, **45**, p.449 [M.A. 7-62]. Xenolithic, olivine Fa_{28-30}, R.A. Binns, Geochimica et Cosmochimica Acta, 1968, **32**, p.299. Metabreccia, T.E. Bunch and D. Stöffler, Contrib. Miner. Petr., 1974, **44**, p.157.
4.3kg Denver, Mus. Nat. Hist.; 3.5kg Washington, U.S. Nat. Mus.; 2kg Chicago, Field Mus. Nat. Hist.; 1.2kg Mainz, Max-Planck-Inst.; 953g Los Angeles, Univ. of California; 850g Tempe, Arizona State Univ.; 441g New York, Amer. Mus. Nat. Hist.;
Specimen(s): [1938,1206], 36.5g. two pieces; [1959,777], 929.5g. slice

Kemis v Krasnojarsk.

Kemiz v Krasnojarsk.

Kendall County 29°24′ N., 98°30′ W.
Kendall County, Texas, U.S.A.
Found 1887, known in this year
Synonym(s): *San Antonio*
Iron. Hexahedrite (IRANOM), brecciated.
A mass of about 21kg was found in Kendall County and acquired by the Naturhist. Hofmus. in Vienna, A. Brezina, Ann. Naturhist. Hofmus. Wien, 1887, **2**, p.115. Described, with an analysis, 5.64 %Ni, E. Cohen, Ann. Naturhist. Hofmus. Wien, 1900, **15**, p.382. Texture and mineralogy of silicate inclusions, T.E. Bunch et al., Contr. Mineral. Petrol., 1970, **25**, p.297. The metal has a low Ni content and may have been reduced, 5.4 %Ni, 70.9 ppm.Ga, 355 ppm.Ge, 1.7 ppm.Ir, J.T. Wasson, Geochimica et Cosmochimica Acta, 1970, **34**, p.957. Description; polycrystalline aggregate of kamacite, silicate and graphite; some samples have been heated artificially, V.F. Buchwald, Iron Meteorites, Univ. of California, 1975, p.717.
11.4kg Vienna, Naturhist. Mus., includes main mass, 8.95kg; 2044g Washington, U.S. Nat. Mus.; 500g Michigan Univ.; 686g Chicago, Field Mus. Nat. Hist.; 500g Budapest, Nat. Mus.; 466g Harvard Univ.; 410g Ottawa, Mus. Geol. Surv.; 380g New York, Amer. Mus. Nat. Hist.; 344g Prague, Nat. Mus.; 123g Tempe, Arizona State Univ.; 51g Berlin, Humboldt Univ.;
Specimen(s): [67015], 556g. slice

Kendelton v Kendleton.

Kendleton 29°27′ N., 96°0′ W.
Fort Bend County, Texas, U.S.A.
Fell 1939, May 2, 1925 hrs
Synonym(s): *Kendelton*
Stone. Olivine-hypersthene chondrite (L4), brecciated.
A shower of stones fell at Kendleton, 44 miles W. by S. of Houston, Texas. Photographs of the trail and observations of the path of the meteor were obtained. Thirteen complete stones and about 15 fragments were recovered, totalling 6937g, the largest 1.6kg, O.E. Monnig, The Sky, New York, 1939, **3** (10), p.6 (11), p.22 [M.A. 7-378]. Description, with details of the phenomena of fall, J.J. King and F.F. Fouts, Univ. Texas Publ., 1940 (3945), p.657 [M.A. 8-194]. Olivine Fa_{25}, B. Mason, Geochimica et Cosmochimica Acta, 1963, **27**, p.1011.
9.6kg Fort Worth, Texas, Monnig Colln.; 1.6kg Houston, in the possession of Mr. J.J. King;

Kenna 33°54′ N., 103°33.2′ W.
Roosevelt County, New Mexico, U.S.A.
Found 1972, approx.
Stone. Achondrite, Ca-poor. Ureilite (AURE).
A single stone of 10.9kg was found. Petrology and mineralogy; olivine $Fa_{20.8-1}$, J.L. Berkley et al., Geochimica et Cosmochimica Acta, 1976, **40**, p.1429. Consortium study, K. Keil, Geochimica et Cosmochimica Acta, 1976, **40**, p.1427.
8kg Mainz, Max-Planck-Inst.; 1260g Washington, U.S. Nat. Mus.; 78g Chicago, Field Mus. Nat. Hist.; 47g Albuquerque, Univ. of New Mexico;
Specimen(s): [1976,M.2], 40.8g. crusted part-slice

Kennard 41°29′ N., 96°10′ W.
Washington County, Nebraska, U.S.A.
Found 1961, about this year
Stone. Olivine-bronzite chondrite (H5).
One stone of 8.2kg was found, Meteor. Bull., 1967 (39), Meteoritics, 1970, **5**, p.91. Olivine Fa_{19}, B. Mason, Geochimica et Cosmochimica Acta, 1967, **31**, p.1100.
156g Washington, U.S. Nat. Mus.;

Kensington v Smith Center.

Kenton County 38°49′ N., 84°36′ W.

Kentucky, U.S.A.
Found 1889
Synonym(s): *Independence, Williamstown*
Iron. Octahedrite, medium (0.9mm) (IIIA).

A mass of 163kg was found on a farm about 8 miles south of Independence, this report includes an analysis, 7.65 %Ni, H.L. Preston, Am. J. Sci., 1892, **44**, p.163. Williamstown, identical to Kenton County, was found in 1892 on a farm 3 miles north of Williamstown. Described, with an analysis, 7.26 %Ni, E.E. Howell, Am. J. Sci., 1908, **25**, p.49. The Williamstown mass weighed 68lb. Newer analysis, 7.38 %Ni, 18.2 ppm.Ga, 35.0 ppm.Ge, 14 ppm.Ir, E.R.D. Scott et al., Geochimica et Cosmochimica Acta, 1973, **37**, p.1957. Description, V.F. Buchwald, Iron Meteorites, Univ. of California, 1975, p.721.

64kg Chicago, Field Mus. Nat. Hist.; 25.5kg New York, Amer. Mus. Nat. Hist., includes 16.8kg of Williamstown; 5.7kg Yale Univ.; 4.2kg Harvard Univ.; 3.9kg Washington, U.S. Nat. Mus.; 3.3kg Vienna, Naturhist. Mus.; 2kg Budapest, Nat. Mus.; 900g Tempe, Arizona State Univ.; 525g Sarajevo, Bosnian Nat. Mus.; 569g Paris, Mus. d'Hist. Nat.; 306g Prague, Nat. Mus.;

Specimen(s): [71527], 2500g. slice, and fragments, 17g.; [1908,183], 838.5g. slice, of Williamstown

Keranouvé v Kernouve.

Keranroué v Kernouve.

Kerhartice v Usti nad Orlici.

Kerilis 48°24′ N., 3°18′ W.

Mäel Pestivien, Callac, Cotes-du-Nord, France
Fell 1874, November 26, 1030 hrs
Synonym(s): *Callac, Mäel Pestivien*
Stone. Olivine-bronzite chondrite (H5).

After detonations, a stone of about 5kg fell, G.A. Daubrée, C. R. Acad. Sci. Paris, 1880, **91**, p.28. Olivine Fa_{19}, B. Mason, Geochimica et Cosmochimica Acta, 1963, **27**, p.1011.

3.8kg Paris, Mus. d'Hist. Nat.; 79g Calcutta, Mus. Geol. Surv. India; 57g Rome, Vatican Colln; 42g Chicago, Field Mus. Nat. Hist.; 25g Vienna, Naturhist. Mus.;

Specimen(s): [71573], 69g.; [54634], 5.75g.

Kermiché v Kermichel.

Kermichel 47°39′ N., 2°46′ W.

Vannes, Morbihan, France
Found 1911, perhaps fell 1903, June 30
Synonym(s): *Kermiché, Limerzel*
Stone. Olivine-hypersthene chondrite (L6).

A stone of about 3kg was ploughed up in 1911 on the farm Kermichel, S. Meunier, C. R. Acad. Sci. Paris, 1912, **154**, p.1739. According to the Marquis de Mauroy, who obtained possession of the mass, a stone was seen to fall, after the usual light and sound phenomena, on 30 June, 1903, but was not found at the time, Marquis de Mauroy, letter of 7 May, 1912 in Min. Dept. BM(NH). Olivine Fa_{24}, B. Mason, Geochimica et Cosmochimica Acta, 1963, **27**, p.1011.

1.4kg Paris, Mus. d'Hist. Nat.; 652g Rome, Vatican Colln; 227g Vienna, Naturhist. Mus.; 70g Yale Univ.; 62g Chicago, Field Mus. Nat. Hist.; 53g New York, Amer. Mus. Nat. Hist.; 42g Washington, U.S. Nat. Mus.;

Specimen(s): [1912,513], 112g.

Kernouve 48°7′ N., 3°5′ W.

Morbihan, France
Fell 1869, May 22, 2200 hrs
Synonym(s): *Cléguérec, Clegueric, Keranouvé, Keranroué, Morbihan, Napoléonsville*
Stone. Olivine-bronzite chondrite (H6), veined.

A conical stone of about 80kg fell, after detonations and appearance of a fireball, de Limur, C. R. Acad. Sci. Paris, 1869, **68**, p.1338. Doubtful analysis, F. Pisani, C. R. Acad. Sci. Paris, 1869, **68**, p.1489., A. Lacroix, Bull. Mus. d'Hist. Nat. Paris, 1927, **33**, p.439. Mineral and bulk analyses, 28.78 % total iron, olivine $Fa_{18.8}$, R. Hutchison et al., Proc. Roy. Soc. London, 1981, **A374**, p.159.

15.5kg Paris, Mus. d'Hist. Nat.; 15kg Nantes, Mus. d'Hist. Nat.; 1.2kg Stockholm, Riksmus.; 1.1kg Washington, U.S. Nat. Mus.; 920g Vienna, Naturhist. Mus.; 520g Berlin, Humboldt Univ.; 363g Harvard Univ.; 124g Chicago, Field Mus. Nat. Hist.; 158g New York, Amer. Mus. Nat. Hist.;

Specimen(s): [44143], 7108g. three pieces, and fragments, 59g; [43400], 1211g.; [1920,297], 131.8g.

Kerulensky 48°15′ N., 115°10′ E. approx.

Choybalsan aymag, Mongolia, [Керуленский]
Synonym(s): *Herlen, Kherlen*
Doubtful..

A bolide was recorded in the Choybalsan district, at Uudiin, 48°15'N., 115°10'E., on the Kerulen river, on 21 March, 1950, at about 0100 hrs., and numerous pieces of a vesicular slaggy mass, totalling 27.78kg, were collected and sent to Ulan-Bator, G.G. Vorobyev and O. Namnandorzh, Метеоритика, 1958, **16**, p.134 [M.A. 14-129]. The description may be similar to that of the problematical Igast object (*q.v.*).

Kesen 38°59′ N., 141°37′ E.

Iwate, Honshu, Japan
Fell 1850, June 13, 0500 hrs
Synonym(s): *Iwate, Okirai*
Stone. Olivine-bronzite chondrite (H4).

A stone of about 135kg fell, after detonations, K. Jimbo, Beitr. Miner. Japan, 1906 (2), p.37, H.A. Ward, Am. J. Sci., 1893, **45**, p.153. Analysis, A. Miyashiro, Jap. J. Geol. Geogr., 1962, **33**, p.73 [M.A. 17-56]. Date of fall and co-ordinates, S. Murayama, priv. comm. Analysis, 27.16 % total iron, H. von Michaelis et al., Earth planet. Sci. Lett., 1969, **5**, p.387. Okirai is part of Kesen, N. Sato, priv. comm., 1983.

106kg Tokyo, Nat. Sci. Mus., main mass, approx. weight; 6kg New York, Amer. Mus. Nat. Hist.; 3kg Chicago, Field Mus. Nat. Hist.; 2105g Washington, U.S. Nat. Mus.; 850g Vienna, Naturhist. Mus.; 698g Harvard Univ.; 736g Tempe, Arizona State Univ.; 423g Yale Univ.; 372g Prague, Nat. Mus.; 356g Paris, Mus. d'Hist. Nat.;

Specimen(s): [71525], 1229g. includes one mass of 1157g, and fragments, 7g.

Keszu v Mocs.

Ketschki v Ryechki.

Ketscki v Ryechki.

Kettree v Khetri.

Keyes 36°43′ N., 102°30′ W.
Cimarron County, Oklahoma, U.S.A.
Found 1939
Synonym(s): *Boise City*
Stone. Olivine-hypersthene chondrite (L6).
A mass of 142kg was found, H.H. and A.D. Nininger, The Nininger Collection of Meteorites, Winslow, Arizona, 1950, p.66. Olivine Fa_{23}, B. Mason, Geochimica et Cosmochimica Acta, 1963, **27**, p.1011. Light noble gas analysis, R.J. Wright et al., J. Geophys. Res., 1973, **78**, p.1308. Cosmic ray exposure age, 18.5 m.y., J.C. Lorin and G. Poupeau, Meteoritics, 1973, **8**, p.410.
1.7kg Washington, U.S. Nat. Mus.; 667 Chicago, Field Mus. Nat. Hist.; 483g Fort Worth, Texas, Monnig Colln.; 59g Tempe, Arizona State Univ.;
Specimen(s): [1959,880], 342g.

Khaipur v Khairpur.

Khairagarh v Kheragur.

Khairpur 29°32′ N., 72°18′ E.
Bahawalpur, Sind, Pakistan
Fell 1873, September 23, 0500 hrs
Synonym(s): *Bhawalpur, Khaipur, Mailsi, Mooltan, Multan*
Stone. Enstatite chondrite (E6).
After the appearance of a cluster of luminous meteors, followed by detonations, a shower of stones fell on both sides of the Sutlej over an area of 16 × 3 miles. Of six stones preserved in the Indian Museums, three weighed about 11lb, 10lb, and 8lb, and the total weight was about 30lb, H.B. Medlicott, J. Asiatic Soc. Bengal, 1874, **43** (2), p.33. Described, with an analysis, G.T. Prior, Min. Mag., 1916, **18**, p.17. The metal phase contains over 1% Si, A.E. Ringwood, Geochimica et Cosmochimica Acta, 1961, **25**, p.1. Discussion, B. Mason, Geochimica et Cosmochimica Acta, 1966, **30**, p.23. Mineralogy, K. Keil, J. Geophys. Res., 1968, **73**, p.6945. Further bulk analysis, 21.1 % total iron, A.A. Moss et al., Min. Mag., 1967, **36**, p.101. Trace element abundances, C.M. Binz et al., Geochimica et Cosmochimica Acta, 1974, **38**, p.1529, P.A. Baedecker and J.T. Wasson, Geochimica et Cosmochimica Acta, 1975, **39**, p.735.
4.9kg Calcutta, Mus. Geol. Surv. India, includes a nearly complete stone of 4.4kg; 500g New York, Amer. Mus. Nat. Hist.; 106g Budapest, Nat. Mus.; 85g Vienna, Naturhist. Mus.; 59.7g Chicago, Field Mus. Nat. Hist.; 27g Washington, U.S. Nat. Mus.;
Specimen(s): [51366], 2716g. includes a nearly complete stone of 2472g, and fragments, 9g; [51189], 201g. a nearly complete stone; [48369], 66g. probably specimen C of Medlicott's list; [1964,704], 210g.; [96265], thin sections (two)

Khandesh district v Bhagur.

Khandesh district v Manegaon.

Khangpur v Khanpur.

Khanpur 25°33′ N., 83°7′ E.
Ghazipur district, Uttar Pradesh, India
Fell 1932, July 8, 1200-1300 hrs
Synonym(s): *Kahrapar, Kakrapar*
Stone. Olivine-hypersthene chondrite, amphoterite (LL5), brecciated.
Many stones fell, but only two pieces, totalling 3698g (the largest 1300g) were collected. Some of the stones fell at Kakrapur, Jaunpur district, and some at Karauli, Benares district. Description, M.S. Krishnan, Rec. Geol. Surv. India, 1934, **68**, p.108 [M.A. 6-12]. Ni and Co in metal, D.W. Sears and H.J. Axon, Meteoritics, 1976, **11**, p.97. Olivine Fa_{29}, B. Mason, Geochimica et Cosmochimica Acta, 1963, **27**, p.1011.
3.2kg Calcutta, Mus. Geol. Surv. India, main masses; 30g Washington, U.S. Nat. Mus.;
Specimen(s): [1933,158], 392g. two pieces, and fragments, 5g; [1954,180], 499g.

Kharauni v Sultanpur.

Kharkov 50°37.5′ N., 35°4.5′ E.
Sumy region, Ukraine, USSR, [Харков], [Жигайловка]
Fell 1787, October 12, 1500 hrs
Synonym(s): *Bobrik, Charkov, Jigalovka, Jigalowka, Lebedin, Zhigailovka, Zhigajlovka, Zigajlovka, Zigajlowka*
Stone. Olivine-hypersthene chondrite (L6), veined.
After detonations, several stones were seen to fall near the villages of Zhigailovka and Lebedin; at least one was preserved but its weight is not given, A. Stoikovitz, Ann. Phys. (Gilbert), 1809, **31**, p.311. History, description, L.A. Kulik, Meteorites of the USSR, Acad. Sci. Moscow, 1935 (2) [M.A. 7-67]. Total weight preserved is about 1.5kg, E.L. Krinov, Астрон. Журнал, 1945, **22**, p.303 [M.A. 9-297]. Olivine Fa_{24}, B. Mason, priv. comm., 1983. Classified as LL, analysis, 18.94 % total iron, V.Ya. Kharitonova, Метеоритика, 1972, **31**, p.116.
858g Moscow, Acad. Sci., main mass; 26g Washington, U.S. Nat. Mus.; 16g Chicago, Field Mus. Nat. Hist.;
Specimen(s): [19966], 437g.

Khatanga v Tunguska.

Khenteisky 49°12′ N., 113°28′ E. approx.
Bayan-dung, Choybalsan aymag, Mongolia, [Хэнтэйский]
Fell 1952, July 10, 2330 hrs
Synonym(s): *Hentey, Khentejsky, Khentey*
Stone.
Four pieces were found, the largest weighing 402.1g, G.G. Vorobyev and O. Namnandorzh, Метеоритика, 1958, **16**, p.134 [M.A. 14-129]. Locality described as southern side of Uldzyn-gol river in Bayan-dung soman, O. Namnandorj, Meteorites of Mongolia, 1980, p.21, English trans.. Co-ordinates are those of Bayan-dung.

Khentejsky v Khenteisky.

Khentey v Khenteisky.

Kheragur 26°57′ N., 77°53′ E.
Bhurtpur, Agra district, Uttar Pradesh, India
Fell 1860, March 28
Synonym(s): *Agra, Bhurtpur, Dhenagur, Khairagarh, Khiragurh*
Stone. Olivine-hypersthene chondrite (L6).
A stone (probably of about 1lb) is said to have fallen, but no details are given as to phenomena or weight, Proc. Asiatic Soc. Bengal, 1860, **29**, p.212, N.S. Maskelyne, Phil. Mag., 1863, **25**, p.446. Olivine Fa_{24}, B. Mason, Geochimica et Cosmochimica Acta, 1963, **27**, p.1011. The place of fall is probably Khairagarh, 26°57'N., 77°53'E., C.A. Silberrad, Min. Mag., 1932, **23**, p.299.
144g Calcutta, Mus. Geol. Surv. India; 23g Vienna, Naturhist. Mus.;
Specimen(s): [34799], 353g.

Kherlen v Kerulensky.

Kherson v Savtschenskoje.

Kherson v Vavilovka.

Khetri 28°1′ N., 75°49′ E.
Shekhawati, Jaipur, Rajasthan, India
Fell 1867, January 19, 0900 hrs
Synonym(s): *Kettree, Saonlod*
Stone. Olivine-bronzite chondrite (H6), brecciated.
After detonations, a shower of stones (about 40) fell. They were mostly pounded to powder by the villagers, and only two fragments appear to have been preserved. Report, with an analysis, D. Waldie, J. Asiatic Soc. Bengal, 1869, **38**, p.252, T. Oldham, Rec. Geol. Surv. India, 1870, **2**, p.101. Two stones were of about 13lb and 4.5lb respectively, J. Anderson, letter of 27 November, 1868 in Min. Dept. BM (NH). Most of the stones fell at Saonlod, 28°1'N., 75°49'E., 3 miles N. of Khetri, C.A. Silberrad, Min. Mag., 1932, **23**, p.302. Very little preserved,
49g Calcutta, Mus. Geol. Surv. India; 15.5g New York, Amer. Mus. Nat. Hist.; 5g Chicago, Field Mus. Nat. Hist.; 4g Vienna, Naturhist. Mus.; 2.4g Tempe, Arizona State Univ.;
Specimen(s): [43060], 13g.

Khiragurh v Kheragur.

Khmelevka 56°45′ N., 75°20′ E.
Omsk region, Federated SSR, USSR, [Хмелевка]
Fell 1929, March 1, 0524 hrs
Synonym(s): *Chmelewka, Hmeljovka*
Stone. Olivine-hypersthene chondrite (L5).
One stone fell in the valley of the Irtish, V.I. Vernadsky, letter of 5 August, 1939 in Min. Dept. BM(NH). Listed, I.S. Astapowitsch, Trans. Roy. Astron. Soc. Canada, 1938, **32**, p.195 [M.A. 7-172]. Several stones fell, but only one (6109g) was recovered. Described, with a micrometric analysis, P.L. Dravert, Метеоритика, 1941, **1**, p.49 [M.A. 9-293]. Analysis, M.I. D'yakonova, Метеоритика, 1964, **25**, p.129. Olivine Fa_{23}, B. Mason, Geochimica et Cosmochimica Acta, 1963, **27**, p.1011.
5950g Moscow, Acad. Sci., main mass;

Khobdo v Adzhi-Bogdo (stone).

Khohar 25°6′ N., 81°32′ E.
Banda district, Uttar Pradesh, India
Fell 1910, September 19, 1300 hrs
Stone. Olivine-hypersthene chondrite (L3).
After detonations, a stone (perhaps more than one) fell at Khohar village. 22 pieces, weighing from 4.8kg to 15g totalling about 9.7kg, were collected and are preserved in Calcutta, Mus. Geol. Surv. India, G. de P. Cotter, Rec. Geol. Surv. India, 1912, **42**, p.274. Analysis, 21.55 % total iron, E. Jarosewich, Geochimica et Cosmochimica Acta, 1966, **30**, p.1261. Unequilibrated, R.T. Dodd et al., Geochimica et Cosmochimica Acta, 1967, **31**, p.921. Mentioned, S.S. Deshmukh and S. Bandyopadhyaya, Rec. Geol. Surv. India, 1966, **94**, p.305 [M.A. 20-136]. Bismuth distribution, references, D.S. Woolum et al., Geochimica et Cosmochimica Acta, 1978 (Suppl. 10), p.1173.
7.84kg Calcutta, Mus. Geol. Surv. India; 230g Paris, Mus. d'Hist. Nat.; 189g New York, Amer. Mus. Nat. Hist.; 92g Washington, U.S. Nat. Mus.; 77g Tempe, Arizona State Univ.;
Specimen(s): [1915,87], 326g.

Khorma v Veliko-Nikolaevsky Priisk.

Khor Temiki 16° N., 36° E. approx.
Gash delta, Kassala, Sudan
Fell 1932, April 8
Synonym(s): *Girger*
Stone. Achondrite, Ca-poor. Aubrite (AUB).
Several stones fell, and about 7lb of fragments and dust were recovered, G.W. Grabham, letters of 11 May, 14 May, 29 September, 13 December, 1933 and 7 June, 1934 in Min. Dept. BM(NH). Described, with an analysis, M.H. Hey and A.J. Easton, Geochimica et Cosmochimica Acta, 1967, **31**, p.1789. Analysis of enstatite, A.M. Reid and A.J. Cohen, Geochimica et Cosmochimica Acta, 1967, **31**, p.661. Deformed pyroxene, reference to gas-rich nature, J.R. Ashworth and D.J. Barber, Contr. Mineral. Petrol., 1975, **49**, p.149.
Main mass, Khartoum, Sudan Geol. Surv.; 90g Washington, U.S. Nat. Mus.; 87g New York, Amer. Mus. Nat. Hist.; 77g Calcutta, Mus. Geol. Surv. India;
Specimen(s): [1934,777], 550g. and fragments, 10.5g; [1934, 778], 286g. and fragments, 2g; [1953,62], 448g.; [1934,780], 138g. fragments and powder; [1934,781], 770g. fragments and powder; [1934,781a], 5g. fragments, and a polished mount.

Khutor Lipowsky v Lipovsky.

Kiangsu v Fengsien-Ku.

Kiangsu Rukao v Min-Fan-Zhun.

Kiel 54°24′ N., 10°9′ E.
Schleswig-Holstein, Germany
Fell 1962, April 26, 1245 hrs, approx. time (U.T.).
Stone. Olivine-hypersthene chondrite (L6).
One stone of 737.6g made a hole in the roof of a house. No acoustic or visual phenomena were recorded, W. Schreyer, Natur und Museum, 1964, **94**, p.118, Meteor. Bull., 1963 (28). Analysis, 22.50 % total iron, H. König, Geochimica et Cosmochimica Acta, 1964, **28**, p.1697. Olivine Fa_{25}, B. Mason, Geochimica et Cosmochimica Acta, 1967, **31**, p.1100.
6.2g Washington, U.S. Nat. Mus.; 4g New York, Amer. Mus. Nat. Hist.;
Specimen(s): [1976,M.5], 1.35g. crusted fragment.

Kielpa 33°36′ S., 136°6′ E.
Jervois County, South Australia, Australia
Found 1948
Stone. Olivine-bronzite chondrite (H5).
A fragment (apparently about one third of the original) of 13.6kg was found 10 miles west of the Kielpa railway siding, Eyre Peninsula. The specimen is a deeply weathered fragment, the remainder of the stone was not found, Meteor. Bull., 1971 (50), Meteoritics, 1971, **6**, p.120, C.J. Barclay and J.B. Jones, J. Geol. Soc. Austr., 1971, **17**, p.221 [M.A. 74-419]. Co-ordinates, analysis, M.J. Fitzgerald, Ph.D. Thesis, Univ. of Adelaide, 1979, p.23. Olivine $Fa_{18.9}$, B. Mason, Rec. Austr. Mus., 1974, **29**, p.169.
Main mass, Adelaide, Univ.; 351g Washington, U.S. Nat. Mus.;
Specimen(s): [1968,273], 275g. end piece

Kiev v Bjelaja Zerkov.

Kiev v Oczeretna.

Kiffa 16°35′ N., 11°20′ W.
The Third Region, Mauritania
Fell 1970, October 23, 1455 hrs, U.T.
Stone. Olivine-bronzite chondrite (H4).
The fall was accompanied by explosions and a white cloud appeared in the sky. Next day almost 1.5kg of fragments were found in a 20cm deep hole in sand. The location is 8 km SW. of Kiffa, Meteor. Bull., 1971 (50), Meteoritics, 1971, **6**, p.114. Description and analysis, 28.33 % total iron, R.S. Clarke, Jr. et al., Smithson. Contrib. Earth Sci., 1975 (14), p.63.
Main mass, Nouakchott, Service Geologue; 25g Washington, U.S. Nat. Mus.; 2.2g Mainz, Max-Planck-Inst.;
Specimen(s): [1972,2], 0.59g.

Kifkakhsyagan 64°24′ N., 172°42′ E. approx.
Chukot Peninsula, Federated SSR, USSR, [Кифкахсяган]
Found 1972
Iron. Octahedrite, medium (1.1mm) (IIIA).
An 18.8kg individual was found lying on a rocky surface on a slope of Mount Kifkakhsyagan. The specimen has a red-brown, badly weathered exterior and is probably very old, Meteor. Bull., 1974 (52), Meteoritics, 1974, **9**, p.112. Description, with an analysis, 7.56 %Ni, O.A. Kirova and M.I. D'yakonova, Метеоритика, 1976, **35**, p.40.
18.5kg Moscow, Acad. Sci.;

Kijima v Fukutomi.

Kijima (1906) 36°51′ N., 138°23′ E.
Shimo Takai, Nagano, Honshu, Japan
Fell 1906, June 15
Synonym(s): *Kizima*
Stone.
Listed, Kwasan Observ. Bull., 1935, p.306 [M.A. 7-173]. A second, larger stone was found in 1908 and is in the possession of Mrs. S. Onozawa, S. Murayama, letter of 6 April, 1962 in Min. Dept. BM(NH). Not to be confused with Kijima = Fukutomi.
282g Nagano, Girl's High School; Specimen, Tokyo, Nat. Sci. Mus.;

Ki-Ki v Coomandook.

Kikino 55° N., 34° E. approx.
Vyazma, Smolensk, Federated SSR, USSR
Fell 1809
Synonym(s): *Viasma, Viazma, Wjasemsk*
Stone. Olivine-bronzite chondrite (H6).
An "ash-coloured" stone fell in the village of Kikino, but no particulars of the fall nor weight are recorded, E. von Eichwald, Arch. Wiss. Kunde Russl., 1847, **5**, p.177. Olivine Fa_{19}, B. Mason, Geochimica et Cosmochimica Acta, 1963, **27**, p.1011.
67g Budapest, Nat. Mus.; 32g Chicago, Field Mus. Nat. Hist.; 27g Moscow, Acad. Sci.; 24g Vienna, Naturhist. Mus.;
Specimen(s): [63622], 24g. and fragments, 3g. Grey material, from the collection of "Taulaubieff" of Petrograd.; [56471], 1g. from the collection of Y.I. Simashko, white material.

Kilbourn 43°35′ N., 89°36′ W.
Columbia County, Wisconsin, U.S.A.
Fell 1911, June 16, 1700 hrs
Stone. Olivine-bronzite chondrite (H5).
After detonations, a stone of 772g fell through the roof to the floor of a barn, penetrating two hemlock boards; description, O.C. Farrington, Field Mus. Nat. Hist. Geol. Ser., 1914, **5** (178), p.10. Olivine Fa_{19}, B. Mason, Geochimica et Cosmochimica Acta, 1963, **27**, p.1011.
87g Vienna, Naturhist. Mus.; 70g Chicago, Field Mus. Nat. Hist.; 70g Harvard Univ.; 50g New York, Amer. Mus. Nat. Hist.; 41g Tempe, Arizona State Univ.; 85g Philadelphia, Acad. Nat. Sci.; 86g Ottawa, Mus. Geol. Surv. Canada;
Specimen(s): [1919,45], 1g. fragments

Killeter 54°40′ N., 7°40′ W.
County Tyrone, Ireland
Fell 1844, April 29, 1530 hrs
Stone. Olivine-bronzite chondrite (H6), veined.
After the appearance of a rapidly moving cloud and detonations, a shower of stones fell over several fields, but only a few fragments were preserved, S. Haughton, Phil. Mag., 1862, **23**, p.47, Rev. J. Love, letter of 20 March, 1847 in Min. Dept. BM(NH). Olivine Fa_{20}, B. Mason, Geochimica et Cosmochimica Acta, 1963, **27**, p.1011.
30g Tübingen, Univ.; 6g Tempe, Arizona State Univ.; 3g Chicago, Field Mus. Nat. Hist.; 2g Budapest, Nat. Mus.; 2g Washington, U.S. Nat. Mus.;
Specimen(s): [90268], 85g.; [90269], 9g.

Kilrea v Bovedy.

Kimble County 30°25′ N., 99°24′ W.
Kimble County, Texas, U.S.A.
Found 1918, recognized 1936
Stone. Olivine-bronzite chondrite (H6).
One mass, broken, total weight 153.8kg, largest mass 132kg, was found, A.D. Nininger, Pop. Astron., Northfield, Minnesota, 1937, **45**, p.449 [M.A. 7-62], H.H. Nininger, letter of 30 May, 1939 in Min. Dept. BM(NH). Described, with an analysis, V.E. Barnes, Univ. Texas Publ. no. 3945, 1940, p.623 [M.A. 9-59]. Olivine Fa_{19}, B. Mason, Geochimica et Cosmochimica Acta, 1963, **27**, p.1011.
4.1kg Washington, U.S. Nat. Mus.; 884g Fort Worth, Texas, Monnig Colln.;
Specimen(s): [1961,393], 559g. three fragments, the largest 444g.

Kimbolton 40°4′17″ S., 175°43′49″ E.
Manawatu district, North Island, New Zealand
Found 1976, March
Stone. Olivine-bronzite chondrite (H4).
A single mass, representing slightly less than half of the stone and weighing 7.5kg, was found in a ploughed field during harrowing. The place of find is 4km WSW. of Kimbolton, Meteor. Bull., 1980 (58), Meteoritics, 1980, **15**, p.235.

Kinas Putts v Gibeon.

Kinejima v Fukutomi.

Kingai 11°38′ N., 24°41′ E.
Darfur Province, Sudan
Fell 1967, November 7, 1400 hrs
Stone. Olivine-bronzite chondrite (H6).
One stone fell, original weight about 450g but only 67.4g were preserved, Meteor. Bull., 1968 (43), Meteoritics, 1970, **5**, p.97.
Main mass, Khartoum, Geol. Surv.;
Specimen(s): [1968,271], 8.9g.

Kingfisher 35°50′ N., 97°56′ W.
Kingfisher County, Oklahoma, U.S.A.
Found 1950, recognized 1951
Stone. Olivine-hypersthene chondrite (L5), black.
A mass of 8.18kg was found, B. Mason, Meteorites, Wiley, 1962, p.240, Cat. Meteor. Arizona State Univ., 1964, p.460. Microprobe analysis of metal, G.J. Taylor and D. Heymann, Geochimica et Cosmochimica Acta, 1970, **34**, p.677.
2.2kg Washington, U.S. Nat. Mus.; 324g Chicago, Field Mus. Nat. Hist.; 185g Los Angeles, Univ. of California; 169g Tempe, Arizona State Univ.; 80g New York, Amer. Mus. Nat. Hist.;
Specimen(s): [1952,139], 75.5g.; [1960,105], 145g. slice

Kingoonya v Naretha.

Kingooya v Naretha.

King Solomon 20°41′58″ S., 139°48′15″ E.
Mary Kathleen, Queensland, Australia
Found 1952, March
Iron.
A mass of unrecorded weight was found about 100 yards from the King Solomon mine, Mary Kathleen, Meteor. Bull., 1976 (54), Meteoritics, 1976, **11**, p.76. Reported, B.R. Houston, Queensland Govt. Min. J., 1971, **72**, p.482.

Kingston 32°54′ N., 107°44′ W.
Sierra County, New Mexico, U.S.A.
Found 1891
Iron. Octahedrite, medium (0.8mm) (IRANOM).
A mass of about 28.5lb was found near the Solitary Mine, Percha Creek, 4 miles north of Kingston. Described, with an analysis, 6.98 %Ni, E.O. Hovey, Ann. New York Acad. Sci., 1912, **22**, p.335. Chemically anomalous, another analysis, 6.88 %Ni, 21.3 ppm.Ga, 58.8 ppm.Ge, 5.1 ppm.Ir, E.R.D. Scott et al., Geochimica et Cosmochimica Acta, 1973, **37**, p.1957. Description; heavily weathered; artificially heated at one end, V.F. Buchwald, Iron Meteorites, Univ. of California, 1975, p.725.
1kg New York, Amer. Mus. Nat. Hist.; 1kg Chicago, Field Mus. Nat. Hist.; 962g Budapest, Nat. Mus.; 913g Tempe, Arizona State Univ.; 541g Vienna, Naturhist. Mus.; 333g Washington, U.S. Nat. Mus.;
Specimen(s): [1912,595], 346.5g. includes a slice of 320g

Kinley 52°2.8′ N., 107°14′ W.
Saskatchewan, Canada
Found 1965, or 1966, recognized 1968
Stone. Olivine-hypersthene chondrite (L6).
A weathered stone of 2kg was ploughed up, Meteor. Bull., 1974 (52), Meteoritics, 1974, **9**, p.112.
880g Mainz, Max-Planck-Inst.; 1535g Ottawa, Geol. Surv. Canada;
Specimen(s): [1973,M.6], 24.5g. sawn fragment

Kinsella 53°12′ N., 111°26′ W. approx.
Echo Lake, Kinsella, Alberta, Canada
Found 1946, May
Iron. Octahedrite, medium (1.0mm) (IIIB).
A single mass of 3.72kg was found. Description, E.R.D. Scott et al., Meteoritics, 1977, **12**, p.425. Analysis, 8.78 %Ni, 20.3 ppm.Ga, 42.0 ppm.Ge, 0.12 ppm.Ir, A. Kracher et al., Geochimica et Cosmochimica Acta, 1980, **44**, p.773.
Main mass, Los Angeles, Univ. of California;

Kiowa v Brenham.

Kiowa County v Brenham.

Kirbyville 30°45′ N., 95°57′ W.
Jasper County, Texas, U.S.A.
Fell 1906, November 12, 1530 hrs
Stone. Achondrite, Ca-rich. Eucrite (AEUC).
A stone of 97.7g fell, A.D. Nininger, Pop. Astron., Northfield, Minnesota, 1937, **45**, p.449 [M.A. 7-62], H.H. Nininger, letter of 30 May, 1939 in Min. Dept. BM(NH). Analysis, B. Mason et al., Smithson. Contrib. Earth Sci., 1979 (22), p.30.
97g Fort Worth, Texas, Monnig Colln.;

Kirghiz Steppes v Karakol.

Kirgisen Steppe v Karakol.

Kirgizskaia Step v Karakol.

Kirin v Jilin.

Kirkland 42°41′35″ N., 122°10′13″ W.
King County, Washington, U.S.A.
Fell 1955, January 17, 1100 hrs, approx. time
Synonym(s): *Seattle*
Pseudometeorite. Iron. Octahedrite.
Two metallic objects of 119.2g and 113.2g are stated to have pierced the dome of an amateur astronomical observatory 2 miles NE. of Kirkland, W.F. Read, Meteoritics, 1963, **2**, p.56. Both masses are fragments of Cañon Diablo, E.P. Henderson, letter of 18 December, 1963, Meteor. Bull., 1964 (29). A fraud, V.F. Buchwald, Iron Meteorites, Univ. of California, 1975, p.727.

Kirkuk v Tauk.

Kis-Gyor 48°1′ N., 20°42′ E.
Miskolcz, Borsod, Hungary
Found 1901, known in this year
Doubtful..
A specimen of 3.61g in the Hungarian Nat. Mus., Budapest, is said to have fallen or been found in 1901, but is not included in the list of Hungarian meteorites. It is not stated whether it is an iron or a stone, L. Tokody and M. Dudich, Magyarorszag Meteoritgyujtemenyei, Budapest, 1951, p.35, 89. Doubtful, V. Zsivny, letters of 4 July, and 5 August, 1952 in Min. Dept. BM(NH).

Kissij 54°52′ N., 50°53′ E.
Chistopol, Kazan, Tatar Republic, USSR,
[Чувашские Киссы]
Found 1899
Synonym(s): *Chuvashskie Kissy, Chuvashskii Kissy, Cuvasskije Kissy, Kissji, Tchivashsky Kissy, Tschuwaschkya, Tschuwaschskije Kissy, Tschuwaschsky Kissy*
Stone. Olivine-bronzite chondrite (H5), black.
A stone of about 5.5kg was ploughed up and was preserved in the Geol. Inst. Kazan Univ., A. Stuckenberg, Sitz-Prot. Naturfor.-Gesell. Kazan, 1900, **32** (188), Neues Jahrb. Min., 1903, **1**, p.212, abs. Analysis, 26.30 % total iron, M.I. D'yakanova, Метеоритика, 1964, **25**, p.129. Olivine Fa_{17}, B. Mason, Geochimica et Cosmochimica Acta, 1963, **27**, p.1011.
3358g Kazan Univ.; 635g Vienna, Naturhist. Mus.; 317g Chicago, Field Mus. Nat. Hist.; 77g Budapest, Nat. Mus.; 33g Berlin, Humboldt Univ.; 28g Moscow, Acad. Sci.; 22g Washington, U.S. Nat. Mus.; 21g Prague, Nat. Mus.;
Specimen(s): [1980,M.3], 41.5g.

Kissji v Kissij.

Kisvarsany 48°10′ N., 22°18′30″ E.
Comitat Szabolcs, Hungary
Fell 1914, May 24
Stone. Olivine-hypersthene chondrite (L).
Thirty stones fell. Mentioned in 1914 in the Nyiregyhaza newspaper Nyirvedek, V. Zsivny, letters of December 1933 and 28 May, 1952 in Min. Dept. BM(NH).. Mentioned, L. Tokody and M. Dudich, Magyarorszag Meteoritgyüjtemenyei, Budapest, 1951, p.35, 89, 99, 102, 103, B. Mauritz et al., Föld. Közl., 1953, **83**, p.138 [M.A. 12-247] . Tokody and Vendl confuse this fall with Nyirabrany (*q.v.*). Olivine Fa_{23}, B. Mason, Geochimica et Cosmochimica Acta, 1967, **31**, p.1100. The specimens that were in the Nat. Mus., Budapest, have been destroyed. Analysis, 20.24 % total iron, H.B. Wiik, Comm. Phys.-Math. Soc. Sci. Fenn., 1969, **34**, p.135.
1549g Nyiregyhaza, Mus.;

Kitano v Gifu.

Kittakittaooloo 28°2′ S., 138°8′ E.
South Australia, Australia
Found 1970
Stone. Olivine-bronzite chondrite (H4).
One mass of 3.6kg was found on a hardened part of a sand dune, about 1 km NW. of Lake Kittakittaooloo, A.F. Williams, Quart. Geol. Notes, Geol. Surv. S. Australia, 1971 (37). Co-ordinates, M.J. Fitzgerald, letter of 11 December, 1972, in Min. Dept. BM(NH). Olivine $Fa_{18.9}$, B. Mason, Rec. Austr. Mus., 1974, **29**, p.169.
Main mass, Adelaide, South Austr. Mus.; 10g Washington, U.S. Nat. Mus.;

Kivesvaara 64°27′ N., 27°34′ E.
Paltramo, Oulu, Finland
Found 1968, May
Stone. Carbonaceous chondrite, type II (CM2).
A mass of 164g was found. Mineral compositions, olivine $Fa_{0.4-22.9}$, orthopyroxene Fs1.5-4.8, Meteor. Bull., 1981 (59), Meteoritics, 1981, **16**, p.195. Description, analysis, 20.75 % total iron, K.A. Kinnunen and R. Saikkonen, Bull. Geol. Soc. Finland, 1983, **55**, p.35.
Specimen.; Helsinki, Geol. Dept. Univ. Helsinki;

Kizima v Kijima (1906).

Klagenfurt 46°38′ N., 14°18′ E.
Carinthia, Austria
Fell 1849, or found in this year
Doubtful. Stone. Olivine-bronzite chondrite?, veined.
Two pieces of 87g and 36g in the Eötvös Lorand University, Budapest, are said to have fallen, or have been found, in 1849, and to have probably come from Klagenfurt. They are described as a veined crystalline bronzite-chondrite, L. Tokody and M. Dudich, Magyarorszag Meteoritgyüjtemenyei, Budapest, 1951, p.90, 97. No other record has been traced, and the fall must be accounted very doubtful.

Klamath Falls 42°10′ N., 121°51′ W.
Klamath County, Oregon, U.S.A.
Found 1952
Synonym(s): *Oregon*
Iron. Octahedrite, medium (0.5mm) (IIIF).
A mass of 37.5lb was found during road building. The finder took a small fragment to a chemist who sent it to H.H. Nininger who confirmed it to be meteoritic, this fragment is now in Tempe (Arizona State Univ.). The finder later sold the main mass to the Institute of Meteoritics, Univ. of New Mexico (see their 1965 catalogue under 'Oregon'). Analysis, 7.24 %Ni, 0.31 %Co., E.F. Lange, The Ore Bin, 1970, **32** (2), p.21, L. LaPaz, Univ. New Mexico Publ. in Meteoritics no. 6, 1969, p.179. Classification and analysis, 8.5 %Ni, 6.79 ppm.Ga, 0.701 ppm.Ge, 0.0059 ppm.Ir, A. Kracher et al., Geochimica et Cosmochimica Acta, 1980, **44**, p.773.
12.96kg Albuquerque, Inst. Meteoritics, Univ. of New Mexico, main mass; 6.2g Tempe, Arizona State Univ.;

Klausemburg v Mocs.

Klausenburg v Mocs.

Klein-Menow v Menow.

Klein-Wenden 51°36′ N., 10°48′ E.
Nordhausen, Erfurt, Germany
Fell 1843, September 16, 1645 hrs
Stone. Olivine-bronzite chondrite (H6).
A stone of about 3.25kg fell, after detonations, Ann. Phys. Chem. (Poggendorff), 1843, **60**, p.157. Analysis, C. Rammelsberg, Ann. Phys. Chem. (Poggendorff), 1844, **62**, p.449. Analysis shows more K than Na and is doubtful, W. Wahl, Geochimica et Cosmochimica Acta, 1950, **1**, p.28. Olivine Fa_{19}, B. Mason, Geochimica et Cosmochimica Acta, 1963, **27**, p.1011.
2.5kg Berlin, Humboldt Univ.; 174g Vienna, Naturhist. Mus.; 67g Washington, U.S. Nat. Mus.; 47g Los Angeles, Univ. of California;
Specimen(s): [35404], 5.5g.

Klondike v Klondike (Gay Gulch).

Klondike v Klondike (Skookum Gulch).

Klondike (Gay Gulch) 63°55′ N., 139°20′ W.
Yukon, North-west Territories, Canada
Found 1901
Synonym(s): *Clondike, Dawson, Gay Gulch, Klondike*
Iron. Octahedrite, plessitic (IRANOM).
One mass of 483g was found in Gay Gulch, Bonanza Creek, in 1901. Once thought to be paired with Klondike (Skookum Gulch), R.A.A. Johnston, Mus. Bull. Geol. Surv. Canada, 1915 (Geol. ser. no. 26), p.1, but is chemically and

structurally distinct, 15.06 %Ni, 6.68 ppm.Ga, 10.7 ppm.Ge, 0.11 ppm.Ir, J.T. Wasson and R. Schaudy, Icarus, 1971, **14**, p.59. Description, under Gay Gulch, V.F. Buchwald, Iron Meteorites, Univ. of California, 1975, p.582.

452g Ottawa, Mus. Geol. Surv. Canada;

Klondike (Skookum Gulch) 63°55′ N., 139°20′ W.

Yukon, North-west Territories, Canada
Found 1905
Synonym(s): *Big Skookum, Clondike, Eldorado Creek, Klondike, Skookum Gulch*
Iron. Ataxite, Ni-rich (IVB).

A mass of about 16kg was found in Pliocene gravels in Skookum Gulch, 10 miles from the find site of the Klondike (Gay Gulch) mass. These finds were first thought to be paired, R.A.A. Johnston, Mus. Bull. Geol. Surv. Canada, 1915 (Geol. Ser. no. 26), p.1. Distinct, analysis, 17.13 %Ni, 0.272 ppm.Ga, 0.056 ppm.Ge, 15 ppm.Ir, R. Schaudy et al., Icarus, 1972, **17**, p.174. Structural description, H.J. Axon and P.L. Smith, Min. Mag., 1972, **38**, p.736. Description, artificially reheated, under Skookum Gulch, V.F. Buchwald, Iron Meteorites, Univ. of California, 1975, p.1137.

Main mass, Ottawa, Mus. Geol. Surv. Canada; 778g New York, Amer. Mus. Nat. Hist.; 715g Chicago, Field Mus. Nat. Hist.;

Specimen(s): [1919,43], 328g. includes a slice, 315g.

Kloster Schefftlar 48°24′ N., 11°44′ E. co-ordinates of Freising

Freising, Bayern, Germany
Fell 1722, June 5, 1530 hrs
Doubtful..

Several stones are said to have fallen, E.F.F. Chladni, Die Feuer-Meteore, Wien, 1819, p.240, but the evidence is not conclusive.

Knaasta v Białystok.

Knasta v Białystok.

Knahyna v Knyahinya.

Kniaginia v Knyahinya.

Knowles 36°54′ N., 100°13′ W.

Beaver County, Oklahoma, U.S.A.
Found 1903
Iron. Octahedrite, medium (0.75mm) (IIIB).

A mass of iron weighing 355lb was found, Ann. Rep. Amer. Mus. Nat. Hist., New York for 1909, 1910, p.39, L.W. MacNaughton, Am. Mus. Novit., 1926 (207). Analysis, 8.56 %Ni, H.B. Wiik and B. Mason, Geochimica et Cosmochimica Acta, 1965, **29**, p.1003. Newer analysis, 9.35 %Ni, 18.5 ppm.Ga, 31.6 ppm.Ge, 0.023 ppm.Ir, E.R.D. Scott et al., Geochimica et Cosmochimica Acta, 1973, **37**, p.1957. Description, V.F. Buchwald, Iron Meteorites, Univ. of California, 1975, p.727.

161kg New York, Amer. Mus. Nat. Hist.;

Specimen(s): [1959,984], 55.5g. and sawings, 3g.

Knoxville v Tazewell.

Knyahinya 48°54′ N., 22°24′ E.

Nagybereszna, Ungvar, Ukraine, USSR
Fell 1866, June 9, 1700 hrs
Synonym(s): *Csillagfalva, Knahyna, Kniaginia, Knyhyna, Kuyahinga, Nagy-Bereszna*
Stone. Olivine-hypersthene chondrite (L5), brecciated.

After appearance of a fireball and detonations, a shower of stones (estimated at over 1000, and of total weight of about 500kg, the largest weighing about 293kg) fell over an area of 2 x 0.75 miles, W. von Haidinger, Sitzungsber. Akad. Wiss. Wien, Math.-naturwiss. Kl., 1866, **54**, p.200, 475. Analysis, W. Wahl and H.B. Wiik, Geochimica et Cosmochimica Acta, 1951, **1**, p.123. Another analysis, 20.15 % total iron, B. Mason and H.B. Wiik, Am. Mus. Novit., 1963 (2154) [M.A. 16-639].

293kg Vienna, Naturhist. Mus., largest mass; 9.7kg Bonn, Univ. Mus.; 8.1kg Paris, Mus. d'Hist. Nat.; 6.5kg Stockholm, Riksmus; 5.25kg Chicago, Field Mus. Nat. Hist.; 3.4kg Budapest, Nat. Mus.; 3kg Philadelphia, Acad. Nat. Sci.; 1.8kg Berlin, Humboldt Univ.; 1575g Washington, U.S. Nat. Mus.; 1kg New York, Amer. Mus. Nat. Hist.; 690g Harvard Univ.; 800g Prague, Nat. Mus.; 489g Los Angeles, Univ. of California; 588g Tempe, Arizona State Univ.; 421g Tübingen, Univ.; 414g Copenhagen, Univ. Geol. Mus.;

Specimen(s): [52213], 6465g. complete stone; [40581], 3980g.; [51606], 249g. complete stone; [41647], 230g. complete stone; [41646], 227.5g. complete stone; [41648], 61g. complete stone; [41650], 17g. complete stone; [42262], 16g. complete stone; [41651], 11g. complete stone; [41653], 2.5g.; [51605], 2.5g. two stones; [1960,494], 29g.; [96266], thin sections (two)

Knyhyna v Knyahinya.

Knyszyn v Białstok.

Kobdo v Adzhi-Bogdo (stone).

Kochi 33°32′42″ N., 133°32′53″ E.

Kochi-Ken, Shikoku, Japan
Fell 1949, November 20, 2000 hrs
Doubtful. Stone.

A small piece of stone, 1.6g, fell, penetrating the window of a house; there remains doubt of its meteoritic origin; its S.G., 3.93, suggests that if it is meteoritic, Kochi is a stony-iron, S. Murayama, Nat. Sci. Mus. Tokyo, 1953, **20** (3-4), p.32 [M.A. 12-611], S. Murayama, Meteoritics, 1955, **1**, p.300, S. Murayama, letter of 6 April, 1962 in Min. Dept. BM(NH).

Specimen, in the possession of K. Komatsu, Kochi.;

Kodaikanal 10°16′ N., 77°24′ E.

Palni Hills, Madura district, Tamil Nadu, India
Found 1898, perhaps fell 8 years previously
Iron. Octahedrite, finest (0.1mm), with silicate inclusions (IIE-ANOM).

A mass of about 35lb was found, T.H. Holland, Proc. Asiatic Soc. Bengal, 1900, p.2. Description, includes data on "weinbergite", since discredited as oligoclase mixture, F. Berwerth, Tschermaks Min. Petr. Mitt., 1906, **25**, p.179. Analysis of metal, 8.22 %Ni, 20.7 ppm.Ga, 65.6 ppm.Ge, 5.2 ppm.Ir, E.R.D. Scott et al., Geochimica et Cosmochimica Acta, 1973, **37**, p.1957. Silicate mineralogy, T.E. Bunch et al., Contr. Mineral. Petrol., 1970, **25**, p.297. Description, detailed summary, refers to Rb/Sr age, 3800 m.y., K/Ar age 3500 m.y., V.F. Buchwald, Iron Meteorites, Univ. of California, 1975, p.728.

4kg Calcutta, Mus. Geol. Surv. India; 844g Paris, Mus. d'Hist. Nat.; 628g Vienna, Naturhist. Mus.; 237g Chicago, Field Mus. Nat. Hist.; 312g Budapest, Nat. Mus.; 283g Washington, U.S. Nat. Mus.; 89g Berlin, Humboldt Univ.;
Specimen(s): [85485], 2308g. and fragments, 27.5g; [1920, 310], 90g. slice

Kofa 33°30′ N., 114° W. approx.
Yuma County, Arizona, U.S.A.
Found 1893
Iron. Octahedrite, plessitic (IRANOM).
A mass of 490g was found, B. Mason, Meteorites, Wiley, 1962, p.228. It is chemically anomalous, analysis, 18.27 % Ni, 4.79 ppm.Ga, 8.61 ppm.Ge, 0.098 ppm.Ir, J.T. Wasson and R. Schaudy, Icarus, 1971, **14**, p.59. Description, well preserved and shock-hardened, V.F. Buchwald, Iron Meteorites, Univ. of California, 1975, p.735.
351g Chicago, Field Mus. Nat. Hist., main mass;
Specimen(s): [1980,M.4], 35.1g.

Kokomo 40°29′ N., 86°22′ W.
Howard County, Indiana, U.S.A.
Found 1862
Synonym(s): *Howard County*
Iron. Ataxite, Ni-rich (IVB).
A mass of about 4lb was found at a depth of about 2 ft, E.T. Cox, Am. J. Sci., 1873, **5**, p.155, J.L. Smith, Am. J. Sci., 1874, **7**, p.391. Analysis, 15.76 %Ni, E. Cohen, Meteoritenkunde, 1905, **3**, p.152. Structure, H.J. Axon and P.L. Smith, Min. Mag., 1972, **38**, p.736. Another analysis, 15.88 %Ni, 0.193 ppm.Ga, 0.031 ppm.Ge, 31 ppm.Ir, R. Schaudy et al., Icarus, 1972, **17**, p.174. Description, some specimens artificially deformed but not reheated, V.F. Buchwald, Iron Meteorites, Univ. of California, 1975, p.736.
308g Harvard Univ.; 62g Chicago, Field Mus. Nat. Hist.; 54g Paris, Mus. d'Hist. Nat.; 35g Budapest, Nat. Mus.; 24g Copenhagen, Univ. Geol. Mus.; 11.7g New York, Amer. Mus. Nat. Hist.; 11g Vienna, Naturhist. Mus.;
Specimen(s): [50807], 38g.; [47240], 7.5g.

Kokstad 30°33′ S., 29°25′ E.
Griqualand East, Cape Province, South Africa
Found 1884
Synonym(s): *Matatiela, Matatiele*
Iron. Octahedrite, coarse (1.4mm) (IIIE).
A mass of 43kg, of jaw-bone shape, was brought to Kokstad at the same time as the Matatiele mass of 298kg, E. Cohen, Ann. South African Mus. Cape Town, 1900, **2**, p.9. The two masses are chemically identical, 8.33 %Ni, 17.4 ppm.Ga, 35.9 ppm.Ge, 0.6 ppm.Ir, E.R.D. Scott et al., Geochimica et Cosmochimica Acta, 1973, **37**, p.1957. Description of the masses; shock-annealed, V.F. Buchwald, Iron Meteorites, Univ. of California, 1975, p.738.
39.87kg Vienna, Naturhist. Mus., includes mass of 38.6kg, of Kokstad; 250g Chicago, Field Mus. Nat. Hist.; 207g Budapest, Nat. Mus., of Kokstad; 298kg Cape Town, South African Mus., previously known as Matatiele;
Specimen(s): [80675], 203g.; [84470], 40g. slice, previously known as Matatiele

Kolapara v Dokachi.

Kolocha v Borodino.

Kolotscha v Borodino.

Komagome
Tokyo, Honshu, Japan, Not given
Fell 1926, April 18
Iron. Octahedrite.
One mass of 238g fell; analysis, 8 %Ni, 0.05 %Co, M. Shima, J. Japan. Assoc. Min. Petr. Econ. Geol., 1965, **54**, p.216 [M.A. 70-542].

Komarinsky v Brahin.

Konia v Adalia.

Konovo 42°31′ N., 26°10′ E.
Nova Zagora, Sliven, Bulgaria
Fell 1931, May 26, 0000-0100 hrs
Stone. Chondrite.
A stone of about 100g fell, N.S. Nikolov, Метеоритика, 1959, **17**, p.93 [M.A. 15-535].
Small fragments, Sofia, Univ. Geol. Inst.;

Kopjes Vlei
29°18′ S., 21°9′ E. co-ordinates of Kenhardt
Kenhardt district, Cape Province, South Africa
Found 1914, known before this year
Synonym(s): *Carnarvon*
Iron. Hexahedrite (IIA).
A mass of elongated shape, consisting of two pieces weighing together about 30lb, was found, E.L. Gill, letter of 3 June, 1926 in Min. Dept. BM(NH). Two pieces, totalling 15.5lb were presented to the South African Mus. in 1914. Metallography, H.J. Axon, Min. Mag., 1968, **36**, p.1139. Analysis, 5.65 %Ni, 59.9 ppm.Ga, 182 ppm.Ge, 3.1 ppm.Ir, J.T. Wasson, Geochimica et Cosmochimica Acta, 1969, **33**, p.859. Highly shocked, some reheating, V.F. Buchwald, Iron Meteorites, Univ. of California, 1975, p.742.
5.6kg Cape Town, South African Mus., includes the main mass, 5.2kg; 169g Vienna, Naturhist. Mus.; 64g Washington, U.S. Nat. Mus.;
Specimen(s): [1950,252], 1028g. and fragments, 15.5g, and filings, 6.5g.

Koraleigh 35°6′ S., 143°24′ E.
New South Wales, Australia
Found 1943
Stone. Olivine-hypersthene chondrite (L6).
A weathered stone of 450g was ploughed up. It was considered similar to the Caroline meteorite, an H5 chondrite, A.B. Edwards and G. Baker, Mem. Nat. Mus. Melbourne, 1943 (13), p.157 [M.A. 9-298], but is distinct. Classification, olivine Fa_{26}, B. Mason, Rec. Austr. Mus., 1974, **29**, p.169.
Main mass, Melbourne, Nat. Mus.; Thin section, Washington, U.S. Nat. Mus.;
Specimen(s): [1983,M.40], 4.2g.

Korgaon v Naoki.

Korrelocking 26° S., 122° E. approx.
Western Australia, Australia
Found 1937
No details of this fall are available, A.L. Coulson, Mem. Geol. Surv. India, 1940, **75**, p.125 [M.A. 8-54], F.C. Leonard, Univ. New Mexico Publ., Albuquerque, 1946 (meteoritics ser. no. 1), p.17.

Koso-sho
China, Not located
Found 1947, known before this year
Pseudometeorite..
A supposed meteorite of this name is noted in the report of an exchange, Rep. U.S. Nat. Mus., 1947, p.44. Discredited, manufactured iron, contains no nickel, B. Mason, Min. Mag., 1969, **37**, p.287.

Kosseir v El Qoseir.

Kossuth County v Forest City.

Köstritz v Pohlitz.

Kota-Kota 13°1′ S., 34°12′ E.
Marimba district, Malawi
Found 1905, known before this year
Synonym(s): *Marimba*
Stone. Enstatite chondrite (E4).
A large stone was said by natives to have fallen near their village some years before 1905 and was held as sacred. A nearly complete stone of 334g (but described by natives as part of the large stone) was brought to Mr. A.J. Swann, the magistrate of the Marimba district. Description, G.T. Prior, Min. Mag., 1914, **17**, p.129, B. Mason, Geochimica et Cosmochimica Acta, 1966, **30**, p.23. Contains zincian daubréelite, K. Keil, Am. Miner., 1968, **53**, p.491. Contains perryite, S.J.B. Reed, Min. Mag., 1968, **36**, p.850. Trace element data, C.M. Binz et al., Geochimica et Cosmochimica Acta, 1974, **38**, p.1579. Mineralogy, K. Keil, J. Geophys. Res., 1968, **73**, p.6945. Bulk analysis, 29.64 % total iron, A.J. Easton and C.J. Elliott, Meteoritics, 1977, **12**, p.409.
8g Vienna, Naturhist. Mus.; 2.7g Chicago, Field Mus. Nat. Hist.; 2.2g New York, Amer. Mus. Nat. Hist.;
Specimen(s): [1905,355], 232g. and fragments, 2.6g.

Kotschki v Ketschki.

Kouga v Kouga Mountains.

Kouga Mountains 33°37′ S., 24°0′ E.
Humansdorf district, Cape Province, South Africa
Found 1903, about
Synonym(s): *Kouga*
Iron. Octahedrite, medium (0.7mm) (IIIB).
A mass of 2586lb was found at Joubert's Kraal, L. Peringuey, letter of 7 January, 1916, in Min. Dept. BM(NH). Analysis, 9.29 %Ni, 18.0 ppm.Ga, 35.3 ppm.Ge, 0.022 ppm.Ir, E.R.D. Scott et al., Geochimica et Cosmochimica Acta, 1973, **37**, p.1957. Description; weathered, mildly shocked, V.F. Buchwald, Iron Meteorites, Univ. of California, 1975, p.744.
1175kg Cape Town, South African Mus.; 177g Vienna, Naturhist. Mus.;
Specimen(s): [1916,60], 288.5g. slice, and filings, 6g.

Koul Bokkeveld v Cold Bokkeveld.

Koursk v Botschetschki.

Koursk v Sevrukovo.

Kraehenberg v Krähenberg.

Krähenberg 49°13′ N., 7°23′ E.
Zweibrücken, Rheinland-Pfalz, Germany
Fell 1869, May 5, 1830 hrs
Synonym(s): *Kraehenberg, Zweibrücken*
Stone. Olivine-hypersthene chondrite, amphoterite (LL5).
After appearance of a cloud and detonations, a stone of about 16.5kg fell. Reported, with analysis, G. von Rath, Ann. Phys. Chem. (Poggendorff), 1869, **137**, p.176, 328, 335, 617. Further report, G. Neumayer, Sitzungsber. Akad. Wiss. Wien, Math.-naturwiss. Kl., 1869, **60**, p.229. The analysis shows more K than Na and is doubtful, W. Wahl, Geochimica et Cosmochimica Acta, 1950, **1**, p.28. Chemical composition and Rb/Sr age of the light and dark portions, W. Kempe and O. Müller, Meteorite Research, ed. P.M. Millman, D. Reidel, Dordrecht-Holland, 1969, p.418. Contains K-rich glass, olivine $Fa_{27.8}$, F. Wlotzka et al., Geochimica et Cosmochimica Acta, 1983, **47**, p.743.
16.5kg Speyer, Germany, Hist. Mus.; 85g Vienna, Naturhist. Mus.; 12g Budapest, Nat. Mus.; 5g Berlin, Humboldt Univ.; 4.4g Washington, U.S. Nat. Mus.;
Specimen(s): [85237], 7.5g.; [1920,311], 1.8g.; [43061], 1.5g.

Krähenholz v Barntrup.

Krakhut v Benares.

Kramer Creek 38°23′33″ N., 104°10′36″ W.
Pueblo County, Colorado, U.S.A.
Found 1966, recognized 1972
Stone. Olivine-hypersthene chondrite (L4).
One mass of 2.3kg was found in a rock garden. About one third of the oriented stone appears to be missing, T.E. Schmidt, Meteoritics, 1973, **8**, p.67, T.E. Schmidt, Earth Sci. (Colorado Springs), 1973, **26**, p.63, Meteor. Bull., 1974 (52), Meteoritics, 1974, **9**, p.113. Description, with an analysis, 20.36 % total iron, olivine $Fa_{21.7}$, E.K. Gibson, Jr. et al., Meteoritics, 1977, **12**, p.95.
Main mass, Possession of T.E. Schmidt, Colorado Springs; 118g Denver, Mus. Nat. Hist.; 71g Chicago, Field Mus. Nat. Hist.; 38g Washington, U.S. Nat. Mus.; 38g Albuquerque, Univ. of New Mexico;
Specimen(s): [1973,M.12], 51.2g. crusted part-slice

Kranganur v Cranganore.

Krasnoiarsk v Krasnojarsk.

Krasnoi-Ugol 54°2′ N., 40°54′ E.
Ryazan, Federated SSR, USSR, [Красный Угол]
Fell 1829, September 9, 1400 hrs
Synonym(s): *Krasnyi Ugol*
Stone. Olivine-hypersthene chondrite (L6).
After detonations, seven stones are said to have fallen, but only two were found, Ann. Phys. Chem. (Poggendorff), 1829, **17**, p.379. Analysis, 23.31 % total iron, M.I. D'yakonova and V.Ya. Kharitonova, Метеоритика, 1961, **21**, p.52. Listed, E.L. Krinov, Каталог Метеоритов Акад. Наук СССР, Москва, 1947 [M.A. 10-511]. Olivine Fa_{24}, B. Mason, Geochimica et Cosmochimica Acta, 1963, **27**, p.1011.
2.35kg Moscow, Acad. Sci., main mass; 61g Berlin, Humboldt Univ.; 11g Vienna, Naturhist. Mus.; 5.5g Washington, U.S. Nat. Mus.;
Specimen(s): [35184], 5g.

Krasnojarsk 54°54′ N., 91°48′ E.

Yeniseisk, Federated SSR, USSR, [Красноярск]
Found 1749
Synonym(s): *Berg Emir, Emir, Kemis, Kemiz, Krasnoiarsk, Krasnoyarsk, Malyi Altai, Malyi Altaj, Medvedeva, Medwedewa, Mount Kemis, Pallace Iron, Pallas Iron*
Stony-iron. Pallasite (PAL).

A mass estimated at about 700kg was discovered in 1749 about 145 miles south of Krasnoyarsk, between the Ubei and Sisim rivers; it was seen by P.S. Pallas in 1772, and was transported to Krasnoyarsk, P.S. Pallas, Reise Russ. Reichs, St. Petersburg, 1776, **3**, p.411, A. Göbel, Bull. Acad. Sci. St.-Petersbourg, 1867, **10**, p.305. Mentioned, G. Thomson, Atti Accad. Sci. Siena, 1808, **9**, p.37. Thomson (=W. Thomson, F.R.S.) studied the etching of the iron with nitric acid and was the first to develop the etch figures commonly called Widmanstätten figures. The fragment mentioned under the name of Malyi Altai, P. Dravert, Sibirske Ogni, 1930 (9), p.122, I.S. Astapowitsch, Trans. Roy. Astron. Soc. Canada, 1938, **32**, p.195 [M.A. 7-172], is a piece of Krasnojarsk, Meteor. Bull., 1959 (13). Co-ordinates, A.I. Eremeeva, Метеоритика, 1980, **39**, p.134. Texture, references, olivine $Fa_{12.2}$, E.R.D. Scott, Geochimica et Cosmochimica Acta, 1977, **41**, p.693. Analysis of metal, 8.9 %Ni, 22 ppm.Ga, 56.6 ppm.Ge, 0.18 ppm.Ir, J.T. Wasson and S.P. Sedwick, Nature, 1969, **222**, p.22.

515kg Moscow, Acad. Sci.; 2.6kg Copenhagen, Univ.; 4kg Vienna, Naturhist. Mus.; 2.7kg Berlin, Humboldt Univ.; 2.6kg Budapest, Nat. Mus. (1.8kg), Eötvö Lorand Univ. (0.8kg); 2.4kg Paris, Mus. d'Hist. Nat.; 2.2kg Tübingen, Univ.; 1.7kg Oxford, Univ.; 1kg New York, Amer. Mus. Nat. Hist.; 1kg Oslo, Min.-Geol. Mus.; 910g Washington, U.S. Nat. Mus.; 800g Chicago, Field Mus. Nat. Hist.; 600g Prague, Nat. Mus.;

Specimen(s): [90234], 2860g.; [90235], 259g.; [36604(2)], 241.5g.; [46982], 30.75g.; [1911,719], 16g.; [36603(1)], 0.5g.; [1959,170], 6.5g.; [1963,772], 1100g.; [1964,713], 170g.; [1969,7], 26g.

Krasnojarsk (iron) v Toubil River.

Krasnoslobodsk v Novo-Urei.

Krasnoyarsk v Krasnojarsk.

Krasnyi Klutsch v Krasnyi Klyuch.

Krasnyi Klyuch 54°20′ N., 56°5′ E.

Ufa, Bashkirskaya ASSR, USSR, [Красный Ключ]
Fell 1946, May 4
Synonym(s): *Krasnyi Klutsch, Rodnik*
Stone. Olivine-bronzite chondrite (H5).

A mass of about 4kg fell 30 km south of Ufa on the Krasnyi Kluch collective farm, near the village of Rodnik. Description and analysis, 27.01 % total iron, O.A. Kirova and M.I. D'yakonova, Метеоритика, 1970, **30**, p.148.

3kg Ufa, Mus.; 617g Moscow, Acad. Sci.;

Krasnyi Ugol v Krasnoi-Ugol.

Kravin v Tabor.

Krawin v Tabor.

Kress 34°21′30″ N., 101°43′48″ W.

Swisher County, Texas, U.S.A.
Found 1951, possible year of find, recognized 1966
Stone. Olivine-hypersthene chondrite (L6).

One mass of 4.3kg was recognized to be meteoritic in 1966; it had possibly been ploughed up in 1951, Meteor. Bull., 1972 (51), Meteoritics, 1972, **7**, p.224, G.I Huss, letter of 5 May, 1971, in Min. Dept. BM(NH). A second stone, Kress (b), is listed, G.I. Huss, Cat. Huss Coll. Meteorites, 1976, p.26. Olivine $Fa_{24.6}$, A.L. Graham, priv. comm., 1981.

2.8kg Mainz, Max-Planck-Inst.; 2.55kg Mainz, Max-Planck-Inst., of Kress (b) stone; 153g Copenhagen, Univ. Geol. Mus.; 113g Washington, U.S. Nat. Mus.;

Specimen(s): [1971,118], 88.5g. crusted slice

Krider 34°28′ N., 103°55′18″ W.

Roosevelt County, New Mexico, U.S.A.
Found 1978
Stone. Olivine-bronzite chondrite (H6).

A single mass of 2386g was ploughed up 9.6 km NE. of Krider, G.I Huss, priv. comm., 1982. Olivine $Fa_{20.4}$, A.L. Graham, priv. comm., 1983.

Specimen(s): [1983,M.23], 139g.

Krimka v Krymka.

Kroonstad v Cronstad.

Kruki v Brahin.

Krukov v Brahin.

Krutikha 56°48′ N., 77°0′ E.

Near Krutikha, Novosibirsk region, Federated SSR, USSR
Fell 1906, or 1907, recognized 1981
Stone. Chondrite.

A single stone of 845.2g fell into a kitchen garden and was recovered at once. Donated by the son of the finder to the Academy of Sciences of the USSR, Moscow, in 1981, Meteor. Bull., 1982 (60), Meteoritics, 1982, **17**, p.95.

845g Moscow, Acad. Sci.;

Krymka 47°50′ N., 30°46′ E.

Nicholayev region, Ukraine, USSR, [Крымка]
Fell 1946, January 21
Synonym(s): *Krimka*
Stone. Olivine-hypersthene chondrite, amphoterite (LL3).

A shower of stones fell, about 25kg, and is in the Ukraine Acad. Sci., E.L. Krinov, Метеоритика, 1949, **5**, p.49 [M.A. 12-105]. Analysis, 19.67 % total iron, M.I. D'yakanova and V.Ya. Kharitonova, Метеоритика, 1960, **18**, p.48 [M.A. 16-447]. A further 25kg of material has since been collected, R.L. Dreizin, Метеоритика, 1958, **16**, p.105 [M.A. 14-129]. Very unequilibrated, LL-group classification, R.T. Dodd and E. Jarosewich, Meteoritics, 1979, **14**, p.380, abs. Experimental study of trace element retention, M. Ikramuddin et al., Geochimica et Cosmochimica Acta, 1977, **41**, p.393.

11.6kg Kiev, Acad. Sci.; 800g Moscow, Acad. Sci.; 136g Washington, U.S. Nat. Mus.; 70g Vienna, Naturhist. Mus.; 48g Chicago, Field Mus. Nat. Hist.;

Specimen(s): [1956,325], 61.5g.

Krzadka 50°22′30″ N., 21°44′ E.
Majdan, Poland
Found 1929, July
Iron. Octahedrite.
A mass of between 2 and 3kg was found in glacial gravel, and was kept in Poznan Univ. until 1944 but is now lost, J. Pokrzywnicki, Acta Geophys. Polon., 1958, **6**, p.84, Urania, 1957 (1), p.16.

Kuangtung Yang-Chiang v Yangchiang.

Kuasti-Knasti v Białystok.

Kuga 34°6′ N., 132°5′ E.
Yamaguchi-Ken, Honshu, Japan
Found 1950, known before this year
Iron. Octahedrite, medium (0.6mm) (IIIB).
A small piece of iron, about 11g, has long been kept in the mineral collection of Yamaguchi University, labelled as a meteorite from Kuga-gun, but its date of fall or find is unknown, S. Murayama, Nat. Sci. Mus. Tokyo, 1953, **20**, p.129 [M.A. 13-80]. A second specimen of about 6kg was found in about 1960, Meteor. Bull., 1963 (28), S. Murayama, letter of 6 April, 1962 in Min. Dept. BM(NH).. Description, S. Murayama, Bull. Nat. Sci. Mus. Tokyo, 1959 (45), p.359. Analysis, 9.65 %Ni, 16.8 ppm.Ga, 25.7 ppm.Ge, 0.03 ppm.Ir, M. Shima et al., Meteoritics, 1979, **14**, p.535, abs.
5.5g Tokyo, Nat. Sci. Mus.; 2.7g Yamaguchi Univ.;

Kukschin 51°9′ N., 31°42′ E.
Nezhin, Chernigov, Ukraine, USSR, [Кукшино]
Fell 1938, June 11, 1400 hrs
Synonym(s): *Kukschino, Kukshin*
Stone. Olivine-hypersthene chondrite (L6).
A mass of 2250g was seen to fall near Kukschin. Description, with an analysis, I.S. Astapovich, Nature, 1939, **143**, p.377 [M.A. 7-271]. Description, with details of the phenomena of fall, L.A. Kulik, Метеоритика, 1941, **2**, p.123 [M.A. 10-174]. Listed, E.L. Krinov, Астрон. Журнал, 1945, **22**, p.303 [M.A. 9-297], R.L. Dreizin, Метеоритика, 1951, **9**, p.64 [M.A. 12-249]. Olivine Fa_{24}, B. Mason, Geochimica et Cosmochimica Acta, 1963, **27**, p.1011. Further analysis, 22.79 % total iron, O.A. Kirova et al., Метеоритика, 1978, **37**, p.87.
1928g Kiev, Acad. Sci. Ukraine; 248g Moscow, Acad. Sci.;

Kukschino v Kukschin.

Kukshin v Kukschin.

Kulak 30°43′52″ N., 66°48′8″ E.
Quetta-Pishin, Baluchistan, Pakistan
Fell 1961, March 25, 1730 hrs
Synonym(s): *Kalak*
Stone. Olivine-hypersthene chondrite (L5).
Two stones of about the same size were found to the east and to the west of the village of Kulak, near Quilla Abdulla, District of Quetta-Pishin, Pakistan. One was lost before October 1970; the other weighed 453.6g. Reported, Times newspaper of 8 April, 1961, Meteor. Bull., 1974 (52), Meteoritics, 1974, **9**, p.103. Description, with microprobe analyses, olivine Fa_{24}, A.L. Graham, Meteoritics, 1973, **8**, p.181.
Main mass, Quetta, Geol. Surv. Pakistan;
Specimen(s): [1972,1], 72.1g. part slice, and fragments, 4.6g.

Kuleschewka v Kuleschovka.

Kuleschovka 50°45′ N., 33°30′ E.
Poltava, Ukraine, USSR, [Кулешовка]
Fell 1811, March 12, 1100 hrs
Synonym(s): *Kuleschewka, Kuleshovka, Poltava (of G. von Blöde)*
Stone. Olivine-hypersthene chondrite (L6), veined.
A stone of over 6kg fell, after detonations, Ann. Phys. (Gilbert), 1811, **38**, p.120, A. Göbel, Bull. Acad. Sci. St.-Petersbourg, 1867, **11**, p.242. Analysis, 22.83 % total iron, V.Ya. Kharitonova and L.D. Barsukova, Метеоритика, 1982, **40**, p.41. Olivine Fa_{25}, B. Mason, Geochimica et Cosmochimica Acta, 1963, **27**, p.1011.
4161g Moscow, Acad. Sci.; 196g Vienna, Naturhist. Mus.; 131g Tempe, Arizona State Univ.; 28g Tübingen, Univ.; 82g Calcutta, Mus. Geol. Surv. India; 24g Chicago, Field Mus. Nat. Hist.; 13g New York, Amer. Mus. Nat. Hist.; 11g Washington, U.S. Nat. Mus.;
Specimen(s): [44774], 32g.; [36610(8)], 9.5g.; [44775], 9g.; [90258], 7.5g.

Kuleshovka v Kuleschovka.

Kuli-schu
China, Not located
Fell 1827, August 30
Doubtful..
A stone is said to have fallen, K.E.A. von Hoff, Ann. Phys. Chem. (Poggendorff), 1830, **18**, p.185, but the evidence is not conclusive.

Kulnine 34°9′ S., 141°47′ E.
Wentworth County, Victoria, Australia
Found 1886, known in this year
Stone. Olivine-hypersthene chondrite (L6).
A mass of 122lb in the South Australian Mus., Adelaide, D.W.P. Corbett, Rec. S. Austr. Mus., 1968, **15**, p.767. Olivine $Fa_{25.4}$, B. Mason, Rec. Austr. Mus., 1974, **29**, p.169.
471g New York, Amer. Mus. Nat. Hist.; 170g Washington, U.S. Nat. Mus.;
Specimen(s): [1973,M.24], 65g.

Kulp 41°7′ N., 45°0′ E.
Kazakh, Azerbaydzhan SSR, USSR, [Кульп]
Fell 1906, March 29
Stone. Olivine-bronzite chondrite (H6).
Two stones were found an hour after they had fallen, one of 3.5kg and the other of 7-8kg, P.N. Chirvinsky, Centralblatt Min., 1923, p.550. Two stones fell, 3.719kg together, E.L. Krinov, Астрон. Журнал, 1945, **22**, p.303 [M.A. 9-297]. Krinov gives the date of fall as March 16, but this appears to be an error due to overlooking the change from Old style to New. Analysis, 28.69 % total iron, M.I. D'yakanova, Метеоритика, 1964, **25**, p.129. Olivine Fa_{17}, B. Mason, Geochimica et Cosmochimica Acta, 1967, **31**, p.1100.
3403g Moscow, Acad. Sci., main mass;

Kumdah 20°23′ N., 45°5′ E.
Saudi Arabia
Found 1973, before this year
Iron.
A mass described as 2-3 feet in diameter was found and a portion removed. The main mass is still in situ. A fragment of 33g was sent to the Smithsonian Institution where its meteoritic nature was confirmed. A completely oxidized iron meteorite, B. Mason, Ms. on Arabian Meteorites, 1981, p.12.

Kumerina 24°55′ S., 119°25′ E.

Western Australia, Australia
Found 1937, February
Iron. Octahedrite, plessitic (IIC).

A mass of 118lb was found near Batthewmurnana Hill, 24°55'S., 119°25'E. Described, with an analysis, 9.55 %Ni, E.S. Simpson, Min. Mag., 1938, **25**, p.160. Classification and analysis, 9.69 %Ni, 36.8 ppm.Ga, 93.4 ppm.Ge, 8.1 ppm.Ir, J.T. Wasson, Geochimica et Cosmochimica Acta, 1969, **33**, p.859. Brief description, V.F. Buchwald, Iron Meteorites, Univ. of California, 1975, p.746.

Main mass, Perth, West Austr. Mus.; 670g Copenhagen, Univ. Geol. Mus.; 339g Washington, U.S. Nat. Mus.;
Specimen(s): [1938,220], 1156g.

Kumisch Choi Cha v Armanty.

Kumys-Tyuya v Armanty.

Kunaschak v Kunashak.

Kunashak 55°47′ N., 61°22′ E.

Chelyabinsk region, Federated SSR, USSR, [Кунашак]
Fell 1949, June 11, 0814 hrs
Synonym(s): *Kunaschak, Kunischak*
Stone. Olivine-hypersthene chondrite (L6).

A shower of about 20 stones fell, total weight over 200kg, the largest stones weighing 120kg, 40kg and 36kg, E.L. Krinov, Метеоритика, 1950, **8**, p.66, I.T. Zotkin and E.L. Krinov, Метеоритика, 1957, **15**, p.51 [M.A. 14-129]. Analysis, 21.10 % total iron, M.I. D'yakanova and V.Ya. Kharitonova, Метеоритика, 1961, **21**, p.52. Olivine Fa_{24}, B. Mason, Geochimica et Cosmochimica Acta, 1963, **27**, p.1011.

135kg Moscow, Acad. Sci., includes the three largest stones; 863g Tempe, Arizona State Univ.; 387g New York, Amer. Mus. Nat. Hist.; 213g Yale Univ.; 175g Washington, U.S. Nat. Mus.; 162g Paris, Mus. d'Hist. Nat.;
Specimen(s): [1956,168], 191g.

Kunersdorf

Germany, Not located
Fell 1591, June 9
Doubtful..

Several large stones are said to have fallen, E.F.F. Chladni, Ann. Phys. (Gilbert), 1815, **50**, p.240, R.P. Greg, Rep. Brit. Assn., 1860, p.53, but the evidence is not conclusive. The exact locality is doubtful; there are eight villages possible; Kunersdorf, Silesia 52°40'N., 14°13'E.; Kunersdorf, Silesia (now Poland), 51°5'N., 17°50'E.; Kunersdorf, Brandenburg (now Poland), 52°22'N., 14°37'E.; Kunersdorf, Brandenburg (now Poland), 52°14'N., 15°14'E.; Kunersdorf, Brandenburg, 51°46'N., 14°12'E.; Kunersdorf, Saxony, 51°18'N., 14°4'E.; Kunnersdorf, Silesia (now Poland), 50°53'N., 15°43'E.; and Kunnersdorf, Saxony, 51°3'N., 14°40'E.

Kunischak v Kunashak.

Kuritawaki-muri v Takenouchi.

Kurgansku v Pesyanoe.

Kurla v Pillistfer.

Kursk v Botschetschki.

Kursk v Sevrukovo.

Kurumi 34°41′ N., 135°7′ E.

Mino, Hyogo, Honshu, Japan
Pseudometeorite..

Two fragments of stone, 8g and 36g, are said to have fallen at noon on 27 May, 1930, F. Watson, Nature, 1938, **141**, p.475 [M.A. 7-87], I. Yamamoto, Kwasan Observ. Bull., 1935, p.306 [M.A. 7-173]. They are fragments of quartz-porphyry, S. Murayama, Nat. Sci. Mus. Tokyo, 1953, **20**, p.153 [M.A. 13-80].

Kushiike 37°3′ N., 138°23′ E.

Naka Kubiki, Niigata, Honshu, Japan
Fell 1920, September 16, 1800 hrs
Synonym(s): *Kusiike*
Stone. Chondrite.

A stone of 4.46kg fell, and is preserved in the primary school at Kushiike, S. Kawai and S. Kanda, The Astron. Herald (Astron. Soc. Japan), 1921, **14**, p.35, S. Iimori and T. Yoshimura, Sci. Papers Inst. Phys. Chem. Res. Tokyo, 1929, **10** (suppl. 9), p.37 [M.A. 5-272], S. Murayama, letter of 6 April, 1962 in Min. Dept. BM(NH). Date of fall given as 20 September, 1920, A.D. Nininger, Pop. Astron., Northfield, Minnesota, 1940, **48**, p.559, Contr. Soc. Res. Meteorites, 1942, **2**, p.227 [M.A. 8-54].

Kusiali 29°41′ N., 78°23′ E.

Kumaun, Uttar Pradesh, India
Fell 1860, June 16, 0500 hrs
Synonym(s): *Kasiali*
Stone. Olivine-hypersthene chondrite (L6).

After detonations, a stone fell on hard rock and was shattered; only a few small fragments were preserved, N.S. Maskelyne, Phil. Mag., 1864, **28**, p.148. The place of fall is Kasiali, 38-40 miles SW. of Pauri, tahsil Lansdowne, district Garwal, C.A. Silberrad, Min. Mag., 1932, **23**, p.299. Olivine Fa_{24}, B. Mason, Geochimica et Cosmochimica Acta, 1963, **27**, p.1011.

0.5g Vienna, Naturhist. Mus.; 0.4g Calcutta, Mus. Geol. Surv. India;
Specimen(s): [35732], 4g.

Kusiike v Kushiike.

Kusnetsovo v Kuznetzovo.

Kusnezowo v Kuznetzovo.

Kutais 44°31′ N., 39°18′ E.

Kutais village, Krasnodar district, USSR
Fell 1977, November 28, 0800 hrs
Stone. Olivine-bronzite chondrite (H5).

A single mass of 23.04g was recovered after it had fallen with a loud whistling noise, Meteor. Bull., 1979 (56), Meteoritics, 1979, **14**, p.168. Brief description, R.L. Khotinok, Метеоритика, 1982, **40**, p.6. Mineralogy, olivine Fa_{19}, A.Ya. Skripnik et al., Метеоритика, 1982, **41**, p.50.

Kuttayi v Kuttippuram.

Kuttippuram 10°50′ N., 76°2′ E.
Ponnani taluq, Malabar district, Kerala, India
Fell 1914, April 6, 0700 hrs
Synonym(s): *Kuttayi, Trekanapuram, Treprangoda*
Stone. Olivine-hypersthene chondrite (L6).
After detonations, a shower of stones fell over four villages; the total weight was about 100lb and the largest stone, which fell in Kuttippuram, weighed about 71lb. Description, J. Coggin Brown, Rec. Geol. Surv. India, 1915, **45**, p.209 [M.A. 1-96]. Mineralogy, analysis, 23.10 % total iron, olivine $Fa_{24.3}$, A. Dube et al., Smithson. Contrib. Earth Sci., 1977 (19), p.71.
38kg Calcutta, Mus. Geol. Surv. India, includes the largest stone, in three pieces; 402g Paris, Mus. d'Hist. Nat.; 130g Tempe, Arizona State Univ.; 120g Washington, U.S. Nat. Mus.; 67g Chicago, Field Mus. Nat. Hist.; 46g Yale Univ.;
Specimen(s): [1915,88], 649g.; [1914,1420], 24.25g.

Kuyahinga v Knyahinya.

Kuznetsovo v Kuznetzovo.

Kuznetzovo 55°12′ N., 75°20′ E.
Tatarsk, Federated SSR, USSR, [Кузнецово]
Fell 1932, May 26, between 1700 and 1800 hrs
Synonym(s): *Kusnetsovo, Kusnezowo, Kuznetsovo*
Stone. Olivine-hypersthene chondrite (L6).
At least eight stones, about 23kg, fell at Kuznetzovo, 12 km NE. of Koloniya station. Three nearly complete stones (largest 2538g) and fragments of two more were preserved (total 4047g); two of the stones fit together. Description, P.L. Dravert, Min. Mag., 1934, **23**, p.509. More than 10 stones fell, totalling about 7kg, but only 4047g have been preserved, E.L. Krinov, Каталог Метеоритов Акад. Наук СССР, Москва, 1947 [M.A. 10-511]. Analysis, 21.42 % total iron, M.I. D'yakonova and V.Ya. Kharitonova, Метеоритика, 1961, **21**, p.52., P.L. Dravert, Метеоритика, 1941, **2**, p.61 [M.A. 9-296]. Olivine Fa_{23}, B. Mason, Geochimica et Cosmochimica Acta, 1963, **27**, p.1011.
3814g Moscow, Acad. Sci., includes largest stone, 2534g;
Specimen(s): [1936,1], 99g.

Kwasli v Białstok.

Kwidzyn v Schwetz.

Kwingauk v Quenggouk.

Kyancutta 33°17′ S., 136° E.
Le Hunte County, South Australia, Australia
Found 1932, June
Iron. Octahedrite, medium (1.1mm) (IIIA).
A mass of 72lb was found 28 miles ESE. of Kyancutta in Buxton Co. Description, with an analysis, 7.3 %Ni, L.J. Spencer, Min. Mag., 1933, **23**, p.329, J.F. Lovering and L.G. Parry, Geochimica et Cosmochimica Acta, 1962, **26**, p.361. Co-ordinates, D.W.P. Corbett, Rec. S. Austr. Mus., 1968, **15**, p.767. Newer analysis, 8.06 %Ni, 19.9 ppm.Ga, 39.5 ppm.Ge, 1.7 ppm.Ir, E.R.D. Scott et al., Geochimica et Cosmochimica Acta, 1973, **37**, p.1957. Description; kamacite shock-hardened, V.F. Buchwald, Iron Meteorites, Univ. of California, 1975, p.747.
Main mass, Canberra, Austr. Nat. Univ.; 1.36kg Sydney, Austr. Mus.; 1.41kg Chicago, Field Mus. Nat. Hist.; 754g Washington, U.S. Nat. Mus.; 750g Perth, West. Austr. Mus.; 464g Tempe, Arizona State Univ.;
Specimen(s): [1932,1564], 350.4g. and filings, 29g.; [1933,6], 19g.

Kybunga 33°54′ S., 138°29′ E.
South Australia, Australia
Found 1956
Synonym(s): *Kybunga I, Kybunga II, Kybunga III*
Stone. Olivine-hypersthene chondrite (L5).
One specimen of 2.76kg was found in 1956, a second mass (Kybunga II) of either 2.26kg or 2.37kg and a third mass (Kybunga III) of 2.86kg were found in 1971, J.M. Scrymgour, letter of 14 July, 1971, D.H. McColl, letter of 30 March, 1972, in Min. Dept. BM(NH). The masses are possibly related to "The Kybunga Daylight Meteor" of 1941, G.F. Dodwell and C. Fenner, Proc. S. Austr. Branch R. geogr. Soc. Austr., 1943, **44**, p.6. Olivine Fa_{25}, B. Mason, Rec. Austr. Mus., 1974, **29**, p.169. Analysis (of the III mass), 20.12 % total iron, M.J. Fitzgerald, Ph.D. Thesis, Univ. of Adelaide, 1979, p.23.
2.4kg Adelaide, Univ.; Largest mass, Adelaide, S. Austr. Mus., on loan from the finder Mr. I.D. Schumacher; 94g Washington, U.S. Nat. Mus.; 8g Chicago, Field Mus. Nat. Hist.;

Kybunga I v Kybunga.

Kybunga II v Kybunga.

Kybunga III v Kybunga.

Kyle 29°58′30″ N., 97°52′ W.
Hayes County, Texas, U.S.A.
Found 1965, recognized in this year
Stone. Olivine-hypersthene chondrite (L6).
One stone of 7.78kg was found on a farm, 3 miles SE. of Kyle, Meteor. Bull., 1972 (51), Meteoritics, 1972, **7**, p.225. Description, analysis, 21.31 % total iron, olivine Fa_{26}, R.V. Fodor et al., Meteoritics, 1971, **6**, p.71.
4.4kg Mainz, Max-Planck-Inst.; 253g Washington, U.S. Nat. Mus.; 105g Tempe, Arizona State Univ.; 88g New York, Amer. Mus. Nat. Hist.;
Specimen(s): [1967,439], 104g. slice

Kylesovice v Opava.

Kyolos v Mocs.

Kyushu 32°2′ N., 130°38′ E.
Kyushu, Japan
Fell 1886, October 26, 1500 hrs
Synonym(s): *Ensigahara, Hanakai, Hanakita, Hishikari, Hisikari, Hishugari, Hislugari, Maeme, Oguchimura, Ogutimura, Oshima, Oynchimura, Satsuma, Shigetome, Sigetome, Simagoe, Torigoe, Yamanomura, Yenshigahara, Yenokigahara*
Stone. Olivine-hypersthene chondrite (L6), veined.
After detonations, a shower of stones fell in the southern part of Kyushu in the provinces of Satsuma and Osumi, Kagoshima prefecture. The largest weighed about 29kg, the smaller were described as "innumerable", K. Jimbo, Beitr. Miner. Japan, 1906 (2), p.43. Described, with an analysis, 22.02 % total iron, B. Mason and H.B. Wiik, Geochimica et Cosmochimica Acta, 1961, **21**, p.272. Newer analysis, 21.22 % total iron, H. von Michaelis et al., Earth planet. Sci. Lett., 1969, **5**, p.387. Mentioned, S. Murayama, letter of 6 April, 1962 in Min. Dept. BM(NH). Olivine Fa_{26}, B. Mason, Geochimica et Cosmochimica Acta, 1963, **27**, p.1011.
3.6kg Tokyo, Sci. Mus.; 8.8kg New York, Amer. Mus. Nat. Hist.; 826g Tokyo, Geol. Surv. Japan; 600g Tokyo, Imp. Univ.; 295g Chicago, Field Mus. Nat. Hist.; 216g Washington, U.S. Nat. Mus.; 193g Tempe, Arizona State

Univ.; 154g Berlin, Humboldt Univ.;
Specimen(s): [80031], 28.8kg. the largest complete stone, "Yenshigahara"; [76809], 1479g. " Oynchimura"; [1905,67], 81g. "Oshima"; [1905,68], 34.5g. " Shigetome"; [68213], 4.1g.

La Baffe v Epinal.

La Bécasse 47°5′ N., 1°45′ E.
Dun le Pöelier, Indre, France
Fell 1879, January 31, 1230 hrs
Synonym(s): *Bécasse, Dun-le-Pöelier*
Stone. Olivine-hypersthene chondrite (L6).
After detonations, a stone of 2.8kg was seen to fall, G.A. Daubrée, C. R. Acad. Sci. Paris, 1879, **89**, p.597. Olivine Fa_{23}, B. Mason, Geochimica et Cosmochimica Acta, 1963, **27**, p.1011.
2.5kg Paris, Mus. d'Hist. Nat., main mass; 102g Calcutta, Mus. Geol. Surv. India; 88g Chicago, Field Mus. Nat. Hist.; 81g Washington, U.S. Nat. Mus.; 67g Vienna, Naturhist. Mus.; 29g New York, Amer. Mus. Nat. Hist.; 23g Berlin, Humboldt Univ.;
Specimen(s): [56467], 19.25g.

La Bella Roca v Bella Roca.

La Bella Rocka v Bella Roca.

Laborel 44°17′ N., 5°35′ E.
Drome, France
Fell 1871, June 14, 2000 hrs
Stone. Olivine-bronzite chondrite (H5).
Two stones, one of about 2kg and the other of 91g, appear to have fallen, but were not discovered (or made known) until 1895, E. Cohen, Ann. Naturhist. Hofmus. Wien, 1896, **11**, p.31, E.A. Wülfing, Die Meteoriten in Samml., Tübingen, 1897, p.193. Olivine Fa_{19}, B. Mason, Geochimica et Cosmochimica Acta, 1963, **27**, p.1011.
750g Grenoble, Faculté des Sciences; 882g Paris, Mus. d'Hist. Nat.; 982g Paris, Collection of Col. L. Vésignié; 236g Vienna, Naturhist. Mus.; 124g Berlin, Humboldt Univ.; 47g New York, Amer. Mus. Nat. Hist.;
Specimen(s): [80999], 283g.

La Caille 43°44′ N., 6°47′ E.
Grasse, Alpes Maritimes, France
Found 1828, recognized in this year
Synonym(s): *Caille, Grasse*
Iron. Octahedrite, medium (1.1mm) (IRANOM).
A mass of 626kg was used for about two centuries as a seat in front of the church of La Caille. It was recognized as meteoritic in 1828, it was said to have been brought from the mountain of Audibergue, about 6 miles SE. of La Caille, C.P. Brard, Séances Publiques Acad. Sci. Bordeaux, 1829, p.39, G.A. Daubrée, C. R. Acad. Sci. Paris, 1867, **64**, p.633. Most of the specimens have been heated to 1000C, but the Paris material appears to be undamaged, V.F. Buchwald, Iron Meteorites, Univ. of California, 1975, p.748. Chemically anomalous, analysis, 9.11 %Ni, 13.7 ppm.Ga, 21.5 ppm.Ge, 9.7 ppm.Ir, J.T. Wasson and R. Schaudy, Icarus, 1971, **14**, p.59.
625.95kg Paris, Mus. d'Hist. Nat.; 370g Vienna, Naturhist. Mus.; 304g Harvard Univ.; 107g Washington, U.S. Nat. Mus.; 101g Chicago, Field Mus. Nat. Hist.; 102g Berlin, Humboldt Univ.;
Specimen(s): [35725], 346g. and a fragment, 13g; [35161], 5g.

La Calle v Feid Chair.

La Charca 20°40′ N., 101°17′ W. approx.
Irapuato, Guanajuato, Mexico
Fell 1878, June 11, 1130 hrs
Synonym(s): *Charca, Irapuato*
Stone. Chondrite.
A stone of 399g fell 5 miles from Irapuato, A. Castillo, Cat. Meteorites Mexique, Paris, 1889, p.13. The stone is said to be in the College of Guanajuato.

Lac Labiche v Vilna.

La Colina 37°20′ S., 61°32′ W.
Gen. Lamadrid dept., Buenos Aires province, Argentina
Fell 1924, March 19, 2330 hrs
Stone. Olivine-bronzite chondrite (H5).
After the appearance of light, and detonations, a stone of 2kg fell on a ranch near La Colina. Described, with an analysis, F. Pastore, Anal. Mus. Nac. Hist. Nat. Buenos Aires, 1925, **33**, p.297 [M.A. 3-389]. Olivine Fa_{17}, B. Mason, Geochimica et Cosmochimica Acta, 1963, **27**, p.1011.
1498g Buenos Aires, Mus. Nac. Hist. Nat.;
Specimen(s): [1962,170], 6.5g.

Ladder Creek 38°37′ N., 101°38′ W.
Greeley County, Kansas, U.S.A.
Found 1937, July
Synonym(s): *Greeley County, Horace (no. 1) (2.9kg), Horace (no. 2) (in part), Tribune*
Stone. Olivine-hypersthene chondrite (L6).
Many fragments totalling 35.1kg were found, E.P. Henderson, letters of 12 May and 3 June, 1939 in Min. Dept. BM(NH). Listed, A.D. Nininger, Pop. Astron., Northfield, Minnesota, 1939, **47**, p.212, H.H. and A.D. Nininger, The Nininger Collection of Meteorites, Winslow, Arizona, 1950, p.67. A 1.108kg fragment in Tempe (Arizona State Univ.) is a hypersthene chondrite and so is ascribed to this fall and not to Horace (*q.v.*). The smaller of the two Horace stones mentioned, B. Mason, Meteoritics, 1969, **4**, p.240, is now assigned to Ladder Creek, as is Tribune. Olivine Fa_{25}, B. Mason, Geochimica et Cosmochimica Acta, 1963, **27**, p.1011.
5.2kg Tempe, Arizona State Univ.; 5.7kg Washington, U.S. Nat. Mus.; 4.7kg New York, Amer. Mus. Nat. Hist.; 440g Chicago, Field Mus. Nat. Hist.;
Specimen(s): [1949,75], 56.2g.; [1959,999], 6349g.

La Encantada v Imilac.

La Escondida 24°20′30″ N., 102°4′30″ W.
near Nuevo Mercurio, Zacatecas, Mexico
Found 1979, March
Stone. Olivine-bronzite chondrite (H5).
A single mass of 8.2g was found about 14 km NNE. of Nuevo Mercurio village during the search for specimens of the Nuevo Mercurio fall (*q.v.*). Olivine Fa_{17}, Meteor. Bull., 1980 (57), Meteoritics, 1980, **15**, p.99.

Lafayette (iron) 39°59′ N., 105°5′ W.
Colorado, U.S.A.
Found 1908, before this year
Iron. Ataxite, Ni-rich?.
A piece weighing 11g in the Geological Collection of the Univ. of Colorado, Boulder, was analysed, 59.4 %Ni, R.M. Butters, West. Chem. Metallurg., 1908, **4**, p.181. It appears to have been all consumed in the analysis, as no specimen is now in the Univ. of Colorado, F. Howland, letter of 9 March, 1927 in Min. Dept. BM(NH). Probably a

pseudometeorite, A. Kracher and J. Willis, Meteoritics, 1981, **16**, p.239.

Lafayette (stone) 40°25′ N., 86°53′ W.

Tippecanoe County, Indiana, U.S.A.
Found 1931, before this year
Synonym(s): *La Fayette, Purdue*
Stone. Achondrite, Ca-rich. Nakhlite (ACANOM).

A mass of about 800g was noticed by O.C. Farrington in 1931 in the geological collections in Purdue Univ., Lafayette. Description, H.H. Nininger, Pop. Astron., Northfield, Minnesota, 1935, **43**, p.404 [M.A. 6-207], B. Mason, Meteorites, Wiley, 1962, p.232. The stone is very fresh and similar to Nakhla; it may be part of that fall. K-Ar age, 1600 m.y., F.A. Podosek and J.C. Huneke, Geochimica et Cosmochimica Acta, 1973, **37**, p.667. Distinct from Nakhla and Governador Valadares, J.L. Berkley et al., Geochimica et Cosmochimica Acta, 1980 (Suppl. 14), p.1089. Petrology, N.Z. Boctor et al., Earth planet. Sci. Lett., 1976, **32**, p.69 [M.A. 77-1999].

637g Washington, U.S. Nat. Mus., main mass; 30g Tempe, Arizona State Univ.; 101g Chicago, Field Mus. Nat. Hist.;
Specimen(s): [1959,755], 35g. slice

La Fayette v Lafayette (stone).

Lageado Ipiranga v Ipiranga.

La Grange 38°24′ N., 85°22′ W.

Oldham County, Kentucky, U.S.A.
Found 1860
Synonym(s): *Oldham County*
Iron. Octahedrite, fine (0.27mm) (IVA).

A mass if 112lb was found. Report, with an analysis, 7.81 % Ni, J.L. Smith, Am. J. Sci., 1861, **31**, p.265. Analysis, and determination of Ga, Au and Pd, E. Goldberg et al., Geochimica et Cosmochimica Acta, 1951, **2**, p.1. Classification and analysis, 7.4 %Ni, 2.07 ppm.Ga, 0.115 ppm.Ge, 2.3 ppm.Ir, R. Schaudy et al., Icarus, 1972, **17**, p.174. Description; cold worked with troilite injected along fissures, V.F. Buchwald, Iron Meteorites, Univ. of California, 1975, p.750.

36.6kg Tempe, Arizona State Univ.; 2.2kg Washington, U.S. Nat. Mus.; 981g Berlin, Humboldt Univ.; 840g Tempe, Arizona State Univ.; 440g Vienna, Naturhist. Mus.; 218g Chicago, Field Mus. Nat. Hist.;
Specimen(s): [34580a], 105g. and fragments, 38g.

La Grange v Bluff.

Laguna de los Manantiales v Laguna Manantiales.

Laguna Manantiales 48°35′ S., 67°25′ W.

Deseado department, Santa Cruz, Argentina
Found 1945
Synonym(s): *Laguna de los Manantiales*
Iron.

One mass of 92kg was found by Sr. A. Naves, L.O. Giacomelli, letters of 30 March, 1960 and 30 October, 1962 in Min Dept. Brit. Mus.

Main mass, held by the finder;
Specimen(s): [1964,67], 2.5g.

Lahrauli 26°47′ N., 82°43′ E.

Basti district, Uttar Pradesh, India
Fell 1955, March 24, 1100 hrs, approx.
Synonym(s): *Basti*
Stone. Achondrite, Ca-poor. Ureilite (AURE).

The original mass is not known but a fragment of 650g was analysed, P.D. Malhotra, Rec. Geol. Surv. India, 1962, **89**, p.479. Reported, Meteor. Bull., 1980 (58), Meteoritics, 1980, **15**, p.237. Description, olivine Fa_{15-21}, N. Bhandari et al., Meteoritics, 1981, **16**, p.185.

650g Calcutta, Mus. Geol. Surv. India; 250g Lucknow, Mus.;
Specimen(s): [1980,M.23], 0.1g.

L'Aigle 48°46′ N., 0°38′ E.

Orne, France
Fell 1803, April 26, 1300 hrs
Synonym(s): *Aigla, Aigle, Ober-Pfalz, Waldau*
Stone. Olivine-hypersthene chondrite (L6), brecciated.

After the appearance of a fireball, followed by detonations, a shower of stones, estimated at 2000-3000 in number and of aggregate weight about 37kg, the largest weighing about 9kg, fell within an area of 6 × 2.5 miles. The detailed report of the phenomena first established beyond doubt the fact of the fall of stones from outer space, J.B. Biot, Mém. Inst. France, 1806, **7** (Histoire), p.224, J.B. Biot, Ann. Phys. (Gilbert), 1804, **16**, p.44. Description, H. Pfahler, Tschermaks Min. Petr. Mitt., 1892, **13**, p.362. Analysis, E.H. von Baumhauer, Arch. Néerland. Sci. Nat. Haarlem, 1872, **7**, p.154. Analysis, olivine Fa_{23}, 21.65 % total iron, R.T. Dodd and E. Jarosewich, Meteoritics, 1981, **16**, p.93.

10.5kg Paris, Mus. d'Hist. Nat.; 4kg Vienna, Naturhist. Mus.; 3.5kg Budapest, Nat. Mus.; 1.9kg Berlin, Humboldt Univ.; 981g Tempe, Arizona State Univ.; 909g Yale Univ.; 755g Chicago, Field Mus. Nat. Hist.; 448g Washington, U.S. Nat. Mus.; 482g Moscow, Acad. Sci.; 378g Tübingen, Univ.; 285g Oxford, Univ. Mus.; 252g Copenhagen, Univ.; 223g Dublin, Nat. Mus.; 177g Perth, West. Austr. Mus.; 168g New York, Amer. Mus. Nat. Hist.;
Specimen(s): [90248], 1018g. complete stone; [90251], 105g.; [90252], 225g.; [33900], 258g.; [33182], 297g. one stone, 145g and a piece, 152g.; [33901], 141g.; [1952,255], 67g.; [90250], 17.5g. complete stone; [94707], 35g. four fragments; [1927, 1291], 3g.; [1950,392], 16g. fragments; [1964,707], 33g. two pieces.; [95738], 17.5g. fragments; [96267], thin section; [1964,569], thin section

Laissac v Favars.

Lajeado Ipiranga v Ipiranga.

Lakangaon 21°52′ N., 76°2′ E.

Nimar, Indore, Madhya Pradesh, India
Fell 1910, November 24, 1800 hrs
Synonym(s): *Lapangaon*
Stone. Achondrite, Ca-rich. Eucrite (AEUC).

After the appearance of a flash of light, followed by a trail of smoke and detonations, a stone fell, portions of which weighing 125g and 87.5g were collected, G. de P. Cotter, Rec. Geol. Surv. India, 1912, **42**, p.275. Analysis, 17.74 % total iron, T.S. McCarthy et al., Meteoritics, 1974, **9**, p.215.

116g Calcutta, Mus. Geol. Surv. India; 10g Washington, U.S. Nat. Mus.; 1.1g New York, Amer. Mus. Nat. Hist.;
Specimen(s): [1915,142], 85g.

Lake Bonney 37°45′ S., 140°18′ E.

South Australia, Australia
Found 1961, October
Stone. Olivine-hypersthene chondrite (L6).

Four stones were found, the largest 1.96kg, total 2.8kg, D.W.P. Corbett, Rec. S. Austr. Mus., 1964, **14** (4). A fifth stone found 42 miles away was originally thought to belong to the same fall but is distinct and is named Nora Creina, D.W.P. Corbett, Min. Mag., 1967, **36**, p.293. Olivine $Fa_{24.5}$, B. Mason, Rec. Austr. Mus., 1974, **29**, p.169. Analysis, 20.63 % total iron, M.J. Fitzgerald, Ph.D. Thesis, Univ. of Adelaide, 1979, p.23.

2534g Adelaide, South Austr. Mus.; Thin section, Washington, U.S. Nat. Mus.;

Lake Brown 31° S., 118°30′ E. approx.

County Avon, Western Australia, Australia
Found 1919
Synonym(s): *Burracoppin, Lake Moore*
Stone. Olivine-hypersthene chondrite (L6).

A stone of 21.5lb was found, Ann. Rep. Geol. Surv. West. Austr. for 1921, 1922, p.53. Description and analysis, G.T. Prior, Min. Mag., 1929, **22**, p.155. A second mass of 13.6kg, originally named Lake Moore, is now ascribed to this fall, G.J.H. McCall, 1st. Suppl. to West. Austr. Mus. Spec. Publ. no. 3, 1968, p.5. Olivine Fa_{25}, B. Mason, Rec. Austr. Mus., 1974, **29**, p.169.

Main mass, Perth, Austr., Geol. Surv. Mus.; 375g Canberra, Austr. Nat. Univ.; 137g New York, Amer. Mus. Nat. Hist.; 0.2g Washington, U.S. Nat. Mus.; 33.6g Tempe, Arizona State Univ., of the Lake Moore stone;
Specimen(s): [1925,1037], 174.5g.

Lake Giles v Mount Dooling.

Lake Grace 33°4′ S., 118°13′ E.

Western Australia, Australia
Found 1956
Stone. Olivine-hypersthene chondrite (L6).

A mass of 10.5kg was found while ploughing 3 miles NNW. of Tarin Rock Siding, Spec. Publ. West. Austr. Mus., 1965 (3), p.37, B. Mason, letter of 4 March, 1964 in Min. Dept. BM(NH). Olivine $Fa_{25.9}$, B. Mason, Rec. Austr. Mus., 1974, **29**, p.169.

Main mass, Adelaide, South Austr. Mus., mass in fragments; 335g Washington, U.S. Nat. Mus.; 20g New York, Amer. Mus. Nat. Hist.;

Lake Labyrinth 30°32′ S., 134°45′ E.

South Australia, Australia
Found 1924, probably fell 1924
Stone. Olivine-hypersthene chondrite, amphoterite (LL6), xenolithic.

Meteorite fragments were found by Mr. B. Austin in 1924, about two weeks after the probable fall. Most material (57lb of fragments) was collected in 1934 by Mr R. Bedford from 8 miles north of Peela well, Wilgena station. The original weight was probably about 75lb and the probable date of fall about 5 February, 1924. Description, L.J. Spencer, Min. Mag., 1936, **24**, p.353, D.W.P. Corbett, Rec. S. Austr. Mus., 1968, **15**, p.776. Analysis, 20.46 % total iron, M.I. D'yakonova, Метеоритика, 1968, **28**, p.131. Olivine $Fa_{29.4}$, B. Mason, Rec. Austr. Mus., 1974, **29**, p.169. Xenolithic, R.A. Binns, Geochimica et Cosmochimica Acta, 1968, **32**, p.299.

Main mass, Canberra, Austr. Nat. Univ.; 1.5kg Adelaide, S. Austr. Mus.; 3.6kg Tempe, Arizona State Univ.; 1.2kg Fort Worth, Texas, Monnig Colln.; 1085g Washington, U.S. Nat. Mus.; 767g New York, Amer. Mus. Nat. Hist.; 450g Perth, West. Austr. Mus.; 326g Sydney, Austr. Mus.; 219g Paris, Mus. d'Hist. Nat.; 113g Chicago, Field Mus. Nat. Hist.;
Specimen(s): [1935,28], 3143g. and fragments, 27g; [1935,29], 84g. and 11 fragments, 83g; [1938,393], 45.2g.; [1938,394], 83g. three fragments

Lake Moore v Lake Brown.

Lake Murray 34°6′ N., 97°0′ W.

Carter County, Oklahoma, U.S.A.
Found 1933, before this year, recognized 1952
Iron. Octahedrite, coarsest (10mm) (IIB).

A 590lb mass with an iron-shale coating up to 6 inches thick was found in a gully, L. LaPaz, Meteoritics, 1953, **1**, p.109 [M.A. 12-359]. The mass found reported as 540kg, probably an error, B. Mason, Meteorites, Wiley, 1962, p.240. Classification and analysis, 6.3 %Ni, 53.9 ppm.Ga, 141 ppm.Ge, 0.018 ppm.Ir, J.T. Wasson, Meteorites, Springer-Verlag, 1974, p.300. Description; cosmically reheated, V.F. Buchwald, Iron Meteorites, Univ. of California, 1975, p.752.

11.2kg Albuquerque, Univ. of New Mexico; Half of the mass, Tucker Tower Mus., Lake Murray State Park; 1kg Tempe, Arizona State Univ.; 80g Washington, U.S. Nat. Mus.;

Lake Okechobee v Okechobee.

Laketon 35°34′ N., 100°40′ W.

Gray County, Texas, U.S.A.
Found 1937
Stone. Olivine-hypersthene chondrite (L6).

Two fragments were found, 3.6kg together, A.D. Nininger, Pop. Astron., Northfield, Minnesota, 1937, **45**, p.449 [M.A. 7-62], H.H. and A.D. Nininger, The Nininger Collection of Meteorites, Winslow, Arizona, 1950, p.68. Olivine Fa_{25}, B. Mason, Geochimica et Cosmochimica Acta, 1963, **27**, p.1011.

953g Tempe, Arizona State Univ.; 205g Los Angeles, Univ. of California; 202g New York, Amer. Mus. Nat. Hist.; 124g Washington, U.S. Nat. Mus.;
Specimen(s): [1952,138], 91g.; [1959,816], 974g.

Lakeview 34°32′ N., 101°42′ W.

Swisher County, Texas, U.S.A.
Found 1970
Stone. Olivine-bronzite chondrite (H4).

A very weathered mass of 1238g was found at the edge of a field 8.4 km SW. of Tulia, olivine $Fa_{19.6}$, Meteor. Bull., 1983 (61), Meteoritics, 1983, **18**, p.81, P.S. Sipiera et al., Meteoritics, 1983, **18**, p.63.

165g Chicago, Field Mus. Nat. Hist.;

Lakewood 32°37′47″ N., 104°21′ W.

Eddy County, New Mexico, U.S.A.
Found 1955, recognized 1966
Stone. Olivine-hypersthene chondrite (L6).

One weathered stone was found at a depth of about 0.8m; fragments totalling 46.5kg were recovered, but the stone was originally much larger, Meteor. Bull., 1970 (49), Meteoritics, 1970, **5**, p.175.

31.6kg Mainz, Max-Planck-Inst., includes main mass, 5.2kg; 581g Fort Worth, Texas, Monnig Colln.; 523g Washington, U.S. Nat. Mus.; 174g Chicago, Field Mus. Nat. Hist.; 96g Vienna, Naturhist. Mus.; 37g Albuquerque, Univ. of New Mexico;
Specimen(s): [1967,440], 33g. slice, and fragments, 1g.

La Lande 34°27′ N., 104°8′ W.
De Baca County, New Mexico, U.S.A.
Found 1933
Stone. Olivine-hypersthene chondrite (L5).
There has been some confusion between this fall and Melrose (a). H.H. Nininger first reported that four stones, totalling 20.382kg had been found on the surface at La Lande, 26 miles west of Melrose, and attributed them to the same fall as Melrose (a), H.H. Nininger, Am. Miner., 1934, **19**, p.370 [M.A. 6-14]. Later he found evidence that the La Lande and Melrose (a) masses were distinct, and that the La Lande find included two quite distinct falls, which he named La Lande and Taiban (*q.v.*). The total known weight of the La Lande fall is 30kg, but it is not clear how many stones this includes. Rare gas contents indicate that some La Lande, Melrose (a) and Taiban individuals belong to the same fall, but that some material ascribed to Taiban is distinct, D. Heymann, Geochimica et Cosmochimica Acta, 1965, **29**, p.1203.
768g Tempe, Arizona State Univ.; 1.5kg Albuquerque, Univ. of New Mexico; 160g Los Angeles, Univ. of California; 158g Washington, U.S. Nat. Mus.; 144g New York, Amer. Mus. Nat. Hist.;
Specimen(s): [1950,142], 63g.; [1959,774], 185g.

Lalitpur 24°27′ N., 78°34′ E. approx.
Jhansi district, Uttar Pradesh, India
Fell 1887, April 7, 1030 hrs
Synonym(s): *Iharaota, Jharaota*
Stone. Olivine-hypersthene chondrite (L6), brecciated, veined.
After detonations, a stone was seen to fall and break into pieces, eight of which, weighing 372g, were recovered, F.R. Mallet, Rec. Geol. Surv. India, 1887, **20**, p.153. As the British Museum (Nat. Hist.) specimen is a nearly complete stone, at least two stones appear to have fallen. Olivine Fa_{24}, B. Mason, Geochimica et Cosmochimica Acta, 1963, **27**, p.1011.
128g Calcutta, Mus. Geol. Surv. India; 29g Vienna, Naturhist. Mus.; 9.5g Chicago, Field Mus. Nat. Hist.;
Specimen(s): [63058], 79.5g. a nearly complete stone

Lamesa 32°53′ N., 101°53′ W.
Dawson County, Texas, U.S.A.
Found 1981
Iron. Octahedrite, fine (0.3mm) (IIICD).
A single mass of 16.9kg was found 16 km north and 3 km west of Lamesa, Meteor. Bull., 1983 (61), Meteoritics, 1983, **18**, p.81. Classification, analysis, paired with Carleton (*q.v.*), 13.7 %Ni, 13.3 ppm.Ga, 11.8 ppm.Ge, 0.041 ppm.Ir, D.J. Malvin et al., priv. comm., 1983.
14kg Watchung, N.J., Dupont Colln;
Specimen(s): [1982,M.22], 448g. slice

Lampa v Cobija.

Lancashire v Appley Bridge.

Lancaster County 40°40′ N., 96°45′ W. approx.
Nebraska, U.S.A.
Found 1903, known in this year
Iron. Octahedrite?.
A mass of 13kg was received by the Nebraska Geol. Surv. in 1903, E.H. Barbour, Rep. Geol. Surv. Nebraska, Lincoln, 1903, **1**, p.184. Its present repository is not known.

Lancé 47°42′ N., 1°4′ E.
Vendome, Loir-et-Cher, France
Fell 1872, July 23, 1720 hrs
Synonym(s): *Authon, Orléans*
Stone. Carbonaceous chondrite, type III (CO3).
After the appearance of a fireball and detonations, a shower of stones fell, of which six were recovered; the total weight was about 51.7kg and the largest stone weighed 47kg, L.M. de Tastes, C. R. Acad. Sci. Paris, 1872, **75**, p.273. Description, R. von Drasche, Tschermaks Min. Petr. Mitt., 1875, p.1. Analysis, 25.58 % total iron, H.B. Wiik, Geochimica et Cosmochimica Acta, 1956, **9**, p.279. Classification, W.R. Van Schmus and J.M. Hayes, Geochimica et Cosmochimica Acta, 1974, **38**, p.47. XRF analysis, 24.65 % total iron, T.S. McCarthy and L.H. Ahrens, Earth planet. Sci. Lett., 1972, **14**, p.97. Perovskite-spinel chondrule, M.J. Frost and R.F. Symes, Min. Mag., 1970, **37**, p.724. Petrology, G. Kurat, Tschermaks Min. Petr. Mitt., 1975, **22**, p.38 [M.A. 76-2705]. Trace element data, G.W. Kallemeyn and J.T. Wasson, Geochimica et Cosmochimica Acta, 1981, **45**, p.1217.
47kg Vienna, Naturhist. Mus., largest stone; 1.226kg Paris, Mus. d'Hist. Nat.; 267g Budapest, Nat. Mus.; 185g Washington, U.S. Nat. Mus.; 164g Chicago, Field Mus. Nat. Hist.; 81g Perth, West. Austr. Mus.; 62g Berlin, Humboldt Univ.;
Specimen(s): [1924,14], 311g. and fragments, 2.5g; [48707], 297g. complete stone; [1920,312], 18g.; [48756], 10.5g.

Lancon 43°45′ N., 5°7′ E.
Bouches du Rhone, France
Fell 1897, June 20, 2030 hrs
Stone. Olivine-bronzite chondrite (H6), veined.
After light and sound phenomena, a stone or stones fell, of total weight probably about 7kg (the De Mauroy Collection in 1909 contained 5.8kg, the largest piece weighing 4.4kg). Description and partial analysis, S. Meunier, C. R. Acad. Sci. Paris, 1900, **131**, p.969. Analysis, 27.2 % total iron, M.I. D'yakonova, Метеоритика, 1968, **28**, p.131. Olivine Fa_{19}, B. Mason, Geochimica et Cosmochimica Acta, 1963, **27**, p.1011.
5.09kg Paris, Mus. d'Hist. Nat.; 1.5kg Rome, Vatican Colln; 232g Vienna, Naturhist. Mus.; 176g Chicago, Field Mus. Nat. Hist.; 79g New York, Amer. Mus. Nat. Hist.; 73g Washington, U.S. Nat. Mus.;
Specimen(s): [83631], 199.5g.; [1920,313], 1g. two fragments

Landes 38°54′ N., 79°11′ W.
Grant County, West Virginia, U.S.A.
Found 1930, approx., recognized 1968
Synonym(s): *Landis*
Iron. Octahedrite, with silicate inclusions (IA).
A 69.8kg individual was found about a mile east of Landes Post Office, co-ordinates given above, Meteor. Bull., 1972 (51), Meteoritics, 1972, **7**, p.225. Silicate mineralogy, Odessa type, T.E. Bunch et al., Meteoritics, 1972, **7**, p.31. Structurally anomalous, analysis of metal, 6.31 %Ni, 88.7 ppm.Ga, 414 ppm.Ge, 2.9 ppm.Ir, J.T. Wasson, Meteorites, Springer-Verlag, 1974, p.297.
33kg Mainz, Max-Planck-Inst.; 1.45kg Chicago, Field Mus. Nat. Hist.; 557g Copenhagen, Univ. Geol. Mus.; 492g Washington, U.S. Nat. Mus.; 155g Vienna, Naturhist. Mus.; 33g New York, Amer. Mus. Nat. Hist.;
Specimen(s): [1971,286], 336.2g. polished part-slice; [1974, M.8], 34.4g. crusted part-slice

Landes v Barbotan.

Landis v Landes.

Landor 25°40′ S., 117° E. approx.
Western Australia, Australia
Found 1931, about
Iron. Octahedrite, fine.
Several masses, totalling 20lb, were found near the head of Wooramel river, Landor sheep station. Description, E.S. Simpson, Min. Mag., 1938, **25**, p.161. Listed, Spec. Publ. West. Austr. Mus., 1965 (3), p.37.
2.6g Perth, West. Austr. Mus.; Main mass, Held by the finder, Mr. Murphy.;

Langeac 45°6′ N., 3°31′ E.
Auvergne, Haute Loire, France
Pseudometeorite..
Mentioned, F. Berwerth, Fortschr. Min. Krist. Petr., 1912, **2**, p.235, P. de Brun, Mém. Proc. verb. Soc. agric. sci. Le Puy, 1902, **11**, p.163-165, and said to have fallen on 13 August, 1900. This fall has been disproved, the specimen being a block of basalt, A. Lacroix, Bull. Mus. d'Hist. Nat. Paris, 1927, **33**, p.572.

Langenpiernitz v Stannern.

Långhalsen 58°51′ N., 16°44′ E.
Vrena, Södermanland, Sweden
Fell 1947, February 6, 1515 hrs, U.T.
Stone. Olivine-hypersthene chondrite (L6).
A stone fell on the ice of Lake Langhalsen and broke into four pieces weighing together 2.3kg. Description, H.O. Gröstrand, Pop. Astron. Tidskr., 1947, p.65. Mentioned, F.E. Wickman, Pop. Astron. Tidskr., 1951, p.63, S. Hjelmqvist, letter of March, 1958, in Min. Dept. BM(NH). Analysis, 21.75 % total iron, H.B. Wiik, Comm. Phys.-Math. Soc. Sci. Fenn., 1969, **34**, p.135. Olivine Fa_{26}, B. Mason, Geochimica et Cosmochimica Acta, 1963, **27**, p.1011.
Main mass, Stockholm, Riksmus.; 76g Washington, U.S. Nat. Mus.; 64g Chicago, Field Mus. Nat. Hist.; 35g Tempe, Arizona State Univ.;

Langres v Chassigny.

Langwarrin v Cranbourne.

Lanton 36°32′ N., 91°48′ W.
Howell County, Missouri, U.S.A.
Found 1932
Iron. Octahedrite, medium (1.1mm) (IIIA).
Four rusted fragments were found, totalling 13.78kg. Description, with an analysis, 8.33 %Ni, J.S. Cullinson and G.A. Muilenburg, J. Geol., 1934, **42**, p.305 [M.A. 6-13]. Classification and analysis, 8.28 %Ni, 20.6 ppm.Ga, 39.3 ppm.Ge, 3.5 ppm.Ir, E.R.D. Scott et al., Geochimica et Cosmochimica Acta, 1973, **37**, p.1957. Description; shock-hardened, V.F. Buchwald, Iron Meteorites, Univ. of California, 1975, p.754.
8.9kg Rolla, Missouri, Missouri Bureau of Geology and Mines; 558g Washington, U.S. Nat. Mus.; 78g Tempe, Arizona State Univ.;
Specimen(s): [1934,986], 15.2g. and fragments, 1.25g.

Lanzenkirchen 47°45′ N., 16°14′ E.
Wiener Neustadt, Nieder Osterreich, Austria
Fell 1925, August 28, 1925 hrs
Synonym(s): *Lauzenkirchen*
Stone. Olivine-hypersthene chondrite (L4).
After appearance of fireball, and detonations, a stone weighing 5kg fell, and was found next day, K. Chudoba, Centralblatt Min., A, 1925, p.373. Another stone of 2kg was found on 7 October, 1925, at Frohsdorf, 2.5 km to the NE., H. Michel, Ann. Naturhist. Hofmus. Wien, 1925, **39**, p.1. Description, and analysis, H. Michel, Tschermaks Min. Petr. Mitt., 1927, **37**, p.16 [M.A. 3-394]. Detailed account of the fall, E. Weinmeister, Ann. Naturhist. Hofmus. Wien, 1933, **46**, p.117 [M.A. 5-403]. Fully illustrated description, light-dark structure; microprobe analysis of minerals, olivine Fa_{24}, G. Kurat and H. Kurzweil, Ann. Naturhist. Hofmus. Wien, 1965, **68**, p.9.
6.7kg Vienna, Naturhist. Mus., includes main mass, 5kg.;

Lapangaon v Lakanagaon.

La Paz v Cañon Diablo.

La Perdida v Campo del Cielo.

La Porte 41°36′ N., 86°43′ W.
La Porte County, Indiana, U.S.A.
Found 1900
Iron. Octahedrite, medium (1.1mm) (IIIA).
Four fragments were found, totalling 14.54kg, the largest 11.56kg, Ann. Rep. Field Mus. Nat. Hist., 1937, **11**, p.212, H.W. Nichols, letters of 1 May and 8 May, 1939, in Min. Dept. BM(NH). Reported, A.D. Nininger, Pop. Astron., Northfield, Minnesota, 1940, **48**, p.556. Description and analysis, 7.29 %Ni, S.K. Roy and R.K. Wyant, Field Mus. Nat. Hist. Geol. Ser., 1950, **7**, p.135 [M.A. 11-271]. Newer analysis, 7.88 %Ni, 21.5 ppm.Ga, 43.1 ppm.Ge, 1.4 ppm.Ir, E.R.D. Scott et al., Geochimica et Cosmochimica Acta, 1973, **37**, p.1957. Description; shocked and annealed, V.F. Buchwald, Iron Meteorites, Univ. of California, 1975, p.755.
10.8kg Chicago, Field Mus. Nat. Hist., includes largest mass; 1.4kg Washington, U.S. Nat. Mus.; 40g Tempe, Arizona State Univ.;
Specimen(s): [1959,952], 43g. and sawings, 2.5g.

La Primitiva 19°55′ S., 69°49′ W.
Santa Catalina, Tarapaca, Chile
Found 1888
Synonym(s): *Angela, Anjela, Oficina Angela, Primitiva, Salitra, Sierra Gorda, Tarapaca*
Iron. Structurally and chemically anomalous (IRANOM).
A mass of 6-8lb was found by a native near the nitrate works of La Primitiva in 1888, E.E. Howell, Proc. Rochester Acad. Sci., 1890, **1**, p.100. In 1903 a mass of 4kg was found embedded in "caliche" at the Angela Nitrate Co. works about 12 km from La Primitiva, and in 1906 and 1911 two other masses, of 1.5kg and 4.3kg respectively, from the same locality were sent to the BM(NH)., G.T. Prior, Min. Mag., 1914, **17**, p.131. The La Primitiva iron described, and analysed, 4.72 %Ni, E. Cohen, Ann. Naturhist. Hofmus. Wien, 1897, **12**, p.122, the Angela iron described, with an analysis, 4.52 %Ni, G.T. Prior, Min. Mag., 1914, **17**, p.131. A further mass of 9kg was found in 1907 on a mountain of the Sierra Gorda, desert of Atacama, Chile, by an Indian. It appears to be a transported fragment of La Primitiva. A sixth mass was discovered in the La Plata Mus. labelled Tarapaca, and weighed 5.4kg. Description; phosphide-rich, V.F. Buchwald, Iron Meteorites, Univ. of California, 1975,

p.756. Structure, H.J. Axon, Prog. Materials Sci., 1968, **13**, p.183. Chemical analysis of La Primitiva and "Tarapaca", 4.90 %Ni, 33.3 ppm.Ga, 37.3 ppm.Ge, 0.039 ppm.Ir, E.R.D. Scott et al., Geochimica et Cosmochimica Acta, 1973, **37**, p.1957.
1221g Los Angeles, Univ. of California, was BM 1906,21; 943g Harvard Univ.; 650g Vienna, Naturhist. Mus.; 135g New York, Amer. Mus. Nat. Hist.;
Specimen(s): [1911,141], 2970g. of Angela; [86831], 4038g. of Angela; [1906,21], 23g. fragments of Angela; [1920,314], 115g. of La Primitiva; [84194], 78g. of La Primitiva; [1927, 77], 8.83kg. main part of the mass found at Sierra Gorda, largest portion weighs 6525g

Laramie County v Silver Crown.

La Rinconada 37°28′ N., 6°1′ W.
Seville, Spain
Fell 1934, April 17, or 19
Synonym(s): *La Rinconda*
Doubtful. Iron.
A mass the size of a chestnut, of iron and iron sulphide, was found after a fire attributed to a meteorite, G.M. Cardoso, Bol. R. Soc. Españ. Hist. Nat. Madrid, 1934, **34**, p.201 [M.A. 7-66]. Very doubtful.

La Rinconda v La Rinconada.

La Rioja v Imilac.

La Rioja v Puerta de Arauco.

Larissa 39°38′ N., 22°25′ E.
Greece
Fell 1706, June 7, 1400-1500 hrs
Doubtful..
A stone of 17lb is said to have fallen, D.A. Soldani, Atti Accad. Sci. Siena, 1808, **9**, p.1, E.F.F. Chladni, Die Feuer-Meteore, Wien, 1819, p.240, but the evidence is not conclusive.

La Rochelle v Chantonnay.

La Rochelle v Esnandes.

Las Adargas v Chupaderos.

La Scarpa v Orvinio.

Lasdany v Lixna.

Lasher Creek 42°50′ N., 74°30′ W. approx.
Montgomery County, New York, U.S.A.
Found 1948
Iron.
A mass of 639.5g was found at Lasher Creek, 5 miles ESE. of Canajoharie (42°50'N., 74°37'W.), K. Servos, letter of 28 January, 1957, in Min. Dept. BM(NH).

La Spezia v Pułtusk.

Las Salinas 23° S., 69°30′ W. approx.
Antofagasta, Chile
Found 1905
Iron. Octahedrite, medium?.
A 3.515kg mass was found in a saltpetre mine near the Antofagasta-Calama railway, 144 km from Antofagasta. May be related to Baquedano (*q.v.*) which is nearby, but the Las Salinas mass has not been properly investigated, V.F. Buchwald, Iron Meteorites, Univ. of California, 1975, p.758.
Mass, Dresden, Mus., possible location;

Las Vegas v Cañon Diablo.

La Touanne v Charsonville.

Laundry East 31°31′ S., 127°8′ E.
Nullarbor Plain, Western Australia, Australia
Found 1967, March
Stone. Olivine-bronzite chondrite (H3).
One almost complete, weathered stone of 43.1g was found 7 miles east of Laundry Rockhole, which is 25 miles north of Madura, G.J.H. McCall and W.H. Cleverly, J. Roy. Soc. West. Austr., 1970, **53**, p.69. Classification, olivine $Fa_{18.5}$, B. Mason, Rec. Austr. Mus., 1974, **29**, p.169.
Main mass, Kalgoorlie, West. Austr. School of Mines; Chip and thin section, Perth, West. Austr. Mus.; 1g Washington, U.S. Nat. Mus.;

Laundry Rockhole 31°32′ S., 127°1′ E.
Nullarbor Plain, Western Australia, Australia
Found 1967, June
Stone. Olivine-bronzite chondrite (H5).
Thirty-two fragments, of total weight 1443g, the largest weighing 1016g, were found at the SE. corner of the fence enclosing the rockhole, G.J.H. McCall and W.H. Cleverly, J. Roy. Soc. West. Austr., 1970, **53**, p.69. Classification, olivine $Fa_{20.4}$, B. Mason, Rec. Austr. Mus., 1974, **29**, p.169.
Main mass, Kalgoorlie, West. Austr. School of Mines; 271g Perth, West Austr. Mus., 27 fragments; 3g Washington, U.S. Nat. Mus.;

Laundry West 31°28′ S., 126°56′ E.
Nullarbor Plain, Western Australia, Australia
Found 1967, March
Stone. Olivine-hypersthene chondrite (L4).
Five angular, badly weathered fragments, of total weight 201.9g, were found 6 miles NW. of Laundry Rockhole (see Laundry East); the fragments were spread over a length of 276 feet, G.J.H. McCall and W.H. Cleverly, J. Roy. Soc. West. Austr., 1970, **53**, p.69. Classification, olivine $Fa_{24.2}$, B. Mason, Rec. Austr. Mus., 1974, **29**, p.169.
141g Perth, West. Austr. Mus.; 53g Kalgoorlie, West. Austr. School of Mines; 2.9g Washington, U.S. Nat. Mus.;

Launton 51°54′ N., 1°7′ W.
Bicester, Oxfordshire, England
Fell 1830, February 15, 1930 hrs
Stone. Olivine-hypersthene chondrite (L6), veined.
After the appearance of a fireball, followed by a triple detonation, a stone of 2lb 5oz was seen to fall, W. Stone, Mag. Nat. Hist. London, 1831, **4**, p.139. Described and analysed, G.T. Prior, Min. Mag., 1916, **18**, p.2. Olivine Fa_{23}, B. Mason, Geochimica et Cosmochimica Acta, 1963, **27**, p.1011.
Specimen(s): [77528], 999g. nearly complete stone, and a fragment, less than a gram

Laurens County 34°30′ N., 82°2′ W.
South Carolina, U.S.A.
Found 1857
Synonym(s): *Laurens Court House*
Iron. Octahedrite, fine (0.3mm) (IRANOM).
A mass of 4.75lb was found in the NW. corner of Laurens County. Description, with an analysis, 13.34 %Ni, W.E. Hidden, Am. J. Sci., 1886, **31**, p.463. Classification and analysis, 12.95 %Ni, 10.5 ppm.Ga, 22.4 ppm.Ge, 7.9 ppm.Ir, J.T. Wasson and R. Schaudy, Icarus, 1971, **14**, p.59. Description, structural classification, V.F. Buchwald, Iron Meteorites, Univ. of California, 1975, p.759.
1.5kg Vienna, Naturhist. Mus.; 81g Budapest, Nat. Mus.; 81g Chicago, Field Mus. Nat. Hist.; 39g New York, Amer. Mus. Nat. Hist.;
Specimen(s): [67448], 61.5g.

Laurens Court House v Laurens County.

Laurentjewka v Lavrentievka.

Lausanne 46°31′ N., 6°38′ E.
Switzerland
Fell 1894, September, 2030 hrs
Doubtful. Iron. Nickel-free.
A mass of about 720g was seen to fall; it contains no nickel and gives a granular structure on etching, P.L. Mercanton, Bull. Soc. Vaud. Sci. Nat., 1946, **63**, p.315 [M.A. 11-446].

Lauzenkirchen v Lanzenkirchen.

La Villa 26°16′18″ N., 97°54′6″ W.
Hidalgo County, Texas, U.S.A.
Found 1956, April
Stone. Olivine-bronzite chondrite (H4).
A weathered mass of 43.5lb was found, Meteor. Bull., 1957 (5), B. Mason, Meteorites, Wiley, 1962, p.243. Olivine Fa_{19}, B. Mason, Geochimica et Cosmochimica Acta, 1963, **27**, p.1011.
Main mass, Edinburg, Texas, Pan-American College; 1.2kg Fort Worth, Texas, Monnig Colln.; 518g Mainz, Max-Planck-Inst.;
Specimen(s): [1982,M.13], 9g.

La Vivionnere v Le Teilleul.

Lavrent'evka v Lavrentievka.

Lavrentievka 52°27′ N., 51°34′ E.
Orenburg, Federated SSR, USSR, [Лаврентьевка]
Fell 1938, January 11, 1430 hrs
Synonym(s): *Laurentjewka, Lavrent'evka, Lawrentjewka*
Stone. Olivine-hypersthene chondrite (L6).
One mass fell near Lavrentievka, and five pieces were recovered. Description, E.L. Krinov, Nature, 1938, **143**, p.624. The total weight recovered is variously given as 781.4g, 794g, 1044g, and 1045g, L.A. Kulik, Метеоритика, 1941, **1**, p.73 [M.A. 9-294], E.L. Krinov, Каталог Метеоритов Акад. Наук СССР, Москва, 1947 [M.A. 10-511]. Olivine Fa_{24}, B. Mason, Geochimica et Cosmochimica Acta, 1963, **27**, p.1011. Analysis, 22.73 % total iron, V.Ya. Kharitonova, Метеоритика, 1968, **28**, p.138.
757g Moscow, Acad. Sci.;

Lawrence 38°58′ N., 95°10′ W.
Douglas County, Kansas, U.S.A.
Found 1928, known in this year
Stone. Olivine-hypersthene chondrite (L6).
A small stone of 515g was found NE. of Lawrence, E.P. Henderson, letters of 12 May and 3 June, 1939, in Min. Dept. BM(NH). Listed, Ann. Rep. U.S. Nat. Mus., 1929, p.90, A.D. Nininger, Pop. Astron., Northfield, Minnesota, 1939, **47**, p.212, H.H. and A.D. Nininger, The Nininger Collection of Meteorites, Winslow, Arizona, 1950, p.68. Olivine Fa_{23}, B. Mason, Geochimica et Cosmochimica Acta, 1963, **27**, p.1011.
328g Washington, U.S. Nat. Mus., main mass; 51g Tempe, Arizona State Univ.;
Specimen(s): [1959,881], 43.5g. slice

Lawrentjewka v Lavrentievka.

Lazarev 71°57′ S., 11°30′ E.
Humboldt Mountains, Antarctica
Found 1961, January 21
Iron. Octahedrite, medium (0.8mm) (IRANOM).
Two fragments, of 8kg and 2kg, were found 3000m above sea-level on a southern spur of the Humboldt Mountains, 35 to 40m from the fringe of the glacial sheet, Meteor. Bull., 1961 (20). Possibly a metal-rich pallasite; bandwidth, V.F. Buchwald, Iron Meteorites, Univ. of California, 1975, p.761. Description, analysis, 9.8 %Ni, 16 ppm.Ga, 47 ppm.Ge, 2.6 ppm.Ir, N.I. Zaslavskaya and G.M. Kolesov, Метеоритика, 1980, **39**, p.64.
8kg Leningrad, City Mus.; 1720g Moscow, Acad. Sci.;

Lazbuddie 34°30′ N., 102°45′ W.
Parmer County, Texas, U.S.A.
Found 1970, recognized 1978
Stone. Olivine-hypersthene chondrite, amphoterite (LL5).
A single mass of 8.6kg was found during ploughing, 21.6 % total iron, olivine $Fa_{26.9}$, P.S. Sipiera et al., Meteoritics, 1983, **18**, p.63.
226g Chicago, Field Mus. Nat. Hist.;
Specimen(s): [1980,M.34], 58.6g. part-slice

Lea Iron v Cleveland.

Leander 30°36′ N., 97°54′ W.
Williamson County, Texas, U.S.A.
Found 1940
Stone. Olivine-hypersthene chondrite (L).
A mass of 760g was found, B. Mason, Meteorites, Wiley, 1962, p.243.
764g Fort Worth, Texas, Monnig Colln.;

Leavenworth County v Tonganoxie.

Lebedin v Kharkov.

Lebedinnyi 55°57′ N., 125°24′ E.
Amur, Federated SSR, USSR, [Лебединыиь Прииск]
Found 1925
Synonym(s): *Lebedinnyi Priisk*
Iron. Octahedrite?.
Found near the head of the Zeya river, a tributary of the Amur, V.I. Vernadsky, letter of 5 August, 1939, in Min. Dept. BM(NH). One mass of 410g was found, E.L. Krinov, Астрон. Журнал, 1945, **22**, p.303 [M.A. 9-297]. The mass was in the collection of N.N. Padurov of Leningrad, but has been lost.

Lebedinnyi Priisk v Lebedinnyi.

Lebjagevka v Pesyanoe.

Leedey 35°53′ N., 99°20′ W.
Dewey County, Oklahoma, U.S.A.
Fell 1943, November 25, 1900 hrs
Synonym(s): *Leedy*
Stone. Olivine-hypersthene chondrite (L6).
A shower of stones fell, and more than 20 were recovered; total known weight about 50kg, H.H. and A.D. Nininger, The Nininger Collection of Meteorites, Winslow, Arizona, 1950, p.68. Analysis, H. König, Geochimica et Cosmochimica Acta, 1964, **28**, p.1697. Olivine Fa_{24}, B. Mason, Geochimica et Cosmochimica Acta, 1963, **27**, p.1011. Further analyses, 22.11 % total iron, E. Jarosewich, Geochimica et Cosmochimica Acta, 1967, **31**, p.1103. 22.60 % total iron, H.B. Wiik, Comm. Phys.-Math. Soc. Sci. Fenn., 1969, **34**, p.135. Rare earth element abundances, discussion, A. Masuda et al., Geochimica et Cosmochimica Acta, 1973, **37**, p.239.
22kg Fort Worth, Texas, Monnig Colln.; 11.7kg Tempe, Arizona State Univ.; 97g Washington, U.S. Nat. Mus.; 91g Chicago, Field Mus. Nat. Hist.; 77g Vienna, Naturhist. Mus.;
Specimen(s): [1949,112], 166g.; [1959,776], 1051g. a slice, and fragments, 7g.

Leeds 46°18′ N., 71°20′ W.
Quebec, Canada
Found 1931, recognized in this year
Iron. Octahedrite, coarse (1.3mm) (IA).
A mass of 1445g is mentioned, H.H. Nininger, Our Stone Pelted Planet, 1933. Determination of Ga, Au and Pd, and partial analysis, 8.08 %Ni, E. Goldberg et al., Geochimica et Cosmochimica Acta, 1951, **2**, p.1. Classification and analysis, 7.99 %Ni, 67.1 ppm.Ga, 241 ppm.Ge, 2.1 ppm.Ir, J.T. Wasson, Icarus, 1970, **12**, p.407. Description; contains silicate inclusions, V.F. Buchwald, Iron Meteorites, Univ. of California, 1975, p.762.
633g Tempe, Arizona State Univ.; 77g Harvard Univ.; 72g Heidelberg, Max.-Planck-Inst.; 58g Ottawa, Geol. Surv. Canada; 32g Los Angeles, Univ. of California;
Specimen(s): [1959,972], 194g. and sawings, 13g.

Leedy v Leedey.

Leeuwfontein 25°40′ S., 28°22′ E.
Engelbrecht drift, Pretoria, Transvaal, South Africa
Fell 1912, June 21, 1400 hrs
Stone. Olivine-hypersthene chondrite (L6).
After detonations a stone of 460g was seen to fall, G.T. Prior, Nature, 1922, **110**, p.757. Description and partial analysis, G.T. Prior, Min. Mag., 1923, **20**, p.136. Listed, C. Frick and E.C.I. Hammerbeck, Bull. Geol. Surv. S. Africa, 1973 (57), p.25. Olivine Fa_{24}, B. Mason, Geochimica et Cosmochimica Acta, 1963, **27**, p.1011.
Main mass, Pretoria, Geol. Surv. Mus.; 72g Heidelberg, Max-Planck-Inst.;
Specimen(s): [1922,769], 135g.

Lefroy 40°44′ S., 146°58′ E.
County Dorset, Tasmania, Australia
Found 1904
Iron.
A minute piece of 3.3 grains was found in alluvial drift 27 miles NW. of Launceston, W.F. Petterd, Proc. Roy. Soc. Tasmania, 1910, p.98.
Main mass, Launceston, Tasmania, Queen Victoria Mus.;

Legnano 45°15′ N., 11°15′ E. approx.
Verona, Veneto, Italy
Fell 1855, August 30
Pseudometeorite. Stone.
A mass of 2006g in the Museum at Trento is labelled as above. It is a terrestrial rock, B. Mason, priv. comm.

Le Gould's Stone 23°25′ S., 148°55′ E.
Leichhardt district, Queensland, Australia
Found 1864, before this year
Synonym(s): *Isaacs River*
Pseudometeorite..
A mass, 10 inches in diameter, which in its fall had broken a tree, was found two days march beyond the Isaacs River, Le Gould, Geol. Mag., 1864, **1**, p.142. Very doubtful, L.J. Spencer, Min. Mag., 1937, **24**, p.452. Discredited, T. Hodge-Smith, Mem. Austr. Mus., 1939 (7), p.32.

Leighton 34°35′ N., 87°30′ W.
Colbert County, Alabama, U.S.A.
Fell 1907, January 12, 2000 hrs
Stone. Olivine-bronzite chondrite (H5), breccia, polymict.
After the appearance of a fireball, and detonations, a stone of 877g was seen to fall 8 miles S. of Leighton. Description, and partial analysis, O.C. Farrington, Field Mus. Nat. Hist. Geol. Ser., 1910, **3** (Publ. 145), p.165. Olivine Fa_{20}, B. Mason, Geochimica et Cosmochimica Acta, 1963, **27**, p.1011. Contains 'light' material with equilibrated olivine, Fa19.5, and 'dark', with olivine Fa5 to Fa40, S.J.B. Reed and J.M. Hall, priv. comm. Gas-rich, J.T. Wasson, Meteorites, Springer-Verlag, 1974, p.104.
508g Chicago, Field Mus. Nat. Hist.; 17g Vienna, Naturhist. Mus.;
Specimen(s): [1910,454], 85.5g. slice, and fragments, 1g

Leikanger 61°16′ N., 6°51′ E.
Sognefjord, Norway
Found 1978, July 22
Stone. Olivine-hypersthene chondrite (L6).
A single mass of 1513g was found on the Myrdalsbreen glacier, north of Leikanger. The co-ordinates are those of Myrdals Lake, Meteor. Bull., 1980 (57), Meteoritics, 1980, **15**, p.99.
1513g Oslo, Min.-Geol. Mus.;

Leland v Forest City.

Lenarco v Lenarto.

Lenarto 49° N., 21° E. approx.
Saros, Slovensko, Czechoslovakia
Found 1814
Synonym(s): *Lenarco, Lenartov, Lenartow, Polen, Saros*
Iron. Octahedrite, medium (1.2mm) (IIIA).
A mass of 108.5kg was found on one of the highest summits of the Carpathians, -. Tehel, Ann. Phys. (Gilbert), 1815, **49**, p.181, near Saros (49°3'N., 21°13'E.). Analysis, 8.58 %Ni, J. Boussingault, C. R. Acad. Sci. Paris, 1872, **74**, p.1288. Newer analysis, 8.85 %Ni, 21.7 ppm.Ga, 43.5 ppm.Ge, 0.33 ppm.Ir, E.R.D. Scott et al., Geochimica et Cosmochimica Acta, 1973, **37**, p.1957. Description; shock-annealed, V.F. Buchwald, Iron Meteorites, Univ. of California, 1975, p.763.
76.5kg Budapest, Nat. Mus.; 3.5kg Tübingen, Univ.; 3.2kg Vienna, Naturhist. Mus.; 677g Chicago, Field Mus. Nat. Hist.; 357g Prague, Nat. Mus.; 541g Berlin, Humboldt Univ.; 208g Washington, U.S. Nat. Mus.; 170g New York, Amer. Mus. Nat. Hist.; 121g Yale Univ.; 139g Tempe, Arizona State Univ.;

Specimen(s): [61305], 1565g.; [61304], 232g.; [90220], 198g.; [33922], 1g. filings

Lenartov v Lenarto.

Lenartow v Lenarto.

Lenorka v Leonovka.

Leon 37°40′ N., 96°46′ W.
Butler County, Kansas, U.S.A.
Found 1943
Stone. Olivine-bronzite chondrite (H5).
One stone of 30kg was found on a farm 7 miles S. of Leon. Olivine Fa_{18}, B. Mason, Geochimica et Cosmochimica Acta, 1967, **31**, p.1100. Reported, Meteor. Bull., 1967 (39), Meteoritics, 1970, **5**, p.91.
17kg Washington, U.S. Nat. Mus.;

Leonovka 52°16′ N., 32°51′ E.
Novgorod-Syeversk district, Chernigov Govt., Ukraine, USSR, [Леоновка]
Fell 1900, August 23, 2100 hrs
Synonym(s): *Lenorka, Leonowka*
Stone. Olivine-hypersthene chondrite (L6).
One mass of 700g fell, P.N. Chirvinsky, Centralblatt Min., 1923, p.548 [M.A. 2-257], E.L. Krinov, Каталог Метеоритов Акад. Наук СССР, Москва, 1947 [M.A. 10-511]. Olivine Fa_{24}, B. Mason, Geochimica et Cosmochimica Acta, 1963, **27**, p.1011.
Main mass, Kiev, Acad. Sci.; 48.3g Moscow, Acad. Sci.; 26g Vienna, Naturhist. Mus.; 1.1g Chicago, Field Mus. Nat. Hist.;

Leonowka v Leonovka.

Leoville 38°48′ N., 100°25′ W.
Decatur County, Kansas, U.S.A.
Found 1961, or 1962, recognized 1965
Stone. Carbonaceous chondrite, type III (CV3).
A 1.6kg mass was ploughed up in 1961 or 1962; a 6.5kg mass was found in a farmyard, Meteor. Bull., 1970 (49), Meteoritics, 1970, **5**, p.175. Classification, W.R. Van Schmus and J.M. Hayes, Geochimica et Cosmochimica Acta, 1974, **38**, p.47. Chemical analysis, 22.02 % total iron, T.S. McCarthy and L.H. Ahrens, Earth planet. Sci. Lett., 1972, **14**, p.97. Magnesium isotopes, references, J.C. Lorin et al., Meteoritics, 1978, **13**, p.537.
705g Mainz, Max-Planck-Inst.; 482g Washington, U.S. Nat. Mus.; 281g Tempe, Arizona State Univ.; 163g Chicago, Field Mus. Nat. Hist.; 133g Vienna, Naturhist. Mus.;
Specimen(s): [1969,144], 195g.

Le Pressoir 47°10′ N., 0°26′ E.
Indre-et-Loir, France
Fell 1845, January 25, 1500 hrs
Synonym(s): *Louans*
Stone. Olivine-hypersthene chondrite (L6).
After detonations, a stone of about 3kg was found next day, G.A. Daubrée, C. R. Acad. Sci. Paris, 1881, **92**, p.984. Olivine Fa_{24}, B. Mason, Geochimica et Cosmochimica Acta, 1963, **27**, p.1011.
169g Paris, Mus. d'Hist. Nat.; 65g Chicago, Field Mus. Nat. Hist.; 15g Vienna, Naturhist. Mus.;
Specimen(s): [1927,7], 3g.

Lerici v Pułtusk.

Leroy
Not known,
Pseudometeorite..
Locality and date unknown. A piece of 102g purchased from E.E. Howell is in Amer. Mus. Nat. Hist., New York, L.W. MacNaughton, Am. Mus. Novit., 1926 (207), p.2. Not meteoritic, B. Mason, Min. Mag., 1964, **33**, p.935.

Leshan 29°36′ N., 103°42′ E.
Sichuan, China
Found 1964, August, or September
Iron. Octahedrite, fine (0.5mm) (IRANOM).
A single mass of 344g was found, D. Bian, Meteoritics, 1981, **16**, p.121. Classification, analysis, 9.5 %Ni, 21.3 ppm.Ga, 68.9 ppm.Ge, 4.1 ppm.Ir, D.J. Malvin et al., priv. comm., 1983.
344g Beijing, Acad. Sin. Inst. Geol.;

Les Ormes 48°21′ N., 3°15′ E.
Yonne, France
Fell 1857, October 1, 1700 hrs
Synonym(s): *Commune des Ormes, Ormes*
Stone. Olivine-hypersthene chondrite (L6).
After the appearance of a fireball and detonations, a stone was seen to fall, of which a piece weighing 125g was presented to the French Academy, L'Institut, Paris, 1857, **25**, p.363. Description, partial analysis, S. Meunier, Bull. Soc. Hist. Nat. Autun, 1892, **5**, p.335. The entry in the register of the BM(NH) specimen states that the stone, weighing about 1lb, fell in the "hameau des Touchards." Olivine Fa_{24}, B. Mason, Geochimica et Cosmochimica Acta, 1963, **27**, p.1011.
76g Paris, Mus. d'Hist. Nat.; 5g Budapest, Nat. Mus.; 1.7g Vienna, Naturhist. Mus.; 1.5g Chicago, Field Mus. Nat. Hist.;
Specimen(s): [27363], 12g.

Lesves 50°22′ N., 4°44′ E.
Namur, Belgium
Fell 1896, April 13, 0730 hrs
Stone. Olivine-hypersthene chondrite (L6).
After detonations, a stone of about 2kg was seen to fall, Nature, 1896, **53**, p.611. Description, with a partial analysis, A.F. Renard, Bull. Acad. Roy. Belg., 1896, **31**, p.654. Olivine Fa_{25}, B. Mason, Geochimica et Cosmochimica Acta, 1963, **27**, p.1011.
166g Vienna, Naturhist. Mus.; 124g Budapest, Nat. Mus.; 88g New York, Amer. Mus. Nat. Hist.; 42g Chicago, Field Mus. Nat. Hist.;
Specimen(s): [81535], 56.7g.

Le Teilleul 48°32′ N., 0°52′ W.
Manche, France
Fell 1845, July 14, 1500 hrs
Synonym(s): *La Vivionnere, Teilleul, Vivionnere*
Stone. Achondrite, Ca-rich. Howardite (AHOW).
After detonations, a stone of 780g fell. Description, with a partial analysis, G.A. Daubrée, C. R. Acad. Sci. Paris, 1879, **88**, p.544. Mentioned, A. Lacroix, Arch. Mus. Hist. Nat. Paris, 1926, **1** (ser. 6), p.15 [M.A. 3-392]. Olivine Fa_{28-42}, C. Desnoyers and D.Y. Jerome, Meteoritics, 1973, **8**, p.342. Analysis, B. Mason et al., Smithson. Contrib. Earth Sci., 1979 (22), p.30.
447g Paris, Mus. d'Hist. Nat.; 104g Vienna, Naturhist. Mus.; 15g Budapest, Nat. Mus.; 12g Chicago, Field Mus. Nat. Hist.;
Specimen(s): [56464], 2g.

Lexington County 34° N., 81°15′ W. approx.
South Carolina, U.S.A.
Found 1880
Iron. Octahedrite, coarse (2.1mm) (IA).
A mass of 10.5lb was found on a farm. Description, with an analysis, 6.08 %Ni, C.U. Shepard, Am. J. Sci., 1881, **21**, p.117. Further analysis, 6.69 %Ni, 85.4 ppm.Ga, 316 ppm.Ge, 2.3 ppm.Ir, J.T. Wasson, Meteorites, Springer-Verlag, 1974, p.297. The 969g end-piece in Chicago (Field Mus. Nat. Hist. no. Me 1114), belongs to group IIIA, with 7.85 %Ni, 21.7 ppm.Ga, 42.6 ppm.Ge, 1.1 ppm.Ir, E.R.D. Scott et al., Geochimica et Cosmochimica Acta, 1973, **37**, p.1957. Description, inclusion-rich, V.F. Buchwald, Iron Meteorites, Univ. of California, 1975, p.765.
3kg Washington, U.S. Nat. Mus.; 210g New York, Amer. Mus. Nat. Hist.; 192g Berlin, Humboldt Univ.; 110g Chicago, Field Mus. Nat. Hist., of gp. IA material; 59g Vienna, Naturhist. Mus.;
Specimen(s): [63629], 201.5g. slice; [54278], 70g. slice.

Lexington County v Ruff's Mountain.

Liangcheng 40°30′ N., 112°30′ E.
Nei Monggol, China
Found 1959
Synonym(s): *Liancheng Daihai*
Iron. Octahedrite, medium (1.3mm) (IIIA).
A single mass of 200kg was found, D. Bian, Meteoritics, 1981, **16**, p.121. Structure, Z. Ouyang et al., Sci. Geol. Sinica, 1964, p.241, Y. Hou et al., Kexue Tongbao, 1964, p.727. Classification, analysis, 7.9 %Ni, 21.6 ppm.Ga, 45.7 ppm.Ge, 0.5 ppm.Ir, D.J. Malvin et al., priv. comm., 1983.
200kg Beijing, Geol. Mus.;

Liangcheng Daihai v Liangcheng.

Libonnez v Juvinas.

Liboschitz v Ploschkovitz.

Lichtenberg 26°9′ S., 26°11′ E.
Barbania farm, Lichtenberg, Transvaal, South Africa
Fell 1973, September 26
Stone. Chondrite.
A single mass of 4kg was seen to fall on Barbania farm, Meteor. Bull., 1980 (58), Meteoritics, 1980, **15**, p.235.

Lichtenfels v Gibeon.

Lick Creek 35°40′ N., 80°15′ W. approx.
Davidson County, North Carolina, U.S.A.
Found 1879
Iron. Hexahedrite (IIA).
A mass of 1.24kg was found. Description, with an analysis, 5.74 %Ni, W.E. Hidden, Am. J. Sci., 1880, **20**, p.324. Heavily weathered, V.F. Buchwald, Iron Meteorites, Univ. of California, 1975, p.767.
990g Vienna, Naturhist. Mus.; 15g Chicago, Field Mus. Nat. Hist.; 11g Berlin, Humboldt Univ.; 9g Washington, U.S. Nat. Mus.;
Specimen(s): [55124], 18.5g. slice

Lider (a) 34°39′18″ N., 101°38′24″ W.
Hale County, Texas, U.S.A.
Found 1972, recognized in this year
Stone. Olivine-hypersthene chondrite (L).
A mass of 2.16kg was found, C.F. Lewis and C.B. Moore, Cat. Meteor. Arizona State Univ., 1976, p.116.
1885g Mainz, Max-Planck-Inst.; 84g Tempe, Arizona State Univ.;

Lider (b) 34°39′18″ N., 101°38′24″ W.
Hale County, Texas, U.S.A.
Found 1972, recognized in this year
Stone. Olivine-bronzite chondrite (H).
A mass of 1.8kg was found, G.I Huss, Cat. Huss Coll. Meteorites, 1976, p.28.
1572g Mainz, Max-Planck-Inst.; 67.7g Tempe, Arizona State Univ.;

Liewipantsun v Yukan.

Liksen v Lixna.

Liksna v Lixna.

Lillaverke 56°39′ N., 15°52′ E.
Kalmar, Smaland, Sweden
Fell 1930, May 11, 1600 hrs
Stone. Olivine-bronzite chondrite (H5), veined.
A mass of 6862g fell, N. Zenzen, Geol. För. Förh. Stockholm, 1930, **52**, p.366 [M.A. 4-417]. Olivine Fa_{18}, B. Mason, Geochimica et Cosmochimica Acta, 1963, **27**, p.1011.
Main mass, Stockholm, Riksmus.; 150g Washington, U.S. Nat. Mus.; 75g Chicago, Field Mus. Nat. Hist.; 73g Mainz, Max-Planck-Inst.;
Specimen(s): [1980,M.5], 22.3g.

Lille v Hainaut.

Lime Creek 31°33′ N., 87°31′ W.
Claiborne, Monroe County, Alabama, U.S.A.
Found 1834
Synonym(s): *Alabama, Claiborne, Clarke County, Limestone Creek*
Iron. Ataxite, Ni-rich (IRANOM).
A mass of irregular shape, 10 inches long by 5 or 6 inches in thickness, was found. Description, with an analysis, C.T. Jackson, Am. J. Sci., 1838, **34**, p.332. Later analysis, 29.99 %Ni, E. Cohen, Meteoritenkunde, 1905, **3**, p.131. The mass was found near Limestone Creek, Claiborne, but the name Lime Creek has been more generally used. Metallography, H.J. Axon and P.L. Smith, Min. Mag., 1972, **38**, p.736. Chemically anomalous, 29.1 %Ni, 15.5 ppm.Ga, 28.5 ppm.Ge, 1.1 ppm.Ir, E.R.D. Scott et al., Geochimica et Cosmochimica Acta, 1973, **37**, p.1957. Description; an 800g piece was broken from the mass which was never relocated; shocked and annealed, V.F. Buchwald, Iron Meteorites, Univ. of California, 1975, p.768.
236g Vienna, Naturhist. Mus.; 130g New York, Amer. Mus. Nat. Hist.; 84g Chicago, Field Mus. Nat. Hist.; 75g Tempe, Arizona State Univ.; 50g Yale Univ.; 22g Harvard Univ.; 2.9g Washington, U.S. Nat. Mus.;
Specimen(s): [35964], 15.5g.; [34583], 3.5g.

Lime Creek v Walker County.

Limerick 52°34′ N., 8°47′ W.
County Limerick, Ireland
Fell 1813, September 10, 0900 hrs
Synonym(s): *Adare, Brasky, Faha*
Stone. Olivine-bronzite chondrite (H5), veined.
A shower of stones fell, after detonations, one of 17lb at Scagh, several smaller stones near Adare, one of 65lb at

Brasky, and another of 24lb at Faha, S. Maxwell, Phil. Mag., 1818, **51**, p.355. Analysis, G.T. Prior, Min. Mag., 1921, **19**, p.167. Analysis of the Brasky mass, H.J. Seymour, Sci. Proc. Roy. Dublin Soc., 1947, **24**, p.157 [M.A. 10-172]. Olivine Fa_{19}, B. Mason, Geochimica et Cosmochimica Acta, 1963, **27**, p.1011.

27.06kg Dublin, Nat. Mus., the Brasky mass, and fragments, 42g; 8.5kg Oxford, Univ. Mus., of Faha; 1kg Tübingen, Univ.; 210g Washington, U.S. Nat. Mus.; 163g Vienna, Naturhist. Mus.; 148g New York, Amer. Mus. Nat. Hist.; 148g Tempe, Arizona State Univ.; 52g Chicago, Field Mus. Nat. Hist.;

Specimen(s): [33910a], 82g.; [1907,15], 47g.; [46973], 22g.; [1935,50], 0.5g.; [1955,115], 101g.; [1964,580], 249g.; [1968, 186], 83g.; [96268], thin section

Limerzel v Kermichel.

Limestone Creek v Lime Creek.

Limousin 45°36′ N., 1°30′ E.

Haute Vienne, France
Fell 1540, April 28
Doubtful..

Two stones, one the size of a barrel, are said to have fallen, E.F.F. Chladni, Die Feuer-Meteore, Wien, 1819, p.212, but the evidence is not conclusive.

Lincoln County 39°22′ N., 103°10′ W.

Colorado, U.S.A.
Found 1937
Synonym(s): *Township no. 7, Township no. 8*
Stone. Olivine-hypersthene chondrite (L6).

A mass of 498g was found in township 7, range 52W., Lincoln Co., E.P. Henderson, letters of 3 June 1939 in Min. Dept. BM(NH). A further mass of about 4kg was also found. Two masses listed, H.H. and A.D. Nininger, The Nininger Collection of Meteorites, Winslow, Arizona, 1950, p.98. Olivine Fa_{24}, B. Mason, Geochimica et Cosmochimica Acta, 1963, **27**, p.1011.

2.2kg Tempe, Arizona State Univ.; 86g New York, Amer. Mus. Nat. Hist.; 345g Denver, Amer. Meteor. Lab.; 71g Washington, U.S. Nat. Mus.; 65g Chicago, Field Mus. Nat. Hist.;

Specimen(s): [1959,882], 1788g. half of the larger stone.

Lincoln County v Petersburg.

Linn County v Marion (Iowa).

Linnville v Linville.

Linville Mountain v Linville.

Linum 52°45′ N., 12°54′ E.

Fehrbellin, Potsdam, Germany
Fell 1854, September 5, 0800 hrs
Synonym(s): *Brandenburg, Fehrbellin*
Stone. Olivine-hypersthene chondrite (L).

After detonations, a stone of 1862g fell, G. Rose, Ann. Phys. Chem. (Poggendorff), 1855, **94**, p.169. Analysis, 22.37 % total iron, H.B. Wiik, Geochimica et Cosmochimica Acta, 1956, **9**, p.279. Olivine Fa_{23}, B. Mason, Geochimica et Cosmochimica Acta, 1963, **27**, p.1011.

1.7kg Berlin, Humboldt Univ.; 4g Vienna, Naturhist. Mus.;

Specimen(s): [86639], 2g.

Linville 35°52′ N., 81°55′ W.

Burke County, North Carolina, U.S.A.
Found 1882
Synonym(s): *Burke County, Linnville, Linville Mountain*
Iron. Ataxite, Ni-rich (IRANOM).

A mass of 442g was found on Linville Mountain, G.F. Kunz, Am. J. Sci., 1888, **36**, p.275. Description, with an analysis, 16.32 %Ni, E. Cohen, Ann. Naturhist. Hofmus. Wien, 1898, **13**, p.145. Classification and analysis, 15.8 %Ni, 7.5 ppm.Ga, 16.1 ppm.Ge, 0.012 ppm.Ir, A. Kracher et al., Geochimica et Cosmochimica Acta, 1980, **44**, p.773. Description, co-ordinates; artificially reheated, V.F. Buchwald, Iron Meteorites, Univ. of California, 1975, p.772.

200g Vienna, Naturhist. Mus.; 66g New York, Amer. Mus. Nat. Hist.; 27g Chicago, Field Mus. Nat. Hist.;

Specimen(s): [67449], 21g.

Linwood 41°26′ N., 96°58′ W.

Butler County, Nebraska, U.S.A.
Found 1940, or 1941
Iron. Octahedrite, coarse (2.8mm), with silicate inclusions (IA).

A mass of 46kg was harrowed up near Linwood. Description, with an analysis of the metal, 5.98 %Ni, E.P. Henderson and S.H. Perry, Proc. U.S. Nat. Mus., 1949, **99**, p.357 [M.A. 10-520]. Determinations of Ga, Au and Pd, and partial analysis, 6.64 %Ni, E. Goldberg et al., Geochimica et Cosmochimica Acta, 1951, **2**, p.1. Silicate inclusions are Odessa-type, T.E. Bunch et al., Contr. Miner. Petrol., 1970, **25**, p.297. Analysis of the metal, 6.4 %Ni, 90.4 ppm.Ga, 374 ppm.Ge, 2.7 ppm.Ir, J.T. Wasson, Geochimica et Cosmochimica Acta, 1970, **34**, p.957. Description; plastically deformed, V.F. Buchwald, Iron Meteorites, Univ. of California, 1975, p.779.

33kg Washington, U.S. Nat. Mus., main mass; 1.6kg Harvard Univ.; 777g Chicago, Field Mus. Nat. Hist.; 1kg New York, Amer. Mus. Nat. Hist.;

Specimen(s): [1959,928], 17.5g.

Lion River v Gibeon.

Lipan Flats v San Angelo.

Lipetsky Khutor v Lipovsky.

Liponnas v Luponnas.

Lipovitz v Oczeretna.

Lipovskii Khutor v Lipovsky.

Lipovsky 49°5′ N., 42°31′ E.

Ust-Medvyeditsk, Federated SSR, USSR, [Липовскиинь Хутор]
Found 1904, known in this year
Synonym(s): *Khutor Lipowski, Lipetsky Khutor, Lipovskii Khutor, Lipowsky Chutor*
Stony-iron. Pallasite (PAL).

A mass of about 3.5kg was found. Description, with an analysis of the metal, 7.78 %Ni, P.N. Chirvinsky, Centralblatt Min., 1922, p.35. Olivine $Fa_{11.5}$, P.R. Buseck and J.I. Goldstein, Bull. Geol. Soc. Amer., 1969, **80**, p.2141.

Main mass, Kharkov, Univ.; 290g Moscow, Acad. Sci.; 34g Tempe, Arizona State Univ.;

Lipno v Lixna.

Lipowsky Chutor v Lipovsky.

Lippe v Barntrup.

Liptó v Nagy-Borové.

Lishui 31°38′ N., 118°59′ E.

Lishui County, Jiangsu, China
Fell 1978, September 10, 1600 hrs
Stone. Olivine-hypersthene chondrite (L5).

A single mass of 498g was recovered, D. Bian, Meteoritics, 1981, **16**, p.119. Analysis, 23.26 % total iron, W. Daode et al., Geochemistry, 1982, **1**, p.186.

490g Nanjing, Purple Mountain Observatory, approx. weight;

Lismore 35°57′ S., 143°20′ E. approx.

Victoria, Australia
Found 1959, November
Iron. Octahedrite, medium (1.0mm) (IIIA?).

One mass of nearly 10kg was found, A.B. Edwards, Proc. Roy. Soc. Victoria, 1960, **72**, p.93, Meteor. Bull., 1961 (21). Present whereabouts not known, R.A. Binns, priv. comm., 1983.

Lissa 50°12′ N., 14°51′ E.

Bunzlau, Bohemia, Czechoslovakia
Fell 1808, September 3, 1530 hrs
Synonym(s): *Bunzlau, Lysa, Lysa on Labe*
Stone. Olivine-hypersthene chondrite (L6), veined.

After detonations, four, possibly five, stones fell, two at Wustra and two at Strataw; the total weight of these four stones was about 10.4kg, the largest weighing about 3kg, K. von Schreibers, Ann. Phys. (Gilbert), 1808, **30**, p.358, M. Reuss, Ann. Chim., 1810, **74**, p.84. Analysis, J. Kokta, Coll. Czech. Chem. Comm., 1937, **9**, p.471 [M.A. 7-173]. Olivine Fa_{23}, B. Mason, Geochimica et Cosmochimica Acta, 1963, **27**, p.1011.

4kg Tübingen, Univ.; 3.8kg Vienna, Naturhist. Mus.; 1.5kg Prague, Nat. Mus.; 697g Berlin, Humboldt Univ.; 222g New York, Amer. Mus. Nat. Hist.; 206g Chicago, Field Mus. Nat. Hist.;

Specimen(s): [76154], 147g.; [35180], 11.5g.; [33732], 10.5g.; [1920,315], 3g.

Litau v Padvarninkai.

Litchenfels v Gibeon.

Little Miami Valley v Brenham.

Little Piney 37°55′ N., 92°5′ W.

Pulaski County, Missouri, U.S.A.
Fell 1839, February 13, 1530 hrs
Synonym(s): *Jefferson City, Missouri, Pine Bluff, Pulaski County*
Stone. Olivine-hypersthene chondrite (L5), xenolithic.

After the appearance of a fireball, followed by detonations, a stone of about 50lb, which had struck an oak tree, was found 2 miles from Pine Bluff and 10 miles from Little Piney, E.C. Herrick, Am. J. Sci., 1839, **37**, p.385, C.U. Shepard, Am. J. Sci., 1848, **6**, p.407. Xenolithic, olivine $Fa_{23.5}$, R.A. Binns, Geochimica et Cosmochimica Acta, 1968, **32**, p.299. Only about 400g known in collections.

78g Washington, U.S. Nat. Mus.; 62g Vienna, Naturhist. Mus.; 38g Tempe, Arizona State Univ.; 24g Yale Univ.; 15g Copenhagen, Univ.; 13g Berlin, Humboldt Univ.; 10.3g New York, Amer. Mus. Nat. Hist.; 8g Harvard Univ.;

Specimen(s): [24005], 98g.

Little River (a) 38°23′ N., 98°1′ W.

Rice County, Kansas, U.S.A.
Found 1967, recognized 1968
Stone. Olivine-bronzite chondrite (H6).

The smaller, 4.4kg, of two stones found in the area, Meteor. Bull., 1970 (49), Meteoritics, 1970, **5**, p.175, originally assigned to a single fall but now distinguished on textural grounds.

1.05kg Mainz, Max-Planck-Inst.; 170g Washington, U.S. Nat. Mus.;

Specimen(s): [1969,3], 72g. part-slice

Little River (b) 38°26.4′ N., 98°4′ W.

Rice County, Kansas, U.S.A.
Found 1965, recognized 1968
Stone. Olivine-bronzite chondrite (H4-5).

The larger, 11.7kg, of two stones found in the area, Meteor. Bull., 1970 (49), Meteoritics, 1970, **5**, p.175, and now distinguished on textural grounds, R. Hutchison, priv. comm., G.I Huss, letter of 23 May, 1974, in Min. Dept. BM (NH).

8.2kg Mainz, Max-Planck-Inst.; 258g Copenhagen, Univ. Geol. Mus.; 131g Chicago, Field Mus. Nat. Hist.; 121g Tempe, Arizona State Univ.;

Specimen(s): [1974,M.11], 167.2g. part-slice

Littlerock 34°31′ N., 117°59′ W.

Palmdale, California, U.S.A.
Found 1979, April
Stone. Olivine-bronzite chondrite (H6).

A single mass of 19.05kg (42lbs) was found on farm land at Littlerock, Meteor. Bull., 1980 (58), Meteoritics, 1980, **15**, p.238.

Livingston (Montana) 45°36′ N., 110°35′ W.

Park County, Montana, U.S.A.
Found 1936, October
Iron. Octahedrite, medium (0.95mm) (IIIA).

A mass of 1.6kg was found in an American indian grave, H.H. and A.D. Nininger, The Nininger Collection of Meteorites, Winslow, Arizona, 1950, p.69, 113. Classification, and analysis, 7.22 %Ni, 19.6 ppm.Ga, 35.4 ppm.Ge, 9.3 ppm.Ir, E.R.D. Scott et al., Geochimica et Cosmochimica Acta, 1973, **37**, p.1957. Description; shock-hardened, V.F. Buchwald, Iron Meteorites, Univ. of California, 1975, p.776.

705g Tempe, Arizona State Univ.; 294g Washington, U.S. Nat. Mus.;

Specimen(s): [1959,953], 424g. and sawings, 15.5g.

Livingston (Tennessee) 36°25′ N., 85°15′ W.

Overton County, Tennessee, U.S.A.
Found 1937
Iron. Octahedrite, medium (IRANOM).

A mass of about 30lb as found, but has been lost, except for fragments totalling 235g. Description, with an analysis, 7.45 %Ni, S.H. Perry and E.P. Henderson, Am. Miner., 1948, **33**, p.639 [M.A. 10-402]. Chemically anomalous, analysis, 6.64 %Ni, 45.3 ppm.Ga, 250 ppm.Ge, 0.73 ppm.Ir, J.T. Wasson, Meteorites, Springer-Verlag, 1974, p.307. Description; carbon-rich, partially recrystallised; annealed, V.F. Buchwald, Iron Meteorites, Univ. of California, 1975, p.778.

123g Washington, U.S. Nat. Mus.; 57g Chicago, Field Mus. Nat. Hist.; 18g Harvard Univ.;

Livingston County v Smithland.

Liweipantsun v Yukan.

Lixna 56°0′ N., 26°26′ E.
Dvinsk, Latvian SSR, USSR, [Ликсна]
Fell 1820, July 12, 1730 hrs
Synonym(s): *Daugavpils, Dünaburg, Lasdany, Liksen, Liksna, Lipno, Lixsna, Uszwalda*
Stone. Olivine-bronzite chondrite (H4), veined.
After the appearance of a fireball (moving from S. to N.), and detonations, a stone of about 40lb was seen to fall at the village of Lasdany; other stones fell into water and were not recovered, J. von Grotthuss, Ann. Phys. (Gilbert), 1821, **67**, p.337, A. Laugier, Ann. Phys. (Gilbert), 1823, **75**, p.264. Description, with a poor analysis, A. Kuhlberg, Arch. Naturk. Liv.-Ehst.-u. Kurlands, Ser. 1, Min. Wiss., Dorpat, 1867, **4**, p.23, 33. Total known weight 5.213kg, E.L. Krinov, Каталог Метеоритов Акад. Наук СССР, Москва, 1947 [M.A. 10-511]. See also, J. Pokrzynicki, Kosmos, Ser. B, 1959, **4**, p.303. Olivine Fa_{20}, B. Mason, Geochimica et Cosmochimica Acta, 1963, **27**, p.1011.
2760g Kiev, Acad. Sci. Ukraine; 269g Vienna, Naturhist. Mus.; 257g Moscow, Acad. Sci.; 73g Chicago, Field Mus. Nat. Hist.; 53g Washington, U.S. Nat. Mus., was BM 54642; 67g Tübingen, Univ.; 64g Berlin, Humboldt Univ.; 49g New York, Amer. Mus. Nat. Hist.; 43g Tempe, Arizona State Univ.;
Specimen(s): [1920,316], 77.5g.; [33735], 5.25g.

Lixsna v Lixna.

Ljungby v Lundsgard.

Llano del Inca v Vaca Muerta.

Llano Estacado v Estacado.

Loch Tay 56°32′ N., 4°10′ W.
Scotland
Fell 1802, September
Doubtful..
Several stones are said to have fallen, E.F.F. Chladni, Die Feuer-Meteore, Wien, 1819, p.268, R.P. Greg, Rep. Brit. Assn., 1860, p.62, T.M. Hall, Min. Mag., 1879, **3**, p.11, but the evidence is not conclusive.

Lockney 34°9′ N., 101°18′ W.
Floyd County, Texas, U.S.A.
Found 1944, recognized in this year
Stone. Olivine-hypersthene chondrite (L6).
A stone of 824g was found, H.H. and A.D. Nininger, The Nininger Collection of Meteorites, Winslow, Arizona, 1950, p.69. Olivine Fa_{23}, B. Mason, Geochimica et Cosmochimica Acta, 1963, **27**, p.1011.
242g Tempe, Arizona State Univ.; 90g Albuquerque, Univ. of New Mexico; 39g Los Angeles, Univ. of California;
Specimen(s): [1959,883], 257g.

Lockport v Cambria.

Locust Grove 33°20′ N., 84°6′ W.
Henry County, Georgia, U.S.A.
Found 1857
Synonym(s): *Henry County*
Iron. Hexahedrite (IIA).
A mass of 10kg was found three days after the appearance of a luminous meteor on 26 July, 1857. Description, with an analysis, 5.57 %Ni, E. Cohen, Sitzungsber. Akad. Wiss. Berlin, 1897, p.76. The mass has been artifically heated to about 1000C.; corroded, probably not an observed fall, V.F. Buchwald, Iron Meteorites, Univ. of California, 1975, p.780. Classification and analysis, 5.55 %Ni, 60.6 ppm.Ga, 180 ppm.Ge, 7.5 ppm.Ir, J.T. Wasson, Meteorites, Springer-Verlag, 1974, p.300.
2.16kg Washington, U.S. Nat. Mus.; 2kg Yale Univ.; 663g Vienna, Naturhist. Mus.; 370g Chicago, Field Mus. Nat. Hist.; 325g New York, Amer. Mus. Nat. Hist.; 251g Tübingen, Univ.; 229g Tempe, Arizona State Univ.; 174g Berlin, Humboldt Univ.;
Specimen(s): [82773], 365g. slice; [1920,317], 192g. slice

Lodhran v Lodran.

Lodi
Milan, Lombardy, Italy
Pseudometeorite..
A 0.5kg meteorite was said to have fallen on 3 June, 1972, La Nazione, Florence, 4 June, 1972. Discredited, G.R. Levi-Donatti, letters of 6 June, 31 July, 1972 in Min. Dept. BM (NH).

Lodran 29°32′ N., 71°48′ E.
Multan, Punjab, Pakistan
Fell 1868, October 1, 1400 hrs
Synonym(s): *Lodhran, Mooltan, Multan*
Stony-iron. Lodranite (LOD).
After detonations, a stone fell 12 miles E. of Lodhran, and a portion of about 1kg was preserved, T. Oldham, Rec. Geol. Surv. India, 1869, **2**, p.20. Description, with an analysis, G. Tschermak, Sitzungsber. Akad. Wiss. Wien, Math.-naturwiss. Kl., 1870, **61**, p.465. The olivines vary in composition, the centres of the grains being the more iron-rich, R.W. Bild and J.T. Wasson, Min. Mag., 1976, **40**, p.721.
690g Calcutta, Mus. Geol. Surv. India; 69g Vienna, Naturhist. Mus.; 22g New York, Amer. Mus. Nat. Hist.; 13g Washington, U.S. Nat. Mus.; 11g Tempe, Arizona State Univ.; 5g Berlin, Humboldt Univ.;
Specimen(s): [44003], 56g.

Loerbeek
Holland
Pseudometeorite..
After a bolide on 1 or 2 September, 1913, a stone was found and classed as meteoritic, Meteor. Bull., 1958 (8). Limestone, K. Keil, Fortschr. Min., 1960, **38**, p.239.

Logan 36°35′ N., 100°12′ W.
Beaver County, Oklahoma, U.S.A.
Found 1918
Stone. Olivine-bronzite chondrite (H5).
Recognized in 1933. One stone of 1315g was found, A.D. Nininger, Pop. Astron., Northfield, Minnesota, 1937, **45**, p.449 [M.A. 7-62], H.H. and A.D. Nininger, The Nininger Collection of Meteorites, Winslow, Arizona, 1950, p.69. Olivine Fa_{19}, B. Mason, Geochimica et Cosmochimica Acta, 1963, **27**, p.1011.
394g Tempe, Arizona State Univ.; 336g Washington, U.S. Nat. Mus.;
Specimen(s): [1959,884], 425.5g.

Logrono v Barea.

Loket v Elbogen.

Loma de la Cosina v Cosina.

Lombard 46°6′ N., 111°24′ W.
Broadwater County, Montana, U.S.A.
Found 1953, recognized in this year
Iron. Hexahedrite (IIA).
A weathered, monocrystalline hexahedrite weighing 7kg was found, V.F. Buchwald, Iron Meteorites, Univ. of California, 1975, p.782. Analysis, 5.59 %Ni, 58.0 ppm.Ga, 174 ppm.Ge, 2.3 ppm.Ir, J.T. Wasson, Geochimica et Cosmochimica Acta, 1969, **33**, p.859.
335g Copenhagen, Univ. Geol. Mus.; 238g New York, Amer. Mus. Nat. Hist.; 201g Tempe, Arizona State Univ.; 100g Tübingen, Univ.; 83g Perth, West. Austr. Mus.; 79g Chicago, Field Mus. Nat. Hist.; 62g Washington, U.S. Nat. Mus.;
Specimen(s): [1964,649], 292.5g.

Lonaconing 39°25′ N., 79°9′ W.
Allegheny County, Maryland, U.S.A.
Found 1888
Synonym(s): *Garrett County, Maryland*
Iron. Octahedrite, coarse (2.0mm) (IIE).
A mass of 45 oz. (1.4kg) was ploughed up in Garrett County, 12 miles S. of the post office of Lonaconing in Allegheny County, A.E. Foote, Am. J. Sci., 1892, **43**, p.64. Possibly found 12 miles north of Lonaconing post office, V.F. Buchwald, Iron Meteorites, Univ. of California, 1975, p.783. Analysis, 9.7 %Ni, 23.5 ppm.Ga, 62.1 ppm.Ge, 0.9 ppm.Ir, J.T. Wasson, Meteorites, Springer-Verlag, 1974, p.301.
727g Paris, Mus. d'Hist. Nat. (127g); Ecole des Mines (600g); 104g Vienna, Naturhist. Mus.; 98g New York, Amer. Mus. Nat. Hist.; 39g Chicago, Field Mus. Nat. Hist.; 24g Washington, U.S. Nat. Mus.;
Specimen(s): [1913,102], 74g.

London 51°31′ N., 0°10′ W.
England
Pseudometeorite..
Several authors have reported a fall of meteorites near Gresham College, London, on 18 May, 1680, E.F.F. Chladni, Die Feuer-Meteore, Wien, 1819, p.239, R.P. Greg, Rep. Brit. Assn., 1860, p.55, but it clear that this, like the Menabilly "stones", was merely a shower of very large hail-stones, E. King, Remarks concerning stones said to have fallen from the clouds.., London, 1796, p.20.

Lone Star 34°15′35″ N., 101°24′30″ W.
Floyd County, Texas, U.S.A.
Found 1965, recognized 1966
Stone. Olivine-bronzite chondrite (H4).
An incomplete stone of 4.3kg was ploughed up, Meteor. Bull., 1972 (51), Meteoritics, 1972, **7**, p.226, G.I Huss, letter of 18 March, 1970, in Min. Dept. BM(NH). Olivine Fa_{19}, R. Hutchison, priv. comm.
1.64kg Mainz, Max-Planck-Inst.; 168g Tempe, Arizona State Univ.; 54g Copenhagen, Univ. Geol. Mus.;
Specimen(s): [1969,230], 101.5g. slice, and fragments, 3g.

Lone Tree 28°15′ N., 91°28′45″ W.
Iowa, U.S.A.
Found 1971
Stone. Olivine-bronzite chondrite (H4).
An orientated, weathered mass of 20.676kg was ploughed up in the SW. quarter of section 18 T77N., R5W., approximately three miles SW. of Lone Tree. Described, with microprobe analyses, olivine Fa_{19}, G.R. McCormick and J.H. Carman, Meteoritics, 1975, **10**, p.67.
Main mass, Los Angeles, Univ. of California;
Specimen(s): [1978,M.15], 1.73kg.

Longchang 29°18′ N., 105°18′ E.
Sichuan, China
Found 1781
Iron. Octahedrite, medium (IVA-ANOM).
A single mass of 158.5g is in Chengdu, College of Geology, D. Bian, Meteoritics, 1981, **16**, p.120. Chemistry and structure, L. Fan, Geological Review, 1979, **25**, p.26. Classification, analysis, 8.6 %Ni, 2.1 ppm.Ga, 0.4 ppm.Ge, 4.0 ppm.Ir, D.J. Malvin et al., priv. comm., 1983.
158.5g Chengdu, Coll. of Geol.;

Long Island 39°56′ N., 99°36′ W.
Phillips County, Kansas, U.S.A.
Found 1891
Synonym(s): *Phillips County*
Stone. Olivine-hypersthene chondrite (L6), veined.
About 3000 pieces, belonging to one stone and weighing together about 1244lb, were found over a small area of about 15-20 ft. long by 6 ft. wide in the corner of Phillips County, three miles W. of the town of Long Island. Description, with an analysis, O.C. Farrington, Field Mus. Nat. Hist. Geol. Ser., 1902, **1** (64), p.283. Mentioned, H.H. Nininger, Trans. Kansas Acad. Sci., 1936, **39**, p.173. Olivine Fa_{25}, B. Mason, Geochimica et Cosmochimica Acta, 1963, **27**, p.1011.
550kg Chicago, Field Mus. Nat. Hist., most of the material, approx. weight; 66.7kg New York, Amer. Mus. Nat. Hist.; 6.1kg Fort Worth, Texas. Monnig Colln.; 2.6kg Washington, U.S. Nat. Mus.; 1.1kg Tempe, Arizona State Univ.; 1kg Budapest, Nat. Mus.; 800g Vienna, Naturhist. Mus.; 500g Prague, Nat. Mus.; 500g Michigan Univ.; 392g Ottawa, Mus. Geol. Surv. Canada; 512g Tübingen, Univ.;
Specimen(s): [81559], 1286g.

Lontolax v Luotolax.

Lookout Hill 30° S., 128°20′ E. approx.
Western Australia, Australia
Found
Stone. Carbonaceous chondrite, type II (CM2).
A single mass of 16.55g was found mid-way between Forrest and Forrest Lakes, Meteor. Bull., 1980 (57), Meteoritics, 1980, **15**, p.100.
Main mass, Perth, West. Austr. Mus.;
Specimen(s): [1977,M.7], 1.1g.

Loomis 40°28′ N., 99°30′ W.
Phelps County, Nebraska, U.S.A.
Found 1933
Stone. Olivine-hypersthene chondrite (L6).
One stone of 3.02kg was found, A.D. Nininger, Pop. Astron., Northfield, Minnesota, 1937, **45**, p.449 [M.A. 7-62], H.H. and A.D. Nininger, The Nininger Collection of Meteorites, Winslow, Arizona, 1950, p.70. Olivine Fa_{23}, B. Mason, Geochimica et Cosmochimica Acta, 1963, **27**, p.1011.
753g Tempe, Arizona State Univ.; 229g Washington, U.S. Nat. Mus.;
Specimen(s): [1959,885], 267g. and fragments, 5.6g.

Loongana Station (iron) v Mundrabilla.

Loongana Station West v Mundrabilla.

Loongana (stone) v Forrest Lakes.

Loop 32°54′ N., 102°17′ W.
Gains County, Texas, U.S.A.
Found 1962, recognized 1964
Stone. Olivine-hypersthene chondrite (L6).
One incomplete stone of 5.6kg was ploughed up, G.I Huss, letter of 18 March, 1970, in Min. Dept. BM(NH). See also Loot (*q.v.*). The name Loop has been used in error for Ashmore, J.R. Craig et al., Meteoritics, 1971, **6**, p.33. Olivine $Fa_{24.6}$, N.Z. Boctor and G. Kullerud, Meteoritics, 1981, **16**, p.61.
2.1kg Mainz, Max-Planck-Inst.; 149g Washington, U.S. Nat. Mus.; 121g Tempe, Arizona State Univ.;
Specimen(s): [1966,273], 112g. crusted slice, and fragments, 0.5g

Loop v Ashmore.

Loot
Stone. Chondrite.
Mentioned without details, P. Ramdohr, Geochimica et Cosmochimica Acta, 1967, **31**, p.1961. Perhaps an error for Loop.

Loreto 26°1′ N., 111°22′ W.
Baja California, Mexico
Found 1896, approx.
Synonym(s): *Loreto Baja*
Iron. Octahedrite, medium (1.2mm) (IIIA).
A mass of 209lb was found, regarded as paired with Morito though they were found 450 miles apart. Description, with an analysis, 7.75 %Ni, E.P. Henderson and S.H. Perry, Meteoritics, 1956, **1**, p.477. Classification and analysis, 7.67 %Ni, 19.3 ppm.Ga, 38.3 ppm.Ge, 3.8 ppm.Ir, E.R.D. Scott et al., Geochimica et Cosmochimica Acta, 1973, **37**, p.1957. No good reason for pairing with Morito, Ir contents differ slightly [Morito 9.2 ppm.Ir], V.F. Buchwald, Iron Meteorites, Univ. of California, 1975, p.784.
89kg Washington, U.S. Nat. Mus., main mass; 620g Fort Worth, Texas, Monnig Colln.; 323g Tempe, Arizona State Univ.; 200g Mainz, Max-Planck-Inst.; 166g Sydney, Austr. Mus.;

Loreto Baja v Loreto.

Los Amates v Toluca.

Los Angeles v Shingle Springs.

Los Guanacos v Campo del Cielo.

Los Lunas 34°48′45″ N., 106°47′30″ W.
Valencia County, New Mexico, U.S.A.
Found 1978, February
Stone. Olivine-bronzite chondrite (H4).
Two small masses totalling 95g were found, Meteor. Bull., 1983 (61), Meteoritics, 1983, **18**, p.81.
53g Chicago, Field Mus. Nat. Hist.; 19g Watchung, Dupont Colln.;

Los Martinez 38° N., 0°50′ W.
Cervera, Murcia, Spain
Fell 1894, May
Stone. Chondrite.
A specimen of 25g is in Madrid (Mus. Cienc. Nat.), M. Faura y Sans, Meteoritos caidos en la Peninsula Iberica, Tortosa, 1922, p.43, L. Fernández Navarro, Bol. R. Soc. Españ. Hist. Nat. Madrid, 1923, **23**, p.229.

Los Reyes 19°16′ N., 97°17′ W.
Mexico state, Mexico
Found 1897
Iron. Octahedrite, medium (0.90mm) (IIIB).
A mass of 43lb was ploughed up about 40 miles E. of Toluca. Description, with an analysis, 7.71 %Ni, O.C. Farrington, Field Mus. Nat. Hist. Geol. Ser., 1902, **1** (64), p.305. The mass has been artificially reheated, distinct from Toluca, V.F. Buchwald, Iron Meteorites, Univ. of California, 1975, p.786. Analysis, 8.71 %Ni, 20.9 ppm.Ga, 40.7 ppm.Ge, 0.12 ppm.Ir, E.R.D. Scott et al., Geochimica et Cosmochimica Acta, 1973, **37**, p.1957.
18.2kg Chicago, Field Mus. Nat. Hist., main mass; 377g Washington, U.S. Nat. Mus.;
Specimen(s): [1983,M.41], 125g.

Los Sauces 29°25′ S., 66°51′ W.
Capital, La Rioja, Argentina
Found 1937
Iron. Octahedrite, medium.
A mass of 997kg was found, L.M. Villar, Cienc. Investig., 1968, **24**, p.302.

Lost City 36°0.5′ N., 95°9′ W.
Cherokee County, Oklahoma, U.S.A.
Fell 1970, January 3, 2014 hrs
Stone. Olivine-bronzite chondrite (H5).
Following observations on the Prairie Photographic network, a search area was delimited and four fragments totalling 17kg were recovered, the first only six days after the fall. Full description, with an analysis, 27.64 % total iron, J. Geophys. Res., 1971, **76**, p.4056-4143. The largest stone, 9.83kg, was sliced and specimens were loaned to investigators by the U.S. Nat. Mus. Cosmogenic nuclides, R.C. Wrigley, J. Geophys. Res., 1971, **76**, p.4124. Cosmic-ray tracks, J.C. Lorin and P. Pellas, Meteoritics, 1975, **10**, p.445, abs.
16kg Washington, U.S. Nat. Mus.;
Specimen(s): [1971,292], 435g. crusted part-slice

Lost Draw 33°17′22″ N., 102°32′35″ W.
Terry County, Texas, U.S.A.
Found 1980
Stone. Chondrite.
A single mass of 309g was found by an artifact hunter, G.I Huss, letter of 26 October, 1983 in Min. Dept. BM(NH).
309g Denver, Amer. Meteorite Lab.;

Lost Lake 37°39′ N., 105°44′ W.
Alamosa County, Colorado, U.S.A.
Found 1931, approx., recognized 1934
Stone. Olivine-hypersthene chondrite (L).
One stone of 11g was found, A.D. Nininger, Pop. Astron., Northfield, Minnesota, 1937, **45**, p.449 [M.A. 7-62], H.H. and A.D. Nininger, The Nininger Collection of Meteorites, Winslow, Arizona, 1950, p.70. Olivine Fa_{25}, B. Mason, Geochimica et Cosmochimica Acta, 1963, **27**, p.1011.
Specimen(s): [1959,886], 11g. complete stone.

Losttown 34°15′ N., 84°30′ W. approx.
Cherokee County, Georgia, U.S.A.
Found 1868
Synonym(s): *Cherokee County*
Iron. Octahedrite, medium (1.0mm) (IID).
A mass of about 6.5lb was ploughed up 2.5 miles SW. of Losttown, C.U. Shepard, Am. J. Sci., 1868, **46**, p.257. The mass appears to have been heated to about 550C., V.F. Buchwald, Iron Meteorites, Univ. of California, 1975, p.787.

Classification and analysis, 10.0 %Ni, 72.7 ppm.Ga, 78.0 ppm.Ge, 18 ppm.Ir, A. Kracher et al., Geochimica et Cosmochimica Acta, 1980, **44**, p.773.
2513g Tempe, Arizona State Univ.; 87g Rome, Vatican Colln; 71g Washington, U.S. Nat. Mus.; 34g Vienna, Naturhist. Mus.;
Specimen(s): [43684], 6.5g.; [1959,962], 39.5 slice

Louans v Le Pressoir.

Louisa County v Staunton.

Louisiana v Red River.

Louisville 38°15′ N., 85°45′ W.
Jefferson County, Kentucky, U.S.A.
Fell 1977, January 31, 1530 hrs
Stone. Olivine-hypersthene chondrite (L6).
After a bright fireball and sonic booms, four stones totalling 1.3kg were recovered within the city of Louisville, Meteor. Bull., 1978 (55), Meteoritics, 1978, **13**, p.341. Description, G. Hunt and T.E. Boone, Trans. Kentucky Acad. Sci., 1978, **39**, p.39 [M.A. 80-4727].
195g Washington, U.S. Nat. Mus.;

Louvain v Tourinnes-la-Grosse.

Lowell 42°41′ N., 71°21′ W.
Massachusetts, U.S.A.
Fell 1846, November 11
Doubtful..
A mass 4 feet in diameter and weighing 442lb is said to have fallen in 1847, G. von Boguslawski, Ann. Phys. Chem. (Poggendorff), 1854, **4** (suppl.), p.377. Listed, R.P. Greg, Phil. Mag., 1854, **8**, p.460, L'Institut, Paris, 1847, **15**, p.8.

Löwenfluss v Gibeon.

Lowick v Łowicz.

Łowicz 52°0′ N., 19°55′ E.
Poland
Fell 1935, March 12, 0052 hrs
Synonym(s): *Lowick*
Stony-iron. Mesosiderite (MES).
A shower of stones fell, from Seligow to Krempa, near Łowicz, and 58 stones, totalling 59kg were found. Described, with an analysis, M. Kobylecki et al., Arch. Min. Tow. Nauk. Warsawa, 1938, **14**, p.1 [M.A. 7-174]. Metal and silicate mineralogy, B.N. Powell, Geochimica et Cosmochimica Acta, 1969, **33**, p.789, Geochimica et Cosmochimica Acta, 1971, **35**, p.5. Analysis of metal, 7.69 %Ni, 15.3 ppm.Ga, 54 ppm.Ge, 3.8 ppm.Ir, J.T. Wasson et al., Geochimica et Cosmochimica Acta, 1974, **38**, p.135. Olivine coronas, classification, references, C.E. Nehru et al., Geochimica et Cosmochimica Acta, 1980, **44**, p.1103.
35 stones, Warsaw, Geol. Mus.; 3.9kg Warsaw, Observ. Astron.; 21kg Cracow, Jagellonian Univ., 10kg in Dept. Min. Petr., 11kg in Astron. Observ.; 327g New York, Amer. Mus. Nat. Hist.; 171g Berlin, Humboldt Univ.; 82g Washington, U.S. Nat. Mus.;
Specimen(s): [1963,793], 61g. complete stone

Loyola
Locality not known
Stone. Olivine-hypersthene chondrite (L5).
A complete stone, initially of 2077g, in Washington, U.S. Nat. Mus. was found unlabelled in the collection of Loyola Univ. (probably Loyola Univ., Chicago; there are smaller universities of the same name in New Orleans and Los Angeles). It is probably a stone from the Homestead shower. Olivine Fa_{23}, B. Mason, Geochimica et Cosmochimica Acta, 1963, **27**, p.1011.
1772g Washington, U.S. Nat. Mus.;

Lozere v Aumieres.

Lua 24°57′ N., 75°9′ E.
Pargana Begu, Rajasthan, India
Fell 1926, June 26, 1600 hrs
Synonym(s): *Dabra*
Stone. Olivine-hypersthene chondrite (L5).
Two stones fell at Lua, one at Dongria, Udaipur State, and one at Dabra, Indore State, a mile or two distant; the fall was first named Dabra. The largest stone (8632g) fell at Lua, and 9241g were recovered in all. Described, A.L. Coulson, Rec. Geol. Surv. India, 1928, **61**, p.318 [M.A. 4-118]. Olivine $Fa_{25.1}$, R.T. Dodd et al., Geochimica et Cosmochimica Acta, 1967, **31**, p.921.
8.6kg Calcutta, Mus. Geol. Surv. India; 41g Tempe, Arizona State Univ.; 25g Washington, U.S. Nat. Mus.;
Specimen(s): [1928,480], 35.6g.

Lubbock 33°36′ N., 101°44′ W.
Lubbock County, Texas, U.S.A.
Found 1938, November 28
Stone. Olivine-hypersthene chondrite (L5).
A weathered stone of 1457.8g, W.A. Waldschmidt, Am. Miner., 1940, **25**, p.528 [M.A. 8-61], A.D. Nininger, Pop. Astron., Northfield, Minnesota, 1940, **48**, p.556, Contr. Soc. Res. Meteorites, 1942, **2**, p.227 [M.A. 8-54]. Olivine Fa_{24}, B. Mason, Geochimica et Cosmochimica Acta, 1963, **27**, p.1011.
390g Washington, U.S. Nat. Mus.; 125g Tempe, Arizona State Univ.; 150g Chicago, Field Mus. Nat. Hist.;
Specimen(s): [1959,887], 141.5g. slice

Lubimowka v Yurtuk.

Lucania 40°30′ N., 16° E. approx.
Basilicata, Italy
Fell 56, B.C.
Doubtful. Iron.
The fall of a mass of iron in Lucania (=Basilicata) is recorded, Pliny, Nat. Hist., II, p.57, E.F.F. Chladni, Die Feuer-Meteore, Wien, 1819, p.180, but the evidence is not conclusive.

Lucé 47°51′ N., 0°29′ E.
Sarthe, France
Fell 1768, September 13, 1630 hrs
Synonym(s): *Sarthe*
Stone. Olivine-hypersthene chondrite (L6), veined.
After detonations a stone of 3.5kg was seen to fall, C. Fougeroux and A.L. Lavoisier, J. Phys. Chim. Hist. Nat., 1772, **2**, p.251. Olivine Fa_{24}, B. Mason, Geochimica et Cosmochimica Acta, 1963, **27**, p.1011.
166g Vienna, Naturhist. Mus.; 24g Budapest, Nat. Mus.; 23g Berlin, Humboldt Univ.; 4g Chicago, Field Mus. Nat. Hist.; 3g Tübingen, Univ.;
Specimen(s): [46009], 3g.; [34590], 1g.; [35159], 1g.

Lucerne Valley 34°30′ N., 116°54′ W.
San Bernardino County, California, U.S.A.
Found 1963, July
Stone. Chondrite.
One complete individual, four nearly complete pieces, and two fragments, total weight 98g, were found on the surface of Lucerne Dry Lake in the area of an ellipse with a major axis of about 2.4km, J.D. Buddhue, Meteoritics, 1964, **2**, p.177, Meteor. Bull., 1964 (29).
Some specimens, Los Angeles, Griffith Observatory;

Lucignano d'Asso v Siena.

Lucky Hill 17°54′ N., 77°38′ W.
Bellevue, St. Elizabeth, Jamaica
Found 1885
Synonym(s): *Jamaica, St. Elizabeth*
Iron. Octahedrite, medium (IIIA).
An oxidised mass of over 45lb was dug up from 2 feet below the surface, Jamaica Gazette (Supplement), October 16, 1886, p.740, J.J. Bowrey, letter of 15 September, 1885 in Min. Dept. BM(NH). Classification, analysis, 7 %Ni, 21 ppm.Ga, 46 ppm.Ge, 0.4 ppm.Ir, E.R.D. Scott and J.T. Wasson, Geochimica et Cosmochimica Acta, 1976, **40**, p.103.
3.2kg London, Geol. Mus.; 196g Vienna, Naturhist. Mus.; 63g Berlin, Humboldt Univ.; 51g Chicago, Field Mus. Nat. Hist.; 40g Washington, U.S. Nat. Mus.;
Specimen(s): [56488], 4640g. rusted fragments

Luhy v Pribram.

Luis Lopez 34°0′ N., 106°58′ W.
Socorro County, New Mexico, U.S.A.
Found 1896
Synonym(s): *Magdalena, Magdaleny*
Iron. Octahedrite, medium (1.2mm) (IIIB).
A mass of about 7kg was picked up about 5 miles SW. of Socorro. Description, with an analysis, 8.17 %Ni, H.L. Preston, Am. J. Sci., 1900, **9**, p.283. Newer analysis, 8.64 % Ni, 20.1 ppm.Ga, 41.9 ppm.Ge, 0.15 ppm.Ir, E.R.D. Scott et al., Geochimica et Cosmochimica Acta, 1973, **37**, p.1957. Description; shock-deformed and annealed, V.F. Buchwald, Iron Meteorites, Univ. of California, 1975, p.789.
3kg Chicago, Field Mus. Nat. Hist.; 476g Harvard Univ.; 362g New York, Amer. Mus. Nat. Hist.; 178g Tübingen, Univ.; 174g Washington, U.S. Nat. Mus.;
Specimen(s): [84548], 426g. slice

Lujan 34°40′ S., 59°22′ W.
Buenos Aires province, Argentina
Found 1878, before this year
Synonym(s): *Villa Lujan*
Iron. Octahedrite?.
A piece of about 50g was found 6 m deep in Quaternary formations below the remains of a megatherium, M. Kantor, Rev. Mus. La Plata, 1920, **25**, p.117. A thoroughly oxidised iron, mainly limonite. Very little preserved. Co-ordinates given as 34°33′S., 59°8′W, and date of find as 1892, L.M. Villar, Cienc. Investig., 1968, **24**, p.302, but it must have been found before that year.
8g Vienna, Naturhist. Mus.; 33g La Plata Mus.; 3g Chicago, Field Mus. Nat. Hist.;

Lumpkin 32°2′ N., 84°46′ W.
Stewart County, Georgia, U.S.A.
Fell 1869, October 6, 1145 hrs
Synonym(s): *Stewart County*
Stone. Olivine-bronzite chondrite (H6).
After detonations, a stone of about 0.75lb was seen to fall 12 miles SW. of Lumpkin. Reported, with an analysis, J.E. Willet, Am. J. Sci., 1870, **50**, p.335. Olivine Fa_{19}, B. Mason, Geochimica et Cosmochimica Acta, 1963, **27**, p.1011.
40g Harvard Univ.; 31g Washington, U.S. Nat. Mus.; 24g New York, Amer. Mus. Nat. Hist.; 22g Vienna, Naturhist. Mus.;
Specimen(s): [47239], 17.5g.

Lunan 24°48′ N., 103°18′ E.
Weize, Lunan County, Yunan, China
Fell 1980, April 4, 1700 hrs
Stone. Olivine-bronzite chondrite (H).
A single mass of 2.52kg fell, D.Bian, Meteoritics, 1981, **16**, p.115. Analysis, 27.88 % total iron, W. Daode et al., Geochemistry, 1982, **1**, p.186.

Lund v Ekeby.

Lundsaur v Lundsgård.

Lundsgård 56°13′ N., 13°2′ E.
Ljungby, Skane, Sweden
Fell 1889, April 3, 2030 hrs
Synonym(s): *Ljungby, Lundsaur, Schonen*
Stone. Olivine-hypersthene chondrite (L6).
After the appearance of a luminous meteor and detonations, a stone of about 11kg was found, E. Svedmark, Geol. För. Förh. Stockholm, 1889, **11**, p.245, Nature, 1889, **40**, p.229. Description, with an analysis, A. Hadding, Geol. För. Förh. Stockholm, 1941, **62**, p.397 [M.A. 8-192]. Olivine Fa_{24}, B. Mason, Geochimica et Cosmochimica Acta, 1963, **27**, p.1011.
9.5kg Stockholm, Riksmus, main mass; 452g Budapest, Nat. Mus.; 147g Vienna, Naturhist. Mus.; 97g Washington, U.S. Nat. Mus.; 94g Harvard Univ.; 92g New York, Amer. Mus. Nat. Hist.; 104g Tempe, Arizona State Univ.; 55g Chicago, Field Mus. Nat. Hist.;
Specimen(s): [69636], 195g.

Luotolax 61°12′ N., 27°42′ E.
Viborg, Finland
Fell 1813, December 13, 1000 hrs, approx. time
Synonym(s): *Lontolax, Sawotaipola*
Stone. Achondrite, Ca-rich. Howardite (AHOW).
After detonations, a shower of stones fell on to the surface of the ice of a lake near the village of Luotolax, and a few were recovered, Scherer's Allg. Nord. Ann. Chem. St. Petersburg, 1818, **1**, p.203, Ann. Phys. (Gilbert), 1821, **67**, p.370. Description, with an analysis, A.E. Arppe, Acta Soc. Sci. Fenn. Helsingfors, 1867, **8**, p.85. Classification, M.B. Duke and L.T. Silver, Geochimica et Cosmochimica Acta, 1967, **31**, p.1637. Analysis, H.B. Wiik, Comm. Phys.-Math. Soc. Sci. Fenn., 1969, **34**, p.135. Co-ordinates, A.A. Yavnel, letter of 22 October, 1975 in Min. Dept. BM(NH).
500g Helsingfors, Univ., minimum weight; 20g Vienna, Naturhist. Mus.; 9g Tübingen, Univ.; 5g Berlin, Humboldt Univ.; 2g Chicago, Field Mus. Nat. Hist.;
Specimen(s): [41105*], 20.5g.

Luponnas 46°13′ N., 5°0′ E.
Ain, France
Fell 1753, September 7, 1300 hrs
Synonym(s): *Ain, Liponnas, Vonnas*
Stone. Olivine-bronzite chondrite (H3-5), brecciated.
After detonations, two stones were found, one of 9kg at Luponnas and the other of 5kg at Pont-de-Vesle, Jerome la Lande, Ann. Phys. (Gilbert), 1803, **13**, p.343. Very little appears to have been preserved. Description, with an analysis, S. Meunier, Bull. Soc. Hist. Nat. Autun, 1892, **5**, p.335. Olivine Fa_{19}, B. Mason, Geochimica et Cosmochimica Acta, 1963, **27**, p.1011.
84g Vienna, Naturhist. Mus.; 54g Paris, Mus. d'Hist. Nat.; 14g Chicago, Field Mus. Nat. Hist.; 5g Tempe, Arizona State Univ.;
Specimen(s): [63923], 6.5g.; [33727], 1g.

Lupton's Iron v Coahuila.

Luray 39°7′ N., 98°41′ W.
Russell County, Kansas, U.S.A.
Found, year not reported
Stone. Chondrite.
Listed, total known weight 861g, Cat. Huss Coll. Meteorites, 1976, p.29. Possibly part of Waldo or Covert.
858g Mainz, Max-Planck-Inst.;

Lusaka 7°13′ S., 29°26′ E.
Zaire
Fell 1951, April, before this time
A small meteorite, which fell before April 1951, is at Elizabethville (=Lubumbashi), Comité Special de Katanga, letter of 27 November, 1952, in Min. Dept. BM(NH).

Lusignan d'Asso v Siena.

Lusignano d'Asso v Siena.

Lusk 42°46′ N., 104°21′ W.
Niabrara County, Wyoming, U.S.A.
Found 1940
Iron. Octahedrite, very oxidized.
An oxidized mass of 46g was found, A.D. Nininger, Pop. Astron., Northfield, Minnesota, 1940, **48**, p.556, Contr. Soc. Res. Meteorites, **2**, p.227 [M.A. 8-54], H.H. and A.D. Nininger, The Nininger Collection of Meteorites, Winslow, Arizona, 1950, p.70.
22.2g Tempe, Arizona State Univ.;
Specimen(s): [1959,924], 23g.

Lutschaunig's Stone 27° S., 70° W. approx.
Atacama, Chile
Found 1861
Synonym(s): *Atacama Desert*
Stone. Olivine-hypersthene chondrite (L6).
A mass of over 100kg was found in a "quebrada" in the desert about 1861 and fragments were brought as silver ore to Mr. A. Lutschaunig's mill at Copiapo to be crushed, A. Lutschaunig, letters of 21 and 25 October, 1889 in Min. Dept. BM(NH). Olivine Fa_{24}, B. Mason, Geochimica et Cosmochimica Acta, 1963, **27**, p.1011. Repository of the large mass unknown.
73g Washington, U.S. Nat. Mus.; 3g Vienna, Naturhist. Mus.;
Specimen(s): [36107], 54.5g.; [44763], 31.5g.; [44764], 6g.

Lysa v Lissa.

Lysa on Labe v Lissa.

Lyubimovka v Yurtuk.

Mabwe-Khoyma v Mabwe-Khoywa.

Mabwe-Khoywa 19° N., 97° E. approx.
Kyebogyi, Kayah, Burma
Fell 1937, September 17, 2245 hrs
Synonym(s): *Mabwe-Khoyma*
Stone. Olivine-hypersthene chondrite (L5).
Listed, A.L. Coulson, Mem. Geol. Surv. India, 1940, **75**, p.139, B. Mason, letters of 4 March and 28 April, 1964, in Min. Dept. BM(NH). Mineralogy, analysis, 21.77 % total iron, olivine Fa_{24}, A. Dube et al., Smithson. Contrib. Earth Sci., 1977 (19), p.71.
536g Calcutta, Mus. Geol. Surv. India; 0.9g New York, Amer. Mus. Nat. Hist.;

Macao v Macau.

Macatuba
Locality not reported
Fell 1951, July 5
Doubtful..
Listed, K. Keil, Fortschr. Min., 1960, **38**, p.239. No details are given and the locality has not been traced in any atlas or gazetteer.

Macau 5°12′ S., 36°40′ W.
Rio Grande do Norte, Brazil
Fell 1836, November 11, 0500 hrs
Synonym(s): *Macao, Macayo*
Stone. Olivine-bronzite chondrite (H5), veined.
After the appearance of a brilliant meteor, followed by detonations, a shower of stones, some said to weigh from 1lb to 80lb, but most the size of doves' eggs, fell near the mouth of the river Assu, killing several cattle, F. Berthou, C. R. Acad. Sci. Paris, 1837, **5**, p.211, P. Partsch, Die Meteoriten, Wien, 1843, p.81, O.A. Derby, Revista do Observatorio, Rio de Janeiro, 1888, p.7. Listed, N. Vidal, Bol. Mus. Nac. Rio de Janeiro, 1936, **12**, p.91 [M.A. 8-376]. Mineralogy, with an analysis, 26.27 % total iron, olivine Fa_{19}, C.B. Gomes et al., Meteoritics, 1977, **12**, p.241, C.B. Gomes et al., Anais Acad. Brasil. Cienc., 1977, **49**, p.575.
476g Rio de Janeiro, Mus. Nac.; 500g Vienna, Naturhist. Mus.; 257g Paris, Mus. d'Hist. Nat.; 84g Chicago, Field Mus. Nat. Hist.; 72g Washington, U.S. Nat. Mus.; 42g Tempe, Arizona State Univ.; 37g Berlin, Humboldt Univ.; 22g New York, Amer. Mus. Nat. Hist.;
Specimen(s): [35176], 3.5g.; [34742], 2.5g.

Macayo v Macau.

Macedonia v Seres.

Macerata v Monte Milone.

Machinga 15°12′44″ S., 35°14′32″ E.
Near Machinga, Southern province, Malawi
Fell 1981, January 22, 1000 hrs
Stone. Olivine-hypersthene chondrite (L6).
A single mass of 93.2kg fell about 7km SW. of Machinga and 20km NNW. of Zomba. Classification, brief description, M.J. Crow et al., Rep. Geol. Surv. Malawi, 1981 (September). Olivine $Fa_{24.5}$, A.L. Graham, priv. comm., 1982.
Specimen(s): [1982,M.24], 56g.

Macibini 28°50′ S., 31°57′ E.
Natal, South Africa
Fell 1936, September 23, 0800 hrs
Stone. Achondrite, Ca-rich. Eucrite (AEUC).
Four complete stones and two fragments totalling 1995g were recovered at Macibini, near Empangeni. Description, with an analysis, S.H. Haughton and F.C. Partridge, Trans. Geol. Soc. S. Africa, 1939, **41**, p.205. Metal analysed, of variable Ni content, 0.4-2.1 %Ni, 1.4-2.2 %Co, J.F. Lovering, Nature, 1964, **203**, p.70. Polymict breccia; pyroxenes exhibit extreme iron enrichment, A.M. Reid, Meteoritics, 1974, **9**, p.398, abs. Olivine Fa_{53-83}, C. Desnoyers and D.Y. Jerome, Meteoritics, 1973, **8**, p.344, abs.
Main mass, Pretoria, Geol. Surv. Mus.; 105g Calcutta, Mus. Geol. Surv. India; 15g Washington, U.S. Nat. Mus.; 9g Yale Univ.;

Mackinney v McKinney.

Macon County v Tombigbee River.

Macquarie River 31°30′ S., 153° E.
New South Wales, Australia
Found 1857
Pseudometeorite..
Listed as a doubtful meteorite, E.A. Wülfing, Die Meteoriten in Samml., Tübingen, 1897, p.402, T. Hodge-Smith, Mem. Austr. Mus., 1939, **7**, p.27. Shown to be a pseudometeorite, O.C. Farrington, Field Mus. Nat. Hist. Geol. Ser., 1914, **5**, p.12, B. Mason, Min. Mag., 1962, **33**, p.68.
9.9g New York, Amer. Mus. Nat. Hist.; 1g Paris, Mus. d'Hist. Nat.; 13g Berlin, Humboldt Univ.;

Madaras v Mezö-Madaras.

Maddur v Muddoor.

Madhipura 25°55′ N., 86°22′ E.
Bhagalpur district, Bihar, India
Fell 1950, May 23, 1500 hrs
Stone. Olivine-hypersthene chondrite (L).
A stone of about 3 inches diameter fell through the roof of a shed. Description, with a partial analysis, D.N. Ojha, Quar. J. Geol. Mining Metall. Soc. India, 1952, **24**, p.169 [M.A. 12-102]. Year of fall given as 1949; 1kg recovered; Mn53 activity, S.K. Bhattacharya et al., Earth planet. Sci. Lett., 1980, **51**, p.45. Rare gas content, K. Gopalan and M.N. Rao, Meteoritics, 1976, **11**, p.131.
Specimen, Patna, Mus.;

Madioen v Ngawi.

Madison County v Duel Hill (1854).

Madison County v Duel Hill (1873).

Madiun 7°45′ S., 111°32′ E. approx.
Java, Indonesia
Fell 1935, June 20
Stone. Olivine-hypersthene chondrite (L6).
A single mass of about 400g was recovered, olivine $Fa_{24.2}$, Meteor. Bull., 1981 (59), Meteoritics, 1981, **16**, p.195. Description (incorrect co-ordinates), C.G.R. Reid and K. Fredriksson, Meteoritics, 1982, **17**, p.27.
Main mass, Bandong, Geol. Surv. Indonesia; 46g Washington, U.S. Nat. Mus.;

Madoc 44°30′ N., 77°28′ W.
Hastings County, Ontario, Canada
Found 1854
Synonym(s): *Hastings County*
Iron. Octahedrite, medium (0.95mm) (IIIA).
A mass of 370lb was found, T.S. Hunt, Am. J. Sci., 1855, **19**, p.417. Description, E. Cohen, Meteoritenkunde, 1905, **3**, p.354. Classification and analysis, 7.52 %Ni, 19.4 ppm.Ir, 36.4 ppm.Ge, 6.8 ppm.Ir, E.R.D. Scott et al., Geochimica et Cosmochimica Acta, 1973, **37**, p.1957. Description; shock-hardened, V.F. Buchwald, Iron Meteorites, Univ. of California, 1975, p.791.
168kg Ottawa, Mus. Geol. Surv. Canada; 301g London, Geol. Mus.; 207g Vienna, Naturhist. Mus.; 179g New York, Amer. Mus. Nat. Hist.; 169g Tempe, Arizona State Univ.;
Specimen(s): [26972], 205.5g.; [1959,167], 0.2g.; [1960,328], 190.5g.

Madrid 40°25′ N., 3°43′ W.
Spain
Fell 1896, February 10, 0930 hrs
Stone. Olivine-hypersthene chondrite (L6), veined.
After the appearance of a luminous meteor, followed by detonations, several small stones, one of which weighed 125g, fell near Madrid; the total weight collected was about 400g, S. Calderon, Bull. Soc. Géol. France, 1896, **24**, p.117. Description, with a partial analysis, A.F. Gredilla, Anal. Soc. Españ. Hist. Nat. Madrid, 1896, **25**, p.223. Olivine Fa_{24}, B. Mason, Geochimica et Cosmochimica Acta, 1963, **27**, p.1011.
144g Madrid, Mus. Cienc. Nat.; 4.5g Tempe, Arizona State Univ.; 3.5g Washington, U.S. Nat. Mus.;
Specimen(s): [81851], 18.5g. a nearly complete stone.

Mäel Pestivien v Kerilis.

Maeme v Kyushu.

Maessing v Mässing.

Mafra 26°10′ S., 49°56′ W.
Santa Catarina, Brazil
Fell 1941
Stone. Olivine-hypersthene chondrite (L3-4), brecciated.
A mass of 600g was collected. Description, J.V. Valarelli and M.R. de Arruda, Ciencia e Cultura, Sao Paulo, 1965, **17**, p.132, A.J. Melfi, Bol. Soc. Brasil. Geol., 1965, **14**, p.41. Olivine Fa_{25}, B. Mason, Geochimica et Cosmochimica Acta, 1967, **31**, p.1100. Petrography, classification as H4 (in error) 26.70 % total iron, G.R. Levi-Donati et al., Meteoritics, 1976, **11**, p.29. L3-4 classification, D.E. Lange et al., Meteoritics, 1979, **14**, p.472, abs.
21g Albuquerque, Univ. of New Mexico; 9g Washington, U.S. Nat. Mus.;

Magadan v Susuman.

Magdalena v Luis Lopez.

Magdaleny v Luis Lopez.

Magdeburg v Erxleben.

Magdeburg 52°8′ N., 11°38′ E.
Magdeburg, Germany
Fell 998, A.D.
Doubtful..
Two stones are said to have fallen, E.F.F. Chladni, Die Feuer-Meteore, Wien, 1819, p.193, but the evidence is not conclusive.

Magetan v Ngawi.

Magnesia 37°52′ N., 27°31′ E.
Aydin, Turkey
Fell 1899
Synonym(s): *Magnissa*
Iron. Octahedrite, medium (0.60mm) (IIICD-ANOM).
The fall of this iron (reported in error as of about 5kg) was witnessed in 1899 by a Turkish peasant in a village near Magnesia (Magnissa), since destroyed. The mass remained in the family until 1918 when it was purchased by a Turkish army officer and ultimately by Col. L. Vésignié, L. Vésignié, letter of 26 April 1939, in Min. Dept. BM(NH). Analysis, 11.0 %Ni, 14.5 ppm.Ga, 22.4 ppm.Ge, 0.18 ppm.Ir, J.T. Wasson, Meteorites, Springer-Verlag, 1974, p.304. The 5kg mass is an end-piece from another mass. Fully illustrated description; shocked but not annealed, V.F. Buchwald, Iron Meteorites, Univ. of California, 1975, p.1401.
4960g Paris, Mus. d'Hist. Nat.;
Specimen(s): [1938,395], 1g.

Magnissa v Magnesia.

Magrour v Umm Ruaba.

Magura 49°20′ N., 19°29′ E.
Arva, Slovensko, Czechoslovakia
Found 1840
Synonym(s): *Arva, Orava, Slanica, Szlanica, Szlanicza*
Iron. Octahedrite, coarse (2.4mm) (IA).
Numerous masses, totalling about 1500kg, were found but most were smelted and only about 150kg were saved, O. Büchner, Die Meteoriten in Samml., Leipzig, 1863, p.168, W. von Haidinger, Ann. Phys. Chem. (Poggendorff), 1844, **61**, p.675. Analysis, 7.08 %Ni, E. Cohen, Ann. Naturhist. Hofmus. Wien, 1900, **15**, p.377. Contains cliftonite, A. Brezina, Ann. Naturhist. Hofmus. Wien, 1889, **4**, p.102. X-ray identification of cohenite in Magura mass, A. Westgren and G. Phragmen, J. Iron Steel Inst., 1924, **109**, p.159 [M.A. 2-517]. Newer analysis, 6.67 %Ni, 94.6 ppm.Ga, 483 ppm.Ge, 3.2 ppm.Ir, J.T. Wasson, Icarus, 1970, **12**, p.407. Co-ordinates, description; heavily shocked, V.F. Buchwald, Iron Meteorites, Univ. of California, 1975, p.792.
45kg Tübingen, Univ.; 25kg Vienna, Naturhist. Mus.; 25.7kg Budapest, Nat. Mus. (24.3kg), Eötvös Lorand Univ. (1.4kg); 10kg Berlin, Humboldt Univ.; 1.5kg Chicago, Field Mus. Nat. Hist.; 1kg New York, Amer. Mus. Nat. Hist.; 982g Prague, Nat. Mus.; 796g Moscow, Acad. Sci.; 1.1kg Yale Univ.; 672g Washington, U.S. Nat. Mus.;
Specimen(s): [33925], 6367g.; [19101c], 1400g.; [19101b], 1217g.

Mahi Kantha v Myhee Caunta.

Mailand v Cereseto.

Mailsi v Khairpur.

Mainardi v Vigarano.

Mainpuri v Chandpur.

Mainz 50°0′ N., 8°16′ E.
Rheinland-Pfalz, Germany
Found 1850, or 1852
Synonym(s): *Mayence*
Stone. Olivine-hypersthene chondrite (L6), veined.
A stone of about 1.5kg was ploughed up. Description, with an analysis, F. Seelheim, Jahrb. Ver. Naturk. Nassau, 1857, p.405. Mineralogy, 20.8 % total iron, olivine Fa_{24}, H. Palme et al., Meteoritics, 1983, **18**, p.abs.
201g Calcutta, Mus. Geol. Surv. India; 119g Vienna, Naturhist. Mus.; 114g New York, Amer. Mus. Nat. Hist.; 55g Tübingen, Univ.; 41g Chicago, Field Mus. Nat. Hist.;
Specimen(s): [36134], 18g.; [34674], 14.5g.; [35151], 1g.

Majorca 39°27′ N., 2°49′ E.
Balearic Isles, Spain
Fell 1935, July 17, 1135 hrs
Synonym(s): *Mallorca*
Iron. Octahedrite?.
One piece of 809g and 15g of fragments were recovered from a depth of 90 cm near the road from Palma to Manacor, G.M. Cardoso, Bol. R. Soc. Españ. Hist. Nat. Madrid, 1935, **35**, p.453 [M.A. 7-66], J.G. Morales, Bol. R. Soc. Españ. Hist. Nat. Madrid, 1936, **36**, p.301 [M.A. 7-270].

Makarewa 46°19′ S., 168°24′ E.
Invercargill, County Southland, New Zealand
Found 1879
Synonym(s): *Invercargill, Makariwa*
Stone. Olivine-hypersthene chondrite (L6), brecciated.
A stone of about 5lb was found in clay about 2.5 feet from the surface. Description, G.H.F. Ulrich, Proc. Roy. Soc. London, 1893, **53**, p.54. Description, with an analysis, L. Fletcher, Min. Mag., 1894, **10**, p.287. Mentioned, G.R. Marriner, Trans. New Zealand Inst., 1910, **42**, p.178. Olivine Fa_{24}, B. Mason, Geochimica et Cosmochimica Acta, 1963, **27**, p.1011.
Small piece, Dunedin, Mus.; Small piece, Wellington, Mus.; 13g New York, Amer. Mus. Nat. Hist.; 1g Vienna, Naturhist. Mus.;
Specimen(s): [76068], 51g. two pieces; [76069], 11.75g.; [76070], Thin sections (four)

Makariwa v Makarewa.

Makedonien v Seres.

Maksimovka v Vavilovka.

Malaga 32°13′ N., 104°0′ W.
Eddy County, New Mexico, U.S.A.
Fell 1933, November, early in the month, at dusk
Stone. Chondrite.
A stone of 150g fell, B. Mason, Meteorites, Wiley, 1962, p.238. An earlier record describes it as a stony-iron and a find rather than a fall, B.H. Wilson, The Mineralogist, Portland, Oregon, 1950, **18**, p.512.

Malakal 9°30′ N., 31°45′ E. approx.
Upper Nile Province, Sudan
Fell 1970, August, a few days before the 15th.
Stone. Olivine-hypersthene chondrite (L5), brecciated.
About 2kg of material, probably from a much larger mass, was brought to Khartoum on 15 August, 1970, A.S. Dawoud and J.R. Vail, Nature, 1971, **229** (Phys. Sci.), p.212.

Reported, Meteor. Bull., 1971 (50), Meteoritics, 1971, **6**, p.112. Irradiation history, P.J. Cressy and L.A. Rancitelli, Meteoritics, 1973, **8**, p.338, abs. Analysis, olivine Fa_{25}, 22.3 % total iron, R.S. Clarke, Jr. et al., Smithson. Contrib. Earth Sci., 1972 (14), p.63.
160g Washington, U.S. Nat. Mus.; 20g Chicago, Field Mus. Nat. Hist.; 18g Mainz, Max-Planck-Inst.;
Specimen(s): [1971,97], 108g. interior fragment

Malampaka 3°8′ S., 33°31′ E.
Mwanza district, Tanzania
Fell 1930, September?, at mid-day hrs
Stone. Olivine-bronzite chondrite (H).
Fell 1 mile W. of Malampaka railway station, F. Oates, Tanganyika Notes and Records, 1941 (12), p.28 [M.A. 8-373], J.R. Harpum, Rec. Geol. Surv. Tanganyika for 1961, 1965, **11**, p.54. Olivine Fa_{19}, B. Mason, Geochimica et Cosmochimica Acta, 1963, **27**, p.1011.
455g Dodoma, Geol. Surv. Tanzania; 15g New York, Amer. Mus. Nat. Hist.;

Maldiak v Maldyak.

Maldyak 63°20′ N., 148°10′ E.
Magadan region, Federated SSR, USSR,
[Мальдяк]
Found 1939
Synonym(s): *Maldiak*
Iron. Octahedrite, medium (1.0mm) (IIIA).
A mass of 992g was found and acquired by the Acad. Sci., Moscow, E.L. Krinov, Метеоритика, 1949, **5**, p.49 [M.A. 12-105]. It has been suggested that it is part of the same fall as Susuman, B.I. Vronsky, Метеоритика, 1960, **19**, p.135 [M.A. 16-638]. The chemical data shows these finds to be distinct. Analysis, 8.95 %Ni, M.I. D'yakonova and V.Ya. Kharitonova, Метеоритика, 1963, **23**, p.42. Further analysis, 9.08 %Ni, 19.8 ppm.Ga, 42.6 ppm.Ge, 0.39 ppm.Ir, E.R.D. Scott and J.T. Wasson, Geochimica et Cosmochimica Acta, 1976, **40**, p.103.
671g Moscow, Acad. Sci.;

Malegaon v Naoki.

Mallenbye v Mellenbye.

Mallorca v Majorca.

Malomhaza v Minnichhof.

Malotas 28°56′ S., 63°14′ W.
Salavina, Santiago del Estero, Argentina
Fell 1931, June 22, 0440 hrs
Synonym(s): *Salavina*
Stone. Olivine-bronzite chondrite (H5).
A shower of some thousands of stones fell over an area of 5 × 2.5km, NE. of Malotas, dept. Salavina, Description, J. Olsacher, Rev. Univ. Nac. Cordoba, 1931, **18** (9-10), p.430. Another description, with an analysis, E.H. Ducloux, Anal. Mus. Argentino Cienc. Nat., 1942, **40**, p.129 [M.A. 10-179]. Olivine Fa_{19}, B. Mason, Geochimica et Cosmochimica Acta, 1963, **27**, p.1011.
206g Washington, U.S. Nat. Mus.; 140g New York, Amer. Mus. Nat. Hist.;
Specimen(s): [1962,174], 8.5g.

Malpas 53°1′ N., 2°51′ W.
Cheshire, England
Fell 1813, in the summer
Doubtful..
Many stones are said to have fallen, Ann. Phil., 1813, **2**, p.396, but the evidence is not conclusive.

Malvern 29°27′ S., 26°46′ E.
Orange Free State, South Africa
Fell 1933, November 20-30
Stone. Achondrite, Ca-rich. Howardite (AHOW).
A stone of 807g was found on Malvern Farm, 25 km SSW. of Thaba Nchu, on 30 November 1933, C. Frick and E.C.I. Hammerbeck, Bull. Geol. Surv. S. Africa, 1973 (57), p.26. Although chemically like the howardites, H. von Michaelis et al., Earth planet. Sci. Lett., 1969, **5**, p.387, this stone has suffered shock-melting, glassy spheres are present, C. Desnoyers and D.Y. Jerome, Geochimica et Cosmochimica Acta, 1977, **41**, p.81.
Main mass, Bloemfontein, Nat. Mus.; 19g Washington, U.S. Nat. Mus.;
Specimen(s): [1974,M.6], 9.8g. four fragments

Malyi Altai v Krasnojarsk.

Malyi Altaj v Krasnojarsk.

Mamra v Mamra Springs.

Mamra Springs 45°13′ N., 62°5′ E.
Kazakhstan SSR, USSR, [Мамра]
Fell 1927, May 5-15, at night
Synonym(s): *Mamra*
Stone. Olivine-hypersthene chondrite (L6).
A stone of about 1kg fell in the night of 5 May, L.A. Kulik, C. R. Acad. Sci. URSS, 1929 (ser. A), p.81 [M.A. 4-261], or 15 May, I.S. Astapowitsch, Trans. Roy. Astron. Soc. Canada, 1938, **32**, p.195 [M.A. 7-172]. The date seems to be uncertain, but somewhere between 5 and 15 May, V.I. Vernadsky, letter of 5 August, 1939, in Min. Dept. BM(NH). Only 58g have been preserved, date given as 15 May, L.G. Kvasha and A.Ya. Skripnik, Метеоритика, 1978, **37**, p.195. Olivine Fa_{24}, B. Mason, Geochimica et Cosmochimica Acta, 1963, **27**, p.1011.
56.3g Moscow, Acad. Sci.;

Manbazar pargana v Manbhoom.

Manbhoom 23°3′ N., 86°42′ E.
Purulia, West Bengal, India
Fell 1863, December 22, 0900 hrs
Synonym(s): *Cossipore, Manbazar pargana, Manboom*
Stone. Olivine-hypersthene chondrite, amphoterite (LL6).
After detonations, several stones, the largest weighing about 1.5kg, were found near the villages of Govindpur, Pandra, and Cossipore, W. von Haidinger, Sitzungsber. Akad. Wiss. Wien, Math.-naturwiss. Kl., 1864, **50**, p.241. Description, with an analysis, 19.11 % total iron, B. Mason and H.B. Wiik, Geochimica et Cosmochimica Acta, 1964, **28**, p.533. Description, olivine Fa_{31}, K. Fredriksson et al., Origin and Distribution of the Elements, ed. L.H. Ahrens, Pergamon, 1968, p.457. Phosphate minerals analysed, W.R. Van Schmus and P.H. Ribbe, Geochimica et Cosmochimica Acta, 1969, **33**, p.637.
661g Calcutta, Mus. Geol. Surv. India; 600g New York, Amer. Mus. Nat. Hist.; 190g Vienna, Naturhist. Mus.; 51g Tempe, Arizona State Univ.; 15g Washington, U.S. Nat. Mus.;
Specimen(s): [36284], 143g.

Manboom v Manbhoom.

Mandiari v Sultanpur.

Mandiga v Bencubbin.

Manegaon 20°58′ N., 76°6′ E.
Bhusawal, East Khandesh district, Maharashtra, India
Fell 1843, June 29, 1530 hrs
Synonym(s): *Khandesh district, Manegaum*
Stone. Achondrite, Ca-poor. Diogenite (ADIO).
After detonations, an oblong stone, about 15 inches long and 5 inches in diameter, was seen to fall; it was broken into pieces and only a few ounces were preserved, J. Abbott, J. Asiatic Soc. Bengal, 1844, **13**, p.880. Description, analysis, N.S. Maskelyne, Phil. Trans. Roy. Soc. London, 1870, **160**, p.211. Further analysis, 11.57 % total iron, T.S. McCarthy et al., Meteoritics, 1974, **9**, p.215. Mineral chemistry, K. Fredriksson, Meteoritics, 1982, **18**, p.141.
25g Calcutta, Mus. Geol. Surv. India; 2g Vienna, Naturhist. Mus.; 1g Chicago, Field Mus. Nat. Hist.;
Specimen(s): [33759], 11g.; [1951,508], 1g.

Manegaum v Manegaon.

Mangwendi 17°39′ S., 31°36′ E.
Mashonaland, Zimbabwe
Fell 1934, March 7, 1245 hrs
Synonym(s): *Salisbury*
Stone. Olivine-hypersthene chondrite, amphoterite (LL6), brecciated.
A mass, probably of about 60lb, fell on the right bank of the Shawanoya river; one large piece of 49lb 3oz and ten fragments totalling 3lb 1oz were collected. Another stone is said to have fallen at Magaya, 17 miles E. of St. Paul's mission. Description, with an analysis, 20.77 % total iron, B. Lightfoot et al., Min. Mag., 1935, **24**, p.1. Xenolithic, R.A. Binns, Geochimica et Cosmochimica Acta, 1968, **32**, p.299. Chondrule with skeletal chromite figured (under Salisbury), P. Ramdohr, Geochimica et Cosmochimica Acta, 1967, **31**, p.1961. Olivine Fa_{29}, B. Mason, Geochimica et Cosmochimica Acta, 1963, **27**, p.1011.
387g New York, Amer. Mus. Nat. Hist.; 52g Chicago, Field Mus. Nat. Hist.;
Specimen(s): [1934,839], 21.465kg. includes main mass, 20.8kg, and fragments, 30g.

Mani v Toluca.

Manitouwabing 45°26′24″ N., 79°52′32″ W.
Ontario, Canada
Found 1962, November, fell not earlier than the summer of 1949
Iron. Octahedrite, medium (1mm) (IIIA).
One mass of about 39kg was found, Meteor. Bull., 1963 (26), R. Knox, Meteoritics, 1964, **2**, p.279. Possibly related to Madoc (*q.v.*), V.F. Buchwald, Iron Meteorites, Univ. of California, 1975, p.797.
Main mass, Toronto, Univ.; 19g Ottawa, Mus. Geol. Surv. Canada;

Manlai 44°20′ N., 106°30′ E.
Omnogov aymag, Mongolia, [Манлай]
Found 1954, known for "several decades"
Synonym(s): *Manlay*
Iron. Octahedrite, fine, or Ni-rich ataxite.
A mass of 95×40×25 cm was found, G.G. Vorobyev and O. Namnandorzh, Метеоритика, 1958, **16**, p.135 [M.A. 14-130]. The mass weighed 166.8kg, O. Namnandorj, Meteorites of Mongolia, 1980, p.17, English trans.
166kg Ulan-Bator, Central Mus.;

Manpur Urf Baghai Kalan v Merua.

Mantala v Muraid.

Mantos Blancos 23°27′ S., 70°7′ W. approx.
Atacama, Chile
Found 1876
Synonym(s): *Antofagasta, Mount Hicks*
Iron. Octahedrite, fine (0.35mm) (IVA).
A mass of 10.3kg was found on the SE. side of Mount Hicks, about 40 miles from Antofagasta. Description, with an analysis, 8.83 %Ni, L. Fletcher, Min. Mag., 1889, **8**, p.257. New analysis, 8.88 %Ni, 2.42 ppm.Ga, 0.133 ppm.Ge, 0.91 ppm.Ir, R. Schaudy et al., Icarus, 1972, **17**, p.174. Description, V.F. Buchwald, Iron Meteorites, Univ. of California, 1975, p.798.
399g Vienna, Naturhist. Mus.; 356g Washington, U.S. Nat. Mus.; 58g Chicago, Field Mus. Nat. Hist.;
Specimen(s): [53323], 8720g. main mass, and fragments, 14g and filings and turnings, 92g.

Manyc v Manych.

Manych 45°49′ N., 44°38′ E.
Stavropol region, Federated SSR, USSR, [Маныч]
Fell 1951, October 20, 1530 hrs, Moscow time
Synonym(s): *Manyc, Manytsch*
Stone. Olivine-hypersthene chondrite, amphoterite (LL3).
One stone of 1.86kg was found in 1951, the other half, 1695g, in 1952, V.G. Gnilovskoi, Природа, 1952, **41**, p.100 [M.A. 12-611], E.L. Krinov, Метеоритика, 1955, **12**, p.29. Analysis, 20.43 % total iron, M.I. D'yakonova, Метеоритика, 1964, **25**, p.129. Mean composition of unequilibated olivines, $Fa_{24.9}$, R.T. Dodd et al., Geochimica et Cosmochimica Acta, 1967, **31**, p.921. Classification as LL3, R.T. Dodd and E. Jarosewich, Meteoritics, 1979, **14**, p.380, abs.
1870g Moscow, Acad. Sci., main mass of 1.86kg stone; Smaller fragment, Stavropol, Mus.;

Manytsch v Manych.

Mapleton 42°11′ N., 95°43′ W.
Monona County, Iowa, U.S.A.
Found 1939, June 17
Iron. Octahedrite, medium (1.0mm) (IIIA).
A mass of 49kg was found, H.W. Nichols, letter of 11 October 1939, H.W. Nichols, Rocks and Minerals, 1939, **14**, p.320.. Description, with an analysis, B.H. Wilson, Pop. Astron., Northfield, Minnesota, 1944, **52**, p.392 [M.A. 9-300]. Another analysis. 7.5 %Ni, S.K. Roy and R.K. Wyant, Field Mus. Nat. Hist. Geol. Ser., 1949, **7**, p.113 [M.A. 11-139]. Further analysis, 7.73 %Ni, 20.3 ppm.Ga, 40.6 ppm.Ge, 1.4 ppm.Ir, E.R.D. Scott et al., Geochimica et Cosmochimica Acta, 1973, **37**, p.1957. Slightly corroded, shock-hardened, V.F. Buchwald, Iron Meteorites, Univ. of California, 1975, p.799.

31kg Chicago, Field Mus. Nat. Hist., main mass, approx. weight; 3.7kg Washington, U.S. Nat. Mus.;
Specimen(s): [1980,M.6], 3.58kg.; [1959,954], 15.5g. and sawings, 1.5g.

Marathon v Peña Blanca Spring.

Marburg 50°49′ N., 8°46′ E.
Marburg-an-der-Lahn, Hessen, Germany
Found 1906
Stony-iron. Pallasite (PAL).
A 3kg mass, found beside the Lahn river at Marburg, was mostly lost during an air-raid. One piece of 110g survived. Analysis of metal, 11.0 %Ni, olivine $Fa_{12.5}$, P.R. Buseck and J.I. Goldstein, Bull. Geol. Soc. Amer., 1969, **80**, p.2141. Textural study, E.R.D. Scott, Geochimica et Cosmochimica Acta, 1977, **41**, p.693.
47g Mainz, Max-Planck-Inst.; Specimen, Heidelberg, Max-Planck Inst.; 7.5g Tempe, Arizona State Univ.;

Mardan 34°14′ N., 72°5′ E.
Takhti-Bhai, North-west Frontier, Pakistan
Fell 1948, May 8, 1200 hrs
Stone. Olivine-bronzite chondrite (H5).
One stone of over 4.5kg fell near Abazo Dheri police station, H. Crookshank, letter of 10 July, 1954, in Min. Dept. BM (NH). Olivine Fa_{19}, B. Mason, Geochimica et Cosmochimica Acta, 1963, **27**, p.1011.
2380g Quetta, Geol. Surv. Pakistan;
Specimen(s): [1954,222], 2170g.

Mar del Plata 38°0′ S., 57°32′ W.
Buenos Aires province, Argentina
Pseudometeorite..
A mass is said to have fallen in November 1947, K. Keil, Fortschr. Min., 1960, **38**, p.241. It is a sedimentary rock.

Marengo v Homestead.

Maria Elena (1928) v North Chile.

Maria Elena (1935) 22°20′ S., 69°40′ W.
Antofagasta, Chile
Found 1935, known before this year
Synonym(s): *Chile*
Iron. Octahedrite, fine (0.30mm) (IVA).
One mass of 15.5kg was found. Description, with an analysis, 7.63 %Ni, V.B. Meen, Am. J. Sci., 1941, **239**, p.412 [M.A. 8-196]. Determination of Ga, Au and Pd, and a partial analysis, 7.71 %Ni, E. Goldberg et al., Geochimica et Cosmochimica Acta, 1951, **2**, p.1. Classification and analysis, 7.64 %Ni, 1.72 ppm.Ga, 0.096 ppm.Ge, 3.1 ppm.Ir, R. Schaudy et al., Icarus, 1972, **17**, p.174. Shocked and annealed several times, V.F. Buchwald, Iron Meteorites, Univ. of California, 1975, p.800.
11kg Washington, U.S. Nat. Mus., includes main mass; 326g New York, Amer. Mus. Nat. Hist.; 294g Tempe, Arizona State Univ., Specimen no. 388.1, labelled Chile; 327g Bloomfield Hills, Cranbrook Inst. Sci.; End piece, Antofagasta, in possession of Mr. Coope;
Specimen(s): [1959,955], 40.5g.

Maria Linden
South Africa
Fell 1925, April 15
Stone. Olivine-hypersthene chondrite (L4).
A 114g stone is in the collection of meteorites in the Naturhist. Mus., Vienna. Little is known of its past history, A.L. Graham, letter of 24 May, 1974, in Min. Dept. BM (NH).
114g Vienna, Naturhist. Mus.;

Mariaville 42°43′ N., 99°23′ W.
Rock County, Nebraska, U.S.A.
Fell 1898, October 16, 0000? hrs
Synonym(s): *Rock County*
Iron. Octahedrite?.
A mass of 340g is said to have fallen after the appearance of light and detonations, E.H. Barbour, Rep. Geol. Surv. Nebraska, Lincoln, 1903, **1**, p.184, H.H. Nininger, Pop. Astron., Northfield, Minnesota, 1937, **45**, p.562 considers that this was a find, not a witnessed fall. Repository of mass not known.

Maricopa 33°15′ N., 112°3′ W.
Maricopa County, Arizona, U.S.A.
Found 1980, May 18
Stone. Olivine-bronzite chondrite (H).
A single mass of 50g was found, olivine Fa_{18}, Meteor. Bull., 1981 (59), Meteoritics, 1981, **16**, p.195.
18g Tempe, Arizona State Univ.;

Maridi 4°40′ N., 29°15′ E.
Equatoria, Sudan
Fell 1941, 1500 hrs, before this year
Synonym(s): *Esu, Meridi*
Stone. Olivine-bronzite chondrite (H6).
A stone of 3.2kg fell at the above co-ordinates, near the source of the Sueh River, Li Rangu, 30 miles SW. of Meridi, G. Andrew, letters of 5 October, 1942 and 30 December, 1945 in Min Dept. Brit. Mus. Olivine Fa_{19}, B. Mason, Geochimica et Cosmochimica Acta, 1963, **27**, p.1011.
Main mass, Khartoum, Mus. Geol. Surv. Sudan; Thin section, Washington, U.S. Nat. Mus., under 'Esu';
Specimen(s): [1946,142], 145g. and a fragment, 2g.

Marilia 22°15′ S., 49°56′ W.
São Paulo, Brazil
Fell 1971, October 5, 1700 hrs, approx. time
Stone. Olivine-bronzite chondrite (H4).
A shower of meteorites was seen to fall by many witnesses. At least seven stones were recovered, the total weight was estimated at about 2.5kg. Description, P.E. Avanzo et al., Meteoritics, 1973, **8**, p.141. Reported, Meteor. Bull., 1974 (52), Meteoritics, 1974, **9**, p.104. Analysis, 26.16 % total iron, M. Shima et al., Meteoritics, 1974, **9**, p.199.
780g Salvador, Brazil, Inst. Geocienc., largest individual.; 200g Modena, Inst. Min.; 57g Albuquerque, Univ. of New Mexico;
Specimen(s): [1982,M.5], 38g.

Marimba v Kota-Kota.

Marion (Iowa) 41°54′ N., 91°36′ W.
Linn County, Iowa, U.S.A.
Fell 1847, February 25, 1445 hrs
Synonym(s): *Hartford, Iowa, Linn County*
Stone. Olivine-hypersthene chondrite (L6), veined.
After detonations, a stone of about 2.5lb was seen to fall 9 miles due S. of Marion, and two specimens were found later,

one of about 40lb, and the other of about 20lb, C.U. Shepard, Am. J. Sci., 1847, **4**, p.288, 429, C.U. Shepard, Am. J. Sci., 1851, **11**, p.38. Analysis, C.F. Rammelsberg, Monatsber. Akad. Wiss. Berlin, 1870, p.457. Listed, G.S. Dillé, Proc. Iowa Acad. Sci., 1928, **35**, p.225 [M.A. 5-157]. Olivine Fa_{24}, B. Mason, Geochimica et Cosmochimica Acta, 1963, **27**, p.1011. Ar-Ar age, 4400 m.y., G. Turner, Meteorite Research, ed. P.M. Millman, D. Reidel, Dordrecht-Holland, 1969, p.407.

11kg Tempe, Arizona State Univ.; 1.7kg Washington, U.S. Nat. Mus.; 1kg Yale Univ.; 495g Calcutta, Mus. Geol. Surv. India; 445g Edinburgh, Roy. Scottish Mus.; 423g Tübingen, Univ.; 379g Harvard Univ.; 263g Vienna, Naturhist. Mus.; 228g New York, Amer. Mus. Nat. Hist.; 222g Cedar Rapids, Iowa, Coe College Mus.;

Specimen(s): [23384], 749g. and fragments, 36g; [25464], 137g.; [21545], 2.5g.; [96269], thin section

Marion (Kansas) 38°22′ N., 97°2′ W. approx.

Marion County, Kansas, U.S.A.
Found 1955
Stone. Olivine-hypersthene chondrite (L5).

2.89kg were found, B. Mason, Meteorites, Wiley, 1962, p.233. Olivine Fa_{24}, B. Mason, Geochimica et Cosmochimica Acta, 1963, **27**, p.1011.

2.45kg Tempe, Arizona State Univ., main mass; 133g Chicago, Field Mus. Nat. Hist.; 32g New York, Amer. Mus. Nat. Hist.;

Specimen(s): [1959,1026], 58g.

Marialahti v Marjalahti.

Mar'inka 47°54′ N., 37°30′ E. approx.

Near Donetsk, Ukraine, USSR
Found 1976
Iron.

A single mass of 144g was found in coal-bearing Carboniferous strata dated at 285-340 m.y., Meteor. Bull., 1982 (60), Meteoritics, 1982, **17**, p.95.

Marjahlts v Marjalahti.

Marjalachti v Marjalahti.

Marjalahti 61°30′ N., 30°30′ E. approx.

Viipuri, Karelia ASSR, USSR
Fell 1902, June, 1, 2200 hrs
Synonym(s): *Marialahti, Marjahlts, Marjalachti, Marjalhai*
Stony-iron. Pallasite (PAL).

After the appearance of a luminous meteor followed by detonations, a stone of about 45kg was seen to fall and was broken into pieces. Description, with an analysis, L.H. Borgström, Bull. Comm. Geol. Finlande, 1903 (14), p.45. Analysis of the metal, B. Mason, Am. Mus. Novit., 1963 (2163) [M.A. 16-640]. Another analysis of the metal, 7.86 % Ni, 23.0 ppm.Ga, 54.7 ppm.Ge, 1.6 ppm.Ir, J.T. Wasson, Meteorites, Springer-Verlag, 1974, p.306. Olivine Fa_{12}, P.R. Buseck and J.I. Goldstein, Bull. Geol. Soc. Amer., 1969, **80**, p.2141. Co-ordinates and locality, A.A. Yavnel, letter of 22 October, 1975 in Min. Dept. BM(NH). Texture, references, E.R.D. Scott, Geochimica et Cosmochimica Acta, 1977, **41**, p.693.

Main mass, Helsinki, Univ.; 1.87kg Vienna, Naturhist. Mus.; 1.25kg Chicago, Field Mus. Nat. Hist.; 467g Moscow, Acad. Sci.; 452g Washington, U.S. Nat. Mus.; 272g Tempe, Arizona State Univ.; 172g Harvard Univ.; 165g Budapest, Nat. Mus.; 60g New York, Amer. Mus. Nat. Hist.;

Specimen(s): [86584], 2990g.; [1920,318], 103g. slice

Marjalhai v Marjalahti.

Markovka 52°24′ N., 79°48′ E. approx.

Kluchevskoj district, Altay region, USSR, [Марковка]
Found 1967, spring
Stone. Olivine-bronzite chondrite (H4).

An 8.8kg individual was ploughed up from 20-30 cm depth, Meteor. Bull., 1969 (48), Meteoritics, 1970, **5**, p.108. Possibly fell in the autumn of 1966. Co-ordinates, A.A. Yavnel, letter of 7 August, 1975 in Min. Dept. BM(NH). Classification, L.G. Kvasha and A.Ya. Skripnik, Метеоритика, 1978, **37**, p.195. Analysis, 25.48 % total iron, O.A. Kirova et al., Метеоритика, 1975, **34**, p.57.

1750g Moscow, Acad. Sci.;

Marlow 34°36′ N., 97°55′ W.

Stephens County, Oklahoma, U.S.A.
Found 1936, February
Stone. Olivine-hypersthene chondrite (L).

A weathered fragment of about 0.75lb was found on a farm 5.75 miles WSW. of Marlow, F.C. Leonard, Meteoritics, 1954, **1**, p.185 [M.A. 13-53], D. Hoffleit, Sky and Telescope, 1955, **14**, p.191, Meteor. Bull., 1958 (8). Olivine Fa_{24}, B. Mason, Geochimica et Cosmochimica Acta, 1963, **27**, p.1011.

Main mass, Los Angeles, Univ. of California;

Marmande 44°30′ N., 0°9′ E.

Lot-et-Garonne, France
Fell 1848, July 4
Synonym(s): *Montignac*
Stone. Olivine-hypersthene chondrite (L5).

A fragment, labelled as from a stone of 3kg that fell on the above date, was found amongst the effects of Col. Gabalda (label with specimen in BM(NH) collection). Listed, R.P. Greg, Phil. Mag., 1862, **24**, p.540. Regarded as doubtful, A. Lacroix, Bull. Mus. d'Hist. Nat. Paris, 1927, **33**, p.444 [M.A. 3-534]. Olivine Fa_{24}, B. Mason, Geochimica et Cosmochimica Acta, 1963, **27**, p.1011.

25g Vienna, Naturhist. Mus.; 25g Rome, Vatican Colln.; 2g Chicago, Field Mus. Nat. Hist.;

Specimen(s): [35158], 4.75g.

Marmaros v Borkut.

Marmoros v Borkut.

Marokhaza v Mocs.

Maros v Mezö-Madaras.

Marradal v Morradal.

Marsala 37°47′ N., 12°25′ E.

Trapani, Sicily, Italy
Fell 1834, December 15, or 10
Doubtful..

A stone of 15lb, described as " yellowish, spheroidal, very hard and solid", is said to have fallen, G. von Boguslawski, Ann. Phys. Chem. (Poggendorff), 1854, **4** (suppl.), p.34, but the evidence is not conclusive.

Marshall County v Plymouth.

Marshall County 37° N., 88°15′ W.
Marshall County, Kentucky, U.S.A.
Found 1860, described in this year
Iron. Octahedrite, medium (1.2mm) (IIIA).
A piece from a mass said to weigh 15lb was analysed, 8.72 %Ni, J.L. Smith, Am. J. Sci., 1860, **30**, p.240. Another analysis, 7.92 %Ni, 21.1 ppm.Ga, 44.3 ppm.Ge, 2.6 ppm.Ir, E.R.D. Scott et al., Geochimica et Cosmochimica Acta, 1973, **37**, p.1957. The mass has been heated or forged, V.F. Buchwald, Iron Meteorites, Univ. of California, 1975, p.803.
1.9kg Tempe, Arizona State Univ.; 254g Harvard Univ.; 172g Paris, Mus. d'Hist. Nat.; 135g Copenhagen, Univ. Geol. Mus.; 73g Vienna, Naturhist. Mus.; 72g Berlin, Humboldt Univ.; 67g Washington, U.S. Nat. Mus.; 45g Tübingen, Univ.;
Specimen(s): [34581], 80.5g.; [1959,956], 196.5g. a slice, and sawings, 4g.

Marshfield v Seymour.

Marsland 42°27′ N., 103°18′ W.
Dawes County, Nebraska, U.S.A.
Found 1933
Stone. Olivine-bronzite chondrite (H5).
A mass of 2250g was found, H.H. Nininger, The Mines Mag., Golden, Colorado, 1933, **23** (8), p.9 [M.A. 5-405]. However a stone of 4.6kg is also recorded, A.D. Nininger, Pop. Astron., Northfield, Minnesota, 1937, **45**, p.449 [M.A. 7-62]. This is confirmed by later reports, H.H. and A.D. Nininger, The Nininger Collection of Meteorites, Winslow, Arizona, 1950, p.71. Olivine Fa_{16}, B. Mason, Geochimica et Cosmochimica Acta, 1963, **27**, p.1011.
742g Tempe, Arizona State Univ.; 258g Harvard Univ.; 243g Washington, U.S. Nat. Mus.; 240g Paris, Mus. d'Hist. Nat.; 129g New York, Amer. Mus. Nat. Hist.; 115g Los Angeles, Univ. of California;
Specimen(s): [1959,831], 523g.

Mart 31°30′ N., 96°53′ W.
McLennan County, Texas, U.S.A.
Found 1898
Iron. Octahedrite, fine (0.38mm) (IVA).
An oval mass of 15.75lb was found on a farm. Description, with an analysis, 9.20 %Ni, G.P. Merrill, Proc. Washington Acad. Sci., 1900, **2**, p.51. Classification and analysis, 9.20 % Ni, 2.16 ppm.Ga, 0.136 ppm.Ge, 0.64 ppm.Ir, R. Schaudy et al., Icarus, 1972, **17**, p.174. Description; shock heated and annealed, V.F. Buchwald, Iron Meteorites, Univ. of California, 1975, p.805.
Main mass, Waco, Texas, Baylor Univ.; 1kg Chicago, Field Mus. Nat. Hist.; 454g Washington, U.S. Nat. Mus.; 450g Tempe, Arizona State Univ.; 202g New York, Amer. Mus. Nat. Hist.;
Specimen(s): [84881], 415g. slice

Martinho Campos v Ibitira.

Maryland v Emmitsburg.

Maryland v Lonaconing.

Maryland v Nanjemoy.

Maryville 35°48′ N., 84°6′ W.
Blount County, Tennessee, U.S.A.
Fell 1983, January 28, 0415 hrs
Stone. Olivine-hypersthene chondrite (L6).
After a fireball and sonic booms, a single crusted mass weighing 1442g was found in a small impact pit, olivine Fa_{23}, SEAN Bulletin, 1983, **8** (3).
923g Washington, U.S. Nat. Mus.;

Masanderan 36°30′ N., 51° E. approx.
Tabaristan, Iran
Fell 852, July, or August; A.D.
Doubtful. Stone.
A large white stone is said to have fallen, S. de Sacy, Ann. Phys. (Gilbert), 1815, **50**, p.293, E.F.F. Chladni, Die Feuer-Meteore, Wien, 1819, p.191, E.J. Holmyard and D.C. Mandeville, Avicennae de congelatione et conglutatione lapidum, Paris, 1927 [M.A. 3-466].

Mascombes 45°22′ N., 1°52′ E.
Correze, France
Fell 1836, January 31, 1300 hrs
Stone. Olivine-hypersthene chondrite (L6).
A stone of about 1kg fell, after detonations, G.A. Daubrée, C. R. Acad. Sci. Paris, 1864, **58**, p.229, O. Büchner, Ann. Phys. Chem. (Poggendorff), 1865, **124**, p.579, A. Lacroix, Bull. Mus. d'Hist. Nat. Paris, 1927, **33**, p.571 [M.A. 3-534]. Olivine Fa_{24}, B. Mason, Geochimica et Cosmochimica Acta, 1963, **27**, p.1011.
420g Paris, Mus. d'Hist. Nat.; 21g Yale Univ.; 15g Chicago, Field Mus. Nat. Hist.;
Specimen(s): [56463], 4.25g.; [41112], 0.75g.

Massa-lubrense 40°36′ N., 14°20′ E.
Napoli, Campania, Italy
Fell 1819, April
Doubtful..
One stone is said to have fallen, E.F.F. Chladni, Ann. Phys. (Gilbert), 1822, **71**, p.359, but the evidence is not conclusive.

Massenya 11°21′ N., 16°9′ E.
Chad
Found 1958, April
Synonym(s): *Brazzaville*
Stone. Olivine-bronzite chondrite (H).
Incomplete description and analysis of Fe-Ni metal and chromite, olivine Fa_{20}, R. Cayé et al., Meteorite Research, ed. P.M. Millman, D. Reidel, Dordrecht-Holland, 1969, p.657. Stone first studied in Brazzaville, P. Vincent, letter of 23 July, 1983 in Min. Dept. BM(NH).
Specimen, Paris, Mus. d'Hist. Nat.;
Specimen(s): [1976,M.8], 4.2g.

Mässing 48°8′ N., 12°37′ E.
Eggenfelden, Bayern, Germany
Fell 1803, December 13, 1030 hrs
Synonym(s): *Eggenfelden, Maessing, St. Nicholas, Sankt Nicolas*
Stone. Achondrite, Ca-rich. Howardite (AHOW).
After detonations a stone of 1.6kg was seen to fall, M. Imhof, Ann. Phys. (Gilbert), 1804, **18**, p.330. Description, analysis, C.W. Gümbel, Sitzungsber. Akad. Wiss. München, Math.-phys. Kl., 1878, **8**, p.32. Petrology, M.B. Duke and L.T. Silver, Geochimica et Cosmochimica Acta, 1967, **31**, p.1637. Further analysis, B. Mason et al., Smithson. Contrib. Earth Sci., 1979 (22), p.30.
22g Berlin, Humboldt Univ.; 22g Paris, Mus. d'Hist. Nat.; 10.2g Tübingen, Univ.; 2.5g Vienna, Naturhist. Mus.; 1.7g

Chicago, Field Mus. Nat. Hist.;
Specimen(s): [35161a], Less than 1g.

Masua

Sardinia, Italy
Found 1967
Iron. Octahedrite, medium.
One weathered mass of at least 1460g was found in Tertiary limestone near the village of Masua, G.I Huss, letter of 7 January, 1972, in Min. Dept. BM(NH).
1.3kg Mainz, Max-Planck-Inst.;
Specimen(s): [1972,6], 53.2g. polished slice

Matad 47°36′ N., 114°45′ E. approx.

Choybalsan aymag, Mongolia, [Матад]
Found 1958, before this year
Stone.
A mass of 189.5g was found, G.G. Vorobyev and O. Namnandorzh, Метеоритика, 1958, **16**, p.134 [M.A. 14-130].
180g Ulan Bator, State Mus.;

Matatiela v Kokstad.

Matatiele v Kokstad.

Mategaon v Naoki.

Mato Grosso v Paranaiba.

Mau v Mhow.

Mauerkirchen 48°11′ N., 13°8′ E.

Ober-Österreich, Austria
Fell 1768, November 20, 1600 hrs
Stone. Olivine-hypersthene chondrite (L6).
After detonations, a stone of about 19kg fell, E.F.F. Chladni, Ann. Phys. (Gilbert), 1803, **15**, p.316. Description, with an analysis, C.W. Gümbel, Sitzungsber. Akad. Wiss. München, Math.-phys. Kl., 1878, **8**, p.16. Olivine Fa_{24}, B. Mason, Geochimica et Cosmochimica Acta, 1963, **27**, p.1011.
2kg Göttingen, Univ.; 590g Vienna, Naturhist. Mus.; 445g Budapest, Nat. Mus.; 219g Berlin, Humboldt Univ.; 146g Chicago, Field Mus. Nat. Hist.; 82g New York, Amer. Mus. Nat. Hist.; 73g Washington, U.S. Nat. Mus.;
Specimen(s): [19967], 288g. and fragments, 1.5g.; [1920,319], 3g.; [1935,51], minute fragment.

Mauléon v Sauguis.

Mauritius 20° S., 57° E. approx.

Indian Ocean,
Fell 1801, December, some time before this date
Synonym(s): *Böttcher Island, Isle de France, Tonnelier*
Stone. Olivine-hypersthene chondrite (L6), brecciated.
After detonations and the appearance of light, three stones, one as large as a melon and two as an orange, fell on the Isle aux Tonneliers, E.F.F. Chladni, Die Feuer-Meteore, Wien, 1819, p.268, J.B.G.M. Bory de St. Vincent, Voyage.. Isles des Mers d'Afrique, 1804, **3**, p.253. Olivine Fa_{26}, B. Mason, Geochimica et Cosmochimica Acta, 1963, **27**, p.1011.
204g Edinburgh, Roy. Scottish Mus.; 6g Chicago, Field Mus. Nat. Hist.; 4g Vienna, Naturhist. Mus.;
Specimen(s): [1965,299], 5.5g.; [1983,M.49], 2g.

Maverick County v Coahuila.

Mayence v Mainz.

Mayday 39°28′28″ N., 96°55′30″ W.

Riley County, Kansas, U.S.A.
Found 1955, July
Stone. Olivine-bronzite chondrite (H), black.
Two stones, together weighing 6.9kg, were found 2 miles SW. of Mayday, and are believed to be part of a much larger stone, W.S. Houston, The Great Plains Observer, 1956, **1** (5), Meteor. Bull., 1957 (5). Olivine Fa_{19}, B. Mason, Geochimica et Cosmochimica Acta, 1963, **27**, p.1011.
7g Fort Worth, Texas, Monnig Colln.; 5g Washington, U.S. Nat. Mus.;

Mayerthorpe 53°46′30″ N., 115°2′ W.

Alberta, Canada
Found 1964
Iron. Octahedrite, coarse (2mm) (IA).
A mass of 8.74kg was found, Meteor. Bull., 1964 (32). A second mass, 3.87kg, was found in 1964, Meteor. Bull., 1965 (33). Classification and analysis, 7.19 %Ni, 75.5 ppm.Ga, 283 ppm.Ge, 2.4 ppm.Ir, J.T. Wasson, Icarus, 1970, **12**, p.407.
8.7kg Alberta, Univ., the first mass found; 3.6kg Ottawa, Mus. Geol. Surv. Canada, the smaller mass;

Mayfield 37°18.5′ N., 97°32.7′ W.

Sumner County, Kansas, U.S.A.
Found 1972
Stone. Olivine-bronzite chondrite (H).
A stone of 38.4kg was discovered by a farmer. He had not noticed the stone before, and assumed that either his father or grandfather had ploughed it up, since it displayed plough-marks; it was thrown under an old fence, Meteor. Bull., 1975 (53), Meteoritics, 1975, **10**, p.151, G.I. Huss, letter of 8 February, 1974 in Min. Dept. BM(NH).
26.9kg Mainz, Max-Planck-Inst.; 260g Los Angeles, Univ. of California; 600g Copenhagen, Univ. Geol. Mus.;
Specimen(s): [1974,M.3], 69.8g. crusted part-slice

Mayo Belwa 8°58′ N., 12°5′ E.

Adamawa Local Authority, Nigeria
Fell 1974, August 3, evening hrs
Stone. Achondrite, Ca-poor. Aubrite (AUB).
After a fireball accompanied by sonic phenomena, a stone of 4.85kg fell. The co-ordinates are those of Adamawa, Smithsonian Institution Centre for Short-lived Phenomena, card no. 1960, event 136-74, 1974. Description and analysis, A.L. Graham et al., Min. Mag., 1977, **41**, p.487 Composition of metal and schreibersite, A.L. Graham, Meteoritics, 1978, **13**, p.235. Contains amphibole, A.W.R. Bevan et al., Min. Mag., 1977, **41**, p.531.
Main mass, Kaduna, Geol. Surv. Nigeria; Specimen, Washington, U.S. Nat. Mus.; Specimen, Paris, Mus. d'Hist. Nat.; Specimen, Moscow, Acad. Sci.; Specimen, Perth, West. Austr. Mus.;
Specimen(s): [1976,M.11], 540g.

Mayodan 36°23′ N., 79°52′ W.

Rockingham County, North Carolina, U.S.A.
Found 1920
Iron. Hexahedrite (IIA).
A mass of 34lb was received in Washington (U.S. Nat. Mus.) in Sept. 1950. Description and analysis, 5.48 %Ni, E.P. Henderson and S.H. Perry, Am. Miner., 1953, **38**, p.1025 [M.A. 12-358]. May belong to Indian Valley; both irons have been shocked and partly recrystallised, V.F. Buchwald, Iron Meteorites, Univ. of California, 1975, p.806. Further

analysis, 5.54 %Ni, 59.3 ppm.Ga, 180 ppm.Ge, 14 ppm.Ir, J.T. Wasson, Geochimica et Cosmochimica Acta, 1969, **33**, p.859. Structure and shock history and 3He loss, V.F. Buchwald, Chem. Erde, 1971, **30**, p.33.

13.2kg Washington, U.S. Nat. Mus.; 559g Tempe, Arizona State Univ.; 210g Moscow, Acad. Sci.;

Mazapil 24°41′ N., 101°41′ W.

Zacatecas, Mexico
Fell 1885, November 27, 2100 hrs
Synonym(s): *Chichimeguilas, Chichimequillas*
Iron. Octahedrite, medium (1.1mm) (IA).

A mass of about 4kg fell during a star-shower. Description, with an analysis, 7.84 %Ni, W.E. Hidden, Am. J. Sci., 1887, **33**, p.223. Chichimeguilas is a fragment of Mazapil, V.F. Buchwald, Iron Meteorites, Univ. of California, 1975, p.808. Another analysis, 8.64 %Ni, 60.2 ppm.Ga, 221 ppm.Ge, 5.5 ppm.Ir, J.T. Wasson, Icarus, 1970, **12**, p.407.

3.55kg Vienna, Naturhist. Mus.; 83g New York, Amer. Mus. Nat. Hist.; 19g Chicago, Field Mus. Nat. Hist.;

Specimen(s): [67451], 14g.

Maziba 1°13′ S., 30° E.

Kabale, Kigezi district, Uganda
Fell 1942, September 24
Synonym(s): *Uganda*
Stone. Olivine-hypersthene chondrite (L6).

A mass of 4975g fell. Description, with an analysis, R.O. Roberts, Uganda J., 1947, **11**, p.42 [M.A. 10-400]. Mentioned, J.S. Albanese, Rocks and Minerals, Peekskill, N.Y., 1950, **25**, p.261. Olivine Fa_{25}, B. Mason, Geochimica et Cosmochimica Acta, 1963, **27**, p.1011.

390g Washington, U.S. Nat. Mus.; 194g Fort Worth, Texas, Monnig Colln.; 37g New York, Amer. Mus. Nat. Hist.;

Mbosi 9°7′ S., 33°4′ E.

Rungwe district, Tanzania
Found 1930
Synonym(s): *M'Bozi, Tanganyika*
Iron. Octahedrite, medium (0.8mm) (IRANOM).

A mass measuring 13.5 × 4 × 4 feet was found. Description, with an analysis, 8.65 %Ni, G.H. Stanley, S. African J. Sci., 1931, **28**, p.88 [M.A. 5-154]. Analysis, 8.69 %Ni, D.R. Grantham and F. Oates, Min. Mag., 1931, **22**, p.487. The weight has been variously estimated, most recently as 16 tons, V.F. Buchwald, Iron Meteorites, Univ. of California, 1975, p.814. Mentioned, J.R. Harpum, Rec. Geol. Surv. Tanganyika for 1961, 1965, **11**, p.54. The main mass remains at the find site and is a protected monument. Classification, a chemically anomalous medium octahedrite, analysis, 8.71 %Ni, 2.54 ppm.Ga, 26.9 ppm.Ge, 6.6 ppm.Ir, E.R.D. Scott et al., Geochimica et Cosmochimica Acta, 1973, **37**, p.1957. Structure, H.J. Axon, Prog. Materials Sci., 1968, **13**, p.185.

779g Washington, U.S. Nat. Mus.; 485g Tempe, Arizona State Univ.; 326g Chicago, Field Mus. Nat. Hist.;

Specimen(s): [1932,1093], 375.7g. and filings, 18.7g.; [1934, 782], 85.7g.; [1931,427], 5.7g.; [1958,1373], 2g.; [1969,146], 27g. rusted fragments

M'Bozi v Mbosi.

McAddo 33°45′ N., 100°56′ W.

Dickens County, Texas, U.S.A.
Found 1935
Synonym(s): *McAdoo*
Stone. Olivine-hypersthene chondrite (L).

One stone of 1.1kg was found, A.D. Nininger, Pop. Astron., Northfield, Minnesota, 1940, **48**, p.556, Contr. Soc. Res. Meteorites, **2**, p.227 [M.A. 8-54]. Olivine Fa_{25}, B. Mason, Geochimica et Cosmochimica Acta, 1963, **27**, p.1011.

957g New York, Amer. Mus. Nat. Hist., main mass; 16g Tempe, Arizona State Univ.;

Specimen(s): [1959,888], 19g.

McAdoo v McAddo.

McAlester 34°56′ N., 95°15′ W.

Pittsburg County, Oklahoma, U.S.A.
Found 1927, about
Doubtful..

Mentioned, without details, H.H. Nininger, Our Stone Pelted Planet, 1933. Not mentioned, B. Mason, Meteorites, Wiley, 1962.

McCamey v Odessa.

McDowell County v Wood's Mountain.

McKinney 33°11′ N., 96°43′ W.

Collin County, Texas, U.S.A.
Found 1870
Synonym(s): *Collin County, Mackinney, Rockport*
Stone. Olivine-hypersthene chondrite (L4), black.

Two stones, the larger weighing 100kg, were found 8 miles west of McKinney, A. Brezina, Ann. Naturhist. Hofmus. Wien, 1895, **10**, p.252. Analysis, 21.88 % total iron, H.B. Wiik, Comm. Phys.-Math. Soc. Sci. Fenn., 1950, **14**, p.14 [M.A. 11-140]. Further analysis, 22.84 % total iron, H. von Michaelis et al., Earth planet. Sci. Lett., 1969, **5**, p.387. Olivine Fa_{24}, B. Mason, Geochimica et Cosmochimica Acta, 1963, **27**, p.1011. Highly shocked, R.T. Dodd and E. Jarosewich, Earth planet. Sci. Lett., 1979, **44**, p.335.

56kg Chicago, Field Mus. Nat. Hist.; 46.5kg Vienna, Naturhist. Mus.; 12.4kg Paris, Mus. d'Hist. Nat.; 6.7kg Rome, Vatican Colln; 5.4kg Tempe, Arizona State Univ.; 5.2kg Washington, U.S. Nat. Mus.; 1kg Schönenwerd, Bally-Prior Mus.; 1023g Copenhagen, Univ. Geol. Mus.; 446g Harvard Univ.; 512g Tübingen, Univ.; 405g Prague, Nat. Mus.; 400g Sarajevo, Bosnian Nat. Mus.; 254g Berlin, Humboldt Univ.; 239g Yale Univ.;

Specimen(s): [1921,439], 999g. end-slice; [76265], 250g. and fragments, 1g.; [1959,771], 1637g. slice

McLean 35°14′ N., 100°36′ W.

Gray County, Texas, U.S.A.
Found 1939
Stone. Olivine-bronzite chondrite (H).

One stone of 4.3kg was found, A.D. Nininger, Pop. Astron., Northfield, Minnesota, 1940, **48**, p.556, Contr. Soc. Res. Meteorites, **2**, p.227 [M.A. 8-54]. Olivine Fa_{19}, B. Mason, Geochimica et Cosmochimica Acta, 1963, **27**, p.1011.

709g Tempe, Arizona State Univ.;

Specimen(s): [1959,889], 899.5g.

McMurchie v Centerville.

Meadow (a) 33°19′53″ N., 102°16′8″ W.
Terry County, Texas, U.S.A.
Found 1975
Stone. Olivine-bronzite chondrite (H5).
A single stone of 1495g was found during ploughing. A second stone of 691g was found in the same area in 1981 but is distinct, Meadow (b), an L6 stone , G.I Huss, letter of 26 October, 1983, in Min. Dept. BM(NH). Olivine $Fa_{18.5}$, A.L. Graham, Priv. comm., 1984.
Main masses of both stones, Denver, Amer. Meteorite Lab.;

Mecca v Kaaba.

Mecherburg v Mühlau.

Mecklenburg County v Monroe.

Medanitos 27°15′ S., 67°30′ W.
Timogasta dept., Catamarca, Argentina
Fell 1953, July 14, 1400 hrs
Synonym(s): *Medanos*
Stone. Achondrite, Ca-rich. Eucrite (AEUC).
Two small stones, 25g and 6g, were recovered after a brilliant bolide, Cat. Vatican Colln. Meteorites, L.O. Giacomelli, letters of 30 March, 1960 and 30 October, 1962, in Min. Dept. BM(NH)., Meteor. Bull., 1958 (10). Classed as a howardite, in error, R.F. Symes and R. Hutchison, Min. Mag., 1970, **37**, p.721. Mineralogically similar to Moore County and Serra de Magé. Rare gas data, D. Heymann et al., Geochimica et Cosmochimica Acta, 1968, **32**, p.1241.
25g Rome, Vatican Colln;
Specimen(s): [1968,3], 4.4g. and powder

Medanos v Medanitos.

Medvedeva v Krasnojarsk.

Medwedewa v Krasnojarsk.

Meekatharra v Dolgunna.

Meers 34°48′ N., 98°35′ W.
Comanche County, Oklahoma, U.S.A.
Found 1913, known in this year
Doubtful. Iron.
Specimens offered for sale from Meers to the U.S. Nat. Mus. in 1913, E.P. Henderson, letter of 3 June 1939, in Min. Dept. BM(NH). Not listed, B. Mason, Meteorites, Wiley, 1962.

Meerut 29°1′ N., 77°48′ E.
Meerut district, Uttar Pradesh, India
Fell 1861, approx.
Stone. Chondrite.
Listed, J.C. Brown, Mem. Geol. Surv. India, 1916, **43**, p.235 [M.A. 1-94].
22g Calcutta, Mus. Geol. Surv. India;

Meester-Cornelis 6°14′ S., 106°53′ E.
Batavia, Java, Indonesia
Fell 1915, June 2, 0600-0615 hrs
Stone. Olivine-bronzite chondrite (H5).
A stone of 24.75kg fell, after detonations, in the village of Duren Sawit. Description, with an analysis, L.J.C. van Es, Jaarb. Mijnwezen Nederland. Oost-Indie, 1918, **47**, p.21, Neues Jahrb. Min., 1922 (2), p.149. Fell 2 August, 1915, S. Darsoprajitno, priv. comm., 1983.
35kg Bandung, Geol. Surv. Mus.;
Specimen(s): [1983,M.29], 7.2g.

Meghei v Mighei.

Meissen 51°9′ N., 13°28′ E.
Dresden, Germany
Fell 1164
Doubtful. Iron.
A large mass of iron is said to have fallen, E.F.F. Chladni, Die Feuer-Meteore, Wien, 1819, p.198, but the evidence is not conclusive. Further, if there was a fall in 1164, it may be identical with the Steinbach stony-irons (*q.v.*).

Mejillones 23°6′ S., 70°30′ W.
Atacama, Chile
Found 1875, before this year
Synonym(s): *Mejillones (1875), Mejillones (1905), Myelenes, Polanko*
Iron. Hexahedrite (IIA).
A mass of 14.5kg was found in a pampa about 9 km from the sea, between the Morro de los Guaneros de Mejillones and Caleta Herradura Grande. Description, with an analysis, 5.8 %Ni, G.P. Merrill, Proc. Nat. Acad. Sci., 1924, **10**, p.309. Another mass reported to be so big that "a cart would be required for its carriage" was found 3 or 4 leagues from the Bay of Mejillones, I. Domeyko, C. R. Acad. Sci. Paris, 1875, **81**, p.597. It appears that this mass has not been located and only the specimen in Paris, Mus. d'Hist. Nat. (and the 2g specimen in Chicago, cut from it) are in collections. Mejillones (1875) and Mejillones (1905) are identical, analysis, 5.53 %Ni, 59.8 ppm.Ga, 177 ppm.Ge, 2.5 ppm.Ir, J.T. Wasson and J.I. Goldstein, Geochimica et Cosmochimica Acta, 1968, **32**, p.329. Shocked, polycrystalline hexahedrite, distinct from North Chile (*q.v.*), V.F. Buchwald, Iron Meteorites, Univ. of California, 1975, p.816.
12.4kg Washington, U.S. Nat. Mus., includes the main mass of the 14.5kg mass, now 12.2kg; 231g New York, Amer. Mus. Nat. Hist.; 151g Paris, Mus. d'Hist. Nat.; 243g Moscow, Acad. Sci.; 236g Madrid,;

Mejillones v Vaca Muerta.

Mejillones (1875) v Mejillones.

Mejillones (1905) v Mejillones.

Melbourne v Cranbourne.

Mellenbye 28°51′ S., 116°17′ E. approx.
Western Australia, Australia
Found 1929, about
Synonym(s): *Mallenbye*
Stone. Olivine-hypersthene chondrite, amphoterite (LL6).
A piece of 337.6g was found, probably 1/4 or less of the original mass, between Mellenbye and Kadji-Kadji homesteads. Description, E.S. Simpson, Min. Mag., 1938, **25**, p.161. Another specimen, of 844g, is reported to have been found in 1923, Spec. Publ. West. Austr. Mus., 1965 (3), p.38. Possibly includes "Yalgoo" (*q.v.*), olivine Fa_{27}, B. Mason, Rec. Austr. Mus., 1974, **29**, p.169.
297g Perth, West. Austr. Mus., main mass;
Specimen(s): [1934,623], 81g. and fragments, 0.5g.

Melrose (a) 34°23′ N., 103°37′ W.

Curry County, New Mexico, U.S.A.
Found 1933, before this year
Stone. Olivine-hypersthene chondrite (L5).

There has been some confusion between this fall and La Lande (*q.v.*). H.H. Nininger reported one stone of 31kg found at Melrose and four, totalling 20.38kg at La Lande, 26 miles to the west of Melrose, and formerly regarded these as identical. More recently, he has concluded that the Melrose (a) stone, 36.4kg, not 31kg, is different from those found at La Lande, the latter are now assigned to two distinct falls, La Lande and Taiban. Description of the Melrose (a) stone, with an analysis, F.G. Hawley, Am. Miner., 1934, **19**, p.370 [M.A. 6-14]. Rare gas content, D. Heymann, Geochimica et Cosmochimica Acta, 1965, **29**, p.1203.

10.7kg Tempe, Arizona State Univ.; 2.78kg Washington, U.S. Nat. Mus.; 2.5kg Albuquerque, Univ. of New Mexico; 2kg Harvard Univ.; 608g New York, Amer. Mus. Nat. Hist.; 1.5kg Chicago, Field Mus. Nat. Hist.; 322g Los Angeles, Univ. of California;

Specimen(s): [1937,385], 41g.; [1959,772], 11226.5g.

Melrose (b) 34°23.9′ N., 103°36.3′ W.

Curry County, New Mexico, U.S.A.
Found 1971
Stone. Achondrite, Ca-rich. Howardite (AHOW).

A crust-free individual weighing 50.5g was found 2.5 miles NE. of Melrose, G.I. Huss, letter of 10 September, 1974, in Min. Dept. BM(NH).

24.3g Mainz, Max-Planck-Inst.; 9g Albuquerque, Univ. of New Mexico;

Specimen(s): [1974,M.22], 6.3g.

Melville v Cape York.

Melville Bay v Cape York.

Melyan v Milena.

Menabilly 50°19′ N., 4°40′ W.

Cornwall, England
Pseudometeorite..

Listed as a meteorite, E.F.F. Chladni, Die Feuer-Meteore, Wien, 1819, p.261 with the reference, E. King, Remarks concerning...from the clouds, London, 1796, but King says quite definitely that this was merely a fall of very large hailstones.

Menindee Lakes 001

33°10′ S., 141°45′ E. approx.

Near Broken Hill, New South Wales, Australia
Found 1969, 1969-1971, also in 1975
Synonym(s): *Popio*
Stone. Olivine-hypersthene chondrite (L6).

A single mass of 747g was found, olivine $Fa_{23.9}$, M.J. Fitzgerald, Ph.D. Thesis, Univ. of Adlaide, 1979, p.244. Fifty-five masses were found in the vicinity of the Menindee Lakes, about 150km SSE. of Broken Hill. Most were found by W.L. Hollmayer in 1969-1971 and further material was found in 1975 by D.H. McColl. All are chondrites and are named Menindee Lakes 001 to Menindee Lakes 055. They range in weight from 1g to 3.35kg, totalling 7.32kg, Meteor. Bull., 1982 (60), Meteoritics, 1982, **17**, p.95.

Most specimens, Canberra, Bureau Min. Resources;

Menindee Lakes 002

32°55′ S., 141°53′ E. approx.

Popio Lakes region, New South Wales, Australia
Found 1969, 1969-1975
Synonym(s): *Twin Wells*
Stone. Olivine-bronzite chondrite (H5).

A mass of about 270g was found, olivine $Fa_{21.1}$, M.J. Fitzgerald, Ph.D. Thesis Univ. of Adelaide, 1979, p.253.

Menindee Lakes 003

32°55′ S., 141°53′ E. approx.

Popio Lakes region, New South Wales, Australia
Found 1969, 1969-1975
Synonym(s): *Nine Mile*
Stone. Olivine-hypersthene chondrite (L6).

A single mass of 172g was found, olivine $Fa_{26.9}$, M.J. Fitzgerald, Ph.D. Thesis, Univ. of Adelaide, 1979, p.238, M.J. Fitzgerald, letter of 24 November, 1980 in Min. Dept. BM(NH).

Specimen, Canberra, Bureau Min. Resources;

Menindee Lakes 004 33°8′ S., 141°43′ E. approx.

Popio Lakes region, New South Wales, Australia
Found 1969, 1969-1975
Synonym(s): *Popilta*
Stone. Olivine-hypersthene chondrite (L6).

A mass of 158.5g was found, olivine $Fa_{25.6}$, M.J. Fitzgerald, Ph.D. Thesis, Univ. of Adelaide, 1976, p.244.

Specimen, Canberra, Bureau Min. Resources;

Menindee Lakes 005

32°55′ S., 141°53′ E. approx.

Popio Lakes region, New South Wales, Australia
Found 1969, 1969-1972
Synonym(s): *Popio Lakes (a)*
Stone. Olivine-bronzite chondrite (H5).

A mass of 72.9g was found, olivine $Fa_{19.8}$, M.J. Fitzgerald, Ph.D. Thesis, Univ. of Adelaide, 1979, p.210.

Specimen, Canberra, Bureau Min. Resources;

Menindee Lakes 006

32°55′ S., 141°53′ E. approx.

Popio Lakes region, New South Wales, Australia
Found 1969, 1969-1975
Synonym(s): *Popio Lakes (b)*
Stone. Olivine-bronzite chondrite (H6).

A stone of 68.2g was found, olivine $Fa_{18.0}$, M.J. Fitzgerald, Ph.D. Thesis, Univ. of Adelaide, 1979, p.211.

Specimen, Canberra, Bureau Min. Resources;

Meno v Menow.

Menow 53°11′ N., 13°9′ E.

FÜrstenburg, Potsdam, Germany
Fell 1862, October 7, 1230 hrs
Synonym(s): *Fürstenberg, Klein-Menow, Meno*
Stone. Olivine-bronzite chondrite (H4).

A stone of about 10.5kg fell, after detonations, near Fürstenberg, Ann. Phys. Chem. (Poggendorff), 1862, **117**, p.637. Ar-Ar age, 4.48 Ga., G. Turner et al., Geochimica et Cosmochimica Acta, 1978 (Suppl. 10), p.989. Mineralogy, metallurgy, analysis, 27.41 % total iron, olivine $Fa_{18.8}$, R. Hutchison et al., Proc. Roy. Soc. London, 1981, **A374**, p.159. I-Xe age, J. Jordon et al., Z. Natur., 1980, **35a**, p.145.

2.7kg Calcutta, Mus. Geol. Surv. India; 500g Berlin, Humboldt Univ.; 695g Chicago, Field Mus. Nat. Hist.; 157g Vienna, Naturhist. Mus.; 110g Washington, U.S. Nat.

Mus.;
Specimen(s): [50928], 1071g. and fragments, 26g; [35181], 0.5g.

Merceditas 26°20′ S., 70°17′ W.
Chañaral, Atacama, Chile
Found 1884
Synonym(s): *Chañaral, Chañarlino, El Chañaralino*
Iron. Octahedrite, medium (1.0mm) (IIIA).
A mass of about 94.5lb was found near a mining camp, 10 to 12 leagues east of Chañaral, E.E. Howell, Proc. Rochester Acad. Sci., 1890, **1**, p.99. Description, with an analysis, 7.33 %Ni, E. Cohen, Ann. Naturhist. Hofmus. Wien, 1900, **15**, p.379. A mass of 11.2 kg also found in 1884 in the interior from Chañaral has been paired with Merceditas, H.A. Ward, Proc. Rochester Acad. Sci., 1906, **4**, p.230, but is distinct and to be paired with Ilimaes (*q.v*), V.F. Buchwald, Iron Meteorites, Univ. of California, 1975, p.675. Analysis, 7.82 %Ni, 19.5 ppm.Ga, 38.9 ppm.Ge, 3.6 ppm.Ir, E.R.D. Scott et al., Geochimica et Cosmochimica Acta, 1973, **37**, p.1957.
11.4kg New York, Amer. Mus. Nat. Hist.; 7.8kg Vienna, Naturhist. Mus.; 2.12kg Budapest, Nat. Mus.; 1.6kg Harvard Univ.; 1kg Chicago, Field Mus. Nat. Hist.; 914g Oslo, Min.-Geol. Mus.; 724g Washington, U.S. Nat. Mus.; 831g Calcutta, Mus. Geol. Surv. India; 710g Sarajevo, Bosnian Nat. Mus.; 177g Berlin, Humboldt Univ.;
Specimen(s): [66742], 1917g.

Meridi v Maridi.

Mern 55°3′ N., 12°4′ E.
Praestö, Sorestrøm, Denmark
Fell 1878, August 29, 1430 hrs
Synonym(s): *Moern, Praestö*
Stone. Olivine-hypersthene chondrite (L6), veined.
A stone of about 4 kg fell after detonations, S. Tromholt, Wochenschr. Astron. Meteor. Geogr. Halle, 1878 (1879), **21**, p.391. Description, A. Brezina and W. Wahl, Mem. Acad. Sci. Denmark, 1909, **6** (3), p.113. Figured, V.F. Buchwald and S. Munck, Analecta geol., 1965, **1**, p.1. Olivine Fa_{24}, B. Mason, Geochimica et Cosmochimica Acta, 1963, **27**, p.1011.
2319g Copenhagen, Univ. Geol. Mus., includes main mass, 2171g; 220g Washington, U.S. Nat. Mus.; 127g Budapest, Nat. Mus.; 110g Vienna, Naturhist. Mus.;
Specimen(s): [1920,321], 73.5g.; [1906,28], 39g.; [1920,320], 4g.

Mertzon 31°16′ N., 100°50′ W.
Irion County, Texas, U.S.A.
Found 1943
Iron. Octahedrite, medium (0.8mm) (IA-ANOM).
Two specimens, 2332g and 1337g, in Washington, U.S. Nat. Mus., and a slice in Chicago, Field Mus. Nat. Hist., are all that are known of this fall, Ann. Rep. U.S. Nat. Mus., 1947, p.43. Chemically anomalous, analysis, 8.98 %Ni, 68 ppm.Ga, 293 ppm.Ge, 2.4 ppm.Ir, J.T. Wasson, Icarus, 1970, **12**, p.407. Description; polycrystalline with dispersed troilite nodules, V.F. Buchwald, Iron Meteorites, Univ. of California, 1975, p.824.
3.3kg Washington, U.S. Nat. Mus.; 186g Chicago, Field Mus. Nat. Hist.;

Meru 0°0′ N., 37°40′ E.
Kenya
Fell 1945, February 2, 1700 hrs
Stone. Olivine-hypersthene chondrite, amphoterite (LL).
A stone of about 13lb fell near Mucheze village, 1 mile east of Marimba, L.S.B. Leakey, letters in Min. Dept. BM(NH). Olivine Fa_{28}, B. Mason, Geochimica et Cosmochimica Acta, 1963, **27**, p.1011.
Main mass, Nairobi, Coryndon Mus.;
Specimen(s): [1947,226], 26g. fragments

Merua 25°29′ N., 81°59′ E.
Allahabad, Uttar Pradesh, India
Fell 1920, August 30, 1115 hrs
Synonym(s): *Chukri, Manpur Urf Baghai Kalan, Mustafabad, Umri*
Stone. Olivine-bronzite chondrite (H5).
After detonations, six stones, weighing together 71.4kg, fell, the largest (weighing 56.7kg) at Merua, two at Manpur Urf Baghai Kalan, and one each at Chukri, Umri, and Mustafabad. Description, G.H. Tipper, Rec. Geol. Surv. India, 1925, **56**, p.345. Olivine Fa_{18}, B. Mason, Geochimica et Cosmochimica Acta, 1963, **27**, p.1011.
66.3kg Calcutta, Mus. Geol. Surv. India; 212g Tempe, Arizona State Univ.; 197g Washington, U.S. Nat. Mus.;
Specimen(s): [1924,134], 2848g. nearly complete stone, and fragments, 23g; [1924,135], 1385g. nearly complete stone, and fragments, 18g.

Mesa Verde Park 37°10′ N., 108°30′ W.
Montezuma County, Colorado, U.S.A.
Found 1922
Synonym(s): *Pipe Shrine House*
Iron. Octahedrite, medium (0.6mm) (IB).
An oxidized mass of 3.5kg was found in the Sun Shrine at the north end of Pipe Shrine House. Description, G.P. Merrill, Proc. U.S. Nat. Mus., 1923, **63** (18), p.1. Analysis, 10.56 %Ni, 53.0 ppm.Ga, 142 ppm.Ge, 1.8 ppm.Ir, J.T. Wasson, Meteorites, Springer-Verlag, 1974, p.299. Polycrystalline; unusual structure, V.F. Buchwald, Iron Meteorites, Univ. of California, 1975, p.826.
3.38kg Washington, U.S. Nat. Mus.;

Mesón de Fierro v Campo del Cielo.

Messina 38°11′ N., 15°34′ E.
Upper Camara Valley, Messina, Sicily, Italy
Fell 1955, July 16, 1307 hrs
Synonym(s): *Camaro, Camaro Superiore*
Stone. Olivine-hypersthene chondrite (L5).
Three broken pieces, only a quarter of the whole mass, weighing 2025g, 338g and 42g, B. Baldanza and G. Labruto, Rend. Soc. Min. Ital., 1956, **12**, p.47 [M.A. 13-358]. Olivine Fa_{24}, B. Mason, Geochimica et Cosmochimica Acta, 1967, **31**, p.1100.
21g Vienna, Naturhist. Mus.; 3.4g Washington, U.S. Nat. Mus.;
Specimen(s): [1983,M.1], 1972g.

Mesquital v San Francisco del Mezquital.

Meteorite Hills A78001 79°41′ S., 155°45′ E.
Darwin Glacier, Antarctica
Found 1978, between December 1978 and January 1979
Stone. Olivine-bronzite chondrite (H4).
A mass of 624g was found, the first of about 28 masses representing a number of individual falls, recovered by the American Antarctic Expedition during the 1978-1979 field

season. All are chondrites of H- or L-group classification, description of some of this material, Antarctic Meteorite Newsletter, 1981, **4**, p.106.

Meteorite Hills A78002

Stone. Olivine-hypersthene chondrite (L6).
A mass of 542g was found.

Meteorite Hills A78003

Stone. Olivine-hypersthene chondrite (L6).
A mass of 1726g was found.

Meteorite Hills A78006

Stone. Olivine-bronzite chondrite (H6).
A mass of 409g was found.

Meteorite Hills A78028

Stone. Olivine-hypersthene chondrite (L6).
A mass of 20.65kg was found.

Metsäkylä 60°39′ N., 27°4′ E.

Finland
Found 1938, autumn
Synonym(s): *Metsaekylae*
Stone. Olivine-bronzite chondrite (H4).
1kg was found, olivine Fa_{18}, B. Mason, Geochimica et Cosmochimica Acta, 1963, **27**, p.1011. Analysis, 23.89 % total iron, H.B. Wiik and B. Mason, Geologi (Helsinki), 1964, p.95.
Main mass, Helsinki, Univ.; Thin section, Washington, U.S. Nat. Mus.; 2.7g Mainz, Max-Planck-Inst.; 2.4g New York, Amer. Mus. Nat. Hist.;

Metsaekylae v Metsäkylä.

Meung sur Loire v Charsonville.

Meuselbach 50°35′ N., 11°6′ E.

Thuringen, Germany
Fell 1897, May 19, 1945 hrs
Stone. Olivine-hypersthene chondrite (L6), veined.
A stone of about 870g was seen to fall after detonations. Description and analysis, G. Linck, Ann. Naturhist. Hofmus. Wien, 1898, **13**, p.103. Olivine Fa_{24}, B. Mason, Geochimica et Cosmochimica Acta, 1963, **27**, p.1011.
Main mass, Rudolstadt,; 58g Vienna, Naturhist. Mus.; 2.2g Chicago, Field Mus. Nat. Hist.;
Specimen(s): [84245], 19.75g.

Mexico v Pampanga.

Mezel 45°46′ N., 3°15′ E.

Clermont-Ferrand, Puy de Dome, France
Fell 1949, January 25, 1956 hrs, U.T.
Stone. Olivine-hypersthene chondrite (L6), veined.
A mass of 1.3kg fell after a notable bolide, H. Dessens, C. R. Acad. Sci. Paris, 1949, **228**, p.813 [M.A. 11-134]. Description, analysis, D. Collier et al., C. R. Acad. Sci. Paris, 1949, **228**, p.1816 [M.A. 11-135]. Olivine Fa_{24}, B. Mason, Geochimica et Cosmochimica Acta, 1963, **27**, p.1011.
998g Paris, Mus. d'Hist. Nat.;
Specimen(s): [1966,279], 177g.

Mezoe-Madaras v Mezö-Madaras.

Mezö-Madaras 46°30′ N., 25°44′ E.

Harghita, Romania
Fell 1852, September 4, 1630 hrs
Synonym(s): *Fekete, Madaras, Maros, Mezoe-Madaras, Mezö-Madarasz, Weiler*
Stone. Olivine-hypersthene chondrite (L3), xenolithic.
After the appearance of a luminous meteor and detonations, a shower of many stones fell, of which the largest weighed about 10kg and the total weight was about 22.7kg, W. Knöpfler, Verh. Mitt. Siebenbürg. Ver. Naturwiss., Hermannstadt, 1853, **4**, p.19. Analysis, C. Rammelsberg, Ber. Deutsch. Geol. Gesell. Berlin, 1871, **23**, p.734. Study of the iron-rich silicates, contains merrihueite, R.T. Dodd et al., Am. Miner., 1966, **51**, p.1177. A polymict breccia, W.R. Van Schmus, Geochimica et Cosmochimica Acta, 1967, **31**, p.2027, R.A. Binns, Geochimica et Cosmochimica Acta, 1968, **32**, p.299. Analysis, 21.6 % total iron, E. Jarosewich, Geochimica et Cosmochimica Acta, 1967, **31**, p.1103. Unequilibrated and unmetamorphosed, G. Kurat, Geochimica et Cosmochimica Acta, 1967, **31**, p.1843, Meteorite Research, ed. P.M. Millman, D. Reidel, Dordrecht-Holland, 1969, p.185. Contains various 'spinels', G. Hoinkes and G. Kurat, Meteoritics, 1973, **8**, p.383, abs. Trace element abundances, R.R. Keays et al., Geochimica et Cosmochimica Acta, 1971, **35**, p.337. Pb isotopic ratios, G.R. Tilton, Earth planet. Sci. Lett., 1973, **19**, p.321.
9kg Vienna, Naturhist. Mus.; 3kg Berlin, Humboldt Univ.; 1.8kg Budapest, Nat. Mus.; 1.7kg Tübingen, Univ.; 613g Moscow, Acad. Sci.; 569g Tempe, Arizona State Univ.; 538g Calcutta, Mus. Geol. Surv. India; 449g Chicago, Field Mus. Nat. Hist.; 335g Washington, U.S. Nat. Mus.; 247g Los Angeles, Univ. of California;
Specimen(s): [33909], 565g. and fragments, 3g; [33188], 65.75g. most of a stone; [90270], 18g.; [96270], thin sections (two)

Mezö-Madarasz v Mezö-Madaras.

Mezquital v San Francisco del Mezquital.

Mhow 25°54′ N., 83°37′ E.

Azamgarh district, Uttar Pradesh, India
Fell 1827, February 16, 1500 hrs
Synonym(s): *Ghazeepore, Mau, Mow*
Stone. Olivine-hypersthene chondrite (L6).
After detonations, four or five stones fell, one of which broke a tree and another wounded a man. The largest stone weighed 3lb, J. Tytler, Edinburgh J. Sci., 1828, **8**, p.172, N. Story Maskelyne, Phil. Mag., 1863, **25**, p.447. Olivine Fa_{24}, B. Mason, Geochimica et Cosmochimica Acta, 1963, **27**, p.1011.
128g Calcutta, Mus. Geol. Surv. India; 24g Vienna, Naturhist. Mus.; 2g Chicago, Field Mus. Nat. Hist.; 1g Berlin, Humboldt Univ.;
Specimen(s): [34806], 163.75g.

Miami 35°40′ N., 100°36′ W.

Roberts County, Texas, U.S.A.
Found 1930, recognized 1937
Stone. Olivine-bronzite chondrite (H5).
One stone, 57.7kg, was found, E.P. Henderson, letter of May 12, 1939 in Min. Dept. BM(NH)., A.D. Nininger, Pop. Astron., Northfield, Minnesota, 1939, **47**, p.213. Olivine Fa_{17}, B. Mason, Geochimica et Cosmochimica Acta, 1963, **27**, p.1011.
29kg Tempe, Arizona State Univ.; 1160g Washington, U.S. Nat. Mus.;
Specimen(s): [1959,1014], 25000g.

Mianchi 34°48′ N., 111°42′ E.
Henan, China
Fell 1980, September 4, 1100 hrs
Stone. Olivine-bronzite chondrite (H5).
A single mass of 1.1kg was recovered, D. Bian, Meteoritics, 1981, **16**, p.119. Analysis, 28.39 % total iron, W. Daode et al., Geochemistry, 1982, **1**, p.186.
1.1kg Beijing, Planetarium;

Michigan Iron v Toluca.

Middle Hoby v Tugalin-Bulin.

Middlesbrough 54°34′ N., 1°10′ W.
Yorkshire, England
Fell 1881, March 14, 1535 hrs
Synonym(s): *Pennyman's Siding, Yorkshire*
Stone. Olivine-hypersthene chondrite (L6).
After detonations (not heard actually at the place of fall), a stone of 3.5lb was seen to fall at Pennyman's Siding on the railway from Middlesbrough to Guisborough, about 1.75 miles from Middlesbrough, A.S. Herschel, Rep. Brit. Assn., 1881, **51**, p.296. Description, with analysis, W. Flight, Phil. Trans. Roy. Soc. London, 1882, **173**, p.896. Olivine Fa_{23}, B. Mason, Geochimica et Cosmochimica Acta, 1963, **27**, p.1011.
Main mass, York, Mus.;
Specimen(s): [54267], 13.9g. and fragments, 6g

Midland 44°45′ N., 79°53′ W.
Simco County, Ontario, Canada
Found 1960, about; recognized 1964
Iron.
One mass, 34g, was found on the shore of Midland Bay, Meteor. Bull., 1971 (50), Meteoritics, 1971, **6**, p.120.
32.9g Ottawa, Geol. Surv. Canada;

Midt-Vaage v Tysnes.

Migei v Mighei.

Mighei 48°4′ N., 30°58′ E.
Olviopol, Kherson, Ukraine, USSR, [Мигеи]
Fell 1889, June 18, 0830 hrs
Synonym(s): *Elisabethpol, Elizavetpol, Meghei, Migei, Migheia, Migheja, Nigheija*
Stone. Carbonaceous chondrite, type II (CM2).
After the " usual light and sound phenomena", a stone, of over 8kg according to the total weight in collections, was seen to fall, Yu.I. Simashko, Nature, 1890, **41**, p.472, abs.. Description and partial analysis, S. Meunier, C. R. Acad. Sci. Paris, 1889, **109**, p.976. Mentioned, A.N. Zavaritsky and L.G. Kvasha, Метеориты СССР, Москва , 1952, E.R. DuFresne and E. Anders, Geochimica et Cosmochimica Acta, 1961, **23**, p.200. Analysis, 21.24 % total iron, H.B. Wiik, Geochimica et Cosmochimica Acta, 1956, **9**, p.279. Analysis, 19.95 % total iron, H. von Michaelis et al., Earth planet. Sci. Lett., 1969, **5**, p.387. Mentioned, G.P. Vdovykin, Метеоритика, 1964, **25**, p.134. Classification, references, W.R. Van Schmus and J.M. Hayes, Geochimica et Cosmochimica Acta, 1974, **38**, p.47. Trace element data, G.W. Kallemeyn and J.T. Wasson, Geochimica et Cosmochimica Acta, 1981, **45**, p.1217.
2.75kg Odessa, Univ.; 1.6kg Chicago, Field Mus. Nat. Hist.; 1590g Moscow, Acad. Sci.; 1.3kg Vienna, Naturhist. Mus.; 620g Washington, U.S. Nat. Mus.;
Specimen(s): [86947], 147g.; [65604], 67g. and fragments, 2g

Migheia v Mighei.

Migheja v Mighei.

Mike 46°14′ N., 17°32′ E.
Somogy, Hungary
Fell 1944, May 3, 1900 hrs
Stone. Olivine-hypersthene chondrite (L).
Four pieces, totalling 224.2g, were recovered, K. Sztrókay, Természettközl, 1947, **2**, p.305, Zentr. Miner. geol. Paläont., 1949 (1), p.120 [M.A. 11-267]. Analysis, 22.29 % total iron, K.J. Sztrókay and A. Földvári, Föld. Közl., 1953, **83**, p.243 [M.A. 12-357, 610]. Olivine Fa_{24}, B. Mason, Geochimica et Cosmochimica Acta, 1963, **27**, p.1011.
139g Budapest, Nat. Mus.; 3.3g New York, Amer. Mus. Nat. Hist.;

Mikenskoi v Grosnaja.

Mikenskoj v Grosnaja.

Mikentskaja v Grosnaja.

Mikentskaia v Grosnaja.

Mikkeli v St. Michel.

Mikmotojima v Mikomotojima.

Mikolawa
Szakad, Hungary
Fell 1837, January 15
Doubtful..
One stone is said to have fallen, G. von Boguslawski, Ann. Phys. Chem. (Poggendorff), 1854, **4** (suppl.), p.356 but the evidence is not conclusive. The exact locality is uncertain; probably near Szalard, 47°13'N., 22°2'E., or near Szalacs, 47°28'N., 22°18'E.

Mikomotojima 34°34′ N., 138°57′ E.
Shizuoka prefecture, Honshu, Japan
1874, before this year
Synonym(s): *Mikmotojima*
Pseudometeorite..
A mass of iron weighing about 200g was originally described as meteoritic, Mitt. deut. Gesell. Natur- Völkerkunde Ostasien, 1874, **1**, p.5. Listed as doubtful, K. Jimbo, Beitr. Miner. Japan, 1906, **2**, p.33, Sci. Rep. Yokohama Univ., 1952 (2, no. 1). Listed as a meteorite, Meteor. Bull., 1958 (8). The specimen has been lost, but it is clear from the original description that it was not meteoritic, S. Murayama, letter of 23 June 1964 in Min. Dept. BM(NH).

Milan
Lombardy, Italy
Fell 1525, June 23
Doubtful..
One stone is said to have fallen in the Castle of Milan; the munitions caught fire and great damage was caused, F. Onofri, Sommario Historico, 1675. The evidence is not conclusive and nothing is preserved, G.R. Levi-Donati, priv. comm. 30 April 1969 in Min. Dept. BM(NH). This is probably the same fall as is listed under the date "circa 1650", E.F.F. Chladni, Die Feuer-Meteore, Wien, 1819, p.230.

Milan v Cereseto.

Milena 46°11′ N., 16°6′ E.

Varazdin, Croatia, Yugoslavia
Fell 1842, April 26, 1500 hrs
Synonym(s): *Businski, Melyan, Miljana, Miljane, Pusinsko Selo, Pusnsko Selo, Varazdin, Warasdin*
Stone. Olivine-hypersthene chondrite (L6).

After appearance of a luminous meteor, followed by detonations, two (or three) stones, each of 5 to 6kg, fell at Pusinsko Selo, 4 miles south of Milena, but were mostly broken up, -. Kocevar, Ann. Phys. Chem. (Poggendorff), 1842, **56**, p.349, F.E. von Rosthorn, Neues Jahrb. Min., 1843, p.79. Olivine Fa_{24}, B. Mason, Geochimica et Cosmochimica Acta, 1963, **27**, p.1011.

900g Vienna, Naturhist. Mus.; 99g Washington, U.S. Nat. Mus.; 83g Harvard Univ.; 81g Tempe, Arizona State Univ.; 71g Zagreb, Mus.; 17g Chicago, Field Mus. Nat. Hist.; 13g New York, Amer. Mus. Nat. Hist.; 9g Berlin, Humboldt Univ.;

Specimen(s): [54815], 127g.; [35403], 20g.; [34594], 0.5g.; [1920,323], 0.5g. fragments

Miljana v Milena.

Miljane v Milena.

Millarville 50°47′50″ N., 114°18′34″ W.

near Calgary, Alberta, Canada
Found 1977, April 28, 0830 hrs
Iron. Structurally anomalous (IVA).

A single mass of 15.636kg was found about 30 km SW. of Calgary during the ploughing of a newly cleared field, Meteor. Bull., 1979 (56), Meteoritics, 1979, **14**, p.168. Classification and analysis, 9.57 %Ni, 2.38 ppm.Ga, 0.144 ppm.Ge, 0.98 ppm.Ir, A. Kracher et al., Geochimica et Cosmochimica Acta, 1980, **44**, p.773.

Main mass, Calgary, Univ. of Alberta; 45g Ottawa, Geol. Surv. Canada;

Specimen(s): [1978,M.23], 50g.

Millbillillie 26°27′ S., 120°22′ E.

Wiluna dist., Western Australia, Australia
Fell 1960, October, day unknown, but about 1pm, recovered 1970
Synonym(s): *Nabberu*
Stone. Achondrite, Ca-rich. Eucrite (AEUC).

A fireball was witnessed in October 1960 and a meteorite appeared to fall on the plain of the Millbillillie and Jundee Stations. No search was initiated at that time but in 1970, the first of at least three specimens was found. The total weight recovered is at least 25.4kg, R.A. Binns, West. Austr. Yearbook, 1973 (12), p.1 [M.A. 74-3318]. Reported, Meteor. Bull., 1972 (51), Meteoritics, 1972, **7**, p.219. Description, M.J. Fitzgerald, Trans. Roy. Soc. South Austr., 1980, **104**, p.201. Analysis, B. Mason et al., Smithson. Contrib. Earth Sci., 1979 (22), p.30.

20kg Perth, West. Aust. Mus., largest stone and 565g; Specimen, collection of C.V. Latz, of "Nabberu"; 241g Chicago, Field Mus. Nat. Hist.; 66g Albuquerque, Univ. of New Mexico;

Specimen(s): [1973,M.8], 250g. crusted part-slice; [1972,493], thin section polished, P1031; [1974,M.7], 20.6g. crusted part-slice of " Nabberu"

Millen 32°50′7″ N., 81°52′26″ W.

Jenkins County, Georgia, U.S.A.
Found 1975
Stone. Olivine-bronzite chondrite (H4).

Two masses were found, one three months after the first, totalling 40.8kg, Meteor. Bull., 1978 (55), Meteoritics, 1978, **13**, p.342.

Specimens, Washington, U.S. Nat. Mus.;

Miller (Arkansas) 35°24′ N., 92°3′ W.

Cleburne County, Arkansas, U.S.A.
Fell 1930, July 13
Stone. Olivine-bronzite chondrite (H5).

A mass of 16.7kg fell, H.H. Nininger, Am. J. Sci., 1932, **23**, p.78 [M.A. 5-156]. Description, C.A. Reeds, Nat. Hist. (New York), 1931, **31**, p.109, B. Mason and H.B. Wiik, Geochimica et Cosmochimica Acta, 1961, **21**, p.266. Olivine Fa_{19}, B. Mason, Geochimica et Cosmochimica Acta, 1963, **27**, p.1011.

Main mass, New York, Amer. Mus. Nat. Hist.; 53g Washington, U.S. Nat. Mus.; 32g Chicago, Field Mus. Nat. Hist.;

Specimen(s): [1959,1039], 9.5g.

Miller (Kansas) 38°38′ N., 96°1′ W.

Lyon County, Kansas, U.S.A.
Found 1950
Stone. Olivine-bronzite chondrite (H4).

A mass of 970g was found, Earth Sci. Digest, 1953, **6**, p.29 [M.A. 12-253], Meteor. Bull., 1958 (8). Olivine Fa_{19}, B. Mason, Geochimica et Cosmochimica Acta, 1963, **27**, p.1011.

82g New York, Amer. Mus. Nat. Hist.; 42g Los Angeles, Univ. of California; 41g Washington, U.S. Nat. Mus.;

Specimen(s): [1960,107], 40g. a slice

Miller's Run v Pittsburg.

Mills 36°13′48″ N., 104°7′8″ W.

Harding County, New Mexico, U.S.A.
Found 1970
Stone. Olivine-bronzite chondrite (H6).

Two large stones weighing 49.48kg and 29.94kg, and four smaller fragments bringing the total weight recovered to 88kg, were found on a ranch. None of the fragments could be fitted together, Meteor. Bull., 1972 (51), Meteoritics, 1972, **7**, p.226, G.I Huss, letter of 5 May 1971 in Min. Dept. BM(NH). Analysis, olivine Fa_{19}, 24.24 % total iron, G.R. Levi-Donati and E. Jarosewich, Meteoritics, 1974, **9**, p.145. Classification as H6, olivine $Fa_{19.8}$, D.E. Lange and K. Keil, Meteoritics, 1976, **11**, p.315.

66.7kg Mainz, Max-Planck-Inst.; 850g Tempe, Arizona State Univ.; 367g Washington, U.S. Nat. Mus.; 273g Copenhagen, Univ. Geol. Mus.; 240g Albuquerque, Univ. of New Mexico; 171g Chicago, Field Mus. Nat. Hist.;

Specimen(s): [1971,119], 75.8g. a crusted polished slice

Milly Milly 26°7′ S., 116°40′ E.

Murchison Goldfield, Western Australia, Australia
Found 1921
Iron. Octahedrite, medium (1.0mm) (IIIA).

A mass of 26.5kg was found, Spec. Publ. West. Austr. Mus., 1965 (3), E.S. Simpson, Min. Mag., 1938, **25**, p.161. Desciption and analysis, 7.84 %Ni, E.S. Simpson, Min. Mag., 1938, **25**, p.161. Analysis, 7.62 %Ni, 19.1 ppm.Ga, 38.6 ppm.Ge, 2.8 ppm.Ir, E.R.D. Scott et al., Geochimica et Cosmochimica Acta, 1973, **37**, p.1957. Description, plastically deformed, V.F. Buchwald, Iron Meteorites, Univ.

of California, 1975, p.827.
24.5kg Perth, West. Austr. Mus.; 283g Mainz, Max-Planck-Inst.; 283g Washington, U.S. Nat. Mus.; 198g New York, Amer. Mus. Nat. Hist.;
Specimen(s): [1973,M.9], 69.4g. a crusted part-slice

Milnesand (iron) 33°38′ N., 103°20′ W.
Roosevelt County, New Mexico, U.S.A.
Found, year not reported
Iron.
Listed, without details, Cat. Dupont Colln., 1981.
326g Watchung, Dupont Colln.;

Milnesand (stone) 33°38′ N., 103°20′ W.
Roosevelt County, New Mexico, U.S.A.
Found, year not reported
Stone. Chondrite.
Listed, without details, Cat. Dupont Colln., 1981.
34g Watchung, Dupont Colln.;

Milwaukee v Trenton.

Minamino 35°4′42″ N., 136°56′ E.
Nagoya-shi, Aichi, Honshu, Japan
Fell 1632, September 27, about midnight hrs
Stone. Chondrite.
After the appearance of a fireball and detonations, a stone fell into a salt-field near the sea shore. It was recovered soon after the fall and presented to a Shinto shrine where it has remained. Its meteoritic nature was confirmed in 1976; its weight, 1.04kg, Meteor. Bull., 1979 (56), Meteoritics, 1979, **14**, p.168.

Minas Geraes v Minas Gerais.

Minas Gerais 18°30′ S., 44°0′ W.
Minas Gerais, Brazil
Found 1888, known in this year
Synonym(s): *Brasil, Minas Geraes*
Stone. Olivine-hypersthene chondrite (L6), veined.
A piece of 1.2kg was found without label among specimens which may have been brought from Minas Gerais, O.A. Derby, Revista do Observatorio, Rio de Janeiro, 1888, p.12, Am. J. Sci., 1888, **36**, p.157. Olivine Fa_{25}, B. Mason, Geochimica et Cosmochimica Acta, 1963, **27**, p.1011.
628g Rio de Janeiro, Mus. Nac., main mass; 373g Chicago, Field Mus. Nat. Hist.; 30g Yale Univ.;
Specimen(s): [1905,434], 62g.; [63235], 3.5g.

Mincy 36°33′ N., 93°6′ W.
Taney County, Missouri, U.S.A.
Found 1857
Synonym(s): *Crawford County, Forsyth, Miney, Newton County, Tanney, Taney County*
Stony-iron. Mesosiderite (MES).
A mass of about 197lb is stated to have fallen in 1857, 11 miles SE. of Forsyth, whence it was taken to a farm in Newton County, Arkansas, G.F. Kunz and J.H. Caswell, Am. J. Sci., 1887, **34**, p.467. Description and analysis, J.L. Smith, Am. J. Sci., 1865, **40**, p.213. Description, G.T. Prior, Min. Mag., 1918, **18**, p.164. Texture and Ni content of metal, B.N. Powell, Geochimica et Cosmochimica Acta, 1969, **33**, p.789. Texture and composition of silicates, B.N. Powell, Geochimica et Cosmochimica Acta, 1971, **35**, p.5. Analysis of metal, 7.2 %Ni, 13.1 ppm.Ga, 52.4 ppm.Ge, 2.5 ppm.Ir, J.T. Wasson et al., Geochimica et Cosmochimica Acta, 1974, **38**, p.135. Classification, olivine coronas, references, C.E. Nehru et al., Geochimica et Cosmochimica Acta, 1980, **44**, p.1103.
39kg Vienna, Naturhist. Mus.; 2.5kg New York, Amer. Mus. Nat. Hist.; 4.2kg Budapest, Nat. Mus.; 2kg Chicago, Field Mus. Nat. Hist.; 527g Rome, Vatican Colln; 402g Prague, Nat. Mus.; 407g Tempe, Arizona State Univ.;
Specimen(s): [68579], 2375g.; [40885], 60g.; [40886], 16g.; [96271], thin section; [1964,574], thin section

Minden 52°18′ N., 8°55′ E.
Nordrhein-Westfalen, Germany
Fell 1379, May 26
Doubtful..
Several stones are said to have fallen, E.F.F. Chladni, Die Feuer-Meteore, Wien, 1819, p.202 but the evidence is not conclusive.

Mineo 37°17′ N., 14°42′ E.
Catania, Sicily, Italy
Fell 1826, May
Doubtful. Stony-iron. Pallasite (PAL).
A luminous object was reported to have fallen at a farm near Mineo after a loud roar was heard, and a mass of iron was recovered from a crater, A. De Gregario, Il Naturalista Siciliano, 1916, **3**, p.129. According to De Gregario a fragment seen by Meunier was hammered and corroded and may not have been meteoritic. Some fragments recovered later were identified as pallasitic, B. Baldanza et al., Meteoriti, Univ. Perugia, 1969.
42g Perugia, Univ.;

Miney v Mincy.

Min-Fan-Zhun 32°20′ N., 120°40′ E.
Wan-Fu-Syan, Dzugao district, Jiangsu, China
Fell 1952, April 1, after 2000 hrs.
Synonym(s): *Jukao, Kiangsu Rukao, Rugao, Rukao*
Stone. Olivine-hypersthene chondrite, amphoterite (LL6).
One stone of 5kg fell, Meteor. Bull., 1958 (10). Description, D. Wang and Z. Ouyang, Geochimica, 1979, p.120, in Chinese [M.A. 80-2085]. Listed with co-ordinates, D. Bian, Meteoritics, 1981, **16**, p.115. Analysis, 21.48 % total iron, W. Daode et al., Geocghemistry, 1982, **1**, p.186. Classification, olivine $Fa_{27.6}$, J.T. Wasson letter of 8 Nov., 1982 in Min. Dept. BM(NH).
5.5kg Nanjing, Purple Mountain Observatory;

Minnesota (iron)
Minnesota, U.S.A.
Found 1950, known before this year
Iron. Octahedrite, medium.
A fragment of 6g from an unknown locality in Minnesota (Catalogue of the Nininger collection, 1950).
Specimen, Tempe, Arizona State Univ.;

Minnesota (stone)
Minnesota, U.S.A.
Found 1950, known before this year
Stone. Olivine-hypersthene chondrite (L).
A fragment of 1.8g from an unknown locality in Minnesota (Catalogue of the Nininger Collection, 1950)., M.L. Karr et al., Cat. Meteor. Arizona State Univ., 1970, p.97.
Specimen, Tempe, Arizona State Univ.;

Minnichhof 47°42′ N., 16°36′ E. approx.
Sopron, Hungary
Fell 1905, May 27, 1045 hrs
Synonym(s): *Malomhaza*
Stone. Chondrite.
Mentioned, F. Berwerth, Fortschr. Min. Krist. Petr., 1912, **2**, p.235. Listed, L. Tokody and M. Dudich, Magyarország Meteoritgyüjteményei, Budapest, 1951, p.38, 81.
517g Budapest, Nat. Mus.; 45g Vienna, Naturhist. Mus.;

Mino v Gifu.

Minsk v Brahin.

Minsk v Zmenj.

Mirzapur 25°41′ N., 83°15′ E.
Ghazipur district, Uttar Pradesh, India
Fell 1910, January 7, 1130 hrs
Stone. Olivine-hypersthene chondrite (L5), brecciated.
After detonations, a stone fell, of which two pieces, weighing respectively 8.3kg and 208.5g were recovered, G. de P. Cotter, Rec. Geol. Surv. India, 1912, **42**, p.272. Mineralogy, analysis, 20.98 % total iron, olivine $Fa_{24.8}$, A. Dube et al., Smithson. Contrib. Earth Sci., 1977 (19), p.71.
8kg Calcutta, Mus. Geol. Surv. India;
Specimen(s): [1915,141], 204.5g. the smaller piece found

Miscolz v Miskolcz.

Miskolcz 48°7′ N., 20°48′ E.
Hungary
Fell 1559
Synonym(s): *Miscolz*
Doubtful. Stone.
Five large stones are said to have fallen, E.F.F. Chladni, Die Feuer-Meteore, Wien, 1819, p.214 but the evidence is not conclusive.

Misshof 56°40′ N., 23°0′ E.
Courland, Latvian SSR, USSR
Fell 1890, April 10, 1530 hrs
Synonym(s): *Baldohn, Mittel-Stuhre*
Stone. Olivine-bronzite chondrite (H5).
After detonations, a stone of about 5.8kg fell. Description, B. Doss, Arbeiten Naturf. Ver. Riga, 1891 (7), p.1, E. Johnson, Arb. Naturf. Ver. Riga, 1891 (7), p.69, Neues Jahrb. Min., 1892, **1**, p.71. Partial analysis by H.B. Wiik, 27.42 % total iron, W. Wahl, Geochimica et Cosmochimica Acta, 1950, **1**, p.28. Olivine Fa_{19}, B. Mason, Geochimica et Cosmochimica Acta, 1963, **27**, p.1011.
2.5kg Riga, Polytech.; 266g Budapest, Nat. Mus.; 260g Chicago, Field Mus. Nat. Hist.; 158g Washington, U.S. Nat. Mus.; 145g Harvard Univ.; 133g New York, Amer. Mus. Nat. Hist.;
Specimen(s): [67591], 134g.; [1920,284], 68g.

Mission 43°19′ N., 100°46′ W.
Todd County, South Dakota, U.S.A.
Found 1949, recognized in this year
Stone. Olivine-hypersthene chondrite (L).
A stone of 12kg was found, H.H. and A.D. Nininger, The Nininger Collection of Meteorites, Winslow, Arizona, 1950, p.72. Olivine Fa_{23}, B. Mason, Geochimica et Cosmochimica Acta, 1963, **27**, p.1011.
10.2g Tempe, Arizona State Univ.;

Missouri v Little Piney.

Missouri v St. Francois County.

Misteca 16°48′ N., 97°6′ W.
Oaxaca, Mexico
Found 1804, known in this year
Synonym(s): *False Yanhuitlan, Oaxaca, Teposcolula*
Iron. Octahedrite, coarse (1.4mm) (IA).
Del Rio mentions La Misteca as a locality for metallic iron, A.M. del Rio, Tablas Mineralogicas, 1804, p.57, L. Fletcher, Min. Mag., 1890, **9**, p.172. A specimen brought away by Karavinsky was presented to the Vienna collection, P. Partsch, Die Meteoriten, Wien, 1843, p.134. Classification, analysis, 8.27 %Ni, 67.8 ppm.Ga, 233 ppm.Ge, 1.6 ppm.Ir, J.T. Wasson, Icarus, 1970, **12**, p.407. Has been confused with Yanhuitlan, particularly since Ward's establishment etched "Misteca" on the Yanhuitlan material they cut and distributed. The whole mass has been heated artificially to 1000C., V.F. Buchwald, Iron Meteorites, Univ. of California, 1975, p.828. Repository of main masses uncertain.
2kg Vienna, Naturhist. Mus.; 1.1kg Budapest, Nat. Mus.; 1231g Berlin, Humboldt Univ.; 487g Harvard Univ.; 344g Washington, U.S. Nat. Mus.; 140g New York, Amer. Mus. Nat. Hist.;
Specimen(s): [35187], 164.5g.; [35173], 148g.; [1959,160], 2g.

Misteca (in part) v Yanhuitlan.

Mitchell County v Waconda.

Mittel-Stuhre v Misshof.

Mixbury 52°0′ N., 1°7′ W.
Oxfordshire, England
Fell 1725, July 3
Doubtful..
A stone of 20lb is said to have fallen, R.P. Greg, Rep. Brit. Assn., 1860, p.56, T.M. Hall, Min. Mag., 1879, **3**, p.7 but the evidence is inconclusive.

Mjelleim 61°44′ N., 5°56′ E.
Hyen, Nordfjord, Norway
Fell 1898, January 24, 1400-1500 hrs
Synonym(s): *Hyen, Mjellum*
Stone. Olivine-bronzite chondrite (H).
A stone of 100.7g was found, but probably a shower of stones fell, J. Schetelig, letter of February 7, 1927 in Min. Dept. BM(NH). Olivine Fa_{19}, B. Mason, Geochimica et Cosmochimica Acta, 1963, **27**, p.1011.
100g Oslo, Min.-Geol. Mus., main mass; 1.1g New York, Amer. Mus. Nat. Hist.;

Mjellum v Mjelleim.

Moab v Cañon Diablo.

Moama 35°57′ S., 144°31′ E.
Womboota, New South Wales, Australia
Found 1940, October, probably fell some months before
Stone. Achondrite, Ca-rich. Eucrite (AEUC).
A well-preserved oriented stone, 3.416kg, was found in a grassy paddock on the farm, "Benarca Park" near Womboota, approximately 17 miles northwest of Moama, Meteor. Bull., 1975 (53), Meteoritics, 1975, **10**, p.140. Full description, with analyses; chemically and texturally similar to Serra de Magé, J.F. Lovering, Meteoritics, 1975, **10**, p.101. This is a coarse-grained unbrecciated eucrite, similar to Moore County, (R. Hutchison, personal observation).

Rare earth element abundances, Nd-Sm age, 4.58 Ga., J. Hamet et al., Geochimica et Cosmochimica Acta, 1978 (Suppl. 10), p.1115.
Main mass, Melbourne, Univ.;
Specimen(s): [1973,M.11], 25.18g. a crusted fragment, and fragments, 2.5g

Mocila v Dubrovnik.

Mociu v Mocs.

Mocs 46°48′ N., 24°2′ E.
Cluj (= Klausenburg, Kolozsvár), Transylvania, Romania
Fell 1882, February 3, 1600 hrs
Synonym(s): *Bare, Cluj, Gjilatelke, Gyulateke, Gyulatelke, Keszu, Klausemburg, Klausenburg, Kyolos, Marokháza, Mociu, Olah Gyéres, Palatka, Vajda-Kamaras, Visa*
Stone. Olivine-hypersthene chondrite (L6), veined.
After appearance of luminous meteor and detonations, a shower of stones fell; the number has been estimated at 3000 and the total weight at about 300kg, the largest stone weighing about 56kg, A. Koch, Sitzungsber. Akad. Wiss. Wien, Math.-naturwiss. Kl., 1882, **85** (1), p.116, A. Koch, Tschermaks Min. Petr. Mitt., 1883, **5**, p.234, G. von Niessl, Sitzungsber. Akad. Wiss. Wien, Math.-naturwiss. Kl., 1884, **89** (2), p.283. Analysis, 21.81 % total iron, B. Mason and H.B. Wiik, Am. Mus. Novit., 1961 (2069). Olivine Fa_{24}, B. Mason, Geochimica et Cosmochimica Acta, 1963, **27**, p.1011. 21.82 % total iron, V.Ya. Kharitonova, Метеоритика, 1965, **26**, p.146. Nature of the metal, H.C. Urey and T. Mayeda, Geochimica et Cosmochimica Acta, 1959, **17**, p.113. Trace element abundances, references, E. Rambaldi, Earth planet. Sci. Lett., 1977, **33**, p.407.
70kg Vienna, Naturhist. Mus.; 42.8kg Cluj, Mus., includes a stone of 35.7kg; 21kg Budapest, Nat. Mus.; 8.25kg Budapest, Eötvös Loránd Univ.; 6kg Chicago, Field Mus. Nat. Hist.; 7kg Washington, U.S. Nat. Mus.; 3.37kg Paris, Mus. d'Hist. Nat.; 1596g Copenhagen, Univ. Geol. Mus.; 1.3kg Harvard Univ.; 1kg Bonn, Univ. Mus.; 1kg New York, Amer. Mus. Nat. Hist.; 1kg Yale Univ.; 672g Edinburgh, Royal Scottish Mus.; 0.5kg Debreczen, Reform College; 615g Prague, Nat. Mus.;
Specimen(s): [54772], 8635g. complete stone from Oláhgyéres; [54773], 4630g. complete stone from Keszu; [54647], 826g. nearly complete stone from Báré; [54648], 276g. complete stone from Báré; [54651], 123.5g. from Báré; [54649], 81g. complete stone from Gyulatelke; [54775], 27.5g. from Vajdakamarás; [54774], 25g. from Palatka; [54776], 20.5g. from Marokháza; [54650], 17.5g. complete stone from Visa; [54777], 5.5g. from Visa; [1964,568], thin section

Moctezuma 29°48′ N., 109°40′ W.
Sonora, Mexico
Found 1889
Iron. Octahedrite, medium (1.3mm) (IA).
Listed, total known weight 1.7kg, H.H. and A.D. Nininger, The Nininger Collection of Meteorites, Winslow, Arizona, 1950, p.73. The main mass, which was in Mexico, appears to have been exchanged and distributed; description; classification, V.F. Buchwald, Iron Meteorites, Univ. of California, 1975, p.831. Analysis, 7.98 %Ni, 67.2 ppm.Ga, 244 ppm.Ge, 2.4 ppm.Ir, J.T. Wasson, Meteorites, Springer-Verlag, 1974, p.298.
362g Tempe, Arizona State Univ.; 357g Chicago, Field Mus. Nat. Hist.; 177g New York, Amer. Mus. Nat. Hist.; 77g Vienna, Naturhist. Mus.; 9g Mexico City, Inst. Geol.;
Specimen(s): [84882], 170g. slice

Moctezuma (of F. Berwerth) v Arispe.

Modena v Albareto.

Modoc no. 2 v Modoc (1948).

Modoc (1905) 38°30′ N., 101°6′ W.
Scott County, Kansas, U.S.A.
Fell 1905, September 2, 2130 hrs
Stone. Olivine-hypersthene chondrite (L6), veined.
After appearance of luminous meteor, followed by detonations, fifteen to twenty stones were found scattered over an area of about 2 miles by 7 miles; the largest stone weighed about 10.75lb and the total weight was about 35lb. Description, O.C. Farrington, Field Mus. Nat. Hist. Geol. Ser., 1907, **3** (6), p.120. Description, analysis, G.P. Merrill, Am. J. Sci., 1906, **21**, p.356. Description, with analysis. 22.42 % total iron, B. Mason and H.B. Wiik, Am. Mus. Novit., 1967 (2280). XRF analysis, 21.60 % total iron, H. von Michaelis et al., Earth planet. Sci. Lett., 1969, **5**, p.387. Olivine Fa_{23}, B. Mason, Geochimica et Cosmochimica Acta, 1963, **27**, p.1011. Sm-Nd age of metamorphism, 4050 m.y., N. Nakamura and M. Tatsumoto, Meteoritics, 1980, **15**, p.334, abs.
20kg New York, Amer. Mus. Nat. Hist., approx.; 5.5kg Washington, U.S. Nat. Mus.; 4kg Chicago, Field Mus. Nat. Hist., approx.; 1.6kg Vienna, Naturhist. Mus.; 776g Tempe, Arizona State Univ.; 439g Michigan, Univ.;
Specimen(s): [1908,260], 2004g. nearly complete stone

Modoc (1948) 38°30′ N., 101°6′ W.
Scott County, Kansas, U.S.A.
Found 1948, recognized in this year
Synonym(s): *Modoc no. 2*
Stone. Olivine-bronzite chondrite (H6).
A weathered stone of 1.8kg was found some years before 1948, H.H. and A.D. Nininger, The Nininger Collection of Meteorites, Winslow, Arizona, 1950, p.73. Olivine Fa_{19}, B. Mason, Geochimica et Cosmochimica Acta, 1967, **31**, p.1100.
850g Tempe, Arizona State Univ.;
Specimen(s): [1959,890], 741g.

Modok v Modoc.

Moenvalle v Toluca.

Moern v Mern.

Mojigasta
Pocho, Cordoba, Argentina
Found 1940, known in this year
Listed, without details, L.M. Villar, Cienc. Investig., 1968, **24**, p.302.

Mokhtuisk v Nochtuisk.

Mokoia 39°38′ S., 174°24′ E.
Taranaki, North Island, New Zealand
Fell 1908, November 26, 1230 hrs
Stone. Carbonaceous chondrite, type III (CV3).
After appearance of moving cloud and detonations, several stones were seen to fall, and two of about 5lb each were recovered. Description; partial analysis, G.R. Marriner, Trans. New Zealand Inst., 1909, **42**, p.176. Description, M.H. Briggs, Nature, 1961, **191**, p.1137 [M.A. 15-537]. Analysis, 24.04 % total iron, H.B. Wiik, Geochimica et

Cosmochimica Acta, 1956, **9**, p.279. Classification, W.R. Van Schmus and J.M. Hayes, Geochimica et Cosmochimica Acta, 1974, **38**, p.47. New analysis, 22.59 % total iron, T.S. McCarthy and L.H. Ahrens, Earth planet. Sci. Lett., 1972, **14**, p.97. Trace element data, G.W. Kallemeyn and J.T. Wasson, Geochimica et Cosmochimica Acta, 1981, **45**, p.1217.

Main masses, Wanganui, Mus.; 148g Tempe, Arizona State Univ.; 118g Paris, Mus. d'Hist. Nat.; 21g New York, Amer. Mus. Nat. Hist.;

Specimen(s): [1910,729], 170g. and fragments, 3g

Molina 38°7′ N., 1°10′ W.

Murcia, Spain

Fell 1858, December 24

Synonym(s): *Murcia*

Stone. Olivine-hypersthene chondrite (L5), brecciated.

A stone of 144kg fell. Description, and analysis, G.A. Daubrée, C. R. Acad. Sci. Paris, 1868, **66**, p.639, A.F. Gredilla, Meteoritos, 1892, p.108. Olivine Fa_{24}, B. Mason, Geochimica et Cosmochimica Acta, 1963, **27**, p.1011.

116.5kg Madrid, Mus. Cienc. Nat., main mass; 1.3kg Washington, U.S. Nat. Mus.; 173g Harvard Univ.; 149g Tempe, Arizona State Univ.;

Specimen(s): [1923,143], 395g.; [41111], 6g.; [1931,487], 7.7g.

Molong 33°17′ S., 148°53′ E.

County Ashburnham, New South Wales, Australia

Found 1912

Stony-iron. Pallasite (PAL).

A mass of about 230lb was found on Ti-Tree Creek, 12 miles NW. of Canoblas. Description and analysis, J.C.H. Mingaye, Rec. Geol. Surv. New South Wales, 1916, **9**, p.161. Olivine $Fa_{11.5}$, P.R. Buseck and J.I. Goldstein, Bull. Geol. Soc. Amer., 1969, **80**, p.2141. Analysis of metal, 9 %Ni, 22.8 ppm.Ga, 62.5 ppm.Ge, 0.19 ppm.Ir, E.R.D. Scott, Geochimica et Cosmochimica Acta, 1977, **41**, p.349. Texture, E.R.D. Scott, Geochimica et Cosmochimica Acta, 1977, **41**, p.693.

Main mass, Sydney, Austr. Mus.; 523g Chicago, Field Mus. Nat. Hist.; 1kg Sydney, Technol. Mus.; 864g Washington, U.S. Nat. Mus.; 488g Fort Worth, Texas, Monnig Colln.; 474g Paris, Mus. d'Hist. Nat.; 411g New York, Amer. Mus. Nat. Hist.; 160g Los Angeles, Univ. of California;

Specimen(s): [1915,144], 1520g. three pieces, and fragments, 454g

Molteno 31°15′ S., 26°28′ E.

Cape Province, South Africa

Fell 1953, April or May, 1700 hrs, approx

Stone. Achondrite, Ca-rich. Howardite (AHOW).

After detonations, a fragment of about 5oz was seen to fall, coming in from due S., about 13 miles NNE. of Molteno (31°24'S., 26°22'E.); the fragment shows both primary and secondary crust. Description and analysis, M.J. Frost, Min. Mag., 1971, **38**, p.89. Olivine $Fa_{59\text{-}70}$, C. Desnoyers and D.Y. Jerome, Meteoritics, 1973, **8**, p.344, abs.

62g Cape Town, Univ.;

Specimen(s): [1966,287], 63g. half the stone and fragment, 0.5g

Molunat v Dubrovnik.

Monahans 31°29′ N., 102°53′ W.

Ward County, Texas, U.S.A.

Found 1938

Iron. Octahedrite, plessitic (IIF).

One mass of 27.9kg was found 7 miles south and 1 mile east of Monahans, (31°50'N., 102°55'W.), A.D. Nininger, Pop. Astron., Northfield, Minnesota, 1939, **47**, p.212, E.P. Henderson, letter of June 3, 1939 in Min. Dept. BM(NH). Description, with analysis, 10.88 %Ni, H.H. Nininger, Pop. Astron., Northfield, Minnesota, 1939, **47**, p.268 [M.A. 7-376]. Description and analysis of the iron shale, 4.99 %NiO, J.D. Buddhue, Pop. Astron., Northfield, Minnesota, 1939, **47**, p.271 [M.A. 7-377]. Structural description, H.J. Axon and P.L. Smith, Min. Mag., 1972, **38**, p.736. Classification and analysis, 10.6 %Ni, 8.9 ppm.Ga, 127 ppm.Ge, 14 ppm.Ir, J.T. Wasson, Geochimica et Cosmochimica Acta, 1969, **33**, p.859. Co-ordinates, description, V.F. Buchwald, Iron Meteorites, Univ. of California, 1975, p.832.

4.6kg Tempe, Arizona State Univ.; 901g Fort Worth, Texas, Monnig Colln.; 873g Washington, U.S. Nat. Mus.; 844g Chicago, Field Mus. Nat. Hist.; 811g Harvard Univ.;

Specimen(s): [1959,910], 5325g. and fragments, 217g and sawings, 168g

Monclar-d'Agenais v Agen.

Monjolo Farm v Ibitira.

Monmouth County v Deal.

Monroe 35°15′ N., 80°30′ W.

Cabarrus County, North Carolina, U.S.A.

Fell 1849, October 31, 1500 hrs

Synonym(s): *Cabarras County, Cabarrus County, Charlotte, Flows, Mecklenburg County*

Stone. Olivine-bronzite chondrite (H4), brecciated.

After detonations, a stone of about 19lb was found near the post-office at Flows, 22 miles east of Charlotte and 18 miles from Monroe, J.H. Gibbon, Am. J. Sci., 1850, **9**, p.143, C.U. Shepard, Am. J. Sci., 1850, **10**, p.127. Description, with a doubtful analysis (no alkalies quoted), G.P. Merrill, Mem. Nat. Acad. Sci. Washington, 1916, **14**, p.20. Mineralogy, analysis, olivine $Fa_{17.9}$, 28.08 % total iron, R. Hutchison et al., Proc. Roy. Soc. London, 1981, **A374**, p.159.

3.9kg Tempe, Arizona State Univ.; 168g Harvard Univ.; 252g Washington, U.S. Nat. Mus.; 232g Yale Univ.; 156g Vienna, Naturhist. Mus.; 127g Berlin, Humboldt Univ.; 115g New York, Amer. Mus. Nat. Hist.; 113g Chicago, Field Mus. Nat. Hist.; 64g Budapest, Nat. Mus.; 99g Copenhagen, Univ.; 39g Paris, Mus. d'Hist. Nat.; 2g Madrid, Mus. Cienc. Nat.;

Specimen(s): [25462], 386g.; [33737], 0.5g.

Monroe County v Forsyth.

Mons 50°27′ N., 3°57′ E.

Belgium

Fell 1186, June 30, or 1187

Doubtful..

Several stones, up to 1lb in weight, are said to have fallen, G. von Boguslawski, Ann. Phys. Chem. (Poggendorff), 1854, **4** (suppl.), p.9 but the evidence is not conclusive.

Montauban v Orgueil.

Monte Alto 14°12′ S., 43°14′ W.
Bahia, Brazil
Found 1888, known in this year
Synonym(s): *Morre Alto*
Doubtful. Iron.
Mentioned, O.A. Derby, Revista do Observatorio, Rio de Janeiro, 1888, p.20. Doubtful, E. de Oliveira, Anais Acad. Brasil. Cienc., 1931, **3**, p.33 [M.A. 5-15].

Monte Colina 29°24′ S., 139°59′ E.
South Australia, Australia
Found 1963
Synonym(s): *Monte Collina*
Stone. Olivine-hypersthene chondrite (L3).
A mass of about 120g was found; very little information about the discovery is known, D.W.P. Corbett, Rec. S. Austr. Mus., 1968, **15**, p.776, B. Mason and H.B. Wiik, Am. Mus. Novit., 1966 (2273). Unequilibrated, L3, and probably part of the Carraweena shower (*q.v.*), B. Mason, Rec. Austr. Mus., 1974, **29**, p.169. Analysis, 19.07 % total iron, M.J. Fitzgerald, Ph.D. Thesis, Univ. of Adelaide, 1979, p.23.
84g Adelaide, Geol. Surv. South Austr.; 10g Adelaide, South Austr. Mus.; 5.1g New York, Amer. Mus. Nat. Hist.;

Monte Collina v Monte Colina.

Monte das Fortes 38°1′ N., 8°15′ W.
Alemtejo province, Portugal
Fell 1950, August 23, late afternoon hrs
Stone. Olivine-hypersthene chondrite (L5).
Five stones were collected along a 2km SW.-NE. line, weights 1834, 1350, 860, 818, and 23 grams. Description and analysis, 21.04 % total iron, E. Jérémine, Comm. Serv. Geol. Portugal, 1954, **35**, p.5, 24 [M.A. 12-609]. Olivine Fa_{24}, B. Mason, Geochimica et Cosmochimica Acta, 1963, **27**, p.1011.
183g Paris, Mus. d'Hist. Nat.;

Monte Milone 43°8′ N., 13°34′ E.
Macerata, Marche, Italy
Fell 1846, May 8, 0915 hrs
Synonym(s): *Fiume Potenza, Macerata, Pollenza, Potenza river*
Stone. Olivine-hypersthene chondrite (L5), brecciated.
After detonations, many stones fell (some in the river Potenza) 8 miles from Macerata; of five stones recovered, the largest weighed about 3kg, L'Institut, Paris, 1846, **14**, p.340, B. Baldanza, Min. Mag., 1965, **35**, p.214. Analysis, olivine $Fa_{25.5}$, 23.36 % total iron, A. Maras et al., Meteoritics, 1979, **14**, p.482, abs. Co-ordinates, G.R. Levi-Donati, priv. comm. 30 April, 1969 in Min. Dept. BM(NH).; X-ray diffraction study, F. Burragato, Periodico Miner., 1967, **36**, p.463 [M.A. 19-210].
2kg Rome,; 326g Milan,; 280g Florence,; 110g Vienna, Naturhist. Mus.; 37g Tempe, Arizona State Univ.; 11g Chicago, Field Mus. Nat. Hist.;
Specimen(s): [33962], 8g.

Montferre 43°23′26″ N., 1°57′45″ E.
Aude, France
Fell 1923, probably summer, night
Stone. Olivine-hypersthene chondrite (L).
During a summer night, 1923, a meteorite was seen to fall near Montferre, but was not located. Years later, a plough struck the mass, which was dug up in 1966. It weighed 149kg, Meteor. Bull., 1972 (51), Meteoritics, 1972, **7**, p.227. The main mass is with finder, Y. Gillet, letter of 13 November, 1972 in Min. Dept. BM(NH).
Specimen(s): [1973,M.13], 25g. a crusted fragment

Montignac v Marmande.

Montlivault 47°38′ N., 1°35′ E.
Loir-et-Cher, France
Fell 1838, July 22, Daytime hrs
Stone. Olivine-hypersthene chondrite (L6).
A stone of about 0.5kg was seen to fall, after detonations, in the valley Cul-de-Four, on the left bank of the Loire, G.A. Daubrée, C. R. Acad. Sci. Paris, 1873, **76**, p.314. Olivine Fa_{24}, B. Mason, Geochimica et Cosmochimica Acta, 1963, **27**, p.1011.
430g Paris, Mus. d'Hist. Nat.; 8g Vienna, Naturhist. Mus.; 4g Chicago, Field Mus. Nat. Hist.;
Specimen(s): [71571], 11g. a slice

Montpelagry v Grazac.

Montrejeau v Ausson.

Montrose 39°4′ N., 79°48′ W.
Randolph County, West Virginia, U.S.A.
Found 1940, before this year
Doubtful. Stone. 'Amathosite'.
A small mass, 5 × 6 × 9cms, of sandstone with a greenish glassy crust is regarded as possibly meteoritic, F.C. Cross, Pop. Astron., Northfield, Minnesota, 1947, **55**, p.96, F.C. Cross, Contr. Meteoritical Soc., 1947, **4**, p.11.

Monturaqui 23°56′ S., 68°17′ W.
Antofagasta, Chile
Found 1965
Iron. Octahedrite, coarse (2.0mm) (I).
A few oxidized fragments, totalling less than 2kg, were found, V.F. Buchwald and P. Graff-Petersen, Analecta geol., 1976 (4), p.49. Details of recovery of material and associated crater, V.F. Buchwald, Iron Meteorites, Univ. of California, 1975, p.1403.
321g Copenhagen, Univ. Geol. Mus.;

Monument Rock v Cañon Diablo.

Monze 15°58′ S., 27°21′ E.
Southern Province, Zambia
Fell 1950, October 5, 0410 hrs
Stone. Olivine-hypersthene chondrite (L6).
A shower of stones fell over a considerable area including the villages of Chizuni (15°58′S., 27°21′E.) and Chiteba, H.J. Lambert, Colonial Geol. Min. Res., London, 1950, **1**, p.354 [M.A. 11-442]. Description, W. Campbell Smith and M.H. Hey, Colonial Geol. Min. Res., London, 1952, **3**, p.52, H.J. Lambert, Northern Rhodesia J., 1951 (4), p.55 [M.A. 12-103]. Olivine Fa_{25}, B. Mason, Geochimica et Cosmochimica Acta, 1963, **27**, p.1011. Analysis, 21.6 % total iron, H. von Michaelis et al., Earth planet. Sci. Lett., 1969, **5**, p.387. Collections of stones are at the Rhodes-Livingstone Museum, Livingstone, Zambia and at the National Museum of Zimbabwe.
208g Tempe, Arizona State Univ.; 265g Washington, U.S. Nat. Mus.; 153g Chicago, Field Mus. Nat. Hist.; 127g Tübingen, Univ.;
Specimen(s): [1951,33-39, 1951,41-71, 1951,73-75, 1951,77, 1951,79-103], 1987g. 64 mostly complete stones, ranging from 67.7g to 7.5g, and fragments; [1953,136-153], 836.3g. 18 stones; [1951,100], thin section; [1951,101], thin sections (two)

Mooltan v Khairpur.

Mooltan v Lodran.

Moonbi 30°55′ S., 151°17′ E.

Moonbi Range, County Inglis, New South Wales, Australia
Found 1892
Iron. Octahedrite, medium (0.55mm) (IIIF).
A mass of 29lb was found about 18 miles from Moonbi, G.W. Card, Rec. Geol. Surv. New South Wales, 1897, **5**, p.49. Description and analysis, 7.87 %Ni, J.C.H. Mingaye, J. and Proc. Roy. Soc. New South Wales, 1893, **27**, p.82. Structural classification and analysis, 7.70 %Ni, 6.84 ppm.Ga, 0.826 ppm.Ge, 1.3 ppm.Ir, R. Schaudy et al., Icarus, 1972, **17**, p.174. Chemical classification, J.T. Wasson, Meteorites, Springer-Verlag, 1974, p.305. The whole mass was heated in a forge to 800-900C., V.F. Buchwald, Iron Meteorites, Univ. of California, 1975, p.834.
6.74kg Sydney, Austr. Mus., main mass, 5.85kg; 286g Washington, U.S. Nat. Mus.;
Specimen(s): [1925,509], 400g. slice; [1927,1292], 191g. fragments, 19.5g, and turnings, 3g

Moorabbie v Moorabie.

Moorabie 30°6′ S., 141°4′ E.

New South Wales, Australia
Found 1965, before this year
Synonym(s): *Moorabbie, Quinyambie*
Stone. Olivine-hypersthene chondrite (L3).
One mass of 14.04kg (31 pounds) was found about 130 miles north of Broken Hill and close to Boolka (*q.v.*), D.H. McColl, letters of 25 January and 6 June, 1972 in Min. Dept. BM(NH). Reported, Meteor. Bull., 1975 (53), Meteoritics, 1975, **10**, p.152. Classification, XRF analysis, 22.68 % total iron, M.J. Fitzgerald, Ph.D. Thesis, Univ. of Adelaide, 1979, p.23.
Main mass, in possession of finder, L. Russell; 1.6kg Adelaide, South Austr. Mus., of Quinyambie;

Mooradabad v Moradabad.

Mooranoppin v Youndegin.

Moore County 35°25′ N., 79°23′ W.

North Carolina, U.S.A.
Fell 1913, April 21, 1700 hrs
Synonym(s): *Graves*
Stone. Achondrite, Ca-rich. Eucrite (AEUC).
A mass of 4lb 2oz fell 3 miles east of Carthage. Description, with analysis and analyses of feldspar, pyroxene and crust, E.P. Henderson and H.T. Davis, Am. Miner., 1936, **21**, p.215 [M.A. 6-396]. Petrographic notes and fabric analysis; a parent body as large as the earth is suggested, H.H. Hess and E.P. Henderson, Am. Miner., 1949, **34**, p.494 [M.A. 10-402; 11-140]. Ba and rare-earth contents, C.C. Schnetzler and J.A. Philpotts, Meteorite Research, ed. P.M. Millman, D. Reidel, Dordrecht-Holland, 1969, p.206. Strontium isotopic ratio, D.A. Papanastassiou and G.J. Wasserburg, Earth planet. Sci. Lett., 1969, **5**, p.361. Petrography, M.B. Duke and L.T. Silver, Geochimica et Cosmochimica Acta, 1967, **31**, p.1637. Contains nuclear tracks, E.A. Carver and E. Anders, Geochimica et Cosmochimica Acta, 1976, **40**, p.935.
2lb Raleigh, North Carolina State Mus.; 460g Washington, U.S. Nat. Mus.; 35g Tempe, Arizona State Univ.;
Specimen(s): [1959,1050], 78g. slice, and fragments, 15g

Mooresfort 52°27′ N., 8°20′ W.

County Tipperary, Ireland
Fell 1810, August, 1200 hrs
Synonym(s): *Tipperary*
Stone. Olivine-bronzite chondrite (H5), xenolithic.
After appearance of moving cloud and sounds like thunder, a stone of 7.75lb was seen to fall, W. Higgins and M.C. Moore, Phil. Mag., 1811, **38**, p.262. Olivine Fa_{19}, B. Mason, Geochimica et Cosmochimica Acta, 1963, **27**, p.1011. Description, analysis, 25.49 % total iron, B. Mason and H.B. Wiik, Am. Mus. Novit., 1966 (2273). Xenolithic, R.A. Binns, Geochimica et Cosmochimica Acta, 1968, **32**, p.299.
1.25kg Dublin, Nat. Mus.; 328g Vienna, Naturhist. Mus.; 316g Tübingen, Univ.; 161g Washington, U.S. Nat. Mus.;
Specimen(s): [19968], 174g.; [61309], 47g.; [90256], 19.5g.

Moorland v Morland.

Moorleah 40°58.5′ S., 145°36′ E.

Wynyard, Tasmania, Australia
Fell 1930, October, 1830 hrs
Stone. Olivine-hypersthene chondrite (L6).
A stone of 8887.5g fell. Description, with analysis, T. Hodge-Smith and R.O. Chalmers, Rec. Queen Vict. Mus., Launceston, Tasmania, 1942, **1**, p.13 [M.A. 8-377]. Olivine $Fa_{24.4}$, B. Mason, Rec. Austr. Mus., 1974, **29**, p.169.
Main mass, Launceston, Queen Vict. Mus.; 0.25lb Sydney, Australian Mus.; 1.4g New York, Amer. Mus. Nat. Hist.; Section, Washington, U.S. Nat. Mus.;

Moorumbunna 28°55′ S., 136°15′ E.

Anna Creek, South Australia, Australia
Found 1943
Iron. Octahedrite, medium (0.9mm) (IIIA).
A much-pitted mass weighing 169lb 14oz was found on Moorumbunna paddock of Anna Creek sheep station, west of Lake Eyre. Description, with an analysis, 8.82 %Ni, A.B. Edwards, Trans. Roy. Soc. South Austr., 1946, **70**, p.348 [M.A. 10-404]. Analysis, 8.98 %Ni, 21.7 ppm.Ga, 44 ppm.Ge, 0.26 ppm.Ir, E.R.D. Scott et al., Geochimica et Cosmochimica Acta, 1973, **37**, p.1957.
70kg Adelaide, Univ., main mass; 580g Adelaide, South Aust. Mus.; 53g Chicago, Field Mus. Nat. Hist.; 45g Los Angeles, Univ. of California;
Specimen(s): [1966,490], 491g. slice

Moradabad 28°47′ N., 78°50′ E. approx.

Moradabad district, Uttar Pradesh, India
Fell 1808
Synonym(s): *Mooradabad*
Stone. Olivine-hypersthene chondrite (L6).
The original weight and details of the fall are unknown; only 70g of fragments were preserved in the collection of the Asiatic Society at Calcutta, J. Asiatic Soc. Bengal, 1859, **28**, p.259, N.S. Maskelyne, Phil. Mag., 1863, **25**, p.449. Exactly where the stone fell is not known, only that it fell on Moradabad district, C.A. Silberrad, Min. Mag., 1932, **23**, p.300. Olivine Fa_{24}, B. Mason, Geochimica et Cosmochimica Acta, 1963, **27**, p.1011.
27g Calcutta, Mus. Geol. Surv. India; 3.4g Budapest, Nat. Mus.; 1g Berlin, Humboldt Univ.; 1g Chicago, Field Mus. Nat. Hist.; 1g Vienna, Naturhist. Mus.;
Specimen(s): [33758], 17g.

Moradabad District v Bahjoi.

Morandi v Vigarano.

Morasko 52°28′ N., 16°54′ E.
Poznan , Poland
Found 1914
Iron. Octahedrite, coarse (2.5mm) (IA).
A mass of 77.5kg was found in 1914, F. Slavik and L.J. Spencer, Min. Mag., 1928, **21**, p.478, A. Laszkiewicz, Mineralogia, Warszawa, 1936, p.195 [M.A. 6-385]. At least 15 more masses, totalling approx. 211kg have since been found, analysis, 6.65 %Ni, J. Pokrzywnicki, Urania, 1955, **26**, p.165, J. Pokrzywnicki, Urania, 1957, **28**, p.232, J. Pokrzywnicki, Acta Geol. Polon., 1955, **3**, p.427, J. Pokrzywnicki, Acta Geophys. Polon., 1956, **4**, p.21, J. Pokrzywnicki, Метеоритика, 1958 (16), p.123 [M.A. 13-79; 14-126]. Eight small craters are associated with the find-sites of several large irons, J. Pokrzywnicki, Studia Geol. Polon., 1964, **15**, p.49. Classification and analysis, paired with Seeläsgen (*q.v.*), 6.56 %Ni, 98.9 ppm.Ga, 496 ppm.Ge, 1.0 ppm.Ir, A. Kracher et al., Geochimica et Cosmochimica Acta, 1980, **44**, p.773. Description; some atmospheric reheating; may be a true crater-forming meteorite, V.F. Buchwald, Iron Meteorites, Univ. of California, 1975, p.836.
72.2kg Poznan , four masses, 61, 4.2, 3.5, and 3.5kg; 88kg Warsaw, Acad. Sci.; 6.4kg Sukhylas, School; 842g Moscow, Acad. Sci.;
Specimen(s): [1963,792], 118g.

Morbihan v Kernouve.

Mordan v Morden.

Morden 30°30′ S., 142°20′ E.
New South Wales, Australia
Found 1922
Synonym(s): *Mordan*
Iron. Hexahedrite? (IA?).
See Cat. Meteorites S. Australian Mus., D.W.P. Corbett, Rec. S. Austr. Mus., 1968, **15**, p.767, G.W. Card, Ms. list of Australian meteorites in Min. Dept. BM(NH). Mentioned, A.R. Alderman, Rec. S. Austr. Mus., 1936, **5**, p.537. Analysis, 6.6 %Ni, 81 ppm.Ga, 329 ppm.Ge, S.J.B. Reed, Meteoritics, 1972, **7**, p.257. The analysis and classification do not seem to be compatible with the structural classification of D.W.P. Corbett (loc. cit.).
5.75lb Adelaide, South Australian Mus.;

Mordvinovka v Pavlograd.

Moreira do Lima v São Julião de Moreira.

Morelos v Chupaderos.

Morelos v Toluca.

Morgan County v Walker County.

Moriarty 34°59′ N., 106°3′ W.
Torrance County, New Mexico, U.S.A.
Found 1975
Stone. Olivine-hypersthene chondrite (L).
A single mass of 330g was found, olivine Fa_{26}, Meteor. Bull., 1981 (59), Meteoritics, 1981, **16**, p.196.

Morito 27°3′ N., 105°26′ W.
Chihuahua, Mexico
Found 1600, known in this year
Synonym(s): *El Morito, Huejuquilla, Humboldt Iron, Parral, San Gregorio, Valle de Allende*
Iron. Octahedrite, medium (1.0mm) (IIIA).
An enormous mass of about 11,000kg (11 tons) was known to the Indians before 1600. History, L. Fletcher, Min. Mag., 1890, **9**, p.124, O.C. Farrington, Mem. Nat. Acad. Sci. Washington, 1915, **13**, p.312. Part of the material has been heated or forged, V.F. Buchwald, Iron Meteorites, Univ. of California, 1975, p.838. Partial analysis, 7.68 %Ni, E. Goldberg et al., Geochimica et Cosmochimica Acta, 1951, **2**, p.1. These authors give the total weight as 11kg, which is incorrect. Further references, E.P. Henderson and S.H. Perry, Meteoritics, 1956, **1**, p.477 [M.A. 13-362], J.C. Haro, Bol. Inst. Geol. Mexico, 1931 (50), p.26, H.H. Nininger, Bol. Inst. Geol. Mexico, 1931 (50), p.84. Distinct from Durango; analysis, 7.38 %Ni, 18.7 ppm.Ga, 35.8 ppm.Ge, 9.2 ppm.Ir, E.R.D. Scott et al., Geochimica et Cosmochimica Acta, 1973, **37**, p.1957.
Main mass, Mexico City, School of Mines; 8.2kg Yale Univ.; 782g Berlin, Humboldt Univ.; 714g Washington, U.S. Nat. Mus.; 580g Tempe, Arizona State Univ.;
Specimen(s): [1959,957], 491g. slice

Morland 39°20′ N., 100°4′ W.
Graham County, Kansas, U.S.A.
Found 1890, approx.
Synonym(s): *Graham County, Moorland*
Stone. Olivine-bronzite chondrite (H6).
A stone of 623lb was found in 1935, 20in. below the surface, H.H. Nininger, J. Geol., 1936, **44**, p.66 [M.A. 6-390]. Description, analysis, H.H. Nininger and F.G. Hawley, Trans. Kansas Acad. Sci., 1936, **39**, p.181. Another stone had been found about 1890 but had broken up through weathering, and when it was recognized as meteoritic in 1935, only six fragments, weighing 3lb, could be recovered, olivine Fa_{19}, B. Mason, Geochimica et Cosmochimica Acta, 1963, **27**, p.1011.
5kg Tempe, Arizona State Univ.; 988g Chicago, Field Mus. Nat. Hist.; 700g Mainz, Max-Planck-Inst.; 600g Albuquerque, Univ. of New Mexico; 298g Washington, U.S. Nat. Mus.; 277g New York, Amer. Mus. Nat. Hist.; 210g Los Angeles, Univ. of California;
Specimen(s): [1937,386], 23.2g. two pieces; [1959,1043], 285.39kg. main mass

Mornans 44°36′ N., 5°8′ W.
Bourdeaux, Drome, France
Fell 1875, September
Synonym(s): *Bourdeaux*
Stone. Olivine-bronzite chondrite (H5), veined.
A stone of about 1.3kg fell, after detonations, J.R. Gregory, Geol. Mag., 1887, **4**, p.553. Olivine Fa_{19}, B. Mason, Geochimica et Cosmochimica Acta, 1963, **27**, p.1011.
32g Washington, U.S. Nat. Mus.; 18g New York, Amer. Mus. Nat. Hist.; 17g Vienna, Naturhist. Mus.;
Specimen(s): [63551], 1015g. the main mass of the stone, 973g, and pieces, 42g

Morradal 62° N., 7°40′ E.
Grjotli, between Skiaker and Stryn, Norway
Found 1892
Synonym(s): *Marradal*
Iron. Ataxite, Ni-rich (IRANOM).
A mass of 2.75kg was found. Description, with an analysis, 18.77 %Ni, E. Cohen, Meteoritenkunde, 1905, **3**, p.122.

Chemically anomalous; analysis, 19.54 %Ni, 46.3 ppm.Ga, 119 ppm.Ge, 0.61 ppm.Ir, J.T. Wasson, Geochimica et Cosmochimica Acta, 1969, **33**, p.859. Description of anomalous structure, H.J. Axon and P.L. Smith, Min. Mag., 1972, **38**, p.736.

2229g Oslo, Univ., main mass; 58g Washington, U.S. Nat. Mus.; 22g Chicago, Field Mus. Nat. Hist.; 22g Vienna, Naturhist. Mus.;

Specimen(s): [1924,146], 92.5g.

Morre Alto v Monte Alto.

Morrill 42°11′ N., 103°56′ W.

Sioux County, Nebraska, U.S.A.
Found 1920, recognized in 1933
Iron. Octahedrite, medium (0.9mm) (IA-ANOM).

A mass of 1387g was found 16 miles north of Morrill, H.H. Nininger, The Mines Mag., Golden, Colorado, 1933, **23** (8), p.9 [M.A. 5-405], A.D. Nininger, Pop. Astron., Northfield, Minnesota, 1937, **45**, p.449 [M.A. 7-62]. Analysis, 8.38 %Ni, 58 ppm.Ga, 296 ppm.Ge, 1.7 ppm.Ir, J.T. Wasson, Icarus, 1970, **12**, p.407. Description; dispersal of graphite crystals, V.F. Buchwald, Iron Meteorites, Univ. of California, 1975, p.845.

246g Tempe, Arizona State Univ.; 87g Washington, U.S. Nat. Mus.;

Specimen(s): [1959,958], 232.5g. slice, and sawings, 24.5g

Morristown 36°12′ N., 83°23′ W.

Hamblen County, Tennessee, U.S.A.
Found 1887
Synonym(s): *East Tennessee, Hamblen County, Safford Meteorite*
Stony-iron. Mesosiderite (MES).

Several masses, weighing together about 36lb, were found about 6 miles WSW. of Morristown. Description and analysis, L.G. Eakins, Am. J. Sci., 1893, **46**, p.283, 482. Description, G.P. Merrill, Am. J. Sci., 1896, **5**, p.149, G.T. Prior, Min. Mag., 1918, **18**, p.169. Petrography, mineral chemistry, B.N. Powell, Geochimica et Cosmochimica Acta, 1969, **33**, p.789, B.N. Powell, Geochimica et Cosmochimica Acta, 1971, **35**, p.5. Analysis of metal, 7.82 %Ni, 13.2 ppm.Ga, 57 ppm.Ge, 3.7 ppm.Ir, J.T. Wasson et al., Geochimica et Cosmochimica Acta, 1974, **38**, p.135. Olivine coronas, classification, references, C.E. Nehru et al., Geochimica et Cosmochimica Acta, 1980, **44**, p.1103.

3.7kg Chicago, Field Mus. Nat. Hist.; 1.8kg Washington, U.S. Nat. Mus.; 1.9kg Budapest, Nat. Mus.; 1.3kg Rome, Vatican Colln; 444g Vienna, Naturhist. Mus.; 396g Paris, Mus. d'Hist. Nat.; 448g Harvard Univ.; 323g New York, Amer. Mus. Nat. Hist.; 164g Berlin, Humboldt Univ.; 167g Tübingen, Univ.; 124g Tempe, Arizona State Univ.; 114g Yale Univ.;

Specimen(s): [81008], 415.5g.; [77095], 142g.

Morro Cavado v Patrimonio.

Morro do Rocio v Santa Catharina.

Morro do Rocio 27° S., 51° W. approx.

Santa Catharina, Brazil
Found, year not reported
Stone. Olivine-bronzite chondrite (H5).

Listed, D. Guimarães, Ms. list of Brazilian meteorites in Min. Dept. BM(NH). Classification, olivine $Fa_{17.8}$, F. Wlotzka and E. Jarosewich, Meteoritics, 1980, **15**, p.387, abs.

359g Belgrade, Mus. Geol. Faculty, 3 specimens; 9.5g Mainz, Max-Planck-Inst.;

Specimen(s): [1928,70] thin section

Morton 33°43′ N., 102°46′ W.

Cochran County, Texas, U.S.A.
Found 1980
Stone. Olivine-bronzite chondrite (H6).

Listed, without details, Cat. Dupont Colln., 1981.

55g Watchung, Dupont Colln.;

Specimen(s): [1982,M.9], 24g.

Morven 44°49′ S., 171°8′ E.

South Canterbury, New Zealand
Found 1925
Stone. Olivine-bronzite chondrite (H5), veined.

A mass of 7100g was found on a farm 4.5 miles south of Morven railway station. Description, with analysis, C.O. Hutton, Min. Mag., 1936, **24**, p.265. Olivine Fa_{18}, B. Mason, Geochimica et Cosmochimica Acta, 1963, **27**, p.1011.

6753g Otago, Univ. Mus., main mass; 112g Washington, U.S. Nat. Mus.; 31g Mainz, Max-Planck-Inst.;

Specimen(s): [1936,93], 228.5g. and fragments, 3.5g

Mosca 37°38′ N., 105°50′ W.

Alamosa County, Colorado, U.S.A.
Found 1942
Stone. Olivine-hypersthene chondrite (L6).

A stone of about 6123g was found 20 miles E. and 2 miles N. of Mosca, C. Wilton, letters of 11 March, 4 April and 9 June, 1963 in Min. Dept. BM(NH). Olivine Fa_{24}, B. Mason, Geochimica et Cosmochimica Acta, 1967, **31**, p.1100.

6136g Washington, U.S. Nat. Mus.;

Moshesh 30°6′ S., 28°43′ E.

Matatiele District, Cape Province, South Africa
Found
Stone. Olivine-bronzite chondrite (H).

A stone weighing 180-225kg was found at Moshesh Location, near Queen's Mercy, of which 5kg are in Natal Museum, Pietermaritzburg. This may be part of the Queen's Mercy fall, C. Frick and E.C.I. Hammerbeck, Bull. Geol. Surv. S. Africa, 1973 (57), p.27. Analysis, 25.19 % total iron, H. von Michaelis et al., Earth planet. Sci. Lett., 1969, **5**, p.387.

5kg Pietermaritzburg, Natal Mus.;

Mosquero 35°45′ N., 103°56′ W.

Harding County, New Mexico, U.S.A.
Found
Stone. Olivine-bronzite chondrite (H4).

A meteorite from this locality is mentioned, C. Wilton, letters of 11 March, 4 April and 9 June, 1963 in Min. Dept. BM(NH). Olivine Fa_{20}, B. Mason, Geochimica et Cosmochimica Acta, 1967, **31**, p.1100.

1586g Washington, U.S. Nat. Mus., main mass; 203g Albuquerque, Univ. of New Mexico;

Mossgiel 33°19′ S., 144°47′ E.

New South Wales, Australia
Found 1967
Stone. Olivine-hypersthene chondrite (L4).

At least 23 fragments weighing over 32kg were found, Meteor. Bull., 1970 (49), Meteoritics, 1970, **5**, p.174. Olivine Fa_{24}, B. Mason, Rec. Austr. Mus., 1974, **29**, p.169. Description, with an analysis, 18.38 % total iron, M.J. Fitzgerald, Trans. Roy. Soc. South Austr., 1979, **103**, p.145. Main mass removed to Canada, in possession of K.D. Collerson.

39g Washington, U.S. Nat. Mus.;
Specimen(s): [1970,29], 33.3g. fragments

Motecka-Nugla v Moti-ka-nagla.

Moteeka-Nugla v Moti-ka-nagla.

Moti-ka-nagla 26°50′ N., 77°20′ E.
Goorda, Biana district, Bharatpur, Rajasthan, India
Fell 1868, December 22, 1700 hrs
Synonym(s): *Bhurtpur, Ghoordha, Gúrdh, Motecka-Nugla, Moteeka-Nugla*
Stone. Olivine-bronzite chondrite (H6).
After the appearance of a luminous meteor (passing from NE. to SW.), and detonations, a shower of stones fell, but only three were found; the largest fragment weighing about 3.25lb was preserved in the Indian Museum, Calcutta, F. Fedden, Cat. Meteor. Indian Mus. Calcutta, 1880, p.26. Mentioned, Political Agent, Bharatpur, extracts from letters of 6 January, 1869 in Min. Dept. BM(NH)., A.C.L. Carlyle, letter of 23 November, 1969 in Min. Dept. BM(NH). Olivine Fa_{19}, B. Mason, Geochimica et Cosmochimica Acta, 1963, **27**, p.1011.
1.1kg Calcutta, Mus. Geol. Surv. India; 121g Paris, Mus. d'Hist. Nat.; 77g Washington, U.S. Nat. Mus.; 75g Vienna, Naturhist. Mus.; 54g New York, Amer. Mus. Nat. Hist.;
Specimen(s): [43332], 389g. and fragments 15.1g

Motpena 31°6′ S., 138°16′ E.
South Australia, Australia
Found 1968
Synonym(s): *Parachilna*
Stone. Olivine-hypersthene chondrite (L6).
A mass of 8.81kg (19lb 17oz) was found about 6 miles WNW. of Parachilna on Lake Torrens Plain, J.M. Scrymgour, letter of 14 July, 1971 in Min. Dept. BM(NH)., Meteor. Bull., 1975 (53), Meteoritics, 1975, **10**, p.153. Classification, olivine $Fa_{25.8}$, B. Mason, Rec. Austr. Mus., 1974, **29**, p.169. Description and figures, S.J.B. Reed and D.H. McColl, Meteoritics, 1975, **10**, p.227. XRF analysis, 21.02 % total iron, M.J. Fitzgerald, Ph.D. Thesis, Univ. of Adelaide, 1979, p.23.
4.71kg Canberra, Austr. Nat. Univ., main mass; 1.36kg Adelaide, South Austr. Mus.;
Specimen(s): [1972,231], 443g. slice

Motta de' Conti v Motta di Conti.

Motta di Conti 45°12′ N., 8°30′ E.
Vercelli, Piemonte, Italy
Fell 1868, February 29, 1100 hrs
Synonym(s): *Casale, Casale Piemonte, Motta de'Conti, Roggia Marcova, Roletta, Piedmont, Villanova, Villanova di Casale, Villanuova, Villanuova di Casale, Villeneuve*
Stone. Olivine-bronzite chondrite (H4).
After appearance of a luminous meteor (moving from NW. to SE.), and detonations, several stones fell between Villanuova and Motta di Conti: one of 1.9kg was seen to fall about 0.25 mile SE. of Villanuova, another of 6.75kg 0.75 mile distant, and a third of about 0.5kg, which broke into fragments at Motta di Conti. Description, P.F. Denza, C. R. Acad. Sci. Paris, 1868, **67**, p.322, B. Baldanza, Min. Mag., 1965, **35**, p.214. Analysis, olivine $Fa_{19.6}$, 28.6 % total iron, G.R. Levi-Donati et al., Meteoritics, 1979, **14**, p.475, abs.. The correct spelling of this locality is Motta de' Conti; amended co-ordinates, G.R. Levi-Donati, priv. comm. received 30 April 1969, in Min. Dept. BM(NH).
6kg Turin, Univ., largest stone; 1.5kg Rome,; 45g Vienna, Naturhist. Mus.; 35g Chicago, Field Mus. Nat. Hist.;
Specimen(s): [1927,8], 5.75g.

Moungar Bou Hadid v Bou Hadid.

Mount Ayliff 30°49′ S., 29°21′ E.
Griqualand East, Cape Province, South Africa
Found 1907, known in this year
Iron. Octahedrite, coarse (1.6mm) (IA).
A mass of about 30lb was found about 1907. Description and analysis, 6.59 %Ni, G.T. Prior, Min. Mag., 1921, **19**, p.163. Analysis, 7.76 %Ni, 70.0 ppm.Ga, 250 ppm.Ge, 1.8 ppm.Ir, J.T. Wasson, Icarus, 1970, **12**, p.407. Description, V.F. Buchwald, Iron Meteorites, Univ. of California, 1975, p.848.
1460g King William's Town Museum, South Africa, main mass; 400g Vienna, Naturhist. Mus.; 397g Paris, Mus. d'Hist. Nat.; 202g New York, Amer. Mus. Nat. Hist.; 95g Washington, U.S. Nat. Mus.;
Specimen(s): [1920,140], 189g. a polished and etched piece; [1920,141], 26g. filings

Mount Baldr 77°35′2″ S., 160°19′35″ E.
Wright Valley, Victoria Land, Antarctica
Found 1976, December 15
Synonym(s): *Mount Baldr (a), Mount Baldr (b), Mount Baldr A76001, Mount Baldr A76002*
Stone. Olivine-bronzite chondrite (H6).
Two masses, 13.782kg and 4.108kg respectively, were found 700 m apart by the U.S.-Japan Joint Expedition to Antarctica during the 1976-1977 field season. First thought to be distinct, the smaller (a), and the larger (b), E. Olsen et al., Meteoritics, 1978, **13**, p.209. Rare gas studies indicate they are from the same fall, H. Weber and L. Schultz, Meteoritics, 1978, **13**, p.658, abs.
3894g Chicago, Field Mus. Nat. Hist.;
Specimen(s): [1978,M.11], 117.1g. of the smaller stone (a); [1978,M.12], 185.9g. of the larger stone (b)

Mount Baldr A76001 v Mount Baldr.

Mount Baldr A76002 v Mount Baldr.

Mount Browne 29°48′ S., 141°42′ E.
County Evelyn, New South Wales, Australia
Fell 1902, July 17, 0930 hrs
Stone. Olivine-bronzite chondrite (H6).
After detonations, a stone of about 25lb was seen to fall, G.W. Card, Rec. Geol. Surv. New South Wales, 1903, **7**, p.218. Analysis, H.P. White, Rec. Geol. Surv. New South Wales, 1903, **7**, p.312. Co-ordinates, olivine $Fa_{18.1}$, B. Mason, Rec. Austr. Mus., 1974, **29**, p.169. Ar-Ar age, 4.4 Ga., G. Turner et al., Geochimica et Cosmochimica Acta, 1978 (Suppl. 10), p.989.
6.24kg Sydney, Austr. Mus., main mass; 330g Chicago, Field Mus. Nat. Hist.; 400g Washington, U.S. Nat. Mus.; 359g Prague, Nat. Mus.; 111g Vienna, Naturhist. Mus.; 55g New York, Amer. Mus. Nat. Hist.;
Specimen(s): [86644], 126g. and fragments, 9g; [1920,325], 53g.

Mount Dooling 29°27′ S., 119°43′ E. approx.
North Yilgarn, Western Australia, Australia
Found 1909
Synonym(s): *Gosnells, Lake Giles*
Iron. Octahedrite, coarse (1-2mm) (IC).
A mass of 69lb was found, E.S. Simpson, Bull. Geol. Surv. Western Australia, 1912 (48), p.83. Listed, Spec. Publ. West.

Austr. Mus., 1965 (3), p.40. A 1.5 kg mass found in 1960 at Gosnells, near Perth, Western Australia and about 400km from Mount Dooling, is interpreted on chemical and structural grounds as a transported mass of the latter, chemically anomalous, analyses quoted, mean values approx., 6.2 %Ni, 55 ppm.Ga, 243 ppm.Ge, J.R. de Laeter, Meteoritics, 1972, 7, p.469. Co-ordinates; description; shocked and recrystallised, V.F. Buchwald, Iron Meteorites, Univ. of California, 1975, p.849. Another mass, 701kg, was found in 1979 about 50km SE. of Diemals, Western Australia; chemically identical to Mount Dooling, J.R. de Laeter, Meteoritics, 1980, **15**, p.149. Classification, analysis, 6.26 %Ni, 52 ppm.Ga, 234 ppm.Ge, 1.2 ppm.Ir, E.R.D. Scott and J.T. Wasson, Geochimica et Cosmochimica Acta, 1976, **40**, p.103.

Main mass, Perth, West. Austr. Mus.; 370g Washington, U.S. Nat. Mus.; 129g Los Angeles, Univ. of California; 113g Mainz, Max-Planck-Inst.;

Specimen(s): [1968,281], 141g.

Mount Dyrring 32°20′ S., 151°12′ E.

Singleton district, County Durham, New South Wales, Australia
Found 1903
Stony-iron. Pallasite (PAL).

Fragments weighing together about 25lb were found 8 miles north of Bridgman, G.W. Card, Rec. Geol. Surv. New South Wales, 1903, **7**, p.218. Analysis, J.C.H. Mingaye, Rec. Geol. Surv. New South Wales, 1903, **7**, p.305. Olivine $Fa_{13.5}$, P.R. Buseck and J.I. Goldstein, Bull. Geol. Soc. Amer., 1969, **80**, p.2141.

2.15kg Sydney, Australian Mus., main mass; 509g Yale Univ.; 672g Washington, U.S. Nat. Mus.; 600g New York, Amer. Mus. Nat. Hist.; 425g Vienna, Naturhist. Mus.; 402g Chicago, Field Mus. Nat. Hist.; 300g Paris, Mus. d'Hist. Nat.; 222g Budapest, Nat. Mus.;

Specimen(s): [86643], 235g.

Mount Edith 22°30′ S., 116°10′ E.

Ashburton district, Western Australia, Australia
Found 1913
Synonym(s): *Duck Creek*
Iron. Octahedrite, medium (0.8mm) (IIIB).

A mass of 161kg, of irregular triangular shape, was found about 80 miles SE. of Onslow. Description, and analysis, 9.45 %Ni, W.M. Foote, Am. J. Sci., 1914, **37**, p.391. Another mass of 364lb, found in 1914, was analysed, 9.18 % Ni, E.S. Simpson, Bull. Geol. Surv. West. Austr., 1916 (67), p.140. Listed, Spec. Publ. West. Austr. Mus., 1965 (3), p.41. Analysis, 9.4 %Ni, 20.1 ppm.Ga, 37.5 ppm.Ge, 0.016 ppm.Ir, E.R.D. Scott et al., Geochimica et Cosmochimica Acta, 1973, **37**, p.1957. Description; shock-hardened, V.F. Buchwald, Iron Meteorites, Univ. of California, 1975, p.850.

Main mass, Perth, Geol. Surv. Mus., from the 364lb mass; 6.7kg Perth, Geol. Surv. Mus., from the first mass; 29.2kg Harvard Univ.; 10kg Chicago, Field Mus. Nat. Hist., approx.; 9.9kg Vienna, Naturhist. Mus.; 8.75kg New York, Amer. Mus. Nat. Hist.; 8kg Philadelphia, Acad. Nat. Sci.; 1.5kg Washington, U.S. Nat. Mus.; 2.75kg Budapest, Nat. Mus.; 5kg Ottawa, Mus. Geol. Surv. Canada;

Specimen(s): [1914,1570], 989g. slice, and a fragment, 29.5g

Mount Egerton 24°46′ S., 117°42′ E.

Gascoyne River, Western Australia, Australia
Found 1941, known in this year
Synonym(s): *Siberia*
Stony-iron. Mesosiderite, anomalous (MESANOM).

Four fragments, totalling 1.7kg, submitted for examination by M.T. Gaffney from a place 12 miles from Mt. Egerton (which has the above lat. and long.), consist of nickel-iron with 6.38% Ni embedded in large crystals of enstatite, one of which measures 8.5 × 5 × 2.8 cm; schreibersite, troilite, and possibly oldhamite are present, H. Bowley, Ann. Rep. Dept. Mines West. Austr. for 1942, 1944, p.76. Description, G.J.H. McCall, Min. Mag., 1965, **35**, p.241. Listed, Spec. Publ. West. Austr. Mus., 1965 (3), p.42. In December 1963, further specimens, totalling 250g, were found. An additional find of thousands of small fragments, of total weight about 20kg, was made in June 1966, 8 miles from the summit of Mount Egerton towards no. 3 well, 24°53'S., 117°38'E. This is thought to be the location of the original find, W.H. Cleverly, J. Roy. Soc. West. Austr., 1968, **51**, p.76. Analysis of enstatite, A.M. Reid and A.J. Cohen, Geochimica et Cosmochimica Acta, 1967, **31**, p.661. Analysis of metal, 6.20 %Ni, 35.2 ppm.Ga, 99.2 ppm.Ge, 1.8 ppm.Ir, J.T. Wasson, Meteorites, Springer-Verlag, 1974, p.307.

144g Perth, Western Austr. Mus.; 70g Kalgoorlie, School of Mines, plus new additional material; 3.7kg Washington, U.S. Nat. Mus.; 103g Chicago, Field Mus. Nat. Hist.;

Specimen(s): [1968,275], 463g. fragments

Mount Elden v Cañon Diablo.

Mount Erin v Ballinoo.

Mount Hicks v Mantos Blancos.

Mount Joy 39°47′ N., 77°13′ W.

Adams County, Pennsylvania, U.S.A.
Found 1887
Synonym(s): *Adams County, Gettysburg*
Iron. Octahedrite, coarsest (10mm) (IIB).

A mass of 847lb was found 5 miles SE. of Gettysburg. Description and analysis, 4.81 %Ni, E.E. Howell, Am. J. Sci., 1892, **44**, p.415. Analysis, 5.79 %Ni, E.P. Henderson, Am. J. Sci., 1941, **239**, p.407 [M.A. 8-195]. Analysis, determination of Ga, Au and Pd, 5.66 %Ni, E. Goldberg et al., Geochimica et Cosmochimica Acta, 1951, **2**, p.1. Further analysis, 5.68 %Ni, 59.1 ppm.Ga, 183 ppm.Ge, 0.46 ppm.Ir, J.T. Wasson, Geochimica et Cosmochimica Acta, 1969, **33**, p.859. Mentioned, R.W. Stone and E.M. Starr, Bull. Pa. Geol. Surv., 1967 (Rep. G2, 1932, rev.), p.10. Description, classification, V.F. Buchwald, Iron Meteorites, Univ. of California, 1975, p.853.

172kg Vienna, Naturhist. Mus.; 20.5kg Chicago, Field Mus. Nat. Hist.; 8.5kg Tempe, Arizona State Univ.; 7.4kg Prague, Nat. Mus.; 4kg Bonn, Univ. Mus.; 3.25kg Washington, U.S. Nat. Mus.; 1.5kg Budapest, Nat. Mus.; 2.5kg Sarajevo, Bosnian Nat. Mus.; 1kg New York, Amer. Mus. Nat. Hist.;

Specimen(s): [83483], 713g. slice, and pieces, 18.75g; [1959, 166], 1.5g.; [1960,490], 46.5g.

Mount Kemis v Krasnojarsk.

Mount Magnet 28°2′ S., 117°58′ E.

Murchison Goldfield, Western Australia, Australia
Found 1916
Synonym(s): *East Mount Magnet*
Iron. Octahedrite, plessitic (IRANOM).

A sickle-shaped mass of 36.5lb was found 6 miles east of Mount Magnet, Ann. Rep. Geol. Surv. West. Austr. for 1916, 1917, p.26. Description, with analysis, 13.56 %Ni, E.S. Simpson, J. Roy. Soc. West. Austr., 1927, **13**, p.37 [M.A. 3-546]. Shows a brecciated structure with the areas outlined by schreibersite. Listed,, Spec. Publ. West. Austr. Mus., 1965 (3), p.44. Chemically anomalous; analysis, 14.56 %Ni, 7.53

ppm.Ga, 5.26 ppm.Ge, 0.012 ppm.Ir, J.T. Wasson and R. Schaudy, Icarus, 1971, **14**, p.59.
835g Perth, Geol. Surv. Mus.; 1.2kg Calcutta, Mus. Geol. Surv. India; 719g Sydney, Australian Mus.; 249g Chicago, Field Mus. Nat. Hist.; 100g Washington, U.S. Nat. Mus.; 52g New York, Amer. Mus. Nat. Hist.;
Specimen(s): [1926,497], 438g. end-piece, and fragments, 10.3g

Mount Morris (New York) 42°42′ N., 77°53′ W.
Livingstone County, New York, U.S.A.
Found 1897
Stone. Olivine-bronzite chondrite (H).
A stone of about 12.5g was found on the Landers farm, 1.5 miles south of Mount Morris, H.P. Whitlock, Bull. New York State Mus., 1913 (164), p.78. Olivine Fa_{17}, B. Mason, Geochimica et Cosmochimica Acta, 1963, **27**, p.1011.
Specimen, Albany, New York State Mus.;

Mount Morris (Wisconsin) 44° N., 89°15′ W.
Waushara County, Wisconsin, U.S.A.
Found 1937, known before this year
Stone. Chondrite, equilibrated (CHANOM).
Three oxidized fragments, totalling 676g, are preserved; nothing is known of their history, E.P. Henderson, letters of May 12 and June 3, 1939 in Min. Dept. BM(NH)., Ann. Rep. U.S. Nat. Mus., 1937, p.47, H.H. and A.D. Nininger, The Nininger Collection of Meteorites, Winslow, Arizona, 1950, p.75. An equilibrated chondrite rich in troilite. Description, and analysis, 19.88 % total iron, olivine Fa_3, A.L. Graham et al., Min. Mag., 1977, **41**, p.201. Relationship to silicates in IAB irons, R.W. Bild, Geochimica et Cosmochimica Acta, 1977, **41**, p.1439. Possibly a silicate-rich fragment of Pine River (*q.v.*), A.W.R. Bevan, priv. comm., 1981.
556g Washington, U.S. Nat. Mus., main mass; 33g Tempe, Arizona State Univ.; 12g New York, Amer. Mus. Nat. Hist.;
Specimen(s): [1959,891], 52g.

Mount Ouray 38°25′ N., 106°13′ W.
Chaffee County, Colorado, U.S.A.
Found 1898
Synonym(s): *Ute Pass*
Iron. Octahedrite, medium (0.8mm) (IID).
A mass of about 2lb was found at an elevation of 10,000 feet on the NE. slope of Mt. Ouray. Description, with an analysis, 7.97 %Ni, C. Palache, Am. J. Sci., 1926, **12**, p.144. Further analysis, 10.13 %Ni, 71.4 ppm.Ga, 84.3 ppm.Ge, 15 ppm.Ir, J.T. Wasson, Meteorites, Springer-Verlag, 1974, p.301. Description, synonymy with Ute Pass, V.F. Buchwald, Iron Meteorites, Univ. of California, 1975, p.856.
483g Harvard Univ.; 120g Chicago, Field Mus. Nat. Hist.; 84g Washington, U.S. Nat. Mus.; 53g New York, Amer. Mus. Nat. Hist.;
Specimen(s): [1927,86], 146g. slice

Mount Ozren v Ozren.

Mount Padbury 25°40′ S., 118°6′ E.
Meekathara, Western Australia, Australia
Found 1964
Stony-iron. Mesosiderite (MES).
272kg of fragments were found, First Suppl. to Spec. Publ. West. Austr. Mus. no. 3, 1968, p.11, G.J.H. McCall and W.H. Cleverly, Nature, 1965, **207**, p.851, W.H. Cleverly, J. Roy. Soc. West. Austr., 1965, **48**, p.55 [M.A. 17-686], G.J.H. McCall, Min. Mag., 1966, **35**, p.1029. Analysis, 6.10 %Ni, 8.9 ppm.Ga, 44.9 ppm.Ge, 2.0 ppm.Ir, J.T. Wasson et al., Geochimica et Cosmochimica Acta, 1974, **38**, p.135. Olivine coronas, classification, references, C.E. Nehru et al., Geochimica et Cosmochimica Acta, 1980, **44**, p.1103.
Main mass, Perth, West. Austr. Mus.; 8.1kg Washington, U.S. Nat. Mus.; 308g Chicago, Field Mus. Nat. Hist.; 300g New York, Amer. Mus. Nat. Hist.;
Specimen(s): [1966,51], 320g. slice; [1967,383], 214.5g. two slices, 123g and 91.5g

Mount Sir Charles 23°50′ S., 134°2′ E.
Northern Territory, Australia
Found 1942
Synonym(s): *Bond Springs*
Iron. Octahedrite, fine (0.3mm) (IVA).
A mass of 22.9kg was found, 7 miles east of Bond Springs Station, D.W.P. Corbett, Rec. S. Austr. Mus., 1975, **15** (4), p.767, Meteor. Bull., 1975 (53), Meteoritics, 1975, **10**, p.153. Classification and analysis, 8.3 %Ni, 2.32 ppm.Ga, 0.126 ppm.Ge, 1.5 ppm.Ir, A. Kracher et al., Geochimica et Cosmochimica Acta, 1980, **44**, p.773.
22.2kg Adelaide, South Austr. Mus., main mass; 21g Washington, U.S. Nat. Mus.;

Mount Stirling v Youndegin.

Mount Tabby v Duchesne.

Mount Vaisi 44°5′ N., 6°52′ E.
Alpes Maritimes, France
Fell 1637, November 29, 1000 hrs
Stone.
A dark coloured stone fell between Guillaumes (lat. and long. above) and Peone. It weighed 38lb and had a specific gravity of about 3.6, D.A. Soldani, Atti Accad. Sci. Siena, 1808, **9**, p.1, E.F.F. Chladni, Die Feuer-Meteore, Wien, 1819, p.225. Although no material is preserved, this fall appears reasonably well authenticated.

Mount Vernon 36°56′ N., 87°24′ W.
Christian County, Kentucky, U.S.A.
Fell 1868, known in this year
Stony-iron. Pallasite (PAL).
A mass of about 351lb was found about 7 miles NE. of Hopkinsville, but was not recognized as meteoritic until 1902, G.P. Merrill, Amer. Geologist, 1903, **31**, p.156. Description and analysis, W. Tassin, Proc. U.S. Nat. Mus., 1905, **28**, p.213. Analysis of metal, 11.5 %Ni, 21.5 ppm.Ga, 49.1 ppm.Ge, 0.14 ppm.Ir, J.T. Wasson and S.P. Sedwick, Nature, 1969, **222**, p.22. Olivine Fa_{12}, P.R. Buseck and J.I. Goldstein, Bull. Geol. Soc. Amer., 1969, **80**, p.2141. Texture, references, E.R.D. Scott, Geochimica et Cosmochimica Acta, 1977, **41**, p.693.
134kg Washington, U.S. Nat. Mus., main mass; 1.1kg Harvard Univ.; 1kg Tempe, Arizona State Univ.; 0.5kg Budapest, Nat. Mus.; 465g Prague, Nat. Mus.; 280g Chicago, Field Mus. Nat. Hist.; 234g Vienna, Naturhist. Mus.; 200g Berlin, Humboldt Univ.;
Specimen(s): [1919,86], 461g. fragments

Mount Zomba v Zomba.

Moustel Pank v Oesel.

Mouza Khoorna v Supuhee.

Mow v Mhow.

Mrirt 33°8′ N., 5°34′ W.
25km NE. of Khenifra, Morocco
Found 1937, or 1938
Iron.
One mass, 79.9kg, was found near Mrirt village. Description, figured, with analysis, 8.08 %Ni, C. Gaudefroy and J. Lucas, Notes Serv. géol. Maroc, 1955, **12**, p.162.
Complete mass, Rabat, Mus. Serv. géol.;

Mtola 11°30′ S., 33°30′ E. approx.
Mzimba district, Malawi
Fell 1944, June 17, 1000 hrs
Stone.
A single stone weighing about 2lb 6oz was recovered. The specimen was in the Geological Survey Mus. but is now lost, M.J. Crow, Geol. Surv. Malawi, 1982 (report MJC/32).

Muckera 30°5′ S., 130°2′ E.
South Australia, Australia
Found 1950, recognized 1972
Stone. Achondrite, Ca-rich. Howardite (AHOW).
The 513g stone was found about 3km SW. of Muckera Rock Hole, some 80km N. of Cook on the Transcontinental Railway, M.J. Fitzgerald, Trans. Roy. Soc. South Austr., 1980, **104**, p.201.
250g Adelaide, Univ., main mass; 129g Adelaide, South Austr. Mus.;

Muchachos v Tucson.

Muddoor 12°38′ N., 77°1′ E.
Mandya, Karnataka, India
Fell 1865, September 21, 0700 hrs
Synonym(s): *Maddur, Mudoor, Mysore*
Stone. Olivine-hypersthene chondrite (L5).
Two stones were seen to fall, after detonations, near Annay Doddi, Maddur taluq; one weighed about 2kg, the other was broken in pieces, L.B. Bowring, Proc. Asiatic Soc. Bengal, 1865, p.195. Analysis, F. Crook, Inaug.-Diss. Göttingen, 1868, p.33. The place of fall is Anedoddi 5 miles north of Maddur, Mysore, C.A. Silberrad, Min. Mag., 1932, **23**, p.302. Olivine Fa_{23}, B. Mason, Geochimica et Cosmochimica Acta, 1967, **31**, p.1100.
1.5kg Calcutta, Mus. Geol. Surv. India; 2kg Bangalore, Govt. Mus.; 286g New York, Amer. Mus. Nat. Hist.; 51g Vienna, Naturhist. Mus.;
Specimen(s): [41019*], 407g.

Mudoor v Muddoor.

Mühlau 47°17′ N., 11°25′ E.
Innsbruck, Tyrol, Austria
Found 1877, approx.
Synonym(s): *Mecherburg*
Stone. Chondrite.
A stone of 5g was found between Weiherburg and Mühlau, probably soon after its fall, A. Brezina, Ann. Naturhist. Hofmus. Wien, 1887, **2**, p.115.
4.3g Vienna, Naturhist. Mus.;

Muizenberg 34°6′ S., 18°28′ E.
Cape Town, Cape Province, South Africa
Found 1880, known since about this time
Synonym(s): *Muizenburg*
Stone. Olivine-hypersthene chondrite (L6).
A stone of 10lb 2.5oz in the South African Mus., Cape Town, is labelled, " Presented by E.J. Dunn, 1929". In a letter of April 4, 1935, in Min. Dept. BM(NH)., Mr. Dunn states that he had this stone over 40 years. See also, K.H. Barnard, letter of 26 May, 1950 in Min. Dept. BM(NH). Analysis, 20.20 % total iron, H. von Michaelis et al., Earth planet. Sci. Lett., 1969, **5**, p.387. Olivine Fa_{24}, B. Mason, Geochimica et Cosmochimica Acta, 1963, **27**, p.1011.
3.4g New York, Amer. Mus. Nat. Hist.;
Specimen(s): [1950,338], 84.5g.

Muizenburg v Muizenberg.

Mukerop v Gibeon.

Muleshoe 34°7′18″ N., 102°41′42″ W.
Bailey County, Texas, U.S.A.
Found 1972, recognized in this year
Stone. Olivine-bronzite chondrite (H4-6).
Total known weight 3.6kg, Cat. Huss Coll. Meteorites, 1976, p.32. Olivine $Fa_{19.7}$, B.D. Dod et al., Meteoritics, 1979, **14**, p.379.
2.1kg Mainz, Max-Planck-Inst.; 63g Albuquerque, Univ. of New Mexico;
Specimen(s): [1975,M.4], 63g. crusted part-slice

Mulga (north) 30°11′ S., 126°22′ E.
Western Australia, Australia
Found 1964
Stone. Olivine-bronzite chondrite (H6).
59 complete stones, total weight 2421g were found, G.J.H. McCall, First Suppl. to West. Austr. Mus. Spec. Publ. no. 3, 1968, p.12. Description, W.H. Cleverly, J. Roy. Soc. West. Austr., 1972, **55**, p.115. Olivine $Fa_{19.3}$, B. Mason, Rec. Austr. Mus., 1974, **29**, p.169.
52 stones, Kalgoorlie, West. Austr. School of Mines; 513g Perth, West. Austr. Mus., six stones; 278g Washington, U.S. Nat. Mus.;

Mulga (south) 30°12′ S., 126°22′ E.
Western Australia, Australia
Found 1963
Synonym(s): *Billygoat Donga III*
Stone. Olivine-bronzite chondrite (H4).
A total of 24 weathered fragments, total weight 894g, were found between 1963 and 1971, 95km NNE. of Haig (*q.v.*), Meteor. Bull., 1975 (53), Meteoritics, 1975, **10**, p.154. Olivine $Fa_{18.4}$, B. Mason, Rec. Austr. Mus., 1974, **29**, p.169. Mentioned, W.H. Cleverly, J. Roy. Soc. West. Austr., 1972, **55**, p.115.
Main mass, Kalgoorlie, West. Austr. School of Mines; 13g Perth, West. Austr. Mus.; 0.4g Washington, U.S. Nat. Mus.;

Mulga (west) 30°11′ S., 126°22′ E. approx.
Western Australia, Australia
Found 1971, December 12
Stone. Carbonaceous chondrite (C5).
A single, crusted individual, weight 169.2g was found 4km north of Billygoat Donga (*q.v.*) and within the strewnfields of the Mulga (north) and Mulga (south) meteorites. Contains magnetite, olivine Fa_{32}, G.J.H. McCall, 2nd. Suppl. to West. Austr. Mus. Spec. Publ. no. 3, 1972, p.18. Reported, Meteor. Bull., 1975 (53), Meteoritics, 1975, **10**, p.154. Recrystallised, classified as C5 or C6, R.A. Binns et al., Meteoritics, 1977, **12**, p.179, abs.
Main mass, Kalgoorlie, West. Austr. School of Mines; Thin section, Perth, West. Austr. Mus.;
Specimen(s): [1978,M.24], 6.8g.

Mulino 45°13′ N., 122°34′ W.
Clackamas County, Oregon, U.S.A.
Fell 1927, May 4
Doubtful. Stone.
A very small stone resembling the small Holbrook stones was obtained by the U.S. Nat. Mus. with the above particulars, but correspondence has failed to reveal any local record of such a fall. Very doubtful, E.P. Henderson, letter of June 3, 1939 in Min. Dept. BM(NH). Omitted, B. Mason, Meteorites, Wiley, 1962, p.240.

Mullaittivu v Mulletiwu.

Mulletiwu 9°20′ N., 80°50′ E.
Northern Province, Sri Lanka
Fell 1795, April 13, 0800 hrs
Synonym(s): *Carnavelpattu, Ceylon, Mullaittivu*
Stone. Olivine-hypersthene chondrite (L).
After detonations a shower of stones fell, E.F.F. Chladni, Die Feuer-Meteore, Wien, 1819, p.262. Description, with partial analysis, S. Meunier, C. R. Acad. Sci. Paris, 1901, **132**, p.501. Olivine Fa_{25}, B. Mason, Geochimica et Cosmochimica Acta, 1963, **27**, p.1011. The place of the fall is Mullaitivu, 9°20'N., 80°50'E., C.A. Silberrad, Min. Mag., 1932, **23**, p.304.
25g Paris, Mus. d'Hist. Nat.;

Multan v Khairpur.

Multan v Lodran.

Mundrabilla 30°47′ S., 127°33′ E.
Nullarbor Plain, Western Australia, Australia
Found 1911
Synonym(s): *Loongana Station (iron), Loogana Station West, Premier Downs*
Iron. Octahedrite, medium (0.6mm) with sulphide and silicate inclusions (IRANOM).
Two masses, 10-12 tons and 4-6 tons, were found in 1966 but the three smaller Premier Downs masses were found in 1911 and 1918. Analysis, 7.79 %Ni, 60 ppm.Ga, 171 ppm.Ge, J.R. de Laeter, Meteoritics, 1972, **7**, p.285. Description, P. Ramdohr and A. El Goresy, Chem. Erde, 1971, **30**, p.269. Further analysis, chemically anomalous, 7.72 %Ni, 59.5 ppm.Ga, 196 ppm.Ge, 0.87 ppm.Ir, J.T. Wasson, Meteorites, Springer-Verlag, 1974, p.307. The larger mass in Perth (Western Aust. Mus.); the smaller mass went to Adelaide, then to Heidelberg for cutting and distribution. Silicate mineralogy, P. Ramdohr et al., Meteoritics, 1975, **10**, p.477, abs. Full description, V.F. Buchwald, Iron Meteorites, Univ. of California, 1975, p.858. Two further masses, totalling about 1640kg, were found in 1979, Meteor. Bull., 1983 (61), Meteoritics, 1983, **18**, p.82.
12tons Perth, West. Austr. Mus.; Specimen, Heidelberg, Max-Planck-Inst.; 217kg Washington, U.S. Nat. Mus.; Specimen, Adelaide,; 263kg Moscow, Acad. Sci.;
Specimen(s): [1973,M.39], 275kg. slice; [1975,M.9], 52kg. prism

Mungen Dusch v Armanty.

Mungindi 28°56′ S., 148°57′ E.
County Benarba, New South Wales, Australia
Found 1897
Iron. Octahedrite, fine (0.40mm) (IIICD).
Two masses, of 62lb and 51lb respectively, were found in Queensland, 3 miles NNE. of Mungindi, New South Wales, G.W. Card, Rec. Geol. Surv. New South Wales, 1897, **5**, p.121. Description, analysis, 10.99 %Ni, E. Cohen, Meteoritenkunde, 1905, **3**, p.268, H.A. Ward, Am. J. Sci., 1898, **5**, p.138. Classification and analysis, 11.5 %Ni, 19.4 ppm.Ga, 22.1 ppm.Ge, 0.47 ppm.Ir, J.T. Wasson and R. Schaudy, Icarus, 1971, **14**, p.59. Description; co-ordinates, V.F. Buchwald, Iron Meteorites, Univ. of California, 1975, p.862.
19.6kg Sydney, Austr. Mus.; 7kg New York, Amer. Mus. Nat. Hist.; 2kg Chicago, Field Mus. Nat. Hist.; 750g Vienna, Naturhist. Mus.; 573g Washington, U.S. Nat. Mus.; 545g Berlin, Humboldt Univ.;
Specimen(s): [83394], 368g. slice

Muonionalusta 67°48′ N., 23°6′ E.
Kiruna, Norrbotten, Sweden
Found 1906
Iron. Octahedrite, fine (0.3mm) (IVA).
A mass of 7.5kg was found 2.5 miles WSW. of Kitkiojärvi in Muonionalusta. Description, with an analysis, 8.02 %Ni, A.G. Högbom, Bull. Geol. Inst. Univ. Uppsala, 1908, **9**, p.229. Measurements of the Widmanstätten figures on two pieces suggest tetragonal rather than cubic symmetry, D. Malmqvist, Bull. Geol. Inst. Univ. Uppsala, 1948, **32**, p.277 [M.A. 10-515]. A second mass of 15kg was found in 1946, and a third of 6.2kg in 1963, Meteor. Bull., 1963 (28), F.E. Wickman, Arkiv Min. Geol., 1964, **3**, p.467. Analysis, 8.42 %Ni, 2.24 ppm.Ga, 0.133 ppm.Ge, 1.6 ppm.Ir, R. Schaudy et al., Icarus, 1972, **17**, p.174. Description; weathered, V.F. Buchwald, Iron Meteorites, Univ. of California, 1975, p.865.
Specimen, Uppsala, Univ. (Geol. Inst.); 15kg Stockholm,; 419g Washington, U.S. Nat. Mus.; 197g Vienna, Naturhist. Mus.; 96g New York, Amer. Mus. Nat. Hist.; 84g Chicago, Field Mus. Nat. Hist.;
Specimen(s): [1911,195], 142.5g. slice

Muraid 24°30′ N., 90°13′ E.
Ghatail sub-division, Mymensingh district, Dacca, Bangladesh
Fell 1924, August 7, 1430 hrs
Synonym(s): *Mantala, Murraid*
Stone. Olivine-hypersthene chondrite (L6).
After detonations, three stones fell but two only, which fitted together, were recovered, one of 2925.5g at Muraid (24°29'40"N., 90°13'5"E.), and the other of 1777.5g at Mantala (24°29'25"N., 90°12'20"E.), a mile distant. Description, G.V. Hobson, Rec. Geol. Surv. India, 1927, **60**, p.143. Olivine Fa_{26}, B. Mason, Geochimica et Cosmochimica Acta, 1967, **31**, p.1100.
4.7kg Calcutta, Mus. Geol. Surv. India; 0.8g New York, Amer. Mus. Nat. Hist.;
Specimen(s): [1983,M.11], 50.3g.

Murchison 36°37′ S., 145°12′ E.
Victoria, Australia
Fell 1969, September 28, 1045-1100 hrs
Stone. Carbonaceous chondrite, type II (CM2).
A shower of stones fell in an area of over 5 square miles. The largest stone weighed about 7kg, and over 100kg were recovered, Meteor. Bull., 1969 (48), Meteoritics, 1970, **5**, p.107. Detailed description and analysis, 20.44, 22.13 % total iron, L.H. Fuchs et al., Smithson. Contrib. Earth Sci., 1973 (10). Rare-earth element abundances, D.L. Showalter et al., Meteoritics, 1972, **7**, p.295. Study of calcium variation in olivines, R. Hutchison and R.F. Symes, Meteoritics, 1972, **7**, p.23. SEM study of olivine crystals in matrix, E. Olsen and L. Grossman, Meteoritics, 1974, **9**, p.243. Study of amino acids, G.E. Pollock et al., Geochimica et Cosmochimica Acta, 1975, **39**, p.1571. Purines and triazines, R. Hayatsu et al., Geochimica et Cosmochimica Acta, 1975, **39**, p.471.

Noble gas study, L. Alaerts et al., Geochimica et Cosmochimica Acta, 1980, **44**, p.189. Trace element data, G.W. Kallemeyn and J.T. Wasson, Geochimica et Cosmochimica Acta, 1981, **45**, p.1217.
49.5kg Chicago, Field Mus. Nat. Hist.; 30kg Washington, U.S. Nat. Mus.; Specimens, Adelaide, Univ.; 7kg Tempe, Arizona State Univ.; 2.4kg Los Angeles, Univ. of California;
Specimen(s): [1970,5], 1187g. almost complete stone; [1970, 6], 151.5g. fragments; [1972,3], 2.2g. fragment

Murchison Downs 26°40′ S., 119° E. approx.

Kyarra district, Murchison Goldfield, Western Australia, Australia
Found 1925
Iron. Octahedrite, fine.
A piece of 33g was found, E.S. Simpson, J. Roy. Soc. West. Austr., 1927, **13**, p.47. Possibly transported Henbury or Boxhole, V.F. Buchwald, Iron Meteorites, Univ. of California, 1975, p.866.
Main mass, Perth, Geol. Surv. Mus.;

Murcia v Cabeza de Mayo.

Murcia v Molina.

Murfreesboro 35°50′ N., 86°25′ W.

Rutherford County, Tennessee, U.S.A.
Found 1847
Synonym(s): *Rutherford County*
Iron. Octahedrite, medium (1.0mm) (IIIA-ANOM).
A mass of about 19lb was found a few miles from Murfreesboro, G. Troost, Am. J. Sci., 1848, **5**, p.351. Description, co-ordinates, V.F. Buchwald, Iron Meteorites, Univ. of California, 1975, p.866. Chemically anomalous; analysis, 7.91 %Ni, 16.7 ppm.Ga, 30.2 ppm.Ge, 2.2 ppm.Ir, E.R.D. Scott et al., Geochimica et Cosmochimica Acta, 1973, **37**, p.1957.
950g Vienna, Naturhist. Mus.; 156g Harvard Univ.; 125g Washington, U.S. Nat. Mus.; 83g Chicago, Field Mus. Nat. Hist.; 73g New York, Amer. Mus. Nat. Hist.; 18g Tübingen, Univ.;
Specimen(s): [22427], 2785g.

Murnpeowie 29°35′ S., 139°54′ E.

South Australia, Australia
Found 1909
Iron. Anomalous (IC).
A mass of 2520lb was found in the Beltana Pastoral Company's Murnpeowie Run, about 16 miles NE. by E. of Mt. Hopeless, L.L. Smith, Am. J. Sci., 1910, **30**, p.264. Probably fell between 1904 and 1909; condition very fresh when found. The place of fall is 29°35'S., 139°54'E., and is 5 miles from the western shore of Lake Callabonna. The iron is a granular metabolite, with well marked Neumann lines and remnants of a coarse octahedrite structure. Description, and analysis, 6.32 %Ni, L.J. Spencer, Min. Mag., 1935, **24**, p.13. Mentioned, W.D. Brentnall and H.J. Axon, J. Iron Steel Inst., 1962, **200**, p.947. Description of structure, H.J. Axon, Min. Mag., 1968, **36**, p.1139. Chemically anomalous; analysis, 6.31 %Ni, 41.8 ppm.Ga, 85.4 ppm.Ge, 1.8 ppm.Ir, J.T. Wasson, Meteorites, Springer-Verlag, 1974, p.307. Description; structurally anomalous, V.F. Buchwald, Iron Meteorites, Univ. of California, 1975, p.867.
1140kg Adelaide, South Austr. Mus., main mass; 191g Washington, U.S. Nat. Mus.; 2.5lb Sydney, Austr. Mus.;
Specimen(s): [1934,52], 773.5g. slice, and filings, 28.5g

Muroc 34°55′ N., 117°50′ W.

Kern County, California, U.S.A.
Found 1936, September 20
Stone. Olivine-hypersthene chondrite (L).
One mass of 18.4g was found on the surface of the desert, 5 miles east of Muroc, in the region called Muroc Dry Lake. Two other stones, apparently not of the same fall, were found only 15 feet away, H.H. Nininger and C.H. Cleminshaw, Pop. Astron., Northfield, Minnesota, 1937, **45**, p.373 [M.A. 7-69]. Olivine Fa_{25}, B. Mason, Geochimica et Cosmochimica Acta, 1963, **27**, p.1011.
Main mass, Los Angeles, Griffith Observatory; 4.3g Los Angeles, Univ. of California; 2.6g Tempe, Arizona State Univ.;

Muroc Dry Lake 34°55′ N., 117°50′ W.

Kern County, California, U.S.A.
Found 1936, September 20
Stone. Olivine-hypersthene chondrite (L6).
Two stones, 165 and 58g, were found on the surface of the desert 5 miles east of Muroc, in the region called Muroc Dry Lake. Another stone, apparently not of the same fall, was found only 15 feet away, H.H. Nininger and C.H. Cleminshaw, Pop. Astron., Northfield, Minnesota, 1937, **45**, p.373 [M.A. 7-69]. Olivine Fa_{25}, B. Mason, Geochimica et Cosmochimica Acta, 1963, **27**, p.1011.
Main masses, Los Angeles, Griffith Observatory; 43g Washington, U.S. Nat. Mus.; 14.1g Tempe, Arizona State Univ.;

Muroshna v Angara.

Murozhnaya v Angara.

Muroznaja v Angara.

Murphy 35°6′ N., 84°2′ W.

Cherokee County, North Carolina, U.S.A.
Found 1899
Iron. Hexahedrite (IIA).
A mass of about 17lb was found 5 miles from Murphy, H.A. Ward, Am. J. Sci., 1899, **8**, p.225. Description, E. Cohen, Meteoritenkunde, 1905, **3**, p.227. Analysis, 5.42 %Ni, 60.5 ppm.Ga, 186 ppm.Ge, 34 ppm.Ir, J.T. Wasson, Meteorites, Springer-Verlag, 1974, p.299. Monocrystalline, cosmically deformed, daubréelite-rich, V.F. Buchwald, Iron Meteorites, Univ. of California, 1975, p.870.
767g Rome, Vatican Colln; 639g Chicago, Field Mus. Nat. Hist.; 582g Washington, U.S. Nat. Mus.; 560g New York, Amer. Mus. Nat. Hist.; 500g Vienna, Naturhist. Mus.; 217g Berlin, Humboldt Univ.; 212g Harvard Univ.; 202g Yale Univ.; 84g Tübingen, Univ.;
Specimen(s): [84552], 1521g.

Murraid v Muraid.

Murray 36°36′ N., 88°6′ W.

Calloway County, Kentucky, U.S.A.
Fell 1950, September 20, 0135 hrs
Synonym(s): *Murray County*
Stone. Carbonaceous chondrite, type II (CM2).
A fireball was seen and a loud explosion heard; thirty seconds later a shower of fragments fell 9 miles E. of Murray, near Wildcat Creek on Kentucky Lake. The total recovered was 7kg, the largest fragment weighing 3.4kg; further material was collected later, to a total of 12.6kg. Description, with analysis, 21.25 % total iron, H.B. Wiik, Geochimica et Cosmochimica Acta, 1956, **9**, p.279, J. Horan,

Meteoritics, 1953, **1**, p.114 [M.A. 12-359], C.P. Olivier, Sky and Telescope, 1954, **13**, p.112, Meteoritics, 1955, **1**, p.247 [M.A. 13-53], B. Mason, Scientific American, 1963, **208** (3), p.44, G.P. Vdovykin, Метеоритика, 1964, **25**, p.134. Composition of olivines and pyroxenes, K. Fredriksson and K. Keil, Meteoritics, 1964, **2**, p.201. XRF analysis, 20.71 % total iron, H. von Michaelis et al., Earth planet. Sci. Lett., 1969, **5**, p.387. Organic compounds, references, B. Nagy, Carbonaceous Meteorites, Elsevier, 1975. Trace element data, G.W. Kallemeyn and J.T. Wasson, Geochimica et Cosmochimica Acta, 1981, **45**, p.1217.

5.8kg Washington, U.S. Nat. Mus.; 3.55kg Tempe, Arizona State Univ.; 267g Vienna, Naturhist. Mus.; 124g Chicago, Field Mus. Nat. Hist.;

Specimen(s): [1966,48], 28.8g.; [1955,225], 1g.; [1971,288], 21.4g. interior, uncrusted fragment; [1971,293], 96g. about half of a crusted stone

Murray County v Murray.

Muskingum County v New Concord.

Mustafabad v Merua.

Muzaffarpur 26°8′ N., 85°32′ E.

Bihar, India
Fell 1964, April 11, 1700 hrs
Synonym(s): *Muzzaffarpur*
Iron. Octahedrite, plessitic (IRANOM).

Two masses, 1092g and 153g, fell at Bahrampur (26°8'N., 85°32'E.) and Man Bishunpur (26°7'N., 85°31'15"E.), 1.6km apart, and about 13km east of Muzzafarpur (the correct spelling), M.V.N. Murthy et al., Current Sci., 1964, **33**, p.403 [M.A. 17-387]. Analysis, 12.03 %Ni, D.R. Das Gupta et al., Geochimica et Cosmochimica Acta, 1969, **33**, p.1298. Further analysis, 14.3 %Ni, 14.9 ppm.Ga, 28.6 ppm.Ge, 0.55 ppm.Ir, D.J. Malvin et al., priv. comm., 1983. Shocked while cold; mixed structure, ataxitic and octahedral, V.F. Buchwald, Iron Meteorites, Univ. of California, 1975, p.871.

1.1kg Calcutta, Geol. Surv. India; 5g Washington, U.S. Nat. Mus.;

Specimen(s): [1983,M.12], 30.4g.

Muzzaffarpur v Muzaffarpur.

Myelenes v Mejillones.

Myersville 28°57′ N., 97°24′10″ W.

De Witt County, Texas, U.S.A.
Found 1969
Stone. Chondrite.

A mass of 3.88kg was found, G.I Huss, Cat. Huss Coll. Meteorites, 1976, p.32. The correct spelling of the place of find is Meyersville.

3.7kg Mainz, Max-Planck-Inst.;

Myhee Caunta 23°3′ N., 72°38′ E.

Ahmadabad, Gujarat, India
Fell 1842, November 30, 1600 hrs
Synonym(s): *Mahi Kantha*
Stone. Chondrite.

After detonations a " number of stones" fell between the villages of Jeetala and Mor Monree; they were broken up by natives and only a fragment was sent to the Bombay Geogr. Soc., H. Giraud, Edinburgh New Phil. J., 1849, **47**, p.55. The place of the fall is probably Mahi Kantha State, near Ahmadabad (which is 23°3'N., 72°38'E.), but the villages mentioned could not be identified, C.A. Silberrad, Min. Mag., 1932, **23**, p.303.

Mysore v Judesegeri.

Mysore v Muddoor.

Nabberu v Millbillillie.

Nadiabondi 12° N., 1° E. approx.

Diapaga, Gourma, Upper Volta
Fell 1956, July 27, 1930 hrs, U.T., approx. time
Stone. Olivine-bronzite chondrite (H5).

The fall of a mass of 3665g was observed; the mass was found 5km W. of Nadiabondi at a depth of 25cm. Description, E. Jérémine et al., C. R. Soc. Savantes, 1959, p.353. Reported, Meteor. Bull., 1958 (7). Olivine Fa_{18}, B. Mason, Geochimica et Cosmochimica Acta, 1963, **27**, p.1011. Contains cristobalite; unequilibrated, M. Christophe-Michel-Levy et al., Bull. Soc. franc. Min. Crist., 1965, **88**, p.122 [M.A. 17-387].

3463g Paris, Mus. d'Hist. Nat.;

Naes v Näs.

Nagai 38°7′18″ N., 140°3′42″ E.

Yamagata Prefecture, Honshu, Japan
Fell 1922, May 30, recognized 1977
Stone. Olivine-hypersthene chondrite (L6).

A single mass of 1.81kg was recovered from a water-covered rice field having fallen in front of three, resting, farmers, Meteor. Bull., 1979 (56), Meteoritics, 1979, **14**, p.169. Description, analysis, 21.0 % total iron, olivine Fa_{25}, S. Murayama et al., Bull. Nat. Sci. Mus. Tokyo, Ser. E, 1978, **1**, p.19. The stone is in the possession of the son of the finder.

Nagareyama 35°51′ N., 139°55′ E.

Saitama, Honshu, Japan
Found 1872
Doubtful..

Mentioned, I. Yamamoto, Kwasan Observ. Bull., 1935, p.306 [M.A. 7-173]., S. Murayama, letter of 6 April, 1962 in Min. Dept. BM(NH).

Nagaria 26°59′ N., 78°13′ E.

Fatehabad pargana, Agra district, Uttar Pradesh, India
Fell 1875, April 24, 0730 hrs
Synonym(s): *Agra, Nageria*
Stone. Achondrite, Ca-rich. Eucrite (AEUC).

A stone of about 26lb fell, but broke into fragments, and only about 20g were preserved, H.B. Medlicott, Proc. Asiatic Soc. Bengal, 1876, p.222, F. Fedden, Cat. Meteor. Indian Mus. Calcutta, 1880, p.16. Analysis, B. Mason et al., Smithson. Contrib. Earth Sci., 1979 (22), p.30.

3g Calcutta, Mus. Geol. Surv. India; 1.4g Chicago, Field Mus. Nat. Hist.; Thin section, Washington, U.S. Nat. Mus.;

Specimen(s): [51368], 8.5g. two fragments; [63879], 4.75g.

Nagaya v Nogoya.

Nageria v Nagaria.

Nagla v Ambapur Nagla.

Nagy-Bereszna v Knyahinya.

Nagy-Borové 49°10′ N., 19°30′ E. approx.
Liptó, Stredoslovensky, Czechoslovakia
Fell 1895, May 9
Synonym(s): *Liptó, Velka-Borové, Velke-Borove*
Stone. Olivine-hypersthene chondrite (L5).
Date of fall recorded, A. Brezina, Ann. Naturhist. Hofmus. Wien, 1896, **10**, p.307. Olivine Fa_{24}, B. Mason, Geochimica et Cosmochimica Acta, 1963, **27**, p.1011.
5.9kg Budapest, Nat. Mus.; 194g Chicago, Field Mus. Nat. Hist., complete stone;
Specimen(s): [84898], 53g.

Nagydévény v Gross-Divina.

Nagy-Divina v Gross-Divina.

Nagy-Diwina v Gross-Divina.

Nagy-Vázsony 46°59′ N., 17°42′ E.
Vezprém, Hungary
Found 1890
Iron. Octahedrite, coarse (1.4mm) (IA).
A mass of about 2kg was found, A. Brezina, Ann. Naturhist. Hofmus. Wien, 1896, **10**, p.284, 356. Classification, analysis, 7.98 %Ni, 68.9 ppm.Ga, 237 ppm.Ge, 2.1 ppm.Ir, J.T. Wasson, Meteorites, Springer-Verlag, 1974, p.298. Mentioned, V.F. Buchwald, Iron Meteorites, Univ. of California, 1975, p.874.
1.25kg Vienna, Naturhist. Mus.; 218g New York, Amer. Mus. Nat. Hist.; 36g Washington, U.S. Nat. Mus.; 36g Budapest, Nat. Mus.; 36g Chicago, Field Mus. Nat. Hist.;
Specimen(s): [67446], 69g.; [1921,20], 38g.

Naifa 19°56′ N., 51°13′ E.
Rub'al Khali, Saudi Arabia
Found 1932, March 4
Synonym(s): *Naifah*
Iron. Octahedrite, medium.
A small piece (8g) was found at Naifa, 110 miles S. by E. of Wabar. Description, L.J. Spencer, Min. Mag., 1933, **23**, p.401. Perhaps a transported piece of Wabar.
Specimen(s): [1932,1140], 7.99g. the entire mass

Naifah v Naifa.

Nainital 29°22′ N., 79°26′ E. approx.
Nainital district, Uttar Pradesh, India
Stone. Olivine-hypersthene chondrite (L).
A single stone weighing 5kg is reported without further details, N. Bhandari et al., Nuclear Tracks, 1980, **4**, p.213.

Nakhla 31°19′ N., 30°21′ E.
Abu Hommos, Alexandria, Egypt
Fell 1911, June 28, 0900 hrs
Synonym(s): *Abdel Malek, El Nakhla el Baharia*
Stone. Achondrite, Ca-rich. Nakhlite (ACANOM).
About 40 stones, of total weight about 40kg, and varying in weight from 1813g to 20g fell, after appearance of cloud and detonations, W.F. Hume, Cairo Scientific Journal, 1911, **5** (59), p.212. Mentioned, J. Ball, Geol. Reports, Surv. Dept. (Egypt), Cairo, 1912 (25). One of the stones killed a dog. Description and analysis, G.T. Prior, Min. Mag., 1912, **16**, p.274. Consists mainly of a green diopside with some highly ferriferous olivine and a little feldspar. New analysis, 16.16 % total iron, T.S. McCarthy et al., Meteoritics, 1974, **9**, p.215. Ar-Ar age, 1300 m.y., F.A. Podosek and J.C. Huneke, Geochimica et Cosmochimica Acta, 1973, **37**, p.667. Rb-Sr age, D.A. Papanastassiou and G.J. Wasserburg, Geophys. Res. Lett., 1974, **1**, p.23, N.H. Gale et al., Earth planet. Sci. Lett., 1975, **26**, p.195. U-Pb "age", R. Hutchison et al., Nature, 1975, **254**, p.678, Nature, 1976, **259**, p.159. Nd-Sm data, N. Nakamura et al., Geochimica et Cosmochimica Acta, 1982, **46**, p.1555. Petrology, references, J.L. Berkley et al., Geochimica et Cosmochimica Acta, 1980 (Suppl. 14), p.1089.
644g Washington, U.S. Nat. Mus.; 602g Berlin, Humboldt Univ.; 500g Vienna, Naturhist. Mus.; 430g Paris, Mus. d'Hist. Nat.; 159g Harvard Univ.; Specimens, Cairo, Geol. Mus.;
Specimen(s): [1911,369], 667g. and fragments, 9.1g; [1911, 370], 156g. and fragments, 17g; [1913,25], 641g.; [1913,26], 313g.; [1913,27], 110g.

Nakhon Pathom 13°44′ N., 100°5′ E.
Don Yai Hom Subdistrict, Thailand
Fell 1923, December 21, 2100 hrs, approx.
Stone. Olivine-hypersthene chondrite (L6).
One stone of 23.2kg fell about 10km SSE. of the town of Nakhom Pathom, olivine Fa_{24}, B. Mason, Geochimica et Cosmochimica Acta, 1967, **31**, p.1100. Description with analyses of light and dark portions, 22.17 and 21.69 % total iron, J.A. Nelen and K. Fredriksson, Smithson. Contrib. Earth Sci., 1972 (9), p.69.
Main mass, Bangkok, Nat. Mus.; 852g Washington, U.S. Nat. Mus.;

Nallah 31°58′ S., 126°15′ E.
near Cocklebiddy, Western Australia, Australia
Found 1968
Stone. Olivine-bronzite chondrite (H).
A crusted stone, weight 4.617g, was found 0.5 mile SE. of Nallah Nallah Rockhole, G.J.H. McCall, 2nd. Suppl. to West. Austr. Mus. Spec. Publ. no. 3, 1972, p.19. Description of shape, like flanged tektite button, figured, G.J.H. McCall and W.H. Cleverly, Min. Mag., 1969, **37**, p.286. Reported, Meteor. Bull., 1975 (53), Meteoritics, 1975, **10**, p.155.
Main mass, Kalgoorlie, West. Austr. School of Mines;

Namaqualand v Gibeon.

Namib Desert 24°45′ S., 15°22′ E.
between Sossusvlei and the Wittenberg, Namibia
Found 1979, April 22
Stone. Olivine-bronzite chondrite (H4).
A single mass of about 1kg was found, Meteor. Bull., 1981 (59), Meteoritics, 1981, **16**, p.196.
Specimen, Washington, U.S. Nat. Mus.;

Nammianthal 12°17′ N., 79°12′ E.
North Arcot district, Tamil Nadu, India
Fell 1886, January 27
Synonym(s): *South Arcot*
Stone. Olivine-bronzite chondrite (H5), veined.
A stone of about 4.5kg was seen to fall after a loud report, H.B. Medlicott, Rec. Geol. Surv. India, 1886, **19**, p.268. The stone fell at Durginammiyandal, 6 miles NE. of Tiruvannamalai, North Arcot district, C.A. Silberrad, Min. Mag., 1932, **23**, p.296. Olivine Fa_{19}, B. Mason, Geochimica et Cosmochimica Acta, 1963, **27**, p.1011.
1.3kg Calcutta, Mus. Geol. Surv. India; 659g Paris, Mus. d'Hist. Nat.; 119g Chicago, Field Mus. Nat. Hist.; 99g Vienna, Naturhist. Mus.;
Specimen(s): [56916], 1620g. and fragments, 11g

Nandan v Nantan.

Nanjemoy 38°25′ N., 77°10′ W.
Charles County, Maryland, U.S.A.
Fell 1825, February 10, 1200 hrs
Synonym(s): *Annapolis, Charles County, Maryland, Port Tobacco*
Stone. Olivine-bronzite chondrite (H6).
After a loud detonation, a stone of about 16.5lb was seen to fall, S.D. Carver and W.D. Harrison, Am. J. Sci., 1825, **9**, p.351. Analysis, G. Chilton, Am. J. Sci., 1826, **10**, p.131. Olivine Fa_{18}, B. Mason, Geochimica et Cosmochimica Acta, 1963, **27**, p.1011.
879g Yale Univ.; 350g Vienna, Naturhist. Mus.; 176g Harvard Univ.; 106g Chicago, Field Mus. Nat. Hist.; 100g Tübingen, Univ.; 79g Washington, U.S. Nat. Mus.; 75g Tempe, Arizona State Univ.;
Specimen(s): [11620], 293g.; [33191], 32g.

Nanseiki 36° N., 118° E. approx.
Shandong, China
Fell 1920, August 13
Synonym(s): *Nanseiseki, Shantung*
Doubtful..
A piece of 9.5g is in Kwasan Observatory, Japan, I. Yamamoto, Kwasan Observ. Bull., 1935, p.306 [M.A. 7-173]. Stony-iron, P.M. Millman, Trans. Roy. Astron. Soc. Canada, 1938, **32**, p.202 [M.A. 7-173]. Pallasite, A.D. Nininger, Pop. Astron., Northfield, Minnesota, 1940, **48**, p.558, Contr. Soc. Res. Meteorites, **2**, p.227 [M.A. 8-54]. Not meteoritic, W. Wahl, Geochimica et Cosmochimica Acta, 1965, **29**, p.177.

Nanseiseki v Nanseiki.

Nantan 25°6′ N., 107°42′ E.
Guangxi, China
Found 1958
Synonym(s): *Dongning, Nandan, Nantan County*
Iron. Octahedrite, medium (1.0mm) (IIICD).
A meteorite shower comprising at least 19 individuals of total weight 9500kg was distributed over 30 square km, J.K. Hsian, Geochim., 1974, p.66 [M.A. 75-0432]. Classification and analysis, 6.8 %Ni, 77 ppm.Ga, 293 ppm.Ge, 1.7 ppm.Ir, A. Kracher et al., Geochimica et Cosmochimica Acta, 1980, **44**, p.773. Co-ordinates, D. Bian, Meteoritics, 1981, **16**, p.120.
1900kg Guiyang, Acad. Sin. Inst. Geochem.;

Nantan County v Nantan.

Nan Yang Pao 35°40′ N., 103°30′ E.
Daohe County, Gansu, China
Fell 1917, July 11, 1200 hrs
Synonym(s): *Daohe, Kansu, Taoho*
Stone. Olivine-hypersthene chondrite (L).
After detonations and appearance of white light, a stone of 117lb (52.9kg) fell, 20 miles south of Chih-nan, Tao-ho prefecture. Description, with an analysis, C.Y. Hsieh, Bull. Geol. Soc. China, 1923, **2**, p.95. Olivine Fa_{24}, B. Mason, Geochimica et Cosmochimica Acta, 1963, **27**, p.1011. Listed with co-ordinates, D. Bian, Meteoritics, 1981, **16**, p.116.
Main mass, Gansu, Provincial Govt. Offices, in 1927; 9g Paris, Mus. d'Hist. Nat.;

Naoki 19°15′ N., 77°0′ E.
Parbhani district, Maharashtra, India
Fell 1928, September 29, 1700 hrs
Synonym(s): *Kawagaon, Korgaon, Malegaon, Mategaon, Purna*
Stone. Olivine-bronzite chondrite (H6).
Two stones, 4920 and 1762g, fell a mile apart at Naoki. Two others, not recovered, fell at Korgaon and Mategaon, A.L. Coulson, Rec. Geol. Surv. India, 1930, **62**, p.444 [M.A. 4-421]. Several stones fell at Kawagaon, one at Malegaon, two at Naoki; one of 10,320g from Kawagaon figured, with description, M.A.R. Khan, J. Osmania Univ. Coll. Hyderabad, 1934, **2**, p.22 [M.A. 6-102]. Olivine Fa_{20}, B. Mason, Geochimica et Cosmochimica Acta, 1963, **27**, p.1011.
5.8kg Calcutta, Mus. Geol. Surv. India; 247g Paris, Mus. d'Hist. Nat.; 227g Washington, U.S. Nat. Mus.; 51g Tempe, Arizona State Univ.;
Specimen(s): [1930,457], 224g.

Napoléonsville v Kernouve.

Naragh 33°45′ N., 51°30′ E.
Iran
Fell 1974, August 18, 1830 hrs
Stone. Olivine-bronzite chondrite (H6).
After a fireball was seen, a 2.7kg stone hit ıhe roof of a school laboratory in the village of Naragh. The hole made in the roof measured 90cm in diameter, olivine $Fa_{19.5}$, Meteor. Bull., 1975 (53), Meteoritics, 1975, **10**, p.136. Description, co-ordinates and analysis, 26.5 % total iron, D. Adib and J.G. Liou, Meteoritics, 1979, **14**, p.257.
Main mass, Tehran, Univ. Geophys. Inst.;
Specimen(s): [1975,M.1], 1.43g. crusted fragment

Nardoo (no.1) 29°32′ S., 143°59′ E.
Wanaaring, New South Wales, Australia
Found 1944
Stone. Olivine-bronzite chondrite (H5).
Two stones, 2.75 and 4.25lb, were found on the surface, 7 miles apart, on Nardoo sheep station, near Wanaaring. They show somewhat different microscopical characters; no.1 is similar to the Elsinora stone found 25 miles NW. of Nardoo, and may belong to the same shower. Description, R.O. Chalmers, Austr. Mus. Mag., 1948, **9**, p.263 [M.A. 10-521]. Olivine $Fa_{18.8}$, B. Mason, Rec. Austr. Mus., 1974, **29**, p.169.
Main mass, Sydney, Austr. Mus.; 75g Washington, U.S. Nat. Mus.;

Nardoo (no.2) 29°30′ S., 144°4′ E.
Wanaaring, New South Wales, Australia
Found 1944
Stone. Olivine-hypersthene chondrite (L6).
Description, R.O. Chalmers, Austr. Mus. Mag., 1948, **9**, p.263 [M.A. 10-521]. Olivine Fa_{25}, B. Mason, Rec. Austr. Mus., 1974, **29**, p.169.
Main mass, Sydney, Austr. Mus.; 109g Washington, U.S. Nat. Mus.;

Narellan 34°3′ S., 150°41′20″ E.
County Cumberland, New South Wales, Australia
Fell 1928, April 8, 1915 hrs
Stone. Olivine-hypersthene chondrite (L6).
A stone of 367.5g fell. Description and analysis, T. Hodge-Smith, Rec. Austr. Mus., 1931, **18**, p.283 [M.A. 5-160]. Olivine Fa_{25}, B. Mason, Geochimica et Cosmochimica Acta, 1963, **27**, p.1011.
189g Sydney, Austr. Mus., main mass; 33g Washington,

U.S. Nat. Mus.;
Specimen(s): [1939,875], 24.5g.

Naretha 31°0′ S., 124°50′ E.

Western Australia, Australia
Found 1915
Synonym(s): *Kingoonya, Kingooya*
Stone. Olivine-hypersthene chondrite (L4).

A stone of 6lb was found, Ann. Rep. Geol. Surv. West. Austr., 1922, p.53. Listed, Spec. Publ. West. Austr. Mus., 1965 (3), p.46. Olivine $Fa_{24.8}$, B. Mason, Rec. Austr. Mus., 1974, **29**, p.169. Includes Kingoonya, a 6lb mass found in about 1927, W.H. Cleverly, Rec. West. Austr. Mus., 1976, **4**, p.101.

0.8kg Kalgoorlie, School of Mines; 0.7kg Perth, Geol. Surv. Mus.; 87g Adelaide, South Austr. Mus., of Kingoonya; 5.8g New York, Amer. Mus. Nat. Hist.; 10g Sydney, Austr. Mus., of Kingoonya;

Narni 42°31′ N., 12°31′ E.

Terni, Umbria, Italy
Fell 921, A.D.

Several black stones are said to have fallen in the Nera river (Benedicti Monachi Sancti Andree in Monte Soracte, manuscript preserved in the Vatican library). In 1957 fourteen lithic fragments were found at that place and they are now preserved in Narni. Incomplete description, C. Merli, Coelum, 1957, **25** (5-6), p.76. Historical notes, G.R. Levi-Donati, Coelum, 1957, **25** (11-12), p.17.

Narrabura v Narraburra.

Narraburra 34°15′ S., 147°42′ E.

Temora, County Bland, New South Wales, Australia
Found 1855
Synonym(s): *Narrabura, Nurraburra, Temora, Yeo Yeo Creek*
Iron. Octahedrite, medium (0.60mm) (IIIB).

A mass of about 71lb was found about 12 miles east of Temora, H.C. Russell, J. and Proc. Roy. Soc. New South Wales, 1890, **24**, p.81. Description, analysis, 9.74 %Ni, A. Liversidge, J. and Proc. Roy. Soc. New South Wales, 1904, **37**, p.234. Mentioned, H.J. Axon, Nature, 1961, **191**, p.1287 [M.A. 15-534]. Classification and analysis, 10.13 %Ni, 16.6 ppm.Ga, 28.7 ppm.Ge, 0.016 ppm.Ir, E.R.D. Scott et al., Geochimica et Cosmochimica Acta, 1973, **37**, p.1957. Shocked to 400-600kb, R.R. Jaeger and M.E. Lipschutz, Nature, 1967, **213**, p.975. Description, V.F. Buchwald, Iron Meteorites, Univ. of California, 1975, p.875.

23kg Sydney, Austr. Mus., main mass; 695g Harvard Univ.; 321g Washington, U.S. Nat. Mus.; 127g Chicago, Field Mus. Nat. Hist.; 57g Tempe, Arizona State Univ.; 52g Vienna, Naturhist. Mus.; 41g New York, Amer. Mus. Nat. Hist.; 19g Berlin, Humboldt Univ.;

Specimen(s): [86997], 1873g. slice; [1927,1264], 100g. and fragments, 4g; [1927,1268], 7g.; [1927,1269], 960g. fragments and turnings; [1916,4], 6.6g. of Temora

Naruna 30°57′ N., 98°16′ W.

Burnet County, Texas, U.S.A.
Found 1935, recognized 1939
Stone. Olivine-bronzite chondrite (H).

One stone of 672g, A.D. Nininger, Pop. Astron., Northfield, Minnesota, 1940, **48**, p.556, Contr. Soc. Res. Meteorites, 1942, **2**, p.227 [M.A. 8-54]. Olivine Fa_{19}, B. Mason, Geochimica et Cosmochimica Acta, 1963, **27**, p.1011.

972g Fort Worth, Texas, Monnig Colln.;

Näs 59°11′ N., 12°13′ E.

Värvik, Dalsland, Sweden
Found 1907
Synonym(s): *Naes, Värvik*
Stone. Olivine-hypersthene chondrite, amphoterite (LL6).

A stone of 375g was found, Cat. Met. Roy. Nat. Hist. Mus. Sweden. Analysis, 20.18 % total iron, B. Mason and H.B. Wiik, Geochimica et Cosmochimica Acta, 1964, **28**, p.533. Olivine Fa_{30}, B. Mason, Geochimica et Cosmochimica Acta, 1963, **27**, p.1011.

Nearly complete stone, Stockholm, Riksmus.; 17g Washington, U.S. Nat. Mus.; 13g Tempe, Arizona State Univ.; 9g Chicago, Field Mus. Nat. Hist.;

Nash County v Castalia.

Nashville v Drake Creek.

Nashville (iron) 35°58′ N., 77°58′ W.

Nash County, North Carolina, U.S.A.
Found 1934, known before this year
Iron. Octahedrite.

One main mass and fragments, total about 18kg, E.P. Henderson, letter of May 12, 1939 in Min. Dept. BM(NH). Mentioned, Ann. Rep. U.S. Nat. Mus., 1934, p.42, Cat. Meteor. Arizona State Univ., 1964, p.1041.

Main mass, Raleigh, North Carolina State Museum; 2574g Washington, U.S. Nat. Mus.; 234g Tempe, Arizona State Univ.;

Nashville (stone) 37°27′ N., 98°25′ W.

Kingman County, Kansas, U.S.A.
Found 1939, before this year
Synonym(s): *Barber County*
Stone. Olivine-hypersthene chondrite (L6).

A mass of 25kg, H.H. and A.D. Nininger, The Nininger Collection of Meteorites, Winslow, Arizona, 1950, p.76, A.D. Nininger, Pop. Astron., Northfield, Minnesota, 1939, **47**, p.213. Olivine Fa_{22}, B. Mason, Geochimica et Cosmochimica Acta, 1963, **27**, p.1011.

23kg Washington, U.S. Nat. Mus., main mass; 9.9kg Mainz, Max-Planck-Inst.;

Specimen(s): [1975,M.5], 109.8g. crusted part-slice

Nassirah 21°44′ S., 165°54′ E.

Noumea, New Caledonia
Fell 1936, July 15, 1630 to 1700 hrs
Synonym(s): *New Caledonia*
Stone. Olivine-bronzite chondrite (H4), veined.

Three fragments, 323g, 17g and 7g were recovered. Description, A. Lacroix, C. R. Acad. Sci. Paris, 1937, **204**, p.625, Bull. Soc. franc. Min. Crist., 1937, **60**, p.226 [M.A. 7-72]. Olivine Fa_{19}, B. Mason, Geochimica et Cosmochimica Acta, 1963, **27**, p.1011.

352g Paris, Mus. d'Hist. Nat.;

Natal

Natal, South Africa, Co-ordinates not reported
Fell?
Stone.

A 1.4g fragment is said to be part of a stone which fell in Natal. The fragment is described as an olivine-enstatite chondrite. It may be part of Moshesh (*q.v.*) which in turn may belong to the Queen's Mercy fall (*q.q.v.*), C. Frick and E.C.I. Hammerbeck, Bull. Geol. Surv. S. Africa, 1973 (57), p.28.

Main mass, Bloemfontein, Nat. Mus.;

Nativitas v Santa Apolonia.

Nativitas Tlaxcala v Santa Apolonia.

Nauheim
Hessen, Germany
Pseudometeorite..
A mass of iron, found at Bad Nauheim, Hesse, in 1826, and described as meteoritic in 1828, is shown to be manufactured iron, E. Cohen, Ann. Naturhist. Hofmus. Wien, 1898, R. Bernges, Wetterauische Gesell. Hanau, 1935 [M.A. 7-547]. It is not clear whether Nauheim near Limburg or Nauheim near Mainz is intended. Analysis, 0.04 %Ni, C.J. Elliott Min. Dept. BM(NH).

Naunhof 51°17′ N., 12°35′ E.
Leipzig, Germany
Fell 1540, between 1540 and 1550
Synonym(s): *Grimma*
Doubtful. Iron.
A large mass fell, E.F.F. Chladni, Die Feuer-Meteore, Wien, 1819, p.212. Probably identical with Steinbach (*q.v.*).

Navajo 35°20′ N., 109°30′ W.
Apache County, Arizona, U.S.A.
Found 1921, July 10
Iron. Octahedrite, coarsest (10mm) (IIB).
A mass of 3306lb was found in 1921 buried in talus, with Indian beads. In 1926 another mass, of 1508lb, was found buried in soil 160ft distant. Analysis, 5.43 %Ni, O.C. Farrington, letter of November 15, 1926 in Min. Dept. BM (NH). Description, S.K. Roy and R.K. Wyant, Field Mus. Nat. Hist. Geol. Ser., 1949, **7** (8), p.113 [M.A. 11-139]. Classification, analysis, 5.50 %Ni, 55.0 ppm.Ga, 180 ppm.Ge, 0.46 ppm.Ir, J.T. Wasson, Meteorites, Springer-Verlag, 1974, p.300. Description; shock-hardened, V.F. Buchwald, Iron Meteorites, Univ. of California, 1975, p.878.
2183kg Chicago, Field Mus. Nat. Hist.; 517g Washington, U.S. Nat. Mus.; 453g Copenhagen, Univ. Geol. Mus.;
Specimen(s): [1980,M.7], 107.9g.

Nawapali 21°15′ N., 83°40′ E.
Sambalpur district, Orissa, India
Fell 1890, June 6, 1800 hrs
Stone. Carbonaceous chondrite, type II (CM2).
After appearance of fireball, three stones appear to have fallen. One which fell in the middle of the village of Nawapali broke into pieces and only three small fragments of about 30g, 20g and 10g were preserved; the other two stones, said to be of the "size of a 9lb shot", were found in a field 500yds distant and appear to have been taken away by the villagers, District Superintendent of Police of Sambalpur, letter of December 28, 1896 in Min. Dept. BM(NH). Analysis, 21.0 % total iron, H.B. Wiik, Geochimica et Cosmochimica Acta, 1956, **9**, p.279. Classification, references, W.R. Van Schmus and J.M. Hayes, Geochimica et Cosmochimica Acta, 1974, **38**, p.47.
73g Calcutta, Mus. Geol. Surv. India; 7g Vienna, Naturhist. Mus.; 1.6g Chicago, Field Mus. Nat. Hist.;
Specimen(s): [82968], 21.5g.

Nazareth v Nazareth (stone).

Nazareth (a) v Nazareth (stone).

Nazareth (b) v Nazareth (iron).

Nazareth (c) 34°35′ N., 102°5′ W.
Castro County, Texas, U.S.A.
Found 1967
Stone. Olivine-hypersthene chondrite (L6).
A specimen of 11.7g is in Albuquerque, Univ. of New Mexico. Listed as Nazareth (b), in error, Cat. Huss Colln. Meteorites, 1976, p.33. Nazareth (b) is a synonym for Nazareth (iron).
3.5kg Mainz, Max-Planck-Inst.; 11.7g Albuquerque, Univ. of New Mexico;

Nazareth (iron) 34°31′36″ N., 102°6′18″ W.
Castro County, Texas, U.S.A.
Found 1968
Synonym(s): *Nazareth (b)*
Iron. Octahedrite, medium (1.0mm) (IIIA).
A mass of 11.31kg was found during the ploughing of virgin land. Description, with figures, 8.75 %Ni, R. Gooley et al., Meteoritics, 1971, **6**, p.93. This iron has a high phosphide (schreibersite) content and has been moderately shocked. New analysis, 9.04 %Ni, 20.3 ppm.Ga, 40.3 ppm.Ge, 0.44 ppm.Ir, J.T. Wasson, Meteorites, Springer-Verlag, 1974, p.303. Shock-hardened, V.F. Buchwald, Iron Meteorites, Univ. of California, 1975, p.879.
6kg Mainz, Max-Planck-Inst.; 315g Washington, U.S. Nat. Mus.; 236g Fort Worth, Texas, Monnig Colln.; 230g Copenhagen, Univ. Geol. Mus.; 190g Chicago, Field Mus. Nat. Hist.; 82g Vienna, Naturhist. Mus.;
Specimen(s): [1973,M.16], 89g. part-slice

Nazareth (stone) 34°30′ N., 102°15′ W. approx.
Castro County, Texas, U.S.A.
Found 1938
Synonym(s): *Nazareth, Nazareth (a)*
Stone. Olivine-bronzite chondrite (H).
One stone of 44g was found, A.D. Nininger, Pop. Astron., Northfield, Minnesota, 1939, **47**, p.213. Olivine Fa_{19}, B. Mason, Geochimica et Cosmochimica Acta, 1963, **27**, p.1011. A further mass, weighing 13.1kg and classified as H6, was found in 1977, P.S. Sipiera et al., Meteoritics, 1983, **18**, p.63.
Specimen(s): [1979,M.1], 152.3g. off the second mass; [1980, M.36], 20.2g. of the second mass

Nebraska v Fort Pierre.

Necajevo v Netschaëvo.

Necevo v Netschaëvo.

Nechaevo v Netschaëvo.

Nedagolla 18°41′ N., 83°29′ E.
Vishakhapatnam district, Andhra Pradesh, India
Fell 1870, January 23, 1900 hrs
Synonym(s): *Nidigullam*
Iron. Chemically and structurally anomalous (IRANOM).
After appearance of a luminous meteor (moving from north to south) and detonations, a mass of about 10lb was seen to fall in a field near the village of Nedagolla, J. Lee Warner, letter of April 1, 1871 in Min. Dept. BM(NH)., G.H. Saxton, Proc. Asiatic Soc. Bengal, 1870, p.64. Description, with an analysis, 6.2 %Ni, E. Cohen, Ann. Naturhist. Hofmus. Wien, 1897, **12**, p.119. Peculiarities in the trace element content and in the metallography, H.J. Axon, Nature, 1962, **196**, p.567. Description of structure induced during pre-atmospheric shock, H.J. Axon, Prog. Materials Sci., 1968, **13**, p.185. Analysis, 6.02 %Ni, 0.665 ppm.Ga,

0.005 ppm.Ge, 3.4 ppm.Ir, R. Schaudy et al., Icarus, 1972, **17**, p.174. Description, V.F. Buchwald, Iron Meteorites, Univ. of California, 1975, p.880.
39g Calcutta, Mus. Geol. Surv. India; 38g Vienna, Naturhist. Mus.; 28g Washington, U.S. Nat. Mus.; 26g Copenhagen, Univ. Geol. Mus.; 18g Harvard Univ.;
Specimen(s): [51184], 4020g. main mass, and fragments, 47g

Nedzed v Wabar.

Needles 34°26′39″ N., 114°49′57″ W.
San Bernadino County, California, U.S.A.
Found 1962
Synonym(s): *Carsons Well*
Iron. Octahedrite, fine (0.47mm) (IID).
A mass of 45.3kg was found in the Turtle Mountains about 50km SSW. of Needles. Although chemically similar to the Wallapai irons of Arizona, it is probably not a transported Wallapai mass. Analysis, 10.3 %Ni, 77 ppm.Ga, 93 ppm.Ge, 4.8 ppm.Ir, J.T. Wasson and J. Kimberlin, Meteoritics, 1969, **4**, p.233. Description, V.F. Buchwald, Iron Meteorites, Univ. of California, 1975, p.882.
27.8kg Los Angeles, Univ. of California; 10kg Washington, U.S. Nat. Mus.; 864g Chicago, Field Mus. Nat. Hist.; 93g Vienna, Naturhist. Mus.;
Specimen(s): [1971,100], 982g. etched slice

Needmore 34°2′41″ N., 102°47′57″ W.
Bailey County, Texas, U.S.A.
Found 1976
Stone. Chondrite.
A single mass of 1793g was found during ploughing, G.I Huss, letter of 26 October, 1983 in Min. Dept. BM(NH).
Main mass, Denver, Amer. Meteorite Lab.;

Neenach 34°48′ N., 118°30′ W.
Los Angeles County, California, U.S.A.
Found 1948
Stone. Olivine-hypersthene chondrite (L6).
A stone of 13.8kg was ploughed up, F.C. Leonard, Griffith Observer, 1953, **17**, p.80 [M.A. 12-253], Meteoritics, 1953, **1**, p.28 [M.A. 12-360].
7.6kg Los Angeles, Univ. of California; 2.2kg Washington, U.S. Nat. Mus.; 310g Tempe, Arizona State Univ.;
Specimen(s): [1966,285], 242g.

Negrillos 19°53′ S., 69°50′ W.
Tarapaca, Chile
Found 1936, known before this year
Iron. Hexahedrite (IIA).
One mass of 28.5kg was found buried in nitrate deposits in the Iquique Pampa, near Negrillos, E.P. Henderson, letter of May 12, 1939 in Min. Dept. BM(NH). Analysis, 5.32 %Ni, E.P. Henderson, Am. Miner., 1941, **26**, p.546. Density, E.P. Henderson and S.H. Perry, Geochimica et Cosmochimica Acta, 1954, **6**, p.221. Negrillos has the highest Ir. content of an iron meteorite and on this basis is distinguishable from North Chile (*q.v.*). Analysis, 5.41 %Ni, 59 ppm.Ga, 179 ppm.Ge, 65 ppm.Ir, J.T. Wasson and J.I. Goldstein, Geochimica et Cosmochimica Acta, 1968, **32**, p.329. Description, co-ordinates, V.F. Buchwald, Iron Meteorites, Univ. of California, 1975, p.883.
15kg Washington, U.S. Nat. Mus.; 252g Moscow, Acad. Sci.;
Specimen(s): [1938,265], 1616g. and fragment, 19g

Nei Monggol v Inner Mongolia.

Nejd v Wabar.

Nejed v Wabar.

Nejed (no. 2)
Found
Iron. Octahedrite, fine.
A specimen in Tempe has 7.78% Ni. and is distinct from Wabar (*q.v.*), C.B. Moore et al., Meteorite Research, ed. P.M. Millman, D. Reidel, Dordrecht-Holland, 1969, p.738. Possibly mis-labelled Gibeon, V.F. Buchwald, Iron Meteorites, Univ. of California, 1975, p.1273.
159g Tempe, Arizona State Univ.;

Nejo 9°30′ N., 35°20′ E.
Walaga Province, Ethiopia
Fell 1970, May 11, 1130 hrs
Synonym(s): *Jarso, Walaga, Wollega*
Stone. Olivine-hypersthene chondrite (L6).
From the sound effects, three objects passed over Jarso, a village 16km W. of Nejo. One object landed near Jarso, made a hole 25cm in diameter and 30cm deep, and broke into pieces, three of which were recovered, of 2.3kg, 139g and 17g. Description and analysis, 22.26 % total iron, R.S. Clarke, Jr. et al., Smithson. Contrib. Earth Sci., 1972 (9), p.67.
183g Washington, U.S. Nat. Mus.;
Specimen(s): [1971,195], 225g. part-crusted fragment, and fragments, 0.7g

Nellore v Yatoor.

Nelson County 37°45′ N., 85°30′ W.
Kentucky, U.S.A.
Found 1856
Iron. Octahedrite, coarsest (1-10mm) (IIIF).
A mass of 161lb was ploughed up. Analysis, 6.11 %Ni, J.L. Smith, Am. J. Sci., 1860, **30**, p.240, C.U. Shepard, Am. J. Sci., 1861, **31**, p.459. Further analysis, 7.02 %Ni, 6.33 ppm.Ga, 0.840 ppm.Ge, 7.9 ppm.Ir, R. Schaudy et al., Icarus, 1972, **17**, p.174. Classification, E.R.D. Scott and J.T. Wasson, Geochimica et Cosmochimica Acta, 1976, **40**, p.103. Description; pre-terrestrial cold deformation, V.F. Buchwald, Iron Meteorites, Univ. of California, 1975, p.885.
25kg Vienna, Naturhist. Mus.; 6.6kg Harvard Univ.; 2.6kg Paris, Mus. d'Hist. Nat.; 1.2kg Washington, U.S. Nat. Mus.; 900g Chicago, Field Mus. Nat. Hist.; 346g Berlin, Humboldt Univ.; 345g New York, Amer. Mus. Nat. Hist.;
Specimen(s): [40872], 3748g. and rusted fragments, 57g; [32046], 237g.

Nenntmannsdorf 50°58′ N., 13°57′ E.
Pirna, Dresden, Germany
Found 1872
Synonym(s): *Pirna*
Iron. Octahedrite, coarsest (10mm) (IIB).
A mass of about 12.5kg was found 2ft below the surface, F.E. Geinitz, Sitzungsber. Naturwiss. Gesell. Isis in Dresden, 1873, p.4, Neues Jahrb. Min., 1876, p.608. Description and analysis, 5.48 %Ni, E. Cohen, Meteoritenkunde, 1905, **3**, p.69. Further analysis, 6.18 %Ni, 58.8 ppm.Ga, 176 ppm.Ge, 0.057 ppm.Ir, A. Kracher et al., Geochimica et Cosmochimica Acta, 1980, **44**, p.773. Description; shocked and annealed, V.F. Buchwald, Iron Meteorites, Univ. of California, 1975, p.888.
11.5kg Dresden, Min. Mus., main mass; 66g Vienna, Naturhist. Mus.; 47g Yale Univ.; 26g Chicago, Field Mus. Nat. Hist.; 15g Washington, U.S. Nat. Mus.;
Specimen(s): [56840], 13g. and fragments, 2g

Neptune Mountains 83°15′ S., 55° W.

Pensacola Mountains, Antarctica
Found 1964
Iron. Octahedrite, coarse (1.9mm) (IA).

One mass of 1070g was found, Meteor. Bull., 1965 (34). Classification and analysis, 7.1 %Ni, 73.9 ppm.Ga, 269 ppm.Ge, 2.0 ppm.Ir, J.T. Wasson, Meteorites, Springer-Verlag, 1974, p.298. Description, V.F. Buchwald, Iron Meteorites, Univ. of California, 1975, p.890.

997g Washington, U.S. Nat. Mus.;

Nerft 56°30′ N., 21°30′ E.

Courland, Latvian SSR, USSR
Fell 1864, April 12, 0445 hrs
Synonym(s): *Poghel, Pohgel, Swajahn*
Stone. Olivine-hypersthene chondrite (L6), veined.

After detonations, two stones fell, one of about 5.5kg at the farmhouse Swajahn, and the other of about 4.75kg at the farmhouse Pohgel, 700yds distant, C. Grewingk and C. Schmidt, Arch. Naturk. Liv.-Ehst.-u. Kurlands, Ser. 1, Min. Wiss., Dorpat, 1864, **3**, p.554. Description, with doubtful analysis (no CaO reported), A. Kuhlberg, Arch. Naturk. Liv.-Ehst.-u. Kurlands, Ser. 1, Min. Wiss., Dorpat, 1867, **4**, p.2. Olivine Fa_{23}, B. Mason, Geochimica et Cosmochimica Acta, 1963, **27**, p.1011. Further analysis, 22.57 % total iron, M.I. D'yakonova, Метеоритика, 1972, **31**, p.119.

5.2kg Tartu, Univ., main mass; 1.5kg Vienna, Naturhist. Mus.; 643g Paris, Mus. d'Hist. Nat.; 340g Moscow, Acad. Sci.; 252g New York, Amer. Mus. Nat. Hist.; 105g Chicago, Field Mus. Nat. Hist.; 71g Washington, U.S. Nat. Mus.; 57g Tübingen, Univ.; 52g Berlin, Humboldt Univ.;

Specimen(s): [39711], 69.5g. from Pohgel; [1920,326], 97g. from Swajahn

Ness City v Ness County (1894).

Ness County (in part) v Beeler.

Ness County (1894) 38°30′ N., 99°36′ W.

Kansas, U.S.A.
Found 1894
Synonym(s): *Kansada, Ness City, Welmannville, Welmanville*
Stone. Olivine-hypersthene chondrite (L6).

The earliest known stone of this fall was a coarsely brecciated one of about 21lb found in "November 1894, about 0.5 mile SW. of Kansada", H.A. Ward, letter of November 14, 1897 in Min. Dept. BM(NH). This stone should probably be assigned to Beeler (*q.v.*), as it is more brecciated than the other stones assigned to Ness County (1894). This confused situation is discussed, W.F. Read, Meteoritics, 1972, **7**, p.417 but no conclusion has yet been made. The next stone was one of 417g found in 1897 in SW. of Ness County, H.L. Ward, Am. J. Sci., 1899, **7**, p.233. Later others were found, including one ploughed up on April 10, 1899, seven miles south and three miles west of Ness City (label of H.A. Ward in Min. Dept. BM(NH).). Altogether 26 stones, varying in weight from 34 to 3467g, and of total weight about 17kg were found, O.C. Farrington, Field Mus. Nat. Hist. Geol. Ser., 1902, **1** (11), p.300. Analysis (showing an improbably large amount of alumina), G.P. Merrill, Am. J. Sci., 1913, **35**, p.517. Olivine Fa_{25}, B. Mason, Geochimica et Cosmochimica Acta, 1963, **27**, p.1011.

22kg Chicago, Field Mus. Nat. Hist.; 14kg New York, Amer. Mus. Nat. Hist., approx.; 6kg Tempe, Arizona State Univ.; 5.8kg Washington, U.S. Nat. Mus.; 4.8kg Los Angeles, Univ. of California; 1.3kg Paris, Mus. d'Hist. Nat.; 692g Harvard Univ.; 695g Michigan, Univ.; 604g Budapest, Nat. Mus.; 667g Illinois, Univ.; 580g Vienna, Naturhist. Mus.; 500g La Plata, Mus.; 570g Copenhagen, Univ. Geol.Mus.;

Specimen(s): [83489], 2008g. off the Kansada stone; [85493], 433.5g. nearly complete stone, and fragments, 35g; [84439], 193g. nearly complete stone of Ness City, and fragments, 4g; [1959,765], 3143g. three fragments, 1054g, 745g and 1344g

Ness County (1938) 38°29′ N., 99°55′ W.

Kansas, U.S.A.
Found 1938, recognized in this year
Stone. Olivine-bronzite chondrite (H4).

One stone of 652g was recognized in 1938 as distinct from the Ness County (1894) fall, A.D. Nininger, Pop. Astron., Northfield, Minnesota, 1939, **47**, p.213. Listed, H.H. and A.D. Nininger, The Nininger Collection of Meteorites, Winslow, Arizona, 1950, p.77. Olivine Fa_{20}, B. Mason, Geochimica et Cosmochimica Acta, 1963, **27**, p.1011.

504g Tempe, Arizona State Univ.;

Specimen(s): [1959,892], 58g. slice

Netschaëvo 54°14′ N., 35°9′ E.

Kaluga, Federated SSR, USSR
Found 1846
Synonym(s): *Necajevo, Necevo, Nechaevo, Netschajewo, Netschjewo, Tula*
Iron. Structurally anomalous octahedrite (1.25mm) (IIE-ANOM).

A mass of about 250kg was found in making a road, and was broken up and used for various purposes, J. Auerbach, Bull. Soc. Nat. Moscou, 1858, **31** (1), p.331. Description, references, V.F. Buchwald, Iron Meteorites, Univ. of California, 1975, p.891. It was re-heated artificially. Contains chondrules, olivine Fa_{14}, E. Olsen and E. Jarosewich, Science, 1971, **174**, p.583. Structural classification and analysis of metal, 8.6 %Ni, 24.8 ppm.Ga, 66 ppm.Ge, 1.8 ppm.Ir, E.R.D. Scott et al., Geochimica et Cosmochimica Acta, 1973, **37**, p.1957. 5858g are known to be preserved, E.L. Krinov, Каталог Метеоритов АИад. Наук СССР, Москва, 1947 [M.A. 10-511].

1kg Vienna, Naturhist. Mus.; 552g Tempe, Arizona State Univ.; 326g Calcutta, Mus. Geol. Surv. India; 318g Tübingen, Univ.; 267g Washington, U.S. Nat. Mus.; 185g Moscow, Acad. Sci.; 180g Chicago, Field Mus. Nat. Hist.; 132g Copenhagen, Univ. Geol. Mus.; 70g Paris, Mus. d'Hist. Nat.;

Specimen(s): [33952], 554g.; [33930], 339g. and fragments 46g, and filings, 16g; [54644], 32.5g.; [33953], 31.7g.

Netschajewo v Netschaëvo.

Neu Granada v Santa Rosa.

Netschjewo v Netschaëvo.

New Almelo 39°40′ N., 100° W.

Norton County, Kansas, U.S.A.
Found 1917, recognized in 1932
Stone. Olivine-hypersthene chondrite (L5), brecciated, veined.

One mass of 3kg was found, H.H. Nininger, Our Stone Pelted Planet, 1933. A second stone of about 1150g has since been found, H.H. Nininger, Trans. Kansas Acad. Sci., 1936, **39**, p.169, The Nininger Collection of Meteorites, Winslow, Arizona, 1950, p.77. Listed, F.C. Leonard, Univ. New Mexico Publ. Albuquerque, 1946 (meteoritics ser. no. 1), p.38. Olivine Fa_{24}, B. Mason, Geochimica et Cosmochimica

Acta, 1963, **27**, p.1011. Analysis, 21.53 % total iron, V.Ya. Kharitonova, Метеоритика, 1969, **29**, p.91.
5kg Tempe, Arizona State Univ.; 1.5kg Washington, U.S. Nat. Mus.; 240g Chicago, Field Mus. Nat. Hist.; 199g Harvard Univ.;
Specimen(s): [1959,1000], 1048g. almost complete stone, and fragments, 15g

New Baltimore 40°0′ N., 78°51′ W.
Somerset County, Pennsylvania, U.S.A.
Found 1922
Synonym(s): *Somerset County*
Iron. Chemically and structurally anomalous (IRANOM).
A mass of about 18kg was ploughed up in a cornfield on the crest of Allegheny Mts., 3 miles NW. of New Baltimore. Description, with an analysis, 6.42 %Ni, G.P. Merrill, Am. J. Sci., 1923, **5**, p.175, Am. J. Sci., **6**, p.262. Perhaps identical with Mount Joy (G.P. Merrill). Description, E.P. Henderson and S.H. Perry, Proc. U.S. Nat. Mus., 1958, **107**, p.339 [M.A. 14-130]. Described and figured, R.W. Stone and E.M. Starr, Bull. Pa. Geol. Surv., 1967 (Rep. G2, 1932, rev.), p.15. Classification and analysis, 6.36 %Ni, 20.4 ppm.Ga, 35.9 ppm.Ge, 10 ppm.Ir, E.R.D. Scott et al., Geochimica et Cosmochimica Acta, 1973, **37**, p.1957. Shock-hardened, sheared and distinct from Mount Joy (*q.v.*) and Pittsburg (*q.v.*), V.F. Buchwald, Iron Meteorites, Univ. of California, 1975, p.894.
13.8kg Harvard Univ.; 3.2kg Washington, U.S. Nat. Mus.; 1.6kg Chicago, Field Mus. Nat. Hist.; 1.4kg Paris, Mus. d'Hist. Nat.;
Specimen(s): [1924,1], 513g. slice, and fragments, 24.5g

Newberry v Ruff's Mountain.

New Caledonia v Nassirah.

New Concord 40°0′ N., 81°46′ W.
Muskingum County, Ohio, U.S.A.
Fell 1860, May 1, 1245 hrs
Synonym(s): *Guernsey County, Muskingum County*
Stone. Olivine-hypersthene chondrite (L6), veined.
After detonations and appearance of fireball, about 30 stones fell over an area of 10 by 3 miles; the largest stone weighed 103lb and the total weight was about 500lb, E.B. Andrews et al., Am. J. Sci., 1860, **30**, p.103, 296. One of the stones killed a colt. Description and analysis, J.L. Smith, Am. J. Sci., 1861, **31**, p.87. Analysis, 21.60 % total iron, B. Mason and H.B. Wiik, Am. Mus. Novit., 1961 (2069). Olivine Fa_{24}, B. Mason, Geochimica et Cosmochimica Acta, 1963, **27**, p.1011. XRF analysis, 23.79 % total iron, H. von Michaelis et al., Earth planet. Sci. Lett., 1969, **5**, p.387.
103lb Marietta College, complete stone; 38.4kg Tempe, Arizona State Univ.; 24.8kg Harvard Univ.; 13.8kg Berlin, Humboldt Univ.; 8.7kg Washington, U.S. Nat. Mus.; 6.3kg Yale Univ.; 4.4kg Chicago, Field Mus. Nat. Hist.; 5kg Washington, U.S. Nat. Mus.; 6kg Philadelphia, Acad. Nat. Sci.; 3kg Tempe, Arizona State Univ.; 2kg New York, Amer. Mus. Nat. Hist., approx.; 1.6kg Vienna, Naturhist. Mus.; 1.55kg Paris, Mus. d'Hist. Nat.; 830g Budapest, Nat. Mus.;
Specimen(s): [47217], 18597g. nearly complete stone (no. 5 of J.L. Smith's list); [34617], 890g. and fragments, 16g; [94706], 37g. three pieces; [96272], Thin section

New Granada v Santa Rosa.

New Ireland v Dyarrl Island.

New Jersey v Deal.

New Leipzig 46°22′ N., 101°57′ W.
Grant County, North Dakota, U.S.A.
Found 1936
Iron. Octahedrite, coarse (2.6mm) (IA).
One mass of 20kg was found, A.D. Nininger, Pop. Astron., Northfield, Minnesota, 1937, **45**, p.449 [M.A. 7-62]. Analysis, 6.88 %Ni, 93.1 ppm.Ga, 445 ppm.Ge, 2.5 ppm.Ir, J.T. Wasson, Meteorites, Springer-Verlag, 1974, p.298. Slightly weathered mass, some pre-terrestrial cracks, V.F. Buchwald, Iron Meteorites, Univ. of California, 1975, p.897.
17kg Washington, U.S. Nat. Mus., main mass; 553 Chicago, Field Mus. Nat. Hist.;
Specimen(s): [1959,937], 19g. slice, and sawings, 0.5g

New Mexico 34°30′ N., 107° W. approx.
New Mexico, U.S.A.
Found 1935, recognized in this year
Iron. Octahedrite, coarsest.
An Indian axe (130g) found in a ruin in New Mexico is of meteoritic iron, H.H. and A.D. Nininger, The Nininger Collection of Meteorites, Winslow, Arizona, 1950, p.78. A single, finger-shaped lamella of kamacite from a coarsest octahedrite, V.F. Buchwald, Iron Meteorites, Univ. of California, 1975, p.899.
124g Tempe, Arizona State Univ., main mass;

New Moore 33°7′ N., 102°7′ W. approx.
Lynn County, Texas, U.S.A.
Found 1972
Stone. Chondrite.
Two stones are listed, New Moore (a) 100.3g, and New Moore (b) 200g found in 1975, Cat. Huss Coll. Meteorites, 1976, p.34.
95.6g Mainz, Max-Planck-Inst., of New Moore (a); 182g Mainz, Max-Planck-Inst., of New Moore (b);

Newport 35°36′ N., 91°16′ W.
Jackson County, Arkansas, U.S.A.
Found 1923
Stony-iron. Pallasite (PAL).
A mass of 5600g was found. Description, H.H. Nininger, Am. J. Sci., 1932, **23**, p.78 [M.A. 5-156]. Analysis of metal, 10.7 %Ni, 17.5 ppm.Ga, 31.2 ppm.Ge, 0.16 ppm.Ir, J.T. Wasson and S.P. Sedwick, Nature, 1969, **222**, p.22. Olivine Fa_{12}, P.R. Buseck and J.I. Goldstein, Bull. Geol. Soc. Amer., 1969, **80**, p.2141. Texture, references, E.R.D. Scott, Geochimica et Cosmochimica Acta, 1977, **41**, p.693.
1.8kg Tempe, Arizona State Univ.; 222g Washington, U.S. Nat. Mus.; 642g Paris, Mus. d'Hist. Nat.; 482g Harvard Univ.; 137g Chicago, Field Mus. Nat. Hist.;
Specimen(s): [1931,159], 200g. and fragments, 29g

Newsom 37°36′ N., 105°50′ W.
Alamosa County, Colorado, U.S.A.
Found 1939
Synonym(s): *Newsome*
Stone. Olivine-hypersthene chondrite (L).
One stone of 892g was found, A.D. Nininger, Pop. Astron., Northfield, Minnesota, 1940, **48**, p.556, Contr. Soc. Res. Meteorites, **2**, p.227 [M.A. 8-54], H.H. and A.D. Nininger, The Nininger Collection of Meteorites, Winslow, Arizona, 1950, p.78. Olivine Fa_{24}, B. Mason, Geochimica et Cosmochimica Acta, 1963, **27**, p.1011.
204g Tempe, Arizona State Univ.; 67g Denver, Mus. Nat. Hist.;
Specimen(s): [1959,893], 252g.

Newsome v Newsom.

Newton v Newtown.

Newton County v Mincy.

Newtown 41°24′ N., 73°19′ W.
Fairfield County, Connecticut, U.S.A.
Found 1925
Synonym(s): *Newton*
Pseudometeorite..
Found on the surface about 5 months after a notable meteor was observed, C.A. Reeds, Bull. Amer. Mus. Nat. Hist., 1937, **73** (6), p.517 [M.A. 7-61]. Not a meteorite, B. Mason, Min. Mag., 1964, **33**, p.935.
216g New York, Amer. Mus. Nat. Hist.;

New Westville 39°48′ N., 84°49′ W.
Preble County, Ohio, U.S.A.
Found 1941
Iron. Octahedrite, fine (0.4mm) (IVA).
A weathered mass of 4.8kg was found. Description with analysis, 9.41 %Ni, E.P. Henderson and S.H. Perry, Smithsonian Misc. Coll., 1946, **104** (17) [M.A. 9-301]. Analysis, 9.45 %Ni, M.I. D'yakonova, Метеоритика, 1958 (16), p.180. Further analysis, 9.36 %Ni, 2.40 ppm.Ga, 0.139 ppm.Ge, 0.55 ppm.Ir, R. Schaudy et al., Icarus, 1972, **17**, p.174. Description, V.F. Buchwald, Iron Meteorites, Univ. of California, 1975, p.900.
3318g Washington, U.S. Nat. Mus., main mass; 206g Chicago, Field Mus. Nat. Hist.;

Ngawi 7°27′ S., 111°25′ E.
Mandioen, Java, Indonesia
Fell 1883, October 3, 1715 hrs
Synonym(s): *Karang Modjo, Madioen, Magetan*
Stone. Olivine-hypersthene chondrite, amphoterite (LL3).
After appearance of fireball and detonations, at least three stones fell. Two stones, one of 1191g and the other of 202g fell at Karang Modjo and Ngawi respectively, E.H. von Baumhauer, Arch. Néerland. Sci. Nat. Haarlem, 1884, **19**, p.175, J. Bosscha, Arch. Néerland. Sci. Nat. Haarlem, 1886, **21**, p.177. Description, analysis, 18.84 % total iron, B. Mason and H.B. Wiik, Am. Mus. Novit., 1966 (2273). Chondrule study, D.E. Lange and J.W. Larimer, Science, 1973, **182**, p.920. Analyses of lithic fragments, R.V. Fodor and K. Keil, Meteoritics, 1975, **10**, p.325.
732g Leiden, Geol. Min. Mus., main mass of Karang Modjo stone and Ngawi stone; 103g Washington, U.S. Nat. Mus.; 45g Vienna, Naturhist. Mus.; 8g Chicago, Field Mus. Nat. Hist.;
Specimen(s): [83396], 35g. another 8g and fragments, 1.12g

N'Goureyma 13°51′ N., 4°23′ W. approx.
Djenne, Ke Macina, Mali
Fell 1900, June 15
Iron. Chemically and structurally anomalous (IRANOM).
A mass of 37.5kg which came into the possession of H. Minod of Geneva in 1901 is said to have fallen and penetrated 1 metre into the clayey soil. Description and analysis, S. Meunier, C. R. Acad. Sci. Paris, 1901, **132**, p.441. Analysis, 9.26 %Ni, E. Cohen, Am. J. Sci., 1903, **15**, p.254. Drastically reheated, H.J. Axon, Prog. Materials Sci., 1968, **13**, p.185. Classification and analysis, 9.26 %Ni, 0.067 ppm.Ga, 0.016 ppm.Ge, 0.058 ppm.Ir, R. Schaudy et al., Icarus, 1972, **17**, p.174. Unique; lowest Ga and Ge contents of any iron meteorite, V.F. Buchwald, Iron Meteorites, Univ. of California, 1975, p.901.
4kg Heidelberg, Max-Planck-Inst.; 1kg Chicago, Field Mus. Nat. Hist.; 1kg Budapest, Nat. Mus., maximum weight; 910g New York, Amer. Mus. Nat. Hist.; 900g Washington, U.S. Nat. Mus.; 707g Paris, Mus. d'Hist. Nat.; 455g Vienna, Naturhist. Mus.; 281g Tübingen, Univ.; 144g Oslo, Min.-Geol. Mus.;
Specimen(s): [85693], 756g. slice, and four pieces 101g

Niagara 48°0′ N., 97°56′ W.
Grand Forks County, North Dakota, U.S.A.
Found 1879
Iron. Octahedrite, coarse (1.4mm) (I).
A small mass of 115g was found 2 miles SE. of Niagara. Description, with an analysis, 7.37 %Ni, H.L. Preston, J. Geol., 1902, **10**, p.518. Description, V.F. Buchwald, Iron Meteorites, Univ. of California, 1975, p.905.
25g Chicago, Field Mus. Nat. Hist.; 19g Yale Univ.; 8g Rome, Vatican Colln; 7g Vienna, Naturhist. Mus.; 3g Berlin, Humboldt Univ.;
Specimen(s): [85428], 16g.

Nickelsville v Dungannon.

Nickolskoye v Nikolskoe.

Nico v Gibeon.

Nicorps 49°2′ N., 1°26′ W.
Normandy, France
Fell 1750, October 11
Stone.
The fall appears to be well authenticated, but no specimen is now preserved, A. Lacroix, Bull. Mus. d'Hist. Nat. Paris, 1927, **33**, p.421. A large stone fell and was broken, the biggest piece weighing about 20lb. It was grey with a black crust. All trace had been lost before 1818, E.F.F. Chladni, Die Feuer-Meteore, Wien, 1819, p.243.

Nidigullam v Nedagolla.

Nieder Finow 52°50′ N., 13°56′ E.
Eberswalde, Frankfurt, Germany
Found 1950, before this year
Iron. Octahedrite, coarse (IA).
A small mass of 287g was found, W. Neben and A. Schüller, Urania, Berlin, 1950, **13**, p.382, F. Heide, Kleine Meteoritenkunde, 2nd edn., 1957, p.130. Analysis, 8.27 %Ni, 72.0 ppm.Ga, 257 ppm.Ge, 2.6 ppm.Ir, J.T. Wasson, Meteorites, Springer-Verlag, 1974, p.299.
Main mass, Berlin, Humboldt Univ.;

Niederreissen 51°4′ N., 11°22′ E.
Buttelstedt, Erfurt, Germany
Fell 1581, July 26, 1300-1400 hrs
Doubtful. Stone?.
One stone of 39lb fell, E.F.F. Chladni, Die Feuer-Meteore, Wien, 1819, p.81, 218, T. von Grotthus, Ann. Physik (Gilbert), 1821, **67**, p.348 but the evidence is not conclusive. The latitude and longitude cited are those of Buttelstedt.

Niger (I) v Niger (C2).

Niger (II) v Niger (LL6).

Niger (III) v Niger (L6).

Niger (C2)
Niger
Found 1969, recognized in this year
Synonym(s): *Niger (I)*
Stone. Carbonaceous chondrite, type II (CM2).
Fragments were sent from Abidjan to Paris in 1969. The C content is 2.75% and water, 13.42%, M. Christophe-Michel-Levy, Meteoritics, 1969, **4**, p.283, abs. Mineral data, petrography, bulk analysis, 20.59 % total iron, C. Desnoyers, Earth planet. Sci. Lett., 1980, **47**, p.223.
Main mass, Paris, Mus. d'Hist. Nat., fragments, 4.95g, and powder, 12g;

Niger (L6)
Niger
Fell 1967, August 1, or found
Synonym(s): *Niger (III)*
Stone. Olivine-hypersthene chondrite (L6).
Either this or Niger (LL6) (*q.v.*) was seen to fall near the village of Koutiaran, Mirria District, Niger. The stone was sent from Abidjan to Paris in 1969. Olivine Fa_{25}, M. Christophe-Michel-Levy, Meteoritics, 1969, **4**, p.283, abs.
3.30g Paris, Mus. d'Hist. Nat., main mass;

Niger (LL6)
Niger
Fell 1967, August 1, or found
Synonym(s): *Niger (II)*
Stone. Olivine-hypersthene chondrite, amphoterite (LL6).
Either this or Niger (L6) (*q.v.*) was seen to fall near the village of Koutiaran, Mirra District, Niger. The stone was sent from Abidjan to Paris in 1969. Olivine Fa_{30}, M. Christophe-Michel-Levy, Meteoritics, 1969, **4**, p.283, abs.
3.30g Paris, Mus. d'Hist. Nat., main mass;

Nigheija v Mighei.

Niho v Nio.

Nihuá v Campo del Cielo.

Nikolaev v Bischtübe.

Nikolaevka 52°27′ N., 78°38′ E.
Pavlodar, Kazakh SSR, USSR, [Николаевка]
Fell 1935, July 11, 0300 hrs
Synonym(s): *Nikolajewka*
Stone. Olivine-bronzite chondrite (H4).
The stone is said to have fallen at 78°38'N., 82°47'E., V.I. Vernadsky, letter of August 5, 1939 in Min. Dept. BM(NH)., but this point is in the Arctic Ocean. The place of fall is mapped as Nikolaevsk (51°6'N., 111°50'E.) in Transbaikal, P.M. Millman, Trans. Roy. Astron. Soc. Canada, 1938, **32**, p.199. Listed, I.S. Astapowitsch, Trans. Roy. Astron. Soc. Canada, 1938, **32**, p.195 [M.A. 7-172]. Neither the latitude and longitude given by Vernadsky nor that given by P.M. Millman is correct, L.A. Kulik, Метеоритика, 1941, **1**, p.73 [M.A. 9-294], E.L. Krinov, Астрон. Журнал, 1945, **22**, p.303 [M.A. 9-297], Каталог Метеоритов Акад. Наук СССР, Москва, 1947 [M.A. 10-511]. The time and date of fall also differ from the earlier accounts. Analysis, 24.57 % total iron, M.I. D'yakonova and V.Ya. Kharitonova, Метеоритика, 1961, **21**, p.51. Olivine Fa_{19}, B. Mason, Geochimica et Cosmochimica Acta, 1963, **27**, p.1011.
3996g Moscow, Acad. Sci., main mass;

Nikolajev v Bischtübe.

Nikolajewka v Nikolaevka.

Nikolskoe 56°7′ N., 37°20′ E.
Solnechnogorsk, Moscow, USSR, [Никольское]
Fell 1954, March 6, 1522 hrs
Synonym(s): *Nickolskoye*
Stone. Olivine-hypersthene chondrite (L4).
6kg of fragments were collected, E.L. Krinov, Метеоритика, 1956, **14**, p.70 [M.A. 13-359], L.G. Kvasha, Метеоритика, 1958, **15**, p.97, V.I. Mikheev and A.I. Kalinin, Метеоритика, 1958, **15**, p.156 [M.A. 14-48]. Analysis, M.I. D'yakonova and V.Ya. Kharitonova, Метеоритика, 1960, **18**, p.48 [M.A. 16-447]. Listed, K. Keil, Fortschr. Min., 1960, **38**, p.249. Olivine Fa_{24}, B. Mason, Geochimica et Cosmochimica Acta, 1963, **27**, p.1011.
3.48kg Moscow, Acad. Sci.; 32g Washington, U.S. Nat. Mus.;
Specimen(s): [1956,326], 48g.

Nilpena 31°5′ S., 138°18′ E.
County Taunton, South Australia, Australia
Found 1975, June 4
Stone. Achondrite, Ca-poor. Ureilite (AURE).
A single mass of 173g was found in the NW. corner of the hundred of Nilpena, north of the Parachilna-Motpena road, Meteor. Bull., 1981 (59), Meteoritics, 1981, **16**, p.196. Description, analysis, A.L. Jaques and M.J. Fitzgerald, Geochimica et Cosmochimica Acta, 1982, **46**, p.893.
110g Canberra, Bureau Min. Res.;
Specimen(s): [1982,M.14], 13g.

Nine Mile v Menindee Lakes 003.

Ningbo 29°54′ N., 121°33′ E.
near Ningbo, Zhejiang, China
Fell 1975, October 4, 1220 hrs
Iron. Octahedrite, fine (0.3mm) (IVA).
A single mass of 14.3kg fell, Meteor. Bull., 1978 (55), Meteoritics, 1978, **13**, p.343. 54Mn activity, S. Song et al., Geochimica, 1976, p.273, in Chinese [M.A. 78-749]. Listed, references, D. Bian, Meteoritics, 1981, **16**, p.122. Analysis, 8.2 %Ni, 2.2 ppm.Ga, 0.13 ppm.Ge, 2.0 ppm.Ir, D.J. Malvin et al., priv. comm., 1983.
14.2kg Beijing, Planetarium;

Nio 34°12′ N., 131°34′ E.
Yoshiki, Yamaguchi, Honshu, Japan
Fell 1897, August 8, 2230 hrs
Synonym(s): *Niho, Niwo*
Stone. Olivine-bronzite chondrite (H).
After appearance of fireball and detonations, two stones of 195g and 253g respectively, and probably others, were found, K. Jimbo, Beitr. Miner. Japan, 1906 (2), p.50. The larger specimen has been lost, S. Murayama, letter of April 6, 1962 in Min. Dept. BM(NH)., S. Murayama, Nat. Sci. Mus. Tokyo, 1953, **20**, p.129 [M.A. 13-80]. Olivine Fa_{19}, B. Mason, Geochimica et Cosmochimica Acta, 1967, **31**, p.1100. Description, analysis, 26.4 % total iron, A. Miyashiro and S. Murayama, Chem. Erde, 1967, **26**, p.219.
192g Tokyo, Nat. Sci. Mus.;

Niro v Verkhne Udinsk.

Niwo v Nio.

Njeschin v Alexandrovsky.

N'Kandhla 28°34′ S., 30°42′ E.
Kwazulu, Natal, South Africa
Fell 1912, August 1, 1330 hrs
Synonym(s): *N'Kandla, Zululand*
Iron. Octahedrite, medium (0.9mm) (IID).
After detonations and appearance of trail of smoke, a mass of about 38lb was seen to fall on the Pokinyoni Hill, near the junction of the Buffalo and Tugela rivers. Description and analysis, 10.68 %Ni, G.H. Stanley, S. African J. Sci., 1914, **10**, p.105. Co-ordinates, C. Frick and E.C.I. Hammerbeck, Bull. Geol. Surv. S. Africa, 1973 (57), p.29. Analysis, 9.96 %Ni, 71.8 ppm.Ga, 83.3 ppm.Ge, 18 ppm.Ir, J.T. Wasson, Geochimica et Cosmochimica Acta, 1969, **33**, p.859. Description, V.F. Buchwald, Iron Meteorites, Univ. of California, 1975, p.906.
146g Washington, U.S. Nat. Mus.; 99g Chicago, Field Mus. Nat. Hist.;
Specimen(s): [1921,17], 16845g. main mass, and pieces, 85g

N'Kandla v N'Kandhla.

Nobleboro v Nobleborough.

Nobleborough 44°5′ N., 69°29′ W.
Lincoln County, Maine, U.S.A.
Fell 1823, August 7, 1630 hrs
Synonym(s): *Nobleboro*
Stone. Achondrite, Ca-rich. Eucrite (AEUC).
After detonations and appearance of cloud, a stone of about 5lb fell and broke into pieces, P. Cleaveland, Am. J. Sci., 1824, **7**, p.170. Classification, howardite, M.B. Duke and L.T. Silver, Geochimica et Cosmochimica Acta, 1967, **31**, p.1637. Classification, eucrite, B. Mason, Geochimica et Cosmochimica Acta, 1967, **31**, p.107. Very little preserved. Analysis, B. Mason et al., Smithson. Contrib. Earth Sci., 1979 (22), p.30.
60g Halle, Univ., in 1897; 14g Chicago, Field Mus. Nat. Hist.; 9.4g Tempe, Arizona State Univ.; 6g Vienna, Naturhist. Mus.; 6g Yale Univ.; 2g Harvard Univ.;
Specimen(s): [35406], less than 1g.

Nochtuisk 59°59′ N., 117°35′ E.
Yakutsk, Federated SSR, USSR, [Нохтуйск]
Found 1876
Synonym(s): *Mokhtuisk, Nochtujsk, Nokhtuisk*
Iron. Octahedrite, coarse.
Only four fragments, weighing 4g, 2g, 1g and 1g respectively, were found in the gold washings and came into the possession of A. Brezina who gives an analysis, 84.90 %Fe, 6.22 %Ni, 0.39 %P, A. Brezina, letter of April 10, 1901 in Min. Dept. BM(NH). The locality is also spelt Mokhtuisk. Corrected latitude and longitude, E.L. Krinov, Астрон. Журнал, 1945, **22**, p.303 [M.A. 9-297].
2g Chicago, Field Mus. Nat. Hist.; 1g Berlin, Humboldt Univ.; 1g Vienna, Naturhist. Mus.;
Specimen(s): [85833], 3.25g. largest piece found

Nochtujsk v Nochtuisk.

Nocoleche 29°52′ S., 144°13′ E.
Wanaaring, County Ularara, New South Wales, Australia
Found 1895, known in this year
Iron. Chemically and structurally anomalous (IC-ANOM).
A mass of about 44lb was found 5 miles SW. of Nocoleche Station, T. Cooksey, Rec. Austr. Mus., 1897, **3**, p.51. Classification and analysis, 6.4 %Ni, 49.3 ppm.Ga, 148 ppm.Ge, 8.2 ppm.Ir, J.T. Wasson, Meteorites, Springer-Verlag, 1974, p.307. Description; anomalous with both granular and Widmanstätten textures; shocked, V.F. Buchwald, Iron Meteorites, Univ. of California, 1975, p.908.
13kg Sydney, Austr. Mus.; 2kg New York, Amer. Mus. Nat. Hist.; 1kg Chicago, Field Mus. Nat. Hist.; 203g Washington, U.S. Nat. Mus.; 93g Copenhagen, Univ. Geol. Mus.; 80g Vienna, Naturhist. Mus.;
Specimen(s): [82747], 685g. slice; [1927,1280], 253g.

Noen v Noyan-Bogdo.

Nogata 33°43′30″ N., 130°45′ E.
Fukuoka Prefecture, Kyushu, Japan
Fell 861, A.D.
Stone. Olivine-hypersthene chondrite (L6).
After detonations and a brilliant flash, a stone fell which was recovered from a hole in the ground the following morning. The stone has been preserved since its fall in a Shinto shrine. A single mass of 472g, description, analysis, olivine $Fa_{25.1}$, 19.45 % total iron, M. Shima et al., Meteoritics, 1983, **18**, p.87.
470g Nogata, Suga Jinja;

Nogaya v Nogoya.

Nogoga v Nogoya.

Nogoya 32°22′ S., 59°50′ W.
Entre Rios, Argentina
Fell 1879, June 30, evening hrs
Synonym(s): *Concepcion, Nagaya, Nogaya, Nogoga*
Stone. Carbonaceous chondrite, type II (CM2).
After appearance of a luminous meteor, a stone of about 4kg fell, C.M.F. Websky, Sitzungsber. Akad. Wiss. Berlin, Math.-naturwiss. Kl., 1882, **1**, p.395, E.H. Ducloux, Anal. Mus. Nac. Hist. Nat. Buenos Aires, 1914, **26**, p.99. Description, G.A. Daubrée, C. R. Acad. Sci. Paris, 1883, **96**, p.1764. Analysis, C. Friedheim, Sitzungsber. Akad. Wiss. Berlin, Math.-naturwiss. Kl., 1888, p.363. Mentioned, E.H. Ducloux, Sitzungsber. Akad. Wiss. Berlin, Math.-naturwiss. Kl., 1888, p.104. Partial analysis, 18.8 % total iron, L.H. Ahrens et al., Earth planet. Sci. Lett., 1969, **6**, p.285. Classification, references, W.R. Van Schmus and J.M. Hayes, Geochimica et Cosmochimica Acta, 1974, **38**, p.47. Trace element data, G.W. Kallemeyn and J.T. Wasson, Geochimica et Cosmochimica Acta, 1981, **45**, p.1217.
1.8kg Berlin, Humboldt Univ.; 356g Buenos Aires, Mus. Argent. Cienc. Natural.; 310g Vienna, Naturhist. Mus.; 210g Paris, Mus. d'Hist. Nat.; 46g Chicago, Field Mus. Nat. Hist.;
Specimen(s): [84191], 24.75g.; [63925], 4g.; [55125], 3g.

Noin v Noyan-Bogdo.

Nokhtuisk v Nochtuisk.

Noon v Arispe.

Nora Creina 37°19′ S., 139°51′ E.
South Australia, Australia
Found 1962
Synonym(s): *Nora Criena*
Stone. Olivine-hypersthene chondrite (L4).
A specimen of 283.5g found at Nora Creina Bay, was at first thought to be part of the Lake Bonney meteorite fall, but subsequent work on the pyroxenes shows it to be distinct, D.W.P. Corbett, Rec. S. Austr. Mus., 1968, **15**, p.778, Min.

Mag., 1967, **36**, p.293. Olivine $Fa_{25.1}$, B. Mason, Rec. Austr. Mus., 1974, **29**, p.169. Analysis, 19.69 % total iron, M.J. Fitzgerald, Ph.D. Thesis, Univ. of Adelaide, 1979, p.23.
250g Adelaide, South Austr. Mus., main mass; 9.9g New York, Amer. Mus. Nat. Hist.; 8.5g Washington, U.S. Nat. Mus.;

Nora Criena v Nora Creina.

Norcateur 39°49′ N., 100°12′ W.
Decatur County, Kansas, U.S.A.
Found 1940, recognized 1948
Stone. Olivine-hypersthene chondrite (L6), veined, brecciated.

One stone of 3.2kg was found, W. Wahl, letter of May 23, 1950 in Min. Dept. BM(NH). Listed, H.H. and A.D. Nininger, The Nininger Collection of Meteorites, Winslow, Arizona, 1950, p.78. Olivine Fa_{25}, B. Mason, Geochimica et Cosmochimica Acta, 1963, **27**, p.1011.
1.4kg Tempe, Arizona State Univ.; 133g Washington, U.S. Nat. Mus.; 66g New York, Amer. Mus. Nat. Hist.; 47g Mainz, Max-Planck-Inst.; 33g Los Angeles, Univ. of California; 31g Chicago, Field Mus. Nat. Hist.;
Specimen(s): [1959,780], 1384g.

Nord-Brabant v Uden.

Nord du Portugal v North Portugal.

Nordheim 28°52′ N., 97°37′ W.
De Witt County, Texas, U.S.A.
Found 1932, recognized in 1936
Iron. Ataxite, Ni-rich (IRANOM).

One mass of 15.15kg was found, A.D. Nininger, Pop. Astron., Northfield, Minnesota, 1937, **45**, p.449 [M.A. 7-62]. The mass was ploughed up 3 miles south of Nordheim. An etched surface shows schlieren-like streaks in 10 directions under suitable lighting, and a minute octahedral structure under high (×230) magnification. Description, with an analysis, 11.69 %Ni, V.E. Barnes, Univ. Texas Publ., 1940 (3945), p.633 [M.A. 8-60]. Classification and analysis, 11.64 %Ni, 0.550 ppm.Ga, 0.644 ppm.Ge, 11 ppm.Ir, R. Schaudy et al., Icarus, 1972, **17**, p.174. Description; not related to the IVB ataxites, V.F. Buchwald, Iron Meteorites, Univ. of California, 1975, p.910.
Main mass, Austin, Texas, Univ. of Texas; 1kg Washington, U.S. Nat. Mus.; 44g Tempe, Arizona State Univ.;

Norfolk 36°54′ N., 76°18′ W.
Norfolk County, Virginia, U.S.A.
Found 1907
Iron. Octahedrite, medium (0.9mm) (IIIA).

A mass of about 23kg, which was later cut into two pieces of 21.6kg and 1.27kg, was seen to fall, L.W. MacNaughton, Am. Mus. Novit., 1926 (207), p.2. The fall is said to have been observed, but this is very doubtful in view of its oxidized condition, H.H. Nininger, Pop. Astron., Northfield, Minnesota, 1937, **45**, p.562 [M.A. 7-69]. Definitely a find, analysis, 7.51 %Ni, H.B. Wiik and B. Mason, Geochimica et Cosmochimica Acta, 1965, **29**, p.1003. Analysis, 7.45 %Ni, 20.2 ppm.Ga, 38.1 ppm.Ge, 10 ppm.Ir, E.R.D. Scott et al., Geochimica et Cosmochimica Acta, 1973, **37**, p.1957. Description, V.F. Buchwald, Iron Meteorites, Univ. of California, 1975, p.912.
Main mass, New York, Amer. Mus. Nat. Hist.; 145g Chicago, Field Mus. Nat. Hist.; 110g Washington, U.S. Nat. Mus.; 85g Los Angeles, Univ. of California;
Specimen(s): [1932,8], 82g. and a fragment, 8g

Norfork 36°13′ N., 92°16′ W.
Baxter County, Arkansas, U.S.A.
Fell 1918, October
Iron. Octahedrite, medium (1.1mm) (IIIA).

A mass of 1050g fell. Description, H.H. Nininger, Pop. Astron., Northfield, Minnesota, 1937, **45**, p.562 [M.A. 7-69]. The fall occurred either at dusk or in darkness, F.E. Wickman, letter of 15 December, 1971 in Min. Dept. BM (NH). Analysis, 7.75 %Ni, 20.3 ppm.Ga, 40.1 ppm.Ge, 3.0 ppm.Ir, E.R.D. Scott et al., Geochimica et Cosmochimica Acta, 1973, **37**, p.1957. Description; the recovered mass is only part of a larger, fallen, body, V.F. Buchwald, Iron Meteorites, Univ. of California, 1975, p.914.
408g Tempe, Arizona State Univ.; 164g Harvard Univ.; 73g Washington, U.S. Nat. Mus.;
Specimen(s): [1959,959], 97g. slice, and sawings, 22g

Norin-Schibir v Norin-Shibir.

Norin-Shibir 51°51′ N., 107°55′ E.
Buryat ASSR, USSR, [Норин–Шибир]
Found 1900, between 1900 and 1937
Synonym(s): *Norin-Schibir*
Iron. Octahedrite, medium.

Fell at 50°51'N., 107°55'E., near Verkne Udinsk, Transbaikal, V.I. Vernadsky, letter of August 5, 1939 in Min. Dept. BM(NH). Norin-Shibir is mapped about 51°N., 29°E., NW. of Kiev, Ukraine, P.M. Millman, Trans. Roy. Astron. Soc. Canada, 1938, **32**, p.198. Listed, I.S. Astapowitsch, Trans. Roy. Astron. Soc. Canada, 1938, **32**, p.195 [M.A. 7-172]. The above latitude and longitude are given, E.L. Krinov, Астрон. Журнал, 1945, **22**, p.303 [M.A. 9-297], Каталог Метеоритов Акад. Наук СССР, Москва, 1947 [M.A. 10-511]. Analysis, 9.0 %Ni, 52 ppm.Ga, 120 ppm.Ge, 6.8 ppm.Ir, N.I. Zaslavskaya, Метеоритика, 1982, **40**, p.54. May be a pallasite, or an iron with silicate inclusions, V.F. Buchwald, Iron Meteorites, Univ. of California, 1975, p.915.
2.8g Moscow, Acad. Sci., small fragment, all that is known;

Norquín 37°43′ S., 70°37′ W.
Neuquen, Argentina
Found 1945, before this year
Iron. Octahedrite, medium (0.8mm) (IIIB?).

A mass of 19.25kg was found. Description, with a poor analysis, 4.52 %Ni, E.H. Ducloux, Notas Mus. La Plata, 1945, **10** (geol. no. 40), p.163 [M.A. 9-302]. Description, V.F. Buchwald, Iron Meteorites, Univ. of California, 1975, p.916.
18.4kg La Plata, Mus., main mass; 58g Harvard Univ.;

Norris v Bennett County.

Norristown 32°31′ N., 82°33′ W.
Emanuel County, Georgia, U.S.A.
Found 1965, or 1966
Iron. Octahedrite, medium (0.65mm) (IIIB).

A 4.2kg individual was uncovered during road working, 2.5 miles west of Norristown, Meteor. Bull., 1978 (55), Meteoritics, 1978, **13**, p.343. Classification and analysis, 9.64 %Ni, 18.2 ppm.Ga, 32.4 ppm.Ge, 0.016 ppm.Ir, E.R.D. Scott et al., Geochimica et Cosmochimica Acta, 1973, **37**, p.1957. Corroded, shock-hardened, V.F. Buchwald, Iron Meteorites, Univ. of California, 1975, p.916.
196g Washington, U.S. Nat. Mus.;

Nörten 51°38′ N., 9°57′ E.

Göttingen, Niedersachsen, Germany
Fell 1580, May 27, 1400 hrs
Doubtful. Stone.
A shower of stones is said to have fallen, E.F.F. Chladni, Die Feuer-Meteore, Wien, 1819, p.217 but the evidence is not conclusive.

North Africa (481)

Fell 481, A.D.
Doubtful..
"Fiery stones fell", E.F.F. Chladni, Ann. Phys. Chem. (Poggendorff), 1826, **8**, p.45 but the evidence is not conclusive.

North Africa (1020)

Fell 1020, July or August
Doubtful..
Many stones are said to have fallen at an unnamed locality in North Africa, and some are said to have killed people, E.F.F. Chladni, Die Feuer-Meteore, Wien, 1819, p.196, S. de Sacy, Ann. Physik (Gilbert), 1815, **50**, p.291 but the evidence is not conclusive. Not improbably hailstones, R.P. Greg, Rep. Brit. Assn., 1860, **30**, p.51.

Northampton 42°19′ N., 72°38′ W.

Hampshire County, Massachusetts, U.S.A.
Found 1963, recognized in this year
Iron. Octahedrite.
The precise location of the find is unknown, L. LaPaz, Cat. Coll. Inst. Meteor. Univ. New Mexico, 1965. Not listed as being in the Univ. of New Mexico Coll., D.E. Lange et al., Spec. Publ. Univ. of New Mexico, 1980 (no. 21).
16.8g Albuquerque, Univ. of New Mexico;

North Chile 23° S., 69° W. approx.

Antofagasta, Chile
Found 1875
Synonym(s): *Buen Huerto, Buey Muerto, Cerros del Buey Muerto, Cerros del Buen Muerto, Chile, Coya Norte, Filomena, Maria Elena (1928), Puripica, Quillagua, Rio Loa, San Martin, Tocopilla, Union*
Iron. Hexahedrite (IIA).
The following eight irons, of total weight 266kg, identical in structure and composition, are now collectively known as North Chile because exact localities are unknown: Coya Norte, Filomena, Puripica, Quillagua, Rio Loa, San Martin, Tocopilla, and Union. Each mass is part of a single kamacite crystal in which rhabdites are irregularly distributed; cosmic and artificial fracturing were controlled by the rhabdite plates. Mean composition, 5.59 %Ni, 58.9 ppm.Ga, 177 ppm.Ge, 3.6 ppm.Ir, V.F. Buchwald, Iron Meteorites, Univ. of California, 1975, p.917. Locality map, composition, and discussion of Chilean hexahedrites, J.T. Wasson and J.I. Goldstein, Geochimica et Cosmochimica Acta, 1968, **32**, p.329. Cooling rates, E. Randich and J.I. Goldstein, Geochimica et Cosmochimica Acta, 1978, **42**, p.221.
22kg Geneva, Mus. d'Hist. Nat., main mass of Union; 11kg Kiel, Univ., of San Martin; 39.8kg Washington, U.S. Nat. Mus., of Filomena, 20kg; of Puripica, 15kg; of Coya Norte, 4.8kg; 2.76kg Paris, Mus. d'Hist. Nat., of Tocopilla;
Specimen(s): [1959,917], 2221g. and a slice, 75.5g, and sawings, 66g, of Coya Norte; [1959,921], 273.5g. and sawings, 4g, of Puripica; [1959,922], 39g. and sawings, 0.5g, of Rio Loa; [1931,13], 2526g. and sawings, 4g, of Tocopilla

North East Reid 30°9′ S., 128°43′ E. approx.

Nullarbor Plain, Western Australia, Australia
Found 1969
Stone. Olivine-hypersthene chondrite (L4).
Two angular, oxidized masses, 31.1g and 7.5g, were found 3 miles NE. of the Reid find, G.J.H. McCall and W.H. Cleverly, J. Roy. Soc. West. Austr., 1970, **53**, p.69. There is some confusion here, for this is listed as H5, B. Mason, Rec. Austr. Mus., 1974, **29**, p.169.
The two main masses, Kalgoorlie, West. Austr. School of Mines; Thin section, Perth, West. Austr. Mus.; 1.9g Washington, U.S. Nat. Mus.;

North Forrest v North West Forrest (H).

North Haig 30°13′ S., 126°13′ E.

Western Australia, Australia
Found 1961
Stone. Achondrite, Ca-poor. Ureilite (AURE).
One mass, 964g, was found 13 miles due N. of Sleeper Camp which is 40 miles N. of Haig, Spec. Publ. West. Austr. Mus., 1965 (3), p.47, G.J.H. McCall, letter of 13 February, 1964 in Min. Dept. BM(NH). It is distinguished from Dingo Pup Donga by its variable mineral composition, olivine Fa_{0-39}, G.J.H. McCall and W.H. Cleverly, Min. Mag., 1968, **36**, p.691. Trace element contents, C.M. Binz et al., Geochimica et Cosmochimica Acta, 1975, **39**, p.1576. Mentioned, J.L. Berkley et al., Geochimica et Cosmochimica Acta, 1980, **44**, p.1579. Contains suessite, iron silicide, K. Keil et al., Am. Mineral., 1982, **67**, p.126.
Main mass, Kalgoorlie, West. Austr. School of Mines; 9.7g Perth, West. Austr. Mus., and a thin section; 3.6g Albuquerque, Univ. of New Mexico;
Specimen(s): [1968,340], 5.5g. fragments

North Inch of Perth v Perth.

North Mandiga v Bencubbin.

North Portugal 41° N., 8° W. approx.

Portugal
Found 1931, known before this year
Synonym(s): *Nord du Portugal*
Iron. (IIB).
A specimen of an iron from a locality in the north of Portugal is in the Inst. Sup. Techn., Lisbon, G. Constanzo, Rev. Quim. pur. apl., Porto, 1931 (ser. 3), p.17. Almost certainly fragments of the weathered São Julião de Moreira meteorite, V.F. Buchwald, Iron Meteorites, Univ. of California, 1975, p.931.

North Reid 30°8′ S., 128°38′ E. approx.

Nullarbor Plain, Western Australia, Australia
Found 1969
Stone. Olivine-hypersthene chondrite, amphoterite (LL5).
Three masses were found. The first of 108.3g, 5 miles NW. of the Reid find, the second, 43.9g, about 7 miles SSW. of the Reid site, and the third, 156.5g, 6 miles SW. of the Reid find (*q.v.*). Each mass is crusted and flight oriented, G.J.H. McCall, 2nd. Suppl. to West. Austr. Mus. Spec. Publ. no. 3, 1972, p.21, G.J.H. McCall and W.H. Cleverly, J. Roy. Soc. West. Austr., 1970, **53**, p.69. Olivine $Fa_{27.4}$, B. Mason, Rec. Austr. Mus., 1974, **29**, p.169.
Main mass, Kalgoorlie, West. Austr. School of Mines; Thin section, Perth, West. Austr. Mus.; 4.7g Washington, U.S. Nat. Mus.;

North Star Bay v Cape York.

Northumberland Island v Cape York.

North West Forrest (E6)

30°36′ S., 127°49′ E. approx.

Nullarbor Plain, Western Australia, Australia
Found 1971
Synonym(s): *North West Forrest (1971)*
Stone. Enstatite chondrite (E6).

300-400 fragments, of total weight 4.4kg, were found in an ellipse about 10 metres long, the result of terrestrial weathering. The site is about 25 miles NW. of Forrest railway station, G.J.H. McCall, 2nd. Suppl. to West. Austr. Mus. Publ. no. 3, 1972, p.22. Pyroxene Fs0.3, B. Mason, Rec. Austr. Mus., 1974, **29**, p.169.

4.4kg Perth, West. Austr. Mus.; 11g Washington, U.S. Nat. Mus.;

North West Forrest (H)

30°46′ S., 128°1′ E. approx.

Nullarbor Plain, Western Australia, Australia
Found 1969, or earlier
Synonym(s): *North Forrest, North West Forrest (1969)*
Stone. Olivine-bronzite chondrite (H4).

A single weathered mass, 238.4g, broken in two, was found about 14km NW. of Forrest railway station. It contains clear or turbid glass, G.J.H. McCall, 2nd. Suppl. to West. Austr. Mus. Spec. Publ. no. 3, 1972, p.22. Olivine $Fa_{19.1}$, B. Mason, Rec. Austr. Mus., 1974, **29**, p.169. North Forrest, a single badly weathered mass of 608.9g, was found in 1969 about 25 miles NW. of Forrest station, G.J.H. McCall and W.H. Cleverly, J. Roy. Soc. West. Austr., 1970, **53**, p.69. Initially classified as L4 but later re-classified as H4, B. Mason, Rec. Austr. Mus., 1974, **29**, p.169.

Main mass, Kalgoorlie, West. Austr. School of Mines; Thin section, Perth, West. Austr. Mus.; 3.8g Washington, U.S. Nat. Mus.;

North West Forrest (1969) v North West Forrest (H).

North West Forrest (1971) v North West Forrest (E6).

Norton County 39°41′ N., 99°52′ W.

Kansas, U.S.A.
Fell 1948, February 18, 1656 hrs
Synonym(s): *Furnas County*
Stone. Achondrite, Ca-poor. Aubrite (AUB).

A shower of stones fell over a considerable area in Norton Co., Kansas, and Furnas Co., Nebraska. Over 100 stones have been recovered, including one of about 1 ton, and one of 131.5lb. It is a friable lime-poor enstatite achondrite (aubrite) with some nickel-iron inclusions, F.C. Leonard, Pop. Astron., Northfield, Minnesota, 1948, **56**, p.434 [M.A. 10-401]. Description and optical data, C.W. Beck and L. LaPaz, Publ. Astron. Soc. Pacific, 1949, **61**, p.63 [M.A. 11-140]. Preliminary notes, L. LaPaz, Science, 1948, **107**, p.543, Pop. Astron., Northfield, Minnesota, 1948, **56**, p.391 [M.A. 10-401]. Consists of enstatite crystals in a groundmass of enstatite and olivine (9%) with small (undetermined) amounts of nickel-iron, 5.82 %Ni, C.W. Beck and L. LaPaz, Am. Miner., 1951, **36**, p.45. Other minerals present, K. Keil and K. Fredriksson, Geochimica et Cosmochimica Acta, 1963, **27**, p.939. Analysis, 1.60 % total iron, H.B. Wiik, Geochimica et Cosmochimica Acta, 1956, **9**, p.279. Further analysis, H. von Michaelis et al., Earth planet. Sci. Lett., 1969, **5**, p.387. Internal Rb/Sr isochron age, 4700 m.y., D. Bogard et al., Earth planet. Sci. Lett., 1967, **3**, p.179. For a statistical study of fragments, B. Lang and K. Liszewska, Meteoritics, 1973, **8**, p.277, Meteoritics, 1974, **9**, p.99. Analysis of enstatite, A.M. Reid and A.J. Cohen, Geochimica et Cosmochimica Acta, 1967, **31**, p.661. Mineralogy, T.R. Watters and M. Prinz, Geochimica et Cosmochimica Acta, 1979 (Suppl. 11), p.1073.

Main mass, Albuquerque, Univ. New Mexico; 1778g Tempe, Arizona State Univ.; 1.7kg Washington, U.S. Nat. Mus.; 956g Los Angeles, Univ. of California; 600g Chicago, Field Mus. Nat. Hist.;

Specimen(s): [1953,8], 5g.; [1959,758], 197g.

Nossi-Bé v Pułtusk.

Nosy Bé v Pułtusk.

Nova Lima 19°59′ S., 43°51′ W.

Minas Gerais, Brazil
Found 1960, known before this year
Pseudometeorite. Iron.

A fragment from the collection of the Escuola de Minas, Engenharia e Metalurgia, Ouro Preto, Minas Gerais, "looks like a meteorite that suffered secondary thermal alteration", W.S. Curvello, letter of February, 1960 in Min. Dept. BM (NH). Pseudometeorite, wrought iron, V.F. Buchwald, Iron Meteorites, Univ. of California, 1975, p.931.

Nova-Petropolis 29°26′ S., 50°55′ W.

Nova-Petropolis County, Rio Grande do Sul, Brazil
Found 1967, May 16
Iron. Octahedrite, medium (1.3mm) (IIIA).

A single mass of 305kg was found by a farmer, H. Grunewaldt, Anais Acad. Brasil. Cienc., 1983, **55** (1). Analysis, 7.8 %Ni, 19.9 ppm.Ga, 36.5 ppm.Ge, 9.4 ppm.Ir, D.J. Malvin et al., priv. comm., 1983.

Main mass, Arrioi do Meio, in private hands; Specimen, Buenos Aires, Mus. Nac.;

Novellara 44°50′ N., 10°45′ E.

Reggio Emilia, Emilia, Italy
Fell 1766, August 15
Doubtful..

The fall is said to have damaged a tree, near which several fragments were found, D. Troili, Della caduta di un sasso dall'aria, Modena, 1766. Listed, B. Baldanza, Min. Mag., 1965, **35**, p.214.

Noventa Vicentina 45°17′30″ N., 11°31′38″ E.

Vicenza province, Veneto, Italy
Fell 1971, May 12
Stone. Olivine-bronzite chondrite (H4).

A single, crusted, cuboidal individual of 177g fell, Meteor. Bull., 1975 (53), Meteoritics, 1975, **10**, p.137.

Novorybinskoe 51°53′ N., 71°15′ E.

Tselinogradskaya oblast, Kazakhstan SSR, USSR, [Новорыбинское]
Found 1937
Iron. Octahedrite, fine (0.3mm) (IVA).

A mass of 3055g was found north of Akmolinsk. Description, P.L. Dravert, Природа, 1939 (2), p.123 [M.A. 7-374]. Listed, L.A. Kulik, Метеоритика, 1941, **1**, p.73 [M.A. 9-294], E.L. Krinov, Астрон. Журнал, 1945, **22**, p.303 [M.A. 9-297], L.G. Kvasha and A.Ya. Skripnik, Метеоритика, 1978, **37**, p.197. Analysis, 9.1 %Ni, 2.45 ppm.Ga, 0.20 ppm.Ge, 0.90 ppm.Ir, J.T. Wasson, Meteorites, Springer-Verlag, 1974, p.306. Briefly figured, A.N. Zavaritsky, Метеоритика, 1954, **11**, p.64 [M.A. 13-48].

2966g Moscow, Acad. Sci.;

Novosibirsk 55°0′ N., 82°54′ E.
Novosibirsk city, Federated SSR, USSR
Found 1978, in the spring
Stone. Olivine-bronzite chondrite (H5-6).
A single mass of 11.41kg was found in sandy soil during excavations, Meteor. Bull., 1981 (59), Meteoritics, 1981, **16**, p.196.

Novo-Urei 54°49′ N., 46°0′ E.
Karamzinka, Mordovsky ASSR, USSR, [Новый-Урей]
Fell 1886, September 4, 0715 hrs
Synonym(s): *Alatyr, Krasnoslobodsk, Novyj Urej, Novyi Urei, Novyi Urey, Nowo-Urei, Urei*
Stone. Achondrite, Ca-poor. Ureilite (AURE).
After appearance of light and detonations, three stones fell, one of 1.9kg on the left bank of the river Alatyr at Karamzinka, another on the right bank at Petrovka which was broken up and lost, and the third in a swamp to the south of the farm Novo-Urei which was also lost, Y.I. Simashko, Cat. Meteorites, St.-Petersbourg, 1891, p.49. Description and analysis, M.I. D'yakonova, Метеоритика, 1964 (25), p.129. Y.I. Simashko says that the correct date of the fall, according to many witnesses, is September 4 (not September 22), and the locality as above and not Krasnoslobodsk, Penza, as usually given. Listed, E.L. Krinov, Каталог Метеоритов Акад. Наук СССР, Москва, 1947 [M.A. 10-511], V.I. Mikheev and A.I. Kalinin, Метеоритика, 1958 (15), p.156 [M.A. 14-48]. Analysis, 15.64 % total iron, H.B. Wiik, Meteoritics, 1972, **7**, p.553. Rare gas content, E. Mazor et al., Geochimica et Cosmochimica Acta, 1970, **34**, p.781. Trace element contents, C.M. Binz et al., Geochimica et Cosmochimica Acta, 1975, **39**, p.1576. Discussion, G.P. Vdovykin, Space Sci. Rev., 1970, **10**, p.483. Mineralogy, J.L. Berkley et al., Geochimica et Cosmochimica Acta, 1980, **44**, p.1579.
Main mass, Leningrad, Mus. Mining Inst.; 463g Moscow, Acad. Sci.; 93g Washington, U.S. Nat. Mus.; 55g Vienna, Naturhist. Mus.; 50g Chicago, Field Mus. Nat. Hist.; 34g Paris, Mus. d'Hist. Nat.;
Specimen(s): [63625], 22g.

Novy-Ergi 58°33′ N., 31°20′ E. approx.
Novgorod region, Federated SSR, USSR, [Новый Ерги]
Fell 1662, December 10, 1600 hrs
Stone.
After appearance of fireball and detonations, a shower of stones fell, and many were preserved in the Kirilo-Belosersky monastery at Novy-Ergi, P.N. Chirvinsky, Centralblatt Min., 1923, p.548 but nothing is now to be found, Meteor. Bull., 1959 (13).

Novyi Urei v Novo-Urei.

Novyi Urey v Novo-Urei.

Novyj Urej v Novo-Urei.

Novy-Projekt 56°0′ N., 22°0′ E.
Novo-Alexandrovsky district, Kovno, Lithuanian SSR, USSR, [Новый Проект]
Fell 1908, April 25, 0200 hrs
Stone. Chondrite.
A stone of 1001g fell in the roadway, P.N. Chirvinsky, Centralblatt Min., 1923, p.550. The present location of this mass is unknown, E.L. Krinov, Каталог Метеоритов Акад. Наук СССР, Москва, 1947 [M.A. 10-511].

Nowo-Urei v Novo-Urei.

Noyan-Bogdo 42°55′ N., 102°28′ E.
Omnogov aymag, Mongolia, [Ноян-Богдо]
Fell 1933, September
Synonym(s): *Noen, Noin, Noyen*
Stone. Olivine-hypersthene chondrite (L6).
220g of fragments were recovered in 1948, L.G. Kvasha, Метеоритика, 1954, **11**, p.81 [M.A. 13-52], G.G. Vorobyev and O. Namnandorzh, Метеоритика, 1958, **16**, p.135 [M.A. 14-130]. Olivine Fa_{23}, B. Mason, Geochimica et Cosmochimica Acta, 1967, **31**, p.1100. Analysis (by L.G. Kvasha), 18.22 % total iron, O. Namnandorj, Meteorites of Mongolia, 1980, p.13, English trans.
48g Ulan-Bator,; 17.9g Moscow, Acad. Sci.;

Noyen v Noyan-Bogdo.

Nuevo Laredo 27°30′ N., 99°30′ W.
Tamaulipas, Mexico
Found 1950
Stone. Achondrite, Ca-rich. Eucrite (AEUC).
One complete stone of 500g was found, C.C. Patterson, Geochimica et Cosmochimica Acta, 1955, **7**, p.151. Apparently concordant U-Pb and Th-Pb age is 4590 m.y., L.T. Silver and M.B. Duke, Meteoritics, 1970, **5**, p.224, abs.. Bulk and mineral analyses, M.B. Duke and L.T. Silver, Geochimica et Cosmochimica Acta, 1967, **31**, p.1637. Rb-Sr data, D.A. Papanastassiou and G.J. Wasserburg, Earth planet. Sci. Lett., 1969, **5**, p.361. Selected elemental abundances, R.A. Schmitt et al., Meteoritics, 1972, **7**, p.131.
132g Washington, U.S. Nat. Mus.; 4.8g New York, Amer. Mus. Nat. Hist.;
Specimen(s): [1980,M.37], 0.26g.

Nuevo Leon v Coahuila.

Nuevo Mercurio 24°18′ N., 102°8′ W.
Zacatecas, Mexico
Fell 1978, December 15, 1850 hrs
Stone. Olivine-bronzite chondrite (H5).
After a fireball, moving NE. to SW, and an explosion, a shower of stones fell over an elliptical area over 10 km in length. Over 300 individuals have been recovered, the total mass being over 5kg, Meteor. Bull., 1980 (57), Meteoritics, 1980, **15**, p.100. Brief description, olivine $Fa_{17.3}$, K. Fredriksson et al., Meteoritics, 1979, **14**, p.400.
111g Fort Worth, Texas, Monnig Colln.;
Specimen(s): [1979,M.2], 136.4g. seven small individuals (a-g); [1981,M.2], 830g. individual

Nuleri 27°50′ S., 123°52′ E.
Central Division, Western Australia, Australia
Found 1902, during or before this year
Iron. Octahedrite, medium (IIIA).
A mass of 120g was found 200 miles east of Mount Sir Samuel. Description and analysis, 5.79 %Ni, E.S. Simpson, Bull. Geol. Surv. West. Austr., 1907 (26), p.24. Mentioned, A. Brezina, Bull. Geol. Surv. West. Austr., 1914 (59), p.213, Spec. Publ. West. Austr. Mus., 1965 (3), p.47. Analysis, 7.44 %Ni, 18.0 ppm.Ga, 36.7 ppm.Ge, 9.3 ppm.Ir, J.T. Wasson, Meteorites, Springer-Verlag, 1974, p.301. Chemically similar to Boxhole (*q.v.*).
93g Perth, Geol. Surv. Mus., main mass; 2g Vienna, Naturhist. Mus.;

Nullarbor 31° S., 132° E. approx.
Nullarbor Plain, South Australia, Australia
Found 1935, before this year
Stone. Olivine-bronzite chondrite (H5).
A mass weighing 40kg was brought to Adelaide between 1926 and 1935. It was probably found near the E-W railway line in South Australia, Meteor. Bull., 1983 (61), Meteoritics, 1983, **18**, p.82.
40kg Adelaide, South Austr. Mus.;

Nullagine 21°52′ S., 120°7′ E.
near Nullagine, Western Australia, Australia
Found 1973
Stone. Olivine-bronzite chondrite (H5).
A single mass of 102g was found during mapping in the Pilbara district of NW. Western Australia, J.D. Lewis, Geol. Surv. West. Austr., Ann. Rept. for 1974, 1975, p.101 [M.A. 77-1997]. Reported, Meteor. Bull., 1978 (55), Meteoritics, 1978, **13**, p.344.

Nulles 41°38′ N., 0°45′ E.
Catalonia, Spain
Fell 1851, November 5, 1730 hrs
Synonym(s): *Barcelona, Tarragona, Vilabella*
Stone. Olivine-bronzite chondrite (H6), brecciated.
After appearance of fireball and detonations, a great number of stones fell between the villages of Nulles, Vilabella and Tarragona, but only a few were preserved. The largest stone weighed about 19.5lb, R.P. Greg, Phil. Mag., 1862, **24**, p.536. Analysis, D.J. Barcells, Lithologia Meteorica, Barcelona, 1854, p.4, 28. Olivine Fa_{19}, B. Mason, Geochimica et Cosmochimica Acta, 1963, **27**, p.1011.
7.8kg Madrid, Mus. Cienc. Nat., main mass; 124g Washington, U.S. Nat. Mus.; 210g Tempe, Arizona State Univ.; 92g Paris, Mus. d'Hist. Nat.;
Specimen(s): [85427], 24g.; [35160], 4.5g.

Numakai 43°20′ N., 141°52′ E.
Ishikari, Hokkaido, Japan
Fell 1925, September 5
Stone. Olivine-bronzite chondrite (H4).
Two stones were seen to fall, but only one of 363g was picked up, I. Yamamoto, Kwasan Observ. Bull., 1935, p.306 [M.A. 7-173], H. Imai, J. Geogr., Tokyo Geogr. Soc., 1926, **38**, p.145, S. Murayama, letter of April 6, 1962 in Min. Dept. BM(NH)., A.D. Nininger, Pop. Astron., Northfield, Minnesota, 1940, **48**, p.559, Contr. Soc. Res. Meteorites, **2**, p.227 [M.A. 8-54]. Analysis, 29.53 % total iron, M. Shima, Meteoritics, 1974, **9**, p.123. Mineralogical study, classification, K. Yagi et al., J. Japan. Assoc. Min. Petr. Econ. Geol., 1976, **71**, p.273 [M.A. 81-0695].
319g in possession of finder, main mass;

Nurraburra v Narraburra.

Nurrah v Sitathali.

Nutwood Downs v Henbury.

Nyaung 21°12′30″ N., 94°55′ E.
Myingyan district, Upper Burma, Burma
Fell 1939, December 24, 1940 hrs
Iron.
A mass of 737.6g fell, M.A.R. Khan, Hyderabad Acad. Studies, 1950, **12** [M.A. 11-441], P. Chatterjee, letter of March 25, 1952 in Min. Dept. BM(NH).
639g Calcutta, Mus. Geol. Surv. India;
Specimen(s): [1983,M.13], 22.7g.

Nyezhin v Alexandrovsky.

Nyirábrany 47°33′ N., 22°1′30″ E.
Hajdu-Bihar, Ungarn, Hungary
Fell 1914, July 17
Stone. Olivine-hypersthene chondrite, amphoterite (LL5).
Mentioned in 1914 in the newspaper Nyirvidék published at Nyiregyháza, V. Zsivny, letters of December 1933 and May 28, 1952 in Min. Dept. BM(NH). This fall has been incorrectly reported as identical with Kisvarsány, L. Tokody and M. Dudich, Magyarország Meteoritgyüjteményei, Budapest, 1951, p.89. Olivine $Fa_{28.7}$, K. Fredriksson et al., Origin and Distribution of the Elements, ed. L.H. Ahrens, Pergamon, 1968, p.457. Mineralogy, analysis, 20.38 % total iron, K.I. Sztrokay et al., Chem. Erde, 1977, **36**, p.287.
1108g Nyiregyháza, Mus., one piece; Thin section, Washington, U.S. Nat. Mus.;

Nyons v Aubres.

Oackley v Oakley (stone).

Oak 31°35′ S., 127°42′ E. approx.
Nullarbor Plain, Western Australia, Australia
Found 1968
Stone. Olivine-hypersthene chondrite (L5).
A partially weathered 75.3g mass was found about 20 miles NW. of Mundrabilla homestead, Eyre Highway, G.J.H. McCall and W.H. Cleverly, J. Roy. Soc. West. Austr., 1970, **53**, p.69. Olivine $Fa_{25.7}$, B. Mason, Rec. Austr. Mus., 1974, **29**, p.169.
Main mass, Kalgoorlie, West. Austr. School of Mines; Thin section, Perth, West. Austr. Mus.; 0.8g Washington, U.S. Nat. Mus.;

Oakley (iron) 42°20′ N., 113°42′ W.
Cassia County, Idaho, U.S.A.
Found 1926
Iron. Octahedrite, coarse (1.4mm) (IIIF).
A mass of about 111kg was found, G.P. Merrill, Am. J. Sci., 1926, **12**, p.532, letter of July 18, 1927 in Min. Dept. BM (NH). Description, with an analysis, 7.04 %Ni, G.P. Merrill, Proc. U.S. Nat. Mus., 1927, **71** (21) [M.A. 3-535]. Classification, analysis, 7.32 %Ni, 7.20 ppm.Ga, 1.13 ppm.Ge, 5.5 ppm.Ir, E.R.D. Scott and J.T. Wasson, Geochimica et Cosmochimica Acta, 1976, **40**, p.103. Description, co-ordinates, V.F. Buchwald, Iron Meteorites, Univ. of California, 1975, p.932.
111kg Washington, U.S. Nat. Mus., main mass;

Oakley (stone) 38°57′ N., 101°1′ W.
Logan County, Kansas, U.S.A.
Found 1895
Synonym(s): *Oackley*
Stone. Olivine-bronzite chondrite (H6).
A stone of 61lb was ploughed up 15 miles SW. of Oakley. Description, H.L. Preston, Am. J. Sci., 1900, **9**, p.410. Olivine Fa_{21}, B. Mason, Geochimica et Cosmochimica Acta, 1963, **27**, p.1011. Analysis, 26.96 % total iron, W. Wahl, Min. Mag., 1950, **29**, p.416. Further analysis, 26.5 % total iron, A.A. Moss et al., Min. Mag., 1967, **36**, p.101.
8.8kg Chicago, Field Mus. Nat. Hist.; 3.8kg Harvard Univ.; 1.6kg Washington, U.S. Nat. Mus.; 750g Budapest, Nat. Mus.; 750g Vienna, Naturhist. Mus.; 733g New York, Amer. Mus. Nat. Hist.; 229g Tempe, Arizona State Univ.; 102g Yale Univ.;
Specimen(s): [84814], 2302g.

Oasis State Park 34°13'42" N., 103°21'30" W.
Roosevelt County, New Mexico, U.S.A.
Found 1968
Stone. Chondrite.
Listed, total known weight 606g, Cat. Huss Coll. Meteor., 1976, p.34.
606g Mainz, Max-Planck-Inst.;

Oaxaca v Misteca.

Oaxaca v Yanhuitlan.

Oberlin 39°48' N., 100°31' W.
Decatur County, Kansas, U.S.A.
Found 1911, approx., recognized 1935
Stone. Olivine-hypersthene chondrite, amphoterite (LL5).
One stone of 5.5lb was found, A.D. Nininger, Pop. Astron., Northfield, Minnesota, 1937, **45**, p.449 [M.A. 7-62]. Mentioned, H.H. Nininger, Trans. Kansas Acad. Sci., 1936, **39**, p.170. Olivine $Fa_{27.2}$, K. Fredriksson et al., Origin and Distribution of the Elements, ed. L.H. Ahrens, Pergamon, 1968, p.457.
1.1kg Tempe, Arizona State Univ.; 195g Washington, U.S. Nat. Mus.; 100g New York, Amer. Mus. Nat. Hist.;
Specimen(s): [1959,779], 483g.

Obernkirchen 52°16' N., 9°6' E.
Bückeberg, Rinteln, Niedersachsen, Germany
Found 1863, known before this year
Synonym(s): *Bückeberg*
Iron. Octahedrite, fine (0.26mm) (IVA).
A mass of about 41kg was found in a quarry on the Bückeberg, 15ft below the surface. Recognized as meteoritic in 1863, W. Wicke and F. Wöhler, Ann. Phys. Chem. (Poggendorff), 1863, **120**, p.509, Nachr. Gesell. Wiss. Göttingen, 1863, p.364. Description, with an analysis, 7.55 %Ni, E. Cohen, Ann. Naturhist. Hofmus. Wien, 1900, **15**, p.366. Analysis, 7.33 %Ni, 1.80 ppm.Ga, 0.092 ppm.Ge, 3.2 ppm.Ir, R. Schaudy et al., Icarus, 1972, **17**, p.174. Description; shocked, partially melted and recrystallised, V.F. Buchwald, Iron Meteorites, Univ. of California, 1975, p.934.
302g Washington, U.S. Nat. Mus.; 283g Harvard Univ.; 184g Chicago, Field Mus. Nat. Hist.; 125g Paris, Mus. d'Hist. Nat.; 90g Yale Univ.; 69g Berlin, Humboldt Univ.; 54g Tempe, Arizona State Univ.;
Specimen(s): [36056], 34.5kg. and fragments, 68g

Oberon Bay 39°4' S., 146°21' E.
Wilson's Promontory, Victoria, Australia
Found 1962, approx.
Stone. Olivine-hypersthene chondrite, amphoterite (LL6).
A mass of 179g was found in a sand-blow away from the coast. Description, classification and analysis olivine $Fa_{25.6}$, 18.43 % total iron, M.J. Fitzgerald, Proc. Roy. Soc. Victoria, 1980, **91**, p.67.
Entire mass, Adelaide, Univ.;

Ober-Pfalz v L'Aigle.

Oborniki 52°40' N., 16°50' E. approx.
Poland
Found 1955, known in this year
Doubtful. Iron.
Two pieces found near Oberniki, 30km NNW. of Poznan, and now lost, perhaps belonged to the same shower as Morasko, J. Pokrzywnicki, Acta Geol. Polon., 1955, **3**, p.427 [M.A. 13-79].

Obritti v Pułtusk.

Obruteza v Owrucz.

Ocatitlan v Toluca.

Ochansk 57°47' N., 55°16' E.
Perm, Federated SSR, USSR, [Оханск]
Fell 1887, August 30, 1300 hrs
Synonym(s): *Ojansk, Okhansk, Oschank, Taborg, Taborsk, Taborskoie Selo, Taborskoje Selo, Tabory, Toborsk*
Stone. Olivine-bronzite chondrite (H4), brecciated.
After appearance of luminous meteor and detonations, a shower of stones of total weight about 500kg, the largest weighing 115kg, fell in the village of Tabory, near Okhansk, G.A. Daubrée, C. R. Acad. Sci. Paris, 1887, **105**, p.987. Description, G.P. Merrill, Mem. Nat. Acad. Sci. Washington, 1919, **14**, p.7. Analysis, 26.19 % total iron, M.I. D'yakonova and V.Ya. Kharitonova, Метеоритика, 1960, **18**, p.48 [M.A. 16-447]. A stone of 180g shows minute pores in the crust, probably due to the escape of gases while the surface was fused. Description, P.L. Dravert, Метеоритика, 1946, **3**, p.63 [M.A. 10-174]. Olivine Fa_{20}, B. Mason, Geochimica et Cosmochimica Acta, 1963, **27**, p.1011. Further analysis, 27.19 % total iron, H.B. Wiik, Geochimica et Cosmochimica Acta, 1956, **9**, p.279. Further analysis, 27.08 % total iron, A.J. Easton and C.J. Elliott, Meteoritics, 1977, **12**, p.409. New data on fall, A.K.Stanyukovich and V.V. Chichmar, Метеоритика, 1977, **36**, p.146.
100kg Kazan, Univ., main mass; 6.615kg Moscow, Acad. Sci.; 13kg Chicago, Field Mus. Nat. Hist.; 4.2kg Vienna, Naturhist. Mus.; 2.6kg Paris, Mus. d'Hist. Nat.; 1.6kg Washington, U.S. Nat. Mus.; 443g Berlin, Humboldt Univ.; 233g New York, Amer. Mus. Nat. Hist.; 190g Harvard Univ.; 162g Oxford, Univ. Mus.;
Specimen(s): [1922,157], 2710g. and fragments, 10g; [1922, 158], 2255g. and fragments, 82g; [63549], 128g. and fragments, 19.5 g; [1964,708], 78.5g.

Octibbeha County v Oktibbeha County.

Ocheretna v Oczeretna.

Ocotitlán v Toluca.

Oczeretna 49°19' N., 31°31' E.
Lipovets, Kiev, Ukraine, USSR, [Очеретна]
Found 1871
Synonym(s): *Kiev, Lipovitz, Ocheretna, Otcheretna, Otscheretnaja*
Stone. Olivine-bronzite chondrite (H4), veined.
A stone of about 130g was obtained by the BM(NH). from Count Ylinski in 1875. A label (signed H. Lazzerini) with it, stated that it was found at Oczeretna in the spring of 1871. The stone is similar to stones of the Pultusk fall and is possibly identical. Listed, E.L. Krinov, Астрон. Журнал, 1945, **22**, p.303 [M.A. 9-297]. Olivine Fa_{19}, B. Mason, Geochimica et Cosmochimica Acta, 1963, **27**, p.1011.
4g Vienna, Naturhist. Mus.; 3g Chicago, Field Mus. Nat. Hist.;
Specimen(s): [48368], 85g.

Odesa v Grossliebenthal.

Odessa (iron) 31°43′ N., 102°24′ W.
Ector County, Texas, U.S.A.
Found 1922, before this year
Synonym(s): *Ector County, Ector County No. 3, McCamey, Penwell*
Iron. Octahedrite, coarse (1.7mm) (IA).
A mass of about 1kg, cut from a larger mass, was described with an analysis, 7.25 %Ni, G.P. Merrill, Am. J. Sci., 1922, **3**, p.335. Determination of Ga, Au and Pd, and partial analysis, 7.24 %Ni, E. Goldberg et al., Geochimica et Cosmochimica Acta, 1951, **2**, p.1. There is a well-developed meteorite crater at the place of fall, and many small pieces of metal (one of 300g) and much iron-shale were found on the crater rim, E.H. Sellards, Bull. Geol. Soc. Amer., 1927, **38**, p.149 [M.A. 5-16], D.M. Barringer, Proc. Acad. Nat. Sci. Philadelphia, 1929, **80**, p.307 [M.A. 4-427]. Description, analysis, 8.73 %Ni, C.W. Beck and L. LaPaz, Pop. Astron., Northfield, Minnesota, 1951, **59**, p.145. The shreibersite was also analysed. Contains silicate inclusions similar to those in Campo del Cielo, Linwood, and Toluca, T.E. Bunch et al., Contr. Miner. Petrol., 1970, **25**, p.297. Contains native copper, H.H. Nininger and G.I. Huss, Meteoritics, 1966, **3**, p.71. McCamey (*q.v.*) belongs to the same fall. Analysis, 7.2 %Ni, 74.7 ppm.Ga, 285 ppm.Ge, 2.2 ppm.Ir, J.T. Wasson, Icarus, 1970, **12**, p.407. Full description, references, V.F. Buchwald, Iron Meteorites, Univ. of California, 1975, p.937. Noble gases, shock effects, G.F. Herzog et al., J. Geophys. Res., 1976, **81**, p.3583 [M.A. 77-2364].
128kg Washington, U.S. Nat. Mus.; 85kg Tempe, Arizona State Univ.; 36.5kg Fort Worth, Texas, Monnig Colln.; 24kg Yale Univ.; 23kg Adelaide, South Austr. Mus.; 17.5kg Michigan Univ.; 15kg Prague, Nat. Mus.; 12.9kg Chicago, Field Mus. Nat. Hist.; 2.0kg Mainz, Max-Planck-Inst.; 1.3kg Harvard Univ.; 2kg Cranbrook Inst. Sci., Michigan; 1.5kg Michigan State College; 1.3kg New York, Amer. Mus. Nat. Hist.; 1.1kg Copenhagen, Univ. Geol. Mus.; 500g Albuquerque, Univ. New Mexico; 700 specimens, Austin, Texas, Univ. Texas;
Specimen(s): [1934,775], 606g.; [1934,776], 24g. shale-ball; [1959,1044], 2792g. complete iron; [1965,396], 308g. pieces; [1966,270], 128g.; [1966,271], 32.5g.; [1965,397], 62g. iron-shale

Odessa (stone) 46°30′ N., 30°46′ E.
Ukraine, USSR, [Одесса]
Found 1960, June
Stone. Olivine-bronzite chondrite (H4).
A stone of 1926g was found on the surface on the outskirts of Odessa, Meteor. Bull., 1964 (29).
1642g Moscow, Acad. Sci., includes main mass, 1268g;
Specimen(s): [1971,199], 16.25g. crusted part-slice

Odessa v Grossliebenthal.

Oesede 52°17′ N., 8°3′ E.
Osnabrück, Niedersachsen, Germany
Fell 1927, December 30, 1130 hrs
Stone. Olivine-bronzite chondrite (H5).
A mass of 3600g fell, but only 1400g were preserved. Description, with an analysis, K. Busz, Veröff. Naturwiss. Verein Osnabrück, 1929, **21**, p.13 [M.A. 4-258]. Olivine Fa_{19}, B. Mason, Geochimica et Cosmochimica Acta, 1967, **31**, p.1100.
600g Münster,; 137g Washington, U.S. Nat. Mus.; 575g Osnabrück, lost in 1945;

Oesel 58°30′ N., 23°0′ E.
Saaremaa, Estonian SSR, USSR, [Эзель]
Fell 1855, May 11, 1530 hrs
Synonym(s): *Estland, Ezel, Kaande, Moustel Pank*
Stone. Olivine-hypersthene chondrite (L6).
After detonations, several stones appear to have fallen on Oesel island and in the sea, but only about 6kg of fragments of one stone were preserved. Description and analysis, A. Göbel, Arch. Naturk. Liv.-Ehst.-u. Kurlands, Ser. 1, Min. Wiss. Dorpat, 1855, **1**, p.447. Olivine Fa_{25}, B. Mason, Geochimica et Cosmochimica Acta, 1963, **27**, p.1011. Analysis (under "Kaande"), 22.58 % total iron, V.Ya. Kharitonova, Метеоритика, 1969, **29**, p.91.
500g Tartu, Univ., approx.; 145g Moscow, Acad. Sci.; 48g Chicago, Field Mus. Nat. Hist.; 38g Vienna, Naturhist. Mus.; 28g Washington, U.S. Nat. Mus.; 21g Berlin, Humboldt Univ.; 20g Tübingen, Univ.; 27g Tempe, Arizona State Univ.;
Specimen(s): [33909a], 15g.; [1920,327], 12g.; [90272], 1.7g.

Ofehértó 47°53′ N., 22°2′ E.
Nyiregyhaza, Szabolcs, Hungary
Fell 1900, July 25
Stone. Olivine-hypersthene chondrite (L).
A stone of 3.75kg fell, F. Berwerth, Verzeichnis der Meteoriten, Naturhist. Hofmus. Wien, 1903, p.72. Olivine Fa_{23}, B. Mason, Geochimica et Cosmochimica Acta, 1963, **27**, p.1011.
3740g Budapest, Nat. Mus., main mass; 10g New York, Amer. Mus. Nat. Hist.;

Ofen 47°48′ N., 18°45′ E.
Budapest, Hungary
Fell 1642, December 2
Doubtful..
A mass (iron or stone?) fell "between Ofen and Gran", E.F.F. Chladni, Die Feuer-Meteore, Wien, 1819, p.100 but the evidence is not conclusive.

Offiglia v Cereseto.

Oficina Angela v La Primitiva.

Ogallala 41°10′ N., 101°40′ W.
Keith County, Nebraska, U.S.A.
Found 1918
Synonym(s): *Brule*
Iron. Octahedrite, coarse (1.6mm) (IA).
A mass of 3300g was ploughed up in 1918, 3.5 miles NE. of Ogallala. Shows a marked heating zone up to 4.5mm wide. Description, with an analysis, 7.93 %Ni, H.H. Nininger, Am. Miner., 1932, **17**, p.221 [M.A. 5-156]. Classification, analysis, 7.85 %Ni, 66.7 ppm.Ga, 266 ppm.Ge, 2.6 ppm.Ir, J.T. Wasson, Icarus, 1970, **12**, p.407. Description, V.F. Buchwald, Iron Meteorites, Univ. of California, 1975, p.942.
1.09kg Tempe, Arizona State Univ.; 118g Washington, U.S. Nat. Mus.; 48g Chicago, Field Mus. Nat. Hist.;
Specimen(s): [1959,960], 1203g.; [1959,961], 49g. sawings

Ogasawara 36° N., 139°30′ E. approx.
Tokyo Fu, Honshu, Japan
Doubtful. Iron.
Mentioned, I. Yamamoto, Kwasan Observ. Bull., 1935, p.306 [M.A. 7-173]. Missing in 1916, S. Murayama, letter of April 6, 1962 in Min. Dept. BM(NH).

Ogg v Ozren.

Ogi 33°17′ N., 130°12′ E.
Saga, Kyushu, Japan
Fell 1741, June 8, 1100 hrs
Synonym(s): *Hizen, Japan*
Stone. Olivine-bronzite chondrite (H6).
After sounds like thunder, four stones fell, one of about 5.6kg, another of 4.6kg, and the two others of about 2kg each. The two largest were for a long time among the offerings annually made in the temple in Ogi, K. Jimbo, Beitr. Miner. Japan, 1906 (2), p.34. Description, analysis, E. Divers, Trans. Asiatic Soc. Japan, 1882, **10**, p.199. Ar data, further analysis, 22.93 % total iron, olivine Fa_{20}, M. Shima et al., Meteoritics, 1981, **16**, p.386, abs.
Largest stone, Tokyo, Nabeshima family; 40g Washington, U.S. Nat. Mus.; 23g Chicago, Field Mus. Nat. Hist.; 22g Vienna, Naturhist. Mus.; 15g New York, Amer. Mus. Nat. Hist.;
Specimen(s): [55256], 4072g. main mass of the second largest stone, and a piece, 144.7g

Oguchimura v Kyushu.

Ogutimura v Kyushu.

Ohaba 46°4′ N., 23°35′ E. approx.
Alba Iulia (= Karlsburg), Transylvania, Romania
Fell 1857, October 11, 0015 hrs
Synonym(s): *Alba Julia, Carlsburg, Karlsburg, Veresegyháza*
Stone. Olivine-bronzite chondrite (H5), veined.
After appearance of fireball, followed by detonations, a stone of 16.25kg was found, J.L. Neugeboren, Verh. Mitt. Siebenbürg. Ver. Naturwiss., Hermannstadt, 1857, **8**, p.229. Description, with an analysis, M. Hörnes, Sitzungsber. Akad. Wiss. Wien, Math.-naturwiss. Kl., 1858, **31**, p.79. Olivine Fa_{20}, B. Mason, Geochimica et Cosmochimica Acta, 1963, **27**, p.1011.
15.75kg Vienna, Naturhist. Mus., main mass; 46g Tempe, Arizona State Univ.; 26g Harvard Univ.; 7g Chicago, Field Mus. Nat. Hist.;
Specimen(s): [55530], 29.75g.; [34588], 10g.

Ohuma 6°45′ N., 8°30′ E.
Igedde district, Benue province, Nigeria
Fell 1963, April 11
Stone. Olivine-hypersthene chondrite (L5).
A stone of about 17lb fell and is in the Geol. Survey of Nigeria, Kaduna. Analysis, 23.4 % total iron, A.A. Moss et al., Min. Mag., 1967, **36**, p.101. Olivine Fa_{25}, B. Mason, Geochimica et Cosmochimica Acta, 1967, **31**, p.1100.
Specimen(s): [1966,53], 162g. and fragments, 18g

Oildale v Cañon Diablo.

Ojansk v Ochansk.

Ojuelos Altos 38°11′ N., 5°24′ W.
Fuente Ovejuna, Cordoba, Spain
Fell 1926, December 10, 0930 hrs
Stone. Olivine-hypersthene chondrite (L6), brecciated.
A mass of 5850g fell 1km from the village. Description and analysis, L. Fernández Navarro, Bol. R. Soc. Españ. Hist. Nat. Madrid, 1929, **29**, p.19 [M.A. 4-259]. Olivine Fa_{25}, B. Mason, Geochimica et Cosmochimica Acta, 1963, **27**, p.1011.
191g Washington, U.S. Nat. Mus.; 73g Mainz, Max-Planck-Inst.; 71g Tempe, Arizona State Univ.; 70g Paris, Mus. d'Hist. Nat.;

Okabe 36°11′ N., 139°13′ E.
Okabe-mura, Saitama, Honshu, Japan
Fell 1958, November 26, 1500 hrs, approx.
Stone. Olivine-bronzite chondrite (H).
One stone of 194g was seen to fall, Meteor. Bull., 1959 (14), Meteoritic Stone and Meteoritic Iron, 1959, **3**, p.7, S. Murayama, letter of 6 April, 1962 in Min. Dept. BM(NH). Olivine Fa_{19}, B. Mason, Geochimica et Cosmochimica Acta, 1967, **31**, p.1100. Analysis, 28.93 % total iron, A. Miyashiro and S. Murayama, Chem. Erde, 1967, **26**, p.219.
Main mass, Tokyo, Nat. Sci. Mus.;
Specimen(s): [1978,M.17], 9g.

Okahandja 21°59′ S., 16°56′ E.
Damaraland, Namibia
Found 1926, known before this year
Synonym(s): *Okehandja*
Iron. Hexahedrite (IIA).
A mass of about 14.5lb was found, A.W. Rogers, letter of 17 November, 1926 in Min. Dept. BM(NH). Classification and analysis, 5.74 %Ni, 56.3 ppm.Ga, 186 ppm.Ge, 10 ppm.Ir, J.T. Wasson, Meteorites, Springer-Verlag, 1974, p.300. Description, slowly cooled, plastically deformed and subsequently shocked, V.F. Buchwald, Iron Meteorites, Univ. of California, 1975, p.944.
Main mass, Pretoria, Geol. Surv. Mus.; 466g Heidelberg, Max-Planck-Inst.;

Okano 35°5′ N., 135°12′ E.
Sasayama, Taki, Hyogo, Honshu, Japan
Fell 1904, April 7, 0635 hrs
Iron. Hexahedrite (IIA).
After the appearance of a fireball and detonations, a mass of 4742g was found, K. Jimbo, Beitr. Miner. Japan, 1906 (2), p.51. Description, and poor analysis, 4.44 %Ni, M. Chikashige and T. Hiki, Mem. Coll. Sci. Eng. Kyoto Univ., 1912, **5** (1), p.1. Analysis, 5.55 %Ni, 59.9 ppm.Ga, 180 ppm.Ge, 11 ppm.Ir, J.T. Wasson, Geochimica et Cosmochimica Acta, 1969, **33**, p.859. Description, a normal hexahedrite, V.F. Buchwald, Iron Meteorites, Univ. of California, 1975, p.946.
3.58kg Kyoto, Univ.; 383g Harvard Univ.; 163g New York, Amer. Mus. Nat. Hist.; 20g Washington, U.S. Nat. Mus.;
Specimen(s): [1923,1016], 302g. slice

Okechobee 26°41′ N., 80°48′ W.
Palm Beach County, Florida, U.S.A.
Found 1916, before this year
Synonym(s): *Lake Okechobee*
Stone. Olivine-hypersthene chondrite (L4).
Fragments weighing about 1kg were brought up in a net some 0.75 miles from the shore, G.P. Merrill, Proc. U.S. Nat. Mus., 1916, **51**, p.525. Olivine Fa_{24}, B. Mason, Geochimica et Cosmochimica Acta, 1963, **27**, p.1011.
968g Washington, U.S. Nat. Mus.; 71g Chicago, Field Mus. Nat. Hist.; 20g Tempe, Arizona State Univ.;

Okehandja v Okahandja.

Okhansk v Ochansk.

Okirai v Kesen.

Okniny 50°50′ N., 25°30′ E.
Krzemieniec, Volhynia, Ukraine, USSR, [Окиниы]
Fell 1834, January 8, 0930 hrs
Stone. Olivine-hypersthene chondrite, amphoterite (LL6), brecciated.
With the usual phenomena, a stone of about 12kg fell, K. Wtorschetzkii, Schrift. Russ. Min. Gesell. St. Petersburg, 1842, **1**, p.lxxii. Listed, E.L. Krinov, Астрон. Журнал, 1945, **22**, p.303 [M.A. 9-297]. Olivine Fa_{27}, B. Mason, Geochimica et Cosmochimica Acta, 1963, **27**, p.1011.
110g Vienna, Naturhist. Mus.; 65g Berlin, Humboldt Univ.; 46.2g Moscow, Acad. Sci.; 11g Tübingen, Univ.; 8g Chicago, Field Mus. Nat. Hist.;
Specimen(s): [35177], 6g.; [35405], 1g.

Oktibbeha County 33°30′ N., 89° W. approx.
Mississippi, U.S.A.
Found 1854, approx.
Synonym(s): *Octibbeha County*
Iron. Anomalous (IB).
A mass of 156g was found in an Indian tumulus. Description and analysis, 59.69 %Ni, W.J. Taylor, Am. J. Sci., 1857, **24**, p.293. Analysis, 62.01 %Ni, E. Cohen, Ann. Naturhist. Hofmus. Wien, 1892, **7**, p.146. Considered as doubtfully meteoritic, J.D. Buddhue, Pop. Astron., Northfield, Minnesota, 1937, **45**, p.106 [M.A. 7-79] but confirmed as meteoritic, with the highest Ni content of metal, 60%, and of schreibersite, (Fe0.7Ni2.3)P, of any meteorite, S.J.B. Reed, Min. Mag., 1972, **38**, p.623. Classification and analysis, 58.5 %Ni, 3.6 ppm.Ga, 9.0 ppm.Ge, 0.026 ppm.Ir, A. Kracher et al., Geochimica et Cosmochimica Acta, 1980, **44**, p.773.
25g Philadelphia, Acad. Nat. Sci.; 13g New York, Amer. Mus. Nat. Hist.; 6g Harvard Univ.; 2g Vienna, Naturhist. Mus.;
Specimen(s): [34595], less than one gram

Olah Gyéres v Mocs.

Oldenburg (1368) 52°57′ N., 8°10′ E.
Niedersachsen, Germany
Fell 1368
Doubtful..
A mass of iron is said to have fallen, E.F.F. Chladni, Die Feuer-Meteore, Wien, 1819, p.201 but the evidence is not conclusive. Perhaps identical to Benthullen (*q.v.*).

Oldenburg (1930) 52°57′ N., 8°10′ E.
Niedersachsen, Germany
Fell 1930, September 10, 1415 hrs
Synonym(s): *Beverbruch, Beverbruck*
Stone. Olivine-hypersthene chondrite (L6).
Two stones fell, one of 4840g at Bissel, 23km south of Oldenburg, co-ordinates as above, and one of 11730g at Beverbruch, 4.4km to the NW., von Buttel-Reepen, Oldenburg Jahrb. Ver. Altertumsk. Landesgesch., Oldenburg, 1930, **34**, p.101 [M.A. 5-403]. Description, E. Preuss, Fortschr. Min., 1949, **28**, p.63 [M.A. 11-267], Sternenwelt, 1950 (7) [M.A. 11-267]. Olivine Fa_{24}, B. Mason, Geochimica et Cosmochimica Acta, 1967, **31**, p.1100. Description, (under "Beverbruch"), P. Ramdohr and A. El Goresy, Meteoritics, 1974, **9**, p.397, abs.
Both stones, Cloppenburg, Mus.;

Oldfield River 33° S., 121° E.
West Point Farm, Ravensthorpe District, Western Australia, Australia
Found 1972, approx.
Stone. Olivine-bronzite chondrite (H5).
A single stone was found in sandy soil on West Point Farm on the Oldfield River. A portion of the stone is held by the owner of the farm, the remainder is in the West. Austr. Mus. Analysis, 28.5 % total iron, Ann. Rep. Govt. Chem. Lab. for 1973, 1973, p.28. Olivine Fa_{16}, R.A. Binns, priv. comm. Mentioned, J.R. de Laeter, letter of 22 August, 1979 in Min. Dept. BM(NH).
150g Perth, West. Austr. Mus.;

Old Fork v Jenny's Creek.

Oldham County v La Grange.

Oldisleben 51°18′ N., 11°11′ E.
Halle, Germany
Fell 1135
Doubtful..
A black stone, the size of a man's head, is said to have fallen, E.F.F. Chladni, Die Feuer-Meteore, Wien, 1819, p.197 but the evidence is not conclusive.

Old Woman 34°28′ N., 115°14′ W.
Old Woman Mountains, San Bernardino County, California, U.S.A.
Found 1976
Iron. Octahedrite, coarsest (10mm) (IIB).
A large single mass of 2753kg was found on the western slopes of the Old Woman Mountains, Meteor. Bull., 1978 (55), Meteoritics, 1978, **13**, p.344. Brief report, J.T. Wasson and J. Willis, Meteoritics, 1976, **11**, p.386, abs. Classification and analysis, 5.71 %Ni, 58.5 ppm.Ga, 190 ppm.Ge, 0.80 ppm.Ir, A. Kracher et al., Geochimica et Cosmochimica Acta, 1980, **44**, p.773.
2753kg Washington, U.S. Nat. Mus.; Specimen, Los Angeles, Univ. of California;

Oliva-Gandia 39°0′ N., 0°2′ W.
Valencia, Spain
Fell 1520, May 26
Stone.
Three stones fell between Gandia and Oliva, S. Calderón, Bol. R. Soc. Españ. Hist. Nat. Madrid, 1906, **6**, p.331. It is doubtful whether any is now preserved. Listed, M. Faura y Sans, Meteoritos caídos en la Península Iberica, Tortosa, 1922, p.6, E.F.F. Chladni, Die Feuer-Meteore, Wien, 1819, p.211.

Olivenza 38°43′ N., 7°4′ W.
Badajoz, Spain
Fell 1924, June 19, 0800 hrs
Stone. Olivine-hypersthene chondrite, amphoterite (LL5).
After appearance of white cloud, with trajectory NNE.-SSW., and detonations, five stones of total weight estimated at 150kg fell and broke into numerous pieces. Description, L. Fernández Navarro, Bol. R. Soc. Españ. Hist. Nat. Madrid, 1924, **24**, p.339. Analysis, F. Raoult, Trab. Mus. Nac. Cienc. Nat. Madrid, 1925 (ser. geol., no. 35), p.1, C. R. Acad. Sci. Paris, 1925, **180**, p.1674. Olivine $Fa_{29.8}$, K. Fredriksson et al., Origin and Distribution of the Elements, ed. L.H. Ahrens, Pergamon, 1968, p.457. Rb-Sr age, 4630 m.y., H.G. Sanz and G.J. Wasserburg, Earth planet. Sci. Lett., 1969, **6**, p.335. Lightly shocked, J.R. Ashworth and D.J. Barber, Earth planet. Sci. Lett., 1975, **27**, p.43. One fragment fell at

Castelo de Vide, just in Portugal, R. de Serpa Pinto, Rev. sismol. geofis. Coimbra, 1932, **3**, p.45 [M.A. 5-297]. Ni and Co contents of metal, D.W. Sears and H.J. Axon, Meteoritics, 1976, **11**, p.97.

Main mass, Madrid, Mus.; Main mass, Badajoz, Mus.; 3.25kg Paris, Mus. d'Hist. Nat.; 1.7kg New York, Amer. Mus. Nat. Hist.; 938g Washington, U.S. Nat. Mus.; 576g Yale Univ.; 452g Tempe, Arizona State Univ.; 238g Chicago, Field Mus. Nat. Hist.;

Specimen(s): [1925,430], 1181g. deeply scooped piece; [1925, 524], 86g.

Ollague v Imilac.

Olmedilla de Alarcón 39°34′ N., 2°6′ W.

Cuenca, Spain

Fell 1929, February 26, 1200 hrs

Stone. Olivine-bronzite chondrite (H5), veined, xenolithic.

Several stones fell, one in Valverdejo district, the others in Olmedilla de Alarcón. Eight were recovered, four of which weighed 14, 13, 2 and 2lb. Description, with an analysis, L. Fernández Navarro, Mem. R. Soc. Españ. Hist. Nat. Madrid, 1929, **15**, p.859. Olivine Fa_{18}, B. Mason, Geochimica et Cosmochimica Acta, 1963, **27**, p.1011. Xenolithic, R.A. Binns, Geochimica et Cosmochimica Acta, 1968, **32**, p.299. In a preliminary note, stones of 30, 7 and 2kg were mentioned, L. Fernández Navarro, Bol. R. Soc. Españ. Hist. Nat. Madrid, 1929, **29**, p.145 [M.A. 4-259].

299g Chicago, Field Mus. Nat. Hist.; 314g Washington, U.S. Nat. Mus.; 233g Harvard Univ.; 99g Mainz, Max-Planck-Inst.;

Specimen(s): [1930,421], 242g. and fragments, 2.5g

Olton 34°10′2″ N., 102°8′1″ W.

Lamb County, Texas, U.S.A.

Found 1948

Stone. Chondrite.

Listed, total known weight 953g, Cat. Monnig Colln. Meteorites, 1983.

953g Fort Worth, Texas, Monnig Colln.;

Omoa 15°48′ N., 87°59′ W.

Honduras

Doubtful..

A meteorite from Omoa is mentioned, H. Heuland, Ann. Phil., 1818, **12**, p.200 but never further described, and probably proved to be a pseudometeorite.

Ootomi v Otomi.

Ooyada v Gifu.

Opava 49°58′ N., 17°54′ E.

Severomoravsky, Czechoslovakia

Found 1925

Synonym(s): *Kylesovice, Troppau*

Iron. Hexahedrite.

Four pieces of 7.4, 5.8, 1.0 and 0.1kg were found 80cm below the surface in a loess deposit on the Kylesovsky hill, F. Drahny, Veda Prírodni, 1926, **7**, p.139. Analysis, 5.6 % Ni, R. Rost, Метеоритика, 1955, **12**, p.54 [M.A. 13-48].

Main masses, Dejvice, Praha Polytechnic; 629g Prague, Nat. Mus.;

Ophir v Illinois Gulch.

Orange River (iron) 30° S., 25° E. approx.

Orange River district, Orange Free State, South Africa

Found 1855

Synonym(s): *Griqualand*

Iron. Octahedrite, medium (1.35mm) (IIIB).

A mass of 328lb, sent by a farmer of the Orange River district, was brought to London in August 1855 and was described, with an analysis, C.U. Shepard, Am. J. Sci., 1856, **21**, p.213. Analysis, 8.46 %Ni, 21.2 ppm.Ga, 43.7 ppm.Ge, 0.12 ppm.Ir, E.R.D. Scott et al., Geochimica et Cosmochimica Acta, 1973, **37**, p.1957. Description, shock-hardened, V.F. Buchwald, Iron Meteorites, Univ. of California, 1975, p.947.

146kg Tempe, Arizona State Univ., main mass; 283g Chicago, Field Mus. Nat. Hist.; 214g Tübingen, Univ.; 120g Washington, U.S. Nat. Mus.;

Specimen(s): [32051], 96g.; [1959,963], 732g. thick slice, and sawings, 10g

Orange River (stone) 29°40′ S., 24°13′ E.

Cape Province, South Africa

Fell 1887, September 8

Synonym(s): *Beaufort*

Doubtful. Stone. Chondrite.

A fragment of 8g of this doubtful stone in Vienna (Naturhist. Mus.), A. Brezina, Die Meteoritensammlung Naturhist. Hofmus. Wien, 1895, p.248.

8g Vienna, Naturhist. Mus.;

Orava v Magura.

Orebro 59°17′ N., 15°13′ E.

central Sweden, Sweden

Fell 1958, October 15

Pseudometeorite. Stone.

A mass of 8kg is said to have fallen, (Wochenpost, 11 July, 1959), K. Keil, Fortschr. Min., 1960, **38**, p.255 but this is unconfirmed. A hoax, F.E. Wickman and A.G. Uddenberg-Anderson, Arkiv Min. Geol., 1968, **4**, p.543.

Oregon v Klamath Falls.

Oregon City v Willamette.

Orgueil 43°53′ N., 1°23′ E.

Montauban, Tarn-et-Garonne, France

Fell 1864, May 14, 2000 hrs

Synonym(s): *Montauban, Orguell*

Stone. Carbonaceous chondrite, type I (CI).

After appearance of a luminous meteor and detonations, about 20 stones, the largest of the size of a man's head, but most as large as a fist, fell over an area of 2 square miles, G.A. Daubrée, C. R. Acad. Sci. Paris, 1864, **58**, p.932, 1065. Analysis, S. Cloez, C. R. Acad. Sci. Paris, 1864, **58**, p.986, C. R. Acad. Sci. Paris, 1864, **59**, p.37. Mentioned, F. Pisani, C. R. Acad. Sci. Paris, 1864, **59**, p.132. H and C isotopes, G. Boato, Geochimica et Cosmochimica Acta, 1954, **6**, p.209. Most of the extensive literature on the carbon compounds and alleged fossils in this meteorite is summarized, with an analysis, 19.47 % total iron, B. Mason, Space Sci. Rev., 1963, **1**, p.621, Ann. New York Acad. Sci., 1963, **108**, p.495, 514, 534, Ann. Biochem. Biophys., 1963, **101**, p.240, Nature, 1963, **198**, p.728, Nature, 1963, **200**, p.565, Nature, 1964, **202**, p.125, 228, J. Amer. Oil Chem. Soc., 1966, **43**, p.189. Texture, M.N. Bass, Meteoritics, 1970, **5**, p.180, abs. Remnent magnetisation, S.K. Banerjee and R.B. Hargraves, Earth planet. Sci. Lett., 1971, **10**, p.392.

Magnetite formation conditions, M.S. Lancet and E. Anders, Geochimica et Cosmochimica Acta, 1973, **37**, p.137. Magnetite and sulphide have very old I-Xe age, R.S. Lewis and E. Anders, Proc. Nat. Acad. Sci., 1975, **72**, p.268. U-Pb-Th systematics, M. Tatsumoto et al., Geochimica et Cosmochimica Acta, 1976, **40**, p.617. D/H ratio, Y. Kolodny et al., Earth planet. Sci. Lett., 1980, **46**, p.149. Pre-solar component, P. Eberhardt et al., Ap. J., 1979, **234**, p.L169. Trace element data, G.W. Kallemeyn and J.T. Wasson, Geochimica et Cosmochimica Acta, 1981, **45**, p.1217.

8.72kg Paris, Mus. d'Hist. Nat.; 370g Prague, Nat. Mus.; 348g Edinburgh, Roy. Scottish Mus.; 183g Washington, U.S. Nat. Mus.; 140g Berlin, Humboldt Univ.; 62g Chicago, Field Mus. Nat. Hist.;

Specimen(s): [36104], 463g. stone, and fragments, 8g; [36273], 131g. and fragments, 2g; [1960,331], 55g. and fragments, 6g; [1920,328], 6g.

Orguell v Orgueil.

Oriang 24° N., 76° E. approx.

Malwa region, Madhya Pradesh, India
Fell 1825, January 16
Doubtful..

A meteorite is said to have fallen and to have killed a man and injured a woman, E.F.F. Chladni, Ann. Phys. Chem. (Poggendorff), 1826, **6**, p.32 but the evidence is not conclusive.

Orimattila 60°35′ N., 25°35′ E.

Mallusjoki, Socken, Finland
Found 1974
Stone. Olivine-bronzite chondrite (H4).

One weathered individual of 1.872kg was found at about 1m depth in moraine gravel, Meteor. Bull., 1975 (53), Meteoritics, 1975, **10**, p.156.

Main mass, Helsinki, Outokumpu Mining Co.; 211g Helsinki, Univ.;

Orléans v Charsonville.

Orléans v Lancé.

Orlovka 56°0′ N., 76°45′ E.

River Ui, Federated SSR, USSR, [Орловка]
Found 1928
Synonym(s): *Orlowka*
Stone. Olivine-bronzite chondrite (H5).

One stone, 40.5kg, was ploughed up near Orlovka, on the river Ui, a tributary of the Irtysh, 120km east of Tara, P.L. Dravert, Trudy miner. Inst., 1931, **1**, p.121 [M.A. 5-154]. Description, with micrometric analysis, P.N. Chirvinsky, Метеоритика, 1948, **4**, p.75 [M.A. 10-516]. Analysis, 27.96 % total iron, M.I. D'yakonova and V.Ya. Kharitonova, Метеоритика, 1960, **18**, p.48 [M.A. 16-447]. Olivine Fa_{19}, B. Mason, Geochimica et Cosmochimica Acta, 1963, **27**, p.1011.

37.3kg Moscow, Acad. Sci., includes main mass, 25.6kg; 127g Washington, U.S. Nat. Mus.; 123g New York, Amer. Mus. Nat. Hist.; 104g Tempe, Arizona State Univ.;

Specimen(s): [1933,272], 53g.

Orlowka v Orlovka.

Ormes v Les Ormes.

Ornana v Ornans.

Ornans 47°7′ N., 6°9′ E.

Doubs, France
Fell 1868, July 11, 1915 hrs
Synonym(s): *Ornana, Salins*
Stone. Carbonaceous chondrite, type III (CO3).

After detonations, a stone of about 6kg fell and broke into two pieces, J. Marcon, Bull. Soc. Géol. France, 1869, **26**, p.92. Description, G.A. Daubrée, Bull. Soc. Géol. France, 1869, **26**, p.95. Analysis, F. Pisani, C. R. Acad. Sci. Paris, 1868, **67**, p.663. Further analysis, 25.83 % total iron, B. Mason, Space Sci. Rev., 1963, **1**, p.621 [M.A. 16-640]. Chosen as representative of one of the two sub-types of type III carbonaceous chondrites, W.R. Van Schmus and J.M. Hayes, Geochimica et Cosmochimica Acta, 1974, **38**, p.47. XRF analysis, 23.97 % total iron, L.H. Ahrens et al., Earth planet. Sci. Lett., 1969, **6**, p.285. Rare gas content, E. Mazor et al., Geochimica et Cosmochimica Acta, 1970, **34**, p.781. Noble gases in separated minerals, B. Srinivasan et al., J. Geophys. Res., 1977, **82**, p.762. Petrography of Ornans-type chondrites, H.Y. McSween, Jr., Geochimica et Cosmochimica Acta, 1977, **41**, p.477. Trace element data, G.W. Kallemeyn and J.T. Wasson, Geochimica et Cosmochimica Acta, 1981, **45**, p.1217.

3.53kg Paris, Mus. d'Hist. Nat.; 89g Yale Univ.; 83g Calcutta, Mus. Geol. Surv. India; 74g Chicago, Field Mus. Nat. Hist.;

Specimen(s): [42474], 925g. and fragments, 3.5g; [1920,329], 1g.

Oro Grande 32°22′18″ N., 106°15′30″ W.

New Mexico, U.S.A.
Found 1971
Stone. Olivine-bronzite chondrite (H5).

A 513g stone was found in a gravel driveway 10 miles W. of Orogrande. The finder, Mr. T.C. Wilton, speculates that the meteorite was brought in with the gravel for the driveway, Meteor. Bull., 1974 (52), Meteoritics, 1974, **9**, p.114. Mineralogical description and analysis, olivine $Fa_{19.3}$, R.V. Fodor et al., Meteoritics, 1972, **7**, p.495.

250g in possession of finder; 230g Albuquerque, Univ. New Mexico; 9g Washington, U.S. Nat. Mus.;

Oroville 39°41′ N., 121°38′ W.

Butte County, California, U.S.A.
Found 1893
Iron. Octahedrite, medium (0.85mm) (IIIB).

A mass of 54lb was found 10 miles N. of Oroville, O.C. Farrington, Mem. Nat. Acad. Sci. Washington, 1915, **13**, p.345. Mentioned, C.S. Bement, letter of October 18, 1894 in Min. Dept. BM(NH)., C.P. Butler, Pacific Discovery, Calif. Acad. Sci., 1964, **17** (5). Analysis, 8.36 %Ni, 20.3 ppm.Ga, 40.7 ppm.Ge, 0.053 ppm.Ir, E.R.D. Scott et al., Geochimica et Cosmochimica Acta, 1973, **37**, p.1957. Description; intensely deformed and fissured, V.F. Buchwald, Iron Meteorites, Univ. of California, 1975, p.949.

46lb San Francisco, Acad. Sci., main mass; 401g Washington, U.S. Nat. Mus.; 315g Chicago, Field Mus. Nat. Hist.; 281g New York, Amer. Mus. Nat. Hist.; 117g Vienna, Naturhist. Mus.;

Specimen(s): [84549], 308g.; [77092], 62g.

Ørsted v Funen.

Ortenau 48°30′ N., 8° E. approx.
Baden-Württemberg, Germany
Fell 1671, February 27, 1200 hrs
Stone.
A mass of 10lb, grey inside, with a black crust, was seen to fall in the district of Ortenau, near Offenburg, L.W. Gilbert, Ann. Physik (Gilbert), 1809, **33**, p.183.

Orvinio 42°8′ N., 12°56′ E.
Rieti, Lazio, Italy
Fell 1872, August 31, 0515 hrs
Synonym(s): *Anticoli Corradi, Canemorto, Gerano, La Scarpa, Pezza del Moleto, Pozzaglia, Rieti, Rom, Roma, Rome, Umbria*
Stone. Olivine-hypersthene chondrite (L6), brecciated, black.
After appearance of fireball and detonations, several stones appear to have fallen, of which six "fragments" were found. The total weight was 3.4kg and the largest piece weighed 1.2kg, P. Keller, Ann. Phys. Chem. (Poggendorff), 1873, **150**, p.171, P. Secchi, C. R. Acad. Sci. Paris, 1872, **75**, p.656. Description, G. Tschermak, Sitzungsber. Akad. Wiss. Wien, Math.-naturwiss. Kl., 1875, **70** (1), p.459. Analysis, L. Sipöcz, Tschermaks Min. Mitt., 1874, p.244. Olivine Fa_{23}, B. Mason, Geochimica et Cosmochimica Acta, 1963, **27**, p.1011. Listed, B. Baldanza, Min. Mag., 1965, **35**, p.214.
750g Rome, Univ.; 625g Vienna, Naturhist. Mus.; 660g Vatican colln.; 96g Paris, Mus. d'Hist. Nat.; 59g Washington, U.S. Nat. Mus.; 50g Tempe, Arizona State Univ.;
Specimen(s): [53930], 57.75g.; [1920,330], 28g.; [48750], 4g.

Oschank v Ochansk.

Oscuro Mountains 33°38′ N., 106°23′ W.
Socorro County, New Mexico, U.S.A.
Found 1895
Iron. Octahedrite, coarse (1.8mm) (IA).
Three pieces of about 3.5, 3.25, and 1.25lb were found in the eastern foothills. Description and analysis, 7.66 %Ni, R.C. Hills, Proc. Colorado Sci. Soc., 1897, **6**, p.30. Analysis, 6.55 %Ni, 91.0 ppm.Ga, 359 ppm.Ge, 2.6 ppm.Ir, J.T. Wasson, Meteorites, Springer-Verlag, 1974, p.297. Cosmically reheated, then annealed at low temperature, V.F. Buchwald, Iron Meteorites, Univ. of California, 1975, p.950.
750g Chicago, Field Mus. Nat. Hist.; 235g New York, Amer. Mus. Nat. Hist.; 102g Washington, U.S. Nat. Mus.; 76g Vienna, Naturhist. Mus.;
Specimen(s): [83981], 494g. slice

Oshima v Kyushu.

Oshkosh 44°4′37″ N., 88°33′46″ W.
Winnebago County, Wisconsin, U.S.A.
Found 1961
Stone. Olivine-bronzite chondrite (H).
About a dozen small fragments were picked up from the surface of a field, the largest being about 2" in diameter. Total weight 144g, W.F. Read, Wisconsin Acad. Review, 1962, **9**, p.152, Meteor. Bull., 1963 (26). Figured, W.F. Read, Meteoritics, 1968, **4**, p.137. Olivine Fa_{18}, B. Mason, Geochimica et Cosmochimica Acta, 1963, **27**, p.1011.
1g New York, Amer. Mus. Nat. Hist.; Remainder of fall, Appleton, Wisconsin, Lawrence College;

Osseo 47°38′ N., 80°5′ W.
Timiskaming, Ontario, Canada
Found 1931
Iron. Octahedrite, coarse (2.8mm) (IA).
A mass of 46.3kg was found. Description and analysis, 6.51 %Ni, J.P. Marble, Am. Miner., 1938, **23**, p.282 [M.A. 7-68, 176]. Description, co-ordinates, V.F. Buchwald, Iron Meteorites, Univ. of California, 1975, p.953. Classification and analysis, 6.44 %Ni, 91.7 ppm.Ga, 450 ppm.Ge, 5.4 ppm.Ir, J.T. Wasson, Icarus, 1970, **12**, p.407.
37kg Washington, U.S. Nat. Mus., main mass; 2kg Ottawa, Mus. Geol. Surv. Canada; 750g Michigan, Univ.; 173g Tempe, Arizona State Univ.; 156g Chicago, Field Mus. Nat. Hist.; 133g Los Angeles, Univ. of California;
Specimen(s): [1959,929], 197.5g. slice, and sawings, 3g

Ostrolenka v Pułtusk.

Ostrzeszów 51°25′ N., 17°55′ E. approx.
Poznán, Poland
Fell 1907, September 3, 0100 hrs, approx.
Doubtful..
A fireball was observed, but nothing was recovered, J. Pokrzywnicki, Studia Geol. Polon., 1964, **15**, p.125.

Otasawian
Alberta, Canada
Found 1907
Doubtful. Iron.
A mass of 8.03kg and a slice of 634g, labelled "Otasawian, Canada", are preserved in Perugia University, B. Baldanza et al., Meteoriti, Univ. Perugia, 1969. Some features indicative of dynamic deformation are figured, B. Baldanza and G. Pialli, Meteorite Research, ed. P.M. Millman, D. Reidel, Dordrecht-Holland, 1969, p.806. Transported Cañon Diablo; a fraud, V.F. Buchwald, Iron Meteorites, Univ. of California, 1975, p.1408.

Otcheretna v Oczeretna.

Otchinjau 16°30′ S., 14° E. approx.
Cunene, Angola
Found 1919
Iron. Octahedrite, fine (0.3mm) (IVA).
A mass of 30kg was found in 1919 by J. d'Almeida, H. Vieira, letters of January 24, 1950 in Min. Dept. BM(NH). Analysis, 7.74 %Ni, D.I. Bothwell Min. Dept. BM(NH). Description and analyses, L. Aires-Barros and R.A. David Gomes, Bol. Serv. Geol. Min. Angola, 1964 (10), p.108 [M.A. 17-758]. Further analysis, 7.82 %Ni, 2.13 ppm.Ga, 0.117 ppm.Ge, 2.6 ppm.Ir, R. Schaudy et al., Icarus, 1972, **17**, p.174. Shock-hardened, unannealed, V.F. Buchwald, Iron Meteorites, Univ. of California, 1975, p.955.
Specimen(s): [1953,61], 1405g. and fragments, 49g and filings, 1g

Oterøy 58°53′ N., 9°24′ E.
Kragerø, Norway
Fell 1928, October 15, 1500 hrs
Stone. Olivine-hypersthene chondrite (L6).
246g of fragments were collected, W.L. Griffin, Meteoritics, 1974, **9**, p.167. Olivine Fa_{25}, B. Mason, Geochimica et Cosmochimica Acta, 1963, **27**, p.1011.
227g Oslo, Min.-Geol. Mus., fragments; 3.7g New York, Amer. Mus. Nat. Hist.;

Otis 38°32′ N., 99°3′ W.

Rush County, Kansas, U.S.A.
Found 1940
Stone. Olivine-hypersthene chondrite (L6).

A mass of 2.6kg was found, W. Wahl, letter of May 23, 1950 in Min. Dept. BM(NH)., H.H. and A.D. Nininger, The Nininger Collection of Meteorites, Winslow, Arizona, 1950, p.80. Olivine Fa_{24}, B. Mason, Geochimica et Cosmochimica Acta, 1963, **27**, p.1011. 40K-40Ar age of separated feldspar, 4230 m.y., is significantly greater than that of pyroxene, olivine and bulk meteorite, 2240-2390 m.y., suggesting diffusion loss of 40Ar from this find, P. Bochsler et al., Meteorite Research, ed. P.M. Millman, D. Reidel, Dordrecht-Holland, 1969, p.857.

1.1kg Tempe, Arizona State Univ.; 215g Washington, U.S. Nat. Mus.; 105g Mainz, Max-Planck-Inst.; 24g New York, Amer. Mus. Nat. Hist.;

Specimen(s): [1959,840], 564.5g.

Otomi 38°24′ N., 140°21′ E.

Kita Murayama, Yamagata, Honshu, Japan
Fell 1867, May 24
Synonym(s): *Ootomi, Otomigo*
Stone. Olivine-bronzite chondrite (H).

A stone of 6.5kg fell, M. Hoshina, J. Geol. Soc. Tokyo, 1925, **32**, p.177, Jap. J. Geol. Geogr., 1926, **4**, p.1 [M.A. 3-257]. Mentioned, S. Murayama, letter of April 6, 1962 in Min. Dept. BM(NH). Olivine Fa_{18}, B. Mason, Geochimica et Cosmochimica Acta, 1967, **31**, p.1100.

Main mass, Kofu, in possession of T. Inuma;

Otomigo v Otomi.

Otscheretnaja v Oczeretna.

Otsego County v Burlington.

Ottawa 38°36′ N., 95°13′ W.

Franklin County, Kansas, U.S.A.
Fell 1896, April 9, 1815 hrs
Stone. Olivine-hypersthene chondrite, amphoterite (LL6), brecciated.

The fall was described in the Ottawa Weekly Times of April 16, 1896. The single stone of 840g which fell passed into the possession of C.S. Bement, O.C. Farrington, Mem. Nat. Acad. Sci. Washington, 1915, **13**, p.347, C.S. Bement, letter of October 18, 1897 in Min. Dept. BM(NH). Analysis, 21.24 % total iron, B. Mason and H.B. Wiik, Am. Mus. Novit., 1961 (2069). Olivine $Fa_{29.3}$, K. Fredriksson et al., Origin and Distribution of the Elements, ed. L.H. Ahrens, Pergamon, 1968, p.457.

222g New York, Amer. Mus. Nat. Hist.; 70g Chicago, Field Mus. Nat. Hist.; 56g Vienna, Naturhist. Mus.; 36g Paris, Mus. d'Hist. Nat.; 12g Washington, U.S. Nat. Mus.;

Specimen(s): [83498], 86.25g. two pieces, 62.75 and 23.5g; [1920,331], 65g.

Ottiglio v Cereseto.

Ottsjö v Föllinge.

Ottsjön v Föllinge.

Otumpa v Campo del Cielo.

Ouallen 24°10′ N., 0°5′ E.

Tanezrouft, Sahara Occidental, Algeria
Found 1936, February
Synonym(s): *Tanesrouft, Tanezrouft*
Stone. Olivine-bronzite chondrite (H).

13 fragments were found in the desert at 24°10'N., 0°5'E., about 135km WSW. of Ouallen, 11 weighing 4849g in one place, and two weighing 325g some hundreds of metres away. Description and analysis, A. Lacroix, C. R. Acad. Sci. Paris, 1936, **203**, p.901 [M.A. 6-395]. Olivine Fa_{20}, B. Mason, Geochimica et Cosmochimica Acta, 1963, **27**, p.1011. Found by T. Monod, February 12, 1936. Account of the finding and photograph of the fragments in situ, T. Monod, Méharées, Paris, 1937, p.289.

2.377kg Paris, Mus. d'Hist. Nat., main mass;

Specimen(s): [1966,280], 7.9g. two part-crusted fragments; [1972,323], 8.31g. part-crusted fragment

Ouaregla v Haniet-el-Beguel.

Oubari 26°48′ N., 13°35′ E.

Fezzan, Libya
Found 1944, September 27
Synonym(s): *Fezzan, Ubari*
Stone. Olivine-hypersthene chondrite, amphoterite (LL6), brecciated.

A broken stone was found 2km from Gabr'on. The fragments total 8kg, the largest being 2156g. Description, with analysis and optical data, E. Jérémine and M. Lelubre, Geochimica et Cosmochimica Acta, 1952, **2**, p.217. Brecciated, W. Wahl, Geochimica et Cosmochimica Acta, 1952, **2**, p.91. Olivine $Fa_{28.0}$, K. Fredriksson et al., Origin and Distribution of the Elements, ed. L.H. Ahrens, Pergamon, 1968, p.457.

7.478kg Paris, Mus. d'Hist. Nat.; 151g Washington, U.S. Nat. Mus.;

Specimen(s): [1953,40], 102g.

Oude v Dyalpur.

Oude v Kaee.

Oufrane 28°17′45″ N., 0°1′30″ E.

Tademait, Algeria
Found 1969
Stone. Olivine-hypersthene chondrite (L).

Two well-preserved stones, 335g and 205g, were found 95m apart, 31km SSW. of the village of Oufrane, which is 57km NNE. of the city of Adrar. Two stones probably fell on the morning of January 10, 1969, a few weeks before their recovery (February 4), Meteor. Bull., 1971 (50), Meteoritics, 1971, **6**, p.121.

500g Paris, Mus. d'Hist. Nat.;

Outpost Nunatak A80301 75°50′ S., 158° E. approx.

Victoria Land, Antarctica
Found 1980
Stone. Olivine-bronzite chondrite (H3).

A single mass of 35g was found about 100km NE. of Allan Hills during the 1980-81 field season in Antarctica, olivine Fa_{17-19}, B. Mason and R.S. Clarke, Jr., Mem. Nat. Inst. Polar Res. Tokyo, 1982 (spec. issue no. 25), p.17. Locality map, L. Schultz et al., Meteoritics, 1981, **16**, p.384.

Ovambo 18° S., 16° E. approx.
Amboland, Namibia
Fell 1900, approx.
Stone. Olivine-hypersthene chondrite (L6).
A 56g stone was found amongst material in Professor Walter Wahl's collection after his death. The available information indicates that after the fall a stone was found in the possession of natives from which a fragment was removed and passed via Finnish missionaries to Prof. Wahl, Meteor. Bull., 1975 (53), Meteoritics, 1975, **10**, p.137.
Main mass, Helsinki, Univ.;

Ovid 40°58′ N., 102°24′ W.
Sedgwick County, Colorado, U.S.A.
Found 1939, February
Stone. Olivine-bronzite chondrite (H6).
One stone of 6.169kg was found, A.D. Nininger, Pop. Astron., Northfield, Minnesota, 1940, **48**, p.556, Contr. Soc. Res. Meteorites, **2**, p.227 [M.A. 8-54], H.H. and A.D. Nininger, The Nininger Collection of Meteorites, Winslow, Arizona, 1950, p.80. Olivine Fa_{20}, B. Mason, Geochimica et Cosmochimica Acta, 1963, **27**, p.1011. A second mass was found subsequently.
1.7kg Tempe, Arizona State Univ.; 612g Chicago, Field Mus. Nat. Hist.; 251g Washington, U.S. Nat. Mus.; 4.59kg Denver, Mus. Nat. Hist., the second mass;
Specimen(s): [1959,992], 1929g.

Oviedo 43°24′ N., 5°52′ W.
Asturias, Spain
Fell 1856, August 5, 1745 hrs
Stone. Olivine-hypersthene chondrite (L6).
After detonations, stones fell of which three fragments of 105, 50 and 50g were recovered. Description and analysis, J.R. de Luanco, Rev. Progr. Ciencias Exactas, Fisicas y Naturales, Madrid, 1867, **17**, p.159. Olivine Fa_{25}, B. Mason, Geochimica et Cosmochimica Acta, 1963, **27**, p.1011.
16g Madrid, Mus. Cien. Nat.; 14g Paris, Mus. d'Hist. Nat.;

Oviedo v Cangas de Onis.

Ovifak 69°30′ N., 53°35′ W. approx.
Disko Island, Greenland
Pseudometeorite..
It is suggested that this and many other occurrences of native iron commonly accepted as of terrestrial origin are in fact meteoritic, L.A. Kulik, Журнал Геофисики, 1937, **7**, p.151 [M.A. 7-172]. Terrestrial origin, H. Löfquist and C. Benedicks, Kungl. Svenska Vetensk. Akad. Handlingar, 1941, **19** (ser. 3, no. 3), p.1.

Ovruch v Owrucz.

Owens Valley 37°28′ N., 118°0′ W.
Inyo County, California, U.S.A.
Found 1913
Iron. Octahedrite, medium (1.2mm) (IIIB).
A mass of 425lb was found 22 miles NE. of Big Pine. Description, with an analysis, 7.65 %Ni, G.P. Merrill, Mem. Nat. Acad. Sci. Washington, 1922, **19**, p.1. Classification and analysis, 8.53 %Ni, 21.5 ppm.Ga, 45.9 ppm.Ge, 0.15 ppm.Ir, E.R.D. Scott et al., Geochimica et Cosmochimica Acta, 1973, **37**, p.1957. Shock-hardened, then cosmically annealed and recrystallised, V.F. Buchwald, Iron Meteorites, Univ. of California, 1975, p.957.
157kg Washington, U.S. Nat. Mus.; 32.6kg New York, Amer. Mus. Nat. Hist.; 430g Tempe, Arizona State Univ.;
Specimen(s): [1924,442], 277g. slice

Owrucz 51°20′ N., 28°50′ E.
Ukraine, USSR, [Овруч]
Fell 1775, or 1776
Synonym(s): *Obruteza, Ovruch*
Stone. Chondrite?.
Several stones fell, and one was preserved for a time but later lost. The actual place of the fall was probably Olenicze, 20 versts NNW. of Ovruch, approx. 51°29'N., 28°38'E., A. Stoikovitz, Ann. Physik (Gilbert), 1809, **31**, p.306, J. Pokrzywnicki, Bull. Soc. Amis Sci. Lettr. Poznan, 1958, **14** (ser. B), p.419.

Oyada v Gifu.

Oynchimura v Kyushu.

Ozona 30°44′ N., 101°18′ W.
Crockett County, Texas, U.S.A.
Found 1929, recognized 1939
Synonym(s): *Crockett County*
Stone. Olivine-bronzite chondrite (H6).
Several fragments, the largest 45kg, and totalling 127.5kg were found, A.D. Nininger, Pop. Astron., Northfield, Minnesota, 1940, **48**, p.556, Contr. Soc. Res. Meteorites, **2**, p.227 [M.A. 8-54]. Listed, F.C. Leonard, Univ. New Mexico Publ. Albuquerque, 1946 (meteoritics ser. no.1), p.47. Olivine Fa_{19}, B. Mason, Geochimica et Cosmochimica Acta, 1963, **27**, p.1011.
48.3kg Chicago, Field Mus. Nat. Hist.; 9kg Tempe, Arizona State Univ.; 1.6kg Washington, U.S. Nat. Mus.;
Specimen(s): [1953,154], 54.8g.

Ozren 44°36′45″ N., 18°25′5″ E.
Bosnia, Yugoslavia
Found 1952
Synonym(s): *Bosna, Bosnia, Mount Ozren, Ogg, Ozren-Bosna, Planina Ozren*
Iron. Octahedrite, coarse (2mm) (IA).
One mass of about 3900g was found, M. Ramovic, Geol. Glasnik, Sarajevo, 1956, p.35. Analysis, 6.73 %Ni, M. Ramovic, Geol. Glasnik Sarajevo, 1958 (Bih-Broj 4), p.273 [M.A. 14-126]. Classification, V.F. Buchwald, Iron Meteorites, Univ. of California, 1975, p.958.
2976g Sarajevo, Nat. Mus., main mass; 17g Washington, U.S. Nat. Mus.;
Specimen(s): [1959,220], 14.9g.

Ozren-Bosna v Ozren.

Pacula 21°3′ N., 99°18′ W.
Jacala, Hidalgo, Mexico
Fell 1881, June 18, afternoon
Synonym(s): *Hidalgo, Jacala, Paculo*
Stone. Olivine-hypersthene chondrite (L6), brecciated.
Three pieces were recovered, weighing together about 3.4kg, and the largest 2kg, A. Castillo, Cat. Météorites Mexique, Paris, 1889, p.12. Olivine Fa_{24}, B. Mason, Geochimica et Cosmochimica Acta, 1963, **27**, p.1011.
1920g Mexico, Inst. Geol.; 266g Vienna, Naturhist. Mus.; 265g New York, Amer. Mus. Nat. Hist.; 179g Chicago, Field Mus. Nat. Hist.; 88g Tempe, Arizona State Univ.; 63g Paris, Mus. d'Hist. Nat.;
Specimen(s): [67455], 28g.; [1959,1031], 35g.

Paculo v Pacula.

Paderborn v Hainholz.

Padvarninkai 55°40′ N., 25° E.
Androniski, Lithuanian SSR, USSR
Fell 1929, February 9, 0045 hrs
Synonym(s): *Andrioniskis, Andronishkis, Andronshkyai, Litau, Padvarninkaj*
Stone. Achondrite, Ca-rich. Eucrite (shergottite) (AEUC).
Eleven stones were found some months after the fall, totalling 3858g, the largest 2128g. Description, with an analysis, K. Slezevicius et al., Lietuvos Univ. Mat.-Gamtos Fak. Darbai, Kaunas, 1930, **5**, p.131 [M.A. 4-419], R. Brauns, Centralblatt Min., 1930 (Abt. A), p.401 [M.A. 4-420]. The above cited latitude and longitude are those of Androniski. Differs texturally and mineralogically from Shergotty and Zagami; Padvarninkai has ferrohypersthene, R.A. Binns, Nature, 1967, **213**, p.1111.
2128g Vilnius, Univ., main mass; 126g Moscow, Acad. Sci.; 87g Prague, Nat. Mus.; 14g Chicago, Field Mus. Nat. Hist.; 5g Tempe, Arizona State Univ.;
Specimen(s): [1931,108], 50g.

Padvarninkaj v Padvarninkai.

Paiksha v Dokachi.

Paint Creek 40°30′ N., 82° W. approx.
Holmes County, Ohio, U.S.A.
Fell 1868?
Doubtful. Iron.
A mass of 20.5lb, said to have been seen to fall in 1868, was presented to Wooster College, Wooster, Ohio, in 1935, R. Ver Steeg, Science, 1935, **81**, p.403. Not mentioned, B. Mason, Meteorites, Wiley, 1962, p.240.

Pakenham v Cranbourne.

Palahatchie 32°19′ N., 89°43′ W.
Rankin County, Mississippi, U.S.A.
Fell 1910, October 17
Synonym(s): *Palchatchie, Pelahatchee*
Stone. Chondrite.
A small stone fell near Palahatchie, G.P. Merrill, Am. J. Sci., 1925, **9**, p.436. The proper spelling of the town is Pelahatchee.
Main mass, Univ. Mississippi;

Palatka v Mocs.

Palchatchie v Palahatchie.

Palenca de Baixo 39° N., 9° W. approx.
Portugal
Fell 1894, July 31
Synonym(s): *Palenca de Taixo, Taixo*
Doubtful. Stone?.
A stone is said to have fallen on the left bank of the Tagus, but nothing has been preserved, S. Meunier, Météorites, 1895, p.78. The accounts available are not conclusive.

Palenca de Taixo v Palenca de Baixo.

Palermo 34°33′ S., 58°26′ W.
Capital Federal, Buenos Aires province, Argentina
Fell 1966, or found
Included in a table, L.M. Villar, Cienc. Investig., 1968, **24**, p.302.

Palézieux v Chervettaz.

Palinch'i v Palinshih.

Palinshih 43°29′ N., 118°37′ E.
Liaoning, China
Fell 1914, July
Synonym(s): *Balin, Palinch'i*
Iron. Octahedrite.
One mass of 18kg fell near Palinshih (= Palinch'i, = Pa-lin-yu-i-ch'i, = Ta-pan-shang). Reported, Meteor. Bull., 1959 (12). Listed, with references, D. Bian, Meteoritics, 1981, **16**, p.120.

Palisades Park v Cañon Diablo.

Pallace Iron v Krasnojarsk.

Pallas Iron v Krasnojarsk.

Palo Blanco Creek 36°30′ N., 104°30′ W. approx.
Colfax County, New Mexico, U.S.A.
Found 1954, approx.
Stone. Achondrite, Ca-rich. Eucrite (AEUC).
Listed, L. LaPaz, Cat. Coll. Inst. Meteor. Univ. New Mexico, 1965. Analysis, B. Mason et al., Smithson. Contrib. Earth Sci., 1979 (22), p.30.
20.5g Albuquerque, Univ. of New Mexico; Thin section, Washington, U.S. Nat. Mus.;

Paloduro 34°54′ N., 101°13′ W.
Armstrong County, Texas, U.S.A.
Found 1935
Synonym(s): *Clarendon*
Iron. Octahedrite, coarse (1.6mm) (IIIE).
A mass of 3kg was found, A.D. Nininger, Pop. Astron., Northfield, Minnesota, 1937, **45**, p.449 [M.A. 7-62], H.H. Nininger, letter of 30 May, 1939 in Min. Dept. BM(NH). Classification, analysis, 9.1 %Ni, 19.7 ppm.Ga, 37.7 ppm.Ge, 0.09 ppm.Ir, D.J. Malvin et al., priv. comm., 1983.
2787g Fort Worth, Texas, Monnig Colln.;

Palolo Valley 21°18′ N., 157°47′ W.
Oahu, Hawaii, U.S.A.
Fell 1949, April 24
Stone. Olivine-bronzite chondrite (H5).
One stone of 682g is in the the Univ. of Hawaii (Geol. Dept.), Meteor. Bull., 1967 (39), Meteoritics, 1970, **5**, p.91. Olivine Fa_{19}, B. Mason, Geochimica et Cosmochimica Acta, 1967, **31**, p.1100.

Pampa de Agua Blanca 24°10′ S., 69°50′ W.
Antofagasta, Chile
Found 1916, known before this year
Stone. Olivine-hypersthene chondrite (L).
10g of much oxidized fragments are in the Field Mus., Chicago, but nothing is known of its history, O.C. Farrington, Cat. Meteor. Field Mus. Nat. Hist. Chicago, 1916, p.287. Olivine Fa_{24}, B. Mason, Geochimica et Cosmochimica Acta, 1963, **27**, p.1011.

Pampa del Infierno 26°41′ S., 61°5′ W.
Avia Terai, Chaco, Argentina
Found 1895
Stone. Olivine-hypersthene chondrite (L6).
A mass of 896g was found in ploughing. Described, with an analysis, E.H. Ducloux, Rev. fac. cienc. quim. Univ. nac. La Plata, 1926, **4**, p.11, E.H. Ducloux, Anal. Soc. Cient.

Argentina, 1929, **107**, p.491 [M.A. 4-424]. Analysis, 22.61 % total iron, T.S. McCarthy et al., Meteoritics, 1974, **9**, p.215. Shocked, contains ringwoodite, classification as L6, olivine Fa_{25}, N.Z. Boctor et al., Geochimica et Cosmochimica Acta, 1982, **46**, p.1903.

Main mass, Buenos Aires, Mus. Nac. Hist. Nat.; 2.5g Paris, Mus. d'Hist. Nat.;

Specimen(s): [1962,162], 3.5g.

Pampa de Tamarugal v Tamarugal.

Pampanga 15°5′ N., 120°42′ E.

Luzon, Philippines
Fell 1859, April 5, 1700 hrs
Synonym(s): *Mexico, Philippine Islands*
Stone. Olivine-hypersthene chondrite (L5), brecciated.

A stone fell near the village of Mexico. The main mass, the original weight of which was 10.5kg, has been lost, M. Selga, Pub. Manila Observ., 1930, **1** (9) [M.A. 4-421]. Description of a 115g piece sent to Paris, G.A. Daubrée, C. R. Acad. Sci. Paris, 1868, **66**, p.637. Olivine Fa_{24}, B. Mason, Geochimica et Cosmochimica Acta, 1963, **27**, p.1011.

93g Paris, Mus. d'Hist. Nat.; 16g Vienna, Naturhist. Mus.; 2g Chicago, Field Mus. Nat. Hist.;

Specimen(s): [1922,242], 7.25g.; [41107], 1.75g.

Panamint Range v Cañon Diablo.

Pan de Azucar 26°30′ S., approx.°30′ W.

Atacama, Chile
Found 1887
Iron. Octahedrite, coarse (2.2mm) (IA).

A mass of about 43lb was found about 67 miles from the port of Pan de Azucar, P.H. Scholberg, letter of 3 May, 1894 to F.H. Butler, in Min. Dept. BM(NH). Analysis, 6.84 %Ni, 82.1 ppm.Ga, 308 ppm.Ge, 2 ppm.Ir, J.T. Wasson, Meteorites, Springer-Verlag, 1974, p.298. Includes Corrizatillo, (*q.v.*), V.F. Buchwald, Iron Meteorites, Univ. of California, 1975, p.960.

210g Chicago, Field Mus. Nat. Hist.; 132g Vienna, Naturhist. Mus.; 101g Washington, U.S. Nat. Mus.;

Specimen(s): [76808], 19280g. and a slice, 86g, and filings, 65g.

Paneth's Iron

No location known
Found 1873, known in this year
Iron. Octahedrite, coarse (1.5mm) (IIIE).

Three slices of a mass, the estimated minimum weight of which is 150kg, were mis-labelled Toluca; all three had passed through the hands of the London dealer/collector John Calvert. The mass was artificially reheated; contains the carbide haxonite. Analysis, 8.98 %Ni, 16.9 ppm.Ga, 34.1 ppm.Ge, 0.37 ppm.Ir, V.F. Buchwald et al., Meteoritics, 1974, **9**, p.307, V.F. Buchwald, Iron Meteorites, Univ. of California, 1975, p.1409.

18.65kg Vienna, Naturhist. Mus., labelled Toluca, specimens A911 and A911a; 15.2kg London, Geol. Mus.; 13.18kg Mainz, Max-Planck-Inst.; Specimen, Los Angeles, Univ. of California;

Specimen(s): [47192], 25.5kg. part-slice, and a fragment, 452g.

Panganur v Punganaru.

Panhandle 35°20′ N., 101°23′ W.

Carson County, Texas, U.S.A.
Found 1969, July 17
Stone. Olivine-bronzite chondrite (H5).

A single stone, weighing 1.36kg, was found by a farmer, Meteor. Bull., 1979 (56), Meteoritics, 1979, **14**, p.169.

1.36kg Houston, Univ. Geol. Dept.;

Pannikin 32°2′30″ S., 126°11′ E.

Nullarbor Plain, Western Australia, Australia
Found 1965
Stone. Olivine-hypersthene chondrite (L6).

Two fragments, 10.4g and 3.2g, were found in a distinct impact pit 3.5 miles east of Cocklebiddy Tank, near the Eyre Highway; they have no crust, G.J.H. McCall, First Suppl. to West. Austr. Mus. Spec. Publ. no. 3, 1968, p.15, G.J.H. McCall, Min. Mag., 1968, **36**, p.691, Meteor. Bull., 1975 (53), Meteoritics, 1975, **10**, p.156. Olivine Fa_{24}, B. Mason, Rec. Austr. Mus., 1974, **29**, p.169.

Main mass, Kalgoorlie, West. Austr. School of Mines; Thin section, Perth, West. Austr. Mus.;

Pantar 8°4′ N., 124°17′ E.

Lanao, Mindanao, Philippines
Fell 1938, June 16, 2045 hrs
Stone. Olivine-bronzite chondrite (H5).

16 stones were recovered; thousands 'as big as corn and rice grains' fell on roofs, H.J. Detrick, Pop. Astron., Northfield, Minnesota, 1946, **54**, p.191 [M.A. 9-298]. Mentioned, H.H. Nininger, Pop. Astron., Northfield, Minnesota, 1946, **54**, p.252 [M.A. 10-175], K. Fredriksson and K. Keil, Geochimica et Cosmochimica Acta, 1963, **27**, p.717. Olivine Fa_{18}, B. Mason, Geochimica et Cosmochimica Acta, 1963, **27**, p.1011. Has light-dark texture, and is gas-rich, H.E. Suess et al., Geochimica et Cosmochimica Acta, 1964, **28**, p.595. Analysis, 26.47 % total iron, H. von Michaelis et al., Earth planet. Sci. Lett., 1969, **5**, p.387.

844g Tempe, Arizona State Univ.; 470g Chicago, Field Mus. Nat. Hist.; 112g Washington, U.S. Nat. Mus.; 20g Vienna, Naturhist. Mus.;

Specimen(s): [1959,894], 53.5g.

Papasquiaro v Bella Roca.

Parachilna v Motpena.

Paracutu

Minas Gerais, Brazil, Co-ordinates not reported
Found
Iron. Octahedrite, coarse (2.6mm) (IA).

Very little information available on this iron. Classification, analysis, 7.54 %Ni, 80.4 ppm.Ga, 320 ppm.Ge, 2.6 ppm.Ir, A. Kracher et al., Geochimica et Cosmochimica Acta, 1980, **44**, p.773.

Specimen, Mainz, Max-Planck-Inst.;

Para de Minas 19°52′ S., 44°37′ W.

Minas Gerais, Brazil
Found 1934, about
Iron. Octahedrite, fine (0.33mm) (IVA).

A mass of 116.3kg is said to have fallen on the ranch Palmital, 12km south of Para de Minas, A. de Oliveira, Serv. Prod. Min. Brasil, Rep. Dir. for 1937, 1938 (31), p.53. Description, analysis, 7.87 %Ni., W.S. Curvello, Bol. Mus. Nac. Rio de Janeiro, 1952 (geol. no. 18) [M.A. 12-255]. Classification, further analysis, 7.99 %Ni, 2.21 ppm.Ga, 0.125 ppm.Ge, 2.3 ppm.Ir, R. Schaudy et al., Icarus, 1972, **17**, p.174. Not an observed fall; weathered, V.F. Buchwald,

Iron Meteorites, Univ. of California, 1975, p.964.
101.9kg Rio de Janeiro, Nat. Mus.; 8kg Sao Paulo, Univ.; 275g Tempe, Arizona State Univ.; 227g Washington, U.S. Nat. Mus.;

Paragould 36°4′ N., 90°30′ W.
Greene County, Arkansas, U.S.A.
Fell 1930, February 17, 0408 hrs
Stone. Olivine-hypersthene chondrite, amphoterite (LL5).
Two stones fell, one of 80lb and one of 820lb, C.C. Wylie, Science, 1930, **72**, p.66, C.C. Wylie, Scientific American, 1931, **144**, p.180 [M.A. 5-12]. Description, analysis, S.K. Roy and R.K. Wyant, Fieldiana, Chicago Nat. Hist. Mus., 1955, **10**, p.283 [M.A. 13-81]. Unequilibrated, mean silicate composition, olivine $Fa_{27.6}$, K. Fredriksson et al., Origin and Distribution of the Elements, ed. L.H. Ahrens, Pergamon, 1968, p.457.
337kg Chicago, Field Mus. Nat. Hist., main mass of the larger stone; 31kg Washington, U.S. Nat. Mus., main mass of the smaller stone; 3.75kg New York, Amer. Mus. Nat. Hist.; 450g Harvard Univ.; 64g Tempe, Arizona State Univ.;
Specimen(s): [1959,820], 61g.

Parambu 6°14′ S., 40°42′ W.
Ceara, Brazil
Fell 1967, July 24, 1900 hrs
Stone. Olivine-hypersthene chondrite, amphoterite (LL5).
After the appearance of a fireball travelling from southwest to northeast, and detonations, a shower of meteorites was seen to fall. A complete individual, 595g, 4 other stones each weighing some hundreds of grams and more than 22 small fragments weighing from 50g to only a few grams each were soon recovered, Meteor. Bull., 1974 (52), Meteoritics, 1974, **9**, p.104. Heavy trace metal content, analysis, 18.95 % total iron, M. Shima et al., Meteoritics, 1974, **9**, p.199. Mineralogical description, olivine Fa_{28}, G.R. Levi-Donati and G.P. Sigholfi, Meteoritics, 1974, **9**, p.1.
117g Albuquerque, Univ. of New Mexico;

Paranaiba 19°8′ S., 51°40′ W.
Sant'Ana, Mato Grosso, Brazil
Fell 1956
Synonym(s): *Cancã, Can-Can, Mato Grosso*
Stone. Olivine-hypersthene chondrite (L6), veined.
A mass of about 100kg fell on the Fazenda Can-Can (or Cancã), near Sant'Ana in the state of Mato Grosso, about 70km NW. of Paranaiba and 80km from the mouth of the Rio Apore. Description, S.E. do Amaral, Bol. Soc. Brasil. Geol., 1962, **11**, p.5, M.R. Arruda, Ciencia e Cultura, Sao Paulo, 1962, **14**, p.154. Full description, analysis, 20.86 % total iron, olivine $Fa_{23.7}$, K. Keil et al., Rev. Brasil. Geocienc., 1977, **7**, p.256.
500g Sao Paulo, Inst. Geocienc.; 322g Washington, U.S. Nat. Mus.; 53g New York, Amer. Mus. Nat. Hist.; 43g Albuquerque, Univ. of New Mexico;

Parish v Vigarano.

Parjabatpur v Bishunpur.

Park 39°6′36″ N., 100°21′42″ W.
Gove County, Kansas, U.S.A.
Found 1969, recognized in this year
Stone. Olivine-hypersthene chondrite (L).
Two masses, totalling 13kg, were found, olivine $Fa_{25.6}$, Meteor. Bull., 1981 (59), Meteoritics, 1981, **16**, p.197.
434g Chicago, Field Mus. Nat. Hist.;

Park Creek v Guffey.

Park Hotel v Bald Eagle.

Parma v Borgo San Donino.

Parma Canyon 43°48′ N., 110°0′ W. approx.
Ada County, Idaho, U.S.A.
Found 1940, April 2
Iron. Octahedrite?.
A mass of 2.15kg was found, F.C. Leonard, Pop. Astron., Northfield, Minnesota, 1947, **55**, p.381, Contr. Meteoritical Soc., **4**, p.58 [M.A. 10-177]. The above latitude and longitude cited by Leonard, are those of Parma, which is in Canyon Co., not Ada Co.; there is evidently some confusion.

Parmallee v Parnallee.

Parnallee 9°14′ N., 78°21′ E.
Madura district, Tamil Nadu, India
Fell 1857, February 28, 1200 hrs
Synonym(s): *Parmallee, Perunali*
Stone. Olivine-hypersthene chondrite, amphoterite (LL3), brecciated.
After detonations, two stones, one of about 134lb and the other of 37lb, were seen to fall, J.L. Cassels, Am. J. Sci., 1861, **32**, p.401. Description, N. Story Maskelyne, Phil. Mag., 1863, **25**, p.438. Unequilibrated, classification and analysis, 18.29 % total iron, R.T. Dodd et al., Geochimica et Cosmochimica Acta, 1967, **31**, p.921. Contains a very large chondrule, R.A. Binns, Min. Mag., 1967, **36**, p.319., R.T. Dodd, Min. Mag., 1969, **37**, p.230. Fine-grained matrix, J.R. Ashworth, Earth planet. Sci. Lett., 1977, **35**, p.25. Pyroxene compositions, R.A. Binns, Min. Mag., 1970, **37**, p.649. Parnallee is said to be 16 miles south of Madura, but no village with this or a similar name can be traced in the district. The place of fall is probably Perunali, Ramnad district, 9°14'N., 78°21'E., 52 miles SSE. of Madura, C.A. Silberrad, Min. Mag., 1932, **23**, p.296.
2.0kg Tempe, Arizona State Univ.; 2.25kg Yale Univ.; 2kg Cleveland, Adelbert College; 1.5kg New York, Amer. Mus. Nat. Hist.; 789g Chicago, Field Mus. Nat. Hist.; 776g Vienna, Naturhist. Mus.; 520g Washington, U.S. Nat. Mus.; 422g Berlin, Humboldt Univ.; 419g Paris, Mus. d'Hist. Nat.; 327g Harvard Univ.; 131g Tübingen, Univ.;
Specimen(s): [34792], 57.8kg. includes main mass of the larger stone, 46.5kg, and fragments, 82g.; [40876], 132g. from the smaller stone; [33893], 10g. and fragments, 3.75g.

Parque v Santa Rosalia.

Parral v Morito.

Parsa 26°12′ N., 85°24′ E.
Muzaffarpur district, Bihar, India
Fell 1942, April 14
Stone. Enstatite chondrite (E4).
Two stones, weighing about 600g and 200g fell near Parsa and Paro respectively, on opposite sides of the Gandak river, Meteor. Bull., 1979 (56), Meteoritics, 1979, **14**, p.170. Description, partial analysis, N. Bhandari et al., Meteoritics, 1980, **10**, p.225.
Specimens, Patna, Mus.;

Pasamonte 36°13′ N., 103°24′ W.
Union County, New Mexico, U.S.A.
Fell 1933, March 24, 0500 hrs
Stone. Achondrite, Ca-rich. Eucrite (AEUC).
75 stones, totalling 3-4kg, the largest under 300g were collected along a track 28 miles long, H.H. Nininger, Pop. Astron., Northfield, Minnesota, 1934, **42**, p.105, 291, Pop. Astron., Northfield, Minnesota, 1936, **44**, p.331, 383 [M.A. 6-105, 6-398]. Description, analysis, W.F. Foshag, Am. J. Sci., 1938, **35**, p.374 [M.A. 7-176]. Petrology, mineralogy and analysis, M.B. Duke and L.T. Silver, Geochimica et Cosmochimica Acta, 1967, **31**, p.1637. Rb-Sr, Sm-Nd, U-Pb ages, references, D.M. Unruh et al., Earth planet. Sci. Lett., 1977, **37**, p.1. XRF analysis, H. von Michaelis et al., Earth planet. Sci. Lett., 1969, **5**, p.387.
1.58kg Washington, U.S. Nat. Mus.; 680g Tempe, Arizona State Univ.; 156g Mainz, Max-Planck-Inst.; 145g Chicago, Field Mus. Nat. Hist.; 144g Los Angeles, Univ. of California; 146g Albuquerque, Univ. of New Mexico; 87g Paris, Mus. d'Hist. Nat.; 61g Harvard Univ.;
Specimen(s): [1937,378], 113g.; [1937,379], 58.5g.; [1937,380], 2.4g.; [1959,756], 518.5g. six individuals

Paso Rio Mayo
Chubut, Argentina, Not located
Found
Not reported.
The name appears in the table, L.M. Villar, Cienc. Investig., 1968, **24**, p.302.

Pastrona v Cereseto.

Patora 20°56′13″ N., 82°3′ E.
Gariaband Tahsil, Raipur district, Madhya Pradesh, India
Fell 1969, October 20, 1100 hrs
Stone. Olivine-bronzite chondrite (H6).
After a bright object was seen in the sky, two stones fell. A 4kg individual fell in the eastern outskirts of Patora village, and a smaller, 375g stone in the vicinity of Sendar village, about 1 mile west of Patora. Both stones were collected on the day of the fall and deposited at the police headquarters at Rajim, 11 miles west of Patora, Meteor. Bull., 1974 (52), Meteoritics, 1974, **9**, p.105. Mineralogical description, olivine $Fa_{19.6}$, A.L. Graham and V.K. Nayak, Meteoritics, 1974, **9**, p.137. The location of the 375g stone is not known.
Main mass, Raipur, Directorate of Geol. and Mines;
Specimen(s): [1973,M.38], 6.8g. crusted fragments

Patos I v Patos de Minas (hexahedrite).

Patos II v Patos de Minas (octahedrite).

Patos de Minas (hexahedrite)
18°35′ S., 46°32′ W.
Minas Gerais, Brazil
Found 1925
Synonym(s): *Corrego Areado, Patos I*
Iron. Hexahedrite (IIA).
A mass of 32kg is said to have fallen in 1925 in the Corrego Areado, near Patos de Minas. Description, with an analysis, 5.29 %Ni, D. Guimarães, Bol. Soc. Brasil. Geol., 1958, **7**, p.33. Mentioned, W.S. Curvello, letter of February, 1960 in Brit. Mus. Nat. Hist. Analysis, 5.36 %Ni, 59.8 ppm.Ga, 170 ppm.Ge, 43 ppm.Ir, A. Kracher et al., Geochimica et Cosmochimica Acta, 1980, **44**, p.773. Weathered, did not fall in 1925, V.F. Buchwald, Iron Meteorites, Univ. of California, 1975, p.965.
Main mass, Belo Horizonte, Geol. Surv. Brasil; 827g Washington, U.S. Nat. Mus.;

Patos de Minas (octahedrite)
18°35′ S., 46°32′ W.
Minas Gerais, Brazil
Found 1925, known before this year
Synonym(s): *Patos II*
Iron. Octahedrite.
A fragment of a badly oxidized octahedrite from Patos de Minas is in the collection of the Escuola de Engenharia de Minas e Metalurgia, Ouro Preto, Minas Gerais, W.S. Curvello, letter of February, 1960 in Min. Dept. BM(NH).

Patricia 32°30′ N., 102°2′ W.
Martin County, Texas, U.S.A.
Found, year not reported
Stone. Olivine-bronzite chondrite (H5).
A mass of 14.9kg was found, olivine $Fa_{18.5}$, T.E. Rodman and E. King, priv. comm., 1983.
14.6kg Odessa, Texas, Library;

Patrimonio 19°32′ S., 48°34′ W.
Conceicão Aparecida, Minas Gerais, Brazil
Fell 1950, August 6, 0905 hrs
Synonym(s): *Alfenas, Conciecão Aparecido, Morro Cavado*
Stone. Olivine-hypersthene chondrite (L6).
At least 20 stones fell, and masses of from 600 to 1800g were recovered, F.C. Leonard, Class. Cat. Meteor., 1956, p.22, K. Keil, Fortschr. Min., 1960, **38**, p.256. Description, analysis, 21.51 % total iron, olivine $Fa_{24.8}$, C.B. Gomes et al., J. Mineral. Recife, 1978, **7**, p.67.
1769g Rio de Janeiro, Mus. Nac.; 54g Sao Paulo, Inst. Geocienc.; 14g Washington, U.S. Nat. Mus.;
Specimen(s): [1982,M.6], 20g.

Patti 38°8′ N., 14°58′ E.
Messina, Sicily, Italy
Fell 1922, approx.
Iron. Octahedrite.
A mass of unknown weight was recovered after the fall, B. Baldanza et al., Meteoriti, Perugia Univ., 1969. The co-ordinates are of the town of Patti. Shock deformation structures figured, B. Baldanza and G. Pialli, Meteorite Research, ed. P.M. Millman, D. Reidel, Dordrecht-Holland, 1969, p.806.
12g Perugia Univ.;

Patwar 23°9′ N., 91°11′ E.
Tippera district, Chittagong, Bangladesh
Fell 1935, July 29, 1420 hrs
Stony-iron. Mesosiderite (MES).
Five masses were recovered, the largest 23kg, totalling 37.35kg, A.L. Coulson, Advance Proc. and Notices Asiatic Soc. Bengal, 1935, **2**, p.94 [M.A. 6-206]. Description, A.L. Coulson, Rec. Geol. Surv. India, 1936, **69**, p.439 [M.A. 6-393]. Analysis, mineralogy, E. Jarosewich and B. Mason, Geochimica et Cosmochimica Acta, 1969, **33**, p.411. Texture, mineralogy, B.N. Powell, Geochimica et Cosmochimica Acta, 1971, **35**, p.5. Analysis of metal, 9.9 %Ni, 9.6 ppm.Ga, 39.3 ppm.Ge, 2.1 ppm.Ir, J.T. Wasson et al., Geochimica et Cosmochimica Acta, 1974, **38**, p.135. Corona structure, references, C.E. Nehru et al., Geochimica et Cosmochimica Acta, 1980, **44**, p.1103.
9.9kg Calcutta, Geol. Surv. India; 698g Tempe, Arizona State Univ.; 592g Washington, U.S. Nat. Mus.; 198g Los Angeles, Univ. of California; 176g New York, Amer. Mus. Nat. Hist.;
Specimen(s): [1937,1479], 396g.

Pau v Beuste.

Paulding County 34° N., 84°48′ W.

Paulding County, Georgia, U.S.A.
Found 1901, approx.
Iron. Octahedrite, coarse.

A mass, of which oxidized fragments weighed 725g, was found; description, with an analysis, 6.34 %Ni, T.L. Watson, Am. J. Sci., 1913, **36**, p.165.

246g Vienna, Naturhist. Mus.; 161g Chicago, Field Mus. Nat. Hist.; 99g New York, Amer. Mus. Nat. Hist.;

Pavel 43°28′ N., 25°31′ E.

Veliko Turnovo, Bulgaria
Fell 1966, February 28, 1400 hrs
Stone. Olivine-bronzite chondrite (H).

The fall of two stones, 2968g and 6.15g, was preceded by detonations. Witnesses observed the fall which occurred a few hundred metres from the village of Pavel, in northern Bulgaria; the trajectory was from west to east, Meteor. Bull., 1966 (36), Meteoritics, 1970, **5**, p.86. Analysis, 26.19 % total iron, D.I. Dimov, Annu. Univ. Sofia, Fac. Geol. Geogr., 1974, **66** (Livre 1, geol.), p.209.

Main mass, Sofia, Astron. Observ. Univ.;

Pavia v Valdinizza.

Pavlodar (pallasite) 51°10′ N., 77°20′ E.

Semipalatinsk, Kazan SSR, USSR, [Павлодар], [Ямышева]
Found 1885
Synonym(s): *Iamysheva, Jamyscheva, Jamyseva, Pawlodar, Samyscheva, Semipalatinsk, Yamuishova, Yamyshev, Yamysheva, Yamyschewa*
Stony-iron. Pallasite (PAL).

A mass of 4.5kg was found near the village of Jamyscheva (Yamyscheva), Bull. Acad. Sci. St.-Petersbourg, 1898, **8** (4), p.xliii. Listed, E.L. Krinov, Каталог Метеоритов Акад. Наук СССР, Москва, 1947 [M.A. 10-511]. Olivine $Fa_{12.5}$, P.R. Buseck and J.I. Goldstein, Bull. Geol. Soc. Amer., 1969, **80**, p.2141. References, analysis of metal, 7.92 %Ni, 23.2 ppm.Ga, 74 ppm.Ge, 5 ppm.Ir, E.R.D. Scott, Geochimica et Cosmochimica Acta, 1977, **41**, p.349.

1.4kg Chicago, Field Mus. Nat. Hist., main mass; 495g Harvard Univ.; 354g Vienna, Naturhist. Mus.; 177g Budapest, Nat. Hist. Mus.; 125g Rome, Vatican Colln; 78g Moscow, Acad. Sci.; 27g Berlin, Humboldt Univ.; 21g New York, Amer. Mus. Nat. Hist.;

Specimen(s): [63550], 58g.; [1920,335], 10g.

Pavlodar (stone) 52°18′ N., 77°2′ E.

Kazakhstan, Kazan SSR, USSR
Fell 1938, May 23, 1340 hrs
Stone. Olivine-bronzite chondrite (H5).

A mass of 120g fell at Pavlodar. Brief description, E.L. Krinov, C. R. (Doklady) Acad. Sci. URSS, 1938, **20**, p.585 [M.A. 7-271]. Locality and date corrected, L.A. Kulik, Метеоритика, 1941, **1**, p.73 [M.A. 9-294]. Listed, L.G. Kvasha and A.Ya. Skripnik, Метеоритика, 1978, **37**, p.178. Total weight given as 142.5g in four fragments, B. Mason, letter of 11 February, 1965 in Min. Dept. BM(NH). Olivine Fa_{18}, B. Mason, Geochimica et Cosmochimica Acta, 1967, **31**, p.1100.

28g Moscow, Acad. Sci.;

Pavlograd 48°32′ N., 35°59′ E.

Ekaterinoslav, Ukraine, USSR, [Мордвиновка]
Fell 1826, May 19
Synonym(s): *Berdjansk, Ekaterinoslav, Jekaterinoslav, Mordvinovka, Pawlograd*
Stone. Olivine-hypersthene chondrite (L6).

A stone of about 40kg is said to have fallen in a field of Frau Sorbinov, K.E.A. von Hoff, Ann. Phys. Chem. (Poggendorff), 1830, **18**, p.185. Analysis, 21.33 % total iron, M.I. D'yakonova and V.Ya. Kharitonova, Метеоритика, 1960, **18**, p.48 [M.A. 16-447]. Whether any specimens in collections really came from such a stone is by no means certain, they are all very similar to Bachmut. However it has been shown from radiation ages that Bachmut and Pavlograd specimens belong to different falls, but some confusion exists, R. Ganapathy and E. Anders, Очерка современной геохимии и аналитической химии. наука, Москва, 1972, p.72. Olivine Fa_{25}, B. Mason, Geochimica et Cosmochimica Acta, 1963, **27**, p.1011.

29.57kg Moscow, Acad. Sci.; 932g Dorpat, Univ.; 681g Budapest, Nat. Mus.; 134g Harvard Univ.; 135g Chicago, Field Mus. Nat. Hist.; 97g Washington, U.S. Nat. Mus.; 52g Vienna, Naturhist. Mus.;

Specimen(s): [90264], 161g.; [1920,334], 1g.; [46011], less than 1g; [46012], 35g. assigned to Bachmut, obtained from J.R. Gregory labelled 'Ekaterinoslav', may belong to Pavlograd, R. Ganapathy and E. Anders, op. cit.

Pavlovka 52°2′ N., 43° E. approx.

Balashov, Sartov region, Federated SSR, USSR, [Павловка]
Fell 1882, August 2, 1700 hrs
Synonym(s): *Pawlowka, Saratov*
Stone. Achondrite, Ca-rich. Howardite (AHOW).

A stone of about 2kg fell, after detonations, F.N. Chernyshev, Z. Deutsch. Geol. Ges., 1883, **35**, p.190. Mentioned, L.A. Kulik, Метеоритика, 1941, **1**, p.73 [M.A. 9-294]. Analysis, B. Mason et al., Smithson. Contrib. Earth Sci., 1979 (22), p.30.

955g Budapest, Nat. Mus., main mass; 148g Chicago, Field Mus. Nat. Hist.; 106g Berlin, Humboldt Univ.; 101g Paris, Mus. d'Hist. Nat.; 74g Vienna, Naturhist. Mus.; 1.9g Moscow, Acad. Sci.;

Specimen(s): [55255], 70g.

Pawlodar v Pavlodar (pallasite).

Pawlograd v Pavlograd.

Pawlowka v Pavlovka.

Peace River 56°8′ N., 117°56′ W.

Alberta, Canada
Fell 1963, March 31, 0435 hrs
Stone. Olivine-hypersthene chondrite (L6).

The fireball travelled N.75°E., detonated at a height of 13 km and broke into two main fragments; the smaller was recovered as a number of pieces broken on impact in the predicted fall area on April 24, after snow had melted. Total weight recovered 45.76kg, J. Roy. Astron. Soc. Canada, 1964, **58**, p.109, Meteor. Bull., 1963 (27, 28). Olivine Fa_{23}, B. Mason, Geochimica et Cosmochimica Acta, 1967, **31**, p.1100. U-Pb measurements, N.H. Gale et al., Nature, 1972, **240** (Phys. Sci.), p.56. Largely out-gassed, G. Turner, Geochimica et Cosmochimica Acta, 1979 (suppl. 11), p.1917. Shocked, contains wadsleyite, G.D. Price et al., Canadian Mineral., 1983, **21**, p.29.

12.3kg Ottawa, Geol. Surv. Canada; 582g Chicago, Field Mus. Nat. Hist.; 551g Washington, U.S. Nat. Mus.; 499g New York, Amer. Mus. Nat. Hist.; 347g Copenhagen, Univ. Geol. Mus.; 82g Tempe, Arizona State Univ.;
Specimen(s): [1967,257], 200g. crusted fragment

Pecklesheim 51°40′ N., 9°15′ E.
Nordrhein-Westfalen, Germany
Fell 1953, March 3, 1430 hrs
Stone. Achondrite, Ca-poor. Unique type (ACANOM).
After a whining noise, the stone, 117.8g, struck the branch of a tree and fell at the feet of workmen. The place of fall is 3.5 km ENE. of Pecklesheim and 12 km N. of Warburg, Meteor. Bull., 1969 (46), Meteoritics, 1970, **5**, p.103. This brecciated stone comprises 90% bronzite, Fs14.4, 2-3% chromite, rare feldspar, An90-94 and An81, Co-rich kamacite and taenite, 4.16% and 1.08% Co respectively, troilite and tridymite, P. Ramdohr and A. El Goresy, Meteoritics, 1969, **4**, p.291, abs.
Main mass, Heidelberg, Max-Planck-Inst.; 2.2g Washington, U.S. Nat. Mus.; 1.9g Tempe, Arizona State Univ.;

Peck's Spring 32° N., 102° W.
Midland County, Texas, U.S.A.
Found 1926
Stone. Olivine-hypersthene chondrite (L5).
A mass of 1600g was found. Description, with an analysis, G.P. Merrill, Proc. U.S. Nat. Mus., 1929, **75** (16) [M.A. 4-262]. The analysis is doubted, alumina being improbably high, W. Wahl, Geochimica et Cosmochimica Acta, 1950, **1**, p.28. Olivine Fa_{25}, B. Mason, Geochimica et Cosmochimica Acta, 1963, **27**, p.1011.
629g Washington, U.S. Nat. Mus., main mass; 19g Tempe, Arizona State Univ.;
Specimen(s): [1959,1045], 17g. slice

Pecora Escarpment 82501
Antarctica
Found 1982, between December 1982 and January 1983
Stone. Achondrite, Ca-rich. Eucrite (AEUC).
A single mass of 54.4g was found by an American party during the 1982-1983 field season in Antarctica.

Pecora Escarpment 82502
Found 1982, between December 1982 and January 1983
Stone. Achondrite, Ca-rich. Eucrite (AEUC).
A single mass of 890g was found.

Peery Meteor v Estherville.

Peetz 40°57′ N., 103°5′ W.
Logan County, Colorado, U.S.A.
Found 1937
Stone. Olivine-hypersthene chondrite (L6).
A mass of 11.5kg was found in NW. 1/4, sect. 30, township 12, range 50W, Logan County, A.D. Nininger, Pop. Astron., Northfield, Minnesota, 1939, **47**, p.212, E.P. Henderson, letter of 3 June, 1939 in Min. Dept. BM(NH). Olivine Fa_{26}, B. Mason, Geochimica et Cosmochimica Acta, 1963, **27**, p.1011. Cooling history, P. Pellas and D. Storzer, Meteoritics, 1979, **14**, p.513, abs.
2.7kg Denver, Mus. Nat. Hist.; 2.4kg Tempe, Arizona State Univ.; 810g Washington, U.S. Nat. Mus.; 726g Chicago, Field Mus. Nat. Hist.; 64g New York, Amer. Mus. Nat. Hist.;
Specimen(s): [1938,1205], 118g.; [1959,1033], 2651g.

Pegu v Quenggouk.

Peine v Imilac.

Pei Xian 34°42′ N., 117° E. approx.
Pei County, Jiangsu, China
Found 1917, before this year
Iron.
A single mass weighing over 400kg is listed, without further references, D. Bian, Meteoritics, 1981, **16**, p.120.
Main mass, Xuzhou, Senior Middle School No. 1;

Pelahatchee v Palahatchie.

Peña Blanca Spring 30°7.5′ N., 103°7′ W.
Marathon, Brewster County, Texas, U.S.A.
Fell 1946, August 2, afternoon hrs
Synonym(s): *Marathon*
Stone. Achondrite, Ca-poor. Aubrite (AUB).
Fell into a murky pond near Marathon; about 70kg were recovered, including masses of 47kg and 13kg. Description, with an analysis, J.T. Lonsdale, Am. Miner., 1947, **32**, p.354 [M.A. 10-177]. Large enstatite crystals, cream-coloured fusion crust, O.E. Monnig, Pop. Astron., Northfield, Minnesota, 1946, **54**, p.483 [M.A. 10-178]. Analysis of enstatite, A.M. Reid and A.J. Cohen, Geochimica et Cosmochimica Acta, 1967, **31**, p.661.
2.2kg Fort Worth, Texas, Monnig Colln.; 457g Washington, U.S. Nat. Mus.; 70g Los Angeles, Univ. of California; 47g Tempe, Arizona State Univ.;
Specimen(s): [1959,759], 65g.

Penkarring Rock v Youndegin.

Pennyman's Siding v Middlesbrough.

Penokee 39°21′ N., 99°55′ W.
Graham County, Kansas, U.S.A.
Found 1947, known before this year
Stone. Olivine-bronzite chondrite (H5).
A stone, initially of 3580g, in Washington, U.S. Nat. Mus. Olivine Fa_{19}, B. Mason, Geochimica et Cosmochimica Acta, 1963, **27**, p.1011.
3552g Washington, U.S. Nat. Mus.;

Pentolina 43°12′ N., 11°10′ E.
Siena, Tuscany, Italy
Fell 1697, January 13, 1700 hrs
Doubtful. Stone?.
Several stones are reported to have fallen at Pentolina and other places near Siena, including one of 13 oz. They were black outside, inside "like an iron ore", D.A. Soldani, Atti Accad. Sci. Siena, 1808, **9**, p.1, E.F.F. Chladni, Die Feuer-Meteore, Wien, 1819, p.239. The evidence of a meteoritic nature is not conclusive.

Penwell v Odessa.

Pep 33°43′48″ N., 102°34′36″ W.
Hockley County, Texas, U.S.A.
Found 1966
Stone. Chondrite.
Listed, total known weight 591g, Cat. Huss Coll. Meteorites, 1976, p.35.
591g Mainz, Max-Planck-Inst.;

Peramiho 10°40′ S., 35°30′ E.
Nguni, Ruvuma, Tanzania
Fell 1899, October 24, 0700 hrs
Stone. Achondrite, Ca-rich. Eucrite (AEUC).
After detonations, and appearance of light, a stone of 165g was found. Description, with an analysis, F. Berwerth, Sitzungsber. Akad. Wiss. Wien, Math.-naturwiss. Kl., 1903, **112**, p.739., J.R. Harpum, Rec. Geol. Surv. Tanganyika for 1961, 1965, **11**, p.54.
142g Vienna, Naturhist. Mus.; 3g Berlin, Humboldt Univ.;

Perpeti 23°19′30″ N., 91°0′ E.
Tippera district, Chittagong, Bangladesh
Fell 1935, May 14, 2300 hrs
Stone. Olivine-hypersthene chondrite (L6).
Fourteen stones were recovered, the largest 6869.8g, totalling 23.474kg. Description, with analyses, A.L. Coulson, Rec. Geol. Surv. India, 1936, **71**, p.123 [M.A. 6-394]. Olivine Fa_{25}, B. Mason, Geochimica et Cosmochimica Acta, 1963, **27**, p.1011.
21.5kg Calcutta, Mus. Geol. Surv. India; 344g Paris, Mus. d'Hist. Nat.; 68g Tempe, Arizona State Univ.; 49g Chicago, Field Mus. Nat. Hist.; 35g Washington, U.S. Nat. Mus.;
Specimen(s): [1937,1478], 1276g. and fragments, 2.5g.

Perry Meteor v Estherville.

Perryville 37°44′ N., 89°51′ W.
Perry County, Missouri, U.S.A.
Found 1906
Iron. Octahedrite, plessitic (0.06mm) (IIC).
A mass of 17.5kg was found. Described, with an analysis, 9.66 %Ni, G.P. Merrill, Proc. U.S. Nat. Mus., 1912, **43**, p.595. Determination of Ga, Au and Pd, E. Goldberg et al., Geochimica et Cosmochimica Acta, 1951, **2**, p.1. Classification and analysis, 9.27 %Ni, 37.0 ppm.Ga, 88.0 ppm.Ge, 11 ppm.Ir, J.T. Wasson, Geochimica et Cosmochimica Acta, 1969, **33**, p.859. Description, V.F. Buchwald, Iron Meteorites, Univ. of California, 1975, p.966.
14kg Washington, U.S. Nat. Mus.; 597g Harvard Univ.; 181g Chicago, Field Mus. Nat. Hist.; 170g Tempe, Arizona State Univ.; 136g New York, Amer. Mus. Nat. Hist.;
Specimen(s): [1959,979], 185g. slice, and sawings, 4g.

Persimmon Creek 35°3′ N., 84°14′ W.
Cherokee County, North Carolina, U.S.A.
Found 1893
Iron. Octahedrite, plessitic, with silicates (IB).
A mass of about 5kg was found. Description, with an analysis, 14.5 %Ni+Co, W. Tassin, Proc. U.S. Nat. Mus., 1904, **27**, p.955. Date of find given as 1903, O.C. Farrington, Cat. Meteor. Field Mus. Nat. Hist. Chicago, 1916, p.288. Classification, analysis of metal, 14.45 %Ni, 34.7 ppm.Ga, 78.3 ppm.Ge, 0.65 ppm.Ir, E.R.D. Scott et al., Geochimica et Cosmochimica Acta, 1973, **37**, p.1957. A member of the Copiapo class of irons with silicate inclusions, T.E. Bunch et al., Contr. Miner. Petrol., 1970, **25**, p.297. Description, co-ordinates, related to carbon-rich group IB irons, V.F. Buchwald, Iron Meteorites, Univ. of California, 1975, p.967.
3519g Washington, U.S. Nat. Mus.; 228g Chicago, Field Mus. Nat. Hist.; 48g Vienna, Naturhist. Mus.;

Perth 56°24′ N., 3°26′ W.
Perthshire, Scotland
Fell 1830, May 17, 1230 hrs
Synonym(s): *North Inch of Perth*
Stone. Olivine-hypersthene chondrite, amphoterite (LL5).
A stone, 7 inches in diameter, fell at Perth in a field known as the North Inch of Perth; only two small pieces weighing together about 2g appear to have been preserved, N. Story Maskelyne, Phil. Mag., 1863, **25**, p.437, Author not known, Rec. Geol. Surv. India, 1868, **1**, p.72. They were in the collection of Dr. Thomson of Glasgow, but passed into the possession of Mr. W. Nevill of Godalming, by whom one was presented to the British Museum and the other to the Geol. Surv. Museum in Calcutta. Olivine Fa_{28}, B. Mason, Geochimica et Cosmochimica Acta, 1963, **27**, p.1011.
Specimen, Calcutta, Geol. Surv. India;
Specimen(s): [34248], 1.5g.

Perugia v Assisi.

Perunali v Parnallee.

Pervomaiskii Poselok v Pervomaisky.

Pervomaisky 56°38′ N., 39°26′ E.
Ivanovo-Vosnesenk district, Vladimir, Federated SSR, USSR, [Первомайский Поселок]
Fell 1933, December 26, 1800 hrs
Synonym(s): *Ivanovo, Pervomaiskii Poselok, Pervomaisky Posyelok, Pervomaysky Poselok, Perwomaiski, Perwomajskij*
Stone. Olivine-hypersthene chondrite (L6).
A shower of stones fell over an area of 4 km by 5 km near Pervomaisky, (56°38'N., 39°26'E.), and at least 66, ranging from 10kg to 30g were recovered, L.A. Kulik, ниманю наблюдателей болидов, Комиссия по метеоритам, Акад. Наук СССР, Москва, 1937 [M.A. 7-62]. 66kg collected, conditions of fall, I.S. Astapowitsch, Метеоритика, 1957, **15** (suppl.), p.3. Analysis, 21.73 % total iron, M.I. D'yakonova and V.Ya. Kharitonova, Метеоритика, 1961, **21**, p.52. Olivine Fa_{24}, B. Mason, Geochimica et Cosmochimica Acta, 1963, **27**, p.1011.
30kg Moscow, Acad. Sci.; 8.8kg Odessa, Univ., second largest stone; 462g Paris, Mus. d'Hist. Nat.; 446g Washington, U.S. Nat. Mus.; 435g Tempe, Arizona State Univ.; 177g New York, Amer. Mus. Nat. Hist.;
Specimen(s): [1971,200], 95.3g. part-crusted fragment

Pervomaisky Posyelok v Pervomaisky.

Pervomaysky Poselok v Pervomaisky.

Perwomaiski v Pervomaisky.

Perwomajskij v Pervomaisky.

Pesqueira v Serra de Magé.

Pesyanoe 55°30′ N., 66°5′ E.
Kurgan, Federated SSR, USSR, [Старое Песьяное]
Fell 1933, October 2, 0600 hrs
Synonym(s): *Lebjagevka, Staroe Pesianoe, Staroe Pesyanoe, Staroje Pesjanoje, Staro-Pesiianoe*
Stone. Achondrite, Ca-poor. Aubrite (AUB).
Co-ordinates of fall site, V.I. Vernadsky, letter of 5 August, 1939 in Min. Dept. BM(NH). Description, with a

micrometric analysis, B.M. Kupletsky and I.A. Ostrovsky, Метеоритика, 1941, **1**, p.59 [M.A. 9-294]. Stones totalling 3393g fell, the largest weighing 905g, E.L. Krinov, Каталог Метеоритов АИад. Наук СССР, Москва, 1947 [M.A. 10-511]. Analysis, M.I. D'yakonova and V.Ya. Kharitonova, Метеоритика, 1960, **18**, p.48 [M.A. 16-447]. Gas-rich, light-dark structure, O. Müller and J. Zähringer, Earth planet. Sci. Lett., 1966, **1**, p.25. Analysis of enstatite, A.M. Reid and A.J. Cohen, Geochimica et Cosmochimica Acta, 1967, **31**, p.661.

1790g Moscow, Acad. Sci., includes main mass; 67g Washington, U.S. Nat. Mus.;

Specimen(s): [1956,322], 36.5g.

Petersburg 35°18′ N., 86°38′ W.

Lincoln County, Tennessee, U.S.A.
Fell 1855, August 5, 1530 hrs
Synonym(s): *Fayetteville, Lincoln County*
Stone. Achondrite, Ca-rich. Howardite (AHOW).

After detonations, a stone of about 4lb was seen to fall, J.M. Safford, Geol. Reconn. Tennessee, Nashville, 1856, p.125. Analyses, J.L. Smith, Am. J. Sci., 1861, **31**, p.264, B. Mason et al., Smithson. Contrib. Earth Sci., 1979 (22), p.30. Classification and details of variable pyroxene composition, M.B. Duke and L.T. Silver, Geochimica et Cosmochimica Acta, 1967, **31**, p.1637.

205g Chicago, Field Mus. Nat. Hist.; 467g Tempe, Arizona State Univ.; 72g Berlin, Humboldt Univ.; 51g Tempe, Arizona State Univ.; 42g Washington, U.S. Nat. Mus.; 26g Vienna, Naturhist. Mus.; 24g Yale Univ.; 7g Harvard Univ.;

Specimen(s): [32053], 42.4g. two pieces; [1920,333], 0.2g.; [1959,757], 8g.

Petropavlovka 48°12′ N., 43°44′ E.

Nizhne Chirskaya, Federated SSR, USSR, [Петропавловка]
Found 1916
Synonym(s): *Petropawlowka*
Stone. Olivine-bronzite chondrite (H4).

One stone was found weighing 1.773kg. Co-ordinates, V.I. Vernadsky, letter of 5 August, 1939 in Min. Dept. BM(NH). Analysis, 26.04 % total iron, M.I. D'yakonova and V.Ya. Kharitonova, Метеоритика, 1960, **18**, p.48 [M.A. 16-447]. Olivine Fa_{16}, B. Mason, Geochimica et Cosmochimica Acta, 1963, **27**, p.1011.

1350g Moscow, Acad. Sci., includes main mass, 950g; 118g Washington, U.S. Nat. Mus.;

Petropavlovsk 53°21′ N., 87°11′ E.

Mrasa River, Tomsk, Federated SSR, USSR, [Петропавловский Прииск]
Found 1841
Synonym(s): *Petropavlovskii Priisk, Petropavlovsky Priisk, Petropawlowsk*
Iron. Octahedrite, medium (1.3mm) (IRANOM).

A mass of about 7kg was found 31.5 feet deep in gold-bearing alluvium, analysis, 6.98 %Ni, A. Ermann, Arch. Wiss. Kunde Russl., 1841, **1**, p.314, 723. Listed, E.L. Krinov, Астрон. Журнал, 1945, **22**, p.303 [M.A. 9-297]. Further analysis, 8.52 %Ni, M.I. D'yakanova, Метеоритика, 1958, **16**, p.180. Classification and analysis, 8.1 %Ni, 21.7 ppm.Ga, 48.6 ppm.Ge, 0.57 ppm.Ir, A. Kracher et al., Geochimica et Cosmochimica Acta, 1980, **44**, p.773. Description, based on the artificially reheated Chicago sample, V.F. Buchwald, Iron Meteorites, Univ. of California, 1975, p.969.

5.9kg Leningrad, Mus. Mining Inst.; 100g Vienna, Naturhist. Mus.; 46g Moscow, Acad. Sci.; 45g Chicago, Field Mus. Nat. Hist.;

Specimen(s): [56472], 12g.

Petropavlovskii Priisk v Petropavlovsk.

Petropavlovsky Priisk v Petropavlovsk.

Petropawlowka v Petropavlovka.

Petropawlowsk v Petropavlovsk.

Petrovskoie-Rasumovskoye v Agricultural College.

Pettiswood 53°32′ N., 7°20′ W.

Westmeath, Ireland
Fell 1779
Stone.

A light-coloured stone, almost white, with a light brown crust, W. Bingley, Gentleman's Mag., 1796, **66**, p.726. The evidence is fairly conclusive that this was a genuine fall, though nothing is now preserved.

Pevensey 34°47′ S., 144°40′ E.

Old Man Plain, County Waradgery, New South Wales, Australia
Found 1868, to 1870
Synonym(s): *Hay*
Stone. Olivine-hypersthene chondrite, amphoterite (LL5).

A stone of 9.5lb was found in a paddock, 10 miles from Hay and 15 miles south of the Murrumbidgee river, C. Anderson, Rec. Austr. Mus., 1913, **10** (5), p.59. Classification, olivine $Fa_{28.5}$, B. Mason, Rec. Austr. Mus., 1974, **29**, p.169.

Main mass, Melbourne, Godfrey collection; 14g Washington, U.S. Nat. Mus.;

Specimen(s): [1983,M.35], 6.6g.

Pezza del Moleto v Orvinio.

Pfullingen 48°28′ N., 9°13′ E.

Tübingen, Baden-Württemberg, Germany
Fell 1904, October 29, 1600 hrs
Doubtful. Iron.

A mass of 7.5kg was said to have fallen in the Echaz valley, P.L. Mercanton, Bull. Soc. Vaud. Sci. Nat., 1929, **57**, p.59 [M.A. 4-418]. Discredited, F. Heide, Chem. Erde, 1954, **17**, p.61.

Phasti v Białstok.

Philadelphia Iron v Cleveland.

Philippine Islands v Pampanga.

Phillips County v Long Island.

Phillips County (pallasite) 40°27′ N., 102°23′ W.

Colorado, U.S.A.
Found 1935
Synonym(s): *Phillips County (1935)*
Stony-iron. Pallasite (PAL).

One mass of 3lb was found, A.D. Nininger, Pop. Astron., Northfield, Minnesota, 1937, **45**, p.449 [M.A. 7-62]. Olivine $Fa_{18.5}$, P.R. Buseck and J.I. Goldstein, Bull. Geol. Soc. Amer., 1969, **80**, p.2141. Analysis of metal, 12.0 %Ni, 17.2 ppm.Ga, 34.4 ppm.Ge, 0.085 ppm.Ir, E.R.D. Scott, Geochimica et Cosmochimica Acta, 1977, **41**, p.349. Texture,

E.R.D. Scott, Geochimica et Cosmochimica Acta, 1977, **41**, p.693.
415g Tempe, Arizona State Univ.; 332g Chicago, Field Mus. Nat. Hist.; 250g Washington, U.S. Nat. Mus.;
Specimen(s): [1959,985], 297.5g. slice

Phillips County (stone) 40°0′ N., 99°15′ W.
Kansas, U.S.A.
Fell 1901, May 9
Synonym(s): *Phillips County (1901)*
Stone. Olivine-hypersthene chondrite (L).
Two stones, of total weight 57.9kg, were seen to fall, Meteor. Bull., 1969 (47), Meteoritics, 1970, **5**, p.106. Olivine Fa_{23}, R. Hutchison, priv. comm.
20.4kg Fort Hays, Kansas State College Mus.;
Specimen(s): [1970,1], 8.2g. part crusted fragment, and a slice

Phillips County (1901) v Phillips County (stone).

Phillips County (1935) v Phillips County (pallasite).

Phnom Penh v Pnompehn.

Phu Hong 11°15′ N., 108°35′ E. approx.
Binh-Chanh, Vietnam
Fell 1887, September 22
Synonym(s): *Phu-Long*
Stone. Olivine-bronzite chondrite (H4), veined.
A nearly spherical stone, 10 cm in diameter and weighing about 500g fell, F.J. Delauney, C. R. Acad. Sci. Paris, 1887, **105**, p.1294. Description, analysis, A. Lacroix, C. R. Acad. Sci. Paris, 1925, **180**, p.1977. Olivine Fa_{18}, B. Mason, Geochimica et Cosmochimica Acta, 1963, **27**, p.1011.
423g Paris, Mus. d'Hist. Nat.; 11g Chicago, Field Mus. Nat. Hist.;

Phulmari 20°8′ N., 75°30′ E.
Aurangabad district, Maharashtra, India
Found 1936, known before this year
Stone.
A mass of 4064g is said to have fallen 20 miles NNE. of Aurangabad, near the village of Phulmari. Description, M.A.R. Khan, Pop. Astron., Northfield, Minnesota, 1936, **35**, p.563 [M.A. 6-395].

Phu-Long v Phu-Hong.

Phum Sambo 12°0′ N., 105°29′ E. approx.
Kompong Cham, Cambodia
Fell 1933, January 9, 1630 hrs
Stone. Olivine-bronzite chondrite (H4).
A mass of 7.8kg and fragments were collected. Description, A. Lacroix, C. R. Acad. Sci. Paris, 1933, **197**, p.565 [M.A. 5-404]. Olivine Fa_{19}, B. Mason, Geochimica et Cosmochimica Acta, 1963, **27**, p.1011.
7.44kg Paris, Mus. d'Hist. Nat., main mass; 127g Chicago, Field Mus. Nat. Hist.;
Specimen(s): [1972,237], 36.18g.

Phuoc-Binh 15°43′ N., 108°6′ E.
Quang Nam, Vietnam
Fell 1941, July 18, 1600 hrs
Stone. Olivine-bronzite chondrite (H).
One stone of 11kg was seen to fall. Description, with an analysis, E. Saurin and F. Nagy, C. R. Acad. Sci. Paris, 1950, **230**, p.2304 [M.A. 11-268].
Main mass, Hanoi, Geol. Surv. Mus.;

Piacenza v Borgo San Donino.

Piancaldoli 44°14′39″ N., 11°30′8″ E.
Florence, Tuscany, Italy
Fell 1968, August 10, 2014 hrs
Stone. Olivine-hypersthene chondrite, amphoterite (LL3).
A fireball was widely observed from Yugoslavia to northern and central Italy. It exploded and broke up. Three fragments, 7.55g, 5.10g and 0.41g, were partly covered with fusion crust. Description, classification, olivine $Fa_{2\text{-}34}$, 19.73 % total iron, M. Carapezza et al., Meteoritics, 1976, **11**, p.195. Microchondrule-bearing clast described, A.E. Rubin et al., Geochimica et Cosmochimica Acta, 1982, **46**, p.1763.
Thin section, Washington, U.S. Nat. Mus.;

Picacho 33°12′ N., 105°0′ W.
Lincoln County, New Mexico, U.S.A.
Found 1952
Iron. Octahedrite, medium (1.0mm) (IIIA).
A mass of 22kg was found 10 miles SE. of Picacho, B. Mason, Meteorites, Wiley, 1962, p.238. Analysis, 7.08 %Ni, 19.1 ppm.Ga, 33.9 ppm.Ge, 19 ppm.Ir, E.R.D. Scott et al., Geochimica et Cosmochimica Acta, 1973, **37**, p.1957. Description, V.F. Buchwald, Iron Meteorites, Univ. of California, 1975, p.971.
12.6kg Mainz, Max-Planck-Inst.; 397g Chicago, Field Mus. Nat. Hist.; 262g Tempe, Arizona State Univ.; 253g Copenhagen, Univ. Geol. Mus.;
Specimen(s): [1974,M.1], 167.3g. slice, polished and etched

Pickarring Rock v Youndegin.

Pickens County 34°30′ N., 84°30′ W.
Georgia, U.S.A.
Found 1908
Stone. Olivine-bronzite chondrite (H6).
A stone of 400g was found; analysis (doubtful, alumina improbably high), S.W. McCallie, Science, 1909, **30**, p.772. Olivine Fa_{20}, B. Mason, Geochimica et Cosmochimica Acta, 1963, **27**, p.1011.
231g Chicago, Field Mus. Nat. Hist.; 67g Washington, U.S. Nat. Mus.;
Specimen(s): [1980,M.9], 27.7g. part-slice

Picote 41°22′ N., 6°14′ W.
Miranda, Traz-os-Montes, Portugal
Fell 1843, September, late in the month
Stone.
Three stones fell, the largest 1125g, but nothing is now preserved, R. de Serpa Pinto, Rev. sismol. geofis. Coimbra, 1932, **3**, p.45 [M.A. 5-297].

Piedade do Bagre 18°56′30″ S., 44°59′ W.
Curvelo, Minas Gerais, Brazil
Found 1922
Iron. Octahedrite, medium (0.7mm) (IRANOM).
A mass of 130lb was found 10 miles SW. of Piedade do Bagre. Description, with an analysis, 7.48 %Ni, L.J. Spencer, Min. Mag., 1930, **22**, p.271 [M.A. 4-422]. Not part of the Curvello fall; has an anomalously small bandwidth, V.F. Buchwald, Iron Meteorites, Univ. of California, 1975, p.973. Chemically anomalous, 7.51 %Ni, 15.1 ppm.Ga, 25.7 ppm.Ge, 11 ppm.Ir, E.R.D. Scott et al., Geochimica et Cosmochimica Acta, 1973, **37**, p.1957.
1557g Chicago, Field Mus. Nat. Hist.; 398g Washington, U.S. Nat. Mus.;
Specimen(s): [1931,257], 57kg. main mass; [1929,261], 70.9g. and fragments, 59g and filings, 105g.

Piedmont v Alessandria.

Piedmont v Cereseto.

Piedmont v Motta di Conti.

Piedmont 44°55′ N., 8°0′ E.
Piemonte, Italy
Fell 1583, March 2
Doubtful..
A stone the size of a pomegranate fell, D.A. Soldani, Atti Accad. Sci. Siena, 1808, **9**, p.1, E.F.F. Chladni, Die Feuer-Meteore, Wien, 1819, p.219. The evidence of meteoritic nature is not conclusive.

Pienza v Siena.

Pierceville (iron) 37°52′ N., 100°40′ W.
Finney County, Kansas, U.S.A.
Found 1917
Iron. Octahedrite (IIIB?).
A very weathered mass of 100kg was found, Trans. Kansas Acad. Sci., 1953, **56**, p.255, Meteor. Bull., 1958 (8). Classification, V.F. Buchwald, Iron Meteorites, Univ. of California, 1975, p.974.
24kg Tempe, Arizona State Univ.; 534g Copenhagen, Univ. Geol. Mus.; 101g Harvard Univ.; 92g Mainz, Max-Planck-Inst.;
Specimen(s): [1959,1053], 50kg. of meteorodes; [1968,183], 78.5g. slice

Pierceville (stone) 37°52′ N., 100°40′ W.
Finney County, Kansas, U.S.A.
Found 1939
Stone. Olivine-hypersthene chondrite (L6).
One stone of 2.125kg was found, A.D. Nininger, Pop. Astron., Northfield, Minnesota, 1940, **48**, p.556, Contr. Soc. Res. Meteorites, **2**, p.227 [M.A. 8-54]. Olivine Fa_{24}, B. Mason, Geochimica et Cosmochimica Acta, 1963, **27**, p.1011.
424g Tempe, Arizona State Univ.; 183g Washington, U.S. Nat. Mus.; 176g Albuquerque, Univ. of New Mexico; 136g Los Angeles, Univ. of California;
Specimen(s): [1959,895], 436.5g.

Pieve v Vigarano.

Pieve-di-Casignano v Borgo San Donino.

Pila v Rancho de la Pila (1882).

Pilawa Gorna v Gnadenfrei.

Pilistvere v Pillistfer.

Piljusa v Jelica.

Pillistfer 58°40′ N., 25°44′ E.
Vohma, Estonian SSR, USSR
Fell 1868, August 8, 1230 hrs
Synonym(s): *Aukoma, Kurla, Pilistvere, Pillistvere, Sawiauk, Wahhe*
Stone. Enstatite chondrite (E6).
After detonations several stones fell, and four were found, at Aukoma, Kurla, Wahhe, and Sawiauk, weighing respectively about 14kg, 7.5kg, 1.5kg and 0.25kg, G. Rose, Monatsber. Akad. Wiss. Berlin, 1863, p.441. The metal contains over 1% Si, A.E. Ringwood, Geochimica et Cosmochimica Acta, 1966, **30**, p.23. Analysis, 27.78 % total iron, E. Jarosewich and B. Mason, Geochimica et Cosmochimica Acta, 1969, **33**, p.411. Mineralogy, I.A. Yudin, Метеоритика, 1972, **31**, p.83, K. Keil, J. Geophys. Res., 1968, **73**, p.6945. Trace element abundances, C.M. Binz et al., Geochimica et Cosmochimica Acta, 1974, **38**, p.1579.
14.7kg Dorpat, Univ.; 1.7kg Vienna, Naturhist. Mus.; 1.2kg Moscow, Acad. Sci.; 151g Tempe, Arizona State Univ.; 120g New York, Amer. Mus. Nat. Hist.; 116g Chicago, Field Mus. Nat. Hist.; 66g Washington, U.S. Nat. Mus.;
Specimen(s): [1920,337], 183g. of Kurla; [56324], 139g. of Aukoma; [36270], 13.5g. of Aukoma

Pillistvere v Pillistfer.

Pima County 32°12′ N., 111°0′ W. approx.
Arizona, U.S.A.
Found 1947, before this year
Iron. Hexahedrite (IIA).
A small mass of 210g, said to have been found near Tucson, Pima County, is described, with an analysis, 5.64 %Ni, E.P. Henderson and S.H. Perry, Proc. U.S. Nat. Mus., 1949, **99**, p.353 [M.A. 10-520]. Further analysis, 5.56 %Ni, 60.3 ppm.Ga, 181 ppm.Ge, 8.9 ppm.Ir, J.T. Wasson, Meteorites, Springer-Verlag, 1974, p.300. The mass is only part of an unidentified iron. A shock recrystallised hexahedrite; fusion crust due to oxy-acetylene cutting, V.F. Buchwald, Iron Meteorites, Univ. of California, 1975, p.974.
150g Washington, U.S. Nat. Mus., main mass;

Pinalta v Campo del Cielo.

Pine Bluff v Little Piney.

Pine Bluffs 41°11′ N., 104°4′ W.
Laramie County, Wyoming, U.S.A.
Found 1935
Stone. Chondrite?.
One stone of 2.7kg was recognized as meteoritic in 1938, A.D. Nininger, Pop. Astron., Northfield, Minnesota, 1939, **47**, p.213.

Pine River 44°8′ N., 89°5′ W.
Waushara County, Wisconsin, U.S.A.
Found 1931
Synonym(s): *Saxeville*
Iron. Octahedrite, medium, (1.2mm), with silicate inclusions (IA-ANOM).
One mass of 3.6kg was found, A.D. Nininger, Pop. Astron., Northfield, Minnesota, 1940, **48**, p.556, Contr. Soc. Res. Meteorites, **2**, p.227 [M.A. 8-54]. Classification, analysis of metal, 7.40 %Ni, 76.9 ppm.Ga, 234 ppm.Ge, 2.6 ppm.Ir, J.T. Wasson, Icarus, 1970, **12**, p.407. Contains Copiapo-type silicates, T.E. Bunch et al., Contr. Miner. Petrol., 1970, **25**, p.297. Description, unusual forms of graphite in kamacite, V.F. Buchwald, Iron Meteorites, Univ. of California, 1975, p.976.
305g Oshkosh, Public Mus.; 297g Washington, U.S. Nat. Mus.; 236g Appleton, Lawrence College; 112g Tempe, Arizona State Univ.; 102g Chicago, Field Mus. Nat. Hist.;
Specimen(s): [1959,989], 126g. and fragments, 15g

Pinnaroo 35°23′ S., 140°55′ E.
County Chandos, South Australia, Australia
Found 1927
Stony-iron. Mesosiderite (MES).
A weathered mass of 39.4kg was ploughed up 9 miles from Pinnaroo. The metal shows Widmanstätten structure, the silicates are olivine with some enstatite and plagioclase, Ab10. Description, with an analysis, A.R. Alderman, Trans. Roy. Soc. South Austr., 1940, **64**, p.109 [M.A. 8-58]. Analysis of metal (Ni-rich for a mesosiderite), 9.66 %Ni, 14.0 ppm.Ga, 51.6 ppm.Ge, 3.2 ppm.Ir, J.T. Wasson et al., Geochimica et Cosmochimica Acta, 1974, **38**, p.135. Olivine composition, references, C.E. Nehru et al., Geochimica et Cosmochimica Acta, 1980, **44**, p.1103.
25.1kg Adelaide, Univ.; 1.13kg Washington, U.S. Nat. Mus.; 183g New York, Amer. Mus. Nat. Hist.;
Specimen(s): [1966,489], 413g.

Piñon 32°40′ N., 105°6′ W.
Otero County, New Mexico, U.S.A.
Found 1928, or 1930
Iron. Ataxite, Ni-rich (IRANOM).
A mass of 17.85kg was found in sect. 22, township 19 south, range 18 east, Chaves County, New Mexico. The date is given as 1928, H.H. Nininger, Our Stone Pelted planet, 1933. Date given as 1930, H.H. Nininger, The Mines Mag., Golden, Colorado, 1933, **23** (8), p.6 [M.A. 5-405]. Description, analysis, 16.32 %Ni, H.H. Nininger, Pop. Astron., Northfield, Minnesota, 1939, **47**, p.155 [M.A. 7-375]. Analysis, 16.58 %Ni, E. Goldberg et al., Geochimica et Cosmochimica Acta, 1951, **2**, p.1. Structural study indicates a partial melting event, H.J. Axon and P.L. Smith, Min. Mag., 1972, **38**, p.736. Chemically anomalous, 15.54 % Ni, 2.32 ppm.Ga, 1.15 ppm.Ge, 15 ppm.Ir, J.T. Wasson and R. Schaudy, Icarus, 1971, **14**, p.59. Description; structurally similar to IVB irons, but has higher Ga and Ge contents, V.F. Buchwald, Iron Meteorites, Univ. of California, 1975, p.978. Contains isotopically anomalous Ag, T. Kaiser and G.J. Wasserburg, Geochimica et Cosmochimica Acta, 1983, **47**, p.43.
12kg Mesilla Park, New Mexico State Univ.; 1.4kg Washington, U.S. Nat. Mus.; 673g Tempe, Arizona State Univ.; 480g Michigan Univ.; 410g Chicago, Field Mus. Nat. Hist.; 88g Perth, West. Austr. Mus.;
Specimen(s): [1959,911], 2007g. and sawings, 138g.

Pinto Mountains 33°42′ N., 116°6′ W.
Riverside County, California, U.S.A.
Found 1954, November
Stone. Olivine-hypersthene chondrite (L6).
A mass of about 39.5lb was found. Description, L. LaPaz, Meteoritics, 1955, **1**, p.295 [M.A. 13-82]. Olivine Fa_{24}, B. Mason, Geochimica et Cosmochimica Acta, 1963, **27**, p.1011. Analysis, 22.27 % total iron, M.I. D'yakonova, Метеоритика, 1968, **28**, p.131.
12.2kg Albuquerque, Univ. of New Mexico; 159g Los Angeles, Univ. of California;
Specimen(s): [1964,703], 17.2g.

Pipe Creek 29°41′ N., 98°55′ W.
Bandera County, Texas, U.S.A.
Found 1887
Synonym(s): *Bandera County, San Antonio*
Stone. Olivine-bronzite chondrite (H6).
A mass of about 30lb was found, 35 miles SW. of San Antonio. Report, with an analysis, A.R. Ledoux, Trans. New York Acad. Sci., 1888-9, **8**, p.186. Olivine Fa_{19}, B. Mason, Geochimica et Cosmochimica Acta, 1963, **27**, p.1011.
4241g Chicago, Field Mus. Nat. Hist.; 460g Vienna, Naturhist. Mus.; 394g New York, Amer. Mus. Nat. Hist.; 312g Washington, U.S. Nat. Mus.; 196g Harvard Univ.; 82g Tempe, Arizona State Univ.;
Specimen(s): [67588], 766g.; [1920,338], 84.6g.; [67216], 56g.

Pipe Shrine House v Mesa Verde Park.

Piprasi v Butsura.

Piprassi v Butsura.

Piquetberg 32°52′ S., 18°43′ E.
Cape Province, South Africa
Fell 1881
Stone. Olivine-bronzite chondrite (H), veined.
37g in Vienna (Naturhist. Mus.), A. Brezina, Ann. Naturhist. Hofmus. Wien, 1887, **2**, p.115. Olivine Fa_{17}, B. Mason, Geochimica et Cosmochimica Acta, 1967, **31**, p.1100.
37g Vienna, Naturhist. Mus.;

Pirapora 17°18′ S., 45°0′ W.
Minas Gerais, Brazil
Found 1888, before this year
Synonym(s): *Angra dos Reis (iron)*
Iron. Hexahedrite (IIA).
A complete individual weighing 6.175kg, showing flight markings, was in a collection presented to Pope Leo XIII in 1888. It was labelled Angra dos Reis, with date of fall 20 January 1869, as for the achondrite, S.J. Salpeter, The Vatican Coll. of Meteorites, 1957. A second mass of 2.56kg was found before 1954, W.S. Curvello, Anais Acad. Brasil. Cienc., 1954, **26** (2), K. Keil, Fortschr. Min., 1960, **38**, p.257 [M.A. 15-351]. Analysis, chemically identical to Angra dos Reis (iron), 5.43 %Ni, 57.8 ppm.Ga, 189 ppm.Ge, 30 ppm.Ir, A. Kracher et al., Geochimica et Cosmochimica Acta, 1980, **44**, p.773. Metallography (of Angra dos Reis (iron)), H.J. Axon, Min. Mag., 1971, **38**, p.94. Description, similarity of Pirapora and Angra dos Reis (iron), V.F. Buchwald, Iron Meteorites, Univ. of California, 1975, p.261. 981.
Main mass, Rome, Vatican Coll., of Angra dos Reis (iron); Specimen, Rio de Janeiro, main mass of Pirapora;
Specimen(s): [1970,26], 26g. of Angra dos Reis (iron)

Pirganj v Pirgunje.

Pirgunje 25°48′ N., 88°27′ E.
Dinajpur district, East Bengal, Bangladesh
Fell 1882, August 29
Synonym(s): *Dinagepur, Pirganj*
Stone. Olivine-hypersthene chondrite (L6), veined.
A stone of 842g, labelled "Pirgunje, 29.8.82", was sent from India to E.A. Pankhurst of Brighton by a man who had no knowledge whatever of it or its antecedents, E.A. Pankhurst, letters of 12, 15 May and 3 November, 1889 in Min. Dept. BM(NH). Olivine Fa_{25}, B. Mason, Geochimica et Cosmochimica Acta, 1963, **27**, p.1011.
28g Chicago, Field Mus. Nat. Hist.; 23g New York, Amer. Mus. Nat. Hist.; 9g Vienna, Naturhist. Mus.;
Specimen(s): [64565], 732g. main mass, and fragments, 16g.

Pirna v Nenntmannsdorf.

Pirot v Dimitrovgrad.

Pirthalla 29°35′ N., 76°0′ E.

Hissar district, Haryana, India
Fell 1884, February 9, 1400 hrs
Stone. Olivine-bronzite chondrite (H6), brecciated.
After detonations, a stone of about 3lb was seen to fall 150 paces from the village of Pirthalla; three pieces, of 510g, 427g, and 224g were preserved, H.B. Medlicott, Rec. Geol. Surv. India, 1885, **18**, p.148. Olivine Fa_{18}, B. Mason, Geochimica et Cosmochimica Acta, 1963, **27**, p.1011.
471g Calcutta, Mus. Geol. Surv. India.; 29g Vienna, Naturhist. Mus.; 2.2g New York, Amer. Mus. Nat. Hist.;
Specimen(s): [56445], 427g. one of the three pieces

Pitts 31°57′ N., 83°31′ W.

Wilcox County, Georgia, U.S.A.
Fell 1921, April 20, 0900 hrs
Iron. Octahedrite, fine (0.2mm), with silicate inclusions (IB).
After the appearance of a brilliant fireball, followed by detonations, four masses, of 57oz, 42.5oz, 30oz, and 2oz, fell over an area of 1 mile by 0.25 mile; description, with an analysis, 6.67 %Ni, S.W. MacCallie, Am. J. Sci., 1922, **3**, p.211. Classification, and analysis of the metal, 12.8 %Ni, 33.0 ppm.Ga, 94.2 ppm.Ge, 0.86 ppm.Ir, J.T. Wasson, Icarus, 1970, **12**, p.407. The silicates are of the Copiapo type, T.E. Bunch et al., Contr. Miner. Petrol., 1970, **25**, p.297. Description, troilite-rich, V.F. Buchwald, Iron Meteorites, Univ. of California, 1975, p.981.
Main mass, In private hands in Georgia; 1022g Washington, U.S. Nat. Mus.; 46g New York, Amer. Mus. Nat. Hist.;

Pittsburg 40°26′ N., 80° W.

Allegheny County, Pennsylvania, U.S.A.
Found
Synonym(s): *Allegheny County, Miller's Run, Pittsburgh*
Iron. Octahedrite, coarse (2.2mm) (IIICD).
A mass said to have been about 292lb was ploughed up on Miller's Run; the main mass was forged into a bar and only a small part (about 600g) has been preserved, B. Silliman, Jr., Proc. Amer. Assoc., 1850, **4**, p.37. Description, analysis, 5.89 %Ni, E. Cohen, Mitt. Naturwiss. Ver. Neu-Vorp. u. Rügen, Greifswald, 1903, **35**, p.4. Further analysis, 6.77 % Ni, E.P. Henderson and S.H. Perry, Proc. U.S. Nat. Mus., 1958, **107**, p.339 [M.A. 12-604]. Described, figured, R.W. Stone and E.M. Starr, Bull. Pa. Geol. Surv., 1967 (Rep. G2, 1932, rev.), p.5. The original weight was possibly 29.2lb, not 292lb as reported in the original reference, V.F. Buchwald, Iron Meteorites, Univ. of California, 1975, p.985. Classification and analysis 6.16 %Ni, 83 ppm.Ga, 359 ppm.Ge, 2.0 ppm.Ir, E.R.D. Scott and J.T. Wasson, Geochimica et Cosmochimica Acta, 1976, **40**, p.103.
158g Yale Univ.; 80g Göttingen, Univ.; 47g Tempe, Arizona State Univ.; 39g Chicago, Field Mus. Nat. Hist.;
Specimen(s): [35418], 199g.

Pittsburgh v Pittsburg.

Plains 33°17′ N., 102°46′ W.

Yoakum County, Texas, U.S.A.
Found 1964
Stone. Olivine-bronzite chondrite (H5).
Two stones, totalling 34.3kg, were ploughed up about 1 mile apart, G.I. Huss, letter of 18 March, 1970 in Min. Dept. BM (NH). Co-ordinates, Cat. Huss Coll. Meteorites, 1976, p.36. Olivine Fa_{16}, R. Hutchison, priv. comm.
10.4kg Mainz, Max-Planck-Inst.; 9.6kg Lubbock, West Texas Mus.; 2kg Tempe, Arizona State Univ.; 363g Washington, U.S. Nat. Mus.; 221g Copenhagen, Univ. Geol. Mus.; 155g Vienna, Naturhist. Mus.; 76g New York, Amer. Mus. Nat. Hist.;
Specimen(s): [1968,181], 85.5g. slice

Plainview (no. 2) v Plainview (1950).

Plainview (1917) 34°7′ N., 101°47′ W.

Hale County, Texas, U.S.A.
Found 1917
Synonym(s): *Death Valley, Hale County*
Stone. Olivine-bronzite chondrite (H5), brecciated.
About 12 stones, totalling about 31kg, were found in 1917 and later, the largest weighing 5kg and the smallest 863g. Description, G.P. Merrill, Proc. U.S. Nat. Mus., 1917, **52**, p.419. About 700kg of this fall have been collected. The Death Valley stone has been assigned to this fall, Cat. Meteor. Arizona State Univ., 1970, p.36. Contains carbonaceous inclusions, analysis, 26.81 % total iron, R.V. Fodor and K. Keil, Geochimica et Cosmochimica Acta, 1976, **40**, p.177. Xenolithic, R.A. Binns, Geochimica et Cosmochimica Acta, 1968, **32**, p.299. Further analysis, 27.65 % total iron, H. von Michaelis et al., Earth planet. Sci. Lett., 1969, **5**, p.387. A regolith meteorite, younger than 3600 m.y., references, K. Keil et al., Earth planet. Sci. Lett., 1980, **51**, p.235 [M.A. 81-3047].
102kg Tempe, Arizona State Univ.; 24.9kg Fort Worth, Texas, Monnig Colln.; 24kg Washington, U.S. Nat. Mus.; 5.7kg Chicago, Field Mus. Nat. Hist.; 5kg Mainz, Max-Planck-Inst.; 4kg New York, Amer. Mus. Nat. Hist.; 2.7kg Yale Univ.; 2.5kg Ann Arbor, Michigan Univ.; 2.32kg Paris, Mus. d'Hist. Nat.; 2.3kg Copenhagen, Univ. Geol. Mus.; 1.9kg Los Angeles, Univ. of California; 1.6kg Harvard Univ.; 1.2kg Cranbrook Inst. Sci.;
Specimen(s): [1919,328], 2793g.; [1919,143], 35g.; [1959,783-788; 1959,790-808], 28kg. 26 mostly complete stones; [1962, 135], 147.5g.; [1976,M.9], 36g.

Plainview (1950) 34°7′ N., 101°47′ W.

Hale County, Texas, U.S.A.
Found 1950, recognized in this year
Synonym(s): *Plainview (no. 2)*
Stone. Olivine-bronzite chondrite (H).
One stone of 2.2kg was recognized as distinct from the other stones of the Plainview shower, H.H. and A.D. Nininger, The Nininger Collection of Meteorites, Winslow, Arizona, 1950, p.132. Olivine Fa_{20}, B. Mason, Geochimica et Cosmochimica Acta, 1963, **27**, p.1011.
1.9kg Tempe, Arizona State Univ., main mass; 2g New York, Amer. Mus. Nat. Hist.;

Plainview (c) 34°7′30″ N., 101°45′24″ W.

Hale County, Texas, U.S.A.
Found 1963
Stone. Chondrite.
A single mass of 631g is listed, Cat. Huss Colln. Meteorites, 1976, p.36.
631g Mainz, Max-Planck-Inst.;

Plainview (d) 34°11′15″ N., 101°42′15″ W.

Hale County, Texas, U.S.A.
Found 1979
Stone. Olivine-hypersthene chondrite (L6).
A single stone was found whose cut fragments totalled about 700g, olivine $Fa_{25.5}$, P. Sipiera et al., Meteoritics, 1983, **18**, p.63.

Planina Ozren v Ozren.

Plantersville 30°42′ N., 96°7′ W.

Grimes County, Texas, U.S.A.
Fell 1930, September 4, 1600 hrs
Stone. Olivine-bronzite chondrite (H6), veined.
A mass of 2085g was recovered. Description, with an analysis, J.T. Lonsdale, Am. Miner., 1937, **22**, p.877 [M.A. 6-397]. In the bulk analysis figures quoted, the Fe present as FeS (2.53%) has been omitted; its inclusion would raise the sum to 100.08%. Olivine Fa_{20}, B. Mason, Geochimica et Cosmochimica Acta, 1963, **27**, p.1011.
1862g Washington, U.S. Nat. Mus., main mass; 54g New York, Amer. Mus. Nat. Hist.;

Pleasanton 38°11′ N., 94°43′ W.

Linn County, Kansas, U.S.A.
Found 1935
Stone. Olivine-bronzite chondrite (H5).
One stone of About 5lb was found, A.D. Nininger, Pop. Astron., Northfield, Minnesota, 1937, **45**, p.449 [M.A. 7-62]. Olivine Fa_{18}, B. Mason, Geochimica et Cosmochimica Acta, 1963, **27**, p.1011.
204g Washington, U.S. Nat. Mus.; 67g Tempe, Arizona State Univ.;
Specimen(s): [1959,762], 68g. slice

Plescowitz v Ploschkovitz.

Ploschkovitz 50°32′ N., 14°7′ E. approx.

Litomerice, Sevrocesky, Czechoslovakia
Fell 1723, June 22, 1400 hrs
Synonym(s): *Bunzlau, Liboschitz, Plescowitz, Ploskovice, Reichstadt*
Stone. Olivine-hypersthene chondrite (L), brecciated.
After detonations and the appearance of a cloud, 33 stones fell, P.M.S. Bigot de Morogues, Mem. sur les Chutes des Pierres, Orléans, 1812, p.86. The only reputed specimen of this fall was the 1.25oz specimen purchased by the Brit. Mus. in 1846 from H. Heuland; it contains less iron than Tabor, N. Story Maskelyne, Phil. Mag., 1863, **25**, p.451. Olivine Fa_{24}, B. Mason, Geochimica et Cosmochimica Acta, 1963, **27**, p.1011.
7.9g New York, Amer. Mus. Nat. Hist.; 4g Chicago, Field Mus. Nat. Hist.; 3g Vienna, Naturhist. Mus.;
Specimen(s): [19973], 25.5g. bought from H. Heuland

Ploskovice v Ploschkovitz.

Plymouth 41°20′ N., 86°19′ W.

Marshall County, Indiana, U.S.A.
Found 1893
Synonym(s): *Marshall County*
Iron. Octahedrite, medium (1.3mm) (IIIA).
A mass measuring 12.5 × 7.4 × 2 inches was ploughed up about 5 miles SW. of Plymouth. In 1872 a larger, pear-shaped, mass about 4 feet long and 3 ft "in its widest diameter" had been found, but was afterwards buried and lost. Description, with an analysis, 8.55 %Ni, H.A. Ward, Am. J. Sci., 1895, **49**, p.53. Further analysis, 8.4 %Ni, 23.1 ppm.Ga, 42.4 ppm.Ge, 0.66 ppm.Ir, E.R.D. Scott et al., Geochimica et Cosmochimica Acta, 1973, **37**, p.1957. Description, cosmically reheated, V.F. Buchwald, Iron Meteorites, Univ. of California, 1975, p.986.
3645g New York, Amer. Mus. Nat. Hist.; 1077g Chicago, Field Mus. Nat. Hist.; 891g Vienna, Naturhist. Mus.; 751g Calcutta, Mus. Geol. Surv. India; 619g Washington, U.S. Nat. Mus.; 565g Harvard Univ.; 427g Budapest, Nat. Mus.; 470g Tempe, Arizona State Univ.;
Specimen(s): [76810], 445g.

Pnompehn 11°35′ N., 104°55′ E.

Cambodia
Fell 1868, June 20-30, 1500 hrs
Synonym(s): *Phnom Penh*
Stone. Olivine-hypersthene chondrite (L6).
Three stones fell, one of which weighed about 2lb, Rep. Brit. Assn., 1869, p.276.The place of fall is now spelt Phnom Penh. Olivine Fa_{24}, B. Mason, Geochimica et Cosmochimica Acta, 1963, **27**, p.1011.
54g Paris, Mus. d'Hist. Nat.; 26g Berlin, Humboldt Univ.; 0.2g Chicago, Field Mus. Nat. Hist.;

Poblazon v Charcas.

Podkamennaya Tunguska v Tunguska.

Poghel v Nerft.

Pohgel v Nerft.

Pohlitz 50°56′ N., 12°8′ E.

Gera, Germany
Fell 1819, October 13, 0800 hrs
Synonym(s): *Gera, Köstritz, Politz*
Stone. Olivine-hypersthene chondrite (L5), veined.
Some days after detonations had been heard, a stone of about 3kg was found between Politz and Langenberg, W.E. Braun, Ann. Phys. (Gilbert), 1819, **63**, p.217. Olivine Fa_{25}, B. Mason, Geochimica et Cosmochimica Acta, 1963, **27**, p.1011.
400g Gera, Mus.; 713g Berlin, Humboldt Univ.; 406g Vienna, Naturhist. Mus.; 340g Budapest, Nat. Mus.; 145g Tübingen, Univ.; 11g Chicago, Field Mus. Nat. Hist.;
Specimen(s): [33961], 87g.

Poinsett Iron v Toluca.

Point of Rocks (iron) 36°30′ N., 104°30′ W. approx.

Colfax County, New Mexico, U.S.A.
Found 1956
Iron. Octahedrite, medium (1.0mm) (IIIA).
Listed, L. LaPaz, Cat. Coll. Inst. Meteor. Univ. New Mexico, 1965. Classification, analysis, 8.4 %Ni, 21.1 ppm.Ga, 41.4 ppm.Ge, 0.46 ppm.Ir, E.R.D. Scott et al., Geochimica et Cosmochimica Acta, 1973, **37**, p.1957.
22g Albuquerque, Univ. of New Mexico;

Point of Rocks (stone) 36°30′ N., 104°30′ W. approx.

Colfax County, New Mexico, U.S.A.
Found 1954, approx.
Stone. Olivine-hypersthene chondrite (L6).
Listed, L. LaPaz, Cat. Coll. Inst. Meteor. Univ. New Mexico, 1965. Olivine $Fa_{25.3}$, D.E. Lange and K. Keil, Meteoritics, 1976, **11**, p.315.
750g Albuquerque, Univ. of New Mexico;

Poitiers v Vouillé.

Pojoaque v Glorieta Mountain.

Pokhra 26°43′ N., 82°40′ E.
Basti district, Gorakhpur, Uttar Pradesh, India
Fell 1866, May 27, 2030 hrs
Synonym(s): *Gorakhpur, Pokra*
Stone. Olivine-bronzite chondrite (H5).
After detonations and the appearance of a fireball, a stone of about 12 oz (350g) fell 6 miles ESE. of Bustee (Basti) at Pokhra, co-ordinates given above, (Calcutta Mus. label with BM(NH). specimen). Mentioned, O. Büchner, Ann. Phys. Chem. (Poggendorff), 1869, **136**, p.458, F. Fedden, Cat. Meteor. Indian Mus. Calcutta, 1880, p.23. Olivine Fa_{18}, B. Mason, Geochimica et Cosmochimica Acta, 1963, **27**, p.1011.
258g Calcutta, Mus. Geol. Surv. India; 26g Vienna, Naturhist. Mus.;
Specimen(s): [41020*], 43g.

Pokra v Pokhra.

Polanko v Mejillones.

Pölau 50°43′ N., 12°29′ E.
Zwickau, Karl-Marx-Stadt, Germany
Fell 1647, February 18, at night hrs
Doubtful..
A stone is said to have fallen, E.F.F. Chladni, Die Feuer-Meteore, Wien, 1819, p.227 but the evidence is inconclusive.

Polen v Lenarto.

Police v Suchy Dul.

Politz v Pohlitz.

Polk County v Fisher.

Pollen 66°20.9′ N., 14°0.9′ E.
Nord-Sjona, Nesna district, Norway
Fell 1942, April 6, 1900 hrs
Stone. Carbonaceous chondrite, type II (CM2).
A stone of 253.6g fell on the Pollen farm, less than 1 metre from the finder in 0.5m of snow; there were no audible phenomena, F.C. Wolff, Geochimica et Cosmochimica Acta, 1963, **27**, p.979, Norges Geol. Unders., 1963, **223**, p.359 [M.A. 16-640]. Classification, variable olivine composition, J.A. Wood, Geochimica et Cosmochimica Acta, 1967, **31**, p.2095. Analysis by H.B. Wiik, wt %; Fe metal 0.1, Ni 0.0, FeS 8.14, SiO2 27.92, TiO2 0.14, Al2O3 2.83, FeO 20.67, MnO 0.216, MgO 19.55, CaO 1.86, Na2O 0.63, K2O 0.30, P2O5 0.30, H2O+ 12.95, H2O- 0.62, C 1.65, Cr2O3 0.45, CoO 0.08, NiO 1.72.
120g Oslo, Min.-Geol. Mus., possibly contaminated;
Specimen(s): [1964,496], 20g.

Pollenza v Monte Milone.

Poltava (of G. von Blöde) v Kuleschovka.

Poltava (of G. von Blöde) v Simbirsk (of P. Partsch).

Poltava (of P. Partsch) v Slobodka.

Polujamki 52°6′ N., 79°42′ E. approx.
Mikhailovsky district, Altay region, USSR, [Полуямки]
Found 1971
Synonym(s): *Poluyamki*
Stone. Olivine-bronzite chondrite (H4).
A mass of 4.5kg was found in 1971 by a tractor driver during autumn sowing. In 1974 two other individuals were found, of 7.8kg and 1.2kg, 300 metres and 4 km respectively from the find-site of the 4.5kg stone, Meteor. Bull., 1976 (54), Meteoritics, 1976, **11**, p.83. Description, analysis, 25.69 % total iron, O.A. Kirova et al., Метеоритика, 1975, **34**, p.57.
13.3kg Moscow, Acad. Sci.; 8.3g Chicago, Field Mus. Nat. Hist.;

Poluyamki v Polujamki.

Pomorze
Poland, Not located
Found 1931, between 1931 and 1939
Doubtful. Iron. Ataxite, Ni-rich?.
A few pieces of iron, of total weight 2kg to 5kg, were found by a man working a quarry for road-stone. The 'iron' was approx. 50/50 Fe/Ni. Two pieces were sent to Gdansk Polytechnic; after the war, neither they nor the original locality could be found, J. Pokrzywnicki, Studia Geol. Polon., 1964, **15**, p.72.

Pomozdino 62°12′ N., 54°10′ E.
Ustkulom, Komi ASSR, USSR, [Помоздино]
Found 1964, summer
Stone. Achondrite, Ca-rich. Eucrite (AEUC).
One complete stone of 327g, apparently a recent fall, was found in a shallow hole, Meteor. Bull., 1964 (31). Full description, figures and analysis, L.G. Kvasha and M.I. D'yakonova, Метеоритика, 1972, **31**, p.109. The high MgO content, 9.96% wt., and a low Fe/Mg ratio indicate that this stone may be similar to Moore County (*q.v.*).
271g Moscow, Acad. Sci.;

Ponca Creek v Ainsworth.

Ponta Grossa 25°5′ S., 50°9′ W.
Parana, Brazil
Fell 1846, April
Doubtful. Stone.
A stone of 667g is said to have fallen, O.A. Derby, Revista do Observatorio, Rio de Janeiro, 1888, p.15. The stone appears to have been lost or was a pseudometeorite, E.A. Wülfing, Die Meteoriten in Samml., Tübingen, 1897, p.404. Very doubtful, E. de Oliveira, Anais Acad. Brasil. Cienc., 1931, **3**, p.33 [M.A. 5-15].

Ponte de Lima v São Julião.

Pont Llyfni v Pontlyfni.

Pontlyfni 53°2′11″ N., 4°19′10″ W.
Gwynedd, Wales
Fell 1931, April 14, 1153 hrs
Synonym(s): *Pont Llyfni*
Stone. Chondrite, anomalous (CHANOM).
The main mass probably fell into the sea. A fragment weighing 5oz was recovered, A. King, J. Brit. Astron. Assoc., 1932, **42**, p.328 [M.A. 5-153]. Composed largely of olivine, Fa1, enstatite Fs2 and rich in troilite, analysis, 33.84

% total iron, A.L. Graham et al., Min. Mag., 1977, **41**, p.201. Trace element analysis, A.M. Davis et al., Earth planet. Sci. Lett., 1977, **35**, p.19. Ar-Ar age, 4520 m.y., G. Turner et al., Geochimica et Cosmochimica Acta, 1978 (suppl. 8), p.989.
18g Chicago, Field Mus. Nat. Hist.; 2.1g Paris, Mus. d'Hist. Nat.;
Specimen(s): [1975,M.6], 107.9g. and fragments, 8.7g.

Pony Creek 31°39′48″ N., 99°57′ W.
Runnels County, Texas, U.S.A.
Found, year not reported
Stone. Chondrite.
A single mass of 4642g was found, Cat. Monnig Colln., 1983.
4642g Fort Worth, Texas, Monnig Colln.;

Poonarunna 27°46′ S., 137°51′ E.
Kalamurina Station, South Australia, Australia
Found 1970
Stone. Olivine-bronzite chondrite (H5).
A single mass of unrecorded weight was found 2km north of the Warburton River and 15km from the Poonarunna oil well. It was too large to be moved by the finder who removed a small chip for investigation which proved its meteoritic nature. The main mass could not be found subsequently, olivine $Fa_{19.9}$, Meteor. Bull., 1981 (59), Meteoritics, 1981, **16**, p.197.

Pooposo 18°20′ S., 66°50′ W. approx.
Pooposo Estate, Ouro, Bolivia
Found 1910, known in this year
Iron. Octahedrite, coarse (2.6mm) (IA).
A mass of this iron was obtained in 1910 by J. Böhm from a missionary who had brought it from Bolivia, R. Koechlin, letter of 23 May, 1922 in Min. Dept. BM(NH). Distinct fron Bolivia (*q.v.*). Classification and analysis, 6.98 %Ni, 78.8 ppm.Ga, 325 ppm.Ge, 2.8 ppm.Ir, A. Kracher et al., Geochimica et Cosmochimica Acta, 1980, **44**, p.773. Brief description, V.F. Buchwald, Iron Meteorites, Univ. of California, 1975, p.989.
7.4kg Vienna, Naturhist. Mus., main mass;
Specimen(s): [1922,243], 26g.

Popilta v Menindee Lakes 004.

Popio v Menindee Lakes 001.

Popio Lakes (a) v Menindee Lakes 005.

Popio Lakes (b) v Menindee Lakes 006.

Poplar Camp v Cranberry Plains.

Poplar Hill v Cranberry Plains.

Portales (a) 34°4′ N., 103°30′ W.
Roosevelt County, New Mexico, U.S.A.
Found 1967
Stone. Olivine-bronzite chondrite (H4).
One well preserved oriented stone, 3.24kg, was found 18.5km SW. of Portales, G.I Huss, letter of 12 February, 1974 in Min. Dept. BM(NH). Olivine Fa_{18}, R. Hutchison, priv. comm., 1976.
3.2kg Mainz, Max-Planck-Inst.;
Specimen(s): [1970,172], 6.6g.

Portales (b) 34°4′ N., 103°30′ W.
Roosevelt County, New Mexico, U.S.A.
Found 1967
Stone. Olivine-hypersthene chondrite (L6).
A heavily weathered fragment, 595.8g, was found 17.7km SW. of Portales, G.I Huss, letter of 12 February, 1974 in Min. Dept. BM(NH). Olivine Fa_{26}, A.L. Graham, priv. comm., 1978.
576g Mainz, Max-Planck-Inst.; 5g Albuquerque, Univ. of New Mexico;
Specimen(s): [1970,173], 3.3g.

Portales (c) 34°6′ N., 103°25′ W.
Roosevelt County, New Mexico, U.S.A.
Found 1967
Stone. Olivine-bronzite chondrite (H4).
A 6kg stone was found 10.4km SW. of Portales; it is more heavily weathered than the Portales (a) stone (*q.v.*), G.I Huss, letter of 12 February, 1974 in Min. Dept. BM(NH).
3.2kg Mainz, Max-Planck-Inst.; 156g Washington, U.S. Nat. Mus.; 95g Albuquerque, Univ. of New Mexico;
Specimen(s): [1970,32], 41.8g.

Portis 39°34′ N., 98°41′ W.
Osborne County, Kansas, U.S.A.
Found 1959, approx.
Stone. Olivine-hypersthene chondrite (L), brecciated.
Listed, total weight not known, Cat. Huss Coll. Meteorites, 1976, p.36.
29.9g Mainz, Max-Planck-Inst.;

Portland Meteor v Washougal.

Port Orford 42°45′ N., 124°30′ W.
Rogue River Mts., Curry County, Oregon, U.S.A.
Found 1859
Synonym(s): *Port Oxford, Rogue River Mts.*
Stony-iron. Pallasite (PAL).
A mass estimated at 10 tons was said to have been found, but only about 30g are known to be preserved, O.C. Farrington, Mem. Nat. Acad. Sci. Washington, 1915, **13**, p.358. Historical note, J.H. Pruett, The Sky, New York, 1939, **3** (11), p.18 [M.A. 7-544]. Mentioned, L. LaPaz, Pop. Astron., Northfield, Minnesota, 1951, **59**, p.101 [M.A. 11-445], E.P. Henderson and H.M. Dole, The Ore Bin, 1964, **26**, p.113 [M.A. 17-594]. Analysis of metal, 10.29 %Ni, E.F. Lange, The Ore Bin, 1970, **32**, p.21. Analysis, 9.43 %Ni, olivine $Fa_{12.5}$, P.R. Buseck and J.I. Goldstein, Bull. Geol. Soc. Amer., 1969, **80**, p.2141. Further analysis of metal, 11.6 %Ni, 17.9 ppm.Ga, 33 ppm.Ge, 0.08 ppm.Ir, E.R.D. Scott, Geochimica et Cosmochimica Acta, 1977, **41**, p.349.
24g Washington, U.S. Nat. Mus.; 3g Vienna, Naturhist. Mus.; Fragment, Calcutta, Mus. Geol. Surv. India;

Port Oxford v Port Orford.

Port Tobacco v Nanjemoy.

Portugal 38°30′ N., 8° W. approx.
Portugal
Fell 1796, February 19
Synonym(s): *San Michele de Mechede, Tasquinha*
Stone.
A stone of 10lb is said to have fallen, E. Howard, Phil. Trans. Roy. Soc. London, 1802, **92**, p.170. Nothing appears to have been preserved. Listed, R. de Serpa Pinto, Rev. sismol. geofis. Coimbra, 1932, **3**, p.45 [M.A. 5-297], E.F.F.

Chladni, Die Feuer-Meteore, Wien, 1819, p.264, M. Faura y Sans, Meteoritos caidos en la Peninsula Iberica, Tortosa, 1922.

Poscente v Rancho Gomelia.

Post 33°7′ N., 101°23′ W.
Garza County, Texas, U.S.A.
Found 1965, about this year
Stone. Olivine-hypersthene chondrite (L), veined, brecciated.
A badly weathered 2kg mass was found. About half the stone has been lost, the other half is in the possession of the finder, Mr. Taylor, Meteor. Bull., 1972 (51), Meteoritics, 1972, **7**, p.229.
20g Houston, in possession of E.A. King, Univ. of Houston;

Potenza River v Monte Milone.

Potosi v Coahuila.

Potosi v Imilac.

Potter 41°14′ N., 103°18′ W.
Cheyenne County, Nebraska, U.S.A.
Found 1941
Stone. Olivine-hypersthene chondrite (L6), brecciated.
261kg of fragments were recovered. Mentioned, with a note that it may be identical to Bushnell or Dix, H.H. Nininger, Contr. Soc. Res. Meteorites, 1942, **3**, p.62. Distinct from Bushnell (H4) and Dix (H6), olivine Fa_{23}, B. Mason, Geochimica et Cosmochimica Acta, 1963, **27**, p.1011.
86kg Chicago, Field Mus. Nat. Hist., main mass; 45.3kg Tempe, Arizona State Univ.; 4.8kg New York, Amer. Mus. Nat. Hist.; 1.9kg Washington, U.S. Nat. Mus.; 805g Harvard Univ.; 706g Los Angeles, Univ. of California; 492g Copenhagen, Univ. Geol. Mus.; 470g Mainz, Max-Planck-Inst.; 182g Vienna, Naturhist. Mus.;
Specimen(s): [1950,223], 50.3g.; [1959,1015], 9760g. and a fragment, 62g; [1971,204], 41.1g. and fragments, 6g

Po-wang Chen 31°25′ N., 118°30′ E.
Bowang, Dangtu County, Anhui, China
Fell 1933, October 23, 1900 hrs
Synonym(s): *Dangtu, Dantu, Tangtu, Tang-t'u*
Stone.
Six stones fell totalling 665g, the largest 357g. Short description, Ko Hsueh, La Science Populaire, 1933, p.352. Listed, references, D. Bian, Meteoritics, 1981, **16**, p.115.
170g Peking, Planetarium, several small fragments;

Powder Mill Creek v Crab Orchard.

Pozo del Cielo v Campo del Cielo.

Pozzaglia v Orvinio.

Prachin v Bohumilitz.

Praestö v Mern.

Prairie Dog Creek 39°38′ N., 100°30′ W.
Decatur County, Kansas, U.S.A.
Found 1893, approx.
Synonym(s): *Decatur County, Dog Creek, Smoky Hill River*
Stone. Olivine-bronzite chondrite (H3).
A stone of 2.9kg was found, H.L. Preston, Am. J. Sci., 1900, **9**, p.412. Description, E. Weinschenk, Tschermaks Min. Petr. Mitt., 1895, **14**, p.471. Classification and mineral compositions, olivine unequilibrated, R.T. Dodd et al., Geochimica et Cosmochimica Acta, 1967, **31**, p.921.
715g New York, Amer. Mus. Nat. Hist.; 314g Washington, U.S. Nat. Mus.; 274g Vienna, Naturhist. Mus.; 215g Budapest, Nat. Mus.; 186g Tübingen, Univ.; 171g Chicago, Field Mus. Nat. Hist.; 50g Yale Univ.;
Specimen(s): [77096], 529g.; [83670], Thin section

Prambachkirchen 48°18′9″ N., 13°56′27″ E.
Eferding, Ober-Österreich, Austria
Fell 1932, November 5, 2155 hrs
Stone. Olivine-hypersthene chondrite (L6).
One stone of 2125g fell. Description, with an analysis, E. Dittler and J. Schadler, Sitzungsber. Akad. Wiss. Wien, Math.-naturwiss. Kl., 1933, **142**, p.213 [M.A. 6-11], J. Schadler and J. Rosenhagen, Jahr. Oberösterr. Musealver. Linz, 1935, **86**, p.102 [M.A. 6-392], J. Schadler, Natur u. Volk, Frankfurt-a-M., 1938, **68**, p.1 [M.A. 7-67].
22g Vienna, Naturhist. Mus.;

Prambanan 7°34′ S., 110°50′ E. approx.
Surakarta, Java, Indonesia
Found 1797, known in this year
Synonym(s): *Brambanan, Sörakarta, Surakarta*
Iron. Octahedrite, finest (0.13mm) (IRANOM).
After detonations and the appearance of a fireball, two masses were found. The smaller has been lost, the larger, described as measuring a metre cube, was brought from Prambanan to Surakarta in 1797; a piece of 0.25kg was sent to Amsterdam in 1866 and described, E.H. von Baumhauer, Arch. Néerland. Sci. Nat. Haarlem, 1866, **1**, p.465. Description, analysis, 9.39 %Ni, E. Cohen, Meteoritenkunde, 1905, **3**, p.308. All of this material has been heated above 900C, V.F. Buchwald, Iron Meteorites, Univ. of California, 1975, p.989. Classification and analysis, 10.1 %Ni, 28.3 ppm.Ga, 190 ppm.Ge, 4.2 ppm.Ir, A. Kracher et al., Geochimica et Cosmochimica Acta, 1980, **44**, p.773.
106g Budapest, Nat. Mus.; 490g Vienna, Naturhist. Mus.; 16g Chicago, Field Mus. Nat. Hist.; 2.3g Washington, U.S. Nat. Mus.;
Specimen(s): [54645], 8.25g.; [43203], 0.5g.

Praskoles v Zebrak.

Praskolesy v Zebrak.

Premier Downs v Mundrabilla.

Preobrazhenka 53°42′ N., 79°30′ E.
Preobrazhenka village, Altai region, Federated SSR, USSR
Found 1949, recognized 1980
Stone. Chondrite.
A mass of 413.9g in Achinsk Museum was recognized as meteoritic in 1980, Meteor. Bull., 1982 (60), Meteoritics, 1982, **17**, p.96.

Pribram 49°40′ N., 14°2′ E.
Stredocesky, Czechoslovakia
Fell 1959, April 7, 1930 hrs, U.T.
Synonym(s): *Drazkov, Hojsis, Kamyk nad Vltavou, Luhy, Prizbram, Velka*
Stone. Olivine-bronzite chondrite (H5).
Four stones, of a possible 19, totalling 5.555kg (the largest 4.25kg) were found after a brilliant meteor, the orbit of which was obtained from simultaneous photographs, Meteor. Bull., 1959 (15), R. Rost, Casopis min. geol., 1959, **5**, p.423, K. Tucek, Bull. Astron. Inst. Czech., 1961, **12**, p.212. Figured, K. Tucek, Метеоритика, 1965, **26**, p.112. Analysis, 25.39 % total iron, V.Ya. Kharitonova, Метеоритика, 1965, **26**, p.146. Further analysis, 26.15 % total iron, R.S. Clarke, Jr. et al., J. Geophys. Res., 1971, **76**, p.4135. Olivine Fa_{20}, B. Mason, Geochimica et Cosmochimica Acta, 1963, **27**, p.1011.
4783g Prague, Nat. Mus.; 140g Washington, U.S. Nat. Mus.;
Specimen(s): [1982,M.18], 5.5g.

Pricetown 39°7′ N., 83°51′ W.
Highland County, Ohio, U.S.A.
Fell 1893, February 13
Synonym(s): *Highland County, Princetown*
Stone. Olivine-hypersthene chondrite (L6).
A stone of about 900g came into the possession of C.S. Bement, but no details of the fall have been published, O.C. Farrington, Mem. Nat. Acad. Sci. Washington, 1915, **13**, p.361, C.S. Bement, letters of 13 February, 1893 and 22 October, 1897 in Min. Dept. BM(NH). Olivine Fa_{24}, B. Mason, Geochimica et Cosmochimica Acta, 1963, **27**, p.1011.
480g New York, Amer. Mus. Nat. Hist., main mass; 59g Vienna, Naturhist. Mus.; 44g Paris, Mus. d'Hist. Nat.;
Specimen(s): [83499], 10.2g.

Prieska v Rateldraai.

Primitiva v La Primitiva.

Prince George 53°50′ N., 122°27′ W.
British Columbia, Canada
Fell 1969, August 21, 0558 hrs, U.T.
Doubtful..
A bright fireball was followed by a seismic event, but a search in difficult terrain revealed nothing, I. Halliday and A.T. Blackwell, Meteoritics, 1971, **6**, p.39.

Prince of Wales's Straits 73°30′ N., 115°0′ W.
Alaska, U.S.A.
Fell 1850, December 3
Doubtful..
A stone or stones are said to have fallen, R.P. Greg, Rep. Brit. Assn., 1860, p.91 but the evidence is not conclusive.

Princetown v Pricetown.

Prizbram v Pribram.

Provence 43°30′ N., 6° E. approx.
Var, France
Fell 1627, November 27
Doubtful..
A stone of 59lb is said to have fallen, R.P. Greg, Rep. Brit. Assn., 1860, p.54 but the evidence is not conclusive.

Providence 38°34′ N., 85°14′ W.
Trimble County, Kentucky, U.S.A.
Found 1903
Iron. Octahedrite, medium (1.2mm) (IIIA).
A mass of 6.804kg was found. Description, with an analysis, 10.3 %Ni, D.M. Young, Pop. Astron., Northfield, Minnesota, 1939, **47**, p.382 [M.A. 7-376]. Classification, analysis, 8.25 %Ni, 20.2 ppm.Ga, 41.5 ppm.Ge, 0.39 ppm.Ir, E.R.D. Scott et al., Geochimica et Cosmochimica Acta, 1973, **37**, p.1957. Description; co-ordinates; shocked and annealed, V.F. Buchwald, Iron Meteorites, Univ. of California, 1975, p.991.
5.49kg Washington, U.S. Nat. Mus., main mass; 299g Harvard Univ.;

Przelazy v Seeläsgen.

Pseudo-Apoala v Durango.

Pueblito de Allende v Allende.

Puente del Zacate
27°52′ N., 101°30′ W. probable location
Monclova district, Coahuila, Mexico
Found 1904, approx.
Iron. Octahedrite, medium (1.0mm) (IIIA).
A mass of 30.79kg in the museum of the Inst. Geol. Mexico labelled Puente del Zacate was described, with an analysis, 7.65 %Ni, H.H. Nininger, Proc. Colorado Mus. Nat. Hist., 1931, **10** (1) [M.A. 5-13]. The bridge of Zacate is also given as the locality of the Coahuila hexahedrites, Bol. Inst. Geol. Mexico, 1923 (40), p.123 there has perhaps been some interchange of labels or specimens. Determination of Ga, Au and Pd, 8.21 %Ni, E. Goldberg et al., Geochimica et Cosmochimica Acta, 1951, **2**, p.1. Further analysis, 8.20 % Ni, 20.6 ppm.Ga, 40.5 ppm.Ge, 1.4 ppm.Ir, E.R.D. Scott et al., Geochimica et Cosmochimica Acta, 1973, **37**, p.1957. Probably not from the Coahuila site; co-ordinates, shock deformation, V.F. Buchwald, Iron Meteorites, Univ. of California, 1975, p.994.
20kg Mexico City, Inst. Geol.; 2.3kg Washington, U.S. Nat. Mus.; 495g Chicago, Field Mus. Nat. Hist.; 191g Tempe, Arizona State Univ.;
Specimen(s): [1952,147], 455g.

Puente-Ladron 34°24′ N., 106°51′ W.
Socorro County, New Mexico, U.S.A.
Found 1944, May 17
Stone. Olivine-hypersthene chondrite (L).
A small stone of 7.673g was picked up near a bridge over the river Puerco, 10 miles from Ladron Peak, H.H. Nininger, Pop. Astron., Northfield, Minnesota, 1944, **52**, p.407, Contr. Soc. Res. Meteorites, 1945, **3**, p.165 [M.A. 9-301], H.H. Nininger, Pop. Astron., Northfield, Minnesota, 1947, **55**, p.325 [M.A. 10-178]. Olivine Fa_{24}, B. Mason, Geochimica et Cosmochimica Acta, 1963, **27**, p.1011. Possibly transported, L. LaPaz, Pop. Astron., Northfield, Minnesota, 1946, **54**, p.95.
7.6g Tempe, Arizona State Univ.;

Puerta de Arauco 28°53′ S., 66°40′ W.
La Rioja, Argentina
Found 1904, middle of the year
Synonym(s): *La Rioja*
Iron. Octahedrite, fine (0.4mm) (IVA).
A mass of 1.5kg was found, M. Kantor, Cat. Meteoritos, Rev. Mus. La Plata, 1920, **25**, p.110 [M.A. 1-146]. Analysis, 6.61 %Ni, E.H. Ducloux, Rev. Mus. La Plata, 1908, **15**,

p.84. Analysis doubtful, more likely about 10%Ni; polycrystalline, and of a rare type, V.F. Buchwald, Iron Meteorites, Univ. of California, 1975, p.996. Mentioned, M.M. Radice, Rev. Mus. La Plata, 1959, **5** (new ser.), p.114.

1.1kg La Plata Mus.; 22g Vienna, Naturhist. Mus.;

Puerto Libertad 29°54′ N., 112°41′ W.

Sonora, Mexico
Found 1973, recognized in this year
Stone. Olivine-hypersthene chondrite (L).

One stone of 66g was found in a dry wash near the beach on the Gulf of California about 3 km WNW. of Puerto Libertad, Meteor. Bull., 1974 (52), Meteoritics, 1974, **9**, p.114.

66g Tempe, Arizona State Univ.;

Pulaski County v Cañon Diablo.

Pulaski County v Little Piney.

Pulivl v Botschetschki.

Pulrose 54°15′ N., 4°30′ W. approx.

Isle of Man
Fell 1813, between this year and 1819
Doubtful..

A fall of pumiceous stones is said to have occurred, E.F.F. Chladni, Ann. Physik (Gilbert), 1821, **68**, p.333, J. Murray, Phil. Mag., 1819, **54**, p.39 but the evidence is not conclusive.

Pulsora 23°22′ N., 75°11′ E.

Ratlam district, Madhya Pradesh, India
Fell 1863, March 16, afternoon
Synonym(s): *Ratlam, Rutlam*
Stone. Olivine-bronzite chondrite (H5), brecciated.

After detonations, three stones fell, one weighing about 1.5lb, W. von Haidinger, Sitzungsber. Akad. Wiss. Wien, Math.-naturwiss. Kl., 1869, **59** (2), p.228. Olivine Fa_{19}, B. Mason, Geochimica et Cosmochimica Acta, 1963, **27**, p.1011, K. Fredriksson et al., Smithson. Contrib. Earth Sci., 1975 (14), p.41.

332g Calcutta, Mus. Geol. Surv. India; 126g Paris, Mus. d'Hist. Nat.; 50g Vienna, Naturhist. Mus.; 5g Chicago, Field Mus. Nat. Hist.; 5g Washington, U.S. Nat. Mus.;

Specimen(s): [43059], 40g.

Pulter

Spain
Doubtful..

Reference is made to a meteorite under this name in Rev. Industr., 1956, vol.11 p.196, Meteor. Bull., 1958 (8). No trace of a place of this name in Spain or elsewhere can be found, and it appears likely that the name is a misprint for Pułtusk, and the location in Spain erroneous.

Pułtusk 52°46′ N., 21°16′ E.

Warsaw, Poland
Fell 1868, January 30, 1900 hrs
Synonym(s): *Gostkowo, La Spezia, Lerici, Nossi-Bé, Nosy Bé, Obritti, Ostrolenka, Spezia, Varsava, Warsaw, Warschau, Warshaw*
Stone. Olivine-bronzite chondrite (H5), veined, brecciated.

After the appearance of a fireball followed by detonations, a shower of stones fell over an area of several square miles between Pułtusk and Ostrolenka on the Narew. The weights of individuals range from 9kg to about a gram, a few were of 4kg and over, 200 of 1kg and over; more than 200kg are preserved in collections, K. Szymanski, Neues Jahrb. Min., 1868, p.326, G. vom Rath, Neues Jahrb. Min., 1869, p.80, W. von Haidinger, Sitzungsber. Akad. Wiss. Wien, Math.-naturwiss. Kl., 1868, **57**, p.405. Analysis, 27.19 % total iron, M.I. D'yakonova and V.Ya. Kharitonova, Метеоритика, 1961, **21**, p.52. Co-ordinates, estimated 2000kg fell, total number of fragments having been about 180000, B. Lang and M. Kowalski, Meteoritics, 1971, **6**, p.149. Xenolithic, olivine $Fa_{18.5}$, R.A. Binns, Geochimica et Cosmochimica Acta, 1968, **32**, p.299. Further analysis, 27.24 % total iron, H. von Michaelis et al., Earth planet. Sci. Lett., 1969, **5**, p.387.

27.98kg Paris, Mus. d'Hist. Nat.; 18.9kg Bonn, Univ.; 15.8kg Vienna, Naturhist. Mus.; 16kg Moscow, Acad. Sci.; 13kg Chicago, Field Mus. Nat. Hist.; 10.9kg Berlin, Humboldt Univ.; 4.9kg Stockholm, Riksmus; 4.8kg Cracow, Jagellonian Univ.; 4.4kg Budapest, Nat. Mus.; 1.5kg Tempe, Arizona State Univ.; 1.18kg Harvard Univ.; 900g Prague, Nat. Mus.; 714g Copenhagen, Univ. Geol. Mus.;

Specimen(s): [42424], 9095g. complete stone; [41676], 3545 complete stone; [41677], 845g. complete stone; [42296], 538g.; [42297], 425g.; [42298], 280.5g.; [41680], 207g. and fragments, 8g; [42301], 158g.; [42302], 145g.; [41679], 139.5g. complete stone; [42309], 111.5g.; [42305], 108g.; [42307], 92g. and fragments, 3g; [42304], 82.5g.; [42310], 76g.; [42311], 74.5g.; [42313 - 42329, 42331 - 42339], 948g. 26 stones from 11g to 64g in weight; [41681 - 41692], 240g. 12 small stones; [64524], 12.5g. two stones; [1927,1289], 121g. and fragments and powder, 23g; [1923,349], 15g.; [90249], 140g.; [1960,493], 30g.; [1964,706], 13.5g.; [85832], 8g. probably Pu±tusk; [1922,298], 132g. from Nossi Be, probably Pu±tusk; [96275], thin section; [1964,570], thin section

Punganaru 13°20′ N., 78°57′ E.

Cuddepah, Andhra Pradesh, India
Fell 1811, November 23
Synonym(s): *Panganur*

A meteorite is said to have fallen at "Panganur" and to be high in iron and nickel, J. Abbott and H. Piddington, J. Asiatic Soc. Bengal, 1844, **13**, p.885 and a 3.5oz specimen presented by Mr Ross of Cuddepah was in the collections of the Society in 1859 (no. 12; the Moradabad fall is no. 1, 2.5oz in 3 fragments, T. Oldham et al., J. Asiatic Soc. Bengal, 1859, **28**, p.261. In 1863 it was suggested that it may be identical to Moradabad, O. Büchner, Die Meteoriten in Samml., Leipzig, 1863, p.30. Listed as a synonym of Moradabad, E.A. Wülfing, Die Meteoriten in Samml., Tübingen, 1897, p.245. The 3.5oz specimen is not listed by T. Oldham among the 7 falls which were represented in the Royal Asiatic Society's collection in 1867, but were not in the Geol. Surv. India collection, Cat. Met. Mus. Geol. Surv. Calcutta, 1867, p.8. Wülfing in 1897 could only trace a few grams from the 2.5oz Moradabad specimen. Wülfing's evidence for equating the two falls remains obscure; the Panganur mass is not listed in 1866, J. Asiatic Soc. Bengal, 1866, **35** (2), p.43 so must have been lost between 1859 and 1866, but there seems no reason to doubt that a meteorite fell at or near Punganaru. It is possible that Gurram Konda (*q.v.*) may be an extant fragment of this fall.

Puquios 27°9′ S., 69°55′ W.

Copiapo, Atacama, Chile
Found 1885
Iron. Octahedrite, medium (0.75mm) (IID).

A mass of about 14.5lb was found near Puquios; description, with an analysis, 9.83 %Ni, E.E. Howell, Am. J. Sci., 1890, **40**, p.224. Structural classification, complex shock

deformation, V.F. Buchwald, Iron Meteorites, Univ. of California, 1975, p.996. Chemical classification, analysis, 10.08 %Ni, 77.0 ppm.Ga, 87.9 ppm.Ge, 13 ppm.Ir, J.T. Wasson, Geochimica et Cosmochimica Acta, 1969, **33**, p.859.
1.4kg Vienna, Naturhist. Mus.; 1.5kg New York, Amer. Mus. Nat. Hist.; 318g Chicago, Field Mus. Nat. Hist.; 208g Harvard Univ.; 90g Washington, U.S. Nat. Mus.;
Specimen(s): [66743], 153g. and a fragment, 9.5g

Pura Dayal v Dyalpur.

Purdue v Lafayette.

Purgatory Peak A77006 77°20′ S., 162°25′ E.
Victoria Valley, Antarctica
Found 1978, January 23
Iron. Octahedrite, coarse (2mm) (IA).
A single mass of 19.068kg was found in moraine about 600m from the terminal face of the Victoria glacier, Meteor. Bull., 1980 (57), Meteoritics, 1980, **15**, p.100. Classification, analysis, 7.27 %Ni, 78.2 ppm.Ga, 245 ppm.Ge, 2.46 ppm.Ir, R.S. Clarke, Jr. et al., Meteoritics, 1980, **15**, p.273, abs.

Puripica v North Chile.

Purna v Naoki.

Purnea v Shikarpur.

Purwa Dayalapur v Dyalpur.

Pusinsko Selo v Milena.

Pusnsko Selo v Milena.

Putinga 29°2′ S., 53°3′ W.
Encantada district, Rio Grande do Sul, Brazil
Fell 1937, August 16, 1630 hrs
Synonym(s): *Encantada*
Stone. Olivine-hypersthene chondrite (L6).
The day after a bolide was observed, fragments of 56kg, 90kg, and 40kg and several of about 10kg were found at various depths about 2 km from Putinga. More stones have been reported, Diario de Noticias, Porto Alegre, August 1 and 19, 1937, E.W. Salpeter, letter of 19 March, 1963, H. Grunewaldt, letters of 21 January and 17 February, 1965 in Min. Dept. BM(NH). Description, with an analysis, 22.70 % total iron, olivine $Fa_{24.3}$, R.F. Symes and R. Hutchison, Min. Mag., 1970, **37**, p.721. Description, analysis, 22.42 % total iron, K. Keil et al., Meteoritics, 1978, **13**, p.165.
45kg Porto Alegre, Mus. Luis Englert, Univ. Federal; Specimen, Pasadena, Calif. Inst. Tech.; Specimen, Hamburg, Min.-techn. Inst.; Specimen, Rio de Janeiro, Mus. Nac.; 4kg in possession of H. Grunewaldt; 962g Washington, U.S. Nat. Mus.; 284g Rome, Vatican Collection; 254g New York, Amer. Mus. Nat. Hist.;
Specimen(s): [1964,34], 18.5g.; [1982,M.7], 33g.

Putiwl v Botschetschki.

Putnam County 33°15′ N., 83°15′ W.
Georgia, U.S.A.
Found 1839
Iron. Octahedrite, fine (0.3mm) (IVA).
A mass of 72lb was found, J.E. Willet, Am. J. Sci., 1854, **17**, p.331. Description, analysis, 7.89 %Ni, E. Cohen, Meteoritenkunde, 1905, **3**, p.343. Further analysis, 7.98 % Ni, 2.17 ppm.Ga, 0.129 ppm.Ge, 2.1 ppm.Ir, R. Schaudy et al., Icarus, 1972, **17**, p.174. Description, heavily weathered, shock hardened, V.F. Buchwald, Iron Meteorites, Univ. of California, 1975, p.999.
12.1kg Tempe, Arizona State Univ.; 2.7kg Washington, U.S. Nat. Mus.; 2.38kg Harvard Univ.; 372g Yale Univ.; 132g Vienna, Naturhist. Mus.;
Specimen(s): [90227], 89.5g.; [90228], 20.5g. and fragments, 2.5g

Qingzhen 26°32′ N., 106°28′ E.
Yongle, Qingzhen County, Guizhou, China
Fell 1976, September 13, 1640 hrs
Stone. Enstatite chondrite (E3).
A bright fireball was seen over western Guizhou (Kweichow) province travelling from SW. to NE. Between September 15 and 31 a search resulted in the recovery of two stones, of total weight 2.6kg, D. Wang and X. Xie, Geochimica, 1977, **4**, p.276 [M.A. 78-3345]. Classification and analysis, 31.65 % total iron, D. Wang and X. Xie, Geochemistry, 1982, **1**, p.69. Reported, Meteor. Bull., 1978 (55), Meteoritics, 1978, **13**, p.345. Details of fall, D. Bian, Meteoritics, 1981, **16**, p.115.
Main mass, Guiyang, Acad. Sinica Inst. Geochem.;

Quairading v Youndegin.

Quarat al Hanish 25°9′ N., 25°35′ E.
Egypt
Found 1979, June 25
Iron. Octahedrite, fine (0.4mm) (IIICD-ANOM).
A single mass of 593g was found in the eastern Libyan desert, in the area of the occurrence of the Libyan desert glass, Meteor. Bull., 1980 (57), Meteoritics, 1980, **15**, p.100. Classification, analysis, 12.8 %Ni, 17.2 ppm.Ga, 29.7 ppm.Ge, 0.86 ppm.Ir, D.J. Malvin et al., priv. comm., 1983.
Specimen(s): [1980,M.38], 49.1g.

Quartz Mountain 37°12′ N., 116°42′ W.
Nye County, Nevada, U.S.A.
Found 1935
Iron. Octahedrite, medium (1.1mm) (IIIA).
A mass of 4832 grams was found, V.P. Gianella, Pop. Astron., Northfield, Minnesota, 1936, **44**, p.448 [M.A. 6-397]. Mentioned, F.C. Leonard, Pop. Astron., Northfield, Minnesota, 1944, **52**, p.512, Contr. Soc. Res. Meteorites, **3**, p.173 [M.A. 9-301]. Classification and analysis, 7.85 %Ni, 18.6 ppm.Ga, 36 ppm.Ge, 4.2 ppm.Ir, E.R.D. Scott and J.T. Wasson, Geochimica et Cosmochimica Acta, 1976, **40**, p.103. Kamacite shocked to epsilon structure, V.F. Buchwald, Iron Meteorites, Univ. of California, 1975, p.1000.
Main mass, Reno, Univ. Nevada; 1.1kg Tempe, Arizona State Univ.; 57g Washington, U.S. Nat. Mus.; 45g New York, Amer. Mus. Nat. Hist.;
Specimen(s): [1959,1034], 17g. slice, and sawings, 1g

Quatahar v Butsura.

Quebracho v Distrito Quebracho.

Quebrada de la Aguada v Aguada.

Quebrada de Vaca Muerta v Vaca Muerta.

Quedlinburg 51°47′ N., 11°8′ E.
Halle, Germany
Fell 1249, July 26
Doubtful..
A shower of stones is said to have fallen at several places in the district of Ballenstädt and Blankenburg, E.F.F. Chladni, Die Feuer-Meteore, Wien, 1819, p.199, but the evidence of meteoritic nature is not conclusive.

Queensland v Gladstone (iron).

Queen's Mercy 30°7′ S., 28°42′ E.
Matatiele, Griqualand East, Cape Province, South Africa
Fell 1925, April 30, 2000 hrs
Stone. Olivine-bronzite chondrite (H6), veined.
After detonations and appearance of a bright light, three stones (at least) fell; one, measuring about 18 × 12 × 9 in. at Queen's Mercy, about 20 miles from Matatiele, where it was broken up by the natives, a smaller stone of 950 grams, about 15 miles from Matatiele, and a small fractured stone of 315 grams found near the large one; described and partially analysed, G.T. Prior, Min. Mag., 1926, **21**, p.190. Olivine Fa_{18}, B. Mason, Geochimica et Cosmochimica Acta, 1963, **27**, p.1011. Ar-Ar age, 4490 m.y., G. Turner et al., Geochimica et Cosmochimica Acta, 1978 (suppl. 10), p.989.
2135g Chicago, Field Mus. Nat. Hist., of the large stone; Main mass, Pitermaritzburg, Natal Mus., of the 950g stone; 424g Washington, U.S. Nat. Mus.; 336g New York, Amer. Mus. Nat. Hist.; 210g Tempe, Arizona State Univ.;
Specimen(s): [1927,1124], 2519g. from the large stone; [1927, 1125], 171.5g. of the small fractured stone; [1926,180], 46.75g. of the 950g stone; [1925,1038], 43g.; [1926,219], 24g.

Quenggouk 17°46′ N., 95°11′ E.
Bassein district, Burma
Fell 1857, December 27, 0230 hrs
Synonym(s): *Bassein, Kwingauk, Pegu*
Stone. Olivine-bronzite chondrite (H4).
After appearance of fireball, and detonations, three stones of 2291, 1909.5, 1844.5 grams respectively were seen to fall, W. von Haidinger, Sitzungsber. Akad. Wiss. Wien, Math.-naturwiss. Kl., 1861, **42**, p.301, W. von Haidinger, ibid., 1862, **44**, p.637. Analysis, 27.98 % total iron, olivine $Fa_{18.5}$, R. Hutchison et al., Proc. Roy. Soc. London, 1981, **A374**, p.159. Fine structure, pyroxenes, J.R. Ashworth, Proc. Roy. Soc. London, 1981, **A374**, p.179. The place of fall is Kwingauk, 17°49'N., 95°11'E., C.A. Silberrad, Min. Mag., 1932, **23**, p.295.
2.3kg Calcutta, Mus. Geol. Surv. India; 1kg New York, Amer. Mus. Nat. Hist.; 503g Vienna, Naturhist. Mus.; 300g Chicago, Field Mus. Nat. Hist.; 311g Tempe, Arizona State Univ.;
Specimen(s): [33764], 654g.; [1920,336], 64g.

Quesa 39°0′ N., 0°40′ W.
Enguera, Valencia, Spain
Fell 1898, August 1, 2100 hrs
Iron. Octahedrite, medium (0.7mm) (IRANOM).
After appearance of fireball and detonations, a mass of about 10.75kg was found; description, F. Berwerth, Ann. Naturhist. Hofmus. Wien, 1909, **23**, p.318. Analysis, 10.75 %Ni, E. Cohen, Meteoritenkunde, 1905, **3**, p.306. Classification and analysis, 11.04 %Ni, 37.3 ppm.Ga, 101 ppm.Ge, 0.08 ppm.Ir, A. Kracher et al., Geochimica et Cosmochimica Acta, 1980, **44**, p.773. Brief description, V.F. Buchwald, Iron Meteorites, Univ. of California, 1975, p.1001.
10.37kg Vienna, Naturhist. Mus., main mass;

Quillagua v North Chile.

Quincay 46°36′ N., 0°15′ E.
Vienne, France
Fell 1851, summer
Stone. Olivine-hypersthene chondrite (L6).
Nothing appears to have been published on the fall of this stone, S. Meunier, Meteorites, Paris, 1884, p.241. Very doubtful; probably fragments of Vouille., A. Lacroix, Bull. Mus. d'Hist. Nat. Paris, 1927, **33**, p.445. Xenolithic chondrite, with moderately recrystallised clasts containing feldspar, and distinct from Vouille (*q.v.*), R.A. Binns, letter of October 19, 1966 in Min. Dept. BM(NH). Only 31g known in collections.
10g Paris, Mus. d'Hist. Nat.; 11g Chicago, Field Mus. Nat. Hist.; 2.5g Vienna, Naturhist. Mus.;
Specimen(s): [56468], 8g.

Quinn Canyon 38°5′ N., 115°32′ W.
Nye County, Nevada, U.S.A.
Found 1908
Synonym(s): *Tonopah*
Iron. Octahedrite, medium (1.1mm) (IIIA).
A mass of 1450kg was found 90 miles east of Tonopah; it is possibly part of the Nevada meteor of 1894 (February 1, 2200hrs), W.P. Jenney, Am. J. Sci., 1909, **28**, p.431. Description, analysis, 7.33 %Ni, O.C. Farrington, Field Mus. Nat. Hist. Geol. Ser., 1910, **3** (Publ. 145), p.165. Mentioned, F.C. Leonard, Pop. Astron., Northfield, Minnesota, 1944, **52**, p.512, Contr. Soc. Res. Meteorites, **3**, p.173 [M.A. 9-301]. Too weathered to be a recent fall, includes Tonopah, V.F. Buchwald, Iron Meteorites, Univ. of California, 1975, p.1003. Classification, analysis, 8.40 %Ni, 20.9 ppm.Ga, 41.5 ppm.Ge, 0.58 ppm.Ir, E.R.D. Scott et al., Geochimica et Cosmochimica Acta, 1973, **37**, p.1957.
1450kg Chicago, Field Mus. Nat. Hist., entire mass; 44g Washington, U.S. Nat. Mus.;

Quinyambie v Moorabie.

Qutahar Bazar v Butsura.

Qutrixpileo v Allende.

Rabbit Flat 20°22′ S., 130°7′ E. approx.
Northern Territory, Australia
Found 1974
Stone. Olivine-bronzite chondrite (H6).
A single stone of 295g was found by a road gang on the road between The Granites and Rabbit Flat, Meteor. Bull., 1978 (55), Meteoritics, 1978, **13**, p.345.
Thin section, Washington, U.S. Nat. Mus.;

Raco 26°40′ S., 65°27′ W.
Tafi, Tucuman, Argentina
Fell 1957, November 17, at night hrs
Synonym(s): *Tucuman*
Stone. Olivine-bronzite chondrite (H5).
A mass of 5kg fell, M.E. Teruggi, Acta Geol. Lilloana, 1963, **4**, p.53, F.R. Siegel, Acta Geol. Lilloana, 1963, **4**, p.147, Meteor. Bull., 1961 (21). Olivine Fa_{18}, B. Mason, Geochimica et Cosmochimica Acta, 1963, **27**, p.1011. K/Ar age, 4400 m.y., R.R. Gonzalez and M.A. Cabrera, Meteoritics, 1971, **6**, p.159.
133g Washington, U.S. Nat. Mus.; 5g Tempe, Arizona State Univ.; 4g Chicago, Field Mus. Nat. Hist.;
Specimen(s): [1959,538], 11.5g.; [1964,70], 10g.

Radford Furnace v Indian Valley.

Raepur v Sitathali.

Rafrüti 47°0′ N., 7°50′ E.
Emmenthal, canton Bern, Switzerland
Found 1886
Iron. Chemically and structurally anomalous (IRANOM).
A mass of 18.2kg was found in 1886, and possibly fell at the end of 1856, E. von Fellenberg, Centralblatt Min., 1900, p.152. Description, analysis, 9.54 %Ni, E. Cohen, Meteoritenkunde, 1905, **3**, p.80. Too weathered to have fallen in 1856, the mass has been reheated to a temperature below 400C, but the original, cosmically recrystallised, structure largely maintained, V.F. Buchwald, Iron Meteorites, Univ. of California, 1975, p.1004. Classification, analysis, 9.0 %Ni, 0.159 ppm.Ga, 0.055 ppm.Ge, 0.007 ppm.Ir, R. Schaudy et al., Icarus, 1972, **17**, p.174.
17.78kg Bern, Naturhist. Mus.; 22g Washington, U.S. Nat. Mus.; 13g Vienna, Naturhist. Mus.; 8g Chicago, Field Mus. Nat. Hist.;
Specimen(s): [1971,296], 284.2g.; [1924,406], 1.7g. turnings

Ragland 34°46′4″ N., 103°33′ W.
Quay County, New Mexico, U.S.A.
Found 1982, recognized in this year
Stone. Olivine-hypersthene chondrite (L3).
A mass of 12.1kg was found in a wheatfield, G.I Huss, letter of 26 October, 1983 in Min. Dept. BM(NH).
12.1kg Denver, Amer. Meteorite Lab.;

Raguli 45°42′ N., 43°42′ E. approx.
Stavropol district, Federated SSR, USSR, [Рагули]
Found 1972
Stone. Olivine-bronzite chondrite (H3-4).
A single stone of 4.239kg was found near Raguli village, L.G. Kvasha and A.Ya. Skripnik, Метеоритика, 1978, **37**, p.200. Reported, Meteor. Bull., 1978 (55), Meteoritics, 1978, **13**, p.346. Description, analysis, 24.5 % total iron, olivine $Fa_{19.0}$, G.V. Baryshnikova et al., Метеоритика, 1982, **40**, p.10.
4233g Moscow, Acad. Sci.;

Railway v South African Railways.

Raipur v Sitalhali.

Rais v Al Rais.

Rakhbah v ar-Rakhbah.

Rakity 51°48′ N., 79°54′ E. approx.
Mikhailovsky district, Altay region, USSR, [Ракиты]
Found 1971
Stone. Olivine-hypersthene chondrite (L3).
A single mass of 11.4kg was found in a field during harvesting; analysis, 20.85 % total iron, O.A. Kirova et al., Метеоритика, 1975, **34**, p.57. Reported, Meteor. Bull., 1976 (54), Meteoritics, 1976, **11**, p.84.
11.4kg Moscow, Acad. Sci.;

Rakovka 52°59′ N., 37°2′ E.
Tula, Federated SSR, USSR, [Раковка]
Fell 1878, November 20, 1500 hrs
Synonym(s): *Rakowka, Tula*
Stone. Olivine-hypersthene chondrite (L6).
One stone about the size of a man's head fell, P. Grigoriev, Z. Deutsch. Geol. Ges., 1880, **32**, p.417. A mass of about 9kg fell, L.G. Kvasha and A.Ya. Skripnik, Метеоритика, 1978, **37**, p.200. Olivine Fa_{24}, B. Mason, Geochimica et Cosmochimica Acta, 1963, **27**, p.1011.
536g Vienna, Naturhist. Mus.; 409g Moscow, Univ.; 387g Moscow, Acad. Sci.; 357g New York, Amer. Mus. Nat. Hist.; 163g Chicago, Field Mus. Nat. Hist.; 92g Paris, Mus. d'Hist. Nat.;
Specimen(s): [53320], 371g.

Rakowka v Rakovka.

Ramnagar 26°27′ N., 82°54′ E.
Azamgarh district, Uttar Pradesh, India
Fell 1940, December 15, 1430 hrs
Stone. Olivine-hypersthene chondrite (L6).
Two pieces were recovered, of 3392g and 374g, M.A.R. Khan, Hyderabad Acad. Studies, 1950, **12** [M.A. 11-441], P. Chatterjee, letter of 25 March, 1952 in Min. Dept. BM(NH). Olivine Fa_{25}, B. Mason, Geochimica et Cosmochimica Acta, 1967, **31**, p.1100.
3.7kg Calcutta, Mus. Geol. Surv. India; 0.8g New York, Amer. Mus. Nat. Hist.;
Specimen(s): [1983,M.15], 50.7g.

Ramos v Bocas.

Rampurhat 24°10′ N., 87°46′ E.
Birbhum district, West Bengal, India
Fell 1916, November 21, 0930 hrs
Stone. Olivine-hypersthene chondrite, amphoterite? (LL?).
After a roaring noise, a stone of 100g was seen to fall at Bouripara, Rampurhat town. Description, H. Walker, Rec. Geol. Surv. India, 1924, **55**, p.136. Probably an amphoterite, olivine Fa_{27}, B. Mason, Geochimica et Cosmochimica Acta, 1967, **31**, p.1100.
99g Calcutta, Mus. Geol. Surv. India; 0.9g New York, Amer. Mus. Nat. Hist.;

Ramsdorf 51°53′ N., 6°56′ E.
Borken, Nordrhein-Westfalen, Germany
Fell 1958, July 26, 1830 hrs
Stone. Olivine-hypersthene chondrite (L6), brecciated.
A mass of 4682g was found, description, R. Mosebach, Natur u. Volk, 1958, **88**, p.329, K. Keil, Fortschr. Min., 1960, **38**, p.258. Has been thermally metamorphosed and mechanically deformed, F. Begemann and F. Wlotzka, Geochimica et Cosmochimica Acta, 1969, **33**, p.1351. Olivine Fa_{25}, B. Mason, Geochimica et Cosmochimica Acta, 1963, **27**, p.1011. Some chondrules and metallic globules formed by shock-melting, F. Wlotzka, Meteorite Research, ed. P.M. Millman, D. Reidel, Dordrecht-Holland, 1969, p.174.
468g Vienna, Naturhist. Mus.; 108g Tempe, Arizona State Univ.; 93g Ottawa, Mus. Geol. Surv. Canada; 47g Tübingen, Univ.; 37g Chicago, Field Mus. Nat. Hist.; 25g Yale Univ.; 20g Washington, U.S. Nat. Mus.;
Specimen(s): [1965,225], 29.5g.

Rana v Dokachi.

Ranchapur 23°59′ N., 87°5′ E.
Jamtara Santhal, Bihar, India
Fell 1917, February 20, 0830 hrs
Synonym(s): *Deoli*
Stone. Olivine-bronzite chondrite (H4).
After detonations, stones fell in the villages of Ranchapur and Deoli. Three fragments of 149.8g, 140.3g, and 0.3g respectively, were obtained from villagers of Ranchapur, and one of 77.8g from the villagers of Deoli. Description, H. Walker, Rec. Geol. Surv. India, 1924, **55**, p.137. Olivine Fa_{19}, B. Mason, Geochimica et Cosmochimica Acta, 1963, **27**, p.1011.
191g Calcutta, Mus. Geol. Surv. India; 59g Washington, U.S. Nat. Mus.; 39g Tempe, Arizona State Univ.;
Specimen(s): [1925,441], 78.5g. nearly complete stone

Ranchito v Bacubirito.

Rancho de la Pila (1834) v Zacatecas (1792).

Rancho de la Pila (1882) 24°7′ N., 104°18′ W.
Durango, Mexico
Found 1882
Synonym(s): *Durango (of P. Partsch), Guadalupe, Karavinsky Iron, Pila*
Iron. Octahedrite, medium (1.0mm) (IIIA).
A mass of 46.5kg was ploughed up at Rancho de la Pila, 9 leagues east of Durango. Report and analysis, 8.35 %Ni, L. Häpke, Abhand. Naturwiss. Ver. Bremen, 1884, **8**, p.513, Abhand. Naturwiss. Ver. Bremen, 1886, **9**, p.358. The fragment from a mass in the plain NE. of Durango acquired in 1834 by Karavinsky for the Vienna collection has been paired with this find, L. Fletcher, Min. Mag., 1890, **9**, p.153, P. Partsch, Die Meteoriten, Wien, 1843, p.113. Chemically distinct from Cacaria (*q.v.*), analysis, 7.93 %Ni, 20.8 ppm.Ga, 42.4 ppm.Ge, 0.70 ppm.Ir, E.R.D. Scott et al., Geochimica et Cosmochimica Acta, 1973, **37**, p.1957. Co-ordinates; problems of mis-identification; description, shock-hardened, V.F. Buchwald, Iron Meteorites, Univ. of California, 1975, p.1006.
Specimen(s): [55253], 44.22kg. main mass, and pieces, 1298g and filings, 74g.

Rancho de la Presa
19°52′ N., 100°49′ W. approx.
Zenapecuaro, Michoacan, Mexico
Fell 1899
Stone. Olivine-bronzite chondrite (H5), brecciated.
An observed fall, original weight 300g, H.H. Nininger, Pop. Astron., Northfield, Minnesota, 1942, **50**, p.388, Contr. Soc. Res. Meteorites, **3**, p.37 [M.A. 8-376]. Listed, O.C. Farrington, Cat. Meteor. Field Mus. Nat. Hist. Chicago, 1916, p.291. Olivine Fa_{19}, B. Mason, Geochimica et Cosmochimica Acta, 1963, **27**, p.1011.
Main mass, Morelia, Michoacan, State Mus.; 63g Mexico City, Inst. Geol.; 36.4g New York, Amer. Mus. Nat. Hist.; 30g Harvard Univ.; 17g Tempe, Arizona State Univ.; 12g Washington, U.S. Nat. Mus.; 3g Chicago, Field Mus. Nat. Hist.;
Specimen(s): [1959,822], 16g.

Rancho Gomelia 24°31′ N., 105°15′ W.
Durango, Mexico
Found 1975
Synonym(s): *Poscente*
Iron. Octahedrite, medium (0.8mm) (IIIB).
Two masses, totalling 15.65kg, were recovered and are in Tempe, (Arizona State Univ.), Meteor. Bull., 1980 (58), Meteoritics, 1980, **15**, p.238. Classification and analysis, 9.7 %Ni, 16.4 ppm.Ga, 28.8 ppm.Ge, 0.013 ppm.Ir, A. Kracher et al., Geochimica et Cosmochimica Acta, 1980, **44**, p.773.
11.65kg Tempe, Arizona State Univ.;

Rancome v Ransom.

Randolf County v Guilford County.

Rangala 25°23′ N., 72°1′ E.
Jodhpur, Rajasthan, India
Fell 1937, December 29, 1000 hrs
Stone. Olivine-hypersthene chondrite (L6), veined.
Description, with an analysis, J.A. Dunn, Rec. Geol. Surv. India, 1939, **34**, p.260. 22 fragments totalling 3224.5g were recovered. Olivine Fa_{25}, B. Mason, Geochimica et Cosmochimica Acta, 1963, **27**, p.1011.
2.7kg Calcutta, Mus. Geol. Surv. India; 29g Tempe, Arizona State Univ.; 26g Washington, U.S. Nat. Mus.;
Specimen(s): [1951,330], 80g.

Ranjitpura v Patwar.

Ransom 38°37′ N., 99°56′ W.
Ness County, Kansas, U.S.A.
Found 1938, recognized in this year
Synonym(s): *Rancome*
Stone. Olivine-bronzite chondrite (H4).
Four stones totalling 15kg were found in NW. 1/4, sect. 25, township 15, range 24, A.D. Nininger, Pop. Astron., Northfield, Minnesota, 1939, **47**, p.213. Listed, H.H. and A.D. Nininger, The Nininger Collection of Meteorites, Winslow, Arizona, 1950, p.88. Additional stones have been found; full, illustrated description, W.F. Read, Meteoritics, 1972, **7**, p.509. Olivine Fa_{19}, B. Mason, Geochimica et Cosmochimica Acta, 1963, **27**, p.1011.
6.4kg Tempe, Arizona State Univ.; 1.82kg Fort Hays, Fort Hays Kansas State College Mus.; 400g Albuquerque, Univ. of New Mexico; 349g Los Angeles, Univ. of California; 343g Chicago, Field Mus. Nat. Hist.; 317g Perth, West. Austr. Mus.; 295g Washington, U.S. Nat. Mus.; 85g New York, Amer. Mus. Nat. Hist.;
Specimen(s): [1949,181], 71.5g.; [1959,825], 3117.5g. an almost complete stone

Rasgata v Santa Rosa.

Rasgrad 43°30′ N., 26°32′ E.
Bulgaria
Fell 1740, October 25, 1200 hrs
Synonym(s): *Razgrad*
Two stones, of 49.5lb and 5lb, fell at Hasargrad (= Rasgrad = Hesargrad), now in Bulgaria, J. von Hammer, Ann. Phys. (Gilbert), 1815, **50**, p.284., N. Bonev, Метеоритика, 1958, **16**, p.143 [M.A. 14-126].

Ras Tanura 26°40′ N., 50°9′ E.
Dakhran, Saudi Arabia
Fell 1961, February 23, 1142 hrs, U.T.
Synonym(s): *Ra's at Tannurah*
Stone. Olivine-bronzite chondrite (H6).
One stone of 6.1g was found immediately after a notable bolide, Meteor. Bull., 1961 (21). Brief description, olivine $Fa_{19.0}$, B. Mason, Smithson. Contrib. Earth Sci., 1977 (19), p.83.
5.9g Washington, U.S. Nat. Mus., main mass;

Ra's at Tannurah v Ras Tanura.

Rateldraai 28°50′ S., 21°8′ E.
Kenhardt County, Cape Province, South Africa
Found 1909
Synonym(s): *Prieska, Ratteldraai*
Iron. Octahedrite, medium (0.9mm) (IIIA).
A mass of 1210lb was found, L. Peringuey, letter of 7 January, 1916 in Min. Dept. BM(NH). Classification and analysis, 7.28 %Ni, 18.5 ppm.Ga, 32.5 ppm.Ge, 12 ppm.Ir, E.R.D. Scott et al., Geochimica et Cosmochimica Acta, 1973, **37**, p.1957. Description; heavily weathered, shock-hardened, V.F. Buchwald, Iron Meteorites, Univ. of California, 1975, p.1009, 1413.
Main mass, Cape Town, South African Mus.; 2.72kg Kimberley, Alexander MacGregor Memorial Mus.; 854g Vienna, Naturhist. Mus.;
Specimen(s): [1916,61], 226g.

Ratlam v Pulsora.

Ratteldraai v Rateldraai.

Ratun v Ratyn.

Ratyn 52°12′ N., 17°59′ E.
Kalisz province, Poland
Fell 1880, August 24, between 1400 and 1500 hrs
Synonym(s): *Ratun*
Stone.
A stone of 2lb fell at Ratyn, between Kolin and Golina and made a hole a foot deep; it has since been lost, J. Pokrzywnicki, Urania, 1955, **26**, p.165 [M.A. 13-79], J. Pokrzywnicki, Acta Geol. Polon., 1955, **3**, p.427, J. Pokrzywnicki, Acta Geophys. Polon., 1956, **4**, p.21.

Ravensthorpe v Oldfield River.

Ravni Njive v Zavid.

Ravni Zavid v Zavid.

Rawlinna (iron) v Haig.

Rawlinna (pallasite) 31°10′ S., 125°16′ E.
Nullarbor Plain, Western Australia, Australia
Found 1959, before this year
Stony-iron. Pallasite (PAL).
A single mass of 74g was found 12 miles SSW. of Rawalinna station on the Trans-Australian railway, Meteor. Bull., 1976 (54), Meteoritics, 1976, **11**, p.85. Listed, G.J.H. McCall, First Suppl. to West. Austr. Mus. Spec. Publ. no. 3, 1968. Olivine $Fa_{16.5}$, P.R. Buseck and J.I. Goldstein, Bull. Geol. Soc. Amer., 1969, **80**, p.2141. An 'anomalous' pallasite; reference to analysis of metal, 13.6 %Ni, E.R.D. Scott, Geochimica et Cosmochimica Acta, 1977, **41**, p.693.
39g Mainz, Max-Planck-Inst.; 13g Tempe, Arizona State Univ.;

Rawlinna (stone) 30°22′ S., 126°5′ E.
Nullarbor Plain, Western Australia, Australia
Found 1952
Stone. Olivine-bronzite chondrite (H5).
Two stones were found; the co-ordinates are those of the find-site of the second, G.J.H. McCall and W.H. Cleverly, J. Roy. Soc. West. Austr., 1970, **53**, p.69. Olivine $Fa_{19.4}$, B. Mason, Rec. Austr. Mus., 1974, **29**, p.169. Mentioned, L. Greenland and J.F. Lovering, Geochimica et Cosmochimica Acta, 1965, **29**, p.843.
136g Tempe, Arizona State Univ.; 107g Perth, West. Austr. Mus.;

Razgrad v Rasgrad.

Reager 39°47′ N., 100°0′ W.
Norton County, Kansas, U.S.A.
Found 1948
Stone. Olivine-hypersthene chondrite (L6).
A stone of 230g was found, B. Mason, Meteorites, Wiley, 1962, p.234. Olivine Fa_{25}, B. Mason, Geochimica et Cosmochimica Acta, 1963, **27**, p.1011.
37g Washington, U.S. Nat. Mus.; 1.3g Los Angeles, Univ. of California;

Rechki v Ryechki.

Reckling Peak A78001 76°16′ S., 159°15′ E.
50 km north of Allan Hills, Antarctica
Found 1978, between December 1978 and January 1979
Synonym(s): *Reckling Peak A78003, Reckling Peak A80202*
Stone. Olivine-hypersthene chondrite (L6).
A single mass of 234.9g was found. Paired with Reckling Peak A78003 (1276g). Petrographic description, contains maskelynite, olivine Fa_{23}, Antarctic Meteorite Newsletter, 1981, **4**, p.110. This is one of four specimens, see following entries, found in the Reckling Peak area during the 1978-1979 field season in Antarctica. Co-ordinates, J.O. Annexstad, letter of 23 May, 1980 in Min. Dept. BM(NH).

Reckling Peak A78002
Stone. Olivine-bronzite chondrite (H4).
A stone of 8.48kg was found, olivine $Fa_{18.5}$, Antarctic Meteorite Newsletter, 1981, **4**, p.111.

Reckling Peak A78003 v Reckling Peak A78001.

Reckling Peak A78004
Stone. Olivine-bronzite chondrite (H4).
A stone of 166g was found; olivine Fa_{17}, Antarctic Meteorite Newsletter, 1981, **4**, p.112.

Reckling Peak A79001 76°16′ S., 159°15′ E.
Near Allan Hills, Antarctica
Found 1979, between December 1979 and January 1980
Stone. Olivine-hypersthene chondrite (L6).
A mass of 3006g was found. Possibly paired with Reckling Peak A78001 and Reckling Peak A78003 (*q.v.*). This mass is one of about 15 found in the Reckling Peak area during the 1979-1980 field season in Antarctica and named Reckling Peak A79001 to Reckling Peak A79015, Antarctic Meteorite Newsletter, 1981, **4**, p.139. Masses from this collection included here are those of over 70g in weight.

Reckling Peak A79002
Stone. Olivine-hypersthene chondrite (L6).
A mass of 203.6g was found. Possibly paired with Reckling Peak A79001, olivine Fa_{24}.

Reckling Peak A79003
Stone. Olivine-bronzite chondrite (H6).
A mass of 182g was found, olivine Fa_{18}.

Reckling Peak A79004
Stone. Olivine-bronzite chondrite (H5).
A mass of 370g was found, olivine Fa_{18}.

Reckling Peak A79008
Stone. Olivine-hypersthene chondrite (L3).
A mass of 73g was found, olivine Fa_{1-29}.

Reckling Peak A79014
Stone. Olivine-bronzite chondrite (H5).
A mass of 77.7g was found, olivine Fa_{18}.

Reckling Peak A79015
Synonym(s): *Reckling Peak A80229, Reckling Peak A80246, Reckling Peak A80258, Reckling Peak A80263*
Stony-iron. Mesosiderite (MESANOM).
A mass of 10.02kg was found. The pyroxene is estimated to be Wo2En73Fs25. Further small masses were found by the 1980-1981 expedition, Antarctic Meteorite Newsletter, 1982, **5** (1). Description, mineralogy, R.S. Clarke, Jr. and B. Mason, 7th Symposium on Antarctic Meteorites, Tokyo, 1982, p.77. Analysis of metal, 10.0 %Ni, 12.9 ppm.Ga, 42.9 ppm.Ge, 0.51 ppm.Ir, D.J. Malvin et al., priv. comm., 1983.

Reckling Peak A80201
76°16′ S., 159°15′ E. approx.
Near Reckling Peak, Antarctica
Found 1980, between December 1980 and January 1981
Stone. Olivine-bronzite chondrite (H6).
A mass of 813g was found, olivine Fa_{19}. This is one of the specimens found in the Reckling Peak area by an American party during the 1980-1980 field season. The material is named Reckling Peak A80201 to Reckling Peak A80268. Most are ordinary chondrites of less than 100g. Reported here are those ordinary chondrites of over 200g in weight and all other types found. Full listing, description, Antarctic Meteorite Newsletter, 1982, **5** (no. 1).

Reckling Peak A80202 v Reckling Peak A78001.

Reckling Peak A80203 v Reckling Peak A80231.

Reckling Peak A80204
Stone. Achondrite, Ca-rich. Eucrite (AEUC).
A mass of 15.4g was found, plagioclase An_{92}.

Reckling Peak A80206 v Reckling Peak A80231.

Reckling Peak A80208 v Reckling Peak A80231.

Reckling Peak A80211 v Reckling Peak A80231.

Reckling Peak A80213 v Reckling Peak A80231.

Reckling Peak A80214 v Reckling Peak A80231.

Reckling Peak A80221 v Reckling Peak A80231.

Reckling Peak A80224
Stone. Achondrite, Ca-rich. Eucrite (AEUC), unbrecciated.
A mass of 8.0g was found, plagioclase An_{89}.

Reckling Peak A80226
Iron. Octahedrite, medium (1.2mm) (IA).
A single mass of 160g was found. Classification, analysis, 8.4 %Ni, 68.4 ppm.Ga, 255 ppm.Ge, 2.1 ppm.Ir, D.J. Malvin et al., priv. comm., 1983.

Reckling Peak A80229 v Reckling Peak A79015.

Reckling Peak A80231
Synonym(s): *Reckling Peak A80203, Reckling Peak A80206, Reckling Peak A80208, Reckling Peak A80211, Reckling Peak A80213, Reckling Peak A80214, Reckling Peak A80221, Reckling Peak A80254, Reckling Peak A80255, Reckling Peak A80262, Reckling Peak A80265, Reckling Peak A80266*
Stone. Olivine-bronzite chondrite (H6).
A mass of 238.1g was found. Synonymous masses of over 30g in weight are Reckling Peak A80206, 46.6g; Reckling Peak A80221, 51.9g; Reckling Peak A80254, 68.5g; Reckling Peak A80262, 32.1g, olivine Fa_{18}, Antarctic Meteorite Newsletter, 1982, **5** (no. 1).

Reckling Peak A80235
Stone. Olivine-hypersthene chondrite, amphoterite (LL6).
A mass of 261g was found, olivine Fa_{30}.

Reckling Peak A80239
Stone. Achondrite, Ca-poor. Ureilite (AURE).
A mass of 5.6g was found, olivine Fa_{16}.

Reckling Peak A80241
Stone. Carbonaceous chondrite, type III (CV3).
A fragment weighing 0.6g was found, olivine $Fa_{0.7-36}$.

Reckling Peak A80246 v Reckling Peak A79015.

Reckling Peak A80254 v Reckling Peak A80231.

Reckling Peak A80255 v Reckling Peak A80231.

Reckling Peak A80258 v Reckling Peak A79015.

Reckling Peak A80259
Stone. Enstatite chondrite (E5).
A mass of 20.2g was found.

Reckling Peak A80262 v Reckling Peak A80231.

Reckling Peak A80263 v Reckling Peak A79015.

Reckling Peak A80265 v Reckling Peak A80231.

Reckling Peak A80266 v Reckling Peak A80231.

Red Deer Hill 53°4′30″ N., 105°50′30″ W.
Prince Albert, Saskatchewan, Canada
Found 1975, May
Stone. Olivine-hypersthene chondrite (L6).
One mass of 1.06kg was found 2.5km north of Red Deer Hill; a second mass, 1.45kg, was found two weeks later about 0.8 km north of the find-site of the first. May be identical to Blaine Lake (*q.v.*), olivine Fa_{26}, Meteor. Bull., 1978 (55), Meteoritics, 1978, **13**, p.346.
11g Ottawa, Geol. Surv. Canada;

Redfields 30°43′ S., 116°30′ E.
Wongan Hills district, Western Australia, Australia
Found 1969, recognized in this year
Synonym(s): *Redlands, Redlands Farm*
Iron. Structurally and chemically anomalous (IRANOM).
A mass of 8.74kg was found at Redfields Farm, about 11 km east of Gabalong. Full description and analysis, 6.65 %Ni, J.R. de Laeter et al., Min. Mag., 1973, **39**, p.30. Further analysis, 6.91 %Ni, 39.3 ppm.Ga, 95.4 ppm.Ge, 0.82 ppm.Ir,

J.T. Wasson, Meteorites, Springer-Verlag, 1974, p.307. Description, complex cooling history, V.F. Buchwald, Iron Meteorites, Univ. of California, 1975, p.1413.

Main mass, Perth, West. Austr. Mus.; 668g In possession of finder, Mrs. Bennett; 580g Perth, West. Austr. Inst. Technol.; 120g Canberra, Austr. Nat. Mus.; 48g Chicago, Field Mus. Nat. Hist.; 23g Vienna, Naturhist. Mus.;

Specimen(s): [1973,M.10], 33.6g.

Redlands v Redfields.

Redlands Farm v Redfields.

Red River 32° N., 95° W. approx.

Texas, U.S.A.

Found 1808

Synonym(s): *Brazos, Cross Timbers, Gibb's Meteorite, Louisiana, Texas*

Iron. Octahedrite, medium (1.0mm) (IIIA).

A mass of about 1635lb and two smaller ones were found by Pawnee Indians about 1808, north by northwest of Natchitoches on the Red River "in lat. 32°7'N., and long. 95°10'W."; the large mass was brought to New York in 1810, C.H., Am. J. Sci., 1824, **8**, p.218. Description, analysis, 8.46 %Ni, B. Silliman Jr. and T.S. Hunt, Am. J. Sci., 1846, **2**, p.370. The place of find cannot be accurately placed; it is believed to be in Smith County, though Johnson County, (32.25°N., 97.25°W.) is also suggested it is certainly a long way from the Red River. Suggested location 33°30'N., 99°30'W.; description, shock-hardened, V.F. Buchwald, Iron Meteorites, Univ. of California, 1975, p.1010. Further analysis, 7.70 %Ni, 19.7 ppm.Ga, 38.5 ppm.Ge, 4.4 ppm.Ir, E.R.D. Scott et al., Geochimica et Cosmochimica Acta, 1973, **37**, p.1957.

784.8kg Yale Univ., main mass; 1.76kg Harvard Univ.; 1.5kg Bonn, Univ.; 1.26kg Vienna, Naturhist. Mus.; 559g Calcutta, Mus. Geol. Surv. India; 615g Tempe, Arizona State Univ.; 247g New York, Amer. Mus. Nat. Hist.; 193g Los Angeles, Univ. of California; 128g Berlin, Humboldt Univ.; 108g Chicago, Field Mus. Nat. Hist.; 82g Tübingen, Univ.; 79g Washington, U.S. Nat. Mus.; 80g Tempe, Arizona State Univ.;

Specimen(s): [24002], 424g.; [46974], 83g.; [34675], 0.5g.

Red River v Wichita County.

Red River County v Deport.

Red Rock 35°25′ N., 117°55′ W.

Near Red Rock Canyon, Kern County, California, U.S.A.

Found 1976, approx.

Iron. Octahedrite, medium (1.1mm) (IIIA).

A single mass of 47.6kg was found and purchased by the University of California, Los Angeles, Meteor. Bull., 1984 (62), Meteoritics, 1984, **19**. Classification, analysis, 7.71 % Ni, 21.6 ppm.Ga, 41.8 ppm.Ge, 2.1 ppm.Ir, D.J. Malvin et al., priv. comm., 1983.

47.6kg Los Angeles, Univ. of California;

Red Willow 40°15′ N., 100°30′ W. approx.

Red Willow County, Nebraska, U.S.A.

Found 1899, approx.

Iron. Octahedrite, medium.

A mass of 2.75kg was found, E.H. Barbour, Rep. Geol. Surv. Nebraska, Lincoln, 1903, **1**, p.184. The present repository of this mass is not known.

Reed City 43°52′ N., 85°31′ W.

Osceola County, Michigan, U.S.A.

Found 1895

Iron. Octahedrite, coarse (1.8mm) (IRANOM).

A mass of about 44lb was ploughed up. Description, with an analysis, 8.18 %Ni, H.L. Preston, Proc. Rochester Acad. Sci., 1903, **4**, p.89. Chemically anomalous, another analysis, 7.35 %Ni, 22.5 ppm.Ga, 55.5 ppm.Ge, 54 ppm.Ir, E.R.D. Scott et al., Geochimica et Cosmochimica Acta, 1973, **37**, p.1957. Not artificially reheated, partially shock-melted, V.F. Buchwald, Iron Meteorites, Univ. of California, 1975, p.1012.

10kg Michigan State Univ., approx. weight; 3kg Chicago, Field Mus. Nat. Hist.; 1.8kg Washington, U.S. Nat. Mus.; 862g Ann Arbor, Michigan Univ.; 600g Ottawa, Mus. Geol. Surv. Canada; 490g Harvard Univ.; 449g Vienna, Naturhist. Mus.; 494g Tempe, Arizona State Univ.; 169g Berlin, Humboldt Univ.; 295g New York, Amer. Mus. Nat. Hist.;

Specimen(s): [86426], 810g. slice, and pieces, 49g

Reichstadt v Ploschkovitz.

Reid 30°11′ S., 128°41′ E.

Nullarbor Plain, Western Australia, Australia

Found 1969

Stone. Olivine-bronzite chondrite (H4).

Four weathered and fractured stones were found, of total weight 144.1g. The largest piece, 90.3g was found, at the given co-ordinates, 48 miles NNE. of Reid railway station; the remainder 10 miles SW. of the initial find, G.J.H. McCall, 2nd. Suppl. to West. Austr. Mus. Spec. Publ. no. 3, 1972, p.25. Classification as H5, olivine $Fa_{18.4}$, B. Mason, Rec. Austr. Mus., 1974, **29**, p.169.

Main mass, Kalgoorlie, West. Austr. School of Mines; 4g Washington, U.S. Nat. Mus.; Thin section, Perth, West. Austr. Mus.;

Reliegos 42°28′30″ N., 5°20′ W.

Leon, Spain

Fell 1947, December 28, 0800 hrs

Stone. Olivine-hypersthene chondrite (L5).

One mass of 17.3kg fell, J.G. de Llarena, letters of 12 December, 1948 and 14 February, 1949 in Min. Dept. BM (NH). Description, with an analysis, J.G. de Llarena and C.R. Arango, Bol. R. Soc. Españ. Hist. Nat. Madrid, 1950, **48**, p.303 [M.A. 11-439]. The high Ni and Co figures are doubtful. Olivine Fa_{23}, B. Mason, Geochimica et Cosmochimica Acta, 1963, **27**, p.1011.

279g New York, Amer. Mus. Nat. Hist.; 61g Mainz, Max-Planck-Inst.; 33g Tempe, Arizona State Univ.; 26g Washington, U.S. Nat. Mus.;

Specimen(s): [1949,76], 415g. and fragments, 1g; [1949,24], 12g. fragments

Rembang 6°44′ S., 111°22′ E.

Java, Indonesia

Fell 1919, August 30

Iron. Octahedrite, fine (IVA).

One mass of 10kg fell, R. Bedford, Nature, 1938, **142**, p.1161 [M.A. 7-269], A.D. Nininger, Pop. Astron., Northfield, Minnesota, 1940, **48**, p.559, A.D. Nininger, Contr. Soc. Res. Meteorites, **2**, p.227 [M.A. 8-54]. Analysis, 8.82 %Ni, 2.25 ppm.Ga, 0.134 ppm.Ge, 1.2 ppm.Ir, E.R.D. Scott and J.T. Wasson, Geochimica et Cosmochimica Acta, 1976, **40**, p.103.

275g Canberra, Austr. Nat. Univ.;

Specimen(s): [1983,M.42], 45g.

Renazzo — 44°46′ N., 11°17′ E.
Ferrara, Emilia, Italy
Fell 1824, January 15, 2030 hrs
Synonym(s): *Arenazzo, Cento, Ferrara*
Stone. Carbonaceous chondrite, type II (CM2).
After the appearance of light, followed by three detonations, several stones fell, of which three were recovered, the largest weighing about 5kg, total 10kg, E.F.F. Chladni, Ann. Phys. Chem. (Poggendorff), 1825, **5**, p.122. Mentioned, B. Baldanza, Min. Mag., 1965, **35**, p.214. Analysis, 24.93 % total iron, B. Mason and H.B. Wiik, Am. Mus. Novit., 1962 (2106) [M.A. 16-639]. Co-ordinates, G.R. Levi-Donati, priv. comm.. Description and analysis of metal grains, J.A. Wood, Icarus, 1967, **6**, p.1. Oxygen isotopic ratios, R.N. Clayton and T.K. Mayeda, Meteoritics, 1977, **12**, p.199, abs. High D/H ratio in organic fraction, Y. Kolodny et al., Earth planet. Sci. Lett., 1980, **46**, p.149.
441g Bologna, Univ.; 100g Vienna, Naturhist. Mus.; 81g Paris, Mus. d'Hist. Nat.; 77g Florence, Univ.; 26g Washington, U.S. Nat. Mus.; 18g Tempe, Arizona State Univ.; 11.3g Harvard Univ.; 13.8g New York, Amer. Mus. Nat. Hist.; 6.7g Tübingen, Univ.; 6.8g Modena, Univ.; 3.6g Chicago, Field Mus. Nat. Hist.;
Specimen(s): [41105], 12g.

Renca — 32°45′ S., 65°17′ W.
San Luis, Argentina
Fell 1925, June 20, 1500 hrs
Stone. Olivine-hypersthene chondrite (L5).
Several fragments fell over an area of several square km; total weight 300g. Description, with an analysis (doubtfully high alkalies reported), E.H. Ducloux, Rev. fac. quim. farm. Univ. nac. La Plata, 1928, **5**, p.111 [M.A. 4-120]. Olivine Fa_{24}, B. Mason, Geochimica et Cosmochimica Acta, 1963, **27**, p.1011.
Main mass, Buenos Aires, Mus. Nac. Hist. Nat.; 5.2g New York, Amer. Mus. Nat. Hist.;
Specimen(s): [1962,179], 7g.

Renqiu — 38°40′ N., 116°8′ E.
Hebei, China
Fell 1916, March 23, 1200 hrs
Stone. Olivine-hypersthene chondrite (L6).
A single stone weighing 300g fell in 1916, it was kept by a family until 1975 when it was recognized as meteoritic. Description, with an analysis, 21.23 % total iron, D. Wang et al., Geochimica, 1977 (4), p.297. Reported, Meteor. Bull., 1978 (55), Meteoritics, 1978, **13**, p.347. Listed, references, weight given as 355g and time of fall as 'noon', D. Bian, Meteoritics, 1981, **16**, p.116.
355g Beijing, Planetarium;

Rensselaer County v Tomhannock Creek.

Repeev Khutor — 48°36′ N., 45°40′ E.
Lower Volga, Astrakhan region, USSR, [Репеев Хутор]
Fell 1933, August 8, 2000 hrs
Synonym(s): *Repeev Khytor, Repeew Chutor*
Iron. Octahedrite, plessitic (IIF).
A mass of 7kg fell at 48°36'N., 45°40'E., near Stalingrad (Tsaritsyn), V.I. Vernadsky, letter of 5 August, 1939 in Min. Dept. BM(NH). See also, I.S. Astapowitsch, Trans. Roy. Astron. Soc. Canada, 1938, **32**, p.195 [M.A. 7-172], A.D. Nininger, Pop. Astron., Northfield, Minnesota, 1937, **45**, p.449 [M.A. 7-62], A.N. Zavaritsky, Метеоритика, 1954, **11**, p.64 [M.A. 13-48]. Chemically anomalous, analysis, 14.3 %Ni, 11.6 ppm.Ga, 193 ppm.Ge, 3.0 ppm. Ir, J.T. Wasson, Meteorites, Springer-Verlag, 1974, p.307. Brief description, V.F. Buchwald, Iron Meteorites, Univ. of California, 1975, p.1014.
12.28kg Moscow, Acad. Sci., main mass;

Repeev Khytor v Repeev Khutor.

Repeew Chutor v Repeev Khutor.

Retchki v Ryechki.

Retschki v Ryechki.

Retscki v Ryechki.

Revelstoke — 51°20′ N., 118°57′ W.
British Columbia, Canada
Fell 1965, March 31, 2147 hrs
Stone. Carbonaceous chondrite, type I (CI).
After a brilliant bolide, a search yielded two small fragments, together weighing about 1g, about 64 km NW. of Revelstoke, Meteor. Bull., 1965 (34). Full description, R.E. Folinsbee et al., Geochimica et Cosmochimica Acta, 1967, **31**, p.1625. Thermomagnetic data, E.E. Larson et al., Earth planet. Sci. Lett., 1974, **21**, p.345.
0.2g Ottawa, Geol. Surv. Canada;

Reventazone v Campo del Cielo.

Rewah State v Sarratola.

Rewari — 28°12′ N., 76°40′ E.
Haryana, India
Fell 1929, July, approx.
Stone. Olivine-hypersthene chondrite (L6).
A single stone weighing 3332g was recovered some time after its fall. Description, with an analysis, 22.40 % total iron, olivine Fa_{23}, S.P. Das Gupta et al., Min. Mag., 1979, **43**, p.423.
1658g Oxford, Univ. Mus.; 397g Calcutta, Geol. Surv. India;
Specimen(s): [1975,M.10], 1276g.

Rhineland — 33°32′ N., 99°37′ W.
Knox County, Texas, U.S.A.
Found 1961
Stone. Olivine-bronzite chondrite (H5).
One stone, 6.24kg, was found during ploughing, Meteor. Bull., 1967 (41), Meteoritics, 1970, **5**, p.95. The stone was loaned to NASA, Manned Spacecraft Center, Houston.

Rhine Valley v Rhine Villa.

Rhine Villa — 34°40′ S., 139°17′ E.
County Sturt, South Australia, Australia
Found 1900
Synonym(s): *Adelaide, Rhine Valley*
Iron. Octahedrite, coarse (1.4mm) (IIIE).
A mass of 3325g was found 50 miles NE. of Adelaide; description, with an analysis, 9.07 %Ni, G.A. Goyder, Trans. Roy. Soc. South Austr., 1901, **25**, p.14. The place of find has since been re-named Cambrai. The main mass was sent to Germany. Co-ordinates, D.H. McColl, letter of 30 March, 1972 in Min. Dept. BM(NH). Classification and analysis, 8.63 %Ni, 18.8 ppm.Ga, 36.3 ppm.Ge, 0.12 ppm.Ir, E.R.D. Scott et al., Geochimica et Cosmochimica Acta, 1973, **37**, p.1957. Description, V.F. Buchwald, Iron

Meteorites, Univ. of California, 1975, p.1016.
Slice, Adelaide, South Austr. Mus.; 333g Vienna, Naturhist. Mus.; 181g Budapest, Nat. Mus.; 124g Chicago, Field Mus. Nat. Hist.; 131g Prague, Nat. Mus.; 118g Copenhagen, Univ. Geol. Mus.; 123g Berlin, Humboldt Univ.; 114g Washington, U.S. Nat. Mus.;
Specimen(s): [85824], 175g. slice, and a fragment, 10g.

Rica Aventura 21°59′ S., 69°37′ W.

Antofagasta, Chile
Found 1910
Iron. Octahedrite, fine (0.27mm) (IVA).
A single mass of 5395g was found. Compositionally distinct from Mantos Blancos (8.89%Ni) and Maria Elena (7.64% Ni), the other Chilean IVA octahedrites, E. Olsen and W. Zeitschel, Meteoritics, 1978, **14**, p.51. Analysis, 9.2 %Ni, 2.29 ppm.Ga, 0.138 ppm.Ge, 0.38 ppm.Ir, A. Kracher et al., Geochimica et Cosmochimica Acta, 1980, **44**, p.773.
5269g Hanau, Zeitschel Colln.; 63g Chicago, Field Mus. Nat. Hist.;

Richa 10° N., 9° E. approx.

Dauda, Nigeria
Found 1960
Iron. Octahedrite, medium (0.5mm) (IID).
A mass of about 1.5kg was found near Dauda mining camp between Bargesh and Richa in the SW. corner of the Jos plateau, R.R.E. Jacobson, letters of 12 September, 1961 and 19 April, 1962 in Min. Dept. BM(NH). The mass has suffered a perculiar oxidative weathering, resulting in the kamacite plates of the structure standing out prominently over the whole surface, V.F. Buchwald, Iron Meteorites, Univ. of California, 1975, p.1018. Analysis, 10.0 %Ni, 78 ppm.Ga, 91 ppm.Ge, 16 ppm.Ir, E.R.D. Scott and J.T. Wasson, Geochimica et Cosmochimica Acta, 1976, **40**, p.103.
Main mass, Kaduna, Geol. Surv. Nigeria;
Specimen(s): [1966,55], 390g. and filings, 10g.

Richardton 46°53′ N., 102°19′ W.

Stark County, North Dakota, U.S.A.
Fell 1918, June 30, 2200 hrs
Stone. Olivine-bronzite chondrite (H5), veined.
After the appearance of a luminous meteor, followed by detonations, several stones fell, totalling 200lb, the largest weighing about 18lb, fell over an area of 9 × 5 miles between Richardton and Mott. Description, with an analysis, T.T. Quirke, J. Geol., 1919, **27**, p.431. Analysis, 29.79 % total iron, B. Mason and H.B. Wiik, Am. Mus. Novit., 1963 (2154) [M.A. 16-639]. Details of orbit, E.A. Fath, Pop. Astron., Northfield, Minnesota, 1945, **53**, p.241 [M.A. 9-301]. Further analysis, 28.97 % total iron, H. von Michaelis et al., Earth planet. Sci. Lett., 1969, **5**, p.387. Another analysis, 25.36 % total iron, A.J. Easton and C.J. Elliott, Meteoritics, 1977, **12**, p.409. Olivine Fa_{19}, B. Mason, Geochimica et Cosmochimica Acta, 1963, **27**, p.1011. U-Pb study, M.C.B. Abranches et al., Earth planet. Sci. Lett., 1980, **46**, p.311. Separated chondrules, ages, N.M. Evensen et al., Earth planet. Sci. Lett., 1979, **42**, p.223.
14.5kg Michigan Univ.; 10.4kg Tempe, Arizona State Univ.; 8.3kg New York, Amer. Mus. Nat. Hist.; 5.1kg Washington, U.S. Nat. Mus.; 3.5kg Chicago, Field Mus. Nat. Hist.; 3.2kg Illinois, Univ.; 2.58kg Harvard Univ.; 1kg Philadelphia, Acad. Nat. Sci.; 1kg Yale Univ.; 389g Los Angeles, Univ. of California;
Specimen(s): [1920,508], 433g. complete stone; [1920,510], 412g. complete stone; [1920,509], 371g. complete stone; [1920,511], 165g. complete stone; [1937,1390], 67.0g. original specimen described by T.T. Quirke; [1937,1658], 124g.; [1959,763], 5324g. complete stone; [1959,764], 3563g. and fragments, 16g

Richland 31°54′ N., 96°24′ W.

Navarro County, Texas, U.S.A.
Found 1951
Iron. Hexahedrite (IIA).
A mass of 30lb was found when an old well was being cleaned out. Listed, F.C. Leonard, Classif. Cat. Meteor., 1956, p.8, 45, 74. Analysis, 5.56 %Ni, E.P. Henderson and O. Monnig, Meteoritics, 1957, **1**, p.459 [M.A. 13-361]. It has been suggested that it is a transported piece of Coahuila, but is chemically distinct. Newer analysis, 5.40 %Ni, 60.6 ppm.Ga, 182 ppm.Ge, 8.2 ppm.Ir, J.T. Wasson, Meteorites, Springer-Verlag, 1974, p.299. Structurally distinct from Coahuila; shock-melted troilite, V.F. Buchwald, Iron Meteorites, Univ. of California, 1975, p.1020.
12kg Dallas, Southern Methodist Univ.; 1kg Washington, U.S. Nat. Mus.;

Richland Springs 31°15′ N., 99°2′ W.

San Saba County, Texas, U.S.A.
Fell 1980, September 20, 2130 hrs
Stone. Chondrite.
After a bright fireball and detonations a search recovered three masses totalling 1.1kg. The largest mass, 0.8kg, was found 18 months after the fall, Meteor. Bull., 1984 (62), Meteoritics, 1984, **19**.
Specimens, Fort Worth, Texas, Mus. Sci. Hist.;

Richmond 37°28′ N., 77°30′ W.

Chesterfield County, Virginia, U.S.A.
Fell 1828, June 4, 0830 hrs
Stone. Olivine-hypersthene chondrite (L5).
After detonations, a stone of about 4lb was seen to fall 7 miles SW. of Richmond, J.H. Cocke, Am. J. Sci., 1829, **15**, p.195. Description, analysis, C.U. Shepard, Am. J. Sci., 1829, **16**, p.191, C.U. Shepard, Am. J. Sci., 1843, **45**, p.102. Further analysis, C. Rammelsberg, Monatsber. Akad. Wiss. Berlin, 1870, p.453. Olivine Fa_{26}, B. Mason, Geochimica et Cosmochimica Acta, 1963, **27**, p.1011.
254g Yale Univ.; 114g Tempe, Arizona State Univ.; 141g Vienna, Naturhist. Mus.; 38.5g Chicago, Field Mus. Nat. Hist.; 24g Berlin, Humboldt Univ.; 23g Tübingen, Univ.; 18g Washington, U.S. Nat. Mus.;
Specimen(s): [24006], 94g.; [35411], 70g.; [15149], 5.5g.

Rich Mountain 33°17′ N., 85°10′ W. approx.

Jackson County, North Carolina, U.S.A.
Fell 1903, June 30, 1400 hrs
Stone. Olivine-hypersthene chondrite (L6), veined.
After the appearance of a fireball, and detonations, a stone (perhaps several) fell, and a fragment of 668g was recovered. Description, with an analysis, G.P. Merrill, Proc. U.S. Nat. Mus., 1907, **32**, p.241. Olivine Fa_{24}, B. Mason, Geochimica et Cosmochimica Acta, 1963, **27**, p.1011.
112g Washington, U.S. Nat. Mus.; 33g New York, Amer. Mus. Nat. Hist.;

Ridgecrest 35°35′ N., 117°34′ W.

San Bernardino County, California, U.S.A.
Found 1958, May
Stone. Chondrite.
One complete stone of 9.7g was found, L.E. Humiston, Meteoritics, 1963, **2**, p.50. Probably from the same fall as Muroc and Muroc Dry Lake, F.C. Leonard, Meteoritics, 1963, **2**, p.52. Co-ordinates, J.T. Wasson, letter of 15 April, 1980 in Min. Dept. BM(NH).

Rieti v Orvinio.

Rifle v Cañon Diablo.

Rincon de Caparrosa v Caparrosa.

Ring Meteorite v Tucson.

Rio Arriba v Glorieta Mountain.

Rio Bunge
Locality not reported
Stone. Olivine-hypersthene chondrite (L).
Listed, with no details other than the source, Nat. Hist. Mus., Buenos Aires, and light noble gas content, J. Zäharinger, Geochimica et Cosmochimica Acta, 1968, **32**, p.209.

Rio Florido v Chupaderos.

Rio Grande do Sul v Santa Barbara.

Rio Loa v North Chile.

Rio Mocoreta 30°37′ S., 57°58′ W.
Corrientes, Argentina
Fell 1844, January
Doubtful..
A large iron, several feet in diameter is said to have fallen, R.P. Greg, Phil. Mag., 1855, **10**, p.14 but the evidence is not conclusive.

Rio Negro 26°6′ S., 49°48′ W.
Parana, Brazil
Fell 1934, September 21, 2030 hrs
Stone. Olivine-hypersthene chondrite (L4).
A stone of 1310g was seen to fall. Description, with a partial analysis, A. Gatterer and J. Junkes, Pontificia Acad. Sci., Comment., Roma, 1940, **4** (6), p.191 [M.A. 10-403]. The report mentions eastern time which suggests that the Rio Negro in question (there are several) is the place with a railway station and post office at 26°6'S., 49°48'W., in Parana State, L.J. Spencer [M.A. 10-403]. Unequilbrated fragment studied, olivine Fa_{25}, R.V. Fodor et al., Rev. Brasil. Geocienc., 1977, **7**, p.45.
517g Rome, Vatican Colln; 66g Washington, U.S. Nat. Mus.; 39g Mainz, Max-Planck-Inst.;
Specimen(s): [1963,791], 50g. slice

Rio San Francisco do Sul v Santa Catharina.

Rittersgrün v Steinbach.

River 30°22′ S., 126°1′ E.
Nullarbor Plain, Western Australia, Australia
Found 1965
Stone. Olivine-hypersthene chondrite (L5).
One complete, flight oriented and crusted stone, weighing 190.5g, was found 15 miles W. of Sleeper Camp (*q.v.*), G.J.H. McCall, First Suppl. to West. Austr. Mus. Spec. Publ. no. 3, 1968, p.16. Figured, G.J.H. McCall and W.H. Cleverly, Min. Mag., 1968, **36**, p.691. Classification, olivine $Fa_{24.7}$, B. Mason, Rec. Austr. Mus., 1974, **29**, p.169.
Main mass, Kalgoorlie, West. Austr. School of Mines; 17g Perth, West. Austr. Mus., and a thin section;

Riverton 50°56′24″ N., 96°59′30″ W.
Riverton, Manitoba, Canada
Found 1960, or 1961, recognized 1968
Stone. Olivine-bronzite chondrite (H5).
One stone of 103.3g was found in a field about 7 km S. of Riverton, olivine $Fa_{20.1}$, Meteor. Bull., 1976 (54), Meteoritics, 1976, **11**, p.86.
22.2g Ottawa, Geol. Surv. Canada; 78.5g Mainz, Max-Planck-Inst.;

Rivolta de Bassi 45°29′ N., 9°31′ E.
Cremona, Lombardy, Italy
Fell 1491, March 22
Stone.
A stone the size of a man's head, light grey with a "burnt" surface, was seen to fall near Crema, E.F.F. Chladni, Die Feuer-Meteore, Wien, 1819, p.204, Phil. Mag., 1826, **67**, p.3, 179. This was clearly a true meteorite. The co-ordinates given are those of Rivolta D'Adda, the modern name of Rivolta de Bassi.

Roa 41°44′ N., 3°56′ W.
Burgos, Spain
Fell 1438
Doubtful. Stone.
Many light, vesicular stones are said to have fallen near Roa, E.F.F. Chladni, Die Feuer-Meteore, Wien, 1819, p.203. Another authority gives the place of fall as Maderuelo, Segovia, (41°29'N., 3°31'W.), M. Faura y Sans, Meteoritos caidos en la Peninsula Iberica, Tortosa, 1922, p.5.

Robertson County v Coopertown.

Robinson Station v Cynthiana.

Roche-Serviere v St. Christophe-la-Chartreuse.

Rochester 41°5′ N., 86°17′ W.
Fulton County, Indiana, U.S.A.
Fell 1876, December 21, 2045 hrs
Synonym(s): *Fulton County*
Stone. Olivine-bronzite chondrite (H6).
After the passage eastward, for over 1000 miles of a cluster of luminous meteors over the United States from Kansas to Ohio, and detonations, a stone of about 12 oz. (340g.) fell 3 miles NW. of Rochester, H.A. Newton, Am. J. Sci., 1877, **13**, p.166, C.U. Shepard, Am. J. Sci., 1877, **13**, p.207, D. Kirkwood, Proc. Amer. Phil. Soc., 1877, **16**, p.592. Description, analysis (doubtful, no K reported), J.L. Smith, Am. J. Sci., 1877, **14**, p.219. Olivine Fa_{20}, B. Mason, Geochimica et Cosmochimica Acta, 1963, **27**, p.1011.
56g Harvard Univ.; 53g Washington, U.S. Nat. Mus.; 11g Berlin, Humboldt Univ.; 14g Yale Univ.; 14g Vienna, Naturhist. Mus.;
Specimen(s): [53289], 8g.; [1920,339], 1.5g.

Rocicky v Brahin.

Rock County v Mariaville.

Rock Creek 34°25′20″ N., 101°29′51″ W.
Swisher County, Texas, U.S.A.
Found 1980
Stone. Olivine-hypersthene chondrite (L4).
A single mass of 1641g was found, Meteor. Bull., 1983 (61), Meteoritics, 1983, **18**, p.82.
Main mass, Chicago, Field Mus. Nat. Hist.; 262g Hanau, Zeitschel Colln.;

Rockhampton 23°23′ S., 150°31′ E.
Port Curtis, Queensland, Australia
Fell 1895, spring, 1600-1700 hrs
Stone.
Three stones, one of 3lb, fell, H. Tryon, Queensland Naturalist, 1910, **1**, p.170, L.J. Spencer, Min. Mag., 1937, **24**, p.451. The specimens appear to have been lost.

Rockingham County v Deep Springs.

Rockingham County v Smith's Mountain.

Rockport v McKinney.

Rockwood v Crab Orchard.

Roda 42°18′ N., 0°33′ E.
Huesca, Spain
Fell 1871, spring
Synonym(s): *Huesca*
Stone. Achondrite, Ca-poor. Diogenite (ADIO).
A stone of about 400g fell. Description, with an analysis, G.A. Daubrée, C. R. Acad. Sci. Paris, 1874, **79**, p.1507. Further description, analysis, A. Lacroix, C. R. Acad. Sci. Paris, 1925, **180**, p.89. Pyroxene determined optically, Fs27, B. Mason, Am. Mus. Novit., 1963 (2155).
80g Paris, Mus. d'Hist. Nat.; 15g Chicago, Field Mus. Nat. Hist.; 12g Vienna, Naturhist. Mus.; 7.5g Rome, Vatican Colln;
Specimen(s): [54664], 7.5g.

Rodach 50°21′ N., 10°48′ E.
Coburg, Bayern, Germany
Fell 1775, September 19, 0900 hrs
Stone.
A stone of 6.5lb, white to grey with a very thin black crust, was seen to fall, but has since been lost, J. Büttner, Ann. Phys. (Gilbert), 1806, **23**, p.93.

Rodeo 25°20′ N., 104°40′ W.
Durango, Mexico
Found 1852
Synonym(s): *El Rodeo*
Iron. Octahedrite, medium (0.6mm) (IID).
A mass of 97lb was found about 7 miles NW. of Rodeo and was used as an anvil for many years; description, O.C. Farrington, Field Mus. Nat. Hist. Geol. Ser., 1905, **3** (Publ. 101), p.1. Description, analysis, 11.27 %Ni, E. Cohen, Meteoritenkunde, 1905, **3**, p.297. Co-ordinates; all of the mass has been heated artificially to 800-900C, V.F. Buchwald, Iron Meteorites, Univ. of California, 1975, p.1022. Further analysis, 10.2 %Ni, 82.1 ppm.Ga, 93.0 ppm.Ge, 8.0 ppm.Ir, J.T. Wasson, Geochimica et Cosmochimica Acta, 1969, **33**, p.859.
28.7kg Chicago, Field Mus. Nat. Hist.; 2kg Washington, U.S. Nat. Mus.; 1.7kg Harvard Univ.; 710g Tempe, Arizona State Univ.; 653g Budapest, Nat. Mus.; 374g Ottawa, Mus. Geol. Surv. Canada; 350g Vienna, Naturhist. Mus.; 319g Prague, Nat. Mus.; 287g Berlin, Humboldt Univ.;
Specimen(s): [1959,974], 768.5g. and sawings, 10.5g; [87036], 409g. slice

Rodnik v Krasnyi Kluch.

Roebourne 22°20′ S., 118° E.
Hamersley Range, Western Australia, Australia
Found 1892
Synonym(s): *Hamersley, Hamersley Range, Hammersley, Hammersley Range*
Iron. Octahedrite, medium (1.1mm) (IIIA).
A mass of 191.5lb was found 200 miles SE. of Roebourne and 8 miles from the Hamersley Range, in lat. 22°20'S., long. 118°E. Report and analysis, 8.33 %Ni, H.A. Ward, Am. J. Sci., 1898, **5**, p.135. Structure, H.J. Axon, Prog. Materials Sci., 1968, **13**, p.185. Analysis, 8.01 %Ni, 21.2 ppm.Ga, 42.4 ppm.Ge, 0.65 ppm.Ir, E.R.D. Scott et al., Geochimica et Cosmochimica Acta, 1973, **37**, p.1957. Description, recrystallised, V.F. Buchwald, Iron Meteorites, Univ. of California, 1975, p.1025.
33.87kg Chicago, Field Mus. Nat. Hist.; 3.2kg Paris, Mus. d'Hist. Nat.; 4kg Washington, U.S. Nat. Mus.; 3.0kg Tempe, Arizona State Univ.; 2.5kg Bonn, Univ.; 2.3kg New York, Amer. Mus. Nat. Hist.; 1.7kg Vienna, Naturhist. Mus.;
Specimen(s): [83314], 960g. four pieces; [1920,340], 93g. slice

Rogers 34°3′54″ N., 103°24′30″ W.
Roosevelt County, New Mexico, U.S.A.
Found 1974
Stone. Olivine-hypersthene chondrite (L).
Listed, total known weight 982g, Cat. Huss Coll. Meteorites, 1976, p.37.
920g Mainz, Max-Planck-Inst.;

Roggia Marcova v Motta di Conti.

Rogue River Mountains v Port Orford.

Rokicky v Brahin.

Rokitskii v Brahin.

Roletta v Motta di Conti.

Rolla (no. 2) v Rolla (1939).

Rolla (no. 3) v Rolla (1941).

Rolla (1936) 37°7′ N., 101°36′ W.
Morton County, Kansas, U.S.A.
Found 1936
Stone. Olivine-bronzite chondrite (H5).
A small stone was found by H.H. Nininger on an Indian camp site near Elkhart, NW. 1/4, sect. 24, township 32, range 40, Morton County. He regarded it as part of the Hugoton fall, E.P. Henderson, letters of 12 May, 3 June, 1939 in Min. Dept. BM(NH). Listed separately from Hugoton, A.D. Nininger, Pop. Astron., Northfield, Minnesota, 1940, **48**, p.556, A.D. Nininger, Contr. Soc. Res. Meteorites, **2**, p.227 [M.A. 8-54]. Nininger has described two further distinct falls from this district, Rolla (1941) and Rolla (1939), H.H. Nininger, Meteoritics, 1970, **5**, p.215. Olivine Fa_{19}, B. Mason, Geochimica et Cosmochimica Acta, 1963, **27**, p.1011.
414g Washington, U.S. Nat. Mus.; 27g Tempe, Arizona State Univ.;

Rolla (1939) 37°7′ N., 101°36′ W.
Morton County, Kansas, U.S.A.
Found 1939
Synonym(s): *Rolla (no. 2)*
Stone. Olivine-bronzite chondrite (H4).
Total known weight 207.5g, H.H. and A.D. Nininger, The Nininger Collection of Meteorites, Winslow, Arizona, 1950, p.132.
153g Tempe, Arizona State Univ.; 10g Washington, U.S. Nat. Mus.;
Specimen(s): [1959,836], 30g.

Rolla (1941) 37°5′ N., 101°36′ W.
Morton County, Kansas, U.S.A.
Found 1941
Synonym(s): *Rolla (no. 3)*
Stone. Olivine-bronzite chondrite (H), black.
Total known weight 200g, H.H. and A.D. Nininger, The Nininger Collection of Meteorites, Winslow, Arizona, 1950, p.132. Further masses were found in 1942 (50g Rolla no. 4) and in 1950 (185g Rolla no. 5).
48g Tempe, Arizona State Univ., Rolla (1941); 45g Tempe, Arizona State Univ., the Rolla no. 4 stone; 174g Tempe, Arizona State Univ., the Rolla no. 5 stone;
Specimen(s): [1949,1032], 68g.

Rom v Orvinio.

Roma v Orvinio.

Rome v Orvinio.

Romero 35°46′ N., 102°57′ W.
Hartley County, Texas, U.S.A.
Found 1938
Stone. Olivine-bronzite chondrite (H4), xenolithic.
One stone of about 17.2kg was found, A.D. Nininger, Pop. Astron., Northfield, Minnesota, 1940, **48**, p.556, A.D. Nininger, Contr. Soc. Res. Meteorites, **2**, p.227 [M.A. 8-54], H.H. and A.D. Nininger, The Nininger Collection of Meteorites, Winslow, Arizona, 1950, p.89. Xenolithic, olivine Fa_{18}, A.L. Graham, Meteoritics, 1981, **16**, p.319, abs.
402g Tempe, Arizona State Univ.;
Specimen(s): [1959,896], 408.5g.

Roosevelt 34°52′ N., 98°57′ W.
Kiowa County, Oklahoma, U.S.A.
Found 1972, recognized in this year
Stone. Olivine-bronzite chondrite (H3).
Total weight 5.2kg, G.I Huss, letter of 6 December, 1974 in Min. Dept. BM(NH).
3.7kg Mainz, Max-Planck-Inst.; 138g Copenhagen, Univ. Geol. Mus.; 59g Tempe, Arizona State Univ.;
Specimen(s): [1975,M.7], 16g. part-slice

Roper River 15° S., 135° E.
Northern Territory, Australia
Found 1953, known before this year
Iron. Octahedrite, medium (0.6mm) (IIIB).
A mass of 14lb was found near the Roper River in lat. 15°S., long. 135°E., G.W. Card, Ms. list of Australian meteorites in Min. Dept. BM(NH). Analysis, 9.8 %Ni, 18.1 ppm.Ga, 33.9 ppm.Ge, 0.04 ppm.Ir, E.R.D. Scott et al., Geochimica et Cosmochimica Acta, 1973, **37**, p.1957.
Main mass, Melbourne, Nat. Mus.;

Roquefort v Barbotan.

Rosamond Dry Lake 34°50′ N., 118°4′ W.
Kern County, California, U.S.A.
Found 1940, November 24
Stone. Olivine-hypersthene chondrite (L).
A chondritic stone was found on the surface of Rosamond Dry Lake, very similar to the stones found at Muroc Dry Lake, 15 miles to the NE., W.T. Whitney, Pop. Astron., Northfield, Minnesota, 1941, **49**, p.387 [M.A. 8-195]. Mentioned, F.C. Leonard, Pop. Astron., Northfield, Minnesota, 1947, **55**, p.381, F.C. Leonard, Contr. Meteoritical Soc., **4**, p.58 [M.A. 10-177]. Olivine Fa_{24}, B. Mason, Geochimica et Cosmochimica Acta, 1963, **27**, p.1011.
111g Chicago, Field Mus. Nat. Hist.; 13.7g Los Angeles, Univ. of California; Main mass, Claremont, California, Pomona College;
Specimen(s): [1980,M.10], 24.2g.

Rosario 14°36′ N., 88°41′ W.
Honduras
Found 1896
Synonym(s): *Honduras*
Iron. Octahedrite, coarse (1.7mm) (IA).
In July 1896 a mass of about 4lb was brought to Mr. E. Schernikov at the Rosario mine by a native who had found it at a ranch called Rosario, about 50 miles from the mine, E. Schernikov, priv. comm., 1922. Analysis, 7.16 %Ni, 89.5 ppm.Ga, 401 ppm.Ge, 1.5 ppm.Ir, J.T. Wasson, Meteorites, Springer-Verlag, 1974, p.298. Brief description, V.F. Buchwald, Iron Meteorites, Univ. of California, 1975, p.1030.
1540g New York, Amer. Mus. Nat. Hist.; 450g Chicago, Field Mus. Nat. Hist.; 190g Washington, U.S. Nat. Mus.; 24g Vienna, Naturhist. Mus.;
Specimen(s): [83614], 118g. four pieces

Rosebud 30°49′ N., 97°3′ W.
Falls County, Texas, U.S.A.
Found 1915, known before this year
Synonym(s): *Glen Rose (stone)*
Stone. Olivine-bronzite chondrite (H), black.
A mass of 54.9kg was found in Milam County, 1.5 miles west of Burlington, and 6 miles south of Rosebud, Falls County, and is said to have fallen about 1907. Description, with an analysis, F.M. Bullard, Am. Miner., 1939, **24**, p.242 [M.A. 7-377]. The Glen Rose stone listed, H.H. Nininger, The Mines Mag., Golden, Colorado, 1933, **23** (8), p.6 [M.A. 5-405] was apparently named in mistake for Rosebud, E.P. Henderson, letter of 3 June, 1939 in Min. Dept. BM(NH). Olivine Fa_{19}, B. Mason, Geochimica et Cosmochimica Acta, 1963, **27**, p.1011.
Main mass, Austin, Texas, Univ. of Texas; 11g Tempe, Arizona State Univ.; 5g Washington, U.S. Nat. Mus.;

Rose City 44°31′ N., 83°57′ W.
Ogemaw County, Michigan, U.S.A.
Fell 1921, October 17, 2300 hrs
Stone. Olivine-bronzite chondrite (H5), brecciated, black.
After the appearance of a brilliant meteor moving NNW. to SSE. over the NE. portion of the Lower Peninsular of Michigan, and detonations, three stones of 3.25lb, 7lb and 13lb respectively, fell about 9 miles NE. of Rose City, and were found next day. Description, with an analysis, E.O. Hovey, J. Am. Mus. Nat. Hist., 1923, **23** (1), p.86, Am. Mus. Novit., 1922 (52), p.1. Mentioned, H.H. Nininger, letter of 11 November, 1933 in Min. Dept. BM(NH). Illustrated description and analysis, 36.4 % total iron, B. Mason and H.B. Wiik, Am. Mus. Novit., 1966 (2272).

Mentioned, H.J. Axon, Prog. Materials Sci., 1968, **13**, p.185, V. del Chamberlain, Meteorites of Michigan, Geol. Surv. Bull., 1968 (5). Vapour-grown crystals in vugs, R.M. Fruland, Meteoritics, 1975, **10**, p.403, abs.

5kg New York, Amer. Mus. Nat. Hist.; 3587g Washington, U.S. Nat. Mus.; 1.4kg Chicago, Field Mus. Nat. Hist.; 847g Michigan Univ.; 684g Harvard Univ.; 247g Bloomfield Hills, Michigan, Cranfield Inst. Sci.; 243g Tempe, Arizona State Univ.;

Specimen(s): [1959,768], 224g. half a stone, and pieces, 16g

Ross's Iron v Cape York.

Roswell v Cañon Diablo.

Round Top 30°2′ N., 96°44′ W.

Fayette County, Texas, U.S.A.
Found 1934, recognized 1937
Stone. Chondrite.

One stone of 7.7kg was found, A.D. Nininger, Pop. Astron., Northfield, Minnesota, 1940, **48**, p.557, A.D. Nininger, Contr. Soc. Res. Meteorites, **2**, p.227 [M.A. 8-54]. Perhaps identical with Bluff or Cedar, F.C. Leonard, Univ. New Mexico Publ., Albuquerque, 1946 (meteoritics ser. no. 1), p.47.

Rousoumousky v Agricultural College.

Rowena 29°48′ S., 148°38′ E.

New South Wales, Australia
Found 1962, January
Stone. Olivine-bronzite chondrite (H6).

A weathered stone was found, broken into a large number of pieces when struck by a plough, 17 miles E. of Rowena. 76.5lb were recovered but a reconstruction of the pieces shows that a lot of it is missing, R.O. Chalmers and B. Mason, Rec. Austr. Mus., 1977, **30**, p.519. Reported, Meteor. Bull., 1965 (33). Olivine $Fa_{19.5}$, B. Mason, Rec. Austr. Mus., 1974, **29**, p.169.

Main mass, Sydney, Austr. Mus.; 78g Washington, U.S. Nat. Mus.;

Rowton 52°46′ N., 2°31′ W.

Wellington, Shropshire, England
Fell 1876, April 20, 1545 hrs
Synonym(s): *Shropshire, Wellington*
Iron. Octahedrite, medium (1.2mm) (IIIA).

After detonations had been heard, a mass of 7.75lb was found in a hole 18 inches deep. Description, with an analysis, 8.58 %Ni, W. Flight, Phil. Trans. Roy. Soc. London, 1882, **173**, p.894. Further analysis, 7.79 %Ni, 20.5 ppm.Ga, 38.1 ppm.Ge, 2.8 ppm.Ir, E.R.D. Scott et al., Geochimica et Cosmochimica Acta, 1973, **37**, p.1957. Description, V.F. Buchwald, Iron Meteorites, Univ. of California, 1975, p.1031.

39g Vienna, Naturhist. Mus.; 32g Washington, U.S. Nat. Mus.; 29g Copenhagen, Univ. Geol. Mus.; 17g Harvard Univ.;

Specimen(s): [50062], 3109g. main mass, and pieces, 70.5g

Roy (no. 2) v Roy (1934).

Roy (1933) 35°57′ N., 104°12′ W.

Harding County, New Mexico, U.S.A.
Found 1933, in the spring
Stone. Olivine-hypersthene chondrite (L5).

A weathered stone was found 11 miles E. of Roy, broken up by the finder, 104lb of fragments were recovered. Description, with an analysis, H.H. Nininger, Pop. Astron., Northfield, Minnesota, 1934, **42**, p.599 [M.A. 6-14], R.E.S. Heineman, Am. Miner., 1935, **20**, p.438 [M.A. 6-104]. Another stone of 5.273kg was found in 1939 and is assigned to Roy (1934) (*q.v.*). Olivine 26.1, D.E. Lange and K. Keil, Meteoritics, 1976, **11**, p.315, abs.

8.6kg Washington, U.S. Nat. Mus.; 6.3kg Tempe, Arizona State Univ.; 1.1kg Calcutta, Mus. Geol. Surv. India; 856g Albuquerque, Univ. of New Mexico; 686g Paris, Mus. d'Hist. Nat.; 493g Chicago, Field Mus. Nat. Hist.; 398g Ottawa, Mus. Geol. Surv. Canada;

Specimen(s): [1959,773], 7108g.; [1937,384], 75g.

Roy (1934) 35°57′ N., 104°12′ W.

Harding County, New Mexico, U.S.A.
Found 1934
Synonym(s): *Roy (no. 2), Tequezquito Creek, Tequezyuito Creek*
Stone. Olivine-hypersthene chondrite (L6).

A small stone, 346g, is very distinct from the other fragments of the Roy fall, H.H. and A.D. Nininger, The Nininger Collection of Meteorites, Winslow, Arizona, 1950, p.90, 117. The 5.273kg stone found in 1939, previously assigned to the Roy (1933) find is now assigned to the Roy (1934) find. Olivine Fa_{25}, D.E. Lange and K. Keil, Meteoritics, 1976, **11**, p.315, abs.

4.5kg Washington, U.S. Nat. Mus.; 169g Tempe, Arizona State Univ.;

Specimen(s): [1959,1028], 169g.

Rozana v Ruschany.

Rozanj v Zavid.

Rozany v Ruschany.

Ruff's Mountain 34°18′ N., 81°24′ W.

Lexington County, South Carolina, U.S.A.
Found 1844
Synonym(s): *Lexington County, Newberry*
Iron. Octahedrite, medium (1.2mm) (IIIA).

A mass of 117lb was found, C.U. Shepard, Am. J. Sci., 1850, **10**, p.128. Analysis, 8.55 %Ni, G.P. Merrill, Mem. Nat. Acad. Sci. Washington, 1919, **14** (4), p.10. Description; annealing history; artificial heating, V.F. Buchwald, Iron Meteorites, Univ. of California, 1975, p.1032. Further analysis, 8.56 %Ni, 21.5 ppm.Ga, 46.9 ppm.Ge, 0.47 ppm.Ir, E.R.D. Scott et al., Geochimica et Cosmochimica Acta, 1973, **37**, p.1957.

23.7kg Tempe, Arizona State Univ.; 4.7kg Washington, U.S. Nat. Mus.; 2.3kg Tempe, Arizona State Univ.; 1.2kg Chicago, Field Mus. Nat. Hist.; 600g Vienna, Naturhist. Mus.; 590g Harvard Univ.; 237g Berlin, Humboldt Univ.;

Specimen(s): [90225], 334g.; [25459], 165g.; [1920,341], 73g. slice

Rugao v Min-Fan-Zhun.

Ruhobobo 01°27′ S., 29°50′ E.

Commune Cyeru, Ruhengeri Prefecture, Rwanda
Fell 1976, October 13, 1630 hrs
Stone. Olivine-hypersthene chondrite (L6).

After detonations and a noise resembling that of a jet aeroplane, a single stone, 465.5g fell, Meteor. Bull., 1978 (55), Meteoritics, 1978, **13**, p.348. Full description, unshocked, olivine $Fa_{23.4}$, H. Klob et al., Meteoritics, 1981, **16**, p.1.

Rukao v Min-Fan-Zhun.

Runa Pocito v Campo del Cielo.

Ruponda v Rupota.

Rupota 10°16′ S., 38°46′ E.

Mtwara, Tanzania
Fell 1949, February 7, 1200 hrs
Synonym(s): *Ruponda*
Stone. Olivine-hypersthene chondrite (L4), xeolithic.
One stone of about 6kg fell at Rupota, 4.5 miles E. of Ruponda, after thunderous noises, it penetrated 18 inches, D.R. Grantham, letter of 14 February, 1950, J.R. Harpum, letter of 22 January, 1958 in Min. Dept. BM(NH). Xenoliths, L6, in L4 matrix, analysis, 21.89 % total iron, E. Jarosewich and B. Mason, Geochimica et Cosmochimica Acta, 1969, **33**, p.411. Mentioned, J.R. Harpum, Rec. Geol. Surv. Tanganyika for 1961, 1965, **11**, p.58.
5765g Dodoma, Tanzania, Geol. Surv.; 975g Washington, U.S. Nat. Mus.;

Ruschany 52°53′ N., 24°53′ E.

Slonim, Belorussiya SSR, USSR, [Рушаны]
Fell 1894, December 7
Synonym(s): *Deretschin, Rozana, Rozany, Slonim*
Doubtful..
A large meteorite is said to have fallen, E.A. Wülfing, Die Meteoriten in Samml., Tübingen, 1897, p.404, J. Pokrzywnicki, Acta Geophys. Polon., 1956, **3**, p.62 [M.A. 13-78]. Included as doubtful, "there is reason to think it may be a pseudometeorite" by., E.L. Krinov, Основы Метеоритики, Москва, 1955, p.340 though definitely classed as a pseudometeorite (though without cited evidence) in the translation, E.L. Krinov, Principles of Meteoritics, 1960, p.473.

Rush County 39°30′ N., 85°30′ W. approx.

Indiana, U.S.A.
Found 1948, recognized in this year
Stone. Olivine-bronzite chondrite (H5).
A stone of 4.3kg was found, H.H. and A.D. Nininger, The Nininger Collection of Meteorites, Winslow, Arizona, 1950, p.90. Olivine Fa_{17}, B. Mason, Geochimica et Cosmochimica Acta, 1963, **27**, p.1011.
4kg Washington, U.S. Nat. Mus.; 139g Chicago, Field Mus. Nat. Hist.; 36g Tempe, Arizona State Univ.;
Specimen(s): [1959,897], 35g.

Rush Creek 38°37′ N., 102°43′ W.

Kiowa County, Colorado, U.S.A.
Found 1938
Stone. Olivine-hypersthene chondrite (L6), brecciated.
Three fragments totalling 9.3kg were found in section 1, township 17, range 47 W., Kiowa County, A.D. Nininger, Pop. Astron., Northfield, Minnesota, 1939, **47**, p.213, E.P. Henderson, letter of 3 June, 1939 in Min. Dept. BM(NH). Olivine Fa_{26}, B. Mason, Geochimica et Cosmochimica Acta, 1963, **27**, p.1011.
3.1kg Washington, U.S. Nat. Mus.; 586g Los Angeles, Univ. of California; 680g Denver, Mus. Nat. Hist.; 493g Tempe, Arizona State Univ.; 464g Chicago, Field Mus. Nat. Hist.;
Specimen(s): [1959,1003], 1968g.

Rushville 39°37′ N., 85°27′ W.

Franklin County, Indiana, U.S.A.
Found 1866
Synonym(s): *Brookville*
Stone. Olivine-hypersthene chondrite (L5), brecciated,
Listed, O.C. Farrington, Mem. Nat. Acad. Sci. Washington, 1915, **13**, p.389.
22g New York, Amer. Mus. Nat. Hist.; 17g Chicago, Field Mus. Nat. Hist.; 7g Harvard Univ.; 3g Berlin, Humboldt Univ.;

Russel Gulch 39°48′ N., 105°30′ W.

Gilpin County, Colorado, U.S.A.
Found 1863
Synonym(s): *Colorado, Gilpin County*
Iron. Octahedrite, medium (0.9mm) (IIIA).
A mass of 29lb was found, description, with an analysis, 7.84 %Ni, J.L. Smith, Am. J. Sci., 1866, **42**, p.218. 7.43 %Ni, E. Cohen, Meteoritenkunde, 1905, **3**, p.360. Not artificially reheated but cosmically deformed, V.F. Buchwald, Iron Meteorites, Univ. of California, 1975, p.1035. Classification and analysis, 7.52 %Ni, 19.1 ppm.Ga, 35.6 ppm.Ge, 7.2 ppm.Ir, E.R.D. Scott et al., Geochimica et Cosmochimica Acta, 1973, **37**, p.1957.
4.97kg New York, Amer. Mus. Nat. Hist.; 1.9kg Harvard Univ.; 502g Berlin, Humboldt Univ.; 273g Chicago, Field Mus. Nat. Hist.; 175g Tempe, Arizona State Univ.; 121g Yale Univ.; 118g Copenhagen, Univ. Geol. Mus.; 85g Denver, Mus. Nat. Hist.; 73g Ottawa, Geol. Surv. Canada; 72g Washington, U.S. Nat. Mus.;
Specimen(s): [40881], 143.5g.; [40882], 102g.; [40883], 46.5g.

Rutherford County v Colfax.

Rutherford County v Murfreesboro'.

Rutlam v Pulsora.

Ryechki 51°8′ N., 34°30′ E.

Sumy district, Kharkov, Ukraine, USSR, [Речки]
Fell 1914, April 9, 1330 hrs
Synonym(s): *Ketschki, Ketscki, Rechki, Retchki, Retschki, Retscki*
Stone. Olivine-hypersthene chondrite (L5).
After detonations, a stone of 7.5kg was seen to fall, and another of about 5.5kg was found about 2 miles to the NW., N.I. Dyakov, Ann. Geol. Min. Russ., 1916, **17**, p.99 [M.A. 2-86]. Analysis, P.N. Chirvinsky, Mem. Soc. Nat. Kiev, 1915, **25**, p.13. Olivine Fa_{23}, B. Mason, Geochimica et Cosmochimica Acta, 1963, **27**, p.1011. Two stones, of 7.5kg and 2.5kg respectively, were seen to fall, and a third stone also probably fell; description, P.N. Chirvinsky, Centralblatt Min., 1923, p.577. Mentioned, E.L. Krinov, Астрон. Журнал, 1945, **22**, p.303 [M.A. 9-297].
3.3kg Kharkov, Univ., main mass; 30.9g Moscow, Acad. Sci.;

Sabetmahet 27°26′ N., 82°5′ E.

near Balrampur, Gonda district, Uttar Pradesh, India
Fell 1855, August 16, evening hrs
Stone. Chondrite.
A stone of about 2.75lb fell and was made an object of worship; only about 3g were secured, H.B. Medlicott, Rec. Geol. Surv. India, 1885, **18**, p.237.
3g Calcutta, Mus. Geol. Surv. India;

Saboriza v Zaborzika.

Saboryzy v Zaborzika.

Sacramento Mountains
32°55′ N., 104°40′ W. approx.
Eddy County, New Mexico, U.S.A.
Found 1890, approx.
Synonym(s): *Badger, Eddy County*
Iron. Octahedrite, medium (1.0mm) (IIIA).
A mass of 523lb was found on the eastern slope of the Sacramento Mountains, 23 miles SW. of Badger, W.M. Foote, Am. J. Sci., 1897, **3**, p.65, H.L. Preston, Am. J. Sci., 1900, **9**, p.284. Determination of Ga, Au and Pd, partial analysis, 8.1 %Ni, E. Goldberg et al., Geochimica et Cosmochimica Acta, 1951, **2**, p.1. Further analysis, 7.82 % Ni, 19.2 ppm.Ga, 36.6 ppm.Ge, 6.7 ppm.Ir, E.R.D. Scott et al., Geochimica et Cosmochimica Acta, 1973, **37**, p.1957. Location unknown; year of find; severe and complex deformation, V.F. Buchwald, Iron Meteorites, Univ. of California, 1975, p.1037.
24.2kg New York, Amer. Mus. Nat. Hist.; 9.9kg Rome, Vatican Colln; 8.4kg Chicago, Field Mus. Nat. Hist.; 6.5kg Washington, U.S. Nat. Mus.; 5kg Stockholm, Riksmus; 4.6kg Yale Univ.; 4.3kg Vienna, Naturhist. Mus.; 2.7kg Ottawa, Mus. Geol. Surv. Canada; 2.7kg Berlin, Humboldt Univ.; 2.7kg Albuquerque, Univ. of New Mexico; 2kg Sydney, Austr. Mus.; 1.8kg Schönenwerd, Bally Mus.; 1.6kg Paris, Mus. d'Hist. Nat.; 1.2kg Harvard Univ.;
Specimen(s): [84267], 14030g. slice; [81871], 13.5g.

Sacramento Peak v Alamogordo.

Safford Meteorite v Morristown.

Sagan 51°32′ N., 14°53′ E.
Zegan, Poland
Fell 1636, March 6, 0600 hrs
Synonym(s): *Dabrowa Łuzycka, Dubrow, Zagan, Zegan*
Stone.
One stone, grey with a black crust, fell in the village of Dubrow, but nothing is now preserved. "Dubrow" cannot be identified with certainty, but is probably Dubraucke (= Dubrawka), 51°47'N., 14°37'E., 27 km NW. of Priebus (= Przewoz)., E.F.F. Chladni, Die Feuer-Meteore, Wien, 1819, p.225, J. Pokrzywnicki, Bull. Soc. Amis Sci. Lettr. Poznan, 1958, **14** (B), p.423. Co-ordinates, J. Pokrzywnicki, Studia Geol. Polon., 1964, **15**, p.1.

Sagauli v Segowlie.

Saginaw 32°51′57″ N., 97°19′23″ W.
Tarrant County, Texas, U.S.A.
Found 1979, July
Iron. Octahedrite, fine to medium.
A single mass of 44.5kg was found embedded in the ground by a survey crew, J.O. Williams, letter of 30 October, 1983 in Min. Dept. BM(NH).
44.5kg Fort Worth, Texas, Mus. Sci. Hist.;

Saharanpur v Akbarpur.

St. Ann 40°27′ N., 100°45′ W.
Frontier County, Nebraska, U.S.A.
Found 1938
Stone. Olivine-bronzite chondrite (H).
One well-oriented stone of 373.5g was found, A.D. Nininger, Pop. Astron., Northfield, Minnesota, 1939, **47**, p.213, H.H. and A.D. Nininger, The Nininger Collection of Meteorites, Winslow, Arizona, 1950, p.91. Olivine Fa_{18}, B. Mason, Geochimica et Cosmochimica Acta, 1963, **27**, p.1011.
171g Tempe, Arizona State Univ.;
Specimen(s): [1959,989], 177g.

St. Caprais-de-Quinsac 44°45′ N., 0°3′ E.
Gironde, France
Fell 1883, January 28, 1445 hrs
Stone. Olivine-hypersthene chondrite (L6).
After the appearance of a "black cloud", and detonations, a stone of 282.5g was seen to fall, (but c.f. 288.5g now in Paris), G. Lespiault and L. Forquignon, C. R. Acad. Sci. Paris, 1883, **97**, p.1022. Olivine Fa_{25}, B. Mason, Geochimica et Cosmochimica Acta, 1963, **27**, p.1011.
288.5g Paris, Mus. d'Hist. Nat.; 33g Vienna, Naturhist. Mus.; 14g Berlin, Humboldt Univ.; 6g New York, Amer. Mus. Nat. Hist.; 4g Chicago, Field Mus. Nat. Hist.;
Specimen(s): [55624], 9g.

St.-Chinian 43°26′ N., 2°57′ E.
Herault, France
Fell 1959, December 25, 1530 hrs
Stone. Olivine-hypersthene chondrite (L6).
Two stones, totalling 134.3g, were found. One, of 77g, was seen to fall and break into six parts near Camprafand, the other, of 57.3g, was seen to fall near Sorteilho, Meteor. Bull., 1960 (19). Brief description, olivine $Fa_{23.5}$, M. Christophe Michel-Levy, Bull. Min., 1978, **101**, p.580 [M.A. 79-2726].

St. Christophe v St. Christophe-la-Chartreuse.

St. Christophe-la-Chartreuse 46°57′ N., 1°30′ W.
Roche-Serviere, Vendée, France
Fell 1841, November 5, evening hrs
Synonym(s): *Bourbon-Vendée, Roche-Serviere, St. Christophe*
Stone. Olivine-hypersthene chondrite (L6).
After the appearance of a fireball, and detonations, a stone of 5.5kg fell, G.A. Daubrée, C. R. Acad. Sci. Paris, 1880, **91**, p.30. Description, analysis, A. Lacroix, Bull. Soc. Sci. Nat. Ouest France, Nantes, ser. 2, 1906, **6**, p.81. Olivine Fa_{25}, B. Mason, Geochimica et Cosmochimica Acta, 1963, **27**, p.1011.
4.96kg Nantes, Mus.; 251g Paris, Mus. d'Hist. Nat.; 7.5g Chicago, Field Mus. Nat. Hist.;

St. Croix County v Hammond.

St. Croix River v Hammond.

St. Denis Westrem 51°3′ N., 3°45′ E. approx.
Ghent, Belgium
Fell 1855, June 7, 1945 hrs
Synonym(s): *Gent, Ghent*
Stone. Olivine-hypersthene chondrite (L6), veined.
A stone of 700g was seen to fall without detonations, F. Duprez, Bull. Acad. Roy. Belg., 1855, **22** (2), p.54, W. von Haidinger, Sitzungsber. Akad. Wiss. Wien, Math.-naturwiss. Kl., 1861, **42**, p.9. Analysis, C. Klement, Bull. Mus. Roy. d'Hist. Nat. Belg., 1886, **4**, p.273. Olivine Fa_{25}, B. Mason, Geochimica et Cosmochimica Acta, 1963, **27**, p.1011.
326g Vienna, Naturhist. Mus.; 189g Ghent Univ., in 1897; 34g Paris, Mus. d'Hist. Nat.; 13g Chicago, Field Mus. Nat. Hist.;
Specimen(s): [34672], 1.5g.

St. Elizabeth v Lucky Hill.

Ste. Marguerite 50°46′ N., 3°0′ E. approx.
Comines, Nord, France
Fell 1962, June 8, between 1730 and 0930 the following day
Synonym(s): *St. Marguerite*
Stone. Olivine-bronzite chondrite (H4).
Six fragments of a mass of 4.95kg were found in an impact pit 45 cm deep. Description, E. Jérémine et al., C. R. Acad. Sci. Paris, 1962, **255**, p.749 [M.A. 16-170]. Olivine Fa_{20}, B. Mason, Geochimica et Cosmochimica Acta, 1963, **27**, p.1011.
4.96kg Paris, Mus. d'Hist. Nat.;
Specimen(s): [1965,393], 1g. fragments

St. Francis Bay 25°4′ S., 14°53′ E.
St. Francis Bay, Namibia
Found 1976, June 6
Stone. Olivine-hypersthene chondrite (L6).
A single mass of 531.6g was found on a sand dune near the sea shore, olivine $Fa_{24.5}$, Meteor. Bull., 1978 (55), Meteoritics, 1978, **13**, p.348.

St. Francois County 37°45′ N., 90°30′ W.
Missouri, U.S.A.
Found 1863, before this year
Synonym(s): *Missouri, South-east Missouri, Südost-Missouri*
Iron. Octahedrite, coarse (2.7mm) (IC).
A specimen weighing about 0.5lb was found in 1863 by B.F. Shumard in the museum of the St. Louis Acad. of Sci. labelled "S.E. Missouri", C.U. Shepard, Am. J. Sci., 1869, **47**, p.233. Later a mass of over 2.5kg was found and became known as St. Francois County. Description, with an analysis, 6.97 %Ni, E. Cohen, Ann. Naturhist. Hofmus. Wien, 1900, **15**, p.369. Classification, analysis, 6.77 %Ni, 49.2 ppm.Ga, 247 ppm.Ge, 0.11 ppm.Ir, E.R.D. Scott and J.T. Wasson, Geochimica et Cosmochimica Acta, 1976, **40**, p.103. The 'S.E. Missouri' specimen is a fragment of Wichita County (*q.v.*), V.F. Buchwald, Iron Meteorites, Univ. of California, 1975, p.1042.
749g Chicago, Field Mus. Nat. Hist.; 641g Vienna, Naturhist. Mus.; 562g Tempe, Arizona State Univ.; 321g Tübingen, Univ.; 314g New York, Amer. Mus. Nat. Hist.; 236g Washington, U.S. Nat. Mus.;
Specimen(s): [1920,353], 4g. St. Francois County

St. Genevieve County 37°58′ N., 90°19′ W.
Missouri, U.S.A.
Found 1888
Synonym(s): *Abancay*
Iron. Octahedrite, medium (0.5mm) (IIIF).
A mass of 539lb was found in the extreme western portion of St. Genevieve County. Report and analysis, 7.98 %Ni, H.A. Ward, Proc. Rochester Acad. Sci., 1901, **4**, p.65. Structure, E. Cohen, Meteoritenkunde, 1905, **3**, p.372. Abancay is a slice of St. Genevieve County; co-ordinates, V.F. Buchwald, Iron Meteorites, Univ. of California, 1975, p.1043. Analysis, 7.68 %Ni, 6.86 ppm.Ga, 0.775 ppm.Ge, 2.0 ppm.Ir, R. Schaudy et al., Icarus, 1972, **17**, p.174. Chemical classification, E.R.D. Scott and J.T. Wasson, Geochimica et Cosmochimica Acta, 1976, **40**, p.103.
106kg Chicago, Field Mus. Nat. Hist.; 54.5kg Tempe, Arizona State Univ.; 7.5kg Vienna, Naturhist. Mus.; 3.7kg New York, Amer. Mus. Nat. Hist.; 669g Los Angeles, Univ. of California; 1.3kg Washington, U.S. Nat. Mus.; 474g Copenhagen, Univ. Geol. Mus.; 257g Berlin, Humboldt Univ.;
Specimen(s): [84861], 6437g. slice; [1959,976], 653.5g. block with etched sides; [1959,968], 79g. slice, and sawings, 1.5g, previously known as Abancay

St. Georges-de-Lévéjac v Aumieres.

St. Germain-du-Pinel 48°1′ N., 1°9′ W. approx.
Vitré, Ille-et-Vilaine, France
Fell 1890, July 4, 1530 hrs
Synonym(s): *St. Germain-en-Puel, Vitré*
Stone. Olivine-bronzite chondrite (H6).
After detonations, a stone of about 4kg fell in two portions 2 miles apart, S. Meunier, C. R. Acad. Sci. Paris, 1912, **154**, p.1741. Olivine Fa_{18}, B. Mason, Geochimica et Cosmochimica Acta, 1963, **27**, p.1011.
1172g Budapest, Nat. Mus.; 686g Vienna, Naturhist. Mus.; 420g Paris, Mus. d'Hist. Nat.; 105g Chicago, Field Mus. Nat. Hist.; 53g Washington, U.S. Nat. Mus.;
Specimen(s): [1912,610], 113g.

St. Germain-en-Puel v St. Germain-du-Pinel.

St. Lawrence 31°44′ N., 101°30′18″ W.
Glasscock County, Texas, U.S.A.
Found 1965
Stone. Olivine-hypersthene chondrite, amphoterite (LL6).
One stone, 2.6kg was found by a boy, Meteor. Bull., 1970 (49), Meteoritics, 1970, **5**, p.175. Classification, B. Mason, Smithson. Contrib. Earth Sci., 1975 (14), p.71.
1.6kg Mainz, Max-Planck-Inst.; 127g New York, Amer. Mus. Nat. Hist.; 100g Washington, U.S. Nat. Mus.; 89g Tempe, Arizona State Univ.;
Specimen(s): [1967,441], 22g.

St. Leo v Willowdale.

St. Louis 38°42′ N., 90°14′ W.
St. Louis County, Missouri, U.S.A.
Fell 1950, December 10, 2307 hrs
Stone. Olivine-bronzite chondrite (H4).
A mass of 1kg hit a car on West Florissant Ave., in St. Louis, F.E. Wickman, letter of 14 May, 1973 in Min. Dept. BM(NH). Olivine Fa_{19}, B. Mason, Geochimica et Cosmochimica Acta, 1963, **27**, p.1011.
Main mass, in possession of E.E. Fitton, St. Louis; 2.4g Washington, U.S. Nat. Mus.; 1.9g Mainz, Max-Planck-Inst.;

St. Marguerite v Ste. Marguerite.

St. Mark's 32°1′ S., 27°25′ E.
St. Mark's Mission Station, Transkei, Cape Province, South Africa
Fell 1903, January 3, 2300 hrs
Synonym(s): *Transkei*
Stone. Enstatite chondrite (E5).
A stone of 13.78kg was seen to fall after the appearance of light and four detonations. Description, with an analysis, E. Cohen, Ann. South African Mus. Cape Town, 1906, **5** (1), p.1. Analysis, 32.43 % total iron, H.B. Wiik, Comm. Phys.-Math. Soc. Sci. Fenn., 1969, **34**, p.135, H. von Michaelis et al., Earth planet. Sci. Lett., 1969, **5**, p.387. Mineralogy, K. Keil, J. Geophys. Res., 1968, **73**, p.6945. Trace element abundances, C.M. Binz et al., Geochimica et Cosmochimica Acta, 1974, **38**, p.1579.
Main mass, Cape Town, South African Mus.; 257g Washington, U.S. Nat. Mus.; 170g New York, Amer. Mus. Nat. Hist.; 83g Calcutta, Mus. Geol. Surv. India; 43g Vienna, Naturhist. Mus.; 32g Tempe, Arizona State Univ.; 28g Berlin, Humboldt Univ.;
Specimen(s): [1916,59], 86.5g. and fragments, 2.5g; [1970, 339], 9.2g. four fragments

St. Mary's County 38°10′ N., 76°23′ W.

Maryland, U.S.A.
Fell 1919, June 20, 1800 hrs
Stone. Olivine-hypersthene chondrite, amphoterite (LL3).
A stone of considerable size was seen to fall about 1 mile from St. Jerome's Creek, near Ridge, St. Mary's County. The main mass has been lost, and only one fragment, 24.25g, has been preserved, A.D. Nininger, Pop. Astron., Northfield, Minnesota, 1939, **47**, p.213, E.P. Henderson, letter of 3 June, 1939 in Min. Dept. BM(NH). Most of the mass is believed to have fallen in the Chesapeake Bay, F.D. Cecil, Sky and Telescope, 1944, **3** (12), p.9 [M.A. 9-300]. Description, analysis, 20.33 % total iron, A.F. Noonan et al., Smithson. Contrib. Earth Sci., 1977 (19), p.96.
19.8g Washington, U.S. Nat. Mus.;

St. Mesmin 48°27′ N., 3°56′ E.

Aube, France
Fell 1866, May 30, 1545 hrs
Stone. Olivine-hypersthene chondrite, amphoterite (LL6), brecciated, polymict.
After the appearance of a fireball, followed by three detonations, three stones were found, of 4.2kg, 2.2kg, and 1.9kg respectively, G.A. Daubrée, C. R. Acad. Sci. Paris, 1866, **62**, p.1305. An LL6 chondrite with olivine-bronzite chondrite, P. Pellas, From Plasma to Planet, ed A. Elvius, (Proc. 21 Nobel Symp.), Wiley, 1972, p.65. Petrology, R.T. Dodd and E. Jarosewich, Meteoritics, 1974, **11**, p.1, R.T. Dodd, Contr. Mineral. Petrol., 1974, **46**, p.129. Olivine Fa_{29}, B. Mason, Geochimica et Cosmochimica Acta, 1963, **27**, p.1011. Petrography of lithic fragments, R.V. Fodor and K. Keil, Meteoritics, 1975, **10**, p.325. Cr-rich olivine, R.T. Dodd et al., Geochimica et Cosmochimica Acta, 1975, **39**, p.1621. Rb-Sr age, J.F. Minster and C.J. Allegre, Nature, 1979, **278**, p.732 [M.A. 80-2076].
1.98kg Paris, Mus. d'Hist. Nat.; 472g Vienna, Naturhist. Mus.; 86g Washington, U.S. Nat. Mus.; 63g Berlin, Humboldt Univ.; 39g Chicago, Field Mus. Nat. Hist.;
Specimen(s): [67590], 68g.; [40991], 27.3g. and fragments, 1.4g; [41427], 7g.

St. Michel 61°39′ N., 27°12′ E.

Mikkeli, Finland
Fell 1910, July 12, 1925 hrs
Synonym(s): *Mikkeli*
Stone. Olivine-hypersthene chondrite (L6).
After the appearance of a fireball followed by detonations, two stones, about 7kg and 10kg respectively were found. Description, with an analysis, L.H. Borgström, Bull. Comm. Geol. Finlande, 1912, **6** (34), p.1. Analysis, 21.71 % total iron, H. von Michaelis et al., Earth planet. Sci. Lett., 1969, **5**, p.387.
8.64kg Helsinki, Univ., main mass; 752g New York, Amer. Mus. Nat. Hist.; 725g Vienna, Naturhist. Mus.; 625g Washington, U.S. Nat. Mus.; 610g Paris, Mus. d'Hist. Nat.; 571g Harvard Univ.; 232g Chicago, Field Mus. Nat. Hist.; 130g Tempe, Arizona State Univ.;
Specimen(s): [1913,145], 953g.

St. Nicholas v Mässing.

Saintonge v Jonzac.

St. Peter 39°24′ N., 100°2′ W.

Graham County, Kansas, U.S.A.
Found 1957, before this year
Stone. Olivine-hypersthene chondrite (L5).
A single mass of 15lb was found, Meteor. Bull., 1957 (5). Olivine Fa_{25}, B. Mason, Geochimica et Cosmochimica Acta, 1963, **27**, p.1011.
Main mass, Fort Hays, Fort Hays Kansas State College; 521g Washington, U.S. Nat. Mus.; 290g Mainz, Max-Planck-Inst.; 40g New York, Amer. Mus. Nat. Hist.;

Saint-Sauveur 43°44′ N., 1°23′ E.

Haute Garonne, France
Fell 1914, July 10, 1400-1500 hrs
Stone. Enstatite chondrite (E4).
After detonations and the appearance of light, a stone of about 14kg was seen to fall, 1.5km south of Saint-Sauveur, -. Mengaud and -. Mourié, C. R. Acad. Sci. Paris, 1923, **177**, p.510. Description, analysis, A. Lacroix, C. R. Acad. Sci. Paris, 1923, **177**, p.561. Mineralogy, B. Mason, Geochimica et Cosmochimica Acta, 1966, **30**, p.23, K. Keil, J. Geophys. Res., 1968, **73**, p.6945.
14kg Toulouse, Mus. d'Hist. Nat.; 323g Paris, Mus. d'Hist. Nat.;
Specimen(s): [1980,M.28], 10.1g. fragments

Saint-Séverin 45°18′ N., 0°14′ E.

Charente, France
Fell 1966, June 27, 1540 hrs
Stone. Olivine-hypersthene chondrite, amphoterite (LL6).
Eight stones totalling 271kg were recovered, Meteor. Bull., 1967 (39, 40), Meteoritics, 1970, **5**, p.89, 94. Charged particle-track studies and ablation loss, Y. Cantelaube et al., Meteorite Research, ed. P.M. Millman, D. Reidel, Dordrecht-Holland, 1969, p.705, D. Lal et al., Meteorite Research, ed. P.M. Millman, D. Reidel, Dordrecht-Holland, 1969, p.275. U-Th-Pb and Rb-Sr study, G. Manhes et al., Earth planet. Sci. Lett., 1978, **39**, p.14 [M.A. 78-4759]. Xenon age, F.A. Podosek, Geochimica et Cosmochimica Acta, 1970, **34**, p.341. Noble gas studies, L. Schultz et al., Meteoritics, 1973, **8**, p.435, abs., P. Signer, Earth planet. Sci. Lett., 1976, **30**, p.191. Analysis, 20.17 % total iron, E. Jarosewich and B. Mason, Geochimica et Cosmochimica Acta, 1969, **33**, p.411.
197kg Paris, Mus. d'Hist. Nat., main mass; 2.4kg Washington, U.S. Nat. Mus.; 267g Chicago, Field Mus. Nat. Hist.; 263g Los Angeles, Univ. of California;
Specimen(s): [1966,493], 146g. slice

St. Vrain 34°19′6″ N., 103°28′18″ W.

Curry County, New Mexico, U.S.A.
Found 1971
Stone. Chondrite.
A single, oriented, mass of 45.5g is listed, Cat. Huss Coll. Meteorites, 1976, p.38.
45.5g Mainz, Max-Planck-Inst.;

Sakauchi 35°40′ N., 136°18′ E.

Ibi, Gifu, Honshu, Japan
Fell 1913, April 13
Synonym(s): *Sakouchi, Sakouti*
Iron. Hexahedrite.
Listed, I. Yamamoto, Kwasan Observ. Bull., 1935, p.306 [M.A. 7-173]. A single mass of 4.18kg, not seen to fall, A.D. Nininger, Pop. Astron., Northfield, Minnesota, 1940, **48**, p.559, Contr. Soc. Res. Meteorites, **2**, p.227 [M.A. 8-54]. The mass was kept in Kyoto Univ. but is now lost, S. Murayama, letter of 6 April, 1962 in Min. Dept. BM(NH).

Classification, S. Murayama, Nat. Sci. Mus. Tokyo, 1960, **27** (3-4).

Sakouchi v Sakauchi.

Sakouti v Sakauchi.

Sakurayama 35°10′ N., 136°55′ E.

Nagoya Shigai, Aichi, Honshu, Japan
Fell 1935, July 7
Doubtful. Iron.

One fragment of 0.84g is known, in the possession of Prof. Mizukawa, I. Yamamoto, Kwasan Observ. Bull., 1935, p.306 [M.A. 7-173], S. Murayama, letter of 6 April, 1962 in Min. Dept. BM(NH).

Salaices 27°0′ N., 105°15′ W.

Chihuahua, Mexico
Found 1971
Stone. Olivine-bronzite chondrite (H4).

One mass of 3.9kg was found 6 km SW. of Salaices and within the Allende strewnfield, about 1.5 km NE. of Rancho Blanco, Meteor. Bull., 1972 (51), Meteoritics, 1972, **7**, p.229. A further mass, 39 fragments totalling 20.6kg, was found in 1981 about 7.2 km WSW. of the original find-site, close to Pueblito de Allende, W. Zeitschel, letter of 29 October, 1981 in Min. Dept. BM(NH).

1203g Tempe, Arizona State Univ.; 395g Washington, U.S. Nat. Mus.;

Specimen(s): [1981,M.4], 1906g. part-slice

Salavina v Malotas.

Salem v Kakangari.

Salem v Smithland.

Salem 44°57′ N., 123°1′ W.

Marion County, Oregon, U.S.A.
Fell 1981, May 11, 0115 hrs, or May 12
Stone. Olivine-hypersthene chondrite, amphoterite (LL4-5).

Fell on the roof of a house in Salem; two masses, totalling 61g, were recovered, Meteor. Bull., 1982 (60), Meteoritics, 1982, **17**, p.96.

Salina 38°59′ N., 111°51′ W.

Sevier County, Utah, U.S.A.
Found 1908, approx.
Iron. Octahedrite, medium.

A number of shale balls, up to 4 or 5lb in weight, some with metallic cores, were found around a small depression, in sect. 36, township 21 S., range 3W., in the Pahvant Mts., 15 miles from Salina, but only one weathered mass of about 235g was preserved, S.H. Perry, Pop. Astron., Northfield, Minnesota, 1939, **47**, p.121 [M.A. 7-378].

221g Washington, U.S. Nat. Mus., main mass; 17g Tempe, Arizona State Univ.;

Saline 39°24′ N., 100°24′ W.

Saline Township, Sheridan County, Kansas, U.S.A.
Found 1901, possibly fell 1898, November 15, 2130 hrs.
Stone. Olivine-bronzite chondrite (H5).

After the appearance of a fireball in 1898, a stone of about 68lb was found three years later. Description, O.C. Farrington, Science, 1902, **16**, p.67, O.C. Farrington, Field Mus. Nat. Hist. Geol. Ser., 1907, **3** (6), p.126. Analysis, O.C. Farrington, Field Mus. Nat. Hist. Geol. Ser., 1911, **3** (9). Further, partial analysis, 27.04 % total iron, W. Wahl and H.B. Wiik, Geochimica et Cosmochimica Acta, 1951, **1**, p.123. Olivine Fa_{18}, B. Mason, Geochimica et Cosmochimica Acta, 1963, **27**, p.1011.

21kg Chicago, Field Mus. Nat. Hist.; 897g Washington, U.S. Nat. Mus.; 643g Calcutta, Mus. Geol. Surv. India; 280g Tempe, Arizona State Univ.; 270g Rome, Vatican Colln; 249g Tübingen, Univ.; 182g Berlin, Humboldt Univ.; 58g Vienna, Naturhist. Mus.;

Specimen(s): [1921,21], 217g.; [86522], 167.5g. and fragments, 5g

Salins v Ornans.

Salisbury v Mangwendi.

Salitra v La Primitiva.

Sall v Kaalijarv.

Salla 66°48′ N., 28°27′ E.

Lappi, Finland
Found 1963
Stone. Olivine-hypersthene chondrite (L6).

Total known weight 2.9kg, H.B. Wiik and B. Mason, Geologi (Helsinki), 1964, p.95 but see weight listed in the Helsinki collection. Olivine Fa_{24}, B. Mason, Geochimica et Cosmochimica Acta, 1967, **31**, p.1100. Analysis, 21.58 % total iron, H.B. Wiik, Comm. Phys.-Math. Soc. Sci. Fenn., 1969, **34**, p.135.

5.366kg Helsinki, Univ., main mass; 318g Washington, U.S. Nat. Mus.; 66g Vienna, Naturhist. Mus.; 39g Mainz, Max-Planck-Inst.; 20g Perugia, Univ.; 13g New York, Amer. Mus. Nat. Hist.; 11g Tempe, Arizona State Univ.;

Salles 46°3′ N., 4°38′ E.

Villefranche, Rhone, France
Fell 1798, March 12, 1800 hrs
Synonym(s): *Villefranche*
Stone. Olivine-bronzite chondrite (H6), veined.

After the appearance of a fireball moving from east to west, a stone of about 20lb fell and buried itself 18 inches in the soil, Marquis de Drée, Phil. Mag., 1803, **16**, p.217, J. Phys. Chim. Hist. Nat., 1802, **56**, p.383. Date of fall perhaps March 8, when a fireball moving in the same direction, and at about the same hour was observed, P. Prevost, J. Phys. Chim. Hist. Nat., 1802, **56**, p.465. Olivine Fa_{18}, B. Mason, Geochimica et Cosmochimica Acta, 1963, **27**, p.1011.

1.52kg Paris, Mus. d'Hist. Nat.; 335g Vienna, Naturhist. Mus.; 146g Budapest, Nat. Hist. Mus.; 101g Chicago, Field Mus. Nat. Hist.; 50g Washington, U.S. Nat. Mus.; 15g Berlin, Humboldt Univ.;

Specimen(s): [90246], 165g.

Salta v Imilac.

Saltillo v Coahuila.

Salt Lake City 40°55′ N., 111°40′ W.

Utah, U.S.A.
Found 1869
Synonym(s): *Echo, Utah*
Stone. Olivine-bronzite chondrite (H5), brecciated.

A stone of 875g was found between Salt Lake City and Echo. Description, with an analysis, E.S. Dana, Am. J. Sci., 1886, **32**, p.226. Olivine Fa_{19}, B. Mason, Geochimica et Cosmochimica Acta, 1963, **27**, p.1011.

631g Yale Univ.; 123g Chicago, Field Mus. Nat. Hist.; 13.8g Paris, Mus. d'Hist. Nat.; 7.9g Vienna, Naturhist. Mus.;
Specimen(s): [64343], 4.5g.; [1973,M.23], 6g.

Salt River 37°57′ N., 85°47′ W.
Bullitt County, Kentucky, U.S.A.
Found 1850, approx.
Synonym(s): *Tocavita*
Iron. Octahedrite, plessitic (IIC).
A mass of about 8lb was found about 20 miles S. of Louisville, and was heated in a forge, B. Silliman, Jr., Proc. Amer. Assoc., 1850, **4**, p.36. Description, analysis, 8.70 % Ni, E. Cohen, Meteoritenkunde, 1905, **3**, p.275. All specimens reheated to 1000C; synonymy of Tocavita, V.F. Buchwald, Iron Meteorites, Univ. of California, 1975, p.1046. Classification, analysis, 10.02 %Ni, 37.8 ppm.Ga, 100 ppm.Ge, 6.3 pppm.Ir, J.T. Wasson, Geochimica et Cosmochimica Acta, 1969, **33**, p.859.
985g Yale Univ.; 272g Harvard Univ.; 170g Tempe, Arizona State Univ.; 112g Washington, U.S. Nat. Mus.; 61g Tübingen, Univ.; 79g Chicago, Field Mus. Nat. Hist.; 61g Univ. Kentucky; 59g New York, Amer. Mus. Nat. Hist.;
Specimen(s): [35416], 431.5g. includes a slice, 61.5g

Saluka v Shalka.

Samelia 25°40′ N., 74°52′ E.
Shahpura, Rajasthan, India
Fell 1921, May 20, 1730 hrs
Synonym(s): *Beshkalai, Beshki, Shahpura, Shimelia*
Iron. Octahedrite, medium (1.1mm) (IIIA).
After the appearance of a fireball moving from south to north and leaving a white trail, loud detonations were heard and three masses fell, one of 1125g in the jungle south of Samelia (25°40'N., 74°52'E.), 7 miles WNW. of Shahpura, another of 587g in the jungle of Beshki, 1.2 miles SSE. of Samelia, and a third, of 749.5g in the village of Beshkalai, in the Banera territory, 5 miles W. of Beshki. Description, L.L. Fermor, Rec. Geol. Surv. India, 1924, **55**, p.327, L.L. Fermor, Rec. Geol. Surv. India, 1931, **65**, p.161 [M.A. 5-10]. Classification and analysis, 8.0 %Ni, 19.5 ppm.Ga, 38.3 ppm.Ge, 2.6 ppm.Ir, E.R.D. Scott and J.T. Wasson, Geochimica et Cosmochimica Acta, 1976, **40**, p.103. Description; shock-hardened, V.F. Buchwald, Iron Meteorites, Univ. of California, 1975, p.1049.
2.1kg Calcutta, Mus. Geol. Surv. India;
Specimen(s): [1927,916], 138.5g. end-piece of Beshkalai mass

Sam's Valley 42°32′ N., 122°52′30″ W.
Jackson County, Oregon, U.S.A.
Found 1894
Iron. Octahedrite, medium (0.7mm) (IIIB).
A mass of 15.25lb was found about 10 miles NW. of Medford. Description, with an analysis, 9.76 %Ni, W.M. Foote, Am. J. Sci., 1915, **39**, p.80. This mass was sliced up and distributed by the Foote Mineral Co. Another mass, 1kg, was found before 1900, description, R.A. Morley, Pop. Astron., Northfield, Minnesota, 1950, **58**, p.236 [M.A. 11-271]. Five individuals may have been found, of which three are positively known, E.F. Lange, The Ore Bin, 1967, **29**, p.145. Further analysis, 9.77 %Ni, 18.4 ppm.Ga, 35.1 ppm.Ge, 0.017 ppm.Ir, E.R.D. Scott et al., Geochimica et Cosmochimica Acta, 1973, **37**, p.1957. Description; some material has been artificially reheated, V.F. Buchwald, Iron Meteorites, Univ. of California, 1975, p.1049.
2.2kg New York, Amer. Mus. Nat. Hist.; 1.1kg Harvard Univ.; 1kg Jacksonville Mus., Oregon; 662g Ottawa, Mus. Geol. Surv. Canada; 501g Chicago, Field Mus. Nat. Hist.; 316g Tempe, Arizona State Univ.;
Specimen(s): [1919,42], 50g. slice, and a fragment, 5.5g

Samyscheva v Pavlodar (pallasite).

San Albano v Valdinizza.

Sanam v as-Sanam.

San Angelo 31°25′ N., 100°21′ W.
Tom Green County, Texas, U.S.A.
Found 1897
Synonym(s): *Lipan Flats*
Iron. Octahedrite, medium (1.0mm) (IIIA).
A mass of 194lb was found 7 miles S. of San Angelo. Description, with an analysis, 7.86 %Ni, H.L. Preston, Am. J. Sci., 1898, **5**, p.269. Further analysis, 7.52 %Ni, 19.4 ppm.Ga, 37.6 ppm.Ge, 8.0 ppm.Ir, E.R.D. Scott et al., Geochimica et Cosmochimica Acta, 1973, **37**, p.1957. Description; shock-deformed, V.F. Buchwald, Iron Meteorites, Univ. of California, 1975, p.1053.
30kg Austin, Univ. of Texas; 5.8kg Chicago, Field Mus. Nat. Hist.; 3.4kg Washington, U.S. Nat. Mus.; 2.3kg Fort Worth, Texas, Monnig Colln.; 1kg New York, Amer. Mus. Nat. Hist.; 977g Rome, Vatican Colln; 903g Tempe, Arizona State Univ.; 547g Los Angeles, Univ. of California; 472g Tübingen, Univ.; 420g Vienna, Naturhist. Mus.; 295g Harvard Univ.; 196g Berlin, Humboldt Univ.; 142g Oslo, Min.-Geol. Mus.;
Specimen(s): [83612], 771g. slice; [1953,1], 151.5g.

San Antonio v Kendall County.

San Antonio v Pipe Creek.

San Bartolome v Chupaderos.

San Bernardino County v Ivanpah.

San Bernardino County v San Emigdio.

San Carlos 35°32′ S., 58°46′ W.
Monte district, Buenos Aires province, Argentina
Found 1942, before this year
Stone. Olivine-bronzite chondrite (H).
A fragment weighing 3600g, part of a larger stone, was found in the surface soil of the San Carlos estate. Description, with an analysis, E.H. Ducloux, Notas Mus. La Plata, 1942, **7** (geol. no 19), p.123 [M.A. 8-376]. Olivine Fa_{18}, B. Mason, Geochimica et Cosmochimica Acta, 1963, **27**, p.1011. Only a little seems to have been preserved.
285g La Plata Mus.; 7.2g New York, Amer. Mus. Nat. Hist.;
Specimen(s): [1964,71], 30g.

San Cerre de Mallorca 39°35′ N., 2°39′ E.
Mallorca, Spain
Fell 1958, or 1959
Doubtful..
A meteorite is said to have fallen on the beach at San Cerre, near Palma, fragments of up to 500kg being strewn over a radius of 100m, Volkswacht, 1959 (31 January), K. Keil, Fortschr. Min., 1960, **38**, p.261. Unconfirmed and very doubtful.

Sancha Estate v Coahuila.

Sanchez Estate v Coahuila.

Sanclerlandia 16°13′ S., 50°18′ W.
Serra do Mangabal, Goias, Brazil
Found 1971, October 5
Iron. Octahedrite, medium (1.0mm) (IIIA).
A single mass of 279kg was found about 500 metres from the Sanclerlandia to Anicuns road, M.R. Ribeiro and A.P. Rodrigues, Bol. Mineral. Recife, 1972, **2**, p.75. Classification and analysis, 7.47 %Ni, 18.6 ppm.Ga, 36.4 ppm.Ge, 7.1 ppm.Ir, A. Kracher et al., Geochimica et Cosmochimica Acta, 1980, **44**, p.773. Co-ordinates of find-site, C.B. Gomes, letter of 28 October 1980 in Min. Dept. BM(NH).
Main mass, Brasilia, Univ. (Dept. of Geosciences); 128g Albuquerque, Univ. of New Mexico;

San Cristobal 23°26′ S., 69°30′ W.
Antofagasta, Chile
Found 1882, approx.
Synonym(s): *Antofagasta*
Iron. Ataxite, Ni-rich (IB).
A mass of about 5kg was found. Description and analysis, 25.6 %Ni, E. Cohen, Meteoritenkunde, 1905, **3**, p.132. 26.4 %Ni, P. Fann, Zeits. Metallk, 1917. Further analysis, 25.0 %Ni, 11.8 ppm.Ga, 25.0 ppm.Ge, 0.32 ppm.Ir, E.R.D. Scott et al., Geochimica et Cosmochimica Acta, 1973, **37**, p.1957. Structure, H.J. Axon and P.L. Smith, Min. Mag., 1972, **38**, p.736, E.R.D. Scott and R.W. Bild, Geochimica et Cosmochimica Acta, 1974, **38**, p.1379, V.F. Buchwald, Iron Meteorites, Univ. of California, 1975, p.1054.
1.3kg Vienna, Naturhist. Mus.; 855g New York, Amer. Mus. Nat. Hist.; 500g Yale Univ.; 136g Washington, U.S. Nat. Mus.; 114g Chicago, Field Mus. Nat. Hist.;
Specimen(s): [86764], 76.5g.; [86763], 58.5g.

Sanderson 30°8′ N., 102°9′ W.
Terrel County, Texas, U.S.A.
Found 1936, March
Iron. Octahedrite, medium (0.7mm) (IIIB).
One mass of 6.8kg was found, A.D. Nininger, Pop. Astron., Northfield, Minnesota, 1940, **48**, p.557, A.D. Nininger, Contr. Soc. Res. Meteorites, **2**, p.227 [M.A. 8-54]. Analysis, 9.69 %Ni, 18.2 ppm.Ga, 35.8 ppm.Ge, 0.021 ppm.Ir, E.R.D. Scott et al., Geochimica et Cosmochimica Acta, 1973, **37**, p.1957. Co-ordinates; deformation history, V.F. Buchwald, Iron Meteorites, Univ. of California, 1975, p.1057.
1.98kg Tempe, Arizona State Univ.; 471g Washington, U.S. Nat. Mus.; 335g Chicago, Field Mus. Nat. Hist.;
Specimen(s): [1959,964], 1708g. and sawings, 60g

Sandia Mountains 35°15′ N., 106°30′ W.
near Albuquerque, New Mexico, U.S.A.
Found 1925
Iron. Octahedrite, coarsest (10mm) (IIB).
Original weight 100lb, only 15kg preserved. Description, with an analysis, 5.75 %Ni, H.H. Nininger, Am. J. Sci., 1929, **18**, p.412 [M.A. 4-263]. Determination of Ga, Au and Pd, 5.94 %Ni, E. Goldberg et al., Geochimica et Cosmochimica Acta, 1951, **2**, p.1. Classification, analysis, 5.87 %Ni, 59 ppm.Ga, 174 ppm.Ge, 0.15 ppm.Ir, J.T. Wasson, Geochimica et Cosmochimica Acta, 1969, **33**, p.859. Description, V.F. Buchwald, Iron Meteorites, Univ. of California, 1975, p.1059.
6.6kg Tempe, Arizona State Univ., includes main mass; 1.67kg Washington, U.S. Nat. Mus.; 773g Harvard Univ.; 700g Albuquerque, Univ. of New Mexico; 460g Chicago, Field Mus. Nat. Hist.;
Specimen(s): [1937,392], 23.3g.; [1959,923], 436.5g.

Sandon v Sierra Sandon.

Sandtown 35°56′ N., 91°38′ W.
Independence County, Arkansas, U.S.A.
Found 1938
Synonym(s): *Joe Wright Mountain II*
Iron. Octahedrite, medium (1.2mm) (IIIA).
The 9.35kg mass previously ascribed to Joe Wright Mountain (*q.v.*), is distinduished as a separate fall on chemical grounds. Analysis, 8.09 %Ni, 21.0 ppm.Ga, 41.4 ppm.Ge, 1.4 ppm.Ir, E.R.D. Scott et al., Geochimica et Cosmochimica Acta, 1973, **37**, p.1957. Co-ordinates; shocked and annealed, V.F. Buchwald, Iron Meteorites, Univ. of California, 1975, p.1060.
7.4kg Chicago, Field Mus. Nat. Hist., main mass; 105g Copenhagen, Univ. Geol. Mus.; 100g Washington, U.S. Nat. Mus.;
Specimen(s): [1980,M.11], 603g. sawn fragment

Sandwich Islands v Honolulu.

San Emigdio 36° N., 119° W. approx.
San Emigdio Mountains, Kern County, California, U.S.A.
Found 1887, known in this year
Synonym(s): *San Bernardino County, San Emigdio Range, Emiglio*
Stone. Olivine-bronzite chondrite (H4).
A stone of about 80lb was found and was crushed to pieces as an ore. Description, with a partial analysis, G.P. Merrill and J.E. Whitfield, Proc. U.S. Nat. Mus., 1888, **11**, p.161. Locality given as Kern County, not San Bernardino County, A.S. Eakle, Minerals of California, 1914, p.22. Olivine Fa_{20}, B. Mason, Geochimica et Cosmochimica Acta, 1963, **27**, p.1011.
490g Washington, U.S. Nat. Mus.; 48g Chicago, Field Mus. Nat. Hist.; 20g Tempe, Arizona State Univ.; 6g Vienna, Naturhist. Mus.; 6g Yale Univ.; 2g Berlin, Humboldt Univ.;
Specimen(s): [1919,88], 33g. 27 fragments

San Emigdio Range v San Emigdio.

San Emiglio v San Emigdio.

San Francisco del Mezquital
23°29′ N., 104°22′ W.
Durango, Mexico
Found 1868, known before this year
Synonym(s): *Mesquital, Mezquital*
Iron. Hexahedrite (IIA).
A mass of about 7.5kg was brought to France from Mexico by Gen. Castlenau. Report and analysis, 5.89 %Ni, G.A. Daubrée, C. R. Acad. Sci. Paris, 1868, **66**, p.573, L. Fletcher, Min. Mag., 1890, **9**, p.154. Description, analysis, 5.46 %Ni, E. Cohen, Meteoritenkunde, 1905, **3**, p.48. Co-ordinates; the mass has been heated artificially to about 800C, V.F. Buchwald, Iron Meteorites, Univ. of California, 1975, p.1062. Further analysis, 5.50 %Ni, 61.4 ppm.Ga, 183 ppm.Ge, 21 ppm.Ir, J.T. Wasson, Meteorites, Springer-Verlag, 1974, p.299.
150g Paris, Mus. d'Hist. Nat.; 70g Vienna, Naturhist. Mus.; 50g Harvard Univ.; 36g New York, Amer. Mus. Nat. Hist.; 24g Chicago, Field Mus. Nat. Hist.;
Specimen(s): [43401], 7085g. main mass, and pieces, 14g

San Francisco do Sul v Santa Catharina.

San Francisco Mountains 35° N., 112° W. approx.
Arizona, U.S.A.
Found 1920, approx.
Iron. Octahedrite, fine (0.2mm) (IVA).
A mass of 1700g was found, evidently a recent fall as it shows a good fresh crust and heating zone. Description, with an analysis, 7.83 %Ni, S.H. Perry, Am. J. Sci., 1934, **28**, p.202 [M.A. 6-13]. Classification and analysis, 7.62 %Ni, 2.09 ppm.Ga, 0.102 ppm.Ge, 3.0 ppm.Ir, R. Schaudy et al., Icarus, 1972, **17**, p.174. Description, V.F. Buchwald, Iron Meteorites, Univ. of California, 1975, p.1064.
1.3kg Washington, U.S. Nat. Mus.; 36g Chicago, Field Mus. Nat. Hist.; 30g Tempe, Arizona State Univ.;
Specimen(s): [1959,980], 28g. and sawings, 1.5g

San Francisco Pass v Barranca Blanca.

San Giovanni d'Asso v Siena.

San Giuliano v Alessandria.

San Giuliano di Alessandria v Alessandria.

San Giuliano Veccio v Alessandria.

San Gregorio v Morito.

Sanguis-Saint-Etienne v Sauguis.

San Jago del Estero v Campo del Cielo.

San José v Heredia.

San José 24°42′ N., 99°5′ W.
Tamaulipas, Mexico
Found 1944, known before this year
Stone. Olivine-bronzite chondrite (H5).
A single, 532g fragment, was recovered, R.S. Clarke, Jr., letter of 13 April, 1973 in Min. Dept. BM(NH). Olivine Fa_{18}, B. Mason, Geochimica et Cosmochimica Acta, 1963, **27**, p.1011.
Main mass, Austin, Texas, Texas Mem. Mus., Univ. of Texas; 142g Washington, U.S. Nat. Mus.;

San Juan Capistrano 33°29′5″ N., 117°39′45″ W.
California, U.S.A.
Fell 1973, March 15, 0000-0400 hrs
Stone. Olivine-bronzite chondrite (H6).
Two partly-crusted interlocking fragments, total weight 56g, were recovered after falling through the aluminium roof of a carport, Meteor. Bull., 1975 (53), Meteoritics, 1975, **10**, p.138. Full description; isotopic data, olivine $Fa_{18.7}$, R.C. Finkel et al., Meteoritics, 1975, **10**, p.61.
18g Washington, U.S. Nat. Mus.;

San Juliano v São Julião de Moreira.

Sankt Nicolas v Mässing.

San Luis 33°20′ S., 66°23′ W.
San Luis, Argentina
Found 1964, known before this year
Doubtful. Stone. Olivine-bronzite chondrite (H).
A specimen from San Luis, supplied by L.O. Giacomelli, is mentioned, B. Mason, letter of 4 March, 1964 in Min. Dept. BM(NH). Perhaps a specimen of Arbol Solo (*q.v.*).

San Luis Potosi v Bocas.

San Luis Potosi v Charcas.

San Martin v North Chile.

San Michele de Mechede v Portugal.

Sano 36°19′ N., 139°35′ E.
Tochigi Prefecture, Honshu, Japan
Found 1925
Doubtful..
Doubtful, I. Yamamoto, Kwasan Observ. Bull., 1935, p.306 [M.A. 7-173].

San Pedro v Imilac.

San Pedro v Vaca Muerta.

San Pedro de Atacama v Imilac.

San Pedro de Atacama v Vaca Muerta.

San Pedro Springs 29°30′ N., 98°30′ W. approx.
Bexar County, Texas, U.S.A.
Found 1887
Stone. Olivine-hypersthene chondrite (L6), veined.
A specimen of 72g is mentioned, C.S. Bement, Catalogue of June, 1894.
49.7g New York, Amer. Mus. Nat. Hist.; 3g Vienna, Naturhist. Mus.; 1g Berlin, Humboldt Univ.;

San Rafael v Grant.

San Sebastiano da Boa Vista v São Sebastião da Boa Vista.

Santa Apolonia 19°13′ N., 98°18′ W.
Nativitas, Tlaxcala, Mexico
Found 1872
Synonym(s): *Apollonia, Nativitas, Nativitas Tlaxcala*
Iron. Octahedrite, medium (0.9mm) (IIIA).
A mass of about 1kg was said to be in Mexico City, H.A. Ward, Cat. Ward-Coonley Coll. Meteorites, Chicago, 1904, p.22. A mass of 1315.6kg was found at Nativitas, Tlaxcala and is referred to Santa Apolonia. Report and analysis, 7.39 %Ni, H.H. Nininger, Proc. Colorado Mus. Nat. Hist., 1931, **10** (1) [M.A. 5-13]. Determination of Ga, Au and Pd, with an analysis, 7.83 %Ni, E. Goldberg et al., Geochimica et Cosmochimica Acta, 1951, **2**, p.1. Further analysis, 7.52 % Ni, 19.5 ppm.Ga, 35.8 ppm.Ge, 8.3 ppm.Ir, E.R.D. Scott et al., Geochimica et Cosmochimica Acta, 1973, **37**, p.1957. Description; shock-hardened, V.F. Buchwald, Iron Meteorites, Univ. of California, 1975, p.1067.
1051kg Mexico City, Inst. Geol.; 9.4kg Tempe, Arizona State Univ.; 5.6kg Washington, U.S. Nat. Mus.; 4kg New York, Amer. Mus. Nat. Hist.; 3.38kg Harvard Univ.; 906g Ann Arbor, Michigan Univ.; 373g Los Angeles, Univ. of California; 211g Chicago, Field Mus. Nat. Hist.;
Specimen(s): [1931,222], 2089g.; [1959,965], 1557; [1959,971], 16.5g. polished plates

Santa Barbara 29°12′ S., 51°52′ W.
Porto Alegre, Rio Grande do Sul, Brazil
Fell 1873, September 26
Synonym(s): *Rio Grande do Sul*
Stone. Olivine-hypersthene chondrite (L4), brecciated.
A stone of about 400g fell, O.A. Derby, Am. J. Sci., 1888, **36**, p.156. Classification, olivine $Fa_{24.6}$, C.B. Gomes et al., Meteoritics, 1977, **12**, p.241, abs., J.L. Berkley et al., Anais Acad. Brasil. Cienc., 1978, **50**, p.191.
43g Fort Worth, Texas, Monnig Colln.; 41g Chicago, Field Mus. Nat. Hist.; 36g Rio de Janeiro,; 2g Vienna, Naturhist. Mus.; Thin section, Albuquerque, Univ. of New Mexico;
Specimen(s): [65494], 1.5g.

Santa Catarina v Santa Catharina.

Santa Catarina Mountains v Tucson.

Santa Catharina 26°13′ S., 48°36′ W.
Santa Catarina, Brazil
Found 1875
Synonym(s): *Morro do Rocio, Rio San Francisco do Sul, San Francisco do Sul, Santa Catarina, Santa Catherina*
Iron. Ataxite, Ni-rich (IRANOM).
Large masses, of total weight about 7000kg, including six weighing respectively 2250kg, 1500kg, 450kg, 375kg, and 300kg, were found in clay overlying granite on the island of São Francisco, -. Lunay, C. R. Acad. Sci. Paris, 1877, **85**, p.84, O.A. Derby, Science, 1892, **20**, p.254. The total weight may have been as much as 25000kg, most of it appears to have been sent to England to be smelted for nickel, E.A. Wülfing, Die Meteoriten in Samml., Tübingen, 1897, p.308. Analysis, 33.97 %Ni, A.A. Damour, C. R. Acad. Sci. Paris, 1877, **84**, p.478. Classification, further analysis, 33.62 %Ni, 5.28 ppm.Ga, 9.07 ppm.Ge, 0.02 ppm.Ir, J.T. Wasson and R. Schaudy, Icarus, 1971, **14**, p.59. Description, references, V.F. Buchwald, Iron Meteorites, Univ. of California, 1975, p.1068.
2559kg Rio de Janeiro, Nat. Mus.; 52.3kg Paris, Mus. d'Hist. Nat.; 45kg Vienna, Naturhist. Mus.; 42.8kg Chicago, Field Mus. Nat. Hist.; 11kg Washington, U.S. Nat. Mus.; 8kg Vienna, Naturhist. Mus., minimum weight; 2.4kg Harvard Univ.; 1.5kg New York, Amer. Mus. Nat. Hist.; 1.8kg Budapest, Nat. Mus.; 1.1kg Tempe, Arizona State Univ.;
Specimen(s): [52283], 5785g. and fragments and filings, 200g; [51834], 24g.

Santa Catherina v Santa Catharina.

Santa Clara 24°28′ N., 103°21′ W.
Near Santa Clara, Durango, Mexico
Found 1976
Iron. Ataxite, Ni-rich (IVB).
A single mass of 63kg was purchased in 1977 by Arizona State Univ., Meteor. Bull., 1982 (60), Meteoritics, 1982, **17**, p.96. Classification and analysis, 17.9 %Ni, 0.22 ppm.Ga, 0.054 ppm.Ge, 18 ppm.Ir, A. Kracher et al., Geochimica et Cosmochimica Acta, 1980, **44**, p.773. Contains isotopically anomalous silver, T. Kaiser and G.J. Wasserburg, Geochimica et Cosmochimica Acta, 1983, **47**, p.43.
Main mass, Tempe, Arizona State Univ.;
Specimen(s): [1983,M.27], 1334g.

Santa Cruz 24°10′ N., 99°20′ W.
Tamaulipas, Mexico
Fell 1939, September 3, 1200 hrs
Stone. Carbonaceous chondrite, type II (CM2).
Several stones fell, total weight unknown, A.D. Nininger, Pop. Astron., Northfield, Minnesota, 1940, **48**, p.558, A.D. Nininger, Contr. Soc. Res. Meteorites, **2**, p.227 [M.A. 8-54]. Analysis, 23.1 % total iron, H.B. Wiik, Geochimica et Cosmochimica Acta, 1956, **9**, p.279. Classification, CM2, trace element analysis, G.W. Kallemeyn and J.T. Wasson, Geochimica et Cosmochimica Acta, 1981, **45**, p.1217.
24g Los Angeles, Univ. of California; 12g Tempe, Arizona State Univ.; Thin section, Washington, U.S. Nat. Mus.;
Specimen(s): [1959,782], 17g. and powder, 0.5g

Santa Fe v Glorieta Mountain.

Santa Fe County v Glorieta Mountain.

Santa Fe I v El Timbu.

Santa Giulietta v Alessandria.

Santa Isabel 33°54′ S., 61°42′ W.
Santa Fe, Argentina
Fell 1924, November 18, 0930 hrs
Stone. Olivine-hypersthene chondrite (L6).
A stone of 5.5kg fell. Description, with an analysis, E.H. Ducloux, Rev. fac. cienc. quim. Univ. nac. La Plata, 1926, **4**, p.21 [M.A. 4-424]. Further description, F. Pastore, Anal. Mus. Nac. Hist. Nat. Buenos Aires, 1925, **33**, p.306 [M.A. 3-389]. Olivine Fa_{24}, B. Mason, Geochimica et Cosmochimica Acta, 1963, **27**, p.1011.
3480g Buenos Aires, Mus. Nac. Hist. Nat.; 39g Paris, Mus. d'Hist. Nat.; 29g Washington, U.S. Nat. Mus.;
Specimen(s): [1962,168], 3.5g.

Sant Albano v Valdinizza.

Santa Luzia 16°16′ S., 47°57′ W.
Goias, Brazil
Found 1921
Synonym(s): *Santa Luzia de Goyaz, Santa Maria*
Iron. Octahedrite, coarsest (5mm) (IIB).
A mass of about 4.5kg was found; analysis, 5.67 %Ni, O.C. Farrington, letter of 15 November, 1926 in Min. Dept. BM (NH). Later a mass of 1890kg was found in 1927 in the Negro Monte ravine, 18 km from Santa Luzia (now Luziania). Description, with a poor analysis, 2.64 %Ni, E. de Oliveira, Anais Acad. Brasil. Cienc., 1931, **3**, p.33 [M.A. 5-15]. Mentioned, N. Vidal, Bol. Mus. Nac. Rio de Janeiro, 1931, **7**, p.9 [M.A. 5-406]. Two other masses, of 18.1kg and 4kg were found on the Upper Santa Maria river, 24 km from Santa Luzia, E.P. Henderson, letter of 12 May, 1939 in Min. Dept. BM(NH). Description, six masses found, analysis, 6.4 %Ni, V.B. Meen, Am. Miner., 1939, **24**, p.598 [M.A. 11-272]. A seventh mass of 1.64kg was found in the National Mus. at Rio de Janeiro and according to N. Vidal was found 100 metres from the main mass, W.S. Curvello, Bol. Mus. Nac. Rio de Janeiro, 1950 (geol. no. 9) [M.A. 11-272]. Further analysis, 6.39 %Ni, 47.9 ppm.Ga, 110 ppm.Ge, 0.010 ppm.Ir, J.T. Wasson, Geochimica et Cosmochimica Acta, 1969, **33**, p.859. Description, V.F. Buchwald, Iron Meteorites, Univ. of California, 1975, p.1072.
1890kg Rio de Janeiro, Nat. Mus.; 23kg Washington, U.S. Nat. Mus.; 3kg Chicago, Field Mus. Nat. Hist., main mass of 4.5kg find; 1.5kg New York, Amer. Mus. Nat. Hist.; 375g Paris, Mus. d'Hist. Nat.; 256g Tempe, Arizona State

Univ.;
Specimen(s): [1959,930], 181g. slice, and sawings, 4g

Santa Luzia de Goyaz v Santa Luzia.

Santa Maria v Santa Luzia.

Santa Rita v Tucson.

Santa Rosa 5°55′ N., 73°0′ W.
Tunja, Boyaca, Colombia
Found 1810
Synonym(s): *Bogota, Colombia, Neu Granada, New Granada, Rasgata, Zipaquira*
Iron. Ataxite, Ni-poor (IC).
A large mass, estimated at 750kg, and smaller pieces were found on the hill of Tocavita, near Santa Rosa, in 1810, and in 1824 the large mass was being used as an anvil in Santa Rosa, M. de Rivero and J.B. Boussingault, Ann. Chim. Phys., 1824, **25**, p.438. In 1874 it was placed on a pillar in the market place where it was seen in 1906 by H.A. Ward, who obtained a portion of 107kg, H.A. Ward, Am. J. Sci., 1907, **23**, p.1. Two other masses of 41kg and 22kg respectively were found in 1810 at the village of Rasgata near the saline of Zipaquira, M. de Rivero and J. Boussingault, loc. cit., p.442. Description, analysis, 6.52 % Ni, E. Cohen, Ann. Naturhist. Hofmus. Wien, 1894, **9**, p.111, E. Cohen, Ann. Naturhist. Hofmus. Wien, 1899, **13**, p.131. Two small specimens formerly attributed to this fall are of a plessitic octahedrite and identified as Tocavita; they are no. Me 1155, Chicago, Field Mus. Nat. Hist., and no. 2220, Tübingen Univ. Tocavita paired with Salt River, V.F. Buchwald, Iron Meteorites, Univ. of California, 1975, p.1048. Full description, history and analysis, 6.63 %Ni, 50.6 ppm.Ga, 222 ppm.Ge, 0.066 ppm.Ir, V.F. Buchwald and J.T. Wasson, Analecta geol., 1968 (3), V.F. Buchwald, Iron Meteorites, Univ. of California, 1975, p.1075.
460kg Bogota, Nat. Mus.; 138kg Bogota, Inst. Geofis.; 99.9kg Chicago, Field Mus. Nat. Hist.; 14kg Harvard Univ.; 5.7kg Washington, U.S. Nat. Mus.; 6.5kg New York, Amer. Mus. Nat. Hist.; 2kg Tübingen, Univ.; 1.5kg Vienna, Naturhist. Mus.; 1.6kg Rio de Janeiro, Nat. Mus.; 1.3kg Berlin, Humboldt Univ.; 1.2kg Vienna, Naturhist. Mus.; 1.1kg Tempe, Arizona State Univ.;
Specimen(s): [1907,132], 876g. slice, from Ward's Santa Rosa, and pieces totalling 50g; [35185], 88g. from G. Rose in 1863 and labelled 'Santa Rosa (Tocavita)'; [61306], 54g. from H. Heuland's specimen of Rasgata; [35409], 6g. from C.U. Shepard and labelled 'Santa Rosa bei Tunga (from Humboldt)'; [90231], Microprobe mount Santa Rosa; [94704], 233.5g. turnings

Santa Rosa v Coahuila.

Santa Rosalia 27°20′ N., 112°20′ W.
Baja California, Mexico
Found 1950, before this year
Synonym(s): *Parque*
Stony-iron. Pallasite (PAL).
A mass of 1631g was found, Cat. Meteor. Arizona State Univ., 1964, p.806. Possibly from near the village of Ignacio, H.H. Nininger, Find a falling star, Eriksson, New York, 1972, p.231. Listed under 'Parque', P.N. Chirvinsky, Pallasites, Moscow, 1969, p.20. Olivine Fa_{12}, P.R. Buseck and J.I. Goldstein, Bull. Geol. Soc. Amer., 1969, **80**, p.2141. References, analysis of metal, 10.1 %Ni, 19.3 ppm.Ga, 41.0 ppm.Ge, 0.17 ppm.Ir, E.R.D. Scott, Geochimica et Cosmochimica Acta, 1977, **41**, p.349.
675g Tempe, Arizona State Univ.; 63g Washington, U.S. Nat. Mus.; 50g Los Angeles, Univ. of California;

Santiago de Chile v Cobija.

Santiago del Estero v Campo del Cielo.

Santiago Papasquiero 24°30′ N., 106° W. approx.
Durango, Mexico
Found 1958, approx.
Iron. Ataxite, anomalous (IRANOM).
One mass of 119.5kg was found about 83 miles SW. of the village of Santiago Papasquiero, Meteor. Bull., 1965 (33). Classification and analysis, 7.48 %Ni, 0.573 ppm.Ga, 0.040 ppm.Ge, 4.0 ppm.Ir, R. Schaudy et al., Icarus, 1972, **17**, p.174. Description; recrystallised kamacite, with dispersed taenite, V.F. Buchwald, Iron Meteorites, Univ. of California, 1975, p.1080.
95kg Tempe, Arizona State Univ.; 1.26kg Chicago, Field Mus. Nat. Hist.; 1.1kg Washington, U.S. Nat. Mus.; 653g Harvard Univ.; 343g New York, Amer. Mus. Nat. Hist.; 155g Copenhagen, Univ. Geol. Mus.;
Specimen(s): [1966,283], 138g.

Santillo v Coahuila.

São João Nepomuceno 21°33′ S., 43°1′ W.
Minas Gerais, Brazil
Found
Iron. Octahedrite, fine, (0.3mm), with silicate inclusions (IVA-ANOM).
A specimen of this iron is in Washington (U.S. Nat. Mus.). Analysis, 8.0 %Ni, 2.16 ppm.Ga, 0.118 ppm.Ge, 2.6 ppm.Ir, A. Kracher et al., Geochimica et Cosmochimica Acta, 1980, **44**, p.773. Reported, Meteor. Bull., 1980 (58), Meteoritics, 1980, **15**, p.239.
Specimen, Washington, U.S. Nat. Mus.;

São Jose do Rio Preto 20°48′36″ S., 49°22′50″ W.
São Paulo, Brazil
Fell 1962, August 14, 0800 hrs
Stone. Olivine-bronzite chondrite (H4).
A mass of 927g fell. Description, with a modal analysis and imperfect chemical analysis, J.M.V. Coutinho and F.M. Arid, Bol. Soc. Brasil. Geol., 1963, **12**, p.75. Petrography, analysis, 26.78 % total iron, olivine $Fa_{18.7}$, C.B. Gomes et al., Naturalia, 1978, **4**, p.25. The correct spelling is São Jose de Rio Preto, F.E. Wickman, letter of 26 June, 1972 in Min. Dept. BM(NH).
7.5g Washington, U.S. Nat. Mus.;

São Julião de Moreira 41°45′58″ N., 8°35′1″ W.
Ponte do Lima, Minho, Portugal
Found 1883, known before this year
Synonym(s): *Moreira do Lima, Ponte de Lima, San Juliano*
Iron. Octahedrite, coarsest (6mm) (IIB).
A mass of 162kg was ploughed up, A. Ben-Saude, Comun. Comm. Trab. Geol. Portugal, Lisbon, 1888, **2** (1), p.14, Neues Jahrb. Min., 1888, **2**, p.371. Analysis, 6.02 %Ni, E. Cohen, Neues Jahrb. Min., 1889, p.213. Classification, further analysis, 6.1 %Ni, 46.2 ppm.Ga, 107 ppm.Ge, 0.013 ppm.Ir, J.T. Wasson, Geochimica et Cosmochimica Acta, 1969, **33**, p.859. Co-ordinates, phosphide-rich, V.F. Buchwald, Iron Meteorites, Univ. of California, 1975, p.1083. History, description, C. Teixeira, Mem. Acad. Cienc. Lisboa, Classe Cienc., 1968, **12**, p.213. The location of the main mass is not known, it was of 159kg and in the Lisbon

Mus. in 1897, it may have been cut and the material distributed.
5kg Vienna, Naturhist. Mus.; 2.7kg Chicago, Field Mus. Nat. Hist.; 1.7kg New York, Amer. Mus. Nat. Hist.; 1.5kg Budapest, Nat. Mus.; 1.8kg Yale Univ.; 1kg Schönenwerd, Bally-Prior Mus.; 2150g Washington, U.S. Nat. Mus.; 815g Harvard Univ.; 563g Copenhagen, Univ. Geol. Mus.; 304g Prague, Nat. Mus.; 660g Paris, Mus. d'Hist. Nat.;
Specimen(s): [83315], 681.5g. and fragments, 35g; [1921,385], 139g.; [68215], 9g.

Saonlod v Khetri.

São Sebastião da Boa Vista 31°6′ S., 54°15′ W. perhaps
Brazil
Fell 1914
Synonym(s): *San Sebastiano da Boa Vista*
Doubtful..
An immense meteorite is said to have fallen, H. Michel, Fortschr. Min. Krist. Petr., 1922, *7*, p.263, Prometheus, 1915, **26**, p.84. São Sebastião is a very common name in Brazil; the place of fall has not been definitely identified.

Saotome 36°30′ N., 137° E.
Toyama, Honshu, Japan
Found 1892
Synonym(s): *Saotomi*
Iron. Octahedrite, fine.
A mass of 10.88kg was found. Perhaps part of Shirahagi, I. Yamamoto, Kwasan Observ. Bull., 1935, p.306 [M.A. 7-173], S. Murayama, letter of 6 April, 1962 in Min. Dept. BM (NH). Almost certainly part of Shirahagi (*q.v.*), S. Murayama, letter of 12 October, 1980 in Min. Dept. BM (NH).
4.22kg Toyama, Science Cultural Centre; 4.6kg privately held;

Saotomi v Saotome.

Sapahi v Supuhee.

Saratov v Pavlovka.

Saratov v Sarepta.

Saratov 52°33′ N., 46°33′ E.
Near Donguz, Penza, Federated SSR, USSR, [Саратов]
Fell 1918, September 6, 1500 hrs
Synonym(s): *Belaya Gora, Donguz, Saratow*
Stone. Olivine-hypersthene chondrite (L4).
After the appearance of a fireball, and detonations, several stones fell at Donguz, 52°35'N., 46°16'E., and Belaya Gora, 52°31'N., 46°49'E., of total weight about 328kg, P.N. Chirvinsky, Centralblatt Min., 1923, p.585, L.A. Kulik, Bull. Acad. Sci. Russ., 1922, **16**, p.391. Listed, I.S. Astapowitsch, Trans. Roy. Astron. Soc. Canada, 1938, **32**, p.195 [M.A. 7-172]. Micrometric analysis, B.M. Kupletsky, Метеоритика, 1941, **2**, p.75 [M.A. 9-296]. Analyses of magnetic and non-magnetic portions, L.A. Selivanov, C. R. (Doklady) Acad. Sci. URSS, 1940, **26**, p.389 [M.A. 8-57]. Contains native copper, I.A. Yudin, Zap. Vsecoiuz Min. Obsh., 1956, **85**, p.403 [M.A. 13-647]. Olivine Fa_{24}, B. Mason, Geochimica et Cosmochimica Acta, 1963, **27**, p.1011. Ar-Ar age, 4440 m.y., G. Turner et al., Geochimica et Cosmochimica Acta, 1978 (suppl. 10), p.989.
159kg Saratov Univ., main mass; 52kg Moscow, Acad. Sci.; 513g Washington, U.S. Nat. Mus.; 509g Los Angeles, Univ. of California; 373g Tempe, Arizona State Univ.; 146g Chicago, Field Mus. Nat. Hist.; 93g Berlin, Humboldt Univ.;
Specimen(s): [1959,169], 250g.

Saratow v Saratov.

Sarbanovac v Soko-Banja.

Sardis 32°56′56″ N., 81°51′54″ W.
Burke County, Georgia, U.S.A.
Found 1940
Iron. Octahedrite, coarse (2.5mm) (IA).
A mass of 800kg was found when ploughing in Jenkins Co., 6.25 miles WSW. of Sardis, Burke County. Description, with an analysis, 6.69 %Ni, E.P. Henderson and C.C. Cooke, Proc. U.S. Nat. Mus., 1942, **92** (3143), p.141. Classification, analysis, 6.58 %Ni, 93.7 ppm.Ga, 400 ppm.Ge, 1.3 ppm.Ir, J.T. Wasson, Meteorites, Springer-Verlag, 1974, p.297. Description; heavily corroded, V.F. Buchwald, Iron Meteorites, Univ. of California, 1975, p.1084.
800kg Washington, U.S. Nat. Mus., main mass; 800g Chicago, Field Mus. Nat. Hist.; 1050g Copenhagen, Univ. Geol. Mus.; 741g Perth, West. Austr. Mus.; 400g Moscow, Acad. Sci.;
Specimen(s): [1966,50], 192g. slice

Sarepta 48°29′ N., 44°49′ E.
Volgograd region, Federated SSR, USSR, [Сарепта]
Found 1854
Synonym(s): *Saratov*
Iron. Octahedrite, coarse (2.2mm) (IA).
A mass of about 14kg was found on the right bank of the Volga, 20 miles from Sarepta, J. Auerbach, Bull. Soc. Nat. Moscou, 1854 (4), p.504. Description, W. von Haidinger, Sitzungsber. Akad. Wiss. Wien, Math.-naturwiss. Kl., 1864, **49** (2), p.497. Analysis, 6.94 %Ni, M.I. D'yakonova, Метеоритика, 1958, **16**, p.180. Further analysis, 6.55 %Ni, 99.9 ppm.Ga, 457 ppm.Ge, 3.4 ppm.Ir, J.T. Wasson, Icarus, 1970, **12**, p.407. Description; pre-atmospheric fissuring, V.F. Buchwald, Iron Meteorites, Univ. of California, 1975, p.1086.
2kg Berlin, Humboldt Univ., main mass; 1.25kg Moscow, Acad. Sci.; 1kg Paris, Mus. d'Hist. Nat.; 750g Vienna, Naturhist. Mus.; 427g Harvard Univ.; 339g Tübingen, Univ.; 286g Chicago, Field Mus. Nat. Hist.; 254g Washington, U.S. Nat. Mus.; 207g Tempe, Arizona State Univ.;
Specimen(s): [33750], 223g.; [36605(3)], 60.5g.

Saricam v Ibrisim.

Saros v Lenarto.

Sarratola 24°24′ N., 81°53′ E.
Sidhi tahsil, Madhya Pradesh, India
Found 1948, known in this year
Synonym(s): *Rewah State*
Stone. Chondrite, veined.
One small stone of 265g was received by the Geol. Surv. of India in September, 1948, P. Chatterjee, letter of 25 March, 1952 with a partial description, in Min. Dept. BM(NH). Mentioned, M.A.R. Khan, Hyderabad Acad. Studies, 1950, **12** [M.A. 11-441].
212g Calcutta, Mus. Geol. Surv. India;

Sarthe v Lucé.

Sasagase 34°43′ N., 137°47′ E.
Hamamatsu-shi, Khizuoka-ken, Honshu, Japan
Fell 1688, February 13, 1200 hrs
Stone. Olivine-bronzite chondrite (H).
A mass of about 695g was seen to fall near the temple Zofukuji, and was kept there; recorded and described, S. Kanda, Tenmon-soho, 1950, **4** (4), p.27. Description, analysis, S. Murayama et al., Jap. J. Geol. Geogr., 1962, **33**, p.239 [M.A. 17-56]. Olivine Fa_{19}, J.T. Wasson, Meteorites, Springer-Verlag, 1974, p.272.
Specimen(s): [1978,M.18], 1.4g.

Saskatchewan v Iron Creek.

Satsuma v Kyushu.

Sauguis 43°9′ N., 0°51′ W.
St. Etienne, Basses-Pyrénées, France
Fell 1868, September 7, 0230 hrs
Synonym(s): *Mauléon, Sauguis-Saint-Etienne, Sauguls*
Stone. Olivine-hypersthene chondrite (L6).
After the appearance of a fireball, followed by detonations, a stone fell into a stream and was broken into pieces, 2kg to 4kg of this was collected. Report and analysis, G.A. Daubrée, C. R. Acad. Sci. Paris, 1868, **67**, p.873. Olivine Fa_{25}, B. Mason, Geochimica et Cosmochimica Acta, 1963, **27**, p.1011.
140g Paris, Mus. d'Hist. Nat.; 13g Chicago, Field Mus. Nat. Hist.; 13g Vienna, Naturhist. Mus.;
Specimen(s): [71572], 11g.; [48758], 4.5g. two fragments

Sauguis-Saint-Etienne v Sauguis.

Sauguls v Sauguis.

Saurette v Apt.

Sautschenskoje v Savtschenskoje.

Savannah 35°10′ N., 88°11′ W.
Hardin County, Tennessee, U.S.A.
Found 1923, approx.
Iron. Octahedrite, medium (1.2mm) (IIIA).
An oxidized, dumb-bell shaped mass of about 135lb (60kg) was found on the high road between Savannah and Cerro Gordo, about 4 miles NE. of Savannah. Description, with an analysis, 7.76 %Ni, G.P. Merrill, Proc. U.S. Nat. Mus., 1923, **63** (18), p.2. Classification and further analysis, 7.99 % Ni, 21.1 ppm.Ga, 44.2 ppm.Ge, 0.6 ppm.Ir, E.R.D. Scott et al., Geochimica et Cosmochimica Acta, 1973, **37**, p.1957. Description; it consists of two parent austenite crystals, V.F. Buchwald, Iron Meteorites, Univ. of California, 1975, p.1088.
30kg Nashville, Geol. Surv. Tennessee; 27kg Washington, U.S. Nat. Mus.;

Savcenskoje v Savtschenskoje.

Savchenskoe v Savtschenskoje.

Savik v Cape York.

Saviksue v Cape York.

Savtschenskoje
Tiraspol, Kherson, Ukraine, USSR, [Савченское]
Fell 1894, July 27, 2000 hrs
Synonym(s): *Cherson, Kherson, Sautschenskoje, Savcenskoje, Savchenskoe, Sawtschenskoje*
Stone. Olivine-hypersthene chondrite, amphoterite (LL4).
After three detonations, a stone of about 2.5kg fell, R. Prendel, Mem. Soc. Nat. Nouvelle Russie, Odessa, 1895, **20** (1), p.49. Analysis, 18.64 % total iron, M.I. D'yakonova and V.Ya. Kharitonova, Метеоритика, 1961, **21**, p.52. Mentioned, L.A. Kulik, Метеоритика, 1941, **1**, p.73 [M.A. 9-294], E.L. Krinov, Каталог Метеоритов Акад. Наук СССР, Москва, 1947 [M.A. 10-511]. Olivine Fa_{28}, B. Mason, Geochimica et Cosmochimica Acta, 1963, **27**, p.1011.
2108g Moscow, Acad. Sci.; 38g Vienna, Naturhist. Mus.; 37g New York, Amer. Mus. Nat. Hist.; 25g Chicago, Field Mus. Nat. Hist.; 13g Budapest, Nat. Hist. Mus.; 8g Berlin, Humboldt Univ.;
Specimen(s): [83490], 62g.

Sawetnoe v Zavetnoe.

Sawetnoje v Zavetnoe.

Sawiauk v Pillistfer.

Sawotaipola v Luotolax.

Saxeville v Pine River.

Sazovice 49°14′ N., 17°34′ E.
Zlin, Moravia, Czechoslovakia
Fell 1934, June 28, 2000 hrs
Stone. Olivine-hypersthene chondrite (L), veined.
A mass of 411.98g fell. Description, with an analysis, J. Kokta, Casopis Moravskeho, Zemskeho Brno, 1937, **30**, p.1, J. Kokta, Coll. Czech. Chem. Comm., 1937, **9**, p.471 [M.A. 7-173].
Main mass, Brno, County Mus., in 1938; 10.5g Prague, Nat. Mus.;

Schaap-Koi v Schaap-Kooi.

Schaap-Kooi 32°5′ S., 21°20′ E.
Fraserburg district, Cape Province, South Africa
Found 1910, known since this year
Stone. Olivine-bronzite chondrite (H4).
A stone of 2.7kg was found on the farm Schaapkooi, C. Frick and E.C.I. Hammerbeck, Bull. Geol. Surv. S. Africa, 1973 (57), p.32. Analysis, 26.32 % total iron, H. von Michaelis et al., Earth planet. Sci. Lett., 1969, **5**, p.387. Olivine Fa_{19}, B. Mason, Geochimica et Cosmochimica Acta, 1963, **27**, p.1011.
5.5lb Cape Town, South African Mus.; 164g Vienna, Naturhist. Mus.;
Specimen(s): [1950,254], 148.5g.

Scheikahr-Stattan v Buschof.

Scheikahr Statten v Buschof.

Schellin 53°21′ N., 15°3′ E. approx.
Stargard, Stettin, Poland
Fell 1715, April 11, 1600 hrs
Synonym(s): *Garz, Skalen, Skalin, Skölen*
Stone. Olivine-hypersthene chondrite (L), veined.
After detonations, two stones fell, one of about 7kg and the other as large as a goose egg, but very little has been preserved, L.W. Gilbert, Ann. Phys. (Gilbert), 1822, **71**, p.213, J. Pokrzywnicki, Urania, 1955 (6), p.165. Olivine Fa_{25}, B. Mason, Geochimica et Cosmochimica Acta, 1963, **27**, p.1011. 341g were in 1897 in the possession of the Bredow family in Wagenitz, Brandenburg.
5g Berlin, Humboldt Univ.; 4.5g Tübingen, Univ.; 2g Vienna, Naturhist. Mus.; 0.5g Chicago, Field Mus. Nat. Hist.;
Specimen(s): [33726], less than 1g

Schenectady 42°51′39″ N., 73°57′1″ W.
Glenville, New York, U.S.A.
Fell 1968, April 12, 2030 hrs
Stone. Olivine-bronzite chondrite (H5).
A crusted mass of 283.3g was recovered two days after its fall, in which it struck the roof of a house, Meteor. Bull., 1968 (44), Meteoritics, 1970, **5**, p.99, R.L. Fleischer et al., Meteoritics, 1969, **4**, p.171, abs. Classification, B. Mason, Smithson. Contrib. Earth Sci., 1975 (14), p.71.
Main mass, Schenectady, Mus.; 30g Washington, U.S. Nat. Mus.;

Scherbanovaz v Soko-Banja.

Schertz v Cañon Diablo.

Schie v Ski.

Schleusingen 50°31′ N., 10°46′ E.
Suhl, Germany
Fell 1552, May 19
Doubtful. Stone.
Many stones fell, one is said to have killed a horse, E.F.F. Chladni, Die Feuer-Meteore, Wien, 1819, p.213. The evidence of meteoritic nature is not conclusive.

Schloin v Grüneberg.

Schobergrund v Gnadenfrei.

Scholakov v Bachmut.

Schoneberg v Schönenberg.

Schonen v Lundsgård.

Schönenberg 48°7′ N., 10°28′ E.
Pfaffenhausen, Schwaben, Bayern, Germany
Fell 1846, December 25, 1400 hrs
Synonym(s): *Schoneberg*
Stone. Olivine-hypersthene chondrite (L6), veined.
After detonations, a stone of about 8kg was seen to fall, L. Landbeck, Jahres Ver. vaterl. Naturk. Württemberg, Stuttgart, 1847, **2**, p.383. Description, with a doubtful analysis, C.W. Gümbel, Sitzungsber. Akad. Wiss. München, Math.-phys. Kl., 1878, **8**, p.40. Olivine Fa_{25}, B. Mason, Geochimica et Cosmochimica Acta, 1963, **27**, p.1011.
135g Paris, Mus. d'Hist. Nat.; 98g Harvard Univ.; 80g Budapest, Nat. Mus.; 37g New York, Amer. Mus. Nat. Hist.; 32g Chicago, Field Mus. Nat. Hist.; 26g Vienna, Naturhist. Mus.;
Specimen(s): [67208], 39.46g. and fragments, 0.85g

Schowtnewi Chutor v Zhovtnevyi.

Schuscha v Indarch.

Schwetz 53°24′ N., 18°27′ E.
Kwidzyn, Poland
Found 1850
Synonym(s): *Kwidzyn, Swiecie, Weichsel*
Iron. Octahedrite, medium (1.0mm) (IIIA).
A mass of about 21.5kg was found, 4 feet below the surface, in making a road, G. Rose, Ann. Phys. Chem. (Poggendorff), 1851, **83**, p.594. Analysis, 7.44 %Ni, 18.3 ppm.Ga, 33.5 ppm.Ge, 11 ppm.Ir, E.R.D. Scott et al., Geochimica et Cosmochimica Acta, 1973, **37**, p.1957. Description; shocked, V.F. Buchwald, Iron Meteorites, Univ. of California, 1975, p.1090.
10kg Berlin, Humboldt Univ.; 942g Los Angeles, Univ. of California; 843g Vienna, Naturhist. Mus.; 676g Tübingen, Univ.; 655g Budapest, Nat. Mus.; 534g Warsaw, Mus. of the Earth, Polish Acad. Sci.; 532g Tempe, Arizona State Univ.; 334g Copenhagen, Univ. Geol. Mus.; 257g Yale Univ.; 165g Washington, U.S. Nat. Mus.; 91g Chicago, Field Mus. Nat. Hist.;
Specimen(s): [33929], 1060g.

Schwiebus v Seeläsgen.

Scott City 38°28′ N., 100°56′ W.
Scott County, Kansas, U.S.A.
Found 1905
Stone. Olivine-bronzite chondrite (H5).
A stone of 135g was found in 1905, and another of about 2kg in 1911, G.P. Merrill, Proc. U.S. Nat. Mus., 1912, **42**, p.295, G.P. Merrill, Am. J. Sci., 1906, **21**, p.360. Olivine Fa_{20}, B. Mason, Geochimica et Cosmochimica Acta, 1963, **27**, p.1011.
935g Chicago, Field Mus. Nat. Hist.; 228g Washington, U.S. Nat. Mus.; 155g Harvard Univ.; 139g New York, Amer. Mus. Nat. Hist.; 54 Paris, Mus. d'Hist. Nat.; 49g Tempe, Arizona State Univ.;
Specimen(s): [1919,144], 124g.

Scottsville 36°46′ N., 86°10′ W.
Allen County, Kentucky, U.S.A.
Found 1867
Synonym(s): *Allen County*
Iron. Hexahedrite (IIA).
A mass of about 10kg was found, J.E. Whitfield, Am. J. Sci., 1887, **33**, p.500. Description, analysis, 5.33 %Ni, E. Cohen, Meteoritenkunde, 1905, **3**, p.218. Further description, V.F. Buchwald, Iron Meteorites, Univ. of California, 1975, p.1091. Further analysis, 5.31 %Ni, 60.4 ppm.Ga, 172 ppm.Ge, 49 ppm.Ir, J.T. Wasson, Geochimica et Cosmochimica Acta, 1969, **33**, p.859.
1.1kg Chicago, Field Mus. Nat. Hist.; 1.5kg Vienna, Naturhist. Mus.; 1042g Washington, U.S. Nat. Mus.; 536g Harvard Univ.; 500g Budapest, Nat. Mus.; 532g New York, Amer. Mus. Nat. Hist.; 490g Rome, Vatican Colln; 123g Tempe, Arizona State Univ.; 85g Yale Univ.;
Specimen(s): [62871], 404g.

Scriba 43°27′ N., 76°26′ W.
Oswego County, New York, U.S.A.
Pseudometeorite..
A mass of 2.7kg was found, C.U. Shepard, Am. J. Sci., 1841, **40**, p.366. It is still occasionally included as a doubtful iron though it was shown to be manufactured iron in 1898, E. Cohen, Ann. Naturhist. Hofmus. Wien, 1898, **13**, p.49.

Scurry 32°30′ N., 101° W. approx.
Mitchell County, Texas, U.S.A.
Found 1937
Stone. Olivine-bronzite chondrite (H5).
About 115kg were found, olivine Fa_{19}, B. Mason, Geochimica et Cosmochimica Acta, 1963, **27**, p.1011.
117kg Fort Worth, Texas, Monnig Colln.; 72g Mainz, Max-Planck-Inst.; 56g Copenhagen, Univ. Geol. Mus.; 52g Tempe, Arizona State Univ.; 26g Washington, U.S. Nat. Mus.;
Specimen(s): [1969,232], 41g.

Seagraves 32°56′ N., 102°35′ W.
Gaines County, Texas, U.S.A.
Found 1962
Stone. Olivine-bronzite chondrite (H4).
Reported, Meteor. Bull., 1967 (39), Meteoritics, 1970, **5**, p.91. Olivine Fa_{18}, B. Mason, Geochimica et Cosmochimica Acta, 1967, **31**, p.1100.
13kg Washington, U.S. Nat. Mus.;

Searsmont 44°22′ N., 69°12′ W.
Waldo County, Maine, U.S.A.
Fell 1871, May 21, 0815 hrs
Synonym(s): *Searsport, Seassport, Waldo County*
Stone. Olivine-bronzite chondrite (H5).
After an explosion, a stone was seen to fall which was broken into fragments on impact, totalling about 12lb; only about 2lb appear to have been preserved. Report and doubtful analysis, C.U. Shepard, Am. J. Sci., 1871, **2**, p.133. Olivine Fa_{19}, B. Mason, Geochimica et Cosmochimica Acta, 1963, **27**, p.1011.
869g Tempe, Arizona State Univ.; 90g Washington, U.S. Nat. Mus.; 50g Harvard Univ.; 45g New York, Amer. Mus. Nat. Hist.; 33g Paris, Mus. d'Hist. Nat.; 28g Vienna, Naturhist. Mus.;
Specimen(s): [44326], 49.8g. two fragments

Searsport v Searsmont.

Seassport v Searsmont.

Seattle v Kirkland.

Sediköy 38°18′ N., 27°8′ E.
Izmir, Turkey
Fell 1917
Stone. Olivine-hypersthene chondrite (L6).
Fragments totalling 240g were in the possession of Prof. A.M. Sayar in 1937, A.M. Sayar, letters of 3 and 11 September, and 26 November, 1937 in Min. Dept. BM(NH). Classification, olivine $Fa_{24.1}$, A.L. Graham, priv. comm., 1982.
Specimen(s): [1970,171], 1g.

Sedlcany v Selcany.

Seelaesgen v Seeläsgen.

Seeläsgen 52°16′ N., 15°33′ E.
Schwiebus (= Swiebodzin), Zielona Gora, Poland
Found 1847, known before this year
Synonym(s): *Brandenburg, Branibor, Przelazy, Seelaesgen, Sulechow, Schwiebus, Züllichau*
Iron. Octahedrite, coarse (3.1mm) (IA).
A mass of about 102kg was found in draining a field, and several years afterwards, in 1847, was recognized as meteoritic, W.G. Schneider, Ann. Phys. Chem. (Poggendorff), 1848, **74**, p.57. Analysis, 6.23 %Ni, C. Rammelsberg, Ann. Phys. Chem. (Poggendorff), 1848, **74**, p.443. The mass has been mildly shocked and cosmically annealed, then artificially heated to 600C; includes Sulechow, V.F. Buchwald, Iron Meteorites, Univ. of California, 1975, p.1094. Classification, further analysis, 6.47 %Ni, 96.8 ppm.Ga, 493 ppm.Ge, 1.1 ppm.Ir, J.T. Wasson, Icarus, 1970, **12**, p.407. Paired with Morasko, A. Kracher et al., Geochimica et Cosmochimica Acta, 1980, **44**, p.773.
15kg Tübingen, Univ.; 6.5kg Vienna, Naturhist. Mus.; 3.7kg Berlin, Humboldt Univ.; 3.5kg Bonn, Univ. Mus.; 1.2kg Moscow, Acad. Sci.; 1kg Chicago, Field Mus. Nat. Hist.; 1kg Budapest, Nat. Mus.; 1kg Prague, Nat. Mus.; 944g Harvard Univ.; 407g Washington, U.S. Nat. Mus.;
Specimen(s): [33927], 8135g.; [23020], 1578g. and fragments, 19.5g; [33197], 70.5g.; [33928], 24g. and sawings, 1.5g

Segowlie 26°45′ N., 84°47′ E.
Bettiah, Champaran district, Bihar, India
Fell 1853, March 4, 1200 hrs
Synonym(s): *Sagauli, Soojoolee*
Stone. Olivine-hypersthene chondrite (L6).
About 30 stones, varying in weight from about 0.5lb to 14lb, were seen to fall, W.S. Sherwill, J. Asiatic Soc. Bengal, 1854, **23**, p.746, T. Oldham et al., J. Asiatic Soc. Bengal, 1859, **28**, p.261, W. von Haidinger, Sitzungsber. Akad. Wiss. Wien, Math.-naturwiss. Kl., 1860, **41**, p.754. Mineralogy, analysis, 21.45 % total iron, olivine $Fa_{25.0}$, A. Dube et al., Smithson. Contrib. Earth Sci., 1977 (19), p.71. The place of fall is Sagouli, C.A. Silberrad, Min. Mag., 1932, **23**, p.294.
4.5kg Calcutta, Mus. Geol. Surv. India; 996g Vienna, Naturhist. Mus.; 184g Chicago, Field Mus. Nat. Hist.; 123g Harvard Univ.; 109g Tempe, Arizona State Univ.;
Specimen(s): [34802], 639g.; [34803], 539g.; [34804], 27.5g.; [96277], thin sections (three)

Seguin 39°22′ N., 100°38′ W.
Sheridan County, Kansas, U.S.A.
Found 1956
Stone. Olivine-bronzite chondrite (H6).
750g were found, olivine Fa_{20}, B. Mason, Geochimica et Cosmochimica Acta, 1963, **27**, p.1011.
49g Washington, U.S. Nat. Mus.;

Seibert 39°18′ N., 102°50′ W.
Kit Carson County, Colorado, U.S.A.
Found 1941
Stone. Olivine-bronzite chondrite (H5).
One stone of 3.5kg was found, F.C. Leonard, Pop. Astron., Northfield, Minnesota, 1947, **55**, p.381, Contr. Meteoritical Soc., **4**, p.58 [M.A. 10-177], H.H. and A.D. Nininger, The Nininger Collection of Meteorites, Winslow, Arizona, 1950, p.94. Olivine Fa_{18}, B. Mason, Geochimica et Cosmochimica Acta, 1963, **27**, p.1011.
1533g Tempe, Arizona State Univ.; 401g Denver, Mus. Nat. Hist.; 139g Chicago, Field Mus. Nat. Hist.; 131g Washington, U.S. Nat. Mus.;
Specimen(s): [1959,899], 1454.5g.

Seidam v Geidam.

Seifersdorf v Grüneberg.

Seifersholz v Grüneberg.

Selakopi 7°14′ S., 107°20′ E.
Near Bandung, Java, Indonesia
Fell 1939, September 26
Stone. Olivine-bronzite chondrite (H5).
A single stone of 1590g is said to have fallen about 30 miles SSW. of Bandung on the above date. The date of fall is the same as that reported for Glanggang (*q.v.*). Distinct from Glanggang, olivine $Fa_{18.2}$, K. Fredriksson and G.S. Peretsman, Meteoritics, 1982, **17**, p.77.

Selcany 49°45′ N., 14°25′ E.
Votice, Pribram, Czechoslovakia
Found 1900
Synonym(s): *Sedlcany*
Iron. Octahedrite, coarse.
One small mass of 20g was found, K. Tucek, Cat. Coll. Meteor., Nat. Mus. Prague, 1968, p.66.
20g Prague, Nat. Mus.;

Seldebourak 22°50′ N., 4°59′ E.
Ahaggar, Algeria
Fell 1947, February 26, 1545 hrs
Stone. Olivine-bronzite chondrite (H5).
A stone fell 55 km W. of Tamanrasset (22°50'N., 5°30'E.), and broke on striking a rock, a fragment of 150g was recovered. Description, with an analysis, E. Jérémine and M. Lelubre, C. R. Acad. Sci. Paris, 1949, **229**, p.425 [M.A. 11-136]. Olivine Fa_{19}, B. Mason, Geochimica et Cosmochimica Acta, 1963, **27**, p.1011.
80g Paris, Mus. d'Hist. Nat.;

Selden 39°32′ N., 100°34′ W. approx.
Sheridan County, Kansas, U.S.A.
Found 1960, recognized in this year
Stone. Chondrite.
A stone of 1.56kg is in Fort Hays, Kansas, Meteor. Bull., 1969 (47), Meteoritics, 1970, **5**, p.105.
1.56kg Fort Hays, Fort Hays Kansas State College;

Seligman 35°17′ N., 112°52′ W.
Coconino County, Arizona, U.S.A.
Found 1949
Iron. Octahedrite, coarse (2.3mm) (IA).
A fresh mass of 2.2kg was found, showing a distinct heating zone; rich in cohenite and schreibersite, H.H. Nininger, letter of 22 May 1950 in Min. Dept. BM(NH). Listed, H.H. and A.D. Nininger, The Nininger Collection of Meteorites, Winslow, Arizona, 1950, p.133, 138. Chemical analysis, 6.69 %Ni, 91 ppm.Ga, 423 ppm.Ge, 3.1 ppm.Ir, J.T. Wasson, J. Geophys. Res., 1968, **73**, p.3207, J.T. Wasson, Icarus, 1970, **12**, p.407. Description, V.F. Buchwald, Iron Meteorites, Univ. of California, 1975, p.1099.
1.2kg Mainz, Max-Planck-Inst., main mass; 274g Tempe, Arizona State Univ.; 147g Washington, U.S. Nat. Mus.;
Specimen(s): [1959,939], 150g.

Selma 32°24′ N., 87°0′ W.
Dallas County, Alabama, U.S.A.
Found 1906
Stone. Olivine-bronzite chondrite (H4).
A stone of 310lb was found; description, G.P. Merrill, Proc. U.S. Nat. Mus., 1907, **32**, p.59. Analysis (doubtful), G.P. Merrill, Mem. Nat. Acad. Sci. Washington, 1916, **14** (1), p.21. Description; weathered and not associated with the meteor of 20 July, 1898; analysis, 22.08 % total iron, B. Mason and H.B. Wiik, Am. Mus. Novit., 1960 (2010).
28kg New York, Amer. Mus. Nat. Hist.; 713g Tempe, Arizona State Univ.; 636g Ottawa, Mus. Geol. Surv. Canada; 555g Calcutta, Mus. Geol. Surv. India; 510g Perth, West. Austr. Mus.;
Specimen(s): [1961,470], 420g.

Semarkona 22°15′ N., 79° E. approx.
Sanjari tahsil, Chhindwara district, Madhya Pradesh, India
Fell 1940, October 26
Stone. Olivine-hypersthene chondrite, amphoterite (LL3).
Two pieces, 423g and 268g, were recovered, M.A.R. Khan, Hyderabad Acad. Studies, 1950, **12** [M.A. 11-441], P. Chatterjee, letter of 25 March, 1952 in Min. Dept. BM(NH). Classification, analysis, 19.09 % total iron, R.T. Dodd et al., Geochimica et Cosmochimica Acta, 1967, **31**, p.921. Classification confirmed, E. Jarosewich and R.T. Dodd, Meteoritics, 1981, **16**, p.83.
386g Calcutta, Mus. Geol. Surv. India; 245g Washington, U.S. Nat. Mus.; 27g New York, Amer. Mus. Nat. Hist.;

Seminole 32°41′ N., 102°37′ W.
Gaines County, Texas, U.S.A.
Found 1961, recognized 1963
Stone. Olivine-bronzite chondrite (H4).
Two fragments totalling 41.1kg were recovered. Full, illustrated description, olivine Fa_{20}, G.I Huss et al., Meteoritics, 1972, **7**, p.463. Two other stones, Seminole (b), total known weight 1183g, and Seminole (c), total known weight 1710g, are also listed, Cat. Huss Coll. Meteorites, 1976, p.39.
26.1kg Mainz, Max-Planck-Inst.; 457g Washington, U.S. Nat. Mus.; 259g Copenhagen, Univ. Geol. Mus.; 253g Harvard Univ.;
Specimen(s): [1965,412], 170.5g.

Seminole Draw (a)
32°43′ N., 102°39′ W. approx.
Gains County, Texas, U.S.A.
Found 1976
Stone. Olivine-hypersthene chondrite (L6).
A single mass of 478g was found, G.I Huss, priv. comm., 1983. Olivine $Fa_{25.9}$, A.L. Graham, 1983.
Specimen(s): [1983,M.25], 35g.

Semipalatinsk v Pavlodar (pallasite).

Sena 41°43′ N., 0°3′ W.
Sarinena, Huesca, Spain
Fell 1773, November 17, 1200 hrs
Synonym(s): *Sigena*
Stone. Olivine-bronzite chondrite (H4), brecciated.
After detonations, a stone of about 4kg was seen to fall, L. Proust, Ann. Phys. (Gilbert), 1806, **24**, p.261. Description, mineralogy, M. Christophe Michel-Levy, Bull. Mineral., 1979, **102**, p.410. Olivine Fa_{17}, B. Mason, Geochimica et Cosmochimica Acta, 1963, **27**, p.1011.
1.7kg Madrid, Mus. Cienc. Nat.; 167g Washington, U.S. Nat. Mus.; 73g Paris, Mus. d'Hist. Nat.; 101g Tempe, Arizona State Univ.; 94g Budapest, Nat. Mus.; 44g New York, Amer. Mus. Nat. Hist.; 29g Vienna, Naturhist. Mus.;
Specimen(s): [86641], less than a gram; [1927,9], 0.25g.

Seneca 39°50′ N., 96°4′ W.
Nemaha County, Kansas, U.S.A.
Found 1936
Stone. Olivine-bronzite chondrite (H4).
Two fragments totalling 1.9kg were found, A.D. Nininger, Pop. Astron., Northfield, Minnesota, 1937, **45**, p.449 [M.A. 7-62]. Description, H.H. Nininger, Trans. Kansas Acad. Sci., 1936, **39**, p.172. Olivine Fa_{19}, B. Mason, Geochimica et Cosmochimica Acta, 1963, **27**, p.1011.
204g Tempe, Arizona State Univ.; 260g Washington, U.S. Nat. Mus.;
Specimen(s): [1959,900], 218g.

Seneca Falls 42°55′ N., 76°47′ W.
Cayuga County, New York, U.S.A.
Found 1850
Synonym(s): *Seneca River*
Iron. Octahedrite, medium (1.1mm) (IIIA).
A mass of about 9lb was found in digging a ditch, O. Root, Am. J. Sci., 1852, **14**, p.439. Description, analysis, C.U. Shepard, Am. J. Sci., 1853, **15**, p.363. Further analysis, 8.25 %Ni, 21.1 ppm.Ga, 42.8 ppm.Ge, 0.25 ppm.Ir, E.R.D. Scott et al., Geochimica et Cosmochimica Acta, 1973, **37**, p.1957. Description; severely shocked, recrystallised to a strange structure, V.F. Buchwald, Iron Meteorites, Univ. of California, 1975, p.1101.
820g Vienna, Naturhist. Mus.; 382g Yale Univ.; 307g Chicago, Field Mus. Nat. Hist.; 74g Paris, Mus. d'Hist. Nat.;
Specimen(s): [26845], 54g.

Seneca River v Seneca Falls.

Seneca Township 41°47′ N., 84°11′ W.
Lenawee County, Michigan, U.S.A.
Found 1923
Iron. Octahedrite, fine (0.28mm) (IVA).
A mass of 11.5kg was found. Description, with an analysis, 11.41 %Ni, G.P. Merrill, Proc. U.S. Nat. Mus., 1927, **72** (4), p.3. Reputed to have fallen in June, 1903, but the corrosion present makes this most unlikely; shocked and annealed, V.F. Buchwald, Iron Meteorites, Univ. of California, 1975, p.1104. Classification, analysis, 8.42 %Ni, 2.17 ppm.Ga, 0.124 ppm.Ge, 1.8 ppm.Ir, R. Schaudy et al., Icarus, 1972, **17**, p.174.
9.37kg Washington, U.S. Nat. Mus., main mass; 189g New York, Amer. Mus. Nat. Hist.; 184g Michigan Univ.; 181g Chicago, Field Mus. Nat. Hist.; 164g Harvard Univ.; 111g Tempe, Arizona State Univ.;
Specimen(s): [1959,1022], 92g. and sawings, 2.5g

Senegal River v Siratik.

Senhadja v Aumale.

Seoni 21°41′2″ N., 79°30′3″ E.
Seoni district, Madhya Pradesh, India
Fell 1966, January 16, 1445 hrs, approx. time
Stone. Olivine-bronzite chondrite (H6).
One mass of about 20kg fell about 1 mile from the village of Khandasa (co-ordinates given above) and was quickly recovered; a smaller stone of about 1kg fell about 1 mile NE. of the nearby village of Piparwani and was also recovered, Meteor. Bull., 1974 (52), Meteoritics, 1974, **9**, p.106. Description, analyses, olivine $Fa_{19.7}$, 26.11 % total iron, T.E. Bunch et al., Meteoritics, 1972, **7**, p.87. Morphology, V.K. Nayak and G.R. Rao, Meteoritics, 1975, **10**, p.115.
Main mass, Seoni, District Headquarters;
Specimen(s): [1973,M.21], 1.68g. fragments

Seres 41°3′ N., 23°34′ E.
Macedonia, Greece
Fell 1818, June
Synonym(s): *Macedonia, Makedonien, Serrai*
Stone. Olivine-bronzite chondrite (H4).
A mass of about 8.5kg fell, but the particulars are unknown, P. Partsch, Die Meteoriten, Wien, 1843, p.75. Analysis, J.J. Berzelius, Ann. Phys. Chem. (Poggendorff), 1829, **16**, p.611. Contains finely granular troilite between troilite and metal, R. Vogel, Chem. Erde, 1966, **25**, p.142. Olivine Fa_{17}, B. Mason, Geochimica et Cosmochimica Acta, 1963, **27**, p.1011.
6.5kg Vienna, Naturhist. Mus.; 116g Tempe, Arizona State Univ.; 40g Chicago, Field Mus. Nat. Hist.;
Specimen(s): [35165], 399.75g.

Sergipe v Bendegó.

Serra de Magé 8°23′ S., 36°46′ W. approx.
Pesqueira, Pernambuco, Brazil
Fell 1923, October 1, 1100 hrs
Synonym(s): *Pesqueira, Sierra de Magi*
Stone. Achondrite, Ca-rich. Eucrite (AEUC).
After the appearance of a fireball travelling from SE., and detonations, a shower of stones fell, of which about 50 were recovered, most of them less than 5 cm in diameter, but some as much as 10 cm. Description, with an analysis, L.J. de Moraes and D. Guimaraes, Bol. Inst. Brasil. Sci., 1926, **2** (1), p.356 [M.A. 3-389]. Illustrated description, unbrecciated eucrite, M.B. Duke and L.T. Silver, Geochimica et Cosmochimica Acta, 1967, **31**, p.1637. Complex irradiation history, low level of 26Al, E.A. Carver and E. Anders, Earth planet. Sci. Lett., 1970, **8**, p.214. Sm-Nd systematics, G.W. Lugmair et al., Meteoritics, 1977, **12**, p.300, abs. Pyroxene chemistry, G.E. Harlow et al., Earth planet. Sci. Lett., 1979, **43**, p.173 [M.A. 79-3983]. Analysis, B. Mason et al., Smithson. Contrib. Earth Sci., 1979 (22), p.30.
1289g Rio de Janeiro, Serv. Geol.; 332g Paris, Mus. d'Hist. Nat.; 25g Washington, U.S. Nat. Mus.; 17g Albuquerque, Univ. of New Mexico; 17g New York, Amer. Mus. Nat. Hist.;
Specimen(s): [1927,1031], 21g. complete stone; [1927,1143], 14.8g. complete stone; [1927,1032], 3.2g.

Serrai v Seres.

Serrania de Varas 24°33′ S., 69°4′ W.
Atacama, Chile
Found 1875
Synonym(s): *Varas*
Iron. Octahedrite, fine (0.3mm) (IVA).
A mass of about 1.5kg was found about 1875. Description, with an analysis, 8 %Ni, L. Fletcher, Min. Mag., 1889, **8**, p.258. Further analysis, 8.00 %Ni, 2.16 ppm.Ga, 0.130 ppm.Ge, 1.8 ppm.Ir, R. Schaudy et al., Icarus, 1972, **17**, p.174. Co-ordinates; shocked, sheared, heated, and recrystallised, V.F. Buchwald, Iron Meteorites, Univ. of California, 1975, p.1107.
14g Vienna, Naturhist. Mus.;
Specimen(s): [53323a], 1469g. main mass, and a fragment, 5g, and turnings, 12g.

Sete Lagoas 19°28′ S., 44°13′ W.
Minas Gerais, Brazil
Fell 1908, December 15
Stone. Olivine-bronzite chondrite (H4).
Four stones, the largest the size of a walnut, totalling 63g, fell, and are preserved at the School of Mines, Ouro Preto,

Brazil, O.C. Farrington, letter of 15 November, 1926 in Min. Dept. BM(NH). Description, E. de Oliveira, Anais Acad. Brasil. Cienc., 1931, **3**, p.53, 55 [M.A. 5-15]. Further description, mineral analyses, olivine $Fa_{19.4}$, C.B. Gomes and K. Keil, Bol. IG. Inst. Geocien. Univ. Sao Paulo, 1977, **8**, p.77.

16g Albuquerque, Univ. of New Mexico; 3.5g New York, Amer. Mus. Nat. Hist.;

Seth Ward 34°16′8″ N., 101°38′50″ W.
Hale County, Texas, U.S.A.
Found 1977
Stone. Olivine-bronzite chondrite (H5).

A single mass was found which was cut into four pieces of total weight 2497g, olivine Fa_{19}, P.S. Sipiera et al., Meteoritics, 1983, **18**, p.63. Possibly part of Plainview (1917).

Specimen, Watchung, Dupont Colln; 26g Chicago, Field Mus. Nat. Hist.;

Setif v Tadjera.

Severny Kolchim 60°30′ N., 57°00′ E.
Krasnovishersk district, Perm region, USSR, [Северныиь Колчим]
Found 1965
Stone. Chondrite, type 3.

A stone of about 2kg was found in loamy soil among the roots of a tree, Meteor. Bull., 1966 (38), Meteoritics, 1970, **5**, p.88.

1.3kg Sverdlovsk, Ural Geol. Mus.; 240g Moscow, Acad. Sci.;

Sevier County v Cosby's Creek.

Sevilla 37°25′ N., 6°0′ W.
Andalucia, Spain
Fell 1862, November 1
Stone. Olivine-hypersthene chondrite, amphoterite (LL4-6), brecciated.

A stone of about 100g fell on November 1, not October 1, as usually stated, O. Büchner, Ann. Phys. Chem. (Poggendorff), 1865, **124**, p.591. Classification on silicate composition, olivine Fa_{28}, B. Mason, Geochimica et Cosmochimica Acta, 1963, **27**, p.1011.

92g Madrid, Mus. Cienc. Nat.; 56g Prague, Nat. Mus.; 15g Berlin, Humboldt Univ.; 7g Tempe, Arizona State Univ.;

Sevriukovo v Sevrukovo.

Sevrukovo 50°37′ N., 36°36′ E.
Belgorod, Kursk, USSR, [Севрюково]
Fell 1874, May 11, 2345 hrs
Synonym(s): *Belgorod, Koursk, Kursk, Sevriukovo, Sevryukovo, Sewrukof, Sewrukowo, Tula*
Stone. Olivine-hypersthene chondrite (L5), black.

After the appearance of a band of light and detonations, a stone of 101kg fell 2 miles E. of the village of Sevrukovo. Description, with an analysis, A. Eberhard, Arch. Naturk. Liv.-Ehst.-u. Kurlands, Ser. 1, Min. Wiss. Dorpat, 1882, **9**, p.115. Analysis, 22.89 % total iron, M.I. D'yakonova and V.Ya. Kharitonova, Метеоритика, 1960, **18**, p.48 [M.A. 16-447]. Mentioned, G.P. Vdovykin, Метеоритика, 1964, **25**, p.134.

92kg Kharkov Mus.; 4.3kg Moscow, Acad. Sci.; 293g Paris, Mus. d'Hist. Nat.; 180g Chicago, Field Mus. Nat. Hist.; 28g Vienna, Naturhist. Mus.; 23g Washington, U.S. Nat. Mus.;

Specimen(s): [54631], 20g.

Sevryukovo v Sevrukovo.

Sewrukof v Sevrukovo.

Sewrukowo v Sevrukovo.

Seymchan 62°54′ N., 152°26′ E. approx.
Magadan district, Federated SSR, USSR
Found 1967
Iron. Octahedrite, coarse (2.0mm) (IIE).

A mass of 272.3kg was found in the bed of a stream flowing into the river Hekandue, a tributary of the Jasachnaja; a second mass, 50kg, was later found, Meteor. Bull., 1968 (42, 43), Meteoritics, 1970, **5**, p.97. The co-ordinates are of the town of Seymchan. Illustrated description, with an analysis, 9.47 %Ni, O.A. Kirova and M.I. D'yakonova, Метеоритика, 1972, **31**, p.104. Further analysis, 9.15 %Ni, 24.6 ppm.Ga, 68.3 ppm.Ge, 0.55 ppm.Ir, E.R.D. Scott and J.T. Wasson, Geochimica et Cosmochimica Acta, 1976, **40**, p.103.

273kg Moscow, Acad. Sci.;

Seymour 37°14′15″ N., 92°47′13″ W.
Webster County, Missouri, U.S.A.
Found 1940, approx., recognized 1963
Synonym(s): *Marshfield*
Iron. Octahedrite, coarse (2.2mm) (IA).

A mass of 57lb was found 6 miles NNW. of Seymour (and SE. of Marshfield), Meteor. Bull., 1964 (29, 31). Description, figured, W.F. Read, Meteoritics, 1964, **2**, p.285. Description, mineral analyses, J.C. Drake, Meteoritics, 1970, **5**, p.19. Analysis, 6.54 %Ni, 89 ppm.Ga, 382 ppm.Ge, 1.7 ppm.Ir, J.T. Wasson, Icarus, 1970, **12**, p.407. Description; may be paired with Jenkins (*q.v.*), V.F. Buchwald, Iron Meteorites, Univ. of California, 1975, p.1110.

5kg Harvard Univ.; 1.3kg Washington, U.S. Nat. Mus.; 495g Chicago, Field Mus. Nat. Hist.;

Specimen(s): [1970,170], 671g. and fragments, 10g

Shafter Lake 32°24′ N., 102°35′ W.
Andrews County, Texas, U.S.A.
Found 1933, about, recognized 1936
Stone. Olivine-bronzite chondrite (H).

One stone of about 6.25lb was found, A.D. Nininger, Pop. Astron., Northfield, Minnesota, 1937, **45**, p.449 [M.A. 7-62], F.C. Leonard, Univ. New Mexico Publ., Albuquerque, 1946 (meteoritics ser. no. 1), p.47. Olivine Fa_{19}, B. Mason, Geochimica et Cosmochimica Acta, 1963, **27**, p.1011.

3kg Fort Worth, Texas, Monnig Colln.;

Shahpur v Futtehpur.

Shahpura v Samelia.

Shaital v Shytal.

Shalfa v ash-Shalfa.

Shalka 23°6′ N., 87°18′ E.
Bishnupur, Bankura district, West Bengal, India
Fell 1850, November 30, early morning
Synonym(s): *Bancoorah, Bankura, Bissempore, Saluka*
Stone. Achondrite, Ca-poor. Diogenite (ADIO).

After detonations, an immense stone, said to measure 3 feet across fell and broke into pieces, and only about 8lb has been preserved, H. Piddington, J. Asiatic Soc. Bengal, 1851, **20**, p.299, W. von Haidinger, Sitzungsber. Akad. Wiss. Wien,

Math.-naturwiss. Kl., 1860, **41**, p.251. Analysis, N.S. Maskelyne and W. Flight, Phil. Trans. Roy. Soc. London, 1871, **161**, p.366. The place of fall is Saluka, 23°6'N., 87°18'E., C.A. Silberrad, Min. Mag., 1932, **23**, p.293. XRF analysis, 12.72 % total iron, H. von Michaelis et al., Earth planet. Sci. Lett., 1969, **5**, p.387. Cosmic ray tracks, R.K. Bull and S.A. Durrani, Earth planet. Sci. Lett., 1976, **32**, p.35. Orthopyroxene analysis, H. Takeda et al., Geochimica et Cosmochimica Acta, 1976 (suppl. 7), p.3535.

2.2kg Calcutta, Mus. Geol. Surv. India; 374g Tempe, Arizona State Univ.; 190g Vienna, Naturhist. Mus.; 87g New York, Amer. Mus. Nat. Hist.; 17.5g Budapest, Nat. Mus.;

Specimen(s): [33761], 965g. three pieces, and fragments, 34g; [32098], 262g.; [1962,156], 8g. fragments; [96278], thin section; [96279], thin sections (three)

Shallowater 33°42′ N., 101°56′ W.

Lubbock County, Texas, U.S.A.
Found 1936, July
Stone. Achondrite, Ca-poor. Aubrite (AUB).

A stone of about 10.25lb was found, H.H. Nininger, letter of 2 March, 1937 in Min. Dept. BM(NH). Description, with an analysis, W.F. Foshag, Am. Miner., 1939, **24**, p.185 [M.A. 7-377]. Weathered and veined with limonite; contains 9% metal. Classified as an aubrite and described, with an analysis of enstatite, W.F. Foshag, Am. Miner., 1940, **25**, p.779 [M.A. 8-60]. Kamacite contains 1% (wt.) Si, J.T. Wasson and C.M. Wai, Geochimica et Cosmochimica Acta, 1970, **34**, p.169. Analysis of enstatite, A.M. Reid and A.J. Cohen, Geochimica et Cosmochimica Acta, 1967, **31**, p.661.

1017g Washington, U.S. Nat. Mus.; 304g Harvard Univ.; 291g Fort Worth, Texas, Monnig Colln.; 289g Chicago, Field Mus. Nat. Hist.; 171g Tempe, Arizona State Univ.;

Specimen(s): [1937,377], 205g. slice; [1959,760], 177g. slice

Shangdu 42°30′ N., 114° E. approx.

Nei Monggol, China
Found 1957
Iron. Octahedrite, coarse (IIIE).

A mass of 247kg is preserved in Shijiazhuang, Hebei Bureau of Geology, references, D. Bian, Meteoritics, 1981, **16**, p.115. Analysis, 6.08 %Ni, Z. Li et al., Geochimica, 1979, p.249.

247kg Shijiazhuang, Hebei Bureau Geol.;

Shantung v Nanseiki.

Shantung Hotse v Hotse.

Shapiyun v Shupiyan.

Sharps 37°50′ N., 76°42′ W.

Richmond County, Virginia, U.S.A.
Fell 1921, April 1
Stone. Olivine-bronzite chondrite (H3).

With a whirring noise, a stone of 1265g was seen to fall. Description, with an analysis, T.L. Watson, Proc. U.S. Nat. Mus., 1923, **64** (2), p.1 [M.A. 2-260]. Description, analysis, 26.26 % total iron, K. Fredriksson, Meteorite Research, ed. P.M. Millman, D. Reidel, Dordrecht-Holland, 1969, p.155. Study of chondrules, R.T. Dodd, Contr. Mineral. Petrol., 1971, **31**, p.201. Undersaturated minerals in olivine, R.T. Dodd, Min. Mag., 1971, **38**, p.451.

401g Washington, U.S. Nat. Mus.; Specimen, Virginia Univ.;

Shassini v Chassigny.

Shaw 39°32′ N., 103°20′ W.

Lincoln County, Colorado, U.S.A.
Found 1937
Stone. Olivine-hypersthene chondrite (L6).

Two fragments weighing 3.7kg were found, A.D. Nininger, Pop. Astron., Northfield, Minnesota, 1937, **45**, p.449 [M.A. 7-62]. Mentioned, Ann. Rep. U.S. Nat. Mus., 1938, p.50. An additional stone of 13.8kg was recovered in 1967, Amer. Meteorite Lab. Catalogue, 1968 (price list of November 25). Description, analysis, olivine Fa_{23}, 16.93 % total iron, K. Fredriksson and B. Mason, Geochimica et Cosmochimica Acta, 1967, **31**, p.1705. Recrystallized at high temperature; assigned to petrologic grade 7, R.T. Dodd et al., Geochimica et Cosmochimica Acta, 1975, **39**, p.1585. Thermal history, E.R.D. Scott and R.S. Rajan, Geochimica et Cosmochimica Acta, 1979 (suppl. 11), p.1031.

2037g Mainz, Max-Planck-Inst.; 1596g Washington, U.S. Nat. Mus.; 528g Tempe, Arizona State Univ.; 202g Copenhagen, Univ. Geol. Mus.; 250g Albuquerque, Univ. of New Mexico; 127g Ottawa, Geol. Surv. Canada; 119g Vienna, Naturhist. Mus.;

Specimen(s): [1955,224], 11.3g.; [1959,761], 232g. slice; [1969, 4], 30.5g.; [1971,205], 24g. fragments

Shelburne 44°3′ N., 80°10′ W.

Grey County, Ontario, Canada
Fell 1904, August 13, 2000 hrs
Stone. Olivine-hypersthene chondrite (L5), veined, brecciated.

After the appearance of a fireball, and detonations, two stones fell, of about 28lb and 13lb respectively. Description, with an analysis, L.H. Borgström, Trans. Roy. Astron. Soc. Canada, 1904, p.69, O.C. Farrington, Field Mus. Nat. Hist. Geol. Ser., 1906, **3** (Publ. 109), p.7, G.T. Prior, Min. Mag., 1919, **18**, p.349. Olivine Fa_{24}, B. Mason, Geochimica et Cosmochimica Acta, 1963, **27**, p.1011.

4.9kg Chicago, Field Mus. Nat. Hist.; 477g Harvard Univ.; 426g Rome, Vatican Colln; 245g Vienna, Naturhist. Mus.; 176g Mainz, Max-Planck-Inst.; 165g Yale Univ.; 155g Berlin, Humboldt Univ.; 93g Ottawa, Geol. Surv. Canada;

Specimen(s): [1905,138], 1670g. and fragments, 24g; [1905, 435], 68.75g.; [1906,31], thin section

Sherghati v Shergotty.

Shergotty 24°33′ N., 84°50′ E.

Gaya, Bihar, India
Fell 1865, August 25, 0900 hrs
Synonym(s): *Behar, Sherghati, Umjhiawar*
Stone. Achondrite, Ca-rich. Eucrite (shergottite) (AEUC).

After detonations, a stone of about 11lb fell, W.C. Costley, Proc. Asiatic Soc. Bengal, 1865, p.194, F. Fedden, Cat. Meteor. Indian Mus. Calcutta, 1880, p.27. Description, mineral analyses, G. Tschermak, Sitzungsber. Akad. Wiss. Wien, Math.-naturwiss. Kl., 1872, **65** (1), p.122. Analysis, E. Ludwig, Tschermaks Min. Petr. Mitt., 1871, p.55. Mineral composition, R.A. Binns, Nature, 1967, **213**, p.1111. Further analysis, 15.03 % total iron, T.S. McCarthy et al., Meteoritics, 1974, **9**, p.215. The place of fall is Shergahti, 24°33'N., 84°50'E., C.A. Silberrad, Min. Mag., 1932, **23**, p.294. Petrogenesis, chronology, C.-Y. Shih et al., Geochimica et Cosmochimica Acta, 1982, **46**, p.2323. Mineralogy, petrology, minor element contents, J.V. Smith and R.L. Hervig, Meteoritics, 1979, **14**, p.121. Experimental petrology, E. Stolper and H.Y. McSween, Jr., Geochimica et Cosmochimica Acta, 1979, **43**, p.1475.

3.6kg Calcutta, Mus. Geol. Surv. India; 270g Washington, U.S. Nat. Mus.; 211g Vienna, Naturhist. Mus.; 37g

Chicago, Field Mus. Nat. Hist.; 46g New York, Amer. Mus. Nat. Hist.;
Specimen(s): [41021], 62g.; [41020], 42.5g. and fragments, 1g

Sherlock v Homestead.

Shibayama 35°45′54″ N., 140°24′36″ E.
Sanbu-gun, Chiba, Honshu, Japan
Found 1969, April
Stone. Olivine-hypersthene chondrite (L6).
A single stone of 235g was found about 50 cm below ground level, Meteor. Bull., 1979 (56), Meteoritics, 1979, **14**, p.170. Description, analysis, 21.41 % total iron, olivine Fa_{23}, M. Shima et al., Meteoritics, 1979, **14**, p.317.

Shields 38°42′ N., 100°21′ W.
Lane County, Kansas, U.S.A.
Found 1962, approx., recognized 1968
Stone. Olivine-bronzite chondrite (H5).
A mass of about 9.78kg was found on a low hill top in an eroded ravine off the Smoky Hill River. The stone may have been carried to this spot, G.I Huss, letter of 16 March, 1973 in Min. Dept. BM(NH). Classification, olivine $Fa_{18.9}$, A.L. Graham, priv. comm., 1981.
6.1kg Mainz, Max-Planck-Inst., includes main mass, 5.5kg; 171g Vienna, Naturhist. Mus.; 66g New York, Amer. Mus. Nat. Hist.;
Specimen(s): [1973,M.18], 86.25g. crusted slice

Shigetome v Kyushu.

Shikarpur 25°51′ N., 87°34′39″ E.
Purnea district, Bihar, India
Fell 1921, August 9, 0900 hrs
Synonym(s): *Purnea*
Stone. Olivine-hypersthene chondrite (L6), veined.
After detonations, a stone of 3679.7g fell at Deari-Shikarpur, Kasba police station. Description, G.V. Hobson, Rec. Geol. Surv. India, 1927, **60**, p.139. Olivine Fa_{25}, B. Mason, Geochimica et Cosmochimica Acta, 1967, **31**, p.1100.
3.6kg Calcutta, Mus. Geol. Surv. India; 0.9g New York, Amer. Mus. Nat. Hist.;
Specimen(s): [1983,M.16], 9.9g.

Shimelia v Samelia.

Shingle Springs 38°40′ N., 120°56′ W.
El Dorado County, California, U.S.A.
Found 1869, or 1870
Synonym(s): *Eldorado County, El Dorado County, Los Angeles*
Iron. Ataxite, Ni-rich (IRANOM).
A mass of about 85lb was found in a field about 0.5 miles from Shingle Springs and was "rescued from the forge of a smith". Description, with an analysis, 17.17 %Ni, B. Silliman, Am. J. Sci., 1873, **6**, p.18. Further description, analysis, 16.69 %Ni, E. Cohen, Meteoritenkunde, 1905, **3**, p.156. Description, references, V.F. Buchwald, Iron Meteorites, Univ. of California, 1975, p.1113. According to H.A. Ward the main mass has been lost. Chemically anomalous, analysis, 16.95 %Ni, 2.06 ppm.Ga, 0.130 ppm.Ge, 2.6 ppm.Ir, R. Schaudy et al., Icarus, 1972, **17**, p.174.
433g Yale Univ.; 141g Vienna, Naturhist. Mus.; 117g Chicago, Field Mus. Nat. Hist.; 110g New York, Amer. Mus. Nat. Hist.; 87g Harvard Univ.; 71g Budapest, Nat. Mus.; 71g Paris, Mus. d'Hist. Nat.; 61g Berlin, Humboldt Univ.; 32g Washington, U.S. Nat. Mus.;
Specimen(s): [51634], 75g.

Shirahagi 36°42′ N., 137°22′ E.
Nakaniikawa, Toyama, Honshu, Japan
Found 1890
Synonym(s): *Shirohagi, Sirohagi*
Iron. Octahedrite, fine (0.3mm) (IVA).
A mass of 22.7kg was found in the bed of the Kamiichikawa river, analysis, 9.30 %Ni, K. Jimbo, Beitr. Miner. Japan, 1906 (2), p.49. Mentioned, S. Murayama, Meteoritics, 1953, **1**, p.99 [M.A. 12-357]. Further analysis, 7.86 %Ni, 2.19 ppm.Ga, 0.120 ppm.Ge, 2.4 ppm.Ir, J.T. Wasson, Meteorites, Springer-Verlag, 1974, p.305. Possibly includes Saotome (*q.v.*), V.F. Buchwald, Iron Meteorites, Univ. of California, 1975, p.1115. Saotome, structurally similar, was found in the same river two years later, S. Murayama, letter of 12 October, 1980 in Min. Dept. BM(NH).
18.2kg Tokyo, Sci. Mus., main mass; 91g Tempe, Arizona State Univ.;
Specimen(s): [1978,M.19], 31.7g.

Shiraiwa 39°30′ N., 140°12′ E. approx.
Senboku, Akita, Honshu, Japan
Found 1915, September
Synonym(s): *Hataya, Siroiwa, Siruiwa*
Stone. Chondrite.
A mass of 950g was found, A.D. Nininger, Pop. Astron., Northfield, Minnesota, 1940, **48**, p.559, A.D. Nininger, Contr. Soc. Res. Meteorites, **2**, p.227 [M.A. 8-54], I. Yamamoto, Kwasan Observ. Bull., 1935, p.306 [M.A. 7-173], P.M. Millman, Trans. Roy. Astron. Soc. Canada, 1938, **32**, p.197 [M.A. 7-173], S. Murayama, letter of 6 April, 1962 in Min. Dept. BM(NH).
373g Kakudate, Akita prefecture, Girls High School;

Shiriya 41°26′ N., 141°28′ E.
Aomori, Honshu, Japan
Fell 1883, October 24
Synonym(s): *Siriya*
Doubtful. Stone?
Doubtful, I. Yamamoto, Kwasan Observ. Bull., 1935, p.306 [M.A. 7-173].

Shirohagi v Shirahagi.

Shohaku 40°19′ N., 126°55′ E.
Neien-gun, Heian-nan-do, Korea
Found 1938, before this year
Iron. Octahedrite, medium.
One mass of 101g was found, S. Kanda, Sci. Rep. Yokohama Univ., 1952 (11), p.104, Meteor. Bull., 1959 (12).
Main mass, possession of S. Kanda;

Showtnjewi Chutor v Zhovtnevyi.

Shrewsbury 39°46′ N., 76°40′ W.
York County, Pennsylvania, U.S.A.
Found 1907
Iron. Octahedrite, medium (1.1mm) (IA).
A mass of about 27lb was ploughed up. Description, with an analysis, 8.80 %Ni, O.C. Farrington, Am. J. Sci., 1910, **29**, p.350. Further description, E.P. Henderson and S.H. Perry, Proc. U.S. Nat. Mus., 1958, **107**, p.339 [M.A. 14-130]. Analysis, 8.42 %Ni, 61.4 ppm.Ga, 204 ppm.Ge, 2.6 ppm.Ir, J.T. Wasson, Icarus, 1970, **12**, p.407. Cosmically annealed, V.F. Buchwald, Iron Meteorites, Univ. of California, 1975, p.1115.
2.1kg Vienna, Naturhist. Mus.; 850g Philadelphia, Penn. Acad. Sci.; 736g Paris, Mus. d'Hist. Nat.; 562g New York, Amer. Mus. Nat. Hist.; 348g Harvard Univ.; 395g

Washington, U.S. Nat. Mus.; 104g Chicago, Field Mus. Nat. Hist.; 90g Yale Univ.;
Specimen(s): [1910,410], 925.5g. and fragments, 9.5g

Shropshire v Rowton.

Shuangyang 43°30′ N., 125°40′ E.
Quayan, Shuangyang County, Jilin, China
Fell 1971, May 25, 1500 hrs, or May 26
Stone. Olivine-bronzite chondrite (H5).
Five pieces were recovered, totalling 3.9kg, the largest weighing 2.0kg, D. Bian, Meteoritics, 1981, **16**, p.117. Analysis, 27.14 % total iron, D. Wang and Z. Ouyang, Geochimica, 1979, p.120, in Chinese [M.A. 80-2085].
3.9kg Beijing, Univ. Geol. Dept.;

Shupiyan 33°43′ N., 74°50′ E.
Kashmir, India
Fell 1912, April
Synonym(s): *Shapiyun*
Stone. Olivine-bronzite chondrite (H6), brecciated.
Two stones of about 4.5kg and 500g appear to have fallen, and were preserved in the Srinagar Museum, J. Coggin Brown, Rec. Geol. Surv. India, 1915, **45**, p.221. Olivine Fa_{19}, B. Mason, Geochimica et Cosmochimica Acta, 1963, **27**, p.1011. The place of fall is Shapiyun, 33°43'N., 74°50'E., 24 miles S. of Srinagar, C.A. Silberrad, Min. Mag., 1932, **23**, p.303. Noble gas data, N. Doshi et al., Meteoritics, 1977, **12**, p.208, abs.
294g Calcutta, Mus. Geol. Surv. India;
Specimen(s): [1915,143], 69g.

Shytal 24°20′ N., 90°10′ E.
Madhupur jungle, Mymensingh district, Tangail, Bangladesh
Fell 1863, August 11, 1200 hrs
Synonym(s): *Dacca, Shaital, Shythal*
Stone. Olivine-hypersthene chondrite (L6), brecciated.
After detonations, a stone of about 7lb was seen to fall, W. von Haidinger, Sitzungsber. Akad. Wiss. Wien, Math.-naturwiss. Kl., 1863, **48**, p.595. Amount and composition of Ni/Fe, G.T. Prior, Min. Mag., 1919, **18**, p.352. Olivine Fa_{25}, B. Mason, Geochimica et Cosmochimica Acta, 1963, **27**, p.1011. There is no such place as Shytal in or near the Madhupur jungle, C.A. Silberrad, Min. Mag., 1932, **23**, p.293.
2.75kg Calcutta, Mus. Geol. Surv. India; 223g Vienna, Naturhist. Mus.; 164g Tempe, Arizona State Univ.; 27g Washington, U.S. Nat. Mus.; 25g New York, Amer. Mus. Nat. Hist.; 22g Chicago, Field Mus. Nat. Hist.; 16g Berlin, Humboldt Univ.;
Specimen(s): [40583], 447g.; [1911,346], 14g.

Shythal v Shytal.

Siberia v Mount Egerton.

Sibley County v Arlington.

Sichote-Alin v Sikhote-Alin.

Sichote-Alinsky v Sikhote-Alin.

Sidney 41°3′ N., 102°54′ W.
Cheyenne County, Nebraska, U.S.A.
Found 1941
Stone. Olivine-hypersthene chondrite (L).
6kg were found, olivine Fa_{24}, B. Mason, Geochimica et Cosmochimica Acta, 1963, **27**, p.1011.
30g Tempe, Arizona State Univ.;

Sidowra v Supuhee.

Siena 43°7′ N., 11°36′ E.
Tuscany, Italy
Fell 1794, June 16, 1900 hrs
Synonym(s): *Cosona, Lucignano d'Asso, Lusignan d'Asso, Lusignano d'Asso, Pienza, San Giovanni d'Asso, Spedalone presso Pienza, Suoma*
Stone. Olivine-hypersthene chondrite, amphoterite (LL5), brecciated.
After appearance of cloud and detonations, a shower of small stones fell the largest weighing about 3.5kg, D. Tata, Ann. Phys. (Gilbert), 1800, **6**, p.156, E. Howard, Phil. Trans. Roy. Soc. London, 1802, p.173. Historical notes, W. Salomon, Mitt. Geschichte Medizin Naturwiss., 1930, **29**, p.135 [M.A. 4-417]. The co-ordinates are of Lucignano d'Asso; the fall was probably between Lucignano and Cosona (43°7'N., 11°35'E.), G.R. Levi-Donati, priv. comm., 1969. Metal content and total iron are high for an LL-group chondrite, analysis, 25.31 % total iron, B. Baldanza et al., Min. Mag., 1969, **37**, p.34, B. Baldanza et al., Rend. Soc. Ital. Miner. Petrol., 1969, **25**, p.173. Mineralogy, olivine $Fa_{28.7}$, G. Kurat et al., Geochimica et Cosmochimica Acta, 1969, **33**, p.765. Analysis of lithic fragments, R. Fodor and K. Keil, Meteoritics, 1975, **10**, p.325.
1719g Bologna, Univ., 985g in Min. Dept., 734g in Geol. Dept.; 185g Vienna, Naturhist. Mus.; 127g Paris, Mus. d'Hist. Nat.; 65g Chicago, Field Mus. Nat. Hist.; 43g Siena, Acad. Fis.; 36g Budapest, Nat. Mus.;
Specimen(s): [33990], 110g.; [90245], 9.5g.; [34589], 8g.; [1960,332], 79g.

Sierra Blanca 27°9′ N., 104°54′ W.
Jimenez, Chihuahua, Mexico
Found 1784
Synonym(s): *Huejuquilla, Jimenez, Villa Nueva*
Iron. Octahedrite, coarse (1.3mm) (I).
Masses of "20, 30 and more hundredweights" were said to have been found, E.F.F. Chladni, Ann. Phys. (Gilbert), 1817, **56**, p.383, H.J. Burkart, Neues Jahrb. Min., 1856, p.278. Only a few hundred grams are known in collections, L. Fletcher, Min. Mag., 1890, **9**, p.149. The above co-ordinates are those of Jimenez. The mass has been heated, very similar to Toluca, V.F. Buchwald, Iron Meteorites, Univ. of California, 1975, p.1117.
230g Dorpat Univ.; 147g Berlin, Humboldt Univ.;
Specimen(s): [35186], 15g.

Sierra County 33° N., 107°30′ W. approx.
New Mexico, U.S.A.
Found 1962
Stone. Olivine-bronzite chondrite (H5).
Listed, L. LaPaz, Cat. Coll. Inst. Meteor. Univ. New Mexico, 1965. Olivine Fa_{19}, B. Mason, Geochimica et Cosmochimica Acta, 1967, **31**, p.1100.
1kg Albuquerque, Univ. of New Mexico;

Sierra de Chaco v Vaca Muerta.

Sierra de Deesa v Copiapo.

Sierra de Deesa v Dehesa.

Sierra de las Adargas v Chupaderos.

Sierra de la Ternera v Ternera.

Sierra de Magi v Serra de Magé.

Sierra Gorda 22°54′ S., 69°21′ W.
Antofagasta, Chile
Found 1898
Iron. Hexahedrite (IIA).
A mass was found at 22°54'S., 69°21'W., on the railway between Calama and Antofagasta, E.P. Henderson, letter of 12 May, 1939 in Min. Dept. BM(NH). Analysis, 5.58 %Ni, E.P. Henderson, Am. Miner., 1941, **26**, p.546. Analysis, determination of Ga, Au and Pd, 5.59 %Ni, E. Goldberg et al., Geochimica et Cosmochimica Acta, 1951, **2**, p.1. Distinct from North Chile (*q.v.*). Analysis, 5.27 %Ni, 57.4 ppm.Ga, 170 ppm.Ge, 43 ppm.Ir, J.T. Wasson, Geochimica et Cosmochimica Acta, 1969, **33**, p.859. Description, shock-melted troilite, V.F. Buchwald, Iron Meteorites, Univ. of California, 1975, p.1119.
17.3kg Washington, U.S. Nat. Mus.; 7kg San Francisco, Ferry Building; 550g Chicago, Field Mus. Nat. Hist.; 315g New York, Amer. Mus. Nat. Hist.;

Sierra Gorda v La Primitiva.

Sierra Sandon 25°10′ S., 69°17′ W.
Taltal, Antofagasta, Chile
Found 1923, approx.
Synonym(s): *Sandon*
Iron. Octahedrite, medium (1.0mm) (IIIA).
A mass of 6.33kg was found at the above co-ordinates at an elevation of 3400 metres. Description, C. Palache, Am. J. Sci., 1926, **12**, p.137. Analysis, 8.55 %Ni, 20.8 ppm.Ga, 43.8 ppm.Ge, 0.28 ppm.Ir, E.R.D. Scott et al., Geochimica et Cosmochimica Acta, 1973, **37**, p.1957. Description; shock-hardened, V.F. Buchwald, Iron Meteorites, Univ. of California, 1975, p.1120.
6.1kg Harvard Univ., main mass; 72g Washington, U.S. Nat. Mus.;
Specimen(s): [1967,381], 67g.

Sigena v Sena.

Sigetome v Kyushu.

Signal Mountain 32°30′ N., 115°30′ W. approx.
Baja California, Mexico
Found 1919, several years before this
Iron. Octahedrite, fine (0.3mm) (IVA).
A mass of 140lb was found. Description, with an analysis, 7.86 %Ni, G.P. Merrill, Proc. U.S. Nat. Mus., 1922, **61**, p.2. Classification, analysis, 7.84 %Ni, 2.11 ppm.Ga, 0.121 ppm.Ge, 2.5 ppm.Ir, R. Schaudy et al., Icarus, 1972, **17**, p.174. Description; weathered, an old fall, V.F. Buchwald, Iron Meteorites, Univ. of California, 1975, p.1122.
58kg New York, Amer. Mus. Nat. Hist.; 41g Washington, U.S. Nat. Mus.;

Signet Iron v Tucson.

Sihote-Alin v Sikhote-Alin.

Sikhote-Alin 46°9′36″ N., 134°39′12″ E.
Maritime Territory, Federated SSR, USSR,
[Сихотэ-Алин]
Fell 1947, February 12, 1038 hrs, U.T.
Synonym(s): *Sichote-Alin, Sichote-Alinsky, Sihote-Alin, Sikhote-Alinskii, Silkhote Alin, Ussuri*
Iron. Octahedrite, coarsest (9mm) (IIB).
A shower of fireballs fell in thick forest in the Sikhote-Alin Mts., 25 miles from Novopoltavka, Maritime Province, producing 106 impact holes, the largest 28 metres across, over an area of 100 × 660 metres, and many fragments, up to 300kg in weight, totalling over 23000kg, were found scattered inside and outside the holes. Description, with an analysis, 5.8 %Ni, V.G. Fesenkov, Астрон. Журнал, 1947, **24**, p.302 [M.A. 10-399]. Description, A.A. Yavnel, Доклады Акад. Наук СССР, 1948, **60**, p.1381 [M.A. 10-399]. Mentioned, E.L. Krinov, Метеоритика, 1949, **5**, p.49, E.L. Krinov and S.S. Fonton, Метеоритика, 1952, **10**, p.74 [M.A. 12-250], A.A. Yavnel, Метеоритика, 1956, **14**, p.87 [M.A. 13-358]. Analysis, 5.87 %Ni, 51.8 ppm.Ga, 161 ppm.Ge, 0.029 ppm.Ir, J.T. Wasson, Geochimica et Cosmochimica Acta, 1969, **33**, p.859. Illustrated descriptions of the 1967-1970 expeditions to the site, E.L. Krinov, Meteoritics, 1971, **6**, p.127, E.L. Krinov, Метеоритика, 1972, **31**, p.62. U distribution, L.I. Genaeva et al., Метеоритика, 1972, **31**, p.137. Cosmogenic nuclides, A.K. Lavrukhina et al., Метеоритика, 1972, **31**, p.151. Fragmentation study, E.L. Krinov, Meteoritics, 1974, **9**, p.255. Magnetic study, E.S. Gorshkov et al., Meteoritics, 1975, **10**, p.9. Description; unannealed, V.F. Buchwald, Iron Meteorites, Univ. of California, 1975, p.1123. Contains carlsbergite, H.J. Axon et al., Min. Mag., 1981, **44**, p.107.
1745kg Moscow, Acad. Sci., principal mass, and other specimens; 4kg Washington, U.S. Nat. Mus.; 4.9kg Copenhagen, Univ. Geol. Mus.; 7kg Albuquerque, Univ. of New Mexico; 3.3kg Tempe, Arizona State Univ.; 3kg Prague, Nat. Mus.;
Specimen(s): [1956,170], 898g.; [1956,171], 714g.; [1974, M.25], 9.35g. turnings

Sikhote-Alinskii v Sikhote-Alin.

Sikkensaare v Tennasilm.

Silkhote Alin v Sikhote-Alin.

Silva v Campo del Cielo.

Silver Bell 32°29′ N., 111°34′ W.
Pima County, Arizona, U.S.A.
Found 1939, known before this year
Iron. Octahedrite, coarsest (5mm) (IIB).
A mass of 5.1kg was found near the settlement of Silver Bell, 60 km NW. of Tucson, Pima County; description, V.F. Buchwald, Iron Meteorites, Univ. of California, 1975, p.1130. Analysis, 6.43 %Ni, 45.6 ppm.Ga, 111 ppm.Ge, 0.012 ppm.Ir, J.T. Wasson, Geochimica et Cosmochimica Acta, 1969, **33**, p.859.
3.3kg Tucson, Univ. of Arizona; 248g Tempe, Arizona State Univ.; 192g Washington, U.S. Nat. Mus.; 89g Chicago, Field Mus. Nat. Hist.;

Silver Crown 41°14′ N., 104°59′ W.
Laramie County, Wyoming, U.S.A.
Found 1887
Synonym(s): *Crow Creek, Laramie County, Wyoming*
Iron. Octahedrite, coarse (2.1mm) (IA).
A mass of 25.6lb was found; erroneous analysis, G.F. Kunz, Am. J. Sci., 1888, **36**, p.276. Description; well preserved, V.F. Buchwald, Iron Meteorites, Univ. of California, 1975, p.1132. Analysis, 6.98 %Ni, 82.0 ppm.Ga, 321 ppm.Ge, 1.6 ppm.Ir, J.T. Wasson, Meteorites, Springer-Verlag, 1974, p.298.
7kg Vienna, Naturhist. Mus.; 176g Chicago, Field Mus. Nat. Hist.; 172g Berlin, Humboldt Univ.; 170g Washington, U.S. Nat. Mus.; 158g Prague, Nat. Mus.; 135g New York, Amer. Mus. Nat. Hist.;
Specimen(s): [67214], 583g.

Silverton (New South Wales) 31°53′ S., 141°12′ E.
County Yancowinna, New South Wales, Australia
Found 1883, approx.
Stone. Olivine-hypersthene chondrite (L6).
A stone of 350.7g was found by R. Bedford in 1933 in the museum at Port Adelaide, South Australia. Description, L.J. Spencer, Min. Mag., 1934, **23**, p.569. Olivine Fa_{25}, B. Mason, Geochimica et Cosmochimica Acta, 1963, **27**, p.1011.
237g Sydney, Austr. Mus.; Specimen, Canberra, Austr. Nat. Univ.; 69g Washington, U.S. Nat. Mus.;
Specimen(s): [1934,50], 32g.

Silverton (Texas) 34°32′ N., 101°11′ W.
Briscoe County, Texas, U.S.A.
Found 1938
Stone. Olivine-bronzite chondrite (H4).
One stone of 1.37kg was found, A.D. Nininger, Pop. Astron., Northfield, Minnesota, 1940, **48**, p.557, A.D. Nininger, Contr. Soc. Res. Meteorites, **2**, p.227 [M.A. 8-54]. Three stones found, 1404g, 762g and 410g, NE. of Silverton (co-ordinates above), olivine $Fa_{18.5}$, P.S. Sipiera et al., Meteoritics, 1983, **18**, p.63.
1.6kg Fort Worth, Texas, Monnig Colln.; 125g Chicago, Field Mus. Nat. Hist., **18**, p.63;

Simagoe v Kyushu.

Simbirsk (of G. von Blöde) v Slobodka.

Simbirsk (of P. Partsch) 54°18′ N., 48°24′ E.
Ulanovsk, Federated SSR, USSR
Found 1838, before this year
Synonym(s): *Poltava (of G. von Blöde)*
Stone. Chondrite, crystalline.
A nearly complete stone of about 1.5kg was in Leningrad, (Mus. Acad. Sci.), in 1846 under the name Poltava, G. von Blöde, Bull. Acad. Sci. St.-Petersbourg, 1848, **6**, p.4, P. Partsch, Die Meteoriten, Wien, 1843, p.46. Not listed, I.S. Astapowitsch, Trans. Roy. Astron. Soc. Canada, 1938, **32**, p.195 [M.A. 7-173]. Identical to Slobodka, Meteor. Bull., 1959 (13).
9g Vienna, Naturhist. Mus.; 0.3g Tempe, Arizona State Univ.;

Simmern 49°59′ N., 7°32′ E.
Hunsrück, Rheinland-Pfalz, Germany
Fell 1920, July 1, 0915 hrs
Synonym(s): *Hochscheid, Hunsrück*
Stone. Olivine-bronzite chondrite (H).
After the appearance of a fireball and detonations, a large number of stones fell over an area of 10 × 2 miles, and three were found, one of 142g at Hochscheid, another of 610g at Götzeroth, and the third of 470g between Hochscheid and Hintzerath, H. Michel, Fortschr. Min. Krist. Petr., 1922, **7**, p.263. Very little preserved. Olivine Fa_{19}, B. Mason, Geochimica et Cosmochimica Acta, 1963, **27**, p.1011.
200g Bonn, Univ.; 71g Tempe, Arizona State Univ.; 600g Berlin, Humboldt Univ.;
Specimen(s): [1972,321], 29.5g.; [1924,327], 7.5g.

Simondium 33°51′ S., 18°57′ E.
Lower Paarl, Cape Province, South Africa
Found 1907
Stony-iron. Mesosiderite (MES).
Two masses each "a foot in diameter" were found, but only about 2.5lb have been preserved. Description, with an analysis, G.T. Prior, Min. Mag., 1910, **15**, p.312, G.T. Prior, Min. Mag., 1918, **18**, p.161, G.T. Prior, Min. Mag., 1921, **19**, p.165. Shock history, A.V. Jain and M.E. Lipschutz, Nature, 1973, **242**, p.26. Olivine coronas, references, C.E. Nehru et al., Geochimica et Cosmochimica Acta, 1980, **44**, p.1103.
500g King William's Town Mus., South Africa; 135g Cape Town, S. African Mus.; 124g New York, Amer. Mus. Nat. Hist.; 46g Chicago, Field Mus. Nat. Hist.;
Specimen(s): [1909,148], 82.5g.; [1909,149], 506.5g. and four tubes of powder, and fragments, 28g; [1920,143], 198.5g. and fragments, 11g

Simonod 46°5′ N., 5°20′ E. approx.
Ain, France
Fell 1835, November 13, 2100 hrs
Synonym(s): *Ain, Belmont*
Doubtful. Stone. Carbonaceous chondrite?.
Original weight not known. History, W. von Haidinger, Sitzungsber. Akad. Wiss. Wien, Math.-naturwiss. Kl., 1867, **55**, p.127. Possibly a pseudometeorite.
0.1g Tübingen, Univ.; 0.1g Tempe, Arizona State Univ.;

Sinai 30°54′ N., 32°29′ E.
Sinai, Egypt
Fell 1916, July 14-17, 1430 hrs
Synonym(s): *Egyptian Meteorite, Kantarah*
Stone. Olivine-hypersthene chondrite (L6).
A stone of 1455g fell 8 miles E. of Kantarah on the desert route to Katia, and another is said to have fallen on the opposite bank of the Suez Canal, H. Wilde, Mem. Manchester Lit. Phil. Soc., 1917, **61** (4). Mentioned, T.A. Coward, letters of 25 January and 10 February, 1923 in Min. Dept. BM(NH). Description, partial analysis, G.T. Prior, Min. Mag., 1923, **20**, p.137. Olivine Fa_{23}, B. Mason, Geochimica et Cosmochimica Acta, 1963, **27**, p.1011.
Main mass, Manchester, Univ.; 137g Paris, Mus. d'Hist. Nat.;
Specimen(s): [1923,1], 99.5g.

Sinaloa v Bacubirito.

Sindhri 26°13′ N., 69°33′ E.

Khipro taluq, Sanghar district, Sind, Pakistan
Fell 1901, June 10, 2300 hrs
Stone. Olivine-bronzite chondrite (H5).

After the appearance of a brilliant fireball travelling from NW. to E, and detonations, a stone of about 4lb which broke into two pieces was seen to fall near Samo Junejo village, and a week later another stone of about 14.5lb was dug up, Director Geol. Surv. India, letter of 10 September, 1902 in Min. Dept. BM(NH). Olivine Fa_{19}, B. Mason, Geochimica et Cosmochimica Acta, 1963, **27**, p.1011. Sindhri could not be located, but Samo Junejo is probably Sumo-juneja, 26°13'N., 69°33'E., C.A. Silberrad, Min. Mag., 1932, **23**, p.297.

3.67kg Calcutta, Mus. Geol. Surv. India; 925g Paris, Mus. d'Hist. Nat.; 647g Chicago, Field Mus. Nat. Hist.; 435g Vienna, Naturhist. Mus.;

Specimen(s): [86146], 1140g. main mass of the larger stone which was seen to fall, and fragments, 13g; [86149], 42.5g.

Singhur 18°19′ N., 73°55′ E.

Poona, Maharashtra, India
Found 1847
Synonym(s): *Singarh*
Stony-iron. Pallasite? (PAL?).

A mass of 31.25lb was found and sent to the Bombay Geogr. Soc. Described with an analysis, 4.24 %Ni, H. Giraud, Edinburgh New Phil. J., 1849, **47**, p.56. Mentioned, W.S. Clarke, Am. J. Sci., 1853, **15**, p.8. The place of find is Sinhgarh, 18°19'N., 73°55'E., C.A. Silberrad, Min. Mag., 1932, **23**, p.295.

Sinhgarh v Singhur.

Sinnai 39°18′ N., 9°12′ E.

Cagliari, Sardinia, Italy
Fell 1956, February 19, early morning hrs
Stone. Olivine-bronzite chondrite (H6).

One stone of about 2kg fell, V. Rossetti and R. Sitzia, Periodico Miner., 1958, **27**, p.179 [M.A. 14-50]. Olivine Fa_{18}, B. Mason, Geochimica et Cosmochimica Acta, 1967, **31**, p.1100.

1230g Cagliari, Univ.; 17g Washington, U.S. Nat. Mus.;

Sioux County 42°35′ N., 103°40′ W.

Nebraska, U.S.A.
Fell 1933, August 8, 1030 hrs
Stone. Achondrite, Ca-rich. Eucrite (AEUC).

20 stones fell, totalling 4.1kg, A.D. Nininger, Pop. Astron., Northfield, Minnesota, 1937, **45**, p.449 [M.A. 7-62]. Mineralogy, analysis, M.B. Duke and L.T. Silver, Geochimica et Cosmochimica Acta, 1967, **31**, p.1637. XRF analysis, H. von Michaelis et al., Earth planet. Sci. Lett., 1969, **5**, p.387. Rb/Sr data, J.L. Birck and C.J. Allegre, Earth planet. Sci. Lett., 1978, **39**, p.37 [M.A. 78-4757]. U-Th-Pb data, M. Tatsumoto et al., Science, 1973, **180**, p.1279. Experimental petrology, E. Stolper, Geochimica et Cosmochimica Acta, 1977, **41**, p.587.

752g Tempe, Arizona State Univ.; 154g Washington, U.S. Nat. Mus.; 37g Paris, Mus. d'Hist. Nat.; 14g Harvard Univ.;

Specimen(s): [1949,113], 153g.; [1959,1029], 659g. half of one of the stones, and fragments, 3.5g

Siratik 14° N., 11° W.

Bambouk, Mali
Found 1716
Synonym(s): *Bambuk, Fouta Senegalais, Senegal River*
Iron. Hexahedrite (IIA).

Large masses of iron were found in 1716 to be used by natives of Siratik and Bambuk for making pots, P. Compagnon, J.J. Schwabe's Allgem. Hist. Reisen Wasser u. Lände, Leipzig, 1748, **2**, p.510. Only about 1.7kg is known. Description, analysis, 5.21 %Ni, E. Cohen, Ann. Naturhist. Hofmus. Wien, 1899, **13**, p.127. Most of the specimens under this name are pseudometeorites, a very small number are altered hexahedrites, V.F. Buchwald, Iron Meteorites, Univ. of California, 1975, p.1134. The two larger pieces [90236-7] in the Brit. Mus. Nat. Hist. collection were brought back from Senegal by Gen. O'Hara and analysed in 1802, 5 %Ni, E. Howard, Phil. Trans. Roy. Soc. London, 1802, **92**, p.211. However, this analysis was in error; these specimens are pseudometeorites. The name 'Siratik', under which the iron is generally known is the title generally given to the kings of the Fouta Toro, A. Lacroix, C. R. Acad. Sci. Paris, 1924, **179**, p.357. Analysis of IIA material, 5.6 %Ni, 59 ppm.Ga, 188 ppm.Ge, 11 ppm.Ir, D.J. Malvin et al., priv. comm., 1983.

436g Vienna, Naturhist. Mus.; 74g Budapest, Nat. Mus.; 212g Tübingen, Univ.; 64g Berlin, Humboldt Univ.;

Specimen(s): [90236], 354.5g. pseudometeorite, contains very little nickel; [90237], 30.5g. pseudometeorite, contains very little nickel; [90238], 10.25g. possibly a pseudometeorite

Siriya v Shiriya.

Sirohagi v Shirahagi.

Siroiwa v Shiraiwa.

Siruiwa v Shiraiwa.

Sitathali 20°55′ N., 82°35′ E.

Raipur district, Madhya Pradesh, India
Fell 1875, March 4, 1100 hrs
Synonym(s): *Nurrah, Raepur, Raipur*
Stone. Olivine-bronzite chondrite (H5).

After detonations, two stones, of about 2lb and 1.5lb respectively, fell 0.75 mile apart, but the two fitted together, H.B. Medlicott, Proc. Asiatic Soc. Bengal, 1876, p.115. Description, analysis, 23.77 % total iron, T.V. Viswanathan et al., Min. Mag., 1971, **38**, p.335. This total iron content is barely consistent with an H-group classification. Olivine Fa_{19}, B. Mason, Geochimica et Cosmochimica Acta, 1963, **27**, p.1011.

766g Calcutta, Mus. Geol. Surv. India; 36g New York, Amer. Mus. Nat. Hist.; 25g Vienna, Naturhist. Mus.; 14g Chicago, Field Mus. Nat. Hist.; 14g Harvard Univ.; 13g Washington, U.S. Nat. Mus.;

Specimen(s): [51186], 649g. includes main mass of the smaller fragment, 593g, and fragments, 2g.

Skalen v Schellin.

Skalin v Schellin.

Skane-Tranas 58° N., 15° E.

Skane, Sweden
Fell 1922, February 12
Doubtful..

A meteorite fell on a frozen lake, making a hole in the ice 20 × 18 cm but nothing was recovered, A. Corlin, Pop. Astron., Northfield, Minnesota, 1939, **47**, p.436, 525 [M.A. 7-540]. The evidence of meteoritic nature is not really conclusive.

Ski 59°44′ N., 10°52′ E.

Akershuus, Oslo, Norway
Fell 1848, December 27, evening hrs
Synonym(s): *Akershuus, Schie*
Stone. Olivine-hypersthene chondrite (L6), veined.

After detonations and the appearance of light, two days later a stone of 850g was found on the ice of a small stream. Description, with an analysis, H.S. Ditten, Ann. Phys. Chem. (Poggendorff), 1855, **96**, p.341. Description, H. Reusch, Neues Jahrb. Min., Beil.-Band, 1886, **4**, p.493. Olivine Fa_{24}, B. Mason, Geochimica et Cosmochimica Acta, 1963, **27**, p.1011.

629g Oslo, Min.-Geol. Mus.; 31g Tübingen, Univ.; 31g Vienna, Naturhist. Mus.; 17g Tempe, Arizona State Univ.; 8.6g Chicago, Field Mus. Nat. Hist.;

Specimen(s): [39712], 4g. two fragments

Skiff 49°15′ N., 111°52′ W.

Alberta, Canada
Found 1966
Stone. Olivine-bronzite chondrite (H4).

A single mass of 3.54kg was found on a farm near Skiff in southern Alberta (NE. 1/4, sect. 31, township 3 range 14W.), olivine Fa_{18}, Meteor. Bull., 1980 (58), Meteoritics, 1980, **15**, p.239. Corrected locality, Meteor. Bull., 1981 (59), Meteoritics, 1981, **16**, p.199.

2.2kg Edmonton, Univ. of Alberta; 1.3kg Ottawa, Geol. Surv. Canada;

Skölen v Schellin.

Skookum Gulch v Klondike (Skookum Gulch).

Slaghek Fabbri v Slaghek's Iron.

Slaghek's Iron

Atacama, Chile, Co-ordinates not reported
Found 1916, before this year
Synonym(s): *Atacama Desert, Slaghek Fabbri*
Iron. Octahedrite, medium (1.3mm) (IIIA).

A mass of 1.9kg was given by G.H. Slaghek-Fabbri to the Techn. Inst. Livorno, and is stated to have come from the Atacama Desert, but there are no details of the place or date of find. Description, with an analysis, 7.89 %Ni, M. Bertolani, Periodico Miner., 1950, **19**, p.127, M. Bertolani, Rend. Soc. Min. Ital., 1950, **6**, p.29 [M.A. 11-271].

Slanica v Magura.

Slavetic 45°41′ N., 15°36′ E.

Zagreb (= Agram), Croatia, Yugoslavia
Fell 1868, May 22, 1030 hrs
Synonym(s): *Jaska*
Stone. Olivine-bronzite chondrite (H4), brecciated.

After the appearance of a cloud, and detonations, several stones appear to have fallen, but only two, of 1583g and 125g respectively, were found, W. von Haidinger, Sitzungsber. Akad. Wiss. Wien, Math.-naturwiss. Kl., 1868, **58** (2), p.162, 943. Olivine Fa_{17}, B. Mason, Geochimica et Cosmochimica Acta, 1963, **27**, p.1011.

1.3kg Vienna, Naturhist. Mus.; 23g Calcutta, Mus. Geol. Surv. India; 11g Chicago, Field Mus. Nat. Hist.;

Specimen(s): [43204], 20.5g.

Sleeper Camp 30°15′ S., 126°20′ E.

Western Australia, Australia
Found 1962
Stone. Olivine-hypersthene chondrite (L6).

A mass of 1.25kg was found, G.J.H. McCall, letter of 13 February, 1964 in Min. Dept. BM(NH). Illustrated description, G.J.H. McCall and W.H. Cleverly, Min. Mag., 1968, **36**, p.691. Olivine $Fa_{24.7}$, B. Mason, Rec. Austr. Mus., 1974, **29**, p.169.

Main mass, Kalgoorlie, West. Austr. School of Mines; 11.6g Perth, West. Austr. Mus.; 3.3g Washington, U.S. Nat. Mus.;

Slobodka 55° N., 35° E.

Yukhov, Kaluga, Federated SSR, USSR,
[Слободка]
Fell 1818, August 10
Synonym(s): *Poltava (of P. Partsch), Simbirsk (of G. von Blöde)*
Stone. Olivine-hypersthene chondrite (L4), veined.

A stone of about 2.75kg fell, A. Göbel, Bull. Acad. Sci. St.-Petersbourg, 1867, **11**, p.246. Description, analysis, M.I. D'yakonova and V.Ya. Kharitonova, Метеоритика, 1961, **21**, p.52. A veined, white, chondrite, E.L. Krinov, Каталог Метеоритов Акад. Наук СССР, Москва, 1947 [M.A. 10-511]. Olivine Fa_{23}, B. Mason, Geochimica et Cosmochimica Acta, 1963, **27**, p.1011.

2455g Moscow, Acad. Sci., includes main mass, 2355g; 132g Berlin, Humboldt Univ.; 90g Vienna, Naturhist. Mus.; 42g Paris, Mus. d'Hist. Nat.; 23g Yale Univ.; 24g Tempe, Arizona State Univ.; 11g Chicago, Field Mus. Nat. Hist.;

Specimen(s): [35182], 21.2g.; [34673], 5g.; [35178], less than 1g

Slobodka (of P. Partsch) 55° N., 35° E.

Kaluga, Federated SSR, USSR
Found 1838, known before this year
Stone. Olivine-hypersthene chondrite (L6), veined.

Mentioned as having fallen on 10 August, 1818, P. Partsch, Die Meteoriten, Wien, 1843, p.55. Some confusion between this and Slobodka.

792g Tübingen, Univ.; 150g Vienna, Naturhist. Mus.; 4g Washington, U.S. Nat. Mus.;

Slonim v Ruschany.

Smith Center 39°50′ N., 99°1′ W.

Smith County, Kansas, U.S.A.
Found 1937
Synonym(s): *Kensington*
Stone. Olivine-hypersthene chondrite (L6).

One stone of 1585g was found in SE. 1/4, sect. 20, township 3, range 15, Smith County, A.D. Nininger, Pop. Astron., Northfield, Minnesota, 1939, **47**, p.213, E.P. Henderson, letter of 3 June, 1939 in Min. Dept. BM(NH)., H.H. and A.D. Nininger, The Nininger Collection of Meteorites, Winslow, Arizona, 1950, p.65. Olivine Fa_{24}, B. Mason, Geochimica et Cosmochimica Acta, 1963, **27**, p.1011.

522g Tempe, Arizona State Univ.; 91g Washington, U.S. Nat. Mus.;

Specimen(s): [1959,1027], 641g. and fragments, 2g

Smith County (Tennessee) v Carthage.

Smith County (Kansas) v Cedar (Kansas).

Smithland 37°8′ N., 88°24′ W.
Livingstone County, Kentucky, U.S.A.
Found 1839, or 1840
Synonym(s): *Livingston County, Salem*
Iron. Ataxite, Ni-rich (IVA).
A large mass was found and partly smelted, but about 5kg have been preserved, G. Troost, Am. J. Sci., 1846, **2**, p.357. Description, E. Cohen, Meteoritenkunde, 1905, **3**, p.101. Structurally anomalous, due to artificial reheating, V.F. Buchwald, Iron Meteorites, Univ. of California, 1975, p.1139. Analysis, 9.23 %Ni, 2.38 ppm.Ga, 0.133 ppm.Ge, 0.89 pm.Ir, R. Schaudy et al., Icarus, 1972, **17**, p.174.
1.5kg Harvard Univ.; 133g Washington, U.S. Nat. Mus.; 110g Vienna, Naturhist. Mus.; 92g Chicago, Field Mus. Nat. Hist.; 72g Paris, Mus. d'Hist. Nat.; 35g New York, Amer. Mus. Nat. Hist.;
Specimen(s): [20645], 2503g. includes largest piece preserved, 2396g, and a polished mount, 3.5g

Smithonia 34°0′ N., 83°10′ W.
Oglethorpe County, Georgia, U.S.A.
Found 1940
Synonym(s): *Elberton, Smithsonia*
Iron. Hexahedrite (IIA).
A mass of 154lb was found near Smithonia, F.C. Leonard, Pop. Astron., Northfield, Minnesota, 1947, **55**, p.102, 167 [M.A. 10-177]. Description, analysis, 5.58 %Ni, S.K. Roy and R.K. Wyant, Field Mus. Nat. Hist. Geol. Ser., 1950, **7** (9), p.129 [M.A. 11-270]. Further analysis, 5.86 %Ni, 65.3 ppm.Ga, 187 ppm.Ge, 34 ppm.Ir, J.T. Wasson, Geochimica et Cosmochimica Acta, 1969, **33**, p.859. Elberton is part of this mass, V.F. Buchwald, Iron Meteorites, Univ. of California, 1975, p.1142.
70kg Chicago, Field Mus. Nat. Hist.; 590g Washington, U.S. Nat. Mus.; 247g Hanau, Zeitschel Colln.;
Specimen(s): [1980,M.12], 127.8g. part-slice

Smith's Mountain 36°25′ N., 80°0′ W.
Rockingham County, North Carolina, U.S.A.
Found 1863, approx.
Synonym(s): *Rockingham County*
Iron. Octahedrite, medium (0.63mm) (IIIB).
A mass of about 11lb was found. Report, with an analysis, 8.74 %Ni, W.C. Kerr, Rep. Geol. Surv. North Carolina, 1875, **1**, p.313 Appendix C, p.56. Analysis, 8.02 %Ni, J.L. Smith, Am. J. Sci., 1877, **13**, p.213. Classification, further analysis, 9.56 %Ni, 17.4 ppm.Ga, 30.6 ppm.Ge, 0.023 ppm.Ir, E.R.D. Scott et al., Geochimica et Cosmochimica Acta, 1973, **37**, p.1957. Kamacite shock-hardened, H.J. Axon, Prog. Materials Sci., 1968, **13**, p.183. Paired with Hopper, J.T. Wasson, Meteorites, Springer-Verlag, 1974, p.304. Description; bandwidth variable, V.F. Buchwald, Iron Meteorites, Univ. of California, 1975, p.1143.
Main mass, Raleigh, North Carolina, State Mus.; 784g Harvard Univ.; 480g New York, Amer. Mus. Nat. Hist.; 459g Chicago, Field Mus. Nat. Hist.; 224g Tübingen, Univ.; 152g Paris, Mus. d'Hist. Nat.;
Specimen(s): [47241], 51.5g.; [47872], 19.5g.

Smithsonia v Smithonia.

Smithsonian Iron
Locality not known
Found 1881, before this year
Iron. Octahedrite, coarsest (10mm) (IIB).
An iron of 3.51kg, unknown provenance, G.P. Merrill, Proc. U.S. Nat. Mus., 1916 (Bull. 94), p.56. Distinct from Coahuila (*q.v.*), which is a hexahdrite. Classification and analysis, 5.64 %Ni, 55.3 ppm.Ga, 165 ppm.Ge, 0.057 ppm.Ir, J.T. Wasson, Meteorites, Springer-Verlag, 1974, p.300. Description; probably an independent fall, V.F. Buchwald, Iron Meteorites, Univ. of California, 1975, p.1145.
3.4kg Washington, U.S. Nat. Mus.;
Specimen(s): [54279], 5.5g.

Smithville 35°59′ N., 85°51′ W.
De Kalb County, Tennessee, U.S.A.
Found 1840
Synonym(s): *Caney Fork, Cany Fork, Caryfort, De Calb County, DeKalb County*
Iron. Octahedrite, coarse (2.2mm) (IA).
A mass of 36lb was found "near the mouth of the Cany Fork" in 1840, G. Troost, Am. J. Sci., 1845, **49**, p.341. Three other masses of about 7lb, 15lb, and 65lb, were found at Smithville in 1892, O.W. Huntington, Proc. Amer. Acad. Arts and Sci., 1894, **21**, p.251. Two further masses, 8lb and 1lb, were found at Berey Cantrell's in 1903, L.C. Glenn, Am. J. Sci., 1904, **17**, p.216. Another mass of 2.9kg was found in January, 1962, W.F. Read, J. Tennessee Acad. Sci., 1963, **38**, p.22, Meteor. Bull., 1963 (26). Analysis, 6.78 %Ni, 86.9 ppm.Ga, 363 ppm.Ge, 2.0 ppm.Ir, J.T. Wasson, Icarus, 1970, **12**, p.407. Description; may include Cookeville (*q.v.*), V.F. Buchwald, Iron Meteorites, Univ. of California, 1975, p.1147.
10kg New York, Amer. Mus. Nat. Hist., approx. weight; 3.7kg Chicago, Field Mus. Nat. Hist.; 3.7kg Washington, U.S. Nat. Mus.; 2.4kg Vienna, Naturhist. Mus.; 2.2kg Harvard Univ.; 1.2kg Berlin, Humboldt Univ.; 1.2kg Tempe, Arizona State Univ.; 260g Paris, Mus. d'Hist. Nat.; 147g Tübingen, Univ.;
Specimen(s): [77093], 1670g. and fragments, 43g; [35791], 4.5g. 'Caryfort'

Smoky Hill River v Prairie Dog Creek.

Smyer 33°35′ N., 102°10′ W.
Hockley County, Texas, U.S.A.
Found 1968, June 1, approx.
Stone. Olivine-bronzite chondrite (H5-6).
A single mass of 3273g was recovered from a cultivated field 4 miles NE. of Smyer, olivine $Fa_{19.7}$, Meteor. Bull., 1980 (57), Meteoritics, 1980, **15**, p.101. Brief description, E.A. King, Meteoritics, 1979, **14**, p.443, abs.

Snyder 32°43′ N., 100°55′ W.
Scurry County, Texas, U.S.A.
Found, year not reported
Stone. Olivine-bronzite chondrite (H3).
A single mass of 16.9kg was found, T.E. Rodman, priv. comm., 1983.
16.9kg Odessa, Texas, Library;

Social Circle 33°42′ N., 83°42′ W.
Walton County, Georgia, U.S.A.
Found 1927, several years before this
Iron. Octahedrite, fine (0.3mm) (IVA).
A mass of 219lb was found. Description, with an analysis, 5.02 %Ni, S.W. McCallie, Am. J. Sci., 1927, **13**, p.360.

Analysis, 7.44 %Ni, E.P. Henderson and S.H. Perry, Am. Miner., 1951, **36**, p.603 [M.A. 11-443]. Shocked then reheated, A.V. Jain and M.E. Lipschutz, Geochimica et Cosmochimica Acta, 1970, **34**, p.883, V.F. Buchwald, Iron Meteorites, Univ. of California, 1975, p.1151. Further analysis, 7.65 %Ni, 1.63 ppm.Ga, 0.092 ppm.Ge, 3.6 ppm.Ir, R. Schaudy et al., Icarus, 1972, **17**, p.174.

70kg Atlanta, Georgia State Mus, approx. weight, main mass; 22kg Washington, U.S. Nat. Mus.; 227g Chicago, Field Mus. Nat. Hist.;

Soheria 27°8′ N., 84°4′ E. approx.

Bagaha district, Champaran, Bihar, India
Fell 1960, January 8, 0200 hrs
Stone. Chondrite.

One small stone of 72.9g was recovered. Description, S.S. Deshmukh, Indian Minerals, 1961, **15**, p.201.

72g Calcutta, Mus. Geol. Surv. India;

Soko-Banja 43°40′ N., 21°52′ E.

Aleksinac, Serbia, Yugoslavia
Fell 1877, October 13, 1400 hrs
Synonym(s): *Aleksinac, Alexinatz, Banja, Blendija, Devica, Device, Dugo Polje, Dungo Polje, Sarbanovac, Scherbanovaz*
Stone. Olivine-hypersthene chondrite, amphoterite (LL4), brecciated.

After the appearance of a fireball, and detonations, a shower of stones, of which about ten were found, fell over an area of 7 × 1 miles; the total weight was about 80kg and the largest stone weighed 38kg, C. Klein, Nachr. Gessell. Wiss. Göttingen, 1879, p.92, E. Doll, Verh. Geol. Reichsanst. Wien, 1877, p.283, S.M. Losanitsch, Ber. Deutsch. Chem. Gesell. Berlin, 1878, **11**, p.96. Analysis, 19.86 % total iron, A.J. Easton and C.J. Elliott, Meteoritics, 1977, **12**, p.409. Xenolithic, olivine Fa_{28}, R.A. Binns, Geochimica et Cosmochimica Acta, 1968, **32**, p.299, R.V. Fodor and K. Keil, Meteoritics, 1975, **10**, p.325.

16kg Belgrade, Nat. Hist. Mus.; 3.5kg Budapest, 1kg Nat. Mus., 2.5kg Eötvös Lorand Univ.; 2.5kg Vienna, Naturhist. Mus.; 1.8kg Paris, Mus. d'Hist. Nat.; 319g Chicago, Field Mus. Nat. Hist.; 337g Washington, U.S. Nat. Mus.; 218g Harvard Univ.; 144g New York, Amer. Mus. Nat. Hist.;

Specimen(s): [51857], 1912g. and fragments, 45g

Sombrerete 23°38′ N., 103°40′ W.

Sombrerete, Zacatecas, Mexico
Found 1958
Iron. Anomalous (IRANOM).

A single mass of about 10kg was found, Meteor. Bull., 1978 (55), Meteoritics, 1978, **13**, p.349. Contains silicate inclusions of albite and o-pyroxene, Fs29. Oxygen isotopic ratios, T.K. Mayeda and R.N. Clayton, Geochimica et Cosmochimica Acta, 1980 (suppl. 14), p.1145. Analysis, 10.0 %Ni, 18.1 ppm.Ga, 11.3 ppm.Ge, 0.073 ppm.Ir, D.J. Malvin et al., priv. comm., 1983.

1kg Washington, U.S. Nat. Mus., approx. weight;

Somerset County v New Baltimore.

Somervell County 32°11′ N., 97°48′ W.

Somervell County, Texas, U.S.A.
Found 1919, approx. recognized 1937
Stony-iron. Pallasite (PAL).

Three masses totalling about 26lb were found, A.D. Nininger, Pop. Astron., Northfield, Minnesota, 1937, **45**, p.449 [M.A. 7-62]. Olivine $Fa_{12.5}$, P.R. Buseck and J.I. Goldstein, Bull. Geol. Soc. Amer., 1969, **80**, p.2141.

Main mass, Fort Worth, Texas, Monnig Colln.; 80g Washington, U.S. Nat. Mus.;

Somesbar 41°23′ N., 123°30′ W.

Siskiyou County, California, U.S.A.
Found 1977
Stone. Olivine-bronzite chondrite (H6).

A single mass of about 60g was found, olivine Fa_{20}, Meteor. Bull., 1981 (59), Meteoritics, 1981, **16**, p.197.

38g Tempe, Arizona State Univ.;

Sone 35°10′ N., 135°20′ E.

Shuchi, Funai, Kyoto Fu, Honshu, Japan
Fell 1866, June 7, 1200 hrs
Stone. Olivine-bronzite chondrite (H).

After detonations, a stone of about 17kg was found, K. Jimbo, Beitr. Miner. Japan, 1906 (2), p.39. Date of fall given as 17 May, 1866, I. Yamamoto, Kwasan Observ. Bull., 1935, p.306 [M.A. 7-173]. Date given discussed, S. Murayama, letter of 6 April, 1962 in Min. Dept. BM(NH). Olivine Fa_{18}, B. Mason, Geochimica et Cosmochimica Acta, 1967, **31**, p.1100. Description, analysis, 27.23 % total iron, A. Miyashiro et al., Bull. Nat. Sci. Mus. Tokyo, 1963, **6**, p.352.

Main mass, held by S. Takagi of Takatsuki, Osaka;

Sonora v Tucson.

Soojoolee v Segowlie.

Soper 34°2′ N., 95°35′ W.

Choctaw County, Oklahoma, U.S.A.
Found 1938
Synonym(s): *Hugo (iron)*
Iron. Anomalous (IRANOM).

One mass, 3700g, was found 6 miles NW. of Hugo, Choctaw Co.; description, with an analysis, 6.21 %Ni, F.C. Wood and C.A. Merritt, Am. Miner., 1939, **24**, p.59 [M.A. 7-273]. Another analysis, 5.66 %Ni, E.P. Henderson and S.H. Perry, Am. Miner., 1948, **33**, p.692 [M.A. 10-520]. Anomalous, both structurally and chemically, analysis, 5.70 %Ni, 9.71 ppm.Ga, 10.8 ppm.Ge, 0.011 ppm.Ir, J.T. Wasson and R. Schaudy, Icarus, 1971, **14**, p.59. Description; dendritic texture formed, then recrystallised, V.F. Buchwald, Iron Meteorites, Univ. of California, 1975, p.1154.

1632g Oklahoma Univ. (Mus. Geol.); 1122g Washington, U.S. Nat. Mus.; 165g Chicago, Field Mus. Nat. Hist.;

Specimen(s): [1959,908], 45.5g. slice, and sawings, 1g.

Sopot 44°25′ N., 23°30′ E.

Craiova, district Dolj, Romania
Fell 1927, April 27, 0000-0100 hrs
Stone. Chondrite.

Eight stones totalling 958g were collected over an area measuring 12 km from east to west, M. Demetrescu, Mem. Muzeuli Regional al Olteniei, Craiova, 1928, **1** (2) [M.A. 5-154].

Sörakarta v Prambanan.

Soroti 1°42′ N., 33°38′ E.

Teso district, Uganda
Fell 1945, September 17
Synonym(s): *Sozoti, Uganda*
Iron. Octahedrite, finest (0.12mm) (IRANOM).

After thunderous noises, fragments of metal fell over a wide area. Four pieces recovered weighed 1000g, 700g, 180g, 170g, and show sharp points and deep hollows, due to the

melting out of the abundant troilite. Description, with an analysis of metal, 12.96 %Ni, E.P. Henderson and S.H. Perry, Proc. U.S. Nat. Mus., 1958, **107**, p.339 [M.A. 14-130]. Mentioned, Ann. Rep. Geol. Surv. Uganda for 1945, 1948, p.38. Troilite-rich, analysis of metal, 12.88 %Ni, 14.1 ppm.Ga, 5.22 ppm.Ge, 0.060 ppm.Ir, J.T. Wasson and R. Schaudy, Icarus, 1971, **14**, p.59. Description; a unique meteorite, V.F. Buchwald, Iron Meteorites, Univ. of California, 1975, p.1156.
1.7kg Entebbe, Geol. Surv. Uganda; 154g Washington, U.S. Nat. Mus.; 61g New York, Amer. Mus. Nat. Hist.;
Specimen(s): [1971,98], 35g. sawn, crusted metal fragment

Sotik
Nairobi, Kenya
Fell 1946, November 19, 2030-2130 hrs
Doubtful..
SW. of Nairobi, a meteor was seen travelling N.; three explosions were heard. The meteor apparently broke into four or five pieces, one of which is said to have struck a hut, but nothing was found, K.P. Oakley, comm. of 19 March, 1968 in Min. Dept. BM(NH).

Sotoville v Tombigbee River.

South African Railways
Cape Province, South Africa, Not well located; possibly in Namibia
Found 1938, before this year
Synonym(s): *Railway*
Iron.
A 47kg mass was found near the main railway line between Windhoek and Cape Town, C. Frick and E.C.I. Hammerbeck, Bull. Geol. Surv. S. Africa, 1973 (57), p.31.
Main mass, Pretoria, Geol. Surv. Mus.; 1kg Heidelberg, Max-Planck Inst.;

South Arcot v Nammianthal.

South Bend 41°39′ N., 86°13′ W.
St. Joseph County, Indiana, U.S.A.
Found 1893
Stony-iron. Pallasite (PAL).
A mass of 5.5lb was ploughed up, 2 miles SE. of South Bend. Description, with an analysis of the iron, 9.35 %Ni, O.C. Farrington, Field Mus. Nat. Hist. Geol. Ser., 1906, **3** (Publ. 109), p.19. Olivine Fa_{12}, P.R. Buseck and J.I. Goldstein, Bull. Geol. Soc. Amer., 1969, **80**, p.2141. Analysis of metal, 9.35 %Ni, 21.2 ppm.Ga, 44.3 ppm.Ge, 0.055 ppm.Ir, J.T. Wasson and S.P. Sedwick, Nature, 1969, **222**, p.22. Trace element data; references, E.R.D. Scott, Geochimica et Cosmochimica Acta, 1977, **41**, p.349.
1292g Chicago, Field Mus. Nat. Hist., main mass; 171g Harvard Univ.; 86g Vienna, Naturhist. Mus.; 35g New York, Amer. Mus. Nat. Hist.;
Specimen(s): [1910,455], 161g. slice

South Byron 43°2′ N., 78°2′ W.
Genesee County, New York, U.S.A.
Found 1915
Iron. Ataxite, Ni-rich (IRANOM).
A mass of nearly 6kg was found about 1/2 mile W. of South Byron, O.C. Farrington, letter of 15 November, 1926 in Min. Dept. BM(NH). Chemically anomalous, but similar to Babb's Mill (Troost's Iron) (*q.v.*), analysis, 17.8 %Ni, 20.0 ppm.Ga, 45.0 ppm.Ge, 28 ppm. Ir, E.R.D. Scott et al., Geochimica et Cosmochimica Acta, 1973, **37**, p.1957. Description, V.F. Buchwald, Iron Meteorites, Univ. of California, 1975, p.1158.
2.8kg Chicago, Field Mus. Nat. Hist.; 465g Washington, U.S. Nat. Mus.; 633g Paris, Mus. d'Hist. Nat.; 168g Tempe, Arizona State Univ.;
Specimen(s): [1959,912], 166.5g. slice, and sawings, 10g

South Canara v Udipi.

South Corston v Strathmore.

South Dahna 22°34′ N., 48°18′ E.
Rub'al Khali, Saudi Arabia
Found 1957
Iron. Octahedrite.
A much-oxidized and broken mass of 275kg was found. Description, with an analysis, D.A. Holm, Am. J. Sci., 1962, **260**, p.303.
142kg Washington, U.S. Nat. Mus.;

South Dixon 41°50′ N., 89°29′ W.
Lee County, Illinois, U.S.A.
Found 1947?
Pseudometeorite..
A mass of 3550g was found. Description, with a partial analysis, B.H. Wilson, The Mineralogist, Portland, Oregon, 1947, **15**, p.63 [M.A. 10-519]. The locality of the find is given as South Dixon, Illinois; this was not traced, but is probably a suburb of Dixon, Lee County. Gabbroic, Meteor. Bull., 1978 (55), Meteoritics, 1978, **13**, p.352.

South-east Missouri v Saint Francois County.

South-east Missouri v Wichita County.

Southern Arizona
Arizona, U.S.A., Locality not known
Found 1947, known in this year
Iron. Octahedrite, coarse (1.5mm) (IA).
A mass of 266g is listed under this name in the catalogue of S.H. Perry's collection, Rep. U.S. Nat. Mus., 1947, p.44. Several coarse octahedrites are known from Arizona and this specimen may belong to one of the known falls. Distinct from Cañon Diablo and Seligman, J.T. Wasson, J. Geophys. Res., 1968, **73**, p.3207. Possibly mislabelled Toluca, V.F. Buchwald, Iron Meteorites, Univ. of California, 1975, p.1161. Analysis, 8.06 %Ni, 66.2 ppm.Ga, 242 ppm.Ge, 1.9 ppm.Ir, J.T. Wasson, Icarus, 1970, **12**, p.407.
193g Chicago, Field Mus. Nat. Hist.;

Southern Michigan
Michigan, U.S.A., Co-ordinates not known
Found
Iron. Octahedrite, medium.
A slice, 49g in Chicago (Field Mus. Nat. Hist.), H. Horback and E.J. Olsen, Fieldiana Geol., 1965, **15** (3).
49g Chicago, Field Mus. Nat. Hist.;

South Forrest v Forrest (b).

South Oman 21°0′ N., 56°40′ E.
Oman
Found 1958, approx.
Stone. Enstatite chondrite (E4).
One weathered stone, about 90g, was found, S.O. Agrell, priv. comm., 1970. Contains kamacite (3.9%Ni) and perryite, S.J.B. Reed, Min. Mag., 1968, **36**, p.850. Analysis, 29.59 % total iron, A.J. Easton, Ph.D. Thesis, 1983.

Main mass, Cambridge, Univ.; 13g Washington, U.S. Nat. Mus.; 6g Tempe, Arizona State Univ.;
Specimen(s): [1966,1], 7.2g.

South Plains 34°16′ N., 101°15′ W.
Floyd County, Texas, U.S.A.
Found 1971, approx.
Stone. Olivine-hypersthene chondrite (L5).
A single mass of 4763g was found in a field, Meteor. Bull., 1980 (57), Meteoritics, 1980, **15**, p.101.

South Queensland v Gladstone (iron).

South Strafford 43°51′ N., 72°21′ W.
Orange County, Vermont, U.S.A.
Found 1942, August
Pseudometeorite. Iron.
A mass of 5lb 2oz was found on Whitcomb Hill, C.G. Doll, Science, 1942, **96**, p.494 [M.A. 8-374]. Cast iron, E.P. Henderson, Am. Miner., 1955, **40**, p.936.

Sowallick Mountains v Cape York.

Soweida 32°43′ N., 36°34′ E.
Egypt
Fell 856, December, A.D.
Doubtful..
Five stones, one of about 3lb, fell, S. de Sacy, Ann. Phys. (Gilbert), 1818, **50**, p.293, E.F.F. Chladni, Die Feuer-Meteore, Wien, 1819, p.192. The evidence of meteoritic nature is not conclusive.

Sozoti v Soroti.

Spearman 36°15′ N., 101°13′ W.
Hansford County, Texas, U.S.A.
Found 1934
Iron. Octahedrite, medium (1.1mm) (IIIA).
One mass, 10.4kg was found, A.D. Nininger, Pop. Astron., Northfield, Minnesota, 1937, **45**, p.449 [M.A. 7-62]. Determination of Ga, Au and Pd, 8.39 %Ni, E. Goldberg et al., Geochimica et Cosmochimica Acta, 1951, **2**, p.1. Further analysis, 8.63 %Ni, 20.2 ppm.Ga, 46.0 ppm.Ge, 0.71 ppm.Ir, E.R.D. Scott et al., Geochimica et Cosmochimica Acta, 1973, **37**, p.1957. Description; shock-hardened, V.F. Buchwald, Iron Meteorites, Univ. of California, 1975, p.1162.
2kg Fort Worth, Texas, Monnig Colln.; 870g Washington, U.S. Nat. Mus.; 750g Chicago, Field Mus. Nat. Hist.; 662g Harvard Univ.; 396g Tempe, Arizona State Univ.;
Specimen(s): [1949,179], 69g.

Spedalone presso Pienza v Siena.

Spezia v Pułtusk.

Spoleto v Collescipoli.

Springbok River v Gibeon.

Springer 36°20′46″ N., 97°11′6″ W.
Carter County, Oklahoma, U.S.A.
Found 1965
Stone. Olivine-bronzite chondrite (H).
Listed, no further details, olivine Fa_{19}, J.T. Wasson, Meteorites, Springer-Verlag, 1974, p.260, 272. Locality, total mass found 8.3kg, Cat. Monnig Colln. Meteorites, 1983, p.20.
8.1kg Fort Worth, Texas, Monnig Colln.;

Springfield 37°23′ N., 102°38′ W.
Baca County, Colorado, U.S.A.
Found 1937
Stone. Olivine-hypersthene chondrite (L6).
Two stones weighing 283g were found, A.D. Nininger, Pop. Astron., Northfield, Minnesota, 1937, **45**, p.449 [M.A. 7-62], H.H. Nininger, Amer. Antiquity, 1938, **4**, p.39 [M.A. 7-272]. Total known weight 3.2kg, W. Wahl, letter of 23 May, 1950 in Min. Dept. BM(NH). Three different falls possibly represented, G.I Huss, letter of 23 April, 1970 in Min. Dept. BM(NH).
1.5kg Tempe, Arizona State Univ.; 517g Denver, Mus. Nat. Hist.; 92g New York, Amer. Mus. Nat. Hist.; 45g Chicago, Field Mus. Nat. Hist.;
Specimen(s): [1959,833], 1005g.

Springlake 34°20′40″ N., 102°13′25″ W.
Lamb County, Texas, U.S.A.
Found 1980, August, approx.
Stone. Olivine-hypersthene chondrite (L6).
Mentioned, co-ordinates, olivine $Fa_{25.5}$, E.J. Olsen, letter of 24 October, 1982 in Min. Dept. BM(NH).
131g Watchung, Dupont Colln; 72g Chicago, Field Mus. Nat. Hist.;

Springwater 52°0′ N., 108°18′ W.
Saskatchewan, Canada
Found 1931
Stony-iron. Pallasite (PAL).
Three masses, of 85lb, 41lb and 23lb, were found near Springwater, 100 miles west of Saskatoon. Description, analysis of metal, 10.72 %Ni, H.H. Nininger, Am. Miner., 1932, **17**, p.396 [M.A. 5-157]. Contains farringtonite, E.R. DuFresne and S.K. Roy, Geochimica et Cosmochimica Acta, 1961, **24**, p.198. Further analysis of metal, 12.6 %Ni, 14.8 ppm.Ga, 31.9 ppm.Ge, 0.069 ppm.Ir, J.T. Wasson and S.P. Sedwick, Nature, 1969, **222**, p.22. Olivine Fa_{18}, P.R. Buseck and J.I. Goldstein, Bull. Geol. Soc. Amer., 1969, **80**, p.2141. Trace element data; references, E.R.D. Scott, Geochimica et Cosmochimica Acta, 1977, **41**, p.349.
17.7kg Tempe, Arizona State Univ.; 2.6kg Washington, U.S. Nat. Mus.; 1.6kg Prague, Nat. Mus.; 1.3kg Los Angeles, Univ. of California; 1.2kg Chicago, Field Mus. Nat. Hist.; 745g Ottawa, Mus. Geol. Surv. Canada; 648g Paris, Mus. d'Hist. Nat.; 570g New York, Amer. Mus. Nat. Hist.;
Specimen(s): [1937,1353], 1387g.; [1959,1017], 20.1kg. half of one of the stones; [1962,136], 98g. slice

Sprucefield v Bovedy.

Squaw Creek 32° N., 98° W. approx.
Somervell County, Texas, U.S.A.
Found
Iron. Anomalous (IIAB).
Listed, without details, analysis, 5.5 %Ni, 59 ppm.Ga, 182 ppm.Ge, 9.7 ppm.Ir, D.J. Malvin et al., priv. comm., 1983.
Main mass, Fort Worth, Texas, Monnig Colln.;

Ssyromolotovo 58°37′ N., 98°56′ E.
Keshma, Krasnoyarsk region, Russian SSR, USSR, [Сыромолотово]
Found 1873
Synonym(s): *Angara, Syromolotov, Syromolotovo*
Iron. Octahedrite, medium (0.9mm) (IIIA).
A mass, stated to be of 196.5kg (but over 216kg is in Moscow, Acad. Sci.), was found in the sand on the left bank of the Angara, A. Göbel, Bull. Acad. Sci. St.-Petersbourg, 1874, **19**, p.544. Analysis, 7.94 %Ni, M.I. D'yakonova, Метеоритика, 1958, **16**, p.180. Place of fall incorrectly mapped, P.M. Millman, Trans. Roy. Astron. Soc. Canada, 1938, **32**, p.199. Further analysis, 7.85 %Ni, 19.6 ppm.Ga, 40.9 ppm.Ge, 3.3 ppm.Ir, J.T. Wasson, Meteorites, Springer-Verlag, 1974, p.302. Description, V.F. Buchwald, Iron Meteorites, Univ. of California, 1975, p.1164.
216.3kg Moscow, Acad. Sci., includes main mass, 214.6kg; 161g Perth, West. Austr. Mus.; 26g Chicago, Field Mus. Nat. Hist.; 7g Vienna, Naturhist. Mus.;
Specimen(s): [66748], 3.75g. slice

Staartje v Uden.

Stade v Bremervörde.

Staelldalen v Ställdalen.

Ställdalen 59°56′ N., 14°57′ E.
Nya Kopparberg, Örebro, Sweden
Fell 1876, June 28, 2330 hrs
Synonym(s): *Staelldalen, Stelldalen*
Stone. Olivine-bronzite chondrite (H5), brecciated.
After the appearance of a fireball, and detonations, 11 stones fell, totalling about 34kg, the largest weighing about 12.5kg, A.E. Nordenskiöld, Geol. För. Förh. Stockholm, 1878, **4**, p.46. Analysis, G. Lindström, Öfvers. Vetensk.-Akad. Förh. Stockholm, 1877 (4), p.35. Olivine Fa_{19}, B. Mason, Geochimica et Cosmochimica Acta, 1963, **27**, p.1011.
23.5kg Stockhholm, Riksmus; 3.5kg Uppsala, Univ.; 500g Budapest, Nat. Mus.; 443g New York, Amer. Mus. Nat. Hist.; 377g Chicago, Field Mus. Nat. Hist.; 255g Oslo, Min.-Geol. Mus.; 242g Vienna, Naturhist. Mus.; 192g Washington, U.S. Nat. Mus.; 87g Paris, Mus. d'Hist. Nat.; 78g Harvard Univ.;
Specimen(s): [51549], 1322g.; [1920,343], 62g.; [1927,1290], 2g.; [1958,713], 215g.

Stannein v Stannern.

Stannern 49°17′ N., 15°34′ E.
Iglau, Jihomoravsky, Czechoslovakia
Fell 1808, May 22, 0600 hrs
Synonym(s): *Cilli, Iglau, Langenpiernitz, Stannein, Stonarov, Stonarow*
Stone. Achondrite, Ca-rich. Eucrite (AEUC).
After detonations, some 200-300 stones fell of which about 66 stones were recovered totalling about 52kg, the largest weighing about 6kg, K. von Schreibers, Ann. Phys. (Gilbert), 1808, **29**, p.225. Analysis, C. Rammelsberg, Ann. Phys. Chem. (Poggendorff), 1851, **83**, p.591, G.P. Merrill, Mem. Nat. Acad. Sci. Washington, 1916, **14** (1), p.22. Full description and bibliography, W. von Englehardt, Beitr. Miner. Petrogr., 1963, **9**, p.65. Rb/Sr data, J.L. Birck and C.J. Allegre, Earth planet. Sci. Lett., 1978, **39**, p.37 [M.A. 78-4757]. Ba and R.E. data, C.C. Schnetzler and J.A. Philpotts, Meteorite Research, ed. P.M. Millman, D. Reidel, Dordrecht-Holland, 1969, p.206. Sm-Nd systematics, G.W. Lugmair and N.B. Schreinin, Meteoritics, 1975, **10**, p.447, abs. Low pressure quenching experiments, E. Stolper, Geochimica et Cosmochimica Acta, 1977, **41**, p.587.
14kg Vienna, Naturhist. Mus.; 4kg Tübingen, Univ.; 2kg Budapest, Nat. Mus.; 1.5kg Berlin, Humboldt Univ.; 1.2kg Prague, Nat. Mus.; 840g Paris, Mus. d'Hist. Nat.; 580g Moscow, Acad. Sci.; 623g Chicago, Field Mus. Nat. Hist.; 500g Budapest, Eötvös Lorand Univ.; 497g Washington, U.S. Nat. Mus.; 309g Copenhagen, Univ. Geol. Mus.; 263g Dublin, Nat. Mus.; 128g Bern, Naturhist. Mus., formerly BM 1921,18; 220g Tempe, Arizona State Univ.;
Specimen(s): [21429], 674g. complete stone; [46970], 157g.; [33892], 237g.; [34374], 139g.; [90255], 192.5g. complete stone; [33731], 10.25g. " Langenpiernitz "

Stannton v Staunton.

Stara Bela v Alt Bela.

Staroe Boriskino v Boriskino.

Staroe Pesianoe v Pesyanoe.

Staroe Pesyanoe v Pesyanoe.

Staroje Pesjanoje v Pesyanoe.

Staro-Pesiianoe v Pesyanoe.

Staroye Boriskino v Boriskino.

Staunton 38°13′ N., 79°3′ W.
Augusta County, Virginia, U.S.A.
Found 1869
Synonym(s): *Augusta County, Folersville, Louisa County, Stannton*
Iron. Octahedrite, coarse (1.6mm) (IIIE).
Three masses, of about 25.5kg, 16.5kg and 1.5kg respectively, were described, with an analysis, 10.24 %Ni, J.W. Mallet, Am. J. Sci., 1871, **2**, p.10. Another mass of about 1kg was also found; description, analysis, 8.85 %Ni, G.F. Kunz, Am. J. Sci., 1887, **33**, p.58. A mass of 152lb and one of 7.1kg, previously assigned to this find are medium octahedrites and now distinguished as Augusta County (*q.v.*), V.F. Buchwald, Iron Meteorites, Univ. of California, 1975, p.1166. Further analysis, 8.21 %Ni, 18.9 ppm.Ga, 36.6 ppm.Ge, 0.11 ppm.Ir, E.R.D. Scott et al., Geochimica et Cosmochimica Acta, 1973, **37**, p.1957.
1.7kg Tempe, Arizona State Univ.; 1.6kg Vienna, Naturhist. Mus.; 1.2kg Washington, U.S. Nat. Mus.; 557g Prague, Nat. Mus.; 220g Harvard Univ.;
Specimen(s): [44761], 1565g.

Staunton No. 4 v Augusta County.

Staunton No. 6 v Augusta County.

Stavrapol v Stavropol.

Stavropol 45°3′ N., 41°59′ E.
Stavrapol region, Federated SSR, USSR, [Ставрополь]
Fell 1857, March 24, 1700 hrs
Synonym(s): *Stavrapol, Stawropol*
Stone. Olivine-hypersthene chondrite (L6).
After detonations, a stone fell measuring 132 × 93 × 66 mm and weighing about 1.5kg. Description, with an analysis, H. Abich, Bull. Acad. Sci. St.-Petersbourg, 1860, **2**, p.404, 436.

Further analysis, 22.46 % total iron, V.Ya. Kharitonova, Метеоритика, 1965, **26**, p.146. Olivine Fa_{24}, B. Mason, Geochimica et Cosmochimica Acta, 1963, **27**, p.1011.

1180g Moscow, Acad. Sci, includes main mass, 921g; 90g Berlin, Humboldt Univ.; 52g Washington, U.S. Nat. Mus.; 47g New York, Amer. Mus. Nat. Hist.; 21g Vienna, Naturhist. Mus.;

Specimen(s): [35216], 22.5g.; [35175], 0.75g.

Stawropol v Stavropol.

Steiermark 47° N., 15° E.

Steiermark, Austria
Fell 1618, August
Doubtful..

A shower of stones fell, up to 3 cwt in weight, E.F.F. Chladni, Die Feuer-Meteore, Wien, 1819, p.220 but the evidence is not conclusive. The fall may have occurred in 1619, July 14, according to a report in a Turkish history book, A. Kizilirmak, letter of 7 January, 1967 in Min. Dept. BM(NH).

Steinbach 50°30′ N., 12°30′ E. approx.

Erzgebirge, Karl-Marx-Stadt, Germany
Found 1724
Synonym(s): *Breitenbach, Eibenstock, Grimma, Johanngeorgenstadt, Rittersgrün*
Iron. Iron with silicate inclusions (IVA-ANOM).

Large masses of iron were said to have fallen at Whitsuntide, 1164, in the region of Meissen, E.F.F. Chladni, Ann. Phys. (Gilbert), 1808, **29**, p.379 and between 1540-1550 a large mass of iron is said to have fallen near Grimma, E.F.F. Chladni, Die Feuer-Meteore, Wien, 1819, p.212. A mass of nearly 1kg which has been known since about 1724 in the Naturalien-Cabinette at Gotha, F. Stromeyer, Göttingische Gelehrte Anzeig, 1824, **3**, p.2082 may possibly be part of the Grimma mass. A mass of about 86.5kg was found at Rittersgrün in 1833 (or 1847), A. Breithaupt, Z. Deutsch. Geol. Ges., 1861, **13**, p.148. In 1861 another mass of about 10.5kg was found at Breitenbach, and was described, with an analysis, N.S. Maskelyne and W. Flight, Phil. Trans. Roy. Soc. London, 1871, **161**, p.359. The Grimma mass was described and analysed, F. Heide, Centralblatt Min., 1923, p.69. Although the Gotha specimen (which has been shown by Heide to belong with the Steinbach masses) has been attributed to the 1540-1550 Naunhof (= Grimma) fall, there is no clear evidence on this point, nor is there any evidence to connect it and the other masses with the Meissen fall of 1164. Analysis, 9.08 %Ni, 2.27 ppm.Ga, 0.132 ppm.Ge, 0.68 ppm.Ir, R. Schaudy et al., Icarus, 1972, **17**, p.174. Mineralogy, pyroxene Fs11, G. Dörfler et al., Tschermaks Min. Petr. Mitt., 1965, **10**, p.413.

1kg Gotha Mus., of Steinbach; 55.5kg Freiberg, Sächs. Bergakad., of Rittersgrün; 4.5kg Berlin, Humboldt Univ.; 3kg Vienna, Naturhist. Mus.; 500g Budapest, Nat. Mus.; 553g Harvard Univ.; 427g Washington, U.S. Nat. Mus.; 226g Chicago, Field Mus. Nat. Hist.;

Specimen(s): [35540], 6427g. includes main mass of Breitenbach, 6230g, and fragments, 14g; [35606], 689g. Rittersgrün; [33956], 130.5g. Steinbach; [34676], 1.7g. Steinbach

Stelldalen v Ställdalen.

Sterling 40°36′ N., 103°11′ W.

Logan County, Colorado, U.S.A.
Found 1900, approx.
Stony-iron. Pallasite (PAL).

A fragment of 679.5g was found, A.D. Nininger, Pop. Astron., Northfield, Minnesota, 1939, **47**, p.213. It may have been found in Nebraska, in sand hills, R.M. Pearl, letter of 11 September, 1973 in Min. Dept. BM(NH).

Stewart County v Lumpkin.

Stockholm v Hessle.

Stolzenau 52°32′ N., 9°3′ E.

Niedersachsen, Germany
Fell 1647, August, 1200 hrs
Stone.

One large stone fell on the heath and broke up. It was black outside and had gold-like specks inside, E.F.F. Chladni, Die Feuer-Meteore, Wien, 1819, p.227. This was clearly a true meteorite.

Stonarov v Stannern.

Stonarow v Stannern.

Stonington 37°17′ N., 102°12′ W.

Baca County, Colorado, U.S.A.
Found 1942, some years before this
Stone. Olivine-bronzite chondrite (H5), veined.

A weathered stone of about 6lb was found. Description, with a partial analysis, J.D. Buddhue, Pop. Astron., Northfield, Minnesota, 1942, **50**, p.97, Contr. Soc. Res. Meteorites, **3**, p.6, J.D. Buddhue, The Mineralogist, Portland, Oregon, 1942, **10**, p.377 [M.A. 8-375]. Olivine Fa_{19}, B. Mason, Geochimica et Cosmochimica Acta, 1963, **27**, p.1011.

336g Fort Worth, Texas, Monnig Colln.; 938g Tempe, Arizona State Univ.; 164g Denver, Mus. Nat. Hist.; 100g Washington, U.S. Nat. Mus.;

Specimen(s): [1959,841], 1031g.

Stonitsa v Borodino.

Stonitza v Borodino.

Stratford 41°12′ N., 73°8′ W.

Fairfield County, Connecticut, U.S.A.
Fell 1974, May 27, 1600 hrs
Stone. Olivine-hypersthene chondrite (L).

A single stone of 50g fell in the city of Stratford after a 'whistling' sound was heard. It made a small dent 1 inch deep in the asphalt of a street. The stone was sent to Los Angeles, Univ. of California, where it was analysed, Meteor. Bull., 1975 (53), Meteoritics, 1975, **10**, p.139.

Main mass, Los Angeles, Univ. of California;

Strathmore 56°35′ N., 3°15′ W.

Perthshire, Scotland
Fell 1917, December 3, 1315 hrs
Synonym(s): *Carsie, Corston, Easter Essendy, Essendy, Keithick, South Corston*
Stone. Olivine-hypersthene chondrite (L).

After the appearance of a brilliant fireball, travelling from SE. to NW., and detonations, four stones fell, three in Perthshire at Easter Essendy (22.25lb), Carsie (2lb 6oz), and Keithick (2.5lb), and one in Forfarshire at South Corston (2lb 5oz), H. Coates, Trans. Perthshire Soc. Nat. Sci., 1920,

7, p.80, R.A. Sampson, Proc. Roy. Soc. Edinburgh, 1918, **38**, p.70. Description, analysis, W.F.P. McLintock and F.R. Ennos, Min. Mag., 1922, **19**, p.323. Further analysis, 20.60 % total iron, A.J. Easton, unpublished in Min. Dept. BM (NH). Olivine Fa_{25}, B. Mason, Geochimica et Cosmochimica Acta, 1963, **27**, p.1011.

22lb Edinburgh, Roy. Scot. Mus., the Easter Essendy stone; 2.5lb Private possession, Keithick stone; 2.4lb Perth, Mus., South Corston stone;

Specimen(s): [1922,793], 1056g. Carsie stone; [1976,M.6], 46g. of Easter Essendy

Stratton 39°13′18″ N., 102°35′24″ W.

Kit Carson County, Colorado, U.S.A.
Found 1964
Stone. Chondrite.

A single mass of 121g was found in the gravel driveway of a farm, Meteor. Bull., 1979 (56), Meteoritics, 1979, **14**, p.170.

89g Mainz, Max-Planck-Inst.;

Stretchleigh 50°23′ N., 3°57′ W.

Ermington, Devonshire, England
Fell 1623, January 10
Stone.

A mass of 23lb fell, "like a stone singed or half burnt for lime", T.M. Hall, Min. Mag., 1879, **3**, p.4, T.M. Hall, Rep. Trans. Devon Ass. Adv. Sci., 1869, **3**, p.75. Appears authentic. Hall gives good reason for believing that the inadequately described Tregnie (Tregony?) fall, dated January 10, 1622, E.F.F. Chladni, Die Feuer-Meteore, Wien, 1819, p.222, R.P. Greg, Rep. Brit. Assn., 1860, **30**, p.48 or 1723, Rep. Brit. Assn., 1867, p.414 is really to be referred to this fall; Strechleigh and Tregony are 42 miles apart.

Strkov v Tabor.

Strkow v Tabor.

Stutsman County v Jamestown.

Sublette 37°30′ N., 100°50′ W.

Haskell County, Kansas, U.S.A.
Found 1952
Stone. Olivine-hypersthene chondrite (L6).

1.3kg were found, olivine Fa_{24}, B. Mason, Geochimica et Cosmochimica Acta, 1963, **27**, p.1011.

239g Los Angeles, Univ. of California; 120g Tempe, Arizona State Univ.; 62g Washington, U.S. Nat. Mus.; 59g New York, Amer. Mus. Nat. Hist.; 50g Harvard Univ.; 41g Mainz, Max-Planck-Inst.;

Success 36°29′ N., 90°40′ W.

Clay County, Arkansas, U.S.A.
Fell 1924, April 18, 0300-0400 hrs
Stone. Olivine-hypersthene chondrite (L6).

One mass of 3.5kg was seen to fall, Meteor. Bull., 1957 (5). Olivine Fa_{24}, B. Mason, Geochimica et Cosmochimica Acta, 1963, **27**, p.1011. Possibly not an observed fall, R.S. Clarke, Jr., letter of 8 May, 1973 in Min. Dept. BM(NH).

3503g Washington, U.S. Nat. Mus.;

Suchy Dul 50°32′17″ N., 16°15′48″ E.

near Police, Nachod district, Czechoslovakia
Fell 1969, September 16, 0715 hrs, U.T.
Synonym(s): *Police*
Stone. Olivine-hypersthene chondrite (L6).

One stone hit the roof of a house and split into two pieces weighing 755.2g and 60.1g. The fall was accompanied by a boom and a whine, Meteor. Bull., 1969 (48), Meteoritics, 1970, **5**, p.107, K. Tucek, Sbor. Nar. Muz. Praze, 1970, **26B**, p.97. Mineralogy, olivine Fa_{25}, P. Jakes, Sbor. Nar. Muz. Praze, 1980, **36B**, p.43.

686g Prague, Nat. Hist. Mus., main mass; 1.8g Washington, U.S. Nat. Mus.;

Südost-Missouri v St. Francois County.

Sulechow v Seeläsgen.

Sultanpur v Dyalpur.

Sultanpur 25°56′ N., 84°17′ E.

Ballia district, Uttar Pradesh, India
Fell 1916, July 10, 1100 hrs
Synonym(s): *Kharauni, Mandiari*
Stone. Olivine-hypersthene chondrite (L6), black.

After detonations, stones fell near three villages, two at Sultanpur, two at Mandiari, 2 miles W. of Sultanpur, and one at Kharauni, 3 miles SSE. of Sultanpur; the total weight found was 1710.5g, the largest piece weighing 586.5g. Description, H. Walker, Rec. Geol. Surv. India, 1924, **55**, p.133. Mineralogy, analysis, 22.44 % total iron, olivine $Fa_{26.2}$, A. Dube et al., Smithson. Contrib. Earth Sci., 1977 (19), p.71.

1.2kg Calcutta, Mus. Geol. Surv. India; 84g New York, Amer. Mus. Nat. Hist.; 71g Washington, U.S. Nat. Mus.;

Specimen(s): [1925,442], 264g.

Sumampa 29°20′ S., 63°21′ W.

Santiago del Estero, Argentina
Pseudometeorite..

Manufactured iron, E.H. Ducloux, Rev. fac. quim. farm. Univ. nac. La Plata, 1928, **5** (1), p.77 [M.A. 4-120].

Summerfield 34°46′ N., 102°25′ W.

Castro County, Texas, U.S.A.
Found 1979, recognized in this year
Synonym(s): *Easter*
Stone. Olivine-hypersthene chondrite (L5).

A single orientated stone of 6.2kg was found approximately equidistant between the towns of Hereford and Summerfield, olivine $Fa_{24.4}$, P.S. Sipiera et al., Meteoritics, 1983, **18**, p.63.

4.5kg Watchung, Dupont Colln.;

Specimen(s): [1983,M.30], 79g.

Summit 34°12′ N., 86°29′ W.

Blount County, Alabama, U.S.A.
Found 1890, known since this year
Synonym(s): *Blount County*
Iron. Octahedrite, coarsest (6mm) (IIB).

A mass of about 1kg was found. Description, with an analysis, 5.62 %Ni, G.F. Kunz, Am. J. Sci., 1890, **40**, p.322. Not artificially reheated, V.F. Buchwald, Iron Meteorites, Univ. of California, 1975, p.1168. Classification, further analysis, 6.56 %Ni, 50.5 ppm.Ga, 115.4 ppm.Ge, 0.025 ppm.Ir, J.T. Wasson, Meteorites, Springer-Verlag, 1974, p.300.

458g Vienna, Naturhist. Mus.; 38g Chicago, Field Mus.

Nat. Hist.; 33g New York, Amer. Mus. Nat. Hist.; 16g Berlin, Humboldt Univ.;
Specimen(s): [67452], 45.7g.

Sumner County v Drake Creek.

Sumter County v Bishopville.

Sungach 44°52′ N., 133°10′ E.
Vladivostok, Federated SSR, USSR, [Сунгач]
Fell 1935, April 10, 0730 hrs
Synonym(s): *Sungatsch*
Stone. Olivine-bronzite chondrite (H5).
One stone of 637g fell, L.A. Kulik, Метеоритика, 1941, **1**, p.73 [M.A. 9-294]. Olivine Fa_{19}, B. Mason, Geochimica et Cosmochimica Acta, 1967, **31**, p.1100. Analysis, 28.00 % total iron, M.I. D'yakonova, Метеоритика, 1964, **25**, p.129.
596g Moscow, Acad. Sci.;

Sungatsch v Sungach.

Suoma v Siena.

Supuhee 26°43′ N., 84°13′ E.
Padrauna, Gorakhpur district, Uttar Pradesh, India
Fell 1865, January 19, 1200 hrs
Synonym(s): *Bubuowly, Goruckpur, Mouza Khoorna, Sapahi, Sidowra*
Stone. Olivine-bronzite chondrite (H6), brecciated.
After detonations, six stones fell, one of about 250g at Sonkhunee, Supuhee, which appears to have been lost; two, of about 3.7kg and 3kg respectively at Mouza Khoorna, near Padrauna, C.C. Drury, copy of report in Min. Dept. BM (NH). and three small stones, one of 145g and two of abour 70g each, at the Bubuowly Indigo Factory, M.M. Brooke, letter of 22 September, 1867 in Min. Dept. BM(NH). The place of fall is Sapahi, 26°43'N., 84°13'E. Portions fell near the villages of Khuria (=Mouza Khoorna) and Babnauli (=Bubuowly), C.A. Silberrad, Min. Mag., 1932, **23**, p.300. One component of this stone is exceptionally rich in Tl, Bi and Ag, J.C. Laul et al., Geochimica et Cosmochimica Acta, 1973, **37**, p.329. Olivine Fa_{19}, B. Mason, Geochimica et Cosmochimica Acta, 1963, **27**, p.1011. Description, analysis of lithic fragments, C.A. Leitch and L. Grossman, Meteoritics, 1976, **11**, p.319, abs.
89g Washington, U.S. Nat. Mus.; 29g Harvard Univ.; 29g Vienna, Naturhist. Mus.; 28g Calcutta, Mus. Geol. Surv. India;
Specimen(s): [39713], 3588g. a nearly complete stone, of Mouza Khoorna; [39714], 206g. and fragments, 12g, of Monza Khoorna; [41049], 145g. complete stone, of Bubuowly; [41050], 55g. a nearly complete stone, of Bubuowly; [41051], 3g. part of a small stone, of Bubuowly

Surakarta v Prambanan.

Surala v Durala.

Surprise Springs 34°10′ N., 115°55′ W.
Bagdad, San Bernardino County, California, U.S.A.
Found 1899
Iron. Octahedrite, coarse (1.4mm) (IA).
A mass of about 1.5kg was found. Description, with an analysis, 7.65 %Ni, E. Cohen, Mitt. Naturwiss. Ver. Neu-Vorp. u. Rügen, Greifswald, 1901, **33**, p.29. Classification, further analysis, 8.12 %Ni, 69.6 ppm.Ga, 265 ppm.Ge, 2.0 ppm.Ir, J.T. Wasson, Icarus, 1970, **12**, p.407. Description, V.F. Buchwald, Iron Meteorites, Univ. of California, 1975, p.1169.
869g Chicago, Field Mus. Nat. Hist., main mass; 130g Vienna, Naturhist. Mus.; 114g Paris, Mus. d'Hist. Nat.;
Specimen(s): [85425], 97.7g.

Suscha v Indarch.

Susquehanna v Bald Eagle.

Susuman 62°43′17″ N., 148°7′49″ E.
Magadan region, Federated SSR, USSR, [Сусуман]
Found 1957, November
Synonym(s): *Magadan*
Iron. Octahedrite, medium (1.0mm) (IIIA).
One mass of 18.9kg was found in alluvial deposits, Meteor. Bull., 1958 (11), A.I. Shulzhenko, Природа, 1959 (5), p.115 [M.A. 14-409], B.I. Vronsky, Метеоритика, 1960, **19**, p.135 [M.A. 16-638]. Not part of Maldyak, O.A. Kirova, Метеоритика, 1962, **22**, p.61. Classification and analysis, 7.86 %Ni, 20.5 ppm.Ga, 41.0 ppm.Ge, 2.2 ppm.Ir, E.R.D. Scott et al., Geochimica et Cosmochimica Acta, 1973, **37**, p.1957. Description; shock-hardened, V.F. Buchwald, Iron Meteorites, Univ. of California, 1975, p.1170.
17.7kg Moscow, Acad. Sci., includes main mass, 12.1kg; 74g Washington, U.S. Nat. Mus.;
Specimen(s): [1971,201], 37g. crusted, part-slice

Sutton 40°36′ N., 97°52′ W.
Clay County, Nebraska, U.S.A.
Found 1964
Stone. Olivine-bronzite chondrite (H5).
One stone of 7.6kg was ploughed up, G.I Huss, letter of 18 March, 1970 in Min. Dept. BM(NH). Olivine Fa_{18}, B. Mason, Geochimica et Cosmochimica Acta, 1967, **31**, p.1100.
3.1kg Mainz, Max-Planck-Inst.; 305g Washington, U.S. Nat. Mus.; 115g Albuquerque, Univ. of New Mexico; 95g New York, Amer. Mus. Nat. Hist.; 106g Tempe, Arizona State Univ.;
Specimen(s): [1966,274], 81.8g.

Suwa 36°2′ N., 138°5′ E.
Nagano, Honshu, Japan
Found 1915, before this year
Iron.
Listed, I. Yamamoto, Kwasan Observ. Bull., 1935, p.306 [M.A. 7-173]. Mentioned, A.D. Nininger, Pop. Astron., Northfield, Minnesota, 1940, **48**, p.559, Contr. Soc. Res. Meteorites, **2**, p.227 [M.A. 8-54].
203g Nagano, Girls' High School;

Suwahib (Adraj) 20°1′ N., 50°56′ E.
Rub'al Khali, Saudi Arabia
Found 1932, February 12
Synonym(s): *Adraj*
Stone. Olivine-hypersthene chondrite (L4).
A mass of 118.1g was found at Adraj, 30 miles from Buwah, was regarded as part of the Suwahib (Buwah) fall, W. Campbell Smith, Min. Mag., 1933, **23**, p.334 but is distinct. Olivine Fa_{23}, B. Mason, Geochimica et Cosmochimica Acta, 1963, **27**, p.1011.
24g Calcutta, Mus. Geol. Surv. India;
Specimen(s): [1932,1232], 90.5g. fragments

Suwahib ('Ain Sala) 19°57′20″ N., 51°2′ E.
Rub'al Khali, Saudi Arabia
Found 1932
Synonym(s): *'Ain Sala*
Stone. Olivine-bronzite chondrite (H6).
A single mass of 106.1g was found by H.St.J. Philby at 'Ain Sala, 25 miles from Buwah, W. Campbell Smith, Min. Mag., 1933, **23**, p.334. Distinct from Suwahib (Buwah) (*q.v.*), olivine Fa_{18}, A.L. Graham, priv. comm., 1976.
Specimen(s): [1932,1231], 100g. three pieces

Suwahib (Buwah) 20°3′20″ N., 51°25′ E.
Rub'al Khali, Saudi Arabia
Found 1931
Synonym(s): *Buwah, Buwaj Suwaihib*
Stone. Olivine-bronzite chondrite (H3), black.
A mass of 241g was found at Buwah, B. Thomas, Geogr. J. London, 1931, **78**, p.236. Description, analysis, W. Campbell Smith, Min. Mag., 1932, **23**, p.43. Petrographically distinct from Suwahib ('Ain Sala) (*q.v.*). Olivine $Fa_{13.5}$, E.R.D. Scott, priv. comm., 1982.
Specimen(s): [1931,428], 221g. main mass

Suwanee Spring 34°57′ N., 107°10′ W.
Valencia County, New Mexico, U.S.A.
Found 1979, July 21
Stone. Olivine-hypersthene chondrite (L5).
A single mass of about 1.2kg was found about 40 km west of Albuquerque, olivine $Fa_{24.6}$, Meteor. Bull., 1980 (58), Meteoritics, 1980, **15**, p.239. Description, mineralogy, A.E. Rubin et al., Meteoritics, 1981, **19**, p.9.

Svonkovje v Zvonkov.

Svonkovoje v Zvonkov.

Swajahn v Nerft.

Swallik v Cape York.

Sweetwater 32°33′ N., 100°25′ W.
Fisher County, Texas, U.S.A.
Found 1961, recognized 1963
Stone. Olivine-bronzite chondrite (H5).
A 1.7kg mass was ploughed up, Meteor. Bull., 1966 (36), Meteoritics, 1970, **5**, p.87. Olivine Fa_{19}, B. Mason, Geochimica et Cosmochimica Acta, 1967, **31**, p.1100.
1.018kg Washington, U.S. Nat. Mus., main mass;
Specimen(s): [1970,79], 5.5g.

Swidnica Gorna v Swindnica Gorna.

Swiecie v Schewtz.

Swindnica Gorna 51°49′ N., 16°20′ E. approx.
Poznan province, Poland
Found 1858, possibly fell 1857
Synonym(s): *Swidnica Gorna*
Doubtful. Stone.
A shower of meteorites are said to have fallen near Wschowa, Poznan, in 1857. One stone (sp. gr. 3.108) was found in 1858 but has since been lost, J. Pokrzywnicki, Urania, 1955, **26**, p.165 [M.A. 13-79], J. Pokrzywnicki, Acta Geol. Polon., 1955, **3**, p.427, J. Pokrzywnicki, Acta Geophys. Polon., 1956, **4**, p.21.

Swonkowe v Zvonkov.

Swonkowoe v Zvonkov.

Swonkowoj v Zvonkov.

Sylacauga 33°14′ N., 86°17′ W.
Talladega County, Alabama, U.S.A.
Fell 1954, November 30, 1300 hrs
Stone. Olivine-bronzite chondrite (H4).
Two stones of 8.5lb and 3.75lb were found 2 miles apart; the larger had fallen through the roof of a house, G.W. Swindel and W.B. Jones, Meteoritics, 1954, **1**, p.125 [M.A. 13-52]. Olivine Fa_{20}, B. Mason, Geochimica et Cosmochimica Acta, 1963, **27**, p.1011.
3.68kg Tuscaloosa, Alabama Mus. Nat. Hist., the larger stone; 1.68kg Washington, U.S. Nat. Mus., of the smaller stone;

Syromolotov v Ssyromolotovo.

Syromolotovo v Ssyromolotovo.

Szadany v Zsadany.

Szlanica v Magura.

Szlanicza v Magura.

Tabarz 50°53′ N., 10°31′ E.
Gotha, Erfurt, Germany
Found 1854
Synonym(s): *Gotha*
Iron. Octahedrite, coarse (2.0mm) (I?).
A mass of iron was said to have been seen to fall by a shepherd on October 18, 1854, but its oxidized surface seems to be incompatible with this, analysis, 5.69 %Ni, W. Eberhard, Ann. Chem. Pharm. Leipzig, 1855, **96**, p.286. Very little preserved. Possibly group I, V.F. Buchwald, Iron Meteorites, Univ. of California, 1975, p.1171.
20g Göttingen, Univ.; 15g Vienna, Naturhist. Mus.; 4.9g Tempe, Arizona State Univ.;
Specimen(s): [35163], 9g.

Tabor 49°24′ N., 14°39′ E.
Jihocesky, Czechoslovakia
Fell 1753, July 3, 2000 hrs
Synonym(s): *Kravin, Krawin, Strkov, Strkow*
Stone. Olivine-bronzite chondrite (H5), brecciated.
After the appearance of light, and detonations, a shower of stones fell the largest weighing 13lb, E.F.F. Chladni, Die Feuer-Meteore, Wien, 1819, p.246, E. Howard, Phil. Trans. Roy. Soc. London, 1802, **92**, p.179. Analysis (MnO figure probably too high), J. Kokta, Coll. Czech. Chem. Comm., 1937, **9**, p.471 [M.A. 7-173]. Olivine Fa_{18}, B. Mason, Geochimica et Cosmochimica Acta, 1963, **27**, p.1011.
4kg Vienna, Naturhist. Mus.; 1.5kg Budapest, 762g, Nat. Mus., 742g, Eötvös Lorand Univ.; 766g Prague, Nat. Mus.; 640g Prague, German Univ., in 1897; 160g Paris, Mus. d'Hist. Nat.; 112g Chicago, Field Mus. Nat. Hist.; 65g Washington, U.S. Nat. Mus., includes 18g specimen formerly BM. 1912,265;
Specimen(s): [90242], 150g.; [1920,344], 7g.; [90243], 3.75g.

Taborg v Ochansk.

Taborsk v Ochansk.

Taborskoie Selo v Ochansk.

Taborskoje Selo v Ochansk.

Tabory v Ochansk.

Tacoma 47°15′ N., 122°25′ W. approx.
Tacoma, Washington, U.S.A.
Found 1925, between 1925 and 1932
Iron. Octahedrite, medium (1.5mm) (IA).
A single mass of 16.7g was found on a farm outside Tacoma (co-ordinates above). Description and analysis, 7.15 %Ni, 73 ppm.Ga, 272 ppm.Ge, 2.3 ppm.Ir, E.R.D. Scott et al., Meteoritics, 1977, **12**, p.425. Chemically very similar to Cañon Diablo, but undeformed.
12.1g Los Angeles, Univ. of California; 2.2g Washington, U.S. Nat. Mus.;

Tacubaya v Toluca.

Tadjera 36°11′ N., 5°25′ E.
Setif, Constantine, Algeria
Fell 1867, June 9, 2230 hrs
Synonym(s): *Setif*
Stone. Olivine-hypersthene chondrite (L5), black.
After the appearance of a fireball (travelling from SE. to NE.) and three detonations, two stones were found which according to weights in collections must have been about 6kg and 3kg respectively, -. Augeraud, C. R. Acad. Sci. Paris, 1867, **65**, p.240. Description, analysis, G.A. Daubrée, C. R. Acad. Sci. Paris, 1868, **66**, p.513. Contains 'black matter', S. Meunier, Bull. Soc. franc. Min. Crist., 1889, **12**, p.76. Illustrated description, analysis, olivine Fa_{25}, 20.94 % total iron, B. Mason and H.B. Wiik, Am. Mus. Novit., 1966 (2272).
6.7kg Paris, Mus. d'Hist. Nat.; 197g New York, Amer. Mus. Nat. Hist.; 132g Vienna, Naturhist. Mus.; 77g Chicago, Field Mus. Nat. Hist.; 69g Washington, U.S. Nat. Mus.;
Specimen(s): [71574], 36g.; [43199], 3.5g.

Tafoya (a) 36°27′ N., 104°6′ W.
Colfax County, New Mexico, U.S.A.
Found 1951, approx.
Stone. Olivine-bronzite chondrite (H4).
Listed, Cat. Coll. Inst. Meteor. Univ. New Mexico, 1965. Two stones found, one classified as H4, Tafoya (a), the other as H5 (Tafoya (b), olivine $Fa_{18.2}$, D.E. Lange and K. Keil, Meteoritics, 1976, **11**, p.315, abs.
22.5g Albuquerque, Univ. of New Mexico;

Tafoya (b) 36°27′ N., 104°6′ W.
Colfax County, New Mexico, U.S.A.
Found 1951, approx.
Stone. Olivine-bronzite chondrite (H5).
Classification, resembles Abbot (*q.v.*), olivine $Fa_{19.2}$, D.E. Lange and K. Keil, Meteoritics, 1976, **11**, p.315, abs.
490g Albuquerque, Univ. of New Mexico;

Taiban 34°27′ N., 104°1′ W.
De Baca County, New Mexico, U.S.A.
Found 1934
Stone. Olivine-hypersthene chondrite (L5), black, veined.
Some of the stones from the La Lande area (*q.v*) prove to be from a distinct fall, total known weight 25kg, H.H. and A.D. Nininger, The Nininger Collection of Meteorites, Winslow, Arizona, 1950, p.97, 118. Olivine Fa_{25}, B. Mason, Geochimica et Cosmochimica Acta, 1963, **27**, p.1011. Rare gas data, proof of distinctness of fall, D. Heymann, Geochimica et Cosmochimica Acta, 1965, **29**, p.1203.
6.7kg Tempe, Arizona State Univ.; 438g Washington, U.S. Nat. Mus.; 302g Albuquerque, Univ. of New Mexico;
Specimen(s): [1959,1018], 7996g.; [1959,1030], 1985g.; [1977, M.3], 103g. of Taiban (b)

Taicang 31°30′ N., 121°5′ E.
Jiangsu, China
Found 1928, April
Stone.
A single mass of unreported weight is listed, D. Bian, Meteoritics, 1981, **16**, p.115.

Taiga (iron) v Toubil River.

Taiga (stone)
Federated SSR, USSR, Not known
Found 1946, approx.
Stone. Olivine-bronzite chondrite (H6).
A specimen in Heidelberg Univ., smaller than an apple, was collected by a German prisoner of war working in the Taiga, P. Ramdohr, letter of 28 June, 1967 in Min. Dept. BM(NH). Olivine Fa_{19}, B. Mason, Geochimica et Cosmochimica Acta, 1963, **27**, p.1011.

Taixo v Palenca de Baixo.

Tajgha v Toubil River.

Tajima v Takenouchi.

Tajka v Toubil River.

Takano v Gifu.

Takenouchi 35°23′ N., 134°54′ E.
Yabu, Hyogo, Honshu, Japan
Fell 1880, February 18, 0530 hrs
Synonym(s): *Iwati, Kuritawaki-mura, Tajima, Takenouti, Toke-uchi-mura, Yosagori*
Stone. Olivine-bronzite chondrite (H).
A stone of about 0.75kg was seen to fall, K. Jimbo, Beitr. Miner. Japan, 1906 (2), p.32, 40. Analysis, A. Lindner, Sitzungsber. Akad. Wiss. Berlin, 1904, p.981. Olivine Fa_{19}, B. Mason, Geochimica et Cosmochimica Acta, 1963, **27**, p.1011. Synonymous with Yosagori and Kuritawaki-mura, J. Geol. Soc. Tokyo, 1895, **2**, p.246, W. Flight, Geol. Mag., 1882, **9**.
Main mass, Tokyo, Nat. Hist. Mus.; 61g Berlin, Humboldt Univ.;
Specimen(s): [86642], 2.5g.

Takenouti v Takenouchi.

Takysie Lake 53°30′ N., 125°30′ W. approx.
British Columbia, Canada
Pseudometeorite..
The brecciated material collected, H.H. Nininger and G.I Huss, Meteoritics, 1967, **3**, p.169 is probably a conglomerate or breccia, R. Hutchison, priv. comm. in Min. Dept. BM (NH).

Talbot Road v De Cewsville.

Taltal v Vaca Muerta.

Tamarugal 20°48′ S., 69°40′ W.

Iquique, Tarapaca, Chile
Found 1903
Synonym(s): *Chile, Inca, El Inca, Pampa de Tamarugal, Tarapaca*
Iron. Octahedrite, medium (1.1mm) (IIIA).

A mass of about 320kg was found in the Pampa de Tamarugal and was called " The Inca" by the finder. Description, with an analysis, 8.20 %Ni, F. Rinne and H.E. Boeke, Neues Jahrb. Min., Festband, 1907, p.227. Further analysis, 8.44 %Ni, 21.6 ppm.Ga, 43.8 ppm.Ge, 0.58 ppm.Ir, E.R.D. Scott et al., Geochimica et Cosmochimica Acta, 1973, **37**, p.1957. Co-ordinates; some specimens in collections as Tarapaca (*q.v.*) are in fact Tamarugal, V.F. Buchwald, Iron Meteorites, Univ. of California, 1975, p.1173.

22kg Vienna, Naturhist. Mus.; 17kg Washington, U.S. Nat. Mus.; 11kg New York, Amer. Mus. Nat. Hist.; 8kg Budapest, Nat. Mus.; 5.4kg Bonn, Univ. Mus.; 4kg Prague, Nat. Mus.; 2.7kg Berlin, Humboldt Univ.; 822g Chicago, Field Mus. Nat. Hist.; 814g Paris, Mus. d'Hist. Nat.; 677g Copenhagen, Univ. Geol. Mus.;

Specimen(s): [1907,1029], 6229g. slice; [1921,5], 598g. slice; [85843], 14g. from A. Brezina, labelled Tarapaca

Tambakwatu 7°45′ S., 112°46′ E.

Tambakwatu village, Purwosari district, Java, Indonesia
Fell 1975, February 14, before midnight
Stone. Olivine-hypersthene chondrite (L6).

A single mass of 10.5kg was seen to fall in the Purwosari district. The mass was later recovered by the Jakarta Planetarium, Meteor. Bull., 1980 (58), Meteoritics, 1980, **15**, p.240. Full description, olivine $Fa_{23.8}$, K. Fredriksson et al., Meteoritics, 1981, **16**, p.77.

Main mass, Jakarta, Planetarium; 700g Washington, U.S. Nat. Mus.;

Tambo Quemado 14°40′ S., 74°30′ W.

Leoncio Prado, Ayacucho, Peru
Found 1950, before this year
Iron. Octahedrite, medium (0.7mm) (IIIB).

A mass of 141kg was brought to Lima in 1950, 8.70 %Ni, A. Freyre, Bol. Inst. Nac. Invest. Form. Min., 1950, **1**, p.143 [M.A. 12-361]. Probably reheated artificially to about 1000C, V.F. Buchwald, Iron Meteorites, Univ. of California, 1975, p.1174. Further analysis, 10.2 %Ni, 17.9 ppm.Ga, 31.5 ppm.Ge, 0.039 ppm.Ir, E.R.D. Scott et al., Geochimica et Cosmochimica Acta, 1973, **37**, p.1957.

Main mass, Lima, Mus.; 970g Washington, U.S. Nat. Mus.; 90g Tempe, Arizona State Univ.; 70g Harvard Univ.;

Specimen(s): [1983,M.47], fragment in plastic

Tamentit 27°43′ N., 0°15′ W.

Tuat, Algeria
Found 1864, known since this year
Synonym(s): *Tuat*
Iron. Octahedrite, medium (1.2mm) (IIIA).

A mass, 0.5 metre in diameter and weighing 510kg, was found lying on the Kasbah heights, G. Rohlf, Petermann's Geogr. Mitth., Gotha, 1865, p.409. It was an object of adoration to natives and is said to have fallen between Noumen Nas and Tittaf towards the end of the 14th century. Description, A. Lacroix, C. R. Acad. Sci. Paris, 1924, **179**, p.360, A. Lacroix, C. R. Acad. Sci. Paris, 1927, **184**, p.1217. Description, analysis, 8.39 %Ni, A. Lacroix, C. R. Acad. Sci. Paris, 1927, **185**, p.313. Mentioned, F.W. Cassirer, Publ. Astron. Soc. Pacific, 1940, **52**, p.13 [M.A. 7-542]. Classification, further analysis, 8.37 %Ni, 20.3 ppm.Ga, 42.7 ppm.Ge, 2.5 ppm.Ir, E.R.D. Scott et al., Geochimica et Cosmochimica Acta, 1973, **37**, p.1957. Description; shock-hardened, V.F. Buchwald, Iron Meteorites, Univ. of California, 1975, p.1177.

504kg Paris, Mus. d'Hist. Nat.; 746g Copenhagen, Univ. Geol. Mus.; 576g Rome, Vatican Colln; 392g Chicago, Field Mus. Nat. Hist.;

Specimen(s): [1966,278], 341g.

Tamir-Tsetserleg 47°27′ N., 101°28′45″ E.

Bulgan somon, Arhangay Aymag, Mongolia
Found 1956, August 27
Stone?.

Fragments totalling 173kg were found 1-2 km SW. of Tsetserleg, O. Namnandorj, Meteorites of Mongolia, 1980, p.23, English trans.

153kg State Central Mus.;

Tanakami v Tanokami.

Tandil 37°22′ S., 59°20′ W.

Buenos Aires province, Argentina
Pseudometeorite..

A mass of 980g is said to have been seen to fall on the road, 3 leagues W. of Tandil, E. Fossa-Mancini, Notas Mus. La Plata, 1948, **13** (geol. no. 49), p.97 [M.A. 10-404]. It is non-magnetic and is almost certainly an iron-ore, V.F. Buchwald, Iron Meteorites, Univ. of California, 1975, p.1180.

Tané 35°26′ N., 136°14′ E.

Higashi Asai, Shiga, Honshu, Japan
Fell 1918, January 25, 1428 hrs
Synonym(s): *Hayami*
Stone. Olivine-hypersthene chondrite (L5).

After an explosion, a stone of 311g was found, K. Niinomi, Nature, 1918, **101**, p.352. A second stone, 594g, fell at Hayami, I. Yamamoto, Kwasan Observ. Bull., **1935**, p.306 [M.A. 7-173]. Olivine Fa_{25}, B. Mason, Geochimica et Cosmochimica Acta, 1963, **27**, p.1011. Analysis, 23.08 % total iron, A. Miyashiro and S. Murayama, Chem. Erde, 1967, **26**, p.219.

One stone, Kwasan Observ.; 594g Washington, U.S. Nat. Mus., the Hayami mass;

Specimen(s): [1978,M.20], 2g.

Tanesrouft v Ouallen.

Taney County v Mincy.

Tanezrouft v Ouallen.

Tanganyika v Mbosi.

Tangtu v Po-wang Chen.

Tang-t'u v Po-wang Chen.

Tannuola v Chinga.

Tannu-Ola v Chinga.

Tanokami Mountain 34°55′ N., 135°58′ E.
Kurimuto, Shiga, Honshu, Japan
Found 1885
Synonym(s): *Tanakami*
Iron. Octahedrite, coarse (1.5mm) (IIIE).
A mass of 170.75kg was found. Description, with an analysis, 8.56 %Ni, Y. Otsuki, J. Geol. Soc. Tokyo, 1900, **7**, p.85, K. Jimbo, Beitr. Miner. Japan, 1906 (2), p.32, 42. Contains carbide minerals typical of chemical group IIIE, classification and analysis, 8.94 %Ni, 18.2 ppm.Ga, 34.6 ppm.Ge, 0.22 ppm.Ir, E.R.D. Scott et al., Geochimica et Cosmochimica Acta, 1973, **37**, p.1957. Description, V.F. Buchwald, Iron Meteorites, Univ. of California, 1975, p.1179.
Main mass, Tokyo, Nat. Hist. Mus.; 148g Chicago, Field Mus. Nat. Hist.; 91g Washington, U.S. Nat. Mus.;
Specimen(s): [1905,69], 174g. slice

Taoho v Nan Yang Pao.

Taonan 45°24′ N., 122°54′ E.
Taonan County, Jilin, China
Fell 1965, February 28, 1200 hrs
Stone. Olivine-hypersthene chondrite (L).
Three stones fell, the largest 2.55kg, totalling 3.85kg, D. Bian, Meteoritics, 1981, **16**, p.115. Analysis, 22.82 % total iron, W. Daode et al., Geochemistry, 1982, **1**, p.186.

Taos v Tucson.

Taranaki v Wairarapa Valley.

Tarapaca v La Primitiva.

Tarapaca v Tamarugal.

Tarapaca 19°58′ S., 69°38′ W.
Tarapaca, Chile
Found 1889
Doubtful. Iron.
Mentioned, F. Berwerth, Ann. Naturhist. Hofmus. Wien, 1903, **18**, p.19. Description, M.M. Radice, Rev. Mus. La Plata, 1959, **5**, p.126. Three different materials have been labelled Tarapaca; a pseudometeorite, pieces of La Primitiva, and pieces of Tamarugal. The 5.4kg mass in the La Plata Mus. is a La Primitiva specimen, V.F. Buchwald, Iron Meteorites, Univ. of California, 1975, p.1174 so also is the 100g specimen in Harvard Univ., E.R.D. Scott et al., Geochimica et Cosmochimica Acta, 1973, **37**, p.1957. The specimen in Tempe, Arizona State Univ. (no. 547.1) is wrought iron, V.F. Buchwald, loc. cit.. The true affinity of the material listed below is not available.
715g La Plata, Mus.; 252g Budapest, Nat. Mus.;

Tarbagatai 51°22′ N., 107°23′ E.
Near Ulan-Ude, Buryat ASSR, USSR, [Тарбагатай]
Found 1912, or fell 10 November, 1912
Stone. Olivine-hypersthene chondrite (L5).
Mentioned, P.L. Dravert, Sibirskne Ogni, 1930 (9), p.122, I.S. Astapowitsch, Trans. Roy. Astron. Soc. Canada, 1938, **32**, p.195 [M.A. 7-172]. One stone of about 370g was found, L.G. Kvasha and A.Ya. Skripnik, Метеоритика, 1978, **37**, p.204. Analysis, 18.96 % total iron, M.I. D'yakonova and V.Ya. Kharitonova, Метеоритика, 1961, **21**, p.52. Olivine Fa_{23}, B. Mason, Geochimica et Cosmochimica Acta, 1963, **27**, p.1011.
300g Moscow, Acad. Sci.;

Tarfa 19°30′ N., 55°30′ E.
Oman
Found 1954
Stone. Olivine-hypersthene chondrite (L6).
Found on a weathered desert limestone surface at the co-ordinates 18°18'N., 58°18'E., near Tarfa, Iraq Petroleum Co., letters of 1954 in Min. Dept. BM(NH). These co-ordinates refer to a locality in the sea. The co-ordinates above are those of Tarfa, Oman, from the Times Atlas. Olivine Fa_{25}, B. Mason, Geochimica et Cosmochimica Acta, 1963, **27**, p.1011.
Specimen(s): [1954,206], 1030g. the stone, in two halves, and fragments, 4.5g

Tarn v Grazac.

Taromaru v Gifu.

Tarragona v Nulles.

Tasmania v Blue Tier.

Tasquinha v Portugal.

Tatahouine 32°57′ N., 10°25′ E.
Foum Tatahouine, Tunisia
Fell 1931, June 27, 0130 hrs
Synonym(s): *Foum Tatahouine, Tataouine*
Stone. Achondrite, Ca-poor. Diogenite (ADIO).
Fragments fell over a radius of 500 m, 4 km NE. of the village; 12kg were collected, mostly in minute fragments. Description, analysis, A. Lacroix, C. R. Acad. Sci. Paris, 1931, **193**, p.305 [M.A. 5-10], A. Lacroix, Bull. Soc. franc. Min. Crist., 1932, **55**, p.101 [M.A. 5-298].
3.5kg Paris, Mus. d'Hist. Nat.; 183g Chicago, Field Mus. Nat. Hist.; 137g Calcutta, Mus. Geol. Surv. India;
Specimen(s): [1931,490], 118g. and fragments, 6g

Tataouine v Tatahouine.

Tathlith 19°23′ N., 43°44′ E.
Saudi Arabia
Fell 1967, October 5, 0515-0530 hrs
Stone. Olivine-hypersthene chondrite (L6).
A single crusted individual was recovered about 15km from Tathlith almost immediately after its fall. From photographs it was estimated to weigh about 2kg. Analysis, 22.55 % total iron, R.S. Clarke, Jr. et al., Smithson. Contrib. Earth Sci., 1975 (14), p.63. Reported, Meteor. Bull., 1968 (43), Meteoritics, 1970, **5**, p.98.
246g Washington, U.S. Nat. Mus.;
Specimen(s): [1980,M.32], 9.43g.

Tatum 33°14′ N., 103°17′ W.
Lea County, New Mexico, U.S.A.
Found 1938
Stone. Olivine-bronzite chondrite (H).
One stone of 1.8kg was found, A.D. Nininger, Pop. Astron., Northfield, Minnesota, 1939, **47**, p.213. Olivine Fa_{18}, B. Mason, Geochimica et Cosmochimica Acta, 1963, **27**, p.1011.
1.7kg Fort Worth, Texas, Monnig Colln.;

Tauk 35°8′ N., 44°27′ E.
Daquq, Iraq
Fell 1929, spring
Synonym(s): *Kirkuk*
Stone. Olivine-hypersthene chondrite (L6).
A mass "as big as a man's head," and estimated to weigh 6 to 7kg, fell near Tawila, 11 km NW. of Tauk (35°8'N., 44°27'E.), W.A. Macfadyen, letters in Min. Dept. BM(NH). A fragment, 205g, of exactly similar appearance was presented to the BM(NH). in 1932, and is said to have fallen at Kirkuk, 24 km N. of Tauk, about 1925 or 1926; it is to be referred to the Tauk fall. Olivine Fa_{25}, B. Mason, Geochimica et Cosmochimica Acta, 1963, **27**, p.1011.
199g Calcutta, Mus. Geol. Surv. India; 194g New York, Amer. Mus. Nat. Hist.;
Specimen(s): [1936,151], 838g. and fragments, 19g, of Tauk; [1932,1094], 194.5g. of Kirkuk

Tauti 46°43′ N., 23°30′ E.
Tauti, Oradea, Romania
Fell 1937, July or August
Stone. Olivine-hypersthene chondrite (L6).
After three explosions and a bright bolide, a 20kg stone fell at Tauti, and two other stones of about 0.5kg each fell at Gura Suvelului, about 6km S. of Tauti, H. Savu, Studii si cercet. Geol., 1959, **2**, p.273, Meteor. Bull., 1979 (56), Meteoritics, 1979, **14**, p.170.

Tawallah Valley 15°42′ S., 135°40′ E.
Borroloola, Northern Territory, Australia
Found 1939
Iron. Ataxite, Ni-rich (IVB).
A mass of 75.75kg was found. Description, with an analysis, 16.90 %Ni, T. Hodge-Smith and A.B. Edwards, Rec. Austr. Mus., 1941, **21**, p.1 [M.A. 8-196]. Mentioned, T. Hodge-Smith, Austr. Mus. Mag., 1942, **7**, p.429 [M.A. 8-377]. Further analysis, 17.06 %Ni, 0.248 ppm.Ga, 0.068 ppm.Ge, 16 ppm.Ir, R. Schaudy et al., Icarus, 1972, **17**, p.174. Description, V.F. Buchwald, Iron Meteorites, Univ. of California, 1975, p.1180.
39kg Canberra, Geol. Surv. Collection; 30kg Sydney, Austr. Mus.; 1.26kg Washington, U.S. Nat. Mus.; 216g New York, Amer. Mus. Nat. Hist.; 152g Moscow, Acad. Sci.; 87g Chicago, Field Mus. Nat. Hist.;
Specimen(s): [1982,M.15], 87g.

Tazewell 36°26′ N., 83°45′ W.
Claiborne County, Tennessee, U.S.A.
Found 1853
Synonym(s): *Claiborne County, East Tennessee, Knoxville*
Iron. Octahedrite, finest (0.045mm) (IIICD).
A mass of about 60lb was ploughed up 10 miles W. of Tazewell, C.U. Shepard, Am. J. Sci., 1854, **17**, p.325. Description, analysis, 14.82 %Ni, J.L. Smith, Am. J. Sci., 1855, **19**, p.153. Further analysis, 16.64 %Ni, 4.69 ppm.Ga, 3.79 ppm.Ge, 0.063 ppm.Ir, J.T. Wasson and R. Schaudy, Icarus, 1971, **14**, p.59. Fine structure, E.R.D. Scott, Geochimica et Cosmochimica Acta, 1973, **37**, p.2283. Description; co-ordinates, V.F. Buchwald, Iron Meteorites, Univ. of California, 1975, p.1182.
9.8kg Tempe, Arizona State Univ.; 1.6kg Washington, U.S. Nat. Mus.; 893g Harvard Univ.; 645g Berlin, Humboldt Univ.; 366g New York, Amer. Mus. Nat. Hist.; 414g Paris, Mus. d'Hist. Nat.; 294g Tübingen, Univ.; 279g Chicago, Field Mus. Nat. Hist.; 380g Budapest, Nat. Mus.; 206g Vienna, Naturhist. Mus.;
Specimen(s): [32047], 277g. and a fragment, 11g

Tchebankol v Chebankol.

Tchervony Koot v Chervony Kut.

Tchinge v Chinga.

Tchivashsky Kissy v Kissij.

Techado 34°32′ N., 108°21′ W.
Catron County, New Mexico, U.S.A.
Iron. Octahedrite, medium (0.6mm) (IIE).
A single mass of 810g was found about 6 km SE. of Techado during the excavation of a prehistoric American Indian pueblo, 8.9 %Ni, 23.2 ppm.Ga, 70.2 ppm.Ge, 4.9 ppm.Ir, D.J. Malvin et al., priv. comm., 1983. Reported, Meteor. Bull., 1984 (62), Meteoritics, 1984, **19**.
Main mass, Albuquerque, Univ. of New Mexico;

Teheran v Veramin.

Teilleul v Le Teilleul.

Tejupilco v Toluca.

Teleutskoje Osero v Barnaul.

Teleutskoje Ozero v Barnaul.

Tell 34°23′ N., 100°24′ W.
Childress County, Texas, U.S.A.
Found 1930, or 1932, recognized 1965
Stone. Olivine-bronzite chondrite (H6).
An individual of about 16.6kg was ploughed up in 1930 or 1932; it was embedded in concrete in late 1964, but after the sample was identified as meteoritic by the American Meteor. Lab., the bulk of the mass was removed from the concrete in 1965, G.I Huss, letter of 18 March, 1970 in Min. Dept. BM (NH). Olivine Fa_{18}, B. Mason, Geochimica et Cosmochimica Acta, 1967, **31**, p.1100.
8.3kg Mainz, Max-Planck-Inst., main mass; 520g Ottawa, Geol. Surv. Canada; 279g Copenhagen, Univ. Geol. Mus.; 230g Los Angeles, Univ. of California; 180g Washington, U.S. Nat. Mus.; 147g Tempe, Arizona State Univ.; 115g Harvard Univ.;
Specimen(s): [1965,413], 70.8g.

Temora v Narraburra.

Temosachic v Huizopa.

Temple 31°7′ N., 97°18′ W.
Bell County, Texas, U.S.A.
Found 1959
Stone. Olivine-hypersthene chondrite (L6).
Known weight 5kg, V.F. Buchwald and S. Munck, Analecta geol., 1965, **1**, p.1. Olivine Fa_{25}, B. Mason, Geochimica et Cosmochimica Acta, 1963, **27**, p.1011.
260g Tempe, Arizona State Univ.; 253g Collection of H.J. Meyers; 186g Ottawa, Geol. Surv. Canada; 169g Copenhagen, Univ. Geol. Mus.; 169g Washington, U.S. Nat. Mus.; 90g Vienna, Naturhist. Mus.; 64g New York, Amer. Mus. Nat. Hist.;
Specimen(s): [1964,653], 54g.

Tendo 38°21′ N., 140°22′24″ E.

Yamagata Prefecture, Honshu, Japan
Found 1910, approx., recognized 1977
Iron. Octahedrite, medium (IIIA).

A single mass of 10.1kg was found by a farmer, Meteor. Bull., 1979 (56), Meteoritics, 1979, **14**, p.171. Classification, analysis, 8.83 %Ni, 18 ppm.Ga, 33.7 ppm.Ge, 0.34 ppm.Ir, M. Shima et al., Meteoritics, 1979, **14**, p.535, abs.

Main mass, in possession of Mr F. Kikuchi;

Tenham 25°44′ S., 142°57′ E.

South Gregory, Queensland, Australia
Fell 1879, in the spring, at night hrs
Synonym(s): *Warbreccan*
Stone. Olivine-hypersthene chondrite (L6), veined.

Description, with an analysis, and details of previous literature, L.J. Spencer, Min. Mag., 1937, **24**, p.437. About 350lb of stones were recovered. The shower was seen to fall over an area of over 12 × 3 miles near Tenham station. New discussion of trajectory, B. Mason, Meteoritics, 1973, **8**, p.1. Size distribution, M.J. Frost, Meteoritics, 1969, **4**, p.217, B. Hellyer, Observatory, 1971, **91**, p.64. Contains ringwoodite, R.A. Binns et al., Nature, 1969, **221**, p.943, A. Putnis and G.D. Price, Nature, 1979, **280**, p.217. See also under Hammond Downs (*q.v.*).

89 stones, Sydney, Austr. Mus.; 5.5kg Washington, U.S. Nat. Mus.; 4.7kg New York, Amer. Mus. Nat. Hist.; 3.6kg Tyrone station, Charleville; 736g Chicago, Field Mus. Nat. Hist.; 600g Mayfield station, Windorah;

Specimen(s): [1905,377], 31953g. and a piece, 25g; of Warbreccan; [1905,378], 29140g. of Warbreccan; [1905,379], 443g. of Warbreccan; [1935,785], 5245g.; [1935,786], 5160g.; [1935,787], 4290g.; [1935,788], 3645g.; [1935,789], 1903g.; [1935,790], 1497g.; [1935,792-882], 25776g. 91 complete stones; [1937,1647-1650], 342g. 4 stones

Teniet-el-Beguel v Haniet-et-Beguel.

Tennant's Iron v Toluca.

Tennasilm 58°2′ N., 26°57′ E.

Pskov, Estonian SSR, USSR
Fell 1872, June 28, 1200 hrs
Synonym(s): *Sikkensaare, Tennasilon, Tennsilom*
Stone. Olivine-hypersthene chondrite (L4), veined.

After the appearance of a cloud, and detonations, a stone of about 28.5kg was found a few days later. It was broken into pieces by gypsies, but most of it was recovered. Description, with an analysis, G. Schilling, Arch. Naturk. Liv.-Ehst.-u. Kurlands, Ser. 1 Min. Wiss. Dorpat, 1882, **9** (2), p.95. Description, analysis, 21.69 % total iron, B. Mason and H.B. Wiik, Am. Mus. Novit., 1965 (2220). Equilibrated, olivine $Fa_{23.5}$, R.T. Dodd et al., Geochimica et Cosmochimica Acta, 1967, **31**, p.921.

4.8kg Washington, U.S. Nat. Mus.; 3.5kg Vienna, Naturhist. Mus.; 3.75kg Reval Mus.; 3kg Dorpat, Mus.; 174g Moscow, Acad. Sci.; 143g Yale Univ.;

Specimen(s): [1913,218], 177g.; [54635], 15.75g.

Tennasilon v Tennasilm.

Teocaltiche 21°26′ N., 102°34′ W.

Jalisco, Mexico
Found 1903
Iron. Octahedrite.

A mass of 10kg was said to be in Mexico City, (Mus. Inst. Geol.), H.A. Ward, Cat. Ward-Coonley Coll. Meteorites, Chicago, 1904, p.25, 89. This mass could not be located in 1968, V.F. Buchwald, Iron Meteorites, Univ. of California, 1975, p.1185.

Tepl v Tepla.

Tepla 49°59′ N., 12°52′ E.

Finsterhölzelries, Marienbad (=Marianske, Lazne), Zapadocesky Kraj, Czechoslovakia
Found 1909, September 18
Synonym(s): *Finsterhölzelries, Fisterhölzelries, Tepl*
Iron. Octahedrite, medium (0.9mm) (IIIB).

Two masses of 14.4kg and 2.6kg were ploughed up. Description, with an analysis, 7.47 %Ni, B. Jezek, Bull. Internat. Acad. Sci. Boheme, 1923, **25**, p.275. There are probably other masses besides the recorded two. General history, K. Tucek, Casopis Narod. Mus. Praha, 1947, **116**, p.1 [M.A. 10-398]. Description, V.F. Buchwald, Iron Meteorites, Univ. of California, 1975, p.1185. Analysis, 10.0 %Ni, 21.1 ppm.Ga, 39.9 ppm.Ge, 0.014 ppm.Ir, D.J. Malvin et al., priv. comm., 1983.

12.5kg Prague, Nat. Mus., includes main mass, 10.7kg; 372g Vienna, Naturhist. Mus.; 42g Chicago, Field Mus. Nat. Hist.;

Specimen(s): [1912,543], 57g.

Teposcolula v Misteca.

Teposcolula v Yanhuitlan.

Tequezquito Creek v Roy (1934).

Terek v Grosnaja.

Ternera 27°20′ S., 69°48′ W. approx.

Atacama, Chile
Found 1891, before this year
Synonym(s): *Galleguillos, Galleguitos, Sierra de la Ternera*
Iron. Ataxite, Ni-rich (IVB).

A mass of 650g was found. Description, with an analysis, 16.22 %Ni, G.F. Kunz and E. Weinschenk, Tschermaks Min. Petr. Mitt., 1891, **12**, p.184. A mass of 1330g, Galleguillos, was also found. It has been artificially heated but is a fragment of Ternera, V.F. Buchwald, Iron Meteorites, Univ. of California, 1975, p.1187. Analysis, 18.13 %Ni, 0.261 ppm.Ga, 0.056 ppm.Ge, 16 ppm.Ir, J.T. Wasson, Meteorites, Springer-Verlag, 1974, p.306.

1.3kg Oslo, Min.-Geol. Mus., of Galleguillos; 550g Berlin, Humboldt Univ.; 31g Washington, U.S. Nat. Mus.; 21g Los Angeles, Univ. of California;

Specimen(s): [86640], 5.5g.

Terni v Collescipoli.

Terranova da Sibari v Terranova di Sibari.

Terranova di Sibari 39°20′ N., 16°20′ E.

Cosenza, Calabria, Italy
Fell 1755, July
Synonym(s): *Terranova da Sibari*
Doubtful..

A stone of 7.5lb fell, it broke up after about 9 years, E.F.F. Chladni, Die Feuer-Meteore, Wien, 1819, p.248. The evidence of meteoritic nature is not conclusive. The correct spelling of the locality is Terranova da Sibari.

Tesic v Tieschitz.

Tesice v Tieschitz.

Texas v Red River.

Texline 36°24′ N., 103°1′ W.

Dallam County, Texas, U.S.A.
Found 1937, July
Stone. Olivine-bronzite chondrite (H5).

Four stones, 26.2kg together, were found, E.P. Henderson, letter of 12 May, 1939 in Min. Dept. BM(NH)., A.D. Nininger, Pop. Astron., Northfield, Minnesota, 1939, **47**, p.214. Olivine Fa_{19}, B. Mason, Geochimica et Cosmochimica Acta, 1963, **27**, p.1011.

3.2kg Fort Worth, Texas, Monnig Colln.; 8.7kg Tempe, Arizona State Univ.; 672g Chicago, Field Mus. Nat. Hist.; 657g New York, Amer. Mus. Nat. Hist.; 646g Washington, U.S. Nat. Mus.; 127g Los Angeles, Univ. of California; 100g Paris, Mus. d'Hist. Nat.;

Specimen(s): [1959,1038], 8168g. half of one of the masses; [1949,180], 19g.

Thackaringa 32°7′ S., 141°5′ E.

County Yancowinna, New South Wales, Australia
Found 1974, March 24
Stone. Olivine-bronzite chondrite (H5).

A single mass of 432.2g was found on Thackeringa station, 8 km east of the state border. A second mass of 6.4g was found nearby in 1975 and may be part of this fall, Meteor. Bull., 1981 (59), Meteoritics, 1981, **16**, p.197.

Thal 33°24′ N., 70°36′ E.

Jhamtanwala, Mianwali tehsil, North-west Frontier, Pakistan
Fell 1950, June
Stone. Olivine-bronzite chondrite (H6).

A stone, weight not stated, fell SSW. of Wan Bachran railway station, "Pakistan Times", 1950 (issue of 21 June), H. Crookshank, letter of 10 July, 1954 in Min. Dept. BM (NH). Olivine Fa_{19}, B. Mason, Geochimica et Cosmochimica Acta, 1963, **27**, p.1011.

Specimen(s): [1954,223], 341.5g. a crusted fragment

Thal von San Giuliano Vecchio v Alessandria.

Thatcher 37°33′ N., 104°9′ W.

Las Animas County, Colorado, U.S.A.
Found
Pseudometeorite. Stone.

Mentioned, C. Wilton, letters of 11 March, 4 April, 9 June 1963 in Min. Dept. BM(NH). Olivine-basalt, Meteor. Bull., 1974 (52), Meteoritics, 1974, **9**, p.120.

Thebes 27°40′ N., 30°47′ E.

Egypt
Synonym(s): *Amun*
Doubtful..

The sacred object of Amun at Thebes was probably a portion of a pear-shaped iron meteorite, G.A. Wainwright, J. Egypt. Archaeol., 1934, **20**, p.139, Z. Ägypt. Sprache, 1935, **71**, p.41 [M.A. 6-99]. Mentioned, F.C. Leonard, Contr. Soc. Res. Meteorites, 1945, **3**, p.224.

Thiel Mountains 85°27′ S., 90° W. approx.

Antarctica
Found 1962, January
Synonym(s): *Horlick Mountains*
Stony-iron. Pallasite (PAL).

Two pieces, 22.7kg and 9.0kg, were found on the surface of a glacier, about 90 metres apart, Meteor. Bull., 1962 (24, 25). Olivine Fa_{13}, P.R. Buseck and J.I. Goldstein, Bull. Geol. Soc. Amer., 1969, **80**, p.2141. Analysis of metal, 10.1 %Ni, 22.2 ppm.Ga, 50.2 ppm.Ge, 0.23 ppm.Ir, J.T. Wasson, Meteorites, Springer-Verlag, 1974, p.306. Texture, references, E.R.D. Scott, Geochimica et Cosmochimica Acta, 1977, **41**, p.693.

28kg Washington, U.S. Nat. Mus.; 278g Chicago, Field Mus. Nat. Hist.;

Specimen(s): [1980,M.14], 63.9 and olivine, 1.26g

Thiel Mountains 82403

Antarctica
Found 1982, between December 1982 and January 1983
Stone. Achondrite, Ca-rich. Eucrite (AEUC).

A single mass of 49.8g was found by an American party during the 1982-1983 field season in Antarctica.

Thomson 33°28′ N., 82°29′ W.

McDuffie County, Georgia, U.S.A.
Found 1888
Stone. Olivine-hypersthene chondrite (L6), veined.

A stone of 218g was found, G.P. Merrill, Smithsonian Misc. Coll., 1909, **52**, p.473. Olivine Fa_{24}, B. Mason, Geochimica et Cosmochimica Acta, 1963, **27**, p.1011.

217g Washington, U.S. Nat. Mus.;

Thoreau 35°17′ N., 108°16′ W.

McKinley County, New Mexico, U.S.A.
Found 1954, before this year
Iron. Octahedrite, coarse (1.8mm) (IA).

Listed, L. LaPaz, Cat. Coll. Inst. Meteor. Univ. New Mexico, 1965. Classification, may be transported Odessa material, analysis, 7.4 %Ni, 73.8 ppm.Ga, 271 ppm.Ge, 1.9 ppm.Ir, J.T. Wasson, Icarus, 1970, **12**, p.407. Co-ordinates; description; similarity to Odessa, V.F. Buchwald, Iron Meteorites, Univ. of California, 1975, p.1190.

294g Washington, U.S. Nat. Mus.; 12g Albuquerque, Univ. of New Mexico;

Thrace 41°40′ N., 25°0′ E.

Greece
Fell 452, A.D.
Doubtful..

Three large stones are said to have fallen, E.F.F. Chladni, Die Feuer-Meteore, Wien, 1819, p.188 but the evidence is not conclusive.

Thule 76°32′ N., 67°33′ W.

Greenland
Found 1955
Iron. Octahedrite, medium (1.2mm) (IIIA).

A 48.6kg mass was found on a nunatak near Thule, north Greenland. originally reported as a new mass of Cape York, F.C. Leonard, Meteoritics, 1955, **1**, p.305 [M.A. 13-82] but is distinct. Illustrated description, V.F. Buchwald and S. Munck, Analecta geol., 1965, **1**, p.1. Analysis, 8.52 %Ni, 19.3 ppm.Ga, 39.8 ppm.Ge, 2.6 ppm.Ir, E.R.D. Scott et al., Geochimica et Cosmochimica Acta, 1973, **37**, p.1957. Severely shocked; terrestrially well preserved, V.F. Buchwald, Iron Meteorites, Univ. of California, 1975, p.1191. Probably part of the Cape York shower, K.H. Esbensen et al., Geochimica et Cosmochimica Acta, 1982, **46**, p.1913.

47.5kg Copenhagen, Univ. Geol. Mus.; 449g Washington, U.S. Nat. Mus.; 169g Tempe, Arizona State Univ.; 137g Moscow, Acad. Sci.;

Specimen(s): [1972,492], 23.3g. part-slice

Thunda 25°42′ S., 143°3′ E.
Windorah, County Grey, Queensland, Australia
Found 1881, known since this year
Synonym(s): *Diamantina, Windorah*
Iron. Octahedrite, medium (1.2mm) (IIIA).
A mass of 137lb was found, A. Liversidge, J. and Proc. Roy. Soc. New South Wales, 1886, **20**, p.73, A. Liversidge, J. and Proc. Roy. Soc. New South Wales, 1888, **22**, p.341. Description, analysis, 8.49 %Ni, E. Cohen, Ann. Naturhist. Hofmus. Wien, 1900, **15**, p.381. Two masses of iron were known to the aboriginals on Githa Creek, 25°42′S., 143°3′E., before 1881. One of these was transported to the Thunda homestead about 15 miles to the north, L.J. Spencer, Min. Mag., 1937, **24**, p.451. Further analysis, 8.08 %Ni, 20.2 ppm.Ga, 38.9 ppm.Ge, 2.2 ppm.Ir, E.R.D. Scott et al., Geochimica et Cosmochimica Acta, 1973, **37**, p.1957. Description, shock hardened, V.F. Buchwald, Iron Meteorites, Univ. of California, 1975, p.1195.
1.5kg Vienna, Naturhist. Mus.; 1.33kg Chicago, Field Mus. Nat. Hist.; 734g Tempe, Arizona State Univ.; 371g Oxford, Univ. Mus.; 209g Washington, U.S. Nat. Mus.;
Specimen(s): [1927,1254], 28.6kg. main mass; [1922,159], 5173g.; [1927,1255], 1066g.; [1927,1257], 613g.; [66594], 396g. slice; [1927,1259], 122.5g.; [1927,1260], 60g.; [1927,1261], 19.5g.; [1927,1262], 11.5g. two microprobe mounts; [1927,1263], 2kg. filings and planings; [1964,566], 152.5g.

Thurlow 44°45′ N., 77°35′ W. approx.
Hastings County, Ontario, Canada
Found 1888
Iron. Octahedrite, medium (0.7mm) (IIIB).
A mass of about 5.5kg was found, G.C. Hoffmann, Am. J. Sci., 1897, **4**, p.325. Description, analysis, 9.92 %Ni, E. Cohen, Meteoritenkunde, 1905, **3**, p.377. Classification, further analysis, 9.9 %Ni, 15.9 ppm.Ga, 27.3 ppm.Ge, 0.017 ppm.Ir, E.R.D. Scott et al., Geochimica et Cosmochimica Acta, 1973, **37**, p.1957. Description; little weathered, V.F. Buchwald, Iron Meteorites, Univ. of California, 1975, p.1197.
403g Washington, U.S. Nat. Mus.; 234g Vienna, Naturhist. Mus.; 197g Chicago, Field Mus. Nat. Hist.; 136g New York, Amer. Mus. Nat. Hist.;
Specimen(s): [83395], 189g. slice

Thurman 39°31′18″ N., 103°10′ W.
Washington County, Colorado, U.S.A.
Found 1965, recognized in this year
Stone. Chondrite.
A single stone of 1959g was found, Meteor. Bull., 1979 (56), Meteoritics, 1979, **14**, p.171.
1667g Mainz, Max-Planck-Inst.;

Tianlin 24°18′ N., 106°6′ E. approx.
Guangxi, China
Found 1956, in the summer
Iron.
Listed; a single mass of 230kg is in Nanning, Guangxi Bureau of Geology, D. Bian, Meteoritics, 1981, **16**, p.115.

Tiberrhamine 28°7′ N., 0°32′ E.
Sahara, Algeria
Found 1967
Stone. Olivine-hypersthene chondrite (L6), veined.
200 stones were found, the largest weighing 40kg, 25kg, 15kg, and 13kg, and lay together at the northern end of a 30 × 18 metres area, total weight is 107kg. Illustrated description, and analysis, 21.11 % total iron, M. Christophe Michel-Levy et al., Bull. Soc. franc. Min. Crist., 1970, **93**, p.114, Meteor. Bull., 1971 (50), Meteoritics, 1971, **6**, p.122.
Main mass, Paris, Mus. d'Hist. Nat.; 800g Los Angeles, Univ. of California; 272g Chicago, Field Mus. Nat. Hist.; 100g Washington, U.S. Nat. Mus.;
Specimen(s): [1972,238], 109g. two fragments; [1972,324], 15.4g. three fragments

Tibooburra 29°26′ S., 142°1′ E.
New South Wales, Australia
Found 1970, approx.
Stone. Carbonaceous chondrite, type III (CV3).
A single mass of 18.6g was found near the town of Tibooburra, Meteor. Bull., 1981 (59), Meteoritics, 1981, **16**, p.198. Classification, analysis, 24.23 % total iron, M.J. Fitzgerald and A.L. Jaques, Meteoritics, 1982, **17**, p.9.

Ticraco Creek v Tieraco Creek.

Tieraco Creek 26°20′ S., 118°20′ E.
North Murchison goldfield, Western Australia, Australia
Found 1922, before this year
Synonym(s): *Ticraco Creek*
Iron. Octahedrite, medium (0.5mm) (IIIB).
A mass of 41.7kg, showing "drill holes", was found near the head of Tieraco Creek, 2000 feet above sea level. Description, with an analysis, 9.66 %Ni, T. Hodge-Smith, Rec. Austr. Mus., 1926, **15**, p.66. Differential oxidation of kamacite and taenite, E.S. Simpson, Min. Mag., 1938, **25**, p.163. Classification, analysis, 10.5 %Ni, 16.2 ppm.Ga, 28.0 ppm.Ge, 0.041 ppm.Ir, E.R.D. Scott et al., Geochimica et Cosmochimica Acta, 1973, **37**, p.1957. Description; unusual corrosion, V.F. Buchwald, Iron Meteorites, Univ. of California, 1975, p.1199.
17.2kg Harvard Univ.; 16.5kg Sydney, Austr. Mus.; 4kg Washington, U.S. Nat. Mus.; 1kg Perth, West. Austr. Mus.;

Tierra Blanca 34°56′ N., 102°1′ W.
Castro County, Texas, U.S.A.
Found 1965, November, before this date
Stone. Achondrite, anomalous (ACANOM).
A single weathered stone of 860g was found about 10 km SW. of Canyon, Randall County by a rancher, Meteor. Bull., 1979 (56), Meteoritics, 1979, **14**, p.171. Description, analysis, 18.74 % total iron, olivine $Fa_{0.5\text{-}8}$, E.A. King et al., Meteoritics, 1981, **16**, p.229.
Specimen(s): [1982,M.12], 5.3g.

Tieschitz 49°36′ N., 17°7′ E.
Prostejov, Jihomoravsky, Czechoslovakia
Fell 1878, July 15, 1345 hrs
Synonym(s): *Tesic, Tesice, Tischtin, Tistin*
Stone. Olivine-bronzite chondrite (H3).
After detonations, a stone of about 28kg was seen to fall. Description with an analysis, A. Makovsky and G. Tschermak, Denkschr. Akad. Wiss. Wien, Math.-naturwiss. Kl., 1879, **39**, p.187. Unequilibrated, classification and analysis, 25.13 % total iron, R.T. Dodd et al., Geochimica et Cosmochimica Acta, 1967, **31**, p.921. Chondrule petrography, G. Kurat, Meteorite Research, ed. P.M. Millman, D. Reidel, Dordrecht-Holland, 1969, p.185, M. Christophe-Michel-Levy, Earth planet. Sci. Lett., 1976, **30**, p.143. Metallography, A.W.R. Bevan and H.J. Axon, Earth planet. Sci. Lett., 1980, **47**, p.353. Mineral chemistry, R. Hutchison et al., Proc. Roy. Soc. London, 1981, **A374**, p.159. Rb/Sr age, J.-F. Minster and C.J. Allegre, Earth planet. Sci. Lett., 1979, **42**, p.333.

27.6kg Vienna, Naturhist. Mus.; 32g Chicago, Field Mus. Nat. Hist.; 31g Washington, U.S. Nat. Mus.;
Specimen(s): [54275], 17g.; [1975,M.11], 61g.

Tilden 38°12′ N., 89°41′ W.
Randolph County, Illinois, U.S.A.
Fell 1927, July 13, 1300 hrs
Stone. Olivine-hypersthene chondrite (L6).
Three stones fell, 110lb, 46lb, and 9lb. Description, C.C. Wylie, Science, 1927, **66**, p.451 [M.A. 4-422]. Further description, with a doubtful analysis, A.R. Crook and O.C. Farrington, Trans. Illinois Acad. Sci., 1930, **22**, p.442 [M.A. 4-422].
Main mass, Springfield, Illinois, Illinois State Mus.; 1.25kg Chicago, Field Mus. Nat. Hist.; 903g Washington, U.S. Nat. Mus.; 990g New York, Amer. Mus. Nat. Hist.;
Specimen(s): [1959,766], 17g.

Tillaberi 14°15′ N., 01°32′ E.
Mari, Niamey, Niger
Fell 1970, April, in the morning hrs, late in the month
Stone. Olivine-hypersthene chondrite (L6).
After three fireballs and associated detonations, two stones were recovered, totalling about 3kg, close to the hamlet of Baye in the village of Mari, 40 km NE. of Tillaberi, olivine Fa_{25}, M. Christophe Michel-Levy and J.-M. Malezieux, Meteoritics, 1976, **11**, p.217, Meteor. Bull., 1979 (56), Meteoritics, 1979, **14**, p.171.
157g Paris, Mus. d'Hist. Nat.;

Timersoi v Timmersoi.

Timmersoi 18°55′ N., 6°15′ E. approx.
Talak, Niger
Found 1966, before this year
Synonym(s): *Timersoi*
Stone. Olivine-hypersthene chondrite (L5).
An inhabitant in a village at approximately 18°55'N., 7°45'E., allowed Mr. C. Hemming, United Nations Locust Control Office, to remove 516g from a meteorite said to have fallen at about 18°55'N., 6°15'E. Illustrated description, with microprobe analyses, olivine Fa_{24}, D.G.W. Smith et al., Meteoritics, 1972, **7**, p.1. Reported, Meteor. Bull., 1971 (50), Meteoritics, 1971, **6**, p.122.
Specimen, Oxford, Univ. Mus.; Specimen, Tucson, Univ. of Arizona;
Specimen(s): [1971,2], 51.6g. slice

Timochin 54°30′ N., 35°12′ E. approx.
Yukhnov, Kaluga, Federated SSR, USSR, [Тимохина]
Fell 1807, March 25, 1500 hrs
Synonym(s): *Iuknov, Juchnow, Timochina, Timokhina, Timoschin*
Stone. Olivine-bronzite chondrite (H5).
After detonations, a stone of about 65.5kg was seen to fall, L.W. Gilbert, Ann. Phys. (Gilbert), 1807, **26**, p.238. Listed, L.G. Kvasha and A.Ya. Skripnik, Метеоритика, 1978, **37**, p.204. Analysis, 27.64 % total iron, V.Ya. Kharitonova, Метеоритика, 1972, **31**, p.116. Olivine Fa_{20}, B. Mason, Geochimica et Cosmochimica Acta, 1963, **27**, p.1011.
48.6kg Moscow, Acad. Sci., main mass, and fragments, 23.5g; 798g Berlin, Humboldt Univ.; 142g Vienna, Naturhist. Mus.; 119g Tübingen, Univ.; 64g Washington, U.S. Nat. Mus.; 55g Chicago, Field Mus. Nat. Hist.;
Specimen(s): [19969], 94.5g.; [373], 54g. two pieces; [35183], 36.5g.; [46010], 8g.

Timochina v Timochin.

Timokhina v Timochin.

Timoschin v Timochin.

Tipperary v Dundrum.

Tipperary v Mooresfort.

Tiree 56°31′ N., 6°50′ W.
Hebrides, Scotland
Synonym(s): *Hebrides*
Doubtful. Stone. Chondrite.
Two small pieces (3.5g), labelled "Tyree, Scotland"(?) and with the date 1808(?) in the catalogue of the W. Nevill collection to which they belong, are in the Geological Mus., London; they resemble Siena and are possibly identical (G.T. Prior).

Tirlemont v Tourinnes-la-Grosse.

Tirnova v Aleppo.

Tirupati 13°38′ N., 79°25′ E.
Chittor district, Andhra Pradesh, India
Fell 1934, March 20, 1830 hrs
Stone. Olivine-bronzite chondrite (H6).
Five pieces, and fragments, totalling 230g, were collected at Sakamvalipalle, 6 miles SE. of Tirupati. Description, M.S. Krishnan, Rec. Geol. Surv. India, 1936, **71**, p.144 [M.A. 6-394]. Olivine Fa_{18}, B. Mason, Geochimica et Cosmochimica Acta, 1963, **27**, p.1011.
143g Calcutta, Mus. Geol. Surv. India; 35g Tempe, Arizona State Univ.; 16g Washington, U.S. Nat. Mus.;
Specimen(s): [1951,329], 29g.

Tischtin v Tieschitz.

Tishomingo 34°15′ N., 96°41′ W.
Johnson County, Oklahoma, U.S.A.
Found 1965
Iron. Ataxite, Ni-rich (IRANOM).
Four masses, totalling about 250kg, were found. The metal comprises martensite and taenite in the volume ratio 79:21, analysis, 32.5 %Ni, R.D. Buchheit et al., Meteoritics, 1967, **3**, p.103, abs., O.E. Monnig, Meteoritics, 1967, **3**, p.120. Co-ordinates are for Tishomingo from the Times Atlas. Structural study, L.K. Ives et al., Geochimica et Cosmochimica Acta, 1978, **42**, p.1051. Analysis, 32.5 %Ni, 0.25 ppm.Ga, 0.088 ppm.Ge, 17 ppm.Ir, A. Kracher et al., Geochimica et Cosmochimica Acta, 1980, **44**, p.773. Description; a unique meteorite, V.F. Buchwald, Iron Meteorites, Univ. of California, 1975, p.1201.
260kg Fort Worth, Texas, Monnig Colln.;

Tistin v Tieschitz.

Tjabe 7°5′ S., 111°32′ E.
Padang, Rembang, Java, Indonesia
Fell 1869, September 19, 2100 hrs
Stone. Olivine-bronzite chondrite (H6).
After the appearance of a fireball, moving from the NE., and a detonation, a stone of about 20kg was seen to fall and was found next day. Description, with a analysis, E.H. von Baumhauer, Arch. Néerland. Sci. Nat. Haarlem, 1871, **6**, p.305. Olivine Fa_{19}, B. Mason, Geochimica et Cosmochimica Acta, 1963, **27**, p.1011. Main mass perhaps still in Java.
500g Budapest, Nat. Mus.; 173g Amsterdam, Univ. Geol.

Inst.; 110g Paris, Mus. d'Hist. Nat.; 69g Chicago, Field Mus. Nat. Hist.; 39g Washington, U.S. Nat. Mus.; 37g Vienna, Naturhist. Mus.; 22g Tempe, Arizona State Univ.;
Specimen(s): [56323], 96g.; [55109], 24g. fragments; [48759], 14g.

Tjerebon — 6°40′ S., 106°35′ E.
Java, Indonesia
Fell 1922, July 10, 2230 hrs
Stone. Olivine-hypersthene chondrite (L5).
After the appearance of a fireball travelling from NE. to SW., and detonations, two stones, 8.5kg and 8kg, were found. Description, with a doubtful analysis, W.F. Gisolf, Jaarb. Mijnwezen in Nederlandsch-Indië, Verhand., 1925, **53**, p.168. Olivine Fa_{24}, B. Mason, Geochimica et Cosmochimica Acta, 1963, **27**, p.1011.
38g Paris, Mus. d'Hist. Nat.; 19.5g New York, Amer. Mus. Nat. Hist.; Thin section, Washington, U.S. Nat. Mus.;
Specimen(s): [1980,M.18], 31.6g. fragments

Tlacotepec — 18°39′ N., 97°33′ W.
Tecamachalco, Puebla, Mexico
Found 1903
Iron. Ataxite, coarse (1.5mm), Ni-rich (IVB).
A mass of 24kg was stated to be in Mexico City, Mus. Inst. Geol., H.A. Ward, Cat. Ward-Coonley Coll. Meteorites, Chicago, 1904, p.25. Description, analysis, 16.23 %Ni, H.H. Nininger, Am. J. Sci., 1931, **22**, p.360 [M.A. 5-14]. Structurally undeformed, H.J. Axon and P.L. Smith, Min. Mag., 1972, **38**, p.736. Further analysis, 15.82 %Ni, 0.195 ppm.Ga, 0.031 ppm.Ge, 24 ppm.Ir, R. Schaudy et al., Icarus, 1972, **17**, p.174. Description, V.F. Buchwald, Iron Meteorites, Univ. of California, 1975, p.1205.
32.6kg Mexico City, Inst. Geol.; 7.4kg Tempe, Arizona State Univ.; 3.9kg Harvard Univ.; 2.7kg New York, Amer. Mus. Nat. Hist.; 2.4kg Washington, U.S. Nat. Mus.; 2.1kg Chicago, Field Mus. Nat. Hist.;
Specimen(s): [1959,213], 8484g. and sawings, 124g

Tobe — 37°12′ N., 103°35′ W.
Las Animas County, Colorado, U.S.A.
Found 1963, before this year
Stone. Olivine-bronzite chondrite (H4).
A stone of 5386g is mentioned, C. Wilton, letters of 11 March, 4 April, 9 June, 1963 in Min. Dept. BM(NH). Olivine Fa_{20}, B. Mason, Geochimica et Cosmochimica Acta, 1967, **31**, p.1100.
5.4kg Washington, U.S. Nat. Mus.;

Toborsk v Ochansk.

Tobychan
Tobychan River, Ojmakonski district, Yakutsk ASSR, USSR, [Тобычан]
Found 1971, August
Iron. Octahedrite (IIE).
A 52.1kg mass was found 1.7 metres below the surface, Meteor. Bull., 1972 (51), Meteoritics, 1972, **7**, p.230. Analysis, 7.91 %Ni, 27.8 ppm.Ga, 75.5 ppm.Ge, 5.7 ppm.Ir, J.T. Wasson, Meteorites, Springer-Verlag, 1974, p.301. Description, G.M. Ivanova and I.K. Kuznetsova, Метеоритика, 1976, **35**, p.47.
40kg Novosibirsk, Siberian Branch, Acad. Sci. USSR, main mass; 1760g Moscow, Acad. Sci.;

Tocavita v Salt River.

Toconao v Imilac.

Tocopilla v North Chile.

Toke-uchi-mura v Takenouchi.

Tokio (a) — 33°13′ N., 102°37′54″ W.
Terry County, Texas, U.S.A.
Found 1974, recognized in this year
Stone. Olivine-bronzite chondrite (H).
A single stone, Tokio (a), of 6.6kg was found at tha above co-ordinates. A second stone, Tokio (b) was found at 33°12′18″N., 102°39′30″W., G.I Huss, Cat. Huss Coll. Meteorites, 1976, p.43. Olivine $Fa_{19.4}$, A.L. Graham, priv. comm., 1980.
4.6kg Mainz, Max-Planck-Inst., of Tokio (a); 553g Mainz, Max-Planck-Inst., of Tokio (b);
Specimen(s): [1976,M.3], 42.6g. crusted part-slice of Tokio (a)

Toluca — 19°34′ N., 99°34′ W.
Xiquipilco, Mexico State, Mexico
Found 1776, known before this year
Synonym(s): *Abert Iron, Albert Iron, Amates, Caparrosa, Hacienda di Mani, Hiquipilco, Ioluca, Ixtlahuaca, Jiquipilco, Los Amates, Mani, Michigan Iron, Moenvalle, Morelos, Ocatitlan, Ocotitlan, Poinsett Iron, Tacubaya, Tejupilco, Tennant's Iron, Xiquipilco, Ziquipilco*
Iron. Octahedrite, coarse (1.4mm) (IA).
Many masses, the largest of 300lb, were found near Xiquipilco and were being forged into agricultural equipment in 1776. Three masses of 220lb, 19.5lb, and 13lb were brought to Germany by G.A. Stein, F. Wöhler, Sitzungsber. Akad. Wiss. Wien, Math.-naturwiss. Kl., 1856, **20**, p.218. Full bibliography, L. Fletcher, Min. Mag., 1890, **9**, p.164. The Abert Iron, a mass of about 456g, was found in the mineral collection of Col. J.J. Abert and presented by his son to the U.S. Nat. Mus., Washington. Partial analysis, determination of Ga, Au and Pd, 8.31 %Ni, E. Goldberg et al., Geochimica et Cosmochimica Acta, 1951, **2**, p.1. Analysis, 8.05 %Ni, M.I. D'yakonova, Метеоритика, 1958, **16**, p.180. Contains Odessa type silicate inclusions, T.E. Bunch et al., Contr. Miner. Petrol., 1970, **25**, p.297. Analysis of metal, 8.07 %Ni, 70.6 ppm.Ga, 246 ppm.Ge, 1.8 ppm.Ir, J.T. Wasson, Icarus, 1970, **12**, p.407. The chemical composition of Amates, Tacubaya and Michigan is within experimental error of that given above, E.R.D. Scott and J.T. Wasson, Geochimica et Cosmochimica Acta, 1976, **40**, p.103. K-Ar age of silicate, 4500 m.y., D. Bogard et al., Earth planet. Sci. Lett., 1968, **3**, p.275. The 'Calvert' iron and B.M. [47192] differ chemically and structurally from Toluca and are distinguished under 'Paneth's Iron' (*q.v.*). Description; includes Tacubaya and Tennant's Iron, V.F. Buchwald, Iron Meteorites, Univ. of California, 1975, p.1209.
535kg Mexico City, Inst. Geol., includes main mass, 400kg; 353kg Fort Worth, Texas, Monnig Colln.; 219kg Chicago, Field Mus. Nat. Hist.; 144kg Tempe, Arizona State Univ.; 150kg Vienna, Naturhist. Mus.; 114kg Hamburg,; 109kg Washington, U.S. Nat. Mus.; 69kg Nantes, Mus. d'Hist. Nat.; 61kg Bonn, Univ. Mus.; 53.5kg Budapest, 47.5kg, Nat. Mus., 6kg Eötvös Lorand Univ.; 43.5kg Stockholm, Riksmus.; 32kg Prague, Nat. Mus.; 21kg Harvard Univ.; 7kg Paris, Mus. d'Hist. Nat.; 9kg New York, Amer. Mus. Nat. Hist.; 7.6kg Naples, Univ.; 6.2kg Yale Univ.; 4kg Schönenwerd, Bally-Prior Mus.; 6kg Albuquerque, Univ. New Mexico; 2kg Copenhagen, Univ. Geol. Mus.; 2.5kg Sydney, Austr. Mus.; 2kg Philadelphia, Acad. Nat. Sci.; 2.2kg Moscow, Acad. Sci.;
Specimen(s): [40221], 74160g.; [35275], 9100g.; [40733], 4545g.; [33913], 1582g.; [87081], 1230g.; [33915], 1166g.;

[33916], 643g.; [33744], 927g. and fragments, 12g; [33747], 472g. seven fragments; [19964], 440g.; [33914], 119.5g.; [19965], 12g.; [63057], 47g. a slice of the Abert Iron; [1959, 942], 29.5g. and sawings, 2g, of Amates; [94705], 117.5g. turnings; [1959,1048], 513g. and fragments, 6.5g; [1959,966], 994g. and fragments, 11.5g, of Tacubaya; [1927,6], 9g. of Tennant's Iron

Tomakovka 47°51′ N., 34°46′ E.

Ekaterinoslav, Ukraine, USSR, [Томаковка]
Fell 1905, January 17, 2130 hrs
Synonym(s): *Tomakowka*
Stone. Olivine-hypersthene chondrite, amphoterite (LL6).
After the appearance of a luminous meteor (moving from west to east), several stones (none weighing more than 500g) fell at the village of Tomakovka. Description, with an analysis, P.N. Chirvinsky, Bull. Soc. franc. Min. Crist., 1921, **44**, p.155. Mentioned, P.N. Chirvinsky, Ann. Inst. Polytech. Novacherkassk, 1930, **14**, p.2 [M.A. 4-421], E.L. Krinov, Астрон. Журнал, 1945, **22**, p.303 [M.A. 9-297], L.G. Kvasha and A.Ya. Skripnik, Метеоритика, 1978, **37**, p.178. Olivine Fa_{30}, B. Mason, Geochimica et Cosmochimica Acta, 1963, **27**, p.1011. Whereabouts of main mass unknown.
66.7g Moscow, Acad. Sci.;

Tomakowka v Tomakovka.

Tomatlan 20°10′ N., 105°13′ W.

Jalisco, Mexico
Fell 1879, September 17, 1630 hrs
Synonym(s): *Bramudor, El Garganitello, Fomatlan, Gargantillo, Jalisco, Tulisca*
Stone. Olivine-bronzite chondrite (H6).
After the appearance of a luminous meteor travelling from SE. to NW., and detonations, two or three stones fell, the largest about 2lb, C.U. Shepard, Am. J. Sci., 1885, **30**, p.105, A. Castillo, Cat. Météorites Mexique, Paris, 1889, p.13. Analysis, G.P. Merrill, Mem. Nat. Acad. Sci. Washington, 1919, **14** (4), p.3. Olivine Fa_{18}, B. Mason, Geochimica et Cosmochimica Acta, 1963, **27**, p.1011.
168g Washington, U.S. Nat. Mus.; 71g New York, Amer. Mus. Nat. Hist.; 43g Vienna, Naturhist. Mus.;
Specimen(s): [63623], 101g.; [63630], 28g.

Tombigbee v Tombigbee River.

Tombigbee River 32°14′ N., 88°12′ W.

Choctaw and Sumter County, Alabama, U.S.A.
Found 1859
Synonym(s): *Auburn, De Sotoville, Macon County, Tombigbee*
Iron. Chemically and structurally anomalous (IRANOM).
Six masses, totalling about 96lb, the largest of 33lb, were found at different times from 1859 to 1886 near De Sotoville. Description, with an analysis, 4.11 %Ni, W.M. Foote, Am. J. Sci., 1899, **8**, p.153. Classification, analysis, 4.3 %Ni, 38.5 ppm.Ga, 62.5 ppm.Ge, 0.021 Ir, E.R.D. Scott et al., Geochimica et Cosmochimica Acta, 1973, **37**, p.1957. Description; large schreibersites present, cosmically shocked, V.F. Buchwald, Iron Meteorites, Univ. of California, 1975, p.1218.
3.9kg Rome, Vatican Colln; 2.2kg Chicago, Field Mus. Nat. Hist.; 2.9kg Washington, U.S. Nat. Mus.; 2.1kg Tempe, Arizona State Univ.; 1.5kg Berlin, Humboldt Univ.; 1kg New York, Amer. Mus. Nat. Hist.; 488g Yale Univ.; 368g Vienna, Naturhist. Mus.; 334g Dublin, Nat. Mus.; 253g Copenhagen, Univ. Geol. Mus.; 41.4g Paris, Mus. d'Hist. Nat.;
Specimen(s): [84646], 7720g. and a polished slice, 10g; [43683], 37g. of Auburn

Tomhannock Creek 42°53′ N., 73°36′ W.

Rensselaer County, New York, U.S.A.
Found 1863, approx.
Synonym(s): *Ironhannock Creek, Rensselaer County*
Stone. Olivine-bronzite chondrite (H5), brecciated.
A stone of about 1.5kg was found, S.C.H. Bailey, Am. J. Sci., 1887, **34**, p.60. Tomhannock Creek is very similar to Yorktown and Homestead, A. Brezina, Ann. Naturhist. Hofmus. Wien, 1896, **10**, p.251. Description, distinct from Homestead, analysis, 26.31 % total iron, B. Mason and H.B. Wiik, Min. Mag., 1960, **32**, p.528. Olivine Fa_{18}, B. Mason, Geochimica et Cosmochimica Acta, 1963, **27**, p.1011.
1.4kg New York, Amer. Mus. Nat. Hist.; 31g Washington, U.S. Nat. Mus.; 24g Chicago, Field Mus. Nat. Hist.; 22g Vienna, Naturhist. Mus.;
Specimen(s): [56157], 17g.

Tomi-naga v Yonozu.

Tomita 34°30′ N., 133°45′ E. approx.

Asakuchi, Okayama, Honshu, Japan
Fell 1916, April 13
Stone. Olivine-hypersthene chondrite (L).
One stone of 600g fell, I. Yamamoto, Kwasan Observ. Bull., 1935, p.306 [M.A. 7-173]. Description, analysis, 23.46 % total iron, A. Miyashiro et al., Jap. J. Geol. Geogr., 1963, **34**, p.63. Olivine Fa_{23}, B. Mason, Geochimica et Cosmochimica Acta, 1967, **31**, p.1100.
Specimen, Tokyo, Nat. Sci. Mus.;
Specimen(s): [1978,M.21], 3.75g.

Tomsk v Demina.

Tomskii v Demina.

Tonganoxie 39°5′ N., 95°7′ W.

Leavenworth County, Kansas, U.S.A.
Found 1886
Synonym(s): *Kansas, Leavenworth County*
Iron. Octahedrite, medium (1.1mm) (IIIA).
A mass of about 26lb was found on a farm one mile west of Tonganoxie, F.H. Snow, Science, 1891, **17**, p.3. Description, analysis, 7.93 %Ni, E.H.S. Bailey, Am. J. Sci., 1891, **42**, p.385. Classification, V.F. Buchwald, Iron Meteorites, Univ. of California, 1975, p.1224. Further analysis, 7.5 %Ni, 19.9 ppm.Ga, 38 ppm.Ge, 3.8 ppm.Ir, D.J. Malvin et al., priv. comm., 1983.
Specimen, Kansas Univ.; 912g Chicago, Field Mus. Nat. Hist.; 349g Rome, Vatican Colln; 343g Vienna, Naturhist. Mus.; 245g Tübingen, Univ.; 231g Washington, U.S. Nat. Mus.; 221g Harvard Univ.;
Specimen(s): [83813], 260g. slice

Tonk 24°39′ N., 76°52′ E.

Rajasthan, India
Fell 1911, January 22, 1555 hrs
Stone. Carbonaceous chondrite, type I (CI).
After the appearance of a brilliant meteor (moving from west to east across the northern sky), and detonations, a shower of small stones fell, of which a total weight of only 7.7g was collected, the largest piece weighing 1.7g. Description, with an analysis, W.A.K. Christie, Rec. Geol. Surv. India, 1914, **44**, p.41. The stones fell at Chhabra, 24°39'N., 76°52'E., Tonk state, C.A. Silberrad, Min. Mag., 1932, **23**, p.303.
2.5g Calcutta, Mus. Geol. Surv. India; 3g Washington, U.S. Nat. Mus.;

Tonnelier v Mauritius.

Tonopah v Quinn Canyon.

Torgau 51°34′ N., 13°3′ E.
Leipzig, Germany
Fell 1561, May 17
Doubtful..
A stone is said to have fallen, E.F.F. Chladni, Die Feuer-Meteore, Wien, 1819, p.215 but the evidence is not conclusive.

Torigoe v Kyushu.

Torre v Assisi.

Torre Assisi v Assisi.

Torreon de Mata 26°50′ N., 105°25′ W.
Chihuahua, Mexico
Found 1983
Stone. Olivine-hypersthene chondrite (L6).
A single, completely crusted individual weighing 610g was found. It was said to have fallen in February, 1981, W. Zeitschel, letter of 12 September, 1983. Olivine $Fa_{26.5}$, A.L. Graham, priv. comm., 1983.
Specimen(s): [1983,M.5], 492g.

Torre presso Assisi v Assisi.

Torrington 42°4′ N., 104°10′ W.
Goshen County, Wyoming, U.S.A.
Fell 1944, September 23, 1230 hrs
Stone. Olivine-bronzite chondrite (H6).
Several stones fell, but only three, totalling 259.1g, were preserved, H.H. and A.D. Nininger, The Nininger Collection of Meteorites, Winslow, Arizona, 1950, p.98, 118. Olivine Fa_{18}, B. Mason, Geochimica et Cosmochimica Acta, 1963, **27**, p.1011.
116g Washington, U.S. Nat. Mus.; 70g Tempe, Arizona State Univ.;
Specimen(s): [1959,901], 48.5g.

Tostado 29°14′ S., 61°46′ W.
Santa Fe, Argentina
Found 1945, before this year
Stone. Olivine-bronzite chondrite (H).
A dark brown stone weighing 22kg was found while ploughing. Description, with an analysis, E.H. Ducloux, Notas Mus. La Plata, 1945, **10** (geol. no. 41), p.165 [M.A. 9-302]. Olivine Fa_{19}, B. Mason, Geochimica et Cosmochimica Acta, 1963, **27**, p.1011.
Specimen(s): [1962,165], 8g.

Touanne v Charsonville.

Toubil v Toubil River.

Toubil River 55°53′ N., 89°6′ E.
Krasnoyarsk, Federated SSR, USSR, [Тубил]
Found 1891
Synonym(s): *Abakan, Krasnojarsk Iron, Taiga (iron), Tajgha, Tajka, Toubil, Tubil*
Iron. Octahedrite, medium (1.2mm) (IIIA).
A mass of about 22kg was found on the river Tubil, 264 miles ("400 versts") from Krasnojarsk, A. Khalponin, Zap. Imp. Miner. Obshch., 1898, **35**, p.233. Identical with Taiga, to which place pieces were brought, A. Brezina, Ann. Naturhist. Hofmus. Wien, 1896, **10**, p.284, 307. Analysis, 7.81 %Ni, M.I. D'yakonova, Метеоритика, 1958, **16**, p.180. All of the mass has been heated and forged, V.F. Buchwald, Iron Meteorites, Univ. of California, 1975, p.1225. Includes Abakan, A.A. Yavnel, Метеоритика, 1969, **29**, p.154. Further analysis, 7.59 %Ni, 19.8 ppm.Ga, 38.1 ppm.Ge, 5.0 ppm.Ir, E.R.D. Scott et al., Geochimica et Cosmochimica Acta, 1973, **37**, p.1957.
13.7kg Leningrad, Mus.Mining Inst., main mass; 300g Moscow, Acad. Sci.; 750g Paris, Mus. d'Hist. Nat.; 547g Chicago, Field Mus. Nat. Hist.; 250g New York, Amer. Mus. Nat. Hist.; 175g Washington, U.S. Nat. Mus.; 135g Berlin, Humboldt Univ.;
Specimen(s): [83955], 454g. and shavings, 7g, of Toubil; [68216], 9g. of Tajgha

Toulon 41°7′ N., 89°48′30″ W.
Stark County, Illinois, U.S.A.
Found 1962, recognized 1964
Stone. Olivine-bronzite chondrite (H5).
Five fragments, which could be fitted together and totalling 1214.5g, were found on farmland 7.6 km NE. of the town of Toulon. Description, olivine $Fa_{17.9}$, E. Olsen and G.I Huss, Meteoritics, 1974, **9**, p.19. Reported, Meteor. Bull., 1974 (52), Meteoritics, 1974, **9**, p.116.
787g Mainz, Max-Planck-Inst.; 53g Chicago, Field Mus. Nat. Hist.; 51g Copenhagen, Univ. Geol. Mus.;

Toulouse 43°36′ N., 1°24′ E.
Haute Garonne, France
Fell 1812, April 10, 2000 hrs
Synonym(s): *Aucamville, Grenade*
Stone. Olivine-bronzite chondrite (H6), veined.
After the appearance of a fireball, followed by three detonations, a small shower of stones fell between La Pradere in the NW. and La Bordette in the SE. About eight stones were found, the largest weighing about 1kg, L.W. Gilbert, Ann. Phys. (Gilbert), 1812, **41**, p.445, L.W. Gilbert, Ann. Phys. (Gilbert), 1812, **42**, p.111, 343. Olivine Fa_{19}, B. Mason, Geochimica et Cosmochimica Acta, 1963, **27**, p.1011.
150g Paris, Mus. d'Hist. Nat.; 29g Berlin, Humboldt Univ.; 22g Chicago, Field Mus. Nat. Hist.; 16g Vienna, Naturhist. Mus.; 14g New York, Amer. Mus. Nat. Hist.; 11g Tübingen, Univ.;
Specimen(s): [63927], 18g.; [90260], 13.5g.

Tounkin 51°44′ N., 102°32′ E.
Tunka, Irkutsk, Federated SSR, USSR, [Тынка]
Fell 1824, February 18, 0700 hrs
Synonym(s): *Irkutsk, Tunka, Tunkin*
Stone. Chondrite.
After detonations, a stone of about 2kg was seen to fall, K.E.A. von Hoff, Ann. Phys. Chem. (Poggendorff), 1832, **24**, p.224. Listed, E.L. Krinov, Астрон. Журнал, 1945, **22**, p.303 [M.A. 9-297].
0.03g Vienna, Naturhist. Mus., only known specimen;

Tourinnes-la-Grosse 50°47′ N., 4°46′ E.
Tirlemont, Belgium
Fell 1863, December 7, 1130 hrs
Synonym(s): *Louvain, Tirlemont*
Stone. Olivine-hypersthene chondrite (L6).
After detonations, two stones fell; one of about 7kg was found in a wood near Opvelp and the other of about 7.5kg was seen to fall at Le Culot sous Tourinnes-la-Grosse, P.A.

Kesselmeyer, Ann. Phys. Chem. (Poggendorff), 1864, **122**, p.186, P.J. van Beneden et al., Bull. Acad. Roy. Belg., 1863, **16**, p.621. Analysis, F. Pisani, C. R. Acad. Sci. Paris, 1864, **58**, p.169. Further analysis, olivine $Fa_{23.7}$, 21.98 % total iron, R.T. Dodd and E. Jarosewich, Meteoritics, 1980, **15**, p.69.
1.3kg Paris, Mus. d'Hist. Nat.; 500g Berlin, Humboldt Univ.; 416g Vienna, Naturhist. Mus.; 245g Chicago, Field Mus. Nat. Hist.; 111g New York, Amer. Mus. Nat. Hist.; 69g Washington, U.S. Nat. Mus.;
Specimen(s): [54814], 143g.; [36274], 60g.

Township no. 7 v Lincoln County.

Township no. 8 v Lincoln County.

Toyah v Davis Mountains.

Transkei v St. Mark's.

Transvaal v Gibeon.

Travis County 30°18′ N., 97°42′ W. approx.
Travis County, Texas, U.S.A.
Found 1889
Synonym(s): *Hill's stone*
Stone. Olivine-bronzite chondrite (H5), black.
A piece of 2.5kg was found. Description, with an analysis, L.G. Eakins, Am. J. Sci., 1890, **39**, p.59. Olivine Fa_{18}, B. Mason, Geochimica et Cosmochimica Acta, 1963, **27**, p.1011.
1480g Washington, U.S. Nat. Mus., main mass; 126g Berlin, Humboldt Univ.; 130g Tempe, Arizona State Univ.; 97g Michigan Univ.; 96g Chicago, Field Mus. Nat. Hist.; 92g Milwaukee Mus.;
Specimen(s): [1959,769], 58g. slice, and fragments, 3g

Tregnie
Devonshire, England, Not well located
Fell 1622, January 10, or 1723
Doubtful..
One large stone is said to have fallen at Tregnie, Devonshire, E.F.F. Chladni, Die Feuer-Meteore, Wien, 1819, p.222 but the evidence is not conclusive and the locality uncertain - possibly Tregony, Cornwall, 50°16'N., 4°54'W. Possibly a reference to the Stretchleigh fall of 10 January, 1623 (*q.v.*), T.M. Hall, Min. Mag., 1879, **3**, p.4.

Trekanapuram v Kuttippuram.

Trentino
near Tenna Apergena, Italy
Pseudometeorite..
An 18.2kg 'meteorite' was said to have fallen and produced a crater 1 metre deep, Smithson. Inst. Center for Short-lived Phenomena, Event 84-71. Not meteoritic, G.R. Levi-Donati, copy of letter of 13 September, 1971 in Min. Dept. BM(NH).

Trenton 43°22′ N., 88°8′ W.
Washington County, Wisconsin, U.S.A.
Found 1858
Synonym(s): *Colorado (of A. Brezina), Milwaukee, Washington County, Wisconsin*
Iron. Octahedrite, medium (1.2mm) (IIIA).
Masses of 60lb, 16lb, 10lb and 8lb were found in 1858, J.L. Smith, Am. J. Sci., 1869, **47**, p.271. Two more masses, 16.2lb and 33lb, were found in 1872, J.A. Lapham, Am. J. Sci., 1872, **3**, p.69. A further mass, about 10lb, was found in 1880, two masses, 6.5lb and 3lb, were said to have been found before 1890. Masses of 413lb, 527lb, and 1.5lb were discovered in 1952, a 9.5lb mass was found in 1964. Illustrated description, W.F. Read and H.O. Stockwell, Wisconsin Acad. Sci. Arts and Lett., 1966, **55**, p.77. Intensely shocked, H.J. Axon, Prog. Materials Sci., 1968, **13**, p.185, V.F. Buchwald, Iron Meteorites, Univ. of California, 1975, p.1229. Analysis, 8.34 %Ni, 20.8 ppm.Ga, 44.5 ppm.Ge, 2.6 ppm.Ir, E.R.D. Scott et al., Geochimica et Cosmochimica Acta, 1973, **37**, p.1957.
391kg Washington, U.S. Nat. Mus.; 22kg Milwaukee Mus.; 4kg Los Angeles, Univ. of California; 5kg Harvard Univ.; 3.6kg Chicago, Field Mus. Nat. Hist.; 7.3kg Wisconsin Univ.; 1.39kg Berlin, Humboldt Univ.; 1.1kg Vienna, Naturhist. Mus.; 1kg Paris, Mus. d'Hist. Nat.; 855g Tempe, Arizona State Univ.; 604g Budapest, Nat. Mus.; 761g New York, Amer. Mus. Nat. Hist.;
Specimen(s): [1921,22], 139g.; [47242], 113g.; [1920,345], 84g.; [53293], 72g.; [43051], 38g.

Trenzano 45°28′ N., 10°0′ E.
Brescia, Lombardy, Italy
Fell 1856, November 12, 1600 hrs
Synonym(s): *Brescia, Chiara, Ferrara, Vilabella*
Stone. Olivine-bronzite chondrite (H6), veined.
After detonations, three stones were said to have fallen, but only two were found, the largest weighing about 9kg, W. von Haidinger, Sitzungsber. Akad. Wiss. Wien, Math.-naturwiss. Kl., 1860, **41**, p.569. Description, analysis, G. Curioni, Atti R. Ist. Lombardo, 1860, **1**, p.457. Listed, B. Baldanza, Min. Mag., 1965, **35**, p.214. Olivine Fa_{19}, B. Mason, Geochimica et Cosmochimica Acta, 1963, **27**, p.1011.
5.5kg Bologna, Univ.; 2kg Vienna, Naturhist. Mus.; 1.5kg Brescia,; 372g New York, Amer. Mus. Nat. Hist.; 285g Washington, U.S. Nat. Mus.; 150g Paris, Mus. d'Hist. Nat.;
Specimen(s): [65603], 93g.; [65574], 55g.; [35157], 8g. powder

Treprangoda v Kuttippuram.

Tres Estacas
Rio Teuco, Chaco, Argentina, Co-ordinates not reported
Stone. Chondrite.
Listed, with no details, L.M. Villar, Cienc. Investig., 1968, **24**, p.302.

Trevlac 39°17′ N., 86°20′ W.
Brown County, Indiana, U.S.A.
Found 1940, before this year
Doubtful. Stone. 'Amathosite'.
A small mass of sandstone coated with a greenish glassy crust is regarded as possibly meteoritic, and named an "amathosite", F.C. Cross, Pop. Astron., Northfield, Minnesota, 1947, **55**, p.96, Contr. Meteoritical Soc., **4**, p.11.

Treysa 50°55′ N., 9°11′ E.
Hessen, Germany
Fell 1916, April 3, 1530 hrs
Iron. Octahedrite, medium (0.9mm) (IIIB-ANOM).
A mass of about 63kg fell, after the appearance of a fireball and detonations, and was found eleven months later from the observations on the path of the meteor, A. Wegener, Schr. Gesell. Beförd. Ges. Naturwiss. Marburg, 1917, **14**, p.1, F. Richarz, Schr. Gesell. Beförd. Ges. Naturwiss. Marburg, 1918, **14**, p.91. Rare gas data, H. Fechtig et al., Geochimica et Cosmochimica Acta, 1960, **18**, p.72. Analysis, 9.1 %Ni, 20.4 ppm.Ga, 43.1 ppm.Ge, 1.2 ppm.Ir, E.R.D. Scott et al.,

Geochimica et Cosmochimica Acta, 1973, **37**, p.1957. Ir content anomalously high for a IIIB iron. Description, V.F. Buchwald, Iron Meteorites, Univ. of California, 1975, p.1232.
Main mass, Marburg,; 240g Vienna, Naturhist. Mus.; 213g New York, Amer. Mus. Nat. Hist.; 195g Washington, U.S. Nat. Mus.; 143g Tübingen, Univ.; 88g Chicago, Field Mus. Nat. Hist.; 84g Ottawa, Geol. Surv. Canada;
Specimen(s): [1923,401], 1173g. slice

Tribune v Ladder Creek.

Trier v Bitburg.

Trifir 20°3′ N., 1°41′ W.
West Sahara, Mali
Found 1956, March 20
Stone. Olivine-hypersthene chondrite (L6).
A stone of about 1kg was found on the loose desert surface, Meteor. Bull., 1957 (4). Olivine Fa_{24}, B. Mason, Geochimica et Cosmochimica Acta, 1967, **31**, p.1100.
865g Paris, Mus. d'Hist. Nat., main mass; 3g Washington, U.S. Nat. Mus.;

Trigueres v Château-Renard.

Triguerre v Château-Renard.

Trinity County v Canyon City.

Trinity County v Glorieta Mountain.

Trojan v Gumoschnik.

Tromay error for Tromøy.

Tromøy 58°28′24″ N., 8°52′ E.
Arendal, Norway
Fell 1950, April 9, 1500 hrs
Stone. Olivine-bronzite chondrite (H).
One stone of 357.4g was seen to fall. Description, T. Vogt, K. Norsk Vidensk. Selsk. Forhandl., 1951, **23**, p.107, Geochimica et Cosmochimica Acta, 1951, **1**, p.127. Olivine Fa_{18}, B. Mason, Geochimica et Cosmochimica Acta, 1963, **27**, p.1011. Analysis, 25.58 % total iron, H.B. Wiik, Geochimica et Cosmochimica Acta, 1956, **9**, p.279.
8.5g New York, Amer. Mus. Nat. Hist.; 0.66g Chicago, Field Mus. Nat. Hist.;

Troost's Iron v Babb's Mill (Troost's Iron).

Troppau v Opava.

Troup 32°10′ N., 95°6′ W.
Smith County, Texas, U.S.A.
Fell 1917, April 26, morning hrs
Stone. Olivine-hypersthene chondrite (L6).
A stone of about 2.25lb fell, after detonations, 3 miles N. of Troup. Description, with a doubtful analysis, J.A. Udden, Proc. U.S. Nat. Mus., 1921, **59**, p.471. Olivine Fa_{25}, B. Mason, Geochimica et Cosmochimica Acta, 1963, **27**, p.1011.
Main mass, Austin, Texas, Texas State Mus.; 314g Fort Worth, Texas, Monnig Colln.; 114g Washington, U.S. Nat. Mus.; 37g Harvard Univ.;

Troy v Bethlehem.

Troyan v Gumoschnik.

Truckton 38°34′53″ N., 104°4′36″ W.
El Paso County, Colorado, U.S.A.
Found 1978, June 1
Stone. Chondrite.
A single mass of 43.4g was found, Meteor. Bull., 1983 (61), Meteoritics, 1983, **18**, p.82.
43g Colorado Springs, Tiara Observ.;

Tryon 41°33′ N., 100°58′ W.
McPherson County, Nebraska, U.S.A.
Found 1934, recognized in this year
Stone. Olivine-hypersthene chondrite (L6).
35lb of fragments were recovered, A.D. Nininger, Pop. Astron., Northfield, Minnesota, 1937, **45**, p.449 [M.A. 7-62]. Olivine Fa_{24}, B. Mason, Geochimica et Cosmochimica Acta, 1963, **27**, p.1011.
5.9kg Tempe, Arizona State Univ.; 2kg Washington, U.S. Nat. Mus.; 446g Yale Univ.; 395g Los Angeles, Univ. of California; 258g New York, Amer. Mus. Nat. Hist.;
Specimen(s): [1937,387], 68g.; [1959,821], 1186g. three fragments

Trysil 61°18′ N., 12°18′ E.
Drevdalen, Norway
Fell 1927, June 21, 0600 hrs
Synonym(s): *Drevdalen*
Stone. Olivine-hypersthene chondrite (L).
A fireball was observed over Moss and Oslo, and exploded at a height of about 38 km over Trysil. A stone of 640g was picked up on Barflo farm, Drevdalen, I. Oftedal, Norsk. Geol. Tids., 1929, **10**, p.461 [M.A. 5-10]. Olivine Fa_{25}, B. Mason, Geochimica et Cosmochimica Acta, 1963, **27**, p.1011. Description of surface features, O.F. Frigstad, Norsk. Geol. Tids., 1971, **51**, p.9.
Main mass, Oslo, Geol. Mus.; 10g New York, Amer. Mus. Nat. Hist.;

Tsarev 48°42′ N., 45°42′ E.
Volgograd district, Federated SSR, USSR
Found 1968, recognized 1979
Stone. Olivine-hypersthene chondrite (L5).
28 masses, totalling 1131.7kg, were found in fields, the largest mass weighed 284kg. The fall may have occurred in 1922, December 6 at 0700 hrs., Meteor. Bull., 1981 (59), Meteoritics, 1981, **16**, p.198. Brief description, R.L. Khotinok, Метеоритика, 1982, **40**, p.6. Analysis, 20.54 % total iron, L.D. Barsukova et al., Метеоритика, 1982, **41**, p.41.

Tschervonny Kut v Chervony Kut.

Tschinga v Chinga.

Tschinge v Chinga.

Tschuwaschskije Kissy v Kissij.

Tschuwashsky Kissy v Kissij.

Tschuwascksya v Kissij.

Tsess v Gibeon.

Tsmen v Zmenj.

Tuan Tuc 9°40′ N., 105°40′ E.

Soc-Trang, Vietnam
Fell 1921, June 30, 1500 hrs
Synonym(s): *Cochin China, Vinh Luoc*
Stone. Olivine-hypersthene chondrite (L6).

After a triple detonation, three stones, one of which weighed about 10.8kg, were seen to fall at the village of Tuan Tuc, in the province of Soc-Trang; another stone, of 2.3kg, fell at the village of Vinh Luoc, in the province of Rach Gia, 40 km distant, C. Jacob, C. R. Acad. Sci. Paris, 1921, **173**, p.1373. Description, analysis, A. Lacroix, C. R. Acad. Sci. Paris, 1925, **180**, p.1977. Further bulk and mineral analyses, olivine $Fa_{23.7}$, 21.02 % total iron, R.T. Dodd and E. Jarosewich, Meteoritics, 1981, **16**, p.93.

Main masses, Service géol. Indochine; 2.1kg Paris, Mus. d'Hist. Nat.;

Tuat v Tamentit.

Tubil v Toubil River.

Tucson 31°51′ N., 110°58′ W.

Pima County, Arizona, U.S.A.
Found 1850, known before this year
Synonym(s): *Ainsa Iron, Arizona, Bartlett Meteorite, Cañada de Hierro, Carleton Iron, Irwin, Irwin-Ainsa Iron, Muchachos, Ring Meteorite, Santa Catarina Mountains, Santa Rita, Signet Iron, Sonora, Taos*
Iron. Ataxite, Ni-rich (IRANOM).

Two large masses, one ring-shaped (the Signet or Irwin-Ainsa Iron) of 688kg and the other (the Carleton Iron) of 287kg known for centuries, had been transported to Tucson from the Puerta de los Muchacos, a pass 20-30 miles south of that town and were used as anvils; first mentioned by J.F. Velasco in 1850. Full bibliography, with analyses, L. Fletcher, Min. Mag., 1890, **9**, p.16. Bibliography for the period 1850-1876, P.J. McGough, Pop. Astron., Northfield, Minnesota, 1943, **51**, p.511, 564, P.J. McGough, Pop. Astron., Northfield, Minnesota, 1944, **52**, p.243. A unique iron with silicate inclusions, containing brezinaite, (Cr3S4), T.E. Bunch and L.H. Fuchs, Am. Miner., 1969, **54**, p.1509. Analysis, 9.45 %Ni, 0.94 ppm.Ga, 0.049 ppm.Ge, 2.1 ppm.Ir, J.T. Wasson, Geochimica et Cosmochimica Acta, 1970, **34**, p.957. The Carleton mass has been heated and forged; different co-ordinates, weights of masses, V.F. Buchwald, Iron Meteorites, Univ. of California, 1975, p.1235. The meteorite was rapidly cooled, G.T. Miyake and J.I. Goldstein, Geochimica et Cosmochimica Acta, 1974, **38**, p.1201.

911kg Washington, U.S. Nat. Mus., the Ring mass and the Carleton mass; 1.1kg Tempe, Arizona State Univ.; 1429g Chicago, Field Mus. Nat. Hist.; 1100g Chicago, Harvard Univ.; 600g Vienna, Naturhist. Mus.; 498g New York, Amer. Mus. Nat. Hist.; 150g Oxford, Univ. Mus.;
Specimen(s): [35195], 256g. three fragments of Carleton; [61922], 137g. two fragments of Irwin-Ainsa; [34604], 10g.; [90229], 5g.; [40880], 2g.

Tucuman v Campo del Cielo.

Tucuman v Raco.

Tugalin-Bulen 45° N., 103° E. approx.

Tugalin-Bulen, Ovor-Hangay aymag, Mongolia
Fell 1967, February 13
Synonym(s): *Middle Hoby*
Stone. Olivine-bronzite chondrite (H6).

Three stones fell, totalling about 10kg, close to Tugalin-Bulen. The co-ordinates are those of the fall-site, Meteor. Bull., 1979 (56), Meteoritics, 1979, **14**, p.172. A brief report in a newspaper of a fall of two pieces totalling 3.92kg in the Middle Hoby district on January 3, 1967 may refer to this fall though the dates differ, Meteor. Bull., 1967 (40).

8g Paris, Mus. d'Hist. Nat.;

Tula v Netschaëvo.

Tula v Rakovka.

Tula v Sevrukovo.

Tule 27°4′ N., 106°16′ W.

Balleza, Chihuahua, Mexico
Found 1889, mentioned in this year
Synonym(s): *El Tule, Huejuquilla*
Pseudometeorite..

A 49g specimen in Mexico City, A. Castillo, Cat. Météorites Mexique, Paris, 1889, p.7 and a specimen, 8g, in Chicago, Field Mus. Nat. Hist. Both are impure cast iron with less than 0.1% Ni, V.F. Buchwald, Iron Meteorites, Univ. of California, 1975, p.1244.

Tulia (a) 34°37′ N., 101°57′ W.

Swisher County, Texas, U.S.A.
Found 1917
Synonym(s): *Avoca*
Stone. Olivine-bronzite chondrite (H3-4), brecciated.

Two stones, 14.9kg and 8.9kg, were ploughed up. Description, considerably oxidized, with an analysis, C. Palache and J.T. Lonsdale, Am. J. Sci., 1927, **13**, p.353. Analysis doubted, alumina too high, W. Wahl, Geochimica et Cosmochimica Acta, 1950, **1**, p.28. Nature of metal, H.C. Urey and T. Mayeda, Geochimica et Cosmochimica Acta, 1959, **17**, p.113. Olivine $Fa_{18.5}$, A.L. Graham, priv. comm., 1982. A large number of stones have been found in the Tulia-Dimmitt area, a number are ascribed to this fall; 27 specimens in the Monnig Colln. totalling 53.9kg, Cat. Monnig Colln., 1983, p.22.

Main part of larger mass, Austin, Texas, Texas Memorial Univ.; 7.2kg Harvard Univ., main part of smaller mass; 53.9kg Fort Worth, Texas, Monnig Colln.; 4.9kg Washington, U.S. Nat. Mus.; 4.3kg Tempe, Arizona State Univ.; 1.98kg Paris, Mus. d'Hist. Nat.; 713g Moscow, Acad. Sci.; 640g Michigan Univ.; 540g Prague, Nat. Mus.; 318g Los Angeles, Univ. of California;
Specimen(s): [1927,83], 298g. and fragments, 5g; [1983,M.32], 1.7g.

Tulia (b) 34°32′ N., 101°42′ W.

Swisher County, Texas, U.S.A.
Found 1917, recognized 1982
Stone. Olivine-hypersthene chondrite (L6).

Some material in collections labelled Tulia is L-group and not H-group material. 2153g obtained from H. Nininger in 1940 by the Field Mus. Nat. Hist., Chicago is L6 material, as is BM 1959,991. These specimens are fairly fresh, olivine $Fa_{25.0}$, A.L. Graham, priv. comm., 1982.

2153g Chicago, Field Mus. Nat. Hist.;
Specimen(s): [1959,991], 2355g.

Tulisca v Tomatlan.

Tulung Dzong 30° N., 90°45′ E. approx.
Tibet
Fell 1944, March 26, 1015 hrs
Stone.
Said to have made a crater 10 feet in diameter; two pieces were brought to Mr. J.D. Auden, M.A.R. Khan, Hyderabad Acad. Studies, 1950, **12** [M.A. 11-441], P. Chatterjee, letter of 25 March, 1952 in Min. Dept. BM(NH). Tulung Dzong is two days march NNW. of Lhasa.
Two pieces, Calcutta, Mus. Geol. Surv. India;

Tunguska 60°54′ N., 101°57′ E.
Krasnoyarskiy Kray, Federated SSR, USSR
Fell 1908, June 30, 0016 hrs, U.T.
Synonym(s): *Khatanga, Podkamennaya Tunguska, Tunguska River, Vanovara*
There appears to be no reason to doubt that an object of considerable size entered the atmosphere, though very little meteoritic material has been recovered, owing mainly to the nature of the ground and the situation of the place of fall. The mass of the fall is estimated at 10000 to 10,000,000 tons from energetic considerations, possibly of "contraterrene" material, L. LaPaz, Pop. Astron., Northfield, Minnesota, 1948, **56**, p.330 [M.A. 10-398]. Shreds of nickel-iron present in soil samples collected in 1927, A.A. Yavnel, Метеоритика, 1959, **17**, p.8 [M.A. 15-534]. General reviews and popular accounts, I.S. Astapovich, Pop. Astron., Northfield, Minnesota, 1938, **46**, p.310 [M.A. 7-176], J.G. Crowther, Sci. Amer., 1931, **144**, p.314 [M.A. 5-17], L.J. Spencer, Geogr. J. London, 1933, **81**, p.227 [M.A. 5-301]. Descriptions of observed phenomena at the date of fall, P.N. Chirvinsky, Centralblatt Min., 1923, p.548 [M.A. 2-257], A.V. Voznesensky, Mirovédénie, 1925, **14**, p.25 [M.A. 3-92], F.J.W. Whipple, Quart. J. Roy. Meteorolog. Soc., 1934, **60**, p.505 [M.A. 6-16]. Accounts of the topography and other phenomena at the place of fall, L.A. Kulik, Pop. Astron., Northfield, Minnesota, 1937, **45**, p.559 [M.A. 7-67], L.A. Kulik, Метеоритика, 1941, **2**, p.119 [M.A. 10-175], L.A. Kulik, Метеоритика, 1955, **13**, p.104 [M.A. 13-50]. Description of magnetic spherules, new data, E.L. Krinov, Giant Meteorites, Pergamon, Oxford, 1966. Experimental simulation, I.T. Zotkin and M.A. Tsikulin, Метеоритика, 1968, **28**, p.114, G.H.S. Jones, Nature, 1977, **267**, p.605. Explosion caused by anti-matter, R.V. Gentry, Nature, 1966, **211**, p.1071, L. Marshall, Nature, **212**, p.1226. Explosion caused by a small 'black hole', A.A. Jackson and M.P. Ryan, Nature, 1973, **245**, p.88, G.L. Wick and J.D. Isaacs, Nature, 1974, **247**, p.139.

Tunguska River v Tunguska.

Tunka v Tounkin.

Tunkin v Tounkin.

Turakina v Wairarapa Valley.

Turgai v Bischtübe.

Turgaj v Bischtübe.

Turin 45°4′ N., 7°41′ E.
Piemonte, Italy
Fell 1782
Doubtful..
One large stone is said to have fallen, E.F.F. Chladni, Die Feuer-Meteore, Wien, 1819, p.256 but the evidence is not conclusive.

Turner Mound v Brenham.

Turra v Atarra.

Turtle River 47°36′ N., 94°46′ W.
Beltrami County, Minnesota, U.S.A.
Found 1953, between 1953 and 1958, recognized 1968
Iron. Octahedrite, medium (1.0mm) (IIIB).
A mass of 22.39kg was ploughed up, G.I Huss, letter of 18 March, 1970 in Min. Dept. BM(NH). Classification and analysis, 8.80 %Ni, 20.5 ppm.Ga, 41.4 ppm.Ge, 0.057 ppm.Ir, E.R.D. Scott et al., Geochimica et Cosmochimica Acta, 1973, **37**, p.1957. Description; shock-hardened, but artificially reheated, V.F. Buchwald, Iron Meteorites, Univ. of California, 1975, p.1245.
10.5kg Mainz, Max-Planck-Inst.; 873g Copenhagen, Univ. Geol. Mus.; 615g Chicago, Field Mus. Nat. Hist.; 387g Washington, U.S. Nat. Mus.; 232g Tempe, Arizona State Univ.; 179g New York, Amer. Mus. Nat. Hist.; 136g Los Angeles, Univ. of California;
Specimen(s): [1969,1], 127.5g. etched slice

Turuma v Duruma.

Tuscon error for Tucson.

Tuva v Chinga.

Tuxtuac 21°40′ N., 103°22′ W.
Zacatecas, Mexico
Fell 1975, October 16, 1620 hrs
Stone. Olivine-hypersthene chondrite, amphoterite (LL5).
Two stones were recovered totalling 4.25kg, olivine Fa_{31}, Meteor. Bull., 1981 (59), Meteoritics, 1981, **16**, p.198.
351g Chicago, Field Mus. Nat. Hist.; 238g Hanau, Zeitschel Colln.;
Specimen(s): [1980,M.39], 1924g.; [1981,M.7], 146g.

Tuzla 44°1′ N., 28°38′ E.
Dobruja, Romania
Stone. Olivine-hypersthene chondrite (L6).
Doubtful, nothing is known as to fall or find. Specimen in Brit. Mus. (Nat. Hist.) obtained from A. Kusche of Munich in 1920. Olivine Fa_{25}, B. Mason, Geochimica et Cosmochimica Acta, 1963, **27**, p.1011.
Specimen(s): [1921,4], 236g.

Twentynine Palms 34°4′30″ N., 116°1′ W.
San Bernardino County, California, U.S.A.
Found 1944, approx.
Stone. Olivine-hypersthene chondrite (L).
A small mass of 100g was found 4-5 km E. and a little N. of Twentynine Palms, Meteor. Bull., 1962 (24). Another mass of 19.6kg was found in 1955 and is an L-group chondrite. The two stones are presumably from the same fall, but there is no definite evidence on this point. Olivine Fa_{26}, B. Mason, Geochimica et Cosmochimica Acta, 1963, **27**, p.1011.
100g Los Angeles, Griffith Observatory, the stone found in 1944; 31g Tempe, Arizona State Univ.; 3.2g New York, Amer. Mus. Nat. Hist.;

Twin City 32°35′ N., 82°1′ W.
Emanuel County, Georgia, U.S.A.
Found 1955
Iron. Ataxite, Ni-rich (IRANOM).
A much-weathered mass of 5.13kg was found, 29.9 %Ni, E.P. Henderson and A.S. Furcron, Georgia Mineral

Newsletter (Georgia Geol. Surv.), 1957, **10** (4), p.137 [M.A. 14-130], Meteor. Bull., 1958 (8), B. Mason, Meteorites, Wiley, 1962, p.231. Chemically anomalous, analysis, 30.06 % Ni, 4.54 ppm.Ga, 7.42 ppm.Ge, 0.015 ppm.Ir, J.T. Wasson and R. Schaudy, Icarus, 1971, **14**, p.59. Description; polycrystalline, V.F. Buchwald, Iron Meteorites, Univ. of California, 1975, p.1247.

3.4kg Washington, U.S. Nat. Mus.;

Twin Wells v Menindee Lakes 002.

Two Buttes (a) 37°38′ N., 102°25′ W.

Baca County, Colorado, U.S.A.
Found 1962, recognized 1968
Stone. Olivine-bronzite chondrite (H5).

One mass, 19.7kg was ploughed up in field, G.I Huss, letters of 18 March, 1970 and 16 March, 1972. Olivine Fa_{19}, R. Hutchison, priv. comm.

4.5kg Mainz, Max-Planck-Inst.; 9.3kg Denver, Mus. Nat. Hist.; 466g Chicago, Field Mus. Nat. Hist.; 191g Tempe, Arizona State Univ.;

Specimen(s): [1969,231], 66g. part-slice

Two Buttes (b) 37°38′ N., 102°25′ W.

Baca County, Colorado, U.S.A.
Found 1970
Stone. Chondrite.

A single mass, 3kg, was found under a fence. Probably distinct from Two Buttes (a), Meteor. Bull., 1974 (52), Meteoritics, 1974, **9**, p.116.

2.3kg Mainz, Max-Planck-Inst., main mass; 507g Denver, Nat. Hist. Mus.;

Tyron error for Tryon.

Tysnes Island 60°0′ N., 5°37′ E.

Hardanger fiord, Norway
Fell 1884, May 20, 2030 hrs
Synonym(s): *Midt-Vaage*
Stone. Olivine-bronzite chondrite (H4), brecciated.

After the appearance of a fireball, and detonations, two stones fell, of 18.95kg and 0.91kg respectively; description, H. Reusch, Neues Jahrb. Min., 1886, **4**, p.473. Analysis, T. Hiortdahl, Nyt Mag. Naturvid., Oslo, 1886, **30**, p.276. Polymict with carbonaceous clasts, K. Keil and R. Fodor, Chem. Erde, 1980, **39**, p.1. Olivine Fa_{20}, B. Mason, Geochimica et Cosmochimica Acta, 1963, **27**, p.1011.

Main mass, Oslo, Min.-Geol. Mus.; 306g Chicago, Field Mus. Nat. Hist.; 104g Washington, U.S. Nat. Mus.; 100g New York, Amer. Mus. Nat. Hist., approx. weight; 93g Vienna, Naturhist. Mus.;

Specimen(s): [62363], 767g. and fragments, 2g; [1920,346], 1g. fragments

Tyumen 57°10′ N., 65°32′ E.

Tyumen region, Federated SSR, USSR, [Тюмен]
Fell 1903, April 20, or 21 or 22
Doubtful. Iron.

A meteorite was observed to fall near Tyumen by P.A. Rossomachin about April 20-22 1903 and a piece of 0.75kg collected and placed in the local museum, but subsequently lost, P.L. Dravert, Природа, 1939 (2), p.123 [M.A. 7-374], Meteor. Bull., 1959 (13).

Ubari v Oubari.

Uberaba 19°49′ S., 48°47′ W.

Minas Gerais, Brazil
Fell 1903, June 29, 1000 hrs
Synonym(s): *Dores do Campo, Dores dos Campos Formosos, Formosas*
Stone. Olivine-bronzite chondrite (H5), veined.

After the appearance of a luminous meteor (travelling from NE. to SW.) and detonations, a stone of about 30-40kg was seen to fall about 50 miles west of Uberaba. Description, E. Hussak, Ann. Naturhist. Hofmus. Wien, 1904, **19**, p.85. Listed, E. de Oliveira, Anais Acad. Brasil. Cienc., 1931, **3**, p.33 [M.A. 5-15]. Analysis, 26.08 % total iron, olivine $Fa_{19.5}$, C.B. Gomes et al., Anais Acad. Brasil. Cienc., 1977, **49**, p.269.

3.5kg Ouro Preto, Escuola de Minas; 1.16kg Paris, Mus. d'Hist. Nat.; 442g Budapest, Nat. Mus.; 430g Vienna, Naturhist. Mus.; 139g Chicago, Field Mus. Nat. Hist.; 103g Washington, U.S. Nat. Mus.; 72g New York, Amer. Mus. Nat. Hist.;

Specimen(s): [87109], 52.5g.; [1927,1144], 31.5g.

Ucera 11°3′ N., 69°51′ W.

Falcon, Venezuela
Fell 1970, January 16, 1900 hrs, approx.
Synonym(s): *Caserio Ucera, Coro*
Stone. Olivine-bronzite chondrite (H5).

A single crusted stone of 4.95kg fell near a house in Caserio Ucera, 50-60 km SSW. of Coro, and was obtained by the Coro police, who reported that the fall was accompanied by a fireball and an explosion, Meteor. Bull., 1971 (50), Meteoritics, 1971, **6**, p.116. Brief description, analysis, 26.43 % total iron, R.S. Clarke, Jr. et al., J. Geophys. Res., 1971, **76**, p.4135. Cosmogenic nuclides, 26Al only 43 dpm/kg, P.J. Cressy, Jr., J. Geophys. Res., 1971, **76**, p.4072. Thermoluminescence, J.E. Vaz, Meteoritics, 1972, **7**, p.77.

205g Washington, U.S. Nat. Mus.;

Specimen(s): [1970,340], 27.8 partly crusted fragment

Udaipur

Rajasthan, India, Co-ordinates not reported
Fell
Stone. Olivine-bronzite chondrite (H4).

A stone weighing about 1.2kg fell. Rare gas data and partial analysis, 28.2 % total iron, K. Gopalan and M.N. Rao, Meteoritics, 1976, **11**, p.131. 2kg recovered, 53Mn activity, S.K. Bhattacharya et al., Earth planet. Sci. Lett., 1980, **51**, p.45.

Main mass, Udaipur, Mus.;

Udei Station 7°57′ N., 8°5′ E.

Benue river, Nigeria
Fell 1927, in the spring
Iron. Octahedrite, medium (0.6mm) with silicate inclusions (IA).

A mass of 226lb was found, but it was reported that at least one other mass fell. The fall was heard by natives, but the exact date cannot be fixed, R.C.Wilson, letters of 5 and 21 May, 1936 in Min. Dept. BM(NH). Account of fall and find, W.N. MacLeod and R. Walls, Rec. Geol. Surv. Nigeria, 1958, p.22. Classed as a Copiapo-type iron with silicate inclusions, analysis of silicates, T.E. Bunch et al., Contr. Miner. Petrol., 1970, **25**, p.297. Analysis of metal, 8.83 %Ni, 60.4 ppm.Ga, 204 ppm.Ge, 0.51 ppm.Ir, J.T. Wasson, Geochimica et Cosmochimica Acta, 1970, **34**, p.957. Texture, composition and cooling rate of metal, B.N. Powell, Geochimica et Cosmochimica Acta, 1969, **33**, p.789, Geochimica et Cosmochimica Acta, 1971, **35**, p.5. Shock loaded to less than 120 kbar pressure, A.V. Jain and M.E.

Lipschutz, Nature, 1973, **242** (Phys. Sci.), p.26. Description, V.F. Buchwald, Iron Meteorites, Univ. of California, 1975, p.1250.
100kg Kaduna, Geol. Surv. Nigeria; 202g Washington, U.S. Nat. Mus.; 110g New York, Amer. Mus. Nat. Hist.;
Specimen(s): [1936,1217], 993g. and a fragment, 4g

Uden 51°39′ N., 5°37′ E.
Noord-Brabant, Holland
Fell 1840, June 12, 1030 hrs
Synonym(s): *Nord-Brabant, Staartje*
Stone. Olivine-hypersthene chondrite, amphoterite (LL6).
After detonations, a stone of 0.71kg was seen to fall, R. von Rees, Ann. Phys. Chem. (Poggendorff), 1843, **59**, p.350. Analysis, E.H. von Baumhauer and F. Seelheim, Ann. Phys. Chem. (Poggendorff), 1862, **116**, p.184. Olivine $Fa_{30.3}$, K. Fredriksson et al., Origin and distribution of the elements, ed. L.H. Ahrens, Pergamon, 1968, p.457.
579g s'Hertogenbosch, Noordbrabantsch Mus.; 6.3g Tempe, Arizona State Univ.; 5.6g Tübingen, Univ.; 3.3g Budapest, Nat. Mus.; 3g Chicago, Field Mus. Nat. Hist.;
Specimen(s): [35155], less than 1g

Uderei v Angara.

Udipi 13°29′ N., 74°47′ E.
South Kanara district, Karnataka, India
Fell 1866, April, 1000 hrs
Synonym(s): *Canara, South Canara, Udipu*
Stone. Olivine-bronzite chondrite (H5), veined.
After detonations, a stone of about 8lb was seen to fall at Yedabettu village (13°29'N., 74°47'E.) in Udipi taluq, Proc. Madras Govt. Public Dept. April 5, 1869 (57), p.5. Olivine Fa_{18}, B. Mason, Geochimica et Cosmochimica Acta, 1963, **27**, p.1011.
88g Vienna, Naturhist. Mus.; 58g New York, Harvard Univ.; 55g Calcutta, Mus. Geol. Surv. India; 21g Chicago, Field Mus. Nat. Hist.; 13g New York, Amer. Mus. Nat. Hist.;
Specimen(s): [43057], 3265g. main mass, and fragments, 12g

Udipu v Udipi.

Uegit 3°49′ N., 43°20′ E. approx.
Alto Guiba, Somalia
Found 1921
Iron. Octahedrite, medium (0.9mm) (IIIA).
A discoid mass of about 252kg was found at Dersa, east of Uegit. Description, with an analysis, 7.11 %Ni, F. Millosevich, Mem. Reale Accad. Lincei, Cl. sci. fis. mat. nat., 1924, **14** (10), p.501. Analysis, 7.4 %Ni, 17 ppm.Ga, 39 ppm.Ge, S.J.B. Reed, Meteoritics, 1972, **7**, p.257. Full description, mineral chemistry, metallography, F. Burragato and V. Faccenda, Meteoritics, 1975, **10**, p.75. Description, co-ordinates, V.F. Buchwald, Iron Meteorites, Univ. of California, 1975, p.1251.
Main mass Rome, Univ. Mus.; 170g Paris, Mus. d'Hist. Nat.;

Ufana 4°16′ S., 35°21′ E.
Mbulu district, Arusha, Tanzania
Fell 1957, August 5, 1820 hrs
Stone. Enstatite chondrite (E6).
Two fragments, 101.1g and 88.1g, were collected. Full account of the fall, J.R. Harpum, Rec. Geol. Surv. Tanganyika for 1961, 1965, **11**, p.54.
Specimen(s): [1962,81], thin section

Uganda v Maziba.

Uganda v Soroti.

Ujimgin v Wu-chu-mu-ch'in.

Ularring 29°58′ S., 120°36′ E.
Western Australia, Australia
Found 1970
Stone. Olivine-hypersthene chondrite (L6).
A single oriented stone, 271.8g, was found 300 metres N. of the Golden Wonder Gold mine, 6 miles S. of the former township of Ularring, G.J.H. McCall, 2nd. Suppl. to West. Austr. Mus. Spec. Publ. no. 3, 1972, p.26. Olivine $Fa_{25.6}$, B. Mason, Rec. Austr. Mus., 1974, **29**, p.169.
Main mass, Kalgoorlie, West. Austr. School of Mines; 1.2g Washington, U.S. Nat. Mus.; Thin section, Perth, West. Austr. Mus.;

Ulmiz 46°56′ N., 7°13′ E.
Murten, canton Freiburg, Switzerland
Fell 1926, December 25, 0650 hrs
Stone. Olivine-hypersthene chondrite (L).
Ten fragments totalling 76.5g were collected, E. Gerber, Mitt. Natur. Gesell. Bern, 1927, p.xi. Description, analysis, E. Hügi, Mitt. Natur. Gesell. Bern, 1929, p.34 [M.A. 4-417].
26g Bern, Nat. Hist. Mus.; 20g Freiburg, Mus. Cant. d'Hist. Nat.;

Ultuna 59°49′ N., 17°40′ E.
Uppsala, Sweden
Found 1944
Doubtful. Stone. Olivine-bronzite chondrite (H).
A small stone was found during excavations in Ultuna in 1944, F.E. Wickman, Pop. Astron. Tidskr., 1951, p.63 [M.A. 12-357], S. Hjelmqvist, letter of March, 1958 in Min. Dept. BM(NH). Probably part of the Hessle shower of 1869 according to K. Fredriksson. Olivine Fa_{18}, B. Mason, Geochimica et Cosmochimica Acta, 1963, **27**, p.1011.
1.9kg Uppsala, Univ.;

Ulysses 37°36′ N., 101°15′ W.
Grant County, Kansas, U.S.A.
Found 1927, approx., recognized 1932
Stone. Olivine-bronzite chondrite (H4).
One stone of 3.9kg was found about 5 miles SE. of Ulysses, A.D. Nininger, Pop. Astron., Northfield, Minnesota, 1937, **45**, p.449 [M.A. 7-62]. Description, H.H. Nininger, Trans. Kansas Acad. Sci., 1936, **39**, p.170. Listed, F.C. Leonard, Univ. of New Mexico Publ., Albuquerque, 1946 (meteoritics ser. no. 1), p.38. Olivine Fa_{17}, B. Mason, Geochimica et Cosmochimica Acta, 1963, **27**, p.1011.
1.65kg Tempe, Arizona State Univ.; 334g Washington, U.S. Nat. Mus.; 151g Chicago, Field Mus. Nat. Hist.;
Specimen(s): [1959,902], 1369g.

Umbala 30°20′ N., 76°20′ E.
Patiala district, Punjab, India
Fell 1822, or 1823
Synonym(s): *Ambala, Umballa*
Stone. Olivine-hypersthene chondrite, amphoterite (LL5).
A stone of 3.5oz, (100g) "fell about 40 miles west of Ambala between the Jumna and Punjab in 1822/3" (original label with specimen in Min. Dept. BM(NH).). Reported, W.S. Atkinson, Proc. Asiatic Soc. Bengal, 1859, **28**, p.260 where, however the date is given erroneously as 1832-3. Olivine Fa_{28}, B. Mason, Geochimica et Cosmochimica Acta, 1963, **27**, p.1011. 40 miles west of Ambala would be 30°20'N., 76°20'E., in Patiala State, C.A. Silberrad, Min. Mag., 1932, **23**, p.303.

35g Calcutta, Mus. Geol. Surv. India; 9.5g Chicago, Field Mus. Nat. Hist.; 3g Vienna, Naturhist. Mus.;
Specimen(s): [34801], 20.5g.

Umballa v Umbala.

Umbarger 34°57′ N., 102°7′3″ W.

Randall County, Texas, U.S.A.
Found 1954, recognized 1979
Stone. Olivine-hypersthene chondrite (L3-6).
A single, highly oxidized, stone of 13kg was found 2.9km WSW. of Umbarger. It has fractures filled with plagioclase and maskelynite, olivine $Fa_{25.4}$, P.S. Sipiera et al., Meteoritics, 1983, **18**, p.63.
837g Chicago, Field Mus. Nat. Hist.;
Specimen(s): [1981,M.3], 214g. slice

Umbria v Orvinio.

Umehara v Gifu.

Um-Hadid 21°41′42″ N., 50°35′48″ E.

Rub'al Khali, Saudi Arabia
Found
Stony-iron. Mesosiderite (MES).
Oxidized fragments up to 1kg in weight were found associated with a crater 10 metres in diameter, F. El Baz and A. El Goresy, Meteoritics, 1971, **6**, p.265, abs. Classification, co-ordinates, 15.4kg collected, B. Mason, Ms. on Arabian Meteorites, 1981.

Umjhiawar v Shergotty.

Umm Hemeima v Umm Ruaba.

Umm Ruaba 13°28′ N., 31°13′ E.

Kordofan, Sudan
Fell 1966, December 27, 1300-1400 hrs
Synonym(s): *Magrour, Umm Hemeima*
Stone. Olivine-hypersthene chondrite (L5).
Following a thundering sound, a bolide was seen to fall. A complete individual, 1.4kg, and a broken fragment, 300g were recovered some 90 km N. of the town of Umm Ruaba, Meteor. Bull., 1967 (40), Meteoritics, 1970, **5**, p.93. Analysis, 22.64 % total iron, A.J. Easton and C.J. Elliott, Meteoritics, 1977, **12**, p.409.
Main mass, Khartoum, Geol. Surv.; 64g Washington, U.S. Nat. Mus.;
Specimen(s): [1967,259], 191g. a partly crusted fragment

Umm Tina 19°1′40″ N., 51°5′ E.

Rub'al Khali, Saudi Arabia
Found 1932, February 20
Stone. Olivine-hypersthene chondrite (L6).
About twelve fragments, totalling 70.2g, were found by H.St.J. Philby. Though badly oxidized, they clearly differ from the Suwahib ('Ain Sala), Suwahib (Adraj) and Suwahib (Buwah) stones from the same district. Description, W. Campbell Smith, Min. Mag., 1933, **23**, p.334. Olivine Fa_{23}, B. Mason, Geochimica et Cosmochimica Acta, 1963, **27**, p.1011.
13.4g New York, Amer. Mus. Nat. Hist.;
Specimen(s): [1932,1233], 56g.

Umri v Merua.

Union v North Chile.

Union County 34°45′ N., 84°0′ W. approx.

Union County, Georgia, U.S.A.
Found 1853
Iron. Octahedrite, coarsest (2.1mm) (IC).
A mass of about 15lb was found, C.U. Shepard, Am. J. Sci., 1854, **17**, p.238. Classification, analysis, 6.12 %Ni, 54.8 ppm.Ga, 245 ppm.Ge, 2.1 ppm.Ir, E.R.D. Scott and J.T. Wasson, Geochimica et Cosmochimica Acta, 1976, **40**, p.103. Similar to Mount Dooling. Description; shock-hardened but not annealed, V.F. Buchwald, Iron Meteorites, Univ. of California, 1975, p.1252.
375g Tempe, Arizona State Univ.; 216g Yale Univ.; 84g Washington, U.S. Nat. Mus.; 77g Paris, Mus. d'Hist. Nat.; 68g Chicago, Field Mus. Nat. Hist.; 58g New York, Amer. Mus. Nat. Hist.; 47g Berlin, Humboldt Univ.;
Specimen(s): [90226], 55g. slice; [1923,348], 43g.; [1959,931], 96g. and sawings, 9g

Unkoku 34°48′ N., 127°0′ E. approx.

Zenranando, South Korea
Fell 1924, September 7
Stone. Chondrite.
A stone of 1kg fell, I. Yamamoto, Kwasan Observ. Bull., 1935, p.306 [M.A. 7-173].
120g Zinsen Observatory;

Un-named 30°29′ S., 126°15′ E.

Nullarbor Plain, Western Australia, Australia
Found 1966
Iron.
An oxidized iron, 39.5g, listed, G.J.H. McCall and W.H. Cleverly, J. Roy. Soc. West. Austr., 1970, **53**, p.69.

Unter-Mässing 49°5′25″ N., 11°20′0″ E.

Bayern, Germany
Found 1920
Iron. Octahedrite, plessitic (IIC).
A mass of 80kg was found at a depth of 1.5 metres, about 2 miles east of Unter-Mässing, near the junction of the roads to Röckenhofen and Oesterberg. Description, with an analysis, 9.93 %Ni, H. Hess, Jahresber. Naturhist. Gesell. Nürnberg, 1920, p.13. Classification, analysis, 9.80 %Ni, 37.1 ppm.Ga, 101 ppm.Ge, 4.4 ppm.Ir, J.T. Wasson, Geochimica et Cosmochimica Acta, 1969, **33**, p.859. Description; well preserved, V.F. Buchwald, Iron Meteorites, Univ. of California, 1975, p.1254.
Main mass, Nürnberg, Nat. Hist. Soc.; 37g Tempe, Arizona State Univ.;

Upper Volta v Guibga.

Ur 30°54′ N., 46°1′ E.

Chaldea, Iraq
Found, prehistoric, about 2500 B.C.
Iron.
During excavations at Ur, fragments of iron were found containing 10.9% Ni and undoubtedly of meteoritic origin, C.H. Desch, Rep. Brit. Assn., 1928, **98**, p.440, G.A. Wainwright, J. Egypt. Archaeol., 1932, **18**, p.3 [M.A. 5-151], C.L. Woolley, The Royal Cemetery, London, p.49, 293, 542. Oxidized metal, perhaps a broken disc, about 10 cm in diameter and 3 mm thick was found corroded on to copper. Parts of the copper are in Philadelphia (Univ. Mus. CBS 17330) and of the iron in Bagdad (Iraq Mus. No. 4187). BM [1972,229] has veins of metal with 28.7-29.6% Ni in an oxidized matrix with 5.0, 6.1% Ni, undoubtedly meteoritic, V.F. Buchwald, letter of 20 December, 1972 in Min. Dept. BM(NH).
Specimen(s): [1972,229], 0.64g.; [1972,230], 0.1g.

Urasaki 34°29′ N., 133°17′ E.
Numakuma, Hiroshima, Honshu, Japan
Fell 1926, April 16
Stone?.
A mass of 1500g fell, I. Yamamoto, Kwasan Observ. Bull., 1935, p.306 [M.A. 7-173].

Urba v Virba.

Urei v Novo-Urei.

Urgailyk-Chinge v Chinga.

Ussuri v Sikhote-Alin.

Usti nad Orlici 49°58′30″ N., 16°22′30″ E.
Kergartice, Gradez Kralove, Czechoslovakia
Fell 1963, June 12, 1358 hrs
Synonym(s): *Kerhartice*
Stone. Olivine-hypersthene chondrite (L6).
One mass of 1269g fell and made a hole 40 cm deep and 30 cm in diameter, Meteor. Bull., 1963 (28). Illustrated description, K. Tucek, Метеоритика, 1965, **26**, p.112. Classification, mineralogy, olivine $Fa_{23.6}$, M. Bukovanska et al., Meteoritics, 1983, **18**, p.223.
1260g Prague, Nat. Mus.;
Specimen(s): [1982,M.19], 9.1g.

Uszwalda v Lixna.

Utah v Salt Lake City.

Ute Creek 36°12′ N., 103°54′ W. approx.
Union County, New Mexico, U.S.A.
Found 1936
Stone. Olivine-bronzite chondrite (H4).
Several fragments, totalling 2kg, were found, A.D. Nininger, Pop. Astron., Northfield, Minnesota, 1937, **45**, p.449 [M.A. 7-62]. Known weight 1kg, W. Wahl, letter of 23 May, 1950 in Min. Dept. BM(NH). Olivine $Fa_{18.8}$, D.E. Lange and K. Keil, Meteoritics, 1976, **11**, p.315, abs. Petrographically similar to Gladstone, G.R. Levi-Donati and E. Jarosewich, Meteoritics, 1974, **9**, p.145.
381g Tempe, Arizona State Univ.; 73g Chicago, Field Mus. Nat. Hist.; 177g Washington, U.S. Nat. Mus.; 58g Albuquerque, Univ. of New Mexico;
Specimen(s): [1959,830], 141g. and fragments, 5g

Ute Pass v Mount Ouray.

Utrecht 52°7′ N., 5°11′ E.
Holland
Fell 1843, June 2, 2000 hrs
Synonym(s): *Blaauw-Kapel*
Stone. Olivine-hypersthene chondrite (L6), veined.
After detonations, a stone of 7kg was seen to fall near Blaauw-Kapel, 3 miles east of Utrecht and three days later a second stone of 2.7kg was found at Loevenhoutje, 2 miles away, R. von Rees, Ann. Phys. Chem. (Poggendorff), 1843, **59**, p.348, E.H. von Baumhauer, Ann. Phys. Chem. (Poggendorff), 1845, **66**, p.465. Olivine Fa_{24}, B. Mason, Geochimica et Cosmochimica Acta, 1963, **27**, p.1011.
6.5kg Budapest, Nat. Mus.; 2.17kg Utrecht, Univ.; 372g Vienna, Naturhist. Mus.; 106g Chicago, Field Mus. Nat. Hist.; 49g Tempe, Arizona State Univ.; 40g Washington, U.S. Nat. Mus.;
Specimen(s): [55108], 116.5g.; [65493], 56.5g.; [54643], 9.5g.

Utzenstorf 47°7′ N., 7°33′ E.
canton Bern, Switzerland
Fell 1928, August 16, 1900 hrs
Stone. Olivine-bronzite chondrite (H5).
Fell on the farm Stigli and was found 11 days later; three pieces were recovered, totalling 3422g, the largest 2764g, E. Gerber, Mitt. Natur. Gesell. Bern, 1928, p.25 [M.A. 4-418]. Description, analysis, H. Huttenlocher and T. Hügi, Mitt. Natur. Gesell. Bern, 2, 1952, **9**, p.67 [M.A. 12-102]. Further analysis, olivine $Fa_{18.3}$, 28.75 % total iron, A.W.R. Bevan and A.J. Easton, Schweiz. mineral. petrogr. Mitt., 1977, **57**, p.169.
2828g Bern, Naturhist. Mus.; 105g Washington, U.S. Nat. Mus.;
Specimen(s): [1963,943], 12g.; [1975,M.14], 24.1g.

Uvalde 29°12′ N., 99°46′ W.
Uvalde County, Texas, U.S.A.
Found 1915, approx., recognized 1938
Stone. Olivine-bronzite chondrite (H), veined.
One mass and fragments, totalling 7.5kg, were recognized as meteoritic, A.D. Nininger, Pop. Astron., Northfield, Minnesota, 1939, **47**, p.214, F.C. Leonard, Univ. New Mexico Publ., Albuquerque, 1946 (meteoritics ser. no. 1), p.48. Olivine Fa_{19}, B. Mason, Geochimica et Cosmochimica Acta, 1963, **27**, p.1011.
8.2kg Fort Worth, Texas, Monnig Colln.; 5.4g New York, Amer. Mus. Nat. Hist.;

Uwet 5°17′ N., 8°15′ E.
Calabar, Cross, Nigeria
Found 1903, known before this year
Iron. Hexahedrite (IIA).
A mass of about 120lb, said to have fallen about 80 years before 1907, was preserved by the natives; in 1908 a piece of about 20lb was sawn off and presented to the Brit. Mus. (Nat. Hist.). Description, with an analysis, 5.78 %Ni, G.T. Prior, Min. Mag., 1914, **17**, p.127. Further analysis, 5.61 % Ni, 62.3 ppm.Ga, 182 ppm.Ge, 2.6 ppm.Ir, J.T. Wasson, Geochimica et Cosmochimica Acta, 1969, **33**, p.859. Description; weathered, so did not fall in 1825; reheated cosmically, V.F. Buchwald, Iron Meteorites, Univ. of California, 1975, p.1255. Cooling rate, E. Randich and J. Goldstein, Geochimica et Cosmochimica Acta, 1978, **42**, p.221.
22.4kg Heidelberg, Max-Planck-Inst.; 1.1kg Washington, U.S. Nat. Mus.; 668g Moscow, Acad. Sci.; 348g Chicago, Field Mus. Nat. Hist.; 152g Harvard Univ.;
Specimen(s): [1908,171], 5038g. four pieces, the largest 2200g

Uwharrie 35°31′ N., 79°58′ W.
Randolph County, North Carolina, U.S.A.
Found 1930
Iron. Octahedrite, medium (1.1mm) (IIIA).
A mass of 72.7kg was found, H.H. Nininger, Our Stone Pelted Planet, 1933, E.P. Henderson, letter of 3 June, 1939 in Min. Dept. BM(NH). Classification and analysis, 7.83 % Ni, 20.6 ppm.Ga, 39.0 ppm.Ge, 3.6 ppm.Ir, E.R.D. Scott et al., Geochimica et Cosmochimica Acta, 1973, **37**, p.1957. Co-ordinates; shocked and annealed; may include Guilford County (*q.v.*)., V.F. Buchwald, Iron Meteorites, Univ. of California, 1975, p.1259.
71kg Raleigh, North Carolina State Mus.; 1287g Washington, U.S. Nat. Mus.;

Vaalbult 29°45′ S., 22°30′ E. approx.
Prieska division, Cape Province, South Africa
Found 1921, before this year
Iron. Octahedrite, coarse (2.0mm) (IA).
A mass of about 26lb was found, L. Peringuey, letter of 11 February, 1921 in Min. Dept. BM(NH). Description, with an analysis, 6.99 %Ni, G.T. Prior, Min. Mag., 1926, **21**, p.188. Further analysis, 6.84 %Ni, 84.3 ppm.Ga, 323 ppm.Ge, 1.7 ppm.Ir, J.T. Wasson, Meteorites, Springer-Verlag, 1974, p.298. Description; may belong to the same fall as Deelfontein (*q.v.*), V.F. Buchwald, Iron Meteorites, Univ. of California, 1975, p.1260.
11kg Cape Town, South African Mus., main mass;
Specimen(s): [1921,274], 157g.

Vaca Muerta 25°45′ S., 70°30′ W. approx.
Taltal, Atacama, Chile
Found 1861, recognized in this year
Synonym(s): *Bomba, Cachinal, Carrisalillo, Carrizalillo, Cerro la Bomba, Chañaral, Chile, Doña Inez, Harvard University, Huanilla, Inca, Janacera Pass, Jarquera, Llano del Inca, Mejillones, Quebrada de Vaca Muerta, San Pedro, San Pedro de Atacama, Sierra de Chaco, Taltal, Vegas i Carrisalillo, Vegas y Carrisalillo*
Stony-iron. Mesosiderite (MES).
Large masses up to 25kg in weight were found before 1864, I. Domeyko, Anal. Univ. Chile, Santiago, 1864, **25**, p.289, I. Domeyko, C. R. Acad. Sci. Paris, 1875, **81**, p.599. In 1888 precisely similar specimens were found on the Llano del Inca, (small pieces totalling 27lb), and at Cerro de Doña Inez, (pieces totalling 16lb), both about 100 miles SE. of Taltal, E.E. Howell, Proc. Rochester Acad. Sci., 1890, **1**, p.93. Discussion of locality, L. Fletcher, Min. Mag., 1889, **8**, p.234. Description, analysis, 6.99 %Ni, G.T. Prior, Min. Mag., 1918, **18**, p.152. For details of the mass examined by Goldschmidt, see Corrizatillo (*q.v.*) which is an iron and distinct from Vaca Muerta. Mean composition of the metal, 8.8 %Ni, 9.6 ppm.Ga, 42.8 ppm.Ge, 2.2 pm.Ir, J.T. Wasson, Geochimica et Cosmochimica Acta, 1974, **38**, p.135. Petrology, mineralogy of metal and silicates, B.N. Powell, Geochimica et Cosmochimica Acta, 1969, **33**, p.789, B.N. Powell, Geochimica et Cosmochimica Acta, 1971, **35**, p.5. Includes Harvard University, R. Bild and J.T. Wasson, Min. Mag., 1976, **40**, p.732. Olivine coronas, classification, references, C.E. Nehru et al., Geochimica et Cosmochimica Acta, 1980, **44**, p.1103.
15.2kg Paris, Mus. d'Hist. Nat.; 8.5kg New York, Amer. Mus. Nat. Hist.; 3.8kg Vienna, Naturhist. Mus.; 2.3kg Paris, Ecole des Mines; 2kg Chicago, Field Mus. Nat. Hist.; 1.4kg Washington, U.S. Nat. Mus.; 503g Harvard Univ.; 700g Tübingen, Univ.; 600g Berlin, Humboldt Univ.; 500g Budapest, Nat. Mus.; 250g Prague, Nat. Mus.;
Specimen(s): [54721], 4355g. and fragments, 8g, of Vaca Muerta; [46200], 1845g. and fragments, 263g, of Vaca Muerta; [52234], 530g. and fragments, 27.5g of Vaca Muerta; [1959,175], 9.5g.; [66203], 46g. of Doña Inez; [66204], 965g. of Doña Inez; [66205-66209, 66211, 66212], 322g. seven pieces, of Llano del Inca; [35124], Thin section; [96280], Thin section

Vago 45°25′ N., 11°8′ E.
Verona, Veneto, Italy
Fell 1688, June 21, at night hrs
Synonym(s): *Caldiero, Verona*
Stone. Olivine-bronzite chondrite (H6).
After detonations and the appearance of a fireball, a shower of stones fell, two of which are said to have weighed about 136kg and 91kg respectively, but only a few grams have been preserved, E.F.F. Chladni, Die Feuer-Meteore, Wien, 1819, p.233. Of the three fragments referred to Vago in the Nat. Hist. Mus., Paris, in 1890, one was a eucrite, one (the largest, 9g) an intermediate chondrite and one, 7g, a spherical chondrite, E.A. Wülfing, Die Meteoriten in Samml., Tübingen, 1897, p.375. Two pieces, 26g and 7g respecitvely, in the Vienna collection in 1921, obtained by A. Brezina from the collection of Count Miniscalchi, are spherical chondrites, these probably came from the original Vago, for the piece of Vago originally in the Moscaldo Mus., Verona passed into the hands of the Miniscalchi family, F. Koechlin, letter of 23 June, 1922 in Min. Dept. BM(NH). Listed, B. Baldanza, Min. Mag., 1965, **35**, p.214. Olivine Fa_{19}, B. Mason, Geochimica et Cosmochimica Acta, 1963, **27**, p.1011.
26g Vienna, Naturhist. Mus.; 7g Paris, Mus. d'Hist. Nat.;
Specimen(s): [1922,241], 6.75g. one of the Vienna specimens

Vajda-Kamaras v Mocs.

Valdavur 11°59′ N., 79°45′ E.
South Arcot district, Tamil Nadu, India
Fell 1944, October 30, 1700-1710 hrs
Synonym(s): *Valudavur*
Stone. Olivine-bronzite chondrite (H).
A stone of 2799g fell, M.A.R. Khan, Hyderabad Acad. Studies, 1950, **12** [M.A. 11-441], P. Chatterjee, letter of March 25, 1952 in Min. Dept. BM(NH). Analysis, A.K. Dey, Rec. Geol. Surv. India, 1959, **86**, p.447 [M.A. 16-171]. Olivine Fa_{20}, B. Mason, Geochimica et Cosmochimica Acta, 1963, **27**, p.1011.
2.5kg Calcutta, Mus. Geol. Surv. India;
Specimen(s): [1974,M.18], 26.05g. crusted fragment

Val di Nizza v Valdinizza.

Valdinizza 44°52′ N., 9°9′ E.
Pavia, Lombardy, Italy
Fell 1903, July 12, 1000 hrs
Synonym(s): *Pavia, San Albano, Sant Albano, Val di Nizza, Valnizza, Varzi*
Stone. Olivine-hypersthene chondrite (L6).
Two stones, of 872.5g and 131.5g, were seen to fall. Description, R. Meli, Boll. Soc. Geol. Ital., 1908, **27**, p.cxxxv. Listed, B. Baldanza, Min. Mag., 1965, **35**, p.214. Olivine Fa_{24}, B. Mason, Geochimica et Cosmochimica Acta, 1963, **27**, p.1011. Description, analysis, 20.83 % total iron, G.R. Levi-Donati and E. Jarosewich, Meteoritics, 1971, **6**, p.1.
850g Washington, U.S. Nat. Mus.; 122g Milan, Mus. Civico di Storia Nat.;
Specimen(s): [1909,31], 2.5g.

Valdinoce 44°4′ N., 12°6′ E.
Forli, Romagna, Italy
Fell 1496, January 26 or 28, 0900 hrs
Synonym(s): *Cesena, Forli*
Stone.
At least five stones fell; the date and time are given as 1497, 26 January, 0400 hrs, M. Vecchiazzani, Historia di Forlimpopuli, 1647, p.188, G.R. Levi-Donati, letter of 12 June, 1972 in Min. Dept. BM(NH). The fall occurred in 1496, E. Breisach, Caterina Sforza, Univ. of Chicago, 1967, p.169, C.M. Botley, letter of 16 September, 1969 in Min. Dept. BM(NH). Nothing is now preserved, E.F.F. Chladni, Phil. Mag., 1826, **67**, p.3, 179. Listed, B. Baldanza, Min. Mag., 1965, **35**, p.214.

Valence v Alais.

Valkeala 61°3′ N., 26°50′ E.
Kymi, Finland
Found 1962, May
Stone. Olivine-hypersthene chondrite (L6).
One stone of about 4kg was found, olivine Fa_{24}, B. Mason, Geochimica et Cosmochimica Acta, 1963, **27**, p.1011. Analysis, 21.22 % total iron, H.B. Wiik and B. Mason, Geologi (Helsinki), 1964, **7**, p.95.
3825g Helsinki, Univ.; 16g Mainz, Max-Planck-Inst.; 9.1g Los Angeles, Univ. of California; 9.4g New York, Amer. Mus. Nat. Hist.;

Valle de Allende v Chupaderos.

Valle de Allende v Morito.

Valle de San Bartolome v Chupaderos.

Valley Wells 35°28′ N., 115°40′ W.
San Bernardino County, California, U.S.A.
Found 1929, June 10
Synonym(s): *Windmill Station*
Stone. Olivine-hypersthene chondrite (L6).
Four weathered fragments, 81.9g, 24.0g, 13.5g and 10.5g were found, C.A. Reeds, Bull. Amer. Mus. Nat. Hist., 1937, **73** (6), p.517 [M.A. 7-61]. Listed, F.C. Leonard, Pop. Astron., Northfield, Minnesota, 1944, **52**, p.513, Contr. Soc. Res. Meteorites, **3**, p.174 [M.A. 9-300]. Co-ordinates, B. Mason, Meteorites, Wiley, 1962, p.229. Olivine Fa_{24}, B. Mason, Geochimica et Cosmochimica Acta, 1963, **27**, p.1011.
78g New York, Amer. Mus. Nat. Hist.; 11g Washington, U.S. Nat. Mus.;
Specimen(s): [1963,531], 21.5g.

Valnizza v Valdinizza.

Valudavur v Valdavur.

Vanovara v Tunguska.

Varano v Borgo San Donino.

Varano de' Marchesi v Borgo San Donino.

Varas v Serrania de Varas.

Varazdin v Milena.

Varea v Barea.

Varpaisjairi v Varpaisjärvi.

Varpaisjärvi 63°18′ N., 27°44′ E.
Kuopio province, Finland
Found 1913
Synonym(s): *Varpaisjairi*
Stone. Olivine-hypersthene chondrite (L6).
A stone of about 1.5 or 2kg was found in Lukkarila village, Varpaisjärvi parish, but was later broken up and mostly lost. A fragment of 118g was recovered for Helsingfors Mus. Description, with an analysis, 21.85 % total iron, W. Wahl and H.B. Wiik, C. R. Soc. Geol. Finlande, 1950, **23**, p.5 [M.A. 11-267]. Olivine Fa_{25}, B. Mason, Geochimica et Cosmochimica Acta, 1963, **27**, p.1011.
Main mass, Helsingfors, Univ.;
Specimen(s): [1951,117], 3.6g. fragments

Varsava v Pułtusk.

Varvik v Näs.

Varzi v Valdinizza.

Vaucluse v Apt.

Vavilovka 46°9′ N., 32°50′ E.
Kherson, Ukraine, USSR, [Вавиловка]
Fell 1876, June 19, 1400 hrs
Synonym(s): *Cherson, Kherson, Maksimovka, Wawilowka*
Stone. Olivine-hypersthene chondrite, amphoterite (LL6).
After detonations, a stone of about 16kg fell about 2 miles from Vavilovka, and was broken up by peasants. Description, with an analysis, R. Prendel, Mem. Soc. Nation. Sci. Natur. Cherbourg, 1877-8, **21**, p.203. Further analysis, 19.91 % total iron, M.I. D'yakonova and V.Ya. Kharitonova, Метеоритика, 1961, **21**, p.52. Olivine $Fa_{30.6}$, K. Fredriksson et al., Origin and Distribution of the Elements, ed. L.H. Ahrens, Pergamon, 1968, p.457. 1932g were preserved, L.G. Kvasha and A.Ya. Skripnik, Метеоритика, 1978, **37**, p.185.
1145g Moscow, Acad. Sci; 90g Budapest, Nat. Mus.; 115g Chicago, Field Mus. Nat. Hist.; 46g Berlin, Humboldt Univ.; 25g New York, Amer. Mus. Nat. Hist.;
Specimen(s): [77498], 8.5g.; [54633], 1.75g.

Vegas i Carrisalillo v Vaca Muerta.

Vegas y Carrisalillo v Vaca Muerta.

Velikoi-Ustyug 60°47′ N., 46°22′ E.
Vologda, Federated SSR, USSR, [Великий Устюг]
Fell 1250, between 1250 and 1350
Doubtful..
A shower of stones is said to have fallen, A. Stoikovitz, Ann. Physik (Gilbert), 1809, **31**, p.306 but the evidence is not conclusive. Very doubtful, Meteor. Bull., 1959 (13).

Velikonikolaevskii Priisk v Veliko-Nikolaevsky Priisk.

Veliko-Nikolaevsky Priisk 53°50′ N., 97°20′ E.
Irkutskaya, Federated SSR, USSR, [Велико-николаевскиинь Прииск]
Found 1902, known in this year
Synonym(s): *Birüssa, Biryusa, Khorma, Velikonikolaevskii Priisk, Velikonikolajevskyi Prüsk*
Iron. Octahedrite, medium (1.2mm) (IIIA).
The mass, formerly called Biryusa, is now re-named Veliko-Nikolaevsky Priisk. A mass of 24.267kg in the Acad. Sci., Moscow, was labelled Biryusa; it had been mentioned by Y.A. Markov in 1902 as coming from the Veliko-Nikolaevsky mine on the Khoma river, a tributary of the Biryusa river, L.A. Kulik, Метеоритика, 1941, **1**, p.73 [M.A. 9-294]. The mass was found in 1902, E.L. Krinov, Астрон. Журнал, 1945, **22**, p.303 [M.A. 9-297]. Analysis, 8.47 %Ni, M.I. D'yakonova, Метеоритика, 1958, **16**, p.180. Classification, further analysis, 8.75 %Ni, 21.5 ppm.Ga, 47.4 ppm.Ge, 0.62 ppm.Ir, J.T. Wasson, Meteorites, Springer-Verlag, 1974, p.303. Illustrated description, V.F. Buchwald, Iron Meteorites, Univ. of California, 1975, p.1261.
20.5kg Moscow, Acad. Sci., includes main mass, 17.64kg;
Specime·(s): [1971,202], 68.08g. part-slice

Velikonikolajevskyi Prüsk v Veliko-Nikolaevsky Priisk.

Velka v Pribram.

Velka-Borové v Nagy-Borové.

Velka-Divina v Gross-Divina.

Velke-Borove v Nagy-Borové.

Venagas v Charcas.

Vengerovo 56°8′ N., 77°16′ E.
Novosibirsk, Federated SSR, USSR, [Венгерово]
Fell 1950, October 11, 1849 hrs
Stone. Olivine-bronzite chondrite (H5).
Several stones fell, but only two, 9.3kg and 1.5kg, were found. Short description, with an analysis, I.A. Yudin, Метеоритика, 1954, **11**, p.89 [M.A. 13-50]. Further description, analysis, 27.54 % total iron, M.I. D'yakonova, Метеоритика, 1972, **31**, p.119. Olivine Fa$_{19}$, B. Mason, Geochimica et Cosmochimica Acta, 1963, **27**, p.1011.
9.5kg Moscow, Acad. Sci.; 34g Washington, U.S. Nat. Mus.;
Specimen(s): [1971,203], 20.4g.

Ventura 34°15′ N., 119°18′ W. approx.
Ventura County, California, U.S.A.
Found 1953
Iron. Octahedrite, fine (0.4mm) (IRANOM).
A 7.7kg mass was found near the high tide level on the rocky beach adjacent to Ventura, J.T. Wasson, letter of 25 May, 1976 in Min. Dept. BM(NH). Classification and analysis, 10.07 %Ni, 14.0 ppm.Ga, 25.0 ppm.Ge, 0.18 ppm.Ir, E.R.D. Scott et al., Geochimica et Cosmochimica Acta, 1973, **37**, p.1957.
Main mass, Los Angeles, Griffith Observatory;

Venus 32°24′ N., 97°5′ W.
Johnson County, Texas, U.S.A.
Found 1960
Stone. Olivine-bronzite chondrite (H).
A single mass of 1013g was found 1.5 miles S. of Venus, 200 yards inside the Johnson County line, during cotton harvesting, G.I Huss, priv. comm., 1983.
Main mass, in the possession of the finder; 89g Fort Worth, Texas, Monnig Colln.;

Vera 29°55′ S., 60°17′ W.
Dept. Vera, Santa Fe, Argentina
Found 1941
Synonym(s): *Calchaqui*
Stone. Olivine-hypersthene chondrite (L4).
A mass of 80kg was found 15 km NE. of the railway station Calchaqui, and was broken up into several parts, F. Carnevali, Revista Min. Geol., 1953, **21** (1), p.29 [M.A. 12-612], Meteor. Bull., 1958 (8), Meteor. Bull., 1961 (21), L.O. Giacomelli, letter of 26 May, 1959 in Min. Dept. BM(NH). Olivine Fa$_{24}$, B. Mason, Geochimica et Cosmochimica Acta, 1963, **27**, p.1011.
4.4kg L.O. Giacomelli's collection; 8kg Santa Fé, Fac. Cienc. Quim.; 50g Washington, U.S. Nat. Mus.; 20g Vienna, Naturhist. Mus.; 7g Tempe, Arizona State Univ.; 6g Chicago, Field Mus. Nat. Hist.;
Specimen(s): [1959,537], 29g.; [1959,71], 7.9g.; [1964,72], 40g.

Veramin 35°20′ N., 51°38′ E.
Karand, Tehran, Iran
Fell 1880, May, three hours before sunset
Synonym(s): *Karand, Teheran*
Stony-iron. Mesosiderite (MES).
After the appearance of a cloud and detonations, a mass of about 54kg was seen to fall "on the 8th of Jamadi-ul-aval A.H. 1298" (i.e. A.D. 1880), H.A. Ward, Am. J. Sci., 1901, **12**, p.453. Description, A. Brezina, Sitzungsber. Akad. Wiss. Wien, Math.-naturwiss. Kl., 1882, **84**, p.277. Analysis of metal, 7.32 %Ni, 14.8 ppm.Ga, 56 ppm.Ge, 3.2 ppm.Ir, J.T. Wasson et al., Geochimica et Cosmochimica Acta, 1974, **38**, p.135. Petrology, mineralogy of metal and silicates, B.N. Powell, Geochimica et Cosmochimica Acta, 1969, **33**, p.789, B.N. Powell, Geochimica et Cosmochimica Acta, 1971, **35**, p.5. Olivine coronas, classification, references, C.E. Nehru et al., Geochimica et Cosmochimica Acta, 1980, **44**, p.1103.
Main mass, Tehran,; 775g Chicago, Field Mus. Nat. Hist.; 125g Harvard Univ.; 88g New York, Amer. Mus. Nat. Hist.; 64g Vienna, Naturhist. Mus.; 52g Washington, U.S. Nat. Mus.;
Specimen(s): [54609], 53.5g.; [84288], 181g. slice

Verchne Cschirskaja v Verkhne Tschirskaia.

Verchneudinsk v Verkhne Udinsk.

Verchne-Udinsk v Verkhne Udinsk.

Veresegyhaza v Ohaba.

Verissimo 19°44′ S., 48°19′ W. approx.
Goias, Brazil
Found
Iron. Octahedrite, medium (0.9mm) (IIIA).
Listed, C.B. Gomes and K. Keil, Rev. Cienc. Cult., 1977, **29**, p.1094. A specimen of this name from the Mus. Nac. Brasil has been analysed, 7.54 %Ni, 18.3 ppm.Ga, 34.9 ppm.Ge, 14 ppm.Ir, A. Kracher et al., Geochimica et Cosmochimica Acta, 1980, **44**, p.773. Description, M.E.F. Vieira, Thesis, Univ. Fed. Fluminense, 1980.
Main mass, privately held; Specimen, Rio de Janeiro, Mus. Nac. Brasil.;

Verkhne Chirskaia v Verkhne Tschirskaia.

Verkhne Chirskaya v Verkhne Tschirskaia.

Verkhne Dnieprovsk 48°38′ N., 34°22′ E.
Ekaterinoslav, Ukraine, USSR, [ерхне Днепровск]
Found 1876
Synonym(s): *Ekaterinoslav, Werchne Dnieprowsk, Werkhne Dnieprowsk*
Iron. Octahedrite, finest (0.05mm) (IIE).
Original weight and details not known. Mentioned in 1882 in Catalog of Y.I. Simashko's Meteorite Collection as found in 1876. Description, E. Cohen, Meteoritenkunde, 1905, **3**, p.264. Not identical to Augustinovka, metallographic study, A.W.R. Bevan et al., Min. Mag., 1979, **43**, p.149. Classification, analysis, 8.78 %Ni, 22.8 ppm.Ga, 70.4 ppm.Ge, 6.1 ppm.Ir, J.T. Wasson, Meteorites, Springer-Verlag, 1974, p.301. The BM(NH) specimen is apparently the only genuine material. The 97g specimen in Chicago, Field Mus. Nat. Hist., is of Augustinovka, A.A. Yavnel, Метеоритика, 1976, **35**, p.111.
Specimen(s): [51183], 23.5g. presented by Prof. Kulibin in 1877

Verkhne Tschirskaia 48°25′ N., 43°12′ E.
Volgograd, Federated SSR, USSR, [Верхне Чирская]
Fell 1843, November 12, 1200 hrs
Synonym(s): *Verkhne Chirskaia, Verchne Cschirskaja, Verkhne Chirskaya, Werchne Tschirskaja, Werkhne Tschirskaia*
Stone. Olivine-bronzite chondrite (H5), veined.
After detonations, a stone of about 8kg was found, -. Borissiak, Bull. Acad. Sci. St.-Petersbourg, 1847, **5**, p.196. Description, analysis, 27.56 % total iron, A.Ya. Skripnik et al., Метеоритика, 1977, **36**, p.59. Olivine Fa_{18}, B. Mason, Geochimica et Cosmochimica Acta, 1963, **27**, p.1011.
7.4kg Kharkov, Mus., main mass; 110g Moscow, Acad. Sci.; 94g Vienna, Naturhist. Mus.; 15g Paris, Mus. d'Hist. Nat.; 14g Chicago, Field Mus. Nat. Hist.;
Specimen(s): [1927,10], 0.25g.

Verkhne Udinsk 54°46′ N., 113°59′ E.
Transbaikal, Federated SSR, USSR, [Верхне Удинск], [Ниро]
Found 1854
Synonym(s): *Niro, Verchneudinsk, Verchne-Udinsk, Vitim, Werkhne Udinsk, Werkne Udinsk, Witim*
Iron. Octahedrite, medium (1.1mm) (IIIA).
A mass of over 18kg was found on the the banks of the river Niro, a tributary of the Vitim, G. Rose, Z. Deutsch. Geol. Ges., 1864, **16**, p.355. Analysis, 7.82 %Ni, M.I. D'yakonova, Метеоритика, 1958, **16**, p.180. Has been heated to 500-800C, V.F. Buchwald, Iron Meteorites, Univ. of California, 1975, p.1264. Listed, under Niro, L.G. Kvasha and A.Ya. Skripnik, Метеоритика, 1978, **37**, p.178. Further analysis, 7.46 %Ni, 19.0 ppm.Ga, 39.8 ppm.Ge, 3.3 ppm.Ir, E.R.D. Scott et al., Geochimica et Cosmochimica Acta, 1973, **37**, p.1957.
2058g Moscow, Acad. Sci.; 1.5kg Stockholm, Riksmus; 727g Chicago, Field Mus. Nat. Hist.; 600g Vienna, Naturhist. Mus.; 563g Berlin, Humboldt Univ.; 396g Tempe, Arizona State Univ.; 330g Paris, Mus. d'Hist. Nat.; 208g New York, Amer. Mus. Nat. Hist.; 133g Tübingen, Univ.;
Specimen(s): [54665], 2233g. main mass; [36012], 640g.

Vernon County 43°30′ N., 91°10′ W. approx.
Vernon County, Wisconsin, U.S.A.
Fell 1865, March 26, 0900 hrs
Synonym(s): *Claywater, Claywater Stone, Wisconsin*
Stone. Olivine-bronzite chondrite (H6), veined.
After the appearance of a fireball and detonations, two stones, 800g and 700g respectively, fell and were found five days later, the first being subsequently lost. Description, with an analysis, J.L. Smith, Am. J. Sci., 1876, **12**, p.207. Olivine Fa_{19}, B. Mason, Geochimica et Cosmochimica Acta, 1963, **27**, p.1011.
123g Harvard Univ.; 65g Paris, Mus. d'Hist. Nat.; 28g Vienna, Naturhist. Mus.; 22g Chicago, Field Mus. Nat. Hist.; 18g Copenhagen, Univ. Geol. Mus.; 9.3g Washington, U.S. Nat. Mus.; 8g Berlin, Humboldt Univ.;
Specimen(s): [50806], 37g.

Verolanuova v Alfianello.

Verona v Vago.

Veseli nad Moravou v Wessely.

Vetluga 57°48′ N., 45°48′ E.
Kostroma district, Federated SSR, USSR, [Ветлуга]
Fell 1949, February 27, day time hrs
Stone. Achondrite, Ca-rich. Eucrite (AEUC).
A single mass of 750g fell in a forest 80 metres from the southern shore of Bobrovo lake, 7 km from the town of Vetluga, Meteor. Bull., 1978 (55), Meteoritics, 1978, **13**, p.350, L.G. Kvasha and A.Ya. Skripnik, Метеоритика, 1978, **37**, p.186. Mineralogy, feldspar An80 (optical determination), A.Ya. Skripnik, Метеоритика, 1980, **39**, p.43. Analysis, V.Ya. Kharitonova and L.D. Barsukova, Метеоритика, 1982, **40**, p.41.
578g Moscow, Acad. Sci.;

Viasma v Kikino.

Viazma v Kikino.

Vicenice 49°13′ N., 15°48′ E.
Jihomoravsky, Czechoslovakia
Found 1911
Iron. Octahedrite, medium.
A mass of 4.37kg was found 0.8m deep in clay, Meteor. Bull., 1964 (31).
4.37kg Brno, Univ.;

Victoria v Cranbourne.

Victoria v Hinojal.

Victoria v Iron Creek.

Victoria West 31°42′ S., 23°45′ E.
Cape Province, South Africa
Found 1860
Iron. Octahedrite, fine (0.2mm) (IRANOM).
A mass of 6.5lb was seen to fall by a Hottentot on a farm 30 miles SW. of Victoria West, G.A. Maeder, copy of letter of 1 February, 1905 in Min. Dept. BM(NH)., J.R. Gregory, Geol. Mag., 1868, **5**, p.532. Description, analysis, 10.14 %Ni, J.L. Smith, Am. J. Sci., 1873, **5**, p.107. Not an observed fall; the mass is corroded and has been heated to about 900C, V.F. Buchwald, Iron Meteorites, Univ. of California, 1975, p.1265. Chemically anomalous, analysis, 11.8 %Ni, 15.3 ppm.Ga, 31.4 ppm.Ge, 0.022 ppm.Ir, E.R.D. Scott et al., Geochimica et Cosmochimica Acta, 1973, **37**, p.1957.
Half the mass, Cape Town, South African Mus.; 250g Calcutta, Mus. Geol. Surv. India; 227g London, Inst. Geol. Sci.; 172g Vienna, Naturhist. Mus.; 83g Paris, Mus. d'Hist. Nat.; 46g Berlin, Humboldt Univ.; 44g Harvard Univ.; 34g Washington, U.S. Nat. Mus.; 17g Chicago, Field Mus. Nat. Hist.;
Specimen(s): [46007], 135g.; [94703], 7.7g.

Vidin v Virba.

View Hill 43°19′12″ S., 172°3′48″ E.
Oxford, Canterbury, New Zealand
Found 1952, approx.
Iron. Octahedrite, medium (0.8mm) (IIIA).
A single mass of 33.6kg was found. Co-ordinates and full, illustrated description, M.J. Frost, Rec. Canterbury Mus., 1967, **8**, p.255 [M.A. 19-43]. Analysis, 8.87 %Ni, 21.0 ppm.Ga, 42.6 ppm.Ge, 0.27 ppm.Ir, E.R.D. Scott et al., Geochimica et Cosmochimica Acta, 1973, **37**, p.1957. Description; insignificantly weathered, V.F. Buchwald, Iron

Meteorites, Univ. of California, 1975, p.1267.
Main mass, Canterbury, N.Z., Mus.; 1.8kg Washington, U.S. Nat. Mus.; 56g Chicago, Field Mus. Nat. Hist.; 52g New York, Amer. Mus. Nat. Hist.;
Specimen(s): [1968,208], 73g. slice

Vigarno v Vigarano.

Vigarano 44°51′ N., 11°24′ E.
Ferrara, Emilia, Italy
Fell 1910, January 22, 2130 hrs
Synonym(s): *Cariani, Ferrana, Mainardi, Morandi, Parish, Pieve, Vigarano Mainarda, Vigarano Pieve, Vigarno*
Stone. Carbonaceous chondrite, type III (CV3).
After detonations, a stone of about 11.5kg was seen to fall, and a month later a second stone of 4.5kg was found; description, A. Rosati, Atti R. Accad. Lincei, Roma, 1910, **19** (1), p.841, A. Rosati, Atti R. Accad. Lincei, Roma, 1910, **19** (2), p.25. Listed, B. Baldanza, Min. Mag., 1965, **35**, p.214. Analysis, 24.71 % total iron, B. Mason, Space Sci. Rev., 1963, **1**, p.621 [M.A. 16-640]. Partial analysis, 22.24 % total iron, T.S. McCarthy and L.H. Ahrens, Earth planet. Sci. Lett., 1972, **14**, p.97. The type of a sub-type of carbonaceous chondrite, W.R. Van Schmus and J.M. Hayes, Geochimica et Cosmochimica Acta, 1974, **38**, p.47. X-ray diffraction study, mainly of magnetite, F. Burragato, Periodico Miner., 1969, **38**, p.327. Chemical mineralogy of a melilite spinel chondrule, M. Christophe-Michel-Levy, Bull. Soc. franc. Min. Crist., 1968, **91**, p.212. Olivine composition, R. Hutchison and R.F. Symes, Meteoritics, 1972, **7**, p.23. Trace element data, G.W. Kallemeyn and J.T. Wasson, Geochimica et Cosmochimica Acta, 1981, **45**, p.1217.
1.4kg Yale Univ.; 1kg Prague, Nat. Mus.; 2kg Washington, U.S. Nat. Mus.; 293g New York, Amer. Mus. Nat. Hist.; 290g Budapest, Nat. Mus.; 220g Paris, Mus. d'Hist. Nat.; 157g Vienna, Naturhist. Mus.; 93g Chicago, Field Mus. Nat. Hist.;
Specimen(s): [1924,15], 784g.; [1911,174], 107g. and fragments, 1g; [1920,347], 50g.

Vigarano Mainarda v Vigarano.

Vigarano Pieve v Vigarano.

Vigarvano error for Vigarano.

Vigavano error for Vigarano.

Vignabona v Borgo San Donino.

Vigo Park 34°41′ N., 101°23′ W.
Swisher County, Texas, U.S.A.
Found 1934
Stone. Olivine-hypersthene chondrite (L4).
A fragment of 35g, from an original weighing about 1.3kg was recognized as meteoritic in 1934, A.D. Nininger, Pop. Astron., Northfield, Minnesota, 1939, **47**, p.214, B. Mason, Meteorites, Wiley, 1962, p.244. Another mass weighing 274g was found 3 km NE. of the original find-site, olivine $Fa_{24.1}$, P.S. Sipiera et al., Meteoritics, 1983, **18**, p.63. Remains of the original mass lost, classification given is that of the 274g mass.
35g Fort Worth, Texas, colln. of O.E. Monnig, the fragment found in 1934;

Vilabella v Nulles.

Vilabella v Trenzano.

Vilanova de Sitjes v Cañellas.

Villa Coronado 26°45′ N., 105°15′ W.
Chihuahua, Mexico
Found 1983
Stone. Olivine-bronzite chondrite (H5).
Three fragments totalling 2.9kg were found, W. Zeitschel, letter of 12 September, 1983 in Min. Dept. BM(NH). Olivine Fa_{18}, A.L. Graham, priv. comm., 1983.
Specimen(s): [1983,M.6], 600g.

Villa di Cella v Borgo San Donino.

Villa Lujan v Lujan.

Villanova v Cañellas.

Villanova v Motta di Conti.

Villanova di Casale v Motta di Conti.

Villa Nueva v Cañellas.

Villa Nueva v Sierra Blanca.

Villanueva del Fresno 38°23′ N., 7°10′ W.
Badojoz, Spain
Pseudometeorite..
A fragment of 132g, described as meteoritic, is clearly manufactured iron, A. Baselga y Recarte, Notas Comun. Inst. Geol. Min. España, 1953 (30), p.35 [M.A. 14-131].

Villanuova v Motta di Conti.

Villanuova di Casale v Motta di Conti.

Villarrica 25°50′ S., 56°30′ W.
Paraguay
Fell 1925, July 20, 1900 hrs
Stone.
A stone measuring 12×7×5.5 cm fell, V. Lolise, letter of 3 August, 1925 in Min. Dept. BM(NH).

Villedieu 47°55′ N., 4°21′ E.
Molesme, Cote-d'Or, France
Found 1890, approx.
Stone. Olivine-bronzite chondrite (H4).
A much oxidized stone of about 14kg was found. Description, A. Lacroix, C. R. Acad. Sci. Paris, 1926, **182**, p.1498. Olivine Fa_{19}, B. Mason, Geochimica et Cosmochimica Acta, 1963, **27**, p.1011.
12.8kg Paris, Mus. d'Hist. Nat.;

Villefranche v Salles.

Villeneuve v Motta di Conti.

Vilna v Zabrodje.

Vilna 54°13′30″ N., 111°41′30″ W.
Alberta, Canada
Fell 1967, February 5, 1855 hrs
Synonym(s): *Lac Labiche*
Stone. Olivine-hypersthene chondrite (L5).
Two small crusted fragments, 48mg and 94mg respectively, were recovered from snow on a small lake 15 km NE. of Vilna, after a bright bolide had been seen. The path of the parent bolide was well defined in a photograph by the auroral all-sky camera operated by the Dominion Observatory at Meanook, Alberta. Full description and mineralogy, olivine Fa_{25}, D.G.W. Smith et al., Meteoritics, 1973, **8**, p.197.
Both fragments, Edmonton, Univ. of Alberta;

Vincent 35°1′ S., 139°55′ E.
County Buccleuch, South Australia, Australia
Found 1926
Stone. Olivine-hypersthene chondrite (L5).
A single mass of 430g was found, approximately 6 miles NNE. of Karoonda, D.W.P. Corbett, Rec. S. Austr. Mus., 1968, **15**, p.767. Olivine $Fa_{24.2}$, B. Mason, Rec. Austr. Mus., 1974, **29**, p.169. Analysis, 21.45 % total iron, M.J. Fitzgerald, Ph.D. Thesis, Univ. of Adelaide, 1979, p.23.
318g Adelaide, South Austr. Mus.; 1.4g New York, Amer. Mus. Nat. Hist.; Thin section, Washington, U.S. Nat. Mus.;

Vinh Luoc v Tuan Tuc.

Virba 43°32′ N., 22°38′ E.
Vidin, Bulgaria
Fell 1873, June 1
Synonym(s): *Belgradjek, Urba, Vidin, Vyrba, Warbe, Widdin, Wirba*
Stone. Olivine-hypersthene chondrite (L6), veined.
After detonations, a stone of 3.6kg fell, G.A. Daubrée, C. R. Acad. Sci. Paris, 1874, **79**, p.276. The stone stated to have fallen "June 2, 1883" in a forest at "Virba, Belgrade Dijk, S. Meunier, C. R. Acad. Sci. Paris, 1893, **117**, p.258 belongs here; a label with the BM(NH) specimen obtained from L. Eger in 1893 says that it fell in the year 1291 (Hedjir) [i.e. 1873] on Wednesday, May 20 (old style), i.e. June 1 (new style), in an oak forest, 1/4 hour from the Belgrade village of Virba. Olivine Fa_{25}, B. Mason, Geochimica et Cosmochimica Acta, 1963, **27**, p.1011.
3040g Budapest, Nat. Mus.; 104g Vienna, Naturhist. Mus.; 74g Paris, Mus. d'Hist. Nat.;
Specimen(s): [70348], 38g. three fragments

Visa v Mocs.

Vishnupur 23°6′ N., 87°26′ E.
Bankura district, West Bengal, India
Fell 1906, December 15, 0930 hrs
Stone. Olivine-hypersthene chondrite, amphoterite (LL6).
After detonations, two stones were seen to fall, one of 670g at Kheraibani village and the other of 1767g at Mathura village, 6.75 miles distant, G. de P. Cotter, Rec. Geol. Surv. India, 1912, **42**, p.266. Olivine Fa_{30}, B. Mason, Geochimica et Cosmochimica Acta, 1963, **27**, p.1011. C and rare gas content, W. Otting and J. Zähringer, Geochimica et Cosmochimica Acta, 1967, **31**, p.1949. K content, 1480 ppm., W. Kaiser and J. Zähringer, Z. Natur., 1965, **20A**, p.963.
1.6kg Calcutta, Mus. Geol. Surv. India; 53g New York, Amer. Mus. Nat. Hist.; 24g Washington, U.S. Nat. Mus.;
Specimen(s): [1983,M.17], 25.7g.

Visuni 25°27′ N., 70°0′ E.
Umarkot, Thar Parkar district, Sind, Pakistan
Fell 1915, January 19, 1200 hrs
Stone. Olivine-bronzite chondrite (H6).
After detonations, a stone of 594g fell, 15 miles ENE. of Umarkot. Description, H. Walker, Rec. Geol. Surv. India, 1916, **47**, p.273. Olivine Fa_{20}, B. Mason, Geochimica et Cosmochimica Acta, 1967, **31**, p.1100.
595g Calcutta, Mus. Geol. Surv. India; 1g New York, Amer. Mus. Nat. Hist.;
Specimen(s): [1983,M.18], 10.8g.

Viterbo 42°25′ N., 12°7′ E.
Lazio, Italy
Fell 1474
Doubtful..
Two large black stones are said to have fallen, E.F.F. Chladni, Ann. Phys. (Gilbert), 1821, **68**, p.332. The evidence is not conclusive.

Vitim v Verkhne Udinsk.

Vitré v St. Germain-du-Pinel.

Vivionnere v Le Teilleul.

Vnorovy v Wessely.

Vonnas v Luponnas.

Vorova v Angara.

Vouillé 46°38′ N., 0°10′ E.
Poitiers, Vienne, France
Fell 1831, May 13, 2300 hrs
Synonym(s): *Poitiers*
Stone. Olivine-hypersthene chondrite (L6), veined.
After the appearance of a fireball (moving from S. to N.), and detonations, a stone of about 20kg was found next day. Description, analysis, G.A. Daubrée, C. R. Acad. Sci. Paris, 1864, **58**, p.226, Anom., Ann. Chim. Phys., 1831, **47**, p.442. Contains strongly recrystallized clasts in which feldspar has been converted to maskelynite, distinct from Quincay (*q.v.*), R.A. Binns, letter of 19 October, 1966 in Min. Dept. BM (NH). Further bulk and mineral analyses, 21.52 % total iron, olivine $Fa_{23.6}$, R.T. Dodd and E. Jarosewich, Meteoritics, 1981, **16**, p.93.
14.278kg Paris, Mus. d'Hist. Nat., main mass; 500g Chicago, Field Mus. Nat. Hist.; 338g Tempe, Arizona State Univ.; 161g Washington, U.S. Nat. Mus.; 175g New York, Amer. Mus. Nat. Hist.; 116g Prague, Nat. Mus.; 90g Vienna, Naturhist. Mus.; 88g Budapest, Nat. Hist. Mus.;
Specimen(s): [35401], 54.5g.; [1920,348], 30g.; [33894], 6.5g.

Vulcan 50°31′ N., 113°8′ W.
Alberta, Canada
Found 1962, April
Stone. Olivine-bronzite chondrite (H6).
One stone of 19kg was found, Meteor. Bull., 1964 (32). Olivine Fa_{20}, B. Mason, Geochimica et Cosmochimica Acta, 1967, **31**, p.1100.
Main mass, Edmonton, Univ. of Alberta; 6.8kg Ottawa, Geol. Surv. Canada; 53g New York, Amer. Mus. Nat. Hist.;
Specimen(s): [1967,258], 880g. slice

Vyrba v Virba.

Vyskovice v Bohumilitz.

Wabar 21°29′59″ N., 50°28′20″ E.

Rub'al Khali, Saudi Arabia
Found 1863
Synonym(s): *al-Hadida, Nedzed, Nejd, Nejed, Nejed (no. 2), Wadee Banee Khaled, Wanee Banee Khaled*
Iron. Octahedrite, medium (0.9mm) (IIIA).

A mass of 131lb, said to have been seen to fall in the Wadee Banee Khaled during a thunderstorm in 1863, was in 1885 obtained from a Persian agent by the BM(NH). Description, with an analysis, 7.40 %Ni, L. Fletcher, Min. Mag., 7, p.179. In 1893 a second mass, of 137lb, much weathered, was similarly obtained by the BM(NH). Another mass of 25lb and other small pieces, 114g together, were found near a series of meteorite craters, H.St.J. Philby, The Empty Quarter, London, 1933, p.365 [M.A. 5-409]. Associated with the iron were found masses of white and black pumiceous silica-glass. The black contains both Fe and Ni, as innumerable minute metallic globules, clearly of meteoritic origin. Description, L.J. Spencer, Min. Mag., 1933, **23**, p.387. Mentioned, D.A. Holm, Am. J. Sci., 1962, **260**, p.303. Further analysis, 7.62 %Ni, 21.3 ppm.Ga, 38.4 ppm.Ge, 6.0 ppm.Ir, E.R.D. Scott et al., Geochimica et Cosmochimica Acta, 1973, **37**, p.1957. Includes Nejed and Nejed (no. 2), V.F. Buchwald, Iron Meteorites, Univ. of California, 1975, p.1273. Two masses, 2200kg found in 1965 and 200kg found in 1966 are on display in Riyadh, R.S. Clarke, Jr. et al., Meteoritics, 1981, **16**, p.303, abs.

48kg Chicago, Field Mus. Nat. Hist., main part of mass found in 1893, Nejed no. 2; 3.3kg Washington, U.S. Nat. Mus.; 1.8kg New York, Amer. Mus. Nat. Hist.;

Specimen(s): [56154], 58160g. main part of the mass obtained in 1885, and pieces, 475g and filings, 153g, of Nejed; [1932, 1136a], 9.5kg.; [1932,1136b], 475g. end-slice, and fragments, 16g; [1932,1136c], 657.5g. shale, and filings, 90g; [1932, 1137], 15.5g. polished, and filings, 1.5g; [1932,1138], 50g. fragments; [1932,1139], 224g. rust; [1932,1141-1166] black and white silica-glass from the craters; [1932,1167-1169] sandstone from near the craters

Waconda 39°20′ N., 98°10′ W.

Mitchell County, Kansas, U.S.A.
Found 1873
Synonym(s): *Mitchell County*
Stone. Olivine-hypersthene chondrite (L6), brecciated.

A stone of about 50kg was found and broken into pieces, one of which weighed 58lb, (26kg), C.U. Shepard, Am. J. Sci., 1876, **11**, p.473. Analysis, G.P. Merrill, Mem. Nat. Acad. Sci. Washington, 1919, **14** (4), p.13. Analysis doubted, no CaO reported, W. Wahl, Geochimica et Cosmochimica Acta, 1950, **1**, p.28. Olivine Fa_{25}, B. Mason, Geochimica et Cosmochimica Acta, 1963, **27**, p.1011.

26kg Tempe, Arizona State Univ.; 5.5kg Vienna, Naturhist. Mus.; 3.0kg Chicago, Field Mus. Nat. Hist.; 850g Harvard Univ.; 1.1kg Washington, U.S. Nat. Mus.; 608g Mainz, Max-Planck-Inst.; 315g Prague, Nat. Mus.; 269g New York, Amer. Mus. Nat. Hist.; 263g Copenhagen, Univ. Geol. Mus.; 139g Dublin, Nat. Mus.;

Specimen(s): [53287], 352g.; [50805], 14g.

Wadee Banee Khaled v Wabar.

Wahhe v Pillistfer.

Wagla v Ambapur Nagla.

Waingaromia 38°15′ S., 178°5′ E.

East Coast province, New Zealand
Found 1915, recognized 1925
Iron. Octahedrite, medium (0.9mm) (IIIA).

A 9.2kg mass, found at Waingaromia 23km NW. of Tolaga Bay, was presented to Canterbury Mus. in 1970, D.R. Gregg, letter of 1 September, 1970 in Min. Dept. BM(NH). Reported, Meteor. Bull., 1971 (50), Meteoritics, 1971, **6**, p.123. Classification and analysis, 9.14 %Ni, 20.9 ppm.Ga, 41.6 ppm.Ge, 0.38 ppm.Ir, A. Kracher et al., Geochimica et Cosmochimica Acta, 1980, **44**, p.773.

Main mass, Christchurch, Canterbury Mus.; 600g Washington, U.S. Nat. Mus.;

Specimen(s): [1976,M.7], 106.6g. slice

Wairarapa Valley 41°19′ S., 175°8′ E.

Wellington, New Zealand
Found 1863
Synonym(s): *Taranaki, Turakina, Wellington*
Stone. Olivine-bronzite chondrite (H5).

Found at Manaia, Waingawa River, near Masterton. Original weight probably about 13lb. Description, with an analysis, M.H. Battey, Min. Mag., 1962, **33**, p.73. Olivine Fa_{18}, B. Mason, Geochimica et Cosmochimica Acta, 1963, **27**, p.1011. Wairarapa is possibly one of the stones of the Taranaki meteorite which fell on 4 December, 1864 0200 hrs., partly in the sea and partly on land at Turakina, after loud detonations, and the appearance of a fireball of which the smoke persisted for two hours, W. von Haidinger, Sitzungsber. Akad. Wiss. Wien, Math.-naturwiss. Kl., 1865, **52** (2), p.151.

Main mass, Auckland, Mus.; 196g Chicago, Field Mus. Nat. Hist.; 102g Paris, Mus. d'Hist. Nat.;

Walaga v Nejo.

Walata v Holetta.

Waldau v L'Aigle.

Waldo 39°6′ N., 98°50′ W.

Russell County, Kansas, U.S.A.
Found 1937
Stone. Olivine-hypersthene chondrite (L6).

One stone of 1.3kg was found, A.D. Nininger, Pop. Astron., Northfield, Minnesota, 1939, **47**, p.214, H.H. and A.D. Nininger, The Nininger Collection of Meteorites, Winslow, Arizona, 1950, p.100. Olivine Fa_{25}, B. Mason, Geochimica et Cosmochimica Acta, 1963, **27**, p.1011.

487g Tempe, Arizona State Univ.; 94g Washington, U.S. Nat. Mus.; 84g Chicago, Field Mus. Nat. Hist.;

Specimen(s): [1959,903], 486.5g. slice

Waldo County v Searsmont.

Waldron Ridge 36°38′ N., 83°50′ W.

Claiborne County, Tennessee, U.S.A.
Found 1887, known in this year
Synonym(s): *Wallens Ridge*
Iron. Octahedrite, coarse (1.6mm) (IA).

A mass found at Waldron (perhaps Wallens) Ridge appears to have been of about 30lb, for a 15lb piece has been described, G.F. Kunz, Am. J. Sci., 1887, **34**, p.475. Description of a 12lb mass, analysis, 6.01 %Ni, A.R. Ledoux, Trans. New York Acad. Sci., 1889, **8**, p.187. Chemically distinct from Cosby's Creek and Greenbrier County (*q.v.*). Classification and analysis, 7.55 %Ni, 74.6 ppm.G 282 ppm.Ge, 2.0 ppm.Ir, J.T. Wasson, Icarus, 1970,

12, p.407. Some samples, perhaps the entire mass, have been heated artifically to about 600C, V.F. Buchwald, Iron Meteorites, Univ. of California, 1975, p.1275.
4.7kg Vienna, Naturhist. Mus.; 427g Chicago, Field Mus. Nat. Hist.; 369g New York, Amer. Mus. Nat. Hist.; 343g Berlin, Humboldt Univ.; 70g Washington, U.S. Nat. Mus.;
Specimen(s): [1922,160], 3196g. two pieces, and a fragment, 6g

Walker County 34° N., 87°10′ W. approx.
Alabama, U.S.A.
Found 1832
Synonym(s): *Alabama, Claiborne, Clairborne, Lime Creek, Morgan County*
Iron. Hexahedrite (IIA).
A mass of about 165lb was found in the NE. corner of Walker County, G. Troost, Am. J. Sci., 1845, **49**, p.344. Description, analysis, 5.30 %Ni, E. Cohen, Meteoritenkunde, 1905, **3**, p.166. Description, kamacite corrosion illustrated, V.F. Buchwald, Iron Meteorites, Univ. of California, 1975, p.1277. Further analysis, 5.46 %Ni, 58.2 ppm.Ga, 189 ppm.Ge, 3.0 ppm.Ir, J.T. Wasson, Meteorites, Springer-Verlag, 1974, p.299.
40kg Tübingen, Univ.; 2.3kg Harvard Univ.; 291g Washington, U.S. Nat. Mus.; 111g Chicago, Field Mus. Nat. Hist.;
Specimen(s): [16867], 22kg. slab; [35413], 14g. and fragments, 2.5g

Walker Township v Grand Rapids.

Walkringen 46°57′ N., 7°38′ E.
Bern, Switzerland
Fell 1698, May 18, 1900-2000 hrs
Synonym(s): *Waltringen*
Doubtful. Stone.
A heavy stone was seen to fall after detonations; it was sent to Bern and was later lost or destroyed, B. Studer, Ann. Phys. Chem. (Poggendorff), 1872, **146**, p.149.

Wallapai 35°48′ N., 113°42′ W. approx.
Mohave County, Arizona, U.S.A.
Found 1927
Synonym(s): *Hualapai*
Iron. Octahedrite, fine (0.4mm) (IID).
Two masses, 306kg and 124kg, somewhat weathered, were found in 1927 in the Wallapai (Hualapai) Indian Reserve, Mohave Co., Arizona. Description, with an analysis, 9.12 % Ni, G.P. Merrill, Proc. U.S. Nat. Mus., 1927, **72** (22), p.1 [M.A. 3-536]. Further analysis, 11.3 %Ni, 82.9 ppm.Ga, 98.3 ppm.Ge, 3.5 ppm.Ir, J.T. Wasson, Geochimica et Cosmochimica Acta, 1969, **33**, p.859. Description; the masses have been affected differently by shock, V.F. Buchwald, Iron Meteorites, Univ. of California, 1975, p.1280.
303kg Washington, U.S. Nat. Mus., main mass of larger mass; 123kg Tucson, Univ.; 527g Tempe, Arizona State Univ.; 499g New York, Amer. Mus. Nat. Hist.;
Specimen(s): [1959,983], 74g. slice, and sawings, 9g

Wallens Ridge v Waldron Ridge.

Walltown 37°19′30″ N., 84°43′0″ W.
Casey County, Kentucky, U.S.A.
Found 1956, or 1957, recognized 1963
Stone. Olivine-hypersthene chondrite (L6).
Several fragments, totalling about 1.6kg were recovered, Meteor. Bull., 1976 (54), Meteoritics, 1976, **11**, p.88. Description, analysis, 23.7 % total iron, W.D. Ehmann and J.R. Busche, Trans. Kentucky Acad. Sci., 1968, **29**, p.5. Olivine Fa_{23}, B. Mason, Geochimica et Cosmochimica Acta, 1967, **31**, p.1100.
1.6kg in possession of J.W. Flanigan; Fragments, Collection of W.D. Ehmann; 59.4g Perth, West. Austr. Mus.; 16.6g Tempe, Arizona State Univ.; 3.7g Washington, U.S. Nat. Mus.;
Specimen(s): [1963,941], 2.8g.; [1977,M.11], 10.3g.

Walters 34°20′ N., 98°18′ W.
Cotton County, Oklahoma, U.S.A.
Fell 1946, July 28, 1545 hrs
Synonym(s): *Walton*
Stone. Olivine-hypersthene chondrite (L6).
A mass of 62lb was found, Rep. U.S. Nat. Mus., 1947, p.64. Date of fall, description, partial analysis, S.K. Roy et al., Fieldiana Geol., 1962, **10**, p.539. Olivine Fa_{25}, B. Mason, Geochimica et Cosmochimica Acta, 1963, **27**, p.1011.
25kg Washington, U.S. Nat. Mus.; 717g Chicago, Field Mus. Nat. Hist.; 241g Tempe, Arizona State Univ.;

Walton v Walters.

Waltringen v Walkringen.

Wanee Banee Khaled v Wabar.

Warasdin v Milena.

Warbe v Virba.

Warbreccan v Tenham.

Warburton Range 26°17′ S., 126°40′ E. approx.
Western Australia, Australia
Found 1963, or 1964
Iron. Ataxite, Ni-rich (IVB).
A mass of 125.5lb was found on sand 12 miles S. of Warburton Mission. Description, with an analysis, 18.21 % Ni, G.J.H. McCall and H.B. Wiik, J. Roy. Soc. West. Austr., 1966, **49**, p.13. Structure, H.J. Axon and P.L. Smith, Min. Mag., 1972, **38**, p.736. Further analysis, 17.8 %Ni, 0.244 ppm.Ga, 0.064 ppm.Ge, 13 ppm.Ir, R. Schaudy et al., Icarus, 1972, **17**, p.174. Description, V.F. Buchwald, Iron Meteorites, Univ. of California, 1975, p.1283.
Main mass, Perth, West. Austr. Mus.; 259g New York, Amer. Mus. Nat. Hist.; 32g Los Angeles, Univ. of California;
Specimen(s): [1966,487], 237g. two pieces, and filings, 5.5g

Wardswell Draw 32°54′12″ N., 102°55′30″ W.
Gaines County, Texas, U.S.A.
Found 1976, recognized in this year
Stone. Olivine-hypersthene chondrite (L6).
A stone of 3kg was found, G.I Huss, priv. comm., 1976. Olivine Fa_{25}, A.L. Graham, priv. comm., 1978.
Specimen(s): [1977,M.4], 94.6g. crusted part-slice

Warialda v Bingera.

Warrenton 38°41′ N., 91°9′ W.
Warren County, Missouri, U.S.A.
Fell 1877, January 3, 0715 hrs
Stone. Carbonaceous chondrite, type III (CO3).
With a whistling noise a stone of about 100lb was seen to strike a tree and break into pieces. Description, with an analysis, J.L. Smith, Am. J. Sci., 1877, **14**, p.222. Further

analysis, 26.18 % total iron, W. Wahl, Min. Mag., 1950, **29**, p.419. Belongs to the Ornans sub-type of the carbonaceous chondrites, W.R. Van Schmus and J.M. Hayes, Geochimica et Cosmochimica Acta, 1974, **38**, p.47. Partial analysis (XRF), 25.43 % total iron, T.S. McCarthy and L.H. Ahrens, Earth planet. Sci. Lett., 1972, **14**, p.97. Trace element content, G.W. Kallemeyn and J.T. Wasson, Geochimica et Cosmochimica Acta, 1981, **45**, p.1217. Only about 1.6kg known in collections.

329g Harvard Univ.; 260g Yale Univ.; 145g Vienna, Naturhist. Mus.; 124g Budapest, Nat. Mus.; 121g Paris, Mus. d'Hist. Nat.; 86g Chicago, Field Mus. Nat. Hist.; 36g Washington, U.S. Nat. Mus.; 66g New York, Amer. Mus. Nat. Hist.;

Specimen(s): [53290], 73g. and fragments, 6g

Warrentown error for Warrenton.

Warsaw v Pułtusk.

Warsaw I 52°14′ N., 21° E.

Poland

Doubtful..

In an official Polish report of 1 February, 1831, two masses, Warsaw I and Warsaw II, of iron are said to have been used by the mint. Evidence for a meteoritic origin is not known; one mass was noted as 'soft', the other 'hard', J. Pokrzywnicki, Acta Geophys. Polon., 1971, **19**, p.235.

Warschau v Pułtusk.

Warshaw v Pułtusk.

Washington v Farmington.

Washington County v Farmington.

Washington County v Trenton.

Washington County 39°42′ N., 103°10′ W.

Colorado, U.S.A.

Found 1927

Synonym(s): *Aricarie, Arickaree, Arickarie*

Iron. Ataxite, Ni-rich (IRANOM).

A very fresh mass of 5750g was found 12 inches down in a wheatfield, possibly fell in 1916. Description, with an analysis, 9.34 %Ni, C. Palache and E.V. Shannon, Am. Miner., 1928, **13**, p.406 [M.A. 4-118]. Another analysis, 9.9 %Ni, R.E. Cech, Geochimica et Cosmochimica Acta, 1962, **26**, p.993. Classification, analysis, 9.96 %Ni, 15.5 ppm.Ga, 20.5 ppm.Ge, 0.067 ppm.Ir, J.T. Wasson and R. Schaudy, Icarus, 1971, **14**, p.59. A unique, troilite-free iron; shear deformed, V.F. Buchwald, Iron Meteorites, Univ. of California, 1975, p.1284.

4.3kg Harvard Univ.; 35g Copenhagen, Univ. Geol. Mus.; 12.8g New York, Amer. Mus. Nat. Hist.;

Washougal 45°35′ N., 122°21′ W.

Clark County, Washington, U.S.A.

Fell 1939, July 2, 0735 hrs

Synonym(s): *Portland Meteor*

Stone. Achondrite, Ca-rich. Howardite (AHOW).

One stone, the size of a tennis ball, was found in NW. 1/4, sect. 8, township 1 N., range 4 E. Brief description, J.H. Pruett, Pop. Astron., Northfield, Minnesota, 1939, **47**, p.500. The mass of the stone was 225g, popular account of the fall, F.C. Leonard, Griffith Observer, 1940, **4**, p.98 [M.A. 8-61]. The place of find is given as the NE. corner of sect. 8, not the NW, A.D. Nininger, Pop. Astron., Northfield, Minnesota, 1940, **48**, p.557, Contr. Soc. Res. Meteorites, **2**, p.227 [M.A. 8-54]. Description of polymict nature, 14.22 % total iron, D.Y. Jérome and M. Christophe-Michel-Levy, Meteoritics, 1972, **7**, p.449. Fission track age, E.A. Carver and E. Anders, Geochimica et Cosmochimica Acta, 1976, **40**, p.467.

Main mass, Eugene, Univ. of Oregon; 10.7g Tempe, Arizona State Univ.;

Specimen(s): [1959,754], 17g. and powder, 1g

Waterfall v Bowden.

Waterville 47°45′ N., 119°51′ W.

Douglas County, Washington, U.S.A.

Found 1927

Iron. Anomalous, troilite-rich iron (IRANOM).

A mass of 75lb was ploughed up on Fachnie farm, F.A. McMillan, The Mineralogist, Portland, Oregon, 1940, **8**, p.223, 239 [M.A. 8-61]. Other smaller masses were also found, one weighing 3.125kg, O.W. Freeman, Northwest Sci., 1948, **22**, p.25. Contains abundant troilite and graphite, together about 15% (by volume) of the mass; co-ordinates; the farm is 16 miles NE. of Waterville, V.F. Buchwald, Iron Meteorites, Univ. of California, 1975, p.1288. Classification and analysis of metal, 7.81 %Ni, 64.8 ppm.Ga, 196 ppm.Ge, 0.30 ppm.Ir, J.T. Wasson, Icarus, 1970, **12**, p.407.

Main mass, Waterville Mus.; 197g Washington, U.S. Nat. Mus.; 118g Tempe, Arizona State Univ.;

Specimen(s): [1959,967], 130g. slice, and sawings, 4g; [1972, 322], 0.107g. troilite

Wathena 39°49′ N., 94°55′ W.

Doniphan County, Kansas, U.S.A.

Found 1939

Iron. Hexahedrite (IIA).

A weathered mass of 566g was found in a ditch near Wathena. Description, with an analysis, 5.56 %Ni, E.P. Henderson and S.H. Perry, Am. Miner., 1949, **34**, p.102 [M.A. 10-519]. Classification, further analysis, 5.51 %Ni, 59.6 ppm.Ga, 184 ppm.Ge, 7.0 ppm.Ir, J.T. Wasson, Meteorites, Springer-Verlag, 1974, p.300. Shock-recrystallised, with some melting, V.F. Buchwald, Iron Meteorites, Univ. of California, 1975, p.1291.

429g Washington, U.S. Nat. Mus., main mass;

Wauneta

Wray County, Colorado, U.S.A., Co-ordinates not reported

Stone.

A mass of 8lb 6oz is listed in the collection of the Texas Observers, Fort Worth, V.E. Barnes, Univ. Texas, 1940 (3945), p.608. Details may be in error, R.M. Pearl, letter of 11 September, 1973 in Min. Dept. BM(NH).

Wawilowka v Vavilovka.

Wayne County v Jenny's Creek.

Wayne County v Wooster.

Wayside 34°48′6″ N., 101°41′ W.

Armstrong County, Texas, U.S.A.

Found 1973

Stone. Olivine-bronzite chondrite (H6).

A stone of 23.6kg was found on newly cultivated ground. It was an old stone having numerous plough marks in its brown crust, Meteor. Bull., 1976 (54), Meteoritics, 1976, **11**, p.89. Co-ordinates reported are of a site in Randall County, olivine Fa_{18}, A.L. Graham, priv. comm., 1978.

14.7kg Mainz, Max-Planck-Inst.; 379g Copenhagen, Univ. Geol. Mus.; 329g Chicago, Field Mus. Nat. Hist.; 314g Los Angeles, Univ. of California;

Specimen(s): [1974,M.12], 75g. crusted part-slice

Weatherford 35°30′ N., 98°42′ W.

Custer County, Oklahoma, U.S.A.
Found 1926
Stony-iron. Mesosiderite (MES).

A 2kg mass was ploughed up in 1926 and recognized as meteoritic in 1940. Description, with an analysis of the metal phase, 6.1 %Ni, C.W. Beck and L. LaPaz, Pop. Astron., Northfield, Minnesota, 1949, **57**, p.450, Contr. Meteoritical Soc., **4**, p.216 [M.A. 11-139]. A polymict breccia of metal, iron-poor silicate and type III carbonaceous chondrite, analysis and illustrated description, B. Mason and J. Nelen, Geochimica et Cosmochimica Acta, 1968, **32**, p.661.

1216g Washington, U.S. Nat. Mus.; 16.3g Tempe, Arizona State Univ.; 5g Albuquerque, Univ. of New Mexico;

Specimen(s): [1959,987], 27g.

Weathersfield v Wethersfield.

Weaver v Weaver Mountains.

Weaver Mountains 34°15′ N., 112°45′ W. approx.

Wickenberg, Maricopa County, Arizona, U.S.A.
Found 1898
Synonym(s): *Weaver*
Iron. Ataxite, Ni-rich (IVB).

A mass of about 85.5lb found in 1898, was in 1904 in the Mus. of the State School of Mines, Arizona, H.A. Ward, Cat. Ward-Coonley Coll. Meteorites, Chicago, 1904, p.27. Analysis, 18.03 %Ni, E.P. Henderson and S.H. Perry, Pop. Astron., Northfield, Minnesota, 1951, **59**, p.263 [M.A. 11-444]. Structural description, troilite probably re-melted and dispersed by shock, H.J. Axon and P.L. Smith, Min. Mag., 1972, **38**, p.736. Further analysis, 16.81 %Ni, 0.233 ppm.Ga, 0.058 ppm.Ge, 17 ppm.Ir, R. Schaudy et al., Icarus, 1972, **17**, p.174. Co-ordinates, description, V.F. Buchwald, Iron Meteorites, Univ. of California, 1975.

28.5kg Tucson, Univ. Arizona; 2.6kg Tempe, Arizona State Univ.; 1kg Chicago, Field Mus. Nat. Hist.; 773g Washington, U.S. Nat. Mus.; 350g New York, Amer. Mus. Nat. Hist.; 97g Berlin, Humboldt Univ.;

Specimen(s): [86946], 155g. slice

Webb 31°45′ S., 127°47′ E. approx.

Nullarbor Plain, Western Australia, Australia
Found 1968, June
Stone. Olivine-hypersthene chondrite (L6).

A single oxidized mass, 410.5g, was found 12 miles NNW. of Mundrabilla Station homestead, G.J.H. McCall and W.H. Cleverly, J. Roy. Soc. West. Austr., 1970, **53**, p.69. Co-ordinates, olivine $Fa_{25.3}$, B. Mason, Rec. Austr. Mus., 1974, **29**, p.169.

376g Kalgoorlie, West. Austr. School of Mines, main mass; 15g Perth, West. Austr. Mus.; 3g Washington, U.S. Nat. Mus.;

Wedderburn 36°26′ S., 143°38′ E. approx.

Victoria, Australia
Found 1951, before this year
Iron. Ataxite, Ni-rich (IIICD).

A complete mass of 210g was found 3 miles NE. of Wedderburn. Reported, with an analysis, 23.95 %Ni, A.B. Edwards, Proc. Roy. Soc. Victoria, 1953, **64** (2), p.73 [M.A. 12-255]. Ablation loss during entry, J.F. Lovering et al., Geochimica et Cosmochimica Acta, 1960, **19**, p.156. Classification, analysis, 22.36 %Ni, 1.51 ppm.Ga, 1.47 ppm.Ge, 0.052 ppm.Ir, J.T. Wasson and R. Schaudy, Icarus, 1971, **14**, p.59. Description, V.F. Buchwald, Iron Meteorites, Univ. of California, 1975, p.1295.

159g Melbourne, Geol. Mus., Mines Dept., Victoria; 17g Sydney, Austr. Mus.;

Weekeroo v Weekeroo Station.

Weekeroo Station 32°16′ S., 139°52′ E.

Mannahill, South Australia, Australia
Found 1924
Synonym(s): *Weekeroo*
Iron. Octahedrite, coarse (2.5mm) (IIE).

A mass of 94.2kg was found at Weekeroo Station. Description, with an analysis, 6.89 %Ni, T. Hodge-Smith, Rec. Austr. Mus., 1932, **18**, p.312 [M.A. 5-159]. Detailed petrology and mineralogy of silicate inclusions, T.E. Bunch et al., Contr. Miner. Petrol., 1970, **25**, p.297. Classification and analysis of metal, 7.51 %Ni, 28.2 ppm.Ga, 67.0 ppm.Ge, 2.8 ppm.Ir, E.R.D. Scott et al., Geochimica et Cosmochimica Acta, 1973, **37**, p.1957. Description; structural history, V.F. Buchwald, Iron Meteorites, Univ. of California, 1975, p.1296. Rb-Sr isochron age, 4400 m.y., D.S. Burnett and G.J. Wasserburg, Earth planet. Sci. Lett., 1967, **2**, p.397. Ar-Ar age, 4540 m.y., S. Niemeyer, Geochimica et Cosmochimica Acta, 1980, **44**, p.33.

47kg Sydney, Austr. Mus.; 11kg Washington, U.S. Nat. Mus.; 6.4kg Chicago, Field Mus. Nat. Hist.; 3.3kg Harvard Univ.; 5.5kg Tempe, Arizona State Univ.; 3.2kg New York, Amer. Mus. Nat. Hist.;

Specimen(s): [1929,196], 4488g. and fragments, 83g

Weichsel v Schwetz.

Wei-hui-fu (a)

Hunan, China, Not reported
Found 1931
Iron. Octahedrite.

A ceremonial bronze broad axe has an oxidized remnant of an iron blade; bulk chemical and microprobe analyses prove that the 'iron' is meteoritic, R.J. Gettens et al., Freer Gallery of Art, 1971 (Occasional paper 4, no. 1), Meteor. Bull., 1972 (51), Meteoritics, 1972, **7**, p.231.

Specimen, Freer Gallery of Art, F.G.A. 34.10;

Wei-hui-fu (b)

Hunan, China, Not reported
Found 1931
Iron. Octahedrite.

A ceremonial bronze dagger-axe has an oxidized remnant of an iron blade; bulk chemical and microprobe analyses indicate that some of the 'iron' originally had a kamacite composition, indicating a meteoritic origin, R.J. Gettens et al., Freer Gallery of Art, 1971 (Occasional paper 4, no. 1), Meteor. Bull., 1972 (51), Meteoritics, 1972, **7**, p.231.

Specimen, Freer Gallery of Art, F.G.A. 34.11;

Weiler v Mezö-Madaras.

Weiyuan 35°16′ N., 104°19′ E.
Weiyuan County, Gansu, China
Found 1978
Stony-iron. Mesosiderite (MES).
A single mass of unreported weight was found, K. Tao, Sci. Geol. Sinica, 1980, p.296 [M.A. 81-1770]. Listed, D. Bian, Meteoritics, 1981, **16**, p.115.
Main mass, Beijing, Acad. Sci. Inst. Geol.;

Weld County v Johnstown.

Weldna error for Weldona.

Weldona 40°21′ N., 103°57′ W.
Morgan County, Colorado, U.S.A.
Found 1934
Stone. Olivine-bronzite chondrite (H4).
One stone of 27.7kg was found, H.H. Nininger, Pop. Astron., Northfield, Minnesota, 1934, **42**, p.341. Olivine Fa_{19}, B. Mason, Geochimica et Cosmochimica Acta, 1963, **27**, p.1011.
9kg Tempe, Arizona State Univ.; 1070g Washington, U.S. Nat. Mus.; 1kg Chicago, Field Mus. Nat. Hist.; 1kg Denver, Mus. Nat. Hist.; 305g Mainz, Max-Planck-Inst.; 295g New York, Amer. Mus. Nat. Hist.;
Specimen(s): [1959,1020], 12417g.

Welland 43°1′ N., 79°13′ W.
Welland County, Ontario, Canada
Found 1888
Iron. Octahedrite, medium (1.2mm) (IIIA).
A mass of about 18lb was ploughed up, about 1.5 miles N. of Welland. Description, with an analysis, E.E. Howell, Proc. Rochester Acad. Sci., 1890, **1**, p.86. Further analysis, 8.77 % Ni, 21.0 ppm.Ga, 46.7 ppm.Ge, 0.29 ppm.Ir, E.R.D. Scott et al., Geochimica et Cosmochimica Acta, 1973, **37**, p.1957. Description, V.F. Buchwald, Iron Meteorites, Univ. of California, 1975, p.1301.
1.5kg Vienna, Naturhist. Mus.; 920g Chicago, Field Mus. Nat. Hist.; 300g Ottawa, Geol. Surv. Canada; 260g New York, Amer. Mus. Nat. Hist.;
Specimen(s): [65971], 466g.; [1982,M.2], 28g.

Wellington v Rowton.

Wellington v Wairarapa Valley.

Wellington 34°57′ N., 100°15′ W.
Collingsworth County, Texas, U.S.A.
Found 1955, approx., recognized about 1968
Stone. Olivine-bronzite chondrite (H5).
A 13.4kg stone was found in flood-plain deposits, Meteor. Bull., 1968 (44), Meteoritics, 1970, **5**, p.100.
Main mass, Houston, in possession of E.A. King; 4.2kg Washington, U.S. Nat. Mus.; 770g Tempe, Arizona State Univ.; 650g New York, Amer. Mus. Nat. Hist.; 489g Los Angeles, Univ. of California;
Specimen(s): [1971,3], 21.7g. drill core, with polished faces

Wellman v Wellman (a).

Wellman (a) 33°2′ N., 102°20′ W.
Terry County, Texas, U.S.A.
Found 1940
Synonym(s): *Wellman*
Stone. Olivine-bronzite chondrite (H5).
One stone of 50.1kg was found, H.H. and A.D. Nininger, The Nininger Collection of Meteorites, Winslow, Arizona, 1950, p.101. A further mass was found in 1963, Meteorite Cat., Amer. Meteorite Lab., 1965, Sept. Olivine Fa_{18}, B. Mason, Geochimica et Cosmochimica Acta, 1963, **27**, p.1011.
26.2kg Tempe, Arizona State Univ.; 3.2kg Mainz, Max-Planck-Inst.; 874g Copenhagen, Univ. Geol. Mus.; 543g Washington, U.S. Nat. Mus.; 523g Chicago, Field Mus. Nat. Hist.; 242g Harvard Univ.; 117g New York, Amer. Mus. Nat. Hist.;
Specimen(s): [1959,1021], 23020g. half of the main mass

Wellman (b) 33°1′30″ N., 102°25′ W.
Terry County, Texas, U.S.A.
Found 1964, recognized in this year
Stone. Chondrite.
Listed, total known weight 343g, Cat. Huss Coll. Meteorites, 1976, p.44.
233g Mainz, Max-Planck-Inst.;

Wellman (c) 33°2′ N., 102°20′ W. approx.
Terry County, Texas, U.S.A.
Found 1964, recognized in this year
Synonym(s): *Wellman no. 3*
Stone. Olivine-bronzite chondrite (H4).
Numerous masses were found totalling over 40kg, the largest weighing 8.4kg, Cat. Huss Coll. Meteorites, 1976, p.45.
40kg Mainz, Max-Planck-Inst.; 630g Tempe, Arizona State Univ.;
Specimen(s): [1969,182], 72g. in three pieces

Wellman (d) 33°1′ N., 102°22′ W.
Terry Couty, Texas, U.S.A.
Found 1966
Stone. Olivine-bronzite chondrite (H).
Listed, total known mass 1615g, Cat. Huss Coll. Meteorites, 1976, p.46.
1438g Mainz, Max-Planck-Inst.;

Wellman (e) 33°40′42″ N., 102°18′54″ W.
Terry County, Texas, U.S.A.
Found 1973, recognized in this year
Stone. Olivine-bronzite chondrite (H4).
A mass of 497g was noticed as distinct from the other Wellman specimens in Chicago, Field Mus. Nat. Hist. A second mass of 462g was found in 1978 at the above co-ordinates, G.I Huss, priv. comm., 1983. Olivine Fa_{20}, A.L. Graham, 1983.
Specimen(s): [1983,M.26], 19g.

Wellman no. 3 v Wellman (c).

Welmannville v Ness County (1894).

Welmanville v Ness County (1894).

Wengerowo v Vengerovo.

Werchne Dnieprowsk v Verkhne Dnieprovsk.

Werchne Tschirskaja v Verkhne Tschirskaia.

Werkhne Dnieprowsk v Verkhne Dnieprovsk.

Werkhne Tschirskaia v Verkhne Tschirskaia.

Werkhne Udinsk v Verkhne Udinsk.

Werkne Udinsk v Verkhne Udinsk.

Wessely 48°57′ N., 17°23′ E.
Hradisch, Jihomoravsky, Czechoslovakia
Fell 1831, September 9, 1530 hrs
Synonym(s): *Veseli nad Moravou, Vnorovy, Znorow*
Stone. Olivine-bronzite chondrite (H5), veined.
After the appearance of a moving cloud and detonations, a stone of about 3.75kg was seen to fall, K. von Schreibers and A.R. von Holger, Z. Phys. u. Math. Wien, 1832, **1**, p.193. This locality appears under the name 'Vnorovy' in the meteorite locality map, K. Tucek, Cat. Coll. Meteor., Nat. Mus. Prague, 1968. Olivine Fa_{18}, B. Mason, Geochimica et Cosmochimica Acta, 1963, **27**, p.1011.
3.67kg Vienna, Naturhist. Mus.; 12g Tübingen, Univ.; 4g Chicago, Field Mus. Nat. Hist.; 2g Berlin, Humboldt Univ.; 6.8g New York, Amer. Mus. Nat. Hist.;
Specimen(s): [63878], 2.75g.; [35729], 0.5g.

Western Arkansas 35° N., 94° W. approx.
Montgomery County, Arkansas, U.S.A.
Found 1890, before this year
Iron. Octahedrite, fine (0.3mm) (IVA).
A mass of 1.75kg in the Canfield collection of minerals was described, with an analysis, 5.12 %Ni, G.P. Merrill, Proc. U.S. Nat. Mus., 1927, **72** (4), p.2. Classification, analysis, 7.62 %Ni, 1.97 ppm.Ga, 0.100 ppm.Ge, 2.8 ppm.Ir, R. Schaudy et al., Icarus, 1972, **17**, p.174. Has been artificially reheated to 800C, V.F. Buchwald, Iron Meteorites, Univ. of California, 1975, p.1304.
1.5kg Washington, U.S. Nat. Mus.; 17g Tempe, Arizona State Univ.;
Specimen(s): [1959,909], 19g. slice, and sawings, 1g

Western Point district v Cranbourne.

Western Port district v Cranbourne.

West Forrest 30°40′ S., 127°50′ E. approx.
Nullarbor Plain, Western Australia, Australia
Found 1971
Stone. Olivine-bronzite chondrite (H5).
A single, fusion crusted, oxidized stone of 170g was found about 10 miles NW. of Forrest Station on the Trans-Australian Railway, G.J.H. McCall, 2nd. Suppl. to West. Austr. Mus. Spec. Publ. no. 3, 1972, p.27. Olivine $Fa_{18.7}$, B. Mason, Rec. Austr. Mus., 1974, **29**, p.169.
Main mass, in the possession of the finder, J. Clohessy; Thin section and cast, Perth, West. Austr. Mus.; 2.3g Washington, U.S. Nat. Mus.;

West Liberty v Homestead.

Weston 41°13′ N., 73°23′ W.
Fairfield County, Connecticut, U.S.A.
Fell 1807, December 14, 0630 hrs
Synonym(s): *Fairfield County*
Stone. Olivine-bronzite chondrite (H4).
After the appearance of a fireball (travelling from N. to S.), and detonations, a shower of several stones fell over an area about 10 miles in length. The total weight was estimated at 330lb and the largest stone, which broke into fragments, at 200lb, B. Silliman and J.L. Kingsley, Trans. Amer. Phil. Soc. Philadelphia, 1809, **6**, p.323, Am. J. Sci., 1869, **47**, p.1. Description, analysis, 26.89 % total iron, B. Mason and H.B. Wiik, Am. Mus. Novit., 1965 (2220). Xenolithic, with unequilibrated host, R.A. Binns, Geochimica et Cosmochimica Acta, 1968, **32**, p.299. Complex irradiation history, L. Schultz et al., Earth planet. Sci. Lett., 1972, **15**, p.403. Contains hydrous silicates, J.R. Ashworth and R. Hutchison, Nature, 1975, **256**, p.714. Comparatively little has been preserved. Mineralogy, petrology, olivine $Fa_{18.4}$, A.F. Noonan and J. Nelen, Meteoritics, 1976, **11**, p.111.
1.2kg Tempe, Arizona State Univ.; 312g Chicago, Field Mus. Nat. Hist.; 211g Harvard Univ.; 195g New York, Amer. Mus. Nat. Hist.; 186g Vienna, Naturhist. Mus.; 181g Washington, U.S. Nat. Mus.; 90g Paris, Mus. d'Hist. Nat.;
Specimen(s): [35410], 722g.; [90254], 114g.; [90253], 26g.; [1920,349], 22g.; [33742], 4.5g.

West Point 33°4′30″ N., 102°2′42″ W.
Lynn County, Texas, U.S.A.
Found 1972
Stone. Olivine-hypersthene chondrite (L).
A single stone of 3.1kg was found, olivine Fa_{25}, Meteor. Bull., 1979 (56), Meteoritics, 1979, **14**, p.172.
1.8kg Mainz, Max-Planck-Inst.;
Specimen(s): [1976,M.4], 31.3g.

West Point (mesosiderite) v Crab Orchard.

West Reid 30°11′ S., 128°40′ E.
Nullarbor Plain, Western Australia, Australia
Found 1969, November
Stone. Olivine-bronzite chondrite (H6).
A single, flight oriented, oxidized, mass of 627.7g was found, G.J.H. McCall and W.H. Cleverly, J. Roy. Soc. West. Austr., 1970, **53**, p.69. Olivine $Fa_{19.6}$, B. Mason, Rec. Austr. Mus., 1974, **29**, p.169.
588g Kalgoorlie, West. Austr. School of Mines; Thin section, Perth, West. Austr. Mus.; 0.8g Washington, U.S. Nat. Mus.;

Wethersfield v Wethersfield (1971).

Wethersfield (1971) 41°42′ N., 72°39′ W.
Hartford County, Connecticut, U.S.A.
Fell 1971, April 8, 0430-1130 hrs, U.T.
Synonym(s): *Wethersfield, Weathersfield*
Stone. Olivine-hypersthene chondrite (L6).
A stone of 350g fell through the roof of a house. Illustrated description, 22.50 % total iron, olivine Fa_{25}, R.S. Clarke, Jr. et al., Smithson. Contrib. Earth Sci., 1975 (14), p.63, Meteor. Bull., 1971 (50), Meteoritics, 1971, **6**, p.115.
290g Washington, U.S. Nat. Mus., main mass;

Wethersfield (1982) 41°42′38″ N., 72°40′25″ W.
Hartford County, Connecticut, U.S.A.
Fell 1982, November 8, 2114 hrs
Stone. Olivine-hypersthene chondrite (L6).
After the appearence of a fire-ball and thunder-like booms, a mass of 2704g and about 52g of fragments were recovered after they had penetrated the roof of a house in Wethersfield. Report, olivine Fa_{25}, SEAN Bulletin, 1982, 7 (10), p.15.

White Sulphur Springs v Greenbrier County.

Whitfield v Cleveland.

Whitfield County v Cleveland.

Whitman 42°2′ N., 101°30′ W.
Grant County, Nebraska, U.S.A.
Found 1937
Synonym(s): *Whittman*
Stone. Olivine-bronzite chondrite (H5).
One stone of 221g was found, A.D. Nininger, Pop. Astron., Northfield, Minnesota, 1939, **47**, p.214, H.H. and A.D. Nininger, The Nininger Collection of Meteorites, Winslow, Arizona, 1950, p.101. Olivine Fa_{19}, B. Mason, Geochimica et Cosmochimica Acta, 1963, **27**, p.1011.
149g Tempe, Arizona State Univ.; 21g Washington, U.S. Nat. Mus.;
Specimen(s): [1959,904], 35g. slice

Whittman v Whitman.

Wichita County 34°4′ N., 98°55′ W.
Wichita County, Texas, U.S.A.
Found 1836, before this year
Synonym(s): *Austin, Brazos, Brazos River, Red River, South-east Missouri, Young County*
Iron. Octahedrite, coarse (2.4mm) (IA).
A mass of 320lb, known to the Comanche Indians for many years, was removed in 1836. Description, B.F. Shumard, Trans. St. Louis Acad. Sci., 1860, **1**, p.622, J.W. Mallett, Am. J. Sci., 1884, **28**, p.285. Analysis, 7.91 %Ni, E. Cohen and E. Weinschenk, Ann. Naturhist. Hofmus. Wien, 1891, **6**, p.153. Classification, further analysis, 6.78 %Ni, 86.9 ppm.Ga, 341 ppm.Ge, 2.1 ppm.Ir, J.T. Wasson, Icarus, 1970, **12**, p.407. Contains krinovite, E. Olsen and L. Fuchs, Science, 1968, **161**, p.786. Description, includes South-east Missouri, V.F. Buchwald, Iron Meteorites, Univ. of California, 1975, p.1305.
Main mass, Austin, Texas Memorial Univ.; 6kg Vienna, Naturhist. Mus.; 3.8kg Chicago, Field Mus. Nat. Hist.; 3kg New York, Amer. Mus. Nat. Hist.; 2251g Washington, U.S. Nat. Mus.; 2.1kg Ottawa, Mus. Geol. Surv. Canada; 1.8kg Harvard Univ.; 1.8kg Schönenwerd, Bally-Prior Mus.; 1.75kg Budapest, Nat. Mus.; 1.3kg Stockholm, Riksmus.; 600g Paris, Mus. d'Hist. Nat.;
Specimen(s): [55825], 1377g.; [34609], 9g.; [35414], 102g. of South-east Missouri

Wickenburg (iron) v Cañon Diablo.

Wickenburg (stone) 33°58′ N., 112°44′ W.
Maricopa County, Arizona, U.S.A.
Found 1940, recognized in this year
Stone. Olivine-hypersthene chondrite (L6), black.
One stone of 9.2kg was found 3 miles W. of Wickenburg, A.D. Nininger, Pop. Astron., Northfield, Minnesota, 1940, **48**, p.557, Contr. Soc. Res. Meteorites, **2**, p.227 [M.A. 8-54]. Olivine Fa_{23}, B. Mason, Geochimica et Cosmochimica Acta, 1963, **27**, p.1011.
2.88kg Tempe, Arizona State Univ.; 1.4kg Chicago, Field Mus. Nat. Hist.; 640g New York, Amer. Mus. Nat. Hist.;
Specimen(s): [1959,1023], 321g. slice

Widdin v Virba.

Wietrzno-Bobrka 49°25′ N., 21°42′ E.
Rzeszowskie Prov., Poland
Found
Iron. Octahedrite, or ataxite?.
An iron hatchet, of weight 376g, was found in a hill fort dated at 700-500 BC. The fort is situated near the Dukla Pass in the Carpathian Mountains. The iron has 8-10%Ni, J. Pokrzywnicki, Acta Geophys. Polon., 1971, **19**, p.235.

Wigan v Appley Bridge.

Wikieup 34°42′ N., 113°36′ W.
Mohave County, Arizona, U.S.A.
Found 1965
Stone. Olivine-bronzite chondrite (H5).
Total known mass, 372g, C.F. Lewis and C.B. Moore, Cat. Meteor. Arizona State Univ., 1976, p.69.
334g Tempe, Arizona State Univ., main mass;

Wilbia 26°27′ S., 131°00′ E.
South Australia, Australia
Found 1965, May
Stone. Olivine-bronzite chondrite (H5).
A 94g mass was found on a sand dune at the eastern end of Wilbia Hill, Musgrave Ranges, South Australia, D.W.P. Corbett, Rec. S. Austr. Mus., 1968, **15**, p.767. Olivine Fa_{19}, B. Mason, Rec. Austr. Mus., 1974, **29**, p.169. Analysis, 23.30 % total iron, M.J. Fitzgerald, Ph.D. Thesis, Univ. of Adelaide, 1979, p.23.
Main mass, Adelaide, South Austr. Mus.; 3.8g Washington, U.S. Nat. Mus.;
Specimen(s): [1973,M.26], 8.5g.

Wilberton error for Wilburton.

Wilburton 37°5′ N., 101°46′ W.
Morton County, Kansas, U.S.A.
Found 1940
Stone. Olivine-hypersthene chondrite (L).
One stone of 207.8g was found in township 35, range 41 W., sect. 6, H.H. and A.D. Nininger, The Nininger Collection of Meteorites, Winslow, Arizona, 1950, p.102. In 1941 a second smaller stone, 48g, was found. Olivine Fa_{25}, B. Mason, Geochimica et Cosmochimica Acta, 1963, **27**, p.1011.
41.2g Tempe, Arizona State Univ.;
Specimen(s): [1959,1040], 23.5g. half of the stone found in 1941

Wild v Gibeon.

Wildara 28°14′ S., 120°51′ E.
Western Australia, Australia
Found 1968, recognized 1969
Stone. Olivine-bronzite chondrite (H5).
A number of fragments totalling about 500kg were found in a creek bed. The material is very oxidized and broken by weathering; two large interlocking masses weighed together 51kg, G.J.H. McCall, 2nd. Suppl. to West. Austr. Mus. Spec. Publ. no. 3, 1972, p.28. Olivine $Fa_{20.7}$, B. Mason, Rec. Austr. Mus., 1974, **29**, p.169.
320kg Perth, West. Austr. Mus.; 2.3kg Kalgoorlie, West. Austr. School of Mines; Thin section, Washington, U.S. Nat. Mus.;
Specimen(s): [1973,M.7], 524g. crusted slice

Wiley 38°9′ N., 102°40′ W.
Prowers County, Colorado, U.S.A.
Found 1938
Iron. Octahedrite, plessitic (IIC).
One mass of 3.5kg was found 4.5 miles NW. of Wiley, A.D. Nininger, Pop. Astron., Northfield, Minnesota, 1939, **47**, p.214, E.P. Henderson, letter of 3 June, 1939 in Min. Dept. BM(NH). Classification and analysis, 11.5 %Ni, 38.8

ppm.Ga, 114 ppm.Ge, 6.2 ppm.Ir, J.T. Wasson, Geochimica et Cosmochimica Acta, 1969, **33**, p.859. Structural description, H.J. Axon and P.L. Smith, Min. Mag., 1972, **38**, p.736. Mildly shocked, figured, V.F. Buchwald, Chem. Erde, 1971, **30**, p.33, V.F. Buchwald, Iron Meteorites, Univ. of California, 1975, p.1309.

912g Tempe, Arizona State Univ.; 324g Michigan Univ.; 285g Chicago, Field Mus. Nat. Hist.; 189g Washington, U.S. Nat. Mus.; 90g Denver, Mus. Nat. Hist.;

Specimen(s): [1959,914], 1089g. and sawings, 46.5g

Wilkanowko v Grüneberg.

Willamette 45°22′ N., 122°35′ W.

Clackamas County, Oregon, U.S.A.
Found 1902
Synonym(s): *Oregon City*
Iron. Octahedrite, medium (1.0mm) (IIIA).

An immense cavernous mass of 14.1 tons was found 2 miles NW. of Willamette. Description, H.A. Ward, Proc. Rochester Acad. Sci., 1904, **4**, p.137, E.O. Hovey, Am. Mus. J., 1906, **6**, p.105. Analysis, 7.62 %Ni, 18.6 ppm.Ga, 37 3 ppm.Ge, 4.7 ppm.Ir, E.R.D. Scott et al., Geochimica et Cosmochimica Acta, 1973, **37**, p.1957. Full history, description; complex shock effects, V.F. Buchwald, Iron Meteorites, Univ. of California, 1975, p.1311.

14.1tons New York, Amer. Mus. Nat. Hist., main mass; 2.7kg Washington, U.S. Nat. Mus.; 2.5kg Tempe, Arizona State Univ.; 2.3kg Chicago, Field Mus. Nat. Hist.; 2kg Los Angeles, Univ. of California; 1.45kg Vienna, Naturhist. Mus.; 1kg Berlin, Humboldt Univ.; 1kg Budapest, Nat. Mus.; 1.5kg Ottawa, Mus. Geol. Surv. Canada; 700g Paris, Mus. d'Hist. Nat.; 693g Prague, Nat. Mus.; 462g Michigan Univ.;

Specimen(s): [86945], 960g.; [1938,303], 123g. iron shale; [1977,M.8], 7.7g. iron shale

Willanueva v Villanueva.

Willard 34°26′ N., 105°47′ W.

Torrence County, New Mexico, U.S.A.
Found 1978
Stone. Olivine-hypersthene chondrite (L6).

A mass of 500g and a further 200-300g of fragments were found 13 miles SE. of Willard, R. Haag, letter of 1 December, 1982 in Min. Dept. BM(NH).

Specimen(s): [1982,M.17], 28g.

Willaroy 30°6′ S., 143°12′ E.

New South Wales, Australia
Found 1970, March 12
Stone. Olivine-bronzite chondrite (H3).

Four fragments were found, the largest weighing 2.54kg, which fitted together to form a single mass of 4.05kg. Description, with an analysis, 25.3 % total iron, olivine $Fa_{10\text{-}19}$, R.O. Chalmers and B. Mason, Rec. Austr. Mus., 1977, **30**, p.519. Reported, Meteor. Bull., 1976 (54), Meteoritics, 1976, **11**, p.92.

Main mass, Sydney, Austr. Mus.; 187g Washington, U.S. Nat. Mus.;

Williamette error for Willamette.

Williamsport v Bald Eagle.

Williamstown v Kenton County.

Willowbar 36°44′ N., 102°12′ W.

Cimarron County, Oklahoma, U.S.A.
Found 1971
Stone. Olivine-hypersthene chondrite (L6).

A mass of 2.07kg was found on farmland between Keyes and the small community of Willowbar. Description, mineral analyses, olivine $Fa_{24.3}$, D.E. Lange et al., Meteoritics, 1973, **8**, p.263. Reported, Meteor. Bull., 1974 (52), Meteoritics, 1974, **9**, p.117.

1816g Tempe, Arizona State Univ.; 145g Chicago, Field Mus. Nat. Hist.;

Willow Creek 43°28′ N., 106°46′ W.

Natrona County, Wyoming, U.S.A.
Found 1914, approx., recognized 1934
Synonym(s): *Central Wyoming*
Iron. Octahedrite, coarse (1.4mm) (IIIE).

A mass of 112.5lb was found, H.H. Nininger, The Mines Mag., Golden, Colorado, 1937, **27**, p.16 [M.A. 7-69]. Analysis, determination of Ga, Au and Pd, 8.75 %Ni, E. Goldberg et al., Geochimica et Cosmochimica Acta, 1951, **2**, p.1. Classification, analysis, 8.76 %Ni, 16.9 ppm.Ga, 36.4 ppm.Ge, 0.054 ppm.Ir, E.R.D. Scott et al., Geochimica et Cosmochimica Acta, 1973, **37**, p.1957. Shock-melted troilite figured, V.F. Buchwald, Iron Meteorites, Univ. of California, 1975, p.1321.

16.9kg Tempe, Arizona State Univ.; 4.8kg Washington, U.S. Nat. Mus.; 1.8kg Chicago, Field Mus. Nat. Hist.; 1.4kg Michigan Univ.;

Specimen(s): [1959,1019], 17.1kg.; [1937,391], 40.5g.

Willowdale 37°32′ N., 98°22′ W.

Kingman County, Kansas, U.S.A.
Found 1951
Synonym(s): *St. Leo*
Stone. Olivine-bronzite chondrite (H4).

A mass of 3kg was found, B. Mason, Meteorites, Wiley, 1962, p.234. Olivine Fa_{20}, B. Mason, Geochimica et Cosmochimica Acta, 1963, **27**, p.1011.

190g New York, Amer. Mus. Nat. Hist.; 74g Washington, U.S. Nat. Mus.; 65g Mainz, Max-Planck-Inst.; 59g Los Angeles, Univ. of California;

Specimen(s): [1965,188], 236g. slice

Wilmot 37°23′ N., 96°52′ W.

Cowley County, Kansas, U.S.A.
Found 1944
Synonym(s): *Cowley County*
Stone. Olivine-bronzite chondrite (H6), veined.

An extensvely weathered mass of 2kg was found in a farmyard, H.H. Nininger, Pop. Astron., Northfield, Minnesota, 1944, **52**, p.42, Contr. Soc. Res. Meteorites, **3**, p.123 [M.A. 9-300], W. Wahl, letter of 23 May, 1950 in Min. Dept. BM(NH). Listed, H.H. and A.D. Nininger, The Nininger Collection of Meteorites, Winslow, Arizona, 1950, p.103. Olivine Fa_{18}, B. Mason, Geochimica et Cosmochimica Acta, 1963, **27**, p.1011.

526g Tempe, Arizona State Univ.; 133g Washington, U.S. Nat. Mus.; 110g Albuquerque, Univ. of New Mexico; 53g Los Angeles, Univ. of California;

Specimen(s): [1959,839], 643g.

Wilson County v Cosby's Creek.

Wilson County v Cross Roads.

Wiluna 26°35′34″ S., 120°19′42″ E.
Wiluna, Western Australia, Australia
Fell 1967, September 2, 2246 hrs
Synonym(s): *Granite Peak*
Stone. Olivine-bronzite chondrite (H5).
After a fireball and sonic phenomena a shower of stones fell, estimated to number 500 to 1000, approximately 5 miles E. of Wiluna township. The distribution was over an ellipse of 4 by 2 miles elongated NW.-SE.; masses ranged from 10kg to 2.2g, over 150kg have been recovered. Full, illustrated description, G.J.H. McCall and P.M. Jeffery, Min. Mag., 1970, **37**, p.880, Meteor. Bull., 1974 (52), Meteoritics, 1974, **9**, p.107. Analysis, 24.26 % total iron, C.J. Elliott, priv. comm., 1973. Olivine $Fa_{19.2}$, B. Mason, Rec. Austr. Mus., 1974, **29**, p.169.
145kg Perth, West. Austr. Mus.; 10.7kg Kalgoorlie, West. Austr. School of Mines; 462g Fort Worth, Texas, Monnig Colln.; 284g Washington, U.S. Nat. Mus.; 135g Harvard Univ.;
Specimen(s): [1968,189a], 906g. and fragments, 3g; [1968, 189b], 352g. complete stone

Wimberley 29°58′ N., 98°07′ W.
Hays County, Texas, U.S.A.
Found 1976
Iron. Octahedrite, medium (1.0mm) (IIIB).
A single mass of about 7.8kg was found approximately 3 km. SW. of Wimberley, Meteor. Bull., 1979 (56), Meteoritics, 1979, **14**, p.172. Analysis, 9.2 %Ni, 21.0 ppm.Ga, 41.3 ppm.Ge, 0.15 ppm.Ir, D.J. Malvin et al., priv. comm., 1983.
Main mass, Fort Worth, Texas, Mus. Sci. Hist.;

Winburg 28°30′ S., 27° E.
Orange Free State, South Africa
Found 1881
Synonym(s): *Doornport*
Iron. Octahedrite, medium (1.3mm) (IC-ANOM).
A mass of about 50kg was said to have been seen to fall at Zeekoegat, Winburg district. Description, with an analysis, 6.91 %Ni, W.A. Douglas Rudge, Proc. Roy. Soc. London, 1914, **A90**, p.19. Co-ordinates, C. Frick and E.C.I. Hammerbeck, Bull. Geol. Surv. S. Africa, 1973 (57). Not an observed fall, V.F. Buchwald, Iron Meteorites, Univ. of California, 1975, p.1325. Classification and analysis, 6.98 % Ni, 51.8 ppm.Ga, 180 ppm.Ge, 0.89 ppm.Ir, A. Kracher et al., Geochimica et Cosmochimica Acta, 1980, **44**, p.773.
40kg Bloemfontein, Nat. Mus.; 3.6kg Copenhagen, Univ. Geol. Mus.;
Specimen(s): [1915,146], 42g.

Windmill Station v Valley Wells.

Windorah v Thunda.

Wingellina 26°3′ S., 128°57′ E.
Western Australia, Australia
Found 1958
Stone. Olivine-bronzite chondrite (H4).
0.2kg were found, B. Mason, Geochimica et Cosmochimica Acta, 1963, **27**, p.1011, Spec. Publ. West. Austr. Mus., 1965 (3), p.1. Olivine $Fa_{18.2}$, B. Mason, Rec. Austr. Mus., 1974, **29**, p.169.
28.5g Adelaide, South Austr. Mus.; Thin section, Washington, U.S. Nat. Mus.;

Winnebago County v Forest City.

Winona 35°12′ N., 111°24′ W.
Coconino County, Arizona, U.S.A.
Found, Prehistoric
Stone. Chondrite, anomalous (CHANOM).
A weathered mass was found in a stone cist in 1928, in the ruins of the Elden pueblo, Winona. It fell to pieces when removed, 24kg of fragments were recovered. Description, L.F. Brady, Pan-Amer. Geol., 1929, **51**, p.287 [M.A. 4-423]. Description, analysis, olivine Fa_5, B. Mason and E. Jarosewich, Geochimica et Cosmochimica Acta, 1967, **31**, p.1097. Similar to silicate inclusions in IAB irons, R.W. Bild, Geochimica et Cosmochimica Acta, 1977, **41**, p.1439.
Main mass, in the possession of the finder, A.J. Townsend, in 1929; 4.5kg Flagstaff, North Arizona Mus.; 1kg Arizona Bureau of Mines; 334g Harvard Univ.; 356g New York, Amer. Mus. Nat. Hist.; 230g Tempe, Arizona State Univ.; 224g Washington, U.S. Nat. Mus.; 127g Los Angeles, Univ. of California; 104g Michigan Univ.;
Specimen(s): [1930,974], 101g.; [1930,975], 19.5g. two fragments; [1959,988], 147.5g.

Winsloe v Cañon Diablo.

Winslow v Cañon Diablo.

Wirba v Virba.

Wisconsin v Hammond.

Wisconsin v Trenton.

Wisconsin v Vernon County.

Witchelina 30° S., 138° E. approx.
South Australia, Australia
Found 1920
Stone. Olivine-bronzite chondrite (H4).
A mass of 3.64kg was found, Meteor. Bull., 1976 (54), Meteoritics, 1976, **11**, p.92. Olivine $Fa_{19.1}$, B. Mason, Rec. Austr. Mus., 1974, **29**, p.169. Analysis, 24.44 % total iron, M.J. Fitzgerald, Ph.D. Thesis, Univ. of Adelaide, 1979, p.23.
1.5kg Adelaide, Univ., main mass; 156g Washington, U.S. Nat. Mus.; 368g Adelaide, South Austr. Mus.;
Specimen(s): [1968,274], 81g. slice

Withrow 47°42′24″ N., 119°49′48″ W.
Douglas County, Washington, U.S.A.
Found 1950, approx.
Iron. Octahedrite, medium (1.2mm) (IIIA?).
A mass of 19.25lb with well preserved ablation surface was found 1 mile west of Withrow. Illustrated description, W.F. Read et al., Meteoritics, 1967, **3**, p.219, 231. Classification, V.F. Buchwald, Iron Meteorites, Univ. of California, 1975, p.1326.
8.7kg Waterville Mus.;

Witim v Verkhne Udinsk.

Witklip Farm 26° S., 30° E.
Carolina district, Transvaal, South Africa
Fell 1918, May 26, 0940 hrs
Stone. Olivine-bronzite chondrite (H5).
After the appearance of a luminous meteor with a " cloudy trail", followed by detonations, a stone fell on the farm Witklip. Only about four fragments totalling 22g appear to have been preserved, Union Observatory, Cape Town, Circular no. 44, January 17, 1919. Description, G.T. Prior,

Min. Mag., 1926, **21**, p.189. Olivine Fa_{18}, B. Mason, Geochimica et Cosmochimica Acta, 1963, **27**, p.1011.
Main mass, Cape Town, South African Mus.;
Specimen(s): [1921,275], 2.5g.

Witsand Farm 28°40′ S., 18°55′ E.
Orange River, Namibia
Fell 1932, December 1, 1700 hrs
Stone. Olivine-hypersthene chondrite, amphoterite (LL4).
One or more stones fell on Witsand farm about 35 miles north of Pofadder on the Orange river. The material was broken up and mostly lost, S.J. Shand, letter of 22 April, 1939 in Min. Dept. BM(NH). Description, with a partial analysis, S.J. Shand, Am. J. Sci., 1942, **240**, p.67 [M.A. 8-374]. Olivine Fa_{27}, B. Mason, Geochimica et Cosmochimica Acta, 1963, **27**, p.1011.
Specimen(s): [1948,295], 74g. main mass; [1934,119], 1.5g.

Wittekrantz 32°30′ S., 23° E. approx.
Beaufort West, Cape Province, South Africa
Fell 1880, December 9, 0800 hrs
Stone. Olivine-hypersthene chondrite (L5).
After the appearance of a moving cloud, and detonations, two stones were seen to fall, the larger weighed about 4.5lb; of the smaller only a broken fragment, weighing about 113g, was preserved. Description, with an analysis, G.T. Prior, Min. Mag., 1913, **17**, p.28, 132. Chondrule with fine-grained chromite figured, P. Ramdohr, Geochimica et Cosmochimica Acta, 1967, **31**, p.1961. Further analysis, 20.65 % total iron, H. von Michaelis et al., Earth planet. Sci. Lett., 1969, **5**, p.387. Olivine Fa_{23}, B. Mason, Geochimica et Cosmochimica Acta, 1963, **27**, p.1011.
Main mass, Cape Town, South African Mus.; 52g Vienna, Naturhist. Mus.; 31g Tempe, Arizona State Univ.;
Specimen(s): [1914,1032], 69.6g. from the smaller stone

Wittens v Eichstädt.

Wittmess v Eichstädt.

Wjasemsk v Kikino.

Wöhler's Iron v Campo del Cielo.

Wolamo 9° N., 39° E. approx.
near Addis Ababa, Ethiopia
Fell 1964, August, before this date
Stone. Chondrite.
Two fragments, 75.5g and 90.9g, were found about 250km SSW. of Addis Ababa. Description, F. Heide, Chem. Erde, 1965, **24**, p.112, Meteor. Bull., 1965 (33).

Wold Cottage 54°8′12″ N., 0°24′48″ W.
Wold Newton, Scarborough, Yorkshire, England
Fell 1795, December 13, 1530 hrs
Synonym(s): *Yorkshire*
Stone. Olivine-hypersthene chondrite (L6).
After detonations heard in adjacent villages, a stone of about 56lb was seen to fall about 0.75 miles SW. of Wold Newton church, E. Howard, Phil. Trans. Roy. Soc. London, 1802, **92**, p.174, Ann. Phys. (Gilbert), 1803, **13**, p.297, Ann. Phys. (Gilbert), **15**, p.318. Mentioned, The Gentleman's Mag., 1797, **67**, p.549, Sowerby, British Mineralogy, 1807, **2**, p.3. Analysis, 22.73 % total iron, A.A. Moss et al., Min. Mag., 1967, **36**, p.101. Olivine Fa_{24}, B. Mason, Geochimica et Cosmochimica Acta, 1963, **27**, p.1011.
102g Vienna, Naturhist. Mus.; 76g Tübingen, Univ.; 61g Harvard Univ.; 57g Washington, U.S. Nat. Mus.; 43g New York, Amer. Mus. Nat. Hist.; 41g Chicago, Field Mus. Nat. Hist.;
Specimen(s): [1073], 19.30kg. main mass, and pieces, 1367g

Wolf Creek 19°18′ S., 127°46′ E.
S. of Hall's Creek, Kimberley, Western Australia, Australia
Found 1947
Iron. Octahedrite, medium (0.85mm) (IIIB).
A large circular crater was first observed from the air in June 1947. Fragments of iron-shale are abundant on the SW. part of the crater rim; they contain 3.5% to 4.5%Ni, some retain a little unaltered metal, Spec. Publ. West. Austr. Mus., 1965 (3), p.52. Further material, 1.3kg found, analysis, 8.6 %Ni, S.R. Taylor, Nature, 1965, **208**, p.944. Further analysis, 9.22 %Ni, 18.4 ppm.Ga, 37.3 ppm.Ge, 0.036 ppm.Ir, E.R.D. Scott et al., Geochimica et Cosmochimica Acta, 1973, **37**, p.1957. Description; a little deformed or fractured, V.F. Buchwald, Iron Meteorites, Univ. of California, 1975, p.1327. Contains pecoraite, a hydrated Ni silicate, G.T. Faust et al., U.S. Geol. Surv. Prof. Paper, 1973 (384C).
387kg Albuquerque, Univ. of New Mexico, oxidised; 350kg Washington, U.S. Nat. Mus., 62g not oxidised; 75kg New York, Amer. Mus. Nat. Hist.; 4.5kg Tempe, Arizona State Univ.; 2.5kg Harvard Univ.;
Specimen(s): [1963,528], 739g.; [1964,748], 1548g. two fragments; [1965,432], 50g. two fragments, metal, from S.R. Taylor

Wolfsegg 48°6′ N., 13°42′ E.
Ober-Österreich, Austria
Found 1886, or earlier
Pseudometeorite. Iron.
A cuboid piece of iron, 785g, was found in Tertiary lignite during underground working in the Wolfsegg mine, S. of Hausruck mountain, A. Gurlt and G.A. Daubrée, C. R. Acad. Sci. Paris, 1886, **103**, p.702. Listed, a pseudometeorite, E. Wülfing, Meteoriten in Samml., Tübingen, 1897, p.407.

Wollaston's Iron v Bendegó.

Wollega v Nejo.

Wonyulgunna 24°55′ S., 120°4′ E.
Bald Hill, Western Australia, Australia
Found 1937, June
Iron. Octahedrite, medium (1.0mm) (IIIB).
A mass of 37.8kg was found on Bald Hill (formerly Wonyulgunna) sheep station, just W. of the 485 mile post on No. 1 rabbit-proof fence. Description, with an analysis, 8.26 %Ni, E.S. Simpson, Min. Mag., 1938, **25**, p.164. Analysis, 8.72 %Ni, 19.5 ppm.Ga, 39.6 ppm.Ge, 0.028 ppm.Ir, E.R.D. Scott et al., Geochimica et Cosmochimica Acta, 1973, **37**, p.1957. Description; shock-hardened, inclusion-rich, V.F. Buchwald, Iron Meteorites, Univ. of California, 1975, p.1329.
36.6kg Perth, West. Austr. Mus.; 260g Sydney, Austr. Mus.; 577g Washington, U.S. Nat. Mus.; 66g Los Angeles, Univ. of California;
Specimen(s): [1938,357], 948g.

Woodbine 42°20′48″ N., 90°10′3″ W.

Jo Davies County, Illinois, U.S.A.
Found 1953, in the spring
Iron. Octahedrite, fine (0.4mm), with silicate inclusions (IB-ANOM).

One mass of 48.2kg was discovered by a farmer while ploughing, Meteor. Bull., 1962 (24), W.F. Read, Trans. Illinois Acad. Sci., 1963, **56** (2), p.75. Classification, analysis, 10.6 %Ni, 36.7 ppm.Ga, 114 ppm.Ge, 1.4 ppm.Ir, J.T. Wasson, Icarus, 1970, **12**, p.407. Contains Copiapo-type silicates, T.E. Bunch et al., Contr. Miner. Petrol., 1970, **25**, p.297. Illustrated description, bulk analysis, B. Mason, Min. Mag., 1967, **36**, p.120. Further description, V.F. Buchwald, Iron Meteorites, Univ. of California, 1975, p.1331. Ar-Ar age, 4570 m.y., S. Niemeyer, Geochimica et Cosmochimica Acta, 1979, **43**, p.1829.

48kg Washington, U.S. Nat. Mus., includes main mass; 1.8kg Chicago, Field Mus. Nat. Hist.; 432g Tempe, Arizona State Univ.; 256g Paris, Mus. d'Hist. Nat.;
Specimen(s): [1971,294], 942g. polished slice

Woodbridge 52°6′ N., 1°19′ E.

Suffolk, England
Fell 1642, August 4, 1630 hrs
Doubtful..

A stone of 4lb is said to have fallen, The Gentleman's Mag., 1796, **66**, p.1007, E.F.F. Chladni, Die Feuer-Meteore, Wien, 1819, p.226, R.P. Greg, Rept. Brit. Assn., 1860, p.54, T.M. Hall, Min. Mag., 1879, **3**, p.6 the evidence is not conclusive.

Wood's Mountain 35°41′ N., 82°11′ W.

McDowell County, North Carolina, U.S.A.
Found 1918
Synonym(s): *McDowell County*
Iron. Octahedrite, fine (0.3mm) (IVA).

One mass of 3017g was found, E.P. Henderson, letter of 12 May, 1939 in Min. Dept. BM(NH). Listed, Ann. Rep. U.S. Nat. Mus., 1938, p.50, A.D. Nininger, Pop. Astron., Northfield, Minnesota, 1939, **47**, p.214. Description, analysis, 8.27 %Ni, S.H. Perry, Am. J. Sci., 1939, **237**, p.569 [M.A. 7-378]. Further analysis, 8.13 %Ni, 2.39 ppm.Ga, 0.143 ppm.Ge, 2.4 ppm.Ir, R. Schaudy et al., Icarus, 1972, **17**, p.174. Cosmically deformed; a weathered mass of 850g was found before 1923, V.F. Buchwald, Iron Meteorites, Univ. of California, 1975, p.1334.

2.1kg Washington, U.S. Nat. Mus., and 322g of the 1923 mass; 400g Raleigh, North Carolina State Mus., of the 1923 mass; 240g Chicago, Field Mus. Nat. Hist.; 79g Harvard Univ.; 68g New York, Amer. Mus. Nat. Hist.; 52g Tempe, Arizona State Univ.;
Specimen(s): [1959,981], 51g. slice and sawings, 1.5g

Woodward County 36°30′ N., 99°30′ W. approx.

Woodward County, Oklahoma, U.S.A.
Found 1923, known before this year
Synonym(s): *Alva*
Stone. Olivine-bronzite chondrite (H4).

A weathered mass of 43kg, and 2.5kg of fragments, were found, E.P. Henderson, letter of 19 April, 1939, W.A. Tarr, letter of 11 April, 1939 in Min. Dept. BM(NH). Listed, Ann. Rep. U.S. Nat. Mus., 1938, p.50, F.C. Leonard, Univ. New Mexico Publ., Albuquerque, 1946 (meteoritics ser. no. 1), p.44. Olivine Fa_{19}, B. Mason, Geochimica et Cosmochimica Acta, 1963, **27**, p.1011.

44kg Washington, U.S. Nat. Mus.; 182g New York, Amer. Mus. Nat. Hist.; 136g Calcutta, Mus. Geol. Surv. India; 69g Tübingen, Univ.; 53g Tempe, Arizona State Univ.;
Specimen(s): [1959,906], 147.5g.

Woolgorong 27°45′ S., 115°50′ E.

Nedlands, Western Australia, Australia
Fell 1960, December 20, 1400 hrs, approx. time and date
Stone. Olivine-hypersthene chondrite (L6).

Seen to fall but not found until July, 1961, 80lb of fragments were recovered. Description, with an analysis, G.J.H. McCall and P.M. Jeffery, J. Roy. Soc. West. Austr., 1964, **47**, p.33, Meteor. Bull., 1965 (33), Spec. Publ. West. Austr. Mus., 1965 (3), p.53. Olivine $Fa_{25.2}$, B. Mason, Rec. Austr. Mus., 1974, **29**, p.169.

Main mass, Perth, West. Austr. Mus.; 700g New York, Amer. Mus. Nat. Hist.; 362g Washington, U.S. Nat. Mus.;
Specimen(s): [1962,385], 290g. and fragments, 31.5g

Wooster 40°46′ N., 81°57′ W.

Wayne County, Ohio, U.S.A.
Found 1858, recognized in this year
Synonym(s): *Wayne County*
Iron. Octahedrite, medium (1.0mm) (IIIA?).

A mass of about 50lb was found in a wood. Description, with an analysis, J.L. Smith, Am. J. Sci., 1864, **38**, p.385. Description, E.P. Henderson and S.H. Perry, Proc. U.S. Nat. Mus., 1958, **107**, p.339 [M.A. 14-130]. The main mass appears to have been lost, only a few grams are known in collections. Description; has been artifically reheated, V.F. Buchwald, Iron Meteorites, Univ. of California, 1975, p.1336.

14g Philadelphia, Acad. Nat. Sci.; 11g Chicago, Field Mus. Nat. Hist.; 9.8g Tempe, Arizona State Univ.; 7g Harvard Univ.; 5g Paris, Mus. d'Hist. Nat.;
Specimen(s): [34584], 3.25g.; [34585], 2g.

Worowo v Angara.

Wray 40°3′ N., 102°12′ W.

Yuma County, Colorado, U.S.A.
Found 1936
Stone. Olivine-bronzite chondrite (H).

One stone of 281.7g found in SW. 1/4, sect. 35, was recognized as meteoritic in 1938, A.D. Nininger, Pop. Astron., Northfield, Minnesota, 1939, **47**, p.214. Olivine Fa_{15}, B. Mason, Geochimica et Cosmochimica Acta, 1963, **27**, p.1011.

167g Tempe, Arizona State Univ.; 24g Chicago, Field Mus. Nat. Hist.;
Specimen(s): [1959,905], 65.5g. slice

Wuchumuchin v Wu-chu-mu-ch'in.

Wu-chu-mu-ch'in 45°30′ N., 118° E.

Hsing-an range, Barin, Nei Monggol, China
Found 1920, September
Synonym(s): *Ujimgin, Wuchumuchin*
Iron. Ataxite.

A mass of 68.86kg was found, west of the Ulchin Gol and about 10 miles north of Barin. It was examined, O. Aochi and K. Owaza, J. Chinese Mining Industry (no. 54), H.T. Chang, Mem. Geol. Surv. China, 1927 (ser. B no. 2), p.372. Listed as an iron, D. Wang and Z. Ouyang, Geochimica, 1979, p.120, in Chinese. Listed, with references, classification, D. Bian, Meteoritics, 1981, **16**, p.115.

68.8kg Dalian, Mus. Nat. Hist.;

Wushe v Wuzhi.

Wushike v Armanty.

Wuzhi 35°8′ N., 113°20′ E.
Henan, China
Fell 1931, June 25, 2300 hrs
Synonym(s): *Wushe*
Stone.
Listed, a single stone of unreported weight, D. Bian, Meteoritics, 1981, **16**, p.116.

Wynella 28°57′ S., 148°8′ E.
Wynella station, Queensland, Australia
Found 1945, known in this year, recognized 1966
Synonym(s): *Dirranbandi*
Stone. Olivine-bronzite chondrite (H4).
A single mass, 40kg, was found, Meteor. Bull., 1968 (42), Meteoritics, 1970, **5**, p.96. Olivine $Fa_{18.7}$, B. Mason, Rec. Austr. Mus., 1974, **29**, p.169.
Main mass, Sydney, Austr. Univ.; 726g Chicago, Field Mus. Nat. Hist.; 431g Washington, U.S. Nat. Mus.; 202g Tempe, Arizona State Univ.;
Specimen(s): [1967,382], 650g. slice, and fragments, 4.5g

Wynyard 51°53′ N., 104°11′ W.
Big Quill Lake, Saskatchewan, Canada
Found 1968, approx.
Stone. Olivine-bronzite chondrite (H5).
A single mass of 3479g was found in a field during ploughing, I. Halliday, priv. comm., 1983. Olivine Fa_{18}, Meteor. Bull., 1984 (62).

Wyoming v Silver Crown.

Xingyang 32°20′ N., 114°19′ E.
Henan, China
Fell 1977, December 1, 1857 hrs
Synonym(s): *Xin Yang*
Stone. Olivine-bronzite chondrite (H5).
Two masses, 48kg and 27.5kg, were recovered; listed with references, D. Bian, Meteoritics, 1981, **16**, p.119. Analysis, 28.82 % total iron, K. Tao et al., Sci. Geol. Sinica, 1979 (3), p.270.
48kg Beijing, Planetarium;

Xinjiang v Armanty.

Xin Yang v Xingyang.

Xinyi 34°22′ N., 118°20′ E.
Jiangsu, China
Found 1975
Stone. Olivine-bronzite chondrite (H5).
A mass of 69kg was found, 26.9 % total iron, Meteor. Bull., 1978 (55), Meteoritics, 1978, **13**, p.351. Olivine Fa_{15}, C. Tzewen et al., Sci. Geol. Sinica, 1975, p.375 [M.A. 76-2704]. Listed, with references, D. Bian, Meteoritics, 1981, **16**, p.119.
Main mass, Beijing, Acad. Sinica Inst. Geol.;

Xiquipilco v Toluca.

Xiquipilco no. 2
Mexico state, Mexico
Found 1949, recognized in this year
Iron. Octahedrite, coarse.
Total known weight about 600g, C.F. Lewis and C.B. Moore, Cat. Meteor. Arizona State Univ., 1976, p.233.
312g Tempe, Arizona State Univ.;

Xi Ujimgin 44°40′ N., 117°30′ E.
Chaidamu, Xi Ujimgin County, Nei Monggol, China
Fell 1980, August 24, 1700 hrs
Stone. Olivine-hypersthene chondrite (L6).
A single mass of 5.9kg fell, D. Bian, Meteoritics, 1981, **16**, p.115. Analysis, 23.53 % total iron, W. Daode et al., Geochemistry, 1982, **1**, p.186.
5.9kg Beijing, Planetarium;

Yaddlethorpe 53°33′ N., 0°39′ W. approx.
Lincolnshire, England
Pseudometeorite..
A small uncrusted stone 27×20×17 mm was said to have fallen on 8 June, 1963. X-ray diffraction indicates that it is composed of goethite, graphite, gehlenite and glass, which are consistent with its being oxidized cast iron with slag inclusions, R.D. Morton and W.A.S. Sarjeant, The Mercian Geologist, 1971, **4**, p.37.

Yafee Mountains
Saudi Arabia
Pseudometeorite..

Yakushima 30°20′ N., 130°30′ E.
Kyushu, Japan
Fell 1902, or found
Doubtful. Stone.
A meteoritic stone of 46.45g in the Hungarian National Mus., Budapest, is said to have come from Yakushima, an island about 40 miles south of Kyushu, and to have fallen or been found in 1902, L. Tokody and M. Dudich, Magyarorszag Meteoritgyüjtemenyei, Budapest, 1951, p.55, 89. Not mentioned, I. Yamamoto, Kwasan Observ. Bull., 1935, p.306 [M.A. 7-173] and seems a very doubtful fall.

Yalgoo 28°23′ S., 116°43′ E.
Western Australia, Australia
Found 1937, known before this year
Stone. Olivine-hypersthene chondrite, amphoterite (LL).
A fragment of 850g was found in 1937 in the store-room of the Western Australian Mus. labelled "portion of a meteorite found near Yalgoo". Similar to Mellenbye, also an LL chondrite, E.S. Simpson, Min. Mag., 1938, **25**, p.165, Spec. Publ. West. Austr. Mus., 1965 (3), p.55. Probably part of Mellenbye, olivine Fa_{27}, B. Mason, Rec. Austr. Mus., 1974, **29**, p.169.
Main mass, Perth, West. Austr. Mus.; 5g New York, Amer. Mus. Nat. Hist.;

Yamanomura v Kyushu.

Yamato (a) v Yamato 6901.

Yamato (b) v Yamato 6902.

Yamato (c) v Yamato 6903.

Yamato (d) v Yamato 6904.

Yamato (e) v Yamato 6905.

Yamato (f) v Yamato 6906.

Yamato (g) v Yamato 6907.

Yamato (h) v Yamato 6908.

Yamato (i) v Yamato 6909.

Yamato (j) v Yamato 7301.

Yamato (k) v Yamato 7305.

Yamato (l) v Yamato 7308.

Yamato (m) v Yamato 7303.

Yamato 6901 71°50′ S., 36°15′ E. approx.
Yamato Mountains, Antarctica
Found 1969
Synonym(s): *Yamato (a)*
Stone. Enstatite chondrite (E3).
An individual of 715g collected by the 1969-70 Japanese Expedition to Antarctica, Meteor. Bull., 1974 (52), Meteoritics, 1974, **9**, p.118. Analysis, 29.75 % total iron, M. Shima et al., Earth planet. Sci. Lett., 1973, **19**, p.246. Full description, T. Nagata, ed., Mem. Nat. Inst. Polar Res., Tokyo, 1975 (Spec. Issue no. 5).

Yamato 6902 71°50′ S., 36°15′ E. approx.
Yamato Mountains, Antarctica
Found 1969
Synonym(s): *Yamato (b)*
Stone. Achondrite, Ca-poor. Diogenite (ADIO).
An individual of 138g collected by the 1969-70 Japanese Expedition to Antarctica, Meteor. Bull., 1974 (52), Meteoritics, 1974, **9**, p.118. Analysis, 11.29 % total iron, M. Shima, Meteoritics, 1974, **9**, p.123. Rare gas content, M. Shima et al., Earth planet. Sci. Lett., 1973, **19**, p.246. Full description, T. Nagata, ed., Mem. Nat. Inst. Polar Res., Tokyo, 1975 (Spec. Issue no. 5). Unbrecciated texture, pyroxene Fs24, A.M. Reid et al., Meteoritics, 1975, **10**, p.479, abs.

Yamato 6903 71°50′ S., 36°15′ E. approx.
Yamato Mountains, Antarctica
Found 1969
Synonym(s): *Yamato (c)*
Stone. Carbonaceous chondrite, type III (CV3).
An individual of 150g was found by the 1969-70 Japanese Expedition to Antarctica, Meteor. Bull., 1974 (52), Meteoritics, 1974, **9**, p.119. Analysis, 24.09 % total iron, M. Shima, Meteoritics, 1974, **9**, p.123. Rare gas content, M. Shima et al., Earth planet. Sci. Lett., 1973, **19**, p.246. Full description, T. Nagata, ed., Mem. Nat. Inst. Polar Res., Tokyo, 1975 (Spec. Issue no. 5).

Yamato 6904 71°50′ S., 36°15′ E. approx.
Yamato Mountains, Antarctica
Found 1969
Synonym(s): *Yamato (d)*
Stone. Olivine-bronzite chondrite (H6).
An individual of 62g was collected by the 1969-70 Japanese Expedition to Antarctica, Meteor. Bull., 1974 (52), Meteoritics, 1974, **9**, p.119. Analysis, 25.41 % total iron, M. Shima, Meteoritics, 1974, **9**, p.123. Rare gas content, M. Shima et al., Earth planet. Sci. Lett., 1973, **19**, p.246. Full description, olivine Fa_{18}, T. Nagata, ed., Mem. Nat. Inst. Polar Res., Tokyo, 1975 (Spec. Issue no. 5).

Yamato 6905 71°50′ S., 36°15′ E. approx.
Yamato Mountains, Antarctica
Found 1969
Synonym(s): *Yamato (e)*
Stone. Chondrite.
Specimens Yamato 6905 to Yamato 6909 are all chondrites, of weight 10-41g, T. Nagata, ed., Mem. Nat. Inst. Polar Res., Tokyo, 1975 (Spec. Issue no. 5).

Yamato 6906
Yamato Mountains, Antarctica
Synonym(s): *Yamato (f)*
Stone. Olivine-bronzite chondrite (H5).
See entry for Yamato 6905.

Yamato 6907
Yamato Mountains, Antarctica
Synonym(s): *Yamato (g)*
Stone. Olivine-bronzite chondrite (H).
See entry for Yamato 6905.

Yamato 6908
Yamato Mountains, Antarctica
Synonym(s): *Yamato (h)*
Stone. Olivine-bronzite chondrite (H5).
See entry for Yamato 6905.

Yamato 6909
Yamato Mountains, Antarctica
Synonym(s): *Yamato (i)*
Stone. Olivine-hypersthene chondrite (L6).
See entry for Yamato 6905.

Yamato 7301 71°50′ S., 35°30′ E.
Southern Yamato Mountains, Antarctica
Found 1973, December 14
Synonym(s): *Yamato (j)*
Stone. Olivine-bronzite chondrite (H4).
An individual of 650g was found, olivine Fa_{20}, 25.13 % total iron, Meteor. Bull., 1979 (56), Meteoritics, 1979, **14**, p.172, K. Yanai, Cat. Yamato Meteorites., Tokyo, 1979, p.5. Mineralogical study, K. Yagi et al., Mem. Nat. Inst. Polar Res., Tokyo, 1978 (8), p.121.

Yamato 7304 72°50′ S., 35°30′ E. approx.
SW. of the Yamato Mountains, Antarctica
Found 1973, December 17
Synonym(s): *Yamato (m)*
Stone. Olivine-hypersthene chondrite (L6).
An individual of 500g was found by the 1973-74 Japanese Antarctic Expedition, olivine Fa_{26}, 22.98 % total iron, Meteor. Bull., 1979 (56), Meteoritics, 1979, **14**, p.173 under Yamato 7303. Listed, K. Yanai, Cat. Yamato Meteorites, Tokyo, 1979, p.5. Mineralogical study, K. Yagi et al., Mem. Nat. Inst. Polar Res., Tokyo, 1978 (8), p.121.

Yamato 7305 71°50′ S., 36°50′ E.
Southern Yamato Mountains, Antarctica
Found 1973, December 21
Synonym(s): *Yamato (k)*
Stone. Olivine-hypersthene chondrite (L6).
An individual of 900g was found by the 1973-74 Japanese Antarctic Expedition. olivine Fa_{26}, 22.64 % total iron, Meteor. Bull., 1979 (56), Meteoritics, 1979, **14**, p.173, K. Yanai, Cat. Yamato Meteorites, Tokyo, 1979, p.6. Mineralogical study, K. Yagi et al., Mem. Nat. Inst. Polar Res., Tokyo (8), p.121.

Yamato 7307 (in error) v Yamato 7308.

Yamato 7308 71°50′ S., 36°30′ E.
Southern Yamato Mountains, Antarctica
Found 1973, December 22
Synonym(s): *Yamato (l), Yamato 7307 (in error)*
Stone. Achondrite, Ca-rich. Howardite (AHOW).
An individual of 480g was found by the 1973-74 Japanese Antarctic Expedition, 13.31 % total iron, Meteor. Bull., 1979 (56), Meteoritics, 1979, **14**, p.174, K. Yanai, Cat. Yamato Meteorites, Tokyo, 1979, p.6. Mineralogical study, under Yamato 7307, M. Miyamoto et al., Mem. Nat. Inst. Polar Res., Tokyo, 1978 (8), p.185.

Yamato 74001 71°50′ S., 35°30′ E. approx.
Yamato Mountains, Antarctica
Found 1974, November
Stone. Olivine-bronzite chondrite (H5).
A mass of 246g was found, the first of a number of recovered specimens. 663 fragments, mainly chondritic, representing a large number of individual falls were found by the 1974-75 Japanese Antarctic Expedition. The specimens range in weight from 0.1g to 5575g and were numbered in the sequence of recovery Yamato 74001 to Yamato 74663, and are so mamed. The majority are chondrites weighing less than 100g. Listed here, with few exceptions, are those ordinary chondrite specimens of mass greater than 500g, achondrites and irons. Full listing of all recovered material, specimen weights, classification and references, K. Yanai, Cat. Yamato Meteorites, Tokyo, 1979. The problems of synonymy are manifold, specimen names apply to masses as found. The above locality information applies to each of the specimens from the 1974-75 expedition.

Yamato 74005 v Yamato 74013.

Yamato 74010 v Yamato 74013.

Yamato 74011 v Yamato 74013.

Yamato 74013
Synonym(s): *Yamato 74005, Yamato 74010, Yamato 74011, Yamato 74031, Yamato 74037, Yamato 74096, Yamato 74097, Yamato 74109, Yamato 74125, Yamato 74126, Yamato 74136, Yamato 74150, Yamato 74151, Yamato 74162, Yamato 74344, Yamato 74347, Yamato 74368, Yamato 74448, Yamato 74546, Yamato 74606, Yamato 74648, Yamato 75001, Yamato 75004, Yamato 75007, Yamato 75014, Yamato 75285*
Stone. Achondrite, Ca-poor. Diogenite (ADIO).
A mass of 2.05kg was found. Masses possibly synonymous over 500g are: Yamato 74097, 2193g; Yamato 74136, 725g; Yamato 74037, 591g. Mineralogy, H. Takeda et al., Mem. Nat. Inst. Polar Res., Tokyo, 1978 (8), p.170. Rb-Sr and Nd-Sm systematics, N. Nakamura, Mem. Nat. Inst. Polar Res., Tokyo, 1979 (15), p.219.

Yamato 74014
Stone. Olivine-bronzite chondrite (H6).
An individual of 2.36kg was found.

Yamato 74031 v Yamato 74013.

Yamato 74037 v Yamato 74013.

Yamato 74044
Stony-iron. Pallasite (PAL).
A mass of 51.8g was found olivine $Fa_{12.3}$, K. Yanai, Cat. Yamato Meteorites, Tokyo, 1979, p.16.

Yamato 74077
Stone. Olivine-hypersthene chondrite (L6).
An individual of 5.57kg was found.

Yamato 74079
Stone. Olivine-bronzite chondrite (H5).
A mass of 620g was found.

Yamato 74080
Stone. Olivine-hypersthene chondrite (L6).
An individual of 536g was found.

Yamato 74094
Stone. Olivine-bronzite chondrite (H6).
A mass of 867g was found.

Yamato 74096 v Yamato 74013.

Yamato 74097 v Yamato 74013.

Yamato 74109 v Yamato 74013.

Yamato 74115
Stone. Olivine-bronzite chondrite (H5).
An individual of 1.04kg was found.

Yamato 74118
Stone. Olivine-hypersthene chondrite (L6).
An individual of 845g was found.

Yamato 74123
Stone. Achondrite, Ca-poor. Ureilite (AURE).
A mass of 69.9g was found, analysis, olivine Fa_{13-23}, K. Yanai, Cat. Yamato Meteorites, Tokyo, 1979, p.31, 188. Mineralogy, H. Takeda et al., Mem. Nat. Inst. Polar Res., Tokyo, 1979 (15), p.54.

Yamato 74125 v Yamato 74013.

Yamato 74126 v Yamato 74013.

Yamato 74130
Stone. Achondrite, Ca-poor. Ureilite (AURE).
A mass of 17.9g was found. Analysis, mineralogy, olivine Fa_{23}, H. Takeda et al., Mem. Nat. Inst. Polar Res., Tokyo, 1979 (15), p.54.

Yamato 74136 v Yamato 74013.

Yamato 74150 v Yamato 74013.

Yamato 74151 v Yamato 74013.

Yamato 74155
Synonym(s): *Yamato 74156*
Stone. Olivine-bronzite chondrite (H4).
Two masses totalling 3788g were found.

Yamato 74156 v Yamato 74155.

Yamato 74159
Synonym(s): *Yamato 75011, Yamato 75015, Yamato 75295, Yamato 75296, Yamato 75307*
Stone. Achondrite, Ca-rich. Eucrite (AEUC).
An individual of 98.2g was found. Other possibly synonymous masses are Yamato 75011, 121g; Yamato 75015, 166g; Yamato 75295, 8.8g; Yamato 75296, 8.6g Yamato 75307, 7.9g. Polymict breccias, mineralogy, H. Takeda et al., Mem. Nat. Inst. Polar Res., Tokyo, 1979 (12), p.82.

Yamato 74162 v Yamato 74013.

Yamato 74190
Stone. Olivine-hypersthene chondrite (L6).
An individual of 3.23kg was found.

Yamato 74191
Stone. Olivine-hypersthene chondrite (L3).
An individual of 1.09kg was found. Unequilibrated. Analysis, K. Yanai, Cat. Yamato Meteorites, Tokyo, 1979, p.188.

Yamato 74193
Stone. Olivine-bronzite chondrite (H4-5).
An individual of 1.81kg was found.

Yamato 74344 v Yamato 74013.

Yamato 74347 v Yamato 74013.

Yamato 74354
Stone. Olivine-hypersthene chondrite (L6).
A fragment of 2.72kg was found.

Yamato 74356
Stone. Achondrite, Ca-rich. Eucrite (AEUC).
An individual of 10g was found. A monomict breccia, H. Takeda et al., Mem. Nat. Inst. Polar Res., Tokyo, 1979 (15), p.54.

Yamato 74362
Stone. Olivine-hypersthene chondrite (L6).
An individual of 4.17kg was found.

Yamato 74364
Stone. Olivine-bronzite chondrite (H4).
An individual of 757g was found.

Yamato 74368 v Yamato 74013.

Yamato 74371
Stone. Olivine-bronzite chondrite (H5).
An individual of 5.06kg was found.

Yamato 74418
Stone. Olivine-bronzite chondrite (H6).
A fragment of 567g was found.

Yamato 74442
Stone. Olivine-hypersthene chondrite, amphoterite (LL4).
An individual of 173.3g was found; analysis, K. Yanai, Cat. Yamato Meteorites, Tokyo, 1979, p.188.

Yamato 74445
Stone. Olivine-hypersthene chondrite (L6).
An individual of 2.29kg was found.

Yamato 74448 v Yamato 74013.

Yamato 74450
Stone. Achondrite, Ca-rich. Eucrite (AEUC).
An individual of 235g was found. Analysis, K. Yanai, Cat. Yamato Meteorites Tokyo, 1979, p.187.

Yamato 74454
Stone. Olivine-hypersthene chondrite (L6).
An individual of 578g was found.

Yamato 74459
Stone. Olivine-bronzite chondrite (H6).
An individual of 1.71kg was found.

Yamato 74546 v Yamato 74013.

Yamato 74605
Stone. Olivine-hypersthene chondrite (L6).
A fragment of 580g was found.

Yamato 74606 v Yamato 74013.

Yamato 74640
Stone. Olivine-bronzite chondrite (H6).
An individual of 1.06kg was found.

Yamato 74641
Stone. Carbonaceous chondrite, type II (CM2).
A specimen of 4.5g was found.

Yamato 74642
Synonym(s): *Yamato 75293*
Stone. Carbonaceous chondrite, type II (CM2).
An individual of 10.6g was found. Analysis, K. Yanai, Cat. Yamato Meteorites, Tokyo, 1979, p.188.

Yamato 74646
Stone. Olivine-hypersthene chondrite, amphoterite (LL6).
An individual of 554g was found. Analysis, K. Yanai, Cat. Yamato Meteorites, Tokyo, 1979, p.188.

Yamato 74647
Stone. Olivine-bronzite chondrite (H5).
An individual of 2.32kg was found.

Yamato 74648 v Yamato 74013.

Yamato 74659
Stone. Achondrite, Ca-poor. Ureilite (AURE).
A fragment of 18.9g was found. Analysis, K. Yanai, Cat. Yamato Meteorites, Tokyo, 1979, p.188. Mineralogy, olivine $Fa_{8.6}$, H. Takeda et al., Mem. Nat. Inst. Polar Res., Tokyo, 1979 (12), p.82.

Yamato 74662
Stone. Carbonaceous chondrite, type II (CM2).
An individual of 150g was found. Analysis, K. Yanai, Cat. Yamato Meteorites, Tokyo, 1979, p.188. Trace element data, G.W. Kallemeyn and J.T. Wasson, Geochimica et Cosmochimica Acta, 1981, **45**, p.1217.

Yamato 75001 v Yamato 74013.

Yamato 75003 71°50′ S., 35°30′ E. approx.
Yamato Mountains, Antarctica
Found 1975, December 1975-January 1976
Stone. Carbonaceous chondrite (C).
A mass of 1.5g was found. 307 pieces, mainly chondritic and weighing less than 100g, were collected by the 1975-76 Japanese Antarctic Expedition. The specimens range in weight from 0.1g to 11.0kg and were numbered in the sequence of recovery, Yamato 75001 to Yamato 75307, and are so named. Listed here are those ordinary chondrite specimens of mass greater than 500g, achondrites and irons. Full listing of samples, weights, classification and references, K. Yanai, Cat. Yamato Meteorites, Tokyo, 1979. The problems of synonymy are manifold, specimen numbers apply to masses as found. The above locality information applies to each of the specimens from the 1975-76 expedition.

Yamato 75004 v Yamato 74013.

Yamato 75007 v Yamato 74013.

Yamato 75011 v Yamato 74159.

Yamato 75014 v Yamato 74013.

Yamato 75015 v Yamato 74159.

Yamato 75028
Stone. Olivine-bronzite chondrite (H3).
An individual of 6.1kg was found. Description, unequilibrated, and analysis, olivine Fa_{14-21}, H. Takeda et al., Mem. Nat. Inst. Polar Res., Tokyo (15), p.54.

Yamato 75031
Iron. Octahedrite, plessitic (IRANOM).
An individual of 60.2g was found. Classification and analysis, 14.2 %Ni, 31.2 ppm.Ga, 232 ppm.Ge, 0.34 ppm.Ir, A. Kracher et al., Geochimica et Cosmochimica Acta, 1980, **44**, p.773.

Yamato 75032
Stone. Achondrite, Ca-poor. Diogenite (ADIO).
An individual of 189g was found. Description; a monomict breccia with minor augite and rare plagioclase; analysis, orthopyroxene Fs33.6, H. Takeda et al., Mem. Nat. Inst. Polar Res., Tokyo, 1979 (8), p.170.

Yamato 75097
Stone. Olivine-hypersthene chondrite (L4).
An individual of 2.57kg was found.

Yamato 75102
Stone. Olivine-hypersthene chondrite (L6).
An individual of 11.0kg was found olivine $Fa_{24.6}$.

Yamato 75105
Iron. Hexahedrite (IIA).
An individual of 19.6g was found. Classification and analysis, 5.62 %Ni, 58.4 ppm.Ga, 170 ppm.Ge, 2.4 ppm.Ir, A. Kracher et al., Geochimica et Cosmochimica Acta, 1980, **44**, p.773.

Yamato 75108
Stone. Olivine-hypersthene chondrite (L4).
A fragment of 590g was found. Yamato 75018 to Yamato 75257 inclusive are probably from the same fall, Y. Matsumoto and M. Hayashi, Mem. Nat. Inst. Polar Res., Tokyo, 1980 (17), p.21.

Yamato 75110
Stone. Olivine-hypersthene chondrite (L4).
A fragment of 706g was found.

Yamato 75258
Stone. Olivine-hypersthene chondrite, amphoterite (LL6).
A mass of 971g was found. Analysis, K. Yanai, Cat. Yamato Meteorites, Tokyo, 1979, p.188.

Yamato 75260
Stone. Carbonaceous chondrite, type II (CM2).
An individual of 4.0g was found.

Yamato 75271
Stone. Olivine-hypersthene chondrite (L4).
An individual of 1.79kg was found.

Yamato 75274
Stony-iron. Mesosiderite (MES).
A mass of 5.1g was found, olivine Fa_4, Y. Matsumoto et al., Eigth Symp. Antarctic Meteorites, Tokyo, 1983, p.6.

Yamato 75285 v Yamato 74013.

Yamato 75293 v Yamato 74642.

Yamato 75295 v Yamato 74159.

Yamato 75296 v Yamato 74159.

Yamato 75299
Stone. Achondrite, Ca-poor. Diogenite (ADIO).
An individual of 9.1g was found.

Yamato 75307 v Yamato 74159.

Yamato 790001 72° S., 35°30′ E. approx.
Yamato Mountains, Antarctica
Found 1979, 1979-1980
Over 3300 meteorite specimens were recovered by the Japanese in the Yamato Mountains area during the 1979-1980 field season in Antarctica. They are named Yamato 790001 to Yamato 79xxxx in order of finding, K. Yanai, Mem. Nat. Inst. Polar Res., Tokyo, 1981 (Spec. issue no. 20), p.1. 3767 specimens collected, total mass about 320kg, includes four irons, one stony-iron, one ureilite, 31 carbonaceous chondrites (one weighing 25kg), 32 diogenites, three howardites and 43 eucrites, K. Yanai and H. Kojima, Eighth Symposium on Antarctic Meteorites, Tokyo, 1983, p.3.

Yamato 790003
Synonym(s): *Yamato 790032, Yamato 790033, Yamato 790034*
Stone. Carbonaceous chondrite, type II (CM2).
A mass of 4.29g was found. Yamato 790032, 6.1g, Yamato 790033, 1.4g and Yamato 790034, 0.3g were found at the same location and are probably part of the same fall, Meteorites News, Tokyo, 1982, **1**, p.1.

Yamato 790032 v Yamato 790003.

Yamato 790033 v Yamato 790003.

Yamato 790034 v Yamato 790003.

Yamato 790122
Stone. Achondrite, Ca-rich. Eucrite (AEUC).
A single mass of 109.5g was found.

Yamato 790260
Stone. Achondrite, Ca-rich. Eucrite (AEUC).
A single stone of 433.9g was found.

Yamato 790266
Stone. Achondrite, Ca-rich. Eucrite (AEUC).
A stone of 208.0g was found.

Yamato 790269
Stone. Olivine-bronzite chondrite (H4).
A mass of 1269g was found, olivine Fa_{18}, Meteorites News, Tokyo, 1982, **1**, p.12.

Yamato 790964
Stone. Olivine-hypersthene chondrite, amphoterite (LL).
A mass of 3335g was found, olivine $Fa_{31.1}$, Meteorites News, Tokyo, 1982, **1**, p.14.

Yamato 790981
Stone. Achondrite, Ca-poor. Ureilite (AURE).
A mass of 213g was found, olivine Fa_{21}, Meteorites News, Tokyo, 1982, **1**, p.15.

Yamato 791209
Stone. Olivine-bronzite chondrite (H5).
A mass of 3288g was found, olivine $Fa_{17.6}$, Meteorites News, Tokyo, 1982, **1**, p.22.

Yamato 791493
Stony-iron. Lodranite (LOD).
Brief petrographic description, olivine $Fa_{11.6}$, K. Yanai and H. Kojima, Meteoritics, 1982, **17**, p.300.

Yamato 791824
Stone. Carbonaceous chondrite, type II (CM2).
A single mass of about 150g was found and kept in cold storage, Meteorites News, Tokyo, 1982, **1**, p.23.

Yamato 793321
Stone. Carbonaceous chondrite, type II (CM2).
A single mass of about 200g was found and kept in cold storage, Meteorites News, Tokyo, 1982, **1**, p.24.

Yamato 8001 71°50′ S., 35°30′ E. approx.
Yamato Mountains, Antarctica
Found 1980
Thirteen specimens were found by the Japanese Antarctic Expedition in the Yamato Mountains region during the 1980-81 field season, K. Yanai and T. Nagata, Meteoritics, 1982, **17**, p.300.

Yamato 81001 71°50′ S., 35°30′ E. approx.
Yamato Mountains, Antarctica
133 meteorite specimens were collected by the Japanese Antarctic Expedition during the 1981-82 field season in the Yamato Mountains area Specimens include 121 chondrites, 2 achondrites and 7 carbonaceous chondrites. The largest mass weighs 10.7kg, Y. Yoshida and K. Sasaki, Eigth Symp. on Antarctic Meteorites, Tokyo, 1983, p.1.

Yambo 1° N., 22°30′ E. approx.
Bosefele, Djolu, Zaire
Fell 1951, October 20, 2100 hrs
Stone. Olivine-bronzite chondrite (H5).
At least 11 stones fell at Yambo, near Bosefele, Djolu, Zaire. Account of fall and analysis, G. Viseur, Bull. Serv. Geol. Congo Belge, 1954 (no. 5), p.36 [M.A. 12-611].
4g Washington, U.S. Nat. Mus.;

Yambo no. 2
Zaire
Stone. Olivine-hypersthene chondrite (L3).
An unequilibrated olivine-hypersthene chondrite, L3, is listed, B. Mason, Smithson. Contrib. Earth Sci., 1975 (14), p.71.
3.2g Washington, U.S. Nat. Mus.;

Yamuishova v Pavlodar (pallasite).

Yamyschewa v Pavlodar (pallasite).

Yamyshev v Pavlodar (pallasite).

Yamysheva v Pavlodar (pallasite).

Yandama 29°45′ S., 141°2′ E.
Big Plain, Lake Stewart station, New South Wales, Australia
Found 1914, known before this year
Stone. Olivine-hypersthene chondrite (L6).
A mass of 5.8kg was acquired by the South Austr. Mus. in 1914. Description, with an analysis, A.R. Alderman, Rec. S. Austr. Mus., 1936, **5**, p.537 [M.A. 7-70]. Similar to Cartoonkana (*q.v.*); tho two could belong to the same fall, olivine $Fa_{25.0}$, B. Mason, Rec. Austr. Mus., 1974, **29**, p.169. Analysis, 20.77 % total iron, M.J. Fitzgerald, Ph.D. Thesis, Univ. of Adelaide, 1979, p.23.
4kg Adelaide, South Austr. Mus.; 377g Washington, U.S. Nat. Mus.; 184g New York, Amer. Mus. Nat. Hist.;
Specimen(s): [1963,526], 276g.

Yang-Chiang v Yangchiang.

Yangchiang 21°50′ N., 111°50′ E.
Guangdong, China
Fell 1954, April 12, 0400 hrs
Synonym(s): *Kuangtung Yang-Chiang, Yang-Chiang, Yangjian, Yangjiang*
Stone. Olivine-bronzite chondrite (H5).
Description, in Chinese with English abstract, and analysis, 28.38 % total iron, C. Tzewen, Acta Geol. Sinica, 1966, **46**, p.64. Listed, references, co-ordinates; a single 20kg mass, D. Bian, Meteoritics, 1981, **16**, p.115.

Yangjian v Yangchiang.

Yangjiang v Yangchiang.

Yanhuitlan 17°32′ N., 97°21′ W.
Oaxaca, Mexico
Found 1825, known before this year
Synonym(s): *Cholula, Goldbach's Iron, Misteca (in part), Oaxaca, Teposcolula*
Iron. Octahedrite, fine (0.3mm) (IVA).
A mass of about 421kg was found by Indians at the foot of the hill Deque-Yacunino and was used as an anvil; in 1825 it was seen by A.F. Morney who analysed a fragment, A.F.

Morney, El Mosaico mexicano, 1840, **3**, p.219, L. Fletcher, Min. Mag., 1890, **9**, p.171. Description, analysis, 7.36 %Ni, E. Cohen, Meteoritenkunde, 1905, **3**, p.316. The whole mass has been artifically reheated to 700-800C, V.F. Buchwald, Iron Meteorites, Univ. of California, 1975, p.1337. Further analysis, 7.49 %Ni, 1.75 ppm.Ga, 0.105 ppm.Ge, 2.7 ppm.Ir, R. Schaudy et al., Icarus, 1972, **17**, p.174.

300kg Mexico City, Mus. de Chopo, approx. weight; 17kg Chicago, Field Mus. Nat. Hist.; 3kg Tempe, Arizona State Univ.; 1.2kg Washington, U.S. Nat. Mus.;

Specimen(s): [1919,142], 296.5g. slice

Yardea 32°27′ S., 135°33′ E.

Gawler Range, South Australia, Australia
Found 1875
Synonym(s): *Gawler Range*
Iron. Octahedrite, coarse (2.0mm) (IA).

A mass of 7.25lb was found 4 miles south of Yardea station, C. Anderson, Rec. Austr. Mus., 1913, **10**, p.66. Classification, analysis, 6.92 %Ni, 88.8 ppm.Ga, 361 ppm.Ge, 4.3 ppm.Ir, A. Kracher et al., Geochimica et Cosmochimica Acta, 1980, **44**, p.773. Description, V.F. Buchwald, Iron Meteorites, Univ. of California, 1975, p.1341.

3.12kg Adelaide, South Austr. Mus., main mass; 73g Chicago, Field Mus. Nat. Hist.;

Yardymlinskii v Yardymly.

Yardymly 38°56′ N., 48°15′ E.

Near Lenkoran, Azerbaydzhan SSR, USSR, [Ярдымлы]
Fell 1959, November 24, 0705 hrs
Synonym(s): *Aroos, Iardymlinskii, Jardymlinsky, Yardymlinskii*
Iron. Octahedrite, coarse (2.2mm) (IA).

Five masses were recovered, of 127.0kg, 11.3kg, 5.83kg, 5.7kg, and 0.36kg, Meteor. Bull., 1960 (16), Meteor. Bull., 1960 (18). Mentioned, G.P. Vdovykin, Метеоритика, 1964, **25**, p.134. Analysis, under Aroos, 6.71 %Ni, 88.2 ppm.Ga, 387 ppm.Ge, 1.6 ppm.Ir, J.T. Wasson, Icarus, 1970, **12**, p.407. U content, L.I. Genaeva et al., Метеоритика, 1972, **31**, p.137. Description; cosmically annealed, V.F. Buchwald, Iron Meteorites, Univ. of California, 1975, p.1343.

133kg Baku, Acad. Sci., approx. weight; 4.5kg Moscow, Acad. Sci.; 81g Washington, U.S. Nat. Mus.;

Yarra Yarra River v Cranbourne.

Yarri 29°27′ S., 121°13′ E. location uncertain

Western Australia, Australia
Found 1908, before this year
Iron. Octahedrite, medium (1.0mm) (IIIA).

A mass of about 1.5kg was found, Cat. Meteor. Arizona State Univ., 1964, p.1062. Illustrated description; co-ordinates and an analysis, 8.06 %Ni, W.H. Cleverly and R.P. Thomas, J. Roy. Soc. West. Austr., 1969, **52**, p.89. Further analysis, 7.77 %Ni, 19.8 ppm.Ga, 38.5 ppm.Ge, 4.0 ppm.Ir, E.R.D. Scott et al., Geochimica et Cosmochimica Acta, 1973, **37**, p.1957. Shock-hardened, V.F. Buchwald, Iron Meteorites, Univ. of California, 1975, p.1347.

1.5kg Kalgoorlie, West. Austr. School of Mines; 210g Tempe, Arizona State Univ.;

Yarroweyah 35°59′ S., 145°35′ E.

Moira, Victoria, Australia
Found 1903
Iron. Hexahedrite (IIA).

A mass of 21lb (9.6kg) was found about 4.5 miles south of Yarroweyah railway station. Description, with an analysis, 4.95 %Ni, R.H. Walcott, Mem. Nat. Mus. Melbourne, 1915 (6), p.47. Further analysis, 5.48 %Ni, 59.3 ppm.Ga, 171 ppm.Ge, 18 ppm.Ir, J.T. Wasson, Geochimica et Cosmochimica Acta, 1969, **33**, p.859.

Main mass, Melbourne, Nat. Mus.; 149g Canberra, Austr. Nat. Univ.;

Specimen(s): [1983,M.36], 35g.

Yatoor 14°18′ N., 79°46′ E.

Nellore, Andhra Pradesh, India
Fell 1852, January 23, 1630 hrs
Synonym(s): *Nellore, Yatur, Yetur*
Stone. Olivine-bronzite chondrite (H5).

After a single detonation, a stone of about 30lb was seen to fall, W. von Haidinger, Sitzungsber. Akad. Wiss. Wien, Math.-naturwiss. Kl., 1861, **44**, p.73, N.S. Maskelyne and V. von Lang, Phil. Mag., 1863, **25**, p.443. Olivine Fa_{19}, B. Mason, Geochimica et Cosmochimica Acta, 1963, **27**, p.1011. The place of fall is Yetur, 14°18′N., 79°46′E., C.A. Silberrad, Min. Mag., 1932, **23**, p.296.

222g Washington, U.S. Nat. Mus.; 202g Vienna, Naturhist. Mus.; 91g Berlin, Humboldt Univ.; 89g Tempe, Arizona State Univ.; 81g Paris, Mus. d'Hist. Nat.; 80g Harvard Univ.; 75g New York, Amer. Mus. Nat. Hist.; 38g Tempe, Arizona State Univ.; 28g Chicago, Field Mus. Nat. Hist.; 27g Yale Univ.;

Specimen(s): [34793], 9.5kg. and five pieces totalling 511g

Yatur v Yatoor.

Yawata v Gifu.

Yayjinna 32°1′30″ S., 126°11′ E.

Western Australia, Australia
Found 1965
Stone. Olivine-hypersthene chondrite (L6).

One complete stone, 262.5g, was found, G.J.H. McCall, First Suppl. to West. Austr. Mus. Spec. Publ. no. 3, 1968. Olivine $Fa_{26.0}$, B. Mason, Rec. Austr. Mus., 1974, **29**, p.169.

Main mass, Kalgoorlie, West. Austr. School of Mines; 14g Washington, U.S. Nat. Mus.;

Ybbsitz 47°57′36″ N., 14°53′24″ E.

Nieder Osterreich, Austria
Found 1977, recognized 1980
Stone. Olivine-bronzite chondrite (H4).

A single mass of about 15kg was collected as a rock sample in 1977 and recognized as meteoritic in 1980, Meteor. Bull., 1980 (58), Meteoritics, 1980, **15**, p.240.

Yelenovka v Elenovka.

Yenberrie 14°15′ S., 132°1′ E.

Northern Territory, Australia
Found 1918
Iron. Octahedrite, coarse (2.1mm) (IA).

A mass of 291lb was found about 20 miles SSE. of Yenberrie. Description, with an analysis, 5.98 %Ni, J.C.H. Mingaye, J. Washington Acad. Sci., 1920, **10**, p.314. Co-ordinates, B. Mason, letter of 14 June, 1974 in Min. Dept. BM(NH). Analysis, 6.72 %Ni, 86.7 ppm.Ga, 312 ppm.Ge,

2.9 ppm.Ir, J.T. Wasson, Meteorites, Springer-Verlag, 1974, p.297. Description, V.F. Buchwald, Iron Meteorites, Univ. of California, 1975, p.1348.
74kg Sydney, Austr. Mus.; 4.5kg Chicago, Field Mus. Nat. Hist.; 3.75kg New York, Amer. Mus. Nat. Hist.; 3.8kg Washington, U.S. Nat. Mus.;
Specimen(s): [1963,527], 599g.

Yenshigahara v Kyushu.

Yeo Yeo Creek v Narraburra.

Yetur v Yatoor.

Yilmia 31°11'30" S., 121°32' E.
Kambalda district, Western Australia, Australia
Found 1969, recognized 1971
Stone. Enstatite chondrite (E5).
In 1969, numerous fragments were found in soil; in 1971 a single large mass with a primary ablation surface, together with fragments totalling 24kg, were found, G.J.H. McCall, 2nd. Suppl. to West. Austr. Mus. Spec. Publ. no. 3, 1972. Illustrated description, mineral chemistry, P.R. Buseck and E.F. Holdsworth, Meteoritics, 1972, **7**, p.429, A. El Goresy and J.F. Lovering, Meteoritics, 1973, **8**, p.31, abs.
Main mass, Perth, West. Austr. Mus.; 587g Chicago, Field Mus. Nat. Hist.; 170g Albuquerque, Univ. of New Mexico; 96g Kalgoorlie, West. Austr. School of Mines; Slice, Tempe, Arizona State Univ.;
Specimen(s): [1972,132], 58.2g. two fragments; [1972,496], 226g. and pieces 8.85g

Yingde 24°12' N., 113°24' approx.
Guangdong, China
Found 1964, possibly found in about 1860
Iron. Octahedrite, fine (0.3mm) (IVA).
A mass of about 300kg is in Guangzhou, Guangdong Mus., D. Bian, Meteoritics, 1981, **16**, p.115. Classification, analysis, 7.7 %Ni, 1.8 ppm.Ga, 0.115 ppm.Ge, 2.9 ppm.Ir, D.J. Malvin et al., priv. comm., 1983.

Yocemento 38°54' N., 99°26' W.
Ellis County, Kansas, U.S.A.
Found 1966
Stone. Olivine-hypersthene chondrite (L4).
A 5.92kg stone was found during ploughing, Meteor. Bull., 1969 (47), Meteoritics, 1970, **5**, p.105. Olivine Fa_{24}, R. Hutchison, priv. comm.
Main mass, Fort Hays, Kansas State College Mus.;
Specimen(s): [1970,2], 6.7g.

Yodze v Jodzie.

Yonatsu v Yonozu.

Yongning 22°45' N., 108°20' E.
Guangxi, China
Found 1971
Iron. Octahedrite, coarsest (3mm) (IA).
A mass of 60kg was found, D. Bian, Meteoritics, 1981, **16**, p.115. Classification, analysis, 6.4 %Ni, 98 ppm.Ga, 490 ppm.Ge, 3.8 ppm.Ir, D.J. Malvin et al., priv. comm., 1983.
Main mass, Nanning, Guanxi Bureau of Geol.;

Yonoosu v Yonozu.

Yonozu 37°45' N., 139° E. approx
Nishi Kambara, Niigata, Honshu, Japan
Fell 1837, July 13, 1600 hrs
Synonym(s): *Echigo, Eschigo, Tomi-naga, Yonatsu, Yonoosu*
Stone. Olivine-bronzite chondrite (H4).
After detonations, a stone of 30.44kg fell. Description, K. Jimbo, Beitr. Miner. Japan, 1906 (2), p.36. Description, analysis, C.W. Beck and R.E. Stevenson, Am. J. Sci., 1951, **249**, p.815 [M.A. 11-441]. Date of fall, S. Murayama, priv. comm. Olivine Fa_{18}, B. Mason, Geochimica et Cosmochimica Acta, 1963, **27**, p.1010.
31.65kg Tokyo, Nat. Mus.; 107g New York, Amer. Mus. Nat. Hist.; 44g Washington, U.S. Nat. Mus.; 18g Tempe, Arizona State Univ.; 11g Vienna, Naturhist. Mus.;
Specimen(s): [1905,70], 34.5g.

York (iron) 40°45' N., 97°30' W.
York County, Nebraska, U.S.A.
Found 1878
Iron. Octahedrite, medium (1.0mm) (IIIA).
A mass of 835g was ploughed up. Report and analysis, 7.38 %Ni, E.H. Barbour, Rep. Geol. Surv. Nebraska, Lincoln, 1903, p.184. Description and classification, V.F. Buchwald, Iron Meteorites, Univ. of California, 1975, p.1350.
795g New York, Amer. Mus. Nat. Hist.; 30g Washington, U.S. Nat. Mus.;

York (stone) 40°52' N., 97°36' W.
York County, Nebraska, U.S.A.
Found 1928, recognized 1936
Stone. Olivine-hypersthene chondrite (L).
A mass of 1440 was found, A.D. Nininger, Pop. Astron., Northfield, Minnesota, 1937, **45**, p.449 [M.A. 7-62]. Olivine Fa_{24}, B. Mason, Geochimica et Cosmochimica Acta, 1963, **27**, p.1011.
1.0kg Mainz, Max-Planck-Inst.; 42g Tempe, Arizona State Univ.;
Specimen(s): [1959,907], 34.5g. slice

Yorkshire 54° N., 1° W. approx.
England
Fell 1360
Doubtful..
A stone or stones are said to have fallen, R.P. Greg, Rept. Brit. Assn., 1860, p.52 but the evidence is not conclusive.

Yorkshire v Middlesbrough.

Yorkshire v Wold Cottage.

Yorktown (New York) 41°17' N., 73°49' W.
Westchester County, New York, U.S.A.
Fell 1869, September
Stone. Olivine-hypersthene chondrite (L5).
A stone of about 250g was found; long confused with Tomhannock Creek but is a distinct fall, B. Mason and H.B. Wiik, Min. Mag., 1960, **32**, p.528. Whereabouts of the main mass not known. Olivine Fa_{24}, B. Mason, Geochimica et Cosmochimica Acta, 1963, **27**, p.1011. The Chicago and BM (NH) specimens have olivine of Fa19.
7g Chicago, Field Mus. Nat. Hist.; 5.5g New York, Amer. Mus. Nat. Hist.;
Specimen(s): [63886], 4g.

Yorktown (Texas) 28°57′ N., 97°24′10″ W.
De Witt County, Texas, U.S.A.
Found 1957, or 1958, recognized 1965
Synonym(s): *Myersville (a)*
Stone. Olivine-bronzite chondrite (H).
A mass of 3.5kg was ploughed up, G.I Huss, letter of 23 May, 1974 in Min. Dept. BM(NH). Olivine Fa_{20}, A.L. Graham, priv. comm., 1978.
1.4kg Mainz, Max-Planck-Inst.;
Specimen(s): [1974,M.13], 86.1g. slice

Yosagori v Takenouchi.

Yoshiki 34°10′ N., 131°27′ E.
Yamaguchi, Honshu, Japan
Fell 1928, June 25
Synonym(s): *Yosiki*
Stone.
Mentioned, I. Yamamoto, Kwasan Observ. Bull., 1935, p.306 [M.A. 7-173]. A piece of 0.12g, formerly in Kwasan Observatory, was missing in 1935.

Yosiki v Yoshiki.

Youanme v Youanmi.

Youanmi 28°30′ S., 118°50′ E.
Western Australia, Australia
Found 1917
Synonym(s): *Youanme*
Iron. Octahedrite, medium (1.1mm) (IIIA).
A mass of 261lb was found, E.S. Simpson, Ann. Prog. Rep. Geol. Surv. West. Austr. for 1917, 1918, p.19. Description, analysis, 8.08 %Ni, E.S. Simpson, Min. Mag., 1938, **25**, p.165. Listed, Spec. Publ. West. Austr. Mus., 1965 (no. 3), p.55. Further analysis, 7.85 %Ni, 21.0 ppm.Ga, 37.7 ppm.Ge, 2.6 ppm.Ir, E.R.D. Scott et al., Geochimica et Cosmochimica Acta, 1973, **37**, p.1957. Description; a well preserved mass, V.F. Buchwald, Iron Meteorites, Univ. of California, 1975, p.1351.
Main mass, Perth, West. Austr. Mus.;
Specimen(s): [1968,279], 195g.

Youndagin v Youndegin.

Youndegin 32°6′ S., 117°43′ E.
Avon, South West Division, Western Australia, Australia
Found 1884
Synonym(s): *Joundegin, Mooranoppin, Mount Stirling, Penkarring Rock, Pickarring Rock, Quairading, Youndagin, Yundagin, Yundegin*
Iron. Octahedrite, coarse (2.3mm) (IA).
In 1884 four pieces, 25.75lb, 24lb, 17.5lb and 6lb, were found 0.75 miles NW. of Penkarring Rock and 70 miles E. of York. Description, with an analysis, 6.46 %Ni, L. Fletcher, Min. Mag., 1887, **7**, p.121, L. Fletcher, Min. Mag., 1899, **12**, p.171. In 1891 a mass of 382.5lb, and in 1892 a still larger mass of 2044lb were found, Nature, 1892-3, **47**, p.90, 469. Another mass, 5789lb, was found SW. of Quairading, in the vicinity of Wamensking Well, and is now assigned to this fall. A further mass of about 203lb was found 25 miles SE. of Mount Stirling, T. Cooksey, Rec. Austr. Mus., 1897, **3**, p.58, 131 and is also assigned to this fall. A mass of 3.5lb was found in about 1893 60 miles E. of York and two further masses, 820g and 742g were found. These three were referred to the Mooranoppin fall, which is now considered to be synonymous with Youndegin, except that this material has been heated, V.F. Buchwald, Iron Meteorites, Univ. of California, 1975, p.1352. Location of find-sites, J.R. de Laeter, Meteoritics, 1973, **8**, p.169. Widely analysed; for example, 6.38 %Ni, 90.8 ppm.Ga, 383 ppm.Ge, 2.0 ppm.Ir, J.T. Wasson, Meteorites, Springer-Verlag, 1974, p.297-298. Contains krinovite, E. Olsen and L. Fuchs, Science, 1968, **161**, p.876.
2626kg Perth, West. Austr. Mus., the mass found at Quairading; 910kg Vienna, Naturhist. Mus.; 146kg Chicago, Field Mus. Nat. Hist.; 92kg Sydney, Austr. Mus., the Mount Stirling mass; 10.9kg Melbourne, Nat. Mus.; 5.7kg Washington, U.S. Nat. Mus.; 3.2kg New York, Amer. Mus. Nat. Hist.; 919g Helsinki,; 1.8kg Tempe, Arizona State Univ.;
Specimen(s): [56150], 9820g. and fragments, 85g, of the first mass; [62964], 2700g. of the fourth, 6lb, mass; [71528], 630g. from the 382lb mass; [1920,350], 73g. etched slice; [86998], 25g. from the 382lb mass; [83613], 1888g. of Mount Stirling; [82748], 225g. and a fragment, 19g, of Mooranoppin

Young County v Wichita County.

Ysleta 31°39′ N., 106°11′ W.
El Paso County, Texas, U.S.A.
Found 1914, before this year
Iron. Anomalous (IRANOM).
A mass of 140.7kg was found, L.W. MacNaughton, Am. Mus. Novit., 1926 (207), p.2. Structurally and chemically anomalous. Figured, with an analysis, 7.66 %Ni, 0.143 ppm.Ga, 0.125 ppm.Ge, 7.0 ppm.Ir, R. Schaudy et al., Icarus, 1972, **17**, p.174. A unique iron, with au .tenite crystals 1-2 cm in diameter indicative of rapid cooling, V.F. Buchwald, Iron Meteorites, Univ. of California, 1975, p.1359.
140kg New York, Amer. Mus. Nat. Hist., main mass; 17g Washington, U.S. Nat. Mus.;

Yudoma 60° N., 140° E. approx.
Yakutia, Federated SSR, USSR, [Юдома]
Found 1946
Synonym(s): *Iudoma, Judoma*
Iron. Octahedrite, fine (0.3mm) (IVA).
One mass of 7.6kg was found, 8 %Ni, G.G. Bergmann, Метеоритика, 1955, **13**, p.128. Brief description, analysis, 9.5 %Ni, 2 ppm.Ga, 0.7 ppm.Ir, N.I. Zaslavskaya, Метеоритика, 1982, **40**, p.34.
7.4kg Yakutsk, Geol. Mus.; 2.8g Moscow, Acad. Sci.;

Yugan v Yukan.

Yukan 28°43′ N., 116°37′ E. approx.
Jiangxi, China
Fell 1931, August 27, 1500 hrs
Synonym(s): *Liewipantsun, Liweipantsun, Yugan*
Stone. Olivine-hypersthene chondrite, amphoterite (LL6), veined.
A shower of stones fell, at Liweipantsun and Chowyantsun, 19 km from Yukan. At least 10 stones were collected, the largest of 8 catties (4.8kg). Description, C.Y. Hsieh, Bull. Geol. Soc. China, 1932, **11**, p.411 [M.A. 5-298]. Mentioned, C.E. Chen, The Universe (Chinese Astron. Soc.), 1931, **2** (6), p.81, in Chinese, C.Y. Hsieh, letter in Min. Dept. BM(NH). Olivine Fa_{29}, B. Mason, Geochimica et Cosmochimica Acta, 1963, **27**, p.1011. Listed, D. Bian, Meteoritics, 1981, **16**, p.115.
Specimen(s): [1933,98], 12.2g. two pieces and fragments 3g

Yundagin v Youndegin.

Yundegin v Youndegin.

Yungay v North Chile.

Yurtuk 47°19′ N., 35°22′ E.
Lubimov, Mikhailov district, Ukraine, USSR, [Юртук]
Fell 1936, April 2, 0100 hrs
Synonym(s): *Iurtuk, Iurtyk, Jurtuk, Lubimowka, Lyubimovka*
Stone. Achondrite, Ca-rich. Howardite (AHOW).
One stone of 509g fell through the roof of a house, while several pieces, one of 51.5g, were picked up outside. Description, with an analysis, L.L. Ivanov, C. R. (Doklady) Acad. Sci. URSS, 1937, **17**, p.371 [M.A. 7-175]. Total mass recovered 1472g, L.G. Kvasha and A.Ya. Skripnik, Метеоритика, 1978, **37**, p.207. Further analysis, B. Mason et al., Smithson. Contrib. Earth Sci., 1979 (22), p.30.
750g Moscow, Acad. Sci., includes main mass, 508g; 40.3g Tempe, Arizona State Univ.; 33g Washington, U.S. Nat. Mus.;
Specimen(s): [1956,321], 21.5g.

Zaborica v Zaborzika.

Zaboritsa v Zaborzika.

Zaboritza v Zaborzika.

Zaborzika 50°17′ N., 27°41′ E.
Zhitomir, Volhynia, Ukraine, USSR, [Заборица]
Fell 1818, April 11
Synonym(s): *Czartorya, Saboriza, Saboryzy, Zaborica, Zaboritsa, Zaboritza, Zaborzyca*
Stone. Olivine-hypersthene chondrite (L6), veined.
A stone of about 4kg fell, but no details are known, A. Laugier, Ann. Phys. (Gilbert), 1823, **75**, p.264, E. von Eichwald, Arch. Wiss. Kunde Russl., Berlin, 1847, **5**, p.178. Date of fall, E.L. Krinov, Cat. Met. Acad. Sci. USSR, Moscow, 1947, p.23. Total mass recovered 3867g, L.G. Kvasha and A.Ya. Skripnik, Метеоритика, 1978, **37**, p.189. Olivine Fa_{25}, B. Mason, Geochimica et Cosmochimica Acta, 1963, **27**, p.1011. Analysis, 22.97 % total iron, M.I. D'yakonova, Метеоритика, 1972, **31**, p.119.
749g Kiev, Acad. Sci.; 407g Moscow, Acad. Sci.; 270g Vienna, Naturhist. Mus.; 107g Tübingen, Univ.; 50g Chicago, Field Mus. Nat. Hist.; 49g Berlin, Humboldt Univ.; 22g Yale Univ.;
Specimen(s): [33183], 8g. three fragments; [56469], 8g.; [54641], 0.75g.

Zaborzyca v Zaborzika.

Zabrodje 55°11′ N., 27°55′ E.
Minsk region, Belorussiya SSR, USSR, [Забродье]
Fell 1893, September 22, two hours before sunset hrs
Synonym(s): *Vilna, Zabrodzie*
Stone. Olivine-hypersthene chondrite (L6), veined.
After the appearance of a cloud moving from NE. to SW., and detonations, a stone of about 3kg fell through the roof of a house, R. Prendel, Mem. Soc. Nat. Nouvelle Russie, Odessa, 1894, **19** (1), p.243. Description, analysis, P.G. Melikov and L.V. Pissarjevsky, Ber. Deutsch. Chem. Gesell. Berlin, 1894, **27** (2), p.1235. Olivine Fa_{25}, B. Mason, Geochimica et Cosmochimica Acta, 1963, **27**, p.1011.
2600g Vilnius, Univ.; 300g Odessa, Univ.; 81g Moscow, Acad. Sci.; 5g Vienna, Naturhist. Mus.; 4g Paris, Mus. d'Hist. Nat.; 4g Berlin, Humboldt Univ.; 4g Chicago, Field Mus. Nat. Hist.; 1.4g New York, Amer. Mus. Nat. Hist.;
Specimen(s): [77497], 3g.

Zabrodzie v Zabrodje.

Zacatecas (1792) 22°49′ N., 102°34′ W.
Zacatecas, Mexico
Found 1792, known before this year
Synonym(s): *Rancho de la Pila (1834)*
Iron. Anomalous (IRANOM).
A large mass of about 1000kg (1 tonne), from time immemorial before 1792, had been in a street of Zacatecas, and was said to have been found near the Quebradilla mine on the western outskirts of the city, L. Fletcher, Min. Mag., 1890, **9**, p.162. Analysis, 5.82 %Ni, H. Müller, Quart. J. Chem. Soc. London, 1860, **11**, p.236. Analysis, 5.98 %Ni, E. Cohen, Ann. Naturhist. Hofmus. Wien, 1897, **12**, p.47. Classification, further analysis, 5.88 %Ni, 83.8 ppm.Ga, 307 ppm.Ge, 2.2 ppm.Ir, J.T. Wasson, Icarus, 1970, **12**, p.407. The main mass is in Mexico City; "a powder aggregate.. sintered at 1000-1100C", V.F. Buchwald, Iron Meteorites, Univ. of California, 1975, p.1361.
780kg Mexico City, Old School of Engineering; 3.9kg Tübingen, Univ.; 2.55kg Bonn, Univ. Mus.; 2kg Vienna, Naturhist. Mus.; 1.4kg Heidelberg, Univ.; 1.3kg Chicago, Field Mus. Nat. Hist.; 1.3kg Paris, Mus. d'Hist. Nat.; 751g Tempe, Arizona State Univ.; 444g Washington, U.S. Nat. Mus.;
Specimen(s): [1977,M.1], 9.5kg. formerly with the Geol. Soc. London; [28296], 2.57kg.; [14201], 718g. and a fragment, 19.5g; [33917], 92g.; [1960,491], 49g. slice; [1964,709], 71g. slice

Zacatecas (1969)
Zacatecas, Mexico, Co-ordinates not known
Found 1969, before this year
Iron. Octahedrite, medium (0.7mm) (IIIB).
A fragment of iron, 6.66kg, of unknown provenance, was purchased in Zacatecas in 1969, Meteor. Bull., 1971 (50), Meteoritics, 1971, **6**, p.124. Classification, analysis, 9.0 %Ni, 20.3 ppm.Ga, 38.8 ppm.Ge, 0.029 ppm.Ir, E.R.D. Scott et al., Geochimica et Cosmochimica Acta, 1973, **37**, p.1957. Shocked and annealed, V.F. Buchwald, Iron Meteorites, Univ. of California, 1975, p.1367.
6.6kg Washington, U.S. Nat. Mus., main mass;

Zaffra 35° N., 94°45′ W. approx.
Le Flore County, Oklahoma, U.S.A.
Found 1919, December
Iron. Octahedrite, coarse (2.5mm) (IIICD-ANOM).
A weathered mass of about 3kg was found, C.A. Reeds, Bull. Amer. Mus. Nat. Hist., 1937, **73**, p.517 [M.A. 7-61]. Analysis, 7.23 %Ni, A.C. Shead, Proc. Oklahoma Acad. Sci., 1922, **2**. Classification and analysis, 7.12 %Ni, 73.2 ppm.Ga, 244 ppm.Ge, 0.061 ppm.Ir, A. Kracher et al., Geochimica et Cosmochimica Acta, 1980, **44**, p.773. Description; shocked and slightly annealed, V.F. Buchwald, Iron Meteorites, Univ. of California, 1975, p.1369.
1.2kg New York, Amer. Mus. Nat. Hist.;
Specimen(s): [1963,529], 180g.

Zagama v Zagami.

Zagami 11°44′ N., 7°5′ E.
Katsina Province, Nigeria
Fell 1962, October 3
Synonym(s): *Zagama*
Stone. Achondrite, Ca-rich. Eucrite (shergottite) (AEUC).
A mass of about 40lb fell about 0.75 miles from Zagami Rock. Analysed by A.J. Easton, Min. Dept. BM(NH).; Mg/(Mg+Fe) 0.48. Composition of pyroxene and plagioclase glass, An52, R.A. Binns, Nature, 1967, **213**, p.1111. Petrogenesis, chronology, C.-Y. Shih et al., Geochimica et Cosmochimica Acta, 1982, **46**, p.2323. Petrology, mineralogy and experimental data, E. Stolper and H.Y. McSween, Geochimica et Cosmochimica Acta, 1979, **43**, p.1475.
Main mass, Kaduna, Geol. Surv. Nigeria;
Specimen(s): [1966,54], 228g. and fragments, 6g

Zagan v Sagan.

Zagrab v Hraschina.

Zagreb v Hraschina.

Zagrebacko zeljezo v Hraschina.

Zaican v Zaisan.

Zaisan 47°30′ N., 83°E′ approx.
Zaisan Lake, Kazakhstan SSR, USSR, [Зайсан]
Fell 1963, December 18, 1100 hrs
Synonym(s): *Zaican, Zaisan Lake, Zaysan*
Stone. Olivine-bronzite chondrite (H5).
A mass of 463g fell on the ice of Lake Zaisan, Meteor. Bull., 1964 (29). Olivine Fa_{18}, B. Mason, Geochimica et Cosmochimica Acta, 1967, **31**, p.1100.
344g Moscow, Acad. Sci., main mass;

Zaisan Lake v Zaisan.

Zaisho 33°42′ N., 133°48′ E.
Kami-gun, Kochi-Ken, Shikoku, Japan
Fell 1898, February 1, 0500 hrs
Stony-iron. Pallasite (PAL).
A mass of 330g was found after the appearance of a fireball, S. Kanda, Sci. Rep. Yokohama Univ., 1952 (2, no. 1), p.97 [M.A. 13-80], S. Murayama, Nat. Sci. Mus. Tokyo, 1953, **20**, p.129 [M.A. 13-80], S. Murayama, letter of 6 April, 1962 in Min. Dept. BM(NH). Olivine Fa_{19}, B. Mason, Am. Mus. Novit., 1963 (2163). Mineralogy, P.R. Buseck and J. Clarke, Meteoritics, 1982, **17**, p.abs.
Main mass, in the possession of S. Goto of Setaya, Tokyo; 8.5g Tokyo, Nat Sci. Mus.;

Zapata County 27° N., 99° W.
Zapata County, Texas, U.S.A.
Found 1930
Iron. Octahedrite.
A piece of Fe-Ni metal about the size of a man's fist is said to have been recovered at a depth of 1525 feet from a borehole in sediments 'of almost certainly Eocene age'; it has since been lost, J.F. Lovering, Nature, 1959, **183**, p.1664.

Zavetnoje v Zavetnoe.

Zavetnoe 47°8′ N., 43°54′ E.
Rostov region, Federated SSR, USSR, [Заветное]
Fell 1952, December 4, 1630 hrs
Synonym(s): *Sawetnoe, Sawetnoje, Zavetnoje*
Stone. Olivine-hypersthene chondrite (L6).
Three fragments were found, two of which (543g and 202g) fit together, S.S. Fonton, Метеоритика, 1955, **13**, p.104 [M.A. 13-50]. Analysis, 22.80 % total iron, M.I. D'yakonova and V.Ya. Kharitonova, Метеоритика, 1960, **18**, p.48 [M.A. 16-447], K. Keil, Fortschr. Min., 1960, **38**, p.271. Olivine Fa_{24}, B. Mason, Geochimica et Cosmochimica Acta, 1963, **27**, p.1011.
702g Moscow, Acad. Sci.;

Zavid 44°24′ N., 19°7′ E.
Zvornik, Bosnia, Yugoslavia
Fell 1897, August 1, 1122 hrs
Synonym(s): *Ravne Njive, Ravni Zavid, Rozanj, Zvornik*
Stone. Olivine-hypersthene chondrite (L6), brecciated.
After the appearance of a fireball (moving from SE. to NW.), and detonations, four stones fell, of about 90kg, 2.5kg, 220g and 48g respectively. Description, F. Berwerth, Wiss. Mitt. Bosnien u. Hercegovina, 1901, **8**, p.409. Analysis, C. Hödlmoser, Tschermaks Min. Petr. Mitt., 1899, **18**, p.513. More stones are thought to have fallen into the river Drina, M. Ramovic, Cat. Meteor. Coll. Yugoslavia, 1965, p.42. Olivine Fa_{24}, B. Mason, Geochimica et Cosmochimica Acta, 1963, **27**, p.1011.
62kg Sarajevo, Bosnian Landesmus.; 3.6kg Vienna, Naturhist. Mus.; 1.6kg Zagreb, Mineral. Petrol. Dept.; 856g New York, Amer. Mus. Nat. Hist.; 830g Budapest, Nat. Mus.; 658g Chicago, Field Mus. Nat. Hist.; 362g Paris, Mus. d'Hist. Nat.; 356g Prague, Nat. Mus.; 234g Washington, U.S. Nat. Mus.; 142g Tempe, Arizona State Univ.;
Specimen(s): [83668], 263g.; [1920,351], 78g.

Zaysan v Zaisan.

Zebrak 49°53′ N., 13°55′ E.
Horovice, Beroun, Stredocesky, Czechoslovakia
Fell 1824, October 14, 0800 hrs
Synonym(s): *Beraun, Horowitz, Hozowitz, Praskoles, Praskolesy*
Stone. Olivine-bronzite chondrite (H5).
After detonations, a stone of about 2kg fell at Praskolesy, C.F.P. v. Martius, Ann. Chim. Phys., 1825, **30**, p.421, E.F.F. Chladni, Ann. Phys. Chem. (Poggendorff), 1826, **6**, p.28. Analysis, J. Kokta, Coll. Czech Chem. Comm., 1937, **9**, p.471 [M.A. 7-173]. Mineralogy, olivine $Fa_{18.3}$, M. Bukovanska, Meteoritics, 1983, **18**, p.abs.
861g Prague, Bohemian Mus.; 450g Vienna, Naturhist. Mus.; 101g New York, Amer. Mus. Nat. Hist.; 73g Washington, U.S. Nat. Mus.; 50g Berlin, Humboldt Univ.; 14g Chicago, Field Mus. Nat. Hist.;
Specimen(s): [76153], 76g.; [1920,352], 11g.; [33736], 7g.

Zegan v Sagan.

Zemaitkiemis 55°18′ N., 25° E.
Ukmerge, Lithuanian SSR, USSR
Fell 1933, February 2, 2033 hrs
Synonym(s): *Dzemajtkemis, Zhemaitkiems*
Stone. Olivine-hypersthene chondrite (L6).
22 stones were collected totalling 44.1kg, L.G. Kvasha and A.Ya. Skripnik, Метеоритика, 1978, **37**, p.188. The largest mass weighed 7258g. Description, with a doubtful analysis, M. Kaveckis, Lietuvos Univ. Mat.-Gamtos Fak.

Darbai, Kaunas, 1935, **9**, p.307 [M.A. 6-207]. Analysis, 22.09 % total iron, V.Ya. Kharitonova, Метеоритика, 1965, **26**, p.146. Olivine Fa_{24}, B. Mason, Geochimica et Cosmochimica Acta, 1963, **27**, p.1011.
Main mass, Vilnius, Univ.; 2.1kg Prague, Nat. Mus.; 1.1kg Moscow, Acad. Sci.; 33g Chicago, Field Mus. Nat. Hist.;
Specimen(s): [1939,247], 636g.

Zenda 42°30′48″ N., 88°29′22″ W.
Walworth County, Wisconsin, U.S.A.
Found 1955
Iron. Octahedrite, medium (0.9mm) (IA-ANOM).
A mass of about 3.7kg was ploughed up, W.F. Read, Wisconsin Acad. Review, 1962, **9** (4), p.152, Meteor. Bull., 1963 (26). Analysis, 8.50 %Ni, 54.7 ppm.Ga, 214 ppm.Ge, 2.1 ppm.Ir, J.T. Wasson, Icarus, 1970, **12**, p.407. Description, V.F. Buchwald, Iron Meteorites, Univ. of California, 1975, p.1370.
3561g Appleton, Wisconsin, Lawrence College; 60g Tempe, Arizona State Univ.;

Zenith 37°58′ N., 98°30′ W.
Stafford County, Kansas, U.S.A.
Pseudometeorite..
A stone of 40lb to 50lb was found in 1901. Description, C.S. Corbett, Am. J. Sci., 1926, **26**, p.495. Consists mainly of anorthite and fayalite, probably a slag, G.T. Prior, priv. comm., 1928. Siliceous residue of the burning of a stack of straw, L. LaPaz, Pop. Astron., Northfield, Minnesota, 1944, **52**, p.195 [M.A. 9-289].

Zerga 20°15′ N., 12°41′ W.
Aouelloul Crater, Adrar, Mauritania
Found 1973, June 26
Stone. Olivine-hypersthene chondrite, amphoterite (LL6).
A single stone of 76g was found just outside the SSE. portion of the crater rim, Meteor. Bull., 1976 (54), Meteoritics, 1976, **11**, p.93. Brief description, rare gas data, R.F. Fudali and P.J. Cressy, Earth planet. Sci. Lett., 1976, **30**, p.262.
76g Washington, U.S. Nat. Mus.;

Zerhamra 29°51′31″ N., 2°38′42″ W.
Sahara, Algeria
Found 1967
Iron. Octahedrite, medium (1.1mm) (IIIA-ANOM).
A mass of 630kg was found about 50 km SW. of Beni-Abbes and 22 km SE. of the oasis of Zerhamra, Meteor. Bull., 1974 (52), Meteoritics, 1974, **9**, p.120. Annealed due to cosmic reheating, V.F. Buchwald, Iron Meteorites, Univ. of California, 1975, p.1371. Analysis, 8.00 %Ni, 18.2 ppm.Ga, 33.5 ppm.Ge, 10 ppm.Ir, J.T. Wasson, Meteorites, Springer-Verlag, 1974, p.302.
629kg Paris, Mus. d'Hist. Nat., main mass; 71g Copenhagen, Univ. Geol. Mus.;
Specimen(s): [1980,M.30], 193.9g. end-piece

Zhemaitkiems v Zemaitkiemis.

Zhigailovka v Kharkov.

Zhigajlovka v Kharkov.

Zhigansk 68°0′ N., 128°18′ E.
About 250 km NE. of Zhigansk, Yakutsk ASSR, USSR
Found 1966, recognized 1981
Iron.
A large mass, estimated to weigh 600-900kg, was found and a 60g piece removed. Its meteoritic nature was confirmed, Meteor. Bull., 1982 (60), Meteoritics, 1982, **17**, p.97.
17.2g Moscow, Acad. Sci.;

Zhmeni v Zmenj.

Zhongxiang 31°12′ N., 112°30′ E. approx.
Zhongxiang County, Hubei, China
Found, year not reported
Iron.
A single mass of 100kg is preserved at Wuhan, Hubei Bureau of Geology, D. Bian, Meteoritics, 1981, **16**, p.115.

Zhovtnevyi 47°35′ N., 37°15′ E.
Pavlovka, Stalino, Ukraine, USSR, [Жовтневый Хутор]
Fell 1938, October 10, night hrs, or October 9
Synonym(s): *Jovtnevy, Schowtnewi Chutor, Showtnjew Chutor, Zhovtnevy, Zhovtnevyi Khutor, Zhovtnevyj, Zovtnevy Hutor*
Stone. Olivine-bronzite chondrite (H5).
Six large and several small fragments were recovered near Zhovtnevyi in 1938, and a further five stones were found in 1939 and 1941. The total mass recovered is 107kg, L.G. Kvasha and A.Ya. Skripnik, Метеоритика, 1978, **37**, p.189. Description, E.L. Krinov, C. R. Acad. Sci. URSS, 1939, **22**, p.436 [M.A. 7-373]. Analysis, 24.93 % total iron, M.I. D'yakonova and V.Ya. Kharitonova, Метеоритика, 1960, **18**, p.48 [M.A. 16-447]. K-Ar age, E.K. Gerling and T.G. Pavlova, Доклады Акад. Наук СССР, 1951, **77**, p.85 [M.A. 11-527]. Olivine Fa_{19}, B. Mason, Geochimica et Cosmochimica Acta, 1963, **27**, p.1011.
71.8kg Moscow, Acad. Sci., includes main mass, 21kg; Specimen, Kharkov, Univ.; 703g Washington, U.S. Nat. Mus.; 308g Tempe, Arizona State Univ.; 240g Paris, Mus. d'Hist. Nat.; 121g Yale Univ.;
Specimen(s): [1956,167], 156g. two fragments

Zhovtnevy v Zhovtnevyi.

Zhovtnevyi Khutor v Zhovtnevyi.

Zhovtnevyj v Zhovtnevyi.

Zhuanghe 39°40′ N., 122°59′ E.
Liaoning, China
Fell 1976, August 18, 2032 hrs
Stone.
Listed, 3.0kg was recovered, D. Bian, Meteoritics, 1981, **16**, p.115.
3.0kg Dalian, Mus. Nat. Hist.;

Zielena Gora v Grüneberg.

Zielona Gora v Grüneberg.

Zigajlovka v Kharkov.

Zigajlowka v Kharkov.

Zindoo v Gyokukei.

Zipaquira v Santa Rosa.

Ziquipilco v Toluca.

Zmeni v Zmenj.

Zmenj 51°50′ N., 26°50′ E.
Minsk, Belorussiya SSR, USSR, [Жмени]
Fell 1858, August
Synonym(s): *Cmien, Minsk, Tsmen, Zhmeni, Zmeni*
Stone. Achondrite, Ca-rich. Howardite (AHOW).
A stone of 246g fell. Description, A. Brezina, Ann. Naturhist. Hofmus. Wien, 1896, **10**, p.240. Analysis, P.G. Melikov, J. Russ. Phys. Chem. Gesell, 1896, **28**, p.114, 299, Neues Jahrb. Min., 1899, **2**, p.31. Mentioned, L.A. Kulik, Метеоритика, 1941, **1**, p.73 [M.A. 9-294], L.G. Kvasha and A.Ya. Skripnik, Метеоритика, 1978, **37**, p.188. Analysis, 13.39 % total iron, H.B. Wiik, Comm. Phys.-Math. Soc. Sci. Fennica, 1969, **34**, p.135.
116g Vienna, Naturhist. Mus., main mass; 25g Moscow, Acad. Sci.; 6.8g New York, Amer. Mus. Nat. Hist.;

Znorow v Wessely.

Zomba 15°11′ S., 35°17′ E.
Southern province, Malawi
Fell 1899, January 25, 0745 hrs
Synonym(s): *Mount Zomba*
Stone. Olivine-hypersthene chondrite (L6).
After the appearance of a luminous meteor and detonations, several stones fell on the slopes of Mt. Zomba over an area of 9×3 miles. Ten stones were found, the largest weighing 5lb 12.5oz (2.6kg) and the total weight was about 7.5kg. Description, with an analysis, L. Fletcher, Min. Mag., 1901, **13**, p.1. Olivine Fa_{23}, B. Mason, Geochimica et Cosmochimica Acta, 1963, **27**, p.1011.
270g Washington, U.S. Nat. Mus.; 202g New York, Amer. Mus. Nat. Hist.; 43g Vienna, Naturhist. Mus.;
Specimen(s): [84357], 553g. and fragments, 3g; [84356], 483.5g. complete stone; [84355], 413g. complete stone; [1908, 48], 297g. complete stone; [84358], 24g. fragments

Zovtnevy Hutor v Zhovtnevyi.

Zsadany 46°56′ N., 21°30′ E.
Bihor district, Romania
Fell 1875, March 31, 1500-1600 hrs
Synonym(s): *Szadany*
Stone. Olivine-bronzite chondrite (H5).
After detonations, a shower of stones fell, of which only nine small ones were found, totalling 552g, the largest weighing 152g, W. Pillitz, Zeits. Anal. Chem. (Fresenius), Wiesbaden, 1879, **18**, p.61. Analysis, E. Cohen, Verh. Naturhist.-Med. Ver. Heideiberg, 1880, **2**, p.154. Olivine Fa_{18}, B. Mason, Geochimica et Cosmochimica Acta, 1963, **27**, p.1011.
223g Budapest, Nat. Mus.; 44g Vienna, Naturhist. Mus.; 14g Paris, Mus. d'Hist. Nat.; 14g Chicago, Field Mus. Nat. Hist.; 13g Washington, U.S. Nat. Mus.;
Specimen(s): [52274], 13.5g. complete stone

Züllichau v Seeläsgen.

Zululand v N'Kandhla.

Zvonkov 50°12′ N., 30°15′ E. approx.
Vasilkov district, Kiev region, Ukraine, USSR, [Звонковое]
Fell 1955, September 2, 1545 hrs
Synonym(s): *Svonkovje, Svonkovoje, Swonkowe, Swonkowoe, Swonkowoj, Zvonkovje, Zvonkovoe*
Stone. Olivine-bronzite chondrite (H6).
Two stones, 1296g and 1272g, were found 4 km apart, V.G. Fesenkov, Метеоритика, 1958, **16**, p.5 [M.A. 14-45], S.P. Rodionov, Метеоритика, 1959, **17**, p.47 [M.A. 15-534]. Olivine Fa_{19}, B. Mason, Geochimica et Cosmochimica Acta, 1963, **27**, p.1011.
587g Moscow, Acad. Sci.; 40g Washington, U.S. Nat. Mus.;

Zvonkovje v Zvonkov.

Zvonkovoe v Zvonkov.

Zvornik v Zavid.

Zweibrücken v Krähenberg.

List of Prepared Sections of Meteorites

Sections of meteorites held in the British Museum (Natural History) collection and including those held in other institutions.

The accepted names only of the meteorites are given; synonyms will be found in the main meteorite list. Included here for the first time are the sections held in other institutions; the nature of the section is indicated (P.M. polished mount, P.T.S. polished thin section, T.S. covered thin section). There are often more than one section of a particular meteorite held in this collection and in other institutions but this is not recorded here.

The names of the other institutions whose holdings are recorded here have been abbreviated as follows:

Field Mus. Nat. Hist.	Field Museum of Natural History, Chicago, USA.
Mus. d'Hist. Nat.	Musée National d'Histoire Naturelle, Paris, France.
Naturhist. Mus.	Naturhistorisches Museum, Vienna, Austria.
U.S. Nat. Mus.	U.S. National Museum, Smithsonian Institution, Washington, D.C., U.S.A.
Univ. New Mexico	Institute of Meteoritics, University of New Mexico, Albuquerque, New Mexico, U.S.A.

Aarhus
P.M.
U.S. Nat. Mus.
P.T.S.
U.S. Nat. Mus.

Abbott
P.M.
Univ. of New Mexico
P.T.S.
Univ. of New Mexico

Abee
P.M.
U.S. Nat. Mus.
P.T.S.
Univ. of New Mexico
B.M.(N.H.)
T.S.
B.M.(N.H.)

Abernathy
P.M.
Field Mus. Nat. Hist.
P.T.S.
U.S. Nat. Mus.
Field Mus. Nat. Hist.
B.M.(N.H.)
T.S.
U.S. Nat. Mus.
B.M.(N.H.)

Acapulco
P.T.S.
U.S. Nat. Mus.
Field Mus. Nat. Hist.

Achilles
P.M.
U.S. Nat. Mus.
P.T.S.
U.S. Nat. Mus.
B.M.(N.H.)

Achiras
P.T.S.
U.S. Nat. Mus.
B.M.(N.H.)
T.S.
B.M.(N.H.)

Acme
P.T.S.
Univ. of New Mexico
T.S.
B.M.(N.H.)

Adalia
P.T.S.
U.S. Nat. Mus.
T.S.
Mus. d'Hist. Nat.

Adams County
P.T.S.
U.S. Nat. Mus.
B.M.(N.H.)
T.S.
U.S. Nat. Mus.
B.M.(N.H.)

ad-Dahbubah
P.M.
U.S. Nat. Mus.
P.T.S.
U.S. Nat. Mus.

Adelie Land
P.M.
U.S. Nat. Mus.
P.T.S.
U.S. Nat. Mus.
T.S.
B.M.(N.H.)

Adhi Kot
P.T.S.
U.S. Nat. Mus.

Admire
P.M.
U.S. Nat. Mus.
Univ. of New Mexico
P.T.S.
Univ. of New Mexico
T.S.
Naturhist. Mus.

Adrian
P.T.S.
U.S. Nat. Mus.
B.M.(N.H.)
T.S.
B.M.(N.H.)

Agen
P.T.S.
U.S. Nat. Mus.
B.M.(N.H.)
T.S.
U.S. Nat. Mus.
Field Mus. Nat. Hist.
Naturhist. Mus.
Mus. d'Hist. Nat.
B.M.(N.H.)

Aggie Creek
P.M.
U.S. Nat. Mus.

Aguada
P.T.S.
U.S. Nat. Mus.

Ahumada
P.M.
U.S. Nat. Mus.

Ainsworth
P.M.
U.S. Nat. Mus.

Aioun el Atrouss
P.T.S.
U.S. Nat. Mus.

Akaba
T.S.
B.M.(N.H.)

Akbarpur
P.T.S.
B.M.(N.H.)
T.S.
Mus. d'Hist. Nat.
B.M.(N.H.)

Akron (1940)
P.T.S.
B.M.(N.H.)
T.S.
B.M.(N.H.)

Akron (1961)
P.T.S.
B.M.(N.H.)
T.S.
B.M.(N.H.)

Alais
P.T.S.
U.S. Nat. Mus.
T.S.
Mus. d'Hist. Nat.
B.M.(N.H.)

Alamogordo
P.T.S.
Univ. of New Mexico
T.S.
B.M.(N.H.)

Alamoso
P.T.S.
U.S. Nat. Mus.

Albareto
T.S.
Field Mus. Nat. Hist.
Naturhist. Mus.
Mus. d'Hist. Nat.
B.M.(N.H.)

Albin (pallasite)
P.M.
U.S. Nat. Mus.

Albin (stone)
- P.M.
  - U.S. Nat. Mus.

Aldsworth
- P.T.S.
  - B.M.(N.H.)
- T.S.
  - B.M.(N.H.)

Aleppo
- P.T.S.
  - U.S. Nat. Mus.
- T.S.
  - Naturhist. Mus.
  - Mus. d'Hist. Nat.
  - B.M.(N.H.)

Alessandria
- P.T.S.
  - U.S. Nat. Mus.
  - B.M.(N.H.)
- T.S.
  - Naturhist. Mus.

Alexander County
- P.M.
  - U.S. Nat. Mus.

Alfianello
- P.M.
  - U.S. Nat. Mus.
  - Field Mus. Nat. Hist.
- P.T.S.
  - U.S. Nat. Mus.
  - Field Mus. Nat. Hist.
  - Naturhist. Mus.
  - Univ. of New Mexico
- T.S.
  - U.S. Nat. Mus.
  - Field Mus. Nat. Hist.
  - Naturhist. Mus.
  - Mus. d'Hist. Nat.
  - B.M.(N.H.)

al-Ghanim (stone)
- P.M.
  - U.S. Nat. Mus.
- P.T.S.
  - U.S. Nat. Mus.

Algoma
- P.M.
  - U.S. Nat. Mus.

Alikatnima
- P.M.
  - U.S. Nat. Mus.

Allegan
- P.M.
  - Field Mus. Nat. Hist.
- P.T.S.
  - U.S. Nat. Mus.
  - B.M.(N.H.)
- T.S.
  - U.S. Nat. Mus.
  - Field Mus. Nat. Hist.
  - Naturhist. Mus.
  - Mus. d'Hist. Nat.
  - B.M.(N.H.)

Allende
- P.M.
  - U.S. Nat. Mus.
- P.T.S.
  - U.S. Nat. Mus.
  - Field Mus. Nat. Hist.
  - Univ. of New Mexico
  - B.M.(N.H.)
- T.S.
  - Field Mus. Nat. Hist.
  - Naturhist. Mus.
  - B.M.(N.H.)

Al Rais
- P.T.S.
  - U.S. Nat. Mus.
- T.S.
  - B.M.(N.H.)

Alta'ameem
- P.T.S.
  - Field Mus. Nat. Hist.

Alt Bela
- P.M.
  - U.S. Nat. Mus.
  - Field Mus. Nat. Hist.

Altonah
- P.M.
  - U.S. Nat. Mus.

Ambapur Nagla
- P.T.S.
  - U.S. Nat. Mus.
- T.S.
  - Naturhist. Mus.
  - Mus. d'Hist. Nat.
  - B.M.(N.H.)

Amber
- P.T.S.
  - B.M.(N.H.)
- T.S.
  - B.M.(N.H.)

Amherst
- P.M.
  - U.S. Nat. Mus.
- P.T.S.
  - U.S. Nat. Mus.

Andover
- P.T.S.
  - U.S. Nat. Mus.
- T.S.
  - B.M.(N.H.)

Andura
- P.T.S.
  - U.S. Nat. Mus.

Angelica
- P.M.
  - U.S. Nat. Mus.

Angers
- T.S.
  - Mus. d'Hist. Nat.
  - B.M.(N.H.)

Angra dos Reis (stone)
- P.T.S.
  - Field Mus. Nat. Hist.
  - U.S. Nat. Mus.
  - Univ. of New Mexico
  - B.M.(N.H.)
- T.S.
  - U.S. Nat. Mus.
  - Naturhist. Mus.
  - Mus. d'Hist. Nat.
  - B.M.(N.H.)

Ankober
- P.M.
  - U.S. Nat. Mus.
- P.T.S.
  - U.S. Nat. Mus.

Anthony
- T.S.
  - U.S. Nat. Mus.
  - B.M.(N.H.)

Antofagasta
- P.M.
  - U.S. Nat. Mus.
- T.S.
  - Mus. d'Hist. Nat.

Apoala
- P.M.
  - U.S. Nat. Mus.

Appley Bridge
- P.M.
  - U.S. Nat. Mus.
  - B.M.(N.H.)
- P.T.S.
  - U.S. Nat. Mus.
  - B.M.(N.H.)
- T.S.
  - U.S. Nat. Mus.
  - B.M.(N.H.)

Apt
- T.S.
  - Field Mus. Nat. Hist.
  - Mus. d'Hist. Nat.
  - B.M.(N.H.)

Arapahoe
- P.T.S.
  - U.S. Nat. Mus.
  - B.M.(N.H.)
- T.S.
  - B.M.(N.H.)

Arcadia
- P.T.S.
  - U.S. Nat. Mus.
- T.S.
  - U.S. Nat. Mus.
  - B.M.(N.H.)

Archie
- P.T.S.
  - B.M.(N.H.)
- T.S.
  - B.M.(N.H.)

Arispe
- P.M.
  - U.S. Nat. Mus.
  - B.M.(N.H.)

Arlington
- P.M.
  - U.S. Nat. Mus.

Arltunga
- P.M.
  - U.S. Nat. Mus.

Armel
- P.T.S.
  - B.M.(N.H.)
- T.S.
  - B.M.(N.H.)

Arriba
- P.T.S.
  - U.S. Nat. Mus.
  - Univ. of New Mexico
- T.S.
  - B.M.(N.H.)

Arroyo Aguiar
- P.T.S.
  - U.S. Nat. Mus.

Artracoona
- P.T.S.
  - U.S. Nat. Mus.
  - B.M.(N.H.)
- T.S.
  - B.M.(N.H.)

Ashdon
- T.S.
  - B.M.(N.H.)

Asheville
- P.M.
  - U.S. Nat. Mus.
- P.T.S.
  - U.S. Nat. Mus.

Ashmore
- P.T.S.
  - U.S. Nat. Mus.
  - Naturhist. Mus.
  - Univ. of New Mexico
  - B.M.(N.H.)
- T.S.
  - B.M.(N.H.)

ash-Shalfah
- P.M.
  - U.S. Nat. Mus.
- P.T.S.
  - U.S. Nat. Mus.

Assam
- P.T.S.
  - B.M.(N.H.)
- T.S.
  - B.M.(N.H.)

Assisi
P.T.S.
U.S. Nat. Mus.
T.S.
Mus. d'Hist. Nat.
B.M.(N.H.)

Aswan
P.M.
U.S. Nat. Mus.

Atarra
P.T.S.
U.S. Nat. Mus.
T.S.
B.M.(N.H.)

Atemajac
P.T.S.
U.S. Nat. Mus.

Athens
P.M.
U.S. Nat. Mus.

Atlanta
P.M.
U.S. Nat. Mus.
P.T.S.
U.S. Nat. Mus.
Field Mus. Nat. Hist.
B.M.(N.H.)
T.S.
B.M.(N.H.)

Atoka
P.M.
U.S. Nat. Mus.
P.T.S.
U.S. Nat. Mus.

Atwood
P.T.S.
B.M.(N.H.)
T.S.
B.M.(N.H.)

Aubres
P.T.S.
B.M.(N.H.)
T.S.
Naturhist. Mus.
Mus. d'Hist. Nat.
B.M.(N.H.)

Augusta County
P.M.
U.S. Nat. Mus.

Augustinovka
P.M.
U.S. Nat. Mus.
Field Mus. Nat. Hist.

Aumale
P.T.S.
U.S. Nat. Mus.
T.S.
Field Mus. Nat. Hist.
Naturhist. Mus.
Mus. d'Hist. Nat.
B.M.(N.H.)

Aumieres
P.T.S.
U.S. Nat. Mus.
T.S.
Mus. d'Hist. Nat.
B.M.(N.H.)

Aurora
P.M.
Univ. of New Mexico
P.T.S.
B.M.(N.H.)
T.S.
B.M.(N.H.)

Ausson
P.T.S.
U.S. Nat. Mus.
Naturhist. Mus.
Univ. of New Mexico
B.M.(N.H.)
T.S.
Field Mus. Nat. Hist.
Naturhist. Mus.
Mus. d'Hist. Nat.
B.M.(N.H.)

Avanhandava
P.T.S.
U.S. Nat. Mus.
Univ. of New Mexico

Avoca (Western Australia)
P.M.
U.S. Nat. Mus.

Awere
P.T.S.
B.M.(N.H.)
T.S.
B.M.(N.H.)

Aztec
P.T.S.
Field Mus. Nat. Hist.
Univ. of New Mexico
B.M.(N.H.)
T.S.
B.M.(N.H.)

Babb's Mill (Troost's Iron)
P.M.
U.S. Nat. Mus.
B.M.(N.H.)

Bachmut
P.T.S.
Naturhist. Mus.
T.S.
Naturhist. Mus.
Mus. d'Hist. Nat.
B.M.(N.H.)

Bacubirito
P.M.
U.S. Nat. Mus.

Bahjoi
P.M.
U.S. Nat. Mus.

Bald Mountain
P.M.
U.S. Nat. Mus.
P.T.S.
U.S. Nat. Mus.
B.M.(N.H.)
T.S.
B.M.(N.H.)

Balfour Downs
P.M.
U.S. Nat. Mus.

Bali
P.T.S.
U.S. Nat. Mus.
Naturhist. Mus.
T.S.
Naturhist. Mus.

Ballinger
P.M.
U.S. Nat. Mus.

Ballinoo
P.M.
U.S. Nat. Mus.
B.M.(N.H.)

Bandong
P.T.S.
U.S. Nat. Mus.
B.M.(N.H.)
T.S.
Naturhist. Mus.
Mus. d'Hist. Nat.
B.M.(N.H.)

Bansur
P.T.S.
U.S. Nat. Mus.

Banswal
P.T.S.
U.S. Nat. Mus.

Baquedano
P.M.
U.S. Nat. Mus.

Barbotan
P.T.S.
U.S. Nat. Mus.
B.M.(N.H.)
T.S.
Field Mus. Nat. Hist.
Naturhist. Mus.
Mus. d'Hist. Nat.

Barea
P.M.
U.S. Nat. Mus.
P.T.S.
U.S. Nat. Mus.
T.S.
Mus. d'Hist. Nat.

Baroti
P.T.S.
U.S. Nat. Mus.
B.M.(N.H.)
T.S.
B.M.(N.H.)

Barranca Blanca
P.M.
U.S. Nat. Mus.

Barratta
P.M.
U.S. Nat. Mus.
P.T.S.
U.S. Nat. Mus.
Field Mus. Nat. Hist.
B.M.(N.H.)
T.S.
Field Mus. Nat. Hist.
Naturhist. Mus.
Mus. d'Hist. Nat.
B.M.(N.H.)

Bartlett
P.M.
U.S. Nat. Mus.

Barwell
P.T.S.
U.S. Nat. Mus.
B.M.(N.H.)
T.S.
B.M.(N.H.)

Barwise
P.T.S.
U.S. Nat. Mus.
B.M.(N.H.)
T.S.
B.M.(N.H.)

Bath
P.M.
U.S. Nat. Mus.
P.T.S.
U.S. Nat. Mus.
Naturhist. Mus.
Field Mus. Nat. Hist.
T.S.
Naturhist. Mus.

Bath Furnace
P.M.
Field Mus. Nat. Hist.
P.T.S.
U.S. Nat. Mus.
Field Mus. Nat. Hist.
T.S.
Field Mus. Nat. Hist.
Naturhist. Mus.
Mus. d'Hist. Nat.

Baxter
P.T.S.
B.M.(N.H.)
T.S.

B.M.(N.H.)

Bear Creek
P.M.
U.S. Nat. Mus.
B.M.(N.H.)

Beardsley
P.M.
U.S. Nat. Mus.
P.T.S.
U.S. Nat. Mus.
T.S.
B.M.(N.H.)

Bear Lodge
P.M.
U.S. Nat. Mus.

Beaver Creek
P.M.
U.S. Nat. Mus.
P.T.S.
U.S. Nat. Mus.
Field Mus. Nat. Hist.
T.S.
Field Mus. Nat. Hist.
Naturhist. Mus.
Mus. d'Hist. Nat.
B.M.(N.H.)

Beddgelert
P.T.S.
U.S. Nat. Mus.
T.S.
B.M.(N.H.)

Beeler
P.M.
B.M.(N.H.)
P.T.S.
Univ. of New Mexico
T.S.
B.M.(N.H.)

Beenham
P.M.
U.S. Nat. Mus.
P.T.S.
Univ. of New Mexico
Field Mus. Nat. Hist.
B.M.(N.H.)
T.S.
B.M.(N.H.)

Bella Roca
P.M.
U.S. Nat. Mus.

Belle Plaine
P.M.
Field Mus. Nat. Hist.
P.T.S.
U.S. Nat. Mus.
Field Mus. Nat. Hist.

Bells
P.T.S.
U.S. Nat. Mus.

Bellsbank
P.M.
U.S. Nat. Mus.

Belly River
P.T.S.
U.S. Nat. Mus.

Belmont
P.T.S.
U.S. Nat. Mus.

Benares
P.M.
Field Mus. Nat. Hist.
P.T.S.
Field Mus. Nat. Hist.
T.S.
Field Mus. Nat. Hist.
Naturhist. Mus.
Mus. d'Hist. Nat.
B.M.(N.H.)

Bencubbin
P.M.
U.S. Nat. Mus.
P.T.S.
U.S. Nat. Mus.
Naturhist. Mus.
T.S.
B.M.(N.H.)

Bendegó
P.M.
U.S. Nat. Mus.
B.M.(N.H.)

Benld
P.T.S.
U.S. Nat. Mus.
T.S.
Field Mus. Nat. Hist.

Bennett County
P.M.
U.S. Nat. Mus.

Benoni
P.T.S.
B.M.(N.H.)
T.S.
B.M.(N.H.)

Benton
P.T.S.
U.S. Nat. Mus.
B.M.(N.H.)
T.S.
B.M.(N.H.)

Béréba
P.T.S.
U.S. Nat. Mus.
Field Mus. Nat. Hist.
T.S.
Mus. d'Hist. Nat.

Berlanguillas
T.S.
Mus. d'Hist. Nat.

B.M.(N.H.)

Beuste
T.S.
Naturhist. Mus.
Mus. d'Hist. Nat.
B.M.(N.H.)

Beyrout
T.S.
Mus. d'Hist. Nat.

Bhagur
T.S.
Naturhist. Mus.
B.M.(N.H.)

Bherai
T.S.
B.M.(N.H.)

Bhola
P.T.S.
U.S. Nat. Mus.

Bholghati
P.T.S.
U.S. Nat. Mus.
T.S.
B.M.(N.H.)

Białystok
P.T.S.
U.S. Nat. Mus.
T.S.
Mus. d'Hist. Nat.
B.M.(N.H.)

Bielokrynitschie
P.T.S.
U.S. Nat. Mus.
Field Mus. Nat. Hist.
T.S.
Field Mus. Nat. Hist.
Naturhist. Mus.
Mus. d'Hist. Nat.
B.M.(N.H.)

Billings
P.M.
U.S. Nat. Mus.

Billygoat Donga
P.T.S.
U.S. Nat. Mus.

Binda
P.M.
U.S. Nat. Mus.
P.T.S.
U.S. Nat. Mus.
T.S.
Mus. d'Hist. Nat.

Bingera
P.M.
U.S. Nat. Mus.
B.M.(N.H.)

Bir Hadi
P.M.
U.S. Nat. Mus.
P.T.S.
U.S. Nat. Mus.

Birni N'konni
T.S.
Mus. d'Hist. Nat.

Bischtübe
P.M.
U.S. Nat. Mus.
B.M.(N.H.)

Bishop Canyon
P.M.
U.S. Nat. Mus.

Bishopville
P.T.S.
U.S. Nat. Mus.
Field Mus. Nat. Hist.
T.S.
Naturhist. Mus.
Mus. d'Hist. Nat.
B.M.(N.H.)

Bishunpur
P.T.S.
U.S. Nat. Mus.
B.M.(N.H.)
T.S.
Mus. d'Hist. Nat.
B.M.(N.H.)

Bjelaja Zerkov
P.T.S.
U.S. Nat. Mus.
T.S.
Mus. d'Hist. Nat.
B.M.(N.H.)

Bjurböle
P.M.
U.S. Nat. Mus.
Field Mus. Nat. Hist.
P.T.S.
U.S. Nat. Mus.
Field Mus. Nat. Hist.
Univ. of New Mexico
B.M.(N.H.)
T.S.
Field Mus. Nat. Hist.
Naturhist. Mus.
B.M.(N.H.)

Black Moshannan Park
P.T.S.
U.S. Nat. Mus.

Black Mountain
P.M.
U.S. Nat. Mus.

Blanket
P.M.
Field Mus. Nat. Hist.
P.T.S.
Field Mus. Nat. Hist.

B.M.(N.H.)
T.S.
Field Mus. Nat. Hist.
B.M.(N.H.)

Blansko
P.T.S.
U.S. Nat. Mus.
T.S.
Field Mus. Nat. Hist.
Mus. d'Hist. Nat.

Bledsoe
P.T.S.
Naturhist. Mus.

Blithfield
P.M.
U.S. Nat. Mus.
P.T.S.
U.S. Nat. Mus.
B.M.(N.H.)
T.S.
B.M.(N.H.)

Bloomington
P.T.S.
Field Mus. Nat. Hist.

Bluff
P.T.S.
U.S. Nat. Mus.
Naturhist. Mus.
T.S.
Field Mus. Nat. Hist.
Naturhist. Mus.
Mus. d'Hist. Nat.
B.M.(N.H.)

Bocas
P.T.S.
U.S. Nat. Mus.
T.S.
Mus. d'Hist. Nat.
B.M.(N.H.)

Bodaibo
P.M.
U.S. Nat. Mus.

Boerne
P.M.
U.S. Nat. Mus.
P.T.S.
U.S. Nat. Mus.
T.S.
B.M.(N.H.)

Bogou
P.M.
U.S. Nat. Mus.

Boguslavka
P.M.
U.S. Nat. Mus.

Bohumilitz
P.M.
U.S. Nat. Mus.

Bolivia
P.M.
U.S. Nat. Mus.

Bolshaya Korta
P.T.S.
B.M.(N.H.)
T.S.
B.M.(N.H.)

Bondoc
P.T.S.
U.S. Nat. Mus.
Field Mus. Nat. Hist.

Bonita Springs
P.M.
U.S. Nat. Mus.
P.T.S.
U.S. Nat. Mus.
B.M.(N.H.)
T.S.
B.M.(N.H.)

Boogaldi
P.M.
U.S. Nat. Mus.
B.M.(N.H.)

Borgo San Donino
T.S.
Naturhist. Mus.
Mus. d'Hist. Nat.
B.M.(N.H.)

Bori
P.T.S.
U.S. Nat. Mus.
T.S.
Mus. d'Hist. Nat.
B.M.(N.H.)

Boriskino
P.T.S.
Field Mus. Nat. Hist.

Borkut
T.S.
Naturhist. Mus.
B.M.(N.H.)

Botschetschki
P.T.S.
U.S. Nat. Mus.

Bovedy
P.T.S.
B.M.(N.H.)
T.S.
B.M.(N.H.)

Bowden
P.T.S.
U.S. Nat. Mus.

Bowesmont
P.T.S.
U.S. Nat. Mus.
T.S.
B.M.(N.H.)

Boxhole
P.M.
U.S. Nat. Mus.
B.M.(N.H.)

Brachina
P.T.S.
U.S. Nat. Mus.

Brady
P.T.S.
U.S. Nat. Mus.

Brahin
P.M.
Field Mus. Nat. Hist.
T.S.
Naturhist. Mus.

Braunau
P.M.
U.S. Nat. Mus.
T.S.
Naturhist. Mus.

Breitscheid
P.M.
U.S. Nat. Mus.

Bremervörde
P.T.S.
U.S. Nat. Mus.
Field Mus. Nat. Hist.
Univ. of New Mexico
B.M.(N.H.)
T.S.
Field Mus. Nat. Hist.
Naturhist. Mus.
Mus. d'Hist. Nat.
B.M.(N.H.)

Brenham
P.M.
U.S. Nat. Mus.
T.S.
Naturhist. Mus.
B.M.(N.H.)

Brewster
P.M.
B.M.(N.H.)
P.T.S.
B.M.(N.H.)
T.S.
B.M.(N.H.)

Bridgewater
P.M.
U.S. Nat. Mus.

Brient
P.T.S.
Field Mus. Nat. Hist.

Briggsdale
P.T.S.
U.S. Nat. Mus.

Briscoe
P.T.S.
U.S. Nat. Mus.
B.M.(N.H.)
T.S.
B.M.(N.H.)

Bristol
P.M.
U.S. Nat. Mus.
B.M.(N.H.)

Britstown
P.M.
U.S. Nat. Mus.

Broken Bow
P.M.
U.S. Nat. Mus.
P.T.S.
U.S. Nat. Mus.
B.M.(N.H.)
T.S.
B.M.(N.H.)

Brownfield (1937)
P.T.S.
U.S. Nat. Mus.
Univ. of New Mexico
B.M.(N.H.)
T.S.
B.M.(N.H.)

Brownfield (1964)
P.T.S.
U.S. Nat. Mus.
B.M.(N.H.)
T.S.
B.M.(N.H.)

Bruderheim
P.M.
Field Mus. Nat. Hist.
P.T.S.
U.S. Nat. Mus.
Field Mus. Nat. Hist.
Naturhist. Mus.
Univ. of New Mexico
B.M.(N.H.)
T.S.
B.M.(N.H.)

Bruno
P.M.
U.S. Nat. Mus.

Budulan
P.T.S.
U.S. Nat. Mus.
Naturhist. Mus.

Bununu
P.M.
U.S. Nat. Mus.
P.T.S.
U.S. Nat. Mus.

Burdett
P.M.
Univ. of New Mexico
P.T.S.
U.S. Nat. Mus.

T.S.
B.M.(N.H.)

Burgavli
P.M.
U.S. Nat. Mus.

Bur-Gheluai
P.T.S.
U.S. Nat. Mus.
B.M.(N.H.)
T.S.
Naturhist. Mus.
Mus. d'Hist. Nat.
B.M.(N.H.)

Burkett
P.M.
U.S. Nat. Mus.

Burlington
P.M.
U.S. Nat. Mus.

Burnabbie
P.T.S.
U.S. Nat. Mus.
B.M.(N.H.)
T.S.
B.M.(N.H.)

Burrika
P.T.S.
U.S. Nat. Mus.

Bursa
P.T.S.
B.M.(N.H.)
T.S.
B.M.(N.H.)

Buschhof
T.S.
U.S. Nat. Mus.
Field Mus. Nat. Hist.
Naturhist. Mus.
Mus. d'Hist. Nat.
B.M.(N.H.)

Bushman Land
P.M.
U.S. Nat. Mus.

Bushnell
P.T.S.
U.S. Nat. Mus.
B.M.(N.H.)
T.S.
B.M.(N.H.)

Bustee
P.T.S.
U.S. Nat. Mus.
B.M.(N.H.)
T.S.
Naturhist. Mus.
Mus. d'Hist. Nat.
B.M.(N.H.)

Butler
P.M.
U.S. Nat. Mus.
B.M.(N.H.)

Butsura
P.T.S.
B.M.(N.H.)
T.S.
Mus. d'Hist. Nat.
B.M.(N.H.)

Cabezo de Mayo
P.T.S.
U.S. Nat. Mus.
Field Mus. Nat. Hist.
T.S.
Naturhist. Mus.
Mus. d'Hist. Nat.
B.M.(N.H.)

Cabin Creek
P.M.
U.S. Nat. Mus.

Cacaria
P.M.
U.S. Nat. Mus.
T.S.
Mus. d'Hist. Nat.
B.M.(N.H.)

Cachari
P.T.S.
U.S. Nat. Mus.
Univ. of New Mexico

Cachiyuyal
P.M.
B.M.(N.H.)

Cacilandia
P.T.S.
U.S. Nat. Mus.

Cadell
P.T.S.
U.S. Nat. Mus.

Calliham
P.T.S.
U.S. Nat. Mus.
Naturhist. Mus.
B.M.(N.H.)
T.S.
B.M.(N.H.)

Cambria
P.M.
U.S. Nat. Mus.
B.M.(N.H.)

Campbellsville
P.M.
U.S. Nat. Mus.
B.M.(N.H.)

Campo del Cielo
P.M.
U.S. Nat. Mus.
B.M.(N.H.)
P.T.S.
U.S. Nat. Mus.

Cañellas
P.T.S.
U.S. Nat. Mus.
T.S.
Mus. d'Hist. Nat.

Cangas de Onis
P.M.
B.M.(N.H.)
P.T.S.
U.S. Nat. Mus.
Field Mus. Nat. Hist.
T.S.
U.S. Nat. Mus.
Field Mus. Nat. Hist.
Mus. d'Hist. Nat.

Canon City
P.T.S.
U.S. Nat. Mus.
Field Mus. Nat. Hist.

Cañon Diablo
P.M.
U.S. Nat. Mus.
Univ. of New Mexico
B.M.(N.H.)

Canton
P.M.
U.S. Nat. Mus.
B.M.(N.H.)

Canyon City
P.M.
U.S. Nat. Mus.

Cape Girardeau
P.M.
Field Mus. Nat. Hist.
P.T.S.
Field Mus. Nat. Hist.
T.S.
B.M.(N.H.)

Cape of Good Hope
P.M.
U.S. Nat. Mus.
B.M.(N.H.)

Caperr
P.M.
U.S. Nat. Mus.
Field Mus. Nat. Hist.

Cape York
P.M.
U.S. Nat. Mus.

Capilla del Monte
T.S.
B.M.(N.H.)

Caratash
T.S.
B.M.(N.H.)

Carbo
P.M.
U.S. Nat. Mus.

Carcote
T.S.
Naturhist. Mus.

Cardanumbi
P.T.S.
U.S. Nat. Mus.

Carlton
P.M.
U.S. Nat. Mus.
B.M.(N.H.)

Caroline
P.T.S.
U.S. Nat. Mus.

Carraweena
P.T.S.
U.S. Nat. Mus.
B.M.(N.H.)
T.S.
B.M.(N.H.)

Carthage
P.M.
U.S. Nat. Mus.

Cartoonkana
P.T.S.
U.S. Nat. Mus.

Casas Grandes
P.M.
U.S. Nat. Mus.
Field Mus. Nat. Hist.

Cashion
P.M.
U.S. Nat. Mus.
P.T.S.
U.S. Nat. Mus.
B.M.(N.H.)
T.S.
B.M.(N.H.)

Casilda
P.T.S.
U.S. Nat. Mus.

Casimiro de Abreu
P.M.
U.S. Nat. Mus.

Castalia
P.T.S.
U.S. Nat. Mus.
Field Mus. Nat. Hist.
Naturhist. Mus.
B.M.(N.H.)
T.S.
Field Mus. Nat. Hist.
Naturhist. Mus.
Mus. d'Hist. Nat.
B.M.(N.H.)

Castine
T.S.
B.M.(N.H.)

Cavour
P.M.
Field Mus. Nat. Hist.
P.T.S.
U.S. Nat. Mus.
Field Mus. Nat. Hist.
B.M.(N.H.)
T.S.
B.M.(N.H.)

Cedar (Kansas)
P.M.
Field Mus. Nat. Hist.
P.T.S.
U.S. Nat. Mus.
T.S.
B.M.(N.H.)

Cedar (Texas)
P.T.S.
U.S. Nat. Mus.
T.S.
Field Mus. Nat. Hist.
B.M.(N.H.)

Cedartown
P.M.
U.S. Nat. Mus.
Field Mus. Nat. Hist.

Cee Vee
P.M.
Univ. of New Mexico
P.T.S.
U.S. Nat. Mus.
B.M.(N.H.)
T.S.
B.M.(N.H.)

Centerville
P.T.S.
U.S. Nat. Mus.

Cereseto
P.T.S.
U.S. Nat. Mus.
T.S.
B.M.(N.H.)

Chainpur
P.T.S.
U.S. Nat. Mus.
B.M.(N.H.)
T.S.
Mus. d'Hist. Nat.
B.M.(N.H.)

Chamberlin
P.T.S.
B.M.(N.H.)
T.S.
B.M.(N.H.)

Chambord
P.M.
Field Mus. Nat. Hist.

Chandakapur
P.T.S.
Naturhist. Mus.
T.S.
Naturhist. Mus.
Mus. d'Hist. Nat.
B.M.(N.H.)

Chandpur
T.S.
Naturhist. Mus.
Mus. d'Hist. Nat.
B.M.(N.H.)

Channing
T.S.
B.M.(N.H.)

Chantonnay
P.T.S.
U.S. Nat. Mus.
T.S.
Field Mus. Nat. Hist.
Mus. d'Hist. Nat.
Naturhist. Mus.
B.M.(N.H.)

Charcas
P.M.
U.S. Nat. Mus.
B.M.(N.H.)

Charlotte
P.M.
U.S. Nat. Mus.

Charsonville
P.M.
Field Mus. Nat. Hist.
P.T.S.
U.S. Nat. Mus.
Field Mus. Nat. Hist.
Naturhist. Mus.
T.S.
Field Mus. Nat. Hist.
Naturhist. Mus.
Mus. d'Hist. Nat.
B.M.(N.H.)

Charwallas
T.S.
B.M.(N.H.)

Chassigny
P.T.S.
U.S. Nat. Mus.
B.M.(N.H.)
T.S.
Naturhist. Mus.
Mus. d'Hist. Nat.
B.M.(N.H.)

Château Renard
P.M.
U.S. Nat. Mus.
P.T.S.
U.S. Nat. Mus.
Field Mus. Nat. Hist.
T.S.
Field Mus. Nat. Hist.
Naturhist. Mus.
Mus. d'Hist. Nat.
B.M.(N.H.)

Chaves
P.T.S.
U.S. Nat. Mus.
T.S.
Mus. d'Hist. Nat.
B.M.(N.H.)

Chebankol
P.M.
U.S. Nat. Mus.

Cherokee Springs
P.T.S.
U.S. Nat. Mus.
B.M.(N.H.)
T.S.
B.M.(N.H.)

Chervettaz
T.S.
Naturhist. Mus.
Mus. d'Hist. Nat.
B.M.(N.H.)

Chervony Kut
P.T.S.
Univ. of New Mexico

Chesterville
P.M.
B.M.(N.H.)

Chico
P.T.S.
U.S. Nat. Mus.
Univ. of New Mexico

Chico Hills
P.M.
Univ. of New Mexico
P.T.S.
Univ. of New Mexico

Chico Mountains
P.M.
U.S. Nat. Mus.

Chicora
P.T.S.
Univ. of New Mexico

Chihuahua City
P.M.
U.S. Nat. Mus.

Chilkoot
P.M.
U.S. Nat. Mus.

Chinautla
P.M.
U.S. Nat. Mus.
P.T.S.
Field Mus. Nat. Hist.

Chinga
P.M.
U.S. Nat. Mus.

Chinguetti
P.T.S.
U.S. Nat. Mus.

Chulafinnee
P.M.
U.S. Nat. Mus.

Chupaderos
P.M.
U.S. Nat. Mus.
Field Mus. Nat. Hist.

Cincinnati
P.M.
U.S. Nat. Mus.
Field Mus. Nat. Hist.

Clareton
P.T.S.
U.S. Nat. Mus.
B.M.(N.H.)
T.S.
B.M.(N.H.)

Clark County
P.M.
U.S. Nat. Mus.
B.M.(N.H.)

Claytonville
P.T.S.
U.S. Nat. Mus.
B.M.(N.H.)
T.S.
B.M.(N.H.)

Cleveland
P.M.
U.S. Nat. Mus.

Clohars
P.T.S.
B.M.(N.H.)
T.S.
Mus. d'Hist. Nat.
B.M.(N.H.)

Clover Springs
P.T.S.
U.S. Nat. Mus.
B.M.(N.H.)
T.S.
B.M.(N.H.)

Clovis (no.1)
P.T.S.
U.S. Nat. Mus.
Naturhist. Mus.

Clovis (no.2)
P.T.S.
U.S. Nat. Mus.
T.S.
Field Mus. Nat. Hist.
B.M.(N.H.)

Coahuila
P.M.
U.S. Nat. Mus.
B.M.(N.H.)
T.S.
Naturhist. Mus.

Cobija
P.T.S.
B.M.(N.H.)
T.S.
Field Mus. Nat. Hist.
Naturhist. Mus.
B.M.(N.H.)

Cockarrow Creek
P.T.S.
U.S. Nat. Mus.

Cockburn
P.T.S.
U.S. Nat. Mus.

Cocklebiddy
P.T.S.
B.M.(N.H.)
T.S.
B.M.(N.H.)

Cocunda
P.T.S.
U.S. Nat. Mus.

Colby (Kansas)
P.T.S.
B.M.(N.H.)
T.S.
B.M.(N.H.)

Colby (Wisconsin)
P.M.
Field Mus. Nat. Hist.
P.T.S.
U.S. Nat. Mus.
Field Mus. Nat. Hist.
T.S.
B.M.(N.H.)

Cold Bay
P.M.
U.S. Nat. Mus.

Cold Bokkeveld
P.T.S.
U.S. Nat. Mus.
Field Mus. Nat. Hist.
Naturhist. Mus.
B.M.(N.H.)
T.S.
Naturhist. Mus.
Mus. d'Hist. Nat.
B.M.(N.H.)

Coldwater (stone)
P.T.S.
U.S. Nat. Mus.
B.M.(N.H.)
T.S.
Field Mus. Nat. Hist.
Mus. d'Hist. Nat.
B.M.(N.H.)

Colfax
P.M.
U.S. Nat. Mus.
Field Mus. Nat. Hist.

Collescipoli
T.S.
U.S. Nat. Mus.
Naturhist. Mus.
Mus. d'Hist. Nat.
B.M.(N.H.)

Colomera
P.M.
U.S. Nat. Mus.
P.T.S.
U.S. Nat. Mus.

Colony
P.T.S.
U.S. Nat. Mus.
B.M.(N.H.)

Comanche (iron)
P.M.
U.S. Nat. Mus.

Concho
P.T.S.
B.M.(N.H.)
T.S.
U.S. Nat. Mus.
B.M.(N.H.)

Conquista
P.M.
Univ. of New Mexico
P.T.S.
Univ. of New Mexico

Coolac
P.M.
B.M.(N.H.)

Coolamon
P.T.S.
U.S. Nat. Mus.

Coolidge
P.T.S.
Field Mus. Nat. Hist.
B.M.(N.H.)
T.S.
B.M.(N.H.)

Coomandook
P.T.S.
U.S. Nat. Mus.

Coonana
P.T.S.
U.S. Nat. Mus.
T.S.
B.M.(N.H.)

Coon Butte
P.T.S.
U.S. Nat. Mus.
T.S.
B.M.(N.H.)

Coopertown
P.M.
U.S. Nat. Mus.

Coorara
P.T.S.
U.S. Nat. Mus.

Cope
P.T.S.
Univ. of New Mexico
T.S.
B.M.(N.H.)

Copiapo
P.M.
U.S. Nat. Mus.
P.T.S.
B.M.(N.H.)
T.S.
Naturhist. Mus.
B.M.(N.H.)

Cortez
T.S.
B.M.(N.H.)

Cosby's Creek
P.M.
U.S. Nat. Mus.

Cosina
T.S.
Mus. d'Hist. Nat.
B.M.(N.H.)

Costilla Peak
P.M.
U.S. Nat. Mus.

Cotesfield
P.T.S.
B.M.(N.H.)
T.S.
B.M.(N.H.)

Covert
P.T.S.
U.S. Nat. Mus.
T.S.
Mus. d'Hist. Nat.
B.M.(N.H.)

Cowra
P.M.
Field Mus. Nat. Hist.
B.M.(N.H.)

Crab Orchard
P.M.
U.S. Nat. Mus.
Field Mus. Nat. Hist.
P.T.S.
U.S. Nat. Mus.
Naturhist. Mus.
T.S.
Field Mus. Nat. Hist.
Naturhist. Mus.
Mus. d'Hist. Nat.
B.M.(N.H.)

Cranberry Plains
P.M.
U.S. Nat. Mus.

Cranbourne
P.M.
U.S. Nat. Mus.

Cranganore
P.T.S.
U.S. Nat. Mus.
T.S.
B.M.(N.H.)

Cratheús (1931)
P.M.
U.S. Nat. Mus.

Credo
P.T.S.
U.S. Nat. Mus.
B.M.(N.H.)
T.S.
B.M.(N.H.)

Crescent
P.T.S.
U.S. Nat. Mus.

Cronstad
P.T.S.
B.M.(N.H.)
T.S.
B.M.(N.H.)

Cross Roads
T.S.
Naturhist. Mus.
B.M.(N.H.)

Crumlin
P.T.S.
B.M.(N.H.)
T.S.
Naturhist. Mus.
B.M.(N.H.)

Cruz del Aire
P.M.
U.S. Nat. Mus.

Cuero
T.S.
B.M.(N.H.)

Culbertson
P.T.S.
B.M.(N.H.)
T.S.
B.M.(N.H.)

Cullison
P.M.
U.S. Nat. Mus.
P.T.S.
Field Mus. Nat. Hist.

B.M.(N.H.)
T.S.
B.M.(N.H.)

Cumberland Falls
P.M.
B.M.(N.H.)
P.T.S.
U.S. Nat. Mus.
Field Mus. Nat. Hist.
B.M.(N.H.)
T.S.
Mus. d'Hist. Nat.
B.M.(N.H.)

Cumpas
P.M.
U.S. Nat. Mus.

Cynthiana
P.M.
U.S. Nat. Mus.
P.T.S.
U.S. Nat. Mus.
B.M.(N.H.)
T.S.
Mus. d'Hist. Nat.
B.M.(N.H.)

Dalgaranga
P.M.
U.S. Nat. Mus.
P.T.S.
U.S. Nat. Mus.

Dalgety Downs
P.M.
U.S. Nat. Mus.
P.T.S.
U.S. Nat. Mus.
Univ. of New Mexico
B.M.(N.H.)
T.S.
B.M.(N.H.)

Dalhart
P.T.S.
Naturhist. Mus.
B.M.(N.H.)
T.S.
B.M.(N.H.)

Dalton
P.M.
U.S. Nat. Mus.

Dandapur
T.S.
Field Mus. Nat. Hist.
Naturhist. Mus.
Mus. d'Hist. Nat.
B.M.(N.H.)

Daniel's Kuil
P.T.S.
U.S. Nat. Mus.
B.M.(N.H.)
T.S.
Mus. d'Hist. Nat.
B.M.(N.H.)

Danville
T.S.
B.M.(N.H.)

Daoura
T.S.
Mus. d'Hist. Nat.

Darmstadt
T.S.
Mus. d'Hist. Nat.

Davis Mountains
P.M.
U.S. Nat. Mus.

Dayton
P.M.
U.S. Nat. Mus.

Deal
P.T.S.
U.S. Nat. Mus.

Deep Springs
P.M.
U.S. Nat. Mus.
B.M.(N.H.)

Dehesa
P.M.
U.S. Nat. Mus.

Delegate
P.M.
U.S. Nat. Mus.

Del Rio
P.M.
U.S. Nat. Mus.

Demina
P.T.S.
U.S. Nat. Mus.
B.M.(N.H.)
T.S.
B.M.(N.H.)

Densmore (1879)
P.T.S.
Field Mus. Nat. Hist.
B.M.(N.H.)
T.S.
B.M.(N.H.)

Densmore (1950)
P.T.S.
B.M.(N.H.)
T.S.
B.M.(N.H.)

Denton County
P.M.
U.S. Nat. Mus.

Denver
P.T.S.
U.S. Nat. Mus.

Deport
P.M.
U.S. Nat. Mus.

Dexter
P.M.
U.S. Nat. Mus.

Dhajala
P.M.
Univ. of New Mexico
P.T.S.
Univ. of New Mexico
U.S. Nat. Mus.
Field Mus. Nat. Hist.

Dhurmsala
P.T.S.
U.S. Nat. Mus.
Field Mus. Nat. Hist.
Naturhist. Mus.
B.M.(N.H.)
T.S.
Field Mus. Nat. Hist.
Naturhist. Mus.
Mus. d'Hist. Nat.
B.M.(N.H.)

Diep River
P.T.S.
U.S. Nat. Mus.
T.S.
Naturhist. Mus.
B.M.(N.H.)

Dimboola
P.T.S.
U.S. Nat. Mus.

Dimitrovgrad
P.M.
U.S. Nat. Mus.

Dimmitt
P.T.S.
U.S. Nat. Mus.
Univ. of New Mexico
Field Mus. Nat. Hist.
T.S.
B.M.(N.H.)

Dingo Pup Donga
P.T.S.
U.S. Nat. Mus.

Dix
P.T.S.
U.S. Nat. Mus.

Djati-Pengilon
P.M.
U.S. Nat. Mus.
P.T.S.
U.S. Nat. Mus.
T.S.
Naturhist. Mus.
Mus. d'Hist. Nat.
B.M.(N.H.)

Djermaia
T.S.
Mus. d'Hist. Nat.

Dokachi
T.S.
Mus. d'Hist. Nat.
B.M.(N.H.)

Dolgovoli
T.S.
Mus. d'Hist. Nat.

Domanitch
T.S.
Mus. d'Hist. Nat.

Doroninsk
T.S.
Mus. d'Hist. Nat.
B.M.(N.H.)

Dosso
T.S.
Mus. d'Hist. Nat.

Douar Mghila
P.T.S.
U.S. Nat. Mus.
B.M.(N.H.)
T.S.
Mus. d'Hist. Nat.
B.M.(N.H.)

Doyleville
P.T.S.
B.M.(N.H.)
T.S.
B.M.(N.H.)

Drake Creek
T.S.
U.S. Nat. Mus.
Naturhist. Mus.
Mus. d'Hist. Nat.
B.M.(N.H.)

Dresden (Kansas)
P.T.S.
U.S. Nat. Mus.
B.M.(N.H.)
T.S.
B.M.(N.H.)

Dresden (Ontario)
P.T.S.
U.S. Nat. Mus.

Drum Mountains
P.M.
U.S. Nat. Mus.

Dubrovnik
P.T.S.
Naturhist. Mus.

Duchesne
P.M.
U.S. Nat. Mus.
Field Mus. Nat. Hist.

Duel Hill (1854)
P.M.
U.S. Nat. Mus.

Dumas (a)
P.T.S.
U.S. Nat. Mus.

Dundrum
T.S.
B.M.(N.H.)

Dungannon
P.M.
U.S. Nat. Mus.

Durala
P.T.S.
B.M.(N.H.)
T.S.
Naturhist. Mus.
B.M.(N.H.)

Durango
P.M.
Field Mus. Nat. Hist.

Duruma
T.S.
Naturhist. Mus.
B.M.(N.H.)

Dwaleni
P.M.
U.S. Nat. Mus.
P.T.S.
U.S. Nat. Mus.
B.M.(N.H.)
T.S.
B.M.(N.H.)

Dwight
P.T.S.
B.M.(N.H.)
T.S.
B.M.(N.H.)

Dyalpur
P.T.S.
U.S. Nat. Mus.
B.M.(N.H.)
T.S.
B.M.(N.H.)

Dyarrl Island
P.T.S.
U.S. Nat. Mus.

Eagle Station
P.M.
U.S. Nat. Mus.
T.S.
Naturhist. Mus.

Edjudina
P.T.S.
U.S. Nat. Mus.

Edmonson
P.T.S.
Naturhist. Mus.
T.S.
B.M.(N.H.)

Edmonton (Canada)
P.M.
U.S. Nat. Mus.

Edmonton (Kentucky)
P.M.
U.S. Nat. Mus.

Efremovka
P.T.S.
U.S. Nat. Mus.
Field Mus. Nat. Hist.
B.M.(N.H.)
T.S.
B.M.(N.H.)

Ehole
P.T.S.
B.M.(N.H.)
T.S.
B.M.(N.H.)

Eichstädt
T.S.
Naturhist. Mus.
B.M.(N.H.)

Ekeby
P.T.S.
U.S. Nat. Mus.

El Burro
P.M.
U.S. Nat. Mus.

Elba
P.T.S.
U.S. Nat. Mus.
B.M.(N.H.)
T.S.
B.M.(N.H.)

Elbogen
P.M.
U.S. Nat. Mus.

Elenovka
P.M.
U.S. Nat. Mus.
P.T.S.
U.S. Nat. Mus.
Univ. of New Mexico
B.M.(N.H.)
T.S.
Mus. d'Hist. Nat.
B.M.(N.H.)

Eli Elwah
P.T.S.
U.S. Nat. Mus.
B.M.(N.H.)
T.S.
B.M.(N.H.)

Elkhart
P.T.S.
B.M.(N.H.)
T.S.
B.M.(N.H.)

Ellemeet
P.T.S.
Univ. of New Mexico
T.S.
Mus. d'Hist. Nat.
B.M.(N.H.)

Ellerslie
P.T.S.
U.S. Nat. Mus.

Ellis County
P.T.S.
Field Mus. Nat. Hist.

Elm Creek
P.M.
U.S. Nat. Mus.
P.T.S.
U.S. Nat. Mus.
T.S.
Naturhist. Mus.
Mus. d'Hist. Nat.
B.M.(N.H.)

El Perdido
P.T.S.
U.S. Nat. Mus.
T.S.
Mus. d'Hist. Nat.

Elsinora
P.M.
U.S. Nat. Mus.
T.S.
B.M.(N.H.)

Emery
P.M.
U.S. Nat. Mus.
P.T.S.
Univ. of New Mexico

Emmaville
P.T.S.
U.S. Nat. Mus.
B.M.(N.H.)

Emmitsburg
P.M.
U.S. Nat. Mus.

Enigma
P.T.S.
U.S. Nat. Mus.

Enon
P.M.
U.S. Nat. Mus.

Ensisheim
P.M.
U.S. Nat. Mus.
Field Mus. Nat. Hist.
P.T.S.
U.S. Nat. Mus.
Field Mus. Nat. Hist.
T.S.
Field Mus. Nat. Hist.
Naturhist. Mus.
Mus. d'Hist. Nat.
B.M.(N.H.)

Épinal
T.S.
Mus. d'Hist. Nat.

Ergheo
P.T.S.
U.S. Nat. Mus.
Naturhist. Mus.
T.S.
Field Mus. Nat. Hist.
Naturhist. Mus.
Mus. d'Hist. Nat.
B.M.(N.H.)

Erie
P.T.S.
U.S. Nat. Mus.

Erxleben
P.T.S.
Field Mus. Nat. Hist.
T.S.
Field Mus. Nat. Hist.
Naturhist. Mus.
Mus. d'Hist. Nat.
B.M.(N.H.)

Esnandes
T.S.
Mus. d'Hist. Nat.

Espiritu Santo
P.M.
Field Mus. Nat. Hist.

Essebi
P.T.S.
U.S. Nat. Mus.

Estacado
P.M.
U.S. Nat. Mus.
P.T.S.
U.S. Nat. Mus.
Field Mus. Nat. Hist.
Naturhist. Mus.
B.M.(N.H.)
T.S.
Naturhist. Mus.
Mus. d'Hist. Nat.
B.M.(N.H.)

Estherville
P.M.
U.S. Nat. Mus.
B.M.(N.H.)
P.T.S.
U.S. Nat. Mus.
B.M.(N.H.)
T.S.
Naturhist. Mus.
Mus. d'Hist. Nat.
B.M.(N.H.)

Etter
P.T.S.
Naturhist. Mus.
Univ. of New Mexico

Eustis
T.S.
U.S. Nat. Mus.

Eva
P.T.S.
U.S. Nat. Mus.
B.M.(N.H.)
T.S.
B.M.(N.H.)

Faith
P.T.S.
U.S. Nat. Mus.
B.M.(N.H.)
T.S.
B.M.(N.H.)

Farley
P.M.
U.S. Nat. Mus.
Univ. of New Mexico
P.T.S.
U.S. Nat. Mus.
Univ. of New Mexico
B.M.(N.H.)
T.S.
B.M.(N.H.)

Farmington
P.M.
U.S. Nat. Mus.
Field Mus. Nat. Hist.
P.T.S.
U.S. Nat. Mus.
T.S.
Field Mus. Nat. Hist.
Naturhist. Mus.
Mus. d'Hist. Nat.
B.M.(N.H.)

Farmville
P.M.
U.S. Nat. Mus.
P.T.S.
U.S. Nat. Mus.

Farnum
P.T.S.
B.M.(N.H.)
T.S.
B.M.(N.H.)

Faucett
P.T.S.
Naturhist. Mus.
B.M.(N.H.)
T.S.
Univ. of New Mexico
B.M.(N.H.)

Favars
T.S.
Mus. d'Hist. Nat.
B.M.(N.H.)

Fayetteville
P.T.S.
U.S. Nat. Mus.

Feid Chair
T.S.
Mus. d'Hist. Nat.

Felix
P.M.
U.S. Nat. Mus.
P.T.S.
U.S. Nat. Mus.
Field Mus. Nat. Hist.
B.M.(N.H.)
T.S.
B.M.(N.H.)

Felt
P.T.S.
Naturhist. Mus.

Fenbark
P.T.S.
U.S. Nat. Mus.

Fenghsien-Ku
T.S.
Mus. d'Hist. Nat.

Ferguson Switch
P.T.S.
B.M.(N.H.)
T.S.
B.M.(N.H.)

Finmarken
P.M.
U.S. Nat. Mus.
T.S.
Naturhist. Mus.

Finney
P.T.S.
U.S. Nat. Mus.
B.M.(N.H.)
T.S.
B.M.(N.H.)

Fisher
P.T.S.
U.S. Nat. Mus.
T.S.
Naturhist. Mus.
B.M.(N.H.)

Fleming
T.S.
B.M.(N.H.)

Floyd
P.T.S.
Univ. of New Mexico
B.M.(N.H.)
T.S.
B.M.(N.H.)

Föllinge
P.M.
U.S. Nat. Mus.

Forest City
P.M.
U.S. Nat. Mus.
Univ. of New Mexico
P.T.S.
Field Mus. Nat. Hist.
Univ. of New Mexico
T.S.
Field Mus. Nat. Hist.
Mus. d'Hist. Nat.
Naturhist. Mus.
B.M.(N.H.)

Forest Vale
P.M.
U.S. Nat. Mus.
P.T.S.
U.S. Nat. Mus.

Forksville
P.T.S.
U.S. Nat. Mus.

Forrest
P.T.S.
U.S. Nat. Mus.

Forrest Lakes
P.T.S.
U.S. Nat. Mus.

Forsbach
P.T.S.
B.M.(N.H.)
T.S.
Naturhist. Mus.

Forsyth
T.S.
Field Mus. Nat. Hist.
Naturhist. Mus.
Mus. d'Hist. Nat.
B.M.(N.H.)

Forsyth County
P.M.
U.S. Nat. Mus.
B.M.(N.H.)

Fort Pierre
P.M.
U.S. Nat. Mus.

Four Corners
P.M.
U.S. Nat. Mus.
P.T.S.
B.M.(N.H.)

Franceville
P.M.
U.S. Nat. Mus.

Frankfort (iron)
P.M.
U.S. Nat. Mus.

Frankfort (stone)
P.T.S.
U.S. Nat. Mus.
Field Mus. Nat. Hist.
T.S.
Naturhist. Mus.
Mus. d'Hist. Nat.
B.M.(N.H.)

Franklin
P.T.S.
B.M.(N.H.)
T.S.
B.M.(N.H.)

Freda
P.M.
U.S. Nat. Mus.

Fremont Butte
P.T.S.
B.M.(N.H.)
T.S.
B.M.(N.H.)

Frenchman Bay
P.T.S.
U.S. Nat. Mus.
T.S.
B.M.(N.H.)

Fukutomi
P.T.S.
U.S. Nat. Mus.
Field Mus. Nat. Hist.
B.M.(N.H.)
T.S.
Naturhist. Mus.
B.M.(N.H.)

Futtehpur
P.T.S.
B.M.(N.H.)
T.S.
Naturhist. Mus.
Mus. d'Hist. Nat.
B.M.(N.H.)

Galapian
T.S.
Mus. d'Hist. Nat.

Galatia
P.T.S.
Univ. of New Mexico

Galim
T.S.
Mus. d'Hist. Nat.

Gambat
T.S.
Mus. d'Hist. Nat.

Garland
P.T.S.
U.S. Nat. Mus.

Garnett
P.T.S.
Field Mus. Nat. Hist.
B.M.(N.H.)
T.S.

U.S. Nat. Mus.
Mus. d'Hist. Nat.
B.M.(N.H.)

Garraf
P.T.S.
B.M.(N.H.)
T.S.
U.S. Nat. Mus.
B.M.(N.H.)

Garrison
P.T.S.
Univ. of New Mexico

Georgetown
P.T.S.
U.S. Nat. Mus.

Ghubara
P.T.S.
U.S. Nat. Mus.
B.M.(N.H.)
T.S.
B.M.(N.H.)

Gibeon
P.M.
U.S. Nat. Mus.
B.M.(N.H.)

Gifu
T.S.
Naturhist. Mus.
Mus. d'Hist. Nat.

Gilgoin
P.M.
U.S. Nat. Mus.
P.T.S.
U.S. Nat. Mus.
Naturhist. Mus.
B.M.(N.H.)
T.S.
Field Mus. Nat. Hist.
Naturhist. Mus.
Mus. d'Hist. Nat.
B.M.(N.H.)

Girgenti
P.M.
U.S. Nat. Mus.
P.T.S.
U.S. Nat. Mus.
Naturhist. Mus.
B.M.(N.H.)
T.S.
Field Mus. Nat. Hist.
Naturhist. Mus.
Mus. d'Hist. Nat.
B.M.(N.H.)

Giroux
P.M.
U.S. Nat. Mus.

Git-Git
P.T.S.
B.M.(N.H.)
T.S.
B.M.(N.H.)

Gladstone (stone)
P.M.
Univ. of New Mexico
P.T.S.
U.S. Nat. Mus.
Univ. of New Mexico
B.M.(N.H.)
T.S.
Mus. d'Hist. Nat.
B.M.(N.H.)

Glasatovo
P.T.S.
B.M.(N.H.)
T.S.
B.M.(N.H.)

Glasgow
P.M.
U.S. Nat. Mus.

Glorieta Mountain
P.M.
U.S. Nat. Mus.

Gnadenfrei
T.S.
Field Mus. Nat. Hist.
B.M.(N.H.)

Goalpara
P.T.S.
U.S. Nat. Mus.
Naturhist. Mus.
Field Mus. Nat. Hist.
B.M.(N.H.)
T.S.
Naturhist. Mus.
B.M.(N.H.)

Gobabeb
P.T.S.
U.S. Nat. Mus.

Goodland
P.M.
U.S. Nat. Mus.
P.T.S.
U.S. Nat. Mus.
B.M.(N.H.)
T.S.
B.M.(N.H.)

Goose Lake
P.M.
U.S. Nat. Mus.
Field Mus. Nat. Hist.

Gopalpur
P.T.S.
B.M.(N.H.)
T.S.
Naturhist. Mus.
Mus. d'Hist. Nat.
B.M.(N.H.)

Gorlovka
P.T.S.
U.S. Nat. Mus.

Grady (1933)
P.T.S.
U.S. Nat. Mus.

Grady (1937)
P.T.S.
U.S. Nat. Mus.
Univ. of New Mexico
B.M.(N.H.)
T.S.
B.M.(N.H.)

Grady (c)
P.T.S.
Naturhist. Mus.
Univ. of New Mexico

Grand Rapids
P.M.
U.S. Nat. Mus.

Granes
T.S.
Mus. d'Hist. Nat.

Grant
P.M.
U.S. Nat. Mus.
Univ. of New Mexico
Field Mus. Nat. Hist.
B.M.(N.H.)

Grant County
P.T.S.
B.M.(N.H.)
T.S.
B.M.(N.H.)

Grassland
P.T.S.
U.S. Nat. Mus.
B.M.(N.H.)
T.S.
B.M.(N.H.)

Great Bear Lake
P.M.
U.S. Nat. Mus.
P.T.S.
U.S. Nat. Mus.

Greenbrier County
P.M.
U.S. Nat. Mus.

Gressk
P.M.
U.S. Nat. Mus.

Gretna
T.S.
B.M.(N.H.)

Grier (b)
P.T.S.
U.S. Nat. Mus.

Grosnaja
P.T.S.
U.S. Nat. Mus.
Field Mus. Nat. Hist.
B.M.(N.H.)
T.S.
Naturhist. Mus.
Mus. d'Hist. Nat.
B.M.(N.H.)

Gross-Divina
P.T.S.
Naturhist. Mus.
T.S.
Naturhist. Mus.
Mus. d'Hist. Nat.

Grossliebenthal
P.T.S.
B.M.(N.H.)
T.S.
Field Mus. Nat. Hist.
Mus. d'Hist. Nat.
B.M.(N.H.)

Grüneberg
P.T.S.
Field Mus. Nat. Hist.
T.S.
Naturhist. Mus.
Mus. d'Hist. Nat.
B.M.(N.H.)

Gruver
P.T.S.
B.M.(N.H.)
T.S.
Mus. d'Hist. Nat.
B.M.(N.H.)

Guareña
P.M.
Field Mus. Nat. Hist.
P.T.S.
U.S. Nat. Mus.
B.M.(N.H.)
T.S.
Field Mus. Nat. Hist.
Mus. d'Hist. Nat.
B.M.(N.H.)

Guffey
P.M.
U.S. Nat. Mus.

Guibga
P.M.
U.S. Nat. Mus.
P.T.S.
U.S. Nat. Mus.

Guidder
P.T.S.
U.S. Nat. Mus.
T.S.
Mus. d'Hist. Nat.

Guilford County
P.M.
U.S. Nat. Mus.

Gumoschnik
T.S.
Mus. d'Hist. Nat.

Gun Creek
P.M.
U.S. Nat. Mus.

Gundaring
P.M.
U.S. Nat. Mus.

Gunnadorah
P.T.S.
U.S. Nat. Mus.

Gurram Konda
T.S.
B.M.(N.H.)

Gütersloh
P.T.S.
U.S. Nat. Mus.
B.M.(N.H.)
T.S.
Mus. d'Hist. Nat.
B.M.(N.H.)

Haig
P.M.
U.S. Nat. Mus.

Hainaut
T.S.
Mus. d'Hist. Nat.

Hainholz
P.M.
U.S. Nat. Mus.
P.T.S.
U.S. Nat. Mus.
Naturhist. Mus.
T.S.
Naturhist. Mus.
Mus. d'Hist. Nat.
B.M.(N.H.)

Hale Center (no. 1)
P.T.S.
B.M.(N.H.)
T.S.
B.M.(N.H.)

Hale Center (no. 2)
P.T.S.
B.M.(N.H.)
T.S.
B.M.(N.H.)

Hallingeberg
P.T.S.
U.S. Nat. Mus.

Hamilton
P.T.S.
U.S. Nat. Mus.

Hamlet
P.T.S.
U.S. Nat. Mus.
Field Mus. Nat. Hist.
B.M.(N.H.)
T.S.
B.M.(N.H.)

Hammond
P.M.
U.S. Nat. Mus.
B.M.(N.H.)

Hammond Downs
P.T.S.
U.S. Nat. Mus.

Haniet-el-Beguel
P.M.
Field Mus. Nat. Hist.

Happy Canyon
P.T.S.
Univ. of New Mexico
U.S. Nat. Mus.
Field Mus. Nat. Hist.

Haraiya
P.T.S.
U.S. Nat. Mus.

Hardwick
P.T.S.
B.M.(N.H.)
T.S.
U.S. Nat. Mus.
B.M.(N.H.)

Haripura
P.T.S.
U.S. Nat. Mus.

Harleton
P.M.
U.S. Nat. Mus.
P.T.S.
U.S. Nat. Mus.
Field Mus. Nat. Hist.

Harriman (Of)
P.M.
Field Mus. Nat. Hist.

Harriman (Om)
P.M.
U.S. Nat. Mus.

Harrison County
P.T.S.
U.S. Nat. Mus.
B.M.(N.H.)
T.S.
Naturhist. Mus.
Mus. d'Hist. Nat.
B.M.(N.H.)

Harrisonville
P.M.
U.S. Nat. Mus.
P.T.S.
Univ. of New Mexico
B.M.(N.H.)
T.S.
B.M.(N.H.)

Hat Creek
P.T.S.
B.M.(N.H.)
T.S.
B.M.(N.H.)

Haven
P.T.S.
U.S. Nat. Mus.

Haverö
P.T.S.
U.S. Nat. Mus.
B.M.(N.H.)

Haviland (stone)
P.T.S.
B.M.(N.H.)
T.S.
B.M.(N.H.)

Hawk Springs
P.T.S.
B.M.(N.H.)
T.S.
B.M.(N.H.)

Hayden Creek
P.M.
Field Mus. Nat. Hist.

Hayes Center
P.T.S.
U.S. Nat. Mus.
B.M.(N.H.)
T.S.
B.M.(N.H.)

Hedeskoga
P.T.S.
U.S. Nat. Mus.

Hedjaz
P.T.S.
U.S. Nat. Mus.
B.M.(N.H.)
T.S.
Mus. d'Hist. Nat.
B.M.(N.H.)

Henbury
P.M.
U.S. Nat. Mus.
Field Mus. Nat. Hist.
B.M.(N.H.)

Hendersonville
T.S.
U.S. Nat. Mus.
Mus. d'Hist. Nat.
B.M.(N.H.)

Heredia
T.S.
Mus. d'Hist. Nat.
B.M.(N.H.)

Hermitage Plains
P.M.
U.S. Nat. Mus.
P.T.S.
U.S. Nat. Mus.
T.S.
B.M.(N.H.)

Hessle
T.S.
Naturhist. Mus.
Mus. d'Hist. Nat.
B.M.(N.H.)

Hex River Mountains
P.M.
U.S. Nat. Mus.
Naturhist. Mus.
B.M.(N.H.)

Hickiwan
P.T.S.
Univ. of New Mexico

Higashi-koen
T.S.
B.M.(N.H.)

High Possil
T.S.
B.M.(N.H.)

Hildreth
P.T.S.
U.S. Nat. Mus.

Hill City
P.M.
U.S. Nat. Mus.

Hinojal
P.T.S.
U.S. Nat. Mus.

Hoba
P.M.
U.S. Nat. Mus.
Naturhist. Mus.
B.M.(N.H.)
T.S.
Naturhist. Mus.

Hobbs
P.T.S.
Univ. of New Mexico

Hökmark
P.T.S.
U.S. Nat. Mus.

Holbrook
P.M.
U.S. Nat. Mus.
P.T.S.
U.S. Nat. Mus.
Field Mus. Nat. Hist.
Naturhist. Mus.
T.S.
Field Mus. Nat. Hist.
Naturhist. Mus.

Mus. d'Hist. Nat.
B.M.(N.H.)

Holland's Store
P.M.
U.S. Nat. Mus.

Holly
P.T.S.
U.S. Nat. Mus.

Holyoke
P.T.S.
Field Mus. Nat. Hist.
B.M.(N.H.)
T.S.
B.M.(N.H.)

Homestead
P.M.
U.S. Nat. Mus.
P.T.S.
U.S. Nat. Mus.
Naturhist. Mus.
B.M.(N.H.)
T.S.
Field Mus. Nat. Hist.
Naturhist. Mus.
Mus. d'Hist. Nat.
B.M.(N.H.)

Honolulu
P.T.S.
U.S. Nat. Mus.
T.S.
Naturhist. Mus.
Mus. d'Hist. Nat.
B.M.(N.H.)

Hope
P.M.
U.S. Nat. Mus.
P.T.S.
Univ. of New Mexico

Hopper
P.M.
U.S. Nat. Mus.

Horace
P.T.S.
U.S. Nat. Mus.
B.M.(N.H.)
T.S.
B.M.(N.H.)

Horse Creek
P.M.
U.S. Nat. Mus.
Field Mus. Nat. Hist.
B.M.(N.H.)

Howe
P.T.S.
B.M.(N.H.)
T.S.
B.M.(N.H.)

Hraschina
P.M.
U.S. Nat. Mus.
Field Mus. Nat. Hist.

Huckitta
P.M.
U.S. Nat. Mus.

Hugoton
P.M.
U.S. Nat. Mus.
P.T.S.
B.M.(N.H.)
T.S.
Mus. d'Hist. Nat.
B.M.(N.H.)

Huizopa
P.M.
U.S. Nat. Mus.

Hungen
T.S.
Naturhist. Mus.

Hvittis
P.M.
U.S. Nat. Mus.
P.T.S.
U.S. Nat. Mus.
B.M.(N.H.)
T.S.
Naturhist. Mus.
B.M.(N.H.)

Ibbenbüren
P.T.S.
U.S. Nat. Mus.
T.S.
Naturhist. Mus.
Mus. d'Hist. Nat.

Ibitira
P.T.S.
U.S. Nat. Mus.
Univ. of New Mexico

Idaho
P.M.
U.S. Nat. Mus.

Ider
P.M.
U.S. Nat. Mus.

Idutywa
T.S.
B.M.(N.H.)

Ilimaes (iron)
P.M.
U.S. Nat. Mus.

Ilinskaya Stanitza
P.M.
U.S. Nat. Mus.

Illinois Gulch
P.M.
B.M.(N.H.)

Imilac
P.M.
U.S. Nat. Mus.
B.M.(N.H.)
T.S.
Naturhist. Mus.

Indarch
P.M.
U.S. Nat. Mus.
P.T.S.
U.S. Nat. Mus.
B.M.(N.H.)
T.S.
Field Mus. Nat. Hist.
Naturhist. Mus.
Mus. d'Hist. Nat.
B.M.(N.H.)

Indianola
P.T.S.
U.S. Nat. Mus.
B.M.(N.H.)

Indian Valley
P.M.
U.S. Nat. Mus.

Indio Rico
P.T.S.
Naturhist. Mus.
T.S.
Naturhist. Mus.
Mus. d'Hist. Nat.

Inman
P.T.S.
U.S. Nat. Mus.
Univ. of New Mexico
T.S.
Univ. of New Mexico

Ioka
P.T.S.
U.S. Nat. Mus.

Ipiranga
P.M.
Univ. of New Mexico
P.T.S.
U.S. Nat. Mus.
Univ. of New Mexico

Iquique
P.M.
U.S. Nat. Mus.

Iredell
P.M.
U.S. Nat. Mus.

Isna
P.T.S.
U.S. Nat. Mus.

Isoulane-n-Amahar
T.S.
Mus. d'Hist. Nat.

Isthilart
T.S.
Mus. d'Hist. Nat.

Itapicuru-Mirim
P.M.
Univ. of New Mexico
P.T.S.
U.S. Nat. Mus.
Univ. of New Mexico

Ivanpah
P.M.
U.S. Nat. Mus.

Ivuna
P.T.S.
U.S. Nat. Mus.

Jackalsfontein
P.T.S.
Naturhist. Mus.
T.S.
Naturhist. Mus.
Mus. d'Hist. Nat.
B.M.(N.H.)

Jackson County
P.M.
U.S. Nat. Mus.

Jajh deh Kot Lalu
P.T.S.
U.S. Nat. Mus.
B.M.(N.H.)
T.S.
B.M.(N.H.)

Jamestown
P.M.
U.S. Nat. Mus.
B.M.(N.H.)

Jamkheir
T.S.
B.M.(N.H.)

Jelica
P.M.
Field Mus. Nat. Hist.
P.T.S.
U.S. Nat. Mus.
Field Mus. Nat. Hist.
Naturhist. Mus.
B.M.(N.H.)
T.S.
Field Mus. Nat. Hist.
Naturhist. Mus.
Mus. d'Hist. Nat.
B.M.(N.H.)

Jemlapur
P.T.S.
U.S. Nat. Mus.
T.S.
B.M.(N.H.)

Jenkins
P.M.
U.S. Nat. Mus.

Jenny's Creek
P.M.
U.S. Nat. Mus.

Jhung
P.T.S.
B.M.(N.H.)
T.S.
Field Mus. Nat. Hist.
Naturhist. Mus.
Mus. d'Hist. Nat.
B.M.(N.H.)

Jiddat al Harasis
P.T.S.
B.M.(N.H.)
T.S.
B.M.(N.H.)

Jilin
P.T.S.
U.S. Nat. Mus.

Jodzie
P.T.S.
U.S. Nat. Mus.
Field Mus. Nat. Hist.

Joe Wright Mountain
P.M.
U.S. Nat. Mus.

Johnson City
P.T.S.
B.M.(N.H.)
T.S.
B.M.(N.H.)

Johnstown
P.T.S.
U.S. Nat. Mus.
Field Mus. Nat. Hist.
T.S.
Mus. d'Hist. Nat.
B.M.(N.H.)

Jonzac
P.T.S.
U.S. Nat. Mus.
T.S.
Naturhist. Mus.
Mus. d'Hist. Nat.
B.M.(N.H.)

Judesegeri
P.T.S.
U.S. Nat. Mus.
T.S.
Naturhist. Mus.
Mus. d'Hist. Nat.

Juvinas
P.M.
B.M.(N.H.)
P.T.S.
U.S. Nat. Mus.
Field Mus. Nat. Hist.
Naturhist. Mus.
Univ. of New Mexico
B.M.(N.H.)
T.S.
Field Mus. Nat. Hist.
Naturhist. Mus.
Mus. d'Hist. Nat.
B.M.(N.H.)

Kaba
P.T.S.
U.S. Nat. Mus.
Naturhist. Mus.
T.S.
B.M.(N.H.)

Kabo
P.M.
U.S. Nat. Mus.
P.T.S.
U.S. Nat. Mus.
B.M.(N.H.)
T.S.
B.M.(N.H.)

Kadonah
T.S.
B.M.(N.H.)

Kaee
P.T.S.
B.M.(N.H.)
T.S.
B.M.(N.H.)

Kagarlyk
P.T.S.
U.S. Nat. Mus.

Kainsaz
P.T.S.
U.S. Nat. Mus.
Field Mus. Nat. Hist.

Kakangari
P.T.S.
U.S. Nat. Mus.
B.M.(N.H.)

Kakowa
T.S.
Mus. d'Hist. Nat.
B.M.(N.H.)

Kalaba
P.T.S.
U.S. Nat. Mus.
T.S.
B.M.(N.H.)

Kaldoonera Hill
P.T.S.
U.S. Nat. Mus.

Kalkaska
P.M.
U.S. Nat. Mus.

Kalumbi
T.S.
Naturhist. Mus.
Mus. d'Hist. Nat.
B.M.(N.H.)

Kamiomi
P.T.S.
U.S. Nat. Mus.

Kamsagar
P.T.S.
U.S. Nat. Mus.

Kandahar (Afghanistan)
P.M.
U.S. Nat. Mus.
P.T.S.
U.S. Nat. Mus.

Kangra Valley
T.S.
B.M.(N.H.)

Kapoeta
P.M.
U.S. Nat. Mus.
P.T.S.
U.S. Nat. Mus.
T.S.
B.M.(N.H.)

Kappakoola
P.T.S.
U.S. Nat. Mus.
T.S.
B.M.(N.H.)

Kaptal-Aryk
P.T.S.
U.S. Nat. Mus.

Karakol
P.T.S.
U.S. Nat. Mus.

Karasburg
P.M.
U.S. Nat. Mus.

Karatu
P.T.S.
U.S. Nat. Mus.
T.S.
B.M.(N.H.)

Karee Kloof
P.M.
U.S. Nat. Mus.

Karewar
T.S.
B.M.(N.H.)

Karkh
P.T.S.
U.S. Nat. Mus.
T.S.
Mus. d'Hist. Nat.

Karloowala
T.S.
B.M.(N.H.)

Karoonda
P.T.S.
U.S. Nat. Mus.
Field Mus. Nat. Hist.
B.M.(N.H.)
T.S.
B.M.(N.H.)

Kaufman
P.T.S.
U.S. Nat. Mus.

Kayakent
P.M.
B.M.(N.H.)

Kearney
P.T.S.
U.S. Nat. Mus.

Kelly
P.M.
U.S. Nat. Mus.
P.T.S.
U.S. Nat. Mus.
B.M.(N.H.)
T.S.
Mus. d'Hist. Nat.

Kendall County
P.M.
U.S. Nat. Mus.
Naturhist. Mus.
P.T.S.
U.S. Nat. Mus.
T.S.
Naturhist. Mus.

Kenna
P.T.S.
U.S. Nat. Mus.
Univ. of New Mexico
T.S.
Univ. of New Mexico

Kenton County
P.M.
U.S. Nat. Mus.

Kerilis
P.T.S.
B.M.(N.H.)
T.S.
Mus. d'Hist. Nat.
B.M.(N.H.)

Kermichel
P.T.S.
Naturhist. Mus.
T.S.
Naturhist. Mus.
Mus. d'Hist. Nat.
B.M.(N.H.)

Kernouve
P.M.
U.S. Nat. Mus.
B.M.(N.H.)
P.T.S.
U.S. Nat. Mus.
Univ. of New Mexico
B.M.(N.H.)

T.S.
Field Mus. Nat. Hist.
Naturhist. Mus.
Mus. d'Hist. Nat.
B.M.(N.H.)

Kesen
P.M.
Univ. of New Mexico
P.T.S.
U.S. Nat. Mus.
Field Mus. Nat. Hist.
Univ. of New Mexico
B.M.(N.H.)
T.S.
U.S. Nat. Mus.
Field Mus. Nat. Hist.
Naturhist. Mus.
Mus. d'Hist. Nat.
B.M.(N.H.)

Khairpur
P.T.S.
U.S. Nat. Mus.
B.M.(N.H.)
T.S.
Field Mus. Nat. Hist.
Mus. d'Hist. Nat.
B.M.(N.H.)

Khanpur
P.T.S.
U.S. Nat. Mus.
B.M.(N.H.)
T.S.
B.M.(N.H.)

Kharkov
P.T.S.
U.S. Nat. Mus.
T.S.
B.M.(N.H.)

Khetri
T.S.
Mus. d'Hist. Nat.
B.M.(N.H.)

Khohar
P.T.S.
U.S. Nat. Mus.
B.M.(N.H.)
T.S.
Mus. d'Hist. Nat.
B.M.(N.H.)

Khor Temiki
P.M.
B.M.(N.H.)
P.T.S.
U.S. Nat. Mus.
B.M.(N.H.)
T.S.
B.M.(N.H.)

Kiel
P.T.S.
U.S. Nat. Mus.

Kielpa
P.T.S.
U.S. Nat. Mus.
B.M.(N.H.)
T.S.
B.M.(N.H.)

Kiffa
P.M.
U.S. Nat. Mus.
P.T.S.
U.S. Nat. Mus.

Kikino
T.S.
Field Mus. Nat. Hist.
Naturhist. Mus.
Mus. d'Hist. Nat.

Kilbourn
T.S.
Field Mus. Nat. Hist.

Killeter
P.T.S.
B.M.(N.H.)
T.S.
B.M.(N.H.)

Kimble County
P.T.S.
U.S. Nat. Mus.
B.M.(N.H.)
T.S.
B.M.(N.H.)

Kingai
P.T.S.
B.M.(N.H.)

Kingfisher
P.T.S.
U.S. Nat. Mus.
B.M.(N.H.)

Kingston
P.M.
U.S. Nat. Mus.
Field Mus. Nat. Hist.

Kirbyville
P.T.S.
U.S. Nat. Mus.

Kissij
P.T.S.
U.S. Nat. Mus.
T.S.
Mus. d'Hist. Nat.

Kisvarsany
P.T.S.
U.S. Nat. Mus.

Kittakittaooloo
P.T.S.
U.S. Nat. Mus.

Klein-Wenden
T.S.
Naturhist. Mus.
B.M.(N.H.)

Klondike (Skookum Gulch)
P.M.
U.S. Nat. Mus.
B.M.(N.H.)

Knowles
P.M.
U.S. Nat. Mus.

Knyahinya
P.M.
U.S. Nat. Mus.
P.T.S.
Naturhist. Mus.
T.S.
Field Mus. Nat. Hist.
Naturhist. Mus.
Mus. d'Hist. Nat.
B.M.(N.H.)

Kochi
P.T.S.
U.S. Nat. Mus.

Kodaikanal
P.M.
U.S. Nat. Mus.
Field Mus. Nat. Hist.
Naturhist. Mus.
B.M.(N.H.)
P.T.S.
U.S. Nat. Mus.
Field Mus. Nat. Hist.
T.S.
Naturhist. Mus.

Kofa
P.M.
U.S. Nat. Mus.

Kopjes Vlei
P.M.
U.S. Nat. Mus.

Koraleigh
P.T.S.
U.S. Nat. Mus.

Kota-Kota
P.T.S.
U.S. Nat. Mus.
B.M.(N.H.)
T.S.
Naturhist. Mus.
B.M.(N.H.)

Krähenberg
P.M.
Univ. of New Mexico
P.T.S.
U.S. Nat. Mus.
Univ. of New Mexico
B.M.(N.H.)
T.S.
Naturhist. Mus.
Mus. d'Hist. Nat.
B.M.(N.H.)

Kramer Creek
P.T.S.
Univ. of New Mexico
U.S. Nat. Mus.

Krasnoi-Ugol
P.T.S.
U.S. Nat. Mus.

Krasnojarsk
P.M.
U.S. Nat. Mus.
T.S.
Naturhist. Mus.

Kress
P.T.S.
B.M.(N.H.)
T.S.
B.M.(N.H.)

Krymka
P.T.S.
U.S. Nat. Mus.
Naturhist. Mus.

Kulak
P.T.S.
B.M.(N.H.)
T.S.
B.M.(N.H.)

Kuleschovka
P.T.S.
U.S. Nat. Mus.
T.S.
Mus. d'Hist. Nat.

Kulnine
P.T.S.
U.S. Nat. Mus.

Kumerina
P.M.
U.S. Nat. Mus.
B.M.(N.H.)

Kunashak
P.T.S.
U.S. Nat. Mus.
T.S.
B.M.(N.H.)

Kusiali
T.S.
B.M.(N.H.)

Kuttippuram
P.T.S.
U.S. Nat. Mus.
T.S.
Mus. d'Hist. Nat.

Kyancutta
P.M.
U.S. Nat. Mus.

Kybunga
P.T.S.
U.S. Nat. Mus.

Kyle
P.T.S.
Univ. of New Mexico
B.M.(N.H.)
T.S.
B.M.(N.H.)

Kyushu
P.M.
Field Mus. Nat. Hist.
T.S.
Naturhist. Mus.
Mus. d'Hist. Nat.
B.M.(N.H.)

La Bécasse
P.T.S.
B.M.(N.H.)
T.S.
Mus. d'Hist. Nat.
B.M.(N.H.)

Laborel
P.T.S.
B.M.(N.H.)
T.S.
Naturhist. Mus.
Mus. d'Hist. Nat.
B.M.(N.H.)

La Caille
P.M.
U.S. Nat. Mus.
Field Mus. Nat. Hist.

La Colina
T.S.
Mus. d'Hist. Nat.

Ladder Creek
P.T.S.
U.S. Nat. Mus.
Univ. of New Mexico

Lafayette (stone)
P.M.
Field Mus. Nat. Hist.
P.T.S.
U.S. Nat. Mus.
Field Mus. Nat. Hist.
T.S.
Field Mus. Nat. Hist.

La Grange
P.M.
U.S. Nat. Mus.

Lahrauli
P.M.
B.M.(N.H.)

L'Aigle
P.M.
U.S. Nat. Mus.
P.T.S.
U.S. Nat. Mus.
T.S.
Field Mus. Nat. Hist.
Naturhist. Mus.
Mus. d'Hist. Nat.
B.M.(N.H.)

Lakangaon
P.T.S.
U.S. Nat. Mus.
B.M.(N.H.)
T.S.
Mus. d'Hist. Nat.
B.M.(N.H.)

Lake Bonney
P.T.S.
U.S. Nat. Mus.

Lake Brown
P.T.S.
U.S. Nat. Mus.
T.S.
B.M.(N.H.)

Lake Grace
P.T.S.
U.S. Nat. Mus.

Lake Labyrinth
P.M.
U.S. Nat. Mus.
Field Mus. Nat. Hist.
P.T.S.
U.S. Nat. Mus.
Univ. of New Mexico
B.M.(N.H.)
T.S.
B.M.(N.H.)

Lake Murray
P.M.
U.S. Nat. Mus.

Laketon
P.T.S.
U.S. Nat. Mus.

Lakewood
P.T.S.
Naturhist. Mus.
Univ. of New Mexico
B.M.(N.H.)
T.S.
B.M.(N.H.)

La Lande
P.M.
U.S. Nat. Mus.
Univ. of New Mexico
P.T.S.
Univ. of New Mexico
T.S.
B.M.(N.H.)

Lalitpur
T.S.
Mus. d'Hist. Nat.
B.M.(N.H.)

Lancé
P.T.S.
U.S. Nat. Mus.
T.S.
Field Mus. Nat. Hist.
P.T.S.
Field Mus. Nat. Hist.
T.S.
Naturhist. Mus.
P.T.S.
Naturhist. Mus.
T.S.
Mus. d'Hist. Nat.
B.M.(N.H.)
P.T.S.
B.M.(N.H.)

Lancon
P.M.
Field Mus. Nat. Hist.
P.T.S.
Field Mus. Nat. Hist.
B.M.(N.H.)
T.S.
Field Mus. Nat. Hist.
Naturhist. Mus.
Mus. d'Hist. Nat.
B.M.(N.H.)

Landes
P.M.
Naturhist. Mus.

La"nghalsen
P.M.
Field Mus. Nat. Hist.
P.T.S.
U.S. Nat. Mus.
Field Mus. Nat. Hist.

Lanton
P.M.
U.S. Nat. Mus.

Lanzenkirchen
P.T.S.
Naturhist. Mus.
T.S.
Naturhist. Mus.

La Porte
P.M.
U.S. Nat. Mus.

La Primitiva
P.M.
U.S. Nat. Mus.
B.M.(N.H.)

Laundry East
P.T.S.
U.S. Nat. Mus.

Laundry Rockhole
P.T.S.
U.S. Nat. Mus.

Laundry West
P.T.S.
U.S. Nat. Mus.

Launton
T.S.
B.M.(N.H.)

Laurens County
P.M.
U.S. Nat. Mus.

Lawrence
P.M.
U.S. Nat. Mus.
P.T.S.
U.S. Nat. Mus.

Leedey
P.M.
Field Mus. Nat. Hist.
P.T.S.
U.S. Nat. Mus.
Field Mus. Nat. Hist.
Naturhist. Mus.
Univ. of New Mexico

Leeds
P.M.
U.S. Nat. Mus.

Leeuwfontain
T.S.
B.M.(N.H.)

Leighton
P.T.S.
B.M.(N.H.)
T.S.
Field Mus. Nat. Hist.
Naturhist. Mus.

Lenarto
P.M.
U.S. Nat. Mus.
B.M.(N.H.)

Leon
P.T.S.
U.S. Nat. Mus.

Leonovka
T.S.
Naturhist. Mus.

Leoville
P.T.S.
U.S. Nat. Mus.
Naturhist. Mus.
Univ. of New Mexico
Field Mus. Nat. Hist.
B.M.(N.H.)
T.S.
B.M.(N.H.)

Le Pressoir
T.S.
Field Mus. Nat. Hist.
Naturhist. Mus.
Mus. d'Hist. Nat.

Les Ormes
P.T.S.
U.S. Nat. Mus.

T.S.
Mus. d'Hist. Nat.
B.M.(N.H.)

Lesves
T.S.
Mus. d'Hist. Nat.

Le Teilleul
P.M.
Field Mus. Nat. Hist.
P.T.S.
Field Mus. Nat. Hist.
T.S.
Naturhist. Mus.
Mus. d'Hist. Nat.

Lexington County
P.M.
U.S. Nat. Mus.

Lick Creek
P.M.
U.S. Nat. Mus.

Lillaverke
P.T.S.
U.S. Nat. Mus.

Lime Creek
T.S.
Naturhist. Mus.

Limerick
P.T.S.
U.S. Nat. Mus.
B.M.(N.H.)
T.S.
Naturhist. Mus.
Mus. d'Hist. Nat.
B.M.(N.H.)

Lincoln County
P.T.S.
U.S. Nat. Mus.
B.M.(N.H.)
T.S.
B.M.(N.H.)

Linville
P.M.
U.S. Nat. Mus.

Linwood
P.M.
U.S. Nat. Mus.

Lissa
T.S.
Field Mus. Nat. Hist.
Naturhist. Mus.
Mus. d'Hist. Nat.
B.M.(N.H.)

Little Piney
P.M.
U.S. Nat. Mus.
P.T.S.
B.M.(N.H.)
T.S.
Mus. d'Hist. Nat.
B.M.(N.H.)

Little River (a)
P.T.S.
B.M.(N.H.)
T.S.
B.M.(N.H.)

Littlerock
P.T.S.
U.S. Nat. Mus.

Livingston (Montana)
P.M.
U.S. Nat. Mus.

Livingston (Tennessee)
P.M.
U.S. Nat. Mus.

Lixna
P.T.S.
U.S. Nat. Mus.
T.S.
Naturhist. Mus.
Mus. d'Hist. Nat.
B.M.(N.H.)

Lockney
P.T.S.
B.M.(N.H.)
T.S.
B.M.(N.H.)

Locust Grove
P.M.
U.S. Nat. Mus.
B.M.(N.H.)

Lodran
P.M.
B.M.(N.H.)
P.T.S.
U.S. Nat. Mus.
T.S.
Naturhist. Mus.
Mus. d'Hist. Nat.
B.M.(N.H.)

Logan
P.T.S.
U.S. Nat. Mus.

Lombard
P.M.
U.S. Nat. Mus.

Lone Star
P.T.S.
B.M.(N.H.)
T.S.
B.M.(N.H.)

Long Island
P.M.
U.S. Nat. Mus.
T.S.
Field Mus. Nat. Hist.
Naturhist. Mus.
Mus. d'Hist. Nat.
B.M.(N.H.)

Loomis
T.S.
B.M.(N.H.)

Loop
P.T.S.
U.S. Nat. Mus.
B.M.(N.H.)
T.S.
B.M.(N.H.)

Loreto
P.M.
U.S. Nat. Mus.

Los Reyes
P.M.
U.S. Nat. Mus.

Lost City
P.M.
U.S. Nat. Mus.
P.T.S.
U.S. Nat. Mus.
B.M.(N.H.)
T.S.
B.M.(N.H.)

Losttown
P.M.
U.S. Nat. Mus.

Łowicz
P.M.
U.S. Nat. Mus.
P.T.S.
U.S. Nat. Mus.

Lua
P.T.S.
U.S. Nat. Mus.

Lubbock
P.T.S.
U.S. Nat. Mus.
B.M.(N.H.)

Lucé
T.S.
Naturhist. Mus.
Mus. d'Hist. Nat.

Lucky Hill
P.M.
U.S. Nat. Mus.

Luis Lopez
P.M.
U.S. Nat. Mus.

Lundsgård
P.T.S.
U.S. Nat. Mus.
T.S.
Mus. d'Hist. Nat.
B.M.(N.H.)

Luotolax
P.T.S.
U.S. Nat. Mus.
Field Mus. Nat. Hist.
T.S.
Naturhist. Mus.
Mus. d'Hist. Nat.
B.M.(N.H.)

Luponnas
T.S.
Mus. d'Hist. Nat.

Lutschaunig's Stone
T.S.
B.M.(N.H.)

Macau
P.M.
Univ. of New Mexico
P.T.S.
Univ. of New Mexico
T.S.
Field Mus. Nat. Hist.
Naturhist. Mus.
Mus. d'Hist. Nat.

Macibini
P.T.S.
U.S. Nat. Mus.

Madoc
P.M.
U.S. Nat. Mus.
Field Mus. Nat. Hist.

Madrid
P.T.S.
U.S. Nat. Mus.
T.S.
Mus. d'Hist. Nat.

Mafra
P.T.S.
U.S. Nat. Mus.
Univ. of New Mexico

Magura
P.M.
U.S. Nat. Mus.
B.M.(N.H.)
T.S.
Naturhist. Mus.

Mainz
T.S.
Mus. d'Hist. Nat.
B.M.(N.H.)

Makarewa
T.S.
Field Mus. Nat. Hist.
Naturhist. Mus.
Mus. d'Hist. Nat.
B.M.(N.H.)

Malakal
P.M.
U.S. Nat. Mus.
P.T.S.

U.S. Nat. Mus.
B.M.(N.H.)
T.S.
B.M.(N.H.)

Malotas
P.M.
U.S. Nat. Mus.
P.T.S.
U.S. Nat. Mus.

Malvern
P.T.S.
U.S. Nat. Mus.

Manbhoom
P.M.
Field Mus. Nat. Hist.
P.T.S.
U.S. Nat. Mus.
Field Mus. Nat. Hist.
T.S.
Mus. d'Hist. Nat.
Naturhist. Mus.
B.M.(N.H.)

Manegaon
T.S.
Mus. d'Hist. Nat.
B.M.(N.H.)

Mangwendi
P.T.S.
U.S. Nat. Mus.
B.M.(N.H.)
T.S.
B.M.(N.H.)

Mantos Blancos
P.M.
U.S. Nat. Mus.
Field Mus. Nat. Hist.
B.M.(N.H.)

Mapleton
P.M.
U.S. Nat. Mus.

Mardan
T.S.
B.M.(N.H.)

Maria Linden
T.S.
Naturhist. Mus.

Maridi
P.M.
U.S. Nat. Mus.
T.S.
B.M.(N.H.)

Marilia
P.M.
Univ. of New Mexico
P.T.S.
Univ. of New Mexico

Marion (Iowa)
P.T.S.
U.S. Nat. Mus.
T.S.
Field Mus. Nat. Hist.
Naturhist. Mus.
B.M.(N.H.)

Marjalahti
P.M.
Field Mus. Nat. Hist.
B.M.(N.H.)
T.S.
Naturhist. Mus.

Marmande
T.S.
Naturhist. Mus.

Marshall County
P.M.
U.S. Nat. Mus.

Mart
P.M.
U.S. Nat. Mus.

Mascombes
T.S.
Mus. d'Hist. Nat.

Mässing
P.T.S.
U.S. Nat. Mus.
T.S.
Naturhist. Mus.
Mus. d'Hist. Nat.
B.M.(N.H.)

Mauerkirchen
P.T.S.
U.S. Nat. Mus.
T.S.
Field Mus. Nat. Hist.
Naturhist. Mus.
Mus. d'Hist. Nat.
B.M.(N.H.)

Mayo Belwa
P.T.S.
B.M.(N.H.)

Mayodan
P.M.
U.S. Nat. Mus.

Mazapil
P.M.
Field Mus. Nat. Hist.

Maziba
P.T.S.
U.S. Nat. Mus.

Mbosi
P.M.
U.S. Nat. Mus.

McKinney
P.T.S.
U.S. Nat. Mus.
Univ. of New Mexico
B.M.(N.H.)
T.S.
Field Mus. Nat. Hist.
Naturhist. Mus.
Mus. d'Hist. Nat.
B.M.(N.H.)

Medanitos
P.T.S.
B.M.(N.H.)

Mejillones
P.M.
U.S. Nat. Mus.
P.T.S.
Field Mus. Nat. Hist.

Melrose (a)
P.M.
U.S. Nat. Mus.
Univ. of New Mexico
P.T.S.
Univ. of New Mexico

Melrose (b)
P.T.S.
Univ. of New Mexico

Menow
T.S.
Naturhist. Mus.
Mus. d'Hist. Nat.
B.M.(N.H.)

Merceditas
P.M.
U.S. Nat. Mus.

Mern
P.M.
U.S. Nat. Mus.
P.T.S.
U.S. Nat. Mus.
T.S.
Naturhist. Mus.
Mus. d'Hist. Nat.

Mertzon
P.M.
U.S. Nat. Mus.

Merua
P.T.S.
U.S. Nat. Mus.
T.S.
B.M.(N.H.)

Messina
P.T.S.
U.S. Nat. Mus.
Naturhist. Mus.
T.S.
Naturhist. Mus.

Metsäkylä
T.S.
U.S. Nat. Mus.

Meuselbach
T.S.
Naturhist. Mus.

Mezel
P.T.S.
B.M.(N.H.)
T.S.
Mus. d'Hist. Nat.
B.M.(N.H.)

Mezö-Madaras
P.T.S.
U.S. Nat. Mus.
Field Mus. Nat. Hist.
Naturhist. Mus.
B.M.(N.H.)
T.S.
Field Mus. Nat. Hist.
Naturhist. Mus.
Mus. d'Hist. Nat.
B.M.(N.H.)

Mhow
T.S.
Mus. d'Hist. Nat.
B.M.(N.H.)

Middlesbrough
T.S.
Naturhist. Mus.
Mus. d'Hist. Nat.
B.M.(N.H.)

Mighei
P.M.
Field Mus. Nat. Hist.
P.T.S.
U.S. Nat. Mus.
Field Mus. Nat. Hist.
T.S.
Naturhist. Mus.
Mus. d'Hist. Nat.
B.M.(N.H.)

Milena
T.S.
Naturhist. Mus.
Mus. d'Hist. Nat.
B.M.(N.H.)

Millarville
P.M.
U.S. Nat. Mus.

Millbillillie
P.T.S.
U.S. Nat. Mus.
Univ. of New Mexico
B.M.(N.H.)

Miller (Arkansas)
P.T.S.
U.S. Nat. Mus.

Miller (Kansas)
P.T.S.
U.S. Nat. Mus.

Mills
P.M.
Univ. of New Mexico

P.T.S.
U.S. Nat. Mus.
Univ. of New Mexico
B.M.(N.H.)
T.S.
B.M.(N.H.)

Milly Milly
P.M.
U.S. Nat. Mus.

Minas Gerais
P.T.S.
U.S. Nat. Mus.
Univ. of New Mexico
T.S.
Mus. d'Hist. Nat.

Mincy
P.T.S.
U.S. Nat. Mus.
Univ. of New Mexico
T.S.
Naturhist. Mus.
Mus. d'Hist. Nat.
B.M.(N.H.)

Mirzapur
T.S.
B.M.(N.H.)

Misshof
T.S.
Field Mus. Nat. Hist.
Naturhist. Mus.
Mus. d'Hist. Nat.

Misteca
P.M.
U.S. Nat. Mus.
Field Mus. Nat. Hist.

Mocs
P.M.
U.S. Nat. Mus.
Field Mus. Nat. Hist.
P.T.S.
U.S. Nat. Mus.
Naturhist. Mus.
B.M.(N.H.)
T.S.
Field Mus. Nat. Hist.
Naturhist. Mus.
Mus. d'Hist. Nat.
B.M.(N.H.)

Modoc (1905)
P.T.S.
U.S. Nat. Mus.
Field Mus. Nat. Hist.
B.M.(N.H.)
T.S.
Field Mus. Nat. Hist.
Naturhist. Mus.
B.M.(N.H.)

Modoc (1948)
T.S.
B.M.(N.H.)

Mokoia
P.T.S.
U.S. Nat. Mus.
T.S.
B.M.(N.H.)

Molina
T.S.
Naturhist. Mus.
Mus. d'Hist. Nat.
B.M.(N.H.)

Molong
P.M.
U.S. Nat. Mus.

Molteno
P.T.S.
B.M.(N.H.)
T.S.
B.M.(N.H.)

Monahans
P.M.
U.S. Nat. Mus.
B.M.(N.H.)

Monroe
P.T.S.
U.S. Nat. Mus.
T.S.
Field Mus. Nat. Hist.
Mus. d'Hist. Nat.
B.M.(N.H.)

Monte Colina
P.T.S.
U.S. Nat. Mus.

Monte das Fortes
T.S.
Mus. d'Hist. Nat.

Monte Milone
P.T.S.
U.S. Nat. Mus.
T.S.
Mus. d'Hist. Nat.
B.M.(N.H.)

Montlivault
T.S.
Mus. d'Hist. Nat.

Monze
T.S.
B.M.(N.H.)

Moonbi
P.M.
U.S. Nat. Mus.
B.M.(N.H.)

Moore County
P.T.S.
U.S. Nat. Mus.
B.M.(N.H.)

Mooresfort
P.T.S.
U.S. Nat. Mus.
B.M.(N.H.)
T.S.
Naturhist. Mus.
Mus. d'Hist. Nat.
B.M.(N.H.)

Moorleah
P.T.S.
U.S. Nat. Mus.

Moradabad
T.S.
B.M.(N.H.)

Morito
P.M.
U.S. Nat. Mus.

Morland
P.M.
Field Mus. Nat. Hist.
P.T.S.
Univ. of New Mexico
B.M.(N.H.)
T.S.
B.M.(N.H.)

Mornans
P.T.S.
B.M.(N.H.)
T.S.
B.M.(N.H.)

Morrill
P.M.
U.S. Nat. Mus.

Morristown
P.M.
U.S. Nat. Mus.
P.T.S.
U.S. Nat. Mus.
Field Mus. Nat. Hist.
T.S.
Naturhist. Mus.
Mus. d'Hist. Nat.
B.M.(N.H.)

Morro do Rocio
T.S.
B.M.(N.H.)

Morven
P.M.
U.S. Nat. Mus.
T.S.
B.M.(N.H.)

Mosca
P.T.S.
U.S. Nat. Mus.

Mosquero
P.T.S.
U.S. Nat. Mus.

Mossgiel
P.T.S.
U.S. Nat. Mus.
B.M.(N.H.)
T.S.
B.M.(N.H.)

Moti-ka-nagla
P.T.S.
B.M.(N.H.)
T.S.
Naturhist. Mus.
Mus. d'Hist. Nat.
B.M.(N.H.)

Motpena
P.T.S.
U.S. Nat. Mus.

Motta di Conti
P.T.S.
U.S. Nat. Mus.
T.S.
Mus. d'Hist. Nat.

Mount Ayliff
P.M.
U.S. Nat. Mus.

Mount Baldr
P.T.S.
U.S. Nat. Mus.

Mount Browne
P.M.
U.S. Nat. Mus.
P.T.S.
U.S. Nat. Mus.
T.S.
Mus. d'Hist. Nat.

Mount Dooling
P.M.
U.S. Nat. Mus.

Mount Dyrring
P.M.
U.S. Nat. Mus.
T.S.
Naturhist. Mus.

Mount Edith
P.M.
U.S. Nat. Mus.
B.M.(N.H.)

Mount Egerton
P.M.
U.S. Nat. Mus.
P.T.S.
U.S. Nat. Mus.

Mount Joy
P.M.
U.S. Nat. Mus.

Mount Magnet
P.M.
U.S. Nat. Mus.
B.M.(N.H.)

Mount Morris (Wisconsin)
P.T.S.
U.S. Nat. Mus.
B.M.(N.H.)
T.S.
B.M.(N.H.)

Mount Ouray
P.M.
U.S. Nat. Mus.

Mount Padbury
P.M.
U.S. Nat. Mus.
P.T.S.
U.S. Nat. Mus.
B.M.(N.H.)

Mount Vernon
P.M.
U.S. Nat. Mus.

Muddoor
P.T.S.
U.S. Nat. Mus.
T.S.
Mus. d'Hist. Nat.
B.M.(N.H.)

Muizenberg
T.S.
B.M.(N.H.)

Muleshoe
P.M.
Field Mus. Nat. Hist.

Mulga (north)
P.T.S.
U.S. Nat. Mus.

Mulga (south)
P.T.S.
U.S. Nat. Mus.

Mulletiwu
T.S.
Mus. d'Hist. Nat.

Mundrabilla
P.M.
U.S. Nat. Mus.
P.T.S.
U.S. Nat. Mus.

Mungindi
P.M.
U.S. Nat. Mus.

Muonionalusta
P.M.
U.S. Nat. Mus.

Murchison
P.M.
Field Mus. Nat. Hist.
P.T.S.
U.S. Nat. Mus.
Field Mus. Nat. Hist.
Naturhist. Mus.
B.M.(N.H.)
T.S.
B.M.(N.H.)

Murfreesboro
P.M.
U.S. Nat. Mus.

Murnpeowie
P.M.
U.S. Nat. Mus.
B.M.(N.H.)

Muroc Dry Lake
P.T.S.
U.S. Nat. Mus.

Murray
P.M.
U.S. Nat. Mus.
P.T.S.
U.S. Nat. Mus.
Naturhist. Mus.
Univ. of New Mexico

Nadiabondi
T.S.
Mus. d'Hist. Nat.

Nagaria
P.T.S.
U.S. Nat. Mus.
T.S.
Mus. d'Hist. Nat.
B.M.(N.H.)

Nagy-Borové
T.S.
B.M.(N.H.)
P.T.S.
B.M.(N.H.)

Nagy-Vázsony
P.M.
U.S. Nat. Mus.

Nakhla
P.T.S.
U.S. Nat. Mus.
Field Mus. Nat. Hist.
B.M.(N.H.)
T.S.
Naturhist. Mus.
Mus. d'Hist. Nat.
B.M.(N.H.)

Nakhon Pathom
P.T.S.
U.S. Nat. Mus.

Nammianthal
T.S.
Mus. d'Hist. Nat.
B.M.(N.H.)

Nanjemoy
T.S.
Mus. d'Hist. Nat.
B.M.(N.H.)

Naoki
T.S.
Mus. d'Hist. Nat.
B.M.(N.H.)

Naragh
P.T.S.
U.S. Nat. Mus.

Nardoo (no.1)
P.M.
U.S. Nat. Mus.
P.T.S.
U.S. Nat. Mus.

Nardoo (no.2)
P.M.
U.S. Nat. Mus.
P.T.S.
U.S. Nat. Mus.

Narellan
P.M.
U.S. Nat. Mus.
P.T.S.
U.S. Nat. Mus.
B.M.(N.H.)
T.S.
B.M.(N.H.)

Naretha
P.T.S.
U.S. Nat. Mus.

Narraburra
P.M.
U.S. Nat. Mus.
Field Mus. Nat. Hist.
B.M.(N.H.)

Näs
P.M.
Field Mus. Nat. Hist.
P.T.S.
U.S. Nat. Mus.

Nassirah
T.S.
Mus. d'Hist. Nat.

Navajo
P.M.
U.S. Nat. Mus.
Field Mus. Nat. Hist.

Nazareth (stone)
P.T.S.
Univ. of New Mexico

Nedagolla
P.M.
U.S. Nat. Mus.
Field Mus. Nat. Hist.
B.M.(N.H.)

Needles
P.M.
U.S. Nat. Mus.

Neenach
P.T.S.
B.M.(N.H.)
T.S.
B.M.(N.H.)

Negrillos
P.M.
U.S. Nat. Mus.
B.M.(N.H.)

Nejed (no. 2)
P.M.
U.S. Nat. Mus.

Nejo
P.M.
U.S. Nat. Mus.
P.T.S.
U.S. Nat. Mus.
B.M.(N.H.)
T.S.
B.M.(N.H.)

Nelson County
P.M.
U.S. Nat. Mus.

Neptune Mountains
P.M.
U.S. Nat. Mus.

Nerft
T.S.
Field Mus. Nat. Hist.
Naturhist. Mus.
Mus. d'Hist. Nat.
B.M.(N.H.)

Ness County (1894)
P.M.
Field Mus. Nat. Hist.
P.T.S.
U.S. Nat. Mus.
B.M.(N.H.)
T.S.
Field Mus. Nat. Hist.
Naturhist. Mus.
Mus. d'Hist. Nat.
B.M.(N.H.)

Ness County (1938)
P.T.S.
B.M.(N.H.)
T.S.
B.M.(N.H.)

Netschaëvo
P.M.
U.S. Nat. Mus.
P.T.S.
Field Mus. Nat. Hist.
B.M.(N.H.)
T.S.
Naturhist. Mus.
B.M.(N.H.)

New Almelo
T.S.
B.M.(N.H.)

New Baltimore
P.M.
U.S. Nat. Mus.

New Concord
P.M.
U.S. Nat. Mus.
P.T.S.
U.S. Nat. Mus.
Field Mus. Nat. Hist.
B.M.(N.H.)
T.S.
Field Mus. Nat. Hist.
Naturhist. Mus.
Mus. d'Hist. Nat.
B.M.(N.H.)

New Leipzig
P.M.
U.S. Nat. Mus.

Newport
P.M.
U.S. Nat. Mus.

New Westville
P.M.
U.S. Nat. Mus.

Ngawi
P.T.S.
U.S. Nat. Mus.
Field Mus. Nat. Hist.
B.M.(N.H.)
T.S.
B.M.(N.H.)

N'Goureyma
P.M.
U.S. Nat. Mus.
B.M.(N.H.)
P.T.S.
U.S. Nat. Mus.

Niagara
P.M.
Field Mus. Nat. Hist.

N'Kandhla
P.M.
U.S. Nat. Mus.

Nobleborough
P.T.S.
Field Mus. Nat. Hist.
T.S.
B.M.(N.H.)

Nocoleche
P.M.
U.S. Nat. Mus.
B.M.(N.H.)

Nogoya
P.M.
Field Mus. Nat. Hist.
P.T.S.
U.S. Nat. Mus.
Field Mus. Nat. Hist.
T.S.
Naturhist. Mus.
Mus. d'Hist. Nat.

Nora Creina
P.T.S.
U.S. Nat. Mus.

Norcateur
T.S.
B.M.(N.H.)

Nordheim
P.M.
U.S. Nat. Mus.

Norfolk
P.M.
U.S. Nat. Mus.

Norfork
P.M.
U.S. Nat. Mus.

North Chile
P.M.
U.S. Nat. Mus.
B.M.(N.H.)

North East Reid
P.T.S.
U.S. Nat. Mus.

North Haig
P.T.S.
U.S. Nat. Mus.

North Reid
P.T.S.
U.S. Nat. Mus.

North West Forrest (E6)
P.T.S.
U.S. Nat. Mus.

North West Forrest (H)
P.T.S.
U.S. Nat. Mus.

Norton County
P.T.S.
U.S. Nat. Mus.
Univ. of New Mexico
Field Mus. Nat. Hist.

Noventa Vicentina
P.T.S.
U.S. Nat. Mus.

Novo-Urei
P.M.
Field Mus. Nat. Hist.
P.T.S.
U.S. Nat. Mus.
Field Mus. Nat. Hist.
T.S.
Naturhist. Mus.
Mus. d'Hist. Nat.

Nuevo Laredo
P.T.S.
U.S. Nat. Mus.

Nuevo Mercurio
P.M.
Univ. of New Mexico
P.T.S.
U.S. Nat. Mus.
Univ. of New Mexico

Nulles
P.T.S.
U.S. Nat. Mus.
T.S.
Mus. d'Hist. Nat.
B.M.(N.H.)

Nyirábrany
P.M.
U.S. Nat. Mus.
P.T.S.
U.S. Nat. Mus.

Oak
P.T.S.
U.S. Nat. Mus.

Oakley (iron)
P.M.
U.S. Nat. Mus.

Oakley (stone)
P.M.
U.S. Nat. Mus.
P.T.S.
U.S. Nat. Mus.
Field Mus. Nat. Hist.
T.S.
Field Mus. Nat. Hist.
Naturhist. Mus.

Oberlin
P.T.S.
U.S. Nat. Mus.

Obernkirchen
P.M.
U.S. Nat. Mus.
B.M.(N.H.)

Ochansk
P.M.
U.S. Nat. Mus.
Univ. of New Mexico
P.T.S.
U.S. Nat. Mus.
Field Mus. Nat. Hist.
Univ. of New Mexico
B.M.(N.H.)
T.S.
Naturhist. Mus.
Mus. d'Hist. Nat.
B.M.(N.H.)

Oczeretna
P.T.S.
B.M.(N.H.)

Odessa (iron)
P.M.
U.S. Nat. Mus.
B.M.(N.H.)

Odessa (stone)
P.T.S.
B.M.(N.H.)
T.S.
B.M.(N.H.)

Oesede
P.T.S.
U.S. Nat. Mus.

Oesel
T.S.
Naturhist. Mus.
B.M.(N.H.)

Ogallala
P.M.
U.S. Nat. Mus.

Ogi
T.S.
Mus. d'Hist. Nat.
B.M.(N.H.)

Ohaba
T.S.
Naturhist. Mus.
Mus. d'Hist. Nat.
B.M.(N.H.)

Ohuma
P.T.S.
U.S. Nat. Mus.
B.M.(N.H.)

Ojuelos Altos
P.T.S.
U.S. Nat. Mus.
T.S.
Mus. d'Hist. Nat.

Okano
P.M.
U.S. Nat. Mus.

Okechobee
P.T.S.
U.S. Nat. Mus.

Okirai
P.T.S.
U.S. Nat. Mus.

Oktibbeha County
P.M.
U.S. Nat. Mus.
B.M.(N.H.)

Oldenburg (1930)
P.T.S.
U.S. Nat. Mus.

Olivenza
P.M.
U.S. Nat. Mus.
Field Mus. Nat. Hist.
P.T.S.
U.S. Nat. Mus.

B.M.(N.H.)
T.S.
Mus. d'Hist. Nat.
B.M.(N.H.)

Ollague
P.M.
U.S. Nat. Mus.
B.M.(N.H.)

Olmedilla de Alarcón
P.T.S.
U.S. Nat. Mus.
B.M.(N.H.)
T.S.
Mus. d'Hist. Nat.

Orange River (iron)
P.M.
U.S. Nat. Mus.

Orgueil
P.T.S.
U.S. Nat. Mus.
T.S.
Mus. d'Hist. Nat.
B.M.(N.H.)

Orimattila
P.T.S.
U.S. Nat. Mus.

Orlovka
P.T.S.
U.S. Nat. Mus.

Ornans
P.M.
Field Mus. Nat. Hist.
P.T.S.
U.S. Nat. Mus.
Field Mus. Nat. Hist.
Naturhist. Mus.
B.M.(N.H.)
T.S.
Mus. d'Hist. Nat.
B.M.(N.H.)

Oro Grande
P.T.S.
U.S. Nat. Mus.
Univ. of New Mexico

Oroville
P.M.
U.S. Nat. Mus.

Orvinio
T.S.
Naturhist. Mus.
B.M.(N.H.)

Oscuro Mountains
P.M.
U.S. Nat. Mus.

Osseo
P.M.
Field Mus. Nat. Hist.

Ottawa
P.M.
Field Mus. Nat. Hist.
P.T.S.
U.S. Nat. Mus.
T.S.
Field Mus. Nat. Hist.
Naturhist. Mus.
B.M.(N.H.)

Oubari
P.T.S.
U.S. Nat. Mus.
T.S.
Mus. d'Hist. Nat.
B.M.(N.H.)

Ovid
P.T.S.
U.S. Nat. Mus.

Oviedo
T.S.
Mus. d'Hist. Nat.

Ozona
P.T.S.
U.S. Nat. Mus.
Field Mus. Nat. Hist.

Ozren
P.M.
U.S. Nat. Mus.

Pacula
P.T.S.
U.S. Nat. Mus.
T.S.
Naturhist. Mus.
Mus. d'Hist. Nat.
B.M.(N.H.)

Padvarninkai
P.T.S.
U.S. Nat. Mus.
Field Mus. Nat. Hist.

Palo Blanco Creek
P.T.S.
Univ. of New Mexico

Palolo Valley
P.T.S.
U.S. Nat. Mus.

Pampanga
T.S.
Naturhist. Mus.
Mus. d'Hist. Nat.

Pan de Azucar
P.M.
U.S. Nat. Mus.
B.M.(N.H.)

Pantar
P.T.S.
Naturhist. Mus.
B.M.(N.H.)
T.S.
Field Mus. Nat. Hist.
B.M.(N.H.)

Para de Minas
P.M.
U.S. Nat. Mus.

Paragould
P.T.S.
Naturhist. Mus.
B.M.(N.H.)
T.S.
Field Mus. Nat. Hist.

Parambu
P.M.
Univ. of New Mexico
P.T.S.
U.S. Nat. Mus.
Univ. of New Mexico

Paranaiba
P.T.S.
U.S. Nat. Mus.
Univ. of New Mexico

Parnallee
P.M.
U.S. Nat. Mus.
P.T.S.
U.S. Nat. Mus.
B.M.(N.H.)
T.S.
Naturhist. Mus.
B.M.(N.H.)

Parsa
P.T.S.
U.S. Nat. Mus.

Pasamonte
P.T.S.
U.S. Nat. Mus.
Univ. of New Mexico
T.S.
Naturhist. Mus.
B.M.(N.H.)

Patora
P.T.S.
B.M.(N.H.)
T.S.
B.M.(N.H.)

Patrimonio
P.T.S.
Univ. of New Mexico
U.S. Nat. Mus.

Patwar
P.M.
U.S. Nat. Mus.
P.T.S.
U.S. Nat. Mus.
T.S.
Mus. d'Hist. Nat.

Pavlodar (pallasite)
P.M.
Field Mus. Nat. Hist.
T.S.
Naturhist. Mus.

Pavlograd
P.T.S.
U.S. Nat. Mus.
T.S.
Naturhist. Mus.
Mus. d'Hist. Nat.
B.M.(N.H.)

Pavlovka
P.T.S.
U.S. Nat. Mus.
T.S.
Naturhist. Mus.
Mus. d'Hist. Nat.

Peace River
P.M.
U.S. Nat. Mus.
P.T.S.
B.M.(N.H.)
T.S.
B.M.(N.H.)

Peck's Spring
T.S.
U.S. Nat. Mus.

Peetz
P.T.S.
U.S. Nat. Mus.
B.M.(N.H.)
T.S.
B.M.(N.H.)

Peña Blanca Spring
P.M.
U.S. Nat. Mus.
P.T.S.
U.S. Nat. Mus.

Penokee
P.M.
U.S. Nat. Mus.
P.T.S.
U.S. Nat. Mus.

Peramiho
T.S.
Naturhist. Mus.

Perpeti
T.S.
Mus. d'Hist. Nat.
B.M.(N.H.)

Perryville
P.M.
U.S. Nat. Mus.

Persimmon Creek
P.M.
U.S. Nat. Mus.
T.S.
Naturhist. Mus.

Perth
T.S.

B.M.(N.H.)

Pervomaisky
P.T.S.
B.M.(N.H.)
T.S.
B.M.(N.H.)

Pesyanoe
P.T.S.
U.S. Nat. Mus.

Petersburg
P.T.S.
U.S. Nat. Mus.
T.S.
Naturhist. Mus.
Mus. d'Hist. Nat.
B.M.(N.H.)

Petropavlovsk
P.M.
Field Mus. Nat. Hist.
P.T.S.
U.S. Nat. Mus.

Pevensey
P.T.S.
U.S. Nat. Mus.

Phillips County (pallasite)
P.M.
U.S. Nat. Mus.

Phillips County (stone)
T.S.
B.M.(N.H.)

Phu Hong
T.S.
Mus. d'Hist. Nat.

Phum Sambo
T.S.
Mus. d'Hist. Nat.

Piancaldoli
P.T.S.
U.S. Nat. Mus.

Pickens County
P.T.S.
U.S. Nat. Mus.

Piedade do Bagre
P.M.
U.S. Nat. Mus.

Pierceville (iron)
P.M.
U.S. Nat. Mus.

Pierceville (stone)
P.T.S.
U.S. Nat. Mus.
Univ. of New Mexico

Pillistfer
P.M.
U.S. Nat. Mus.
P.T.S.
U.S. Nat. Mus.
B.M.(N.H.)
T.S.
U.S. Nat. Mus.
Naturhist. Mus.
Mus. d'Hist. Nat.
B.M.(N.H.)

Pima County
P.M.
U.S. Nat. Mus.

Pine River
P.M.
U.S. Nat. Mus.
P.T.S.
U.S. Nat. Mus.

Pinnaroo
P.M.
U.S. Nat. Mus.
P.T.S.
U.S. Nat. Mus.
B.M.(N.H.)

Piñon
P.M.
U.S. Nat. Mus.

Pinto Mountains
P.T.S.
Univ. of New Mexico

Pipe Creek
T.S.
Field Mus. Nat. Hist.
Naturhist. Mus.
Mus. d'Hist. Nat.
B.M.(N.H.)

Pirgunje
T.S.
B.M.(N.H.)

Pirthalla
T.S.
Mus. d'Hist. Nat.

Pitts
P.M.
U.S. Nat. Mus.

Plains
P.M.
Field Mus. Nat. Hist.
P.T.S.
Naturhist. Mus.
Univ. of New Mexico

Plainview (1917)
P.M.
U.S. Nat. Mus.
P.T.S.
U.S. Nat. Mus.
Naturhist. Mus.
B.M.(N.H.)
T.S.
Mus. d'Hist. Nat.
B.M.(N.H.)

Pleasanton
T.S.
B.M.(N.H.)

Ploschkovitz
T.S.
B.M.(N.H.)

Plymouth
P.M.
U.S. Nat. Mus.

Pnompehn
T.S.
Mus. d'Hist. Nat.

Pohlitz
T.S.
Field Mus. Nat. Hist.
Naturhist. Mus.
Mus. d'Hist. Nat.
B.M.(N.H.)

Point of Rocks (stone)
P.T.S.
Univ. of New Mexico

Pokhra
T.S.
Mus. d'Hist. Nat.
B.M.(N.H.)

Pollen
P.T.S.
Field Mus. Nat. Hist.

Pontlyfni
P.T.S.
B.M.(N.H.)

Portales (a)
P.T.S.
B.M.(N.H.)
T.S.
B.M.(N.H.)

Portales (b)
P.T.S.
Univ. of New Mexico
B.M.(N.H.)
T.S.
B.M.(N.H.)

Portales (c)
P.T.S.
U.S. Nat. Mus.
Univ. of New Mexico
B.M.(N.H.)
T.S.
B.M.(N.H.)

Potter
P.T.S.
Naturhist. Mus.
B.M.(N.H.)
T.S.
Field Mus. Nat. Hist.
B.M.(N.H.)

Prairie Dog Creek
P.T.S.
U.S. Nat. Mus.
T.S.
Naturhist. Mus.
Mus. d'Hist. Nat.
B.M.(N.H.)

Prambachkirchen
T.S.
Naturhist. Mus.

Prambanan
P.M.
U.S. Nat. Mus.
Field Mus. Nat. Hist.

Pribram
P.M.
U.S. Nat. Mus.
P.T.S.
U.S. Nat. Mus.

Pricetown
T.S.
Naturhist. Mus.
Mus. d'Hist. Nat.

Providence
P.M.
U.S. Nat. Mus.

Puente del Zacate
P.M.
U.S. Nat. Mus.

Pulsora
P.T.S.
U.S. Nat. Mus.
B.M.(N.H.)
T.S.
B.M.(N.H.)

Pultusk
P.M.
U.S. Nat. Mus.
P.T.S.
U.S. Nat. Mus.
Naturhist. Mus.
B.M.(N.H.)
T.S.
Field Mus. Nat. Hist.
Naturhist. Mus.
Mus. d'Hist. Nat.
B.M.(N.H.)

Puquios
P.M.
U.S. Nat. Mus.

Putinga
P.T.S.
Univ. of New Mexico
B.M.(N.H.)
T.S.
B.M.(N.H.)

Putnam County
P.M.
U.S. Nat. Mus.

Quartz Mountain
P.M.
U.S. Nat. Mus.

Queen's Mercy
P.T.S.
Field Mus. Nat. Hist.
T.S.
B.M.(N.H.)

Quenggouk
T.S.
U.S. Nat. Mus.
Naturhist. Mus.
Mus. d'Hist. Nat.
B.M.(N.H.)

Quincay
T.S.
Mus. d'Hist. Nat.
B.M.(N.H.)

Quinn Canyon
P.M.
U.S. Nat. Mus.

Rabbit Flat
P.T.S.
U.S. Nat. Mus.

Raco
P.T.S.
U.S. Nat. Mus.

Rafrüti
P.M.
U.S. Nat. Mus.

Rakovka
T.S.
Naturhist. Mus.
Mus. d'Hist. Nat.

Ramsdorf
P.M.
U.S. Nat. Mus.
P.T.S.
U.S. Nat. Mus.
Field Mus. Nat. Hist.
Naturhist. Mus.
T.S.
Naturhist. Mus.
B.M.(N.H.)

Ranchapur
P.T.S.
B.M.(N.H.)
T.S.
B.M.(N.H.)

Rancho de la Pila (1882)
P.M.
U.S. Nat. Mus.

Rancho de la Presa
P.T.S.
U.S. Nat. Mus.
B.M.(N.H.)

Rangala
P.T.S.
U.S. Nat. Mus.

Ras Tanura
P.T.S.
U.S. Nat. Mus.

Red River
P.M.
U.S. Nat. Mus.

Reed City
P.M.
U.S. Nat. Mus.

Reid
P.T.S.
U.S. Nat. Mus.

Reliegos
P.T.S.
U.S. Nat. Mus.
B.M.(N.H.)

Renazzo
P.T.S.
U.S. Nat. Mus.
Naturhist. Mus.
Field Mus. Nat. Hist.
B.M.(N.H.)
T.S.
Mus. d'Hist. Nat.
B.M.(N.H.)

Renca
T.S.
Mus. d'Hist. Nat.

Rhine Villa
P.M.
U.S. Nat. Mus.

Rich Mountain
P.M.
U.S. Nat. Mus.

Richardton
P.T.S.
U.S. Nat. Mus.
Univ. of New Mexico
B.M.(N.H.)
T.S.
Mus. d'Hist. Nat.
B.M.(N.H.)

Richland
P.M.
U.S. Nat. Mus.
P.T.S.
U.S. Nat. Mus.

Richmond
T.S.
Field Mus. Nat. Hist.
Naturhist. Mus.
Mus. d'Hist. Nat.
B.M.(N.H.)

Rio Negro
P.T.S.
U.S. Nat. Mus.
T.S.
Naturhist. Mus.
B.M.(N.H.)

River
P.T.S.
U.S. Nat. Mus.

Rochester
P.T.S.
U.S. Nat. Mus.

Roda
P.M.
Field Mus. Nat. Hist.
P.T.S.
Field Mus. Nat. Hist.
T.S.
Mus. d'Hist. Nat.

Roebourne
P.M.
U.S. Nat. Mus.
B.M.(N.H.)

Rolla (1936)
P.M.
U.S. Nat. Mus.
P.T.S.
U.S. Nat. Mus.

Rolla (1939)
P.T.S.
U.S. Nat. Mus.

Romero
P.T.S.
Arizona State Univ.

Rosamund Dry Lake
P.T.S.
Field Mus. Nat. Hist.

Rosario
P.M.
U.S. Nat. Mus.

Rose City
P.T.S.
U.S. Nat. Mus.
B.M.(N.H.)

Rowena
P.M.
U.S. Nat. Mus.
P.T.S.
U.S. Nat. Mus.

Rowton
P.M.
B.M.(N.H.)

Roy (1933)
P.T.S.
Univ. of New Mexico

Roy (1934)
P.T.S.
U.S. Nat. Mus.
Univ. of New Mexico

Ruff's Mountain
P.M.
U.S. Nat. Mus.
B.M.(N.H.)

Rupota
P.T.S.
U.S. Nat. Mus.

Rush County
P.T.S.
U.S. Nat. Mus.

Rush Creek
P.M.
U.S. Nat. Mus.
T.S.
B.M.(N.H.)

Rushville
P.T.S.
Field Mus. Nat. Hist.
T.S.
Field Mus. Nat. Hist.

Russel Gulch
P.M.
U.S. Nat. Mus.

Sacramento Mountains
P.M.
U.S. Nat. Mus.
B.M.(N.H.)

St. Caprais-de-Quinsac
P.T.S.
U.S. Nat. Mus.

St. Christophe-la-Chartreuse
T.S.
Mus. d'Hist. Nat.

St. Denis Westrem
T.S.
Mus. d'Hist. Nat.

Ste. Marguerite
T.S.
Mus. d'Hist. Nat.

St. Francis Bay
P.T.S.
U.S. Nat. Mus.

St. Francois County
P.M.
U.S. Nat. Mus.

St. Genevieve County
P.M.
U.S. Nat. Mus.

St. Germain-du-Pinel
P.T.S.

U.S. Nat. Mus.
B.M.(N.H.)
T.S.
Naturhist. Mus.
Mus. d'Hist. Nat.
B.M.(N.H.)

St. Lawrence
P.T.S.
B.M.(N.H.)
T.S.
B.M.(N.H.)

St. Louis
P.T.S.
U.S. Nat. Mus.

St. Mark's
P.M.
U.S. Nat. Mus.
P.T.S.
U.S. Nat. Mus.
B.M.(N.H.)
T.S.
Naturhist. Mus.
Mus. d'Hist. Nat.
B.M.(N.H.)

St. Mary's County
P.M.
U.S. Nat. Mus.
P.T.S.
U.S. Nat. Mus.

St. Mesmin
P.T.S.
U.S. Nat. Mus.
U.S. Nat. Mus.
B.M.(N.H.)
T.S.
Naturhist. Mus.
Mus. d'Hist. Nat.

St. Michel
T.S.
Mus. d'Hist. Nat.
Naturhist. Mus.
B.M.(N.H.)

St. Peter
P.M.
U.S. Nat. Mus.

Saint-Sauveur
P.T.S.
U.S. Nat. Mus.
T.S.
Mus. d'Hist. Nat.

Saint-Séverin
P.M.
U.S. Nat. Mus.
P.T.S.
U.S. Nat. Mus.
Field Mus. Nat. Hist.
B.M.(N.H.)
T.S.
Mus. d'Hist. Nat.
B.M.(N.H.)

Salaices
P.M.
U.S. Nat. Mus.
P.T.S.
U.S. Nat. Mus.

Saline
P.M.
U.S. Nat. Mus.
P.T.S.
Field Mus. Nat. Hist.
B.M.(N.H.)
T.S.
Field Mus. Nat. Hist.
Naturhist. Mus.
B.M.(N.H.)

Salla
P.T.S.
U.S. Nat. Mus.
Naturhist. Mus.

Salles
P.T.S.
U.S. Nat. Mus.
T.S.
Mus. d'Hist. Nat.
B.M.(N.H.)

Salta
P.M.
U.S. Nat. Mus.

Salt Lake City
T.S.
Naturhist. Mus.

Salt River
P.M.
U.S. Nat. Mus.

Sam's Valley
P.M.
Field Mus. Nat. Hist.

San Angelo
P.M.
U.S. Nat. Mus.

San Cristobal
P.M.
U.S. Nat. Mus.
T.S.
Naturhist. Mus.

San Emigdio
P.T.S.
Field Mus. Nat. Hist.
T.S.
Mus. d'Hist. Nat.

San Francisco del Mezquital
P.M.
Field Mus. Nat. Hist.
B.M.(N.H.)

San Francisco Mountains
P.M.
U.S. Nat. Mus.

Sanderson
P.M.
U.S. Nat. Mus.

Sandia Mountains
P.M.
U.S. Nat. Mus.
Univ. of New Mexico

San Pedro Springs
T.S.
Mus. d'Hist. Nat.

Santa Apolonia
P.M.
U.S. Nat. Mus.
B.M.(N.H.)

Santa Barbara
P.T.S.
Univ. of New Mexico

Santa Catharina
P.M.
U.S. Nat. Mus.
B.M.(N.H.)

Santa Cruz
P.T.S.
U.S. Nat. Mus.

Santa Isabel
T.S.
Mus. d'Hist. Nat.

Santa Luzia
P.M.
U.S. Nat. Mus.

Santa Rosa
P.M.
U.S. Nat. Mus.
B.M.(N.H.)
T.S.
Naturhist. Mus.

Santiago Papasquiero
P.M.
U.S. Nat. Mus.

São Jose do Rio Preto
P.M.
Univ. of New Mexico
P.T.S.
U.S. Nat. Mus.

São Julião de Moreira
P.M.
U.S. Nat. Mus.

Saratov
P.T.S.
U.S. Nat. Mus.
B.M.(N.H.)
T.S.
B.M.(N.H.)

Sardis
P.M.
U.S. Nat. Mus.

Sarepta
P.M.
U.S. Nat. Mus.

Sauguis
P.T.S.
Field Mus. Nat. Hist.
T.S.
Mus. d'Hist. Nat.
Field Mus. Nat. Hist.

Savannah
P.M.
U.S. Nat. Mus.

Savtschenskoje
P.T.S.
U.S. Nat. Mus.
T.S.
Naturhist. Mus.
Mus. d'Hist. Nat.

Schaap-Kooi
P.T.S.
U.S. Nat. Mus.
T.S.
Naturhist. Mus.

Schenectady
P.T.S.
U.S. Nat. Mus.

Schönenberg
T.S.
Naturhist. Mus.
Mus. d'Hist. Nat.
B.M.(N.H.)

Schwetz
P.M.
U.S. Nat. Mus.

Scott City
T.S.
Mus. d'Hist. Nat.
B.M.(N.H.)

Scottsville
P.M.
U.S. Nat. Mus.

Scurry
P.T.S.
B.M.(N.H.)
T.S.
B.M.(N.H.)

Searsmount
P.T.S.
Field Mus. Nat. Hist.
T.S.
Field Mus. Nat. Hist.
Naturhist. Mus.
Mus. d'Hist. Nat.
B.M.(N.H.)

Seeläsgen
P.M.
U.S. Nat. Mus.
B.M.(N.H.)

T.S.
Naturhist. Mus.

Segowlie
T.S.
Naturhist. Mus.
Mus. d'Hist. Nat.
B.M.(N.H.)

Seguin
P.T.S.
U.S. Nat. Mus.

Seibert
P.T.S.
U.S. Nat. Mus.
Field Mus. Nat. Hist.

Seldebourak
T.S.
Mus. d'Hist. Nat.

Selma
T.S.
Naturhist. Mus.
Mus. d'Hist. Nat.
B.M.(N.H.)

Semarkona
P.M.
U.S. Nat. Mus.
P.T.S.
U.S. Nat. Mus.
Univ. of New Mexico

Seminole
P.T.S.
U.S. Nat. Mus.
B.M.(N.H.)
T.S.
B.M.(N.H.)

Sena
P.M.
U.S. Nat. Mus.
P.T.S.
U.S. Nat. Mus.
T.S.
Mus. d'Hist. Nat.

Seneca
P.T.S.
U.S. Nat. Mus.

Seneca Township
P.M.
U.S. Nat. Mus.
B.M.(N.H.)

Seres
P.T.S.
Naturhist. Mus.
U.S. Nat. Mus.
Field Mus. Nat. Hist.
T.S.
Naturhist. Mus.
Mus. d'Hist. Nat.
B.M.(N.H.)

Serra de Magé
P.T.S.
U.S. Nat. Mus.
T.S.
Mus. d'Hist. Nat.
P.T.S.
Univ. of New Mexico
B.M.(N.H.)

Sete Lagoas
P.M.
Univ. of New Mexico
P.T.S.
Univ. of New Mexico

Sevilla
T.S.
Mus. d'Hist. Nat.

Sevrukovo
P.M.
U.S. Nat. Mus.
T.S.
Naturhist. Mus.
Mus. d'Hist. Nat.

Seymour
P.M.
U.S. Nat. Mus.

Shalka
P.T.S.
U.S. Nat. Mus.
Field Mus. Nat. Hist.
T.S.
Field Mus. Nat. Hist.
Naturhist. Mus.
Mus. d'Hist. Nat.
B.M.(N.H.)

Shallowater
P.M.
U.S. Nat. Mus.
P.T.S.
U.S. Nat. Mus.
T.S.
B.M.(N.H.)

Sharps
P.T.S.
U.S. Nat. Mus.
U.S. Nat. Mus.

Shaw
P.M.
U.S. Nat. Mus.
Field Mus. Nat. Hist.
Univ. of New Mexico
P.T.S.
U.S. Nat. Mus.
U.S. Nat. Mus.
Univ. of New Mexico
B.M.(N.H.)
T.S.
B.M.(N.H.)

Shelburne
T.S.
Field Mus. Nat. Hist.
Naturhist. Mus.
Mus. d'Hist. Nat.
B.M.(N.H.)

Shergotty
P.T.S.
U.S. Nat. Mus.
Univ. of New Mexico
B.M.(N.H.)
T.S.
Naturhist. Mus.
Mus. d'Hist. Nat.
B.M.(N.H.)

Shields
P.T.S.
Naturhist. Mus.

Shingle Springs
P.M.
B.M.(N.H.)

Shirahagi
P.M.
Univ. of New Mexico

Shupiyan
T.S.
B.M.(N.H.)

Shytal
P.T.S.
U.S. Nat. Mus.
T.S.
Mus. d'Hist. Nat.
B.M.(N.H.)

Siena
P.T.S.
U.S. Nat. Mus.
Naturhist. Mus.
T.S.
Naturhist. Mus.
Mus. d'Hist. Nat.
B.M.(N.H.)

Sierra County
P.M.
Univ. of New Mexico
P.T.S.
Univ. of New Mexico

Sierra Gorda
P.M.
U.S. Nat. Mus.

Sikhote-Alin
P.M.
U.S. Nat. Mus.

Silver Bell
P.M.
U.S. Nat. Mus.

Silverton (New South Wales)
P.T.S.
U.S. Nat. Mus.
T.S.
B.M.(N.H.)

Silverton (Texas)
P.M.
Field Mus. Nat. Hist.
P.T.S.
Field Mus. Nat. Hist.

Simbirsk (of P. Partsch)
T.S.
Field Mus. Nat. Hist.

Simondium
P.T.S.
U.S. Nat. Mus.
T.S.
Naturhist. Mus.
B.M.(N.H.)

Sinai
T.S.
Mus. d'Hist. Nat.
B.M.(N.H.)

Sindhri
P.T.S.
U.S. Nat. Mus.
Field Mus. Nat. Hist.
Naturhist. Mus.
B.M.(N.H.)
T.S.
Mus. d'Hist. Nat.
B.M.(N.H.)

Sinnai
P.T.S.
U.S. Nat. Mus.

Sioux County
P.T.S.
U.S. Nat. Mus.
T.S.
B.M.(N.H.)

Siratik
T.S.
B.M.(N.H.)

Sitathali
P.T.S.
U.S. Nat. Mus.
Naturhist. Mus.
T.S.
Naturhist. Mus.
B.M.(N.H.)

Ski
T.S.
Mus. d'Hist. Nat.
B.M.(N.H.)

Slavetic
T.S.
Naturhist. Mus.
Mus. d'Hist. Nat.

Sleeper Camp
P.T.S.
U.S. Nat. Mus.

Slobodka
P.T.S.

U.S. Nat. Mus.
T.S.
Mus. d'Hist. Nat.
B.M.(N.H.)

Smith Center
T.S.
B.M.(N.H.)

Smithland
P.M.
Field Mus. Nat. Hist.
B.M.(N.H.)

Smithonia
P.M.
U.S. Nat. Mus.

Smith's Mountain
P.M.
U.S. Nat. Mus.
B.M.(N.H.)

Smithsonian Iron
P.M.
U.S. Nat. Mus.

Smithville
P.M.
U.S. Nat. Mus.

Social Circle
P.M.
U.S. Nat. Mus.

Soko-Banja
P.T.S.
U.S. Nat. Mus.
B.M.(N.H.)
T.S.
Naturhist. Mus.
Mus. d'Hist. Nat.
B.M.(N.H.)

Somervell County
P.M.
U.S. Nat. Mus.

Sone
P.T.S.
U.S. Nat. Mus.

Soper
P.M.
U.S. Nat. Mus.
B.M.(N.H.)

Soroti
P.T.S.
U.S. Nat. Mus.

South Bend
P.M.
U.S. Nat. Mus.
Field Mus. Nat. Hist.

South Byron
P.M.
U.S. Nat. Mus.

South Dahna
P.M.
U.S. Nat. Mus.
P.T.S.
U.S. Nat. Mus.

Southern Arizona
P.M.
U.S. Nat. Mus.

South Oman
P.T.S.
U.S. Nat. Mus.
B.M.(N.H.)

Spearman
P.M.
U.S. Nat. Mus.
Field Mus. Nat. Hist.

Springfield
P.T.S.
B.M.(N.H.)
T.S.
B.M.(N.H.)

Springwater
P.M.
U.S. Nat. Mus.

Ssyromolotovo
P.M.
Field Mus. Nat. Hist.

Ställdalen
P.T.S.
B.M.(N.H.)
T.S.
Field Mus. Nat. Hist.
Mus. d'Hist. Nat.
B.M.(N.H.)

Stannern
P.M.
U.S. Nat. Mus.
P.T.S.
U.S. Nat. Mus.
Naturhist. Mus.
T.S.
Field Mus. Nat. Hist.
Naturhist. Mus.
Mus. d'Hist. Nat.
B.M.(N.H.)

Stavropol
T.S.
Field Mus. Nat. Hist.
Naturhist. Mus.
Mus. d'Hist. Nat.
B.M.(N.H.)

Steinbach
P.M.
U.S. Nat. Mus.
P.T.S.
U.S. Nat. Mus.
Field Mus. Nat. Hist.
T.S.
Naturhist. Mus.
Mus. d'Hist. Nat.
B.M.(N.H.)

Stonington
P.T.S.
U.S. Nat. Mus.

Sublette
P.T.S.
U.S. Nat. Mus.

Success
P.M.
U.S. Nat. Mus.
P.T.S.
U.S. Nat. Mus.

Suchy Dul
P.T.S.
U.S. Nat. Mus.

Sultaupur
T.S.
B.M.(N.H.)

Supuhee
P.T.S.
U.S. Nat. Mus.
B.M.(N.H.)
T.S.
Mus. d'Hist. Nat.
B.M.(N.H.)

Sutton
P.M.
Univ. of New Mexico
P.T.S.
B.M.(N.H.)
T.S.
B.M.(N.H.)

Suwahib (Adraj)
P.T.S.
U.S. Nat. Mus.
T.S.
B.M.(N.H.)

Suwahib ('Ain Sala)
P.T.S.
B.M.(N.H.)
T.S.
B.M.(N.H.)

Suwahib (Buwah)
P.T.S.
U.S. Nat. Mus.
B.M.(N.H.)
T.S.
B.M.(N.H.)

Sweetwater
T.S.
B.M.(N.H.)

Sylacauga
P.T.S.
U.S. Nat. Mus.

Tabor
P.T.S.
U.S. Nat. Mus.
T.S.
Field Mus. Nat. Hist.
Naturhist. Mus.
Mus. d'Hist. Nat.
B.M.(N.H.)

Tadjera
P.T.S.
U.S. Nat. Mus.
T.S.
Naturhist. Mus.
Mus. d'Hist. Nat.

Tafoya
P.T.S.
Univ. of New Mexico

Taiban
P.M.
Univ. of New Mexico
U.S. Nat. Mus.
P.T.S.
Univ. of New Mexico
B.M.(N.H.)
T.S.
B.M.(N.H.)

Taiga
P.M.
U.S. Nat. Mus.
P.T.S.
U.S. Nat. Mus.

Tambo Quemado
P.M.
U.S. Nat. Mus.

Tamentit
P.M.
Field Mus. Nat. Hist.

Tané
P.T.S.
U.S. Nat. Mus.

Tanokami Mountain
P.M.
U.S. Nat. Mus.

Tarapaca
T.S.
Naturhist. Mus.

Tarfa
T.S.
B.M.(N.H.)

Tatahouine
P.M.
Field Mus. Nat. Hist.
P.T.S.
U.S. Nat. Mus.
B.M.(N.H.)
T.S.
Mus. d'Hist. Nat.
B.M.(N.H.)

Tathlith
P.T.S.
U.S. Nat. Mus.

Tauk
P.T.S.
B.M.(N.H.)
T.S.
B.M.(N.H.)

Tawallah Valley
P.M.
U.S. Nat. Mus.

Tazewell
P.M.
U.S. Nat. Mus.
B.M.(N.H.)

Tell
P.T.S.
B.M.(N.H.)
T.S.
B.M.(N.H.)

Temple
P.T.S.
U.S. Nat. Mus.

Tenham
P.T.S.
U.S. Nat. Mus.
Univ. of New Mexico
U.S. Nat. Mus.
B.M.(N.H.)
T.S.
B.M.(N.H.)

Tennasilm
P.T.S.
Field Mus. Nat. Hist.
T.S.
Naturhist. Mus.
Mus. d'Hist. Nat.

Ternera
P.M.
U.S. Nat. Mus.

Texline
P.T.S.
Univ. of New Mexico
T.S.
B.M.(N.H.)

Thal
P.T.S.
B.M.(N.H.)
T.S.
B.M.(N.H.)

Thiel Mountains
P.M.
U.S. Nat. Mus.

Thoreau
P.M.
U.S. Nat. Mus.

Thule
P.M.
U.S. Nat. Mus.

Thunda
P.M.
U.S. Nat. Mus.
B.M.(N.H.)

Thurlow
P.M.
U.S. Nat. Mus.

Tierra Blanca
P.T.S.
U.S. Nat. Mus.

Tieschitz
P.M.
Field Mus. Nat. Hist.
P.T.S.
U.S. Nat. Mus.
Field Mus. Nat. Hist.
Naturhist. Mus.
Univ. of New Mexico
T.S.
Naturhist. Mus.

Tilden
T.S.
Field Mus. Nat. Hist.

Timmersoi
P.T.S.
B.M.(N.H.)
T.S.
B.M.(N.H.)

Timochin
T.S.
Mus. d'Hist. Nat.
B.M.(N.H.)

Tirupati
P.T.S.
U.S. Nat. Mus.

Tjabe
P.T.S.
U.S. Nat. Mus.
T.S.
Naturhist. Mus.
Mus. d'Hist. Nat.

Tjerebon
T.S.
Mus. d'Hist. Nat.

Tobe
P.T.S.
U.S. Nat. Mus.

Toluca
P.M.
U.S. Nat. Mus.
B.M.(N.H.)
T.S.
Naturhist. Mus.

Tomatlan
P.T.S.
U.S. Nat. Mus.
T.S.
Mus. d'Hist. Nat.

Tombigbee River
P.M.
U.S. Nat. Mus.
B.M.(N.H.)

Tomhannock Creek
P.T.S.
Field Mus. Nat. Hist.
T.S.
Field Mus. Nat. Hist.
Naturhist. Mus.
Mus. d'Hist. Nat.
B.M.(N.H.)

Tonganoxie
P.M.
U.S. Nat. Mus.

Torrington
P.T.S.
U.S. Nat. Mus.

Toubil River
P.M.
U.S. Nat. Mus.

Toulon
P.M.
Field Mus. Nat. Hist.
P.T.S.
Field Mus. Nat. Hist.

Toulouse
P.T.S.
Field Mus. Nat. Hist.
T.S.
Mus. d'Hist. Nat.
B.M.(N.H.)

Tourinnes-la-Grosse
P.T.S.
U.S. Nat. Mus.
Field Mus. Nat. Hist.
T.S.
Field Mus. Nat. Hist.
Naturhist. Mus.
Mus. d'Hist. Nat.
B.M.(N.H.)

Travis County
T.S.
B.M.(N.H.)

Trenton
P.M.
U.S. Nat. Mus.

Trenzano
P.T.S.
Field Mus. Nat. Hist.
T.S.
Field Mus. Nat. Hist.
Naturhist. Mus.
Mus. d'Hist. Nat.
B.M.(N.H.)

Treysa
P.M.
U.S. Nat. Mus.

Trifir
P.T.S.
U.S. Nat. Mus.

Tryon
P.T.S.
U.S. Nat. Mus.
T.S.
Mus. d'Hist. Nat.

Tuan Tuc
T.S.
Mus. d'Hist. Nat.

Tucson
P.M.
U.S. Nat. Mus.
B.M.(N.H.)
T.S.
Naturhist. Mus.

Tulia
P.T.S.
U.S. Nat. Mus.
Univ. of New Mexico
B.M.(N.H.)
T.S.
B.M.(N.H.)

Turtle River
P.M.
U.S. Nat. Mus.

Tuxtuac
P.T.S.
Field Mus. Nat. Hist.
B.M.(N.H.)

Tuzla
T.S.
B.M.(N.H.)

Twin City
P.M.
U.S. Nat. Mus.

Two Buttes (a)
P.T.S.
B.M.(N.H.)
T.S.
B.M.(N.H.)

Tysnes Island
P.M.
U.S. Nat. Mus.
P.T.S.
U.S. Nat. Mus.
Field Mus. Nat. Hist.
B.M.(N.H.)
T.S.
Field Mus. Nat. Hist.
Naturhist. Mus.
Mus. d'Hist. Nat.
B.M.(N.H.)

Uberaba
P.M.
Univ. of New Mexico
P.T.S.
Field Mus. Nat. Hist.

Univ. of New Mexico
T.S.
Naturhist. Mus.
Mus. d'Hist. Nat.

Ucera
P.M.
U.S. Nat. Mus.
P.T.S.
U.S. Nat. Mus.

Udei Station
P.M.
U.S. Nat. Mus.
P.T.S.
B.M.(N.H.)
T.S.
B.M.(N.H.)

Uden
P.T.S.
U.S. Nat. Mus.

Udipi
P.T.S.
Field Mus. Nat. Hist.
B.M.(N.H.)
T.S.
Mus. d'Hist. Nat.
B.M.(N.H.)

Ufana
P.T.S.
U.S. Nat. Mus.
T.S.
B.M.(N.H.)

Ularring
P.T.S.
U.S. Nat. Mus.

Ulysses
P.T.S.
U.S. Nat. Mus.
T.S.
Mus. d'Hist. Nat.

Umbala
T.S.
B.M.(N.H.)

Um-Hadid
P.T.S.
U.S. Nat. Mus.

Umm Ruaba
T.S.
B.M.(N.H.)

Umm Tina
T.S.
B.M.(N.H.)

Union County
P.M.
U.S. Nat. Mus.

Ur
P.M.
B.M.(N.H.)

Ute Creek
P.M.
Univ. of New Mexico
P.T.S.
Univ. of New Mexico
U.S. Nat. Mus.
T.S.
B.M.(N.H.)

Utrecht
T.S.
Naturhist. Mus.
Mus. d'Hist. Nat.

Utzensdorf
P.T.S.
U.S. Nat. Mus.
B.M.(N.H.)

Uwet
P.M.
U.S. Nat. Mus.

Uwharrie
P.M.
U.S. Nat. Mus.

Vaalbult
T.S.
B.M.(N.H.)

Vaca Muerta
P.M.
U.S. Nat. Mus.
P.T.S.
U.S. Nat. Mus.
Field Mus. Nat. Hist.
B.M.(N.H.)
T.S.
Field Mus. Nat. Hist.
Naturhist. Mus.
Mus. d'Hist. Nat.
B.M.(N.H.)

Vago
T.S.
B.M.(N.H.)

Valdinizza
P.T.S.
U.S. Nat. Mus.
T.S.
Mus. d'Hist. Nat.

Valley Wells
T.S.
B.M.(N.H.)

Varpaisjärvi
T.S.
B.M.(N.H.)

Vavilovka
P.T.S.
U.S. Nat. Mus.
Field Mus. Nat. Hist.
T.S.
Naturhist. Mus.
Mus. d'Hist. Nat.

Vengerovo
P.T.S.
B.M.(N.H.)
T.S.
B.M.(N.H.)

Vera
T.S.
B.M.(N.H.)

Veramin
P.M.
Field Mus. Nat. Hist.
P.T.S.
U.S. Nat. Mus.
T.S.
Mus. d'Hist. Nat.
B.M.(N.H.)

Verkhne Dnieprovsk
P.M.
Field Mus. Nat. Hist.

Verkhne Tschirskaia
P.M.
U.S. Nat. Mus.
T.S.
Mus. d'Hist. Nat.

Verkhne Udinsk
P.M.
B.M.(N.H.)

Vernon County
P.T.S.
U.S. Nat. Mus.

Victoria West
P.M.
U.S. Nat. Mus.

View Hill
P.M.
U.S. Nat. Mus.

Vigarano
P.M.
B.M.(N.H.)
P.T.S.
U.S. Nat. Mus.
Field Mus. Nat. Hist.
B.M.(N.H.)
T.S.
Naturhist. Mus.
Mus. d'Hist. Nat.
B.M.(N.H.)

Villedieu
T.S.
Mus. d'Hist. Nat.

Vincent
P.T.S.
U.S. Nat. Mus.

Virba
P.T.S.
Field Mus. Nat. Hist.
T.S.
Mus. d'Hist. Nat.

Vishnupur
P.T.S.
U.S. Nat. Mus.

Vouillé
T.S.
Mus. d'Hist. Nat.
P.T.S.
U.S. Nat. Mus.
Field Mus. Nat. Hist.
T.S.
B.M.(N.H.)

Vulcan
P.T.S.
U.S. Nat. Mus.
B.M.(N.H.)
T.S.
B.M.(N.H.)

Wabar
P.M.
U.S. Nat. Mus.
B.M.(N.H.)

Waconda
P.M.
U.S. Nat. Mus.
P.T.S.
U.S. Nat. Mus.
Field Mus. Nat. Hist.
T.S.
Field Mus. Nat. Hist.
Naturhist. Mus.
Mus. d'Hist. Nat.
B.M.(N.H.)

Wairarapa Valley
T.S.
Mus. d'Hist. Nat.

Waldo
P.T.S.
U.S. Nat. Mus.

Waldron Ridge
P.M.
U.S. Nat. Mus.
B.M.(N.H.)

Walker County
P.M.
B.M.(N.H.)

Walltown
P.T.S.
U.S. Nat. Mus.

Walters
P.M.
U.S. Nat. Mus.
T.S.
Field Mus. Nat. Hist.

Warrenton
P.M.
Field Mus. Nat. Hist.
P.T.S.
U.S. Nat. Mus.
Field Mus. Nat. Hist.

Naturhist. Mus.
T.S.
Naturhist. Mus.
B.M.(N.H.)

Waterville
P.M.
U.S. Nat. Mus.
B.M.(N.H.)

Wathena
P.M.
U.S. Nat. Mus.

Weatherford
P.M.
U.S. Nat. Mus.
P.T.S.
U.S. Nat. Mus.
Univ. of New Mexico

Weaver Mountains
P.M.
U.S. Nat. Mus.

Webb
P.T.S.
U.S. Nat. Mus.

Weekeroo Station
P.M.
U.S. Nat. Mus.
Field Mus. Nat. Hist.

Weldona
P.T.S.
U.S. Nat. Mus.
T.S.
Mus. d'Hist. Nat.

Welland
P.M.
U.S. Nat. Mus.

Wellington
P.T.S.
B.M.(N.H.)
T.S.
B.M.(N.H.)

Wellman
P.M.
U.S. Nat. Mus.
P.T.S.
U.S. Nat. Mus.
Univ. of New Mexico
T.S.
B.M.(N.H.)

Wellman no. 3
P.T.S.
B.M.(N.H.)
T.S.
B.M.(N.H.)

Wessely
P.T.S.
U.S. Nat. Mus.

Western Arkansas
P.M.
U.S. Nat. Mus.

West Forrest
P.T.S.
U.S. Nat. Mus.

West Reid
P.T.S.
U.S. Nat. Mus.

Weston
P.T.S.
U.S. Nat. Mus.
Field Mus. Nat. Hist.
B.M.(N.H.)
T.S.
Field Mus. Nat. Hist.
Mus. d'Hist. Nat.
Naturhist. Mus.
B.M.(N.H.)

Wethersfield
P.M.
U.S. Nat. Mus.
P.T.S.
U.S. Nat. Mus.

Whitman
P.T.S.
U.S. Nat. Mus.

Wichita County
P.M.
U.S. Nat. Mus.

Wickenburg (stone)
P.T.S.
B.M.(N.H.)

Wilbia
P.T.S.
U.S. Nat. Mus.

Wildara
P.T.S.
U.S. Nat. Mus.

Wiley
P.M.
U.S. Nat. Mus.

Willamette
P.M.
U.S. Nat. Mus.
Field Mus. Nat. Hist.
B.M.(N.H.)

Willaroy
P.M.
U.S. Nat. Mus.
P.T.S.
U.S. Nat. Mus.

Willow Creek
P.T.S.
U.S. Nat. Mus.

Willowdale
P.T.S.
U.S. Nat. Mus.

Wilmot
T.S.
B.M.(N.H.)

Wiluna
P.T.S.
U.S. Nat. Mus.
B.M.(N.H.)
T.S.
B.M.(N.H.)

Wingellina
P.T.S.
U.S. Nat. Mus.

Witchelina
P.T.S.
U.S. Nat. Mus.
B.M.(N.H.)
T.S.
B.M.(N.H.)

Witklip Farm
T.S.
B.M.(N.H.)

Witsand Farm
T.S.
B.M.(N.H.)

Wittekrantz
T.S.
Naturhist. Mus.
B.M.(N.H.)

Wold Cottage
P.T.S.
U.S. Nat. Mus.
B.M.(N.H.)
T.S.
B.M.(N.H.)

Wolf Creek
P.M.
U.S. Nat. Mus.

Wonyulgunna
P.M.
U.S. Nat. Mus.

Wood's Mountain
P.M.
U.S. Nat. Mus.

Woodbine
P.M.
U.S. Nat. Mus.

Woodward County
P.M.
U.S. Nat. Mus.

Woolgorong
P.T.S.
U.S. Nat. Mus.
T.S.
B.M.(N.H.)

Wynella
P.T.S.
U.S. Nat. Mus.
B.M.(N.H.)
T.S.
B.M.(N.H.)

Yambo
P.T.S.
U.S. Nat. Mus.

Yandama
P.T.S.
U.S. Nat. Mus.

Yanhuitlan
P.M.
U.S. Nat. Mus.

Yardymly
P.M.
U.S. Nat. Mus.

Yatoor
P.T.S.
B.M.(N.H.)
T.S.
Mus. d'Hist. Nat.
B.M.(N.H.)

Yayjinna
P.T.S.
U.S. Nat. Mus.

Yenberrie
P.M.
U.S. Nat. Mus.

Yilmia
P.T.S.
Univ. of New Mexico
T.S.
B.M.(N.H.)

Yocemento
P.M.
B.M.(N.H.)
T.S.
B.M.(N.H.)

Yonozu
P.T.S.
Univ. of New Mexico

York (iron)
P.M.
U.S. Nat. Mus.

Yorktown (New York)
T.S.
B.M.(N.H.)

Youanmi
P.M.
U.S. Nat. Mus.

Youndegin
P.M.

U.S. Nat. Mus.
B.M.(N.H.)

Yukan
P.T.S.
B.M.(N.H.)

Yurtuk
P.T.S.
U.S. Nat. Mus.

Zaborzika
P.T.S.
U.S. Nat. Mus.
T.S.
Mus. d'Hist. Nat.
B.M.(N.H.)

Zabrodje
T.S.
Naturhist. Mus.
Mus. d'Hist. Nat.

Zacatecas (1792)
P.M.
U.S. Nat. Mus.
T.S.
Naturhist. Mus.

Zacatecas (1969)
P.M.
U.S. Nat. Mus.

Zagami
P.T.S.
B.M.(N.H.)
T.S.
B.M.(N.H.)

Zavid
P.T.S.
Field Mus. Nat. Hist.
B.M.(N.H.)
T.S.
Field Mus. Nat. Hist.
Naturhist. Mus.
Mus. d'Hist. Nat.

Zebrak
T.S.
B.M.(N.H.)

Zemaitkiemis
T.S.
Mus. d'Hist. Nat.
B.M.(N.H.)

Zerga
P.T.S.
U.S. Nat. Mus.

Zhovtnevyi
P.T.S.
U.S. Nat. Mus.
B.M.(N.H.)

Zmenj
P.T.S.
Naturhist. Mus.

Zomba
P.T.S.
B.M.(N.H.)
T.S.
B.M.(N.H.)

Zsadany
P.T.S.
U.S. Nat. Mus.

Zvonkov
P.T.S.
U.S. Nat. Mus.

Catalogue of Meteorite Craters

Craters are classified here into three broad categories:

Authenticated craters are true explosion craters from which meteorite material has been recovered, or for which an origin by explosive meteorite impact is undoubted (mere impact holes are excluded).

Doubtful craters are either cryptoexplosion craters which lack conclusive evidence of an origin by meteorite impact or structures (some associated with meteorites) for which a truly explosive origin may be in doubt.

Discredited craters are those for which there is conclusive evidence against an origin by explosive meteorite impact.

Classification is stated in each case and it should be noted that this differs from the convention used in the third edition (1966).

The references quoted below refer to the craters only; descriptions of the generating meteorites (where they have been recovered) are in the main list of meteorites, to which cross-references are given in some instances.

Several bibliographies and lists of meteorite craters, authenticated or suspected, are available: E. O'Connell (A catalog of meteorite craters and related features with a guide to the literature, The Rand Corporation, Santa Monica, 1965); Th. Monod (Contribution à l'établissment d'une liste d'accidents circulaires d'origine météoritique (reconnue, possible ou supposée), cryptoexplosive, etc., Univ. de Dakar, Ifan, Dakar, 1965; Catalogues et Documents no. 18). In the following list these bibliographies are referred to shortly as O'Connell, 1965, and Monod, 1965. A comprehensive bibliography of the literature to 1968 is presented by J. H. Freeberg (Terrestrial impact structures — a bibliography U.S. Geological Survey Bulletin, 1966, no. 1220) and a supplement (*ibid.*, 1969, no. 1320). More recently, two further lists have been published; P. B. Robertson, A bibliography of Canadian impact sites (Earth Physics Branch, Dept. Energy Mines and Resources, Canada, 1975, and J. Classen (Meteoritics, 1977, *12*, p. 61). See also R.A.F. Grieve and P. B. Robertson (Icarus, 1979, *38*, p. 212) and P. B. Robertson and R. A. F. Grieve (Journal Royal Astronomical Society, Canada, 1975, *69*, p. 1).

I. T. Zotkin and V. L. Tsvetkov (Astronomicheskii Vestnik, 1970, *4*, p. 55); B. S. Zeylik and E. Yu. Seytmuratova (Izvestia Akademii Nauk Kazakh. SSR., Ser. Geol., 1975, no. 1, p. 62); A. I. Dabizha (Meteoritika, 1975, *34*, p. 88); V. L. Masaitis (Sovetskaya Geologiya, 1975, *11*, p. 52) and V. L. Masaitis *et al.*, (Doklady Akademii Nauk SSRR., Ser. Geol. 1978, *240*, p. 1191 — English translation, Doklady Earth Science Section, 1978, *240*, p. 91) list numerous structures of possible impact origin on the Russian Platform. Unfortunately, only the English translation of Zotkin and Tsvetkov's paper was available (Solar System Studies, 1970, *4*, p. 44) and so the Cyrillic transliterations of crater names referred to therein are excluded. See also, V. V. Fedynskiy and L. P. Khrianina (Astronomicheskii Vestnik, 1976, *10*, p. 81).

In addition to references cited under individual craters, accounts of the effects of large meteorite impact, including crater formation and a wide range of associated phenomena, are to be found in the selection of papers listed below and the references therein; for literature published before 1960 readers are referred to the third edition of the Catalogue of Meteorites (1966).

References

Ahrens, T. J. and O'Keefe, J. D. Journal of Geophysical Research, 1983, *88*, Suppl., p. A799.

Austin, M. G., Thomsen, J. M., Ruhl, S. F., Orphal, D. L. and Schultz, P. H. Geochimica et Cosmochimica Acta, 1980, Suppl. 14, p. 2325.

Baldwin, R. B. Icarus, 1971, *14*, p. 36.

Bucher, W. H. Nature, 1963, *197*, p. 1241.

—— American Journal of Science, 1963, *261*, p. 597.

Carstens, H. Contributions to Mineralogy and Petrology, 1975, *50*, p. 145.

Carter, N. L. Science, 1968, *160*, p. 526.

Chapman, D. R. Journal of Geophysical Research, 1971, *76*, p. 6309.

Croft, S. K. Geochimica et Cosmochimica Acta, 1980, Suppl. 14, p. 2347.

Currie, K. L. In *Geology and Economic Minerals of Canada*, ed. R. J. W. Douglas. Geological Survey of Canada Economic Geology Report, 1970, no. 1, p. 135.

Dabizha, A. I. and Fedynskiy, V. V. Meteoritika, 1977, *36*, p. 113.

Dachille, F. Bulletin of the South Carolina Academy Science, 1962, *261*, p. 650.

David, E. Earth and Planetary Science Letters, 1966, *1*, p. 75.

Dence, M. R. Geological Association of Canada, Special Paper 10, 1972, p. 7.

Dietz, R. S. Nature, 1963, *197*, p. 39.

—— American Journal of Science, 1963, *261*, p. 650.

Engelhardt, W. von Naturwissenschaften, 1974, *61*, p. 413.

—— Naturwissenschaften, 1974, *61*, p. 389.

Florenski, P. V. and Dikov, Yu. P. Geochemistry International, 1981, *18*, p. 92.

French, B. M. Bulletin Volcanologique, 1970, *34*, p. 466.

—— and Short, N. M., eds. *Shock Metamorphism of Natural Materials.* Mono Book Corp., Baltimore, Md., 1968, 644 pp.

Gallant, R. C. Nature, 1963, *197*, p. 38.

Ganapathy, R, Science 1980, *209*, p. 921.

—— Science, 1982, *216*, p. 885.

Grieve, R. A. F. and Dence, M. R. Icarus, 1979, *38*, p. 230.

Gurov, E. P., Val'ter, A. A. and Rakitskaya, R. B. International Geology Review, 1980, *22*, p. 329.

Hodge, P. W. and Wright, F. W. Nature 1970, *225*, p. 717.

Holsapple, K. A. Geochimica et Cosmochimica Acta, 1980, Suppl. 14, p. 2379.

—— and Schmidt, R. M. Geochimica et Cosmochimica Acta, 1979, Suppl. 11, p. 2757.

—— Journal of Geophysical Research, 1982, *87*, p. 1849.

Hörz, F. Proceedings of the meeting on Meteorite Impact and Volcanism. *In* Journal of Geophysical Research, 1971, *76*, p. 5381.

Jones, G. H. S. Nature, 1978, *273*, p. 211.

Kaula, W. M. Transactions of the American Geophysical Union, 1971, *52*, IUGG 1–4.

Kellaway, G. A. and Durrance, E. M. Nature, 1978, *273*, p. 75.
Khryanina, L. P. Doklady Academy Science USSR, Earth Science Sections, 1980, *238*, p. 245.
—— and Ivanov, O. P. Doklady Akademii Nauk SSR, 1977, *233*, p. 457.
Kieffer, S. W. and Simonds, C. H. Reviews of Geophysics and Space Physics, 1980, *18*, p. 143.
King, E. A. Naturwissenschaften, 1977, *64*, p. 379.
Krinov, E. L. *Giant Meteorites*, Pergamon Press, New York 1966, 387 pp.
Lyttleton, R. A. Nature, 1973, *245*, p. 144.
McCall, G. J. H., ed. *Astroblemes*, Dowden Hutchinson and Ross, Stronsburg, 1979, Benchmark papers in Geology, *50*.
Melosh, H. J. and Gaffney, E. S. Journal of Geophysical Research, 1983, *88*, Suppl., p. A830.
Middlehurst, B. M. and Kuiper, G. P., eds. *The Moon Meteorites and Comets*, University of Chicago Press, Chicago, 1963, 810 pp.
Millman, P. M. Nature, 1971, *232*, p. 161.
—— and Dence, M. R. Proceedings 24th. International Geological Congress, 1972, section 15.
Mizutani, H., Kawakami, S., Takagi, Y., Kato, M. and Kumazawa, M. Journal of Geophysical Research, 1983, *88*, Suppl., p. A835.
Mullins, W. W. Icarus, 1978, *33*, p. 624.
Murray, J. B. and Guest, J. E. Modern Geology, 1970, *1*, p. 145.
Neukum, G., König, B., Fechtig, H. and Störzer, D. Geochimica et Cosmochimica Acta, 1975, Suppl. 6, p. 2597.
Nordyke, M. D. Journal of Geophysical Research, 1961, *66*, p. 3439.
—— ed. Proceedings of the Geophysical Laboratory, Lawrence Radiation Laboratory Cratering Symposium, Publ. University of California no. URCL-6438, 1961.
O'Keefe, J. D. and Ahrens, T. J. Science, 1977, *198*, p. 1249.
Orphal, D. L., Borden, W. F., Larson, S. A. and Schultz, P. H. Geochimica et Cosmochimica Acta, 1980, Suppl. 14, p. 2309.
Palme, H., Grieve, R. A. F. and Wolf, R. Geochimica et Cosmochimica Acta, 1981, *45*, p. 2417.
Pike, R. J. Nature (Physical Science), 1971, *234*, p. 56.
—— Geochimica et Cosmochimica Acta, 1980, Suppl. 14, p. 2159.
Remo, J. L. and Skalafuris, A. L. Journal of Geophysical Research, 1969, *73*, p. 3727.
Roddy, D. J., Pepin, R. O. and Merrill, R. B., eds. *Impact and Explosion Cratering — planetary and terrestrial implications*, Pergamon Press, New York, 1977, 1301 pp.
—— Schuster, S. H., Kreyenhagen, K. N. and Orphal, D. L. Geochimica et Cosmochimica Acta, 1980, Suppl. 14, p. 2275.
Saul, J. M. Nature, 1978, *271*, p. 345.
Schmidt, R. M. Geochimica et Cosmochimica Acta, 1980, Suppl. 14, p. 2099.
Schneider, E. and Wagner, G. A. Earth and Planetary Science Letters, 1976, *32*, p. 40.
Schultz, P. H. and Merrill, R. B., eds. *Multi-Ring Basins*, Geochimica et Cosmochimica Acta, 1981, Suppl. 15, 295 pp.
Settle, M. Icarus, 1980, *42*, p. 1.
Shaw, H. F. and Wasserburg, G. J. Earth and Planetary Science Letters, 1982, *60*, p. 155.
Shoemaker, E. M. Report 21st International Geological Congress, 1960, part 18, p. 418.
Short, N. M. *Planetary Geology*, Prentice Hall, New Jersey, 1975, 361 pp.
Simonds, C. H., Warner, J. L., Phinney, W. C. and McGee, P. E. Geochimica et Cosmochimica Acta, 1976, Suppl. 7, p. 2509.
Stanyukovich, A. K. Meteoritika, 1975, *34*, p. 83.
Stöffler, D., Dence, M. R., Graup, G. and Abadian, M. Geochimica et Cosmochimica Acta, 1974, Suppl. 5, p. 137.
Urey, H. C. Nature, 1973, *242*, p. 32.
Violet, C. E. Journal of Geophysical Research, 1961, *66*, p. 3461.
Whipple, H. E., ed. Geological problems in lunar research. Annals of the New York Academy of Science, 1965, *123*, p. 367.

Aflou 34°0′ N., 2°3′ E.
Algeria
Discredited.
A meteoritic origin has been suggested for an oval depression (3×5 km) situated about 80 km WNW of Laghouat and 12 km SSW of Aflou, P. Marks et al., Proc. K. ned. Akad. Wet., 1972, vol.75, p.348. A detailed survey (P. Lambert et al., Meteoritics, 1980, vol.15, p.157) indicates that the feature is due to erosion and/or dissolution in the underlying limestone and gypsum formations..

Afton Craters 32°5′ N., 106°50′ W.
New Mexico, U.S.A.
Synonym(s): *Hunt's Hole, Kilbourne Hole, Stehling's Crater*
Discredited.
Three holes (Kilbourne Hole, Philipp's Hole and Hunt's Hole or Stehling's Crater), two crater-like, are almost certainly volcanic (O'Connell, 1965).

Agheir 19°25′ N., 11°30′ W.
Mauritania
Synonym(s): *Hofrat Aghreydh*
Discredited.
"Grande dépression qui ressemble a un cratére" (Monod, 1965). Co-ordinates, J. Classen, Meteoritics, 1977, vol.12, p. 61.

Agnak Island Craters 67°30′ N., 108°0′ W.
Franklin District, North-west Territories, Canada
Doubtful.
A group of five small craters, 100-500 feet in diameter and with raised rims may be meteoritic or collapsed pingoes, K. L. Currie, Annal. New York Acad. Sci., 1965, vol.123, p. 915.

Akchoky 47°42′ N., 72°23′ E.
Kazakhstan, USSR, [Акщокы]
Doubtful.
A circular structure approximately, 10 km in total diameter is possibly of meteoritic origin, B.S. Zeylik and E.Yu. Seytmuratova, Izv. Acad. Nauk Kazakhstan SSR, Ser. Geol., 1975, vol.1, p.62.

Alexeyevskoye 55.3° N., 50.1° E.
Tatar A.S.S.R., USSR
Doubtful.
A circular structure, 40 metres in diameter may be meteoritic, I.T. Zotkin and V.L. Tsvetkov Astron. Vestnik, 1970, vol.4, p.55.

Al Hadida v Wabar.

Al Umchaimin 32°41′ N., 39°35′ E.
Iraq
Doubtful.
A circular depression, inadequately examined (O'Connell, 1965). Co-ordinates, J. Classen, Meteoritics, 1977, vol.12, p. 61.

Amak Island 55°44′ N., 163°9′ W.
Aleutians, Alaska, U.S.A.
Doubtful.
A circular crater, 70 yards across and 50 yards deep, may perhaps be meteoritic (L. LaPaz Pop. Astron., Northfield Minnesota, 1947, vol.55, p.156; O'Connell, 1965). Brief description, rim to rim diameter, 160 metres W.A. Cassidy and E.G. Lidiak, abs. Meteoritics, 1980, vol.15, p.271.

Amami 28°25′ N., 129°38′ E.
Kagoshima, Japan
Doubtful.
Two craters, up to 2.4 km in diameter, are listed, J. Classen, Meteoritics, 1977, vol.12, p.61.

Amguid 26°5′ N., 4°23′30″ E.
Algeria
Doubtful.
A circular depression observed from the air (O'Connell, 1965; Monod, 1965) Full description with petrological evidence of shock metamorphism indicating an origin by meteorite impact; amended co-ordinates, P. Lambert et al., Meteoritics, 1980, vol.15, p.157. Diameter, 450 metres .

Amurskii 48° N., 132.7° E.
Khabarovsk Territory, USSR
Doubtful.
A lake, 175 metres in diameter may be of meteoritic origin, I.T. Zotkin and V.L. Tsvetkov, Astron. Vestnik, 1970, vol.4, p.55.

Aouelloul 20°15′ N., 12°41′ W.
Adrar, Mauritania
Synonym(s): *Hofrat Aouelloul*
Authenticated.
A crater near Chinguetti, 250 metres in diameter, was at first regarded as an explosion crater by its discoverer, but a small amount of highly siliceous glass containing 2.5% Fe and 0.02% Ni suggests a meteoritic origin. The Tenoumer crater, and two smaller craters at Temimichât-Ghallaman and Richat (all in the same district), are very similar and probably of the same origin. See T. Monod, Méharées, Paris, 1937, pp.145, 160, 188; Compt. Rend. Acad. Sci. Paris, 1950, vol.231, p.202; T. Monod, and A. Pourquié, Bull. Inst. Franç. Afrique Noire, 1951, vol.13, p.293; A. Allix, Rev. Géogr., Lyon, 1951, vol.26, p.357. Description and analyses of the silica glass and of the country rock (sandstone), W. Campbell Smith and M.H. Hey, Bull. Inst. Franç. Afrique Noire, 1952, no.14, p.762; Bull. Direct. Mines, Afrique Occ. Franç., 1952, no.15, p.443. Ni content W.D. Ehmann, Geochimica Acta, 1962, vol. 26, p. 489. See also Monod, 1965 and O'Connell, 1965. Further studies of the crater and silica glass, E.C.T. Chao et al., abs. in Trans. Amer. Geophys. Union, 1966, vol.47, p.144. Discovery of Fe-Ni spherules in silica glass, E.C.T. Chao et al., Science, 1966, vol.154, p. 759 and 765. J.A. O'Keefe supports an extraterrestrial origin for the silica glass (Meteoritics, 1969, vol.4, p.200). Fission-track age (3.3 m.y.) from glasses, D. Storzer, Meteoritics, 1971, vol.6, p.319. Detailed description, with gravity data, R.F. Fudali and W.A. Cassidy, Meteoritics, 1972, vol.7, p.51.
Specimen(s) : [1950,241-244] silica glass; [1951,320] silica glass; [1955,116] silica glass; [1971,298], 4.7g. glass

Araguainha Dome 16°46′ S., 52°59′ W.
Mato Grosso, Brazil
Doubtful.
A circular structure, 40 km in diameter consisting of a central uplift 10 km in diameter surrounded by an annular ring syncline, for which both igneous and meteoritic origins have been proposed. Shock metamorphism and shattercones, R.S. Dietz and B.M. French, abs. in Meteoritics, 1973, vol.8, p.345, and Nature, 1973 vol. 244, p.561.

Archangel'skoe 57.8° N., 46.3° E.
Gorki Province, USSR
Doubtful.
A circular lake, 90 metres in diameter and covered with peat, may be of meteoritic origin, I.T. Zotkin and V.L. Tsvetkov, Astron. Vestnik, 1970, vol.4, p.55.

Arizona v Meteor Crater.

Arnhem v Eastern Arnhem Land.

Arnhem Land v Eastern Arnhem Land.

Ashanti Crater v Lake Bosumtwi.

Aukstadvaris 54.5° N., 24.5° E.
Lithuanian S.S.R., USSR
Doubtful.
An origin by meteoritic impact has been suggested for a depression, 185 metres in diameter and 65 metres deep, I.T. Zotkin and V.L. Tsvetkov, Astron. Vestnik, 1970, vol.4, p. 55.

Avalon Peninsula 47°32′ N., 52°57′ W.
Newfoundland, Canada
Doubtful.
Shattercone occurrence but no related impact structure, D. W. Roy and R.H. Hansman, Bull. Geol. Soc. Amer., 1971, vol.82, p.3183. Possibly related to Holyrood Bay (*q . v* .).

Baghdad 33°20′ N., 44°25′ E.
Iraq
Doubtful.
Two oval depressions seen from the air (O'Connell, 1965). Co-ordinates, J.H. Freeberg, U.S. Geol. Surv. Bull., no. 1220, 1966, p. 22.

Balkhash 47° N., 73° E.
Kazakhstan, USSR
Doubtful.
A circular ridge in flat country and some 2 km in diameter is of doubtful meteoritic origin, I.T. Zotkin and V.L. Tsvetkov, Astron. Vestnik, 1970, vol.4, p.55.

Balkhash Lake 45.75° N., 77.5° E.
Kazakhstan, USSR, [Прибалхашско-Илийская]
Doubtful.
A huge, ovoid depression trending WNW, with major axis, 720 km (includes hilly rim), is listed and figured by B.S. Zeylik and E.Yu Seytmuratova, Izv. Akad. Nauk Kazakhstan SSR, Ser. Geol., 1975, vol.1, p.62.

Baluchistan v Gwarkuk.

Baraba 54.5° N., 77.5° E.
Novosibirsk, USSR
Doubtful.
Several small, circular lakes in steppe are of doubtful meteoritic origin, I.T. Zotkin and V.L. Tsvetkov, Astron. Vestnik, 1970, vol.4, p.55.

Barringer Crater v Meteor Crater.

Basra 30°20′ N., 47°40′ E.
Iraq
Doubtful.
A crater-like feature seen from the air (O'Connell, 1965).

Bass Strait 40° S., 146° E.
Australia
Discredited.
Bass Strait, between the Australian mainland and Tasmania, has been proposed as a cometary impact site; see references in J.H. Freeberg, U.S. Geol. Surv. Bull., 1969, no.1320, p.13.

Bay of Campechy 20° N., 94° W.
Mexico
Doubtful.
A meteoritic origin for this feature has been suggested (R. Gallant, Bombarded Earth, London (Baker), 1964; Monod, 1965).

Berdichev 50° N., 28.5° E.
Zhitomir Province, Ukraine, USSR
Doubtful.
Several lakes approximately, 30 metres in diameter are of doubtful meteoritic origin, I.T. Zotkin and V.L. Tsvetkov, Astron. Vestnik, 1970, vol.4, p.55.

Berezovo 63° N., 65° E.
Tyumen Province, USSR
Doubtful.
A lake, 250 metres in diameter is of doubtful meteoritic origin, I.T. Zotkin and V.L. Tsvetkov, Astron. Vestnik, 1970, vol.4, p.55.

Beyenchime-Salata 71° N., 122° E. approx
Siberia, USSR, [Беенчиме-Салаатинская]
Doubtful.
A circular structure, 8 km in diameter is possibly of meteorite impact origin; shock metamorphism, V.L. Masytis, Sov. Geol., 1975, no.11, p.52; but compare L.Ya. Pinchuk, In Kimberlite Vulcanism Northeast of the Siberian Platform, Leningrad, 1971.

Big Lake 64°52′ N., 112°57′ W.
MacKenzie district, North-west Territories, Canada
Doubtful.
Mentioned in list of possible impact structures, P.B. Robertson and R.A.F. Grieve, Journ. Roy. Astron. Soc. Canada 1975, vol.69, p.1.

Bishoftu Craters 8.75° N., 39° E.
Addis Ababa, Ethiopia
Discredited.
A meteoritic origin for a group of 14 craters has been suggested, but their volcanic origin seems definite (Monod, 1965).

Boltyschskij v Boltysh.

Boltysh
Kirovograd oblast, Ukraine, USSR, [Болтиськая Котловина]
Synonym(s): *Boltyschskij, Boltyshka, Boltyshskiy*
Doubtful.
A circular depression, with central uplift surrounded by breccias, 20-25 km in diameter in Precambrian basement rocks with a Mesozoic-Cenozoic sedimentary infill, for which meteoritic, volcanic and tectonic origins have been proposed

(V.A. Golubev et al., Dop. Akad. Nauk. URSR., 1974, Ser. B, no.1, p.10; Yu.Yu. Yurk et al., ibid., 1974, no.3, p.244; V. L. Masaitis, Meteoritika, 1974, vol.33, p.64). Mineralogical and petrographic description of country rocks with arguments supporting meteoritic origin, Yu.Yu. Yurk et al., Sov. Geol., 1975, vol.2, p.138 (trans- Int. Geol. Rev., 1976, vol.18, p.196); but compare Yu.B. Bass et al., Razv. Okhr. Nedr, 1967, no.9, p.11. General description, 70 m.y. age, V. L. Masaitis, Sov. Geol. 1975, No.11, p.52 (English Trans. Int. Geol. Rev. 1976, vol.18, p.1249.

Boltyshka v Boltysh.

Boltyshskiy v Boltysh.

Boriskovo 56° N., 36.5° E.

Moscow, USSR
Doubtful.
Several "conical" depressions, 10 metres in diameter, are of doubtful meteoritic origin, I.T. Zotkin and V.L. Tsvetkov, Astron. Vestnik, 1970, vol.4, p.55.

Borly 47°5′ N., 74°45′ E.

Kazakhstan, USSR, [Борлы]
Doubtful.
A structure, 150-200 metres across is possibly of meteoritic origin; magnetic Fe-Ni spherules have been found, B.S. Zeylik and E.Yu Seytmuratova, Izv. Acad. Nauk Kazakhstan S.S.R., Geol., 1975, vol.1, p.62.

Bosumtwi v Lake Bosumtwi.

Boxhole 22°37′ S., 135°12′ E.

Plenty River, Northern Territory, Australia
1937, June, recognized in this year
Authenticated.
A crater, 570-575 feet across and 30-52 feet deep, was found in gravelly alluvium. Masses of meteoritic iron have been found outside the craters, but no silica-glass has been observed (C.T. Madigan, Trans. Roy. Soc. South Australia, 1937, vol.61, p.187; Min. Mag., 1940, vol.25, p.481). Microprobe analyses of metallic spherules, P.W. Hodge and F.W. Wright, Meteoritics, 1973, vol.8, p.315. Additional reference, J.H. Freeberg, U.S. Geol. Surv. Bull., 1969, no. 1320, p.14. See also main meteorite list.

BP structure v Gebel Dalma.

Brenham v Haviland.

Brent Crater 46°4′ N., 78°29′ W.

Nipissing County, Ontario, Canada
1951, May, discovered
Authenticated.
A circular depression, 2 miles across, seen from the air, proved on detailed seismic, gravitational, and magnetic study to be underlain by a lens of breccia 1000 feet below the present surface; the breccia appears to have been subjected to sudden intense heat. The feature is estimated to be of Ordovician age (P.M. Millman et al., Contr. Dominion Observ., Ottawa, 1960, vol.24, no.4). See also O'Connell, 1965 and Monod, 1965. Unmetamorphosed sedimentary infill, G.P. Lozej and F.W. Beals, Can. Journ. Earth Sci., 1975, vol.12, p.606. Correlation of thermal resistivity of country rocks and shock metamorphic grade, A.E. Beck et al., Can. Journ. Earth Sci., 1976, vol.13, p.929. Petrochemistry of melt rocks, R.A.F. Grieve, Meteoritics, 1978, vol.13, p.484. See also R.A.F. Grieve, Geochim. Cosmochim. Acta, 1978, suppl. 10, vol.2, p.2579. For additional references, J.H. Freeberg, U.S. Geol. Surv. Bull., 1966, no. 1220, p.33, and 1969, no.1320, p.14.

K-Ar ages of country rocks and breccias (approx. 350m. y.); details of coring in the crater, M. Shafiqullah et al., Earth planet. Sci. Lett., 1968, vol.5, p.148. Compare K-Ar age of 426ł20 m.y. for the impact, J.B. Hartung et al., abs. in Meteoritics, 1969, vol.4, p.183; see also J.B. Hartung, Journ. Geophys. Res., 1971, vol.76, p.5437. Structure, M.R. Dence, abs. in Can. Mineral., 1969, vol.10, p.131. K.L. Currie (Can. Journ. Earth Sci., 1971, vol.8, p.481) considers the structure to be an alkaline carbonatite igneous complex. Projectile an L or LL chondrite? H. Palme et al., Geochim. cosmochim. Acta, 1981, vol.45, p.2417.

Brukkaros 25°50′ N., 18° E.

Namibia
Doubtful.
A crater, 2 km in diameter with breccia is of doubtful meteoritic origin; listed without further details by J. Classen, Meteoritics, 1977, vol.12, p.61.

Bushveld 25° S., 29° E. centered on

South Africa
Doubtful.
An origin by meteorite impact, contemporaneous with the Vrederfort Ring, has been suggested for the Bushveld igneous complex (R.S. Dietz, abs. in Geol. Soc. Amer. for 1962, Spec. Paper 73, p.35; W. Hamilton, Geol. Soc. South Africa, Spec. Publ.1, 1970, p.367). Absence of shock metamorphic features, B.M. French and R.B. Hargraves, Journ. Geology, 1971, vol.79, p.616. Detailed description with evidence supporting impact origin, R.C. Rhodes, Geology, 1975, vol.3, p.549.

Butare Crater 2°36′ S., 29°45′ E.

Rwanda
Synonym(s): *Rwanda*
Doubtful.
A circular structure observed in aerial photographs and for which an origin by meteorite impact has been proposed, M. Dehousse, Bull. Acad. Roy. de Belgique, Cl. Sci., Sér.5, 1966, vol.52, p.76. See also M.-E. Denaeyer and J. Gérards, C. R. Acad. Sci., Paris, 1973, ser.D, vol.277, p.1837..

Cabrerolles v Hérault Craters.

Campo del Cielo 27°28′ S., 61°30′ W.

Gran Chaco Gualamba, Argentina
1933, recognized
Authenticated.
A group of small depressions ("hoyos" or "pozos"; one is 70 metres diameter and 5 metres deep) in Campo del Cielo are probably meteorite craters and associated with the Campo del Cielo (Otumpa) irons (L.J. Spencer, Geogr. Journ. London, 1933, vol.81, p.227. They were formerly regarded as of artificial origin (J.J. Nagéra, Dirección Gen. Minas, Geol. Hidrol. Argentina, 1926, publ. no.19. See also numerous references in O'Connell, 1965, and W.A. Cassidy et al., Science, 1965, vol.149, p.1055.

Additional meteorite find, of estimated weight 18 tons, W. A. Cassidy, abs. in Meteoritics, 1970, vol.5, p.187; entry trajectory and impact energy considerations, W.A. Cassidy and M.E. Renard, ibid., p.187. Co-ordinates of the 20 known craters, A. Romana and W.A. Cassidy, Meteoritics, 1973, vol.8, p.430. For additional references, J.H. Freeberg, U.S. Geol. Surv. Bull., 1966, no.1220, p.33, and 1969, no.1320, p. 14. See also main meteorite list.

Cañon Diablo v Meteor Crater.

Canyon Diablo v Meteor Crater.

Carolina "bays"

Carolina, U.S.A.
Discredited.

Meteoritic origin has been proposed for a number of shallow elliptical depressions on the coastal plain of North and South Carolina (F.A. Melton and W. Schriever, Journ. Geol. Chicago, 1933, vol.41, p.52; C.C. Wylie, Pop. Astron., Northfield, Minnesota, 1933, vol.41, p.410; F.A. Melton, Discovery, London, 1934, vol.15, p.151; G.R. MacCarthy, Bull. Geol. Soc. Amer., 1936, vol.48, p.1211; F.A. Melton, Proc. Geol. Soc. Amer., 1938, (for 1937), p.312, and Bull. Geol. Soc. Amer., 1938, vol.49, p.1954; W.F. Prouty, Science, New York, 1938, vol.88, p.475; Bull. Geol. Soc. Amer., 1952, vol.63, p.167, but the evidence is by no means satisfactory, and various other origins have been suggested (C.W. Cooke, Journ. Geol. Chicago, 1934, vol.42, p.88; F. Watson, Pop. Astron., Northfield Minnesota, 1936, vol.44, p. 2; D. Johnson, Science, New York 1936, vol.84, p.15; P.M. Millman, Journ. Roy. Astron. Soc., Canada, 1936, vol.30, p. 57; H.T. Odum, Amer. Journ. Sci., 1952, vol.250, p.263.

Many further papers have been published on the origin of these depressions; D. Johnson, The origin of the Carolina Bays, New York, 1942; C.W. Cooke, Amer. Journ. Sci., 1943, vol.241, p.589; J.C. Campbell, Journ. Geol. Chicago, 1945, vol.53, p.66, and Pop. Astron. Northfield, Minnesota, 1945, vol.53, p.388; E.B. Bailey, Nature, London, 1944, vol. 154, p.383; B. Barringer, Pop. Astron. Northfield, Minnesota, 1947, vol.55, p.215; C. Grant, ibid., 1948, vol.56, p.511; W.F. Prouty, ibid., 1948, vol.56, p.499, and 1950, vol. 58, p.17; F.A. Melton, Journ. Geol. Chicago, 1950, vol.58, p. 128; W. Schriever, Trans. Amer. Geophys. Union, 1951, vol. 32, p.87; C.W. Cooke, U.S. Geol. Surv. Prof. Paper no.254-1, 1954; D.J. Hagar in H.T.V. Smith, Final Report on Contract Nonr-2242, Geol. Dept. Univ. Massachusetts, Amherst, 1960.

Carswell Lake 58°27′ N., 109°30′ W.

Saskatchewan, Canada
Doubtful.

A circular area, 18 miles in diameter observed from the air in 1957 shows deformed strata tilted 90 degrees away from the centre and brecciation. C.S. Beals et al., (Contr. Dominion Observ. Ottawa, 1960, vol.24, no.4; Current Science, Bangalore, 1960, vol.29, pp 205, 249) favour a meteoritic origin but W.F. Fahrig (Bull. Geol. Surv. Canada, 1961, no.68) does not. See also O'Connell, 1965.

Literature with arguments for and against a meteoritic origin, J.H. Freeberg, U.S. Geol. Surv. Bull., 1969, no.1320, p.15. Features of shock metamorphism, K.L. Currie, in Shock Metamorphism of Natural Materials, eds. B.M. French and N.M. Short, Mono Book Corp., Baltimore, 1968, p.379. Chemical studies (K.L. Currie and M. Shafiquliah, Nature, 1968, vol.218, p.457) argue against an impact origin. See also K.L. Currie, Geol. Surv. Can. Paper 67-32, 1969. Radiometric age, 485 ± 50 m.y., P.B. Robertson and R.A.F. Grieve, Journ. Roy. Astron. Soc. Canada, 1975, vol.69, p.1. Planar elements in quartz, P. Lambert and M. Pagel, C. R. Acad. Sci. Paris, 1977, vol.284, p.1623.

Chaglan Toushtou v Murgab Craters.

Charlevoix Structure 47°32′ N., 70°18′ W.

Quebec, Canada
Synonym(s): *La Malbaie structure*
Doubtful.

A semicircular depression, 35 km in diameter, truncated to the south and east by the St. Lawrence River. Central peak, shattercones and evidence of shock metamorphism, P.B. Robertson, Meteoritics, 1968, vol.4, p.89; also J. Rondot, Can. Journ. Earth Sci., 1968, vol.5, p.1305. Breccia-dykes and mylonization, J. Rondot, abs. in Meteoritics, 1969, vol.4, p.291 and Can. Journ. Earth Sci., 1970, vol.7, p.1194. Thermoluminescence studies, D.J. McDougall, Meteoritics, 1970, vol.5, p.75. Shattercones, D.W. Roy and J. Rondot, abs. ibid., p.219. Breccias, J. Rondot, abs. in Meteoritics, 1971, vol.6, p.307; and impactites, J. Rondot, Journ. Geophys. Res., 1971, vol.76, p.5414, also Proc. 24th Internat. Geol. Cong., 1972, Section 15, p.140. Shock metamorphic zones P.B. Robertson, EOS 1974, vol.55, p.336 and Bull. Geol. Soc. Amer. 1975, vol.86, p.1630; origin and evolution, D.W. Roy, EOS, 1974, vol.55 p.336. Palaeomagnetism, R.B. Hargraves and D.W. Roy, Can. Journ. Earth Sci., 1974, vol.11, p.854. See also, D.W. Roy, abs. in Meteoritics, 1975, vol.10, p.481. Age, 360 ± 25 m.y., P. B. Robertson and R.A.F. Grieve, Journ. Roy. Astron. Soc. Canada, 1975, vol.69, p.1. Fluid inclusion studies, M. Pagel and B. Poty, Fortschr. Min., 1975, vol.52 (Special issue), p. 479. Comparison with Siljan Ring, J. Rondot, Bull. Geol. Inst., Univ. Uppsala, N-S-6, 1975, p.85. Mosaicism in plagioclase, M.J. Walawender, Can. Journ. Earth Sci., 1977, vol.14, p.74. Planar elements in quartz, P. Lambert and M. Pagel, Compt. Rendus Acad. Sci. Paris, 1977, vol.284, p. 1623. Seismic studies, D.W. Roy and R. DuBerger, Can. J. Earth Sci., 1983, vol.20, p.1613.

Charron Lake 52°44′ N., 95°15′ W.

Manitoba, Canada
Doubtful.

Listed as a possible meteorite crater, P.B. Robertson, in the Observer's handbook, ed. J.R. Percy, Roy. Astron. Soc. Canada, 1975, p.77.

Chassenon Structure 45°50′ N., 0°56′ E.

Haut-Vienne district, France
Synonym(s): *Rochechouart Structure*
Doubtful.

A poorly defined structure, minimum diameter 10 km situated 42 km west of Limoges and for which both volcanic and meteoritic origins have been proposed. Breccias with shock metamorphic features, F. Kraut et al., Meteoritics, 1969, vol.4, p.190. Minimum K-Ar age, 150-170 m.y., F. Kraut and B.M. French, abs. in Meteoritics, 1970, vol.5, p. 206. Breccias cover 280 square km, P. Lambert, Bull. B.R.G. M., 1974, vol.1. p.153. Palaeomagnetic data, J. Pohl and H. Soffel, abs. in Meteoritics, 1971, vol.6, p.299. Fission-track age, 210 ± 100 m.y., from crater glass, D. Störzer, abs. in Meteoritics, 1971, vol.6, p.319. See also J.H. Freeberg, U.S. Geol. Surv. Bull., 1969, no.1320, p.15.

Shatter cones, D. Sorel, C. R. Acad. Sci. Paris, Sér D, 1977, sér.D, vol.284, p.2087 (with English summary). Identification of the projectile, M-J. Janssens, et al., J. Geophys Res., 1977, vol.82, p.750. See also, F. Kraut and B. M. French, J. Geophys. Res., 1971, vol.76, p.5407. F. Kraut and J. Becker, C. R. Acad. Sci. Paris, Sér D, 1974, vol.278, p.2893; P. Lambert, Earth planet. Sci. Lett., 1977, vol.35, p. 258. Planar elements in quartz, P. Lambert and M. Pagel, C. R. Acad. Sci. Paris, Sér D, 1977, vol.284, p.1623. Gravity measurements, J. Pohl et al., abs. in Meteoritics, 1978, vol. 13, p.601. Search for coesite and stishovite, P. Lambert, abs. in Meteoritics, 1978, vol.13, p.530. Discovery of Fe-Cr-Ni

metallic residues in rocks from the floor of the crater, W. Horn and A. El Goresy, abs. in Meteoritics, 1979, vol.14, p. 424 and Bull. Minéral., 1981, vol.104, p.587.
Arguments in favour of an origin by "clustered" impact, P. Lambert, Abs. in Meteoritics, 1982, vol.17, p.240.

Chemsiya t Dome v Semsiyyât Dome.

Chik 55° N., 82.5° E.
Novosibirsk Province, USSR
Doubtful.
A lake, 75 metres in diameter and 10 metres deep, situated on an elevation, is possibly of impact origin, I.T. Zotkin and V.L. Tsvetkov, Astron. Vestnik, 1970, vol.4, p.55.

Chinga 51 0° N., 94°0′ E.
Tuvinskaya, USSR, [чинге]
Authenticated.
The numerous irons (Ni-rich ataxite) from Chinga have the "torn" appearance characteristic of irons from meteorite craters. At the time they were found, the only known meteorite crater was Meteor Crater, Arizona, and it is possible that these Chinga irons came from a crater that was not observed, or simply not recognized as such, at the time they were collected; or they may have come from a crater since destroyed by erosion (many of the masses are highly oxidized) (E.L. Krinov, Catalogue of Meteorites, Acad. Nauk USSR, Moscow, 1947, pp.57, 82.
Results of the 1963 expedition to the site, B.I. Vronskii and I.T. Zotkin, Meteoritika, 1968, vol.28, p.125. See also J. H. Freeberg, U.S. Geol. Surv. Bull., 1966, no.1220, p.37.

Christmas Canyon Dome v Upheaval Dome.

Chubb crater 61°17′ N., 73°40′ W.
Ungava district, Quebec, Canada
Synonym(s): *New Quebec Crater, Ungava*
Authenticated.
A circular crater, 2 miles in diameter was located in July 1950; although no meteoritic material has yet been found there is no sign of volcanic action, but every evidence of a violent explosion, and there seems no reason to doubt that the crater is truly meteoritic (V.B. Meen, Journ. Roy. Astron. Soc. Canada, 1950, vol.44, p.169; Globe and Mail, Toronto, Aug.7 and Aug.8, 1950; Ward's Nat. Sci. Bull., 1950, vol.24, no.2, p.19; Nat. Geogr. Mag., Washington, 1952, vol.101, p.1; V.B. Meen, Journ. Roy. Astron. Soc. Canada, 1957, vol.51, p.137; J.M. Harrison, Journ. Roy. Astron. Soc. Canada, 1954, vol.48, p.16; P. Millman, Publ. Dominion Observ., Ottawa, 1956, vol.18, no.4, p.61; E.M. Shoemaker, U.S. Geol. Surv. Astrogeologic Studies, Semiannual Progress Report, Feb-Aug., 1961, p.74), but K. L. Currie does not agree (Nature, 1964, vol.201, p.385; Meteoritics, 1964, vol.2, p.93), nor do L. LaPaz (Meteoritics, 1954, vol.1, p.228) and F.C. Leonard (ibid., p.229). See also O'Connell, 1965 for further references.
Co-ordinates and additional references, J.H. Freeberg, U.S. Geol. Surv. Bull., 1966, no.1220, p.60, and 1969, no.1320, p. 25. Radiometric age, less than 5 m.y., P.B. Robertson and R. A.F. Grieve, Journ. Roy. Astron. Soc. Canada, 1975, vol.69, p.1.

Clayton's Craters v Libyan Desert Craters.

Clearwater Lakes 56°10′ N., 74°20′ W.
Ungava district, Quebec, Canada
Doubtful.
Two circular lakes, 16 miles , 19 miles . in diameter, are possibly of meteoritic origin (C.S. Beals et al., Journ. Roy. Astron. Soc. Canada, 1956, vol.50, pp.203 and 250; C.S. Beals et al., Current Science, Bangalore, 1960, vol.29, pp.205 and 249; R.W. Tanner, Journ. Roy. Astron. Soc. Canada, 1963, vol.57, p.109; I. Halliday and A.A. Griffin, Meteoritics, 1964, vol.2, p.79; D.B. McIntyre, Progr. Amer. Geophys. Union, Los Angeles meeting, 1961, p.55; M.R. Dence, et al., Journ. Roy. Astron. Soc. Canada, 1965, vol.59, p.1), but compare S.H. Kranck and G.W. Sinclair, Bull. Geol. Surv. Canada, 1963, no.100, also K.L. Currie, Meteoritics, 1964, vol.2, p.93. See also, O'Connell, 1965.
Metamorphic evidence suggests an impact origin, M.R. Dence et al., Journ. Roy. Astron. Soc. Canada, 1965, vol.59, p.13, but compare K.L. Currie and M. Shafiqullah, Nature, 1968, vol.218, p.457, and H.H. Bostock, Geol. Surv. Canada Bull., 1969, no.178. Fission track age from glass 287 ± 43 m.y., D. Störzer, Meteoritics, 1971, vol.6, p.319. Immiscible fluids in crater glass, M.R. Dence et al., Contr. Mineral. Petrol., 1974, vol.46, p.81. See J.H. Freeberg, U.S. Geol. Surv. Bull., 1966, no. 1220, p.37, and 1969, no.1320, p.15, for other references.
Field geology, structure and bulk chemistry of crater rocks, C.H. Simonds et al., Geochim. Cosmochim. Acta, Suppl. 10, 1978, vol.2, p.2633. Petrology of West Clearwater crater, W.C. Phinney et al., Geochim. Cosmochim. Acta, Suppl. 10, 1978, vol.2, p.2659. Meteoritic component and impact melt composition, R.A.F. Grieve, Geochim. Cosmochim. Acta, 1978, vol.42, p.429. Impact melts at East Clearwater crater, H. Palme and R.A.F. Grieve, abs. Meteoritics, 1979, vol.13, p.595. Distribution of volatile and siderophile elements in impact melts of East Clearwater indicating a C-chondrite impacting body, H. Palme et al., Geochim. cosmochim. Acta Suppl.11, 1979, p.2465 see also, R.A.F. Grieve et al., Contr. Min. Petr., 1980, vol.75, p.187. Rb-Sr dating of impact melts at East Clearwater, W.V. Reimold, et al., Contr. Min. Petr., 1981, vol.76, p.73. Comparison with Puchezh-Katunki Disturbance, V.M. Gordin et al., Phys. Earth Planet. Inter., 1979, vol.20, p.1.

Colonia Crater
Sao Paulo, Brazil
Doubtful.
R.S. Dietz suspects that this may be meteoritic, abs. in Meteoritics, 1969, vol.4, p.269.

Conception Bay v Holyrood Bay.

Confolent
Haute-Loire, France
Discredited.
"Meandre abandonné, avec au centre une butte de roche en place" (A. Cailleux, in Monod, 1965).

Coon Butte v Meteor Crater.

Crater Elegante 32°40′ N., 112°55′ W.
Sonora, Mexico
Discredited.
A crater near Sonoyta was thought by A.O. Kelly (Sci. Monthly, Amer. Assoc. Adv. Sci., 1952, vol.74, p.291) to be of meteoritic origin, but this claim was withdrawn (A.O. Kelly, Sky and Telescope, 1953, vol.12, p.90). See also O'Connell, 1965. Co-ordinates, J.H. Freeberg, U.S. Geol. Surv. Bull., 1966, no.1220, p.38.

Crater Mound v Meteor Crater.

Crestone 37°54′ N., 105°39′ W.
Saguache County, Colorado, U.S.A.
Discredited.
Meteoritic origin was suggested for a regular elliptical bowl in sand by R. Barringer (Progr. Meteoritical Soc. meeting, Ottawa, 1963) but rejected by U.B. Marvin (Progr. Meteoritical Soc. meeting Ottawa, 1964, p.12). The crater morphology re-examined; lack of evidence of impact or explosion indicates that it is not meteoritic or cometary, U.B. Marvin and T.C. Marvin, Meteoritics, 1966, vol.3, p.1.

Crooked Creek Structure 37°50′ N., 91°23′ W.
St. Louis County, Missouri, U.S.A.
Doubtful..
A crater, 3-4 miles in diameter in Ordovician strata is accepted by many authors as meteoritic in origin, on the basis of intense brecciation and shattercones directed upward. See O'Connell, 1965, and J.H. Freeberg, U.S. Geol. Surv. Bull., 1966, no.1220, p.38, and 1969, no.1320, p.16, for literature. Shock metamorphism, R.S. Dietz and P. Lambert, abs. in Meteoritics, 1980, vol.15, p.281.

Colluma Crater 18°32′ S., 68°5′ W.
Bolivia
Discredited.
A poorly studied crater, perhaps a collapsed dome, see J.H. Freeberg, U.S. Geol. Surv. Bull., 1969, no.1320, p.16.

Daiet el Maadna v Talemzane Crater.

Dalgaranga 27°43′ S., 117°15′ E.
Western Australia, Australia
1923, Noticed
Authenticated.
A crater, 75 feet across and 15 feet deep, was found near Dalgaranga, and shows signs of explosive action, while in and around the crater many small fragments of meteoritic iron were found (E.S. Simpson, Min. Mag., 1938, vol.25, p. 157). The crater and meteoritic material, of a mesosiderite, have been studied in detail by H.H. Nininger and G. I Huss (Min. Mag., 1960, vol.32, p.619) and by G.J.H. McCall (ibid., 1965, vol.35, p.476). See also O'Connell, 1965. For further references and distribution of material see main meteorite list.

Darwin Crater v Mount Darwin Crater.

Darwin glass v Mount Darwin Crater.

Debra Zeit Craters v Bishoftu Craters.

Decaturville Dome 37°54′ N., 92°43′ W.
Camden and Laclede Counties, Missouri, U.S.A.
Doubtful.
A typical cryptoexplosion structure with brecciation and shattercones, 3.7 miles in diameter and of late Cambrian or early Ordovician age. Opinions differ as to its origin (see O'Connell, 1965 and J.H. Freeberg, U.S. Geol. Surv. Bull., 1966, no.1220, p.39, and 1969, no.1320, p.16 for literature). Style and sequence of deformation, T.W. Offield and H.A. Pohn, Meteoritics, 1971, vol.6, p.296. See also T.W. Offield and H.A. Pohn, In Impact and Explosion Cratering, eds D.J. Roddy, R.O. Pepin and R.B. Merrill, Pergamon Press (New York), 1977, p.321.

Deep Bay 56°24′ N., 102°59′ W.
Saskatchewan, Canada
Authenticated.
A bay of Reindeer Lake is of circular outline and is much deeper than the rest of the lake. Generally accepted as a fossil meteorite crater on the basis of intense shattering of the rocks, concentric fractures and gravitational anomalies around the bay, and magnetic anomalies (M.J.S. Innes, Journ. Roy. Astron. Soc. Canada, 1957, vol.51, p.236; Journ. Geophys. Res., 1961, vol.66, p.2225; M.J.S. Innes et al., Publ. Dominion Observ., Ottawa, 1964; C.S. Beals et al., Contr. Dominion Observ., Ottawa, 1960, vol.4, no.4; Current Science, Bangalore, 1960, vol.29, pp.205 and 249). For additional references, O'Connell, 1965, and J.H. Freeberg, U. S. Geol. Surv. Bull., 1966, no.1220, p.39 and 1969, no.1320, p.16. Seismic and magnetic investigations of the crater, C.W. Sander et al., Journ. Roy. Astron. Soc. Canada, 1964, vol.58, p.1; see also W.K. Hartmann, Icarus, 1965, vol.4, p.157. Age 100 ± 50 m.y., P.B. Robertson and R.A.F. Grieve, Journ. Roy. Astron. Soc. Canada, 1975, vol.69, p.1.

Des Plaines Disturbance 42°2′ N., 87°56′ W.
Cook County, Illinois, U.S.A.
Doubtful.
An area of negative gravity anomaly with "high-angle faulting of rock of two ages not explainable by cryptovolcanic theory" has been included by R.B. Baldwin (The Measure of the Moon, Univ. Chicago Press, 1963) as a possible meteorite crater. Co-ordinates and additional references, J.H. Freeberg, U.S. Geol. Surv. Bull., 1966, no. 1220, p.40. See also O'Connell, 1965.

Dirranbandi 28°35′ S., 148°10′ E.
Queensland, Australia
Doubtful.
Several craters may be meteoritic, see J.H. Freeberg, U.S. Geol. Surv. Bull., 1969, no.1320, p.16.

Dogubayazid 39°32′ N., 44°14′ E.
Turkey
Doubtful.
A circular structure, 35 metres in diameter and 30 metres deep, for which an impact origin has been proposed, W. Sander, Studien über Meteoritenkrater, Sirius, Heppenheim, 1975.

Dolgoe 54° N., 37.9° E.
Tula Province, USSR
Doubtful.
Six small lakes some 30 metres in diameter and situated on a watershed, are of doubtful meteoritic origin, I.T. Zotkin and V.L. Tsvetkov, Astron. Vestnik, 1970, vol.4, p.55.

Double Punchbowl v Henbury.

Duckwater 38°7′ N., 115°7′ W.
Nye County, Nevada, U.S.A.
Discredited.
A circular crater, 225 feet in diameter and 10-15 feet deep, may perhaps be of meteoritic origin (J.S. Rinehart and C.T. Elvey, Contr. Meteoritical Soc., 1951, vol.5, p.44), but is probably a limestone sink (literature, see O'Connell, 1965).

Dumas
Saskatchewan, Canada
Doubtful.
A buried area of disturbed rocks identified on seismic traces is possibly a meteorite impact crater of Late Cretaceaous age, H.B. Sawatzky, In Impact and Explosion Cratering, eds. D.J. Roddy, R.O. Pepin and R.B. Merrill, Pergamon Press (New York), 1977, p.461.

Dycus Structure 36°22′ N., 85°45′ W.
Jackson County, Tennessee, U.S.A.
Doubtful.
Meteoritic origin is suggested for a circular disturbed area in Ordovician strata with upwardly directed shattercones (literature in O'Connell, 1965). Additional references, J.H. Freeberg, U.S. Geol. Surv. Bull., 1966, no.1220, p.40.

Dzhaus Craters 39.2° N., 67.3° E.
Tadzhik S.S.R., USSR
Doubtful.
Several small depressions approximately 7 metres in diameter were found in 1950 after the appearance of a bolide; they might therefore be of meteoritic origin, D.I. Kravtzev, Priroda, 1951, vol.40, no.7, p.41, and I.T. Zotkin and V.L. Tsvetkov, Astron. Vestnik, 1970, vol.4, p.55.

Dzhezkazgan 47° N., 70° E.
Kazakhstan, USSR, [Джезказган]
Doubtful.
A large, circular, disturbed area of total diameter about 300 km and with a region of central uplift may be of impact origin, B.S. Zeylik and E. Yu Seytmuratova, Izvestia Akad. Nauk Kazakhstan S.S.R., Ser. Geol., 1975, vol.1, p.62.

Dzioua v Guerrara.

Eagle Butte 49°42′ N., 110°30′ W.
Alberta, Canada
Doubtful.
A disturbed area, 10 km in diameter and 30-40 m.y. old, may be meteoritic, P.B. Robertson and R.A.F. Grieve, Journ. Roy. Astron. Soc. Canada, 1975, vol.69, p.1, but compare T.B. Haites and H. van Hees, Journ. Alberta Soc. Petrol. Geologists, 1962, vol.10, p.511. Bore hole and electric log data for the structure, H.B. Sawatzky, In Impact and Explosion Cratering, eds. D.J. Roddy, R.O. Pepin and R.B. Merrill, Pergamon Press (New York), 1977, p.461.

Eastern Pamir Craters v Murgab Craters.

Eastern Arnhem Land 13°10′ S., 135°40′ E.
Northern Territory, Australia
Synonym(s): *Arnhem, Arnhem Land*
Doubtful.
A feature that may be a meteorite crater has been reported in the Koolatong River area near Blue Mud Bay; its nature is still uncertain (letter from the Official Secretary, Australia House, London, of April 6, 1950, in Min. Dept.BM(NH).). See also, J.H. Freeberg, U.S. Geol. Surv. Bull., 1966, no. 1220, p.22.

Elbow Structure 50°58′ N., 106°45′ W.
Saskatchewan, Canada
Doubtful.
This structure, described by G. DeMille, Journ. Alberta Soc. Petrol. Geologists, 1960, vol.8, p.154, is included in a list of possible meteorite craters due to W.H. Bucher (Monod, 1965, p.35). An infilled erosion channel cut by a thrust fault, T.B. Haites and H. van Hees, Journ. Alberta Soc. Petrol. Geologists, 1962, vol.10, p.511. Compare P.B. Robertson and R.A.F. Grieve, Journ. Roy. Astron. Soc. Canada, 1975, vol. 69, p.1; this latter paper gives an age of 70-80 m.y.

Elgegytgyn v Lake Elgytkhyn.

Elgygytgyn v Lake Elgytkhyn.

Ellef Ringnes Island 78.5° N., 102.5° W.
Franklin district, North-west Territory, Canada
Doubtful.
Four crater-like structures with regular outlines and raised rims noted on aerial photographs may be meteoritic but are probably ring dykes and laccoliths (I.C. Brown, Amer. Journ. Sci., 1951, vol. 249, p.785). Described, P.B. Robertson and R.A.F. Grieve, Journ. Roy. Astron. Soc. Canada, 1975, vol.69, p.1.

El Mouilah 33°51′ N., 2°3′ E.
Algeria
Discredited.
An almost circular depression, 5×4 km with a central dome situated about 80 km west of Laghouat has been investigated, and discredited as a possible impact structure; general geological considerations indicate a recent collapse structure, P. Lambert et al., Meteoritics, 1980, vol.15, p.157.

El Mreiti 23.5° N., 6.5° W.
Mauritania
Doubtful.
A depression 700 metres in diameter observed from the air (Monod, 1965).

Erg Ghech
Algeria
Synonym(s): *Erg Touat*
Doubtful.
Several small shallow craters were observed (Monod, 1965).

Erg Touat v Erg Ghech.

Ernur 56.8° N., 47.6° E.
Mari A.S.S.R., USSR
Doubtful.
A circular lake, 125 metres in diameter, may be of impact origin, I.T. Zotkin and V.L. Tsvetkov, Astron. Vestnik, 1970, vol.4, p.55.

Estherville
Emmet County, Iowa, U.S.A.
Discredited.
The impact pit (12 ft. diameter, 5 ft. deep) formed by this meteorite (see main list) is included by O'Connell (1965) in her bibliography of meteorite craters and related structures.

Esthonia v Kaalijärv.

Eyre Peninsula 34° S., 136° E.
South Australia, Australia
Synonym(s): *Lake Hamilton Craters, Weepra Park Depressions*
Discredited.
Depressions in calcareous sandstone are probably not meteoritic; they are interpreted as sink holes and collapsed caverns (A.F. Wilson, Proc. Roy. Geogr. Soc. Australasia, South Australian Branch, 1947, vol.48, p.25.

Faugéres v Hérault Craters.

Firth v The Firth.

Flynn Creek Disturbance 36°16′ N., 85°37′ W.
Jackson County, Tennessee, U.S.A.
Doubtful.
A nearly circular structure in Devonian strata, with intense brecciation and shattercones, has been interpreted as meteoritic, but opinion seems about equally divided on this question (literature, see O'Connell, 1965). Additional references, J.H. Freeberg, U.S. Geol. Surv. Bull., 1966, no. 1220, p.41, and 1969, no.1320, p.17. Pre-impact conditions and cratering process, D.J. Roddy, In Impact and Explosion Cratering, eds. D.J. Roddy, R.O. Pepin and R.B. Merrill, Pergamon Press (New York), 1977, p.277. Structural deformation revealed by deep drilling, D.J. Roddy, Geochim. Cosmochim. Acta, 1979, Suppl.11, vol.3, p.2519.

Fontaine Lake 59°42′ N., 106°25′ W.
Saskatchewan, Canada
Doubtful.
Possible impact origin, P.B. Robertson and R.A.F. Grieve, Journ. Roy. Astron. Soc. Canada, 1975, vol.69, p.1.

Foum Teguemtour v Foum Teguentour.

Foum Teguentour 26°14.5′ N., 2°57′ E.
Ahnet, central Sahara, Algeria
Synonym(s): *Foum Teruemtour, Foum Teguemtour, Foum Teruentour*
Doubtful.
Meteoritic origin has been tentatively suggested for a disturbance, 8 km in diameter in Devonian strata (Monod, 1965; J.F. McHone and R.S. Dietz, abs. in Meteoritics, 1978, vol.13, p.557). A detailed survey of the structure indicates a diapiric rather than meteoritic origin, P. Lambert et al., Meteoritics, 1981, vol.16, p.203.

Foum Teruemtour v Foum Teguentour.

Foum Teruentour v Foum Teguentour.

Franktown Crater 45°3′ N., 76°4′ W.
Ontario, Canada
Discredited.
A roughly circular area for which a meteoritic origin has been suggested but since withdrawn; probably remains of an old lake (literature in O'Connell, 1965 and Monod, 1965). Evidence from drill cores taken from the crater suggests that it is not meteoritic, M.R. Dence et al., in Shock Metamorphism of Natural Materials, eds. B.M. French and N.M. Short, Mono, Baltimore, 1968, p.339.

Freetown 8°20′ N., 13°20′ W.
Sierra Leone
Doubtful.
On the basis of echo soundings and magnetic measurements an impact origin has been proposed for the Freetown layered basic igneous complex, D.C. Krause, Nature, 1963, vol.200, p.1280.

Frombork 54°20′ N., 19°41′ E.
Vistula Lagoon, Poland
Authenticated.
An oval crater, approximately 100 metres in diameter from around which magnetic fragments and spherules have been collected. Full description, H. Korpikiewica, Meteoritics, 1980, vol.15, p.63.

Furnas County 40°0′ N., 99°50′ W.
Nebraska, U.S.A.
Discredited.
The impact hole of the Norton County meteorite (see main list) is included by O'Connell (1965) in her list of meteorite craters and related features.

Gallouédec Crater 21°0′ N., 15°40′ W.
Mauritania
Doubtful.
Observed on aerial photographs in 1962 and not further examined (Monod, 1965).

Garet El-Lefet 25°0′ N., 16°30′ E.
Fezzan, Libya
Doubtful.
Meteoritic origin has been suggested for a feature 3-4 km in diameter (Monod, 1965).

Gebel Dalma 25°19′ N., 24°20′ E.
Southern Cyrenaica, Libya
Synonym(s): *BP structure*
Doubtful.
A circular structure, 2.8 km in diameter in Mesozoic strata 165 km northeast of the Kufra Oasis and 75 miles ESE of Gebel Dalma is possibly of meteoritic origin, A.J. Martin, Nature, 1969, vol.223, p.940; an association with Libyan Desert Glass is suggested. Description of shock metamorphic features, amended co-ordinates and suggestion that the feature had a simultaneous origin with the "Oasis structure" 80 km to the south, B.M. French et al., Bull. Geol. Soc. Amer., 1974, vol.85, p.1425.

Glasford 40°22′ N., 89°48′ W.
Peoria County, Illinois, U.S.A.
Doubtful.
Meteoritic origin is suggested for a circular disturbed area with severe brecciation (T.C. Buschbach and R. Ryan, Geol. Soc. Amer. Spec. paper 73 (1962), p.126). See also, J.H. Freeberg, U.S. Geol. Surv. Bull., 1966, no.1220, p.42.

Glover Bluff 43°58.2′ N., 89°32.3′ W.
Marquette County, Wisconsin, U.S.A.
Discredited.
Meteoritic origin has been suggested for a locally disturbed and uplifted area, but faulting and glacial action appear sufficient causes (O'Connell, 1965, with literature). Additional references, J.H. Freeberg, U.S. Geol. Surv. Bull., 1966, no.1220, p.42. Co-ordinates and the discovery of shattercones, W.F. Read, Meteoritics, 1983, vol.18, p.241.

Goat Paddock 18°20′ S., 126°40′ E.
Kimberley district, Western Australia, Australia
Doubtful.
A structure, 5 km in diameter excavated in Proterozoic sedimentary rocks is possibly meteoritic; shattercones, brecciation and melt rocks. The crater is of Eocene age, or older, D. Milton et al., abs. Meteoritics, 1980, vol.15, p.333. Full description, J.E. Harms et al., Nature, 1980, vol.286, p. 704.

Gonamski 56.3° N., 126.8° E.
Yakut A.S.S.R., USSR
Doubtful.
A depression, 25 metres in diameter may be of meteoritic origin, I.T. Zotkin and V.L. Tsvetkov, Astron. Vestnik, 1970, vol.4, p.55.

Gosses Bluff 23°50′ S., 132°18′ E.
Northern Territory, Australia
Doubtful.
A circular cryptoexplosion structure approximately 22 km in diameter, with raised rim and well developed shattercones; references in J.H. Freeberg, U.S. Geol. Surv. Bull., 1969, no. 1320, p.18, e.g. P.J. Cook, Journ. Geol., 1968, vol.76, p.123. Gravity investigations revealed a symmetrical, circular gravity low, of radius 10.8 km, centred at focus of shatter cones, B.C. Barlow, B.M.R. Journ. Austral. Geol. Geophys., 1979, vol.4, p.323.

Gourma 1°19′ N., 15°17′ W. approx.
Mali
Doubtful.
About 20 small craters, 20-200 metres in diameter and 1-3 metres deep, are possibly meteoritic. Localities lie between 1°16'N.-1°22'N. and 15°5'W.-15°29'W. (Monod, 1965).

Gow Lake 56°27′ N., 104°29′ W.
Saskatchewan, Canada
Doubtful.
A circular lake roughly 4 km in diameter is of doubtful meteritic origin. Evidence of an origin by meteoritic impact including large central island and shocked and brecciated country rocks, M.D. Thomas and M.J.S. Innes, Can. Journ. Earth Sci., 1977, vol.14, p.1788. Age less than 600 m.y., P.B. Robertson and R.A.F. Grieve, Journ. Roy. Astron. Soc. Canada, 1975, vol.69, p.1. The impacting body was probably an iron meteorite, R. Wolf et al., Geochim. Cosmochim. Acta, 1980, vol.44, p.1015. The crater is the smallest known with central uplift, and has an age of less than 200 m.y.

Guerrara 33°19′ N., 5°17′ E.
Algeria
Synonym(s): *Dzioua*
Doubtful.
A dozen depressions with vertical sides, for which a meteoritic origin was suggested; this is, however, doubtful in view of the absence of any raised rim and the presence of gypsum in the tertiary substratum (Monod, 1965).

Gulf of St. Lawrence
Canada 47°6′ N., 63°3′ W. centre of area
Doubtful.
A meteoritic origin has been suggested for an area, 180 miles in diameter, bounded in part by the coastlines of New Brunswick and Nova Scotia, and a relation to tektite fields in the U.S.A. has also been put forward (literature; O'Connell, 1965; J.H. Freeberg, U.S. Geol. Surv. Bull., 1966, no.1220, p.42 and 1969, no.1320, p.19. Arguments in favour of an impact origin, J.B. Hartung, Abs. in Meteoritics, 1982, vol.17, p.228.

Gusev 48.5° N., 40.3° E.
Federated SSR, USSR, [Гусевская]
Doubtful.
An ellipsoidal basin, 3 km in diameter and 400 metres deep, with explosion breccias, is possibly of meteorite impact origin and contemporaneous with the Kamensk structure situated 1 km to the southwest, V.L. Masaitis, Sov. Geol., 1975, no.11, p.52. English translation, Int. Geol. Rev. 1976, vol.18, p. 1249.

Gwarkuh 28°30′ N., 60°40′ E.
Baluchistan, Iran
Discredited.
A supposed meteorite crater near Gwarkuh (Gauhar Kuh) is certainly not of meteoritic origin (L.J. Spencer, Geogr. Journ., London, 1933, vol.81, p.227; Monod, 1965, with additional literature).

Hagens Fjord 81°45′ N., 28°15′ W.
Greenland
Doubtful.
More than 50 circular areas of varying size observed from the air and partly explored on the ground may be of meteoritic origin (K. Ellitsgaard-Rasmussen, Medd. Dansk. Geol. Foren., 1954, vol.12, p.433).

Hartney 49°24′ N., 100°40′ W.
Manitoba, Canada
Doubtful.
A disturbed area approximately, 6 km in diameter for which an impact origin has been suggested, P.B. Robertson and R. A.F. Grieve, Journ. Roy. Astron. Soc. Canada, 1975, vol.69, p.1. Anomalous stratigraphy, T.B. Haites and H. van Hees, Journ. Alberta Soc. Petrol. Geologists, 1962, vol.10, p.511. See also H.B. Sawatzky, Bull. Amer. Assoc. Petrol. Geologists, 1975, vol.59, p.694 and Impact and Explosion Cratering, eds D.J. Roddy, R.O. Pepin and R.B. Merrill, Pergamon Press (New York), 1977, p.461.

Haughton Dome 75°22′ N., 89°40′ W.
Franklin district, Devon Island, North-west Territories, Canada
Doubtful.
A circular depression, 17 km in diameter, with raised rim of overall diameter 35 km and for which an impact origin has been suggested. See M.R. Dence, Proc. 24th Int. Geol. Cong. Section 15 1972, p.77. Shattercones and other evidence of shock metamorphism, P.B. Robertson and G.D. Mason, Nature, 1975, vol.255, p.393; see also P.B. Robertson and R. A.F. Grieve, abs. in Meteoritics, 1975, vol.10, p.480. Description; Miocene-Pliocene age, P.B. Robertson and R.A. F. Grieve, abs. Meteoritics, 1978, vol.13, p.615; T. Frisch and R. Thorsteinsson, Arctic, 1978, vol.31, p.108. Shock metamorphism in sillimanite, P.B. Robertson and A.G. Plant, Contr. Min. Petrol., 1981, vol.78, p.12.

Haviland 37°34′57″ N., 99°9′49″ W.
Kiowa County, Kansas, U.S.A.
1882, formed before
Synonym(s): *Brenham*
Authenticated.
The site of the fall of the Brenham pallasite has been excavated, and a large number of oxidized, limonitic fragments, many still containing unaltered metal and olivine, were found scattered throughout a funnel-shaped volume in the swampy depression (H.H. Nininger and J.D. Figgins, Proc. Colorado Mus. Nat. Hist., 1933, vol.12, no.3, p.9; H. H. Nininger, Pop. Astron., Northfield, Minnesota, 1938, vol. 46, p.110). It is not clear whether this represents a true explosion crater or is merely a large penetration hole in soft ground. See also O'Connell, 1965 and J.H. Freeberg, U.S. Geol. Surv. Bull., 1966, no.1220, p.43, for additional references. Location and co-ordinates, P.W. Hodge, Meteoritics, 1979, vol.14, p.233.
Specimen(s) : [1962,141] nickeliferous nodules

Hemauer Pulk 49°5′ N., 11°47′ E.

Bayern, Germany

Doubtful.

Fourteen small craters, the largest with diameter, 2 km may be of impact origin and related to Nördlinger Ries, E. Rutte, Geoforum, 1971, vol.7, p.84; see also E. Rutte, Oberrhein. geol. Abh., 1974, vol.23, p.66.

Hebron Crater v Labrador Crater.

Henbury 24°34′ S., 133°10′ E.

McDonnell Ranges, Northern Territory, Australia

Found 1931

Synonym(s): *Double Punchbowl*

Authenticated.

The group of well-defined craters (the largest oval, 220 by 120 yards, and 40-50 feet deep) at Henbury, described by A. R. Alderman (Min. Mag., 1932, vol.23, p.19; Rec. South Australian Mus., 1932, vol.4, p.561), are one of the best authenticated meteorite craters, and their shape and the condition of their walls clearly indicate an explosive origin. They are associated with many masses of meteoritic iron and with dark-brown nickeliferous silica glass, described by L.J. Spencer (with analyses by M.H. Hey, Min. Mag., 1933, vol. 23, p.387). See also L.J. Spencer, Nature, London, 1932, vol. 129, p.781; A. Troller, La Nature, Paris, 1932, no.2885, p.58; L.J. Spencer, Geogr. Journ. London, 1933, vol.81, p.227; A. R. Alderman, Ann. Rep. Smithsonian Inst., Washington, 1933 (for 1932), p.223, (descriptions of craters); E. Preuss, Die Naturwiss., 1934, vol.22, p.480, and Chem. Erde, 1935, vol.9, p.365 (spectrum analyses of the silica glass and sandstone); and P.S. Goel and T.P. Kohman, Science, 1962, vol.136, p.875 (terrestrial age from carbon isotope data).

Further evidence that these are true explosion craters is given by J.M. Rayner (Australian Journ. Sci., 1938, vol.1, p. 93; Rept. Australian New Zealand Assoc. Adv. Sci., 1939, vol.24, p.72). See also P.W. Hodge, Smithsonian Contribution to Astrophysics, 1965, vol.8, no.8, p.199, 16 photos. Numerous additional references J.H. Freeberg, U.S. Geol. Surv. Bull., 1966, no.1220, p.44 and 1969, no.1320, p. 19. Crater glass has fission-track age less than 10,000 years, D. Störtzer, abs. in Meteoritics, 1971, vol.6, p.319. See also main meteorite list. Metallic spherules in impactites, R.V. Gibbons et al., Geochim. Cosmochim. Acta, 1976, Suppl.7, p.863.

Specimen(s) : Iron, silica glass, and country rock (sandstone), see main meteorite list.

Hérault Craters 43°32′ N., 3°8′ E.

France

Synonym(s): *Cabrerolles, Faugeres, Le Clot, Montagne Noire*

Discredited.

Six craterlike depressions near the towns of Cabrerolles and Faugeres, ranging from 45 to 220 m. in diameter (the largest, Le Clot, having a raised rim) have been widely accepted as meteoritic in origin (literature in O'Connell, 1965), but this hypothesis is rejected by C.S. Beals (Meteoritics, 1964, vol.2, p.85; Monod, 1965, p.10). Additional references J.H. Freeberg, U.S. Geol. Surv. Bull., 1966, no.1220, p.45.

Hicks Dome

Harding County, Illinois, U.S.A.

Doubtful.

Mentioned by W.H. Bucher (Nature, 1963, vol.197, p.1241; further literature, see Monod, 1965).

Hofrat Aghreydh v Agheir.

Hofrat Aouelloul v Aouelloul.

Holleford Crater 44°28′ N., 76°38′ W.

Lanark County, Ontario, Canada

Authenticated.

A circular depression 1.5 miles in diameter in Precambrian or early Palaeozoic strata. Concentric gravitational anomalies and an underlying lens of breccia have been observed, and coesite is present (C.S. Beals et al., Sky and Telescope, 1956, vol.15, p.296; C.S. Beals, ibid., 1957, vol.16, p.2; Publ. Dominion Observ., Ottawa, 1960, vol.24, p.117; C.S. Beals et al., Current Science, Bangalore, 1960, vol.29, pp.205 and 249; T.E. Bunch, Science, 1963, vol.142, p.379; I. Halliday and A.A. Griffin, Meteoritics, 1964, vol.2, p.79; M.J.S. Innes, Journ. Geophys. Res., 1961, vol.66, p.2225).

Structure, D.P. Gold, abs., in Proc. North-east Meeting, Geol. Soc. Amer., 1968, p.29; sedimentary rocks, B.E. St. John, Can. Journ. Earth Sci., 1968, vol.5, p.935. Magnetic survey, J.F. Clark, Proc. Geol. Assoc. Canada, 1969, vol.20, p.24; and P. Andrieux and J.F. Clark, Can. Journ. Earth Sci., 1969, vol.6, p.1325. Additional references, J.H. Freeberg, U.S. Geol. Surv. Bull., 1966, no.1220, p.46, and 1969, no.1320, p.19. Age, 550 ± 100 m.y., P.B. Robertson and R.A.F. Grieve, Journ. Roy. Astron. Soc. Canada, 1975, vol. 69, p.1.

Holyrood Bay

Northeast Avalon Peninsula, Newfoundland, Canada

Synonym(s): *Conception Bay*

Doubtful.

An area of disturbed basement rocks with breccias, shattercones and other shock metamorphic features, is suggested as an eroded meteorite impact structure, W.v. Engelhardt and J. Walzebuck, abs. Meteoritics, 1978, vol.13, p.449. The occurrence of shattercones listed under Avalon Peninsula (*q . v .*) may be related to Holyrood Bay. For a description of the geology of the Conception Bay area of the Avalon Peninsula (including Holyrood Bay) with evidence supporting an origin by meteorite impact see W.v. Engelhardt, Naturwissenschaften, 1975, vol.62, p.234. Geophysical interpretation of the geology of the area, H.G. Miller, Can. J. Earth Sci., 1983, vol.20, p.1421.

Hoshikubo 26°16′ N., 127°47′ E.

Okinawa, Japan

Doubtful.

A crater 600 metres in diameter may be meteoritic; listed with no further details by J. Classen, Meteoritics, 1977, vol. 12, p.61.

Hoshinoko-Zan 35°44′ N., 133°14′ E.

Hiroshima, Japan

Doubtful.

A small crater may be meteoritic; listed without further details, J. Classen, Meteoritics, 1977, vol.12, p.61.

Howell Structure 35°15′ N., 86°35′ W.

Lincoln County, Tennessee, U.S.A.

Doubtful.

A circular disturbed area 1.5 miles in diameter is "tentatively considered as an example of the cryptovolcanic structures" (K.E. Born and C.W. Wilson, Jr., Journ. Geol., 1939, vol.47, p.371). See also O'Connell, 1965.

Hudson Bay Arc 57°54′ N., 80°2′ W. centre
Canada
Doubtful.
The east shore of Hudson Bay forms a circular arc of 138 miles radius, with all sediments dipping inwards. Regarded by C.S. Beals et al., Current Science, Bangalore, 1960, vol. 29, pp.205, 249 as of meteoritic origin.
Support both for and against an impact origin, C.S. Beals, in Science, History and Hudson Bay, 1968, vol.2, eds. C.S. Beals and D.A. Shenstone, Dept. Energy and Mines, Res., Ottawa, p.985. But compare R.A. Gibb, Earth planet. Sci. Lett., 1971, vol.10, p.365, who refutes an impact origin and suggests instead rifting of an Archaean protocontinent. A reinvestigation by R.S. Dietz and J.P. Barringer indicates that an impact origin is unlikely (Meteoritics, 1973, vol.8, p. 28).

Hungarian Plain 47° N., 21° E.
Hungary
Discredited.
J.Kaljuvee [= J.Kalkun] has suggested that the Hungarian plain represents a gigantic meteorite crater, rimmed by the Transylvanian Alps (Die Grossprobleme der Geologie, Tallinn, 1933). Meteoritic origin is discredited by F. Heide (Kleine Meteoritenkunde, Berlin, 1957).

Hunt's Hole v Afton craters.

Idritsa 56.3° N., 28.8° E.
Pskov Province, USSR
Doubtful.
A circular lake is of doubtful meteoritic origin, I.T. Zotkin and V.L. Tsvetkov, Astron Vestnik, 1970, vol.4, p.55.

Ile Rouleau 50°41′ N., 73°53′ W.
Mistassini Lake, Quebec, Canada
Synonym(s): *Mistassini Island*
Doubtful.
A circular structure 1 km in diameter, with evidence of brecciation and with shattercones may represent the central portion of a partly submerged impact structure; full description, J.-L. Caty et al., Can. Journ. Earth Sci., 1976, vol.13, p.824.

Il'inets v Il'intsy.

Il'intsy 49°8′ N., 29°11′ E.
Vinnitsa Oblast, Ukraine, USSR, [Ильинцы]
Synonym(s): *Il'inets*
Doubtful.
An origin by meteorite impact has been suggested for a lens 5 km in diameter, of breccias and suevites in the Ukrainian Shield; petrographic indications of shock metamorphism from borehole samples, A.A. Val'ter and V.A. Ryabenko, Geol. Journ., Acad. Sci. Ukrainian SSR., 1973, vol.33, no.6, p.142. Planar elements in biotite, E.P. Gurov, Zap. Vses. Min. Obshch., 1977, vol.106, p.715. Description V.L. Masaitis, Sov. Geol., 1975, no.11, p.52. English translation, Int. Geol. Rev. 1976, vol.18, p.1249.

Illumetsa v Ilumetsa Craters.

Ilumetsa Craters 58°0′ N., 27°14′ E.
Estonia, USSR
Synonym(s): *Illumetsa*
Authenticated.
Three small craters (two 80 m. and 50 m. in diameter, the third 19 m. by 28 m.) appear to be typical explosion craters, though no meteoritic material has been recovered (A. Aaloé, Meteopitika, 1960, vol.18, p.26).

Intaly 49° N., 69°55′ E.
Kazakhstan, USSR, [Ынталы]
Doubtful.
A depression, approximately 20 km in diameter, is possibly of meteoritic origin, B.S. Zeylik and E.Yu. Seytmuratova, Izvestia Akad. Nauk Kazakhstan S.S.R., Ser. Geol., 1975, vol.1, p.62.

Ishim v Tengiz.

Janisjärvi 61°58′ N., 30°55′ E.
Karelia, USSR
Synonym(s): *Yanis'yarvi*
Doubtful.
A structure 13-17 km in diameter is listed as a possible impact crater by M.R. Dence, Proc. 24th. Internat. geol. Cong., Section 15, 1972, p.77. Description of central impactites and breccias in middle of flooded basin in Proterozoic schists, V.L. Masaitis et al., Meteoritika, 1976, vol.35, p.103. Impactites 700 m.y. old; shattercones are present, V.L. Masaytis, Sov. Geol., 1975, no.11, p.52 (English translation: Int. Geol. Rev. 1976, vol.18, p.1249). Shock metamorphism of minerals in country rocks, V.J. Feldman et al., in Cosmic Mineralogy (Proc. 11th IMA Meeting, Novosibirsk), ed. A.V. Sidorenko et al., Acad. Sci. USSR., 1980, p.86.

Jeptha Knob 38°6′ N., 85°6′ W.
Shelby County, Kentucky, U.S.A.
Doubtful.
A nearly circular disturbed area 2 miles in diameter in Silurian strata is regarded by a few authors as meteoritic in origin, but by others as non-meteoritic (literature in O'Connell, 1965).
Petrological and geophysical data, and discussion of the origin, C.R. Seeger, Amer. Journ. Sci., 1968, vol.266, p.630. Additional references, J.H. Freeberg, U.S. Geol. Surv. Bull., 1966, no.1220, p.47, and 1969, no.1320, p.20.

Kaalijarv 58°24′ N., 22°40′ E.
Saaremaa (= Oesel), Estonian SSR, USSR
1928, Recognized
Synonym(s): *Esthonia, Oesel, Saarema Island Craters, Sall Craters*
Authenticated.
After much discussion as to its origin (I. Reinwaldt and A. Luha, Sitzungsber. Naturfors. Gesell. Univ. Tartu, 1928, vol. 35, p.30; E. Kraus et al., Gerlands Beiträge zur Geophysik, 1928, vol.20, p.312; A.G. Ingalls, Scient. American, New York, 1928, vol.139, p.45; P.N. Chirvinsky, Mém. Soc. Russe Min., 1931, ser.2, vol.60, p.135; L.J. Spencer, Geogr. Journ., London, 1933, vol.81, p.227; C. Fisher, Nat. Hist., Amer. Mus. Nat. Hist., 1936, vol.38, p.292; W. Kranz, Gerlands Beiträge Geophysik, 1937, vol.51, p.50), the discovery in 1937 of fragments of meteoritic iron affords definite evidence of the meteoritic origin of this crater, in agreement with the occurrence of tilted, brecciated and powdered rock (dolomite) in the crater walls (I.A. Reinwald, Natur und Volk, Frankfurt-a-M., 1938, vol.68, p.16; Publ. Geol. Inst.

Univ. Tartu, 1939, no.55; L.J. Spencer, Min. Mag., 1937, vol.25, p.75). Besides the main crater, which measures about 100 metres across and 15.5 metres deep and contains a lake, five smaller craters have been recognized.

General accounts of the craters, see also: E.L. Krinov, Izv. Acad. Nauk SSSR., Ser. Geogr. Geofiz., 1945, vol.9, no.4, p. 409; Priroda, 1960, vol.49, p.55; Amer. Journ. Sci., 1961, vol.259, p.430; I.A. Reinwald, Tartu Ulikooli juures oleva Loodusuurijate Seltsi Aruandad, 1939, vol.45, p.1; idem, Meteoritika, 1946, vol.3, p.46; A. Aaloé, Meteoritika, 1958, vol.16, p.108.

Shattercones in main crater, R.S. Dietz, abs. in Meteoritics, 1967, vol.3, p.108. Thermoluminesence studies, V.G. Maksenkov and A.A. Nikulova, Meteoritika, 1968, vol. 28, p.51. Additional references, J.H. Freeberg, U.S. Geol. Surv. Bull., 1966, no.1220, p.48.

Specimen(s) : Meteoritic iron and country rock (dolomite), see main meteorite list.

Kaibsko-Chuiskaya 46° N., 72°30′ E.
Kazakhstan, USSR, [Каибско-Чуйская]
Doubtful.
A large circular structure 550 km in total diameter, with central uplift region and for which an impact origin is suggested, B.S. Zeylikk and E.Yu. Seytmuratova, Izvestia Akad. Nauk Kazakhstan S.S.R., Ser. Geol., 1975, vol.1, p.62.

Kai-Imu-Hoku 20°54′ N., 156°54′ W.
Lanai Island, Hawaii, U.S.A.
Discredited.
Possibly the impact pit of another stone of the Honolulu fall (q.v.); see K.P. Emory, Bull. Bernice Bishop Mus., 1924, no. 12, p.29; J.D. Buddhue, Pop. Astron., 1947, vol.55, p.553. An aerial and land survey revealed no evidence of a crater, B. Barringer, Meteoritics, 1968, vol.4, p.57.

Kakiattukallak Lake 57°42′ N., 71°40′ W.
Quebec, Canada
Doubtful.
Listed as possible impact structure, 6 km in diameter, P.B. Robertson and R.A.F. Grieve, Journ. Roy. Astron. Soc. Canada, 1975, vol.69, p.1.

Kalkkop 32°30′ S., 24°35′ E.
South Africa
Doubtful.
Meteoritic origin has been suggested for a circular feature, 640 metres in diameter; literature, J.H. Freeberg, U.S. Geol. Surv. Bull., 1966, no.1220, p.50. Listed as a " possible" impact crater, M.R. Dence, Proc. 24th Internat. Geol. Cong., Section 15, 1972, p.77.

Kaluga 49° N., 24°22′ E.
Ukraine, USSR, [Калужская]
Synonym(s): *Kalushkaya*
Doubtful.
A circular structure proved by drilling, about 15 km in diameter in crystalline basement rocks buried below Devonian and Carboniferous strata is possibly of meteorite impact origin; breccias and diaplectic glasses as evidence of shock metamorphism, V.L. Masaitis, Sov. Geol., 1975, no.11, p.52 (English Trans.: Int. Geol. Rev., 1976, vol.18, p.1249). See also, V.V. Fedynski and L.P. Chrianin, Astron. Vestnik, 1976, vol.10, p.81.

Kalushkaya v Kaluga.

Kamensk 48.4° N., 40.2° E.
Federated SSR, USSR, [Каменская]
Doubtful.
A circular structure, 25 km in diameter, with central uplift 400 metres, in Carboniferous- Triassic rocks buried below Danian-Paleocene rocks is possibly of meteorite impact origin, V.L. Masaitis, Sov. Geol., 1975, no.11, p.52 (English translation: Int. Geol. Rev. 1976, vol.18, p.1249).

Kara 69° N., 64.3° E. approx.
Pay-Khoy, USSR, [Карская]
Synonym(s): *Karskay*
Doubtful.
A circular basin, 50 km in diameter in folded Palaeozoic rocks, with basic intrusives, is almost completely covered by Quaternary sediments. Breccias and suevites are present, the former containing late Cretaceous rocks not represented elsewhere in the area. A central uplift of breccia is 7-8 km across, V.L. Masaitis, Sov. Geol. 1975, no.11, p.52 and M.A. Maslov, Meteoritika, 1977, vol.36, p.123. Nickel and cobalt enrichment of country rocks indicates an iron meteorite impacting body, V.I. Fel'dman, abs. in Lunar and Planetary Science 10, 1979, p.382. Coesite in country rocks, S.A. Vishnevskii, Dokl. Earth Sci. section, 1977, vol.232, p.154.

Karachev 53.2° N., 35.1° E.
Orlovsk Province, USSR
Doubtful.
A round lake is possibly of meteoritic origin, I.T. Zotkin and V.L. Tsvetkov, Astron. Vestnik, 1970, vol.4, p.55.

Karaganda 49.8° N., 73.2° E.
Karaganda Province, USSR
A small depression, 80 metres in diameter and situated on a small hill may be meteoritic, I.T. Zotkin and V.L. Tsvetkov, Astron. Vestnik, 1970, vol.4, p.55.

Karla 55.5° N., 48.5° E.
Tatar SSR, USSR
Doubtful.
A circular structure about 10 km in diameter in Carboniferous rocks, with explosion breccia, about 500 metres thick, which includes Miocene fragments. One area of breccia extends 9 km northeast from the main field. A partial cover of Pliocene-Quaternary sediments indicates a late Miocene to early Pliocene age, V.L. Masaitis et al., Dokl. Akad. Nauk SSSR, 1976, vol.230, p.174.

Karskay v Kara.

Kasaba 53.8° N., 61.4° E.
Chelyabinsk Province, USSR
Doubtful.
Three lakes about 1 km in diameter are of doubtful meteoritic origin, I.T. Zotkin and V.L. Tsvetkov, Astron. Vestnik, 1970, vol.4, p.55.

Kasarik 49.9° N., 38.3° E.
Luganskoe Province, USSR
Doubtful.
A depression 700 metres in diameter and 20 metres deep, is of doubtful meteoritic origin, I.T. Zotkin and V.L. Tsvetkov, Astron. Vestnik, 1970, vol.4, p.55.

Katen 47° N., 136.4° E.
Khabarovsk, USSR
Doubtful.
A circular lake 40 metres in diameter, is possibly of meteoritic origin, I.T. Zotkin and V.L. Tsvetkov, Astron. Vestnik, 1970, vol.4, p.55.

Keeley Lake 54°54′ N., 108°8′ W.
Saskatchewan, Canada
Doubtful.
A circular lake 8 miles in diameter "with features suggesting possible ancient crater"; references in O'Connell, 1965. "Doubtful" according to C.S. Beals (in Monod, 1965).

Kentland Disturbance 40°45′ N., 87°24′ W.
Newton County, Indiana, U.S.A.
Doubtful.
A circular disturbed area 1100 yards in diameter with well-defined magnetic anomaly, brecciation, coesite, and upwardly-directed shattercones (R.S. Dietz, Science, 1947, vol.105, p.42; A.J. Cohen, T.E. Bunch and A.M. Reid, Science, 1961, vol.134, p.1624; see also references cited by O'Connell, 1965 and J.H. Freeberg, U.S. Geol. Surv. Bull., 1966, no.1220, p.50, and 1969, no.1320, p.20). Structural study, R.T. Laney and W.R. Van Schmus, Geochim. Cosmochim. Acta, 1978, Suppl.10, vol.2, p.2609.
Specimen(s) : [1973,M.28-31] breccias and limestone with shattercones

Khatanga fall v Tunguska river.

Kilbourne Hole v Afton craters.

Kilmichael Structure 33°30′ N., 89°33′ W.
Montgomery County, Mississippi, U.S.A.
Doubtful.
A roughly circular "cryptoexplosion structure" is assigned meteoritic origin by M.D. Butler (Journ. Mississippi Acad. Sci., 1962, vol.8, p.51; see also O'Connell, 1965). Additional rreferences, J.H. Freeberg, U.S. Geol. Surv. Bull., 1966, no. 1220, p.51, and 1969,no.1320, p.20.

Kobrin 52.2° N., 24.2° E.
Brest Province, USSR
Doubtful.
A lake 250 metres in diameter is of doubtful meteoritic origin, I.T. Zotkin and V.L. Tsvetkov, Astron. Vestnik, 1970, vol.4, p.55.

Köfels 47° N., 11° E.
Oetztal, Austria
Doubtful.
It has been suggested that the widening of the Oetztal near Köfels, where scattered blocks and pumiceous material occur in shattered gneiss, represents a meteorite crater 3-4 km across (F.E. Suess, Neues Jahrb. Min., 1936, Abt. A., vol.72, p.98; O. Stutzer, Zeits. Deut. Geol. Gesell., 1936, vol.88, p. 525), but the "pumice" contains only 0.001% of nickel and a volcanic origin is not excluded (W. Schmidt, Zentr. Min., Abt. A, 1937, p.221; W. Hammer, Verh. Geol. Bundesanst. Wien, 1937, p.195; O. Hackl, ibid., p.269; F. Heide, Naturwiss., 1938, vol.26, p.495; W. Kranz, Neues Jahrb. Min., 1938, Abt. B, Beil.-Bd. 80, p.113. See also F.E. Suess, Verhandl. III Internat. Quatär-Konferenz, Wien, 1938, p. 167; Suess holds this feature to be due to a meteorite in late glacial times. See also references in O'Connell, 1965. C.S. Beals (in Monod, 1965) considers a meteoritic origin "doubtful".
A reinvestigation of "pumiceous" material, G. Kurat and W. Richter, abs. in Meteoritics, 1969, vol.4, p.192. Fission-track age of crater glass 8000‡6000 years, D. Störzer et al., abs. in Meteoritics, 1971, vol.6, p.319, also Earth planet. Sci. Lett., 1971, vol.12, p.238. Additional references, J.H. Freeberg, U.S. Geol. Surv. Bull., 1966, no.1220, p.51, and 1969, no.1320, p.20. " Pumice" generated by a landslide, Th. Erismann et al., Tschermaks Min. Petr. Mitt., 1977, vol.24, p.67.

Kogram
Aldanian Shield, USSR, [Kограм]
Doubtful.
An area of eroded disturbed rocks is possibly of meteorite impact origin, E.P. Gurov et al., Dokl. Akad. Nauk SSSR, 1983, vol.270, p.172.

Konder 57.5° N., 134.8° E.
Khabarovsk Territory, USSR
Doubtful.
A circular mountain with a ring shaped ridge, approximately 9 km in diameter; has been investigated and a meteoritic origin proposed, I.T. Zotkin and V.L. Tsvetkov, Astron. Vestnik, 1970, vol.4, p.55.

Konus 48°33′ N., 76°36′ E.
Kazakhstan, USSR, [Конус]
Doubtful.
Five craters up to 250 metres in diameter are associated with breccias and glasses, B.S. Zeylik and E. Yu. Seytmuratova, Izvestia Akad. Nauk Kazakhstan S.S.R., Ser. Geol., 1975, vol.1, p.62.

Kozhva 65° N., 56.5° E.
Komi ASSR, USSR
Doubtful.
An oval lake 100 metres in diameter around which trees have been overturned; of doubtful meteoritic origin, I.T. Zotkin and V.L. Tsvetkov, Astron. Vestnik, 1970, vol.4, p. 55.

Kursk 51°45′ N., 36°14′ E. approx.
USSR
Doubtful.
A circular structure, 5 km in diameter, with central uplift, explosion breccias and diaplectic glass of Triassic age is of suggested impact origin, V.L. Masaitis et al., Dokl. Akad. Nauk SSSR, Ser. Geol., 1978, vol.240, no.5, p.1191 (English translation: p.91).

Kyardla 59°0′ N., 22°42′ E.
Estonian SSR, USSR
Synonym(s): *Kärdla*
Doubtful.
A structure 4 km in diameter, with explosion breccia and diaplectic glass is listed by V.L. Masaitis et al., Dokl. Akad. Nauk SSSR, Ser. Geol., 1978, vol.240, no.5, p.1191 (English translation: p.91). Co-ordinates are of "Kärla" in the Times Atlas.

Labrador Crater 58°2′ N., 64°2′ W.

Labrador, Canada

Synonym(s): *Hebron Crater, Merewether Crater, Wetherbee Crater*

Doubtful.

A slightly oval crater 700 feet . across is widely regarded as probably meteoritic in origin (V.B. Meen, Proc. Geol. Assoc. Canada, 1957, vol.9, p.49; see also O'Connell, 1965). Additional references, J.H. Freeberg, U.S. Geol. Surv. Bull., 1966, no.1220, p.57, and 1969, no.1320, p.24.

Labynkyr 62.5° N., 143° E.

Yakut ASSR, USSR

Doubtful.

A large circular depression 60 km in diameter and containing a lake, is possibly an astrobleme, I.T. Zotkin and V.L. Tsvetkov, Astron. Vestnik, 1970, vol.4, p.55.

Lac Bouchet 44°56′ N., 3°47′ E.

Haute Loire, France

Doubtful.

"Il s'agit d'un lac de cratére" (Monod, 1965, p.34). Co-ordinates, J. Classen, Meteoritics, 1977, vol.12, p.72.

Lac Chatelain 60°15′ N., 74°36′ W.

Quebec, Canada

Doubtful.

A shallow depression 180 feet in diameter with a raised rim composed of angular blocks; may be a collapsed pingoe, K.L. Currie, Annals New York Acad. Sci., 1965, vol.123 p.915.

Lac Couture 60°8′ N., 75°20′ W.

Quebec, Canada

Doubtful.

A circular lake 16 km in diameter in the Ungava peninsula, observed on aerial survey photographs (references in O'Connell, 1965). Additional references, J.H. Freeberg, U.S. Geol. Surv. Bull., 1966, no.1220, p.53, and 1969, no.1320, p. 20. Ar-Ar age of impact melt rocks (410-430 m.y.), R.J. Bottomley et al., abs. in Meteoritics, 1978, vol.13, p.395.

Lac La Moinerie 57°26′ N., 66°36′ W.

Quebec, Canada

Doubtful.

A possible meteoritic origin has been proposed for this circular lake 8 km in diameter, W. von Engelhardt, Die Naturwissenschaften, 1974, vol.61, p.413; Ar-Ar age of impact melt rocks (380-410 m.y.), R.J. Bottomley et al., abs. in Meteoritics, 1978, vol.13, p.395.

Lac Menihek v Menihek Lakes.

Lac Roz 54°49′ N., 72°55′ W.

Quebec, Canada

Doubtful.

Listed as possible impact structure, P.B. Robertson and R.A. F. Grieve, Journ. Roy. Astron. Soc. Canada, 1975, vol.69, p. 1.

Lac St. Jean v Lake St. John.

Lago di Tremorgio 46°28′ N., 8°43′ E.

Canton Ticino, Switzerland

Doubtful.

A circular depression 1360 metres in diameter containing a lake is possibly of meteorite impact origin; shattercones, deformation and evidence of cataclasis, K. Bächtiger, Experientia, 1976, vol.32, p.1102. See also, K. Bächtiger, Schweiz. Min. Petr. Mitt., 1976, vol.56, p.545; K. Bächtiger, Meteoritics, 1977, vol.12, p.169. However, F. Bianconi (Schweiz. Min. Petr. Mitt., 1977, vol.57, p.435) argues that the structure was formed mainly by glacial action with some karst effects.

Laguna Guatavita 4°59′ N., 73°47′ W.

Cordillera Orientale, Colombia

Discredited.

A circular crater 700 metres in diameter and 125 metres deep, containing a lake, is possibly of meteoritic origin, H. Raasveldt, Misc. Publ. Inst. Geol. Nacional, Bogotá, 1954. R.S. Dietz and J.F. McHone suggest that it is probably a salt collapse structure, Meteoritics, 1972, vol.7, p.303, and ibid., 1973, vol.8, p.27.

Lake Bosumtwi 6°32′ N., 1°24′ W.

Ashanti, Ghana

Doubtful.

A meteoritic origin has been suggested for the crater Lake Bosumtwi, Ashanti, 6.5 miles across, 1150 feet deep (M. Maclaren, Geogr. Journ., London, 1931, vol.78, p.270), but it is almost certainly of volcanic origin (L.J. Spencer, ibid., 1933, vol.81, p.227; N.R. Junner, Rep. Geol. Surv. Gold Coast, 1933, p.4 and Bull. Geol. Surv. Gold Coast, 1937, no. 8, p.5; H.P.T. Rohleder, Geogr. Journ. London, 1936, vol.87, p.51. Coesite has been found (J. Littler, Spec. Paper Geol. Soc. Amer., 1961, no.68, p.218), and there is a very considerable literature both supporting and rejecting meteoritic origin (see references in O'Connell, 1965, and Monod, 1965, and in particular S.O. Bampo, Nature, 1963, vol.198, p.1150, and A.F.G. Smit, ibid., 1964, vol.203, p. 179). A connection with the Ivory Coast tektites has been suggested.

Petrographic studies of shocked rocks and minerals, E.C.T. Chao, N. Jb. Miner. Abh., 1968, vol.108, p.209. Fission-track age 1.03 ± 0.02 m.y., D. Störzer, abs. in Meteoritics, 1971, vol. 6, p.319. Additional references J.H. Freeberg, U.S. Geol. Surv. Bull., 1966, no.1220, p.53, and 1969, no.1320, p.21. Recent description of the crater, W.B. Jones et al., Geol. Soc. Amer. Bull., 1981, Part 1, vol.92, p.342. Chemical evidence for an iron meteorite impact, H. Palme et al., Geochim. Cosmochim. Acta, 1978, vol.42, p.313.

Specimen(s) : [1933,250]; [1934,1017] volcanic agglomerates; [1934,1018] volcanic agglomerates; [1965,184] iron oxides containing traces of Ni

Lake Constance 47°35′ N., 9°25′ E.

Switzerland

Doubtful.

An origin by impact synchronous with Nördlinger Ries is suggested, F. Hoffmann, Ecologae geol. Helv., 1973, vol.66, p.83.

Lake Dellen 61°50′ N., 16°45′ E.

Sweden

Doubtful.

A roughly circular disturbed area 17 km in diameter is regarded by K. Fredriksson and F.E. Wickman (Svensk Naturvetenskap., 1963, p.121) as probably meteoritic.

Lake Elgytkhyn 67°29′ N., 172°4′ E.

Northeastern Siberia, USSR

Synonym(s): *Elgegytgyn, Elgygytgyn*

Doubtful.

A large circular crater 18 km in diameter containing a lake for which an origin by meteorite impact has been proposed

(I.A. Nekrasov and P.A. Raudonis, Priroda, 1963, p.102; O'Connell, 1965). Recent description, R.S. Dietz and J.F. McHone, Geology, 1976, vol.4, p.391; but compare R.J. Pike, ibid., 1977, vol.5, p.262 and R.S. Dietz and J.F. McHone, ibid., p.263. Possibly the source of the australasian tektites, R.S. Dietz, Meteoritics, 1977, vol.12, p.145. Further description and Russian literature, E.P. Gurov, et al., Dokl. Akad. Nauk SSSR, 1978, vol.240, p.1407 (English translation, p.103). Fission track dating of crater glasses, 4. 24 m.y., indicates that it was not the source of the Australasian or North American tektite fields, D. Störzer and G.A. Wagner, abs. in Meteoritics, 1979, vol.14, p.541. Impactites and glass bombs from the crater, E.P. Gurov et al., Dokl. Akad. Nauk SSSR, ser. geol., 1980, no.1, p.54.

Lake Hamilton Craters v Eyre Peninsula Craters.

Lake Humeln 57°24′ N., 16°12′ E.

Sweden
Synonym(s): *Lake Hummeln*
Doubtful.
A circular depression, 1 km in diameter and up to 60 metres deep in the southern part of an otherwise shallow lake, is regarded by K. Fredriksson and F.E. Wickman (Svensk Naturvetenskap., 1963, p.121) as probably meteoritic. "Only general geological arguments favour a meteoritic impact origin for the crater", N.B. Svensson, abs. in Meteoritics, 1969, vol.4, p.208. Additional references, J.H. Freeberg, U.S. Geol. Surv. Bull., 1969, no.1320, p.22.

Lake Hummeln v Lake Humeln.

Lake Lappajärvi 63°10′ N., 23°40′ E.

Finland
Doubtful.
Both terrestrial and meteoritic origins have been suggested for this depression containing a lake; deformation in quartz grains from country rocks is taken as supporting impact, N. B. Svensson, Nature, 1968, vol.217, p.438; but compare J. McCall, ibid., vol.218, p.1152. Fe-Ni metal in impact melt rocks, S. Fregerslev and H. Carstens, Contr. Min. Petr., 1976, vol.55, p.255; carbonaceous chondrite proposed as the impacting projectile, E. Göbel, et al., Z. Naturforsch., 1980, vol.35a, p.197. Crystallization of impact melt rocks, U. Maerz and D. Stöffler, abs. in Meteoritics, 1979, vol.14, p. 480. Isotope, major and trace element chemistry of impact melt, W.V. Reimold and D. Stöffler, abs. in Meteoritics, 1979, vol.14, p.526. See also, M. Lehtinen, Bull. Geol. Surv. Finland, 1976, vol.282, p.1, and H. Palme, Geochim. Cosmochim Acta, Suppl.14, 1980, p.481. Petrography, Rb-Sr, major and trace element geochemistry of impact melt (age 77 m.y.) and basement rocks, W.U. Reimold, Geochim. Cosmochim. Acta, 1982, vol.46, p.1203.

Lake Michikamau 54°34′ N., 64°27′ W.

Labrador, Canada
Doubtful.
A circular crater 3.5 miles in diameter, is mentioned in several lists of possible meteorite craters (see O'Connell, 1965, and P.B. Robertson and R.A.F. Grieve, Journ. Roy. Astron. Soc. Canada, 1975, vol.69, p.1.

Lake Leveque v Mecatina Crater.

Lake Mien 56°25′ N., 14°55′ E.

Sweden
Doubtful.
A lake 6 km in diameter is perhaps a fossil meteorite crater (K. Fredriksson and F.E. Wickman, Svensk Naturvetenskap., 1963, p.121). Coesite is present (N.B. Svensson and F.E. Wickman, Nature, 1965, vol.205, p.1202). Fission-track age of crater glasses, 92 ± 6m.y., D. Störzer, abs. in Meteoritics, 1971, vol.6, p.319. Additional references, J.H. Freeberg, U.S. Geol. Surv. Bull., 1966, no.1220, p.55, and 1969, no.1320, p. 22. K-Ar dating and Sr-isotopic composition of rhyolitic rocks, E. Welin, Geol. Fören. Förh., 1975, vol.97, p.307. Ar-Ar age (120 m.y.) of melt rocks, R.J. Bottomley et al., abs. in Meteoritics, 1977, vol.12, p.182 and Contr. Mineral. Petrol., 1978, vol.68, p.79. Nickel, Co and Cr enrichment of impact melts indicates a stony meteorite impacting body, H. Palme, et al., abs. in Lunar and Planetary Science 11, 1980, p.848.

Lake Sääksjärvi

Finland
Doubtful.
A basin of disturbed rocks (containing a lake) for which an origin by meteorite impact has been suggested. Description and chemical composition of shock metamorphic rocks, H. Papunen, Bull. Geol. Soc. Finland, 1973, no.45, p.29. Trace element studies of the country rocks indicate a stony-iron impactor, H. Palme et al., abs. In Lunar and Planetary Science XI, Lunar and Planetary Inst., Houston, 1980, p.848. See also H. Palme, Geochim. Cosmochim. Acta, 1980, Suppl. 14, p.481.

Lake St. John 48°37′ N., 72° W.

Quebec, Canada
Synonym(s): *Lac St. Jean*
Discredited.
Geophysical surveys of this circular structure indicate that it is not meteoritic, M.R. Dence et al., in Shock Metamorphism of Natural Materials, eds. B.M. French and N.M. Short, Mono, Baltimore, 1968, p.339. Compare P.B. Robertson and R.A.F. Grieve, Journ. Roy. Astron. Soc. Canada, 1975, vol.69, p.1. Palaeomagnetism of anorthosites and related rocks in the area, K.L. Buchan et al., Can. J. Earth Sci., 1983, vol.20, p.246.

Lake St. Martin 51°43′ N., 98°30′ W.

Manitoba, Canada
Doubtful.
A broad, shallow depression with central uplift 4 miles in diameter is possibly an impact structure of Permian or Triassic age, K.L. Currie, Nature, 1970, vol.226, p.839, and Geol. Surv. Canada, Paper 70-1, 1970, Part A, p.111. Age, 225Ŧ40m.y., P.B. Robertson and R.A.F. Grieve, Journ. Roy. Astron. Soc. Canada, 1975, vol.69, p.1. Petrology of impactites, C.H. Simonds and P.E. McGee, Geochim. Cosmochim. Acta, 1979, Suppl.11, p.2493.

Lake Siljan v Siljan Ring.

Lake Wanapitei 46°45′ N., 80°44′ W.

Ontario, Canada
Doubtful.
The lake is situated approximately 40 km north-east of Sudbury; the northern part of the lake is roughly circular and 8.6 km in diameter and may be of impact origin, M.R. Dence and J. Popelar, Geol. Assoc. Canada, 1972, Spec. Paper 10, p.117. Electromagnetic survey, J.F. Clark, abs. in Meteoritics, 1973, vol.8, p.24. Occurrence of coesite, M.R.

Dence et al., Earth planet. Sci. Lett., 1974, vol.22, p.118. Age, 37 ±2m.y., P.B. Robertson and R.A.F. Grieve, Journ. Roy. Astron. Soc. Canada, 1975, vol.69, p.1. Rb and Sr isotopic composition of rocks and glasses; K-Ar age 37 m.y., S.R. Winzer et al., Geochimica Cosmochim. Acta, 1976, vol. 40, p.51.
Specimen(s) : [1978,M.27], 105g. of breccia

La Malbaie Structure v Charlevoix Structure.

La Sauvetat 44°52′ N., 1°31′ E.
Puy-de-Dome, France
Doubtful.
Mentioned as a possible meteorite crater by R. Gallant, Bombarded Earth, London, 1964; Monod, 1965. Coordinates, J. Classen, Meteoritics, 1977, vol.12, p.72.

Le Clot v Hérault Craters.

Lepsinsk 45.5° N., 80.7° E.
Alma-Ata Province, Kazakhstan, USSR
Doubtful.
A crater 550 metres in diameter for which an impact origin is possible, I.T. Zotkin and V.L. Tsvetkov, Astron. Vestnik, 1970, vol.4, p.55.

Lesser Antilles
West Indies
Doubtful.
The arc of the Lesser Antilles is perhaps part of a meteorite crater 950 km in diameter, R. Gallant, Bombarded Earth, London, 1964; Monod, 1965.

Lianozovo 55.9° N., 37.6° E.
Moscow Province, USSR
Doubtful.
A "conical" depression 50 metres in diameter is of doubtful impact origin, I.T. Zotkin and V.L. Tsvetkov, Astron. Vestnik, 1970, vol.4, p.55.

Libyan Desert Craters 22°18′ N., 25°30′ E.
Egypt
Synonym(s): *Clayton's Craters*
Discredited.
A group of seven craters, 0.25-0.5 miles in diameter, are usually assumed to be of volcanic origin, but have been held to be meteoritic and to be associated with the Libyan glass (q.v.). See references in O'Connell, 1965, and in Monod, 1965.

Breakdown of zircon to baddeleyite observed in Libyan Desert Glass is evidence of impact origin, B. Kleinmann, Earth planet. Sci. Lett., 1968, vol.5, p.497. Fission-track age 28.5 ± 2.3m.y., D. Störzer, abs. in Meteoritics, 1971, vol.6, p. 319. Gas content of bubbles in the glasses, O. Müller and W. Gentner, Earth planet. Sci. Lett., 1968, vol.4, p.406.

Limestone Mountain Disturbance
Houghton County, Michigan, U.S.A.
Doubtful.
Mentioned by R. Gallant (Bombarded Earth, London, 1964; Monod, 1965) as possibly meteoritic.

Liverpool Crater 12°24′ S., 134°3′ E.
Northern Territory, Australia
Doubtful.
A circular structure, 1.6 km in diameter and with raised rim, may be of impact origin; rocks show microscopic evidence of shock metamorphism, R. Brett et al., Meteoritics, 1970, vol. 5, p.184.

Logoysk 54°8′ N., 27°42′ E.
Belorussian SSR, USSR
Doubtful.
A structure 17 km in diameter, with explosion breccias, impactites and diaplectic glass, is listed, V.L. Masaitis et al., Dokl. Akad. Nauk SSSR, Ser. Geol. 1978, vol.240, no.5, p. 1191 (English translation, p.91). Coesite and stishovite from the crater, E.P. Gurov et al., Dokl. Akad. Nauk SSSR, 1980, tom 24, no.2, p.168.

Lonar Lake 19°59′ N., 76°34′ E.
Maharashtra, India
Doubtful.
A circular crater, 2000 yards in diameter with underlying breccia lens and raised rim, unrelated to the surrounding geology, is probably of meteoritic origin (see literature in O'Connell, 1965, and Monod, 1965). Discovery of impact glasses, V.K. Nayak, Earth planet. Sci. Lett., 1972, vol.14, p. 1. Results of coring and trenching in the crater, K. Fredriksson et al., abs. in Meteoritics, 1973, vol.8, p.34 and ibid., p.35. Additional references, J.H. Freeberg, U.S. Geol. Surv. Bull., 1966, no.1220, p.55, and 1969, no.1320, p.23.

For recent descriptions see K. Fredriksson et al., Science, 1973, vol.180, p.862; K. Fredriksson et al., Smithsonian Contr. Earth Sci., 1979, no.22, p.1. Magnetic studies of country rocks, S.M. Cisowski, abs. in Meteoritics, 1975, vol. 10, p.383. Deposition of ejecta, D.J. Milton, et al., abs, in Meteoritics, 1975, vol.10, p.456; chemical study of impact glass and basalt, W.B. Stroube, Jr., and W.D. Ehmann, abs. in ibid., 1976, vol.11, p.371; D. Milton and A. Dube, abs. ibid., 1977, vol.12, p.311. Experimental analogues to shocked basalt, S.W. Kieffer et al., Geochim. Cosmochim. Acta, 1976, Suppl.7, p.1391. Chemical study of impact glasses, W. B. Stroube et al., Meteoritics, 1978, vol.13, p.201. Meteoritic contamination of crater glasses, J.W. Morgan, Geochim. Cosmochim. Acta, 1978, Suppl.10, p.2713.

Macamic Lake 48°52′ N., 79°1′ W.
Quebec, Canada
Synonym(s): *Makamik Lake*
Discredited.
A circular area 1 mile in diameter, with low rim (C.S. Beals et al., Journ. Roy. Astron. Soc. Canada, 1956, vol.50, pp. 203, 250); "gravitational observations discount meteoritic origin" (O'Connell, 1965).

Madera Mountain v Sierra Madera.

Makamik Lake v Macamic Lake.

Malha 15°6′ N., 26°15′ E.
Sudan
Discredited.
H.H. Nininger (Pop. Astron., Northfield, Minnesota, 1943, vol.51, p.96) mentions a feature shown on aerial photographs as a possible meteorite crater, but L.J. Spencer points out that this is the known volcanic explosion crater of Malha (Sudan Geogr. Journ., 1933, vol.82, p.116; Quart. Journ. Geol. Soc., 1935, vol.91, p.360). Additional references, J.H. Freeberg, U.S. Geol. Surv. Bull., 1966, no.1220, p.56.

Manicouagan Lake 51°23′ N., 68°42′ W.

Quebec, Canada

Doubtful.

A circular domed area, 40 miles in diameter for which meteoritic origin has been suggested. Shattercones, R.S. Dietz, Meteoritics, 1966, vol.3, p.27. Chemical argument against impact origin, K.L. Currie and M. Shafiqullah, Nature, 1968, vol.218, p.457. Refraction of shock wave, D. W. Roy, abs. in Meteoritics, 1969, vol.4, p.292. Study of anorthosites, U. Dworak, Contr. Mineral. Petrol., 1969, vol. 24, p.306. Shocked scapolite, S.H. Wolfe and F. Hörz, abs. in Meteoritics, 1969, vol.4, p.299 and Amer. Mineral., 1970, vol.55, p.1313. Fission-track age of crater glass, 200±30m.y., D. Störzer, abs. in Meteoritics, 1971, vol.6, p.319; K-Ar ages, S.H. Wolfe, Journ. Geophys. Res., 1971, vol.76, p.5424. Geology and petrology, K.L. Currie, Geol. Surv. Canada Bull., 1972, no.198, p.130, but compare J.G. Murtaugh, Proc. 24th Internat. Geol. cong., 1972, Section 15, p.133. Additional references, O'Connell, 1965; Monod, 1965; J.H. Freeberg, U.S. Geol. Surv. Bull., 1966, no.1220, p.56, and 1969, no.1320, p.23.

Petrology, structure and origin, R.J. Floran et al., Geophys. Res. Lett., 1976, vol.3, p.49; crater morphology, R. J. Floran and M.R. Dence, Geochim. Cosmochim. Acta, Suppl.7, 1976, p.2845; M.R. Dence, NASA Spec. Publ., 1977, vol.380, p.175. See also, R.J. Floran et al., J. Geophys. Res., 1978, vol.83, p.2737; R.A.F. Grieve and R.J. Floran, ibid., p.2761; Rb-Sr isochron age (214±5 m.y.), B.-M. Jahn et al., ibid., p.2799; petrogenesis of melt rocks, C.H. Simonds et al., ibid., p.2773; thermal history, P.I.K. Onorato et al., ibid., p.2789; gravity study, J.F. Sweeny, ibid., p.2809; central magnetic anomaly, R.L. Coles and J.F. Clark, ibid., p.2805. An alternative model for the impact structure, D.L. Orphal and P.H. Schultz, Geochim. Cosmochim. Acta, 1978, Suppl. 10, vol.2, p.2695. See also, D.L. Orphal and P.H. Schultz, abs. in Meteoritics, 1978, vol.13, p.591; R.A.F. Grieve and J. W. Head, abs. in Meteoritics, 1981, vol.16, p.320.

Physical properties, crystallization and structural and genetic implications of diaplectic labradoritic glass in country rocks, J. Arndt et al., Physics and Chemistry of Minerals, 1982, vol.8, p.230. Analysis of the original dimensions and form of the crater, R.A.F. Grieve and J.W. Head III, J. Geophys. Res., 1983, vol.88 Suppl., p.A807.

Manson structure 42°35′ N., 94°31′ W.

Calhoun County, Iowa, U.S.A.

Discredited.

A large circular disturbed area concealed by glacial drift (W. H. Bucher, Amer. Journ. Sci., 1963, vol.261, p.597; R.S. Dietz, ibid., p.650). Additional references, J.H. Freeberg, U. S. Geol. Surv. Bull., 1966, no.1220, p.57.

Matam 15°40′ N., 13°18′ W. Approx.

Senegal

Doubtful.

Several circular structures have been observed on aerial photographs (Monod, 1965). Co-ordinates, R.W. Barringer, Meteoritics, 1967, vol.3, p.151.

Mazoula 28°24′ N., 7°19′ E.

Algeria

Discredited.

A circular multi-ring feature, 800 metres in diameter located in Cretaceaous rocks about 102 km ENE of Zaouia El Khala (Fort Flatters) has been investigated, and discredited as a possible meteorite crater; the structure may be a fossil reef or salt diapir, P. Lambert et al., Meteoritics, 1981, vol.16, p. 203.

McIntosh Bay 52°35′ N., 94°5′ W.

Ontario, Canada

Doubtful.

Listed as possible impact structure, P.B. Robertson and R.A. F. Grieve, Journ. Roy. Astron. Soc. Canada, 1975, vol.69, p. 1.

Mecatina Crater 50°50′ N., 59°22′ W.

Quebec, Canada

Synonym(s): *Lake Leveque*

Doubtful.

A crater-like feature, 2 miles in diameter with Lake Leveque in the centre, is a " possible ancient crater filled with sediments later transformed into gneiss" (C.S. Beals et al., Current Science, Bangalore, 1960, vol.29, pp.205 and 249; see also O'Connell, 1965);R. Gallant (Bombarded Earth, London, 1964), frontispiece. Crater sedimentation, C.S. Beals and A. Hitchin, Pub. Dominion Obs., 1971, vol.39, no.4. Additional references, J.H. Freeberg, U.S. Geol. Surv. Bull., 1966, no.1220, p.57, and 1969, no.1320, p.24.

Meen Lake 64°58′ N., 87°40′ W.

Keewatin district, North-west Territories, Canada

Doubtful.

Listed as possible impact structure, P.B. Robertson and R.A. F. Grieve, Journ. Roy. Astron. Soc. Canada, 1975, vol.69, p. 1.

Mejaouda 22°43′ N., 7°18′ W.

Mauritania

Doubtful.

A possible crater 3 km in diameter observed on aerial photographs (Monod, 1965).

Melville Island 76°40′ N., 109°0′ W.

Franklin district, North-west Territories, Canada

Doubtful.

Two nearly circular structures with rims raised above the surrounding country and a central dome; both possibly ring dykes (I.C. Brown, Amer. Journ. Sci., 1951, vol.249, p.785; O'Connell, 1965; P.B. Robertson and R.A.F. Grieve, Journ. Roy. Astron. Soc. Canada, 1975, vol.69, p.1).

Mendorf 48°50′ N., 11°36′ E.

Bayern, Germany

Doubtful.

A circular structure 2.5 km in diameter is possibly of impact origin, E. Rutte, Geoforum, 1971, vol.7, p.84.

Menihek Lakes 54° N., 67° W.

Labrador, Canada

Synonym(s): *Lac Menihek*

Doubtful.

Two circular features, one 3 miles in diameter at 53°42'N. 66°40'W., the other 2.5 miles in diameter at 54°19'N. 67°10'W., were observed on aerial survey photographs; not further studied (literature see O'Connell, 1965; Monod, 1965; J.H. Freeberg, U.S. Geol. Surv. Bull., 1966, no.1220, p.57).

Merewether Crater v Labrador Crater.

Meteor Crater 35°3′ N., 111°2′ W.

Coconino County, Arizona, U.S.A.

1905, recognized

Synonym(s): *Arizona Crater, Barringer Crater, Cañon Diablo, Canyon Diablo, Coon Butte, Crater Mound, Winslow Crater*

Authenticated.

The meteoritic origin of Meteor Crater (or Coon Butte) and its connection with the Cañon Diablo irons is generally recognized; much work has been done trying to locate the supposed main mass in the bottom of the crater, but it is generally believed that the mass would be broken up and ejected. See G.P. Merrill, Publ. Astron. Soc. Pacific, San Francisco, 1920, vol.32, p.259; F.L. Thurmond, Eng. Mining Journ., New York, 1926, vol.122, p.817; D.M. Barringer, Proc. Acad. Nat. Sci. Philadelphia, 1925, vol.76, p.275; H.L. Fairchild, Science, New York, 1929, vol.69, p.485; 1930, vol. 72, p.463 and 1931, vol.73, p.66; F.S. Dellenburgh ibid., 1931, vol.73, p.38; P.N. Chirvinsky, Mém. Soc. Russe Min., 1931, ser.2, vol.60, p.135; L.J. Spencer, Geogr. Journ. London, 1933, vol.81, p.227; J.J. Jakosky, Eng. Mining Journ., New York, 1932, vol.133, p.392; E. Blackwelder, Science, New York, 1932, vol.76, p.557. F.M. Brown, ibid., 1933, vol.77, p.239; H.P.T. Rohleder, Zeits. Deut. Geol. Gesell., 1933, vol.85, p.463; E. Öpik, Publ. Observ. Astron., Univ. Tartu, 1936, vol.28, no.6; O. Stutzer, Zeits. Deut. Geol. Gesell., 1936, vol.88, p.510; W.F. Bingham, Pan-Amer. Geol., 1937, vol.68, p.196; H. Lundberg, Bull. Geol. Soc. Amer., 1938, vol.49, p.1953.

Soil two miles from the crater shows no trace of Ni (J.D. Buddhue, Pop. Astron., Northfield Minnesota, 1945, vol.53, p.287; Contr. Soc. Res. Meteorites, vol.3, p.203); but in contradiction to this H.H. Nininger (Science, Amer. Assoc. Adv. Sci., 1951, vol.113, p.755) found minute metallic spheres in concentrations up to 100 grams per cubic foot over an area of 100 sq. miles. They are highly enriched in Nickel (Ni = 17%) and clearly correspond to the Wabar spheres. The heating effects are less than at Wabar and Henbury, suggesting a lower velocity (H.H. Nininger, Pop. Astron., 1947, vol.55, p.103). J.D. Buddhue (ibid., 1948, vol. 56, p.387) has made a grain size analysis of the crushed sandstone. Aerial photographs from 1400 feet show that the crater is not truly round, but has a squarish outline (W.W. Zimmerman, ibid., 1948, vol. 56, p.496). H.H. Nininger has made a new survey of the area around the crater and plotted the distribution of the masses found; he finds that the Widmanstetter structure is indistinct or absent in masses of small size found on the crater rim, but quite distinct in small masses found 2 to 3 miles from the crater and in larger masses found on the surrounding plain (Pop. Astron., Northfield, Minnesota, 1949, vol.57, pp.17, 333, and 1950, vol.58, p.169). He believes that the fall was not a single mass, but a swarm of irons, one main mass and at least three distinct smaller ones (H.H. Nininger, Scientific Monthly, 1951, vol.72, p.75), and has given a detailed explanation of the form of the crater and the distribution of the metal and silica glass, based on this theory.

Two estimates have been made of the mass of the originating meteorite - 500 to 15,000 tons, and 25,000 tons (C.C. Wylie, Pop. Astron., Northfield, Minnesota, 1943, vol. 51, pp.97, 158, and 200; compare L. LaPaz, ibid., p.339). Suggestions have again been made that the crater is not of meteoritic origin, see N.H. Darton, Bull. Geol. Soc. Amer., 1945, vol.56, p.1154, and D. Hager, Pop. Astron., Northfield, Minnesota, 1949, vol.57, p.457 and Bull. Amer. Assoc. Petroleum Geol., 1953, vol.37, p.821; compare H.H. Nininger, Amer. Journ. Sci., 1948, vol.246, p.101.

Silica glass has been found in the crater, grading from completely fused material to unaltered sandstone (A.F. Rodgers, Amer. Journ. Sci., 1930, ser.5, vol.19, p.195). A small amount of a transparent silica-glass containing metallic spheres has been found on the crater rim, in pieces up to 0.5 inch across (H.H. Nininger, Scientific Monthly, 1951, vol.72, p.75), also friable scoriaceous nodules ("bombs") have been found abundantly (H.H. Nininger, Amer. Journ. Sci., 1954, vol.252, p.277). Coesite (E.C.T. Chao et al., Science, 1960, vol.132, p.220) and stishovite (E.C.T. Chao et al., Journ. Geophys. Res., 1962, vol.67, p.419) have been found at Meteor Crater. The mechanics and energetics of impact are discussed by E.M. Shoemaker (Impact mechanics at Meteor Crater, Arizona, U.S. Atomic Energy Commiss., 1959; Bull. Geol. Soc. Amer., 1959, vol.70, p.1748; Rept. 21st Internat. Geol. Congr., 1960, Part 18, p.418; Amer. Journ. Sci., 1963, vol.261, p.668), by R.L. Bjork (Journ. Geophys. Res., 1961, vol.66, p.3379), and by M.D. Nordyke (ibid., p.3439).

Method for determining the residual meteoritical mass, H. L. Crowson, abs. in Meteoritics, 1969, vol.4, p.163. Morphology of impact bombs, W.R. Greenwood and D.A. Morrison, ibid., p.182. Bedded white sands possibly representing shock disaggregated Coconino Sandstone, J.F. McCauley and H. Masursky, ibid., p.196. Recent drilling in rim, D.J. Roddy et al., Meteoritics, 1971, vol.6, p.306. Chemistry of metallic spheroids, W.R. Kelly et al., Geochimica Acta, 1974, vol.38, p.533. Additional references, O'Connell, 1965; Monod, 1965; J.H. Freeberg, U.S. Geol. Surv. Bull., 1966, no.1220, p.22, and 1969, no.1320, p.12.

Results of rim drilling with thickness, structural uplift, depth, volume, and mass balance calculations, D.J. Roddy et al., Geochim. Cosmochim. Acta, 1975, Suppl.6, p.2621. Gravity and magnetic investigations, R.D. Regan and W.J. Hinze, J. Geophys. Res., 1975, vol.80, p.776. Seismic refraction studies, H.D. Ackermann et al., ibid., p.765. Shock processes in porous quartzite, S.W. Kieffer et al., Contr. Min. Petr., 1976, vol.59, p.41. General geological considerations, D.J. Roddy Geochim. Cosmochim. Acta, 1978, Suppl.10, p.3891. Computer simulation of the impact event, J.B. Bryan et al., Geochim. Cosmochim. Acta, 1978, Suppl.10, p.3931. See also, D.J. Roddy et al., Geochim. Cosmochim. Acta, 1980, Suppl.14, p.2275. Shock induced luminesence, D.J. Roddy et al., Meteoritics, 1980, vol.15, p. 356. See also "A checklist of published references to Barringer Meteorite Crater, Arizona 1891-1970", D.J. Briley and C.B. Moore, Publication no.15, Center for Meteorite Studies, Arizona State Univ., Tempe.

Thermoluminescence of shocked rocks, S.R. Sutton, Abs. in Meteoritics, 1982, vol.17, p.284. Energy of formation, R. M. Scmidt, Geochim. Cosmochim. Acta, Suppl.14, 1980, p. 2099. Computer code simulations of the formation of the crater, D.J. Roddy et al., Geochim. Cosmochim. Acta, Suppl.14, 1980, p.2275.

Specimen(s) : [1930,2] silica glass; [1930,3] silica glass; [1933, 322] silica glass; [1959,70] silica glass; [1959,218] scoriaceous nodules (bombs); [1962,155] magnetic grains; [1934,54] unaltered sandstone; [1964,35-36] sediment excavated from the crater floor; [1973,M.32-35] Coconino sandstone from the crater floor. For specimens of the meteorites, see main meteorite list under Cañon Diablo

Michlifen Craters 32° N., 3° E.

Azron, Morocco

Doubtful.

Two craters, one 800 by 600 m., the other 1.9 km in diameter: "on pouvait penser a une dépression karstique ou a un cratere creusé par la chute d'un météorite" (M. Gigout, Bull. Soc. Sci. Nat. Phys. Maroc, 1956 (for 1955), vol.35, p. 3.

Middlesboro 36°37′ N., 83°44′ W.

Bell County, Kentucky, U.S.A.

Doubtful.

A meteoritic origin has been proposed for a circular depression 4 miles in diameter with central uplift " eye" 0.25 miles in diameter, K.J. Englund and J.B. Roen, U.S. Geol. Surv. Prof. Paper 450-E, 1962, Art.184, pp.E20-22. Highly disturbed underlying strata; shattercones in central uplift, R. S. Dietz, Meteoritics, 1966, vol.3, p.27. See also, J.H. Freeberg, U.S. Geol. Surv. Bull., 1966, no.1220, p.58. Mentioned by Monod, 1965 as very doubtful.

Mishina Gora 58.7° N., 28.0° E. approx

Pskov, Rossiya, USSR, [Мишиногорская]

Synonym(s): *Mishinogorsk*

Doubtful.

A rounded, cup-shaped basin 2.5 km in diameter, situated 25 km southeast of Gdov is possibly of meteorite impact origin; allogenic breccia infill, plagioclase glass and other evidence of shock metamorphism, V.L. Masaitis, Sov. Geol., 1975, no. 11, p.52 (English translation: Int. Geol. Rev., 1976, vol.18, p. 1249).

Mistassini Island v Ile Rouleau.

Mistastin Lake 55°52′ N., 63°22′ W.

Labrador, Canada

Doubtful.

A roughly elliptical depression about 20 km in diameter with central uplift for which both cryptovolcanic and meteorite impact origins have been proposed. Geological features and cryptovolcanic argument, K.L. Currie, Nature, 1968, vol.220, p.776; but compare F.C. Taylor and M.R. Dence, Can. Journ. Earth Sci., 1969, vol.6, p.39. Magnetic study, K.L. Currie and A. Larochelle, Earth planet. Sci. Lett., 1969, vol. 6, p.309. Maskelynite, K.L. Currie, Min. Mag., 1971, vol.38, p.511; see also K.L. Currie, Geol. Surv. Canada Bull., 1971, no. 207. Impact melting, R.A.F. Grieve, abs. in EOS, 1974, vol.55, p.336, also Bull. Geol. Soc. Amer., 1975, vol.86, p. 1617. Age, 40±3m.y., P.B. Robertson and R.A.F. Grieve, Journ. Roy. Astron. Soc. Canada, 1975, vol.69, p.1.

Age of the crater, E.K. Mak et al., Earth planet. Sci. Lett., 1976, vol.31, p.345. Strontium isotopes and trace element chemistry of impact melts, M. Marchand and J.H. Crocket, Geochim. Cosmochim. Acta, 1977, vol.41, p.1487. The impacting meteorite was possibly an iron, R. Wolf et al., Geochim. Cosmochim. Acta, 1980, vol.44, p.1015.

Specimen(s) : [1973,M.15] 5 rock fragments of total weight 92.2 grams.

Mizarayskaya

Lithuanian SSR, USSR

Doubtful.

A structure 5 km in diameter, with explosion breccia, shattercones and diaplectic glass, is listed, V.L. Masaitis et al., Dokl. Akad. Nauk. SSSR, Ser. Geol., 1978, vol.240, no.5, p.1191 (English translation, p.91).

Mogol 57.5° N., 108.5° E.

Irkutsk Province, USSR

Doubtful.

Several " conical" depressions 40 metres in diameter are situated in ejected material and may be the product of meteorite impact, I.T. Zotkin and V.L. Tsvetkov, Astron. Vestnik, 1970, vol.4, p.55.

Montagne Noire v Hérault Craters.

Monturaqui 23°57′ S., 68°17′ W.

Atacama Desert, Antofagasta, Chile

Authenticated.

A Pleistocene to Recent crater, about 370 metres in diameter. Iron shale fragments indicate that the projectile was a coarse octahedrite, group IA, and similar to Cañon Diablo. Full description with references, V.F. Buchwald, Handbook of Iron Meteorites, Universities of California and Arizona State, 1975, vol.3, p.1403 (vol.1, especially p.243, for references). Metallic spherules in impactites, R.V. Gibbons et al., Geochim. Cosmochim. Acta, 1976, Suppl.7, p.863.

Mora County Explosion Crater

New Mexico, U.S.A. 36° N., 105° W. approx

Synonym(s): *New Mexico, Northern New Mexico*

Doubtful.

A small crater discovered from the air in 1948 in north-eastern New Mexico has a diameter of 30 feet depth 3 feet, and has angular rock fragments on the rim; it is perhaps meteoritic (L. LaPaz, Pop. Astron., Northfield, Minnesota, 1949, vol.57, p.136). "In photograph, crater resembles surface burst" (O'Connell, 1965). Additional references, J.H. Freeberg, U.S. Geol. Surv. Bull., 1966, no.1220, p.59.

Morasko 52°29′ N., 16°54′ E.

Poznan, Poland

Doubtful.

Eight small craters, the largest 60 metres in diameter, are closely associated with the find-sites of several large irons, J. Pokrzywnicki, Studia Geol. Polon., 1964, vol.15, p.49, with English summary, p.139. The craters are probably impact pits rather than explosion craters. Recent description, J. Classen, Meteoritics, 1978, vol.13, p.245; see also, H. Korpikiewicz, ibid., p.311. Paired with Seeläsgen, A. Kracher et al., Geochim. Cosmochim. Acta, 1980, vol.44, p. 773.

Mosquito Gulf 9° N., 81.5° W.

Central America

Doubtful.

Mentioned by R. Gallant (Bombarded Earth, London, 1964; Monod, 1965).

Mount Darwin v Mount Darwin Crater.

Mount Darwin Crater 42°15′ S., 145°36′ E.

Tasmania, Australia

1933, recognized in this year

Synonym(s): *Darwin Crater, Darwin glass, Mount Darwin*

Authenticated.

A circular depression 1 km in diameter is situated 5 miles southeast of Mt. Darwin. The crater was possibly associated with the event which produced Darwin glass; R.J. Ford, Earth planet. Sci. Lett., 1972, vol.16, p.228. Darwin glass, formerly classed with the tektites, is generally recognized as a typical meteorite crater glass (L.J. Spencer, Nature, London, 1933, vol.131, p.117, and vol.132, p.571; and Min. Mag., 1937, vol.24, p.505; H. Conder, Indust. Australian and Mining Standard, 1934, vol.89, p.329; H. Michel, Fortschr. Min. Krist. Petr., 1939, vol.23, p.cxliii; T. Hodge-Smith, Australian Meteorites, Mem. Australian Mus., 1939, no.7). The Darwin glass contains coesite (A.M. Reid and A.J. Cohen, Journ. Geophys. Res., 1962, vol.67, p.1654). Germanium content, A.J. Cohen, Rept. 21st Internat. Geol. Congr. Norden., 1960, Part 1, p.30; K-Ar age, W. Gentner and J. Zähringer, Zeits. Naturforsch., 1960, vol.15A, p.93. See also S.R. Taylor and M. Solomon, Nature, 1962, vol.196, p.124; R.L. Fleischer and R.B. Price, 2nd Internat.

Symposium on Tektites, 1963 (abs., p.13) and Geochimica Acta, 1964, vol.28, p.753. Fission-track age of glass, 0.70± 0.01m.y., D. Störzer, abs. in Meteoritics, 1971, vol.6, p.319. The age is in agreement with ages obtained for the interior and rim of the crater, and for the southeast Asian-Australasian tektites, W. Gentner et al., Meteoritics, 1973, vol.8, p.37, and W. Gentner et al., Earth planet. Sci Lett., 1973, vol.20, p.204. Full description, R.F. Fudali, and R.J. Ford, Meteoritics, 1979, vol.14, p.283.
Specimen(s) : [1925,1089]; [1926,437]; [1927,1191]; [1934, 123]; [1939,278-280]; [1946,271]; [1950,414-416]; [1962,143-146] silica glass

Mount Doreen — 23° S., 133° E.

Northern Territory, Australia
Doubtful.
A very doubtful complex of four craters within an outer one; literature see O'Connell, 1965.

Murgab Craters — 38°6′ N., 74°17′ E.

Tadzhikistan, USSR
Synonym(s): *Chaglan Toushtou, Eastern Pamir Craters, Pamir Craters*
Doubtful.
A pair of craters, 260 feet in diameter and 33 feet deep in limestone, with thin deposits containing iron on their walls, are possibly meteoritic (Science, 1929, vol.69, no. 1786, suppl. p.xii; A.N. Bakarev, Meteoritika, 1954, no.11, p.183; ibid., 1956, no.14; D. Hoffleit, Sky and Telescope, 1952, vol. 12, p.8). Additional references, J.H. Freeberg, U.S. Geol. Surv. Bull., 1966, no.1220, p.59.

Nastapoka Arc v Hudson Bay Arc.

Nebiewale caldera — 10°35′ N., 1°40′ W.

Ghana
Discredited.
Included by R. Barringer (Meteoritics, 1963, vol.2, p.169) in a list of meteorite craters, but is rejected by A.F.J. Smit (in Monod, 1965, p.24).

New Mexico v Mora County Explosion Crater.

Newporte

Renville County, North Dakota, U.S.A.
Doubtful.
A circular depressed structure 3.2 km in diameter in early oil bearing Palaeozoic sediments is possibly of meteorite impact origin, R.R. Donofrio, Journ. Petr. Geology, 1981, vol.3, p.279; but compare J.H. Clement and T.E. Mayhew, Oil and Gas Journ., 1979, vol.77, p.165, who favour an origin by "localized late Precambrian - early Palaeozoic differential vertical - basement faulting".

New Quebec Crater v Chubb Crater.

Nicholson Lake — 62°40′ N., 102°41′ W.

MacKenzie district, North-west Territory, Canada
Doubtful.
An impact origin has been suggested for this deeply eroded structure with irregular outline; brecciated and fused country rock, and shattercones are present, M.R. Dence et al., in Shock Metamorphism of Natural Materials, eds. B.M. French and N.M. Short, Mono, Baltimore, 1968, p.339; see also P.B. Robertson et al., ibid., p.433. Age, less than 450 m. y., P.B. Robertson and R.A.F. Grieve, Journ. Roy. Astron. Soc. Canada, 1975, vol.69, p.1. Chemical evidence from country rocks indicates an achondritic impacting body, R. Wolf et al., Geochim. Cosmochim. Acta, 1980, vol.44, p. 1015.

Nördlinger Ries — 48°53′ N., 10°37′ E.

Bayern, Germany
Synonym(s): *Ries Basin, Rieskessel*
Doubtful.
A "large and nearly circular (21 by 24 km) disturbed area with extensive brecciation and scattering of large blocks of limestone, many of them outside the basin itself". Coesite and stishovite have been found, also shattercones. Meteoritic impact and cryptovolcanic explosion have both been advanced as causes, and there is a very extensive literature, summarized by Monod, 1965; O'Connell, 1965, and J.H. Freeberg, U.S. Geol. Surv. Bull., 1966, no.1220, p.66 and ibid., 1969, no.1320, p.26. A connection with the moldavites has been suggested.

Fission-track age of glass, 14.6±0.1 m.y., D. Störzer, abs. in Meteoritics, 1971, vol.6, p.319; see also D. Störzer and W. Gentner, ibid., 1970, vol.5, p.225; and W. Gentner, ibid., 1971, vol.6, p.274. Suggestion that a triple cratering event occurred, P. Rauser et al., Meteoritics, 1971, vol.6, p.304. Composition of glasses, V. Stähle, Earth planet. Sci. Lett., 1973, vol.18, p.385; see also D. Stöffler, ibid., 1970, vol.10, p.115, and V. Stähle, ibid., 1975, vol.25, p.71. Review, J.G. Dennis, Journ. Geophys. Res., 1971, vol.76, p.5394.

There is a considerable recent literature of which the following is only a selection; cratering mechanics, impact metamorphism and distribution of ejected masses, D. Stöffler, Fortschr. Min., 1974, Beiheft vol.52, p.109; see also E. Rutte, Oberrhein geol. Abh., 1974, vol.23, p.97; origin, W. von. Engelhardt, Fortschr. Min., 1975, special issue, vol. 52, p.375 and D. Stöffler, ibid., p.385; results of research drilling at the crater, many papers in Geol. Bavarica, 1977, vol.75; see also, E.R. Padovani et al., Geochim. Cosmochim. Acta, 1978, Suppl.10, p.2731. Significance of armalcolite in glasses, A. El Goresy, and E.C.T. Chao, Earth planet. Sci. Lett., 1976, vol.30, p.200; search for meteoritic material at the crater, J.W. Morgan et al., Geochim. Cosmochim. Acta, 1979, vol.43, p.803; discovery of Fe-Cr-Ni veinlets below the crater, A. El Goresy and E.C.T. Chao, Earth planet. Sci. Lett., 1976, vol.31, p.330; on the basis of trace element data from country rocks it has been suggested that the impacting projectile was aubritic, J.W. Morgan et al., abs. in Meteoritics, 1977, vol.12, p.319. Evidence for the coincidence of a geomagnetic reversal with the impact event, J. Pohl, abs. in Meteoritics, 1978, vol.13, p.600; cratering models, E.C.T. Chao, Geol. Jahrb., 1977, A.43, p.3; but compare J. Pohl, abs. in Meteoritics, 1979, vol.14, p.520; corrected impact mechanical data, E. David, abs. in Meteoritics, 1979, vol.14, p.377; age and intensity of thermal events by fission track analysis, D.S. Miller and G.A. Wagner, Earth planet. Sci. Lett., 1979, vol.43, p.351; possibility of terrestrial chondrules, G. Graup, Earth planet. Sci. Lett., 1981, vol.55, p.407; petrology of suevite and conclusions on crater formation, W. von. Engelhardt and G. Graup, abs. in Meteoritics, 1981, vol.16, p.311. For summary and discussion of the available data (to 1977) on the structure, see J. Pohl et al., In Impact and Explosion Cratering, eds. D.J. Roddy, R.O. Pepin and R.B. Merrill, Pergamon Press (New York), 1977, p.343, but compare E.C.T. Chao and J.E. Minkin, ibid., p.405. Results of shallow drilling in the Bunte breccia impact deposits, F. Hörz et al., ibid., p.425.
Specimen(s) : [1968,2], 8.25kg. suevite from the eastern rim of the crater.; [1982,M.26] fragments of belemnite from shocked rocks; [1982,M.27] impact melted rock; [1982,M.29] bentonitic clay containing glass shards

North Caspian Sea 44° N., 53° E.
Kazakhstan, USSR
Discredited.
An impact origin has been suggested for this WNW trending depression with major axis some 800 km long, B.S. Zeylik and E.Yu. Seytmuratova, Izvestia Akad. Nauk Kazakhstan S.S.R., Ser. Geol., 1975, vol.1, p.62.

Northern New Mexico v Mora County Explosion Crater.

Novki 56.4° N., 41.1° E.
Vladimir Province, USSR
Doubtful.
Eight dry "conical" depressions 40 metres in diameter are of doubtful meteoritic origin, I.T. Zotkin and V.L. Tsvetkov, Astron. Vestnik, 1970, vol.4, p.55.

Novo Gurovka 54.1° N., 56.1° E.
Bachkir A.S.S.R., USSR
Doubtful.
Two "conical" depressions some 10 metres in diameter are surrounded by ejected stones, but are of doubtful impact origin, I.T. Zotkin and V.L. Tsvetkov, Astron. Vestnik, 1970, vol.4, p.55.

Nyika Plateau Crater 10°35′ S., 33°43′ E. approx
Malawi
Doubtful.
A crater 80 metres in diameter and 6 metres deep, reportedly was formed early in 1959. No meteorite material has been recovered and available evidence neither proves nor precludes an impact origin, D.J. Mossman, Meteoritics, 1972, vol.7 p.71.

Oasis structure 24°35′ N., 24°24′ E.
Libya
Doubtful.
A circular structure, 11.5 km in diameter in Nubian sandstones is possibly of meteoritic origin and contemporaneous with "Gebel Dalma" (q.v.) 80 km to the north; shock metamorphic features, B.M. French et al., Bull. Geol. Soc. Amer., 1974, vol.85, p.1425.

Obolon′ 49°36′ N., 32°51′ E.
Poltava oblast, Ukraine, USSR, [Оболонская]
Doubtful.
An origin by meteorite impact has been suggested for a buried circular basin of disturbed rocks, 14-15 km in diameter with central uplift of crystalline basement rocks (Yu.Yu. Yurk et al., Sov. Geol., 1975, no.2, p.138). Shock metamorphic evidence from borehole samples, V.L. Masaitis, et al., Dokl. Akad. Nauk SSSR., 1976, vol.230, p.174 and A. A. Val′ter et al., ibid., 1977, vol.232, p.170. Planar elements in biotite, E.P. Gurov, Zap. Vses. Min. Obshch., 1977, vol. 106, p.715.

Odessa 31°43′ N., 102°24′ W.
Ector County, Texas, U.S.A.
1929, Recognized
Authenticated.
A typical explosion crater of about 530 feet in diameter and 18 feet depth occurs in limestone; meteoritic iron and shale were found around the crater (D.M. Barringer, Proc. Acad. Nat. Sci., Philadelphia, 1929, vol.80, p.307; L.J. Spencer, Geogr. Journ., London, 1933, vol.81, p.227; H.H. Nininger, Pop. Astron., Northfield, Minnesota, 1934, vol.42, p.46; O.E. Monnig and R. Brown, ibid., 1935, vol.43, p.34; H.H. Nininger, The Sky, Amer. Mus. Nat. Hist., 1939, vol.3, no.4, p.6; compare E.H. Sellards, Bull. Geol. Soc. Amer., 1927, vol.38, p.149). There are two smaller craters.
Further excavations, with recovery of more meteorites, see E.H. Sellards and V.E. Barnes, Pop. Astron., Northfield, Minnesota, 1943, vol. 51, p.224. Additional references; O'Connell, 1965; Monod, 1965; J.H. Freeberg, U.S. Geol. Surv. Bull., 1966, no.1220, p.62, and 1969, no.1320, p.25. Meteoritic material from soil near the crater, P.W. Hodge, abs. in Meteoritics, 1979, vol.14, p.422.
Specimen(s) : See main list of meteorites.

Oesel v Kaalijärv.

Ogni 51.8° N., 83.5° E.
Altai Territory, USSR
Doubtful.
A circular crater 140 metres in diameter and 25 metres deep, situated on a mountain slope, is possibly meteoritic, I.T. Zotkin and V.L. Tsvetkov, Astron. Vestnik, 1970, vol.4, p. 55.

Old Fort Rae 62°40′ N., 115°49′ W.
MacKenzie district, North-west Territories, Canada
Doubtful.
Mentioned as possibly of meteoritic origin, C.S. Beals, Journ. Roy. Astron. Soc. Canada, 1956, vol.50, p.203.

Oman Ring 19°55′ N., 56°58′ E.
Arabian Peninsula
Doubtful.
A circular, double ring structure with overall diameter of 6 km is possibly of meteorite impact origin, R.S. Dietz et al., abs. in Meteoritics, 1975, vol.10, p.393.

Ostashkov 57.2° N., 33.2° E.
Kalinin Province, USSR
Doubtful.
A lake 75 metres in diameter and with a horseshoe shaped wall, is possibly of impact origin, I.T. Zotkin and V.L. Tsvetkov, Astron. Vestnik, 1970, vol.4, p.55.

Otnos 53° N., 35° E.
Orlovsk Province, USSR
Doubtful.
Four lakes, 50 metres in diameter are of doubtful meteoritic origin, I.T. Zotkin and V.L. Tsvetkov, Astron. Vestnik, 1970, vol.4, p.55.

Padun 60° N., 53.5° E.
Kirov Province, USSR
Doubtful.
A circular lake 100 metres in diameter is possibly of meteoritic origin, I.T. Zotkin and V.L. Tsvetkov, Astron. Vestnik, 1970, vol.4, p.55.

Pamir Craters v Murgab Craters.

Panamint Crater 36°6′ N., 117°22′ W.
Inyo County, California, U.S.A.
Discredited.
A crater 230 ft in diameter by 40 ft deep is similar to known meteorite craters in general appearance, but is probably merely a solution collapse pit (L.E. Humiston et al., Progr. 24th meeting Meteoritical Soc., 1961, p.16; R.S. Dietz and E. C. Buffington, Meteoritics, 1964, vol.2, p.179).

Paragould 36°4′ N., 90°30′ W.
Greene County, Arkansas, U.S.A.
Discredited.
The impact pit of this meteorite (q.v.) is included by O'Connell (1965) in her " Catalog of meteorite craters and related features".

Paris 48.8° N., 2.5° E.
Sucy-en-Brie and Alentours, France
Doubtful.
Small lakes are possibly meteorite craters; listed with co-ordinates, R.W. Barringer, Meteoritics, 1967, vol.3, p.151.

Parry Sound 45°22′30″ N., 79°56′ W.
Ontario, Canada
Doubtful.
A roughly circular structure about 1.4 miles in diameter and containing two lakes may be a fossil meteorite crater, F.K. McKean, Meteoritics, 1964, vol.2, p.243.

Patomskii Crater v Perevoz Crater.

Perevoz Crater 59°0′ N., 116°0′ E.
Irkutsk, Siberia, USSR
Synonym(s): *Patomskii Crater*
Doubtful.
A circular crater 86 metres in diameter, resembling known meteorite craters (A.M. Portnov, Priroda, 1962, p.102; Meteoritika, 1964, vol.25, p.194). Additional references, J.H. Freeberg, U.S. Geol. Surv. Bull., 1966, no.1220, p.65, and 1969, no.1320, p.25.

Pfahldorf Basin 48°57′ N., 11°22′ E.
West Germany
Doubtful.
Several craters, the largest about 2.5 km in diameter, are possibly of impact origin, H. Illies, Oberrhein geol. Abh., 1969, vol.18, p.1.

Pilot Lake 60°17′ N., 111°1′ W.
MacKenzie district, North-west Territories, Canada
Doubtful.
A lake with a roughly square outline approximately 5.5 km across and 68 metres deep may be meteoritic, M.R. Dence et al., in Shock Metamorphism of Natural Materials, eds. B.M. French and N.M. Short, Mono, Baltimore, 1968, p.339; see also N.M. Short and T.E. Bunch, ibid., p.255. Age, less than 600 m.y., P.B. Robertson and R.A. F. Grieve, Journ. Roy. Astron. Soc. Canada, 1975, vol.69, p.1.

Piute County Explosion Pits
Utah, U.S.A. 38°12′ N., 112°6′ W.
Synonym(s): *Southern Utah Craters*
Discredited.
Four small (4 ft diameter) shallow craters regarded as possibly meteoritic (R.N. Thomas, Sky and Telescope, 1952, vol.11, p.300) are in fact the result of dynamite explosions (D. Hoffleit, ibid., 1954, vol.13, p.157).

Podkamennaye Tunguska v Tunguska.

Popigai Basin 71° N., 111° E.
Siberia, USSR
Synonym(s): *Popigay*
Doubtful.
A circular basin 70-80 km in diameter for which both volcanic and impact origins have been suggested; description with analyses of country rocks, V.L. Masaitis et al., Meteoritics, 1972, vol.7, p.39. Compare M.M. Polyakov and A.I. Trukhalev, Izvestia Akad. Nauk SSSR., ser. geol., 1974, no.4, p.85. See also V.L. Masaitis et al., Dokl. Acad. Sci. USSR, Earth Sci. Sect., 1971, vol.197, p.105 (translation from Dokl. Akad. Nauk SSSR., 1971, vol.197, p.1390.

Coesite found at the crater, V.L. Masaitis et al., Zap. Vses, Min. Obshch., 1974, vol.103, p.122; Nickel bearing iron sulphides and native nickel in suevites, V.L. Masaitis and A. G. Sysoev, ibid., 1975, vol.104, p.204; Graphite in country rocks, S.A. Vishnevskii and N.A. Pal'chik, Geol. Geofiz., 1975, p.67.

Description of tagamites, T.V. Selivanovskaya, Meteoritika, 1977, vol.36, p.131. Comparison with Ries suevites, ibid., p. 135. See also, A.I. Raikhlin and M.S. Maschack, ibid., p.140. Possible source of bediasite tektites, R.S. Dietz, Meteoritics, 1977, vol.12, p.145; he quotes an age of Popigai of 30 m.y., from V.L. Masaitis, Sov. Geol., 1976, vol.11, p.52 (see English translation, Int. Geol. Rev., 1976, vol.18, p.1249). Stishovite from cataclastic breccias, S.A. Vishnevskii et al., Dokl. Akad. Nauk SSSR, 1975, vol.221, p.1167 (English trans.: Dokl. Acad. Sci. USSR, Earth Sci. Section, 1975, vol. 221, p.167).

Popigay v Popigai.

Poplar Bay 50°22′ N., 95°47′ W.
Manitoba, Canada
Doubtful.
An origin by meteorite impact has been suggested for this circular lake, 3 km in diameter which forms the eastern extremity of Lac du Bonnet (q.v.); aeromagnetic and gravimetric data, D.L. Trueman, Can. Journ. Earth Sci., 1976, vol.13, p.1608.

Pretoria saltpan 25°25′ S., 28°5′ E.
South Africa
Discredited.
A meteoritic origin has been suggested for the Pretoria saltpan (H.P.T. Rohleder, Zeits. Deut. Geol. Gesell., 1933, vol.85, p.463; Geol. Mag., London, 1933, vol.70, p.489), but is very improbable. Compare P.A. Wagner, Mem. Geol. Surv. South Africa, 1922, no.20.

Comparison of gravity profile data with a hypothetical impact crater model indicates that the saltpan is not meteoritic in origin, but is a true cryptovolcano, R.F. Fudali and D.P. Gold, Meteoritics, 1973, vol.8, p.36. Additional references, J.H. Freeberg, U.S. Geol. Surv. Bull., 1969, no. 1320, p.25.

Puchezh-Balakhna Disturbance v Puchezh-Katunki Disturbance.

Puchezh-Katunki Disturbance 56°30′ N., 43° E.
Gorki Province, USSR
Synonym(s): *Puchezh-Balakhna Disturbance*
Doubtful.
A roughly elliptical structure 60 by 100 km of controversial origin; the structure is characterised by intensely crushed and deformed rocks and may be of meteoritic origin, L. Firsov (trans. by S.W. Kieffer), Meteoritics, 1973, vol.8, p.223. Co-ordinates, J.H. Freeberg, U.S. Geol. Surv. Bull., 1969, no. 1320, p.25.

Geophysical and geochemical aspects of the disturbance, V.M. Gordin et al., Phys. Earth Planet. Inter., 1979, vol.20, p.1.

Quillagua 21°30′ N., 69°20′ W. Approx
Antofagasta, Chile
Doubtful.
A chain of small craters or impact pits arranged in three main groups aligned in a 40×15 km distribution ellipse trending NNE of the Quillagua Oasis are probably of meteorite impact origin; described, with discovery of abundant silica glass and small amounts of iron, A. Thomas, Geol. Rundschau, 1969, vol.58, p.903. The location is coincident with the find sites of the North Chilean hexahedrites (A. Bevan pers. obs.). See co-ordinates in V.F. Buchwald, Iron Meteorites, Univ. of California, 1975, p.917, but the masses are mildly deformed which seems inconsistent with a major crater-forming iron.

Rajasthan v Ramgargh.

Ramgargh 25°20′ N., 76°37′ E.
Rajasthan, India
Synonym(s): *Rajasthan*
Doubtful.
An impact origin has been suggested for an annular ridge approximately 3 km in diameter and standing on a flat plain. The site is located 350 km SSW of New Delhi and near the border between Rajasthan and Madhya Pradesh, A.R. Crawford, Nature, 1972, vol.237, p.96. See also, H.S. Sharma, Nature, 1973, vol.242, p.39; M.S. Balasundaram and A. Dube, ibid., p.40. Recent description, N. Ahmad, et al., Current Sci. India, 1974, vol.43, p.598.

Randecker Maar
Germany
Discredited.
Included by R. Gallant (Bombarded Earth, London, 1964, plate 15) as of meteoritic origin; Monod (1965) comments: "On voit mal pourquoi ce maar pourrait, plutôt que n'importe quel autre, justifier une hypothese météoritique".

Red Wing Creek
North Dakota, U.S.A.
Doubtful.
A structure about 5 miles in diameter has a central dome which is oil-producing; shattercones.. have been recognized in the cores..", H.B. Sawatzky, Bull. Amer. Assoc. Petrol. Geol., 1975, vol.59, p.694. Deformed Mississippian-Triassic strata are overlain and underlain by undeformed formations, R.L. Brenan et al., Wyoming Geol. Assoc. Earth Sci. Bull., 1975, vol.8, p.1. See also H.B. Sawatzky, In Impact and Explosion Cratering, eds. D.J. Roddy, R.O. Pepin and R.B. Merrill, Pergamon Press (New York), 1977, p.461.

Reitz ring
South Africa
Doubtful.
The arcuate NW rim of the upper Witwatersrand sedimentary basin is regarded as the remnant of an original circular structure 400 km in diameter, possibly produced by early meteorite impact, R.B. Hargraves and A.O. Fuller, Precambrian Research, 1981, vol.14(2), p.99.

Richât Dome 20°5′ N., 11°20′ W.
Mauritania
Doubtful.
A circular disturbed area 50 km in diameter with central dome and some breccia, aligned along a straight fault with the Tenoumer and Temimichat craters, for which both volcanic and meteoritic origins have been proposed (literature in Monod, 1965); coesite is present (A. Cailleux et al., C.R. Acad. Sci. Paris, 1964, vol.258, p.5488). Lack of evidence for cryptoexplosive activity, absence of shattercones and of megascopic and microscopic evidence of shock metamorphism indicate that the crater is not of impact origin, R.S. Dietz et al., abs. in Meteoritics, 1969, vol.4, p. 164. Additional references, J.H. Freeberg, U. S. Geol. Surv. Bull., 1966, no.1220, p.65, and 1969, no.1320, p.26.

Ries basin v Nördlinger Ries.

Rieskessel v Nördlinger Ries.

Riga 57° N., 24.5° E.
Latvian S.S.R., USSR
Doubtful.
A depression approximately 300 metres in diameter is possibly of meteoritic origin, I.T. Zotkin and V.L. Tsvetkov, Astron. Vestnik, 1970, vol.4, p.55.

Rochechouart Structure v Chassenon Structure.

Rogozhino 52.5° N., 39.5° E.
Lipetsk Province, USSR
Doubtful.
A depression 16 metres in diameter is of doubtful meteoritic origin, I.T. Zotkin and V.L. Tsvetkov, Astron. Vestnik, 1970, vol.4, p.55.

Rotchegda 62.5° N., 43.5° E.
Archangel'sk Province, USSR
Doubtful.
A "conical" depression 10 meters in diameter is of doubtful meteoritic origin, I.T. Zotkin and V.L. Tsvetkov, Astron. Vestnik, 1970, vol.4, p.55.

Roter Kamm Crater 27°45′ S., 16°17′ E.
Namibia
Doubtful.
A crater 1.5 miles in diameter resembles known meteorite craters, R.S. Dietz, Progr. 27th meeting Meteoritical Soc., 1964, p.7. Morphology, and explosive energy considerations, R.S. Dietz, Meteoritics, 1965, vol.2, p.311, and ibid., 1966, vol.3, p.31. Description, petrology of crater rocks, magnetic and gravimetric surveys, R.F. Fudali, Meteoritics, 1973, vol. 8, p.245.

Rotmistrovka 49°09′ N., 31°44′ E. approx
Cherkassy oblast, USSR, [Ротмистровская]
A circular basin, 5 km in diameter, of disturbed Proterozoic rocks covered by a Cretaceous sedimentary infill has been suggested as a fossil meteorite impact crater of late Juassic age; breccias, glasses and other evidence of shock metamorphism from borehole samples, V.L. Masaitis et al., Dokl. Akad. Nauk SSSR., 1976, vol.230, p.174; V.L. Masaitis, Sov. Geol., 1975, no.11, p.52 (English trans.: Int. Geol. Rev., 1976, vol.18, p.1249).

Rub' al Khali Craters v Wabar.

Rwanda v Butare.

Ryazan′ 54° N., 40° E.
Ryazan′ Province, USSR
Doubtful.
A "conical" depression 20 metres in diameter at the centre of a small mountain is possibly of impact origin, I.T. Zotkin and V.L. Tsvetkov, Astron. Vestnik, 1970, vol.4, p.55.

Saal 48°48′ N., 11°58′ E.
Bayern, Germany
Doubtful.
A small circular structure may be associated with the Nördlinger Ries impact, E. Rutte, Z. dt. geol. Ges., 1975, vol.126, p.183.

Saarema Island Craters v Kaalijärv.

Saint Georges Bay
Argentina
Doubtful.
Included by R. Gallant (Bombarded Earth, London, 1964; Monod, 1965) on a map of metorite craters.

St.-Imier 47°10′ N., 7° E.
St. Imier, Bern, Switzerland
Doubtful.
A crater, 400 metres in diameter on the slope of a limestone mountain in the Swiss Jura is of doubtful meteorite impact origin; shattercones, discovery of small iron particles, and thermoluminescence data for country rock, Von F. Hofmann and K. Bächtiger, Ecologae Geol. Helv., 1976, vol.69, p.177.

St. Magnus Bay 60°25′ N., 1°34′ W.
Shetland Islands, Scotland
Doubtful.
A roughly circular bay about 11 km in diameter for which an origin by meteorite impact has been suggested, D. Flinn, Proc. Geol. Soc. London, 1970, no.1663, p.131; also A.W. Sharp, "The Moon", Dordrecht-Holland, 1971, vol.2, p.144. Sub-bottom profiling of the bay, A.G. McKay, Scottish Journ. Geol., 1974, vol.10, p.31.

Sall Craters v Kaalijärv.

Samborombon Bay
Argentina
Doubtful.
Included by R. Gallant (Bombarded Earth, London, 1964; Monod, 1965) on map of meteorite craters.

Sanar 60.4° N., 106.2° E.
Irkutsk Province, USSR
Doubtful.
A "well" 50 metres in diameter is of doubtful meteoritic origin, I.T. Zotkin and V.L. Tsvetkov, Astron. Vestnik, 1970, vol.4, p.55.

Sault-Aux-Cochons 49°17′ N., 70°5′ W.
Quebec, Canada
Doubtful.
A large circular feature observed only on aerial survey photographs that suggests a " basin filled with ancient sediments dipping towards center" is included by C.S. Beals et al., (Current Sci., 1960, vol.29, pp.205 and 249) among probable meteorite craters. Literature, J.H. Freeberg, U.S. Geol. Surv. Bull., 1966, no.1220, p.71.

Sausthal 48°58′ N., 11°52′ E.
Bayern, West Germany
Doubtful.
A circular structure 1 km in diameter is possibly of meteoritic origin, E. Rutte, Geoforum, 1971, vol.7, p.84, and Oberrhein. geol. Abh., 1974, vol.23, p.97.

Savonoski Crater 58°31′48″ N., 154°55′31″ W.
Alaska, U.S.A.
Doubtful.
An elliptical basin about 1700 feet in diameter and containing a lake; both volcanic and impact origins have been suggested; described and figured, B.M. French et al., Meteoritics, 1972, vol.7, p.97.

Sayan Crater v Udjei Bowl.

Sea of Japan
Doubtful.
Included by R. Gallant (Bombarded Earth, London, 1964; Monod, 1965) as a possible meteorite crater.

Semsiyyât Dome 21°1′ N., 11°50′ W.
Mauritania
Synonym(s): *Chemsiyat Dome*
Doubtful.
A circular disturbed area with central dome, 5 km in diameter, may be of meteoritic origin (Monod, 1965; see also O'Connell, 1965). Similarities with the Richât Dome lead to the conclusion that the Semsiyya†t Dome is not an impact structure, R.S. Dietz et al., abs. in Meteoritics, 1969, vol.4, p.164.

Serpent Mound 39°2′ N., 83°25′ W.
Adams County, Ohio, U.S.A.
Doubtful.
A nearly circular disturbed area, 4 miles in diameter, to which both meteoritic and cryptovolcanic origins have been assigned. Shattercones and coesite have been observed (A.J. Cohen et al., Science, 1961, vol.134, p.1624). Literature, see O'Connell, 1965, Monod, 1965, and J.H. Freeberg, U.S. Geol. Surv. Bull., 1966, no.1220, p.71, and 1969, no.1320, p. 30.

Serra da Cangalha Structure 8°5′ S., 46°52′ W.
Goias, Brazil
Discredited.
An eroded structure 12 km in diameter known from aerial photographs. Thought to be an alkaline complex, but R.S. Dietz (abs. in Meteoritics, 1969, vol.4, p.269) suggests that it may be of meteoritic origin. No evidence of shock metamorphism, R.S. Dietz and B.M. French, abs. in Meteoritics, 1973, vol.8, p.345, and Nature, 1973, vol.244, p. 561.

Severo-Kounradskaya 47°9′ N., 75°14′ E.
Kazakhstan, USSR, [Северо–Коунрадская]
Doubtful.
A 15 km by 6 km area of disturbed rocks associated with a magnetic anomaly may be of impact origin, B.S. Zeylik and E.Yu. Seytmuratova, Izvestia Akad. Nauk Kazakhstan S.S. R., Ser. Geol., 1975, vol.1, p.62.

Shunak 47°10′ N., 72°45′ E. approx.
Kazakhstan, USSR, [Шынак]
Doubtful.
A circular structure, 2.5 km in diameter in crystalline rocks is possibly of meteoritic origin, V.I. Feld'man et al., Meteoritika, 1979, vol.38, p.99; depth diameter relationship, R.A.F. Grieve et al., Multi-ring Basins, Geochim. Cosmochim. Acta, 1981, Suppl.15, p.37. Listed as having explosion breccia and diaplectic glass, V.L. Masaitis et al., Dokl. Akad. Nauk. SSSR, Ser. Geol., 1978, vol.240, no.5, p. 1191 (English translation: p.91). See also D.M. Borisenko

and V.N. Levin, Dokl. Earth Sci. section, 1980, vol.237, p. 124.

Sichote-Alinsk v Sikhote-Alin.

Sierra Madera 30°36′ N., 102°55′ W.

Pecos County, Texas, U.S.A.
Synonym(s): *Madera Mountain*
Doubtful.

A roughly circular disturbed area about 8 miles in diameter, with underlying breccia lens about 1.5 miles wide and 0.5 mile deep; shattercones are present (J.D. Boon and C.C. Albritton, Jr., Field and Lab., 1937, vol.5, p.53; R.S. Dietz, Pop. Astron., 1946, vol.54, p.465; R.E. Eggleton and E.M. Shoemaker, Spec. Paper Geol. Soc. Amer., 1962, no.68, p. 169; E.M. Shoemaker and R.E. Eggleton, U.S. Geol. Surv. Astrogeolog. Studies, Ann. Rept. for 1962-3, 1964, p.106; J. R. Van Lopik and R.A. Geyer, Science, 1963, vol.142, p.45). Additional references, J.H. Freeberg, U.S. Geol. Surv. Bull., 1966, no.1220, p.72, and 1969, no.1320, p.30.

A bowl or funnel-shaped body, 8 miles in diameter and 6000 to 8000 feet deep; general geology, intrusive breccias and shattercones, H.G. Wiltshire et al., U.S. Geol. Surv. Prof. Paper 599-H, 1972, 42pp.

Sikhote-Alin 46°9.6′ N., 134°39.2′ E.

Primorskiy Kray, USSR
Synonym(s): *Sichote-Alinsk, Ussuri Fall*
Doubtful.

While most of the craters associated with this meteorite shower are simple impact pits, a few of the largest may be true explosion craters. Literature see main meteorite list, and J.H. Freeberg, U.S. Geol. Surv. Bull., 1966, no.1220, p.72.

X-ray study of meteoritic dust, N.I. Zaslavskaya, Meteoritika, 1968, vol.28, p.142. Fragmentation of the meteorite body, E.L. Krinov, Meteoritics, 1974, vol.9, p.255.

Siljan Ring 61°5′ N., 15°0′ E.

Sweden
Synonym(s): *Lake Siljan*
Doubtful.

A complex crater with central dome and surrounding annular depression formed by the outline of several arc-form lakes is possibly a fossil meteorite crater (K. Fredriksson and F.E. Wickman, Svensk Naturvetenskap., 1963, p.121). Country rocks shock-metamorphosed, N.-B. Svensson, Nature (Phys.Sci.), 1971, vol.229, p.90. Additional references, J.H. Freeberg, U.S. Geol. Surv. Bull., 1966, no. 1220, p.55, and 1969, no.1320, p.23; see also R.S. Dietz, abs. in Meteoritics, 1970, vol.5, p.192. Evidence supporting meteorite impact origin, P. Thorslund and C. Auton, Bull. Geol. Inst. Univ. Uppsala, 1975, vol.6 (N.S.), p.69. For a comparison with the Charlevoix Structure (q.v.) see J. Rondot, Bull. Geol. Inst., Univ. Uppsala, ibid., p.85. Examination of the granite/sandstone contact, S. Helmqvist, ibid., 1977, vol.7, p.67. Ar-Ar age of structure (362 m.y.), R. J. Bottomley et al., Contr. Mineral. Petrol., 1978, vol.68, p. 79. Preliminary determination of the impact centre from the orientation of the axes of shattercones, F.E. Wickman, Geologiska Föreningens i Stockholm Förhandlingar, 1980, vol.102, p.105.

Sithylemenkat Lake 66°7′ N., 151°23′ W.

Alaska, U.S.A.
Doubtful.

An origin by meteorite impact has been suggested for a bowl shaped depression with radial features in the rim, 12.4 km in diameter containing a lake. The structure was recognized using Landsat imagery; aeromagnatic survey and ground geochemical survey showing Ni anomaly support an impact origin, P. Jan Cannon, Science, 1977, vol.196, p.1322.

Skeleton Lake 45°16′ N., 79°27′ W.

Muskoka district, Ontario, Canada
Doubtful.

A nearly circular lake is interpreted as the eroded remnant of an impact crater of Palaeozoic age recently exhumed from beneath an Ordovician cover. Lake is 3.5 km in diameter. Airborne and surface magnetic surveys, J.F. Clark, abs. in Meteoritics, 1970, vol.5, p.188. Structure and age, E.D. Waddington and M.R. Dence, Can. Journ. Earth Sci., 1979, vol.16, p.256.

Slate Islands 48°40′ N., 87° W.

Ontario, Canada
Doubtful.

The islands form a circle 7 km in diameter and are thought to represent the central uplift of a complex impact structure, P.B. Robertson and R.A.F. Grieve, abs. in Meteoritics, 1975, vol.10, p.480. Remanent magnetization of country rocks, H. C. Halls, Nature, 1975, vol.255, p.692.

Full description with evidence supporting impact origin, H. C. Halls and R.A.F. Grieve, Can. Journ. Earth Sci., 1976, vol.13, p.1301. Variations in shock deformation, R.A.F. Grieve and P.B. Robertson, Contr. Min. Petr., 1976, vol.58, p.37. Structural analysis of shattercones, R.M. Stesky and H. C. Halls, EOS (Amer. Geophys. Union Trans.), 1978, vol.59, p.228.

Argument against an impact origin, R.P. Sage, Geol. Soc. Amer. Bull., 1978, vol.89, p.1529; but compare H.C. Halls, ibid., 1979, part 1, vol.90, p.1084 and P.B. Robertson and R. A.F. Grieve, ibid., p.1087; see also R.P. Sage, ibid., 1980, vol.91, p.313, 315. Structural analysis of shattercones, R.M. Stesky and H.C. Halls, Can. J. Earth Sci., 1983, vol.20, p.1.

Sobolevskii 46.3° N., 137.9° E.

Primorye Territory, USSR
Doubtful.

A crater 51 metres in diameter and 8 metres deep is surrounded by overturned trees and may be of impact origin, I.T. Zotkin and V.L. Tsvetkov, Astron. Vestnik, 1970, vol.4, p.55. Listed as containing "pulverised meteorite material", V. L. Masaitis etal., Dokl. Akad. Nauk SSSR, 1978, vol.240, no.5, p.1191 (English translation: p.91). Description, L.P. Khryanina, Internat. Geol. Rev., 1981, vol.23, no.1, p.1.

South Balkhash Lake 45° N., 75°30′ E.

Kazakhstan, USSR, [Южно-Прибапхашская]
Doubtful.

A roughly circular structure 380 km in total diameter associated with a magnetic anomaly is possibly of impact origin, B.S. Zeylik and E.Yu. Seytmuratova, Izvestia Akad. Nauk Kazakhstan S.S.R., Ser. Geol., 1975, vol.1, p.62.

Southern Utah Craters v Piute County Explosion Pits.

Spornyi 51.8° N., 139° E.

Khabarovsk Territory, USSR
Doubtful.

Several small " conical" depressions on a slope are of doubtful impact origin, I.T. Zotkin and V.L. Tsvetkov, Astron. Vestnik, 1970, vol.4, p.55.

Srednekan 62.5° N., 152.7° E.
Magadan Province, USSR
Doubtful.
A circular lake 50 metres in diameter may be of impact origin, I.T. Zotkin and V.L. Tsvetkov, Astron. Vestnik, 1970, vol.4, p.55.

Stehling's Crater v Afton Craters.

Steen River 59°31′ N., 117°37′ W.
Alberta, Canada
Doubtful.
An area of disturbed crystalline rocks lying below undisturbed Lower Cretaceous strata and for which an impact origin has been suggested, M.A. Carrigy, in Shock Metamorphism of Natural Materials, eds. B.M. French and N.M. Short, Mono, Baltimore, 1968, p.367; also S.R. Winzer, Proc. 24th Internat. Geol. Congress, Section 15, 1972, p.148. Additional reference, J.H. Freeberg, U.S. Geol. Surv. Bull., 1969, no.1320, p.31. Age, 97±5m.y., P.B. Robertson and R.A.F. Grieve, Journ. Roy. Astron. Soc. Canada, 1975, vol.69, p.1.

Steinheim 48°36′ N., 10°33′ E.
Bayern, Germany
Doubtful.
A meteoritic origin has been suggested for the Steinheim basin, 2.5 km across and 80 metres deep (J. Kaljuvee, Die Grossprobleme der Geologie. Tallinn, 1933; H.P.T. Rohleder, Zeits. Deut. Geol. Gesell., 1933, vol.85, p.463; Geol. Mag., London, 1933, vol.70, p.489; W. Kranz, Petermanns Geogr. Mitt., 1937, vol.83, p.198; O. Stutzer, Zeits. Deut. Geol. Gesell., 1936, vol.88, p.510; R.S. Dietz, Pop. Astron., Northfield, Minnesota, 1946, vol.54, p.465).

See also V. Vand, Pennsylvania State Univ. Min. Indust. Bull., vol.32, p.1; and further references in Monod, 1965, O'Connell, 1965, and J.H. Freeberg, U.S. Geol. Surv. Bull., 1969, no.1320, p.31. Full description, W. Reiff, In Impact and Explosion Cratering, eds. D.J. Roddy, R.O. Pepin and R.B. Merrill, Pergamon Press (New York), 1977, p.309.
Specimen(s) : [1982,M.28], 187g. deformed rock

Strangways Crater 15°12′ S., 133°35′ E.
Northern Territory, Australia
Doubtful.
An impact origin has been suggested for a circular structure 16 km in diameter with possible central uplift and abundant evidence of shock metamorphism, R. Brett et al., Meteoritics, 1970, vol.5, p.184. Preliminary results of a recent survey, J. Ferguson et al., abs. in Meteoritics, 1978, vol.13, p.459. Trace element studies of country rocks indicate that the crater may have been formed by the impact of an olivine-rich achondrite, and melt rocks appear to contain approximately 3 wt. per cent of projectile material, J.W. Morgan and G.A. Wandless, J. Geophys. Res., 1983, vol.88 Suppl., p.A819.

Sudbury Structure 46°20′ N., 81°10′ W.
Ontario, Canada
Doubtful.
A meteoritic origin has been suggested for this area (R.S. Dietz, Journ. Geol., 1964, vol.72, p.412; Amer. Journ. Sci., 1963, vol.261, p.650). Petrographic evidence of impact origin, B.M. French, in Shock Metamorphism of Natural Materials, eds. B.M. French and N.M. Short, Mono, Baltimore, 1968, p.383; distribution of shock metamorphic features, B.M. French, abs. in Meteoritics, 1969, vol.4, p.173; relation between meteorite impact and igneous petrogenesis, B.M. French, Bull. Volc., 1970, vol.34, p.466. The nickel sulphide ores may be meteoritic, R.S. Dietz, abs. in Meteoritics, 1970, vol.5, p.191; also review, abs. in Meteoritics, 1971, vol.6, p. 259, and in New Developments in Sudbury Geology, ed. J.V. Guy-Bray, Geol. Assoc. Canada Spec. Paper no.10, 1972, p. 29; structure of Sudbury basin, M.R. Dence, ibid., p.7; shock metamorphism, B.M. French, ibid., p.19. Melting considerations, B.M. French, Proc. 24th Internat. Geol. Congress, Section 15, 1972, p.125. Rb-Sr study of nickeliferous irruptive, R.W. Hurst and G.W. Wetherill, abs. in EOS, 1974, vol.55, p.466. See also F.W. Beals et al., Can. Journ. Earth Sci., 1975, vol.12, p.629. Additional references, J.H. Freeberg, U.S. Geol. Surv. Bull., 1966, no.1220, p.77, and 1969, no.1320, p.31.

Geological considerations of the structure, S.J. Brocoum and I.W.D. Dalziel, Bull. Geol. Soc. Amer., 1974, vol.85, p. 1571. Origin of the Sudbury structure, F.W. Beals and G.P. Lozej, Can. Journ. Earth. Sci., 1976, vol.13, p.179. Meteorite impact theory reviewed, J.S. Stevenson and L.S. Stevenson, Geoscience Canada, 1980, vol.7, p.103; but compare D.H. Rousell, ibid., 1981, vol.8, p.167.

Surgut 61.2° N., 73.6° E.
Tyumen Province, USSR
Doubtful.
A " conical" depression 50 metres in diameter may be of impact origin, I.T. Zotkin and V.L. Tsvetkov, Astron. Vestnik, 1970, vol.4, p.55.

Svetloyar 56.7° N., 45.1° E.
Gorkii Province, USSR
Doubtful.
A circular lake 450 metres in diameter is possibly of meteoritic origin, I.T. Zotkin and V.L. Tsvetkov, Astron. Vestnik, 1970, vol.4, p.55.

Svyatoe Ozero 55.8° N., 38.5° E.
Moscow Province, USSR
Doubtful.
A circular lake 100 metres in diameter may be of impact origin, I.T. Zotkin and V.L. Tsvetkov, Astron. Vestnik, 1970, vol.4, p.55.

Swan Valley Craters
Idaho, U.S.A.
Discredited.
Three craters 80, 28 and 10 metres in diameter are probably collapse structures due to solution of underlying limestones, R. Greeley et al., abs. in Meteoritics, 1969, vol.4, p.182.

Sym 60.4° N., 88.4° E.
Krasnoyarsk Territory, USSR
Doubtful.
Several craters about 80 metres in diameter are possibly due to meteorite impact, I.T. Zotkin and V.L. Tsvetkov, Astron. Vestnik, 1970, vol.4, p.55.

Tabun Khara Obo Crater 44°6′ N., 109°36′ E.
Dornogovi, Mongolia
Doubtful.
A 1.3 km diameter sand-filled circular crater in the eastern Gobi Desert is possibly of meteorite impact origin, described O.D. Suetenko and L.M. Shkerin, Astron. Vestnik, 1970, vol. 4, p.261 and L.M. Shkerin, Meteoritika, 1976, vol.35, p.97. See also, J.F. McHone and R.S. Dietz, abs. in Meteoritics, 1976, vol.11, p.332.

Tademaît 27°36′ N., 5°7′ E.
In Salah, Sahara, Algeria
Synonym(s): *Tin Bider*
Doubtful.
A circular, multi-ring structure 6 km in diameter for which a meteoritic origin has been suggested (see Monod, 1965). Full description with evidence supporting an origin by meteorite impact, P. Lambert et al., Meteoritics, 1981, vol. 16, p.203.

Tagil 57.8° N., 60° E.
Sverdlovsk Province, USSR
Doubtful.
An oval lake (Lake Bezdonnoe) some 250 metres across and 50 metres deep may be of impact origin, I.T. Zotkin and V. L. Tsvetkov, Astron. Vestnik, 1970, vol.4, p.55.

Talemzane 33°19′ N., 4°2′ E.
Algeria
Synonym(s): *Daiet el Maadna*
Doubtful.
A circular crater 1.7 km with raised rim of upturned strata, coesite, and shattercones (R. Karpoff, Meteoritics, 1953, vol. 1, p.31; L.F. Brady, Sky and Telescope, 1954, vol.13, p.297; further references in Monod, 1965, and O'Connell, 1965).
Full description with evidence supporting an origin by meteorite impact 0.5-3 m.y. ago; amended co-ordinates, P. Lambert et al., Meteoritics, 1980, vol.15, p.157.

Talkuduksk 46°24′ N., 73°10′ E.
Kazakhstan, USSR, [Талкудукская]
Doubtful.
A structure 500 metres in diameter is listed by B.S. Zeylik and E.Yu. Seytmuratova, Izvestia Akad. Nauk Kazakhstan S.S.R., Ser. Geol., 1975, vol.1, p.62.

Tasaral 46°29′ N., 73°42′ E.
Kazakhstan, USSR, [Тектониты , Тасарал-Кызылеспинский]
Doubtful.
An oval structure about 6.5 km by 2 km is associated with breccia and possibly glass, B.S. Zeylik and E.Yu. Seytmuratova, Izvestia Akad. Nauk Kazakhstan S.S.R., Ser. Geol., 1975, vol.1, p.62.

Tektite Crater v Wilkes Land.

Temimichât-Ghallaman Crater
Mauritania 24°15′ N., 9°39′ W.
Doubtful.
A crater in granitic rock, apparently of explosive origin, aligned on a fault with Tenoumer and Richât; diameter about 500 meters T. Monod, 19th Internat. Geol. Congr., fasc. 20, 1954, p.85, and other references in Monod, 1965, and O'Connell, 1965.
A reinvestigation and gravimetric reconnaissance of the crater, which may not be of impact origin, R.F. Fudali and W.A. Cassidy, Meteoritics, 1972, vol.7, p.51.

Tengiz 52° N., 70° E. approx
Kazakhstan, USSR
Synonym(s): *Ishim*
Doubtful.
A depression approximately 350 km in diameter which contains the Tengiz lake is possibly of impact origin, B.S. Zeylik and E Yu. Seytmuratova, Izvestia Akad. Nauk Kazakhstan S.S.R., Ser. Geol., 1975, vol.1, p.62.

Tenoumer 22°55′30″ N., 10°24′ W.
Mauritania
Doubtful.
This crater, 1800 metres across and 100 metres deep, with a flat bottom, and also two smaller craters, one 195 km to the NE. and the other 200 km to the SW., are regarded by J. Richard-Molard as explosion craters on a NE-SW fault (Compt. Rend. Acad. Sci. Paris, 1948, vol.227, p.213; Rev. Géogr., Lyon, 1949, vol.24, p.309) but as pointed out by L.J. Spencer, they bear a close resemblance to known meteorite craters, and the similar crater at Aouelloul in the same region is associated with silica glass very similar to the Darwin glass. See also A. Allix, Rev. Géogr., Lyon, 1951, vol.26, p.357; W. Campbell-Smith and M.H. Hey, Bull. Inst. Franç. Afrique Noire, 1952, vol.14, p.762; Monod, 1965; O'Connell, 1965.
Reinvestigation and gravity survey of the crater, R.F. Fudali and W.A. Cassidy, Meteoritics, 1972, vol.7, p.51. See also T. Monod and C. Pomerol, Soc. Géol. France Bull., ser. 7, vol.8, no.2, p.165. Genesis of melt rocks, R.F. Fudali, J. Geophys. Res., 1974, vol.79, p.2115.See also, S.R. Winzer et al., abs. in Meteoritics 1977, vol.12, p.389.

Ternovka v Ternovska.

Ternovska
Krivoi Rog, Ukraine, USSR, [Терновская]
Synonym(s): *Ternovka, Ternovskaia*
Doubtful.
A circular structure 6 km in diameter for which an origin by meteorite impact has been suggested. Description, shattercones, G.K. Ermenko and V.M. Iakovlev, Dokl. Akad. Nauk SSSR, 1980, vol.253, p.449. See also V.L. Masaitis et al., Dokl. Akad. Nauk SSSR, 1980, vol.255, p. 709. Stishovite in country rocks, E.P. Gurov, Min. Zhurn., 1982, vol.4 (2), p.75.

Ternovskaia v Ternovska.

The Firth 60°30′ N., 0°58′ W.
Shetland Islands, Scotland
Synonym(s): *Firth*
Doubtful.
A deep embayment is possibly a fossil meteorite crater, D. Flinn, Proc. Geol. Soc. London, 1970, no.1663, p.131.

Tibesti 21°30′ N., 17°30′ E.
Chad
Doubtful.
An impact origin has been suggested for a circular structure of about 18 km in diameter seen on photographs from the Gemini-IV space flight, see J.H. Freeberg, U.S. Geol. Surv. Bull., 1969, no.1320, p.32.

Tierra Del Fuego and Falkland Islands Arc
Doubtful.
Meteoritic origin is suggested by R. Gallant (Bombarded Earth, London, 1964; Monod, 1965).

Tiffin 41°48′ N., 91°40′ W.
Johnson County, Iowa, U.S.A.
Discredited.
An oval depression, 100 by 75 feet across and 8 feet deep, in glacial gravels is possibly of meteoritic origin (C.C. Wylie, Pop. Astron., Northfield, Minnesota, 1937, vol.45, p.445; 1938, vol.46, p.221; J.D. Buddhue, ibid., 1938, vol.46, p.222).
Appears to be merely a swirl pit formed during floods (O'Connell, 1965).

Tiksi 71.7° N., 128.4° E.

Yakut A.S.S.R., USSR

Doubtful.

An origin by meteorite impact has been suggested for a crater of diameter 10 metres I.T. Zotkin and V.L. Tsvetkov, Astron. Vestnik, 1970, vol.4, p.55.

Tin Bider v Tademai †t.

Tindouf

Morocco

Doubtful.

A circular feature 3-4 km in diameter may be of meteoritic origin; see Monod, 1965.

Tobys' 63.3° N., 53.1° E.

Komi A.S.S.R., USSR

Doubtful.

Thirteen "conical" depressions and circular lakes, some 50 metres in diameter are possibly of impact origin, I.T. Zotkin And V.L. Tsvetkov, Astron. Vestnik, 1970, vol.4, p.55.

Tokrauskaya 47°44′ N., 75°29′ E.

Kazakhstan, USSR, [Токрауская]

Doubtful.

A roughly circular structure 220-250 km in diameter displaying shattercones, glass and breccias, may be of impact origin, B.S. Zeylik and E.Yu. Seytmuratova, Izvestia Akad. Nauk Kazakhstan S.S.R., Ser. Geol., 1975, vol.1, p.62.

Trafalgar Bay 48°19′ N., 90°39′ W.

Ontario, Canada

Doubtful..

Listed as possible meteorite impact structure, P.B. Robertson and R.A.F. Grieve, Journ. Roy. Astron. Soc. Canada, 1975, vol.69, p.1.

Tsepochkino 57.2° N., 50° E.

Kirov Province, USSR

Doubtful.

Seven "conical" depressions 80 metres in diameter are of doubtful impact origin, I.T. Zotkin and V.L. Tsvetkov, Astron. Vestnik, 1970, vol.4, p.55.

Tunguska 60°54′ N., 101°57′ E.

Yeniseisk, Siberia, USSR, [Тунгуска]

1908, June 30, 0016 U.T. hrs, formed at this time

Synonym(s): *Podkamennaye Tunguska, Tunguska River*

Authenticated.

A large number of holes up to 50 metres across and 4 metres deep were found in swampy ground, but owing to the nature of the ground no meteoritic material has been recovered. Around the impact site the forest had been destroyed; at an average radius of 15-20 km, there is a zone where the trees remain standing, but beyond this, over a zone 30-40 km in diameter, all the trees had fallen radially outwards; farther out, a proportion of the trees had been felled, and some damage extended out to some 120 km.

Additional observations at the time of the fall, see L.A. Kulik, Meteoritika, 1941, vol.2, p.119. It is suggested that an area of dead forest in the basin of the river Ket, about 100 km north of the place of fall of the Tunguska meteorite, may be due to a fragment of the same fall (P.L. Dravert, Meteoritika, 1948, vol.4, p.112). The energy required to produce the air and earthquake waves is calculated as 5×10^{23} ergs, corresponding to the kinetic energy of a mass of 10^{7} tons with a velocity of 10 km/sec, or of 10^{5} tons at 100 km per sec. It is suggested that as no meteoritic material has been found, the meteorite must have been "contraterrene"; a much smaller mass totally annihilated, would of course give the same amount of energy (L. LaPaz, Pop. Astron., Northfield, Minnesota, 1948, vol.56, p.330).

Non-meteoritic origin has been suggested by P.N. Chirvinsky (Mém. Soc. Russe Min., 1931, ser.2, vol.60, p. 135).

The bolide probably exploded before impact, and it has been suggested (V.G. Fesenkov, Soviet Astron., 1962, vol.5, p.441; Meteoritika, 1964, vol.25, p.163; Priroda, 1962, vol.8, p.24) that it was the head of a comet, possibly the Pons-Winnecke.

See also E.L. Krinov, Chem. Erde, 1958, vol.19, no.3, p. 207, and translation, S. and B. Barringer, Internat. Geol. Review, 1960, vol.2, p.8; K.P. Florenskiy, Meteoritika, 1963, vol.23, p.3, preliminary results from the 1961 combined Tunguska meteorite expedition.

See also V.G. Fesenkov, Meteoritika, 1968, vol.28, p.107; I.T. Zotkin and M.A. Tsikulin, ibid., p.114. X-ray study of meteoritic dust, N.I. Zaslavskaya, ibid., p.142 and I.A. Yudin et al., ibid., p.158. Further discussion on shape of impacting wave, I.T. Zotkin, ibid., 1972, vol.31, p.35. A.A. Jackson and M.P. Ryan (Nature, 1973, vol.245, p.88) suggest that the event was caused by a "black hole", but compare G. L. Wick and J.D. Isaacs, Nature, 1974, vol.247, p.139, and W. Beasley and B.A. Tinsley, ibid., vol.250, p.555.

Numerous additional references, see main meteorite list, also Monod, 1965; O'Connell, 1965; J.H. Freeberg, U.S. Geol. Surv. Bull., 1966, no.1220, p.78, and 1969, no.1320, p. 32. High explosive analogue, Nature, 1977, vol.267, p.605. Analyses of silicate spherules and associated metallic phases, B.P. Glass, abs in Meteoritics, 1976, vol.11, p.287. Experiment to measure antimatter content, H. Cranell, Nature, 1974, vol.248, p.396. Composition of silicate spherules near impact point, Yu.A. Dolgov et al., Dokl. Acad. Sci. USSR, Earth Sci. sect., 1971, vol.200, p.212 (translated from Dokl. Akad. Nauk SSSR, 1971, vol.200, p. 201).

Specimen(s) : [1932,1580] basalt; [1932,1581] basalt

Tunguska River v Tunguska.

Tvären Bay 58°46′ N., 17°25′ E.

Sweden

Doubtful.

A circular depression, 2 km in diameter, in the middle of the shallow Tvären Bay is included as possibly of meteoritic origin by K. Fredriksson and F.E. Wickman (Svensk Naturvetenskap., 1963, p.121).

Tverdovo 53.2° N., 34.7° E.

Smolensk Province, USSR

Doubtful.

A dried out circular lake 100 metres in diameter may be of meteoritic origin, I.T. Zotkin and V.L. Tsvetkov, Astron. Vestnik, 1970, vol.4, p.55.

Tyuptyalir 69.9° N., 124.9° E.

Yakut A.S.S.R., USSR

Doubtful.

Two lakes 300 metres in diameter are of doubtful impact origin, I.T. Zotkin and V.L. Tsvetkov, Astron. Vestnik, 1970, vol.4, p.55.

Ubehebe Craters 36°59′ N., 117°32′ W.
California, U.S.A.
Discredited.
Two craters, one 200 ft. in diameter, the other 500 ft, are apparently explosion craters, but there does not appear to be any evidence of meteoritic origin (O'Connell, 1965).

Udjei Bowl 53°45′ N., 93°10′ E.
Sayan Mts., Krasnoyarsk Province, USSR
Synonym(s): *Sayan Crater, Western Sayan Crater*
Discredited.
Meteoritic origin is postulated for a large oval depression with a raised central oval structure in clayey soil (M.V. Voroshilov, Priroda, 1962, vol.3, p.107).

Um-Hadid 21°30′ N., 50°40′ E. approx
Saudi Arabia
Authenticated.
A small crater less than 10 metres in diameter has been found in the region of the Wabar craters. A number of deeply weathered and fractured fragments of iron meteorite, the largest weighing about 1 kg, were found. Silica glass is present, F. El-Baz and A. El Goresy, Meteoritics, 1971, vol. 6, p.265. See also main meteorite list and under Wabar, to which Um-Hadid is probably related.

Ungava Bay 60° N., 67°20′ W.
Quebec, Canada
Doubtful.
Three sides of the bay form an irregular circular area; the feature is included in lists of possible meteorite craters by C. S. Beals et al., (references, see O'Connell, 1965; Monod, 1965).

Ungava Crater v Chubb Crater.

Unnamed Lake 64°58′ N., 87°41′ W.
North-west Territories, Canada
Doubtful.
Listed with co-ordinates, M.R. Dence, Proc. 24th Internat. Geol. Cong., Section 15, 1972, p.77.

Upheaval Dome 38°26′ N., 109°54′ W.
San Juan County, Utah, U.S.A.
Synonym(s): *Christmas Canyon Dome*
Discredited.
A circular crater-like disturbed area has been assigned possible meteoritic origin, but is most probably a collapsed salt dome (references see O'Connell, 1965, and J.H. Freeberg, U.S. Geol. Surv. Bull., 1966, no.1220, p.82). Listed as "possible" impact crater, M.R. Dence, Proc. 24th Internat. Geol. Cong.,Section 15, 1972, p.77.

Ussuri Fall v Sikhote-Alin.

Ust′-Ozernyi 58.9° N., 87.7° E.
Krasnoyarsk Territory, USSR
Doubtful.
A crater 70 metres in diameter and 10 metres deep may be of impact origin, I.T. Zotkin and V.L. Tsvetkov, Astron. Vestnik, 1970, vol.4, p.55.

Ust′-Vikhorevo 56.7° N., 101.4° E.
Irkutsk Province, USSR
Doubtful.
A " conical" depression 10 metres in diameter may be an impact crater, I.T. Zotkin and V.L. Tsvetkov, Astron. Vestnik, 1970, vol.4, p.55.

Uzhur 55.2° N., 90.3° E.
Krasnoyarsk Territory, USSR
Doubtful.
A depression is of doubtful impact origin, I.T. Zotkin and V. L. Tsvetkov, Astron. Vestnik, 1970, vol.4, p.55.

Veevers Crater 22°58′6″ S., 125°22′7″ E.
Western Australia, Australia
Doubtful.
A circular structure, 80 metres in diameter and with raised rim, situated between the Great Sandy and Gibson Deserts is possibly of meteorite impact origin, A.N. Yeates et al., BMR. J. Austr. Geol. Geophys., 1976, vol.1, p.77.

Versailles Disturbance 38°2′ N., 84°45′ W.
Wood County, Kentucky, U.S.A.
Doubtful.
Included as a cryptoexplosion structure by D.F.B. Black (U. S. Geol. Surv. Prof. Paper, 1964, no.501, p.B9). Co-ordinates from J.H. Freeberg, U.S. Geol. Surv. Bull., 1966, no.1220, p. 82, and ibid., 1969, no.1320, p.34. Geophysical investigation is consistent with an impact origin for the stucture, C.R. Seeger, Bull. Geol. Soc. Amer., 1972, vol.83, p.3515.

Viewfield 49°35′ N., 103°4′ W.
Saskatchewan, Canada
Doubtful.
A bowl-shaped depression 1.5 miles in diameter is possibly an old meteorite crater, H.B. Sawatzky, J. Canadian Soc. Explor. Geophysicists, 1972, vol.8, p.22 and Bull. Amer. Assoc. Petrol. Geologists, 1975, vol.59, p.694. See also, H.B. Sawatzky, J. Canadian Soc. Exploration Geophysicists, 1974, vol.10, p.23 and Impact and Explosion Cratering, eds D.J. Roddy, R.O. Pepin and R.B. Merrill, Pergamon Press (New York), 1977, p.461.

Vinogradovo 55.5° N., 38° E.
Moscow Province, USSR
Doubtful.
Five depressions 150 metres in diameter are possibly of meteoritic origin, I.T. Zotkin and V.L. Tsvetkov, Astron. Vestnik, 1970, vol.4, p.55.

Vitim 53.5° N., 112.5° E.
Buryat-Mongolian A.S.S.R., USSR
Doubtful.
A circular lake 200 metres in diameter may be of impact origin, I.T. Zotkin and V.L. Tsvetkov, Astron. Vestnik, 1970, vol.4, p.55.

Vostochnoye 53.5° N., 68.2° E.
Kokchetav Province, USSR
Doubtful.
A circular crater 200 metres in diameter and 35 metres deep and with a lake, is of doubtful meteoritic origin, I.T. Zotkin and V.L. Tsvetkov, Astron. Vestnik, 1970, vol.4, p.55.

Voyevodskoye 53.8° N., 85.6° E.
Altai Territory, USSR
Doubtful.
Several "conical" depressions 40 metres in diameter are of doubtful impact origin, I.T. Zotkin and V.L. Tsvetkov, Astron. Vestnik, 1970, vol.4, p.55.

Vredefort Ring 27° S., 27°22′ E.

Orange Free State, South Africa

Doubtful.

J.D. Boon and C.C. Albritton suggested that this structure might be meteoritic (Field and Laboratory, Contr. Sci. Dept. Southern Methodist Univ., Dallas, Texas, 1936, vol.5, p.1[M. A. 6-399]; ibid., 1938, vol.6, p.44 [M.A. 7-178], and this idea is elaborated by R.A. Daly (Journ. Geol., Chicago, 1947, vol. 55, p.125).

More recent studies show intense shattering of rocks, with shattercones; the origin of the structure is still in dispute (references, Monod, 1965; O'Connell, 1965; J.H. Freeberg, U.S. Geol. Surv. Bull., 1966, no.1220, p.82, and 1969, no. 1320, p.34.

Occurrence of coesite and stishovite, J.E.J. Martini, Nature, 1972, vol.272, p.715, but compare P.A. Lilly, ibid., 1979, vol.277, p.495. Composition of spherules and other features on shattercones, see N.C. Gay, Science, 1976, vol. 194, p.724 and N.C. Gay et al., Earth Planet. Sci. Lett., 1978, vol.41, p.372. Shattercones post-date overturning and formation of fault breccia, apparently precluding an impact origin, C.J. Simpson, J. Geophys. Res., 1981, vol.86, p.10701.

Vyapryayskaya

Lithuanian SSR, USSR

Doubtful.

A structure 8 km in diameter, with explosion breccias and shattercones, is listed by V.L. Masaitis et al., Dokl. Akad. Nauk SSSR, Ser. Geol., 1978, vol.240, no.5, p.1191 (English translation, p.91).

Wabar 21°29′59″ N., 50°28′20″ E.

Rub′ al Khali, Saudi Arabia

1932, February, recognized

Synonym(s): *Al-Hadida, Rub'al Khali Craters*

Authenticated.

Two craters were found, 100 × 100 and 55 × 40 metres across, the larger one 10.5 metres deep but partly filled with drifted sand, and there are indications of several others buried under the sand. Iron meteorites were found in association with the craters. There is abundant silica-glass in the walls of the craters, which were formed in a very pure sandstone; the silica glass is in part white and pumiceous, in part black, and contains numerous minute spherules of metallic nickel-iron (Ni=8.8%; Fe:Ni=10.4), body centred cubic (a = 2.856 A.) with indistinct martensitic structure suggesting rapid cooling. Chemical analysis shows the silica glass to have the same composition as the sandstone except for the loss of volatiles and the addition of iron and nickel in the ratio 15:1 (the Fe:Ni ratio of the associated iron meteorites is 12.6). These data clearly support the theory that meteorite craters are of an explosive nature, originating from the kinetic energy of the falling mass and involving fusion and even vaporization of metal and rock.

The craters are described by H. St. J. Philby (The Empty Quarter, London, 1933; Geogr. Journ. London, 1933, vol.81, p.1; Journ. Roy. Central Asian Soc. London, 1932, vol.19, p. 569) and L.J. Spencer (Nature, 1932, vol.129, p.781; Geogr. Journ. London, 1933, vol.81, p.227), and the meteorites, sandstone, and silica glass described by L.J. Spencer, with analyses by M.H. Hey (Min. Mag., 1933, vol.23, p.387) and examined spectrographically by E. Preuss (Chem. Erde, 1935, vol.9, p.365).

Coesite has been found (E.C.T. Chao et al., Science, 1961, vol.133, p.882). Fission-track age of crater glass, approximately 6400 years, D. Störzer, abs. in Meteoritics, 1971, vol.6, p.319; see also F. El-Baz and A. EL Goresy, ibid., p.265. Additional references; Monod, 1965; O'Connell, 1965; J.H. Freeberg, U.S. Geol. Surv. Bull., 1969, no.1320, p. 34.

Metallic spherules in impactites, R.V. Gibbons, et al., Geochim. Cosmochim Acta, 1976, Suppl.7, vol.1, p.863.

Specimen(s) : Iron, iron-shale, sandstone, and silica glass, see main meteorite list.

Watterson Lake 65°13′ N., 99°23′ W.

Keewatin district, North-west Territories, Canada

Doubtful.

Listed as possible impact structure, P.B. Robertson and R.A. F. Grieve, Journ. Roy. Astron. Soc. Canada, 1975, vol.69, p. 1.

Weddell Sea

Antarctica

Doubtful.

Meteoritic origin is suggested by R. Gallant (Bombarded Earth, London, 1964; Monod, 1965).

Weepra Park Depressions v Eyre Peninsula Craters.

Wells Creek Basin 36°23′ N., 87°40′ W.

Tennessee, U.S.A.

Doubtful.

Five craters, from 2 by 3 miles oval to 375 ft. diameter, occur in a line; there is extreme brecciation and shattercones indicating explosive force from above (C.W. Wilson, Jr., Bull. Geol. Soc. Amer., 1953, no.64, p.753; W.H. Bucher, Nature, 1963, vol.197, p.1241; and Amer. Journ. Sci., 1963, vol.261, p.597; R.S. Dietz, Journ. Geol., 1959, vol.67, p.496; Science, 1960, vol.131, p.1781. Impact origin is supported by the orientation and distribution of shattercones, R.G. Stearns et al., in Shock Metamorphism of Natural Materials, eds. B. M. French and N.M. Short, Mono, Baltimore, 1968, p.323; see also C.W. Wilson et al., Geol. Soc. Amer. Spec. Paper 101, 1968, p.241. Additional references, O'Connell, 1965; J. H. Freeberg, U.S. Geol. Surv. Bull., 1969, no.1320, p.34.

Western Sayan Crater v Udjei Bowl.

West Hawk Lake 49°46′ N., 95°12′ W.

Manitoba, Canada

Doubtful.

A roughly circular lake 3 miles in diameter is included by C. S. Beals et al., (Current Science, Bangalore, 1960, vol.29, pp. 205 and 249) in a list of probable meteorite craters; see also I. Halliday and A.A. Griffin, Journ. Geophys. Res., 1963, vol.68, p.5297; Meteoritics, 1964, vol.2, p.79. A magnetic survey indicates a crater 2.7 km in diameter, J.F. Clark, abs. in Meteoritics, 1969, vol.4, p.268. See also M.R. Dence et al., in Shock Metamorphism of Natural Materials, eds. B.M. French and N.M. Short, Mono, Baltimore, 1968, p.339; full description, with petrography, N.M. Short, Bull. Geol. Soc. Amer., 1970, vol.81, p.609. Age, 100±50 m.y., P.B. Robertson and R.A.F. Grieve, Journ. Roy. Astron. Soc. Canada, 1975, vol.69, p.1. Additional references, J.H. Freeberg, U.S. Geol. Surv. Bull., 1966, no.1220, p.85, and 1969, no.1320, p.35. See also I. Halliday and A.A. Griffin, Journ. Roy. Astron. Soc. Canada, 1966, vol.60, p.59.

Wetherbee Crater v Labrador Crater.

Wilbarger County 34°10′ N., 99°13′ W.

Texas, U.S.A.

Doubtful..

O.E. Monnig (Meteoritics, 1963, vol.2, p.71) suggests that this upthrust may be of meteoritic origin.

Wilkes Land — 71° S., 140° E.

Antarctica
Synonym(s): *Tektite Crater*
Doubtful.

An area 150 miles in diameter with negative gravitational anomaly is suggested as a meteorite crater (buried under the ice) and as the source of the australites by V.E. Barnes (Sci. American, 1961, vol.205, no.5, p.58), A.J. Cohen (3rd Internat. Space Symposium, 1962), and R.A. Schmidt (Science, 1962, vol.138, p.443). See also R.A. Schmidt, Sci. American, 1962, vol.206, no.2, p.12, and V.E. Barnes, ibid., p.12. Evidence supporting an impact origin, J.G. Weihaupt, J. Geophys. Res., 1976, vol.81, p.5651; but compare C.R. Bently, ibid., 1979, vol.84, p.5681.

Willenhofen — 49°14′ N., 11°42′ E.

Bayern, Germany
Doubtful.

Several small craters may be of meteoritic origin, E. Rutte, Geoforum, 1971, vol.7, p.84, and Oberrhein. geol. Abh., 1974, vol.23, p.97.

Wiltshire Crater — 50°2′ N., 2°12′ W.

England
Discredited..

A small hole, 8 feet in diameter, was reported as due to the fall of a meteorite; no meteoritic material was found, and the hypothesis of an ice meteorite was advanced (P. Moore, New Scientist, 1963, vol.19, p.304; W.S. Houston, ibid., p.567). It is quite clear that there was no meteorite.

Winkler — 39°29′ N., 96°49′ W.

Riley County, Kansas, U.S.A.
Discredited.

A circular depression, 2500 feet in diameter filled with loess, is regarded as a meteorite crater by W.S. Houston (Progr. 24th meeting Meteoritical Soc., 1961). See also R. Barringer, Meteoritics, 1964, vol.2, p.169. Petrographic study, drilling, and magnetic survey failed to yield evidence of meteoritic impact. The crater was probably formed by kimberlite emplacement followed by some limestone solution, D.G. Brookins, abs. in Meteoritics, 1969, vol.4, p.263.

Winslow Crater v Meteor Crater.

Wipfelsfurt — 48°51′ N., 11°50′ E.

Bayern, Germany
Doubtful.

A circular structure dissected by the River Danube is possibly an impact crater; shattercones, E. Rutte, Geoforum, 1971, vol.7, p.84, also Oberrhein geol. Abh., 1974, vol.23, p. 97.

Wolf Creek — 19°18′ S., 127°46′ E.

Wyndham, Kimberley, Western Australia, Australia
Authenticated.

A large circular crater was first observed from the air in June 1947; it is 2800 feet across and 160 feet deep, with a rim of shattered sandstone tilted outwards and rising 60 to 100 feet above the surroundings; the level floor, 70 feet below the surrounding land, is covered with porous gypsum and sand, and some trees are growing on the floor. The rim is of angular quartzite blocks, and there are no signs of volcanic activity. Fragments of iron shale are abundant on the SW. part of the crater rim; they contain 3.5-4.5% Ni (earlier analyses gave 1.9%), and some still contain a little metal. See H.H. Holmes, Walkabout, Australian Geogr. Mag., 1948, vol.14, no.13, p.10 [M.A. 10-523];. F. Reeves and R.O. Chambers, Australian Journ. Sci., 1949, vol.11, p. 154 (the latitude is given here as 19 degrees 10 minutes S); F. Reeves and N.B. Sauve, Rocks and Minerals, 1949, vol. 24, p.592; D.J. Guppy and R.S. Matheson, Journ. Geol., Chicago, 1950, vol.58, p.30.

For recent literature, see O'Connell, 1965 and J.H. Freeberg, U.S. Geol. Surv. Bull., 1966, no.1220, p.85, and 1969, no.1320, p.35. Gravity investigation; depth inferred for crater floor below rim-crest is 140-150 metres; rim-crest is 40-50 metres above the level of the gently undulating surface of the country rock, R.F. Fudali, J. Geol., 1979, vol.87, p.55. *Specimen(s)* : see main meteorite list.

Yanis'yarvi v Janisjärvi.

Zapadno-Akkuduksk — 47°1′ N., 73°41′ E.

Kazakhstan, USSR, [Западно-Аккудукская]
Doubtful.

A circular structure about 18 km across is listed by B.S. Zeylik and E.Yu. Seytmuratova, Izvestia Akad. Nauk Kazakhstan S.S.R., Ser. Geol., 1975, vol.1, p.62.

Zapadno-Karkaralinsk — 49°25′ N., 75° E.

Kazakhstan, USSR, [Западно-Каркаралинская]
Doubtful.

A structure about 200 metres in diameter is listed by B.S. Zeylik and E.Yu. Seytmuratova, Izvestia Akad. Nauk Kazakhstan S.S.R., Ser. Geol., 1975, vol.1, p.62.

Zeleny Gay v Zelenyy Gay.

Zelenyy Gay

Kirovograd oblast, Ukraine, USSR, [Зеленогаыск]
Synonym(s): *Zeleny Gay*
Doubtful.

A circular structure, approx. 1.3 km in diameter, in crystalline basement rocks beneath Cenozoic sedimentary cover, may be of meteorite impact origin and contemporaneous with the Boltysh (q.v.) and Rotmistrovka (qq.v.) structures with which it lies in line; gravity anomaly and petrographic evidence of shock metamorphism in borehole samples, A.A. Val'ter et al., Dokl. Akad. Nauk SSSR., 1976, vol.229, p.160. Planar elements in biotite, E.P. Gurov, Zap. Vses. Min. Obshch., 1977, vol.106, p.715.

Zhamanshin — 49° N., 59° E.

Aktyubinsk Province, USSR
Doubtful.

A structure approximately 15 km in diameter, with associated fused glass, for which an impact origin is suggested, I.T. Zotkin and V.L. Tsvetkov, Astron. Vestnik, 1970, vol.4, p.55. Listed with co-ordinates, M.R. Dence, Proc. 24th Internat. Geol. Cong., Section 15, 1972, p.77.

Extensive recent literature of which the following is a selection; discovery of tektites, P.V. Florensky, Meteoritika, 1977, vol.36, p.120; analyses of crater glasses and comparison with tektites, W.D. Ehmann et al., abs. in Meteoritics, 1977, vol.12, p.212. Geology and petrography of country rocks, P.V. Florensky et al., abs. in Meteoritics, ibid., p.227; and K. Fredriksson et al., ibid., p.229; see also V.L. Bouska et al., Meteoritics, 1981, vol.16, p.171; P.W. Florensky, Chem. Erde, 1977, vol.36, p.83; B.P. Glass, Geology, 1979, vol.7, p.351; P.V. Florensky, Izv. Akad. Nauk SSSR., ser. geol. 1975, vol.10, p.73; Astron. Vestnik, 1975, vol.9, no.4, p.237; J.A. Philpotts et al., abs. in Meteoritics, 1977, vol.12, p.338; P.V. Florensky et al.,

Meteoritika, 1979, vol.38, p.86; P.V. Florensky et al., Astron. Vestnik, 1979, vol.13, p.178.

Full description see Zhamanshin Meteorite Crater by P.V. Florensky and A.N. Dabizha, Nauka Press, Moscow, 1980, 127pp. For specimens see tektite list.

Zhuan-Tobe 47°8′ N., 73°51′ E.

Kazakhstan, USSR, [Жуан–Тобе]

Doubtful..

A circular structure 15-20 km in diameter is mentioned, B.S. Zeylik and E.Yu. Seytmuratova, Izvestia Akad. Nauk Kazakhstan S.S.R., Ser. Geol., 1975, vol.1, p.62.

Catalogue of Tektites in the British Museum (Natural History) Collection

TEKTITE

MOLDAVITE

Pisek, Bohemia, Czechoslovakia
Specimen(s): [63655], 45g. 3 spec.

Moldauthein, Bohemia, Czechoslovakia
Specimen(s): [32948], 23g.; [91606], 10g.; [91608], 29g.; [91609], 48g. 3 spec.; [93483], less than 1g. faceted

Radomilice, Bohemia, Czechoslovakia
Specimen(s): [1922,1150], 54g. 20 frag.; [1933,218], 56g. 4 spec.; [1950,408], 21g. 2 spec.

Malovice, Bohemia, Czechoslovakia
Specimen(s): [1933,215], 13g.

Netolice, Bohemia, Czechoslovakia
Specimen(s): [1933,217], 80g. 5 spec.; [1950,400], 18g.

Chrastany, Bohemia, Czechoslovakia
Specimen(s): [1933,211], 56g. 4 spec.

Dolni Chrastany, Bohemia, Czechoslovakia
Specimen(s): [1950,403], 7g. 2 spec.

Lhenice, Bohemia, Czechoslovakia
Specimen(s): [1933,214], 18g.; [1950,393], 3g.; [1950,394], 8g.

Locenice, Bohemia, Czechoslovakia
Specimen(s): [1977,M.5 (a-h)], 40g. 8 spec.

Trebanice, Bohemia, Czechoslovakia
Specimen(s): [1933,220], 14g.

Zahori, Bohemia, Czechoslovakia
Specimen(s): [86551], 90g. 2 spec.

Habri, Bohemia, Czechoslovakia
Specimen(s): [1927,122], 3g.

Ceske Budejovice, Bohemia, Czechoslovakia
Specimen(s): [1925,1086], 15g.; [1925,1088], 18g.; [1927,118], 15g.; [1927,121], 9g.; [1950,395], 2g.; [1950,396], 18g. 3 spec.

Dobrkovska Lhotka, Bohemia, Czechoslovakia
Specimen(s): [1979,M.3], 10.5g.

Vrabce and Slavce, Bohemia, Czechoslovakia
Specimen(s): [1933,219], 21g. 3 spec.; [1933,221], 12g.; [1935,1158], 108g. 19 spec.; [1950,399], 24g. 6 spec.; [1950,404], 2g.; [1950,405], 16g. 7 spec.; [1950,406], 6g.; [1961,212], 13g.; [1961,213], 18g.; [1962,140], 6g.

Vodice, Bohemia, Czechoslovakia
Specimen(s): [1971,297], 3g.

Koroseky, Bohemia, Czechoslovakia
Specimen(s): [1925,1087], 6g. 2 spec.; [1933,212], 32g. 4 spec.; [1935,1157], 10g.; [1950,398], 7g.; [1950,401], 15g. 4 spec.; [1950,402], 10g. 4 spec.

Kroclov, Bohemia, Czechoslovakia
Specimen(s): [1933,213], 16g.

Nechov, Bohemia, Czechoslovakia
Specimen(s): [1933,216], 21g. 3 spec.; [1950,397], 5g.

Localities unknown, Bohemia, Czechoslovakia
Specimen(s): [91607], 11g.; [1911,277], 54g. 7 spec.; [1927,1410], 5g. faceted; [1935,263], 29g. 4 spec.; [1974,M.10], 0.95g. 2 spec., faceted

Trebic, Moravia, Czechoslovakia
Specimen(s): [1925,1085], 15g.; [1927,119], 24g.; [1937,120], 20g.

Teruvky, Moravia, Czechoslovakia
Specimen(s): [1933,223], 17g.

Teltsch, Moravia, Czechoslovakia
Specimen(s): [1925,629], 61g. 2 spec.

Skryje, Moravia, Czechoslovakia
Specimen(s): [1933,222], 22g.

Dukovany, Moravia, Czechoslovakia
Specimen(s): [1950,407], 7g.

Locality not known, Moravia, Czechoslovakia
Specimen(s): [1906,129], less than 0.5g faceted; [1950,417], 3g. two polished drops

INDOMALAYSIANITE

Kaliosso, Solo, Java, Indonesia
Specimen(s): [1937,1638], 23g. 4 spec.; [1937,1639], 33g.; [1937,1640], 76g. 7 spec.; [1937,1641], 14g. 11 frag.; [1937, 1642], 21g. 5 spec.; [1937,1643], 27g. 9 spec.; [1937,1644], 7g. 4 spec.; [1937,1645], 24g. 6 spec.

Billiton Island, Indonesia
Specimen(s): [1925,1084], 100g. 7 spec.; [1933,439], 19g. 4 spec.; [1938,104], 34g.; [1938,388], 3g. 3 spec.; [1958,580], 14g.; [1958,705], 7g.; [1958,706], 7g.; [1938,707], 7g.; [1958, 708], 8g.; [1959,36], 20g.; [1959,37], 16g.

Locality unknown, Indonesia
Specimen(s): [1927,128], 4g.; [1927,129], 7g.; [1927,130], 2g.; [1927,131], 7g.; [1927,132], 34g.; [1927,133], 14g.; [1927,134], 15g.; [1927,135], 12g.; [1927,136], 19g.

Batu Gajah, Perak, Malaysia
Specimen(s): [1927,117], 24g.

Gambang and dist., Pahang, Malaysia
Specimen(s): [1923,52], 41g. 3 spec.; [1934,635], 144g. 8 spec.; [1934,663], 65g. 5 spec.; [1949,135], 32g.; [1951,312], 29g.

Butir, Brunei Bay, Sabah, Malaysia
Specimen(s): [1968,184], 17g.

Locality not known, Malaya, Malaysia
Specimen(s): [1926,679], 368g. in three pieces; [1926,680], 317g.

INDOCHINITE

Kan Luang Dong, Nakhon Phanom province, Thailand
Specimen(s): [1963,944], 160g. 10 spec.

Sukhotai, Thailand
Specimen(s): [1983,M.3], 62g.

Phang Daeng, Nakhon Phanom province, Thailand
Specimen(s): [1963,737], 203g. 11 spec.

Khorat plateau, Nakhom Rachasima province, Thailand
Specimen(s): [1970,78], 764g.

Plain of Jars, Laos
Specimen(s): [1972,7 (a-g)], 137.5g. 7 spec.

Locality not known, Thailand
Specimen(s): [1964,654], 1016g. 17 specs.; [1964,655], 1008g. 41 specs.; [1964,656], 969g. 361 specs.

northern Cambodia, Cambodia
Specimen(s): [1933,100], 34g. 5 spec.

Dalat, Vietnam
Specimen(s): [1950,409], 24g. 8 spec.; [1964,749], 16g.; [1964, 750], 43g.; [1964,751], 115g.; [1964,752], 65g.; [1964,753], 42g. 3 spec.; [1964,754], 59g. 5 spec.; [1964,755], 84g. 4 spec.; [1964,756], 40g. 5 spec.; [1964,757], 138g. 14 spec.; [1964,758], 36g. 15 spec.; [1965,263], 25g. 2 spec.; [1965, 264], 9g. 2 spec.; [1965,265], 30g. 4 spec.; [1973,M.19 (a-c)], 63g. 3 spec.

Lang Bian plateau, Vietnam
Specimen(s): [1933,101], 133g. 6 spec.

South-East Asia, exact locality unknown,
Specimen(s): [1967,57], 115g.; [1975,M.15], 185g.

Tan-Hai Island, Kwangchow-Wan, China
Specimen(s): [1933,99], 186g. 6 spec.

PHILIPPINITE

Manila, Luzon, Philippine Islands
Specimen(s): [1958,709], 10g.; [1958,710], 11g.; [1958,711], 18g.; [1958,712], 23g.; [1959,38], 55g.; [1959,39], 32g.; [1959, 40], 13g.; [1969,233], 80g.; [1969,234], 90g. 5 spec.; [1969, 235], 85g. 7 spec.; [1969,236], 39g. 4 spec.

Anda, Cabarruyan Island, Pangasinan province, Luzon, Philippine Islands
Specimen(s): [1959,503], 12g.; [1967,420], 70g. 6 spec.

Busuanga Island, Palawan, Philippine Islands
Specimen(s): [1934,774], 74g. 13 spec.

Babuyan, Luzon, Philippine Islands
Specimen(s): [1966,276], 26g. 4 spec.

Salapan, Luzon, Philippine Islands
Specimen(s): [1966,277], 10g. 4 spec.

Santa Mesa, Luzon, Philippine Islands
Specimen(s): [1966,275], 14g. 4 spec.

Pagrayanan, Isabela province, Luzon, Philippine Islands
Specimen(s): [1967,419], 148g. 10 spec.

Cubao, Rizal province, Luzon, Philippine Islands
Specimen(s): [1967,421], 45g. 4 spec.

Coco Grove, Province Camarines, Luzon, Philippine Islands
Specimen(s): [1982,M.8], 457g.

Exact locality unknown, Luzon, Philippine Islands
Specimen(s): [1967,418], 24g.

BEDIASITE

Grimes County, Texas, U.S.A.
Specimen(s): [1939,877], 9g.; [1939,878], 30g.; [1939,879], 6g.; [1939,880], 13g.; [1939,881], 10g.; [1939,882], 18g.; [1939,883], 14g.; [1939,884], 9g.; [1939,885], 16g.; [1939,886], 10g.; [1943,38], 43g.; [1943,39], 14g.; [1943,40], 13g.; [1943,41], 14g.; [1943,42], 20g.; [1943,43], 15g.; [1962,139], 34g. 4 spec.

Lee county, Texas, U.S.A.
Specimen(s): [1957,682], 2g.; [1968,341], 176g. 22 spec.; [1967,58], 91g. 11 spec.

Somerville, Texas, U.S.A.
Specimen(s): [1974,M.9], 0.45g.

GEORGIA TEKTITE

Dodge County, Georgia, U.S.A.
Specimen(s): [1973,M.36 (a-d)], 32.9g. 4 spec.; [1976,M.13], 19.9g.

AUSTRALITE

Peake Station, Ashburton River, Western Australia, Australia
Specimen(s): [1925,1079], 10g.

Katanning, Western Australia, Australia
Specimen(s): [1925,1078], 7g.

Kurrawang, Western Australia, Australia
Specimen(s): [1935,260b], 8g.

Lake Ballard, Western Australia, Australia
Specimen(s): [1972,318], 17.4g.

300 miles north of Coolgardie, Western Australia, Australia
Specimen(s): [1935,261], 20g.

Coolgardie and district, Western Australia, Australia
Specimen(s): [1925,1069], 28g. 3 spec.; [1925,1070], 6g. 5 spec.; [1925,1071], 2g.; [1926,356], 8g.; [1926,406], 2g.; [1926,413], 6g.

Dry Lakes, near Coolgardie, Western Australia, Australia
Specimen(s): [1926,342], 1g.; [1926,366], 1g.; [1926,386], 4g.; [1926,402], 3g.; [1926,425], 1g.; [1926,426], 1g.

Kalgoorlie and district, Western Australia, Australia
Specimen(s): [85544], 54g. 15 spec.; [1916,372], 180g. 8 spec.; [1925,1072], 11g. 3 spec.

Bulong, Western Australia, Australia
Specimen(s): [1925,1065], 27g. 5 spec.; [1925,1066], 12g. 3 spec.; [1925,1067], 28g. 3 spec.; [1925,1068], 13g.; [1949,249], 7g.; [1949,250], 5g.; [1949,251], 4g.; [1949,252], 1g.; [1949,253], 1g.; [1949,254], less than 0.5g.; [1951,246], 2g.

Kanowna, Western Australia, Australia
Specimen(s): [1926,350], 14g.; [1926,392], 26g.

Kurnalpi, Western Australia, Australia
Specimen(s): [1935,260a], 12g.

Kurawah, Western Australia, Australia
Specimen(s): [1935,260b], 8g.

Broad Arrow, Western Australia, Australia
Specimen(s): [1925,1075], 2g.

Feysville, Western Australia, Australia
Specimen(s): [1925,1073], 8g.; [1925,1074], 4g.

Nolans, Western Australia, Australia
Specimen(s): [1935,262], 7g. 2 spec.

Edjudina, Western Australia, Australia
Specimen(s): [1935,259], 6g.; [1946,272], 50g. 6 spec.

Goongarrie, Western Australia, Australia
Specimen(s): [1935,258], 16g.

Nullarbor Plain, Western Australia, Australia
Specimen(s): [1927,127], 17g. 44 frag.; [1936,1300], 2g.; [1936,1301], 3g.; [1936,1302], 3g.; [1936,1303], 1g.; [1936,1304], 1g.; [1936,1305], 1g.; [1936,1306], 1g.; [1936,1307], 1g.; [1936,1308], 3g.; [1936,1309], 2g.; [1936,1310], 3g. 2 spec.; [1936,1311], 3g. 3 spec.; [1936,1312], 2g. 3 spec.; [1936,1313], 1g. 3 spec.; [1936,1314], less than 0.5g. 2 spec.; [1936,1315], 3g. 3 spec.; [1936,1316], 2g.; [1936,1317], 3g. 2 spec.; [1936,1318], 2g. 2 spec.; [1936,1319], 1g. 2 spec.;

[1936,1320], 2g.; [1936,1321], 3g. 2 spec.; [1936,1322], 2g. 2 spec.; [1936,1323], less than 0.5g.; [1936,1324], 2g.; [1936, 1325], 1g.; [1936,1326], 1g.; [1936,1327], 3g. 2 spec.; [1936, 1328], 2g. 2 spec.; [1936,1329], 2g. 2 spec.; [1936,1330], 2g. 2 spec.; [1936,1331], 1g. 2 spec.; [1982,M.25], 3.2g. three fragments

Exact locality unknown, Western Australia, Australia
Specimen(s): [1927,1161], 23g. 14 spec.

Localities unknown, Western Australia, Australia
Specimen(s): [1925,1076], 13g. 6 spec.; [1925,1077], 6g.; [1928,67], 44g. 15 spec.; [1939,214], 8g.; [1939,215], 7g.

Charlotte Waters, Northern Territory, Australia
Specimen(s): [1919,304], 22g. 3 spec.; [1926,337], 3g.; [1926, 358], 8g.; [1926,380], 28g.; [1926,381], 27g.; [1926,396], 17g.; [1926,398], 12g.; [1926,405], 3g.; [1926,408], 2g.; [1926,412], 7g.; [1926,423], 2g.; [1936,1332], 43g. 6 spec.

150 miles north of Oodnadatta, Northern Territory, Australia
Specimen(s): [1926,391], 13g.; [1926,397], 14g.; [1926,399], 12g.

Blood's Creek, South Australia, Australia
Specimen(s): [1966,37], 15g. 4 spec.

Alpara, near Pedirka, South Australia, Australia
Specimen(s): [1966,41], 68g. 4 spec.

Macumba Creek, South Australia, Australia
Specimen(s): [1966,40], 4g. 2 spec.; [1966,42], 36g. 2 spec.

Lake Eyre, South Australia, Australia
Specimen(s): [1935,257], 3g.

Warrina, South Australia, Australia
Specimen(s): [1935,256], 28g. 2 spec.

near Anna Creek, South Australia, Australia
Specimen(s): [1925,1056], 25g. 7 spec.; [1925,1057], 5g. 4 spec.; [1925,1058], 120g. 9 spec.; [1925,1059], 70g. 5 spec.; [1925,1060], 5g.; [1925,1061], 6g.; [1925,1062], 8g.; [1925, 1063], 27g.; [1925,1064], 36g. 4 spec.

William Creek, South Australia, Australia
Specimen(s): [1938,1182], 14g.; [1938,1183], 11g.; [1938, 1184], 28g.

Mungerannie, South Australia, Australia
Specimen(s): [1938,1185], 6g.; [1938,1186], 22g.; [1938,1187], 3g.; [1938,1188], 3g.; [1938,1189], 1g.; [1938,1190], 1g.; [1938,1191], 11g.; [1938,1192], 3g.

Lake Eyre dist., exact localities unknown, South Australia, Australia
Specimen(s): [1925,1050], 450g. 16 spec.; [1925,1051], 430g. 28 spec.; [1925,1052], 300g. 52 spec.; [1925,1053], 102g. 39 spec.; [1925,1054], 183g. 25 spec.; [1926,378], 6g.

Serpentine Lakes, South Australia, Australia
Specimen(s): [1966,38], 5g.; [1966,39], 10g. 5 spec.

Nullarbor Plain, South Australia, Australia
Specimen(s): [1921,604], 3g. 3 spec.

Ooldea, South Australia, Australia
Specimen(s): [1948,189], less than 0.5g.

Wellington, Murray River, South Australia, Australia
Specimen(s): [1935,252], 9g.

Maralinga, South Australia, Australia
Specimen(s): [1977,M.10], 1.86g. 2 spec.

near Abminga, South Australia, Australia
Specimen(s): [1972,317 (a-e)], 76.85g. 5 spec.

Mount Dare, South Australia, Australia
Specimen(s): [1972,500 (a-i)], 119.8g. 9 spec.

Curley Creek, Kangaroo Island, South Australia, Australia
Specimen(s): [1926,347], 44g.

Exact locality unknown, South Australia, Australia
Specimen(s): [1925,1055], 90g. fragments

Broken Hill and district, New South Wales, Australia
Specimen(s): [1935,253], 20g.; [1935,254], 28g.; [1935,255], 1g.

Uralla, Harding County, New South Wales, Australia
Specimen(s): [1927,1298], 14g.; [1927,1299], 16g. 3 spec.; [1927,1300], 19g. 12 spec.

near Craigie, Wellesley County, New South Wales, Australia
Specimen(s): [1926,344], 77g.

The Mallee, Victoria, Australia
Specimen(s): [1926,321], 6g.; [1926,354], 12g.; [1926,357], 7g.; [1926,418], 3g.

Lake Buloke, Victoria, Australia
Specimen(s): [1926,376], 9g.

Edenhope, Victoria, Australia
Specimen(s): [1927,1164], 21g.

Balmoral, Victoria, Australia
Specimen(s): [1927,1165], 52g.

Stony Creek, near Hall's Gap, Victoria, Australia
Specimen(s): [1926,330], 4g.; [1926,436b], 3g. fragments

Mount William, Victoria, Australia
Specimen(s): [1926,308], less than 0.5g.; [1926,309], less than 0.5g.; [1926,310], less than 0.5g.; [1926,311], less than 0.5g.; [1926,312], less than 0.5g.; [1926,313], 0.5g.; [1926,314], 0.5g.; [1926,315], less than 0.5g.; [1926,318], 8g.; [1926,319], 6g.; [1926,320], 6g.; [1926,322], 6g.; [1926,323], 7g.; [1926, 324], 6g.; [1926,325], 5g.; [1926,327], 4g.; [1926,329], 5g.; [1926,332], 3g.; [1926,333], 3g.; [1926,334], 2g.; [1926,335], 2g.; [1926,336], 2g.; [1926,339], 1g.; [1926,340], 1g.; [1926, 341], 1g.; [1926,343], 75g.; [1926,345], 57g.; [1926,348], 32g.; [1926,351], 11g.; [1926,352], 11g.; [1926,355], 8g.; [1926,359], 3g.; [1926,361], 3g.; [1926,362], 2g.; [1926,363], 2g.; [1926, 365], 1g.; [1926,367], 1g.; [1926,368], less than 0.5g.; [1926, 369], 6g.; [1926,370], 7g.; [1926,371], 6g.; [1926,372], 2g.; [1926,373], 1g.; [1926,377], 9g.; [1926,379], 33.; [1926,382], 21g.; [1926,383], 4g.; [1926,385], 3g.; [1926,387], 4g.; [1926, 388], 3g.; [1926,390], 15g.; [1926,394], 22g.; [1926,395], 16g.; [1926,401], 13g.; [1926,403], 2g.; [1926,404], 4g.; [1926,410], 13g.; [1926,411], 14g.; [1926,417], 3g.; [1926,420], 2g.; [1926, 421], 3g.; [1926,422], 4g.; [1926,436a], 6g. fragments

Hamilton, Victoria, Australia
Specimen(s): [1926,317], 9g.; [1926,326], 4g.; [1926,328], 4g.; [1926,331], 4g.; [1926,338], 1g.; [1926,349], 25g.; [1926,353], 9g.; [1926,374], 32g.; [1926,375], 37g.; [1926,407], 2g.; [1926, 415], 4g.; [1926,416], 1g.; [1926,424], 1g.

Condah Swamp, Victoria, Australia
Specimen(s): [1926,346], 41g.

South Portland, Victoria, Australia
Specimen(s): [1926,364], 2g.

Woolsthorpe, Victoria, Australia
Specimen(s): [1927,1163], 32g.

near Warrnambool, Victoria, Australia
Specimen(s): [1927,1162], 59g.

Port Campbell district, Victoria, Australia
Specimen(s): [1958,33], 3g.; [1958,35], 1g.; [1958,36], less than 0.5g.; [1958,37], less than 0.5g.; [1958,38], 2g.; [1958, 39], 3g. 2 spec.; [1980,M.31], 28g. 36 frag.

Hexham, Victoria, Australia
Specimen(s): [1926,393], 90g.

Rokewood, Victoria, Australia
Specimen(s): [1926,389], 1g.

Linton, Victoria, Australia
Specimen(s): [1926,316], less than 0.5g.

Napoleons, Victoria, Australia
Specimen(s): [1926,384], 4g.

Daylesford, Victoria, Australia
Specimen(s): [1926,360], 2g.

Exact localities unknown, Victoria, Australia
Specimen(s): [1927,1166], 34g.; [1927,1167], 173g.

Monkira, Queensland, Australia
Specimen(s): [1973,M.20 (a-c)], 16.47g.

Tasmania, Australia
Specimen(s): [33970], 7g. 3 spec.; [1950,414b], 1g.; [1950,415 (3)], 2g.

Localities unknown, Australia
Specimen(s): [1971,301], 1.35g.; [1926,400], 15g.; [1934,997], 4g. 4 spec.; [1925,1080], 8g. 4 spec.; [1925,1081], 3g.; [1925, 1082], 7g.; [1950,410], 46g. 7 spec.; [1950,411], 45g.; [1950, 412], 10g.; [1950,413], 17g. 4 spec.

IVORY COAST TEKTITE

Kongoti, Ivory Coast
Specimen(s): [1969,173], 10g.

Locality not known, Ivory Coast
Specimen(s): [1983,M.27], 7.9g.

IRGHIZITE

Zhamanshin, Aktyubinsk, USSR
Specimen(s): [1983,M.39], 1.8g.

OTHER NATURAL GLASSES

LIBYAN DESERT GLASS

Libyan Desert, Egypt
Description, with figures, P.A. Clayton and L.J. Spencer, Min. Mag., 1934, 23, p.501. The BM(NH) also holds 6.8kg of duplicate material and 16 thin sections.

Specimen(s): [1933,200], 318g.; [1933,224], 2279g.; [1933, 225], 260g.; [1933,226], 161g.; [1933,227], 122g.; [1933,228], 90g.; [1933,229], 10.57g. faceted; [1933,342], 2g.; [1934,651], 129g.; [1934,652], 103g.; [1934,653], 39g.; [1934,654], 116g. 4 pieces; [1935,70], 6030g.; [1935,71], 2985g.; [1935,72], 2890g.; [1935,73], 2303g.; [1935,74], 1232g.; [1935,75], 1181g.; [1935, 76], 1028g.; [1935,77], 1001g.; [1935,78], 999g.; [1935,79], 755g. 2 pieces; [1935,80], 708g.; [1935,81], 707g.; [1935,82], 473g.; [1935,83], 427g.; [1935,84], 405g.; [1935,85], 363g.; [1935,86], 362g.; [1935,87], 291g.; [1935,88], 266g.; [1935,89], 240g.; [1935,90], 199g.; [1935,91], 188g.; [1935,92], 187g.; [1935,93], 157g.; [1935,94], 157g.; [1935,95], 85g.; [1935,96], 85g.; [1935,97], 441g.; [1935,98], 374g.; [1935,99], 327g.; [1935,100], 326g.; [1935,101], 317g.; [1935,102], 284g.; [1935, 103], 276g.; [1935,104], 192g.; [1935,105], 257g.; [1935,106], 263g.; [1935,107], 90g.; [1935,108], 743g. 7 worked flakes; [1935,109], 415g. 48 worked flakes; [1935,110], 333g. 127 flakes; [1935,111], 28g. 8 pieces; [1935,112], 311g. 5 pieces; [1935,113], 19g. 3 pieces; [1935,114], 18g. 3 pieces; [1935, 115], 87g. 29 pieces; [1935,116], 37g.; [1935,117], 147g. 6 pieces; [1935,118], 137g.; [1935,119], 116g.; [1935,120], 215g. 5 pieces, 1 worked flake; [1935,121], 248g. 4 pieces; [1935, 122], 43g. 2 pieces; [1935,123], 40g.; [1935,124], 28g. 5 pieces; [1935,125], 17g.; [1935,126], 93g. 6 pieces; [1935,127], 132g. 2 pieces; [1935,128], 424g.; [1935,129], 50g.; [1935, 130], 110g.; [1935,131], 34g.; [1935,132], 62g.; [1935,133], 168g.; [1935,134], 75g.; [1935,135], 55g.; [1935,136], 64g.; [1935,137], 29g.; [1935,138], 23g.; [1935,139], 54g.; [1935, 140], 52g.; [1935,141], 19g.; [1935,142], 5g.; [1935,143], 85g.; [1935,144], 142g.; [1935,145], 43g.; [1935,146], 4g. 3 frag.; [1935,147], 135g.; [1935,148], 106g.; [1950,418], 60g. 2 pieces, and frags.; [1959,714], 1076g.; [1959,715], 1379g.; [1959,716], 161.3g. faceted; [1959,717], 26.24g. faceted; [1959,718], 35.54g. faceted; [1959,719], 14.15g. faceted; [1959,720], 10.59g. cabochon; [1959,721], 3.34g. cabochon; [1962,261], 5g.

IGAST GLASS 57°50′ N., 26°16′ E.

Igast, Latvian SSR, USSR
Fell 1855, May 17, 1800 hrs

A pumaceous glassy object of 35g was reported to have fallen at Igast, accompanied by detonations. Description, analysis (chemically similar to the moldavites from Radomilice), C. Grewingk and C. Schmidt, Archiv. Naturk. Liv-Ehst- und Kurlands, Min. Sci., 1st. ser., 1864, 3, p.421. A 500g object was reported later, A. Göbel, Bull. Acad. Imp. Sci. St.-Petersbourg, Ser. 3, 1867, 2, p.291. This material is artificial and many Igast specimens have come from it, H. Michel, Ann. k.k. Hofmus. Wien, Min. Sci., 1st. ser., 1913, 27, p.6. The original material is a vesicular glass with quartz grains, likened to a pseudoscoria, 1864 analysis quoted, P.D. Lowman, Jr. and J.A. O'Keefe, Nature, 1966, 209, p.67.

15g Dorpat, Univ., in 1897, genuine material; Specimen, Paris, Mus. d'Hist. Nat.;

Specimen(s): [36271], 1.5g.

AOUELLOUL GLASS

Aouelloul, Mauretania

See Aouelloul, under Meteorite Craters.

DARWIN GLASS

Mount Darwin, Tasmania, Australia

See Mount Darwin, under Meteorite Craters.